	3A (13)	4A (14)	5A (15)	6A (16)	7A (17)	8A (18)
						Helium 2 **He** 4.0026
	Boron 5 **B** 10.811	Carbon 6 **C** 12.011	Nitrogen 7 **N** 14.0067	Oxygen 8 **O** 15.9994	Fluorine 9 **F** 18.9984	Neon 10 **Ne** 20.1797
2B (12)	Aluminum 13 **Al** 26.9815	Silicon 14 **Si** 28.0855	Phosphorus 15 **P** 30.9738	Sulfur 16 **S** 32.066	Chlorine 17 **Cl** 35.4527	Argon 18 **Ar** 39.948
Zinc 30 **Zn** 65.38	Gallium 31 **Ga** 69.723	Germanium 32 **Ge** 72.61	Arsenic 33 **As** 74.9216	Selenium 34 **Se** 78.96	Bromine 35 **Br** 79.904	Krypton 36 **Kr** 83.80
Cadmium 48 **Cd** 112.411	Indium 49 **In** 114.818	Tin 50 **Sn** 118.710	Antimony 51 **Sb** 121.760	Tellurium 52 **Te** 127.60	Iodine 53 **I** 126.9045	Xenon 54 **Xe** 131.29
Mercury 80 **Hg** 200.59	Thallium 81 **Tl** 204.3833	Lead 82 **Pb** 207.2	Bismuth 83 **Bi** 208.9804	Polonium 84 **Po** (208.98)	Astatine 85 **At** (209.99)	Radon 86 **Rn** (222.02)
Copernicium 112 **Cn** (285)	Ununtrium 113 **Uut** Discovered 2004	Ununquadium 114 **Uuq** Discovered 1999	Ununpentium 115 **Uup** Discovered 2004	Ununhexium 116 **Uuh** Discovered 1999	Ununseptium 117 **Uus** Discovered 2010	Ununoctium 118 **Uuo** Discovered 2002

Terbium 65 **Tb** 158.9254	Dysprosium 66 **Dy** 162.50	Holmium 67 **Ho** 164.9303	Erbium 68 **Er** 167.26	Thulium 69 **Tm** 168.9342	Ytterbium 70 **Yb** 173.054	Lutetium 71 **Lu** 174.9668
Berkelium 97 **Bk** (247.07)	Californium 98 **Cf** (251.08)	Einsteinium 99 **Es** (252.08)	Fermium 100 **Fm** (257.10)	Mendelevium 101 **Md** (258.10)	Nobelium 102 **No** (259.10)	Lawrencium 103 **Lr** (262.11)

Standard Colors for Atoms in Molecular Models

 carbon atoms

 hydrogen atoms

 oxygen atoms

 nitrogen atoms

 chlorine atoms

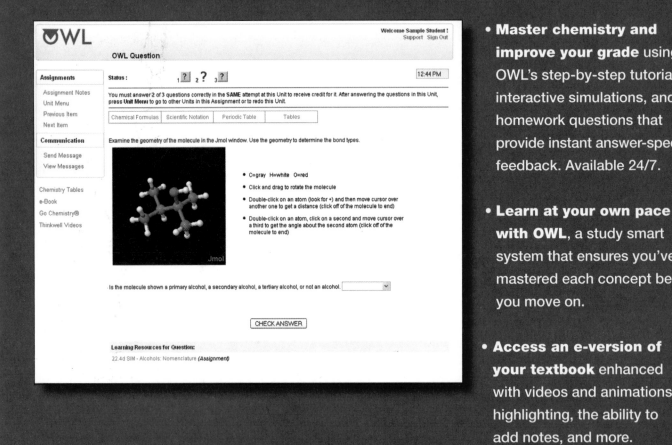

EIGHTH EDITION
HYBRID

CHEMISTRY
& Chemical Reactivity

John C. Kotz

State University of New York
College at Oneonta

Paul M. Treichel

University of Wisconsin–Madison

John R. Townsend

West Chester University of Pennsylvania

BROOKS/COLE
CENGAGE Learning

Australia • Brazil • Japan • Korea • Mexico • Singapore • Spain • United Kingdom • United States

BROOKS/COLE
CENGAGE Learning™

Chemistry & Chemical Reactivity, **Eighth Edition, Hybrid**

John C. Kotz, Paul M. Treichel, John R. Townsend

Publisher: Mary Finch

Executive Editor: Lisa Lockwood

Senior Developmental Editor: Peter McGahey

Assistant Editor: Elizabeth Woods

Editorial Assistant: Krista Mastroianni

Senior Media Editor: Lisa Weber

Media Editor: Stephanie Van Camp

Senior Marketing Manager: Nicole Hamm

Marketing Coordinator: Julie Stefani

Marketing Communications Manager: Linda Yip

Content Project Manager: Teresa L. Trego

Design Director: Rob Hugel

Art Director: John Walker

Print Buyer: Rebecca Cross

Rights Acquisitions Specialist: Dean Dauphinais

Production Service: Graphic World Inc.

Text Designer: Jeanne Calabrese

Photo Researcher: Bill Smith Group/Scott Rosen

Text Researcher: Sue Howard

Copy Editor: Graphic World Inc.

Illustrators: Patrick Harman/Graphic World Inc.

OWL producers: Stephen Battisti, Cindy Stein, David Hart (Center for Educational Software Development, University of Massachusetts, Amherst)

Cover Designer: Riezebos Holzbaur/Tim Heraldo

Cover Image: Joanna Aizenberg Rights-Managed

Compositor: Graphic World Inc.

For product information and technology assistance, contact us at
Cengage Learning Customer & Sales Support, 1-800-354-9706.
For permission to use material from this text or product, submit all requests online at **www.cengage.com/permissions.**
Further permissions questions can be emailed to
permissionrequest@cengage.com

Library of Congress Control Number: 2010938984

ISBN-13: 978-1-111-57498-7

ISBN-10: 1-111-57498-7

Brooks/Cole
20 Davis Drive
Belmont, CA 94002-3098
USA

Cengage Learning is a leading provider of customized learning solutions with office locations around the globe, including Singapore, the United Kingdom, Australia, Mexico, Brazil, and Japan. Locate your local office at **www.cengage.com/global**

Cengage Learning products are represented in Canada by Nelson Education, Ltd.

To learn more about Brooks/Cole, visit **www.cengage.com/brookscole**

Purchase any of our products at your local college store or at our preferred online store **www.CengageBrain.com**

Printed in the United States of America
2 3 4 5 6 7 14 13 12 11

BRIEF CONTENTS

PART ONE THE BASIC TOOLS OF CHEMISTRY

1   Basic Concepts of Chemistry   1

    Let's Review: The Tools of Quantitative Chemistry   20

2   Atoms, Molecules, and Ions   42

3   Chemical Reactions   92

4   Stoichiometry: Quantitative Information about Chemical Reactions   132

5   Principles of Chemical Reactivity: Energy and Chemical Reactions   172

PART TWO THE STRUCTURE OF ATOMS AND MOLECULES

6   The Structure of Atoms   204

7   The Structure of Atoms and Periodic Trends   232

8   Bonding and Molecular Structure   260

9   Bonding and Molecular Structure: Orbital Hybridization and Molecular Orbitals   308

10   Carbon: Not Just Another Element   338

PART THREE STATES OF MATTER

11   Gases and Their Properties   384

12   Intermolecular Forces and Liquids   416

13   The Chemistry of Solids   444

14   Solutions and Their Behavior   472

PART FOUR THE CONTROL OF CHEMICAL REACTIONS

15   Chemical Kinetics: The Rates of Chemical Reactions   506

16   Principles of Chemical Reactivity: Equilibria   548

17   Principles of Chemical Reactivity: The Chemistry of Acids and Bases   576

18   Principles of Chemical Reactivity: Other Aspects of Aqueous Equilibria   618

19   Principles of Chemical Reactivity: Entropy and Free Energy   660

20   Principles of Chemical Reactivity: Electron Transfer Reactions   690

PART FIVE THE CHEMISTRY OF THE ELEMENTS AND THEIR COMPOUNDS

21   The Chemistry of the Main Group Elements   734

22   The Chemistry of the Transition Elements   782

23   Nuclear Chemistry   820

APPENDICES

A   Using Logarithms and Solving Quadratic Equations   A-2

B   Some Important Physical Concepts   A-6

C   Abbreviations and Useful Conversion Factors   A-9

D   Physical Constants   A-13

E   A Brief Guide to Naming Organic Compounds   A-15

F   Values for the Ionization Energies and Electron Attachment Enthalpies of the Elements   A-18

G   Vapor Pressure of Water at Various Temperatures   A-19

H   Ionization Constants for Aqueous Weak Acids at 25 °C   A-20

I   Ionization Constants for Aqueous Weak Bases at 25 °C   A-22

J   Solubility Product Constants for Some Inorganic Compounds at 25 °C   A-23

K   Formation Constants for Some Complex Ions in Aqueous Solution at 25 °C   A-25

L   Selected Thermodynamic Values   A-26

M   Standard Reduction Potentials in Aqueous Solution at 25°C   A-32

N   Answers to Chapter Opening Questions and Case Study Questions   A-36

O   Answers to Check Your Understanding Questions   A-47

P   Answers to Review & Check Questions   A-63

CONTENTS

Preface xvii

PART ONE THE BASIC TOOLS OF CHEMISTRY
....

1 Basic Concepts of Chemistry 1

Gold! 1

1.1 **Chemistry and Its Methods** 2
Hypotheses, Laws, and Theories 3
A Closer Look: Careers in Chemistry 4
Goals of Science 5
Dilemmas and Integrity in Science 5

1.2 **Sustainability and Green Chemistry** 5
A Closer Look: Principles of Green Chemistry 6

1.3 **Classifying Matter** 6
States of Matter and Kinetic-Molecular Theory 7
Matter at the Macroscopic and Particulate Levels 8
Pure Substances 8
Mixtures: Homogeneous and Heterogeneous 9

1.4 **Elements** 10
A Closer Look: Element Names and Symbols 11

1.5 **Compounds** 12

1.6 **Physical Properties** 13
Extensive and Intensive Properties 14

1.7 **Physical and Chemical Changes** 15

1.8 **Energy: Some Basic Principles** 16
Case Study: CO_2 in the Oceans 17
Conservation of Energy 18
....
CHAPTER GOALS REVISITED 19
KEY EQUATION 19

Let's Review: The Tools of Quantitative Chemistry 20

Copper 20

1 **Units of Measurement** 21
Temperature Scales 21
Length, Volume, and Mass 23
A Closer Look: Energy and Food 25
Energy Units 25

2 **Making Measurements: Precision, Accuracy, Experimental Error, and Standard Deviation** 26
Experimental Error 27
Standard Deviation 28

3 **Mathematics of Chemistry** 29
Exponential or Scientific Notation 29
Significant Figures 31

4 **Problem Solving by Dimensional Analysis** 35
Case Study: Out of Gas! 36

5 **Graphs and Graphing** 37

6 **Problem Solving and Chemical Arithmetic** 38

2 Atoms, Molecules, and Ions 42

The Periodic Table, the Central Icon of Chemistry 42

2.1 **Atomic Structure—Protons, Electrons, and Neutrons** 43

2.2 **Atomic Number and Atomic Mass** 44
Atomic Number 44
Relative Atomic Mass and the Atomic Mass Unit 44
Mass Number 44

2.3 **Isotopes** 46
Isotope Abundance 46
Determining Atomic Mass and Isotope Abundance 46

2.4 **Atomic Weight** 47
Case Study: Using Isotopes: Ötzi, the Iceman of the Alps 50

2.5 **The Periodic Table** 50
Developing the Periodic Table 50
A Closer Look: The Story of the Periodic Table 51
Features of the Periodic Table 53
A Brief Overview of the Periodic Table and the Chemical Elements 54

2.6 **Molecules, Compounds, and Formulas** 58
Formulas 58
Molecular Models 60

2.7 **Ionic Compounds: Formulas, Names, and Properties** 61
Ions 61
Formulas of Ionic Compounds 65
Names of Ions 66
Properties of Ionic Compounds 68

2.8 **Molecular Compounds: Formulas and Names** 70

2.9 **Atoms, Molecules, and the Mole** 72
Atoms and Molar Mass 72
A Closer Look: Amedeo Avogadro and His Number 73
Molecules, Compounds, and Molar Mass 74

2.10 **Describing Compound Formulas** 77
Percent Composition 77
Empirical and Molecular Formulas from Percent Composition 79
Determining a Formula from Mass Data 81

Case Study: Mummies, Bangladesh, and the Formula of Compound 606 84

Determining a Formula by Mass Spectrometry 84

A Closer Look: Mass Spectrometry, Molar Mass, and Isotopes 85

2.11 Hydrated Compounds 86

CHAPTER GOALS REVISITED 88

KEY EQUATIONS 89

3   Chemical Reactions 92

Black Smokers and Volcanoes 92

3.1 Introduction to Chemical Equations 93
A Closer Look: Antoine Laurent Lavoisier, 1743–1794 94

3.2 Balancing Chemical Equations 96

3.3 Introduction to Chemical Equilibrium 98

3.4 Aqueous Solutions 101
Ions and Molecules in Aqueous Solutions 101
Solubility of Ionic Compounds in Water 104

3.5 Precipitation Reactions 105
Predicting the Outcome of a Precipitation Reaction 106
Net Ionic Equations 108

3.6 Acids and Bases 110
Acids and Bases: The Arrhenius Definition 110
Acids and Bases: The Brønsted–Lowry Definition 112
A Closer Look: The Hydronium Ion—The H^+ Ion in Water 113
Reactions of Acids and Bases 114
A Closer Look: Sulfuric Acid 115
Oxides of Nonmetals and Metals 116

3.7 Gas-Forming Reactions 118

3.8 Oxidation–Reduction Reactions 119
Oxidation-Reduction Reactions and Electron Transfer 120
Oxidation Numbers 121
A Closer Look: Are Oxidation Numbers "Real"? 122
Recognizing Oxidation–Reduction Reactions 123

3.9 Classifying Reactions in Aqueous Solution 126
Case Study: Killing Bacteria with Silver 126

CHAPTER GOALS REVISITED 129

4   Stoichiometry: Quantitative Information about Chemical Reactions 132

The Chemistry of Pyrotechnics 132

4.1 Mass Relationships in Chemical Reactions: Stoichiometry 133

4.2 Reactions in Which One Reactant Is Present in Limited Supply 137
A Stoichiometry Calculation with a Limiting Reactant 137

4.3 Percent Yield 141

4.4 Chemical Equations and Chemical Analysis 142
Quantitative Analysis of a Mixture 143
Case Study: Green Chemistry and Atom Economy 144
Determining the Formula of a Compound by Combustion 145

4.5 Measuring Concentrations of Compounds in Solution 149
Solution Concentration: Molarity 149
Preparing Solutions of Known Concentration 151
A Closer Look: Serial Dilutions 154

4.6 pH, a Concentration Scale for Acids and Bases 154

4.7 Stoichiometry of Reactions in Aqueous Solution 157
Solution Stoichiometry 157
Titration: A Method of Chemical Analysis 158
Standardizing an Acid or Base 160
Determining Molar Mass by Titration 161
Titrations Using Oxidation–Reduction Reactions 162
Case Study: How Much Salt Is There in Seawater? 163

4.8 Spectrophotometry 164
Case Study: Forensic Chemistry: Titrations and Food Tampering 165
Transmittance, Absorbance, and the Beer–Lambert Law 165
Spectrophotometric Analysis 167

CHAPTER GOALS REVISITED 169

KEY EQUATIONS 170

5   Principles of Chemical Reactivity: Energy and Chemical Reactions 172

Energy and Your Diet 172

5.1 Energy: Some Basic Principles 173
Systems and Surroundings 174
Directionality and Extent of Transfer of Heat: Thermal Equilibrium 174
A Closer Look: What Is Heat? 175

5.2 Specific Heat Capacity: Heating and Cooling 176
Quantitative Aspects of Energy Transferred as Heat 178

5.3 Energy and Changes of State 180

5.4 The First Law of Thermodynamics 183
A Closer Look: P–V Work 185
Enthalpy 186
State Functions 186

Contents **v**

5.5 **Enthalpy Changes for Chemical Reactions** 188

5.6 **Calorimetry** 190
Constant Pressure Calorimetry, Measuring ΔH 190
Constant Volume Calorimetry, Measuring ΔU 192

5.7 **Enthalpy Calculations** 194
Hess's Law 194
Energy Level Diagrams 195
Standard Enthalpies of Formation 197
Enthalpy Change for a Reaction 198
A Closer Look: Hess's Law and Equation 5.6 199

5.8 **Product- or Reactant-Favored Reactions and Thermodynamics** 199
Case Study: The Fuel Controversy—Alcohol and Gasoline 200

CHAPTER GOALS REVISITED 202
KEY EQUATIONS 203

PART TWO THE STRUCTURE OF ATOMS AND MOLECULES

6 The Structure of Atoms 204

Fireworks 204

6.1 **Electromagnetic Radiation** 205

6.2 **Quantization: Planck, Einstein, Energy, and Photons** 207
Planck's Equation 207
Einstein and the Photoelectric Effect 209
Energy and Chemistry: Using Planck's Equation 209

6.3 **Atomic Line Spectra and Niels Bohr** 210
The Bohr Model of the Hydrogen Atom 211
The Bohr Theory and the Spectra of Excited Atoms 213

6.4 **Particle–Wave Duality: Prelude to Quantum Mechanics** 216
Case Study: What Makes the Colors in Fireworks? 217

6.5 **The Modern View of Electronic Structure: Wave or Quantum Mechanics** 219
Quantum Numbers and Orbitals 220
Shells and Subshells 221

6.6 **The Shapes of Atomic Orbitals** 222
s Orbitals 222
p Orbitals 223
d Orbitals 224
A Closer Look: More about H Atom Orbital Shapes and Wavefunctions 225
f Orbitals 226

6.7 **One More Electron Property: Electron Spin** 226
The Electron Spin Quantum Number, m_s 226
A Closer Look: Paramagnetism and Ferromagnetism 227
Diamagnetism and Paramagnetism 227

CHAPTER GOALS REVISITED 228
A Closer Look: Quantized Spins and MRI 229
KEY EQUATIONS 230

7 The Structure of Atoms and Periodic Trends 232

Rubies and Sapphires—Pretty Stones 232

7.1 **The Pauli Exclusion Principle** 233

7.2 **Atomic Subshell Energies and Electron Assignments** 235
Order of Subshell Energies and Assignments 235
Effective Nuclear Charge, Z^* 236

7.3 **Electron Configurations of Atoms** 237
Electron Configurations of the Main Group Elements 239
Electron Configurations of the Transition Elements 242
A Closer Look: Orbital Energies, Z^*, and Electron Configurations 244

7.4 **Electron Configurations of Ions** 245
A Closer Look: Questions about Transition Element Electron Configurations 246

7.5 **Atomic Properties and Periodic Trends** 247
Atomic Size 247
Ionization Energy 249
Electron Attachment Enthalpy and Electron Affinity 252
Trends in Ion Sizes 254

7.6 **Periodic Trends and Chemical Properties** 255
Case Study: Metals in Biochemistry and Medicine 257

CHAPTER GOALS REVISITED 258

8 Bonding and Molecular Structure 260

Chemical Bonding in DNA 260

8.1 **Chemical Bond Formation** 261

8.2 **Covalent Bonding and Lewis Structures** 262
Valence Electrons and Lewis Symbols for Atoms 262
Lewis Electron Dot Structures and the Octet Rule 264
Drawing Lewis Electron Dot Structures 265
A Closer Look: Useful Ideas to Consider When Drawing Lewis Electron Dot Structures 267
Predicting Lewis Structures 267

8.3 **Atom Formal Charges in Covalent Molecules and Ions** 270
A Closer Look: Comparing Oxidation Number and Formal Charge 271

8.4 **Resonance** 272
A Closer Look: Resonance 273
A Closer Look: A Scientific Controversy—Are There Double Bonds in Sulfate and Phosphate Ions? 275

8.5 **Exceptions to the Octet Rule** 276
Compounds in Which an Atom Has Fewer Than Eight Valence Electrons 276
Compounds in Which an Atom Has More Than Eight Valence Electrons 277
Molecules with an Odd Number of Electrons 278
Case Study: Hydroxyl Radicals, Atmospheric Chemistry, and Hair Dyes 279

8.6 **Molecular Shapes** 280
Central Atoms Surrounded Only by Single-Bond Pairs 280
Central Atoms with Single-Bond Pairs and Lone Pairs 282
Multiple Bonds and Molecular Geometry 284

8.7 **Bond Polarity and Electronegativity** 287
Charge Distribution: Combining Formal Charge and Electronegativity 289

8.8 **Bond and Molecular Polarity** 291
A Closer Look: Visualizing Charge Distributions and Molecular Polarity—Electrostatic Potential Surfaces and Partial Charge 294

8.9 **Bond Properties: Order, Length, and Energy** 297
Bond Order 297
Bond Length 298
Bond Dissociation Enthalpy 299
Case Study: Ibuprofen, A Study in Green Chemistry 301
A Closer Look: DNA—Watson, Crick, and Franklin 303

8.10 **DNA, Revisited** 304
CHAPTER GOALS REVISITED 305
KEY EQUATIONS 307

9 **Bonding and Molecular Structure: Orbital Hybridization and Molecular Orbitals** 308
The Noble Gases: Not So Inert 308

9.1 **Orbitals and Theories of Chemical Bonding** 309

9.2 **Valence Bond Theory** 310
The Orbital Overlap Model of Bonding 310
Hybridization of Atomic Orbitals 312
Multiple Bonds 319
Benzene: A Special Case of π Bonding 323

9.3 **Molecular Orbital Theory** 324
Principles of Molecular Orbital Theory 325
A Closer Look: Molecular Orbitals for Molecules Formed from p-Block Elements 331
Electron Configurations for Heteronuclear Diatomic Molecules 331
Resonance and MO Theory 332
Case Study: Green Chemistry, Safe Dyes, and Molecular Orbitals 334
A Closer Look: Three-Center Bonds and Hybrid Orbitals with d Orbitals 335
CHAPTER GOALS REVISITED 336
KEY EQUATION 337

10 **Carbon: Not Just Another Element** 338
The Food of the Gods 338

10.1 **Why Carbon?** 339
Structural Diversity 339
Isomers 340
A Closer Look: Writing Formulas and Drawing Structures 341
Stability of Carbon Compounds 342
A Closer Look: Chirality and Elephants 343

10.2 **Hydrocarbons** 343
Alkanes 343
Alkenes and Alkynes 349
A Closer Look: Flexible Molecules 349
Aromatic Compounds 353

10.3 **Alcohols, Ethers, and Amines** 357
A Closer Look: Petroleum Chemistry 358
Alcohols and Ethers 358
Properties of Alcohols 361
Amines 362

10.4 **Compounds with a Carbonyl Group** 364
Case Study: An Awakening with L-DOPA 364
Aldehydes and Ketones 366
Carboxylic Acids 367
A Closer Look: Glucose and Other Sugars 367
Esters 369
Amides 370

10.5 **Polymers** 373
Classifying Polymers 373
Addition Polymers 373
Condensation Polymers 377
A Closer Look: Copolymers and the Book Cover 377
A Closer Look: Copolymers and Engineering Plastics for Lego Bricks and Tattoos 378
A Closer Look: Green Chemistry: Recycling PET 379
Case Study: Green Adhesives 381
CHAPTER GOALS REVISITED 382

PART THREE STATES OF MATTER

11 Gases and Their Properties 384

The Atmosphere and Altitude Sickness 384

11.1 Gas Pressure 386
A Closer Look: Measuring Gas Pressure 387

11.2 Gas Laws: The Experimental Basis 387
Boyle's Law: The Compressibility of Gases 387
The Effect of Temperature on Gas Volume: Charles's Law 389
Combining Boyle's and Charles's Laws: The General Gas Law 391
Avogadro's Hypothesis 392
A Closer Look: Studies on Gases—Robert Boyle and Jacques Charles 394

11.3 The Ideal Gas Law 394
The Density of Gases 395
Calculating the Molar Mass of a Gas from P, V, and T Data 397

11.4 Gas Laws and Chemical Reactions 398

11.5 Gas Mixtures and Partial Pressures 400

11.6 The Kinetic-Molecular Theory of Gases 403
Molecular Speed and Kinetic Energy 403
A Closer Look: The Earth's Atmosphere 404
Kinetic-Molecular Theory and the Gas Laws 407

11.7 Diffusion and Effusion 408
A Closer Look: SCUBA Diving—An Application of the Gas Laws 410

11.8 Nonideal Behavior of Gases 410
Case Study: What to Do with All of That CO_2? More on Green Chemistry 412

CHAPTER GOALS REVISITED 413
KEY EQUATIONS 413

12 Intermolecular Forces and Liquids 416

Geckos Can Climb Up der Waals 416

12.1 States of Matter and Intermolecular Forces 417

12.2 Interactions between Ions and Molecules with a Permanent Dipole 418

12.3 Interactions between Molecules with a Dipole 420
Dipole–Dipole Forces 420
A Closer Look: Hydrated Salts 421
Hydrogen Bonding 422
Hydrogen Bonding and the Unusual Properties of Water 424
Case Study: Hydrogen Bonding & Methane Hydrates: Opportunities and Problems 426

12.4 Intermolecular Forces Involving Nonpolar Molecules 427
Dipole-Induced Dipole Forces 427
London Dispersion Forces: Induced Dipole-Induced Dipole Forces 428
A Closer Look: Hydrogen Bonding in Biochemistry 429

12.5 A Summary of van der Waals Intermolecular Forces 431

12.6 Properties of Liquids 432
Case Study: A Pet Food Catastrophe 433
Vaporization and Condensation 433
Vapor Pressure 436
Vapor Pressure, Enthalpy of Vaporization, and the Clausius–Clapeyron Equation 438
Boiling Point 439
Critical Temperature and Pressure 439
Surface Tension, Capillary Action, and Viscosity 439
A Closer Look: Supercritical CO_2 and Green Chemistry 442

CHAPTER GOALS REVISITED 442
KEY EQUATIONS 443

13 The Chemistry of Solids 444

Lithium and "Green Cars" 444

13.1 Crystal Lattices and Unit Cells 445
A Closer Look: Packing Oranges and Marbles 449

13.2 Structures and Formulas of Ionic Solids 452
Case Study: High-Strength Steel and Unit Cells 454

13.3 Bonding in Metals and Semiconductors 456
Semiconductors 458

13.4 Bonding in Ionic Compounds: Lattice Energy 460
Lattice Energy 460
Calculating a Lattice Enthalpy from Thermodynamic Data 461

13.5 The Solid State: Other Types of Solid Materials 463
Molecular Solids 463
Network Solids 463
Amorphous Solids 463

13.6 Phase Changes Involving Solids 464
Melting: Conversion of Solid into Liquid 464
Case Study: Graphene—The Hottest New Network Solid 465
Sublimation: Conversion of Solid into Vapor 467

13.7 Phase Diagrams 468
Water 468
Phase Diagrams and Thermodynamics 468
Carbon Dioxide 468

CHAPTER GOALS REVISITED 470

14 Solutions and Their Behavior 472

Survival at Sea 472

14.1 Units of Concentration 474

14.2 The Solution Process 476
Liquids Dissolving in Liquids 477
A Closer Look: Supersaturated Solutions 478
Solids Dissolving in Water 478
Enthalpy of Solution 479
Enthalpy of Solution: Thermodynamic Data 481

14.3 Factors Affecting Solubility: Pressure and Temperature 482
Dissolving Gases in Liquids: Henry's Law 482
Temperature Effects on Solubility: Le Chatelier's Principle 483
Case Study: Exploding Lakes and Diet Cokes 485

14.4 Colligative Properties 486
Changes in Vapor Pressure: Raoult's Law 486
Boiling Point Elevation 487
Freezing Point Depression 490
Osmotic Pressure 491
A Closer Look: Reverse Osmosis for Pure Water 493
Colligative Properties and Molar Mass Determination 494
A Closer Look: Osmosis and Medicine 496
Colligative Properties of Solutions Containing Ions 496

14.5 Colloids 499
Types of Colloids 500
Surfactants 501

CHAPTER GOALS REVISITED 502
KEY EQUATIONS 503

PART FOUR THE CONTROL OF CHEMICAL REACTIONS

15 Chemical Kinetics: The Rates of Chemical Reactions 506

Where Did the Indicator Go? 506

15.1 Rates of Chemical Reactions 507

15.2 Reaction Conditions and Rate 512

15.3 Effect of Concentration on Reaction Rate 513
Rate Equations 514
The Order of a Reaction 514
The Rate Constant, k 515
Determining a Rate Equation 516

15.4 Concentration–Time Relationships: Integrated Rate Laws 519
First-Order Reactions 519
Second-Order Reactions 521
Zero-Order Reactions 522
Graphical Methods for Determining Reaction Order and the Rate Constant 522
Half-Life and First-Order Reactions 523

15.5 A Microscopic View of Reaction Rates 527
Collision Theory: Concentration and Reaction Rate 527
Collision Theory: Temperature and Reaction Rate 528
Collision Theory: Activation Energy 528
A Closer Look: Reaction Coordinate Diagrams 530
Collision Theory: Activation Energy and Temperature 530
Collision Theory: Effect of Molecular Orientation on Reaction Rate 530
The Arrhenius Equation 531
Effect of Catalysts on Reaction Rate 533

15.6 Reaction Mechanisms 535
Molecularity of Elementary Steps 536
Rate Equations for Elementary Steps 537
Molecularity and Reaction Order 537
Reaction Mechanisms and Rate Equations 538
Case Study: Enzymes—Nature's Catalysts 540

CHAPTER GOALS REVISITED 544
KEY EQUATIONS 545

16 Principles of Chemical Reactivity: Equilibria 548

Dynamic and Reversible! 548

16.1 Chemical Equilibrium: A Review 549

16.2 The Equilibrium Constant and Reaction Quotient 550
Writing Equilibrium Constant Expressions 552
A Closer Look: Activities and Units of K 553
A Closer Look: Equilibrium Constant Expressions for Gases—K_c and K_p 554
The Meaning of the Equilibrium Constant, K 554
The Reaction Quotient, Q 555

16.3 Determining an Equilibrium Constant 558

16.4 Using Equilibrium Constants in Calculations 561
Calculations Where the Solution Involves a Quadratic Expression 562

16.5 More about Balanced Equations and Equilibrium Constants 566

16.6 Disturbing a Chemical Equilibrium 568

Effect of the Addition or Removal of a Reactant or
Product 569

Effect of Volume Changes on Gas-Phase Equilibria 571

Effect of Temperature Changes on Equilibrium
Composition 572

Case Study: Applying Equilibrium Concepts—The Haber–
Bosch Ammonia Process 574

CHAPTER GOALS REVISITED 574

KEY EQUATIONS 575

**17 Principles of Chemical Reactivity:
The Chemistry of Acids and Bases** 576

Aspirin Is Over 100 Years Old! 576

17.1 Acids and Bases: A Review 577

**17.2 The Brønsted-Lowry Concept of Acids and Bases
Extended** 578

Conjugate Acid–Base Pairs 580

17.3 Water and the pH Scale 580

Water Autoionization and the Water Ionization
Constant, K_w 581

The pH Scale 583

Calculating pH 583

17.4 Equilibrium Constants for Acids and Bases 584

K_a Values for Polyprotic Acids 587

Logarithmic Scale of Relative Acid Strength, pK_a 588

Relating the Ionization Constants for an Acid and Its
Conjugate Base 588

17.5 Acid–Base Properties of Salts 589

17.6 Predicting the Direction of Acid–Base Reactions 591

17.7 Types of Acid–Base Reactions 594

The Reaction of a Strong Acid with a Strong Base 594

The Reaction of a Weak Acid with a Strong Base 594

The Reaction of a Strong Acid with a Weak Base 595

The Reaction of a Weak Acid with a Weak Base 595

17.8 Calculations with Equilibrium Constants 596

Determining K from Initial Concentrations and
Measured pH 596

What Is the pH of an Aqueous Solution of a Weak Acid
or Base? 597

Case Study: Would You Like Some Belladonna Juice in
Your Drink? 604

17.9 Polyprotic Acids and Bases 605

**17.10 Molecular Structure, Bonding, and Acid–Base
Behavior** 607

Acid Strength of the Hydrogen Halides, HX 607

Comparing Oxoacids: HNO_2 and HNO_3 607

A Closer Look: Acid Strengths and Molecular
Structure 608

Why Are Carboxylic Acids Brønsted Acids? 609

Why Are Hydrated Metal Cations Brønsted Acids? 610

Why Are Anions Brønsted Bases? 611

17.11 The Lewis Concept of Acids and Bases 611

Cationic Lewis Acids 612

Molecular Lewis Acids 614

Molecular Lewis Bases 614

CHAPTER GOALS REVISITED 616

KEY EQUATIONS 616

**18 Principles of Chemical Reactivity: Other
Aspects of Aqueous Equilibria** 618

Nature's Acids 618

18.1 The Common Ion Effect 619

18.2 Controlling pH: Buffer Solutions 622

General Expressions for Buffer Solutions 624

Preparing Buffer Solutions 626

How Does a Buffer Maintain pH? 628

18.3 Acid–Base Titrations 630

Titration of a Strong Acid with a Strong Base 630

Case Study: Take a Deep Breath 631

Titration of a Weak Acid with a Strong Base 632

Titration of Weak Polyprotic Acids 636

Titration of a Weak Base with a Strong Acid 636

pH Indicators 638

18.4 Solubility of Salts 640

The Solubility Product Constant, K_{sp} 641

Relating Solubility and K_{sp} 642

A Closer Look: Minerals and Gems—The Importance of
Solubility 643

A Closer Look: Solubility Calculations 645

Solubility and the Common Ion Effect 646

The Effect of Basic Anions on Salt Solubility 649

18.5 Precipitation Reactions 651

K_{sp} and the Reaction Quotient, Q 651

Case Study: Chemical Equilibria in the Oceans 652

K_{sp}, the Reaction Quotient, and Precipitation
Reactions 653

18.6 Equilibria Involving Complex Ions 655

18.7 Solubility and Complex Ions 656

CHAPTER GOALS REVISITED 658

KEY EQUATIONS 659

19 Principles of Chemical Reactivity: Entropy and Free Energy 660

Hydrogen for the Future? 660

19.1 Spontaneity and Energy Transfer as Heat 661

19.2 Dispersal of Energy: Entropy 663
A Closer Look: Reversible and Irreversible Processes 664

19.3 Entropy: A Microscopic Understanding 664
Dispersal of Energy 664
Dispersal of Matter: Dispersal of Energy Revisited 666
A Summary: Entropy, Entropy Change, and Energy Dispersal 668

19.4 Entropy Measurement and Values 668
Standard Entropy Values, S^o 668
Determining Entropy Changes in Physical and Chemical Processes 670

19.5 Entropy Changes and Spontaneity 671
In Summary: Spontaneous or Not? 674

19.6 Gibbs Free Energy 676
The Change in the Gibbs Free Energy, ΔG 676
Gibbs Free Energy, Spontaneity, and Chemical Equilibrium 677
A Summary: Gibbs Free Energy ($\Delta_r G$ and $\Delta_r G^o$), the Reaction Quotient (Q) and Equilibrium Constant (K), and Reaction Favorability 679
What Is "Free" Energy? 679

19.7 Calculating and Using Free Energy 680
Standard Free Energy of Formation 680
Calculating $\Delta_r G^o$, the Free Energy Change for a Reaction Under Standard Conditions 680
Free Energy and Temperature 682
Case Study: Thermodynamics and Living Things 683
Using the Relationship between $\Delta_r G^o$ and K 685

CHAPTER GOALS REVISITED 686
KEY EQUATIONS 687

20 Principles of Chemical Reactivity: Electron Transfer Reactions 690

Battery Power 690

20.1 Oxidation–Reduction Reactions 692
Balancing Oxidation–Reduction Equations 692

20.2 Simple Voltaic Cells 699
Voltaic Cells with Inert Electrodes 702
Electrochemical Cell Notations 703

20.3 Commercial Voltaic Cells 704
Primary Batteries: Dry Cells and Alkaline Batteries 705
Secondary or Rechargeable Batteries 706
Fuel Cells 708

20.4 Standard Electrochemical Potentials 709
Electromotive Force 709
Measuring Standard Potentials 709
A Closer Look: EMF, Cell Potential, and Voltage 711
Standard Reduction Potentials 711
Tables of Standard Reduction Potentials 712
Using Tables of Standard Reduction Potentials 714
Relative Strengths of Oxidizing and Reducing Agents 715
A Closer Look: An Electrochemical Toothache 717

20.5 Electrochemical Cells under Nonstandard Conditions 717
The Nernst Equation 717
Case Study: Manganese in the Oceans 718

20.6 Electrochemistry and Thermodynamics 721
Work and Free Energy 721
E^o and the Equilibrium Constant 722

20.7 Electrolysis: Chemical Change Using Electrical Energy 725
Electrolysis of Molten Salts 725
Electrolysis of Aqueous Solutions 727
A Closer Look: Electrochemistry and Michael Faraday 730

20.8 Counting Electrons 730

CHAPTER GOALS REVISITED 731
KEY EQUATIONS 732

PART FIVE THE CHEMISTRY OF THE ELEMENTS AND THEIR COMPOUNDS

21 The Chemistry of the Main Group Elements 734

Carbon and Silicon 734

21.1 Element Abundances 735

21.2 The Periodic Table: A Guide to the Elements 736
Valence Electrons 736
Ionic Compounds of Main Group Elements 736
Molecular Compounds of Main Group Elements 737

21.3 Hydrogen 740
Chemical and Physical Properties of Hydrogen 740
A Closer Look: Hydrogen, Helium, and Balloons 741
Preparation of Hydrogen 742

21.4 The Alkali Metals, Group 1A 743
Preparation of Sodium and Potassium 744
Properties of Sodium and Potassium 744
A Closer Look: The Reducing Ability of the Alkali Metals 746
Important Lithium, Sodium, and Potassium Compounds 746

21.5 The Alkaline Earth Elements, Group 2A 748
Properties of Calcium and Magnesium 749
Metallurgy of Magnesium 749
A Closer Look: Alkaline Earth Metals and Biology 750
Calcium Minerals and Their Applications 750
A Closer Look: Of Romans, Limestone, and Champagne 751
Case Study: Hard Water 752

21.6 Boron, Aluminum, and the Group 3A Elements 753
Chemistry of the Group 3A Elements 753
Boron Minerals and Production of the Element 753
Metallic Aluminum and Its Production 754
Boron Compounds 756
Aluminum Compounds 757

21.7 Silicon and the Group 4A Elements 758
Silicon 758
Silicon Dioxide 759
Silicate Minerals with Chain and Ribbon Structures 760
Silicates with Sheet Structures and Aluminosilicates 760
Silicone Polymers 761
Case Study: Lead, Beethoven, and a Mystery Solved 762

21.8 Nitrogen, Phosphorus, and the Group 5A Elements 763
Properties of Nitrogen and Phosphorus 763
Nitrogen Compounds 764
Case Study: A Healthy Saltwater Aquarium and the Nitrogen Cycle 765
A Closer Look: Making Phosphorus 767
Hydrogen Compounds of Phosphorus and Other Group 5A Elements 768
Phosphorus Oxides and Sulfides 768
Phosphorus Oxoacids and Their Salts 770

21.9 Oxygen, Sulfur, and the Group 6A Elements 772
Preparation and Properties of the Elements 772
Sulfur Compounds 773
A Closer Look: Snot-tites and Sulfur Chemistry 775

21.10 The Halogens, Group 7A 775
Preparation of the Elements 775
Fluorine Compounds 777
Chlorine Compounds 778
··
CHAPTER GOALS REVISITED 780

22 The Chemistry of the Transition Elements 782

Memory Metal 782

22.1 Properties of the Transition Elements 784
Electron Configurations 785
Oxidation and Reduction 785

Periodic Trends in the *d*-Block: Size, Density, Melting Point 786
A Closer Look: Corrosion of Iron 787

22.2 Metallurgy 789
Pyrometallurgy: Iron Production 790
Hydrometallurgy: Copper Production 791

22.3 Coordination Compounds 792
Complexes and Ligands 792
Formulas of Coordination Compounds 795
Naming Coordination Compounds 797
A Closer Look: Hemoglobin 798

22.4 Structures of Coordination Compounds 800
Common Coordination Geometries 800
Isomerism 800

22.5 Bonding in Coordination Compounds 804
The *d* Orbitals: Ligand Field Theory 804
Electron Configurations and Magnetic Properties 806

22.6 Colors of Coordination Compounds 809
Color 809
The Spectrochemical Series 810
Case Study: Accidental Discovery of a Chemotherapy Agent 813

22.7 Organometallic Chemistry: Compounds with Metal–Carbon Bonds 813
Carbon Monoxide Complexes of Metals 813
The Effective Atomic Number Rule and Bonding in Organometallic Compounds 814
Ligands in Organometallic Compounds 815
Case Study: Ferrocene—The Beginning of a Chemical Revolution 816
··
CHAPTER GOALS REVISITED 817

23 Nuclear Chemistry 820

A Primordial Nuclear Reactor 820

23.1 Natural Radioactivity 821

23.2 Nuclear Reactions and Radioactive Decay 822
Equations for Nuclear Reactions 822
Radioactive Decay Series 823
Other Types of Radioactive Decay 825

23.3 Stability of Atomic Nuclei 827
The Band of Stability and Radioactive Decay 827
Nuclear Binding Energy 829

23.4 Rates of Nuclear Decay 832
Half-Life 832
Kinetics of Nuclear Decay 833
Radiocarbon Dating 835

23.5 Artificial Nuclear Reactions 837
A Closer Look: The Search for New Elements 838

23.6 Nuclear Fission 840

23.7 Nuclear Fusion 842

23.8 Radiation Health and Safety 843
Units for Measuring Radiation 843
Radiation: Doses and Effects 843
A Closer Look: What Is a Safe Exposure? 845

23.9 Applications of Nuclear Chemistry 845
Nuclear Medicine: Medical Imaging 845
A Closer Look: Technetium-99m 846
Nuclear Medicine: Radiation Therapy 847
Analytical Methods: The Use of Radioactive Isotopes as Tracers 847
Analytical Methods: Isotope Dilution 847
Space Science: Neutron Activation Analysis and the Moon Rocks 848
Food Science: Food Irradiation 848
Case Study: Nuclear Medicine and Hyperthyroidism 849
....
CHAPTER GOALS REVISITED 850
KEY EQUATIONS 850

A Appendices A-1

A **Using Logarithms and Solving Quadratic Equations** A-2

B **Some Important Physical Concepts** A-6

C **Abbreviations and Useful Conversion Factors** A-9

D **Physical Constants** A-13

E **A Brief Guide to Naming Organic Compounds** A-15

F **Values for the Ionization Energies and Electron Attachment Enthalpies of the Elements** A-18

G **Vapor Pressure of Water at Various Temperatures** A-19

H **Ionization Constants for Aqueous Weak Acids at 25 °C** A-20

I **Ionization Constants for Aqueous Weak Bases at 25 °C** A-22

J **Solubility Product Constants for Some Inorganic Compounds at 25 °C** A-23

K **Formation Constants for Some Complex Ions in Aqueous Solution at 25 °C** A-25

L **Selected Thermodynamic Values** A-26

M **Standard Reduction Potentials in Aqueous Solution at 25 °C** A-32

N **Answers to Chapter Opening Questions and Case Study Questions** A-36

O **Answers to Check Your Understanding Questions** A-47

P **Answers to Review & Check Questions** A-63

Index/Glossary I-1

Go Chemistry Modules [go Chemistry]

Go Chemistry® modules are mini video lectures prepared by the book author, John C. Kotz, that may include animations, problems, or eFlashcards for quick review of key concepts. They play on video iPods, iPhones, iPads, personal video players, and iTunes and are correlated to the text by annotations in the margin. If you are using OWL, Go Chemistry is included in the Cengage *YouBook*. You can download two sample modules and purchase modules individually or as a set at **www.cengagebrain.com** (ISBN 0-495-38228-0).

Chapter 1	Basic Concepts of Chemistry	Module 1	The Periodic Table
Chapter 2	Atoms, Molecules, and Ions	Module 2	Predicting Ion Charges
		Module 3	Names to Formulas of Ionic Compounds
		Module 4	The Mole
Chapter 3	Chemical Reactions	Module 5	Predicting the Water Solubility of Ionic Compounds
		Module 6	Writing Net Ionic Equations
Chapter 4	Stoichiometry: Quantitative Information about Chemical Reactions	Module 7	Simple Stoichiometry
		Module 8a	Stoichiometry and Limiting Reactants (Part 1)
		Module 8b	Stoichiometry and Limiting Reactants (Part 2)
		Module 9a	pH (Part 1)
		Module 9b	pH (Part 2)
Chapter 5	Principles of Chemical Reactivity: Energy and Chemical Reactions	Module 10	Thermochemistry and Hess's Law
Chapter 7	The Structure of Atoms and Periodic Trends	Module 11	Periodic Trends
Chapter 8	Bonding and Molecular Structure	Module 12	Drawing Lewis Electron Dot Structures
		Module 13	Molecular Polarity
Chapter 9	Bonding and Molecular Structure: Orbital Hybridization and Molecular Orbitals	Module 14	Hybrid Atomic Orbitals
Chapter 10	Carbon: Not Just Another Element	Module 15	Naming Organic Compounds
Chapter 11	Gases and Their Properties	Module 16	Gas Laws and the Kinetic Molecular Theory
Chapter 12	Intermolecular Forces and Liquids	Module 17	Intermolecular Forces
Chapter 13	The Chemistry of Solids	Module 18	The Solid State
Chapter 14	Solutions and Their Behavior	Module 19	Colligative Properties
Chapter 15	Chemical Kinetics: The Rates of Chemical Reactions	Module 20	Chemical Kinetics
Chapter 16	Principles of Chemical Reactivity: Equilibria	Module 21	Chemical Equilibrium
Chapter 17	Principles of Chemical Reactivity: The Chemistry of Acids and Bases	Module 22	Equilibrium: pH of a Weak Acid
Chapter 18	Principles of Chemical Reactivity: Other Aspects of Aqueous Equilibria	Module 23	Understanding Acid–Base Buffers
Chapter 19	Principles of Chemical Reactivity: Entropy and Free Energy	Module 24	Gibbs Free Energy and Equilibrium
Chapter 20	Principles of Chemical Reactivity: Electron Transfer Reactions	Module 25	Oxidation–Reduction Reactions

Thank you for purchasing the hybrid edition of *Chemistry & Chemical Reactivity*. This trimmer version of the book includes no end-of-chapter questions. Instead, because you are using OWL (Online Web Learning), all end-of-chapter questions are assignable online in OWL, along with tutorials, mastery questions, simulations, visualizations, and the new, enhanced, and customizable Cengage *YouBook*. Your students can print questions from OWL with correct answers and valuable problem-solving feedback to help guide them to superior understanding and mastery of concepts.

John C. Kotz

The reduced page count in this book allows us to offer it at a lower price to students, as well as lessen the environmental impact of this edition and still provide a useful print product.

The authors of this book have many years of experience teaching general chemistry and other areas of chemistry at the college level. Although we have been at different institutions, both large and small, during our careers, we share several goals. One is to provide a broad overview of the principles of chemistry, the reactivity of the chemical elements and their compounds, and the applications of chemistry. To reach that goal with our students, we have tried to show the close relation between the observations chemists make of chemical and physical changes in the laboratory and in nature and the way these changes are viewed at the atomic and molecular level.

Another of our goals has been to convey a sense that chemistry not only has a lively history but is also dynamic, with important new developments occurring every year. Furthermore, we want to provide some insight into the chemical aspects of the world around us. Indeed, a major objective of this book is to provide the tools needed for you to function as a chemically literate citizen. Learning about the chemical world is just as important as understanding some basic mathematics and biology and as important as having an appreciation for history, music, and literature. For example, you should know what materials are important to our economy, some of the reactions in plants and animals and in our environment, and the role that chemists play in protecting the environment. In this regard, one growing area of chemistry, highlighted throughout this edition, is "green" or sustainable chemistry.

These goals have been translated into *Chemistry & Chemical Reactivity*, a book that has been used by more than 1 million students in its first seven editions. The first edition had a copyright date of 1987, and the copyright date for this edition is 2012. So, **this is the 25th anniversary of the book**. It is its silver (Ag) anniversary!

Looking back over the previous editions, we can see how the book has changed. There have been many new and exciting additions to the content of the book. In addition, there have been significant advances in the technology of communicating information, and we have tried to take advantage of those new approaches. A desire to make the book even better for our students has been the impetus behind the preparation of each new edition. With this edition, you will see a new approach to problem solving, new ways to describe contemporary uses of chemistry, new technologies, and improved integration with existing technologies.

Emerging Developments in Content Usage and Delivery: OWL, Go Chemistry®, and the Cengage *YouBook*

Our ongoing challenge as authors and educators is to use multimedia to engage students and to help them reach a higher level of conceptual understanding. More than 15 years ago we incorporated electronic media into this text with the first edition of our *Interactive General Chemistry* CD, a learning tool used by thousands of students worldwide.

As technology has advanced, we have made major changes in our integrated media program. Through several editions we redesigned the media so that students can interact with simulations, tutorials, active figures, and end-of-chapter questions, first through the *Interactive General Chemistry* CD and then with **OWL (Online Web Learning)**. OWL, which was developed at the University of Massachusetts, has been used by hundreds of thousands of students in the past few years.

More recently, we developed and integrated **Go Chemistry** tutorial videos into the seventh edition and more fully into this new edition. These tutorials are 5- to 10-minute mini lectures on topics such as solving equilibrium problems, features of the periodic table, naming compounds, polar molecules, writing net ionic equations, and identifying intermolecular forces.

What's New in This Edition

1. All **Example problems** in the book illustrate a **NEW approach to problem solving**. Each Example problem is broken down into the following categories: Problem, What Do You Know?, Strategy, Solution, Think About Your Answer, and Check Your Understanding. The "Check Your Understanding" questions are largely a revision of the Exercises from previous editions. Being included in the Example format should make them a more useful tool.

2. **NEW Interactive Examples in OWL** allow students to work approximately 70 examples from the book multiple times in slightly different versions to encourage thinking their way through the example instead of passively reading through to the solution.

3. At the end of almost every section in a chapter there are **NEW multiple choice Review & Check questions**. These are meant to be done in a few minutes to check the understanding of the section. In the Cengage *YouBook* these questions are interactive quizzes with feedback. (These questions could also be used in class by instructors to assess student understanding through electronic student response systems.)

4. **Strategy Maps** are a **NEW** feature of this edition. There are approximately 60 maps accompanying Example problems throughout the book. These are visual representations of the pathways to solving problems.

5. Except for Chapter 1, each chapter has **NEW extended Study Questions called *Applying Chemical Principles***. These help students apply principles learned across several chapters to real-world problems. Topics include the discovery of the noble gases, the discovery of elements on the sun, antacids, gunpowder, the rare earth elements, dating meteorites, and lighter-than-air ships.

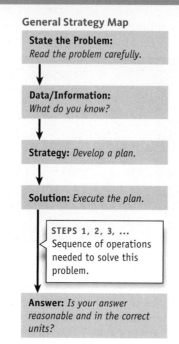

General Strategy Map

State the Problem: *Read the problem carefully.*

↓

Data/Information: *What do you know?*

↓

Strategy: *Develop a plan.*

↓

Solution: *Execute the plan.*

STEPS 1, 2, 3, … Sequence of operations needed to solve this problem.

Answer: *Is your answer reasonable and in the correct units?*

6. There are 2210 end-of-chapter **Study Questions** in the book in addition to the Check Your Understanding, Review & Check, and Applying Chemical Principles questions. Over 1900 of these questions are available in OWL, more than double the number of questions available in OWL in the previous edition.

7. Another **NEW** feature is a discussion of the Principles of **Green Chemistry** are noted in Chapter 1. This is followed by 10 articles on green chemistry throughout the book. See, for example, an explanation of atom economy (page 144), the synthesis of ibuprofen (page 301), and lithium and green cars (page 448). The development of this NEW feature was assisted by Professor Michael Cann of the University of Scranton, a green chemistry authority.

8. This edition of the book is also available as a Cengage *YouBook*, a digital textbook. This includes all the same content as the print book, but it also has clickable videos, animations, Guided Tours of figures with tutorials, three-dimensional molecular models, and quick quiz Review & Check questions. Those who choose the Cengage *YouBook* will also have access to strategy maps with audio/video explanations done by one of the authors or by Salman Khan, who has published hundreds of video tutorials on the Internet in many areas of science.

9. There are 10 **NEW chapter opening stories**. See, for example, essays on gold (page 1), energy and diet (page 172), rubies and sapphires (page 232), chocolate (page 338), and green cars (page 444).

10. A total of **17 NEW Case Studies** have been added. These include the story of Ötzi, the Iceman of the Alps (page 50); free radicals and hair dye (page 279); methane hydrates and the Gulf of Mexico oil spill (page 414); a pet food catastrophe (page 433); and exploding lakes and Diet Cokes (page 485).

11. *Reorganization/addition/revision of material:*
   - A short introduction to energy has been moved from Chapter 5 to Chapter 1, and the units used in thermochemistry are introduced in the *Let's Review* portion of Chapter 1. This will assist instructors who wish to use this book in an "atoms-first" approach.
   - The material on metallic bonding and semiconductors has been moved from the materials interchapter into the chapter on solid-state chemistry (Chapter 13).
   - A short discussion on activities has been added to the equilibrium chapter (Chapter 16).
   - Many of the illustrations have been updated and/or redone.
   - New Study Questions have been added to a number of the chapters.

In addition, an entirely new *digital textbook*—the Cengage *YouBook*—has been developed for this edition. The Cengage *YouBook* is a fully electronic, full-color version of the book with extensive interactivity. You can use it with your desktop or laptop computer, to read the text, access useful databases, watch videos of chemical reactions, take a Guided Tour of a book figure, and much more.

Audience for *Chemistry & Chemical Reactivity* and OWL

The textbook (both as a printed book and the Cengage *YouBook* digital version) and OWL are designed for introductory courses in chemistry for students interested in further study in science, whether that science is chemistry, biology, engineering, geology, physics, or related subjects. Our assumption is that students beginning this course have had some preparation in algebra and in general science. Although undeniably helpful, a previous exposure to chemistry is neither assumed nor required.

Philosophy and Approach of the *Chemistry & Chemical Reactivity* Program

We have had several major, but not independent, objectives since the first edition of the book. The first was to write a book that students would enjoy reading and that would offer, at a reasonable level of rigor, chemistry and chemical principles in a format and organization typical of college and university courses today. Second, we wanted to convey the utility and importance of chemistry by introducing the properties of the elements, their compounds, and their reactions as early as possible and by focusing the discussion as much as possible on these subjects. Finally, with the **Go Chemistry** modules and complete integration of **OWL,** we have incorporated electronic tools to bring students to a higher level of conceptual understanding.

Flame colors by salts of boron, sodium, and strontium.

© Cengage Learning/Charles D. Winters

The American Chemical Society has been urging educators to put "chemistry" back into introductory chemistry courses. We agree wholeheartedly. Therefore, we have tried to describe the elements, their compounds, and their reactions as early and as often as possible by:

- Bringing **material on the properties of elements and compounds** as early as possible into the Examples and Study Questions (and especially the *Applying Chemical Principles* questions) and to introduce new principles using realistic chemical situations.
- Using numerous **color photographs** of the elements and common compounds, of chemical reactions, and of common laboratory operations and industrial processes.
- Introducing each chapter with a **problem in practical chemistry**—for example, a short discussion of the energy in common foods or the source of lithium in car batteries—that is relevant to the chapter.
- Using numerous *Case Studies* and introducing new *Applying Chemical Principles* study questions that delve into practical chemistry.

General Organization of the Book

Chemistry & Chemical Reactivity has two broad themes: *Chemical Reactivity* and *Bonding and Molecular Structure*. The chapters on *Principles of Reactivity* introduce the fac-

tors that lead chemical reactions to be successful in converting reactants to products. Under this topic there is a discussion of common types of reactions, the energy involved in reactions, and the factors that affect the speed of a reaction. One reason for the enormous advances in chemistry and molecular biology in the last several decades has been an understanding of molecular structure. Therefore, sections of the book on *Principles of Bonding and Molecular Structure* lay the groundwork for understanding these developments. Particular attention is paid to an understanding of the structural aspects of such biologically important molecules as DNA.

Flexibility of Chapter Organization

A glance at the introductory chemistry texts currently available shows that there is a generally common order of topics used by educators. With only minor variations, we have followed that order. That is not to say that the chapters in our book cannot be used in some other order. We have written this book to be as flexible as possible. An example is the **flexibility of covering the behavior of gases** (Chapter 11). It has been placed with chapters on liquids, solids, and solutions (Chapters 12–14) because it logically fits with those topics. However, it can easily be read and understood after covering only the first four chapters of the book.

Similarly, chapters on atomic and molecular structure (Chapters 6–9) could be used in an **atoms-first approach** before the chapters on stoichiometry and common reactions (Chapters 3 and 4). To facilitate this, we have moved an introduction to energy and its units to Chapter 1.

Also, the chapters on chemical equilibria (Chapters 16–18) can be covered before those on solutions and kinetics (Chapters 14 and 15).

Organic chemistry (Chapter 10) is often left to one of the final chapters in chemistry textbooks. However, we believe the importance of organic compounds in biochemistry and in consumer products means that material should be presented earlier in the sequence of chapters. Therefore, it follows the chapters on structure and bonding because organic chemistry illustrates the application of models of chemical bonding and molecular structure. However, one can use the remainder of the book without including this chapter.

The order of topics in the text was also devised to introduce as early as possible the background required for the laboratory experiments usually performed in introductory chemistry courses. For this reason, chapters on chemical and physical properties, common reaction types, and stoichiometry begin the book. In addition, because an understanding of energy is so important in the study of chemistry, energy and its units are introduced in Chapter 1 and thermochemistry is introduced in Chapter 5.

Organization and Purposes of the Sections of the Book

Part One: The Basic Tools of Chemistry

Crystals of fluorite, CaF_2.

The basic ideas and methods that are the basis of all chemistry are introduced in Part One. Chapter 1 defines important terms, and the accompanying *Let's Review* section reviews units and mathematical methods. Chapter 2 introduces atoms, molecules, and ions, and the most important organizational device in chemistry, the periodic table. In Chapters 3 and 4 we begin to discuss the principles of chemical reactivity and to introduce the numerical methods used by chemists to extract quantitative information from chemical reactions. Chapter 5 is an introduction to the energy involved in chemical processes. The supplemental chapter *The Chemistry of Fuels and Energy Sources* follows Chapter 5 and uses the concepts developed in the preceding chapters.

Part Two: The Structure of Atoms and Molecules

The goal of this section is to outline the current theories of the arrangement of electrons in atoms (Chapters 6 and 7). This discussion is tied closely to the arrangement of elements in the periodic table and to periodic properties. In Chapter 8 we discuss the details of chemical bonding and the properties of these bonds. In addition, we show how to derive the three-dimensional structure of simple molecules. Finally, Chapter 9 considers the major theories of chemical bonding in more detail.

This part of the book is completed with a discussion of organic chemistry (Chapter 10), primarily from a structural point of view.

This section includes the interchapter on *Milestones in the Development of Chemistry and the Modern View of Atoms and Molecules*. It also includes *The Chemistry of Life: Biochemistry* to provide an overview of some of the most important aspects of biochemistry.

Part Three: States of Matter

The behavior of the three states of matter—gases, liquids, and solids—is described in Chapters 11–13. The discussion of liquids and solids is tied to gases through the description of intermolecular forces in Chapter 12, with particular attention given to liquid and solid water.

In Chapter 14 we describe the properties of solutions, intimate mixtures of gases, liquids, and solids.

Designing and making new materials with useful properties is one of the most exciting areas of modern chemistry. Therefore, this section includes the interchapter *The Chemistry of Modern Materials.*

Part Four: The Control of Chemical Reactions

This section is wholly concerned with the *Principles of Reactivity*. Chapter 15 examines the rates of chemical processes and the factors controlling these rates. Next, we move to Chapters 16–18, which describe chemical reactions at equilibrium. After an introduction to equilibrium in Chapter 16, we highlight the reactions involving acids and bases in water (Chapters 17 and 18) and reactions leading to slightly soluble salts (Chapter 18). To tie together the discussion of chemical equilibria, we again explore thermodynamics in Chapter 19. As a final topic in this section we describe in Chapter 20 chemical reactions involving the transfer of electrons and the use of these reactions in electrochemical cells.

The Chemistry of the Environment supplemental chapter is at the end of Part Four. This chapter uses ideas from kinetics and chemical equilibria, in particular, as well as principles described in earlier chapters in the book.

Part Five: The Chemistry of the Elements and Their Compounds

Although the chemistry of the various elements is described throughout the book, Part Five considers this topic in a more systematic way. Chapter 21 is devoted to the chemistry of the main group elements, whereas Chapter 22 is a discussion of the transition elements and their compounds. Finally, Chapter 23 is a brief discussion of nuclear chemistry.

Features of the Book

Several years ago a student of one of the authors, now an accountant, shared an interesting perspective with us. He said that, while general chemistry was one of his hardest subjects, it was also the most useful course he had taken because it taught him how to solve problems. We were gratified by this perspective. We have always thought that, for many students, an important goal in general chemistry was not only to teach students chemistry but also to help them learn critical thinking and problem-solving skills. Many of the features of the book are meant to support those goals.

Problem-Solving Approach: Organization and Strategy Maps

Worked-out examples are an essential part of each chapter. To better assist students in following the logic of a solution, these problems are now organized around the following outline:

PROBLEM

This is the statement of the problem.

WHAT DO YOU KNOW?

You outline what information you have and begin to think about a solution.

STRATEGY

You combine the information available with the objective to devise a pathway to solution.

SOLUTION

You work through the steps, both logical and mathematical, to the answer.

THINK ABOUT YOUR ANSWER

You ask if the answer is reasonable or what it means.

CHECK YOUR UNDERSTANDING

This is a similar problem for you to try. Solutions to the problems are in Appendix O.

For many problems, a **strategy map** can be a useful tool in solving the problem. For example, on pages 38–40, we ask how thick the oil layer would be if you spread a given mass of oil on the surface of water in a dish. The density of the oil is also given. To help see the logic of the problem, the Example is accompanied by the strategy map given here.

There are approximately 60 strategy maps in the book accompanying Example problems. Many of the strategy maps in the Cengage *YouBook* digital textbook are accompanied by an audio explanation.

Review & Check: New Quick Review Questions

We have also added new, multiple choice questions at the end of almost every section. Students can check their understanding of the section by attempting these brief questions. Answers to the questions are in

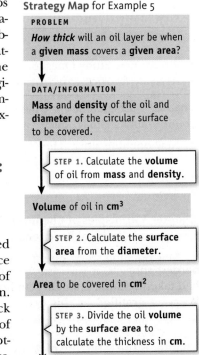

Strategy Map for Example 5

PROBLEM

How thick will an oil layer be when a **given mass** covers a **given area?**

DATA/INFORMATION

Mass and **density** of the oil and **diameter** of the circular surface to be covered.

STEP 1. Calculate the **volume** of oil from **mass** and **density**.

Volume of oil in **cm³**

STEP 2. Calculate the **surface area** from the **diameter**.

Area to be covered in **cm²**

STEP 3. Divide the oil **volume** by the **surface area** to calculate the thickness in **cm**.

Thickness of oil layer in **cm**.

Appendix P. In the Cengage *YouBook* digital textbook the questions are clickable so you can quickly check to see if you know the correct answer.

Chapter Goals/Revisited

The learning goals for each chapter are listed on the first page of each chapter and then are revisited on the last page. There the revisited goals are given in more detail. These goals are of great use in studying. Students can go through the goals and ask themselves if they have met each one. Furthermore, specific end-of-chapter Study Questions are listed that help students determine if they have met those goals.

End-of-Chapter Study Questions

There are 50 to over 100 Study Questions for each chapter. They are grouped as follows:

Practicing Skills

These questions are grouped by the topic covered by the questions.

General Questions

There is no indication regarding the type of question.

In the Laboratory

These are problems that may be encountered in a laboratory experiment on the chapter material.

Summary and Conceptual Questions

These questions use concepts from the current chapter as well as preceding chapters.

Applying Chemical Principles

These questions are preceded by a short description giving the background necessary to address the problem.

Study Questions have been available in the OWL Online Web Learning system for the last two editions. In this edition, we have more than doubled the number available. The OWL system now has over 1900 of the roughly 2200 Study Questions in the book.

Finally, note that some questions are marked with a small red triangle (▲). These are meant to be somewhat more challenging than other questions.

Boxed Essays

As in the seventh edition, we continue to include boxed essays titled *A Closer Look* (for a more in-depth look at relevant material), and *Problem Solving Tips*. We have added or revised a number of the *Case Studies*, some of which deal with "green" or sustainable chemistry.

A Word About the Cengage *YouBook*

The Cengage *YouBook* is an interactive digital version of the complete book and retains the paging integrity of the printed textbook. Either the digital version or the printed version of the book can be used in class.

The Cengage *YouBook* has clickable videos of reactions, Guided Tours of book figures, three-dimensional molecular models, searchable databases of chemical information, and the end-of-section Review & Check questions have clickable answers.

A unique feature of the Cengage *YouBook* is the **audio versions of the strategy maps**. For many of the maps in the book, one of the book authors will give a step-by-step audio explanation of the Example problem and explain some of the details of the strategy and solution.

Some of the audio strategy maps have been done by Salman Khan, who has recently been recognized around the world for his online tutorials, not only in chemistry but also in biology, linear algebra, geometry, trigonometry, statistics, pre-calculus, economics, money and banking, finance, and others. The web address for these tutorials is

www.khanacademy.org

All of Khan's tutorials are free and are on YouTube. The Cengage *YouBook* will link to those videos that Mr. Khan has done specifically for *Chemistry & Chemical Reactivity*. However, students should also explore all of the others for chemistry and for other areas of interest.

Salman Khan of the Khan Academy.

Changes for the Eighth Edition

Significant additions to the book, such as a new problem-solving format, strategy maps, essays on green chemistry, and *Applying Chemical Principles* problems, have been outlined in the section on "What's New." In addition, we have produced many new photos and new illustrations and have continually tried to improve the writing throughout. The following chapter-by-chapter listing indicates more specific changes from the seventh edition of the book to this edition.

Chapter 1 Basic Concepts of Chemistry

- New opening story: Gold!
- New opening section: we now outline the forensic investigation of the Iceman of the Alps.
- New *Closer Look:* Careers in Chemistry. Features a former student who is now a forensic chemist.
- New Section 1.2, Sustainability and Green Chemistry. Green chemistry is a theme used throughout the book.
- New *Closer Look:* Principles of Green Chemistry.
- New *Closer Look:* Element Names and Symbols.
- New Section 1.8: Energy: Some Basic Principles. Introduction to energy moved into this chapter from Chapter 5.
- New *Case Study:* CO_2 in the Oceans.
- Twelve new or revised Study Questions (out of 46).

Let's Review: The Tools of Quantitative Chemistry

- New: Energy units introduced. (This was in Chapter 5 in the seventh edition.)

- Strategy maps introduced into the book for the first time.
- Six new or revised Study Questions (out of 67).

Chapter 2 Atoms, Molecules, and Ions

- New *Case Study:* Using Isotopes: Ötzi, the Iceman of the Alps.
- New *Case Study:* Mummies, Bangladesh, and the Formula of Compound 606.
- Fourteen new Study Questions (out of 165).
- *Applying Chemical Principles:* Argon—An Amazing Discovery.

Chapter 3 Chemical Reactions

- New figure (Figure 3.9) to predict the species present in aqueous solution.
- Updated *Closer Look:* Sulfuric Acid.
- Twelve new Study Questions (out of 93).
- *Applying Chemical Principles:* Superconductors.

Chapter 4 Stoichiometry: Quantitative Information about Chemical Reactions

- New opening story: The Chemistry of Pyrotechnics.
- New *Case Study:* Green Chemistry and Atom Economy.
- Three new Study Questions (out of 139).
- *Applying Chemical Principles:* Antacids.

Chapter 5 Principles of Chemical Reactivity: Energy and Chemical Reactions

- New opening story: Energy and Your Diet.
- Section on basic principles of heat (pp. 209–211 in seventh edition) moved to Chapter 1 Let's Review.
- Ten new Study Questions (out of 110).
- *Applying Chemical Principles:* Gunpowder.

Chapter 6 The Structure of Atoms

- New opening story: Fireworks.
- Boxed essay on orbitals rewritten and slightly expanded.

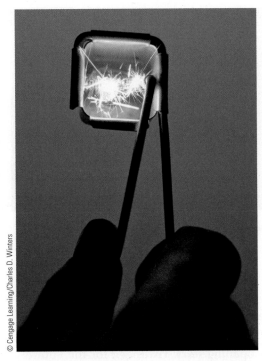

A device used to ignite gas burners. Depends on rare earth elements.

- Section on introduction to wave mechanics rewritten.
- Three new Study Questions (out of 84).
- *Applying Chemical Principles:* Chemistry of the Sun.

Chapter 7 The Structure of Atoms and Periodic Trends

- New opening story: Rubies and Sapphires—Pretty Stones.
- Rewritten discussion of effective nuclear charge with new figures.
- Expanded discussion of configurations of transition metal electron configurations, especially Cr and Cu.
- New *Closer Look:* Orbital Energies, Z^*, and Electron Configurations.
- Clarified relation of electron attachment enthalpy and electron affinity.
- *Applying Chemical Principles:* The Not-So-Rare Earths.

Chapter 8 Bonding and Molecular Structure

- New *Closer Look:* A Scientific Controversy—Are There Double Bonds in Sulfate and Phosphate Ions?
- New *Case Study:* Hydroxyl Radicals, Atmospheric Chemistry, and Hair Dyes.
- Electrostatic potential maps were introduced in the seventh edition. We have slightly enlarged the use of these figures for this edition and are using the industry standard software to create them.
- New *Case Study:* Ibuprofen, A Study in Green Chemistry.
- Eleven new or revised Study Questions (out of 96).
- *Applying Chemical Principles:* Linus Pauling and Electronegativity.

Chapter 9 Bonding and Molecular Structure: Orbital Hybridization and Molecular Orbitals

- Updated discussion of molecular orbital theory.
- New *Case Study:* Green Chemistry, Safe Dyes, and Molecular Orbitals.
- New *Closer Look:* Three-Center Bonds and Hybrid Orbitals with *d* Orbitals.
- Thirteen new or revised Study Questions (out of 80).
- *Applying Chemical Principles:* Probing Molecules with Photoelectron Spectroscopy.

Chapter 10 Carbon: Not Just Another Element

- New opening story: The Food of the Gods.
- New *Case Study:* An Awakening with L-DOPA.
- New *Closer Look:* Copolymers and Engineering Plastics for Lego Bricks and Tattoos.
- New *Closer Look:* Green Chemistry: Recycling PET.
- New *Case Study:* Green Adhesives.
- Deleted boxed essays on fats and oils, biofuels, dyes, and "super diapers," but some information on fats and oils was incorporated into the biochemistry interchapter.
- Fifteen new or revised Study Questions (out of 109).
- *Applying Chemical Principles:* Biodiesel—An Attractive Fuel for the Future?

Chapter 11 Gases and Their Properties

- New *Closer Look:* SCUBA Diving—An Application of the Gas Laws.
- New *Case Study:* What to Do with All of That CO_2? More on Green Chemistry.
- Six new or rewritten Study Questions (out of 108).
- *Applying Chemical Principles:* The Goodyear Blimp.

White phosphorus.

Chapter 12 Intermolecular Forces and Liquids

- New opening story: Geckos Can Climb Up der Waals.
- New *Case Study:* Hydrogen Bonding & Methane Hydrates: Opportunities and Problems. This is about the 2010 oil release in the Gulf of Mexico.
- New *Case Study:* A Pet Food Catastrophe.
- New *Closer Look:* Supercritical CO_2 and Green Chemistry.
- Seven new or rewritten Study Questions (out of 67).
- *Applying Chemical Principles:* Chromatography.

Chapter 13 The Chemistry of Solids

- New opening story: Lithium and "Green Cars."
- New *Case Study:* High-Strength Steels and Unit Cells.
- New Section 13.3, Bonding in Metals and Semiconductors. This was moved to Chapter 13 from the Interchapter on materials.
- New *Case Study:* Graphene—The Hottest New Network Solid.
- Eighteen new or rewritten Study Questions (out of 62).
- *Applying Chemical Principles:* Tin Disease.

Chapter 14 Solutions and Their Behavior

- New opening story: Survival at Sea.
- New *Case Study:* Exploding Lakes and Diet Cokes.
- New *Closer Look:* Reverse Osmosis for Pure Water.
- Four new or rewritten Study Questions (out of 106).
- *Applying Chemical Principles:* Distillation.

Chapter 15 Chemical Kinetics: The Rates of Chemical Reactions

- Four new Study Questions (out of 88).
- *Applying Chemical Principles:* Kinetics and Mechanisms: A 70-Year-Old Mystery Solved.

Chapter 16 Principles of Chemical Reactivity: Equilibria

- New *Closer Look:* Activities and Units of K.
- Eight new or rewritten Study Questions (out of 72).
- *Applying Chemical Principles:* Trivalent Carbon.

Chapter 17 Principles of Chemical Reactivity: The Chemistry of Acids and Bases

- New *Case Study:* Would You Like Some Belladonna Juice in Your Drink?
- Rearranged the sections on Polyprotic Acids and Bases (Section 17.9) and Molecular Structure, Bonding, and Acid–Base Behavior (Section 17.10).
- Some of the 121 Study Questions were reorganized.
- *Applying Chemical Principles:* The Leveling Effect, Nonaqueous Solvents, and Superacids.

Chapter 18 Principles of Chemical Reactivity: Other Aspects of Aqueous Equilibria

- New opening story: Nature's Acids.
- New *Case Study:* Chemical Equilibria in the Oceans.
- Several new and modified Study Questions (out of 112).

Samples of cobalt metal.

- *Applying Chemical Principles:* Everything That Glitters. . . .

Chapter 19 Principles of Chemical Reactivity: Entropy and Free Energy

- New opening story: Hydrogen for the Future?
- Given the key role of thermodynamics in chemistry, this chapter was carefully revised. Sections on free energy were reorganized.
- Three modified or new Study Questions (out of 84).
- *Applying Chemical Principles:* Are Diamonds Forever?

Chapter 20 Principles of Chemical Reactivity: Electron Transfer Reactions

- New opening story: Battery Power.
- Problem Solving Tip on balancing equations for reactions in basic solution was revised.
- Fourteen new Study Questions (out of 103).
- *Applying Chemical Principles:* Sacrifice!

Chapter 21 The Chemistry of the Main Group Elements

- Six new Study Questions (out of 106).
- *Applying Chemical Principles:* van Arkel Triangles and Bonding.

Chapter 22 The Chemistry of the Transition Elements

- *Applying Chemical Principles:* Green Catalysts.

Chapter 23 Nuclear Chemistry

- The *Closer Look:* The Search for New Elements was updated with the newest discoveries.
- Study Questions were reorganized.
- *Applying Chemical Principles:* The Age of Meteorites.

Alternate Versions
Cengage *YouBook* with OWL

The Cengage *YouBook* is a Flash-based, interactive, and customizable eBook version available with OWL. The instructor text edit feature allows for modification of the narrative by adding notes, re-ordering entire sections and chapters, and hiding content. Additional assets include animated figures, video clips, and student highlighting and note tools. See the OWL description below for more details.

Chemistry & Chemical Reactivity, Eighth Edition, Hybrid Version with OWL
ISBN-10: 1-111-57498-7; ISBN-13: 978-111-57498-7

This briefer version of *Chemistry & Chemical Reactivity* does not contain the end-of-chapter problems, which can be assigned in OWL. Access to OWL and the Cengage *YouBook* is packaged with the hybrid version.

Supporting Materials

☾WL
OWL for General Chemistry

Instant Access OWL with Cengage *YouBook* (6 months)
ISBN-10: 1-111-30524-2; ISBN-13: 978-1-111-30524-6

Instant Access OWL with Cengage *YouBook* (24 months)
ISBN-10: 1-111-30521-8; ISBN-13: 978-1-111-30521-5

By Roberta Day and Beatrice Botch of the University of Massachusetts, Amherst, and William Vining of the State University of New York at Oneonta.

OWL Online Web Learning offers more assignable, gradable content (including end-of-chapter questions specific to this textbook) and more reliability and flexibility than any other system. OWL's powerful course management tools allow instructors to control due dates, number of attempts, and whether students see answers or receive feedback on how to solve problems. OWL includes the Cengage *YouBook*, a Flash-based eBook that is interactive and customizable. It features a text edit tool that allows instructors to modify the textbook narrative as needed. With the Cengage *YouBook*, instructors can quickly re-order entire sections and chapters or hide any content they don't teach to create an eBook that perfectly matches their syllabus. Instructors can further customize the Cengage *YouBook* by publishing web links. Additional media assets include animated figures, video clips, highlighting, notes, and more.

Developed by chemistry instructors for teaching chemistry, OWL is the only system specifically designed to support **mastery learning**, where students work as long as they need to master each chemical concept and skill. OWL has already helped hundreds of thousands of students master chemistry through a wide range of assignment types, including tutorials, interactive simulations, and algorithmically generated homework questions that provide instant, answer-specific feedback.

OWL is continually enhanced with online learning tools to address the various learning styles of today's students such as:

- **Quick Prep** review courses that help students learn essential skills to succeed in General and Organic Chemistry
- **Jmol** molecular visualization program for rotating molecules and measuring bond distances and angles
- **Go Chemistry®** mini video lectures on key concepts that students can play on their computers or download to their video iPods, smart phones, or personal video players
- For this text, OWL includes **How Do I Solve It** problem-solving exercises, new **Interactive Example** assignments, as well as parameterized end-of-chapter questions and **Student Self Assessment** questions

In addition, when you become an OWL user, you can expect service that goes far beyond the ordinary. For more information or to see a demo, please contact your Cengage Learning representative or visit us at **www.cengage.com/owl.**

FOR THE INSTRUCTOR

Supporting instructor materials are available to qualified adopters. Please consult your local Cengage Learning, Brooks/Cole representative for details. Visit **login.cengage.com** and search for this book to access the Instructor's Companion Site, where you can

- See samples of materials
- Request a sample copy
- Locate your local representative
- Download digital files of the ExamView test bank and other helpful materials for instructors and students

PowerLecture Instructor's CD/DVD Package with JoinIn® and ExamView®
ISBN-10: 1-111-42719-4; ISBN-13: 978-1-111-42719-1

This digital library and presentation tool that includes:

- **PowerPoint® lecture slides**, written specially for *Chemistry & Chemical Reactivity,* which instructors can customize by importing their own lecture slides or other materials.
- Image libraries that contain **digital files for all text art, most photographs, all numbered tables, and multimedia animations** in a variety of digital formats. These files can be used to print transparencies, create your own PowerPoint slides, and supplement your lectures.
- Digital files of the complete **Instructor's Resource Manual** and **ExamView test bank.**
- Sample chapters from the **Student Solutions Manual** and **Study Guide.**
- **ExamView testing software** that enables you to create, deliver, and customize tests using the more than

1250 test bank questions written specifically for this text by David Treichel, Nebraska Wesleyan University.

- **JoinIn student response (clicker) questions** written for this book for use with the classroom response system of the instructor's choice.

Instructor's Companion Site

Go to **login.cengage.com** and search for this book to access the Instructor's Companion site, which has resources such as a Blackboard version of ExamView.

Instructor's Resource Manual

by John Vincent, The University of Alabama

ISBN-10: 1-111-42697-X; ISBN-13: 978-1-111-42697-2

Available both on the PowerLecture Instructor's Resource DVD and in print, this comprehensive resource contains worked-out solutions to *all* end-of-chapter Study Questions and features ideas for instructors on how to fully utilize resources and technology in their courses. It provides questions for electronic response (clicker) systems, suggests classroom demonstrations, and emphasizes good and innovative teaching practices.

Transparencies

ISBN-10: 1-111-57489-8; ISBN-13: 978-1-111-57489-5

A collection of 150 full-color transparencies of key images selected from the text by the authors. The PowerLecture Instructor's Resource DVD also gives instructors access to all text art and many photos to help in preparing transparencies for material not present in this set.

FOR THE STUDENT

Visit CengageBrain.com

At **www.cengagebrain.com** you can access additional course materials as well as purchase Cengage products, including those listed below. Search by ISBN using the list below or find this textbook's ISBN on the back cover of your book. Instructors can log in at **login.cengage.com**.

Student Companion Site

This site includes a glossary, flashcards, an interactive periodic table, and samples of the Study Guide and Student Solutions Manual, which are all accessible from **www.cengagebrain.com**.

ᴕWL
Quick Prep for General Chemistry
Instant Access OWL Quick Prep for General Chemistry (90 days)
ISBN-10: 0-495-56030-8; ISBN-13: 978-0-495-56030-2

Quick Prep is a self-paced online short course that helps students succeed in general chemistry. Students who completed Quick Prep through an organized class or self-study averaged almost a full letter grade higher in their subsequent general chemistry course than those who did not. Intended to be taken prior to the start of the semester, Quick Prep is appropriate for both underprepared students and for students who seek a review of basic skills and concepts. Quick Prep features an assessment quiz to focus students on the concepts they need to study to be prepared for general chemistry. Quick Prep is approximately 20 hours of instruction delivered through OWL with no textbook required and can be completed at any time in the student's schedule. Professors can package a printed access card for Quick Prep with the textbook or students can purchase instant access at **www.cengagebrain.com**. To view an OWL Quick Prep demonstration and for more information, visit **www.cengage.com/chemistry/quickprep**.

Go Chemistry® for General Chemistry
ISBN-10: 0-495-38228-0; ISBN-13: 978-0-495-38228-7

Pressed for time? Miss a lecture? Need more review? Go Chemistry for General Chemistry is a set of 27 downloadable mini video lectures, accessible via the printed access card packaged with your textbook or available for purchase separately. Developed by one of this book's authors, Go Chemistry helps you quickly review essential topics—whenever and wherever you want! Each video contains animations and problems and can be downloaded to your computer desktop or portable video player (e.g., iPod or iPhone) for convenient self-study and exam review. Selected Go Chemistry videos have e-Flashcards to briefly introduce a key concept and then test student understanding with a series of questions. The Cengage *YouBook* in OWL contains Go Chemistry. Professors can package a printed access card for Go Chemistry with the textbook. Students can enter the

ISBN above at www.cengagebrain.com to download two free videos or to purchase instant access to the 27-video set or individual videos.

CengageBrain.com App

Now students can prepare for class anytime and anywhere using the CengageBrain.com application developed specifically for the Apple iPhone® and iPod touch®, which allows students to access free study materials—book-specific quizzes, flashcards, related Cengage Learning materials and more—so they can study the way they want, when they want to . . . even on the go. For more information about this complementary application, please visit www.cengagebrain.com. Available on the iTunes App Store.

Student Solutions Manual

by Alton J. Banks, North Carolina State University

ISBN-10: 1-111-42698-8; ISBN-13: 978-1-111-42698-9

Improve your performance at exam time with this manual's detailed solutions to the blue-numbered end-of-chapter Study Questions found in the text. This comprehensive guide helps you develop a deeper intuitive understanding of chapter material through constant reinforcement and practice. Solutions match the problem-solving strategies used in the text. Sample chapters are available for review on the PowerLecture Instructor's DVD and on the student companion website, which is accessible from www.cengagebrain.com.

Study Guide

by Michael J. Moran and John R. Townsend, West Chester University of Pennsylvania

ISBN-10: 1-111-42699-6; ISBN-13: 978-1-111-42699-6

With learning tools explicitly linked to the Chapter Goals introduced in each chapter, this guide helps ensure that you are well prepared for class and exams. It includes chapter overviews, key terms and definitions, expanded commentary, study tips, worked-out examples, and direct references back to the text. Sample chapters are available for review on the student companion website, which is accessible from www.cengagebrain.com.

Essential Math for Chemistry Students, Second Edition

by David W. Ball, Cleveland State University

ISBN-10: 0-495-01327-7; ISBN-13: 978-0-495-01327-3

This short book is intended to help you gain confidence and competency in the essential math skills you need to succeed in general chemistry. Each chapter focuses on a specific type of skill and has worked-out examples to show how these skills translate to chemical problem solving. The book includes references to the OWL learning system where you can access online algebra skills exercises.

Survival Guide for General Chemistry with Math Review, Second Edition

by Charles H. Atwood, University of Georgia

ISBN-10: 0-495-38751-7; ISBN-13: 978-0-495-38751-0

Intended to help you practice for exams, this survival guide shows you how to solve difficult problems by dissecting them into manageable chunks. The guide includes three levels of proficiency questions—A, B, and minimal—to quickly build confidence as you master the knowledge you need to succeed in your course.

For the Laboratory
CENGAGE LEARNING Brooks/Cole Lab Manuals

Cengage Learning offers a variety of printed manuals to meet all general chemistry laboratory needs. Visit www.cengage.com/chemistry for a full listing and description of these laboratory manuals and laboratory notebooks. All of our lab manuals can be customized for your specific needs.

Signature Labs . . . for the Customized Laboratory

Signature Labs is Cengage Learning's digital library of tried-and-true labs that help you take the guesswork out of running your chemistry laboratory. Select just the experiments you want from hundreds of options and approaches. Provide your students with only the experiments they will conduct and know you will get the results you seek. Visit www.signaturelabs.com to begin building your manual today.

ACKNOWLEDGMENTS

Preparing this new edition of *Chemistry & Chemical Reactivity* took over two years of continuous effort. As in our work on the first seven editions, we have had the support and encouragement of our families and of wonderful friends, colleagues, and students.

A Special Acknowledgment: John Vondeling

The publication of this book, with its copyright of 2012, marks its 25th anniversary. The first edition had a copyright date of 1987 and was the result of a collaboration with our publisher, the late John Vondeling of Saunders College Publishing. John literally invented many of the practices now considered standard among companies publishing college science textbooks. He was responsible for publishing some of the classic textbooks in chemistry, physics, and astronomy. The success of this book in its various editions owes much to John. It may not have become a reality without his confidence that it would be a worthwhile addition to the library of science texts. He was also a good friend, and we miss his humanity, his sense of humor, and his colorful personality.

CENGAGE LEARNING Brooks/Cole

The seventh edition of this book was published by the Brooks/Cole group of Cengage Learning, and we continue with the same excellent team we have had in place for the previous several years.

The seventh edition of the book was very successful, in large part owing to the work of Lisa Lockwood as our executive editor. She has an excellent sense of the market and has helped guide the new edition.

Peter McGahey has been our the Development Editor since he joined us to work on the fifth edition. Peter is blessed with energy, creativity, enthusiasm, intelligence, and good humor. He is a trusted friend and confidant who cheerfully answers our many questions during almost-daily phone calls.

No book can be successful without proper marketing. Nicole Hamm was a great help in marketing the seventh edition, and she is back in that role for this edition. She is knowledgeable about the market and has worked tirelessly to bring the book to everyone's attention.

Our team at Brooks/Cole is completed with Teresa Trego, Production Manager; Lisa Weber and Stephanie VanCamp, Media Editors; Julie Stefani, Marketing Coordinator; and Elizabeth Woods, Assistant Editor. Schedules are very demanding in textbook publishing, and Teresa has helped to keep us on schedule. We certainly appreciate her organizational skills. Lisa was involved in the development of the Go Chemistry modules and our expanded use of OWL. Stephanie handles the Cengage *YouBook*, and Liz also oversees the development of the ancillaries, such as the PowerPoint notes that accompany text.

Megan Greiner of Graphic World Inc. guided the book through its almost year-long production, and Andy Vosburgh of that company was enormously helpful in getting us through the use of new software. Scott Rosen of the Bill Smith Studio directed the photo research for the book and was successful in filling our sometimes offbeat requests for particular photos.

Photography, Art, and Design

Most of the color photographs for this edition were beautifully created by Charles D. Winters. He produced several dozen new images for this book, always with a creative eye. Charlie's work gets better with each edition. We have worked with Charlie for more than 20 years and have become close friends. We listen to his jokes, both new and old—and always forget them.

When the fifth edition was being planned, we brought in Patrick Harman as a member of the team. Pat designed the first edition of our *Interactive General Chemistry* CD (published in the 1990s), and we believe its success is in no small way connected to his design skill. For the fifth, sixth, and seventh editions of the book, Pat went over almost every figure, and almost every word, to bring a fresh perspective to ways to communicate chemistry. Once again he has worked on designing and producing new illustrations for this edition, and his creativity is obvious in their clarity. In addition, he has been an enormous help in designing and producing media for the *YouBook*. Pat has also become a good friend, and we share interests not only in books but in music.

Other Collaborators

We have been fortunate to have a number of other colleagues who have played valuable roles in this project.

- Bill Vining (State University of New York, Oneonta) was a co-author of the *Interactive GeneralChemistry* CD and has authored many of the media assets in OWL. He has been a friend for many years and recently took the place of one of the authors on the faculty at SUNY–Oneonta. Bill has again applied his considerable energy and creativity in preparing many more OWL questions with tutorials and some of the assets in the Cengage *YouBook*.

- Barbara Stewart (University of Maine) authored the new Interactive Examples in OWL.

- Alton Banks (North Carolina State University) has also been involved for a number of editions preparing the *Student Solutions Manual*. Alton has been very helpful in ensuring the accuracy of the Study Question answers in the book, as well as in their respective manuals.

- Michael Moran (West Chester University) has again updated and revised the *Study Guide* for this text. Our textbook has had a history of excellent study guides, and this manual follows that tradition.

- Jay Freedman was the development editor for the first edition of the book and played an important role in its success. For several editions of the book Jay has also done a masterful job compiling the index/glossary for this edition.

- Donnie Byers (Johnson County Community College) has been a long-time user of the book and a member of our Advisory Board. For this edition she coordinated the revisions of the end-of-chapter changes for the international edition.

- David Treichel (Nebraska Wesleyan University) wrote the *Applying Chemical Principles* problems, did the accuracy review, and developed the test bank.

- John Vincent (The University of Alabama–Tuscaloosa) wrote the *Instructor's Resource Manual*, did several reviews, and did an accuracy review.

- Barbara Mowrey, York College of Pennsylvania, also was an accuracy reviewer of the page proofs.

- Michael C. Cann (University of Scranton) helped identify ways in which green chemistry content could be incorporated.

- Salman Khan worked with us on developing audio/video tutorials of some of the book's Example problems. He is making significant contributions to education in the United States and elsewhere. See his website (**www.khanacademy.org**) for the extensive list of free tutorials he has done in all fields of science as well as in mathematics, economics, and finance.

- John Emsley (University of Cambridge) wrote the interchapter on the history of the development of chemistry.

- Jeffrey J. Keaffaber (University of Florida) wrote the Case Study *A Healthy Aquarium and the Nitrogen Cycle* in Chapter 21.

- Eric Scerri (University of California, Los Angeles) wrote the *A Closer Look: The Story of the Periodic Table* in Chapter 2.

Reviewers for the Eighth Edition

As we began to develop the new edition we had helpful input from the following reviewers:

- Gerald M. Korenowski, Rensselaer Polytechnic Institute

- Robert LaDuca, Michigan State University

- Jeffrey Mack, California State University, Sacramento

- Armando Rivera, East Los Angeles College

- Daniel Williams, Kennesaw State University

- Steven Wood, Brigham Young University

- Roger A. Hinrichs, Weill Cornell Medical College in Qatar, reviewed the energy interchapter

- Leonard Interrante, Rensselaer Polytechnic Institute, reviewed the materials interchapter

- Trudy E. Thomas-Smith, SUNY College at Oneonta, reviewed the environment interchapter

- John Vincent, The University of Alabama, reviewed the environment interchapter

Advisory Board for the Eighth Edition

As the new edition was being planned, this board listened to some of our ideas and made other suggestions. We hope to continue our association with these energetic and creative chemical educators.

- Donnie Byers, Johnson County Community College

- Elizabeth Dorland, Washington University of St. Louis

- Michael Finnegan, Washington State University

- Greg Gellene, Texas Tech University

- Milton Johnson, University of South Florida

- Jeffrey Mack, California State University, Sacramento

- Sara McIntosh, Rensselaer Polytechnic Institute

- MaryKay Orgill, University of Nevada, Las Vegas

- Don Siegel, Rutgers University

- Eric Simanek, Texas A&M University

Reviewers for the Seventh Edition

- Gerald M. Korenowski, Rensselaer Polytechnic Institute

- Robert L. LaDuca, Michigan State University

- Jeffrey Alan Mack, California State University, Sacramento

- Armando M. Rivera-Figueroa, East Los Angeles College

- Daniel J. Williams, Kennesaw State University

- Steven G. Wood, Brigham Young University

- Roger A. Hinrichs, Weill Cornell Medical College in Qatar (reviewed the Energy interchapter)

- Leonard Fine, Columbia University (reviewed the Materials interchapter)

Advisory Board for the Seventh Edition

As the seventh edition was being planned, this board listened to some of our ideas and made other suggestions. We hope to continue our association with these energetic and creative chemical educators.

- Donnie Byers, Johnson County Community College

- Sharon Fetzer Gislason, University of Illinois, Chicago

- Adrian George, University of Nebraska

- George Grant, Tidewater Community College, Virginia Beach Campus

- Michael Hampton, University of Central Florida

- Milton Johnston, University of South Florida

- Jeffrey Alan Mack, California State University, Sacramento

- William Broderick, Montana State University

- Shane Street, University of Alabama

- Martin Valla, University of Florida

John C. Kotz, a State University of New York Distinguished Teaching Professor, Emeritus, at the College at Oneonta, was educated at Washington and Lee University and Cornell University. He held National Institutes of Health post-doctoral appointments at the University of Manchester Institute for Science and Technology in England and at Indiana University. He has coauthored three text-books in several editions (*Inorganic Chemistry, Chemistry & Chemical Reactivity,* and *The Chemical World*) and the *General ChemistryNow CD.* His research in inorganic chemistry and electrochemistry also has been published. He was a Fulbright Lecturer and Research Scholar in Portugal in 1979 and a Visiting Professor there in 1992. He was also a Visiting Professor at the Institute for Chemical Education (University of Wisconsin, 1991–1992), at Auckland University in New Zealand (1999), and at Potchefstroom University in South Africa in 2006. He has been an invited speaker on chemical education at conferences in South Africa, New Zealand, and Brazil. He also served four years as a mentor for the U.S. National Chemistry Olympiad Team. He has received several awards, among them a State University of New York Chancellor's Award (1979), a National Catalyst Award for Excellence in Teaching (1992), the Estee Lecturership at the University of South Dakota (1998), the Visiting Scientist Award from the Western Connecticut Section of the American Chemical Society (1999), the Distinguished Education Award from the Binghamton (NY) Section of the American Chemical Society (2001), the SUNY Award for Research and Scholarship (2005), and the Squibb Lecturership in Chemistry at the University of North Carolina-Asheville (2007). He may be contacted by email at kotzjc@oneonta.edu.

Left to right: Paul Treichel, John Townsend, and John Kotz.

Paul M. Treichel, received his B.S. degree from the University of Wisconsin in 1958 and a Ph.D. from Harvard University in 1962. After a year of postdoctoral study in London, he assumed a faculty position at the University of Wisconsin–Madison. He served as department chair from 1986 through 1995 and was awarded a Helfaer Professorship in 1996. He has held visiting faculty positions in South Africa (1975) and in Japan (1995). Retiring after 44 years as a faculty member in 2007, he is currently Emeritus Professor of Chemistry. During his faculty career he taught courses in general chemistry, inorganic chemistry, organometallic chemistry, and scientific ethics. Professor Treichel's research in organometallic and metal cluster chemistry and in mass spectrometry, aided by 75 graduate and under-graduate students, has led to more than 170 papers in scientific journals. He may be contacted by email at treichel@chem.wisc.edu.

John R. Townsend, Professor of Chemistry at West Chester University of Pennsylvania, completed his B.A. in Chemistry as well as the Approved Program for Teacher Certification in Chemistry at the University of Delaware. After a career teaching high school science and mathematics, he earned his M.S. and Ph.D. in biophysical chemistry at Cornell University, where he also received the DuPont Teaching Award for his work as a teaching assistant. After teaching at Bloomsburg University, he joined the faculty at West Chester University, where he coordinates the chemistry education program for prospective high school teachers and the general chemistry lecture program for science majors. He has been the university supervisor for more than 50 prospective high school chemistry teachers during their student teaching semester. His research interests are in the fields of chemical education and biochemistry. He may be contacted by email at jtownsend@wcupa.edu.

The cover photograph, titled "Save Our Earth, Let's Go Green," is from the research of Professor Joanna Aizenberg, Boaz Pokroy, and Sung Hoon Kang. Professor Aizenberg holds a joint appointment at Harvard University in the Department of Chemistry and Chemical Biology and the Department of Materials Science. This electron microscope photograph, showing hairlike fibers of epoxy resin assembling around a 2-μm polystyrene sphere, was the first place winner in the 2009 International Science and Engineering Visualization Challenge sponsored by *Science* magazine.

For this study, these scientists created a regular array of hairlike epoxy fibers, anchored at one end to a horizontal base. In water and other solvents, these epoxy pillars stand straight up and do not interact with each other. As the solvent evaporates, however, the fibers self-organize, clumping together in a helical pattern, a result of the attractive intermolecular forces in a process called capillary action. If polystyrene spheres are suspended in the liquid, the fibers wind around the sphere. Aizenberg said her group is now trying to make the process reversible, which would allow its possible use in drug release or self-cleaning materials. For example, she envisions polymer fingers that grab dust particles or floating

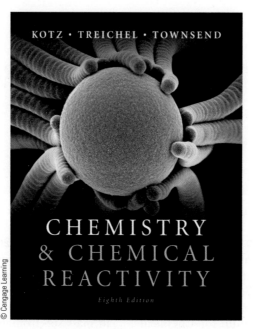

KOTZ • TREICHEL • TOWNSEND

CHEMISTRY & CHEMICAL REACTIVITY

Eighth Edition

© Cengage Learning

bacteria, later to release them so the contaminants can be washed away.

Aizenberg said the image "also brings to mind our collaborative effort to hold up the planet and keep it running." The judges in the photo competition liked both the image and the message. We also liked this photo because it portrays a dynamic research area and the importance of chemistry in the broader arena of science.

Regarding her research in general, Aizenberg said "We try to identify biological systems that have unusual structures that are naturally optimized to make extremely sophisticated, efficient, and highly potent devices and materials." Then the group uses the underlying biological design "to fabricate a new generation of self-assembling and adaptive materials based on biological architectures."

You will encounter further research of Professor Aizenberg on sea sponges in the Let's Review section of Chapter 1. See Chapter 10 for more information on the polymers involved, and see Chapter 12 for a discussion on intermolecular forces. The image originally appeared in *Science*, 19 February 2010 (Vol. 327. no. 5968, pp. 954–955). The website for Professor Aizenberg's research is **www.seas.harvard.edu/aizenberg_lab**.

Registering for OWL

1. To use OWL for the first time, go to **www.cengage.com/owl**, choose your course from the red box shown below, and then choose **Register**.

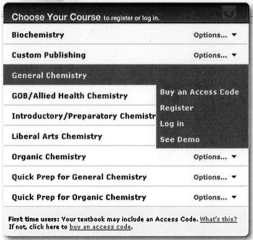

2. On the next User Login and Student Registration page, choose your textbook, making sure that you have chosen the correct title, author, and edition.

3. Follow the prompts in OWL to set up your account and register the access code from the OWL card bound into this book. (If the card is missing, you can purchase instant access to OWL at CengageBrain.com.)

4. When you see the User Login page similar to this one, bookmark it. This is where you'll log in to OWL in the future, using the login name and password that you used during registration.

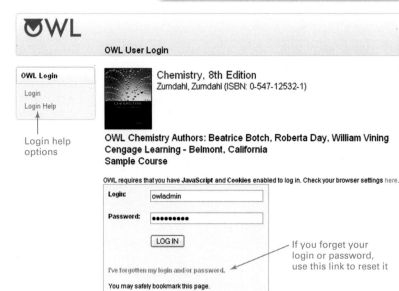

Logging in to OWL

After you register, if you forget to bookmark your OWL login page, you can always log in from **www.cengage.com/owl** and follow these steps:

1. Choose your course and then choose **Log In**.
2. Choose your textbook and then choose your school.
3. Click the blue arrow under **User Login Page**. On the next screen, enter the login name and password you chose during registration.

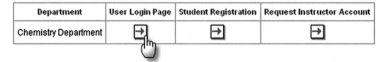

Department	User Login Page	Student Registration	Request Instructor Account
Chemistry Department	→	→	→

4. For Technical Support, to download a Student Access Code Guide and a Student Quick Start Guide, or to search our FAQs, please visit **www.cengage.com/chemistry/owlsupport**.

Using OWL

OWL Introductory Assignments

Every OWL course begins with Introductory Assignments that may be identical or similar to the ones shown below. These teach you the basics of using OWL and help you correctly set up your browser. We strongly recommend that you work through these assignments at the beginning of each semester. If you have trouble with your browser during the semester, you can go to the Introductory Assignments in your Past Due Assignments view and redo the browser tests.

Folder/Assignment
▼ 🗀 Introduction to OWL
▶ 🗀 1 - Navigation, Messages, and Browsers
▶ 🗀 2 - Question Modes and Types
▶ 🗀 3 - Scientific Notation
▶ 🗀 4 - MarvinSketch

OWL Information Menu Bar

At the top of each question page, you can click **Chemical Formulas** or **Scientific Notation** on the information menu bar.

Chemical Formulas	Scientific Notation	Periodic Table	Tables

Periodic Table and **Tables** on the left menu take you to additional resources in OWL. You can also access these resources by choosing **Chemistry Tables** from the left navigation menu on most OWL screens.

Answer Formatting in OWL

Like most online homework systems, when you answer questions in OWL, your responses must be correctly formatted in order to be graded. For some answers, you must enter superscripts and subscripts in your answer. Proper answer formatting is especially important when entering chemical formulas, mathematical equations, and units of measurement. Improper formatting will lead to otherwise correct answers being marked wrong.

Be sure to complete the **Intro to OWL** assignments to learn the basics of answer formatting and setting up your browser correctly. Answer formatting basics are summarized in the **Chemical Formulas** screen below (accessible from the information menu bar on any question page):

To Write a Chemical Formula in OWL

Enclose **subscripts** with *underscores* _.	Enclose **superscripts** with *carats* ^.
The *underscore* key is next to the number zero on the keyboard.	The *carat* key is the number six on the keyboard.
H_2_O = H_2O	Cr^3+^ = Cr^{3+}
Combined: SO_4_^2-^ = SO_4^{2-}	

Ions	Unit Charge Ions
Write the **number first** and then the charge.	Do **not** include the number one in unit charge ions.
N^3-^ = N^{3-} Ca^2+^ = Ca^{2+}	Na^+^ = Na^+ Cl^-^ = Cl^-

Using the Chemical Formula Input

The chemical formula input box displays the superscripts and subscripts as you enter the formula. There are 3 ways to use the input box.

- **Keyboard:** Use the keyboard to enter underscores and carats on your own.
- **Buttons after:** Enter the formula without underscores or carats, then highlight each superscript and/or subscript, click the appropriate subscript or superscript button, and the underscores or carats will be filled in automatically.
- **Button during:** Use the subscript or superscript buttons to enter the underscores and carats while you type the formula.

Using the Chemical Formula Input Tool

Use this tool to help correctly format your answers with subscripts and superscripts.* Use the help 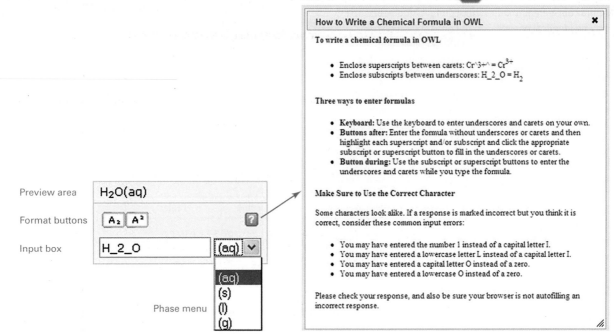 button if you need assistance.

Click A_2 to subscript text and click A^2 to superscript text.
1. For example, to write the chemical formula for water, type a capital "H" in the input box.
2. Click A_2 and then type "2."
3. Click A_2 again to set the text format back to normal font. Note that you press the button before AND after you enter the character to be formatted.
4. Type a capital "O."

Check the Preview Area to see your entry before submitting it.

Help for Common Errors

If your answer is incorrect, click **Why** to display rollover help for common errors such as the one below, where a zero was entered instead of a capital letter O.

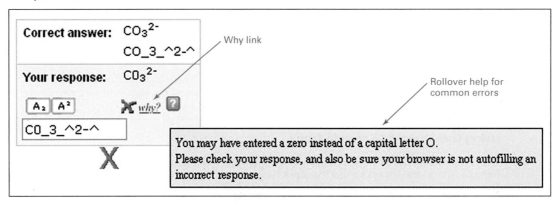

*If you are familiar with previous versions of OWL, you can still format your answers with underscores (Shift-minus) for subscripts and carats (Shift-6) for superscripts.

Other common errors:

- You may have entered the number 1 instead of a capital letter I.
- You may have entered a lowercase letter L instead of a capital letter I.
- You may have entered a lowercase letter O instead of a zero.

Please check your response, and also be sure your browser is not automatically filling in an incorrect response.

Scientific Notation in OWL

The basics of formatting scientific notation in OWL appear in the **Scientific Notation** screen below (accessible from the information menu bar on any question page):

SCIENTIFIC NOTATION IN OWL

OWL uses a similar format to the "EE" or "EXP" format found on most calculators.

6.02×10^{23} should be entered as **6.02E23**

Where "E23" represents "$\times 10^{23}$"

DO NOT confuse SCIENTIFIC NOTATION with entering SUPERSCRIPTS!

Use "E" notation for **NUMERICAL VALUES**. 6.02E23 **This is a numerical value.**

Use "carat" notation for **TEXT** input. Cu^2+^ **This is text.**

Examples:

Number to Enter	OWL Format	Calculator Format
6.02×10^{23}	6.02E23	6.02 EE 23 or 6.02 EXP 23
0.000000454	4.54E-7	4.54 EE -7 or 4.54 EXP -7

Note:
Also do not confuse this notation with the irrational number "e" = 2.718282 which forms the base for the natural log, ln. On calculators this is usually the e^x or **inv ln** key.

- High levels of copper and arsenic were found in the Iceman's hair. These observations, combined with the discovery that his ax was nearly pure copper, led the investigators to conclude he had been involved in copper smelting.
- One fingernail was still present on his body. Amazingly, scientists could conclude from its appearance that he had been sick three times in the 6 months before he died and his last illness had lasted for 2 weeks.
- Australian scientists took samples of blood residues from his stone-tipped knife, his arrows, and his coat. Using techniques developed to study ancient DNA, they found the blood came from four different individuals. In fact, the blood on one arrow tip was from two different individuals, suggesting that the man had killed two different people. Perhaps he had killed one person, retrieved the arrow, and used it to kill another.

The many different methods used to reveal the life of the Iceman and his environment are used by scientists around the world, including present-day forensic scientists to study accidents and crimes. As you study chemistry and the chemical principles in this book, keep in mind that many areas of science depend on chemistry and that many different careers in the sciences are available. As one example of a scientific career, you can read on page 4 about the experience in forensic chemistry of a former student of one of the authors.

1.1   Chemistry and Its Methods

Chemistry is about change. It was once only about changing one natural substance into another—wood and oil burn, grape juice turns into wine, and cinnabar (Figure 1.1), a red mineral from the earth, ultimately changes into shiny quicksilver (mercury) upon heating. The emphasis was largely on finding a recipe to carry out the desired transformation with little understanding of the underlying structure of the materials or explanations for why particular changes occurred. Chemistry is still about change, but now chemists focus on the change of one pure substance, whether natural or synthetic, into another and on understanding that change (Figure 1.2). As you will see, in modern chemistry, we now picture an exciting world of submicroscopic atoms and molecules interacting with each other. We have also developed ways to predict whether or not a particular reaction may occur.

Although chemistry is endlessly fascinating—at least to chemists—why should you study chemistry? Each person probably has a different answer, but many students take a chemistry course because someone else has decided it is an

FIGURE 1.1 Cinnabar and mercury.

(a) The red crystals of cinnabar are the chemical compound mercury(II) sulfide.

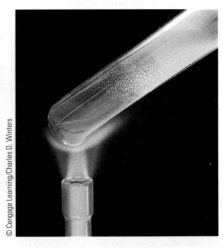

(b) Heating cinnabar in air changes it into orange mercury(II) oxide, which, on further heating, decomposes to the elements mercury (the droplets seen on the inside of the test tube wall) and oxygen gas.

Solid sodium, Na

Chlorine gas, Cl₂

Photos: © Cengage Learning/Charles D. Winters

Sodium chloride solid, NaCl

FIGURE 1.2 Forming a chemical compound. Sodium chloride, table salt, can be made by combining sodium metal (Na) and yellow chlorine gas (Cl₂). The result is a crystalline solid, common salt. (The tiny spheres show how the atoms are arranged in the substances. In the case of the salt crystal, the spheres represent electrically charged sodium and chlorine ions.)

important part of preparing for a particular career. Chemistry is especially useful because it is central to our understanding of disciplines as diverse as biology, geology, materials science, medicine, physics, and some branches of engineering. In addition, chemistry plays a major role in the economy of developed nations, and chemistry and chemicals affect our daily lives in a wide variety of ways. Furthermore, a course in chemistry can help you see how a scientist thinks about the world and how to solve problems. The knowledge and skills developed in such a course will benefit you in many career paths and will help you become a better informed citizen in a world that is becoming technologically more complex—and more interesting.

Hypotheses, Laws, and Theories

As scientists we study questions of our own choosing or ones that someone else poses in the hope of finding an answer or discovering some useful information. When the Iceman was discovered, there were many questions that scientists could try to answer, such as where he was from. They knew he was likely from an area on the border of what is now Austria and Italy. As you will learn later (page 50), the type of oxygen present in water differs slightly from place to place, so testing the oxygen in the Iceman's body might help locate his home. That is, the scientists formed a **hypothesis**, a tentative explanation or prediction based on experimental observations.

After formulating one or more hypotheses, scientists perform experiments designed to give results that confirm or invalidate these hypotheses. In chemistry this usually requires that both quantitative and qualitative information be collected. **Quantitative** information is numerical data, such as the temperature at which a chemical substance melts or its mass (Figure 1.3). **Qualitative** information, in contrast, consists of nonnumerical observations, such as the color of a substance or its physical appearance.

In the case of the Iceman, scientists have assembled a great deal of qualitative and quantitative information on his body, his clothing, and his weapons. Among this was information from recent studies on his tooth enamel and bones. Different oxygen isotopes in these body parts showed that the Iceman must have consumed water from a relatively small location within what is now Italy (page 50). The mystery of his place of origin was solved.

Careers in Chemistry

Your course in general chemistry is just the beginning, and this may be the time when you are just starting to have a sense of your future career direction. This course may interest you in learning even more about chemistry as you prepare for a career in chemistry, but it may also serve as one of the first steps to prepare you for a career in other areas of science, medicine, engineering, or even one as seemingly far from chemistry as finance. One area of science that draws on many others is forensic science, and the photo here is of Caitlin Kilcoin, a forensic chemist and former student of one of this book's authors.

While taking General Chemistry, Ms. Kilcoin became interested in forensic chemistry, so, after her undergraduate studies, she earned a M.S. degree in forensic science at Virginia Commonwealth University in Richmond, Virginia. As part of that program she worked as an intern in Las Vegas, Nevada.

After finishing her advanced degree, she began working for the Department of General Services of the State of Virginia in the Chemical Terrorism Unit. There she worked under the Food Emergency Response Network, a program established by the Food and Drug Administration. This work involved testing food samples from customs agencies

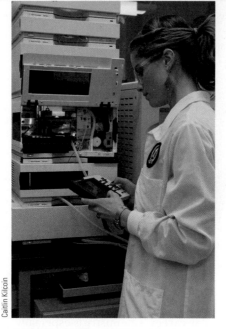

Caitlin Kilcoin

Caitlin Kilcoin, forensic chemist and former student of one of this book's authors.

during the time when companies in China were found to be contaminating food stuffs with the chemical compound melamine (▶ Chapter 12, page 433). Another aspect of

that work was testing food for other toxins. In addition, her laboratory was responsible for testing unknown samples that were submitted as evidence in legal cases. These included suspicious powders, adulterated food samples, and sometimes arson samples.

Ms. Kilcoin told us that deciding that you would like to work in the field of chemistry is the easy part. Where it becomes more involved is defining exactly which path you want to take within the field. "But you should take time to figure out whether you're the type of person who likes to have an established set of procedures or one who likes the research aspect of chemistry and is not afraid of the unknowns." Either way, she stresses the importance of obtaining an advanced degree. Within a highly competitive field, this degree will help you stand out among other applicants. "As a chemist working in the field, I found that I always need to be prepared for the unexpected. Just when I think that people can't get any more outlandish in their criminal attempts, another sample is submitted to the lab for analysis that tops the one before it. The best part is that they probably always think they're going to get away with it. However, I think science has proven that that is not always the case."

The analysis using oxygen isotopes to determine the Iceman's home can only be done because it is well known that oxygen isotopes in water vary with altitude in predictable ways. That is, the variation in isotope composition with location can be deemed a law of science. After numerous experiments by many scientists over an extended period of time, an hypothesis may become a **law**—a concise verbal or mathematical statement of a behavior or a relation that seems always to be the same under the same conditions.

We base much of what we do in science on laws because they help us predict what may occur under a new set of circumstances. For example, we know from experience

FIGURE 1.3 Qualitative and quantitative observations. A new substance is formed by mixing two known substances in solution, and several observations can be made. *Qualitative* observations: Before mixing, one solution is light green and the other is colorless. After mixing, the solution is colorless but there is now a bright red solid in the beaker. *Quantitative* observations: 100 mL of one solution is mixed with 10 mL of the other. Although not shown here, the mass of the solid after mixing is 3.6 g.

© Cengage Learning/Charles D. Winters

that if the chemical element sodium comes in contact with water, a violent reaction occurs and new substances are formed (Figure 1.4), and we know that the mass of the substances produced in the reaction is exactly the same as the mass of sodium and water used in the reaction. That is, *mass is always conserved in chemical reactions.*

Once enough reproducible experiments have been conducted and experimental results have been generalized as a law or general rule, it may be possible to conceive a theory to explain the observation. A **theory** is a well-tested, unifying principle that explains a body of facts and the laws based on them. It is capable of suggesting new hypotheses that can be tested experimentally.

Sometimes nonscientists use the word *theory* to imply that someone has made a guess and that an idea is not yet substantiated. But, to scientists, a theory is based on carefully determined and reproducible evidence. Theories are the cornerstone of our understanding of the natural world at any given time. Remember, though, that theories are inventions of the human mind. Theories can and do change as new facts are uncovered.

FIGURE 1.4 The metallic element sodium reacts with water.

Goals of Science

The sciences, including chemistry, have several goals. Two of these are prediction and control. We do experiments and seek generalities because we want to be able to predict what may occur under other sets of circumstances. We also want to know how we might control the outcome of a chemical reaction or process.

Two further goals are understanding and explaining. We know, for example, that certain elements such as sodium will react vigorously with water. But why should this be true? To explain and understand this, we turn to theories such as those developed in Chapters 7 and 8.

Dilemmas and Integrity in Science

You may think research in science is straightforward: Do experiments, collect information, and draw a conclusion. But, research is seldom that easy. Frustrations and disappointments are common enough, and results can be inconclusive. Experiments sometimes contain some level of uncertainty, and spurious or contradictory data can be collected. For example, suppose you do an experiment expecting to find a direct relation between two experimental quantities. You collect six data sets. When plotted on a graph, four of the sets lie on a straight line, but two others lie far away from the line. Should you ignore the last two points? Or should you do more experiments when you know the time they take will mean someone else could publish their results first and thus get the credit for a new scientific principle? Or should you consider that the two points not on the line might indicate that your original hypothesis is wrong and that you will have to abandon a favorite idea you have worked on for a year? Scientists have a responsibility to remain objective in these situations, but sometimes it is hard to do.

It is important to remember that scientists are human and therefore subject to the same moral pressures and dilemmas as any other person. To help ensure integrity in science, some simple principles have emerged over time that guide scientific practice:

- Experimental results should be reproducible. Furthermore, these results should be reported in the scientific literature in sufficient detail that they can be used or reproduced by others.
- Conclusions should be reasonable and unbiased.
- Credit should be given where it is due.

1.2 Sustainability and Green Chemistry

The world's population is about 6.8 billion people, with about 7 million added every month. Each of these new persons needs shelter, food, and medical care, and each uses increasingly scarce resources like fresh water and energy. And each produces

Principles of Green Chemistry

Paul Anastas and John Warner enunciated the principles of green chemistry in their book *Green Chemistry: Theory and Practice* (Oxford, 1998). Among these are the ones stated below. As you read *Chemistry & Chemical Reactivity*, we will remind you of these principles, and others, and how they can be applied.

- "It is better to prevent waste than to treat or clean up waste after it is formed."
- New pharmaceuticals or consumer chemicals are synthesized, that is, made, by a large number of chemical processes. Therefore, "synthetic methods should be designed to maximize the incorporation of all materials used in the final product."
- Where possible, synthetic methods "should be designed to use and generate substances that possess little or no toxicity to human health or the environment."

- "Chemical products should be designed to [function effectively] while still reducing toxicity."
- "Energy requirements should be recognized for their environmental and economic impacts and should be minimized. Synthetic methods should be conducted at ambient temperature and pressure."
- Raw materials "should be renewable whenever technically and economically practical."
- "Chemical products should be designed so that at the end of their function, they do not persist in the environment or break down into dangerous products."
- Substances used in a chemical process should be chosen to minimize the potential for chemical accidents, including releases, explosions, and fires.

© Cengage Learning/Charles D. Winters

A student performs a "green" synthesis. In this experiment a chemical compound is made using a new method that avoids toxic chemicals.

by-products in the act of living and working that can affect our environment. With such a large population, these individual effects can add up to cause large consequences for our planet. The focus of scientists, planners, and politicians is increasingly turning to a concept of "sustainable development."

James Cusumano, a chemist and former president of a chemical company, said that "On one hand, society, governments, and industry seek economic growth to create greater value, new jobs, and a more enjoyable and fulfilling lifestyle. Yet, on the other, regulators, environmentalists, and citizens of the globe demand that we do so with *sustainable development*—meeting today's global economic and environmental needs while preserving the options of future generations to meet theirs. How do nations resolve these potentially conflicting goals?" This is even more true now than it was in 1995 when Dr. Cusumano made this statement in the *Journal of Chemical Education*.

Much of the increase in life expectancy and quality of life, at least in the developed world, is derived from advances in science. But we have paid an environmental price for it, with increases in gases such as nitrogen oxides and sulfur oxides in the atmosphere, acid rain falling in many parts of the world, and waste pharmaceuticals entering the water supply. Among many others, chemists are seeking answers to this dilemma, and one response has been to practice *green chemistry*.

This concept of green chemistry began to take root more than 10 years ago and is now beginning to lead to new ways of doing things and to lower pollutant levels. By the time you finish this book, you will have been introduced to most, if not all, the underlying principles of green chemistry. As you can see in "A Closer Look: Principles of Green Chemistry," they are simple ideas. The challenge is to put them into practice.

1.3 Classifying Matter

This chapter begins our discussion of how chemists think about science in general and about matter in particular. After looking at a way to classify matter, we will turn to some basic ideas about elements, atoms, compounds, and molecules and describe how chemists characterize these building blocks of matter.

States of Matter and Kinetic-Molecular Theory

An easily observed property of matter is its **state**—that is, whether a substance is a solid, liquid, or gas (Figure 1.5). You recognize a material as a solid because it has a rigid shape and a fixed volume that changes little as temperature and pressure change. Like solids, liquids have a fixed volume, but a liquid is fluid—it takes on the shape of its container and has no definite shape of its own. Gases are fluid as well, but the volume of a gas is determined by the size of its container. The volume of a gas varies more than the volume of a liquid with changes in temperature and pressure.

At low enough temperatures, virtually all matter is found in the solid state. As the temperature is raised, solids usually melt to form liquids. Eventually, if the temperature is high enough, liquids evaporate to form gases. Volume changes typically accompany changes in state. For a given mass of material, there is usually a small increase in volume on melting—water being a significant exception—and then a large increase in volume occurs upon evaporation.

The **kinetic-molecular theory of matter** helps us interpret the properties of solids, liquids, and gases. According to this theory, all matter consists of extremely tiny particles (atoms, molecules, or ions) that are in constant motion.

- In solids these particles are packed closely together, usually in a regular array. The particles vibrate back and forth about their average positions, but seldom do particles in a solid squeeze past their immediate neighbors to come into contact with a new set of particles.
- The particles in liquids are arranged randomly rather than in the regular patterns found in solids. Liquids and gases are fluid because the particles are not confined to specific locations and can move past one another.
- Under normal conditions, the particles in a gas are far apart. Gas molecules move extremely rapidly and are not constrained by their neighbors. The molecules of a gas fly about, colliding with one another and with the container walls. This random motion allows gas molecules to fill their container, so the volume of the gas sample is the volume of the container.

An important aspect of the kinetic-molecular theory is that the higher the temperature, the faster the particles move. The energy of motion of the particles (their **kinetic energy**, page 16) acts to overcome the forces of attraction between particles. A solid melts to form a liquid when the temperature of the solid is raised to the point at which the particles vibrate fast enough and far enough to push one another out of the way and move out of their regularly spaced positions. As the temperature increases even more, the particles move even faster until finally they can escape the clutches of their comrades and enter the gaseous state. *Increasing temperature corresponds to faster and faster motions of atoms and molecules*, a general rule you will find useful in many future discussions.

● **Water—Changes in Volume on Freezing.** Water is an exception to the general statement that a given mass of a substance has a smaller volume as a solid than as a liquid. Water is unique in that, for a given mass, the volume *increases* on changing from a liquid to a solid, that is, its density decreases. (See Table 1.2 on page 13.)

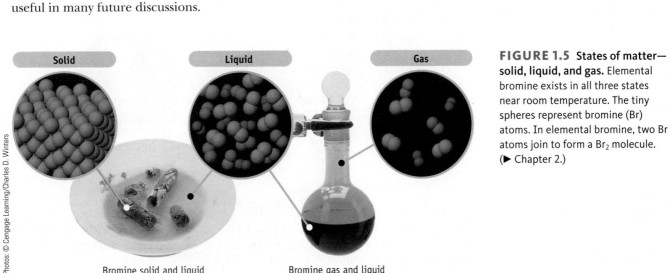

Photos: © Cengage Learning/Charles D. Winters

Solid   **Liquid**   **Gas**

Bromine solid and liquid   Bromine gas and liquid

FIGURE 1.5 States of matter—solid, liquid, and gas. Elemental bromine exists in all three states near room temperature. The tiny spheres represent bromine (Br) atoms. In elemental bromine, two Br atoms join to form a Br_2 molecule. (▶ Chapter 2.)

Matter at the Macroscopic and Particulate Levels

The characteristic properties of gases, liquids, and solids are observed by the unaided human senses. They are determined using samples of matter large enough to be seen, measured, and handled. Using such samples, we can determine, for example, what the color of a substance is, whether it dissolves in water, whether it conducts electricity, and if it reacts with oxygen. Observations such as these generally take place in the **macroscopic** world of chemistry (Figure 1.6). This is the world of experiments and observations.

Now let us move to the level of atoms, molecules, and ions—a world of chemistry we cannot see. Take a macroscopic sample of material and divide it, again and again, past the point where the amount of sample can be seen by the naked eye, past the point where it can be seen using an optical microscope. Eventually you reach the level of individual particles that make up all matter, a level that chemists refer to as the **submicroscopic** or **particulate** world of atoms and molecules (Figures 1.5 and 1.6).

Chemists are interested in the structure of matter at the particulate level. Atoms, molecules, and ions cannot be "seen" in the same way that one views the macroscopic world, but they are no less real. Chemists imagine what atoms must look like and how they might fit together to form molecules. They create models to represent atoms and molecules (Figures 1.5 and 1.6)—where tiny spheres are used to represent atoms—and then use these models to think about chemistry and to explain the observations they have made about the macroscopic world.

It has been said that chemists carry out experiments at the macroscopic level, but they think about chemistry at the particulate level. They then write down their observations as "symbols," the formulas (such as H_2O for water or NH_3 for ammonia molecules) and drawings that signify the elements and compounds involved. This is a useful perspective that will help you as you study chemistry. Indeed, one of our goals is to help you make the connections in your own mind among the symbolic, particulate, and macroscopic worlds of chemistry.

Pure Substances

A chemist looks at a glass of drinking water and sees a liquid. This liquid could be the pure chemical compound water. More likely, though, the liquid is a homogeneous mixture of water and dissolved substances—that is, a **solution**. Specifically, we can classify a sample of matter as being either a pure substance or a mixture (Figure 1.7).

A pure substance has a set of unique properties by which it can be recognized. Pure water, for example, is colorless and odorless. If you want to identify a

FIGURE 1.6 Levels of matter. We observe chemical and physical processes at the macroscopic level. To understand or illustrate these processes, scientists often imagine what has occurred at the particulate atomic and molecular levels and write symbols to represent these observations. A beaker of boiling water can be visualized at the particulate level as rapidly moving H_2O molecules. The process is symbolized by the chemical equation H_2O (liquid) → H_2O (gas).

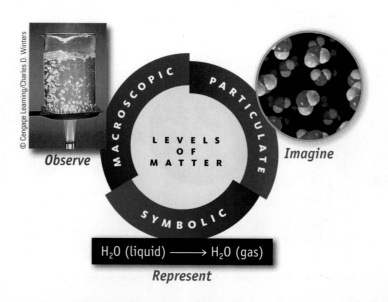

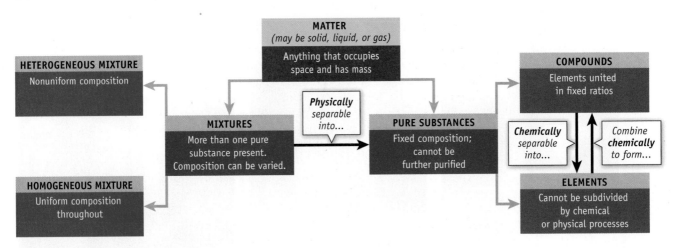

FIGURE 1.7 Classifying matter.

substance conclusively as water, you would have to examine its properties carefully and compare them against the known properties of pure water. Melting point and boiling point serve the purpose well here. If you could show that the substance melts at 0 °C and boils at 100 °C at atmospheric pressure, you can be certain it is water. No other known substance melts and boils at precisely these temperatures.

A second feature of a pure substance is that it cannot be separated into two or more different species by any physical technique at ordinary temperatures. If it could be separated, our sample would be classified as a mixture.

Mixtures: Homogeneous and Heterogeneous

A mixture consists of two or more pure substances that can be separated by physical techniques. Sand on the beach is a **heterogeneous** mixture of solids and liquids (Figure 1.8a), a mixture in which the uneven texture of the material can be detected. However, there are heterogeneous mixtures that may appear completely uniform but on closer examination are not (Figure 1.8b). Milk, for example, appears smooth in texture to the unaided eye, but magnification would reveal fat and protein globules within the liquid. In a heterogeneous mixture the properties in one region are different from those in another region.

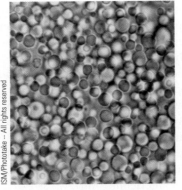

(a) A sample of beach sand is clearly heterogeneous. Each particle of sand and shells can be seen with the naked eye, and the sand and other particles can be separated.

(b) A sample of milk may look homogeneous, but examination with an optical microscope shows it is, in fact, a heterogeneous mixture of liquids and suspended particles (fat globules).

(c) A homogeneous mixture, here consisting of salt in water. The model shows that salt in water consists of separate, electrically charged particles (ions), but the particles cannot be seen with an optical microscope.

FIGURE 1.8 Heterogeneous and homogeneous mixtures.

(a) A laboratory setup. A beaker full of muddy water is passed through a paper filter, and the mud and dirt are removed.

(b) A water treatment plant uses filtration to remove suspended particles from the water.

FIGURE 1.9 Purifying a heterogeneous mixture by filtration.

A **homogeneous** mixture consists of two or more substances in the same phase (Figure 1.8c). No amount of optical magnification will reveal a homogeneous mixture to have different properties in different regions. Homogeneous mixtures are often called **solutions**. Common examples include air (mostly a mixture of nitrogen and oxygen gases), gasoline (a mixture of carbon- and hydrogen-containing compounds called *hydrocarbons),* and a soft drink in an unopened container.

When a mixture is separated into its pure components, the components are said to be **purified**. Efforts at separation are often not complete in a single step, however, and repetition almost always gives an increasingly pure substance. For example, soil particles can be separated from water by filtration (Figure 1.9). When the mixture is passed through a filter, many of the particles are removed. Repeated filtrations will give water a higher and higher state of purity. This purification process uses a property of the mixture, its clarity, to measure the extent of purification. When a perfectly clear sample of water is obtained, all of the soil particles are assumed to have been removed.

1.4 Elements

Module 1: The Periodic Table covers concepts in this section.

Passing an electric current through water can decompose it to gaseous hydrogen and oxygen (Figure 1.10a). Substances like hydrogen and oxygen that are composed of *only one type of atom* are classified as **elements**. Currently 118 elements are known. Of these, only about 90—some of which are illustrated in Figure 1.10—are found in nature. The remainder have been created by scientists. *Names and symbols for each element are listed in the tables at the front and back of this book.* Carbon (C), sulfur (S), iron (Fe), copper (Cu), silver (Ag), tin (Sn), gold (Au), mercury (Hg), and lead (Pb) were known to the early Greeks and Romans and to the alchemists of ancient China, the Arab world, and medieval Europe. However, many other elements—such as aluminum (Al), silicon (Si), iodine (I), and helium (He)—were not discovered until the 18th and 19th centuries. Finally, scientists in the 20th and 21st centuries have made elements that do not exist in nature, such as technetium (Tc), plutonium (Pu), and americium (Am).

The table inside the front cover of this book, in which the symbol and other information for the elements are enclosed in a box, is called the **periodic table**. We will describe this important tool of chemistry in more detail beginning in Chapter 2.

An **atom** is the smallest particle of an element that retains the characteristic chemical properties of that element. Modern chemistry is based on an understanding and exploration of nature at the atomic level (▶ Chapters 6 and 7).

Element Names and Symbols

The stories behind some of the names of the elements are fascinating. Many obviously have names and symbols with Latin or Greek

Erich Lessing/Art Resource, NY

Nicolaus Copernicus (1473–1543). The most recently named element, 112, was named for Copernicus.

origins. Examples include helium (He), from the Greek word *helios* meaning "sun," and lead, whose symbol, Pb, comes from the Latin word for "heavy," *plumbum.* More recently discovered elements have been named for their place of discovery or place of significance. Examples include americium (Am), californium (Cf), scandium (Sc), europium (Eu), francium (Fr), polonium (Po), and many others.

A number of elements are named for their discoverers or famous scientists: curium (Cm), einsteinium (Es), fermium (Fm), mendeleevium (Md), nobelium (No), seaborgium (Sg), and meitnerium (Mt), among others. The most recently named element, element 112, was given its official name, copernicium (Cn), only in 2010. It was named after Nicolaus Copernicus (1473–1543), who first proposed that Earth and the

other planets orbited the sun. Some say his work was the beginning of the scientific revolution.

What do you suppose is the origin of the names of the elements uranium, plutonium, and neptunium?

When writing the symbol for an element notice that only the first letter of an element's symbol is capitalized. For example, cobalt is Co, not CO. The notation CO represents the chemical compound carbon monoxide. Also note that the element name is not capitalized, except at the beginning of a sentence.

To learn more about element names and the periodic table, there are several excellent references on the Internet.
• www.ptable.com
• www.pubs.acs.org/cen/80th/elements .html

REVIEW & CHECK FOR SECTION 1.4

Using the periodic table inside the front cover of this book:

1. What is the symbol for the element sodium?

   (a) S   (b) Na   (c) So   (d) Sm

2. What is the name of the element with the symbol Si?

   (a) silver   (b) sulfur   (c) selenium   (d) silicon

● **Review and Check** At the end of almost all sections of each chapter there will be a set of multiple-choice questions to check your understanding. Answers to these questions are in Appendix P. In the YouBook version, you can click on an answer.

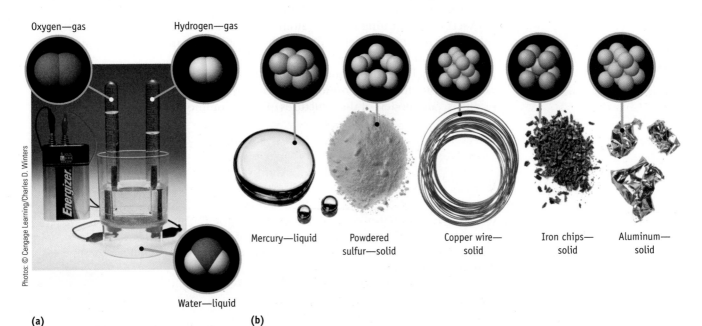

Oxygen—gas   Hydrogen—gas

Water—liquid

Mercury—liquid   Powdered sulfur—solid   Copper wire—solid   Iron chips—solid   Aluminum—solid

Photos: © Cengage Learning/Charles D. Winters

(a)   (b)

FIGURE 1.10 Elements. (a) Passing an electric current through water produces the elements hydrogen (test tube on the right) and oxygen (test tube on the left). **(b)** Chemical elements can often be distinguished by their color and their state at room temperature.

(a) The material in the dish is a mixture of iron chips and sulfur. The iron can be removed easily by using a magnet.

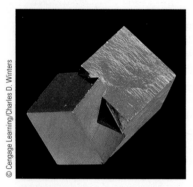

(b) Iron pyrite is a chemical compound composed of iron and sulfur. It is often found in nature as perfect, golden cubes.

FIGURE 1.11 Mixtures and compounds.

1.5 Compounds

A pure substance like sugar, salt, or water, which is composed of two or more different elements held together by **chemical bonds**, is referred to as a **chemical compound**. Even though only 118 elements are known, there appears to be no limit to the number of compounds that can be made from those elements. More than 54 million compounds are now known, with many thousands added to the list each year.

When elements become part of a compound, their original properties, such as their color, hardness, and melting point, are replaced by the characteristic properties of the compound. Consider common table salt (sodium chloride), which is composed of two elements (see Figure 1.2):

- Sodium is a shiny metal that reacts violently with water. Its solid state structure has sodium atoms tightly packed together.
- Chlorine is a light yellow gas that has a distinctive, suffocating odor and is a powerful irritant to lungs and other tissues. The element is composed of Cl_2 molecules in which two chlorine atoms are tightly bound together.
- Sodium chloride, or common salt ($NaCl$), is a colorless, crystalline solid composed of sodium and chlorine ions bound tightly together. Its properties are completely unlike those of the two elements from which it is made.

It is important to distinguish between a mixture of elements and a chemical compound of two or more elements. Pure metallic iron and yellow, powdered sulfur (Figure 1.11a) can be mixed in varying proportions. In the chemical compound iron pyrite (Figure 1.11b), however, there is no variation in composition. Not only does iron pyrite exhibit properties peculiar to itself and different from those of either iron or sulfur, or a mixture of these two elements, but it also has a definite percentage composition by mass (46.55% Fe and 53.45% S). Thus, two major differences exist between a mixture and a pure compound: A compound has distinctly different characteristics from its parent elements, and it has a definite percentage composition (by mass) of its combining elements.

Some compounds—such as table salt, $NaCl$—are composed of **ions**, which are electrically charged atoms or groups of atoms (▶ Chapter 2). Other compounds—such as water and sugar—consist of **molecules**, the smallest discrete units that retain the composition and chemical characteristics of the compound.

The composition of any compound is represented by its **chemical formula**. In the formula for water, H_2O, for example, the symbol for hydrogen, H, is followed by a subscript 2, indicating that two atoms of hydrogen occur in a single water molecule. The symbol for oxygen appears without a subscript, indicating that one oxygen atom occurs in the molecule.

As you shall see throughout this book, molecules can be represented with models that depict their composition and structure. Figure 1.12 illustrates the names, formulas, and models of the structures of a few common molecular compounds.

NAME	Water	Methane	Ammonia	Carbon dioxide
FORMULA	H_2O	CH_4	NH_3	CO_2
MODEL				

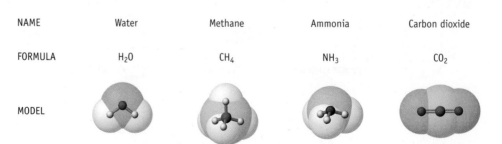

FIGURE 1.12 Names, formulas, and models of some common molecular compounds. Models of molecules appear throughout this book. In such models, C atoms are gray, H atoms are white, N atoms are blue, and O atoms are red.

FIGURE 1.13 Physical properties. An ice cube and a piece of lead can be differentiated easily by their physical properties (such as density, color, and melting point).

1.6 Physical Properties

You recognize your friends by their physical appearance: their height and weight and the color of their eyes and hair. The same is true of chemical substances. You can tell the difference between an ice cube and a cube of lead of the same size not only because of their appearance (one is clear and colorless, and the other is a lustrous metal) (Figure 1.13) but also because one is more dense (lead) than the other (ice). Properties such as these, which can be observed and measured without changing the composition of a substance, are called **physical properties**. The chemical elements in Figure 1.10, for example, clearly differ in terms of their color, appearance, and state (solid, liquid, or gas). Physical properties allow us to classify and identify substances. Table 1.1 lists a few physical properties of matter that chemists commonly use.

Density, the ratio of the mass of an object to its volume, is a physical property useful for identifying substances.

$$\text{Density} = \frac{\text{mass}}{\text{volume}} \qquad (1.1)$$

For example, you can readily tell the difference between an ice cube and a cube of lead of identical size because lead has a high density, 11.35 g/cm^3 (11.35 grams per cubic centimeter), whereas ice has a density slightly less than 0.917 g/cm^3. An ice cube with a volume of 16.0 cm^3 has a mass of 14.7 g, whereas a cube of lead with the same volume has a mass of 182 g.

The **temperature** of a sample of matter often affects the numerical values of its properties. Density is a particularly important example. Although the change in water density with temperature seems small (Table 1.2), it affects our environment profoundly. For example, as the water in a lake cools, the density of the water

● **Units of Density** As described on page 21 the decimal system of units in the sciences is called the Systemé International d'Unites; often referred to as SI units. The SI unit of mass is the kilogram and the SI unit of length is the meter. Therefore, the SI unit of density is kg/m^3. In chemistry, the more commonly used unit is g/cm^3. To convert from kg/m^3 to g/cm^3, divide by 1000.

● **Calculations Involving Density and Mathematics Review** See *Let's Review* beginning on page 20 for a review of some of the mathematics used in introductory chemistry.

● **Temperature Scales** Scientists use the Celsius (°C) and Kelvin scales (K) for temperature. (▶ page 22.)

Table 1.1 Some Physical Properties

Property	Using the Property to Distinguish Substances
Color	Is the substance colored or colorless? What is the color, and what is its intensity?
State of matter	Is it a solid, liquid, or gas? If it is a solid, what is the shape of the particles?
Melting point	At what temperature does a solid melt?
Boiling point	At what temperature does a liquid boil?
Density	What is the substance's density (mass per unit volume)?
Solubility	What mass of substance can dissolve in a given volume of water or other solvent?
Electric conductivity	Does the substance conduct electricity?
Malleability	How easily can a solid be deformed?
Ductility	How easily can a solid be drawn into a wire?
Viscosity	How easily will a liquid flow?

Table 1.2
Temperature Dependence of Water Density

Temperature (°C)	Density of Water (g/cm^3)
0 (ice)	0.917
0 (liq water)	0.99984
2	0.99994
4	0.99997
10	0.99970
25	0.99707
100	0.95836

FIGURE 1.14 Temperature dependence of physical properties. **(a)** *Change in density with temperature.* Ice cubes were placed in the right side of the tank and blue dye in the left side. The water beneath the ice is cooler and denser than the surrounding water, so it sinks. The convection current created by this movement of water is traced by the dye movement as the denser, cooler water sinks. **(b)** *Temperature and calibration.* Laboratory glassware is calibrated for specific temperatures. The pipet will deliver and the volumetric flask will contain the specified volume at the indicated temperature.

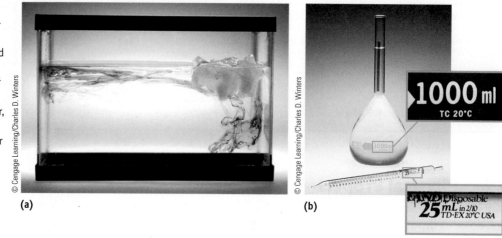

(a)   (b)

increases and the denser water sinks (Figure 1.14a). This continues until the water temperature reaches 3.98 °C, the point at which water has its maximum density (0.999973 g/cm³). If the water temperature drops further, the density decreases slightly and the colder water floats on top of water at 3.98 °C. If water is cooled below about 0 °C, solid ice forms. Water is unique among substances in the universe: Its solid form is less dense than its liquid form, so ice floats on water.

The volume of a given mass of liquid changes with temperature, so its density does as well. This is the reason laboratory glassware used to measure precise volumes of solutions always specifies the temperature at which it was calibrated (Figure 1.14b).

Extensive and Intensive Properties

Extensive properties depend on the amount of a substance present. The mass and volume of the samples of elements in Figure 1.10, or the amount of energy transferred as heat from burning gasoline, are extensive properties, for example.

In contrast, **intensive properties** do *not* depend on the amount of substance. A sample of ice will melt at 0 °C, no matter whether you have an ice cube or an iceberg.

The mass and volume of an object are extensive properties because they depend on the amount of sample present. Interestingly, the density of the object, the quotient of mass and volume, is an intensive property. The density of gold, for example, is the same (19.3 g/cm³ at 20 °C) whether you have a flake of pure gold or a solid gold ring.

Intensive properties are often useful in identifying a material. For example, the temperature at which a material melts (its melting point) is often so characteristic that it can be used to identify the solid (Figure 1.15).

FIGURE 1.15 A physical property used to distinguish compounds. Aspirin and naphthalene both are white solids at 25 °C. One way to tell them apart is by a difference in physical properties. At the temperature of boiling water, 100 °C, naphthalene is a liquid (left), whereas aspirin is a solid (right).

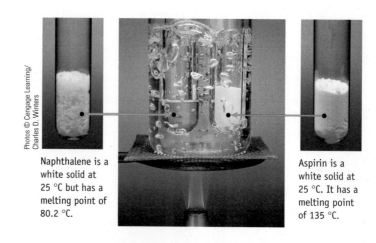

Naphthalene is a white solid at 25 °C but has a melting point of 80.2 °C.

Aspirin is a white solid at 25 °C. It has a melting point of 135 °C.

1.7 Physical and Chemical Changes

Changes in physical properties are called **physical changes**. In a physical change the identity of a substance is preserved even though it may have changed its physical state or the gross size and shape of its pieces. A physical change does not result in a new chemical substance being produced. The substances (atoms, molecules, or ions) present before and after the change are the same. An example of a physical change is the melting of a solid (Figure 1.15). In the case of ice melting, the molecules present both before and after the change are H_2O molecules. Their chemical identity has not changed; they are now simply able to flow past one another in the liquid state instead of being locked in position in the solid.

A physical property of hydrogen gas (H_2) is its low density, so a balloon filled with H_2 floats in air (Figure 1.16). Suppose, however, that a lighted candle is brought up to the balloon. When the heat causes the skin of the balloon to rupture, the hydrogen combines with the oxygen (O_2) in the air, and the heat of the candle sets off a chemical reaction, producing water, H_2O. This reaction is an example of a

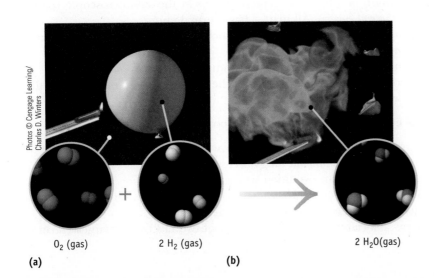

O_2 (gas)   2 H$_2$ (gas)   2 H$_2$O(gas)

(a)   **(b)**

FIGURE 1.16 A chemical change—the reaction of hydrogen and oxygen. (a) A balloon filled with molecules of hydrogen gas and surrounded by molecules of oxygen in the air. (The balloon floats in air because gaseous hydrogen is less dense than air.) **(b)** When ignited with a burning candle, H_2 and O_2 react to form water, H_2O.

Photos © Cengage Learning/Charles D. Winters

chemical change, in which one or more substances (the **reactants**) are transformed into one or more different substances (the **products**).

A chemical change at the particulate level is illustrated by the reaction of hydrogen and oxygen molecules to form water molecules.

$$2 \ H_2(gas) + O_2(gas) \longrightarrow 2 \ H_2O(gas)$$

Reactants   Products

The representation of the change using chemical formulas is called a **chemical equation**. It shows that the substances on the left (the reactants) produce the substances on the right (the products). As this equation shows, there are four atoms of H and two atoms of O before *and* after the reaction, but the molecules before the reaction are different from those after the reaction.

A **chemical property** indicates whether and sometimes how readily a material undergoes a chemical change with another material. For example, a chemical property of hydrogen gas is that it reacts vigorously with oxygen gas.

REVIEW & CHECK FOR SECTION 1.7

When camping in the mountains, you boil a pot of water on a campfire to make tea. Which of the following is a chemical change?

(a)   The water boils.

(b)   The campfire wood burns.

(c)   The tea dissolves in the hot water.

(d)   The pot melts from the heat of the fire.

1.8 Energy: Some Basic Principles

● **Units of Energy** Energy in chemistry is measured in units of joules. See *Let's Review* (page 20) and Chapter 5 (page 172) for calculations involving energy units.

Energy, which is a crucial part of many chemical and physical changes, is defined as the capacity to do work. You do work against the force of gravity when carrying yourself and hiking equipment up a mountain. The energy to do this is provided by the food you have eaten. Food is a source of chemical energy—energy stored in chemical compounds and released when the compounds undergo the chemical reactions of metabolism in your body. Chemical reactions almost always either release or absorb energy.

Energy can be classified as kinetic or potential. **Kinetic energy** is energy associated with motion, such as:

- The motion of atoms, molecules, or ions at the submicroscopic (particulate) level (*thermal energy*). All matter has thermal energy.
- The motion of macroscopic objects such as a moving tennis ball or automobile (*mechanical energy*).
- The movement of electrons in a conductor (*electrical energy*).
- The compression and expansion of the spaces between molecules in the transmission of sound (*acoustic energy*).

Potential energy results from an object's position and includes:

- Energy possessed by a ball held above the floor and by water at the top of a water wheel (*gravitational energy*) (Figure 1.17a).
- Energy stored in an extended spring.
- Energy stored in fuels (*chemical energy*) (Figure 1.17b). Almost all chemical reactions involve a change in chemical energy.
- Energy associated with the separation of two electrical charges (*electrostatic energy*) (Figure 1.17c).

CO₂ in the Oceans

"Over the past 200 years, the oceans have absorbed approximately 550 billion tons of CO_2 from the atmosphere, or about a third of the total amount of anthropogenic emissions over that period." This amounts to about 22 million tons *per day*. This statement was made by R. A. Feely, a scientist at the National Oceanographic and Atmospheric Administration, in connection with studies on the effects of carbon dioxide on ocean chemistry.

The amount of CO_2 dissolved in the oceans is of great concern and interest to oceanographers because it affects the pH of the water, that is, its level of acidity. This in turn can affect the growth of marine systems such as corals and sea urchins and microscopic coccolithophores (single-cell phytoplankton).

Clown fish. The larvae of the clown fish are affected by high levels of CO_2 in the ocean.

Image copyright David Mckee, 2010. Used under license from Shutterstock.com

Sea urchins. The growth of spines by sea urchins depends on the level of CO_2 in the water. (left) Normal growth at the current CO_2 level in the ocean (400 ppm). (right) Growth at a level of 2850 ppm (an extreme level used for lab studies).

Image courtesy of Justin B. Ries, Department of Marine Sciences, University of North Carolina at Chapel Hill

 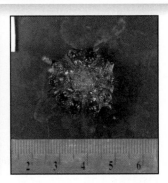

Recent studies have indicated that, in water with a high CO_2 content, the spines of sea urchins are greatly impaired, the larvae of orange clown fish lose their homing ability, and the concentrations of calcium, copper, manganese, and iron in sea water are affected, sometimes drastically.

Questions:

1. Much has been written about CO_2. What is its name?
2. Give the symbols for the four metals mentioned in this article.
3. Of the four metals mentioned here, which is the most dense? The least dense? (Use an Internet tool such as www.ptable.com to find this information.)
4. The spines of the sea urchin, corals, and coccolithophores all are built on the compound $CaCO_3$. What elements are involved in this compound? Do you know its name?

Answers to these questions are available in Appendix N.

Reference:

"Off-Balance Ocean: Acidification from absorbing atmospheric CO_2 is changing the ocean's chemistry," Rachel Petkewich, *Chemical and Engineering News*, February 23, 2009, page 56.

Bruce Roberts/Photo Researchers, Inc.

NASA/Glenn Research Center (GRC)

Image copyright Andriano, 2010. Used under license from Shutterstock.com

(a) Water at the top of a water wheel represents stored, or potential, energy. As water flows over the wheel, its potential energy is converted to mechanical energy.

(b) Chemical potential energy of the fuel and oxygen is converted to thermal and mechanical energy in a jet engine.

(c) Lightning converts electrostatic energy into radiant and thermal energy.

FIGURE 1.17 Energy and its conversion.

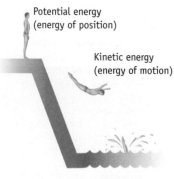

Potential energy
(energy of position)

Kinetic energy
(energy of motion)

Heat and work
(thermal and
mechanical energy)

FIGURE 1.18 The law of energy conservation. The diver's potential energy is converted to kinetic energy, which is then transferred to the water, illustrating the law of conservation of energy.

Potential energy and kinetic energy can be interconverted. For example, as water falls over a waterfall, its potential energy is converted into kinetic energy. Similarly, kinetic energy can be converted into potential energy: The kinetic energy of falling water can turn a turbine to produce electricity, which can then be used to convert water into H_2 and O_2 by electrolysis. Hydrogen gas contains stored chemical potential energy because it can be burned to produce heat and light or electricity.

Conservation of Energy

Standing on a diving board, you have considerable potential energy because of your position above the water. Once you dive off the board, some of that potential energy is converted into kinetic energy (Figure 1.18). During the dive, the force of gravity accelerates your body so that it moves faster and faster. Your kinetic energy increases and your potential energy decreases. At the moment you hit the water, your velocity is abruptly reduced and much of your kinetic energy is transferred to the water as your body moves it aside. Eventually you float to the surface, and the water becomes still again. If you could see them, however, you would find that the water molecules are moving a little faster in the vicinity of your entry into the water; that is, the kinetic energy of the water molecules is slightly higher.

This series of energy conversions illustrates the **law of conservation of energy**, which states that *energy can neither be created nor destroyed.* Or, to state this law differently, *the total energy of the universe is constant.* The law of conservation of energy summarizes the results of many experiments in which the amounts of energy transferred have been measured and in which the total energy content has been found to be the same before and after an event.

Let us examine this law in the case of a chemical reaction, the reaction of hydrogen and oxygen to form water (see Figure 1.16). In this reaction, the reactants (hydrogen and oxygen) have a certain amount of energy associated with them. When they react, some of this energy is released to the surroundings. If we were to add up all of the energy present before the reaction and all of the energy present after the reaction, we would find that the energy has only been redistributed; the total amount of energy in the universe has remained constant. Energy has been conserved.

REVIEW & CHECK FOR SECTION 1.8

1. Which of the following has the highest thermal energy?

    (a) 1.0 g of ice at 0 °C

    (b) 1.0 g of liquid water at 25 °C

    (c) 1.0 g of liquid water at 100 °C

    (d) 1.0 g of water vapor at 100 °C

2. Which of the following is an incorrect statement?

    (a) A mixture of H_2 and O_2 has lower chemical potential energy than H_2O.

    (b) Input of energy is required to split a H_2 molecule into two H atoms.

    (c) Water at 30 °C has a higher thermal energy than the same amount of water at 20 °C.

    (d) In the reaction of H_2O and Na (see Figure 1.4), chemical potential energy is converted to thermal energy.

CHAPTER GOALS REVISITED

Now that you have studied this chapter, you should ask whether you have met the chapter goals. In particular, you should be able to

Understand the differences between hypotheses, laws, and theories
a. Recognize the difference between a hypothesis and a theory and how laws are established (Section 1.1).

Be aware of the principles of green chemistry (Section 1.2)

Apply the kinetic-molecular theory to the properties of matter
a. Understand the basic ideas of the kinetic-molecular theory (Section 1.3).

Classify matter
a. Recognize the different states of matter (solids, liquids, and gases) and give their characteristics (Section 1.3). **Study Question: 31.**
b. Appreciate the difference between pure substances and mixtures and the difference between homogeneous and heterogeneous mixtures (Section 1.3). **Study Questions: 21, 29, 30.**
c. Recognize the importance of representing matter at the macroscopic level and at the particulate level (Section 1.3). **Study Questions: 25, 26.**

Recognize elements, atoms, compounds, and molecules
a. Identify the name or symbol for an element, given its symbol or name, respectively (Section 1.4). **Study Questions: 1–4, and Go Chemistry Module 1.**
b. Use the terms *atom, element, molecule, and compound* correctly (Sections 1.4 and 1.5). **Study Questions: 5, 6.**

Identify physical and chemical properties and changes
a. List commonly used physical properties of matter (Sections 1.5 and 1.6).
b. Identify several physical and chemical properties of common substances (Sections 1.5–1.7). **Study Questions: 7–10, 17, 18, 20.**
c. Relate density to the volume and mass of a substance (Section 1.6). **Study Questions: 15, 16, 27, 28, 33, 35, 39, 42, 43.**
d. Explain the difference between chemical and physical changes (Section 1.7). **Study Questions: 8, 23, 24, 45.**
e. Understand the difference between extensive and intensive properties and give examples of them (Section 1.6). **Study Questions: 15, 16.**

Describe various forms of energy
a. Identify types of potential and kinetic energy (Section 1.8). **Study Question: 13.**
b. Recognize and apply the law of conservation of energy (Section 1.8).

Key Equation

Equation 1.1 (page 13) Density: In chemistry the common unit of density is g/cm³, whereas kg/m³ is commonly used in geology and oceanography.

$$\text{Density} = \frac{\text{mass}}{\text{volume}}$$

Let's Review

The Tools of Quantitative Chemistry

John C. Kotz

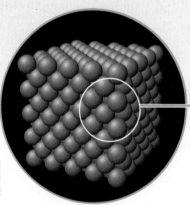
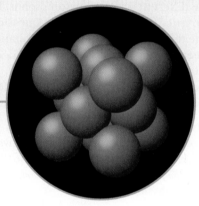

Copper Copper (Cu) is the 26th most abundant element in the Earth's crust (not too different from its near neighbors nickel and zinc in the periodic table). Copper and its minerals are widely distributed, and obtaining the metal from its ores is relatively easy. As a result, elemental copper is used around the world for many useful items, from cooking pots to electric wires. The photo (above left) shows large copper pots on sale in a market in southwestern China.

Pure copper (often called *native copper*) is found in nature, but more commonly it is found combined with other elements in minerals such as cuprite, azurite, or malachite. Copper metal is relatively soft but, when combined in a ratio of about 2 to 1 with tin, it forms bronze. Bronze was important in early civilizations and gave its name to an epoch of human development, the Bronze Age, which started around 3000 BC and lasted until about 1000 BC. (The Iceman you learned about on page 1 lived just before the Bronze Age and had an ax made of copper.) The development of bronze was significant because bronze is stronger than copper and can be shaped into a sharper edge. This improved the cutting edges of plows and weapons, thus giving cultures that possessed bronze advantages over those that did not.

Copper is now used in electric wiring because it is a good conductor of electricity, and it is used in cooking pots because it conducts heat well. It is also described as one of the "coinage metals" (along with silver and gold) because it has been used in coins for centuries. And most gold jewelry is a mixture of gold with some copper.

Compounds of copper are common, and copper is one of the eight essential metals in our bodies, where it is needed for some enzymes to use oxygen more effectively. Fortunately, it is found in common foods (in meats such as lamb, duck, pork, and beef, and in nuts such as almonds and walnuts). The average person has about 72 mg of copper in his or her body.

The figure (above right) shows what happens as we zoom into copper at the particulate level. We begin to see atoms arranged in a regular array, or *lattice*, as chemists call it. Zooming in even closer, we see the smallest repeating unit of the crystal.

You can learn more about copper and its properties by answering Study Questions 60 and 61 at the end of this *Let's Review* section.

At its core, chemistry is a quantitative science. Chemists make measurements of, among other things, size, mass, volume, time, and temperature. Scientists then manipulate that information to search for relationships among properties and to provide insight into the molecular basis of matter.

This section reviews the units used in chemistry, briefly describes the proper treatment of numerical data, and reviews some mathematical skills you will need in chemical calculations. After studying this section you should be able to:

- Use the common units for measurements in chemistry and make unit conversions (such as from liters to milliliters).
- Express and use numbers in exponential or scientific notation.
- Incorporate quantitative information into an algebraic expression and solve that expression.
- Read information from graphs.
- Prepare and interpret graphs of numerical information, and, if a graph produces a straight line, find the slope and equation of the line.
- Recognize and express uncertainties in measurements.

OWL

Sign in to OWL at **www.cengage.com/owl** to view tutorials and simulations, develop problem-solving skills, and complete online homework assigned by your professor.

1 Units of Measurement

Doing chemistry requires observing chemical reactions and physical changes. We make qualitative observations—such as changes in color or the evolution of heat—and quantitative measurements of temperature, time, volume, mass, and length or size. To record and report measurements, the scientific community has chosen a modified version of the **metric system**. This decimal system, used internationally in science, is called the Système International d'Unités (International System of Units), abbreviated **SI**.

All SI units are derived from base units, and these are listed in Table 1. Larger and smaller quantities are expressed by using appropriate prefixes with the base unit (Table 2). The nanometer (nm), for example, is 1 billionth of a meter. That is, it is equivalent to 1×10^{-9} m (meter). Dimensions on the nanometer scale are common in chemistry and biology because, for example, a typical molecule is about 1 nm across and a bacterium is about 1000 nm in length. The prefix *nano-* is also used in the name for a new area of science, *nanotechnology*, which involves the synthesis and study of materials having this tiny size.

Temperature Scales

Two temperature scales are commonly used in scientific work: Celsius and Kelvin (Figure 1). The Celsius scale is generally used worldwide for measurements in the laboratory. When calculations incorporate temperature data, however, the Kelvin scale must almost always be used.

Table 1 **The Seven SI Base Units**

Measured Property	Name of Unit	Abbreviation
Mass	Kilogram	kg
Length	Meter	m
Time	Second	s
Temperature	Kelvin	K
Amount of substance	Mole	mol
Electric current	Ampere	A
Luminous intensity	Candela	cd

● **The Kilogram, a New Standard Needed?** Unlike the second and the meter, the kilogram is defined by a physical object: a block of platinum-iridium alloy in a building in Paris, France. The block has been mysteriously losing mass, so there is great interest in the scientific community to find a better way to define the kilogram. One suggestion is that it be defined as the mass of 5.0184515×10^{25} atoms of the carbon-12 isotope.

● Common Conversion Factors

1000 g = 1 kg
1×10^9 nm = 1 m
10 mm = 1 cm
100 cm = 10 dm = 1 m
1000 m = 1 km
Conversion factors for SI units are given in Appendix C and inside the back cover of this book.

Table 2 Selected Prefixes Used in the Metric System

Prefix	Abbreviation	Meaning	Example
Giga-	G	10^9 (billion)	1 gigahertz = 1×10^9 Hz
Mega-	M	10^6 (million)	1 megaton = 1×10^6 tons
Kilo-	k	10^3 (thousand)	1 kilogram (kg) = 1×10^3 g
Deci-	d	10^{-1} (tenth)	1 decimeter (dm) = 1×10^{-1} m
Centi-	c	10^{-2} (one hundredth)	1 centimeter (cm) = 1×10^{-2} m
Milli-	m	10^{-3} (one thousandth)	1 millimeter (mm) = 1×10^{-3} m
Micro-	μ	10^{-6} (one millionth)	1 micrometer (μm) = 1×10^{-6} m
Nano-	n	10^{-9} (one billionth)	1 nanometer (nm) = 1×10^{-9} m
Pico-	p	10^{-12}	1 picometer (pm) = 1×10^{-12} m
Femto-	f	10^{-15}	1 femtometer (fm) = 1×10^{-15} m

● Lord Kelvin

William Thomson (1824–1907), known as Lord Kelvin, was a professor of natural philosophy at the University in Glasgow, Scotland, from 1846 to 1899. He was best known for his study of heat and work, from which came the concept of the absolute temperature scale.

E. F. Smith Collection/Van Pelt Library/University of Pennsylvania

The Celsius Temperature Scale

The size of the Celsius degree is defined by assigning zero as the freezing point of pure water (0 °C) and 100 as its boiling point (100 °C). You may recognize that a comfortable room temperature is around 20 °C and your normal body temperature is 37 °C. We find that the warmest water we can stand to immerse a finger in is about 60 °C.

The Kelvin Temperature Scale

William Thomson, known as Lord Kelvin (1824–1907), first suggested the temperature scale that now bears his name. The Kelvin scale uses the same size unit as the Celsius scale, but it assigns zero as the lowest temperature that can be achieved, a point called **absolute zero**. Many experiments have found that this limiting temperature is −273.15 °C (−459.67 °F). *Kelvin units and Celsius degrees are the same size.*

FIGURE 1 Comparison of Fahrenheit, Celsius, and Kelvin scales. The reference, or starting point, for the Kelvin scale is absolute zero (0 K = −273.15 °C), which has been shown theoretically and experimentally to be the lowest possible temperature.

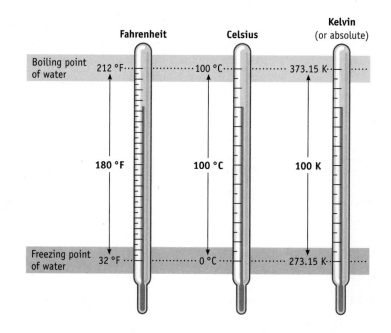

Thus, the freezing point of water is reached at 273.15 K; that is, 0 °C = 273.15 K. The boiling point of pure water is 373.15 K. Temperatures in Celsius degrees are readily converted to kelvins, and vice versa, using the relation

$$T \text{ (K)} = \frac{1 \text{ K}}{1 \text{ °C}} (T \text{ °C} + 273.15 \text{ °C}) \qquad (1)$$

Thus, a common room temperature of 23.5 °C is

$$T \text{ (K)} = \frac{1 \text{ K}}{1 \text{ °C}} (23.5 \text{ °C} + 273.15 \text{ °C}) = 296.7 \text{ K}$$

There are two important things to notice here:

- Converting from degrees Celsius to kelvins simply requires adding 273.15 to the temperature in Celsius, and converting from kelvins to degrees Celsius requires subtracting 273.15 from the value in kelvins.
- The degree symbol (°) is not used with Kelvin temperatures. The name of the unit on this scale is the *kelvin* (not capitalized), and such temperatures are designated with a capital K.

● **Temperature Conversions and Significant Figures** When converting 23.5 °C to kelvins, adding 273.15 gives 296.65. However, the rules of "significant figures" (page 33) tell us that the sum or difference of two numbers can have no more decimal places than the number with the fewest decimal places. Thus, we round the answer to 296.7 K, a number with one decimal place.

Length, Volume, and Mass

The meter is the standard unit of *length*, but objects observed in chemistry are frequently smaller than 1 meter. Measurements are often reported in units of centimeters (cm), millimeters (mm), or micrometers (μm) (Figure 2), and objects on the atomic and molecular scale have dimensions of nanometers (nm; 1 nm = 1×10^{-9} m) or picometers (pm; 1 pm = 1×10^{-12} m) (Figure 3).

To illustrate the range of dimensions used in science, let us look at a recent study of the glassy skeleton of a sea sponge. The sea sponge in Figure 2a is about 20 cm long and a few centimeters in diameter. A closer look (Figure 2b) shows more detail of the lattice-like structure. Scientists at Bell Laboratories found that each strand of the lattice is a ceramic-fiber composite of silica (SiO_2) and protein less than 100 μm in diameter (Figure 2c). These strands are composed of "spicules," which, at the nanoscale level, consist of silica nanoparticles just a few nanometers in diameter (Figure 2d).

Photos courtesy of Joanna Aizenberg, Bell Laboratories. Reference: J. Aizenberg, et al., Science, Vol 309, pages 275–278, 2005.

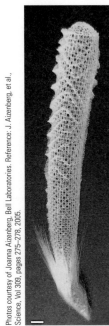

(b) Fragment of the structure showing the square grid of the lattice with diagonal supports. Scale bar = 1 mm.

(c) Scanning electron microscope (SEM) image of a single strand showing its ceramic-composite structure. Scale bar = 20 μm.

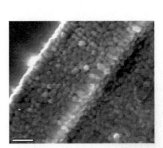

(d) SEM image of the surface of a strand showing that is it composed of nanoscale spheres of hydrated silica. Scale bar = 500 nm.

(a) Photograph of the glassy skeleton of a sea sponge, *Euplectella*. Scale bar = 5 cm.

FIGURE 2 Dimensions in chemistry and biology. These photos are from the research of Professor Joanna Aizenberg, now at Harvard University. Another image from her research on new materials is featured on the cover of this book.

FIGURE 3 Dimensions in the molecular world. Dimensions on the molecular scale are often given in terms of nanometers (1 nm = 1×10^{-9} m) or picometers (1 pm = 1×10^{-12} m). Here, the distance between C atoms in diamond is 0.154 nm or 154 pm. An older but often-used non-SI unit is the Ångstrom unit (Å), where 1 Å = 1.0×10^{-10} m. The C–C distance in diamond would be 1.54 Å.

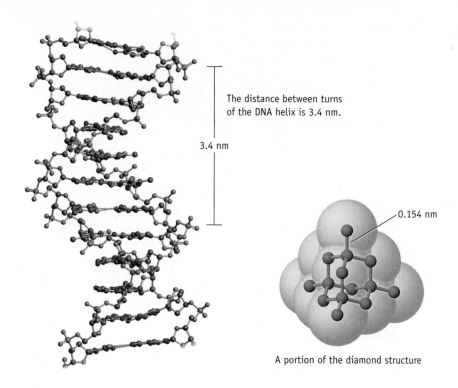

The distance between turns of the DNA helix is 3.4 nm.

3.4 nm

0.154 nm

A portion of the diamond structure

EXAMPLE 1   Distances on the Molecular Scale

Problem The distance between an O atom and an H atom in a water molecule is 95.8 pm. What is this distance in nanometers (nm)?

What Do You Know? You know the interatomic O–H distance and the relationships of the metric units.

95.8 pm

Strategy You can solve this problem by knowing the relationship or conversion factor between the units in the information you are given (picometers) and the desired units (meters or nanometers). (For more about conversion factors and their use in problem solving, see page 35.) There is no conversion factor given in Table 2 to change nanometers to picometers directly, but relationships are listed between meters and picometers and between meters and nanometers. Therefore, we first convert picometers to meters, and then we convert meters to nanometers.

$$\text{picometers} \xrightarrow{\;x\; ^m/_{pm}\;} \text{meters} \xrightarrow{\;y\; ^{nm}/_m\;} \text{nanometers}$$

Solution Using the appropriate conversion factors (1 pm = 1×10^{-12} m and 1 nm = 1×10^{-9} m), we have

$$95.8 \text{ pm} \times \frac{1 \times 10^{-12} \text{ m}}{1 \text{ pm}} = 9.58 \times 10^{-11} \text{ m}$$

$$9.58 \times 10^{-11} \text{ m} \times \frac{1 \text{ nm}}{1 \times 10^{-9} \text{ m}} = \boxed{9.58 \times 10^{-2} \text{ nm}} \text{ or } 0.0958 \text{ nm}$$

Think about Your Answer A nanometer is a larger unit than a picometer, so the same distance expressed in nanometers should have a smaller numerical value. Our answer is in line with this. Notice how the units cancel to leave an answer whose unit is that of the numerator of the conversion factor. The process of using units to guide a calculation is called *dimensional analysis*. It is explored further on pages 35–37.

Check Your Understanding

The C–C distance in diamond (Figure 3) is 0.154 nm. What is this distance in picometers (pm)? In centimeters (cm)?

Energy and Food

The U.S. Food and Drug Administration (FDA) mandates that nutritional data, including energy content, be included on almost all packaged food. The Nutrition Labeling and Education Act of 1990 requires that the total energy from protein, carbohydrates, fat, and alcohol be specified. How is this determined? Initially, the method used was calorimetry. In this method (described in Section 5.6), a food product is burned, and the energy transferred as heat in the combustion is measured. Now, however, energy contents are estimated using the Atwater system. This specifies the following average values for energy sources in foods:

   1 g protein = 4 kcal (17 kJ)
   1 g carbohydrate = 4 kcal (17 kJ)
   1 g fat = 9 kcal (38 kJ)
   1 g alcohol = 7 kcal (29 kJ)

Because carbohydrates may include some indigestible fiber, the mass of fiber is subtracted from the mass of carbohydrate when calculating the energy from carbohydrates.

As an example, one serving of cashew nuts (about 28 g) has

   14 g fat = 126 kcal
   6 g protein = 24 kcal
   7 g carbohydrates − 1 g fiber = 24 kcal
   Total = 174 kcal (728 kJ)

A value of 170 kcal is reported on the package.

You can find data on more than 6000 foods at the Nutrient Data Laboratory website (www.ars.usda.gov/ba/bhnrc/ndl).

Nutrition Facts

Serving Size 1 cup (30g)
   Children Under 4 - ¾ cup (20g)
Servings Per Container About 19
   Children Under 4 - About 28

Amount Per Serving	Cheerios	with ½ cup skim milk	Cereal for Children Under 4
Calories	110	150	70
Calories from Fat	15	20	10

	% Daily Value**		
Total Fat 2g*	3%	3%	1g
Saturated Fat 0g	0%	3%	0g
Polyunsaturated Fat 0.5g			0g
Monounsaturated Fat 0.5g			0g
Cholesterol 0mg	0%	1%	0mg
Sodium 210mg	9%	12%	140mg
Potassium 200mg	6%	12%	130mg

Energy and food labels. All packaged foods must have labels specifying nutritional values, with energy given in Calories (where 1 Cal = 1 kilocalorie).

Chemists often use glassware such as beakers, flasks, pipets, graduated cylinders, and burets, which are marked in volume units (Figure 4). The SI unit of volume is the cubic meter (m^3), which is too large for everyday laboratory use. Chemists usually use the liter (L) or the milliliter (mL) for volume measurements. One liter is equivalent to the volume of a cube with sides equal to 10 cm [$V = (0.1 \text{ m})^3 = 0.001 \text{ m}^3$].

$$1 \text{ liter (L)} = 1000 \text{ cm}^3 = 1000 \text{ mL} = 0.001 \text{ m}^3$$

Because there are exactly 1000 mL (= 1000 cm^3) in a liter, this means that

$$1 \text{ mL} = 0.001 \text{ L} = 1 \text{ cm}^3$$

The units *milliliter and cubic centimeter* (or "cc") *are interchangeable.* Therefore, a flask that contains exactly 125 mL has a volume of 125 cm^3.

Although not widely used in the United States, the cubic decimeter (dm^3) is a common unit in the rest of the world. A length of 10 cm is called a decimeter (dm). Because a cube 10 cm on a side defines a volume of 1 liter, *a liter is equivalent to a cubic decimeter:* $1 \text{ L} = 1 \text{ dm}^3$. Products in Europe, Africa, and other parts of the world are often sold by the cubic decimeter.

The *deciliter, dL,* which is exactly equivalent to 0.100 L or 100 mL, is widely used in medicine. For example, standards for concentrations of environmental contaminants are often set as a certain mass per deciliter. The state of Massachusetts recommends that children with more than 10 micrograms (10×10^{-6} g) of lead per deciliter of blood undergo further testing for lead poisoning.

Finally, when chemists prepare chemicals for reactions, they often take given masses of materials. The *mass* of a body is the fundamental measure of the quantity of matter, and the SI unit of mass is the kilogram (kg). Smaller masses are expressed in grams (g) or milligrams (mg).

$$1 \text{ kg} = 1000 \text{ g } and \text{ 1 g} = 1000 \text{ mg}$$

Energy Units

When expressing energy quantities, most chemists (and much of the world outside the United States) use the **joule** (J), the SI unit. The joule is related directly to the units used for mechanical energy: 1 J equals 1 kg · m^2/s^2. Because the joule is inconveniently small for most uses in chemistry, the kilojoule (kJ), equivalent to 1000 joules, is often the unit of choice.

© Cengage Learning/Charles D. Winters

FIGURE 4 Some common laboratory glassware. Volumes are marked in units of milliliters (mL). Remember that 1 mL is equivalent to 1 cm^3.

● **James Joule** The joule is named for James P. Joule (1818–1889), the son of a wealthy brewer in Manchester, England. The family wealth and a workshop in the brewery gave Joule the opportunity to pursue scientific studies. Among the topics that Joule studied was the issue of whether heat was a massless fluid. Scientists at that time referred to this idea as the caloric hypothesis. Joule's careful experiments showed that heat and mechanical work are related, providing evidence that heat is not a fluid.

● **Review & Check** Answers to the Review & Check questions are in Appendix P. In the Cengage *YouBook*, click on an answer to see if you are correct.

To give you some feeling for joules, suppose you drop a six-pack of soft-drink cans, each full of liquid, on your foot. Although you probably will not take time to calculate the kinetic energy at the moment of impact, it is between 4 J and 10 J.

The calorie (cal) is an older energy unit. It is defined as the energy transferred as heat that is required to raise the temperature of 1.00 g of pure liquid water from 14.5 °C to 15.5 °C. A kilocalorie (kcal) is equivalent to 1000 calories. The conversion factor relating joules and calories is

$$\text{1 calorie (cal)} = \text{4.184 joules (J)}$$

The dietary Calorie (with a capital C) is often used in the United States to represent the energy content of foods. The dietary Calorie (Cal) is equivalent to the kilocalorie or 1000 calories. Thus, a breakfast cereal that gives you 100.0 Calories of nutritional energy per serving provides 100.0 kcal or 418.4 kJ.

REVIEW & CHECK FOR SECTION 1

1. Liquid nitrogen boils at 77 K. What is this temperature in Celsius degrees?

   (a) 350 °C   (b) −196 °C   (c) 77 °C   (d) −109 °C

2. A square platinum sheet has sides 2.50 cm long and a thickness of 0.25 mm. What is the volume of the platinum sheet (in cm³)?

   (a) 16 cm³   (b) 630 cm³   (c) 0.16 cm³   (d) 160 cm³

3. A standard wine bottle has a volume of 750. mL. What volume, in liters, does this represent?

   (a) 0.750 L   (b) 0.00750 L   (c) 7.50 L   (d) 75.0 L

4. One U.S. gallon is equivalent to 3.7865 L. What is the volume in liters of a 2.0-quart carton of milk? (There are 4 quarts in a gallon.)

   (a) 19 L   (b) 3.8 L   (c) 1.9 L   (d) 0.95 L

5. A circulated U.S. quarter has a mass of 5.59 g. Express this mass in milligrams.

   (a) 5.59×10^3 mg   (b) 5.59 mg   (c) 55.9 mg   (d) 559 mg

6. An environmental study of a river found a pesticide present to the extent of 0.02 microgram per liter of water. Express this amount in grams per liter.

   (a) 2 g/L   (b) 0.02 g/L   (c) 2×10^3 g/L   (d) 2×10^{-8} g/L

7. The label on a cereal box indicates that one serving (with skim milk) provides 251 Cal. What is this energy in kilojoules (kJ)?

   (a) 251 kJ   (b) 59.8 kJ   (c) 1050 kJ   (d) 4380 kJ

---

2 Making Measurements: Precision, Accuracy, Experimental Error, and Standard Deviation

The **precision** of a measurement indicates how well several determinations of the same quantity agree. This is illustrated by the results of throwing darts at a target. In Figure 5a, the dart thrower was apparently not skillful, and the precision of the dart's placement on the target is low. In Figures 5b and 5c, the darts are clustered together, indicating much better consistency on the part of the thrower—that is, greater precision.

Accuracy is the agreement of a measurement with the accepted value of the quantity. Figure 5c shows that our thrower was accurate as well as precise—the average of all shots is close to the targeted position, the bull's eye.

Figure 5b shows it is possible to be precise without being accurate—the thrower has consistently missed the bull's eye, although all the darts are clustered precisely around one point on the target. This is analogous to an experiment with some flaw

● **Accuracy and NIST** The National Institute of Standards and Technology (NIST) is an important resource for the standards used in science. Comparison with NIST data is a test of the accuracy of the measurement (see www.nist.gov).

(a) Poor precision and poor accuracy

(b) Good precision and poor accuracy

(c) Good precision and good accuracy

FIGURE 5 **Precision and accuracy.**

(either in design or in a measuring device) that causes all results to differ from the correct value by the same amount.

The accuracy of a result in the laboratory is often expressed in terms of percent error, whereas the precision is expressed as a standard deviation.

Experimental Error

If you measure a quantity in the laboratory, you may be required to report the error in the result, the difference between your result and the accepted value,

Error in measurement = experimentally determined value − accepted value

or the **percent error.**

$$\text{Percent error} = \frac{\text{error in measurement}}{\text{accepted value}} \times 100\%$$

only 1 significant figure

● **Percent Error** Percent error can be positive or negative, indicating whether the experimental value is too high or too low compared to the accepted value. In Example 2, Student B's percent error is −0.2%, indicating it is 0.2% lower than the accepted value.

EXAMPLE 2   Precision, Accuracy, and Error

Problem A coin has an "accepted" diameter of 28.054 mm. In an experiment, two students measure this diameter. Student A makes four measurements of the diameter of the coin using a precision tool called a micrometer. Student B measures the same coin using a simple plastic ruler. The two students report the following results:

Student A	Student B
28.246 mm	27.9 mm
28.244	28.0
28.246	27.8
28.248	28.1

What is the average diameter and percent error obtained in each case? Which student's data are more accurate?

What Do You Know? You know the data collected by the two students and want to compare them with the "accepted" value by calculating the percent error.

Strategy For each set of values, we calculate the average of the results and then compare this average with 28.054 mm.

Solution The average for each set of data is obtained by summing the four values and dividing by 4.

Average value for Student A = 28.246 mm
Average value for Student B = 28.0 mm

Although Student A has four results very close to one another (and so of high precision), Student A's result is less accurate than that of Student B. The average diameter for Student A differs from the "accepted" value by 0.192 mm and has a percent error of 0.684%:

$$\text{Percent error} = \frac{28.246 \text{ mm} - 28.054 \text{ mm}}{28.054 \text{ mm}} \times 100\% = \boxed{0.684\%}$$

Student B's measurement has a percent error of only about $\boxed{-0.2\%}$.

Think about Your Answer Although Student A had less accurate results than Student B, they were more precise; the standard deviation for Student A is 2×10^{-3} (calculated as described below), in contrast to Student B's larger value (standard deviation = 0.14). Possible reasons for the error in Student A's result are incorrect use of the micrometer or a flaw in the instrument.

Check Your Understanding

Two students measured the freezing point of an unknown liquid. Student A used an ordinary laboratory thermometer calibrated in 0.1 °C units. Student B used a thermometer certified by NIST (National Institute of Standards and Technology) and calibrated in 0.01 °C units. Their results were as follows:

Student A: −0.3 °C, 0.2 °C, 0.0 °C, −0.3 °C

Student B: −0.02 °C, +0.02 °C, 0.00 °C, +0.04 °C

Calculate the average value for A and B and, knowing the liquid was water (and using kelvins for temperature), calculate the percent error for each student. Which student has the smaller error?

Standard Deviation

Laboratory measurements can be in error for two basic reasons. First, "determinate" errors are caused by faulty instruments or human errors such as incorrect record keeping. Second, so-called indeterminate errors arise from uncertainties in a measurement where the cause is not known and cannot be controlled by the lab worker. One way to judge the indeterminate error in a result is to calculate the standard deviation.

The **standard deviation** of a series of measurements is *equal to the square root of the sum of the squares of the deviations for each measurement from the average, divided by one less than the number of measurements.* It has a precise statistical significance: Assuming a large number of measurements is used to calculate the average, 68% of the values collected are expected to be within one standard deviation of the value determined, and 95% are within two standard deviations.

Suppose you carefully measured the mass of water delivered by a 10-mL pipet. (A pipet containing a green solution is shown in Figure 4.) For five attempts at the measurement (shown in column 2 of the following table), the standard deviation is found as follows. First, the average of the measurements is calculated (here, 9.984). Next, the deviation of each individual measurement from this value is determined (column 3). These values are squared, giving the values in column 4, and the sum of these values is determined. The standard deviation is then calculated by dividing this sum by the number of determinations minus 1 (= 4) and taking the square root of the result.

Determination	Measured Mass (g)	Difference between Measurement and Average (g)	Square of Difference
1	9.990	0.006	4×10^{-5}
2	9.993	0.009	8×10^{-5}
3	9.973	−0.011	12×10^{-5}
4	9.980	−0.004	2×10^{-5}
5	9.982	−0.002	0.4×10^{-5}

$$\text{Average mass} = 9.984 \text{ g}$$

$$\text{Sum of squares of differences} = 26 \times 10^{-5}$$

$$\text{Standard deviation} = \sqrt{\frac{26 \times 10^{-5}}{4}} = 0.008$$

Based on this calculation, it would be appropriate to represent the measured mass as 9.984 ± 0.008 g. This would tell a reader that if this experiment were repeated, a majority of the values would fall in the range from 9.976 g to 9.992 g.

REVIEW & CHECK FOR SECTION 2

Two students were assigned to determine the mass of a sample of an unknown liquid. Student A used an ordinary laboratory balance that could determine mass to ± 0.01 g. Student B used an analytical balance that could measure mass to ± 0.1 mg. Each made four measurements, giving the following results:

Student A: 8.19 g, 8.22 g, 8.21 g, 8.25 g

Student B: 8.2210 g, 8.2210 g, 8.2209 g, 8.2210 g

1.   What is the standard deviation for Student A?

(a)   0.2 g   (b)   0.02 g   (c)   0.03 g   (d)   0.04 g

2.   Which student is more precise?

(a)   A   (b)   B

FIGURE 6 Lake Otsego. Cooperstown, NY, is located by this beautiful lake.

John C. Katz

3   Mathematics of Chemistry

Exponential or Scientific Notation

Lake Otsego in northern New York is also called *Glimmerglass,* a name suggested by James Fenimore Cooper (1789–1851), the great American author and an early resident of the village now known as Cooperstown. Extensive environmental studies have been done along this lake (Figure 6), and some quantitative information useful to chemists, biologists, and geologists is given in the following table:

Lake Otsego Characteristics	Quantitative Information
Area	2.33×10^7 m^2
Maximum depth	505 m
Dissolved solids in lake water	2×10^2 mg/L
Average rainfall in the lake basin	1.02×10^2 cm/year
Average snowfall in the lake basin	198 cm/year

All of the data collected are in metric units. However, some data are expressed in **fixed notation** (505 m, 198 cm/year), whereas other data are expressed in **exponential**, or **scientific**, **notation** (2.33×10^7 m^2). Scientific notation is a way of presenting very large or very small numbers in a compact and consistent form that simplifies calculations. Because of its convenience, scientific notation is widely used in sciences such as chemistry, physics, engineering, and astronomy (Figure 7).

In scientific notation a number is expressed as the product of two numbers: $N \times 10^n$. N is the *digit term* and is a number between 1 and 9.9999. . . . The second number, 10^n, the *exponential term*, is some integer power of 10. For example, 1234 is written in scientific notation as 1.234×10^3, or 1.234 multiplied by 10 three times:

$$1234 = 1.234 \times 10^1 \times 10^1 \times 10^1 = 1.234 \times 10^3$$

FIGURE 7 **Exponential numbers in astronomy.** The spiral galaxy M-83 is 3.0×10^6 parsecs away from Earth and has a diameter of 9.0×10^3 parsecs. One parsec (pc), a unit used in astronomy, is equivalent to 206265 AU (astronomical units), where 1 AU is 1.496×10^8 km. What is the distance between Earth and M-83 in kilometers (km)?

W. Keel, U. Alabama/NASA

Conversely, a number less than 1, such as 0.01234, is written as 1.234×10^{-2}. This notation tells us that 1.234 should be divided twice by 10 to obtain 0.01234:

$$0.01234 = \frac{1.234}{10^1 \times 10^1} = 1.234 \times 10^{-1} \times 10^{-1} = 1.234 \times 10^{-2}$$

When converting a number to scientific notation, notice that the exponent n is positive if the number is greater than 1 and negative if the number is less than 1. The value of n is the number of places by which the decimal is shifted to obtain the number in scientific notation:

$$1\,2\,3\,4\,5. = 1.2345 \times 10^4$$

(a) Decimal shifted four places to the left. Therefore, n is positive and equal to 4.

$$0.0\,0\,1\,2 = 1.2 \times 10^{-3}$$

(b) Decimal shifted three places to the right. Therefore, n is negative and equal to 3.

If you wish to convert a number in scientific notation to one using fixed notation (that is, not using powers of 10), the procedure is reversed:

$$6\,.\,2\,7\,3 \times 10^2 = 627.3$$

(a) Decimal point moved two places to the right because n is positive and equal to 2.

$$0\,0\,6.273 \times 10^{-3} = 0.006273$$

(b) Decimal point shifted three places to the left because n is negative and equal to 3.

Two final points should be made concerning scientific notation. First, be aware that calculators and computers often express a number such as 1.23×10^3 as 1.23E3 or 6.45×10^{-5} as 6.45E-5. Second, some electronic calculators can readily convert numbers in fixed notation to scientific notation. If you have such a calculator, you may be able to do this by pressing the EE or EXP key and then the "=" key (but check your calculator manual to learn how your device operates).

In chemistry, you will often have to use numbers in exponential notation in mathematical operations. The following five operations are important:

- *Adding and Subtracting Numbers Expressed in Scientific Notation*
  When adding or subtracting two numbers, first convert them to the same powers of 10. The digit terms are then added or subtracted as appropriate:

  $$(1.234 \times 10^{-3}) + (5.623 \times 10^{-2}) = (0.1234 \times 10^{-2}) + (5.623 \times 10^{-2})$$
  $$= 5.746 \times 10^{-2}$$

- *Multiplication of Numbers Expressed in Scientific Notation*
  The digit terms are multiplied in the usual manner, and the exponents are added. The result is expressed with a digit term with only one nonzero digit to the left of the decimal place:

  $$(6.0 \times 10^{23}) \times (2.0 \times 10^{-2}) = (6.0)(2.0 \times 10^{23-2}) = 12 \times 10^{21} = 1.2 \times 10^{22}$$

- *Division of Numbers Expressed in Scientific Notation*
  The digit terms are divided in the usual manner, and the exponents are subtracted. The quotient is written with one nonzero digit to the left of the decimal in the digit term:

  $$\frac{7.60 \times 10^3}{1.23 \times 10^2} = \frac{7.60}{1.23} \times 10^{3-2} = 6.18 \times 10^1$$

● **Comparing the Earth and a Plant Cell—Powers of 10**
Earth = 12,760,000 meters wide
   = 12.76 million meters
   = 1.276×10^7 meters
Plant cell = 0.00001276 meter wide
      = 12.76 millionths of a meter
      = 1.276×10^{-5} meters

Using Your Calculator

You will be performing a number of calculations in general chemistry, most of them using a calculator. Many different types of calculators are available, but this problem-solving tip describes several of the kinds of operations you will need to perform on a typical calculator. Be sure to consult your calculator manual for specific instructions to enter scientific notation and to find powers and roots of numbers.

1. Scientific Notation

When entering a number such as 1.23×10^{-4} into your calculator, you first enter 1.23 and then press a key marked EE or EXP (or something similar). This enters the "$\times 10$" portion of the notation for you. You then complete the entry by keying in the exponent of the number, 4. (To change the exponent from $+4$ to -4, press the "$+/-$" key.)

A common error made by students is to enter 1.23, press the multiply key ($\times$), and then key in 10 before finishing by pressing EE or EXP followed by -4. This gives you an entry that is 10 times too large.

2. Powers of Numbers

Electronic calculators often offer two methods of raising a number to a power. To square a number, enter the number and then press the x^2 key. To raise a number to any power, use the y^x (or similar key such as ^). For example, to raise 1.42×10^2 to the fourth power:

1. Enter 1.42×10^2.
2. Press y^x.
3. Enter 4 (this should appear on the display).
4. Press =, and 4.0659×10^8 appears on the display.

3. Roots of Numbers

A general procedure for finding any root is to use the y^x key. For a square root, x is 0.5 (or 1/2), whereas x is 0.3333 (or 1/3) for a cube root, 0.25 (or 1/4) for a fourth root, and so on. For example, to find the fourth root of 5.6×10^{-10}:

1. Enter the number.
2. Press the y^x key.
3. Enter the desired root. Because we want the fourth root, enter 0.25.
4. Press =. The answer here is 4.9×10^{-3}.

To make sure you are using your calculator correctly, try these sample calculations:

1. $(6.02 \times 10^{23})(2.26 \times 10^{-5})/367$
   (Answer = 3.71×10^{16})
2. $(4.32 \times 10^{-3})^3$
   (Answer = 8.06×10^{-8})
3. $(4.32 \times 10^{-3})^{1/3}$
   (Answer = 0.163)

- *Powers of Numbers Expressed in Scientific Notation*
  When raising a number in exponential notation to a power, treat the digit term in the usual manner. The exponent is then multiplied by the number indicating the power:

$$(5.28 \times 10^3)^2 = (5.28)^2 \times 10^{3 \times 2} = 27.9 \times 10^6 = 2.79 \times 10^7$$

- *Roots of Numbers Expressed in Scientific Notation*
  Unless you use an electronic calculator, the number must first be put into a form in which the exponent is exactly divisible by the root. For example, for a square root, the exponent should be divisible by 2. The root of the digit term is found in the usual way, and the exponent is divided by the desired root:

$$\sqrt{3.6 \times 10^7} = \sqrt{36 \times 10^6} = \sqrt{36} \times \sqrt{10^6} = 6.0 \times 10^3$$

Significant Figures

In most experiments, several kinds of measurements must be made, and some can be made more precisely than others. It is common sense that a result calculated from experimental data can be no more precise than the least precise piece of information that went into the calculation. This is where the rules for significant figures come in. **Significant figures** are the digits in a measured quantity that were observed with the measuring device.

Determining Significant Figures

Suppose we place a new U.S. dime on the pan of a standard laboratory balance such as the one pictured in Figure 8 and observe a mass of 2.265 g. This number has four significant figures or digits because all four numbers are observed. However, you will learn from experience that the final digit (5) is somewhat uncertain because you may notice the balance readings can change slightly and give masses of 2.264, 2.265, and 2.266, with the mass of 2.265 observed most of the time. Thus,

FIGURE 8 Standard laboratory balance and significant figures. Such balances can determine the mass of an object to the nearest milligram. Thus, an object may have a mass of 13.456 g (13456 mg, five significant figures), 0.123 g (123 mg, three significant figures), or 0.072 g (72 mg, two significant figures).

© Cengage Learning/Charles D. Winters

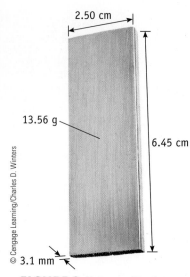

2.50 cm

13.56 g

6.45 cm

3.1 mm

FIGURE 9 Data used to determine the density of a metal.

of the four significant digits (2.265) the last (5) is uncertain. In general, *in a number representing a scientific measurement, the last digit to the right is taken to be inexact.* Unless stated otherwise, it is common practice to assign an uncertainty of ± 1 to the last significant digit.

Suppose you want to calculate the density of a piece of metal (Figure 9). The mass and dimensions were determined by standard laboratory techniques. Most of these data have two digits to the right of the decimal, but they have different numbers of significant figures.

Measurement	Data Collected	Significant Figures
Mass of metal	13.56 g	4
Length	6.45 cm	3
Width	2.50 cm	3
Thickness	3.1 mm = 0.31 cm	2

The quantity 0.31 cm has two significant figures. The 3 in 0.31 is exactly known, but the 1 is uncertain. That is, the thickness of the metal piece may have been as small as 0.30 cm or as large as 0.32 cm, but it is most likely 0.31 cm. In the case of the width of the piece, you found it to be 2.50 cm, where 2.5 is known with certainty, but the final 0 is uncertain. There are three significant figures in 2.50.

When you read a number in a problem or collect data in the laboratory (Figure 10), how do you determine which digits are significant?

First, is the number an exact number or a measured quantity? If it is an exact number, you don't have to worry about the number of significant figures. For example, there are exactly 100 cm in 1 m. We could add as many zeros after the decimal place as we want, and the expression would still be true. Using this number in a calculation will not affect how many significant figures you can report in your answer.

If, however, the number is a measured value, you must take into account significant figures. The number of significant figures in our data above is clear, with the possible exception of 0.31 and 2.50. Are the zeroes significant?

1.   *Zeroes between two other significant digits are significant.* For example, the zero in 103 is significant.

The 10-mL graduated cylinder is marked in 0.1-mL increments. Graduated cylinders are not considered precision glassware, so, at best, you can expect no more than two significant figures when reading a volume.

A 50-mL buret is marked in 0.10-mL increments, but it may be read with greater precision (0.01 mL).

A volumetric flask is meant to be filled to the mark on the neck. For a 250-mL flask, the volume is known to the nearest 0.12 mL, so the flask contains 250.0 ± 0.1 mL when full to the mark (four significant figures).

A pipet is like a volumetric flask in that it is filled to the mark on its neck. For a 20-mL pipet the volume is known to the nearest 0.02 mL.

FIGURE 10 Glassware and significant figures.

2. *Zeroes to the right of a nonzero number, and also to the right of a decimal place, are significant.* For example, in the number 2.50 cm, the zero is significant.

3. *Zeroes that are placeholders are not significant.* There are two types of numbers that fall under this rule.

    (a) The first are decimal numbers with zeroes that occur *before* the first nonzero digit. For example, in 0.0013, only the 1 and the 3 are significant; the zeroes are not. This number has two significant figures.

    (b) The second are *numbers with trailing zeroes* that must be there to indicate the magnitude of the number. For example, the zeroes in the number 13,000 may or may not be significant; it depends on whether they were measured or not. To avoid confusion with regard to such numbers, *we shall assume in this book that trailing zeroes are significant when there is a decimal point to the right of the last zero.* Thus, we would say that 13,000 has only two significant figures but that 13,000. has five. We suggest that the best way to be unambiguous when writing numbers with trailing zeroes is to use scientific notation. For example 1.300×10^4 clearly indicates four significant figures, whereas 1.3×10^4 indicates two and 1.3000×10^4 indicates five.

Using Significant Figures in Calculations

When doing calculations using measured quantities, we follow some basic rules so that the results reflect the precision of all the measurements that go into the calculations. *The rules used for significant figures in this book are as follows:*

Rule 1. When adding or subtracting numbers, the number of decimal places in the answer is equal to the number of decimal places in the number with the fewest digits after the decimal.

0.122 decimal places	
+ 1.9	1 decimal place
+10.925	3 decimal places
12.945	3 decimal places

The sum should be reported as 12.9, a number with one decimal place, because 1.9 has only one decimal place.

Rule 2. In multiplication or division, the number of significant figures in the answer is determined by the quantity with the fewest significant figures.

$$\frac{0.01208}{0.0236} = 0.512, \text{ or in scientific notation, } 5.12 \times 10^{-1}$$

Because 0.0236 has only three significant digits, while 0.01208 has four, the answer should have three significant digits.

Rule 3. When a number is rounded off, the last digit to be retained is increased by one only if the following digit is 5 or greater.

Full Number	Number Rounded to Three Significant Digits
12.696	12.7
16.349	16.3
18.35	18.4
18.351	18.4

● **Zeroes and Common Laboratory Mistakes** We often see students find the mass of a chemical on a balance and fail to write down trailing zeroes. For example, if you find the mass is 2.340 g, the final zero is significant and must be reported as part of the measured value. The number 2.34 g has only three significant figures and implies the 4 is uncertain, when in fact the balance reading indicated the 4 is certain.

● **Calculations and Significant Figures** Keep in mind that there is one rule for addition/subtraction and another rule for multiplication/division. A common error that students make is to learn only one rule and use it in all circumstances.

Now let us apply these rules to calculate the density of the piece of metal in Figure 9.

$$\text{Length} \times \text{width} \times \text{thickness} = \text{volume}$$

$$6.45 \text{ cm} \times 2.50 \text{ cm} \times 0.31 \text{ cm} = 5.0 \text{ cm}^3$$

$$\text{Density} = \frac{\text{mass (g)}}{\text{volume (cm}^3)} = \frac{13.56 \text{ g}}{5.0 \text{ cm}^3} = 2.7 \text{ g/cm}^3$$

The calculated density has two significant figures because *a calculated result can be no more precise than the least precise data used,* and here the thickness has only two significant figures.

One last word on significant figures and calculations: When working problems, you should do the calculation with all the digits allowed by your calculator and round off only at the end of the calculation. *Rounding off in the middle of a calculation can introduce errors.*

EXAMPLE 3   Using Significant Figures

Problem An example of a calculation you will do later in the book (▶Chapter 11) is

$$\text{Volume of gas (L)} = \frac{(0.120)(0.08206)(273.15 + 23)}{(230/760.0)}$$

Calculate the final answer to the correct number of significant figures.

What Do You Know? You know the rules for determining the number of significant figures for each number in the equation.

Strategy First decide on the number of significant figures represented by each number and then apply Rules 1–3.

Solution

Number	Number of Significant Figures	Comments
0.120	3	The trailing 0 is significant.
0.08206	4	The first 0 to the immediate right of the decimal is not significant.
273.15 + 23 = 296	3	23 has no decimal places, so the sum can have none.
230/760.0 = 0.30	2	230 has two significant figures because the last zero is not significant. In contrast, there is a decimal point in 760.0, so there are four significant digits. The quotient will have only two significant digits.

Multiplication and division gives 9.63 L. However, analysis shows that one of the pieces of information is known to only two significant figures. Therefore, we should report the volume of gas as 9.6 L, a number with two significant figures.

Think about Your Answer Be careful when you add or subtract two numbers because it is easy to make significant figure errors when doing so.

Check Your Understanding

What is the result of the following calculation?

$$x = \frac{(110.7 - 64)}{(0.056)(0.00216)}$$

REVIEW & CHECK FOR SECTION 3

1. How many significant figures are indicated by 2.33×10^3?

   (a) 1 (b) 2 (c) 3 (d) 4

2. How many significant figures are indicated by 200?

   (a) 1 (b) 2 (c) 3

3. What is the sum of 10.26 g and 0.063 g?

   (a) 10.323 g (b) 10.32 g (c) 10.3 g (d) 10. g

4. What is the product of 10.26 cm and 0.063 cm?

   (a) 0.6464 cm^2 (b) 0.646 cm^2 (c) 0.65 cm^2 (d) 0.6 cm^2

4 Problem Solving by Dimensional Analysis

Figure 9 illustrated the data that were collected to determine the density of a piece of metal. The thickness was measured in millimeters, whereas the length and width were measured in centimeters. To find the volume of the sample in cubic centimeters, we first had to have the length, width, and thickness in the same units, so we converted the thickness to centimeters.

$$3.1\ \cancel{\text{mm}} \times \frac{1\ \text{cm}}{10\ \cancel{\text{mm}}} = 0.31\ \text{cm}$$

Here, we multiplied the number we wished to convert (3.1 mm) by a *conversion factor* (1 cm/10 mm) to produce the result in the desired unit (0.31 cm). Notice that units are treated like numbers. Because the unit "mm" was in both the numerator and the denominator, dividing one by the other leaves a quotient of 1. The units are said to "cancel out." This leaves the answer in centimeters, the desired unit.

This approach to problem solving is often called **dimensional analysis** (or sometimes the **factor-label method**). It is a general problem-solving approach that uses the dimensions or units of each value to guide us through calculations.

A **conversion factor** expresses the equivalence of a measurement in two different units (1 cm ≡ 10 mm; 1 g ≡ 1000 mg; 12 eggs ≡ 1 dozen; 12 inches ≡ 1 foot). Because the numerator and the denominator describe the same quantity, the conversion factor is equivalent to the number 1. Therefore, multiplication by this factor does not change the measured quantity, only its units. A conversion factor is always written so that it has the form "new units divided by units of original number."

$$\underset{\substack{\uparrow\\ \text{Quantity to express in new units}}}{\text{Number in original unit}} \times \underbrace{\left[\frac{\text{new unit}}{\text{original unit}}\right]}_{\text{Conversion factor}} = \underset{\substack{\uparrow\\ \text{Quantity now expressed in new units}}}{\text{new number in new unit}}$$

• **Using Conversion Factors and Doing Calculations** As you work problems in this book and read Example problems, notice that proceeding from given information to an answer very often involves a series of multiplications. That is, we multiply the given data by a conversion factor, multiply the answer of that step by another factor, and so on, to the answer.

EXAMPLE 4 Using Conversion Factors and Dimensional Analysis

Problem Oceanographers often express the density of sea water in units of kilograms per cubic meter. If the density of sea water is 1.025 g/cm^3 at 15 °C, what is its density in kilograms per cubic meter?

What Do You Know? You know the density in a unit involving mass in grams and volume in cubic centimeters. Each of these has to be changed to its equivalent in kilograms and cubic meters, respectively.

● **Who Is Right—You or the Book?**
If your answer to a problem in this book does not quite agree with the answers in Appendix N through P, the discrepancy may be the result of rounding the answer after each step and then using that rounded answer in the next step. This book follows these conventions:

(a) Final answers to numerical problems result from retaining several digits more than the number required by significant figures throughout the calculation and rounding only at the end.

(b) In Example problems, the answer to each step is given to the correct number of significant figures for that step, but more digits are carried to the next step. The number of significant figures in the final answer is dictated by the number of significant figures in the original data.

Strategy To simplify this problem, break it into three steps. First, change the mass in grams to kilograms. Next, convert the volume in cubic centimeters to cubic meters. Finally, calculate the density by dividing the mass in kilograms by the volume in cubic meters.

Solution First convert the mass in grams to a mass in kilograms.

$$1.025 \; \cancel{g} \times \frac{1 \; kg}{1000 \; \cancel{g}} = 1.025 \times 10^{-3} \; kg$$

No conversion factor is available in one of our tables to directly change units of cubic centimeters to cubic meters. You can find one, however, by cubing (raising to the third power) the relation between the meter and the centimeter:

$$1 \; cm^3 \times \left(\frac{1 \; m}{100 \; cm} \right)^3 = 1 \; \cancel{cm^3} \times \left(\frac{1 \; m^3}{1 \times 10^6 \; \cancel{cm^3}} \right) = 1 \times 10^{-6} \; m^3$$

Therefore, the density of sea water is

$$\text{Density} = \frac{1.025 \times 10^{-3} \; kg}{1 \times 10^{-6} \; m^3} = \boxed{1025 \; kg/m^3}$$

Think about Your Answer Densities in units of kg/m^3 can often be large numbers. For example, the density of platinum is 21,450 kg/m^3, but dry air has a density of only 1.204 kg/m^3.

Check Your Understanding

The density of gold is 19,320 kg/m^3. What is this density in g/cm^3?

Out of Gas!

On July 23, 1983, a new Boeing 767 jet aircraft was flying at 26,000 ft from Montreal to Edmonton as Air Canada Flight 143. Warning buzzers sounded in the cockpit. One of the world's largest planes was now a glider—the plane had run out of fuel!

How did this modern airplane, having the latest technology, run out of fuel? A simple mistake had been made in calculating the amount of fuel required for the flight!

Like all Boeing 767s, this plane had a sophisticated fuel gauge, but it was not working properly. The plane was still allowed to fly, however, because there is an alternative method of determining the quantity of fuel in the tanks. Mechanics can use a stick, much like the oil dipstick in an automobile engine, to measure the fuel level in each of the three tanks. The mechanics in Montreal read the dipsticks, which were calibrated in centimeters, and translated those readings to a volume in liters. According to this, the plane had a total of 7682 L of fuel.

Pilots always calculate fuel quantities in units of mass because they need to know the total mass of the plane before take-off. Air Canada pilots had always calculated the quantity of fuel in pounds, but the new 767's fuel consumption was given in kilograms. The pilots knew that 22,300 kg of fuel was required for the trip. If 7682 L of fuel

remained in the tanks, how much had to be added? This involved using the fuel's density to convert 7682 L to a mass in kilograms. The mass of fuel to be added could then be calculated, and that mass converted to a volume of fuel to be added.

The First Officer of the plane asked a mechanic for the conversion factor to do the volume-to-mass conversion, and the mechanic replied "1.77." Using that number, the First Officer and the mechanics calculated that 4917 L of fuel should be added. But later calculations showed that this is only about one fourth of the required amount of fuel! Why? Because no one thought about the units of the number 1.77. They realized later that 1.77 has units of pounds per liter and not kilograms per liter.

Out of fuel, the plane could not make it to Winnipeg, so controllers directed them to the town of Gimli and to a small airport abandoned by the Royal Canadian Air Force. After gliding for almost 30 minutes, the plane approached the Gimli runway. The runway, however, had been converted to a race course for cars, and a race was underway. Furthermore, a steel barrier had been erected across the runway. Nonetheless, the pilot managed to touch down very near the end of the runway. The plane sped down the concrete strip; the nose wheel collapsed; several tires blew—and

the plane skidded safely to a stop just before the barrier. The Gimli glider had made it! And somewhere an aircraft mechanic is paying more attention to units on numbers.

Questions:
1. What is the fuel density in units of kg/L?
2. What mass and what volume of fuel should have been loaded? (1 lb = 453.6 g) (See Study Question 62, page 49.)

Answers to these questions are available in Appendix N.

© Wayne Glowacki/Winnipeg Free Press, July 23, 1987, reproduced with permission.

The Gimli glider. After running out of fuel, Air Canada Flight 143 glided 29 minutes before landing on an abandoned airstrip at Gimli, Manitoba, near Winnipeg.

5 Graphs and Graphing

In a number of instances in this text, graphs are used when analyzing experimental data with a goal of obtaining a mathematical equation that may help us predict new results. The procedure used will often result in a straight line, which has the equation

$$y = mx + b$$

In this equation, y is usually referred to as the dependent variable; its value is determined from (that is, is dependent on) the values of x, m, and b. In this equation, x is called the independent variable, and m is the slope of the line. The parameter b is the y-intercept—that is, the value of y when $x = 0$. Let us use an example to investigate two things: (1) how to construct a graph from a set of data points and (2) how to derive an equation for the line generated by the data.

A set of data points to be graphed is presented in Figure 11. We first mark off each axis in increments of the values of x and y. Here, our x-data are within the range from -2 to 4, so the x-axis is marked off in increments of 1 unit. The y-data fall within the range from 0 to 2.5, so we mark off the y-axis in increments of 0.5. Each data point is marked as a circle on the graph.

After plotting the points on the graph (round circles), we draw a straight line that comes as close as possible to representing the trend in the data. (Do not just connect the dots!) Because there is always some inaccuracy in experimental data, the straight line we draw is unlikely to pass exactly through every point.

To identify the specific equation corresponding to our data, we must determine the y-intercept *(b)* and slope *(m)* for the equation $y = mx + b$. The y-intercept is the point at which $x = 0$ and thus is the point where the line intersects the y-axis. (In Figure 11, $y = 1.87$ when $x = 0$). The slope is determined by selecting two points *on the line* (marked with squares in Figure 11) and calculating the difference in values

● **Determining the Slope with a Computer Program—Least-Squares Analysis** Generally, the easiest method of determining the slope and intercept of a straight line (and thus the line's equation) is to use a program such as Microsoft Excel or Apple's Numbers. These programs perform a "least-squares" or "linear regression" analysis and give the best straight line based on the data. (This line is referred to in Excel or Numbers as a trendline.)

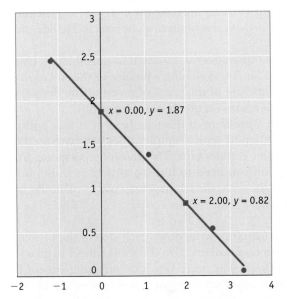

Experimental data

x	y
3.35	0.0565
2.59	0.520
1.08	1.38
-1.19	2.45

Using the points marked with a square, the slope of the line is:

$$\text{Slope} = \frac{\Delta y}{\Delta x} = \frac{0.82 - 1.87}{2.00 - 0.00} = -0.525$$

FIGURE 11 Plotting data.
Data for the variable x are plotted on the horizontal axis (abscissa), and data for y are plotted on the vertical axis (ordinate). The slope of the line, m in the equation $y = mx + b$, is given by $\Delta y / \Delta x$. The intercept of the line with the y-axis (when $x = 0$) is b in the equation. Using Microsoft Excel with these data and doing a linear regression analysis, we find $y = -0.525x + 1.87$.

of y ($\Delta y = y_2 - y_1$) and x ($\Delta x = x_2 - x_1$). The slope of the line is then the ratio of these differences, $m = \Delta y/\Delta x$. Here, the slope has the value -0.525. With the slope and intercept now known, we can write the equation for the line

$$y = -0.525x + 1.87$$

and we can use this equation to calculate y-values for points that are not part of our original set of x–y data. For example, when $x = 1.50$, we find $y = 1.08$.

REVIEW & CHECK FOR SECTION 5

To find the mass of 32 jelly beans, we weighed several samples of beans.

Number of Beans	Mass (g)
5	12.82
11	27.14
16	39.30
24	59.04

Plot these data with the number of beans on the horizontal or x-axis and the mass of beans on the vertical or y-axis.

1. What is the approximate slope of the line?

    (a) 2.4   (b) 0.42   (c) 3.2   (d) 4.8

2. Use your graph or equation of a straight line to calculate the mass of 32 jelly beans.

    (a) 150 g   (b) 100 g   (c) 13 g   (d) 78 g

---

6   Problem Solving and Chemical Arithmetic

Some of the calculations in chemistry can be complex. Students frequently find it helpful to follow a definite plan of attack as illustrated in examples throughout this book.

Step 1: State the Problem. Read it carefully—and then read it again.

Step 2: What Do You Know? Determine specifically what you are trying to determine and what information you are given. What key principles are involved? What information is known or not known? What information might be there just to place the question in the context of chemistry? Organize the information to see what is required and to discover the relationships among the data given. Try writing the information down in table form. If the information is numerical, be sure to include units.

Step 3: Strategy. One of the greatest difficulties for a student in introductory chemistry is picturing what is being asked for. Try sketching a picture of the situation involved. For example, we sketched a picture of the piece of metal whose density we wanted to calculate and put the dimensions on the drawing (page 32).

Develop a plan. Have you done a problem of this type before? If not, perhaps the problem is really just a combination of several simpler ones you have seen before. Break it down into those simpler components. Try reasoning backward from the units of the answer. What data do you need to find an answer in those units? Drawing a strategy map such as that shown to the left may help you in planning how you will go about solving the problem.

Step 4: Solution. Execute the plan. Carefully write down each step of the problem, being sure to keep track of the units on numbers. (Do the units cancel to give you the answer in the desired units?) Don't skip steps. Don't do anything except the simplest steps in your head. Students often say they got a problem wrong because

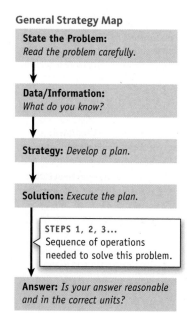

General Strategy Map

State the Problem:
Read the problem carefully.

↓

Data/Information:
What do you know?

↓

Strategy: *Develop a plan.*

↓

Solution: *Execute the plan.*

↓

STEPS 1, 2, 3...
Sequence of operations
needed to solve this problem.

↓

Answer: *Is your answer reasonable and in the correct units?*

● **Strategy Maps** Many Example problems in this book are accompanied by a Strategy Map that outlines a route to a solution. In the YouBook many of these maps are accompanied by an audio/video explanation and a tutorial.

they "made a stupid mistake." Your instructors—and book authors—also make them, and it is usually because they don't take the time to write down the steps of the problem clearly.

Step 5: Think about Your Answer. Ask yourself whether the answer is reasonable and if you obtained an answer in the correct units.

Step 6: Check Your Understanding. In this text each Example is followed by another problem for you to try. (The solutions to those questions are in Appendix O.) When doing homework Study Questions, try one of the simpler *Practicing Skills* questions to see if you understand the basic ideas.

The steps we outline for problem solving are ones that many students have found to be successful, so try to consciously follow this scheme. But also be flexible. The "What Do You Know?" and "Strategy" steps often blend into a single set of ideas.

⬥WL   **INTERACTIVE EXAMPLE 5**   Problem Solving

Problem A mineral oil has a density of 0.875 g/cm³. Suppose you spread 0.75 g of this oil over the surface of water in a large dish with an inner diameter of 21.6 cm. How thick is the oil layer? Express the thickness in centimeters.

What Do You Know? You know the mass and density of the oil and the diameter of the surface to be covered.

Strategy It is often useful to begin solving such problems by sketching a picture of the situation.

This helps you recognize that the solution to the problem is to find the volume of the oil on the water. If you know the volume, then you can find the thickness because

$$\text{Volume of oil layer} = (\text{thickness of layer}) \times (\text{area of oil layer})$$

So, you need two things: (1) the volume of the oil layer and (2) the area of the layer. The volume can be found using the mass and density of the oil. The area can be found because the oil will form a circle, which has an area equal to $\pi \times r^2$ (where r is the radius of the dish).

Solution First, calculate the volume of oil. The mass of the oil layer is known, so combining the mass of oil with its density gives the volume of the oil used:

$$0.75 \text{ g} \times \frac{1 \text{ cm}^3}{0.875 \text{ g}} = 0.86 \text{ cm}^3$$

Next, calculate the area of the oil layer. The oil is spread over a circular surface, whose area is given by

$$\text{Area} = \pi \times (\text{radius})^2$$

The radius of the oil layer is half its diameter (= 21.6 cm) or 10.8 cm, so

$$\text{Area of oil layer} = (3.142)(10.8 \text{ cm})^2 = 366 \text{ cm}^2$$

With the volume and the area of the oil layer known, the thickness can be calculated.

$$\text{Thickness} = \frac{\text{volume}}{\text{area}} = \frac{0.86 \text{ cm}^3}{366 \text{ cm}^2} = \boxed{0.0023 \text{ cm}}$$

Strategy Map for Example 5

PROBLEM
How thick will an oil layer be when a **given mass** covers a **given area**?

↓

DATA/INFORMATION
Mass and **density** of the oil and **diameter** of the circular surface to be covered.

↓

STEP 1. Calculate the **volume** of oil from **mass** and **density**.

↓

Volume of oil in **cm³**

STEP 2. Calculate the **surface area** from the **diameter**.

↓

Area to be covered in **cm²**

STEP 3. Divide the oil **volume** by the **surface area** to calculate the thickness in **cm**.

↓

Thickness of oil layer in **cm**

Think about Your Answer In the volume calculation, the calculator shows 0.857143. . . . The quotient should have two significant figures because 0.75 has two significant figures, so the result of this step is 0.86 cm^3. In the area calculation, the calculator shows 366.435. . . . The answer to this step should have three significant figures because 10.8 has three. When these interim results are combined in calculating thickness, however, the final result can have only two significant figures. Remember that premature rounding can lead to errors.

Check Your Understanding

A particular paint has a density of 0.914 g/cm^3. You need to cover a wall that is 7.6 m long and 2.74 m high with a paint layer 0.13 mm thick. What volume of paint (in liters) is required? What is the mass (in grams) of the paint layer?

REVIEW & CHECK FOR SECTION 6

A spool of copper wire has a mass of 4.26 kg. How long is the wire in the spool if the wire has a diameter of 2.2 mm? (The density of copper is 8.96 g/cm^3.)

(a)   280 m   (b)   220 m   (c)   130 m   (d)   14 m

OWL Questions and problems for this chapter are available in OWL. Use the OWL access card that came with this textbook to access assigned questions and problems for this chapter.

2 Atoms, Molecules, and Ions

TABELLE II.

REIHEN	GRUPPE I. — R^2O	GRUPPE II. — RO	GRUPPE III. — R^2O^3	GRUPPE IV. RH^4 RO^2	GRUPPE V. RH^3 R^2O^5	GRUPPE VI. RH^2 RO^3	GRUPPE VII. RH R^2O^7	GRUPPE VIII. — RO^4
1	H=1							
2	Li=7	Be=9,4	B=11	C=12	N=14	O=16	F=19	
3	Na=23	Mg=24	Al=27,3	Si=28	P=31	S=32	Cl=35,5	
4	K=39	Ca=40	—=44	Ti=48	V=51	Cr=52	Mn=55	Fe=56, Co=59, Ni=59, Cu=63.
5	(Cu=63)	Zn=65	—=68	—=72	As=75	Se=78	Br=80	
6	Rb=85	Sr=87	?Yt=88	Zr=90	Nb=94	Mo=96	—=100	Ru=104, Rh=104, Pd=106, Ag=108.
7	(Ag=108)	Cd=112	In=113	Sn=118	Sb=122	Te=125	J=127	
8	Cs=133	Ba=137	?Di=138	?Ce=140	—	—		— — — —
9	(—)	—	—	—	—	—	—	
10	—	—	?Er=178	?La=180	Ta=182	W=184	—	Os=195, Ir=197, Pt=198, Au=199.
11	(Au=199)	Hg=200	Tl=204	Pb=207	Bi=208	—	—	
12	—	—	—	Th=231	—	U=240	—	— — — —

Photos © Cengage Learning/Charles D. Winters

The Periodic Table, the Central Icon of Chemistry

Nineteenth-century chemists such as Newlands, Chancourtois, Mayer, and others devised ways to organize the chemistry of the elements with varying degrees of success. However, it was Dmitri Mendeleev in 1870 who first truly recognized the periodicity of the chemistry of the elements, who proposed the first periodic table, and who used this to predict the existence of yet-unknown elements.

Mendeleev placed the elements in a table in order of increasing atomic weight. In doing so Li, Be, B, C, N, O, and F became the first row of the table. The next element then known, sodium (Na), had properties quite similar to those of lithium (Li), so Na began the next row of the table. As additional elements were added in order of increasing atomic weight, elements with similar properties fell in columns or groups.

If you compare the periodic table published by Mendeleev in 1871 (shown here) with the table in the front of this book, you will see that many elements are missing in the 1871 table. Mendeleev's genius was that he recognized there must be yet-undiscovered elements, so he left a place for them in the table (marking the empty places with a —). For example, Mendeleev concluded that "Gruppe IV" was missing an element between silicon (Si) and tin (Sn) and marked its position as "— = 72." He called the missing element *eka-silicon* and predicted the element would have an atomic weight of 72 and a density of 5.5 g/cm³. Based on this and other predictions, chemists knew what to look for in mineral samples, and soon many of the missing elements were discovered.

Questions:

1. What is eka-silicon, and how close were Mendeleev's predictions to the actual values for this element?
2. How many of the missing elements can you identify?

Answers to these questions are available in Appendix N.

CHAPTER OUTLINE

2.1 Atomic Structure–Protons, Electrons, and Neutrons

2.2 Atomic Number and Atomic Mass

2.3 Isotopes

2.4 Atomic Weight

2.5 The Periodic Table go Chemistry

2.6 Molecules, Compounds, and Formulas

2.7 Ionic Compounds: Formulas, Names, and Properties go Chemistry

2.8 Molecular Compounds: Formulas and Names

2.9 Atoms, Molecules, and the Mole go Chemistry

2.10 Describing Compound Formulas

2.11 Hydrated Compounds

CHAPTER GOALS

See Chapter Goals Revisited (page 88) for Study Questions keyed to these goals.

- Describe atomic structure and define atomic number and mass number.

- Understand the nature of isotopes and calculate atomic masses from isotopic masses and abundances.

- Know the terminology of the periodic table.

- Interpret, predict, and write formulas for ionic and molecular compounds.

- Name ionic and molecular compounds.

- Understand some properties of ionic compounds.

- Explain the concept of the mole and use molar mass in calculations.

- Derive compound formulas from experimental data.

The chemical elements are forged in stars and, from these elements, molecules such as water and ammonia are made in outer space. These simple molecules and much more complex ones such as DNA and hemoglobin are found on earth. To comprehend the burgeoning fields of molecular biology as well as all modern chemistry, we have to understand the nature of the chemical elements and the properties and structures of molecules. This chapter begins our exploration of the chemistry of the elements, the building blocks of chemistry, and the compounds they form.

2.1 Atomic Structure—Protons, Electrons, and Neutrons

Around 1900 a series of experiments done by scientists in England such as Sir Joseph John Thomson (1856–1940) and Ernest Rutherford (1871–1937) established a model of the atom that is still the basis of modern atomic theory. Atoms themselves are made of subatomic particles, three of which are important in chemistry: electrically positive protons, electrically negative electrons, and, in all except one type of hydrogen atom, electrically neutral neutrons. The model places the more massive protons and neutrons in a very small nucleus (Figure 2.1), which contains all the positive charge and almost all the mass of an atom. Electrons, with a much smaller mass than protons or neutrons, surround the nucleus and occupy most of the volume. In a neutral atom, the number of electrons equals the number of protons.

The chemical properties of elements and molecules depend largely on the electrons in atoms. We shall look more carefully at their arrangement and how they influence atomic properties in Chapters 6 and 7. In this chapter, however, we first want to describe how the composition of the atom relates to its mass and then to the mass of compounds. This is crucial information when we consider the quantitative aspects of chemical reactions in later chapters.

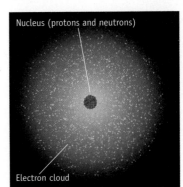

Nucleus (protons and neutrons)

Electron cloud

FIGURE 2.1 The structure of the atom. All atoms contain a nucleus with one or more protons (positive electric charge) and, except for one type of H atom, neutrons (no charge). Electrons (negative electric charge) are found in space as a "cloud" around the nucleus. In an electrically neutral atom, the number of electrons equals the number of protons. Note that this figure is not drawn to scale. If the nucleus were really the size depicted here, the electron cloud would extend over 200 m. The atom is mostly empty space!

43

2.2 Atomic Number and Atomic Mass

Atomic Number

All atoms of a given element have the same number of protons in the nucleus. Hydrogen is the simplest element, with one nuclear proton. All helium atoms have two protons, all lithium atoms have three protons, and all beryllium atoms have four protons. The number of protons in the nucleus of an element is given by its **atomic number**, which is generally indicated by the symbol **Z**.

Currently known elements are listed in the periodic table inside the front cover of this book and on the list inside the back cover. The integer number at the top of the box for each element in the periodic table is its atomic number. A sodium atom (Na), for example, has an atomic number of 11, so its nucleus contains 11 protons. A uranium atom (U) has 92 nuclear protons and $Z = 92$.

Relative Atomic Mass and the Atomic Mass Unit

| Copper |
| 29 ·······Atomic number |
| Cu ······Symbol |

With the quantitative work of the great French chemist Antoine Laurent Lavoisier (1743–1794), chemistry began to change from medieval alchemy to a modern field of study (page 94). As 18th- and 19th-century chemists tried to understand how the elements combined, they carried out increasingly quantitative studies aimed at learning, for example, how much of one element would combine with another. Based on this work, they learned that the substances they produced had a constant composition, so they could define the relative masses of elements that would combine to produce a new substance. At the beginning of the 19th century, John Dalton (1766–1844) suggested that the combinations of elements involve atoms and proposed a relative scale of atom masses. Apparently for simplicity, Dalton chose a mass of 1 for hydrogen on which to base his scale.

The atomic mass scale has changed since 1800, but like the 19th-century chemists, we still use *relative* masses with the standard today being carbon. A carbon atom having six protons and six neutrons in the nucleus is assigned a mass value of exactly 12. From chemical experiments and physical measurements, we know an oxygen atom having eight protons and eight neutrons has 1.33291 times the mass of carbon, so it has a relative mass of 15.9949. Masses of atoms of other elements are assigned in a similar manner.

Masses of fundamental atomic particles are often expressed in **atomic mass units (u)**. *One atomic mass unit, 1 u, is one-twelfth of the mass of an atom of carbon with six protons and six neutrons.* Thus, such a carbon atom has a mass of exactly 12 u. The atomic mass unit can be related to other units of mass using the conversion factor 1 atomic mass unit (u) $= 1.661 \times 10^{-24}$ g.

Mass Number

● **How Small Is an Atom?** The radius of the typical atom is between 30 and 300 pm (3×10^{-11} m to 3×10^{-10} m). To get a feeling for the incredible smallness of an atom, consider that 1 cm³ contains about three times as many atoms as the Atlantic Ocean contains teaspoons of water.

Because proton and neutron masses are so close to 1 u, while the mass of an electron is only about 1/2000 of this value (Table 2.1), the approximate mass of an atom can be estimated if the number of neutrons and protons is known. The sum of the number of protons and neutrons for an atom is called its **mass number** and is given the symbol **A**.

$$A = \text{mass number} = \text{number of protons} + \text{number of neutrons}$$

For example, a sodium atom, which has 11 protons and 12 neutrons in its nucleus, has a mass number of 23 ($A = 11$ p $+ 12$ n). The most common atom of uranium has 92 protons and 146 neutrons, and a mass number of $A = 238$. Using this information, we often symbolize atoms with the following notation:

$$\text{Mass number} \rightarrow {}_{Z}^{A}X \leftarrow \text{Element symbol}$$
$$\text{Atomic number} \rightarrow$$

The subscript Z is optional because the element's symbol tells us what the atomic number must be. For example, the atoms described previously have

Table 2.1 Properties of Subatomic Particles*

| Particle | Mass | | Charge | Symbol |
	Grams	Atomic Mass Units		
Electron	9.109383×10^{-28}	0.0005485799	1−	$_{-1}^{0}e$ or e^{-}
Proton	1.672622×10^{-24}	1.007276	1+	$_{1}^{1}p$ or p^{+}
Neutron	1.674927×10^{-24}	1.008665	0	$_{0}^{1}n$ or n

*These values and others in the book are taken from the National Institute of Standards and Technology website at http://physics.nist.gov/cuu/Constants/index.html

the symbols $_{11}^{23}$Na or $_{92}^{238}$U, or just ^{23}Na or ^{238}U. In words, we say "sodium-23" or "uranium-238."

EXAMPLE 2.1 Atomic Composition

Problem What is the composition of an atom of phosphorus with 16 neutrons? What is its mass number? What is the symbol for such an atom? If the atom has an actual mass of 30.9738 u, what is its mass in grams? Finally, what is the mass of this phosphorus atom relative to the mass of a carbon atom with a mass number of 12?

What Do You Know? You know the name of the element and the number of neutrons. You also know the actual mass, so you can determine its mass relative to carbon-12.

Strategy The symbol for phosphorus is P. You can look up the atomic number (which equals the number of protons) for this element on the periodic table. The mass number is the sum of the number of protons and neutrons. The mass of the atom in grams can be obtained from the mass in atomic mass units using the conversion factor $1 u = 1.661 \times 10^{-24}$ g. The relative mass of an atom of P compared to ^{12}C can be determined by dividing the mass of the P atom in atomic mass units by the mass of a ^{12}C atom, 12.0000 u.

Solution A phosphorus atom has 15 protons and, because it is electrically neutral, also has 15 electrons. A phosphorus atom with 16 neutrons has a mass number of 31.

$$\text{Mass number} = \text{number of protons} + \text{number of neutrons} = 15 + 16 = 31$$

The atom's complete symbol is $_{15}^{31}$P.

$$\text{Mass of one } ^{31}\text{P atom} = (30.9738 \text{ u}) \times (1.661 \times 10^{-24} \text{ g/u}) = 5.145 \times 10^{-23} \text{ g}$$

Mass of ^{31}P relative to the mass of an atom of ^{12}C: 30.9738/12.0000 = 2.58115

Think about Your Answer Because phosphorus has an atomic number greater than carbon's, you expect its relative mass to be greater than 12.

Check Your Understanding

1. What is the mass number of an iron atom with 30 neutrons?

2. A nickel atom with 32 neutrons has a mass of 59.930788 u. What is its mass in grams?

3. How many protons, neutrons, and electrons are in a ^{64}Zn atom?

REVIEW & CHECK FOR SECTION 2.2

1. The mass of an atom of manganese is 54.9380 u. How many neutrons are contained in one atom of this element?

    (a) 25 (b) 30 (c) 29 (d) 55

2. An atom contains 12 neutrons and has a mass number of 23. Identify the element.

    (a) C (b) Mg (c) Na (d) Cl

FIGURE 2.2 Ice made from "heavy water." Water containing ordinary hydrogen ($_1^1$H, protium) forms a solid that is less dense ($d = 0.917$ g/cm³ at 0 °C) than liquid H_2O ($d = 0.997$ g/cm³ at 25 °C), so it floats in the liquid. (Water is unique in this regard. The solid phase of virtually all other substances sinks in the liquid phase of that substance.) Similarly, "heavy ice" (D_2O, deuterium oxide) floats in "heavy water." D_2O-ice is denser than liquid H_2O, however, so cubes made of D_2O sink in liquid H_2O.

2.3 Isotopes

In only a few instances (for example, aluminum, fluorine, and phosphorus) do all atoms in a naturally occurring sample of a given element have the same mass. Most elements consist of atoms having several different mass numbers. For example, there are two kinds of boron atoms, one with a mass of about 10 (^{10}B) and a second with a mass of about 11 (^{11}B). Atoms of tin can have any of 10 different masses. Atoms with the same atomic number but different mass numbers are called **isotopes**.

All atoms of an element have the same number of protons—five in the case of boron. To have different masses, isotopes must have different numbers of neutrons. The nucleus of a ^{10}B atom ($Z = 5$) contains five protons and five neutrons, whereas the nucleus of a ^{11}B atom contains five protons and six neutrons.

Scientists often refer to a particular isotope by giving its mass number (for example, uranium-238, ^{238}U), but the isotopes of hydrogen are so important that they have special names and symbols. All hydrogen atoms have one proton. When that is the only nuclear particle, the isotope is called *protium*, or just "hydrogen." The isotope of hydrogen with one neutron, $_1^2$H, is called *deuterium*, or "heavy hydrogen" (symbol = D). The nucleus of radioactive hydrogen-3, $_1^3$H, or *tritium* (symbol = T), contains one proton and two neutrons.

The substitution of one isotope of an element for another isotope of the same element in a compound sometimes can have an interesting effect (Figure 2.2). This is especially true when deuterium is substituted for hydrogen because the mass of deuterium is double that of hydrogen.

Isotope Abundance

A sample of water from a stream or lake will consist almost entirely of H_2O where the H atoms are the ^{1}H isotope. A few molecules, however, will have deuterium (^{2}H) substituted for ^{1}H. We can predict this outcome because we know that 99.985% of all hydrogen atoms on earth are ^{1}H atoms. That is, the abundance of ^{1}H atoms is 99.985%.

$$\text{Percent abundance} = \frac{\text{number of atoms of a given isotope}}{\text{total number of atoms of all isotopes of that element}} \times 100\% \quad \text{(2.1)}$$

The remainder of naturally occurring hydrogen is deuterium, whose abundance is only 0.015% of the total hydrogen atoms. Tritium, the radioactive ^{3}H isotope, occurs naturally in only trace amounts.

Consider again the two isotopes of boron. The boron-10 isotope has an abundance of 19.91%; the abundance of boron-11 is 80.09%. Thus, if you could count out 10,000 boron atoms from an "average" natural sample, 1991 of them would be boron-10 atoms and 8009 of them would be boron-11 atoms.

Determining Atomic Mass and Isotope Abundance

The masses of isotopes and their abundances are determined experimentally using a mass spectrometer (Figure 2.3). A gaseous sample of an element is introduced into the evacuated chamber of the spectrometer, and the atoms or molecules of the sample are converted to positively charged particles (called *ions*). The cloud of ions forms a beam as they are attracted to negatively charged plates within the instrument. As the ions stream toward the negatively charged detector, they fly through a magnetic field, which causes the paths of the ions to be deflected. The extent of deflection depends on particle mass: The less massive ions are deflected more, and the more massive ions are deflected less. The ions, now separated by mass, are detected at the end of the chamber. Chemists using modern instruments can measure isotopic masses to as many as nine significant figures.

Except for carbon-12, whose mass is defined to be exactly 12 u, isotopic masses do not have integer values. However, the isotopic masses are always very close to the mass numbers for the isotope. For example, the mass of an atom of boron-11 (^{11}B, 5 protons and 6 neutrons) is 11.0093 u, and the mass of an atom of iron-58 (^{58}Fe, 26 protons and 32 neutrons) is 57.9333 u.

● **Atomic Masses of Some Isotopes**

Atom	Atomic Mass (u)
^{4}He	4.0092603
^{13}C	13.003355
^{16}O	15.994915
^{58}Ni	57.935346
^{60}Ni	59.930788
^{79}Br	78.918336
^{81}Br	80.916289
^{197}Au	196.966543
^{238}U	238.050784

● **Isotopic Masses and the Mass Defect** Actual masses of atoms are always less than the sum of the masses of the subatomic particles composing that atom. This is called the *mass defect* and the reason for it is discussed in Chapter 23.

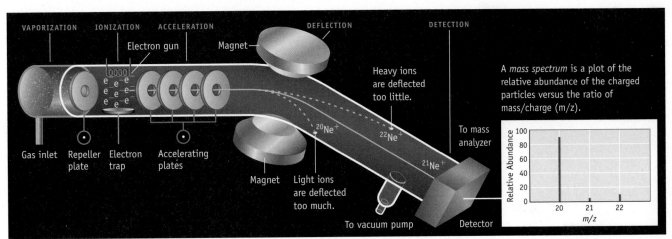

1. A sample is introduced as a vapor into the ionization chamber.

There it is bombarded with high-energy electrons that strip electrons from the atoms or molecules of the sample.

2. The resulting positive particles are accelerated by a series of negatively charged accelerator plates into an analyzing chamber.

3. This chamber is in a magnetic field, which is perpendicular to the direction of the beam of charged particles.

The magnetic field causes the beam to curve. The radius of curvature depends on the mass and charge of the particles (as well as the accelerating voltage and strength of the magnetic field).

4. Here, particles of ^{21}Ne$^+$ are focused on the detector, whereas beams of ions of ^{20}Ne$^+$ and ^{22}Ne$^+$ (of lighter or heavier mass) experience greater and lesser curvature, respectively, and so fail to be detected.

By changing the magnetic field, charged particles of different masses can be focused on the detector to generate the observed spectrum.

FIGURE 2.3 Mass spectrometer. A mass spectrometer will separate ions of different mass and charge in a gaseous sample of ions. The instrument allows the researcher to determine the accurate mass of each ion, whether the ions are composed of individual atoms, molecules, or molecular fragments.

REVIEW & CHECK FOR SECTION 2.3

Silver has two isotopes, one with 60 neutrons (percent abundance = 51.839%) and the other with 62 neutrons. What is the symbol of the isotope with 62 neutrons, and what is its percent abundance?

(a) $^{107}_{47}$Ag, 51.839%   (b) $^{107}_{47}$Ag, 48.161%   (c) $^{109}_{47}$Ag, 51.839%   (d) $^{109}_{47}$Ag, 48.161%

2.4 Atomic Weight

Every sample of boron has some atoms with a mass of 10.0129 u and others with a mass of 11.0093 u. The **atomic weight** of the element, the average mass of a representative sample of boron atoms, is somewhere between these values. For boron, for example, the atomic weight is 10.81. If isotope masses and abundances are known, the atomic weight of an element can be calculated using Equation 2.2.

$$\text{Atomic weight} = \left(\frac{\text{\% abundance isotope 1}}{100}\right)(\text{mass of isotope 1})$$
$$+ \left(\frac{\text{\% abundance isotope 2}}{100}\right)(\text{mass of isotope 2}) + \ldots$$

(2.2)

For boron with two isotopes (^{10}B, 19.91% abundant; ^{11}B, 80.09% abundant), we find

$$\text{Atomic weight} = \left(\frac{19.91}{100}\right) \times 10.0129 + \left(\frac{80.09}{100}\right) \times 11.0093 = 10.81$$

Equation 2.2 gives an average mass, weighted in terms of the abundance of each isotope for the element. As illustrated by the data in Table 2.2, *the atomic weight of an element is always closer to the mass of the most abundant isotope or isotopes.*

● **Atomic Mass, Relative Atomic Mass, and Atomic Weight** The atomic mass is the mass of an atom at rest. The relative atomic mass, also known as the atomic weight or average atomic weight, is the average of the atomic masses of all of the element's isotopes. The term *atomic weight* is slowly being phased out in favor of "relative atomic mass."

Table 2.2 Isotope Abundance and Atomic Weight

Element	Symbol	Atomic Weight	Mass Number	Isotopic Mass	Natural Abundance (%)
Hydrogen	H	1.00794	1	1.0078	99.985
	D*		2	2.0141	0.015
	T†		3	3.0161	0
Boron	B	10.811	10	10.0129	19.91
			11	11.0093	80.09
Neon	Ne	20.1797	20	19.9924	90.48
			21	20.9938	0.27
			22	21.9914	9.25
Magnesium	Mg	24.3050	24	23.9850	78.99
			25	24.9858	10.00
			26	25.9826	11.01

*D = deuterium; †T = tritium, radioactive.

The atomic weight of each stable element is given in the periodic table inside the front cover of this book. In the periodic table, each element's box contains the atomic number, the element symbol, and the atomic weight. For unstable (radioactive) elements, the atomic weight or mass number of the most stable isotope is given in parentheses.

© Cengage Learning/Charles D. Winters

Elemental bromine. Bromine is a deep orange-red, volatile liquid at room temperature. It consists of Br_2 molecules in which two bromine atoms are chemically bonded together. There are two, stable, naturally occurring isotopes of bromine atoms: ^{79}Br (50.69% abundance) and ^{81}Br (49.31% abundance).

EXAMPLE 2.2   Calculating Atomic Weight from Isotope Abundance

Problem Bromine has two naturally occurring isotopes. One has a mass of 78.918338 u and an abundance of 50.69%. The other isotope has a mass of 80.916291 u and an abundance of 49.31%. Calculate the atomic weight of bromine.

What Do You Know? You know the mass and abundance of each of the two isotopes.

Strategy The atomic weight of any element is the weighted average of the masses of the isotopes in a representative sample. To calculate the atomic weight, multiply the mass of each isotope by its percent abundance divided by 100 (Equation 2.2).

Solution

Atomic weight of bromine = $(50.69/100)(78.918338) + (49.31/100)(80.916291) = $ 79.90 u

Think about Your Answer You can quickly estimate the atomic weight from the data given. There are two isotopes, mass numbers of 79 and 81, in approximately equal abundance. From this, we would expect the average mass to be about 80, midway between the two mass numbers. The calculation bears this out.

Check Your Understanding

Verify that the atomic weight of chlorine is 35.45, given the following information:

^{35}Cl mass = 34.96885; percent abundance = 75.77%

^{37}Cl mass = 36.96590; percent abundance = 24.23%

EXAMPLE 2.3   Calculating Isotopic Abundances

Problem Antimony, Sb, has two stable isotopes: ^{121}Sb, 120.904 u, and ^{123}Sb, 122.904 u. What are the relative abundances of these isotopes?

What Do You Know? You know the masses of the two isotopes of the element and know their weighted average, the atomic weight, is 121.760 u (see the periodic table).

Strategy You can predict that the lighter isotope (^{121}Sb) must be the more abundant because the atomic weight is closer to 121 than to 123. To calculate the abundances recognize there are two unknown but related quantities, and you can write the following expression (where the fractional abundance of an isotope is the percent abundance of the isotope divided by 100)

$$\text{Atomic weight} = 121.760$$

$$= (\text{fractional abundance of } ^{121}\text{Sb})(120.904) +$$

$$(\text{fractional abundance of } ^{123}\text{Sb})(122.904)$$

or

$$121.760 = x(120.904) + y(122.904)$$

where x = fractional abundance of ^{121}Sb and y = fractional abundance of ^{123}Sb. Because you know that the fractional abundances of the isotopes must equal 1, $x + y = 1$, and you can solve the equations simultaneously for x and y.

Solution Because y = fractional abundance of ^{123}Sb = $1 - x$, you can make a substitution for y.

$$121.760 = x(120.904) + (1 - x)(122.904)$$

Expanding this equation, you have

$$121.760 = 120.904x + 122.904 - 122.904x$$

Finally, solving for x, you find

$$121.760 - 122.904 = (120.904 - 122.904)x$$

$$x = 0.5720$$

The fractional abundance of ^{121}Sb is 0.5720 and its percent abundance is 57.20%. This means that the percent abundance of ^{123}Sb must be 42.80%.

Think about Your Answer The result confirms your initial inference that the lighter isotope is the more abundant of the two.

Check Your Understanding

Neon has three stable isotopes, one with a small abundance. What are the abundances of the other two isotopes?

^{20}Ne, mass = 19.992435; percent abundance = ?

^{21}Ne, mass = 20.993843; percent abundance = 0.27%

^{22}Ne, mass = 21.991383; percent abundance = ?

A sample of the metalloid antimony. The element has two stable isotopes, ^{121}Sb and ^{123}Sb.

REVIEW & CHECK FOR SECTION 2.4

1. Which is the more abundant isotope of copper, ^{63}Cu or ^{65}Cu?

    (a) ^{63}Cu   (b) ^{65}Cu

2. Which of the following is closest to the observed abundance of ^{71}Ga, one of two stable gallium isotopes (^{69}Ga and ^{71}Ga)?

    (a) 60%   (b) 40%   (c) 20%   (d) 70%

CASE STUDY

Using Isotopes: Ötzi, the Iceman of the Alps

In 1991 a hiker in the Alps on the Austrian-Italian border found the well-preserved remains of an approximately 46-year-old man, now nicknamed "The Iceman," who lived about 5200 years ago (page 1). Studies using isotopes of oxygen, strontium, lead, and argon, among others, have helped scientists paint a detailed picture of the man and his life.

The ^{18}O isotope of oxygen can give information on the latitude and altitude in which a person was born and raised. Oxygen in biominerals such as teeth and bones comes primarily from ingested water. The important fact is that there is a variation in the amount of ^{18}O water ($H_2{}^{18}O$) that depends on how far inland the watershed is found and on its altitude. As rain clouds move inland, water based on ^{18}O will be "rained out" before $H_2{}^{16}O$. The lakes and rivers on the northern side of the Alps are known to have a lower ^{18}O content than those on the southern side of the mountains. On the northern side precipitation originates in the cooler, and more distant, Atlantic Ocean. On the southern side, the precipitation comes from the closer and warmer Mediterranean Sea. The ^{18}O content of the teeth and bones of the Iceman was found to be relatively high and characteristic of the watershed south of the Alps. He had clearly been born and raised in that area.

The relative abundance of isotopes of heavier elements also varies slightly from place to place and in their incorporation into different minerals. Strontium, a member of Group 2A along with calcium, is incorporated into teeth and bones. The ratio of strontium isotopes, $^{87}Sr/^{86}Sr$, and of lead isotopes, $^{206}Pb/^{204}Pb$, in the Iceman's teeth and bones was characteristic of soils from a narrow region of Italy south of the Alps, which established more clearly where he was born and lived of his life.

The investigators also looked for food residues in the Iceman's intestines. Although a few grains of cereal were found, they also located tiny flakes of mica believed to have broken off stones used to grind grain and that were therefore eaten when the man ate the grain. They analyzed these flakes using argon isotopes, ^{40}Ar and ^{39}Ar, and found their signature was like that of mica in an area south of the Alps, thus establishing where he lived in his later years.

The overall result of the many isotope studies showed that the Iceman lived thousands of years ago in a small area about 10–20 kilometers west of Merano in northern Italy.

For details of the isotope studies, see W. Müller, et al., *Science*, Volume 302, October 13, 2003, pages 862–866.

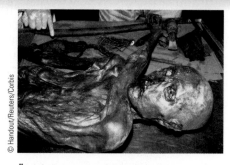

© Handout/Reuters/Corbis

Ötzi the Iceman. A well-preserved mummy of a man who lived in northern Italy about 5000 years ago.

Questions:

1. How many neutrons are there in atoms of ^{18}O? In each of the two isotopes of lead?
2. ^{14}C is a radioactive isotope of carbon that occurs in trace amounts in all living materials. How many neutrons are in a ^{14}C atom?
3. The ratio $^{87}Sr/^{86}Sr$ in the Iceman study was in the range of 0.72. How does this compare with the ratio calculated from average abundances (^{87}Sr = 7.00% and ^{86}Sr = 9.86%)?

Answers to these questions are available in Appendix N.

2.5   The Periodic Table

The periodic table of elements is one of the most useful tools in chemistry. Not only does it contain a wealth of information, but it can also be used to organize many of the ideas of chemistry. It is important to become familiar with its main features and terminology.

Developing the Periodic Table

● **About the Periodic Table** For more information on the periodic table we recommend the following:

- American Chemical Society (pubs .acs.org/cen/8oth/elements.html).
- www.ptable.com
- J. Emsley: *Nature's Building Blocks— An A–Z Guide to the Elements*, New York, Oxford University Press, 2001.
- E. Scerri, *The Periodic Table*, New York, Oxford University Press, 2007.

Although the arrangement of elements in the periodic table is now understood on the basis of atomic structure [▶ Chapters 6 and 7], the table was originally developed from many experimental observations of the chemical and physical properties of elements and is the result of the ideas of a number of chemists in the 18th and 19th centuries.

In 1869, at the University of St. Petersburg in Russia, Dmitri Ivanovitch Mendeleev (1834–1907) was pondering the properties of the elements as he wrote a textbook on chemistry. On studying the chemical and physical properties of the elements, he realized that, if the elements were arranged in order of increasing atomic mass, elements with similar properties appeared in a regular pattern. That is, he saw a **periodicity** or periodic repetition of the properties of elements. Mendeleev organized the known elements into a table by lining them up in horizontal rows in order of increasing atomic mass (page 42). Every time he came to an element with properties similar to one already in the row, he started a new row. For example, the elements Li, Be, B, C, N, O, and F were in a row. Sodium was the next element then

Module 1: The Periodic Table covers concepts in this section.

A CLOSER LOOK

The Story of the Periodic Table

by Eric R. Scerri, UCLA

Dmitri Mendeleev was probably the greatest scientist produced by Russia. The youngest of 14 children, he was taken by his mother on a long journey, on foot, in order to enroll him into a university. However, several attempts initially proved futile because, as a Siberian, Mendeleev was barred from attending certain institutions. His mother did succeed in enrolling him in a teacher training college, thus giving Mendeleev a lasting interest in science education, which contributed to his eventual discovery of the periodic system that essentially simplified the subject of inorganic chemistry.

Statue of Dmitri Mendeleev and a periodic table. This statue and mural are at the Institute of Metrology in St. Petersburg, Russia.

After completing a doctorate, Mendeleev headed to Germany for a postdoctoral fellowship and then returned to Russia, where he set about writing a book aimed at summarizing all of inorganic chemistry. It was while writing this book that he identified the organizing principle with which he is now invariably connected—the periodic system of the elements.

More correctly, though, the periodic system was developed by Mendeleev, as well as five other scientists, over a period of about 10 years, after the Italian chemist Cannizzaro had published a consistent set of atomic weights in 1860. It appears that Mendeleev was unaware of the work of several of his co-discoverers, however.

In essence, the periodic table groups together sets of elements with similar properties into vertical columns. The underlying idea is that if the elements are arranged in order of increasing atomic weights, there are approximate repetitions in their chemical properties after certain intervals. As a result of the existence of the periodic table, students and even professors of chemistry were no longer obliged to learn the properties of all the elements in a disorganized fashion. Instead, they could concentrate on the properties of representative members of the eight columns or groups in the early short-form periodic table, from which they could predict properties of other group members.

Mendeleev is justly regarded as the leading discoverer of the periodic table since he continued to champion the finding and drew out its consequences to a far greater extent than any of his contemporaries. First, he accommodated the 65 or so elements that were known at the time into a coherent scheme based on ascending order of atomic weight while also reflecting chemical and physical similarities. Next, he noticed gaps in his system, which he reasoned would eventually be filled by elements that had not yet been discovered. In addition, by judicious interpolation between the properties of known elements, Mendeleev predicted the nature of a number of completely new elements. Within a period of about 20 years, three of these elements—subsequently called gallium, scandium, and germanium—were isolated and found to have almost the exact properties that Mendeleev had predicted.

What is not well known is that about half of the elements that Mendeleev predicted were never found. But given the dramatic success of his early predictions, these later lapses have largely been forgotten.

Eric Scerri, *The Periodic Table: Its Story and Its Significance,* Oxford University Press, New York, 2007.

---

known; because its properties closely resembled those of Li, Mendeleev started a new row. As more and more elements were added to the table, new rows were begun, and elements with similar properties (such as Li, Na, and K) were placed in the same vertical column.

An important feature of Mendeleev's table—and a mark of his genius—was that he left an empty space in a column when he believed an element was not known but should exist and have properties similar to the elements above and below it in his table. He deduced that these spaces would be filled by undiscovered elements. For example, he left a space between Si (silicon) and Sn (tin) in Group 4A for an element he called *eka-silicon*. Based on the progression of properties in this group, Mendeleev was able to predict the properties of the missing element. With the discovery of germanium (Ge) in 1886, Mendeleev's prediction was confirmed.

In Mendeleev's table the elements were ordered by increasing mass. A glance at a modern table, however, shows that, if some elements (such as Ni and Co, Ar and K, and Te and I) were ordered by mass and not chemical and physical properties, they would be reversed in their order of appearance. Mendeleev recognized these discrepancies and simply assumed the atomic masses known at that time were inaccurate—not a bad assumption based on the analytical methods then in use. In fact, his order is correct and what was wrong was his assumption that element properties were a function of their mass.

● **Mendeleev and Atomic Numbers** Mendeleev developed the periodic table based on atomic masses. The concept of atomic numbers was not known until after the development of the structure of the atom in the early 20th century.

In 1913 H. G. J. Moseley (1887–1915), a young English scientist working with Ernest Rutherford (1871–1937), bombarded many different metals with electrons in a cathode-ray tube and examined the x-rays emitted in the process. Moseley realized the wavelength of the x-rays emitted by a given element was related in a precise manner to the positive charge in the nucleus of the atoms of the element and that

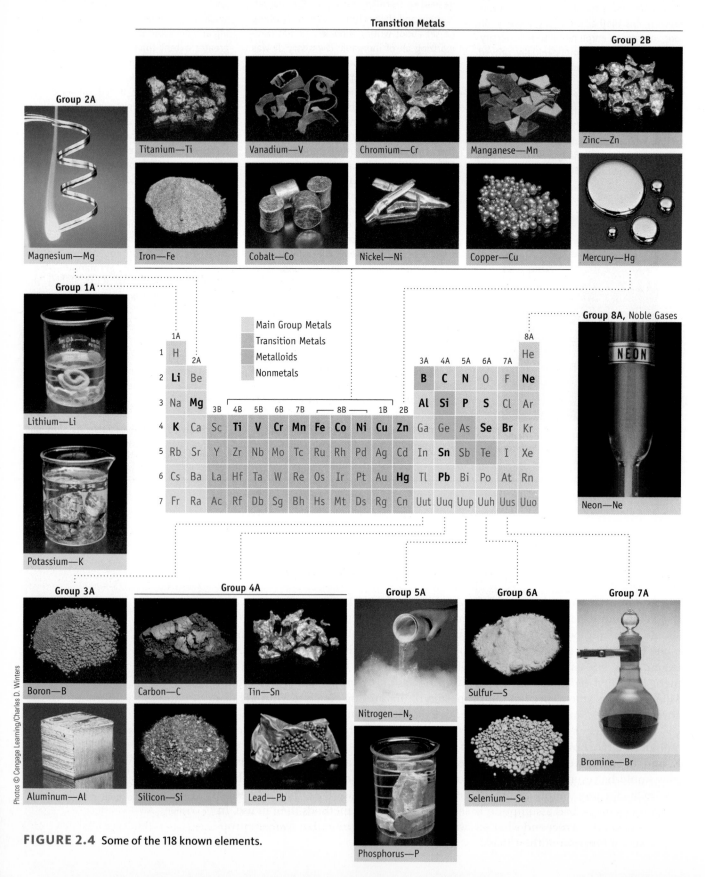

FIGURE 2.4 Some of the 118 known elements.

this provided a way to experimentally determine the atomic number of a given element. Indeed, once atomic numbers could be determined, chemists recognized that organizing the elements in a table by increasing atomic number corrected the inconsistencies in Mendeleev's table. The **law of chemical periodicity** is now stated as *the properties of the elements are periodic functions of atomic number.*

Features of the Periodic Table

The main organizational features of the periodic table are the following:

- Elements are arranged so that those with similar chemical and physical properties lie in vertical columns called **groups** or **families**. The periodic table commonly used in the United States has groups numbered 1 through 8, with each number followed by the letter A or B. The A groups are often called the **main group elements** and the B groups are the **transition elements**.
- The horizontal rows of the table are called **periods**, and they are numbered beginning with 1 for the period containing only H and He. For example, sodium, Na, in Group 1A, is the first element in the third period. Mercury, Hg, in Group 2B, is in the sixth period (or sixth row).

The periodic table can be divided into several regions according to the properties of the elements. On the table inside the front cover of this book, elements that behave as *metals* are indicated in purple, *nonmetals* are indicated in yellow, and elements called *metalloids* appear in green. Elements gradually become less metallic as one moves from left to right across a period, and the metalloids lie along the metal-nonmetal boundary. Some elements are shown in Figure 2.4.

You are probably familiar with many properties of **metals** from your own experience (Figure 2.5a). At room temperature and normal atmospheric pressure metals are solids (except for mercury), can conduct electricity, are usually ductile (can be drawn into wires) and malleable (can be rolled into sheets), and can form alloys (mixtures of one or more metals in another metal). Iron (Fe) and aluminum (Al) are used in automobile parts because of their ductility, malleability, and low cost relative to other metals. Copper (Cu) is used in electric wiring because it conducts electricity better than most other metals.

The **nonmetals** lie to the right of a diagonal line that stretches from B to Te in the periodic table and have a wide variety of properties. Some are solids (carbon, sulfur, phosphorus, and iodine). Five elements are gases at room temperature (hydrogen, oxygen, nitrogen, fluorine, and chlorine). One nonmetal, bromine, is a liquid at room temperature (Figure 2.5b). With the exception of carbon in the form

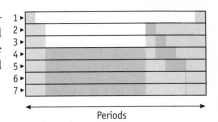

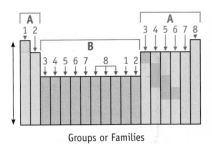

Periods

Groups or Families

● **Periods and Groups in the Periodic Table** One way to designate periodic groups is to number them 1 through 18 from left to right. This method is generally used outside the United States. The system predominant in the United States labels main group elements as Groups 1A–8A and transition elements as Groups 1B–8B. This book uses the A/B system.

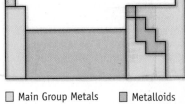

☐ Main Group Metals   ☐ Metalloids
☐ Transition Metals   ☐ Nonmetals

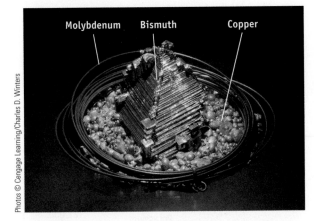

(a) **Metals** Molybdenum (Mo, wire), bismuth (Bi, center object), and copper (Cu) are metals. Metals can be generally drawn into wires and conduct electricity.

FIGURE 2.5 Metals, nonmetals, and metalloids.

(b) **Nonmetals** Only about 18 elements can be classified as nonmetals. Here are orange-red liquid bromine and purple solid iodine.

(c) **Metalloids** Only 6 elements are generally classified as metalloids or semimetals. This photograph shows solid silicon in various forms, including a wafer that holds printed electronic circuits.

of graphite, nonmetals do not conduct electricity, which is one of the main features that distinguishes them from metals.

The elements next to the diagonal line from boron (B) to tellurium (Te) have properties that make them difficult to classify as metals or nonmetals. Chemists call them **metalloids** or, sometimes, **semimetals** (Figure 2.5c). You should know, however, that chemists often disagree about which elements fit into this category. We will define a metalloid as an element that has some of the physical characteristics of a metal but some of the chemical characteristics of a nonmetal; we include only B, Si, Ge, As, Sb, and Te in this category. This definition reflects the ambiguity in the behavior of these elements. Antimony (Sb), for example, conducts electricity as well as many elements that are true metals. Its chemistry, however, resembles that of a nonmetal such as phosphorus.

A Brief Overview of the Periodic Table and the Chemical Elements

Elements in the leftmost column, **Group 1A**, are known as the **alkali metals** (except H). All the alkali metals are solids at room temperature and all are reactive. For example, they react with water to produce hydrogen and alkaline solutions (Figure 2.6). Because of their reactivity, these metals are only found in nature combined in compounds (such as NaCl) (◄ Figure 1.2), never as free elements.

The second group in the periodic table, **Group 2A**, is also composed entirely of metals that occur naturally only in compounds. Except for beryllium (Be), these elements react with water to produce alkaline solutions, and most of their oxides (such as lime, CaO) form alkaline solutions; hence, they are known as the **alkaline earth metals**. Magnesium (Mg) and calcium (Ca) are the seventh and fifth most abundant elements in the earth's crust, respectively (Table 2.3). Calcium, one of the important elements in teeth and bones, occurs naturally in vast limestone deposits. Calcium carbonate ($CaCO_3$) is the chief constituent of limestone and of corals, sea shells, marble, and chalk. Radium (Ra), the heaviest alkaline earth element, is radioactive and is used to treat some cancers by radiation.

Aluminum, an element of great commercial importance, is in **Group 3A** (Figure 2.4). This element and three others of Group 3A (gallium, indium, and thallium) are metals, whereas boron (B) is a metalloid. Aluminum (Al) is the most

● **Alkali and Alkaline** The word *alkali* comes from the Arabic language. Ancient Arabian chemists discovered that ashes of certain plants, which they called al-qali, gave water solutions that felt slippery and burned the skin. These ashes contain compounds of Group 1A elements that produce alkaline (basic) solutions.

● **Placing H in the Periodic Table** Where to place H? Tables often show it in Group 1A even though it is clearly not an alkali metal. However, in its reactions it forms a 1+ ion just like the alkali metals. For this reason, H is often placed in Group 1A.

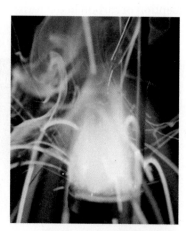

(a) Cutting sodium Cutting a bar of sodium with a knife is like cutting a stick of cold butter.

(b) Potassium reacts with water. When an alkali metal such as potassium is treated with water, a vigorous reaction occurs, giving an alkaline solution and hydrogen gas, which burns in air.

FIGURE 2.6 Properties of the alkali metals.

Photos © Cengage Learning/Charles D. Winters

Table 2.3 **The 10 Most Abundant Elements in the Earth's Crust**

Rank	Element	Abundance (ppm)*
1	Oxygen	474,000
2	Silicon	277,000
3	Aluminum	82,000
4	Iron	41,000
5	Calcium	41,000
6	Sodium	23,000
7	Magnesium	23,000
8	Potassium	21,000
9	Titanium	5,600
10	Hydrogen	1,520

*ppm = parts per million = g per 1000 kg.

abundant metal in the earth's crust at 8.2% by mass. It is exceeded in abundance only by the nonmetal oxygen and metalloid silicon. Not surprisingly, these three elements are found combined in common materials such as clays and minerals. Boron occurs in the mineral borax, a compound used as a cleaning agent, antiseptic, and flux for metal work.

As a metalloid, boron has a different chemistry than the other elements of Group 3A, all of which are metals. Nonetheless, all form compounds with analogous formulas such as BCl_3 and $AlCl_3$, and this similarity marks them as members of the same periodic group.

Thus far all the elements we have described, except boron, have been metals. Beginning with **Group 4A**, however, the groups contain more and more nonmetals. In Group 4A there is a nonmetal, carbon (C), two metalloids, silicon (Si) and germanium (Ge), and two metals, tin (Sn) and lead (Pb) (Figure 2.4). Because of the change from nonmetallic to metallic behavior, more variation occurs in the properties of the elements of this group than in most others. Nonetheless, there are similarities. For example, these elements form compounds with analogous formulas such as CO_2, SiO_2, GeO_2, and PbO_2.

Carbon is the basis for the great variety of chemical compounds that make up living things. It is also found in Earth's atmosphere as CO_2, on the surface of the Earth in carbonates like limestone and coral (calcium carbonate, $CaCO_3$), and in coal, petroleum, and natural gas—the fossil fuels.

One interesting aspect of the chemistry of the nonmetals is that a particular element can often exist in several different and distinct forms, called **allotropes**, each having its own properties. Carbon has many allotropes, the best known of which are graphite and diamond. Graphite consists of flat sheets in which each carbon atom is connected to three others (Figure 2.7a). Because the sheets of carbon atoms cling only weakly to one another, one layer can slip easily over another. This explains why graphite is soft, is a good lubricant, and is used in pencil lead. [Pencil "lead" is not the element lead (Pb) but a composite of clay and graphite that leaves a trail of graphite on the page as you write.]

In diamond each carbon atom is connected to four others at the corners of a tetrahedron, and this extends throughout the solid (Figure 2.7b). This structure

Liquid gallium. Bromine and mercury are the only elements that are liquids under ambient conditions. Gallium (melting point = 29.8 °C) and cesium (melting point = 28.4 °C) melt slightly above room temperature. Here gallium melts when held in the hand.

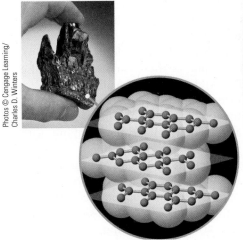

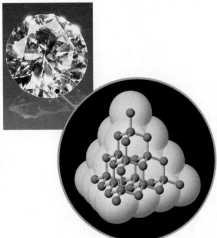

(a) Graphite. Graphite consists of layers of carbon atoms. Each carbon atom is linked to three others to form a sheet of six-member, hexagonal rings.

(b) Diamond. In diamond the carbon atoms are also arranged in six-member rings, but the rings are not planar because each C atom is connected tetrahedrally to four other C atoms.

(c) Buckyballs. A member of the family called buckminsterfullerenes, C_{60} is an allotrope of carbon. Sixty carbon atoms are arranged in a spherical cage that resembles a hollow soccer ball. Notice that each six-member ring shares an edge with three other six-member rings and three five-member rings. Chemists call this molecule a "buckyball." C_{60} is a black powder; it is shown here in the tip of a pointed glass tube.

FIGURE 2.7 The allotropes of carbon.

FIGURE 2.8 Compounds containing silicon. Ordinary clay, sand, and many gemstones are based on compounds of silicon and oxygen. Here clear, colorless quartz and dark purple amethyst lie in a bed of sand. All are silicon dioxide, SiO_2. The different colors are due to impurities.

causes diamonds to be extremely hard, denser than graphite ($d = 3.51$ g/cm^3 for diamond versus $d = 2.22$ g/cm^3 for graphite), and chemically less reactive. Because diamonds are not only hard but are excellent conductors of heat, they are used on the tips of metal- and rock-cutting tools.

In the late 1980s another form of carbon was identified as a component of black soot, the stuff that collects when carbon-containing materials are burned in a deficiency of oxygen. This substance is made up of molecules with 60 carbon atoms arranged as a spherical "cage" (Figure 2.7c). You may recognize that the surface is made up of five- and six-member rings and resembles a hollow soccer ball. The shape also reminded its discoverers of an architectural dome conceived several decades ago by the American philosopher and engineer, R. Buckminster Fuller. This led to the official name of the allotrope, buckminsterfullerene, although chemists often simply call these molecules "buckyballs."

Oxides of silicon are the basis of many minerals such as clay, quartz, and beautiful gemstones like amethyst (Figure 2.8). Tin and lead have been known for centuries because they are easily smelted from their ores. Tin alloyed with copper makes bronze, which was used in ancient times in utensils and weapons. Lead has been used in water pipes and paint, even though the element is toxic to humans.

Nitrogen in **Group 5A** occurs naturally in the form of the diatomic molecule N_2 (Figures 2.9) and makes up about three-fourths of earth's atmosphere. It is also incorporated in biochemically important substances such as chlorophyll, proteins, and DNA. Scientists have long sought ways to make compounds from atmospheric nitrogen, a process referred to as "nitrogen fixation." Nature accomplishes this easily in some prokaryotic organisms, but severe conditions (high temperatures, for example) must be used in the laboratory and in industry to cause N_2 to react with other elements (such as H_2 to make ammonia, NH_3, which is widely used as a fertilizer).

Phosphorus is also essential to life. It is an important constituent in bones, teeth, and DNA. The element glows in the dark if it is in the air (owing to its reaction with O_2), and its name, based on Greek words meaning "light-bearing," reflects this. This element also has several allotropes, the most important being white (Figure 2.4) and red phosphorus. Both forms of phosphorus are used commercially. White phosphorus ignites spontaneously in air, so it is normally stored under water. When it does react with air, it forms P_4O_{10}, which can react with water to form phosphoric acid (H_3PO_4), a compound used in food products such as soft drinks. Red phosphorus is used in the striking strips on match books. When a match is struck, potassium chlorate in the match head mixes with some red phosphorus on the striking strip, and the friction is enough to ignite this mixture.

As with Group 4A, we again see nonmetals (N and P), metalloids (As and Sb), and a metal (Bi, Figure 2.5a) in Group 5A. In spite of these variations, they also form analogous compounds such as the oxides N_2O_5, P_4O_{10} and As_2O_5.

Oxygen, which constitutes about 20% of earth's atmosphere and which combines readily with most other elements, is at the top of **Group 6A**. Most of the energy that powers life on earth is derived from reactions in which oxygen combines with other substances.

Sulfur has been known in elemental form since ancient times as brimstone or "burning stone" (Figure 2.10). Sulfur, selenium, and tellurium are often referred to collectively as *chalcogens* (from the Greek word, *khalkos*, for copper) because most copper ores contain these elements. Their compounds can be foul-smelling and poisonous; nevertheless, sulfur and selenium are essential components of the

FIGURE 2.9 Elements that Exist as Diatomic or Triatomic Molecules. Seven of the known elements exist as diatomic, or two-atom, molecules. Oxygen has an additional allotrope, ozone, with three O atoms in each molecule.

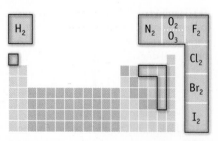

human diet. By far the most important compound of sulfur is sulfuric acid (H_2SO_4), which is manufactured in larger amounts than any other compound.

As in Group 5A, the second- and third-period elements of Group 6A have different structures. Like nitrogen, oxygen is also a diatomic molecule (Figure 2.9). Unlike nitrogen, however, oxygen has an allotrope, the triatomic molecule ozone, O_3. Sulfur, which can be found in nature as a yellow solid, has many allotropes, the most common of which consists of eight-member, crown-shaped rings of sulfur atoms (Figure 2.10).

Polonium, the radioactive element in Group 6A, was isolated in 1898 by Marie and Pierre Curie, who separated a small amount from tons of a uranium-containing ore and named it for Madame Curie's native country, Poland.

In Group 6A we once again observe a variation of properties. Oxygen, sulfur, and selenium are nonmetals, tellurium is a metalloid, and polonium is a metal. Nonetheless, there is a family resemblance in their chemistries. All of oxygen's fellow group members form oxygen-containing compounds (SO_2, SeO_2, and TeO_2), and all form sodium-containing compounds (Na_2O, Na_2S, Na_2Se, and Na_2Te).

At the far right of the periodic table are two groups composed entirely of nonmetals. The **Group 7A** elements—fluorine, chlorine, bromine, iodine, and radioactive astatine—are nonmetals and all exist as diatomic molecules (Figure 2.9). At room temperature fluorine (F_2) and chlorine (Cl_2) are gases. Bromine (Br_2) is a liquid and iodine (I_2) is a solid, but bromine and iodine vapor are clearly visible over the liquid or solid (Figure 2.5b).

The Group 7A elements are among the most reactive of all elements, and all combine violently with alkali metals to form salts such as table salt, NaCl (◄ Figure 1.2). The name for this group, the **halogens**, comes from the Greek words *hals*, meaning "salt," and *genes*, for "forming." The halogens react not only with metals but also with most nonmetals.

The **Group 8A** elements—helium, neon, argon, krypton, xenon, and radioactive radon—are the least reactive elements (Figure 2.11). All are gases, and none is abundant on Earth or in the Earth's atmosphere (although argon is the third most abundant gas in dry air at 0.9%). Because of this, they were not discovered until the end of the 19th century. Helium, the second most abundant element in the universe after hydrogen, was detected in the sun in 1868 by analysis of the solar spectrum but was not found on Earth until 1895. (The name of the element comes from the Greek word for the sun, *helios*.)

It was long believed that none of the Group 8A elements would combine chemically with any other element. However, in 1962, a compound of xenon and fluorine (XeF_4) was first prepared, and this opened the way to the preparation of a number of other such compounds. A common name for this group, the **noble gases**, denotes their general lack of reactivity. For the same reason they are sometimes called the *inert gases* or, because of their low abundance, the *rare gases*.

© Cengage Learning/Charles D. Winters

FIGURE 2.10 Sulfur. The most common allotrope of sulfur consists of S atoms arranged in eight-member, crown-shaped rings.

● **Special Group Names** Some groups have common and widely used names (see Figure 2.4).

Group 1A: Alkali metals

Group 2A: Alkaline earth metals

Group 7A: Halogens

Group 8A: Noble gases

© Cengage Learning/Charles D. Winters

FIGURE 2.11 The noble gases. This kit is sold for detecting the presence of radioactive radon in the home. Neon gas is used in advertising signs, and xenon-containing headlights are increasingly popular on automobiles.

Lanthanides. If you use a Bunsen burner in the lab, you may light it with a "flint" lighter. The flints are composed of iron and "mischmetal," a mixture of lanthanide elements, chiefly Ce, La, Pr, and Nd with traces of other lanthanides. (The word *mischmetal* comes from the German for "mixed metals.") It is produced by the electrolysis of a mixture of lanthanide oxides.

Stretching between Groups 2A and 3A is a series of elements called the **transition elements**. These fill the B-groups (1B through 8B) in the fourth through the seventh periods in the center of the periodic table. All are metals (Figure 2.4), and 13 of them are in the top 30 elements in terms of abundance in the earth's crust (Table 2.3). Most occur naturally in combination with other elements, but a few—copper (Cu), silver (Ag), gold (Au), and platinum (Pt)—are much less reactive and so can be found in nature as pure elements.

Virtually all of the transition elements have commercial uses. They are used as structural materials (iron, titanium, chromium, copper); in paints (titanium, chromium); in the catalytic converters in automobile exhaust systems (platinum and rhodium); in coins (copper, nickel, zinc); and in batteries (manganese, nickel, zinc, cadmium, mercury).

A number of the transition elements play important biological roles. For example, iron, a relatively abundant element (Table 2.3), is the central element in the chemistry of hemoglobin, the oxygen-carrying component of blood.

Two rows at the bottom of the table accommodate the **lanthanides** [the series of elements between the elements lanthanum ($Z = 57$) and hafnium ($Z = 72$)] and the **actinides** [the series of elements between actinium ($Z = 89$) and rutherfordium ($Z = 104$)]. Some lanthanide compounds are used in color television picture tubes, uranium ($Z = 92$) is the fuel for atomic power plants, and americium ($Z = 95$) is used in smoke detectors.

REVIEW & CHECK FOR SECTION 2.5

1. Which of the following elements is a metalloid?

    (a) Ge   (b) S   (c) Be   (d) Al

2. What is the symbol for the element in the third period and the fourth group?

    (a) Si   (b) Sc   (c) V   (d) N

3. What is the most abundant element in the Earth's crust?

    (a) Fe   (b) C   (c) O   (d) N

4. What is the name given to elements that exist in different forms, such as graphite, diamond, and buckyballs?

    (a) isotopes   (b) isomers   (c) allotropes   (d) nonmetals

---

2.6   Molecules, Compounds, and Formulas

A molecule is the smallest identifiable unit into which some pure substances like sugar and water can be divided and still retain the composition and chemical properties of the substance. Such substances are composed of identical molecules consisting of two or more atoms bound firmly together. For example, atoms of aluminum, Al, combine with molecules of bromine, Br_2, to produce a molecule of the compound aluminum bromide, Al_2Br_6 (Figure 2.12).

$$2 \; Al(s) + 3 \; Br_2(\ell) \rightarrow Al_2Br_6(s)$$

aluminum + bromine → aluminum bromide

To describe this chemical change (or chemical reaction) on paper, the composition of each element and compound is represented by a symbol or formula. Here one molecule of Al_2Br_6 is composed of two Al atoms and six Br atoms.

Formulas

For molecules more complicated than water, there is often more than one way to write the formula. For example, the formula of ethanol (also called ethyl alcohol) can be represented as C_2H_6O (Figure 2.13). This **molecular formula** describes the

(a) Solid aluminum and (in the beaker) liquid bromine.

(b) When the aluminum is added to the bromine, a vigorous chemical reaction occurs.

(c) The reaction produces white, solid aluminum bromide, Al_2Br_6.

FIGURE 2.12 Reaction of the elements aluminum and bromine.

composition of ethanol molecules—two carbon atoms, six hydrogen atoms, and one atom of oxygen occur per molecule—but it gives us no structural information. Structural information—how the atoms are connected and how the molecule fills space—is important, however, because it helps us understand how a molecule can interact with other molecules, which is the essence of chemistry.

To provide some structural information, it is useful to write a **condensed formula**, which indicates how certain atoms are grouped together. For example, the condensed formula of ethanol, CH_3CH_2OH (Figure 2.13), informs us that the molecule consists of three "groups": a CH_3 group, a CH_2 group, and an OH group. Writing the formula as CH_3CH_2OH also shows that the compound is not dimethyl ether, CH_3OCH_3, a compound with the same molecular formula but with a different structure and distinctly different properties.

That ethanol and dimethyl ether are different molecules is further apparent from their **structural formulas** (Figure 2.13). This type of formula gives us an even higher level of structural detail, showing how all of the atoms are attached within a molecule. The lines between atoms represent the chemical bonds that hold atoms together in this molecule [▶ Chapter 8].

● **Writing Formulas** When writing molecular formulas of organic compounds (compounds with C, H, and other elements) the convention is to write C first, then H, and finally other elements in alphabetical order. For example, acrylonitrile, a compound used to make consumer plastics, has the condensed formula CH_2CHCN. Its molecular formula would be C_3H_3N.

● **Standard Colors for Atoms in Molecular Models** The colors listed here are used for molecular models in this book and are generally used by chemists.

NAME	MOLECULAR FORMULA	CONDENSED FORMULA	STRUCTURAL FORMULA	MOLECULAR MODEL
Ethanol	C_2H_6O	CH_3CH_2OH		
Dimethyl ether	C_2H_6O	CH_3OCH_3		

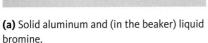

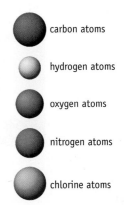

carbon atoms

hydrogen atoms

oxygen atoms

nitrogen atoms

chlorine atoms

FIGURE 2.13 Four approaches to showing molecular formulas. Here the two molecules have the same molecular formula. However, once they are written as condensed or structural formulas, or illustrated with a molecular model, it is clear that these molecules are different.

FIGURE 2.14 **Ice.** Snowflakes are six-sided structures, reflecting the underlying structure of ice. Ice consists of six-sided rings formed by water molecules, in which each side of a ring consists of two O atoms and an H atom.

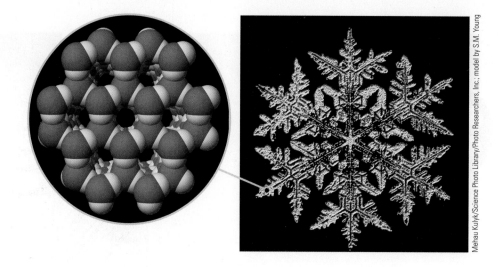

Mehau Kulyk/Science Photo Library/Photo Researchers, Inc.; model by S.M. Young

Molecular Models

Molecular structures are often beautiful in the same sense that art is beautiful, and there is something intrinsically beautiful about the pattern created by water molecules assembled in ice (Figure 2.14).

More important, however, is the fact that the physical and chemical properties of a molecular compound are often closely related to its structure. For example, two well-known features of ice are related to its structure. The first is the shape of ice crystals: The sixfold symmetry of macroscopic ice crystals also appears at the particulate level in the form of six-sided rings of hydrogen and oxygen atoms. The second is water's unique property of being less dense when it is solid than when it is liquid. The lower density of ice, which has enormous consequences for earth's climate, results from the fact that molecules of water are not packed together tightly in ice.

Because molecules are three dimensional, it is often difficult to represent their shapes on paper. Certain conventions have been developed, however, that help represent three-dimensional structures on two-dimensional surfaces. Simple perspective drawings are often used (Figure 2.15).

Several kinds of molecular models exist. In the **ball-and-stick model**, spheres, usually in different colors, represent the atoms, and sticks represent the bonds holding them together. These models make it easy to see how atoms are attached to one another. Molecules can also be represented using **space-filling models**. These models are more realistic because they offer a better representation of relative sizes of atoms and their proximity to each other when in a molecule. A disadvantage of pictures of space-filling models is that atoms can often be hidden from view.

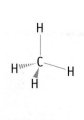

Simple perspective drawing

© Cengage Learning/Charles D. Winters

Plastic model

Ball-and-stick model

Space-filling model

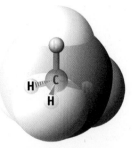

All visualizing techniques represent the same molecule.

FIGURE 2.15 Ways of depicting a molecule, here the methane (CH_4) molecule.

REVIEW & CHECK FOR SECTION 2.6

Cysteine, whose molecular model and structural formula are illustrated here, is an important amino acid and a constituent of many living things. What is its molecular formula?

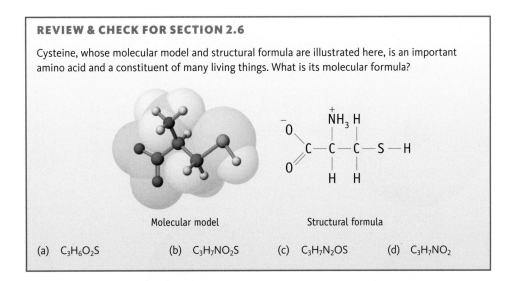

Molecular model   Structural formula

(a) $C_3H_6O_2S$ (b) $C_3H_7NO_2S$ (c) $C_3H_7N_2OS$ (d) $C_3H_7NO_2$

2.7   Ionic Compounds: Formulas, Names, and Properties

The compounds you have encountered so far in this chapter are **molecular compounds**—that is, compounds that consist of discrete molecules at the particulate level. **Ionic compounds** constitute another major class of compounds. They consist of **ions**, atoms, or groups of atoms that bear a positive or negative electric charge. Many familiar compounds are composed of ions (Figure 2.16). Table salt, or sodium chloride (NaCl), and lime (CaO) are just two. To be able to recognize ionic compounds and to write formulas for these compounds, it is important to know the formulas and charges of common ions. You also need to know the names of ions and be able to name the compounds they form.

Modules 2: Predicting Ion Charges and 3: Names to Formulas of Ionic Compounds cover concepts in this section.

Ions

Atoms of many elements can gain or lose electrons in the course of a chemical reaction. To be able to predict the outcome of chemical reactions [▶ Chapter 3], you need to know whether an element will likely gain or lose electrons and, if so, how many.

Cations

If an atom loses an electron (which is transferred to an atom of another element in the course of a reaction), the atom now has one fewer negative electrons than it has positive protons in the nucleus. The result is a positively charged ion called a **cation**

FIGURE 2.16 Some common ionic compounds.

Common Name	Name	Formula	Ions Involved
Calcite	Calcium carbonate	$CaCO_3$	Ca^{2+}, CO_3^{2-}
Fluorite	Calcium fluoride	CaF_2	Ca^{2+}, F^-
Gypsum	Calcium sulfate dihydrate	$CaSO_4 \cdot 2\,H_2O$	Ca^{2+}, SO_4^{2-}
Hematite	Iron(III) oxide	Fe_2O_3	Fe^{3+}, O^{2-}
Orpiment	Arsenic sulfide	As_2S_3	As^{3+}, S^{2-}

Hematite, Fe_2O_3 Calcite, $CaCO_3$

Gypsum, $CaSO_4 \cdot 2\,H_2O$ Fluorite, CaF_2 Orpiment, As_2S_3

© Cengage Learning/Charles D. Winters

FIGURE 2.17 Ions. A lithium-6 atom is electrically neutral because the number of positive charges (three protons) and negative charges (three electrons) are the same. When it loses one electron, it has one more positive charge than negative charge, so it has a net charge of 1+. We symbolize the resulting lithium cation as Li⁺. A fluorine atom is also electrically neutral, having nine protons and nine electrons. A fluorine atom can acquire an electron to produce an F⁻ anion. This anion has one more electron than it has protons, so it has a net charge of 1−.

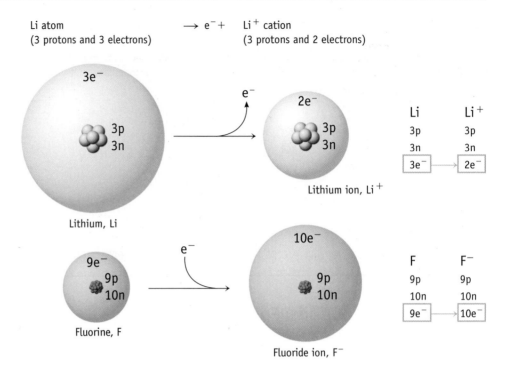

(Figure 2.17). (The name is pronounced "cat´-i-on.") Because it has an excess of one positive charge, we write the cation's symbol as, for example, Li⁺:

$$\text{Li atom} \rightarrow e^- + \text{Li}^+ \text{ cation}$$

(3 protons and 3 electrons)   (3 protons and 2 electrons)

Anions

Conversely, if an atom gains one or more electrons, there is now one or more negatively charged electrons than protons. The result is an **anion** (pronounced "ann´-i-on") (Figure 2.17).

$$\text{O atom} + 2\,e^- \rightarrow \text{O}^{2-} \text{ anion}$$

(8 protons and 8 electrons)   (8 protons and 10 electrons)

Here the O atom has gained two electrons, so we write the anion's symbol as O^{2-}.

How do you know whether an atom is likely to form a cation or an anion? It depends on whether the element is a metal or a nonmetal.

- Metals generally lose electrons in their reactions to form cations.
- Nonmetals frequently gain one or more electrons to form anions in their reactions.

Monatomic Ions

Monatomic ions are single atoms that have lost or gained electrons. As indicated in Figure 2.18, *metals typically lose electrons to form monatomic cations, and nonmetals typically gain electrons to form monatomic anions.*

How can you predict the number of electrons gained or lost? Like lithium in Figure 2.18, *metals of Groups 1A–3A form positive ions having a charge equal to the group number of the metal.*

Group	Metal Atom	Electron Change		Resulting Metal Cation
1A	Na (11 protons, 11 electrons)	−1	→	Na⁺ (11 protons, 10 electrons)
2A	Ca (20 protons, 20 electrons)	−2	→	Ca²⁺ (20 protons, 18 electrons)
3A	Al (13 protons, 13 electrons)	−3	→	Al³⁺ (13 protons, 10 electrons)

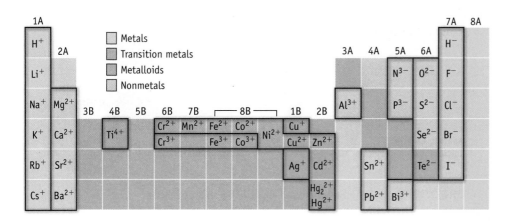

FIGURE 2.18 Charges on some common monatomic cations and anions. Metals usually form cations and nonmetals usually form anions. (The boxed areas show ions of identical charge.)

Transition metals (B-group elements) also form cations. Unlike the A-group metals, however, no easily predictable pattern of behavior occurs for transition metal cations. In addition, transition metals often form several different ions. An iron-containing compound, for example, may contain either Fe^{2+} or Fe^{3+} ions. Indeed, 2+ and 3+ ions are typical of many transition metals (Figure 2.18).

● **Writing Ion Formulas** When writing the formula of an ion, the charge on the ion must be included.

Group	Metal Atom	Electron Change		Resulting Metal Cation
7B	Mn (25 protons, 25 electrons)	−2	→	Mn^{2+} (25 protons, 23 electrons)
8B	Fe (26 protons, 26 electrons)	−2	→	Fe^{2+} (26 protons, 24 electrons)
8B	Fe (26 protons, 26 electrons)	−3	→	Fe^{3+} (26 protons, 23 electrons)

Nonmetals often form ions having a negative charge equal to the group number of the element minus 8. For example, nitrogen is in Group 5A, so it forms an ion having a 3− charge because a nitrogen atom can gain three electrons.

Group	Nonmetal Atom	Electron Change		Resulting Nonmetal Anion
5A	N (7 protons, 7 electrons)	+3	→	N^{3-} (7 protons, 10 electrons) Charge = 5 − 8 = −3
6A	S (16 protons, 16 electrons)	+2	→	S^{2-} (16 protons, 18 electrons) Charge = 6 − 8 = −2
7A	Br (35 protons, 35 electrons)	+1	→	Br^- (35 protons, 36 electrons) Charge = 7 − 8 = −1

Notice that hydrogen appears at two locations in Figure 2.18. The H atom can either lose or gain electrons, depending on the other atoms it encounters.

Electron lost: H (1 proton, 1 electron) → H^+ (1 proton, 0 electrons) + e^-

Electron gained: H (1 proton, 1 electron) + e^- → H^- (1 proton, 2 electrons)

Finally, the noble gases *very* rarely form monatomic cations and never form monatomic anions in chemical reactions.

Ion Charges and the Periodic Table

The metals of Groups 1A, 2A, and 3A form ions having 1+, 2+, and 3+ charges (Figure 2.18); that is, their atoms lose one, two, or three electrons, respectively. *For Group 1A and 2A metals and aluminum, the number of electrons remaining on the cation is the same as the number of electrons in an atom of the noble gas that precedes it in the periodic table.* For example, Mg^{2+} has 10 electrons, the same number as in an atom of the noble gas neon (atomic number 10).

An atom of a nonmetal near the right side of the periodic table would have to lose a great many electrons to achieve the same number as a noble gas atom of lower atomic number. (For instance, Cl, whose atomic number is 17, would have to lose 7 electrons to have the same number of electrons as Ne.) If a nonmetal atom were to gain just a few electrons, however, it would have the same number as a noble gas

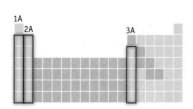

Group 1A, 2A, 3A metals form M^{n+} cations where n = group number.

Main group metals.

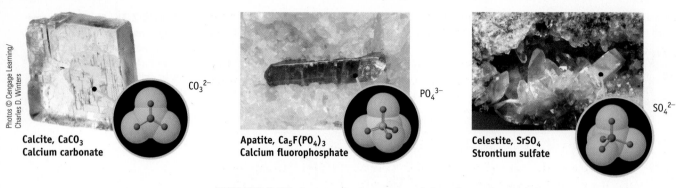

Calcite, CaCO₃
Calcium carbonate

Apatite, Ca₅F(PO₄)₃
Calcium fluorophosphate

Celestite, SrSO₄
Strontium sulfate

FIGURE 2.19 Common ionic compounds based on polyatomic ions.

atom of higher atomic number. For example, an oxygen atom has eight electrons. By gaining two electrons per atom it forms O^{2-}, which has 10 electrons, the same number as neon. *Anions having the same number of electrons as the noble gas atom succeeding it in the periodic table are commonly observed in chemical compounds.*

Polyatomic Ions

Polyatomic ions are made up of two or more atoms, and the collection has an electric charge (Figure 2.19 and Table 2.4). For example, carbonate ion, CO_3^{2-}, a common polyatomic anion, consists of one C atom and three O atoms. The ion has two

Table 2.4 Formulas and Names of Some Common Polyatomic Ions

Formula	Name	Formula	Name
Cation: Positive Ion			
NH_4^+	Ammonium ion		
Anions: Negative Ions			
Based on a Group 4A element		**Based on a Group 7A element**	
CN^-	Cyanide ion	ClO^-	Hypochlorite ion
$CH_3CO_2^-$	Acetate ion	ClO_2^-	Chlorite ion
CO_3^{2-}	Carbonate ion	ClO_3^-	Chlorate ion
HCO_3^-	Hydrogen carbonate ion (or bicarbonate ion)	ClO_4^-	Perchlorate ion
$C_2O_4^{2-}$	Oxalate ion		
Based on a Group 5A element		**Based on a transition metal**	
NO_2^-	Nitrite ion	CrO_4^{2-}	Chromate ion
NO_3^-	Nitrate ion	$Cr_2O_7^{2-}$	Dichromate ion
PO_4^{3-}	Phosphate ion	MnO_4^-	Permanganate ion
HPO_4^{2-}	Hydrogen phosphate ion		
$H_2PO_4^-$	Dihydrogen phosphate ion		
Based on a Group 6A element			
OH^-	Hydroxide ion		
SO_3^{2-}	Sulfite ion		
SO_4^{2-}	Sulfate ion		
HSO_4^-	Hydrogen sulfate ion (or bisulfate ion)		

units of negative charge because there are two more electrons (a total of 32) in the ion than there are protons (a total of 30) in the nuclei of one C atom and three O atoms.

The ammonium ion, NH_4^+, is a common polyatomic cation. In this case, four H atoms surround an N atom, and the ion has a 1+ electric charge. This ion has 10 electrons, but there are 11 positively charged protons in the nuclei of the N and H atoms (seven and one each, respectively).

Formulas of Ionic Compounds

Compounds are electrically neutral; that is, they have no net electric charge. Thus, in an ionic compound the numbers of positive and negative ions must be such that the positive and negative charges balance. In sodium chloride, the sodium ion has a 1+ charge (Na^+) and the chloride ion has a 1− charge (Cl^-). These ions must be present in a 1 : 1 ratio, and so the formula is NaCl.

The gemstone ruby is largely the compound formed from aluminum ions (Al^{3+}) and oxide ions (O^{2-}) (but the color comes from a trace of Cr^{3+} ions.) Here the ions have positive and negative charges that are of different absolute value. To have a compound with the same number of positive and negative charges, two Al^{3+} ions [total charge $= 2 \times (3+) = 6+$] must combine with three O^{2-} ions [total charge $= 3 \times (2-) = 6-$] to give a formula of Al_2O_3.

Calcium is a Group 2A metal, and it forms a cation having a 2+ charge. It can combine with a variety of anions to form ionic compounds such as those in the following table:

Ruby, Al_2O_3. Gems called *rubies* are largely composed of Al^{3+} and O^{2-} ions with a trace of Cr^{3+} ion. It is the chromium(III) ions that give the gem its color.

© Cengage Learning/Charles D. Winters

Compound	Ion Combination	Overall Charge on Compound
$CaCl_2$	$Ca^{2+} + 2\ Cl^-$	$(2+) + 2 \times (1-) = 0$
$CaCO_3$	$Ca^{2+} + CO_3^{2-}$	$(2+) + (2-) = 0$
$Ca_3(PO_4)_2$	$3\ Ca^{2+} + 2\ PO_4^{3-}$	$3 \times (2+) + 2 \times (3-) = 0$

In writing formulas of ionic compounds, the convention is that *the symbol of the cation is given first, followed by the anion symbol.* Also notice the use of parentheses when more than one polyatomic ion of a given kind is present [as in $Ca_3(PO_4)_2$]. (None, however, are used when only one polyatomic ion is present, as in $CaCO_3$.)

● **Balancing Ion Charges in Formulas** Aluminum, a metal in Group 3A, loses three electrons to form the Al^{3+} cation. Oxygen, a nonmetal in Group 6A, gains two electrons to form an O^{2-} anion. Notice that in the compound formed from these ions the charge on the cation is the subscript on the anion, and vice versa.

$$2\ Al^{3+} + 3\ O^{2-} \rightarrow Al_2O_3$$

This often works well, but be careful. The subscripts of Ti^{4+} and O^{2-} are reduced to the simplest ratio in TiO_2 (1 Ti to 2 O, rather than 2 Ti to 4 O).

$$Ti^{4+} + 2\ O^{2-} \rightarrow TiO_2$$

EXAMPLE 2.4   Ionic Compound Formulas

Problem For each of the following ionic compounds, write the symbols for the ions present and give the relative number of each: (a) Li_2CO_3, and (b) $Fe_2(SO_4)_3$.

What Do You Know? You know the formulas of the ionic compounds, the predicted charges on monatomic ions (Figure 2.18), and the formulas and charges of polyatomic ions (Table 2.4).

Strategy Divide the formula of the compound into the cations and anions. To accomplish this you will have to recognize, and remember, the composition and charges of common ions.

Solution

(a) Li_2CO_3 is composed of two lithium ions, Li^+, for each carbonate ion, CO_3^{2-}. Li is a Group 1A element and always has a 1+ charge in its compounds. Because the two 1+ charges balance the negative charge of the carbonate ion, the latter must be 2−.

(b) $Fe_2(SO_4)_3$ contains two iron(III) ions, Fe^{3+}, for every three sulfate ions, SO_4^{2-}. The way to recognize this is to recall that sulfate has a 2− charge. Because three sulfate ions are present (with a total charge of 6−), the two iron cations must have a total charge of 6+. This is possible only if each iron cation has a charge of 3+.

Think about Your Answer The charges predicted are in line with those in Figure 2.18 and Table 2.4. Remember that the formula for an ion must include its composition and its charge. Formulas for ionic compounds are always written with the cation first and then the anion, but ion charges are not included.

Check Your Understanding

(a)   Give the number and identity of the constituent ions in each of the following ionic compounds: NaF, $Cu(NO_3)_2$, and $NaCH_3CO_2$.

(b)   Iron, a transition metal, forms ions having 2+ and 3+ charges. Write the formulas of the compounds formed between chloride ions and these two different iron cations.

EXAMPLE 2.5   Ionic Compound Formulas

Problem Write formulas for ionic compounds composed of an aluminum cation and each of the following anions: (a) fluoride ion, (b) sulfide ion, and (c) nitrate ion.

What Do You Know? You know the names of the ions involved, the predicted charges on monatomic ions (Figure 2.18), and the names of polyatomic ions (Table 2.4).

Strategy First decide on the formula of the Al cation and the formula of each anion. Combine the Al cation with each type of anion to form electrically neutral compounds.

Solution An aluminum cation is predicted to have a charge of 3+ because Al is a metal in Group 3A.

(a)   Fluorine is a Group 7A element. The charge of the fluoride ion is predicted to be 1− (from 7 − 8 = 1−). Therefore, we need 3 F^- ions to combine with one Al^{3+}. The formula of the compound is AlF_3.

(b)   Sulfur is a nonmetal in Group 6A, so it forms a 2− anion. Thus, we need to combine two Al^{3+} ions [total charge is 6+ = 2 × (3+)] with three S^{2-} ions [total charge is 6− = 3 × (2−)]. The compound has the formula Al_2S_3.

(c)   The nitrate ion has the formula NO_3^- (see Table 2.4). The answer here is therefore similar to the AlF_3 case, and the compound has the formula $Al(NO_3)_3$. Here we place parentheses around NO_3 to show that three polyatomic NO_3^- ions are involved.

Think about Your Answer The most common error students make is not knowing the correct charge on an ion.

Check Your Understanding

Write the formulas of all neutral ionic compounds that can be formed by combining the cations Na^+ and Ba^{2+} with the anions S^{2-} and PO_4^{3-}.

Names of Ions
Naming Positive Ions (Cations)

With a few exceptions (such as NH_4^+), the positive ions described in this text are metal ions. Positive ions are named by the following rules:

1.   For a monatomic positive ion (that is, a metal cation) the name is that of the metal plus the word "cation." For example, we have already referred to Al^{3+} as the aluminum cation.

2.   Some cases occur, especially in the transition series, in which a metal can form more than one type of positive ion. In these cases the charge of the ion is indicated by a Roman numeral in parentheses immediately following the ion's name. For example, Co^{2+} is the cobalt(II) cation, and Co^{3+} is the cobalt(III)' cation.

Finally, you will encounter the ammonium cation, NH_4^+, many times in this book and in the laboratory. Do not confuse the ammonium cation with the ammonia molecule, NH_3, which has no electric charge and one less H atom.

● **"-ous" and "-ic" Endings** An older naming system for metal ions uses the ending -ous for the ion of lower charge and -ic for the ion of higher charge. For example, there are cobaltous (Co^{2+}) and cobaltic (Co^{3+}) ions. In addition, this older system sometimes uses the root of the Latin name of some elements in the names of their ions. For example, the Latin name for iron is *ferrum*, and this system calls the iron cations the ferrous (Fe^{2+}) and ferric (Fe^{3+}) ions. We do not use this system in this book, but some chemical manufacturers continue to use it.

Naming Negative Ions (Anions)

There are two types of negative ions: those having only one atom (*monatomic*) and those having several atoms (*polyatomic*).

1. A monatomic negative ion is named by adding *-ide* to the stem of the name of the nonmetal element from which the ion is derived (Figure 2.20). The anions of the Group 7A elements, the halogens, are known as the fluoride, chloride, bromide, and iodide ions and as a group are called **halide ions**.
2. Polyatomic negative ions are common, especially those containing oxygen (called **oxoanions**). The names of some of the most common oxoanions are given in Table 2.4. Although most of these names must simply be learned, some guidelines can help. For example, consider the following pairs of ions:

NO_3^- is the nitrate ion, whereas NO_2^- is the nitrite ion.

SO_4^{2-} is the sulfate ion, whereas SO_3^{2-} is the sulfite ion.

The oxoanion having the greater number of oxygen atoms is given the suffix -ate, and the oxoanion having the smaller number of oxygen atoms has the suffix -ite. For a series of oxoanions having more than two members, the ion with the largest number of oxygen atoms has the prefix *per-* and the suffix *-ate*. The ion having the smallest number of oxygen atoms has the prefix *hypo-* and the suffix *-ite*. The chlorine oxoanions are the most commonly encountered example.

ClO_4^-	*Perchlorate* ion
ClO_3^-	*Chlorate* ion
ClO_2^-	*Chlorite* ion
ClO^-	*Hypochlorite* ion

Oxoanions that contain hydrogen are named by adding the word "hydrogen" before the name of the oxoanion. If two hydrogens are in the anion, we say "dihydrogen." Many hydrogen-containing oxoanions have common names that are used as well. For example, the hydrogen carbonate ion, HCO_3^-, is called the bicarbonate ion.

Ion	Systematic Name	Common Name
HPO_4^{2-}	Hydrogen phosphate ion	
$H_2PO_4^-$	Dihydrogen phosphate ion	
HCO_3^-	Hydrogen carbonate ion	Bicarbonate ion
HSO_4^-	Hydrogen sulfate ion	Bisulfate ion
HSO_3^-	Hydrogen sulfite ion	Bisulfite ion

Names of Ionic Compounds

The name of an ionic compound is built from the names of the positive and negative ions in the compound. The name of the positive cation is given first, followed by the name of the negative anion. Examples of ionic compound names are given below.

Ionic Compound	Ions Involved	Name
$CaBr_2$	Ca^{2+} and $2\ Br^-$	Calcium bromide
$NaHSO_4$	Na^+ and HSO_4^-	Sodium hydrogen sulfate
$(NH_4)_2CO_3$	$2\ NH_4^+$ and CO_3^{2-}	Ammonium carbonate
$Mg(OH)_2$	Mg^{2+} and $2\ OH^-$	Magnesium hydroxide
$TiCl_2$	Ti^{2+} and $2\ Cl^-$	Titanium(II) chloride
Co_2O_3	$2\ Co^{3+}$ and $3\ O^{2-}$	Cobalt(III) oxide

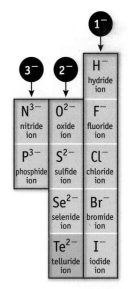

FIGURE 2.20 Names and charges of some common monatomic ions.

● **Naming Oxoanions**

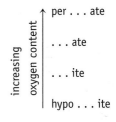

● **Names of Compounds Containing Transition Metal Cations** Be sure to notice that the charge on a transition metal cation is indicated by a Roman numeral and is included in the name.

Formulas for Ions and Ionic Compounds

Writing formulas for ionic compounds takes practice, and it requires that you know the formulas and charges of the most common ions. The charges on monatomic ions are often evident from the position of the element in the periodic table, but you simply have to remember the formula and charges of polyatomic ions, especially the most common ones such as nitrate, sulfate, carbonate, phosphate, and acetate.

If you cannot remember the formula of a polyatomic ion or if you encounter an ion you have not seen before, you may be able to figure out its formula. For example, suppose you are told that $NaCHO_2$ is sodium formate. You know that the sodium ion is Na^+, so the formate ion must be the remaining portion of the compound; it must have a charge of $1-$ to balance the $1+$ charge on

the sodium ion. Thus, the formate ion must be CHO_2^-.

Finally, when writing the formulas of ions, you *must* include the charge on the ion (except in the formula of an ionic compound). Writing Na when you mean sodium ion is incorrect. There is a vast difference in the properties of the element sodium (Na) and those of its ion (Na^+).

Properties of Ionic Compounds

When a particle having a negative electric charge is brought near another particle having a positive electric charge, there is a force of attraction between them (Figure 2.21). In contrast, there is a repulsive force when two particles with the same charge—both positive or both negative—are brought together. These forces are called **electrostatic** forces, and the force of attraction (or repulsion) between ions is given by **Coulomb's law** (Equation 2.3):

charge on + and − ions⟍   charge on electron

$$\text{Force} = -k\,\frac{(n^+\text{e})(n^-\text{e})}{d^2} \tag{2.3}$$

proportionality constant   distance between ions

where, for example, n^+ is $+3$ for Al^{3+} and n^- is -2 for O^{2-}. Based on Coulomb's law, the force of attraction between oppositely charged ions increases

- as the ion charges (n^+ and n^-) increase. Thus, the attraction between ions having charges of $2+$ and $2-$ is greater than that between ions having $1+$ and $1-$ charges (Figure 2.21).
- as the distance between the ions becomes smaller (Figure 2.21).

Ionic compounds do not consist of simple pairs or small groups of positive and negative ions. The simplest ratio of cations to anions in an ionic compound is represented by its formula, but an ionic solid consists of millions upon millions of ions arranged in an extended three-dimensional network called a **crystal lattice**. A

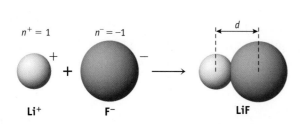

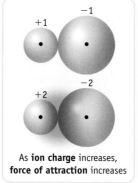

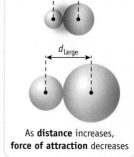

(a) Ions such as Li^+ and F^- are held together by an electrostatic force. Here a lithium ion is attracted to a fluoride ion, and the distance between the nuclei of the two ions is *d*.

(b) Forces of attraction between ions of opposite charge increase with increasing ion charge and decrease with increasing distance (*d*).

FIGURE 2.21 Coulomb's law and electrostatic forces.

Is a Compound Ionic?

Students often ask how to know whether a compound is ionic. Here are some useful guidelines.

1. Most metal-containing compounds are ionic. So, if a metal atom appears in the formula of a compound and especially when it is the element listed first, a good first guess is that it is ionic. (There are interesting exceptions, but few come up in introductory chemistry.) It is helpful in this regard to recall trends in metallic

behavior: All elements to the left of a diagonal line running from boron to tellurium in the periodic table are metallic.

2. If there is no metal in the formula, it is likely that the compound is not ionic. The exceptions here are compounds composed of polyatomic cations based on nonmetals (e.g., NH_4Cl or NH_4NO_3).

3. Learn to recognize the formulas of polyatomic ions (see Table 2.4). Chemists write the formula of ammonium nitrate as

NH_4NO_3 (not as $N_2H_4O_3$) to alert others to the fact that it is an ionic compound composed of the common polyatomic ions NH_4^+ and NO_3^-.

As an example of these guidelines, you can be sure that $MgBr_2$ (Mg^{2+} with Br^-) and K_2S (K^+ with S^{2-}) are ionic compounds. On the other hand, the compound CCl_4, formed from two nonmetals, C and Cl, is not ionic.

---

portion of the lattice for NaCl, illustrated in Figure 2.22, represents a common way of arranging ions for compounds that have a 1:1 ratio of cations to anions.

Ionic compounds have characteristic properties that can be understood in terms of the charges of the ions and their arrangement in the lattice. Because each ion is surrounded by oppositely charged nearest neighbors, it is held tightly in its allotted location. At room temperature each ion can move just a bit around its average position, but considerable energy must be added before an ion can escape the attraction of its neighboring ions. Only if enough energy is added will the lattice structure collapse and the substance melt. Greater attractive forces mean that ever more energy—and higher and higher temperatures—is required to cause melting. Thus, Al_2O_3, a solid composed of Al^{3+} and O^{2-} ions, melts at a much higher temperature (2072 °C) than NaCl (801 °C), a solid composed of Na^+ and Cl^- ions.

Most ionic compounds are "hard" solids. That is, the solids are not pliable or soft. The reason for this characteristic is again related to the lattice of ions. The nearest neighbors of a cation in a lattice are anions, and the force of attraction makes the lattice rigid. However, a blow with a hammer can cause the lattice to break cleanly along a sharp boundary. The hammer blow displaces layers of ions just enough to cause ions of like charge to become nearest neighbors, and the repulsion between these like-charged ions forces the lattice apart (Figure 2.23).

FIGURE 2.22 Sodium chloride. A crystal of NaCl consists of an extended lattice of sodium ions and chloride ions in a 1:1 ratio.

FIGURE 2.23 Ionic solids.

(a) An ionic solid is normally rigid owing to the forces of attraction between oppositely charged ions. When struck sharply, however, the crystal can cleave cleanly.

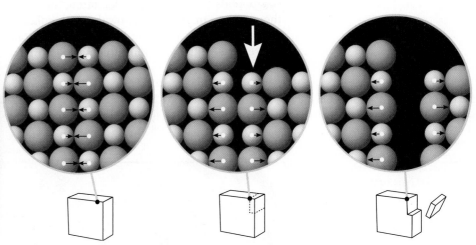

(b) When a crystal is struck, layers of ions move slightly, and ions of like charge become nearest neighbors. Repulsions between ions of similar charge cause the crystal to cleave.

REVIEW & CHECK FOR SECTION 2.7

1. What is the most likely charge on an ion of barium?

    (a) 2−   (b) 2+   (c) 3+   (d) 1−

2. When gallium forms an ion, it

    (a) loses 3 electrons   (c) loses 2 electrons

    (b) gains 3 electrons   (d) gains 2 electrons

3. The name of the compound $(NH_4)_2S$ is

    (a) nitrohydrogen sulfide   (c) ammonium sulfur

    (b) ammonium sulfide   (d) ammonia sulfide

4. The formula of barium acetate is

    (a) $Ba(CH_3CO_2)_2$ (c) $BaMnO_4$

    (b) $BaCH_3CO_2$ (d) $BaCO_3$

5. The name of the compound with the formula V_2O_3 is

    (a) vanadium(III) oxide   (c) divanadium trioxide

    (b) vanadium oxide   (d) vanadium trioxide

6. Which should have the higher melting point, MgO or NaCl?

    (a) MgO   (b) NaCl

2.8 Molecular Compounds: Formulas and Names

Many familiar compounds are not ionic, they are molecular: the water you drink, the sugar in your coffee or tea, or the aspirin you take for a headache.

Ionic compounds are generally solids, whereas molecular compounds can range from gases to liquids to solids at ordinary temperatures (Figure 2.24). As size and molecular complexity increase, compounds generally exist as solids. We will explore some of the underlying causes of these general observations in Chapter 12.

Some molecular compounds have complicated formulas that you cannot, at this stage, predict or even decide if they are correct. However, there are many simple compounds you will encounter often, and you should understand how to name them and, in many cases, know their formulas.

Let us look first at molecules formed from combinations of two nonmetals. These "two-element" or **binary compounds** of nonmetals can be named in a systematic way.

FIGURE 2.24 Molecular compounds. Ionic compounds are generally solids at room temperature. In contrast, molecular compounds can be gases, liquids, or solids. The molecular models are of caffeine (in coffee), water, and citric acid (in lemons).

Hydrogen forms binary compounds with all of the nonmetals except the noble gases. For compounds of oxygen, sulfur, and the halogens, the H atom is generally written first in the formula and is named first. The other nonmetal is named as if it were a negative ion.

Compound	Name
HF	Hydrogen fluoride
HCl	Hydrogen chloride
H_2S	Hydrogen sulfide

Although there are exceptions, *most binary molecular compounds are a combination of nonmetallic elements from Groups 4A–7A with one another or with hydrogen.* The formula is generally written by putting the elements in order of increasing group number. When naming the compound, the number of atoms of a given type in the compound is designated with a prefix, such as "di-," "tri-," "tetra-," "penta-," and so on.

Compound	Systematic Name
NF_3	Nitrogen trifluoride
NO	Nitrogen monoxide
NO_2	Nitrogen dioxide
N_2O	Dinitrogen monoxide
N_2O_4	Dinitrogen tetraoxide
PCl_3	Phosphorus trichloride
PCl_5	Phosphorus pentachloride
SF_6	Sulfur hexafluoride
S_2F_{10}	Disulfur decafluoride

Finally, many of the binary compounds of nonmetals were discovered years ago and have common names.

Compound	Common Name	Compound	Common Name
CH_4	Methane	N_2H_4	Hydrazine
C_2H_6	Ethane	PH_3	Phosphine
C_3H_8	Propane	NO	Nitric oxide
C_4H_{10}	Butane	N_2O	Nitrous oxide ("laughing gas")
NH_3	Ammonia	H_2O	Water

● **Formulas of Binary Nonmetal Compounds Containing Hydrogen** Simple hydrocarbons (compounds of C and H) such as methane (CH_4) and ethane (C_2H_6) have formulas written with H following C, and the formulas of ammonia and hydrazine have H following N. Water and the hydrogen halides, however, have the H atom preceding O or the halogen atom. Tradition is the only explanation for such irregularities in writing formulas.

● **Hydrocarbons** Compounds such as methane, ethane, propane, and butane belong to a class of hydrocarbons called *alkanes.*

methane, CH_4

propane, C_3H_8

ethane, C_2H_6

butane, C_4H_{10}

REVIEW & CHECK FOR SECTION 2.8

1. What is the formula for dioxygen difluoride?

   (a) O_2F_2 (b) OF   (c) O_2F (d) OF_2

2. Compound names and formulas are listed below. Which name is incorrect?

   (a) carbon disulfide, CS_2 (c) boron trifluoride, BF_3

   (b) nitrogen pentaoxide, N_2O_5 (d) sulfur tetrafluoride, SF_4

3. The name of the compound with the formula N_2F_4 is

   (a) nitrogen fluoride   (c) dinitrogen tetrafluoride

   (b) dinitrogen fluoride   (d) nitrogen tetrafluoride

4. The name of the compound with the formula P_4O_{10} is

   (a) phosphorus oxide   (c) tetraphosphorus oxide

   (b) tetraphosphorus decaoxide   (d) phosphorus decaoxide

Module 4: The Mole covers concepts in this section.

2.9 Atoms, Molecules, and the Mole

One of the most exciting aspects of chemical research is the discovery of some new substance, and part of this process of discovery involves quantitative experiments. When two chemicals react with each other, we want to know how many atoms or molecules of each are used so that formulas can be established for the reaction's products. To do so, we need some method of counting atoms and molecules. That is, we must discover a way of connecting the macroscopic world, the world we can see, with the particulate world of atoms, molecules, and ions. The solution to this problem is to define a unit of matter that contains a known number of particles. That chemical unit is the mole.

The **mole** (abbreviated mol) is the SI base unit for measuring an *amount of a substance* (◄ Table 1, page 21) and is defined as follows:

> A **mole** is the amount of a substance that contains as many elementary entities (atoms, molecules, or other particles) as there are atoms in exactly 12 g of the carbon-12 isotope.

● **The "Mole"** The term *mole* was introduced about 1895 by Wilhelm Ostwald (1853–1932), who derived the term from the Latin word *moles*, meaning a "heap" or a "pile."

The key to understanding the concept of the mole is recognizing that *one mole always contains the same number of particles, no matter what the substance.* One mole of sodium contains the same number of atoms as one mole of iron and the same as the number of molecules in one mole of water. How many particles? Many, many experiments over the years have established that number as

$$1 \text{ mole} = 6.0221415 \times 10^{23} \text{ particles}$$

This value is known as **Avogadro's number** (symbolized by N_A) in honor of Amedeo Avogadro, an Italian lawyer and physicist (1776–1856) who conceived the basic idea (but never determined the number).

Atoms and Molar Mass

● **A Difference Between the Terms** *Amount* **and** *Quantity* The terms "amount" and "quantity" are used in a specific sense by chemists. The amount of a substance is the number of moles of that substance. In contrast, quantity refers, for example, to the mass or volume of the substance. See W. G. Davies and J. W. Moore. *Journal of Chemical Education*, Vol. 57, p. 303, 1980. See also http://physics.nist.gov

The mass in grams of one mole of any element (6.0221415×10^{23} atoms of that element) is the **molar mass** of that element. Molar mass is conventionally abbreviated with a capital italicized M and has units of grams per mole (g/mol). *An element's molar mass is the amount in grams numerically equal to its atomic weight.* Using sodium and lead as examples,

Molar mass of sodium (Na) = mass of 1.000 mol of Na atoms

= 22.99 g/mol

= mass of 6.022×10^{23} Na atoms

Molar mass of lead (Pb) = mass of 1.000 mol of Pb atoms

= 207.2 g/mol

= mass of 6.022×10^{23} Pb atoms

Figure 2.25 shows the relative sizes of a mole of some common elements. Although each of these "piles of atoms" has a different volume and different mass, each contains 6.022×10^{23} atoms.

The mole concept is the cornerstone of quantitative chemistry. It is essential to be able to convert from moles to mass and from mass to moles. Dimensional analysis, which is described in *Let's Review*, page 35, shows that this can be done in the following way:

MASS ⟷ MOLES CONVERSION	
Moles to Mass	*Mass to Moles*
Moles × $\dfrac{\text{grams}}{1 \text{ mol}}$ = grams	Grams × $\dfrac{1 \text{ mol}}{\text{grams}}$ = moles
↑	↑
molar mass	1/molar mass

FIGURE 2.25 **One mole of common elements.** (*Left to right*) Sulfur powder, magnesium chips, tin, and silicon. (*Above*) Copper beads.

Copper
63.546 g

Sulfur
32.066 g

Magnesium
24.305 g

Tin
118.71 g

Silicon
28.086 g

© Cengage Learning/Charles D. Winters

For example, what mass, in grams, is represented by 0.35 mol of aluminum? Using the molar mass of aluminum (27.0 g/mol), you can determine that 0.35 mol of Al has a mass of 9.5 g.

$$0.35 \text{ mol Al} \times \frac{27.0 \text{ g Al}}{1 \text{ mol Al}} = 9.5 \text{ g Al}$$

Molar masses of the elements are generally known to at least four significant figures. The convention followed in calculations in this book is to use a value of the molar mass with at least one more significant figure than in any other number in the problem. For example, if you weigh out 16.5 g of carbon, you use 12.01 g/mol for the molar mass of C to find the amount of carbon present.

$$16.5 \text{ g C} \times \frac{1 \text{ mol C}}{12.01 \text{ g C}} = 1.37 \text{ mol C}$$
↑
Note that four significant figures are used in the
molar mass, but there are three in the sample mass.

Using one more significant figure for the molar mass means the accuracy of this value will not affect the accuracy of the result.

A CLOSER LOOK

Amedeo Avogadro and His Number

Amedeo Avogadro, conte di Quaregna, (1776–1856) was an Italian nobleman and a lawyer. In about 1800, he turned to science and was the first professor of mathematical physics in Italy.

Avogadro did not himself propose the notion of a fixed number of particles in a chemical unit. Rather, the number was named in his honor because he had performed experiments in the 19th century that laid the groundwork for the concept.

Just how large is Avogadro's number? One mol of unpopped popcorn kernels would cover the continental United States to a depth of about 9 miles. One mole of pennies divided equally among every man, woman, and child in the United States would allow each person to pay off the national debt ($11.6 trillion or 11.6×10^{12} dollars) and there would still be about $8 trillion left over per person! A mole of pennies does not go as far as it used to!

Is the number a unique value like π? No. It is fixed by the definition of the mole as exactly 12 g of carbon-12. If one mole of carbon were defined to have some other mass, then Avogadro's number would have a different value.

E. F. Smith Collection/
Van Pelt Library/
University of Pennsylvania

Lead. A 150-mL beaker containing 2.50 mol or 518 g of lead.

EXAMPLE 2.6   Mass, Moles, and Atoms

Problem Consider two elements in the same vertical column of the periodic table, lead and tin.

(a)   What mass of lead, in grams, is equivalent to 2.50 mol of lead (Pb)?

(b)   What amount of tin, in moles, is represented by 36.6 g of tin (Sn)? How many atoms of tin are in the sample?

What Do You Know? You know the amount of lead and the mass of tin. You also know, from the periodic table in the front of the book, the molar masses of lead (207.2 g/mol) and tin (118.7 g/mol). For part (b) Avogadro's number is needed (and is given in the text and inside the back cover of the book).

Strategy The molar mass is the quantity in grams numerically equal to the atomic weight of the element.

Part (a) Multiply the amount of Pb by the molar mass.

Part (b) Multiply the mass of tin by (1/molar mass). To determine the number of atoms, multiply the amount of tin by Avogadro's number.

Solution

(a)   Convert the amount of lead in moles to mass in grams.

$$2.50 \text{ mol Pb} \times \frac{207.2 \text{ g}}{1 \text{ mol Pb}} = \boxed{518 \text{ g Pb}}$$

(b)   Convert the mass of tin to the amount in moles,

$$36.6 \text{ g Sn} \times \frac{1 \text{ mol Sn}}{118.7 \text{ g Sn}} = \boxed{0.308 \text{ mol Sn}}$$

and then use Avogadro's number to find the number of atoms in the sample.

$$0.308 \text{ mol Sn} \times \frac{6.022 \times 10^{23} \text{ atoms Sn}}{1 \text{ mol Sn}} = \boxed{1.86 \times 10^{23} \text{ atoms Sn}}$$

Think about Your Answer To find a mass from an amount of substance, use the conversion factor (mass/mol) (= molar mass). To find the amount from mass of a substance, use the conversion factor (mol/mass) (= 1/molar mass). You can sometimes catch a mistake in setting up a conversion factor upside down if you think about your answers. For example, if we had made this mistake in part b, we would have calculated that there was less than one atom in 0.308 mol of Sn, clearly an unreasonable answer.

Check Your Understanding

The density of gold, Au, is 19.32 g/cm³. What is the volume (in cubic centimeters) of a piece of gold that contains 2.6×10^{24} atoms? If the piece of metal is a square with a thickness of 0.10 cm, what is the length (in centimeters) of one side of the piece?

Tin. A sample of tin having a mass of 36.6 g (or 1.86×10^{23} atoms).

Molecules, Compounds, and Molar Mass

The formula of a compound tells you the type of atoms or ions in the compound and the relative number of each. For example, one molecule of methane, CH_4, is made up of one atom of C and four atoms of H. But suppose you have Avogadro's number of C atoms (6.022×10^{23}) combined with the proper number of H atoms. For CH_4 this means there are $4 \times 6.022 \times 10^{23}$ H atoms per mole of C atoms. What masses of atoms are combined, and what is the mass of this many CH_4 molecules?

C	+	4 H	⟶	CH₄
6.022×10^{23} C atoms		$4 \times 6.022 \times 10^{23}$ H atoms		6.022×10^{23} CH₄ molecules
= 1.000 mol of C		= 4.000 mol of H atoms		= 1.000 mol of CH₄ molecules
= 12.01 g of C atoms		= 4.032 g of H atoms		= 16.04 g of CH₄ molecules

Because we know the number of moles of C and H atoms, we can calculate the masses of carbon and hydrogen that combine to form CH_4. It follows that the mass of CH_4 is the sum of these masses. That is, 1 mol of CH_4 has a mass equal to the mass of 1 mol of C atoms (12.01 g) plus 4 mol of H atoms (4.032 g). Thus, the *molar mass, M*, of CH_4 is 16.04 g/mol. The molar masses of some substances are:

Molar and Molecular Masses

Substance	Molar Mass, *M* (g/mol)	Average Mass of One Molecule (g/molecule)
O_2	32.00	5.314×10^{-23}
NH_3	17.03	2.828×10^{-23}
H_2O	18.02	2.992×10^{-23}
$NH_2CH_2CO_2H$ (glycine)	75.07	1.247×10^{-22}

Ionic compounds such as NaCl do not exist as individual molecules. Thus, for ionic compounds we write the simplest formula that shows the relative number of each kind of atom in a "formula unit" of the compound, and the molar mass is calculated from this formula (*M* for NaCl = 58.44 g/mol). To differentiate substances like NaCl that do not contain molecules, chemists sometimes refer to their *formula weight* instead of their molar mass.

Figure 2.26 illustrates 1-mol quantities of several common compounds. To find the molar mass of any compound, you need only to add up the atomic masses for each element in the compound, taking into account any subscripts on elements. As an example, let us find the molar mass of aspirin, $C_9H_8O_4$. In one mole of aspirin there are 9 mol of carbon atoms, 8 mol of hydrogen atoms, and 4 mol of oxygen atoms, which add up to 180.2 g/mol of aspirin.

$$\text{Mass of C in 1 mol } C_9H_8O_4 = 9 \text{ mol C} \times \frac{12.01 \text{ g C}}{1 \text{ mol C}} = 108.1 \text{ g C}$$

$$\text{Mass of H in 1 mol } C_9H_8O_4 = 8 \text{ mol H} \times \frac{1.008 \text{ g H}}{1 \text{ mol H}} = 8.064 \text{ g H}$$

$$\text{Mass of O in 1 mol } C_9H_8O_4 = 4 \text{ mol O} \times \frac{16.00 \text{ g O}}{1 \text{ mol O}} = 64.00 \text{ g O}$$

$$\text{Total mass of 1 mol of } C_9H_8O_4 = \text{molar mass of } C_9H_8O_4 = 180.2 \text{ g}$$

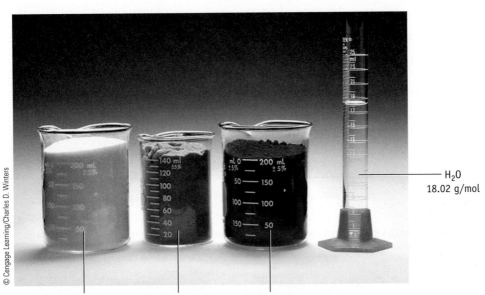

H_2O
18.02 g/mol

Aspirin, $C_9H_8O_4$
180.2 g/mol

Copper(II) chloride dihydrate, $CuCl_2 \cdot 2\ H_2O$
170.5 g/mol

Iron(III) oxide, Fe_2O_3
159.7 g/mol

FIGURE 2.26 One-mole quantities of some compounds. The second compound, $CuCl_2 \cdot 2H_2O$, is called a *hydrated compound* because water is associated with the $CuCl_2$ (see page 86). Thus, one "formula unit" consists of one Cu^{2+} ion, two Cl^- ions, and two water molecules. The molar mass is the sum of the mass of 1 mol of Cu, 2 mol of Cl, and 2 mol of H_2O.

© Cengage Learning/Charles D. Winters

● **Aspirin Formula** Aspirin has the molecular formula $C_9H_8O_4$ and a molar mass of 180.2 g/mol. Aspirin is the common name of the compound acetylsalicylic acid.

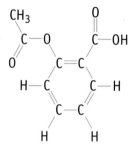

As was the case with elements, it is important to be able to convert between amounts (moles) and mass (grams). For example, if you take 325 mg (0.325 g) of aspirin in one tablet, what amount of the compound have you ingested? Based on a molar mass of 180.2 g/mol, there is 0.00180 mol of aspirin per tablet.

$$0.325 \text{ g aspirin} \times \frac{1 \text{ mol aspirin}}{180.2 \text{ g aspirin}} = 0.00180 \text{ mol aspirin}$$

Using the molar mass of a compound it is possible to determine the number of molecules in any sample from the sample mass and to determine the mass of one molecule. For example, the number of aspirin molecules in one tablet is

$$0.00180 \text{ mol aspirin} \times \frac{6.022 \times 10^{23} \text{ molecules}}{1 \text{ mol aspirin}} = 1.08 \times 10^{21} \text{ molecules}$$

and the mass of one molecule is

$$\frac{180.2 \text{ g aspirin}}{1 \text{ mol aspirin}} \times \frac{1 \text{ mol aspirin}}{6.022 \times 10^{23} \text{ molecules}} = 2.99 \times 10^{-22} \text{ g/molecule}$$

⬮WL   INTERACTIVE EXAMPLE 2.7   Molar Mass and Moles

Problem You have 16.5 g of oxalic acid, $H_2C_2O_4$.

(a)   What amount is represented by 16.5 g of oxalic acid?

(b)   How many molecules of oxalic acid are in 16.5 g?

(c)   How many atoms of carbon are in 16.5 g of oxalic acid?

What Do You Know? You know the mass and formula of oxalic acid. The molar mass of the compound can be calculated based on the formula.

Strategy The strategy is outlined in the strategy map.

•   The molar mass is the sum of the masses of the component atoms.

•   Part (a) Use the molar mass to convert mass to amount.

•   Part (b) Use Avogadro's number to calculate the number of molecules from the amount.

•   Part (c) From the formula you know there are two atoms of carbon in each molecule.

Solution

(a)   *Moles represented by 16.5 g*

Let us first calculate the molar mass of oxalic acid:

$$2 \text{ mol C per mol } H_2C_2O_4 \times \frac{12.01 \text{ g C}}{1 \text{ mol C}} = 24.02 \text{ g C per mol } H_2C_2O_4$$

$$2 \text{ mol H per mol } H_2C_2O_4 \times \frac{1.008 \text{ g H}}{1 \text{ mol H}} = 2.016 \text{ g H per mol } H_2C_2O_4$$

$$4 \text{ mol O per mol } H_2C_2O_4 \times \frac{16.00 \text{ g O}}{1 \text{ mol O}} = 64.00 \text{ g O per mol } H_2C_2O_4$$

Molar mass of $H_2C_2O_4$ = 90.04 g per mol $H_2C_2O_4$

Now calculate the amount in moles. The molar mass (expressed here in units of 1 mol/90.04 g) is used in all mass-mole conversions.

$$16.5 \text{ g } H_2C_2O_4 \times \frac{1 \text{ mol}}{90.04 \text{ g } H_2C_2O_4} = \boxed{0.183 \text{ mol } H_2C_2O_4}$$

Strategy Map 2.7

PROBLEM

Find **amount of oxalic acid** in a given mass. Then find **number of molecules** and **number of C atoms** in the sample.

↓

DATA/INFORMATION KNOWN

• **Mass** of sample
• **Formula** of compound
• **Avogadro's number**

↓

STEP 1. Calculate **molar mass** of oxalic acid.

↓

Molar mass of oxalic acid **(g/mol)**

↓

STEP 2. Use molar mass to calculate **amount** (multiply **mass by 1/molar mass**).

↓

Amount (mol) of oxalic acid

↓

STEP 3. Multiply by **Avogadro's number**.

↓

Number of molecules

↓

STEP 4. Multiply by number of **C atoms** per molecule.

↓

Number of **C atoms** in sample

(b) *Number of molecules*

Use Avogadro's number to find the number of oxalic acid molecules in 0.183 mol of $H_2C_2O_4$.

$$0.183 \; \text{mol} \times \frac{6.022 \times 10^{23} \; \text{molecules}}{1 \; \text{mol}} = 1.10 \times 10^{23} \; \text{molecules}$$

(c) *Number of C atoms*

Because each molecule contains two carbon atoms, the number of carbon atoms in 16.5 g of the acid is

$$1.10 \times 10^{23} \; \text{molecules} \times \frac{2 \; \text{C atoms}}{1 \; \text{molecule}} = 2.21 \times 10^{23} \; \text{C atoms}$$

Think about Your Answer The mass of oxalic acid was 16.5 g, much less than the mass of a mole, so check to make sure your answer reflects this. The number of molecules of the acid should be many fewer than in one mole of molecules.

Check Your Understanding

If you have 454 g of citric acid ($H_3C_6H_5O_7$), what amount (moles) does this represent? How many molecules? How many atoms of carbon?

REVIEW & CHECK FOR SECTION 2.9

1. What is the molar mass of calcium nitrate?

   (a) 150.08 g/mol

   (b) 102.08 g/mol

   (c) 164.09 g/mol

   (d) 68.09 g/mol

2. Which of the following contains the largest number of molecules?

   (a) 24 g of O_2 (b) 4.0 g of H_2 (c) 19 g of F_2 (d) 28 g of N_2

3. Which of the following has the largest mass?

   (a) 0.50 mol Na   (b) 0.20 mol K   (c) 0.40 mol Ca   (d) 0.60 mol Mg

4. Which quantity represents the largest amount (moles) of the indicated substance?

   (a) 6.02×10^{23} molecules of H_2O

   (b) 22 g of CO_2

   (c) 3.01×10^{23} molecules of Br_2

   (d) 30 g of HF

5. How many atoms of oxygen are contained in 16 g of O_2?

   (a) 6.0×10^{23} atoms of O

   (b) 3.0×10^{23} atoms of O

   (c) 1.5×10^{23} atoms of O

   (d) 1.2×10^{24} atoms of O

2.10 Describing Compound Formulas

Given a sample of an unknown compound, how can its formula be determined? The answer lies in *chemical analysis,* a major branch of chemistry that deals with the determination of formulas and structures.

Percent Composition

Any sample of a pure compound always consists of the same elements combined in the same proportion by mass. This means molecular composition can be expressed in at least three ways:

1. in terms of the number of atoms of each type per molecule or per formula unit—that is, by giving the formula of the compound

2. in terms of the mass of each element per mole of compound

3.   in terms of the mass of each element in the compound relative to the total mass of the compound—that is, as a mass percent

Suppose you have 1.0000 mol of NH_3 or 17.031 g. This mass of NH_3 is composed of 14.007 g of N (1.0000 mol) and 3.0237 g of H (3.0000 mol). If you compare the mass of N to the total mass of compound, 82.246% of the total mass is N (and 17.755% is H).

$$\text{Mass of N per mole of } NH_3 = \frac{1 \text{ mol N}}{1 \text{ mol } NH_3} \times \frac{14.007 \text{ g N}}{1 \text{ mol N}} = 14.007 \text{ g N/1 mol } NH_3$$

$$\text{Mass percent N in } NH_3 = \frac{\text{mass of N in 1 mol } NH_3}{\text{mass of 1 mol } NH_3} \times 100\%$$

$$= \frac{14.007 \text{ g N}}{17.031 \text{ g } NH_3} \times 100\%$$

$$= 82.244\% \text{ (or 82.244 g N in 100.000 g } NH_3)$$

$$\text{Mass of H per mole of } NH_3 = \frac{3 \text{ mol H}}{1 \text{ mol } NH_3} \times \frac{1.0079 \text{ g H}}{1 \text{ mol H}}$$

$$= 3.0237 \text{ g H/1 mol } NH_3$$

$$\text{Mass percent H in } NH_3 = \frac{\text{mass of H in 1 mol } NH_3}{\text{mass of 1 mol } NH_3} \times 100\%$$

$$= \frac{3.0237 \text{ g H}}{17.031 \text{ g } NH_3} \times 100\%$$

$$= 17.755\% \text{ (or 17.755 g H in 100.000 g } NH_3)$$

● **Molecular Composition** Molecular composition can be expressed as a percent (mass of an element in a 100-g sample × 100%). For example, NH_3 is 82.244% N. Therefore, it has 82.244 g of N in 100.000 g of compound.

82.244% of NH_3 mass is **nitrogen**.

17.755% of NH_3 mass is **hydrogen**.

These values represent the mass percent of each element, or percent composition by mass. They tell you that in a 100.00-g sample there are 82.244 g of N and 17.755 g of H.

EXAMPLE 2.8   Using Percent Composition

Problem What is the mass percent of each element in propane, C_3H_8? What mass of carbon is contained in 454 g of propane?

What Do You Know? You know the formula of propane. You will need the atomic weights of C and H to calculate the mass percent of each element.

Strategy

*   Calculate the molar mass of propane.

*   From the formula, you know there are 3 moles of C and 8 moles of H per mole of C_3H_8. Determine the mass of C and H represented by these amounts. The percent of each element is the mass of the element divided by the molar mass.

*   The mass of C in 454 g of C_3H_8 is obtained by multiplying this mass by the % C/100, (that is, by the decimal fraction of C in the compound).

Solution

(a)   The molar mass of C_3H_8 is 44.097 g/mol.

(b)   Mass percent of C and H in C_3H_8:

$$\frac{3 \text{ mol C}}{1 \text{ mol } C_3H_8} \times \frac{12.01 \text{ g C}}{1 \text{ mol C}} = 36.03 \text{ g C/1 mol } C_3H_8$$

$$\text{Mass percent of C in } C_3H_8 = \frac{36.03 \text{ g C}}{44.097 \text{ g } C_3H_8} \times 100\% = \boxed{81.71\% \text{ C}}$$

$$\frac{8 \text{ mol H}}{1 \text{ mol } C_3H_8} \times \frac{1.008 \text{ g H}}{1 \text{ mol H}} = 8.064 \text{ g H/1 mol } C_3H_8$$

$$\text{Mass percent of H in } C_3H_8 = \frac{8.064 \text{ g H}}{44.097 \text{ g } C_3H_8} \times 100\% = \boxed{18.29\% \text{ H}}$$

(c)   Mass of C in 454 g of C_3H_8:

$$454 \text{ g } C_3H_8 \times \frac{81.71 \text{ g C}}{100.0 \text{ g } C_3H_8} = 371 \text{ g C}$$

Think about Your Answer   Once you know the percent C in the sample, you could calculate the percent H from it using the formula %H = 100% − %C.

Check Your Understanding

1.   Express the composition of ammonium carbonate, $(NH_4)_2CO_3$, in terms of the mass of each element in 1.00 mol of compound and the mass percent of each element.

2.   What is the mass of carbon in 454 g of octane, C_8H_{18}?

Empirical and Molecular Formulas from Percent Composition

Now let us consider the reverse of the procedure just described: using relative mass or percent composition data to find a molecular formula. Suppose you know the identity of the elements in a sample and have determined the mass of each element in a given mass of compound by chemical analysis [▶ Section 4.4]. You can then calculate the relative amount (moles) of each element, which is also the relative number of atoms of each element in the formula of the compound. For example, for a compound composed of atoms of A and B, the steps from percent composition to a formula are as follows:

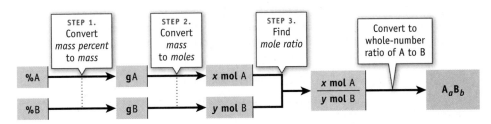

As an example, let us derive the formula for hydrazine, a compound used to remove oxygen from water in heating and cooling systems. It is composed of 87.42% N and 12.58% H and is a close relative of ammonia.

Step 1: *Convert mass percent to mass.* The mass percentages of N and H in hydrazine tell us there are 87.42 g of N and 12.58 g of H in a 100.00-g sample.

Step 2: *Convert the mass of each element to moles.* The amount of each element in the 100.00-g sample is

$$87.42 \text{ g N} \times \frac{1 \text{ mol N}}{14.007 \text{ g N}} = 6.241 \text{ mol N}$$

$$12.58 \text{ g H} \times \frac{1 \text{ mol H}}{1.0079 \text{ g H}} = 12.48 \text{ mol H}$$

Step 3: *Find the mole ratio of elements.* Use the amount (moles) of each element in the 100.00-g of sample to find the amount of one element relative to the other. To do this divide the larger amount by the smaller amount. For hydrazine, this ratio is 2 mol of H to 1 mol of N,

$$\frac{12.48 \text{ mol H}}{6.241 \text{ mol N}} = \frac{2.00 \text{ mol H}}{1.00 \text{ mol N}} \longrightarrow NH_2$$

showing that there are 2 mol of H atoms for every 1 mol of N atoms in hydrazine. Thus, in one molecule, two atoms of H occur for every atom of N; that is, the formula is NH_2. This simplest, whole-number atom ratio of atoms in a formula is called the **empirical formula**.

● **Deriving a Formula**   Percent composition gives the mass of an element in 100 g of a sample. However, in deriving a formula, any amount of sample is appropriate if you know the mass of each element in that sample mass.

Percent composition data allow us to calculate the atom ratios in a compound. A *molecular formula*, however, must convey two pieces of information: (1) the relative numbers of atoms of each element in a molecule (the atom ratios) and (2) the total number of atoms in the molecule. For hydrazine there are twice as many H atoms as N atoms, so the molecular formula could be NH_2. Recognize, however, that percent composition data give only the simplest possible ratio of atoms in a molecule. The empirical formula of hydrazine is NH_2, but the true molecular formula could be NH_2, N_2H_4, N_3H_6, N_4H_8, or any other formula having a 1:2 ratio of N to H.

To determine the molecular formula from the empirical formula, the molar mass must be obtained from experiment. For example, experiments show that the molar mass of hydrazine is 32.0 g/mol, twice the formula mass of NH_2, which is 16.0 g/mol. Thus, the molecular formula of hydrazine is two times the empirical formula of NH_2, that is, N_2H_4.

⬡WL INTERACTIVE EXAMPLE 2.9 Calculating a Formula from Percent Composition

Problem Many soft drinks contain sodium benzoate as a preservative. When you consume the sodium benzoate, it reacts with the amino acid glycine in your body to form hippuric acid, which is then excreted in the urine. Hippuric acid has a molar mass of 179.17 g/mol and is 60.33% C, 5.06% H, and 7.82% N; the remainder is oxygen. What are the empirical and molecular formulas of hippuric acid?

What Do You Know? You know the mass percent of C, H, and N. The mass percent of oxygen is not known but is obtained by difference. You also know the molar mass and will need atomic weights of these four elements for the calculation.

Strategy Assume the mass percent of each element is equivalent to its mass in grams, and convert each mass to moles. The ratio of moles gives the empirical formula. The mass of a mole of compound having the calculated empirical formula is compared with the actual, experimental molar mass to find the true molecular formula.

Solution The mass of oxygen in a 100.0-g sample of hippuric acid is

$$100.00 \text{ g} = 60.33 \text{ g C} + 5.06 \text{ g H} + 7.82 \text{ g N} + \text{mass of O}$$

$$\text{Mass of O} = 26.79 \text{ g O}$$

The amount of each element is

$$60.33 \text{ g C} \times \frac{1 \text{ mol C}}{12.011 \text{ g C}} = 5.023 \text{ mol C}$$

$$5.06 \text{ g H} \times \frac{1 \text{ mol H}}{1.008 \text{ g H}} = 5.02 \text{ mol H}$$

$$7.82 \text{ g N} \times \frac{1 \text{ mol N}}{14.01 \text{ g N}} = 0.558 \text{ mol N}$$

$$26.79 \text{ g O} \times \frac{1 \text{ mol O}}{15.999 \text{ g O}} = 1.674 \text{ mol O}$$

To find the mole ratio, the best approach is to base the ratios on the smallest number of moles present—in this case, nitrogen.

$$\frac{\text{mol C}}{\text{mol N}} = \frac{5.023 \text{ mol C}}{0.558 \text{ mol N}} = \frac{9.00 \text{ mol C}}{1.00 \text{ mol N}} = 9 \text{ mol C/1 mol N}$$

$$\frac{\text{mol H}}{\text{mol N}} = \frac{5.02 \text{ mol C}}{0.558 \text{ mol N}} = \frac{9.00 \text{ mol C}}{1.00 \text{ mol N}} = 9 \text{ mol H/1 mol N}$$

$$\frac{\text{mol O}}{\text{mol N}} = \frac{1.674 \text{ mol O}}{0.558 \text{ mol N}} = \frac{3.00 \text{ mol O}}{1.00 \text{ mol N}} = 3 \text{ mol O/1 mol N}$$

Now we know there are 9 mol of C, 9 mol of H, and 3 mol of O per 1 mol of N. Thus, the empirical formula is $C_9H_9NO_3$.

The experimentally determined molar mass of hippuric acid is 179.17 g/mol. This is the same as the empirical formula weight, so the molecular formula is $C_9H_9NO_3$.

Strategy Map 2.9

PROBLEM

Determine **empirical** and **molecular formulas** based on *known composition* and *known molar mass*.

↓

DATA/INFORMATION KNOWN

- Molar mass
- Percent composition

STEP 1. Assume each **atom %** is equivalent to **mass in grams** in **100 g** sample.

Mass of each element in a **100 g** sample of the compound

STEP 2. Use **atomic weight** of each element to calculate **amount** of each element in 100 g sample (multiply **mass** by **mol/g**).

Amount (mol) of each element in **100 g** sample

STEP 3. Divide the *amount of each element* by the *amount of the element present in the least amount*.

Whole-number ratio of the *amount of each element* to the *amount of element present in the least amount* = **empirical formula**

STEP 4. Divide *known* **molar mass** by **empirical formula mass**.

Molecular formula

Think about Your Answer There is another approach to finding the molecular formula here. Knowing the percent composition of hippuric acid and its molar mass, you could calculate that in 179.17 g of hippuric acid there are 108.06 g of C (8.997 mol of C), 9.06 g of H (8.99 mol of H), 14.01 g N (1.000 mol N), and 48.00 g of O (3 mol of O). This gives us a molecular formula of $C_9H_9NO_3$. However, you must recognize that *this approach can only be used when you know* **both** the percent composition and the molar mass.

Check Your Understanding

1. What is the empirical formula of naphthalene, $C_{10}H_8$?

2. The empirical formula of acetic acid is CH_2O. If its molar mass is 60.05 g/mol, what is the molecular formula of acetic acid?

3. Isoprene is a liquid compound that can be polymerized to form natural rubber. It is composed of 88.17% carbon and 11.83% hydrogen. Its molar mass is 68.11 g/mol. What are its empirical and molecular formulas?

4. Camphor is found in camphor wood, much prized for its wonderful odor. It is composed of 78.90% carbon and 10.59% hydrogen. The remainder is oxygen. What is its empirical formula?

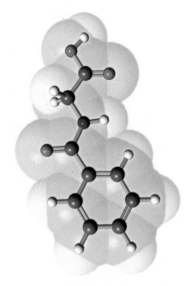

Hippuric Acid, $C_9H_9NO_3$ This substance, which can be isolated as white crystals, is found in the urine of humans and of herbivorous animals.

Determining a Formula from Mass Data

The composition of a compound in terms of mass percent gives us the mass of each element in a 100.0-g sample. In the laboratory we often collect information on the composition of compounds slightly differently. We can

1. Combine known masses of elements to give a sample of the compound of known mass. Element masses can be converted to amounts (moles), and the ratio of amounts gives the combining ratio of atoms—that is, the empirical formula. This approach is described in Example 2.10.

2. Decompose a known mass of an unknown compound into "pieces" of known composition. If the masses of the "pieces" can be determined, the ratio of moles of the "pieces" gives the formula. An example is a decomposition such as

$$Ni(CO)_4(\ell) \rightarrow Ni(s) + 4\ CO(g)$$

The masses of Ni and CO can be converted to moles, whose 1:4 ratio would reveal the formula of the compound. We will describe this approach in Chapter 4.

⭕WL   **INTERACTIVE EXAMPLE 2.10**   Formula of a Compound from Combining Masses

Problem Oxides of virtually every element are known. Bromine, for example, forms several oxides when treated with ozone. Suppose you allow 1.250 g of bromine, Br_2, to react with ozone and obtain 1.876 g of Br_xO_y. What is the formula of the product?

What Do You Know? You know that you begin with a given mass of bromine and that all of the bromine becomes part of bromine oxide of unknown formula. You also know the mass of the product, and because you know the mass of Br in this product, you can determine the mass of O in the product.

Strategy

• The mass of oxygen is determined as the difference between the product mass and the mass of bromine used.

• Calculate the amounts of Br and O from the masses of each element.

• Find the lowest whole number ratio between the moles of Br and moles of O. This defines the empirical formula.

Strategy Map 2.10

PROBLEM
Determine **empirical formula** based on **masses of combining elements**.

↓

DATA/INFORMATION KNOWN
• Mass of one element in binary compound
• Mass of product

> **STEP 1.** Find *mass of second element* by **difference**.

Mass of each element in a sample of the compound

> **STEP 2.** Use **atomic weight** of each element to calculate **amount** of each element in sample (multiply mass by mol/g).

Amount (mol) of each element in sample

> **STEP 3.** Divide the *amount of each element* by the *amount of the element present* **in the least amount**.

Whole-number ratio of the *amount of each element* to the *amount of element present in the least amount* = **empirical formula**

Finding Empirical and Molecular Formulas

- The experimental data available to find a formula may be in the form of percent composition or the masses of elements combined in some mass of compound. No matter what the starting point, the first step is always to convert masses of elements to moles.
- Be sure to use *at least* three significant figures when calculating empirical formulas. Using fewer significant figures can give a misleading result.

- When finding atom ratios, always divide the larger number of moles by the smaller one.
- Empirical and molecular formulas can differ for molecular compounds. In contrast, there is no "molecular" formula for an ionic compound; all that can be recorded is the empirical formula.

- Determining the molecular formula of a compound after calculating the empirical formula requires knowing the molar mass.
- When *both* the percent composition and the molar mass are known for a compound, the alternative method mentioned in *Think about Your Answer* in Example 2.9 could be used.

Solution You already know the mass of bromine in the compound, so you can calculate the mass of oxygen in the compound by subtracting the mass of bromine from the mass of the product.

$$1.876 \text{ g product} - 1.250 \text{ g Br}_2 = 0.626 \text{ g O}$$

Next, calculate the amount of each reactant. Notice that, although Br_2 was the reactant, we need to know the amount of Br in the product.

$$1.250 \text{ g Br}_2 \times \frac{1 \text{ mol Br}_2}{159.81 \text{ g}} = 0.007822 \text{ mol Br}_2$$

$$0.007822 \text{ mol Br}_2 \times \frac{2 \text{ mol Br}}{1 \text{ mol Br}_2} = 0.01564 \text{ mol Br}$$

$$0.626 \text{ g O} \times \frac{1 \text{ mol O}}{16.00 \text{ g O}} = 0.0391 \text{ mol O}$$

Find the ratio of moles of O to moles of Br:

$$\text{Mole ratio} = \frac{0.0391 \text{ mol O}}{0.01564 \text{ mol Br}} = \frac{2.50 \text{ mol O}}{1.00 \text{ mol O}}$$

The atom ratio is 2.5 mol O/1.0 mol Br. However, atoms combine in the ratio of small whole numbers, so we double this to give a ratio of 5 mol O to 2 mol Br. Thus, the product is dibromine pentaoxide, Br_2O_5.

Think about Your Answer The whole number ratio of 5 : 2 was found by realizing that 2.5 = 2 1/2 = 5/2. This problem rests on the principle of the conservation of matter. All of the bromine used in the reaction is part of the product.

Check Your Understanding

Gallium oxide, Ga_xO_y, forms when gallium is combined with oxygen. Suppose you allow 1.25 g of gallium (Ga) to react with oxygen and obtain 1.68 g of Ga_xO_y. What is the formula of the product?

EXAMPLE 2.11   Determining a Formula from Mass Data

Problem Tin metal (Sn) and purple iodine (I_2) combine to form orange, solid tin iodide with an unknown formula.

$$\text{Sn metal} + \text{solid I}_2 \longrightarrow \text{solid Sn}_x\text{I}_y$$

Weighed quantities of Sn and I_2 are combined, where the quantity of Sn is more than is needed to react with all of the iodine. After Sn_xI_y has been formed, it is isolated by filtration. The mass of excess tin is also determined. The following data were collected:

Mass of tin (Sn) in the original mixture	1.056 g
Mass of iodine (I_2) in the original mixture	1.947 g
Mass of tin (Sn) recovered after reaction	0.601 g

What is the empirical formula of the tin iodide obtained?

(a) Weighed samples of tin (left) and iodine (right).

(b) The tin and iodine are heated in a solvent.

(c) The hot reaction mixture is filtered to recover unreacted tin.

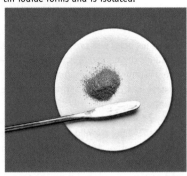

(d) When the solvent cools, solid, orange tin iodide forms and is isolated.

Photos © Cengage Learning/Charles D. Winters

Determining a formula. The formula of a compound composed of tin and iodine can be determined by finding the mass of iodine that combines with a given mass of tin.

What Do You Know? You know the mass of iodine used in the reaction. The mass of tin used in this reaction is needed, and it can be determined from the mass information given. You need the atomic weights of tin and iodine.

Strategy

- The mass of tin consumed in the reaction is determined by subtracting the mass of tin recovered from the initial mass of tin.

- Calculate the amounts of Sn and I from the masses of each element.

- Find the lowest whole number ratio between the moles of Sn and moles of I. This defines the empirical formula.

Solution First, let us find the mass of tin that combined with iodine:

Mass of Sn in original mixture	1.056 g
Mass of Sn recovered	− 0.601 g
Mass of Sn combined with 1.947 g I_2	0.455 g

Now convert the mass of tin to the amount of tin.

$$0.455 \text{ g Sn} \times \frac{1 \text{ mol Sn}}{118.7 \text{ g Sn}} = 0.00383 \text{ mol Sn}$$

No I_2 was recovered; it all reacted with Sn. Therefore, 0.00383 mol of Sn combined with 1.947 g of I_2. Because we want to know the amount of I that combined with 0.00383 mol of Sn, we calculate the amount of I from the mass of I_2.

$$1.947 \text{ g } I_2 \times \frac{1 \text{ mol } I_2}{253.81 \text{ g } I_2} \times \frac{2 \text{ mol I}}{1 \text{ mol } I_2} = 0.01534 \text{ mol I}$$

Finally, we find the ratio of moles.

$$\frac{\text{mol I}}{\text{mol Sn}} = \frac{0.01534 \text{ mol I}}{0.00383 \text{ mol Sn}} = \frac{4.01 \text{ mol I}}{1.00 \text{ mol Sn}} = \frac{4 \text{ mol I}}{1 \text{ mol Sn}}$$

There are four times as many moles of I as moles of Sn in the sample. Therefore, there are four times as many atoms of I as atoms of Sn per formula unit. The empirical formula is SnI_4.

Think about Your Answer This is very similar to Example 2.10.

Check Your Understanding

Analysis shows that 0.586 g of potassium metal combines with 0.480 g of O_2 gas to give a white solid having a formula of K_xO_y. What is the empirical formula of the compound?

Mummies, Bangladesh, and the Formula of Compound 606

The Iceman of the Alps lived over 5000 years ago (page 50), but his body was preserved in the ice for thousands of years. Other civilizations deliberately mummified their dead, and one curious case has been studied by archeologists for some years.

Over 6800 years ago, in what is now the nation of Chile, a child died at an age of about 6 months. After removing the head and internal organs, someone modeled a head of clay and placed it on the torso. The cause of death for this child, and of others whose mummified remains have been found in this region, was apparently arsenic poisoning.

Arsenic occurs naturally in geological formations in many parts of the world, including the mountains of Chile. Rain and snowmelt flow from the mountains, carrying dissolved minerals, among them arsenic-containing salts. In one region of Chile, a

region in which the mummies were found, the river water has 860 mg of arsenic per liter, 86 times the World Health Organization's guideline.

Another region of the world greatly affected by arsenic salts in the ground water is Bangladesh. Here the arsenic is carried from the mountains of Tibet into the flood plains of Bangladesh by the great Brahmaputra River. Enormous efforts have been made to alleviate the problem with some success.

The practice in medicine for some centuries has been to find compounds that are toxic to certain organisms but not so toxic that the patient is harmed. In the early part of the 20th century, Paul Ehrlich set out to find just such a compound that would cure syphilis, a venereal disease that was rampant at the time. He screened hundreds of compounds, and found that his 606th compound was effective: an arsenic-containing drug now called salvarsan. It was used for some years for syphilis treatment until penicillin was discovered in the 1930s.

Salvarsan was a forerunner of the modern drug industry. Interestingly, what chemists long thought to be a single compound was in fact a mixture of compounds. Question 2 below will lead you to the molecular formula for both of them.

For more about arsenic and its effect in the environment, see H. Pringle, *Science*, 29 May 2009, Volume 324, page 1130 and Y. Bhattacharjee, *Science*, 23 March 2007, Volume 315, page 1659.

A sample of orpiment, a common arsenic-containing mineral (As_2S_3). The name of the element is thought to come from the Greek word for this mineral, which was long favored by 17th century Dutch painters as a pigment.

Questions:

1. Enargite is an arsenic-containing ore. It has 19.024% As, 48.407% Cu, and 32.569% S. What is the empirical formula of the ore?

2. Salvarsan was long thought to be a single substance. Recently, however, a mass spectrometry study of the compound shows it to be a mixture of two molecules with the same empirical formula. Each has the composition 39.37% C, 3.304% H, 8.741% O, 7.652% N, and 40.932% As. One has a molar of mass of 549 g/mol and the other has a molar mass of 915 g/mol. What are the molecular formulas of the compounds?

Answers to these questions are available in Appendix N.

The hands of a woman in Bangladesh. The spots are the symptom of arsenocosis caused by ingesting arsenic-containing water.

Determining a Formula by Mass Spectrometry

We have described chemical methods of determining a molecular formula, but there are many instrumental methods as well. One of them is *mass spectrometry* (Figure 2.27). We introduced this technique earlier when discussing the existence of isotopes and their relative abundance (Figure 2.3). If a compound can be turned into a vapor, the vapor can be passed through an electron beam in a mass spectrometer where high energy electrons collide with the gas phase molecules. These high-energy collisions cause the molecule to lose electrons and turn the molecules into positive ions. These ions usually break apart or fragment into smaller pieces. As illustrated in Figure 2.27 the cation created from ethanol ($CH_3CH_2OH^+$) fragments (losing an H atom) to give another cation ($CH_3CH_2O^+$), which further fragments. A mass spectrometer detects and records the masses of the different particles. Analysis of the spectrum can help identify a compound and can give an accurate molar mass.

Mass Spectrometry, Molar Mass, and Isotopes

Bromobenzene, C_6H_5Br, has a molar mass of 157.010 g/mol. Why, then, are there two prominent lines at mass-to-charge ratios (m/Z) of 156 and 158 in the mass spectrum of the compound (when $Z = +1$)? The answer shows us the influence of isotopes on molar mass.

Bromine has two, naturally occurring isotopes, ^{79}Br and ^{81}Br. They are 50.7% and 49.3% abundant, respectively. What is the mass of C_6H_5Br based on each isotope? If we use the most abundant isotopes of C and H (^{12}C and ^{1}H), the mass of the molecule having the ^{79}Br isotope, $C_6H_5{}^{79}Br$, is 156. The mass of the molecule containing the ^{81}Br isotope, $C_6H_5{}^{81}Br$, is 158.

The calculated molar mass of bromobenzene is 157.010, a value derived from the atomic masses of the elements. The molar mass reflects the abundances of all of the isotopes. In contrast, the mass spectrum has a line for each possible combination of isotopes. This also explains why there are also small lines at the mass-to-charge ratios of 157 and 159. They arise from various combinations of ^{1}H, ^{12}C, ^{13}C, ^{79}Br, and ^{81}Br atoms. In fact careful analysis of such patterns can identify a molecule unambiguously.

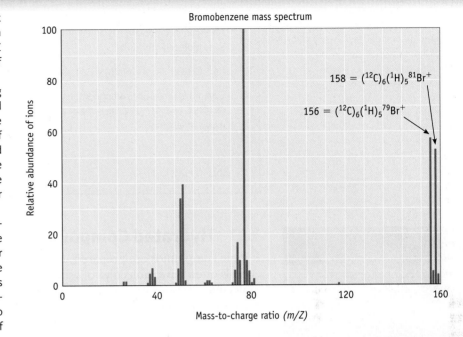

Bromobenzene mass spectrum

$158 = (^{12}C)_6(^{1}H)_5{}^{81}Br^+$

$156 = (^{12}C)_6(^{1}H)_5{}^{79}Br^+$

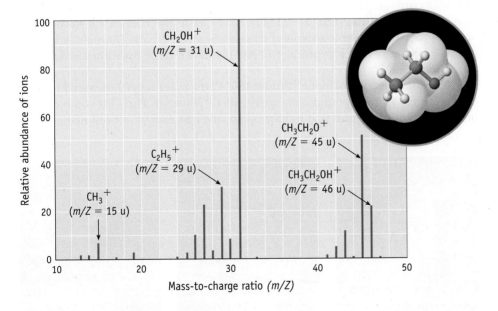

FIGURE 2.27 Mass spectrum of ethanol, CH_3CH_2OH. A prominent peak or line in the spectrum is the "parent" ion ($CH_3CH_2OH^+$) at mass 46. (The parent ion is the heaviest ion observed.) The mass designated by the peak for the parent ion confirms the formula of the molecule. Other peaks are for "fragment" ions. This pattern of lines can provide further, unambiguous evidence of the formula of the compound. (The horizontal axis is the mass-to-charge ratio of a given ion. Because almost all observed ions have a charge of $Z = +1$, the value observed is the mass of the ion.) (See "A Closer Look: Mass Spectrometry, Molar Mass, and Isotopes.")

REVIEW & CHECK FOR SECTION 2.10

1. Which of the following hydrocarbons has the highest percentage of carbon?

    (a) methane, CH_4

    (b) ethane, C_2H_6

    (c) propane, C_3H_8

    (d) butane, C_4H_{10}

2. Which of the following organic compounds has the highest percentage of carbon?

   (a) benzene, C_6H_6 (c) ethanol, C_2H_5OH

   (b) ethane, C_2H_6 (d) acetic acid, CH_3CO_2H

3. An organic compound has an empirical formula CH_2O and a molar mass of 180 g/mol. What is its molecular formula?

   (a) CH_2O (b) $C_6H_{12}O_6$ (c) $C_2H_4O_2$ (d) $C_5H_{10}O_5$

4. Eugenol is the major component in oil of cloves. It has a molar mass of 164.2 g/mol and is 73.14% C and 7.37% H; the remainder is oxygen. What are the empirical and molecular formulas of eugenol?

   (a) C_5H_6O, C_5H_6O (c) C_6H_5O, $C_{12}H_{10}O_2$

   (b) C_5H_6O, $C_{10}H_{12}O_2$ (d) C_3H_6O, $C_6H_{12}O_2$

FIGURE 2.28 Gypsum wall-board. Gypsum is hydrated calcium sulfate, $CaSO_4 \cdot 2H_2O$.

2.11 Hydrated Compounds

If ionic compounds are prepared in water solution and then isolated as solids, the crystals often have molecules of water trapped in the lattice. Compounds in which molecules of water are associated with the ions of the compound are called **hydrated compounds**. The beautiful blue copper(II) compound in Figure 2.26, for example, has a formula that is conventionally written as $CuCl_2 \cdot 2\ H_2O$. The dot between $CuCl_2$ and $2\ H_2O$ indicates that 2 mol of water are associated with every mole of $CuCl_2$; it is equivalent to writing the formula as $CuCl_2(H_2O)_2$. The name of the compound, copper(II) chloride dihydrate, reflects the presence of 2 mol of water per mole of $CuCl_2$. The molar mass of $CuCl_2 \cdot 2\ H_2O$ is 134.5 g/mol (for $CuCl_2$) plus 36.0 g/mol (for $2\ H_2O$) for a total molar mass of 170.5 g/mol.

Hydrated compounds are common. The walls of your home may be covered with wallboard, or "plaster board" (Figure 2.28) These sheets contain hydrated calcium sulfate, or gypsum ($CaSO_4 \cdot 2\ H_2O$), as well as unhydrated $CaSO_4$, sandwiched between paper. Gypsum is a mineral that can be mined. Now, however, it is more commonly obtained as a byproduct in the manufacture of hydrofluoric acid and phosphoric acid.

If gypsum is heated between 120 and 180 °C, the water is partly driven off to give $CaSO_4 \cdot \frac{1}{2}\ H_2O$, a compound commonly called "plaster of Paris." If you have ever broken an arm or leg and had to have a cast, the cast may have been made of this

FIGURE 2.29 Dehydrating hydrated cobalt(II) chloride, $CoCl_2 \cdot 6H_2O$. $CoCl_2 \cdot 6H_2O$ also makes a good "invisible ink." A solution of cobalt(II) chloride in water is red, but if you write on paper with the solution it cannot be seen. When the paper is warmed, however, the cobalt compound dehydrates to give the deep blue anhydrous compound, and the writing becomes visible.

Cobalt(II) chloride hexahydrate [$CoCl_2 \cdot 6H_2O$] is a deep red compound.

When it is heated, the compound loses the water of hydration and forms the deep blue compound $CoCl_2$.

compound. It is an effective casting material because, when added to water, it forms a thick slurry that can be poured into a mold or spread out over a part of the body. As it takes on more water, the material increases in volume and forms a hard, inflexible solid. These properties also make plaster of Paris a useful material for artists, because the expanding compound fills a mold completely and makes a high-quality reproduction.

Hydrated cobalt(II) chloride is the red solid in Figure 2.29. When heated it turns purple and then deep blue as it loses water to form anhydrous $CoCl_2$; "anhydrous" means a substance without water. On exposure to moist air, anhydrous $CoCl_2$ takes up water and is converted back into the red hydrated compound. It is this property that allows crystals of the blue compound to be used as a humidity indicator. You may have seen them in a small bag packed with a piece of electronic equipment.

There is no simple way to predict how much water will be present in a hydrated compound, so it must be determined experimentally. Such an experiment may involve heating the hydrated material so that all the water is released from the solid and evaporated. Only the anhydrous compound is left. The formula of hydrated copper(II) sulfate, commonly known as "blue vitriol," is determined in this manner in Example 2.12.

FIGURE 2.30 Heating a hydrated compound. The formula of a hydrated compound can be determined by heating a weighed sample enough to cause the compound to release its water of hydration. Knowing the mass of the hydrated compound before heating and the mass of the anhydrous compound after heating, we can determine the mass of water in the original sample.

◉WL INTERACTIVE EXAMPLE 2.12 Determining the Formula of a Hydrated Compound

Problem You want to know the value of x in blue, hydrated copper(II) sulfate, $CuSO_4 \cdot x\ H_2O$, that is, the number of water molecules for each unit of $CuSO_4$. In the laboratory you weigh out 1.023 g of the solid. After heating the solid thoroughly in a porcelain crucible (Figure 2.30), 0.654 g of nearly white, anhydrous copper(II) sulfate, $CuSO_4$, remains.

$$1.023 \text{ g } CuSO_4 \cdot x\ H_2O + \text{heat} \rightarrow 0.654 \text{ g } CuSO_4 + ? \text{ g } H_2O$$

What Do You Know? You know the mass of the copper sulfate sample including water (before heating) and with no water (after heating). Therefore, you know the mass of water in the sample and the mass of $CuSO_4$.

Strategy To find x you need to know the amount of H_2O per mole $CuSO_4$.

- Determine the mass of water released on heating the hydrated compound.
- Calculate the amount (moles) of $CuSO_4$ and H_2O from their masses and molar masses.
- Determine the smallest whole number ratio (amount H_2O/amount $CuSO_4$).

Solution Find the mass of water.

Mass of hydrated compound	1.023 g
− Mass of anhydrous compound, $CuSO_4$	−0.654
Mass of water	0.369 g

Next convert the masses of $CuSO_4$ and H_2O to moles.

$$0.369 \text{ g } H_2O \times \frac{1 \text{ mol } H_2O}{18.02 \text{ g } H_2O} = 0.0205 \text{ mol } H_2O$$

$$0.654 \text{ g } CuSO_4 \times \frac{1 \text{ mol } CuSO_4}{159.6 \text{ g } CuSO_4} = 0.00410 \text{ mol } CuSO_4$$

The value of x is determined from the mole ratio.

$$\frac{0.0205 \text{ mol } H_2O}{0.00410 \text{ mol } CuSO_4} = \frac{5.00 \text{ mol } H_2O}{1.00 \text{ mol } CuSO_4}$$

The water-to-$CuSO_4$ ratio is 5:1, so the formula of the hydrated compound is $CuSO_4 \cdot 5\ H_2O$. Its name is copper(II) sulfate pentahydrate.

Strategy Map 2.12

PROBLEM

Determine **formula of hydrated salt** based on *masses of water and dehydrated salt*.

↓

DATA/INFORMATION KNOWN

- Mass of sample before and after heating to dehydrate

STEP 1. Find *masses of salt and water* by **difference**.

Mass of salt and water in a sample of the hydrated compound

STEP 2. Use **molar mass** of salt and water to calculate **amount** of each in sample (multiply **mass** by **mol/g**).

Amount (mol) of salt and water in sample

STEP 3. Divide the *amount of water* by the *amount of the dehydrated salt*.

Formula = *ratio of the amount of water to the amount of salt* in dehydrated sample

Think about Your Answer The ratio of the amount of water to the amount of $CuSO_4$ is a whole number. This is almost always the case with hydrated compounds.

Check Your Understanding

Hydrated nickel(II) chloride is a beautiful green, crystalline compound. When heated strongly, the compound is dehydrated. If 0.235 g of $NiCl_2 \cdot x\ H_2O$ gives 0.128 g of $NiCl_2$ on heating, what is the value of x?

REVIEW & CHECK FOR SECTION 2.11

Epsom salt is $MgSO_4 \cdot 7H_2O$. When heated to 70 to 80 °C it loses some, but not all, of its water of hydration. Suppose you heat 2.465 g of Epsom salt to 75 °C and find that the mass is now only 1.744 g. What is the formula of the slightly dehydrated salt?

(a) $MgSO_4 \cdot 6H_2O$ (c) $MgSO_4 \cdot 4H_2O$

(b) $MgSO_4 \cdot 5H_2O$ (d) $MgSO_4 \cdot 3H_2O$

---

 and

Sign in at **www.cengage.com/owl** to:
- View tutorials and simulations, develop problem-solving skills, and complete online homework assigned by your professor.
- For quick review and exam prep, download Go Chemistry mini lecture modules from OWL (or purchase them at **www.cengagebrain.com**)

Access **How Do I Solve It?** tutorials on how to approach problem solving using concepts in this chapter.

CHAPTER GOALS REVISITED

Now that you have studied this chapter, you should ask whether you have met the chapter goals. In particular, you should be able to

Describe atomic structure and define atomic number and mass number
a. Describe electrons, protons, and neutrons, and the general structure of the atom (Section 2.1). **Study Questions: 1, 7.**
b. Understand the relative mass scale and the atomic mass unit (Section 2.2).

Understand the nature of isotopes and calculate atomic masses from isotopic masses and abundances
a. Define isotope and give the mass number and number of neutrons for a specific isotope (Sections 2.2 and 2.3). **Study Questions: 5, 6, 8, 14, 97.**
b. Do calculations that relate the atomic weight (atomic mass) of an element and isotopic abundances and masses (Section 2.4). **Study Questions: 16–18, 98, 158.**

Know the terminology of the periodic table
a. Identify the periodic table locations of groups, periods, metals, metalloids, nonmetals, alkali metals, alkaline earth metals, halogens, noble gases, and the transition elements (Section 2.5). **Study Questions: 24, 25, 29, 103, and Go Chemistry Module 1.**
b. State similarities and differences in properties of some of the common elements of a group (Section 2.5). **Study Questions: 28, 30.**

Interpret, predict, and write formulas for ionic and molecular compounds
a. Recognize and interpret molecular formulas, condensed formulas, and structural formulas (Section 2.6). **Study Questions: 31, 33.**
b. Recognize that metal atoms commonly lose one or more electrons to form positive ions, called cations, and nonmetal atoms often gain electrons to form negative ions, called anions (Figure 2.17).
c. Predict the charge on a metal cation: for Groups 1A, 2A, and 3A ions, the charge is equal to the group number in which the element is found in the periodic table (M^{n+}, n = Group number) (Section 2.7); for transition metal cations, the charges are often 2+ or 3+, but other charges are observed. **Study Questions: 35, 37, 43, and Go Chemistry Module 2.**

d. Recognize that the negative charge on a single-atom or monatomic anion, X^{n-}, is given by n = Group number − 8 (Section 2.7). **Study Questions: 37, 40.**

e. Write formulas for ionic compounds by combining ions in the proper ratio to give no overall charge (Section 2.7). **Study Questions: 41, 43, 45, 47.**

Name ionic and molecular compounds

a. Give the names or formulas of polyatomic ions, knowing their formulas or names, respectively (Table 2.4 and Section 2.7). **Study Questions: 37, 38.**

b. Name ionic compounds and simple binary compounds of the nonmetals (Sections 2.7 and 2.8). **Study Questions: 49, 57–60, 128, and Go Chemistry Module 3.**

Understand some properties of ionic compounds

a. Understand the importance of Coulomb's law (Equation 2.3), which describes the electrostatic forces of attraction and repulsion of ions. Coulomb's law states that the force of attraction between oppositely charged species increases with electric charge and with decreasing distance between the species (Section 2.7). **Study Question: 56.**

Explain the concept of the mole and use molar mass in calculations

a. Understand that the molar mass of an element is the mass in grams of Avogadro's number of atoms of that element (Section 2.9). **Study Questions: 61, 62, 65, 105, and Go Chemistry Module 4.**

b. Know how to use the molar mass of an element and Avogadro's number in calculations (Section 2.9). **Study Questions: 65, 68, 106, 107.**

c. Understand that the molar mass of a compound (often called the molecular weight) is the mass in grams of Avogadro's number of molecules (or formula units) of a compound (Section 2.9). For ionic compounds, which do not consist of individual molecules, the sum of atomic masses is often called the formula mass (or formula weight).

d. Calculate the molar mass of a compound from its formula and a table of atomic weights (Section 2.9). **Study Questions: 69, 71, 73.**

e. Calculate the number of moles of a compound that is represented by a given mass, and vice versa (Section 2.9). **Study Questions: 73–75, 117.**

Derive compoud formulas from experimental data

a. Express the composition of a compound in terms of percent composition (Section 2.10). **Study Questions: 79, 81, 113.**

b. Use percent composition or other experimental data to determine the empirical formula of a compound (Section 2.10). **Study Questions: 83, 85, 89, 91, 129.**

c. Understand how mass spectrometry can be used to find a molar mass (Section 2.10). **Study Question: 159.**

d. Use experimental data to find the number of water molecules in a hydrated compound (Section 2.11) **Study Questions: 153, 157.**

Key Equations

Equation 2.1 (page 46) Percent abundance of an isotope.

$$\text{Percent abundance} = \frac{\text{number of atoms of a given isotope}}{\text{total number of atoms of all isotopes of that element}} \times 100\%$$

Equation 2.2 (page 47) Calculate the atomic weight from isotope abundances and the exact atomic mass of each isotope of an element.

$$\text{Atomic weight} = \left(\frac{\%\ \text{abundance isotope 1}}{100}\right)(\text{mass of isotope 1})$$
$$+ \left(\frac{\%\ \text{abundance isotope 2}}{100}\right)(\text{mass of isotope 2}) + \ldots$$

Equation 2.3 (page 68) **Coulomb's Law,** the force of attraction between oppositely charged ions.

$$\underset{\substack{\uparrow \\ \text{proportionality constant}}}{\text{Force} = -k} \frac{\overbrace{(n^+e)}^{\text{charge on + and − ions}} \overbrace{(n^-e)}^{\text{charge on electron}}}{\underset{\substack{\downarrow \\ \text{distance between ions}}}{d^2}}$$

OWL Questions and problems for this chapter are available in OWL. Use the OWL access card that came with this textbook to access assigned questions and problems for this chapter.

3 Chemical Reactions

© Ralph White/Corbis

John C. Kotz

Metal sulfides from a black smoker and a volcano. *(above)* A "black smoker." This one was photographed deep in the Pacific Ocean along the East Pacific Rise. *(right)* The water in this waterfall (on the island of St. Lucia in the Caribbean) flows from a volcano. The sides of the rock wall are streaked with iron(III) hydroxide. The water in the pool and the rock wall are also blackened with insoluble sulfides of copper(II), iron(III), and manganese(II).

Black Smokers and Volcanoes

In 1977, scientists were exploring the junction of two of the tectonic plates that form the floor of the Pacific Ocean. There they found thermal springs gushing a hot, black soup of minerals. Seawater seeps into cracks in the ocean floor and, as it sinks deeper into the earth's crust, the water is superheated to between 300 °C and 400 °C by the magma of the earth's core. This superhot water dissolves minerals in the crust and is pushed back to the surface. When this hot water, now laden with dissolved metal cations and rich in anions such as sulfide and sulfate, gushes through the surface, it cools, and metal sulfates such as calcium sulfate and sulfides—such as those of copper, manganese, iron, zinc, and nickel—precipitate. Many metal sulfides are black, and the plume of material coming from the sea bottom looks like black "smoke;" thus, the vents have been called "black smokers." The solid sulfides and other minerals settle around the edges of the vent on the sea floor and eventually form a "chimney" of precipitated minerals.

You can see the same deposits of metal sulfides around steam vents from volcanoes and in the water flowing away from a volcano or steam vent.

Scientists were amazed to discover that the vents were surrounded by peculiar animals living in the hot, sulfide-rich environment. Because black smokers are under hundreds of feet of water and sunlight does not penetrate to these depths, the animals have developed a way to live without the energy from sunlight. In a terrestrial environment, plants use the energy of the sun to create organic molecules by the process of photosynthesis. In the lightless ecosystem deep in the ocean, energy is derived from the oxidation of sulfides. With this source of energy, microbes are able to make the organic molecules that are the basis of life.

Question:

Write balanced, net ionic equations for the reactions of Fe^{2+} and Bi^{3+} with H_2S and for Ca^{2+} with sulfate ions.

The answer to this question is available in Appendix N.

CHAPTER OUTLINE

3.1 Introduction to Chemical Equations

3.2 Balancing Chemical Equations

3.3 Introduction to Chemical Equilibrium

3.4 Aqueous Solutions go Chemistry

3.5 Precipitation Reactions go Chemistry

3.6 Acids and Bases

3.7 Gas-Forming Reactions

3.8 Oxidation–Reduction Reactions

3.9 Classifying Reactions in Aqueous Solution

CHAPTER GOALS

See Chapter Goals Revisited (page 125) for Study Questions keyed to these goals.

- Balance equations for simple chemical reactions.

- Understand the nature and characteristics of chemical equilibria.

- Understand the nature of ionic substances dissolved in water.

- Recognize common acids and bases and understand their behavior in aqueous solution.

- Recognize the common types of reactions in aqueous solution.

- Write chemical equations for the common types of reactions in aqueous solution.

- Recognize common oxidizing and reducing agents and identify oxidation–reduction reactions.

Chemical reactions are at the heart of chemistry. We begin a chemical reaction with one set of materials and end up with different materials. Just reading this sentence involves an untold number of chemical reactions in your body. Indeed, every activity of living things depends on carefully regulated chemical reactions. Our objective in this chapter is to introduce you to the symbolism used to represent chemical reactions and to describe several types of common chemical reactions.

3.1 Introduction to Chemical Equations

When a stream of chlorine gas, Cl_2, is directed onto solid phosphorus, P_4, the mixture bursts into flame, and a chemical reaction produces liquid phosphorus trichloride, PCl_3 (Figure 3.1). We can depict this reaction using a **balanced chemical equation**.

$$\underbrace{P_4(s) + 6\ Cl_2(g)}_{\text{reactants}} \longrightarrow \underbrace{4\ PCl_3(\ell)}_{\text{product}}$$

In a chemical equation, the formulas for the **reactants** (the substances combined in the reaction) are written to the left of the arrow and the formulas of the **products** (the substances produced) are written to the right of the arrow. The physical states of reactants and products can also be indicated. The symbol (s) indicates a solid, (g) a gas, and (ℓ) a liquid. A substance dissolved in water, that is, in an *aqueous* solution, is indicated by (aq).

In the 18th century, the French scientist Antoine Lavoisier (1743–1794) introduced the **law of conservation of matter**, which states that *matter can neither be created nor destroyed.* This means that if the total mass of reactants is 10 g, and if the reaction completely converts reactants to products, you must end up with 10 g of products. This also means that if 1000 atoms of a particular element are contained in the reactants, then those 1000 atoms must appear in the products in some fashion. *Atoms are conserved in chemical reactions.*

OWL

Sign in to OWL at **www.cengage.com/owl** to view tutorials and simulations, develop problem-solving skills, and complete online homework assigned by your professor.

go Chemistry

Download mini lecture videos for key concept review and exam prep from OWL or purchase them from **www.cengagebrain.com**

● **Information from Chemical Equations** The same number of atoms must exist after a reaction as before it takes place. However, these atoms are arranged differently. In the phosphorus/chlorine reaction, for example, the P atoms were in the form of P_4 molecules before reaction but appear in the PCl_3 molecules after reaction.

$$P_4(s) + 6\ Cl_2(g) \longrightarrow 4\ PCl_3(\ell)$$

REACTANTS   PRODUCT

FIGURE 3.1 Reaction of solid white phosphorus with chlorine gas. The product is liquid phosphorus trichloride.

A CLOSER LOOK

Antoine Laurent Lavoisier, 1743–1794

On Monday, August 7, 1774, the Englishman Joseph Priestley (1733–1804) isolated oxygen. (The Swedish chemist Carl Scheele [1742–1786] also discovered the element, perhaps in 1773, but did not publish his results until later.) Priestley accomplished this by heating mercury(II) oxide, HgO, which caused the oxide to decompose to mercury and oxygen.

$$2\ HgO(s) \rightarrow 2\ Hg(\ell) + O_2(g)$$

Priestley did not immediately understand the significance of the discovery, but he mentioned it to the French chemist Antoine Lavoisier in October 1774. One of Lavoisier's contributions to science was his recognition of the importance of exact scientific measurements and of carefully planned experiments, and he applied these methods to the study of oxygen. From this work, Lavoisier proposed that oxygen was an element, that it was one of the constituents of the compound water, and that burning involved a reaction with oxygen. He also mistakenly came to believe Priestley's gas was present in all acids, so he named it "oxygen" from the Greek words meaning "to form an acid."

In other experiments, Lavoisier observed that the heat produced by a guinea pig when exhaling a given amount of carbon dioxide is similar to the quantity of heat produced by burning carbon to give the same amount of carbon dioxide. From these and other experiments he concluded that, "Respiration is a combustion, slow it is true, but otherwise perfectly similar to that of charcoal." Although he did not understand the details of the process, this was an important step in the development of biochemistry.

Lavoisier was a prodigious scientist, and the principles of naming chemical substances that he introduced are still in use today. Furthermore, he wrote a textbook in which he applied the principles of the conservation of matter to chemistry, and he used the idea to write early versions of chemical equations.

Because Lavoisier was an aristocrat, he came under suspicion during the Reign of Terror of the French Revolution. He was an investor in the Ferme Générale, the infamous tax-collecting organization in 18th-century France. Tobacco was a monopoly product of the Ferme Générale, and it was common to cheat the purchaser by adding water to the tobacco, a practice that Lavoisier opposed. Nonetheless, because of his involvement with the Ferme, his career was cut short by the guillotine on May 8, 1794, on the charge of "adding water to the people's tobacco."

Lavoisier and his wife, as painted in 1788 by Jacques-Louis David. Lavoisier was then 45, and his wife, Marie Anne Pierrette Paulze, was 30. (The Metropolitan Museum of Art, Purchase, Mr. and Mrs. Charles Wrightsman gift, in honor of Everett Fahy, 1997. Photograph © 1989 The Metropolitan Museum of Art.)

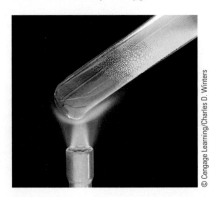

The decomposition of red mercury(II) oxide. The decomposition reaction gives mercury metal and oxygen gas. The mercury is seen as a film on the surface of the test tube.

When applied to the reaction of phosphorus and chlorine, the law of conservation of matter tells us that 1 molecule of phosphorus, P_4 (with 4 phosphorus atoms), and 6 diatomic molecules of Cl_2 (with 12 atoms of Cl) will produce four molecules of PCl_3. Because each PCl_3 molecule contains 1 P atom and 3 Cl atoms, the four PCl_3 molecules are needed to account for 4 P atoms and 12 Cl atoms in the product. The equation is balanced; the same number of P and Cl atoms appear on each side of the equation.

$$\underbrace{\overset{6 \times 2\ =}{\underset{12\ \text{Cl atoms}}{}} \quad \overset{4 \times 3\ =}{\underset{12\ \text{Cl atoms}}{}}}$$

$$\underbrace{P_4(s) + 6\ Cl_2(g)}_{4\ \text{P atoms}} \longrightarrow \underbrace{4\ PCl_3(\ell)}_{4\ \text{P atoms}}$$

Next, consider the balanced equation for the reaction of iron and and chlorine (Figure 3.2). In this case, there are two iron atoms and six chlorine atoms on each side of the equation.

$$\underset{\text{stoichiometric coefficients}}{2\ Fe(s) + 3\ Cl_2(g) \longrightarrow 2\ FeCl_3(s)}$$

The numbers in front of the formulas in balanced chemical equations are required by the law of conservation of matter. They can be read as a number of atoms or molecules (2 atoms of Fe and 3 molecules of Cl_2), or as formula units (2 formula units of the ionic compound $FeCl_3$). They can refer equally well to amounts of reactants and products: 2 mol of solid iron combine with 3 mol of chlorine gas to produce 2 mol of solid $FeCl_3$. The relationship between the quantities of chemical reactants and products is called **stoichiometry** (pronounced "stoy-key-AHM-uh-tree") (▶ Chapter 4), and the coefficients in a balanced equation are the **stoichiometric coefficients**.

REVIEW & CHECK FOR SECTION 3.1

The reaction of aluminum with bromine is shown on page 59. The equation for the reaction is

$$2\ Al(s) + 3\ Br_2(\ell) \rightarrow Al_2Br_6(s)$$

1. What are the stoichiometric coefficients in this equation?

    (a) 1, 3, 4   (b) 2, 3, 1   (c) 1, 6, 8   (d) 1, 1, 1

2. If you were to use 8000 atoms of Al, how many molecules of Br_2 are required to consume the Al completely?

    (a) 5333   (b) 8000   (c) 12,000   (d) 20,000

FIGURE 3.2 The reaction of iron and chlorine. Here, hot iron gauze is inserted into a flask containing chlorine gas. The heat from the reaction causes the iron gauze to glow, and brown iron(III) chloride forms.

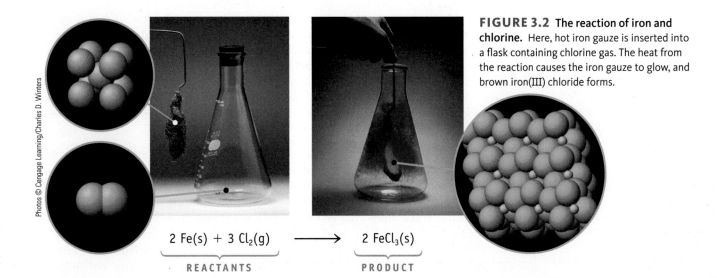

Photos © Cengage Learning/Charles D. Winters

$$2\ Fe(s) + 3\ Cl_2(g) \longrightarrow 2\ FeCl_3(s)$$

REACTANTS   PRODUCT

FIGURE 3.3 Reactions of a metal and two nonmetals with oxygen.

(a) Reaction of iron and oxygen to give iron(III) oxide, Fe_2O_3.

(b) Reaction of sulfur (in the spoon) with oxygen.

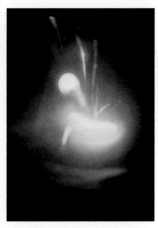

(c) Reaction of phosphorus and oxygen to give tetraphosphorus decaoxide, P_4O_{10}.

3.2　Balancing Chemical Equations

● The Importance of Balanced Chemical Equations Balanced chemical equations are fundamentally important for understanding the quantitative basis of chemistry. *You must always begin with a balanced equation before carrying out a quantitative study of a chemical reaction.*

In a balanced chemical equation, the same number of atoms of each element appears on each side of the equation. Many chemical equations can be balanced by trial and error, and this is the method that will be used for now, but there are also more systematic methods to balance some chemical equations (▶ Section 20.1).

One general class of chemical reactions is the reaction of metals or nonmetals with oxygen to give oxides of the general formula M_xO_y. For example, iron reacts with oxygen to give iron(III) oxide (Figure 3.3a).

$$4\ Fe(s) + 3\ O_2(g) \rightarrow 2\ Fe_2O_3(s)$$

The nonmetals sulfur and oxygen react to form sulfur dioxide (Figure 3.3b),

$$S(s) + O_2(g) \rightarrow SO_2(g)$$

and phosphorus, P_4, reacts vigorously with oxygen to give tetraphosphorus decaoxide, P_4O_{10} (Figure 3.3c).

$$P_4(s) + 5\ O_2(g) \rightarrow P_4O_{10}(s)$$

The equations written above are balanced. The same number of iron, sulfur, or phosphorus atoms and oxygen atoms occurs on each side of these equations.

The **combustion**, or burning, of a fuel in oxygen is accompanied by the evolution of energy as heat. You are familiar with combustion reactions such as the burning of octane, C_8H_{18}, a component of gasoline, in an automobile engine:

$$2\ C_8H_{18}(\ell) + 25\ O_2(g) \rightarrow 16\ CO_2(g) + 18\ H_2O(g)$$

In all combustion reactions, some or all of the elements in the reactants end up as oxides, compounds containing oxygen. When the reactant is a hydrocarbon (a compound that contains only C and H, such as octane), the products of complete combustion are always carbon dioxide and water.

So far, we have given you the balanced chemical equations for the reactions we have considered, but often this will not be the case; you will have to balance the equations. When balancing chemical equations, there are two important things to remember.

- Formulas for reactants and products must be correct or the equation is meaningless. Once the correct formulas for the reactants and products have been determined, the subscripts in their formulas cannot be changed to balance equations. Changing the subscripts changes the identity of the substance. For example, you cannot change CO_2 to CO to balance an equation; carbon monoxide, CO, and carbon dioxide, CO_2, are different compounds.

- Chemical equations are balanced using stoichiometric coefficients. The entire chemical formula of a substance is multiplied by the stoichiometric coefficient.

As an example of equation balancing, let us write the balanced equation for the complete combustion of propane, C_3H_8.

Step 1. *Write correct formulas for the reactants and products.*

$$C_3H_8(g) + O_2(g) \xrightarrow{\text{unbalanced equation}} CO_2(g) + H_2O(g)$$

Here propane and oxygen are the reactants, and carbon dioxide and water are the products.

Step 2. *Balance the C atoms.* In combustion reactions such as this it is usually best to balance the carbon atoms first and leave the oxygen atoms until the end (because the oxygen atoms are often found in more than one product). In this case three carbon atoms are in the reactants, so three must occur in the products. Three CO_2 molecules are therefore required on the right side:

$$C_3H_8(g) + O_2(g) \xrightarrow{\text{unbalanced equation}} 3\ CO_2(g) + H_2O(g)$$

Step 3. *Balance the H atoms.* Propane, the reactant, contains 8 H atoms. Each molecule of water has two hydrogen atoms, so four molecules of water account for the required eight hydrogen atoms on the right side:

$$C_3H_8(g) + O_2(g) \xrightarrow{\text{unbalanced equation}} 3\ CO_2(g) + 4\ H_2O(g)$$

Step 4. *Balance the O atoms.* Ten oxygen atoms are on the right side ($3 \times 2 = 6$ in CO_2 plus $4 \times 1 = 4$ in H_2O). Therefore, five O_2 molecules are needed to supply the required ten oxygen atoms:

$$C_3H_8(g) + 5\ O_2(g) \rightarrow 3\ CO_2(g) + 4\ H_2O(g)$$

Step 5. *Verify that the number of atoms of each element is balanced.* The equation shows three carbon atoms, eight hydrogen atoms, and ten oxygen atoms on each side.

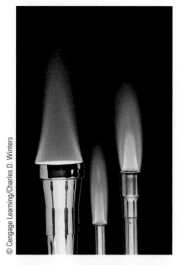

A combustion reaction. Here, propane, C_3H_8, burns to give CO_2 and H_2O. These simple oxides are always the products of the complete combustion of a hydrocarbon.

© Cengage Learning/Charles D. Winters

⬙WL   INTERACTIVE EXAMPLE 3.1   Balancing an Equation for a Combustion Reaction

Problem Write the balanced equation for the combustion of ammonia gas (NH_3) to give water vapor and nitrogen monoxide gas.

What Do You Know? You know the correct formulas and/or names for the reactants (NH_3 and oxygen, O_2) and the products (H_2O and nitrogen monoxide, NO). Also, you know their states.

Strategy First write the unbalanced equation. Next balance the N atoms, then the H atoms, and finally the O atoms.

Solution

Step 1. *Write correct formulas for the reactants and products.* This is a combustion reaction, so the ammonia is reacting with O_2. The unbalanced equation for the combustion is

$$NH_3(g) + O_2(g) \xrightarrow{\text{unbalanced equation}} NO(g) + H_2O(g)$$

Step 2. *Balance the N atoms.* There is one N atom on each side of the equation. The N atoms are in balance, at least for the moment.

$$NH_3(g) + O_2(g) \xrightarrow{\text{unbalanced equation}} NO(g) + H_2O(g)$$

Step 3. *Balance the H atoms.* There are three H atoms on the left and two on the right. To have the same number on each side (6), let us use two molecules of NH_3 on the left and three molecules of H_2O on the right (which gives us six H atoms on each side).

$$2\ NH_3(g) + O_2(g) \xrightarrow{\text{unbalanced equation}} NO(g) + 3\ H_2O(g)$$

Strategy Map 3.1

PROBLEM

Balance the equation for the reaction of **NH₃** and **O₂**

↓

DATA/INFORMATION

The **formulas** of the **reactants** and **products** are given

↓

STEP 1. Balance **N** atoms.

↓

N atoms balanced but overall equation not balanced

STEP 2. Balance **H** atoms.

↓

N and H atoms balanced but overall equation not balanced

STEP 3. Balance **O** atoms. Best left to final step.

↓

N, H, and O atoms balanced.
Overall equation now balanced.

Notice that when we balance the H atoms, the N atoms are no longer balanced. To bring them into balance, let us use 2 NO molecules on the right.

$$2\ NH_3(g) + O_2(g) \xrightarrow{\text{unbalanced equation}} 2\ NO(g) + 3\ H_2O(g)$$

Step 4. *Balance the O atoms.* After Step 3, there is an even number of O atoms (two) on the left and an odd number (five) on the right. Because there cannot be an odd number of O atoms on the left (O atoms are paired in O_2 molecules), multiply each coefficient on both sides of the equation by 2 so that an even number of oxygen atoms (10) now occurs on the right side:

$$4\ NH_3(g) + O_2(g) \xrightarrow{\text{unbalanced equation}} 4\ NO(g) + 6\ H_2O(g)$$

Now the oxygen atoms can be balanced by having five O_2 molecules on the left side of the equation:

$$4\ NH_3(g) + 5\ O_2(g) \xrightarrow{\text{balanced equation}} 4\ NO(g) + 6\ H_2O(g)$$

Step 5. *Verify the result.* Four N atoms, 12 H atoms, and 10 O atoms occur on each side of the equation.

Think about Your Answer   An alternative way to write this equation is

$$2\ NH_3(g) + {}^5/_2\ O_2(g) \rightarrow 2\ NO(g) + 3\ H_2O(g)$$

where a fractional coefficient has been used. This equation is correctly balanced and will be useful under some circumstances. In general, however, we balance equations with whole-number coefficients.

Check Your Understanding

(a) Butane gas, C_4H_{10}, can burn completely in air [use $O_2(g)$ as the other reactant] to give carbon dioxide gas and water vapor. Write a balanced equation for this combustion reaction.

(b) Write a balanced chemical equation for the complete combustion of liquid tetraethyl lead, $Pb(C_2H_5)_4$ (which was used until the 1970s as a gasoline additive). The products of combustion are $PbO(s)$, $H_2O(g)$, and $CO_2(g)$. (The PbO was a significant environmental pollutant!)

REVIEW & CHECK FOR SECTION 3.2

The (unbalanced) equation describing the oxidation of propanol is

$$2\ C_3H_7OH(\ell) + \underline{\ \ }\ O_2(g) \rightarrow 6\ CO_2(g) + \underline{\ \ }\ H_2O(g)$$

Determine the two missing stoichiometric coefficients.

(a) 4.5 O_2, 4 H_2O (b) 9 O_2, 8 H_2O (c) 5 O_2, 4 H_2O (d) 5 O_2, 7 H_2O

Dr. Arthur N. Palmer

FIGURE 3.4 Cave chemistry.
Calcium carbonate stalactites cling to the roof of a cave, and stalagmites grow up from the cave floor. The chemistry producing these formations is a good example of the reversibility of chemical reactions.

3.3 Introduction to Chemical Equilibrium

To this point, we have treated chemical reactions as proceeding in one direction only, with reactants being converted *completely* to products. Nature, however, is more complex than this. Chemical reactions are reversible, and many reactions lead to incomplete conversion of reactants to products.

The formation of stalactites and stalagmites in a limestone cave is an example of a system that depends on the reversibility of a chemical reaction (Figure 3.4). Stalactites and stalagmites are made chiefly of calcium carbonate, a mineral found in underground deposits in the form of limestone, a leftover from ancient oceans. If water seeping through the limestone contains dissolved CO_2, a reaction occurs in which the mineral dissolves, giving an aqueous solution of $Ca(HCO_3)_2$.

$$CaCO_3(s) + CO_2(aq) + H_2O(\ell) \rightarrow Ca(HCO_3)_2(aq)$$

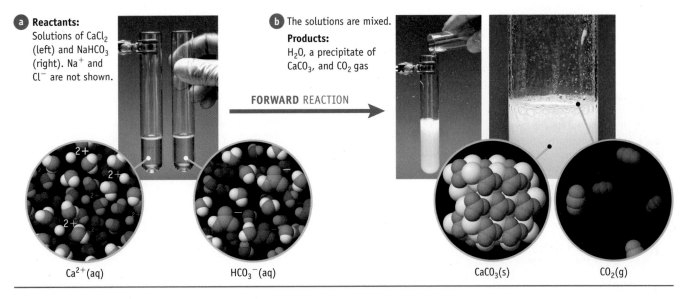

a **Reactants:**
Solutions of $CaCl_2$ (left) and $NaHCO_3$ (right). Na^+ and Cl^- are not shown.

Ca^{2+}(aq) HCO_3^-(aq)

b The solutions are mixed.
Products:
H_2O, a precipitate of $CaCO_3$, and CO_2 gas

FORWARD REACTION

$CaCO_3$(s) CO_2(g)

Equilibrium equation:

$$Ca^{2+}(aq) + 2\ HCO_3^-(aq) \rightleftharpoons CaCO_3(s) + CO_2(g) + H_2O(\ell)$$

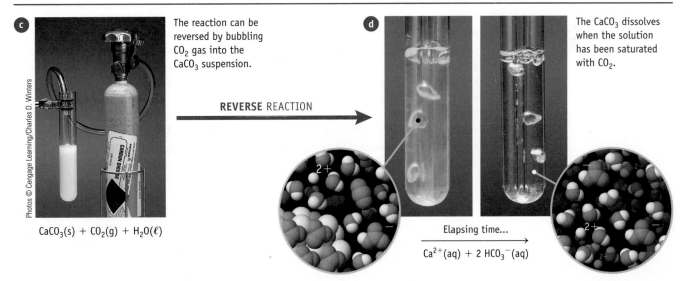

c
The reaction can be reversed by bubbling CO_2 gas into the $CaCO_3$ suspension.

Photos © Cengage Learning/Charles D. Winters

$CaCO_3$(s) + CO_2(g) + $H_2O(\ell)$

REVERSE REACTION

d
The $CaCO_3$ dissolves when the solution has been saturated with CO_2.

Elapsing time...
Ca^{2+}(aq) + 2 HCO_3^-(aq)

FIGURE 3.5 **The reversibility of chemical reactions.** The experiments here demonstrate the reversibility of chemical reactions. (*top*) Solutions of $CaCl_2$ (a source of Ca^{2+} ions) and $NaHCO_3$ (a source of HCO_3^- ions) are mixed **(a)** and produce a precipitate of $CaCO_3$ and CO_2 gas **(b)**. (*bottom*) If CO_2 gas is bubbled into a suspension of $CaCO_3$ **(c)**, the reverse of the reaction displayed in the top panel occurs. That is, solid $CaCO_3$ and gaseous CO_2 produce Ca^{2+} and HCO_3^- ions **(d)**.

When the mineral-laden water reaches a cave, the reverse reaction occurs, with CO_2 being evolved into the cave and solid $CaCO_3$ being deposited.

$$Ca(HCO_3)_2(aq) \rightarrow CaCO_3(s) + CO_2(g) + H_2O(\ell)$$

Cave chemistry can be done in a laboratory (Figure 3.5) using reactions that further demonstrate the reversibility of the reactions involved.

Another example of a reversible reaction is the reaction of nitrogen with hydrogen to form ammonia gas, a compound made industrially in enormous quantities and used directly as a fertilizer and in the production of other fertilizers.

$$N_2(g) + 3\ H_2(g) \rightarrow 2\ NH_3(g)$$

Nitrogen and hydrogen react to form ammonia, but, under the conditions of the reaction, the product ammonia also breaks down into nitrogen and hydrogen in the reverse reaction.

$$2\ NH_3(g) \rightarrow N_2(g) + 3\ H_2(g)$$

Let us consider what would happen if we mixed nitrogen and hydrogen in a closed container under the proper conditions for the reaction to occur. At first, N_2 and H_2 react to produce some ammonia. As the ammonia is produced, however,

FIGURE 3.6 The reaction of N₂ and H₂ to produce NH₃. N₂ and H₂ in a 1:3 mixture react to produce some NH₃. As the reaction proceeds, the rate or speed of NH₃ production slows, as does the rate of consumption of N₂ and H₂. Eventually, the amounts of N₂, H₂, and NH₃ no longer change. At this point, the reaction has reached equilibrium. Nonetheless, the forward reaction to produce NH₃ continues, as does the reverse reaction (the decomposition of NH₃).

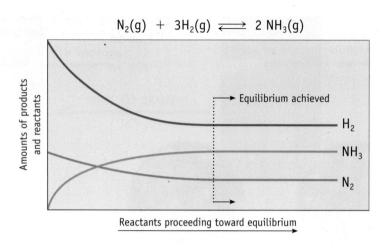

$$N_2(g) \ + \ 3H_2(g) \ \rightleftharpoons \ 2 \ NH_3(g)$$

Equilibrium achieved

H₂

NH₃

N₂

Amounts of products and reactants

Reactants proceeding toward equilibrium

some NH₃ molecules decompose to re-form nitrogen and hydrogen in the reverse reaction (Figure 3.6). At the beginning of the process, the forward reaction to give NH₃ predominates, but, as the reactants are consumed, the rate (or speed) of the forward reaction progressively slows. At the same time, the reverse reaction speeds up as the amount of ammonia increases. Eventually, the rate of the forward reaction will equal the rate of the reverse reaction. At this point, no further *macroscopic* change is observed; the amounts of nitrogen, hydrogen, and ammonia in the container stop changing but the forward and reverse reactions continue. We say the system has reached **chemical equilibrium**. The reaction vessel will contain all three substances: nitrogen, hydrogen, and ammonia. Because both the forward and reverse processes are still occurring (but at equal rates), we refer to this state as a **dynamic equilibrium**. We represent a system at dynamic equilibrium by writing a double arrow symbol ($\rightleftharpoons$) connecting the reactants and products.

$$N_2(g) \ + \ 3 \ H_2(g) \ \rightleftharpoons \ 2 \ NH_3(g)$$

Chemical reactions always proceed spontaneously toward equilibrium. A reaction will never proceed spontaneously in a direction that takes a system further from equilibrium. A key question that arises is "When a reaction reaches equilibrium, will the reactants be converted largely to products or will most of the reactants still be present?" The answer will depend on the nature of the substances involved, the temperature, and other factors, and that is the subject of later chapters (▶ Chapters 16–18). For the present, though, it is useful to define **product-favored reactions** as *reactions in which reactants are completely or largely converted to products when equilibrium is reached.* The combustion reactions we have been studying are examples of reactions that are product-favored at equilibrium. In fact, most of the reactions that we shall study in the rest of this chapter are product-favored reactions at equilibrium. We usually write the equations for reactions that are very product-favored using only the single arrows we have been using up to this point.

The opposite of a product-favored reaction is one that is **reactant-favored** at equilibrium. Such reactions lead to the conversion of only a small amount of the reactants to products. An example of such a reaction is the ionization of acetic acid in water where only a tiny fraction of the acid ionizes to produce ions.

● **Quantitative Description of Chemical Equilibrium** As you shall see in Chapters 16–18, the extent to which a reaction is product-favored can be described by a mathematical expression called the *equilibrium constant expression*. Each chemical reaction has a numerical value for the equilibrium constant, symbolized by K. Product-favored reactions have large values of K; small K values indicate reactant-favored reactions. For the ionization of acetic acid in water, $K = 1.8 \times 10^{-5}$.

$$CH_3CO_2H(aq) \ + \ H_2O(\ell) \ \rightleftharpoons \ CH_3CO_2^-(aq) \ + \ H_3O^+(aq)$$

acetic acid   water   acetate ion   hydronium ion

As you shall see below, acetic acid is an example of a large number of acids called "weak acids" because the reaction is so reactant-favored at equilibrium that only a few percent of the molecules react with water to form ionic products.

REVIEW & CHECK FOR SECTION 3.3

Which of the following statements about chemical equilibrium is not true?

(a) At equilibrium the rates of the forward and reverse reactions are equal.

(b) There is no observable change in a chemical system at equilibrium.

(c) Chemical reactions always proceed toward equilibrium.

(d) All chemical equilibria are product-favored.

3.4 Aqueous Solutions

Many of the reactions you will study in your chemistry course and almost all of the reactions that occur in living things are carried out in solutions in which the reacting substances are dissolved in water. In Chapter 1, we defined a **solution** as a homogeneous mixture of two or more substances. One substance is generally considered the **solvent**, the medium in which another substance—the **solute**—is dissolved. The remainder of this chapter is an introduction to some of the types of reactions that occur in **aqueous solutions**, solutions in which *water* is the solvent. To understand these reactions, it is important first to understand something about the behavior of compounds dissolved in water.

Module 5: Predicting the Water Solubility of Ionic Compounds covers concepts in this section.

Ions and Molecules in Aqueous Solutions

Dissolving an ionic solid requires separating each ion from the oppositely charged ions that surround it in the solid state. Water is especially good at dissolving ionic compounds because each water molecule has a positively charged end and a negatively charged end (Figure 3.7). When an ionic compound dissolves in water, each negative ion becomes surrounded by water molecules with the positive ends of water molecules pointing toward it, and each positive ion becomes surrounded by the negative ends of several water molecules.

The water-encased ions produced by dissolving an ionic compound are free to move about in solution. Under normal conditions, the movement of ions is random, and the cations and anions from a dissolved ionic compound are dispersed uniformly throughout the solution. However, if two **electrodes** (conductors of electricity such as copper wire) are placed in the solution and connected to a battery, ion movement is no longer random. Positive cations move through the solution to the negative electrode and negative anions move to the positive electrode (Figure 3.8). If a light bulb is inserted into the circuit, the bulb lights, showing that ions are available to conduct charge in the solution just as electrons conduct charge in the wire part of the circuit. Compounds whose aqueous solutions conduct electricity are called **electrolytes**. *All ionic compounds that are soluble in water are electrolytes.*

For every mole of NaCl that dissolves, 1 mol of Na^+ and 1 mol of Cl^- ions enter the solution.

$$NaCl(s) \rightarrow Na^+(aq) + Cl^-(aq)$$

100% Dissociation → strong electrolyte

Because the solute has dissociated (broken apart) completely into ions, the solution will be a good conductor of electricity. Substances whose solutions are good electrical conductors owing to the presence of ions are **strong electrolytes** (Figure 3.8). The ions into which an ionic compound will dissociate are given by the compound's name, and the relative amounts of these ions are given by its formula. For example,

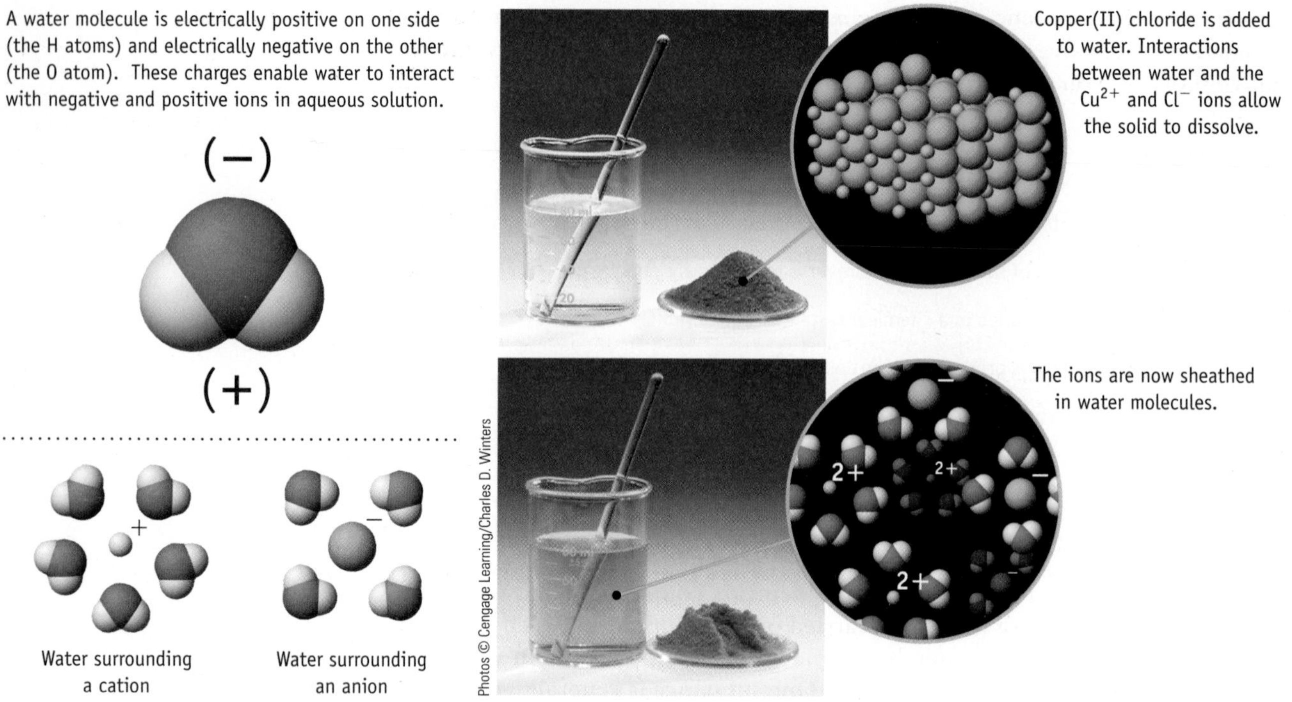

A water molecule is electrically positive on one side (the H atoms) and electrically negative on the other (the O atom). These charges enable water to interact with negative and positive ions in aqueous solution.
(−)
(+)
+
−
Water surrounding a cation
Water surrounding an anion
Copper(II) chloride is added to water. Interactions between water and the Cu2+ and Cl− ions allow the solid to dissolve.
The ions are now sheathed in water molecules.
2+
2+
2+
−
Photos © Cengage Learning/Charles D. Winters

(a) Water molecules are attracted to both positive cations and negative anions in aqueous solution.

(b) When an ionic substance dissolves in water, each ion is surrounded by water molecules.

FIGURE 3.7 Water as a solvent for ionic substances.

as we have seen, sodium chloride yields sodium ions (Na^+) and chloride ions (Cl^-) in solution in a 1:1 ratio. The ionic compound barium chloride, $BaCl_2$, is also a strong electrolyte. In this case there are two chloride ions for each barium ion in solution.

$$BaCl_2(s) \rightarrow Ba^{2+}(aq) + 2\ Cl^-(aq)$$

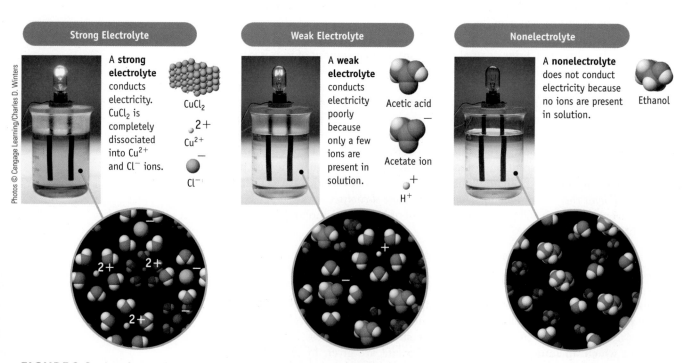

Strong Electrolyte
A strong electrolyte conducts electricity. CuCl2 is completely dissociated into Cu2+ and Cl− ions.
CuCl2
2+
Cu2+
−
Cl−
Weak Electrolyte
A weak electrolyte conducts electricity poorly because only a few ions are present in solution.
Acetic acid
−
Acetate ion
+
H+
Nonelectrolyte
A nonelectrolyte does not conduct electricity because no ions are present in solution.
Ethanol
Photos © Cengage Learning/Charles D. Winters

FIGURE 3.8 Classifying solutions by their ability to conduct electricity.

Notice that the two chloride ions per formula unit are present as two separate particles in solution; they are not present as diatomic particles, Cl_2^{2-}. In yet another example, the ionic compound barium nitrate yields barium ions and nitrate ions in solution. For each Ba^{2+} ion in solution, there are two NO_3^- ions.

$$Ba(NO_3)_2(s) \rightarrow Ba^{2+}(aq) + 2\ NO_3^-(aq)$$

Notice that the polyatomic ion stays together as one unit, NO_3^-, but that the two nitrate ions are separate ions in solution.

Compounds whose aqueous solutions do not conduct electricity are called **nonelectrolytes**. The solute particles present in these aqueous solutions are molecules, not ions. *Most molecular compounds that dissolve in water are nonelectrolytes.* For example, when the molecular compound ethanol (C_2H_5OH) dissolves in water, each molecule of ethanol stays together as a single unit. We do not get ions in the solution.

$$C_2H_5OH(\ell) \rightarrow C_2H_5OH(aq)$$

Other examples of nonelectrolytes are sucrose ($C_{12}H_{22}O_{11}$) and antifreeze (ethylene glycol, $HOCH_2CH_2OH$).

Some molecular compounds (strong acids, weak acids, and weak bases) (Section 3.6), however, react with water to form ions and thus are electrolytes. Gaseous hydrogen chloride, a molecular compound, reacts with water to form ions, and the solution is referred to as hydrochloric acid.

$$HCl(g) + H_2O(\ell) \rightarrow H_3O^+(aq) + Cl^-(aq)$$

This reaction is very product-favored. Each molecule of HCl produces ions in solution, so hydrochloric acid is a strong electrolyte.

Some molecular compounds are **weak electrolytes**. When these compounds dissolve in water only a small fraction of the molecules forms ions; the majority remains intact. These aqueous solutions are poor conductors of electricity (Figure 3.8). As described on page 100, the interaction of acetic acid with water is very reactant-favored. In vinegar, an aqueous solution of acetic acid, about 40 molecules in every 10,000 molecules of acetic acid are ionized to form acetate ($CH_3CO_2^-$) and hydronium (H_3O^+) ions. Thus, aqueous acetic acid is a weak electrolyte.

$$CH_3CO_2H(aq) + H_2O(\ell) \rightleftharpoons CH_3CO_2^-(aq) + H_3O^+(aq)$$

Figure 3.9 summarizes whether a given type of solute will be present in aqueous solution as ions, molecules, or a combination of ions and molecules.

● **Dissolving Halides** When an ionic compound with halide ions dissolves in water, the halide ions are released into aqueous solution. Thus, $BaCl_2$ produces two Cl^- ions for each Ba^{2+} ion (and not Cl_2 or Cl_2^{2-} ions).

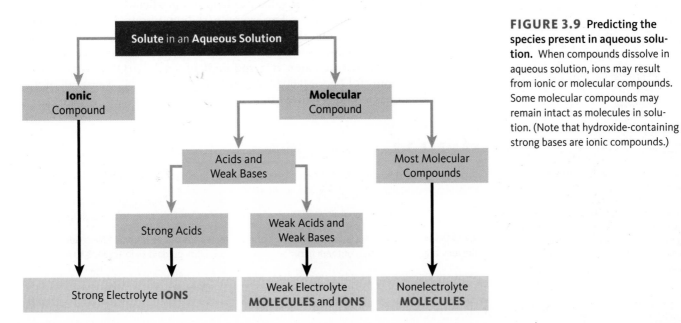

FIGURE 3.9 Predicting the species present in aqueous solution. When compounds dissolve in aqueous solution, ions may result from ionic or molecular compounds. Some molecular compounds may remain intact as molecules in solution. (Note that hydroxide-containing strong bases are ionic compounds.)

SILVER COMPOUNDS

AgNO₃   AgCl   AgOH

(a) Nitrates are generally soluble, as are chlorides (except AgCl). Hydroxides are generally not soluble.

SULFIDES

(NH₄)₂S   CdS   Sb₂S₃   PbS

(b) Sulfides are generally not soluble (exceptions include salts with NH₄⁺ and Na⁺).

HYDROXIDES

Photos © Cengage Learning/
Charles D. Winters

NaOH   Ca(OH)₂   Fe(OH)₃   Ni(OH)₂

(c) Hydroxides are generally not soluble, except when the cation is a Group 1A metal.

SOLUBLE COMPOUNDS	EXCEPTIONS
Almost all salts of Na^+, K^+, NH_4^+ Salts of nitrate, NO_3^- chlorate, ClO_3^- perchlorate, ClO_4^- acetate, $CH_3CO_2^-$	
Almost all salts of Cl^-, Br^-, I^-	Halides of Ag^+, Hg_2^{2+}, Pb^{2+}
Salts containing F^-	Fluorides of Mg^{2+}, Ca^{2+}, Sr^{2+}, Ba^{2+}, Pb^{2+}
Salts of sulfate, SO_4^{2-}	Sulfates of Ca^{2+}, Sr^{2+}, Ba^{2+}, Pb^{2+}, Ag^+

INSOLUBLE COMPOUNDS	EXCEPTIONS
Most salts of carbonate, CO_3^{2-} phosphate, PO_4^{3-} oxalate, $C_2O_4^{2-}$ chromate, CrO_4^{2-} sulfide, S^{2-}	Salts of NH_4^+ and the alkali metal cations
Most metal hydroxides and oxides	Alkali metal hydroxides and $Ba(OH)_2$ and $Sr(OH)_2$

FIGURE 3.10 Guidelines to predict the solubility of ionic compounds. If a compound contains one of the ions in the column on the left in the top chart, it is predicted to be at least moderately soluble in water. There are exceptions, which are noted at the right. Most ionic compounds formed by the anions listed at the bottom of the chart are poorly soluble (with the exception of compounds with NH_4^+ and the alkali metal cations).

Solubility of Ionic Compounds in Water

Many ionic compounds are soluble in water, but some dissolve only to a small extent; still others are essentially insoluble. Fortunately, we can make some general statements about which ionic compounds are water-soluble. In this chapter, we consider solubility as an "either-or" question, referring to those materials that are soluble beyond a certain extent as "soluble" and to those that do not dissolve to that extent as "insoluble." To get a better idea of the amounts that will actually dissolve in a given quantity of water, we could do an experiment or perform a calculation that uses the concept of equilibrium (▶ Chapter 18).

Figure 3.10 lists broad guidelines that can help you predict whether a particular ionic compound is soluble in water. For example, sodium nitrate, $NaNO_3$, contains an alkali metal cation, Na^+, and the nitrate anion, NO_3^-. The presence of either of these ions ensures that the compound is soluble in water. In contrast, calcium hydroxide is poorly soluble in water. If a spoonful of solid $Ca(OH)_2$ is added to 100 mL of water, only 0.17 g, or 0.0023 mol, will dissolve at 10 °C. Nearly all of the $Ca(OH)_2$ remains as a solid (Figure 3.10c).

● **Solubility Guidelines** Observations such as those shown in Figure 3.10 were used to create the solubility guidelines. Note, however, that these are general guidelines and not rules followed under all circumstances. There are exceptions, but the guidelines are a good place to begin. See B. Blake, *Journal of Chemical Education*, Vol. 80, pp. 1348–1350, 2003.

EXAMPLE 3.2   Solubility Guidelines

Problem   Predict whether the following ionic compounds are likely to be water-soluble. For soluble compounds, list the ions present in solution.

(a) KCl   (b) $MgCO_3$ (c) Fe_2O_3 (d) $Cu(NO_3)_2$

What Do You Know? You know the formulas of the compounds but need to be able to identify the ions that make up each of them in order to use the solubility guidelines in Figure 3.10.

Strategy Decide the probable water solubility based on the solubility guidelines (Figure 3.10). Soluble compounds will dissociate into their respective ions in solution.

Solution

(a) KCl is composed of K^+ and Cl^- ions. The presence of *either* of these ions means that the compound is likely to be soluble in water. The solution contains K^+ and Cl^- ions dissolved in water.

$$KCl(s) \rightarrow K^+(aq) + Cl^-(aq)$$

(The solubility of KCl is about 35 g in 100 mL of water at 20 °C.)

(b) Magnesium carbonate is composed of Mg^{2+} and CO_3^{2-} ions. Salts containing the carbonate ion usually are insoluble, unless combined with an ion like Na^+ or NH_4^+. Therefore, $MgCO_3$ is predicted to be insoluble in water. (The solubility of $MgCO_3$ is less than 0.2 g/100 mL of water.)

(c) Iron(III) oxide is composed of Fe^{3+} and O^{2-} ions. Oxides are soluble only when O^{2-} is combined with an alkali metal ion; Fe^{3+} is a transition metal ion, so Fe_2O_3 is insoluble.

(d) Copper(II) nitrate is composed of Cu^{2+} and NO_3^- ions. Nitrate salts are soluble, so this compound dissolves in water, giving ions in solution as shown in the equation below.

$$Cu(NO_3)_2(s) \rightarrow Cu^{2+}(aq) + 2\ NO_3^-(aq)$$

Think about Your Answer For chemists, a set of simple guidelines like this is useful. However, it is possible to get accurate solubility information for many compounds in chemical resource books or databases.

Check Your Understanding

Predict whether each of the following ionic compounds is likely to be soluble in water. If it is soluble, write the formulas of the ions present in aqueous solution.

(a) $LiNO_3$ (b) $CaCl_2$ (c) CuO (d) $NaCH_3CO_2$

REVIEW & CHECK FOR SECTION 3.4

1. Which of the compounds listed below is water-soluble?

     (a) $Ba(NO_3)_2$ (b) CuS (c) $Fe_3(PO_4)_2$ (d) $Mg(OH)_2$

2. Calcium nitrate dihydrate is dissolved in water. What ions are present in solution?

     (a) Ca^{2+} and $(NO_3)_2^{2-}$ (c) Ca and $2\ NO_3$

     (b) Ca^{2+} and $2\ NO_3^-$ (d) Ca^+ and NO_3^-

3. A 1-g sample of each of the four compounds listed below was placed in a beaker with 100 mL of water. The mixture was then tested to see if it conducted an electric current. Which solution was a weak conductor of electricity?

     (a) ethanol   (b) acetic acid   (c) NaCl   (d) NaOH

3.5 Precipitation Reactions

Now that you can predict whether compounds will yield ions or molecules when they dissolve in water and whether ionic compounds are soluble or insoluble in water, we can begin to discuss the chemical reactions that occur in aqueous solutions. We will introduce you to four major categories of reactions in aqueous solution: precipitation, acid–base, gas-forming, and oxidation–reduction reactions. As you learn about these reactions, it will be useful to look for patterns that allow you to predict the reaction products. You will notice that precipitation, acid–base, and many gas-forming (but not oxidation–reduction) reactions are **exchange reactions** (sometimes called double dis-

Module 6: Writing Net Ionic Equations covers concepts in this section.

FIGURE 3.11 Precipitation of silver chloride. Mixing aqueous solutions of silver nitrate and potassium chloride produces white, insoluble silver chloride, AgCl.

(b) Initially, the Ag^+ ions (silver color) and Cl^- ions (green) are widely separated.

(c) Ag^+ and Cl^- ions approach and form ion pairs.

(d) As more and more Ag^+ and Cl^- ions come together, a precipitate of solid AgCl forms.

(a) Mixing aqueous solutions of silver nitrate and potassium chloride produces white, insoluble silver chloride, AgCl.

placement, double replacement, or metathesis reactions) in which *the ions of the reactants change partners.*

$$\overset{\frown}{A^+B^-} \; + \; \underset{\smile}{C^+D^-} \; \longrightarrow \; A^+D^- \; + \; C^+B^-$$

For example, aqueous solutions of silver nitrate and potassium chloride react to produce solid silver chloride and aqueous potassium nitrate (Figure 3.11a).

$$AgNO_3(aq) + KCl(aq) \rightarrow AgCl(s) + KNO_3(aq)$$

This particular reaction is an example of the first type of reaction that we shall examine: a precipitation reaction. A **precipitation reaction** produces a water-insoluble solid product, known as a **precipitate**. The reactants in such reactions are generally water-soluble ionic compounds. When these substances dissolve in water, they dissociate to give the appropriate cations and anions. If the cation from one reactant can form an insoluble compound with the anion from the other reactant, precipitation occurs. In Figure 3.11, both silver nitrate and potassium chloride are water-soluble ionic compounds. When combined in water, they undergo an *exchange reaction* to produce insoluble silver chloride and soluble potassium nitrate.

$$AgNO_3(aq) + KCl(aq) \rightarrow AgCl(s) + KNO_3(aq)$$

Reactants	**Products**
$Ag^+(aq) + NO_3^-(aq)$	Insoluble AgCl(s)
$K^+(aq) + Cl^-(aq)$	$K^+(aq) + NO_3^-(aq)$

Predicting the Outcome of a Precipitation Reaction

Many combinations of positive and negative ions give insoluble substances (see Figure 3.10). For example, the solubility guidelines indicate that most compounds containing the chromate ion are not soluble (alkali metal chromates and ammonium chromate are exceptions). Thus, we can predict that yellow, solid lead(II) chromate will precipitate when a water-soluble lead(II) compound is combined with a water-soluble chromate compound (Figure 3.12a).

$$Pb(NO_3)_2(aq) + K_2CrO_4(aq) \rightarrow PbCrO_4(s) + 2\ KNO_3(aq)$$

Reactants	**Products**
$Pb^{2+}(aq) + 2\ NO_3^-(aq)$	Insoluble $PbCrO_4$(s)
$2\ K^+(aq) + CrO_4^{2-}(aq)$	$2\ K^+(aq) + 2\ NO_3^-(aq)$

Similarly, we know from the solubility guidelines that almost all metal sulfides are insoluble in water (Figure 3.12b). If a solution of a soluble metal compound comes in contact with a source of sulfide ions, the metal sulfide precipitates.

$$Pb(NO_3)_2(aq) + (NH_4)_2S(aq) \rightarrow PbS(s) + 2\ NH_4NO_3(aq)$$

Reactants	**Products**
$Pb^{2+}(aq) + 2\ NO_3^-(aq)$	Insoluble PbS(s)
$2\ NH_4^+(aq) + S^{2-}(aq)$	$2\ NH_4^+(aq) + 2\ NO_3^-(aq)$

Photos © Cengage Learning/Charles D. Winters

(a) **(b)** **(c)** **(d)**

FIGURE 3.12 Precipitation reactions. Many ionic compounds are insoluble in water. Guidelines for predicting the solubilities of ionic compounds are given in Figure 3.10. **(a)** $Pb(NO_3)_2$ and K_2CrO_4 produce yellow, insoluble $PbCrO_4$ and soluble KNO_3. **(b)** $Pb(NO_3)_2$ and $(NH_4)_2S$ produce black, insoluble PbS and soluble NH_4NO_3. **(c)** $FeCl_3$ and $NaOH$ produce orange, insoluble $Fe(OH)_3$ and soluble $NaCl$. **(d)** $AgNO_3$ and K_2CrO_4 produce orange, insoluble Ag_2CrO_4 and soluble KNO_3. (See Example 3.3.)

In yet another example, the solubility guidelines indicate that with the exception of the alkali metal cations (and Sr^{2+} and Ba^{2+}), all metal cations form insoluble hydroxides. Thus, water-soluble iron(III) chloride and sodium hydroxide react to give insoluble iron(III) hydroxide (Figures 3.10c and 3.12c).

$$FeCl_3(aq) + 3\ NaOH(aq) \rightarrow Fe(OH)_3(s) + 3\ NaCl(aq)$$

Reactants	**Products**
$Fe^{3+}(aq) + 3\ Cl^-(aq)$	Insoluble $Fe(OH)_3(s)$
$3\ Na^+(aq) + 3\ OH^-(aq)$	$3\ Na^+(aq) + 3\ Cl^-(aq)$

EXAMPLE 3.3   Writing the Equation for a Precipitation Reaction

Problem Is an insoluble product formed when aqueous solutions of potassium chromate and silver nitrate are mixed? If so, write the balanced equation.

What Do You Know? Names of the two reactants are given. You should recognize that this is an exchange reaction, and you will need the information on solubilities in Figure 3.10.

Strategy

- Determine the formulas from the names of the reactants.

- Determine the formulas for the cations and anions making up the reactants.

- Write formulas for the products in this reaction by exchanging cations and anions and determine whether either product is insoluble using information in Figure 3.10.

- Write and balance the equation.

Solution The reactants are silver nitrate and potassium chromate. The possible products of the exchange reaction are silver chromate (Ag_2CrO_4) and potassium nitrate (KNO_3). Based on the solubility guidelines, we know that silver chromate is an insoluble compound (chromates are insoluble except for those with Group 1A metals), whereas potassium nitrate is soluble in water. A precipitate of silver chromate will therefore form if these reactants are mixed.

$$2\ AgNO_3(aq) + K_2CrO_4(aq) \rightarrow Ag_2CrO_4(s) + 2\ KNO_3(aq)$$

Think about Your Answer This reaction is illustrated in Figure 3.12d.

Check Your Understanding

In each of the following cases, does a precipitation reaction occur when solutions of the two water-soluble reactants are mixed? Give the formula of any precipitate that forms, and write a balanced chemical equation for the precipitation reactions that occur.

(a) sodium carbonate and copper(II) chloride

(b) potassium carbonate and sodium nitrate

(c) nickel(II) chloride and potassium hydroxide

© Cengage Learning/Charles D. Winters

Black tongue, a precipitate of Bi_2S_3. Pepto-Bismol™ has anti-diarrheal, antibacterial, and antacid effects in the digestive tract, and has been used for over 100 years as an effective remedy. However, some people find their tongues blackened after taking this over-the-counter medicine. The active ingredient in Pepto-Bismol is bismuth subsalicylate. (It also contains pepsin, zinc salts, flavoring, and salol, a compound related to aspirin.) The tongue-blackening comes from the reaction of bismuth ions with traces of sulfide ions found in saliva to form black Bi_2S_3. The discoloration is harmless and lasts only a few days.

Net Ionic Equations

We have seen that when aqueous solutions of silver nitrate and potassium chloride are mixed, insoluble silver chloride forms, leaving potassium nitrate in solution (Figure 3.11). The balanced chemical equation for this process is

$$AgNO_3(aq) + KCl(aq) \rightarrow AgCl(s) + KNO_3(aq)$$

We can represent this reaction in another way by writing an equation in which we show that the *soluble* ionic compounds are present in solution as dissociated ions. An aqueous solution of silver nitrate contains Ag^+ and NO_3^- ions, and an aqueous solution of potassium chloride contains K^+ and Cl^- ions. In the products, the potassium nitrate is present in solution as K^+ and NO_3^- ions. The silver chloride, however, is insoluble and thus is not present in the solution as dissociated ions. It is shown in the equation by its entire formula, AgCl.

$$\underset{\text{before reaction}}{Ag^+(aq) + NO_3^-(aq) + K^+(aq) + Cl^-(aq)} \longrightarrow \underset{\text{after reaction}}{AgCl(s) + K^+(aq) + NO_3^-(aq)}$$

This type of equation is called a **complete ionic equation**.

The K^+ and NO_3^- ions are present in solution both before and after reaction, so they appear on both the reactant and product sides of the complete ionic equation. Such ions are often called **spectator ions** because they do not participate in the net reaction; they only "look on" from the sidelines. Little chemical information is lost if the equation is written without them, so we can simplify the equation to

$$Ag^+(aq) + Cl^-(aq) \rightarrow AgCl(s)$$

● **Net Ionic Equations** All chemical equations, including net ionic equations, must be balanced. The same number of atoms of each kind must appear on both the product and reactant sides. In addition, the sum of positive and negative charges must be the same on both sides of the equation.

The balanced equation that results from leaving out the spectator ions is the **net ionic equation** for the reaction.

Leaving out the spectator ions does not mean that K^+ and NO_3^- ions are unimportant in the $AgNO_3 + KCl$ reaction. Indeed, Ag^+ ions cannot exist alone in solution; a negative ion, in this case NO_3^-, must be present to balance the positive charge of Ag^+. Any anion will do, however, as long as it forms a water-soluble compound with Ag^+. Thus, we could have used $AgClO_4$ instead of $AgNO_3$. Similarly, there must be a positive ion present to balance the negative charge of Cl^-. In this case, the positive ion present is K^+ in KCl, but we could have used NaCl instead of KCl. The net ionic equation would have been the same.

Finally, notice that there must always be a *charge balance* as well as a mass balance in a balanced chemical equation. Thus, in the $Ag^+ + Cl^-$ net ionic equation, the cation and anion charges on the left add together to give a net charge of zero, the same as the zero charge on AgCl(s) on the right.

Writing Net Ionic Equations

Net ionic equations are commonly written for chemical reactions in aqueous solution because they describe the actual chemical species involved in a reaction. To write net ionic equations we must know which compounds exist as ions in solution.

1. Strong acids, strong bases, and soluble salts exist as ions in solution. Examples include the acids HCl and HNO_3, a base such as NaOH, and salts such as NaCl and $CuCl_2$.
2. All other species should be represented by their complete formulas. Weak acids such as acetic acid (CH_3CO_2H) exist in aqueous solutions primarily as molecules.

(See Section 3.6.) Insoluble salts such as $CaCO_3(s)$ or insoluble bases such as $Mg(OH)_2(s)$ should not be written in ionic form, even though they are ionic compounds.

The best way to approach writing net ionic equations is to follow precisely a set of steps.

1. Write a complete, balanced equation. Indicate the state of each substance (aq, s, ℓ, g).
2. Next rewrite the whole equation, writing all strong acids, strong bases, and soluble salts as ions. (Consider only species labeled "(aq)" in this step.)
3. Some ions may remain unchanged in the reaction (the ions that appear in the equation both as reactants and products). These "spectator ions" are not part of the chemistry that is going on, and you can cancel them from each side of the equation.
4. Like full chemical equations, net ionic equations must be balanced. The same number of atoms appears on each side of the arrow, and the sum of the ion charges on the two sides must also be equal.

◯WL   **INTERACTIVE EXAMPLE 3.4**   Writing and Balancing Net Ionic Equations

Problem Write a balanced, net ionic equation for the reaction of aqueous solutions of $BaCl_2$ and Na_2SO_4.

What Do You Know? The formulas for the reactants are given. You should recognize that this is an exchange reaction, and that you will need the information on solubilities in Figure 3.10.

Strategy Follow the strategy outlined in Problem Solving Tip 3.1.

Solution

Step 1. In this *exchange reaction*, the Ba^{2+} and Na^+ cations exchange anions (Cl^- and SO_4^{2-}) to give $BaSO_4$ and $NaCl$. Now that the reactants and products are known, we can write an equation for the reaction. To balance the equation, we place a 2 in front of the $NaCl$.

$$BaCl_2 + Na_2SO_4 \rightarrow BaSO_4 + 2\ NaCl$$

Step 2. Decide on the solubility of each compound (Figure 3.10). Compounds containing sodium ions are always water-soluble, and those containing chloride ions are almost always soluble. Sulfate salts are also usually soluble, one important exception being $BaSO_4$. We can therefore write

$$BaCl_2(aq) + Na_2SO_4(aq) \rightarrow BaSO_4(s) + 2\ NaCl(aq)$$

Step 3. Identify the ions in solution. All soluble ionic compounds dissociate to form ions in aqueous solution.

$$BaCl_2(s) \quad \rightarrow \quad Ba^{2+}(aq) \quad + \quad 2\ Cl^-(aq)$$
$$Na_2SO_4(s) \quad \rightarrow \quad 2\ Na^+(aq) \quad + \quad SO_4^{2-}(aq)$$
$$NaCl(s) \quad \rightarrow \quad Na^+(aq) \quad + \quad Cl^-(aq)$$

This results in the following complete ionic equation:

$$Ba^{2+}(aq) + 2\ Cl^-(aq) + 2\ Na^+(aq) + SO_4^{2-}(aq) \rightarrow BaSO_4(s) + 2\ Na^+(aq) + 2\ Cl^-(aq)$$

Step 4. Identify and eliminate the spectator ions (Na^+ and Cl^-) to give the net ionic equation.

$$Ba^{2+}(aq) + SO_4^{2-}(aq) \rightarrow BaSO_4(s)$$

Think about Your Answer Notice that the sum of ion charges is the same on both sides of the equation. On the left, $2+$ and $2-$ give zero; on the right the charge on $BaSO_4$ is also zero.

Check Your Understanding

In each of the following cases, aqueous solutions containing the compounds indicated are mixed. Write balanced net ionic equations for the reactions that occur.

(a) $CaCl_2 + Na_3PO_4$

(b)   iron(III) chloride and potassium hydroxide (a similar reaction is in Figure 3.12c)

(c)   lead(II) nitrate and potassium chloride

REVIEW & CHECK FOR SECTION 3.5

1.   Which of the following compounds is not soluble in water?

   (a) $Mg(NO_3)_2$ (b) Ag_2S (c) $(NH_4)_2SO_4$ (d) Na_2CO_3

2.   Aqueous solutions of the two reactants listed below are mixed. In which solutions will a precipitate form?

   (1) $NaI(aq) + Pb(NO_3)_2(aq)$ (3) $H_2SO_4(aq) + BaCl_2(aq)$

   (2) $CaCl_2(aq) + Na_2CO_3(aq)$ (4) $NaOH(aq) + AgNO_3(aq)$

   (a) 1, 2, and 3   (b) 2 and 3   (c) 1, 2, and 4   (d) 1, 2, 3, and 4

Precipitation reaction. The reaction of barium chloride and sodium sulfate produces insoluble barium sulfate and water-soluble sodium chloride.

© Cengage Learning/Charles D. Winters

Strategy Map 3.4

PROBLEM
Write **balanced net ionic equation** for the reaction of $BaCl_2 + Na_2SO_4$.

↓

DATA/INFORMATION
The **formulas** of the **reactants** are given

STEP 1. Decide on **products** and then write **complete balanced equation.**

Complete balanced equation with *reactants* and *products*

STEP 2. Decide if each reactant and product is **solid**, **liquid**, **gas**, or **dissolved in water.**

Complete balanced equation with **indication of state** of each reactant and product

STEP 3. Identify ions in solution.

Complete ionic equation with reactants and products **dissociated into ions** if appropriate

STEP 4. Eliminate spectator ions.

Balanced net ionic equation

3. What is the net ionic equation for the reaction of $AgNO_3(aq)$ and $Na_2CO_3(aq)$?

   (a) $Ag^+(aq) + CO_3^{2-}(aq) \rightarrow AgCO_3(s)$

   (b) $Na^+(aq) + NO_3^-(aq) \rightarrow NaNO_3(s)$

   (c) $2\,Ag^+(aq) + CO_3^{2-}(aq) \rightarrow Ag_2CO_3(s)$

   (d) $AgNO_3(aq) + Na_2CO_3(aq) \rightarrow 2\,NaNO_3(aq) + Ag_2CO_3(s)$

3.6 Acids and Bases

Acids and bases are two important classes of compounds. You may already be familiar with some common properties of acids. They produce bubbles of CO_2 gas when added to a metal carbonate such as $CaCO_3$ (Figure 3.13a), and they react with many metals to produce hydrogen gas (H_2) (Figure 3.13b). Although tasting substances is *never* done in a chemistry laboratory, you have probably experienced the sour taste of acids such as acetic acid in vinegar and citric acid (commonly found in fruits and added to candies and soft drinks). Acids and bases have some related properties. Solutions of acids or bases, for example, can change the colors of vegetable pigments (Figure 3.13c). You may have seen acids change the color of litmus, a dye derived from certain lichens, from blue to red. If an acid has made blue litmus paper turn red, adding a base reverses the effect, making the litmus blue again. Thus, acids and bases seem to be opposites. A base can neutralize the effect of an acid, and an acid can neutralize the effect of a base. Table 3.1 lists common acids and bases.

Over the years, chemists have examined the properties, chemical structures, and reactions of acids and bases and have proposed different definitions of the terms *acid* and *base*. In this section, we shall examine the two most commonly used definitions, one proposed by Svante Arrhenius (1859–1927) and another proposed by Johannes N. Brønsted (1879–1947) and Thomas M. Lowry (1874–1936).

Acids and Bases: The Arrhenius Definition

The Swedish chemist Svante Arrhenius made a number of important contributions to chemistry, but he is perhaps best known for studying the properties of solutions of salts, acids, and bases. In the late 1800s, Arrhenius proposed that these compounds dissolve in water and ultimately form ions. This theory of electrolytes predated any

(a) A piece of coral (mostly $CaCO_3$) dissolves in acid to give CO_2 gas.

(b) Zinc reacts with hydrochloric acid to produce zinc chloride and hydrogen gas.

(c) The juice of a red cabbage is normally blue-purple. On adding acid, the juice becomes more red. Adding base produces a yellow color.

FIGURE 3.13 Some properties of acids and bases.

Table 3.1 **Common Acids and Bases**

Strong Acids (Strong Electrolytes) *		Soluble Strong Bases	
HCl	Hydrochloric acid	LiOH	Lithium hydroxide
HBr	Hydrobromic acid	NaOH	Sodium hydroxide
HI	Hydroiodic acid	KOH	Potassium hydroxide
HNO_3	Nitric acid	$Ba(OH)_2$	Barium hydroxide
$HClO_4$	Perchloric acid	$Sr(OH)_2$	Strontium hydroxide
H_2SO_4	Sulfuric acid		
Weak Acids (Weak Electrolytes) *		**Weak Base (Weak Electrolyte) ***	
HF	Hydrofluoric acid	NH_3	Ammonia
H_3PO_4	Phosphoric acid		
H_2CO_3	Carbonic acid		
CH_3CO_2H	Acetic acid		
$H_2C_2O_4$	Oxalic acid		
$H_2C_4H_4O_6$	Tartaric acid		
$H_3C_6H_5O_7$	Citric acid		
$HC_9H_7O_4$	Aspirin		

*The electrolytic behavior refers to aqueous solutions of these acids and bases.

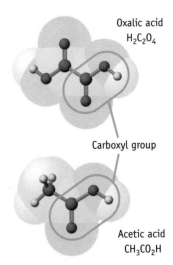

Oxalic acid
$H_2C_2O_4$

Carboxyl group

Acetic acid
CH_3CO_2H

● **Weak Acids** Common acids and bases are listed in Table 3.1. There are numerous other weak acids and bases, and many of these are natural substances. Many of the natural acids contain CO_2H groups. (The H of this group is lost as H^+.) The CO_2H group is circled in the figures showing the structures of oxalic and acetic acids, two naturally occurring weak acids.

knowledge of the composition and structure of atoms and was not well accepted initially. With a knowledge of atomic structure, however, we now take it for granted.

The Arrhenius definitions for acids and bases derive from his theory of electrolytes and focuses on formation of H^+ and OH^- ions in aqueous solutions.

- An acid is a substance that, when dissolved in water, increases the concentration of hydrogen ions, H^+, in solution.

$$HCl(g) \rightarrow H^+(aq) + Cl^-(aq)$$

- A base is a substance that, when dissolved in water, increases the concentration of hydroxide ions, OH^-, in the solution.

$$NaOH(s) \rightarrow Na^+(aq) + OH^-(aq)$$

- The reaction of an acid and a base produces a salt and water. Because the characteristic properties of an acid are lost when a base is added, and vice versa, acid–base reactions were logically described as resulting from the combination of H^+ and OH^- to form water.

$$HCl(aq) + NaOH(aq) \rightarrow NaCl(aq) + H_2O(\ell)$$

Arrhenius further proposed that acid strength was related to the extent to which the acid ionized. Some acids such as hydrochloric acid (HCl) and nitric acid (HNO_3) ionize completely in water; they are strong electrolytes, and we now call them **strong acids**. Other acids such as acetic acid and hydrofluoric acid are incompletely ionized; they are weak electrolytes and are **weak acids**. Weak acids exist in solution primarily as acid molecules, and only a fraction of these molecules ionize to produce $H^+(aq)$ ions along with the appropriate anion.

Water-soluble compounds that contain hydroxide ions, such as sodium hydroxide (NaOH) and potassium hydroxide (KOH), are strong electrolytes and **strong bases**.

Aqueous ammonia, $NH_3(aq)$, is a weak electrolyte. Even though it does not have an OH^- ion as part of its formula, it does produce ammonium ions and hydroxide ions from its reaction with water and so is a base (Figure 3.14). The fact that this is a weak electrolyte indicates that this reaction with water to form ions is reactant-favored at equilibrium. Most of the ammonia remains in solution in molecular form.

© Cengage Learning/Charles D. Winters

FIGURE 3.14 Ammonia, a weak electrolyte. Ammonia, NH_3, interacts with water to produce a very small number of NH_4^+ and OH^- ions per mole of ammonia molecules. (The name on the bottle, ammonium hydroxide, is misleading. The solution consists almost entirely of NH_3 molecules dissolved in water. It is better referred to as "aqueous ammonia.")

Although the Arrhenius theory is still used to some extent and is interesting in an historical context, modern concepts of acid–base chemistry such as the Brønsted–Lowry theory have gained preference among chemists.

Acids and Bases: The Brønsted–Lowry Definition

In 1923, Brønsted in Copenhagen (Denmark) and Lowry in Cambridge (England) independently suggested a new concept of acid and base behavior. They viewed acids and bases in terms of the transfer of a proton (H^+) from one species to another, and they described all acid–base reactions in terms of equilibria. The Brønsted–Lowry theory expanded the scope of the definition of acids and bases and helped chemists make predictions of product- or reactant-favorability based on acid and base strength. We will describe this theory here qualitatively; a more complete discussion will be given in Chapter 17.

The main concepts of the Brønsted–Lowry theory are the following:

- *An acid is a proton donor.* This is similar to the Arrhenius definition.
- *A base is a proton acceptor.* This definition includes the OH^- ion but it also broadens the number and type of bases.
- *An acid–base reaction involves the transfer of a proton from an acid to a base to form a new base and a new acid. The reaction is written as an equilibrium reaction and the equilibrium favors the weaker acid and base.* This allows the prediction of product- or reactant-favored reactions based on acid and base strength.

From the point of view of the Brønsted–Lowry theory, the behavior of acids such as HCl or CH_3CO_2H in water is written as an acid–base reaction. Both species (both Brønsted acids) donate a proton to water (a Brønsted base) forming $H_3O^+(aq)$, the hydronium ion. Hydrochloric acid, HCl(aq), is a strong electrolyte because it ionizes completely in aqueous solution; it is classified as a strong acid.

Hydrochloric acid, a strong acid. 100% ionized. Equilibrium strongly favors products.

$$HCl(aq) \;+\; H_2O(\ell) \;\rightleftharpoons\; H_3O^+(aq) \;+\; Cl^-(aq)$$

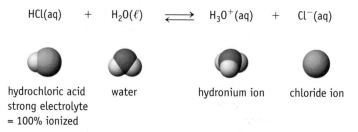

| hydrochloric acid | water | hydronium ion | chloride ion |

strong electrolyte
= 100% ionized

In contrast, CH_3CO_2H is a weak electrolyte, evidence that it is only weakly ionized in water and is therefore a weak acid.

Acetic acid, a weak acid, << 100% ionized. Equilibrium favors reactants.

$$CH_3CO_2H(aq) \;+\; H_2O(\ell) \;\rightleftharpoons\; H_3O^+(aq) \;+\; CH_3CO_2^-(aq)$$
weak Brønsted acid

The different extent of ionization for these two acids relates to their acid strength. Hydrochloric acid is a strong acid, and the equilibrium strongly favors the products. In contrast, acetic acid is a weak acid and its reaction with water is reactant-favored.

Sulfuric acid, a *diprotic acid* (an acid capable of transferring two H^+ ions), reacts with water in two steps. The first step strongly favors products, whereas the second step is reactant-favored.

Strong acid: $\quad H_2SO_4(aq) + H_2O(\ell) \;\rightleftharpoons\; H_3O^+(aq) \;+\; HSO_4^-(aq)$
　　　　　　　sulfuric acid　　　　　　　　　　hydronium ion　　hydrogen
　　　　　　　100% ionized　　　　　　　　　　　　　　　　　　sulfate ion

Weak acid: $\quad HSO_4^-(aq) + H_2O(\ell) \;\rightleftharpoons\; H_3O^+(aq) \;+\; SO_4^{2-}(aq)$
　　　　　　　hydrogen sulfate ion　　　　　　hydronium ion　　sulfate ion
　　　　　　　<100% ionized

A CLOSER LOOK

The Hydronium Ion—The H⁺ Ion in Water

The H⁺ ion is a hydrogen atom that has lost its electron. Only the nucleus, a proton, remains. Because a proton is only about 1/100,000 as large as the average atom or ion, water molecules can approach closely, and the proton and water molecules are strongly attracted. In fact, the H⁺ ion in water is better represented as H_3O^+, called the **hydronium ion**. This ion is formed by combining H⁺ and H_2O. Experiments also show that other forms of the ion exist in water, one example being $[H_3O(H_2O)_3]^+$.

There will be instances when, for simplicity, we will use H⁺(aq). However, in this book we will usually use the H_3O^+(aq) symbol to represent the hydrogen ion in water. Thus, what we call hydrochloric acid is better represented as a solution of H_3O^+ and Cl^-.

hydronium ion
H_3O^+(aq)

chloride ion
Cl^-(aq)

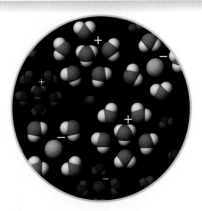

Hydronium ions in solution. When HCl ionizes in aqueous solution, it produces the hydronium ion, H_3O^+, and the chloride ion, Cl^-.

Ammonia, a weak base, reacts with water to produce OH⁻(aq) ions. The reaction is reactant-favored at equilibrium.

Ammonia, a weak base, << 100% ionized. Equilibrium favors reactants.

$$NH_3(aq) + H_2O(\ell) \rightleftharpoons NH_4^+(aq) + OH^-(aq)$$

ammonia, base
weak electrolyte
< 100% ionized

water

ammonium
ion

hydroxide ion

Some species are described as **amphiprotic**, that is, they can function either as acids or as bases depending on the reaction. In the examples above, water functions as a base in reactions with acids (it accepts a proton) and as an acid in its reaction with ammonia (where it donates a proton to ammonia, forming the ammonium ion.)

EXAMPLE 3.5   Brønsted Acids and Bases

Problem Write a balanced net ionic equation for the reaction that occurs when the cyanide ion, CN⁻, accepts a proton (H⁺) from water to form HCN. Is CN⁻ a Brønsted acid or base?

What Do You Know? You know the formulas of the ion (CN⁻) and molecule (H_2O) involved as reactants and one of the products, (HCN). You also know a proton transfer occurs from water to CN⁻.

Strategy As it is a proton transfer, you should move an H⁺ ion from H_2O to CN⁻ to give the products.

Solution

$$H_2O(\ell) + CN^-(aq) \rightleftharpoons OH^-(aq) + HCN(aq)$$

In this reaction the water is the Brønsted acid and the CN⁻ ion is the Brønsted base.

Think about Your Answer The CN⁻ ion is a proton acceptor and so is a Brønsted base.

Check Your Understanding

(a) Write a balanced equation for the reaction that occurs when H_3PO_4, phosphoric acid, donates a proton to water to form the dihydrogen phosphate ion.

(b) Write a net ionic equation showing the dihydrogen phosphate ion acting as a Brønsted acid in a reaction with water. Write another net ionic equation showing the dihydrogen phosphate ion acting as a Brønsted base in a reaction with water. What term is used to describe a species such as the dihydrogen phosphate ion that can act both as an acid or as a base?

Reactions of Acids and Bases

Acids and bases in aqueous solution usually react to produce a salt and water. For example (Figure 3.15),

$$HCl(aq) \ + \ NaOH(aq) \ \longrightarrow \ H_2O(\ell) \ + \ NaCl(aq)$$
hydrochloric acid   sodium hydroxide   water   sodium chloride

The word "salt" has come into the language of chemistry to describe any ionic compound whose cation comes from a base (here Na^+ from NaOH) and whose anion comes from an acid (here Cl^- from HCl). The reaction of any of the acids listed in Table 3.1 with any of the hydroxide-containing bases listed there produces a salt and water.

Hydrochloric acid and sodium hydroxide are strong electrolytes in water (Figure 3.15 and Table 3.1), so the complete ionic equation for the reaction of HCl(aq) and NaOH(aq) should be written as

$$\underbrace{H_3O^+(aq) + Cl^-(aq)}_{\text{from HCl(aq)}} + \underbrace{Na^+(aq) + OH^-(aq)}_{\text{from NaOH(aq)}} \longrightarrow 2\ \underbrace{H_2O(\ell)}_{\text{water}} + \underbrace{Na^+(aq) + Cl^-(aq)}_{\text{from salt}}$$

Because Na^+ and Cl^- ions appear on both sides of the equation, the *net ionic equation* is just the combination of the ions H_3O^+ and OH^- to give water.

$$H_3O^+(aq) + OH^-(aq) \rightarrow 2\ H_2O(\ell)$$

This is always the net ionic equation when a strong acid reacts with a strong base.

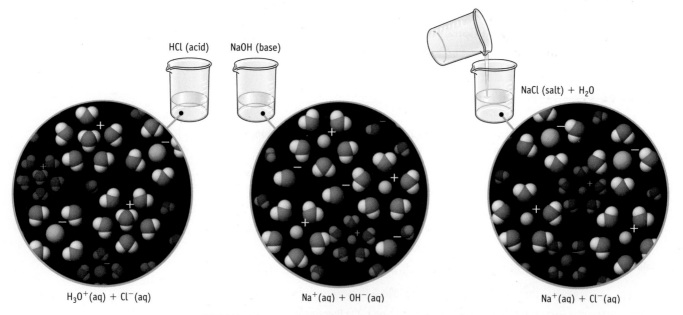

HCl (acid)   NaOH (base)   NaCl (salt) + H_2O

$H_3O^+(aq) + Cl^-(aq)$ $Na^+(aq) + OH^-(aq)$ $Na^+(aq) + Cl^-(aq)$

FIGURE 3.15 An acid–base reaction, HCl and NaOH. On mixing, the H_3O^+ and OH^- ions combine to produce H_2O, whereas the ions Na^+ and Cl^- remain in solution.

Sulfuric Acid

For many years sulfuric acid has been the chemical produced in the largest quantity in the United States (as well as in many other industrialized countries). About 40–50 billion kilograms (40–50 million metric tons) is made annually in the United States. The acid is so important to the economy of industrialized nations that some economists have said sulfuric acid production is a measure of a nation's industrial strength.

Sulfuric acid is a colorless, syrupy liquid with a density of 1.84 g/mL and a boiling point of 337 °C. It has several desirable properties that have led to its widespread use: it is generally less expensive to produce than other acids, is a strong acid, and can be handled in steel containers. It reacts readily with many organic compounds to produce useful products and reacts readily with lime (CaO), the least expensive and most readily available base, to give calcium sulfate, a compound used to make wall board for the construction industry.

The first step in the industrial preparation of sulfuric acid is the production of sulfur dioxide. This comes from the combustion of sulfur in air,

$$S(s) + O_2(g) \rightarrow SO_2(g)$$

Sulfur. Much of the sulfur used in the U.S. used to be mined, but it is now largely a by-product from natural gas and oil-refining processes. It takes about 1 ton of sulfur to make 3 tons of the acid.

Some products that require sulfuric acid for their manufacture or use.

or from the SO_2 produced in smelting sulfur-containing copper, nickel, or other metal ores. The SO_2 is then combined with more oxygen, in the presence of a catalyst (a substance that speeds up the reaction), to give sulfur trioxide,

$$2 SO_2(g) + O_2(g) \rightarrow 2 SO_3(g)$$

which can give sulfuric acid when dissolved in water.

$$SO_3(g) + H_2O(\ell) \rightarrow H_2SO_4(aq)$$

Currently, over two thirds of the production is used in the fertilizer industry. The remainder is used to make pigments, explosives, pulp and paper, detergents, and as a component in storage batteries.

"The Acid Touch," *Chemical and Engineering News*, April 14, 2008, p. 27.

Reactions between strong acids and strong bases are called **neutralization reactions** because, on completion of the reaction, the solution is neutral if exactly the same amounts (number of moles) of the acid and base are mixed; that is, it is neither acidic nor basic. The other ions (the cation of the base and the anion of the acid) remain unchanged. If the water is evaporated, however, the cation and anion form a solid salt. In the example above, NaCl can be obtained. If nitric acid, HNO_3, and NaOH were allowed to react, the salt sodium nitrate, $NaNO_3$, (and water) would be obtained.

$$HNO_3(aq) + NaOH(aq) \rightarrow H_2O(\ell) + NaNO_3(aq)$$

If acetic acid and sodium hydroxide are mixed, the following reaction will take place.

$$CH_3CO_2H(aq) + NaOH(aq) \rightarrow H_2O(\ell) + NaCH_3CO_2(aq)$$

Because acetic acid is a weak acid and ionizes to such a small extent (Figure 3.8), the molecular species is the predominant form in aqueous solutions. In ionic equations, therefore, acetic acid is shown as molecular $CH_3CO_2H(aq)$. The *complete ionic equation* for this reaction is

$$CH_3CO_2H(aq) + Na^+(aq) + OH^-(aq) \rightarrow H_2O(\ell) + Na^+(aq) + CH_3CO_2^-(aq)$$

The only spectator ions in this equation are the sodium ions, so the *net ionic equation* is

$$CH_3CO_2H(aq) + OH^-(aq) \rightarrow H_2O(\ell) + CH_3CO_2^-(aq)$$

© Cengage Learning/Charles D. Winters

Reaction of gaseous HCl and NH₃.
Open dishes of aqueous ammonia and hydrochloric acid were placed side by side. When gas molecules of NH_3 and HCl escape from solution to the atmosphere and encounter one another, we observe a cloud of solid ammonium chloride, NH_4Cl.

EXAMPLE 3.6　　Net Ionic Equation for an Acid–Base Reaction

Problem Ammonia, NH_3, is one of the most important chemicals in industrial economies. Not only is it used directly as a fertilizer but it is the raw material for the manufacture of nitric acid. As a base, it reacts with acids such as hydrochloric acid. Write a balanced, net ionic equation for this reaction.

What Do You Know? The reactants are $NH_3(aq)$ and $HCl(aq)$. A proton will transfer from the acid to the base.

Strategy Follow the general strategy for writing net ionic equations as outlined in Problem Solving Tip 3.1.

Solution A proton transfers from HCl to NH_3, a weak Brønsted base, to form the ammonium ion, NH_4^+. This positive ion must have a negative counterion from the acid, Cl^-, so the reaction product is NH_4Cl, and the complete balanced equation is

$$NH_3(aq) \quad + \quad HCl(aq) \quad \rightarrow \quad NH_4Cl(aq)$$
$$\text{ammonia} \qquad \text{hydrochloric acid} \qquad \text{ammonium chloride}$$

To write the net ionic equation, start with the fact that hydrochloric acid is a strong acid and produces H_3O^+ and Cl^- ions in water and that NH_4Cl is a soluble, ionic compound. On the other hand, ammonia is a weak base and so is predominantly present in the solution as the molecular species, NH_3.

$$NH_3(aq) + H_3O^+(aq) + Cl^-(aq) \rightarrow NH_4^+(aq) + Cl^-(aq) + H_2O(\ell)$$

Eliminating the spectator ion, Cl^-, we have

$$NH_3(aq) + H_3O^+(aq) \rightleftharpoons NH_4^+(aq) + H_2O(\ell)$$

Think about Your Answer The net ionic equation shows that the important aspect of the reaction between the weak base ammonia and the strong acid HCl is the transfer of an H^+ ion from the acid to the NH_3. Any strong acid could be used here (HBr, HNO_3, $HClO_4$, H_2SO_4) and the net ionic equation would be the same. Also notice that, even though H_2O is not in the overall balanced equation, it is present in the net ionic equation because the H^+ from the acid is present in the initial solution as H_3O^+.

Check Your Understanding

Write the balanced, overall equation and the net ionic equation for the reaction of magnesium hydroxide with hydrochloric acid. (*Hint:* Think about the solubility guidelines.)

CO_2

SO_2

SO_3

NO_2

Some common nonmetal oxides that form acids in water.

Oxides of Nonmetals and Metals

Each acid shown in Table 3.1 has one or more H atoms in the molecular formula that ionize in water to form H_3O^+ ions. There are, however, less obvious compounds that form acidic solutions. Oxides of nonmetals, such as carbon dioxide and sulfur trioxide, have no H atoms but react with water to produce H_3O^+ ions. Carbon dioxide, for example, dissolves in water to a small extent, and some of the dissolved molecules react with water to form the weak acid, carbonic acid. This acid then ionizes to a small extent to form the hydronium ion, H_3O^+, and the hydrogen carbonate (bicarbonate) ion, HCO_3^-.

$$CO_2(g) \quad + \quad H_2O(\ell) \quad \rightleftharpoons \quad H_2CO_3(aq)$$

$$H_2CO_3(aq) \quad + \quad H_2O(\ell) \quad \rightleftharpoons \quad HCO_3^-(aq) \quad + \quad H_3O^+(aq)$$

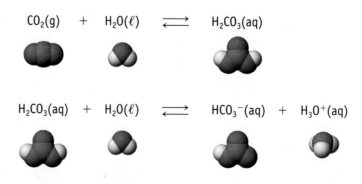

The HCO_3^- ion can also function as an acid, ionizing to produce H_3O^+ and the carbonate ion, CO_3^{2-}.

$$HCO_3^-(aq) \ + \ H_2O(\ell) \ \rightleftharpoons \ CO_3^{2-}(aq) \ + \ H_3O^+(aq)$$

These reactions are important in our environment and in the human body. Carbon dioxide is normally found in small amounts in the atmosphere, so rainwater is always slightly acidic. In the human body, carbon dioxide is dissolved in body fluids where the HCO_3^- and CO_3^{2-} ions perform an important "buffering" action (▶ Chapter 18).

Oxides like CO_2 that can react with water to produce H_3O^+ ions are known as **acidic oxides**. Other acidic oxides include those of sulfur and nitrogen. For example, sulfur dioxide, SO_2, from human and natural sources, can react with oxygen to give sulfur trioxide, SO_3, which then reacts with water to form sulfuric acid.

$$2\ SO_2(g) + O_2(g) \longrightarrow 2\ SO_3(g)$$
$$SO_3(g) + H_2O(\ell) \longrightarrow H_2SO_4(aq)$$

Nitrogen dioxide, NO_2, reacts with water to give nitric and nitrous acids.

$$2\ NO_2(g) + H_2O(\ell) \rightarrow \underset{\text{nitric acid}}{HNO_3(aq)} + \underset{\text{nitrous acid}}{HNO_2(aq)}$$

Oxides of metals are called **basic oxides** because they give basic solutions if they dissolve appreciably in water. Perhaps the best example is calcium oxide, CaO, often called *lime*, or *quicklime*. Almost 20 billion kg of lime is produced annually in the United States for use in the metals and construction industries, in sewage and pollution control, in water treatment, and in agriculture. This metal oxide reacts with water to give calcium hydroxide, commonly called *slaked lime*. Although only slightly soluble in water (about 0.2 g/100 g H_2O at 10 °C), $Ca(OH)_2$ is widely used in industry as a base because it is inexpensive.

$$\underset{\text{lime}}{CaO(s)} + H_2O(\ell) \rightarrow \underset{\text{slaked lime}}{Ca(OH)_2(s)}$$

● **Acid Rain** Oxides of sulfur and nitrogen are the major source of the acid in what is called *acid rain*. These acidic oxides arise from the burning of fossil fuels such as coal and gasoline. The gaseous oxides mix with water and other chemicals in the troposphere, and the rain that falls is more acidic than if it contained only dissolved CO_2. When the rain falls on areas that cannot easily tolerate this greater-than-normal acidity, serious environmental problems can occur.

REVIEW & CHECK FOR SECTION 3.6

1. Which of the following is a strong acid in aqueous solution?

    (a) HF   (b) H_2CO_3 (c) CH_3CO_2H (d) H_2SO_4

2. The hydrogen phosphate ion is amphiprotic. Identify the product formed from this ion when it reacts with a base.

    (a) H_3PO_4 (b) $H_2PO_4^-$ (c) HPO_4^{2-} (d) PO_4^{3-}

3. What is the net ionic equation for the reaction of acetic acid and sodium hydroxide?

    (a) $H_3O^+(aq) \ + \ OH^-(aq) \rightarrow 2\ H_2O(\ell)$

    (b) $Na^+(aq) \ + \ CH_3CO_2^-(aq) \rightarrow NaCH_3CO_2(aq)$

    (c) $CH_3CO_2H(aq) \ + \ OH^-(aq) \rightarrow H_2O(\ell) \ + \ CH_3CO_2^-(aq)$

    (d) $CH_3CO_2H(aq) \ + \ NaOH(aq) \rightarrow H_2O(\ell) \ + \ NaCH_3CO_2(aq)$

4. Identify the basic oxide from those listed below.

    (a) SO_2 (b) BaO   (c) P_4O_{10} (d) CO_2

FIGURE 3.16 Dissolving limestone (calcium carbonate, CaCO₃) in vinegar. Notice the bubbles of CO₂ rising from the surface of the limestone. This reaction shows why vinegar can be used as a household cleaning agent. It can be used, for example, to clean the calcium carbonate deposited from hard water.

Table 3.2 **Gas-Forming Reactions**

Metal carbonate or hydrogen carbonate + acid → metal salt + CO₂(g) + H₂O(ℓ)
$Na_2CO_3(aq) + 2\ HCl(aq) \rightarrow 2\ NaCl(aq) + CO_2(g) + H_2O(\ell)$
$NaHCO_3(aq) + HCl(aq) \rightarrow NaCl(aq) + CO_2(g) + H_2O(\ell)$
Metal sulfide + acid → metal salt + H₂S(g)
$Na_2S(aq) + 2\ HCl(aq) \rightarrow 2\ NaCl(aq) + H_2S(g)$
Metal sulfite + acid → metal salt + SO₂(g) + H₂O(ℓ)
$Na_2SO_3(aq) + 2\ HCl(aq) \rightarrow 2\ NaCl(aq) + SO_2(g) + H_2O(\ell)$
Ammonium salt + strong base → metal salt + NH₃(g) + H₂O(ℓ)
$NH_4Cl(aq) + NaOH(aq) \rightarrow NaCl(aq) + NH_3(g) + H_2O(\ell)$

3.7 Gas-Forming Reactions

Several different chemical reactions lead to gas formation (Table 3.2), and the most common are those that produce CO_2. All metal carbonates (and bicarbonates) react with acids to produce carbonic acid, H_2CO_3, which in turn decomposes rapidly to carbon dioxide and water. For example, the reaction of calcium carbonate and hydrochloric acid is:

$$CaCO_3(s) + 2\ HCl(aq) \rightarrow CaCl_2(aq) + H_2CO_3(aq)$$

$$H_2CO_3(aq) \rightarrow H_2O(\ell) + CO_2(g)$$

Overall reaction: $CaCO_3(s) + 2\ HCl(aq) \rightarrow CaCl_2(aq) + H_2O(\ell) + CO_2(g)$

If the reaction is done in an open beaker, most of the CO_2 gas bubbles out of the solution.

Calcium carbonate is a common residue from hard water in home heating systems and cooking utensils. Washing with vinegar is a good way to clean the system or utensils because the insoluble calcium carbonate is turned into water-soluble calcium acetate in the following gas-forming reaction (Figure 3.16).

$$2\ CH_3CO_2H(aq) + CaCO_3(s) \rightarrow Ca(CH_3CO_2)_2(aq) + H_2O(\ell) + CO_2(g)$$

What is the net ionic equation for this reaction? Acetic acid is a weak acid, and calcium carbonate is insoluble in water. Therefore, the reactants are simply $CH_3CO_2H(aq)$ and $CaCO_3(s)$. On the products side, calcium acetate is water-soluble and so is present in solution as aqueous calcium and acetate ions. Water and carbon dioxide are molecular compounds, so the complete and net ionic equation is

$$2\ CH_3CO_2H(aq) + CaCO_3(s) \rightarrow Ca^{2+}(aq) + 2\ CH_3CO_2^-(aq) + H_2O(\ell) + CO_2(g)$$

There are no spectator ions in this reaction.

Have you ever made biscuits or muffins? As you bake the dough, it rises in the oven (Figure 3.17) because a gas-forming reaction occurs between an acid and baking soda, sodium hydrogen carbonate (bicarbonate of soda, $NaHCO_3$). One acid used for this purpose is tartaric acid, a weak acid found in many foods. The net ionic equation for a typical reaction is

$$\underset{\text{tartaric acid}}{H_2C_4H_4O_6(aq)} + \underset{\text{hydrogen carbonate ion}}{HCO_3^-(aq)} \longrightarrow \underset{\text{hydrogen tartrate ion}}{HC_4H_4O_6^-(aq)} + H_2O(\ell) + CO_2(g)$$

In dry baking powder, the acid and $NaHCO_3$ are kept apart by using starch as a filler. When mixed into the moist batter, however, the acid and sodium hydrogen carbonate dissolve, come into contact, and react to produce CO_2, which causes the dough to rise.

FIGURE 3.17 Muffins rise because of a gas-forming reaction. The acid and sodium bicarbonate in baking powder produce carbon dioxide gas. The acid used in many baking powders is CaHPO₄, but tartaric acid and NaAl(SO₄)₂ are also common. (Aqueous solutions containing the aluminum ion are acidic.)

Recognizing Gas-Forming Reactions

How can you recognize that a particular reaction will lead to gas formation? After you predict the products of the exchange reaction, be alert for certain products:

(a) H_2CO_3: This will decompose into carbon dioxide gas and water.

(b) H_2SO_3: This will decompose into sulfur dioxide gas and water.

(c) H_2S: This is a gaseous product already.

(d) If NH_4^+ and OH^- ions are produced, they will form ammonia gas and water.

EXAMPLE 3.7   Gas-Forming Reactions

Problem Write a balanced equation for the reaction that occurs when nickel(II) carbonate is treated with sulfuric acid.

What Do You Know? You know the names of the reactants and therefore their formulas. You should recognize that the reaction of a metal carbonate with an acid is a gas-forming reaction (CO_2 is formed in these reactions).

Strategy

• Write the formulas for the reactants.

• Determine the products of the reaction and their formulas.

• Write and balance the equation.

Solution The reactants are $NiCO_3$ and H_2SO_4, and the products of the reaction are $NiSO_4$, CO_2 and H_2O. The complete, balanced equation is

$$NiCO_3(s) + H_2SO_4(aq) \longrightarrow NiSO_4(aq) + H_2O(\ell) + CO_2(g)$$

Think about Your Answer The products in this reaction were determined by first exchanging cations (Ni^{2+} and $2\,H^+$) and anions (CO_3^{2-} and SO_4^{2-}). This exchange reaction is then followed by a second reaction in which one product, H_2CO_3, decomposes to give CO_2 and H_2O.

Check Your Understanding

Barium carbonate, $BaCO_3$, is used in the brick, ceramic, glass, and chemical manufacturing industries. Write a balanced equation that shows what happens when barium carbonate is treated with nitric acid. Give the name of each of the reaction products.

REVIEW & CHECK FOR SECTION 3.7

Which of the following compounds, when treated with acid, results in gas evolution?

(a) NH_4Cl (b) Na_2CO_3 (c) Na_2SO_4 (d) Na_3PO_4

(a) Iron ore, which is largely Fe_2O_3, is reduced to metallic iron with carbon (C) or carbon monoxide (CO) in a blast furnace. Here iron(III) oxide is reduced to Fe by the reducing agent C or CO. The C or CO is oxidized to CO_2.

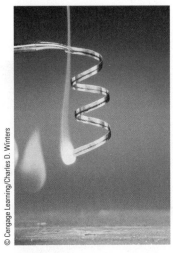

(b) Burning magnesium metal in air produces magnesium oxide. The magnesium is oxidized by the oxidizing agent O_2. Oxygen is reduced by the reducing agent Mg.

FIGURE 3.18 Oxidation–reduction reactions.

3.8 Oxidation–Reduction Reactions

The terms *oxidation* and *reduction* come from reactions that have been known for centuries. Ancient civilizations learned how to change metal oxides and sulfides into the metal, that is, how to "reduce" ore to the metal. A modern example is the reduction of iron(III) oxide with carbon monoxide to give iron metal (Figure 3.18a).

Fe₂O₃ loses oxygen and is reduced.

$$Fe_2O_3(s) + 3\,CO(g) \longrightarrow 2\,Fe(s) + 3\,CO_2(g)$$

CO is the reducing agent. It gains oxygen and is oxidized.

In this reaction carbon monoxide is the agent that brings about the reduction of iron ore to iron metal, so carbon monoxide is called the **reducing agent**.

When Fe_2O_3 is reduced by carbon monoxide, oxygen is removed from the iron ore and added to the carbon monoxide. The carbon monoxide, therefore, is "oxidized" by the addition of oxygen to give carbon dioxide. *Any process in which oxygen is added to another substance is an oxidation.* In the reaction of oxygen with magnesium, for example (Figure 3.18b), oxygen is the **oxidizing agent** because it is the agent responsible for the oxidation of magnesium.

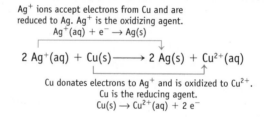

Mg combines with
oxygen and is oxidized.

$$2\ Mg(s) + O_2(g) \longrightarrow 2\ MgO(s)$$

O_2 is the oxidizing agent.

Oxidation–Reduction Reactions and Electron Transfer

The concept of oxidation–reduction reactions has been extended from these and similar examples to a vast number of other reactions, many of which do not even involve oxygen. Rather than concentrate on whether oxygen is gained or lost, let us look at what is going on with electrons during the course of the reaction. All oxidation and reduction reactions can be accounted for by considering them to occur by means of a transfer of electrons between substances. When a substance accepts electrons, it is said to be reduced because there is a reduction in the numerical value of the charge on an atom of the substance. In the net ionic equation for the reaction of a silver salt with copper metal, positively charged Ag^+ ions accept electrons from copper metal and are reduced to uncharged silver atoms (Figure 3.19).

Ag^+ ions accept electrons from Cu and are
reduced to Ag. Ag^+ is the oxidizing agent.
$$Ag^+(aq) + e^- \rightarrow Ag(s)$$

$$2\ Ag^+(aq) + Cu(s) \longrightarrow 2\ Ag(s) + Cu^{2+}(aq)$$

Cu donates electrons to Ag^+ and is oxidized to Cu^{2+}.
Cu is the reducing agent.
$$Cu(s) \rightarrow Cu^{2+}(aq) + 2\ e^-$$

Pure copper
wire

Copper wire in dilute $AgNO_3$
solution after several hours

Blue color due to
Cu^{2+} ions formed
in redox reaction

Silver crystals
formed after
several weeks

FIGURE 3.19 The oxidation of copper metal by silver ions. A clean piece of copper wire is placed in a solution of silver nitrate, $AgNO_3$. Over time, the copper reduces Ag^+ ions, forming silver crystals, and the copper metal is oxidized to copper ions, Cu^{2+}. The blue color of the solution is due to the presence of aqueous copper(II) ions.

Because copper metal supplies the electrons and causes Ag^+ ions to be reduced, Cu is the reducing agent.

When a substance *loses electrons*, the numerical value of the charge on an atom of the substance increases. The substance is said to have been **oxidized**. In our example, copper metal releases electrons on going to Cu^{2+}, so the metal is oxidized. For this to happen, something must be available to accept the electrons from copper. In this case, Ag^+ is the electron acceptor, and its charge is reduced to zero in silver metal. Therefore, Ag^+ is the "agent" that causes Cu metal to be oxidized; that is, Ag^+ is the *oxidizing agent*.

In every oxidation–reduction reaction, one reactant is reduced (and therefore is the oxidizing agent) and one reactant is oxidized (and therefore is the reducing agent). We can show this by dividing the general redox reaction $X + Y \rightarrow X^{n+} + Y^{n-}$ into two parts or *half-reactions*:

Half Reaction	Electron Transfer	Result
$X \rightarrow X^{n+} + n\,e^-$	X transfers electrons to Y	X is oxidized to X^{n+}. X is the reducing agent.
$Y + n\,e^- \rightarrow Y^{n-}$	Y accepts electrons from X	Y is reduced to Y^{n-}. Y is the oxidizing agent.

In the reaction of magnesium and oxygen, O_2 is reduced because it gains electrons (four electrons per molecule) on going to two oxide ions. Thus, because O_2 is involved in the oxidation of magnesium, it is the oxidizing agent.

Mg releases 2 e^- per atom. Mg is oxidized to Mg^{2+} and is the reducing agent.

$$2\ Mg(s) + O_2(g) \longrightarrow 2\ MgO(s)$$

O_2 gains 4 e^- per molecule to form 2 O^{2-}. O_2 is reduced and is the oxidizing agent.

In the same reaction, magnesium is the reducing agent because it releases two electrons per atom on being oxidized to the Mg^{2+} ion (and so two Mg atoms are required to supply the four electrons required by one O_2 molecule). All redox reactions can be analyzed in a similar manner.

The observations outlined so far lead to several important conclusions:

- If one substance is oxidized, another substance in the same reaction must be reduced. For this reason, such reactions are called *oxidation–reduction reactions*, or **redox reactions** for short.
- The reducing agent is itself oxidized, and the oxidizing agent is reduced.
- Oxidation is the opposite of reduction. For example, the removal of oxygen is reduction and the addition of oxygen is oxidation. The gain of electrons is reduction and the loss of electrons is oxidation.

Oxidation Numbers

How can you tell an oxidation–reduction reaction when you see one? How can you tell which substance has gained or lost electrons and so decide which substance is the oxidizing (or reducing) agent? Sometimes it is obvious. For example, if an uncombined element becomes part of a compound (Mg becomes part of MgO, for example), the reaction is definitely a redox process. If it's not obvious, then the answer is to *look for a change in the oxidation number of an element in the course of the reaction*. The **oxidation number** of an atom in a molecule or ion is defined as the charge an atom has, *or appears to have*, as determined by the following guidelines for assigning oxidation numbers.

1. **Each atom in a pure element has an oxidation number of zero.** The oxidation number of Cu in metallic copper is 0, and it is 0 for each atom in I_2 and S_8.
2. **For monatomic ions, the oxidation number is equal to the charge on the ion.** You know that magnesium forms ions with a 2+ charge (Mg^{2+}); the oxidation number of magnesium in this ion is therefore +2.

• **Balancing Equations for Redox Reactions** The notion that a redox reaction can be divided into a reduction portion and an oxidation portion will lead us to a method of balancing more complex equations for redox reactions described in Chapter 20.

• **Writing Charges on Ions** Conventionally, charges on ions are written as (number, sign), whereas oxidation numbers are written as (sign, number). For example, the oxidation number of the Cu^{2+} ion is +2 and its charge is 2+.

A CLOSER LOOK

Are Oxidation Numbers "Real"?

Do oxidation numbers reflect the actual electric charge on an atom in a molecule or ion? With the exception of monatomic ions such as Cl^- or Na^+, the answer is no.

Oxidation numbers assume that the atoms in a molecule are positive or negative ions, which is not true. For example, in H_2O, the H atoms are not H^+ ions and the O atoms are not O^{2-} ions. This is not to say, however, that atoms in molecules do not bear an electric charge of any kind. Calculations on water indicate the O atom

has a charge of about 0.4– (or 40% of the electron charge) and the H atoms are each about 0.2+.

So why use oxidation numbers? Oxidation numbers provide a way of dividing up the electrons among the atoms in a molecule or polyatomic ion. Because the distribution of electrons changes in a redox reaction, we use this method as a way to decide whether a redox reaction has

occurred and to distinguish the oxidizing and reducing agents.

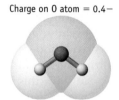

Charge on O atom = 0.4–

Charge on each H atom = 0.2+

• **Peroxides** In peroxides, the oxidation number of oxygen is −1. For example, in hydrogen peroxide (H_2O_2), each hydrogen atom has an oxidation number of +1. To balance this, each oxygen must have an oxidation number of −1. A 3% aqueous solution of H_2O_2 is sometimes used as an antiseptic because it is a good oxidizing agent.

3. **When combined with another element, fluorine always has an oxidation number of −1.**

4. **The oxidation number of O is −2 in most compounds.** The exceptions to this rule occur
   (a) when oxygen is combined with fluorine (where oxygen takes on a positive oxidation number),
   (b) in compounds called *peroxides* (such as Na_2O_2) and *superoxides* (such as KO_2) in which oxygen has an oxidation number of −1 and −1/2, respectively.

5. **Cl, Br, and I have oxidation numbers of −1 in compounds, except when combined with oxygen and fluorine.** This means that Cl has an oxidation number of −1 in NaCl (in which Na's oxidation number is +1, as predicted by the fact that it is an element of Group 1A). In the ion ClO^-, however, the Cl atom has an oxidation number of +1 (and O has an oxidation number of −2; see Guideline 4).

6. **The oxidation number of H is +1 in most compounds.** The key exception to this guideline occurs when H forms a binary compound with a metal. In such cases, the metal forms a positive ion and H becomes a hydride ion, H^-. Thus, in CaH_2 the oxidation number of Ca is +2 (equal to the group number) and that of H is −1.

7. **The algebraic sum of the oxidation numbers for the atoms in a neutral compound must be zero; in a polyatomic ion, the sum must be equal to the ion charge.** For example, in $HClO_4$ the H atom is assigned +1 and each O atom is assigned −2. This means the Cl atom must be +7.

EXAMPLE 3.8   Determining Oxidation Numbers

Problem Determine the oxidation number of the indicated element in each of the following compounds or ions:

(a) aluminum in aluminum oxide, Al_2O_3 (c) sulfur in the sulfate ion, SO_4^{2-}

(b) phosphorus in phosphoric acid, H_3PO_4 (d) each Cr atom in the dichromate ion, $Cr_2O_7^{2-}$

What Do You Know? Correct formulas for each species are given.

Strategy Follow the guidelines in the text, paying particular attention to Guidelines 4, 6, and 7, above.

Solution

(a) Al_2O_3 is a neutral compound. Assuming that oxygen has its usual oxidation number of −2, we can solve the following algebraic equation for the oxidation number of aluminum.

Net charge on Al_2O_3 = sum of oxidation numbers for two Al atoms + three O atoms

$$0 = 2(x) + 3(-2) \text{ and so } x = +3$$

The oxidation number of Al must be +3, in agreement with its position in the periodic table.

(b) H_3PO_4 has an overall charge of 0. If each of the oxygen atoms has an oxidation number of −2 and each of the H atoms is +1, then we can determine the oxidation number of phosphorus as follows:

Net charge on H_3PO_4 = sum of oxidation numbers for three H atoms + one P atom + four O atoms

$$0 = 3(+1) + (x) + 4(-2) \text{ and so } x = +5$$

The oxidation number of phosphorus in this compound is +5.

(c) The sulfate ion, SO_4^{2-}, has an overall charge of 2−. Oxygen is assigned its usual oxidation number of −2.

Net charge on SO_4^{2-} = sum of oxidation number of one S atom + four O atoms

$$-2 = (x) + 4(-2) \text{ and so } x = +6$$

The sulfur in this ion has an oxidation number of +6.

(d) The net charge on the $Cr_2O_7^{2-}$ ion is 2−. Oxygen is assigned its usual oxidation number of −2.

Net charge on $Cr_2O_7^{2-}$ = sum of oxidation numbers for two Cr atoms + seven O atoms

$$-2 = 2(x) + 7(-2) \text{ and so } x = +6$$

The oxidation number of each chromium in this polyatomic ion is +6.

Think about Your Answer Notice that in each of these examples, the oxidation number matched the number of the group in which the element is found. This is often (but not always) the case.

Check Your Understanding

Assign an oxidation number to the underlined atom in each ion or molecule.

(a) $\underline{Fe}_2O_3$ (b) $H_2\underline{S}O_4$ (c) $\underline{C}O_3^{2-}$ (d) $\underline{N}O_2^+$

Recognizing Oxidation–Reduction Reactions

You can always tell whether a reaction involves oxidation and reduction by assessing the oxidation number of each element and noting whether any of these numbers change in the course of the reaction. In many cases, however, this will not be necessary. For example, it will be obvious that a redox reaction has occurred if an uncombined element is converted to a compound or if a well-known oxidizing or reducing agent is involved (Table 3.3).

Like oxygen (O_2), the halogens (F_2, Cl_2, Br_2, and I_2) are oxidizing agents in their reactions with metals and nonmetals. An example is the reaction of chlorine with sodium metal (◀ Figure 1.2).

Na releases 1 e⁻ per atom.
Oxidation number increases.
Na is oxidized to Na^+ and is the reducing agent.

$$2 \text{ Na(s)} + \text{Cl}_2(g) \longrightarrow 2 \text{ NaCl(s)}$$

Cl_2 gains 2 e⁻ per molecule.
Oxidation number decreases by 1 per Cl.
Cl_2 is reduced to Cl^- and is the oxidizing agent.

NO$_2$ gas

Copper metal oxidized to green Cu(NO$_3$)$_2$

FIGURE 3.20 The reaction of copper with nitric acid. Copper (a reducing agent) reacts vigorously with concentrated nitric acid (an oxidizing agent) to give the brown gas NO$_2$ and a deep blue-green solution of copper(II) nitrate.

Table 3.3 **Common Oxidizing and Reducing Agents**

Oxidizing Agent	Reaction Product	Reducing Agent	Reaction Product
O$_2$, oxygen	O^{2-}, oxide ion or O combined in H$_2$O or other molecule	H$_2$, hydrogen	H$^+$(aq), hydrogen ion or H combined in H$_2$O or other molecule
Halogen, F$_2$, Cl$_2$, Br$_2$, or I$_2$	Halide ion, F$^-$, Cl$^-$, Br$^-$, or I$^-$	M, metals such as Na, K, Fe, and Al	M^{n+}, metal ions such as Na$^+$, K$^+$, Fe^{2+} or Fe^{3+}, and Al^{3+}
HNO$_3$, nitric acid	Nitrogen oxides* such as NO and NO$_2$	C, carbon (used to reduce metal oxides)	CO and CO$_2$
Cr$_2$O$_7$$^{2-}$, dichromate ion	Cr^{3+}, chromium(III) ion (in acid solution)		
MnO$_4$$^-$, permanganate ion	Mn^{2+}, manganese(II) ion (in acid solution)		

*NO is produced with dilute HNO$_3$, whereas NO$_2$ is a product of concentrated acid.

A chlorine molecule ends up as two Cl$^-$ ions, each atom having acquired an electron (from a Na atom). Thus, the oxidation number of each Cl atom has decreased from 0 to -1. This means Cl$_2$ has been reduced and therefore is the oxidizing agent.

Figure 3.20 illustrates the chemistry of another excellent oxidizing agent, nitric acid, HNO$_3$. Here, copper metal is oxidized to give copper(II) nitrate, and the nitrate ion is reduced to the brown gas NO$_2$. The net ionic equation for the reaction is

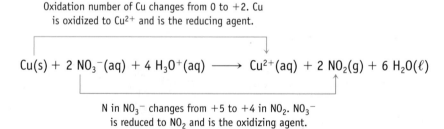

Oxidation number of Cu changes from 0 to +2. Cu is oxidized to Cu^{2+} and is the reducing agent.

$$Cu(s) + 2\ NO_3^-(aq) + 4\ H_3O^+(aq) \longrightarrow Cu^{2+}(aq) + 2\ NO_2(g) + 6\ H_2O(\ell)$$

N in NO$_3$$^-$ changes from +5 to +4 in NO$_2$. NO$_3$$^-$ is reduced to NO$_2$ and is the oxidizing agent.

Nitrogen has been reduced from +5 (in the NO$_3$$^-$ ion) to +4 (in NO$_2$); therefore, the nitrate ion in acid solution is the oxidizing agent. Copper metal is the reducing agent; each metal atom has given up two electrons to produce the Cu^{2+} ion.

In the reactions of sodium with chlorine and copper with nitric acid, the metals are oxidized. This is typical of metals, which characteristically lose electrons in chemical reactions. In yet another example of this, aluminum metal, a good reducing agent, is capable of reducing iron(III) oxide to iron metal in a reaction called the *thermite reaction* (Figure 3.21).

$$\underset{\text{oxidizing agent}}{Fe_2O_3(s)} +\ \underset{\text{reducing agent}}{2\ Al(s)} \rightarrow 2\ Fe(\ell) + Al_2O_3(s)$$

Such a large quantity of energy is evolved as heat in the reaction that the iron produced is in the molten state.

Tables 3.3 and 3.4 may help you organize your thinking as you look for oxidation–reduction reactions and use their terminology.

FIGURE 3.21 Thermite reaction. Here, Fe$_2$O$_3$ is reduced by aluminum metal to produce iron metal and aluminum oxide.

Table 3.4 Recognizing Oxidation–Reduction Reactions

	Oxidation	Reduction
In terms of oxidation number	Increase in oxidation number of an atom	Decrease in oxidation number of an atom
In terms of electrons	Loss of electrons by an atom	Gain of electrons by an atom
In terms of oxygen	Gain of one or more O atoms	Loss of one or more O atoms

EXAMPLE 3.9   Oxidation–Reduction Reaction

Problem For the reaction of the iron(II) ion with permanganate ion in aqueous acid,

$$5\ Fe^{2+}(aq) + MnO_4^-(aq) + 8\ H_3O^+(aq) \rightarrow 5\ Fe^{3+}(aq) + Mn^{2+}(aq) + 12\ H_2O(\ell)$$

decide which atoms are undergoing a change in oxidation number and identify the oxidizing and reducing agents.

What Do You Know? The equation given here is balanced (but, for practice, you might verify this). From the reactants and products (MnO_4^- is a common oxidizing agent; iron changes from Fe^{2+} to Fe^{3+}) you can quickly decide this is an oxidation–reduction reaction.

Strategy Determine the oxidation number of the atoms in each molecule or ion in the equation and identify which atoms change oxidation number.

Solution The Mn oxidation number in MnO_4^- is +7, and it decreases to +2 in the product, the Mn^{2+} ion. Thus, the MnO_4^- ion has been reduced and is the oxidizing agent (Table 3.3).

$$5\ Fe^{2+}(aq) + MnO_4^-(aq) + 8\ H_3O^+(aq) \rightarrow 5\ Fe^{3+}(aq) + Mn^{2+}(aq) + 12\ H_2O(\ell)$$
$$+2+7,\ -2+1,\ -2+3+2+1,\ -2$$

The oxidation number of iron has increased from +2 to +3, so each Fe^{2+} ion has lost one electron upon being oxidized to Fe^{3+} (Table 3.4). This means the Fe^{2+} ion is the reducing agent.

Think about Your Answer If one of the reactants in a redox reaction is a simple substance (here Fe^{2+}), it usually is obvious whether its oxidation number has increased or decreased. Once a species has been established as having been reduced (or oxidized), you know another species has undergone the opposite process. It is also helpful to recognize common oxidizing and reducing agents (Table 3.3).

KMnO₄(aq) oxidizing agent

Fe²⁺(aq) reducing agent

The reaction of iron(II) ion and permanganate ion. The reaction of purple permanganate ion (MnO_4^-, the oxidizing agent) with the iron(II) ion (Fe^{2+}, the reducing agent) in acidified aqueous solution gives the nearly colorless manganese(II) ion (Mn^{2+}) and the iron(III) ion (Fe^{3+}).

Check Your Understanding

The following reaction occurs in a device for testing the breath of a person for the presence of ethanol. Identify the oxidizing and reducing agents, the substance oxidized, and the substance reduced.

$$3\ CH_3CH_2OH(aq) + 2\ Cr_2O_7^{2-}(aq) + 16\ H_3O^+(aq) \rightarrow 3\ CH_3CO_2H(aq) + 4\ Cr^{3+}(aq) + 27\ H_2O(\ell)$$

ethanol   dichromate ion;   acetic acid   chromium(III) ion;
                 orange-red   green

REVIEW & CHECK FOR SECTION 3.8

1. What is the oxidation number of Cr in $(NH_4)_2Cr_2O_7$?

    (a) 0   (b) +3   (c) +6   (d) +12

2. Which of the following reactions are oxidation–reduction reactions?

    (1) $NaOH(aq) + HNO_3(aq) \rightarrow NaNO_3(aq) + H_2O(\ell)$

    (2) $Cu(s) + Cl_2(g) \rightarrow CuCl_2(s)$

    (3) $Na_2CO_3(aq) + 2\ HClO_4(aq) \rightarrow CO_2(g) + H_2O(\ell) + 2\ NaClO_4(aq)$

    (4) $2\ S_2O_3^{2-}(aq) + I_2(aq) \rightarrow S_4O_6^{2-}(aq) + 2\ I^-(aq)$

    (a) all of them   (b) 2, 3, and 4   (c) 2 and 4   (d) 2 only

FIGURE 3.22 A gas-forming reaction. An Alka-Seltzer tablet contains an acid (citric acid) and sodium hydrogen carbonate ($NaHCO_3$), the reactants in a gas-forming reaction.

3. In which of the manganese compounds below does Mn have the highest oxidation number?

   (a) MnO_2 (b) Mn_2O_3 (c) $KMnO_4$ (d) $MnCl_2$

3.9 Classifying Reactions in Aqueous Solution

One goal of this chapter has been to explore the most common types of reactions that can occur in aqueous solution. This helps you decide, for example, that a gas-forming reaction occurs when an Alka-Seltzer tablet (containing citric acid and $NaHCO_3$) is dropped into water (Figure 3.22).

$$\underset{\text{citric acid}}{H_3C_6H_5O_7(aq)} + \underset{\text{hydrogen carbonate ion}}{HCO_3^-(aq)} \longrightarrow$$

$$\underset{\text{dihydrogen citrate ion}}{H_2C_6H_5O_7^-(aq)} + H_2O(\ell) + CO_2(g)$$

We have examined four types of reactions in aqueous solution: precipitation reactions, acid–base reactions, gas-forming reactions, and oxidation–reduction reactions. Three of these four (precipitation, acid–base, and gas-forming) fall into the category of exchange reactions.

Precipitation Reactions (Figure 3.12): Ions combine in solution to form an insoluble reaction product.

Overall Equation

$$Pb(NO_3)_2(aq) + 2\ KI(aq) \rightarrow PbI_2(s) + 2\ KNO_3(aq)$$

Net Ionic Equation

$$Pb^{2+}(aq) + 2\ I^-(aq) \rightarrow PbI_2(s)$$

CASE STUDY

Killing Bacteria with Silver

There is a washing machine on the market that injects silver ions into the wash water, the purpose being to kill bacteria in the wash water. The advertisement for the machine tells us that 100 quadrillion silver ions are injected into a wash load. How does this machine work? Is 100 quadrillion silver ions a lot? Does the silver kill bacteria?

The washing machine works by using electrical energy to oxidize silver metal to give silver ions.

$$Ag(s) \rightarrow Ag^+(aq) + e^-$$

This is a simple electrolysis procedure. (More about that in Chapter 20.) And 100 quadrillion silver ions? This is 100×10^{15} ions.

Do silver ions act as a bacteriocide? There is plenty of medical evidence for this property. In fact, when you were born, the physician or nurse may have put drops of a very dilute silver nitrate solution in your eyes to treat neonatal conjunctivitis. And severely burned patients are treated with silver sulfadiazine ($C_{10}H_9AgN_4O_2S$) to prevent bacterial or fungal infections.

Silver ions as a bacteriocide in dental floss.

The use of silver to prevent infections has a long history. Phoenicians kept wine, water, and vinegar in silver vessels. Early settlers in America put silver coins into water barrels. And you might have been born with a "silver spoon in your mouth." Babies in wealthier homes, who were fed from silver spoons and used a silver pacifier, were found to be healthier.

The historical uses of silver carry over to modern society. One can buy many different kinds of silver-containing water purifiers for the home, and dental floss coated with silver nitrate is available.

Although silver does have health benefits, beware of fraudulent claims. For example, consuming large amounts of "colloidal silver" (nanosized particles of silver suspended in water) is claimed to have health-giving properties. One person who tried this, Stan Jones, ran for the U.S. Congress in 2002 and 2006. From consuming silver, he acquired argyria, a medically irreversible condition in which the skin turns a gray-blue color.

Questions:
1. How many moles of silver are used in a wash cycle?
2. What mass of silver is used?

Answers to these questions are available in Appendix N.

Acid–Base Reactions (Figures 3.13 and 3.15): Water is a product of many acid–base reactions, and the cation of the base and the anion of the acid form a salt.

Overall Equation for the Reaction of a Strong Acid and a Strong Base

$$HNO_3(aq) + KOH(aq) \rightarrow HOH(\ell) + KNO_3(aq)$$

Net Ionic Equation for the Reaction of a Strong Acid and a Strong Base

$$H_3O^+(aq) + OH^-(aq) \rightarrow 2\,H_2O(\ell)$$

Overall Equation for the Reaction of a Weak Acid and a Strong Base

$$CH_3CO_2H(aq) + NaOH(aq) \rightarrow NaCH_3CO_2(aq) + HOH(\ell)$$

Net Ionic Equation for the Reaction of a Weak Acid and a Strong Base

$$CH_3CO_2H(aq) + OH^-(aq) \rightarrow CH_3CO_2^-(aq) + H_2O(\ell)$$

Gas-Forming Reactions (Figures 3.16 and 3.22): The most common examples involve metal carbonates and acids but others exist (Table 3.2). One product with a metal carbonate is always carbonic acid, H_2CO_3, most of which decomposes to H_2O and CO_2. Carbon dioxide is the gas in the bubbles you see during these reactions.

Overall Equation:

$$CuCO_3(s) + 2\,HNO_3(aq) \rightarrow Cu(NO_3)_2(aq) + CO_2(g) + H_2O(\ell)$$

Net Ionic Equation

$$CuCO_3(s) + 2\,H_3O^+(aq) \rightarrow Cu^{2+}(aq) + CO_2(g) + 3\,H_2O(\ell)$$

Oxidation–Reduction Reactions (Figure 3.19): These reactions are *not* ion exchange reactions. Rather, electrons are transferred from one material to another.

Overall Equation

$$Cu(s) + 2\,AgNO_3(aq) \rightarrow Cu(NO_3)_2(aq) + 2\,Ag(s)$$

Net Ionic Equation

$$Cu(s) + 2\,Ag^+(aq) \rightarrow Cu^{2+}(aq) + 2\,Ag(s)$$

These four types of reactions usually are easy to recognize, but keep in mind that a reaction may fall into more than one category. For example, barium hydroxide reacts readily with sulfuric acid to give barium sulfate and water, a reaction that is both a precipitation and an acid–base reaction.

$$Ba(OH)_2(aq) + H_2SO_4(aq) \rightarrow BaSO_4(s) + 2\,H_2O(\ell)$$

⏻WL   INTERACTIVE EXAMPLE 3.10   Types of Reactions

Problem Complete and balance each of the following equations and classify each as a precipitation, acid–base, or gas-forming reaction.

(a) $Na_2S(aq) + Cu(NO_3)_2(aq) \rightarrow$

(b) $Na_2SO_3(aq) + HCl(aq) \rightarrow$

(c) $HClO_4(aq) + NaOH(aq) \rightarrow$

What Do You Know? You know the formulas for the reactants and that these are all exchange reactions.

Strategy Map 3.10a

PROBLEM
Write equation for the reaction of **Na₂S** and **Cu(NO₃)₂** and decide on *reaction type*

↓

DATA/INFORMATION
The **formulas** of the **reactants** are given

↓

> STEP 1. Decide on **products** and then write **complete balanced equation.**

↓

Complete balanced equation with *reactants* and *products*

↓

> STEP 2. Decide if each reactant and product is **solid, liquid, gas,** or **dissolved in water.**

↓

Complete balanced equation with **indication of state** of each reactant and product

↓

> STEP 3. Decide on **reaction type.**

↓

One product is *insoluble in water,* so this is a **precipitation reaction.**

Strategy

• Recognize that these are exchange reactions. The products of each reaction can be found by exchanging cations and anions between the two reactants.

• Write and balance each equation.

• To determine the kind of reaction examine the reactants and products. Look specifically for common acids and bases, for a product that is insoluble, and for anions that react with acid to give a gas (CO_3^{2-}, S^{2-}, SO_3^{2-}).

Solution

(a) The products of the exchange reaction are predicted to be $NaNO_3$ and CuS. The first of these is water-soluble, but the second is an insoluble salt. Thus, this is a precipitation reaction. The balanced chemical equation is

$$Na_2S(aq) + Cu(NO_3)_2(aq) \rightarrow 2\ NaNO_3(aq) + CuS(s)$$

(b) The products of the exchange reaction are predicted to be NaCl and H_2SO_3. The H_2SO_3 should immediately alert us to the fact that this is a gas-forming reaction because it will decompose into $SO_2(g)$ and $H_2O(\ell)$ (Table 3.2). The balanced equation is

$$Na_2SO_3(aq) + 2\ HCl\ (aq) \rightarrow 2\ NaCl(aq) + SO_2\ (g) + H_2O(\ell)$$

(c) The products of the exchange reaction are predicted to be $NaClO_4$ and H_2O, a salt and water; this is an acid–base reaction. The balanced equation is

$$HClO_4(aq) + NaOH(aq) \rightarrow NaClO_4(aq) + H_2O(\ell)$$

Think about Your Answer As practice, try writing the net ionic equations for each of the preceding reactions. The answers are:

(a) $S^{2-}(aq) + Cu^{2+}(aq) \rightarrow CuS(s)$

(b) $SO_3^{2-}(aq) + 2\ H_3O^+(aq) \rightarrow 3\ H_2O(\ell) + SO_2(g)$

(c) $H_3O^+(aq) + OH^-(aq) \rightarrow 2\ H_2O(\ell)$

Check Your Understanding

Classify each of the following reactions as a precipitation, acid–base, gas-forming, or oxidation–reduction reaction. Predict the products of the reaction, and then balance the completed equation. Write the net ionic equation for each.

(a) $CuCO_3(s) + H_2SO_4(aq) \rightarrow$

(b) $Ga(s) + O_2(g) \rightarrow$

(c) $Ba(OH)_2(s) + HNO_3(aq) \rightarrow$

(d) $CuCl_2(aq) + (NH_4)_2S(aq) \rightarrow$

REVIEW & CHECK FOR SECTION 3.9

Sometimes a reaction can fall in more than one category. Into what category (or categories) does the reaction of $Ba(OH)_2(aq) + H_3PO_4(aq)$ fit?

(a) acid–base and oxidation–reduction

(b) oxidation–reduction

(c) acid–base and precipitation

(d) precipitation

CHAPTER GOALS REVISITED

Now that you have studied this chapter, you should ask if you have met the chapter goals. In particular, you should be able to

Balance equations for simple chemical reactions
a. Understand the information conveyed by a balanced chemical equation (Section 3.1).
b. Balance simple chemical equations (Section 3.2). **Study Questions: 1, 3, 5, 51, 60.**

Understand the nature and characteristics of chemical equilibria
a. Recognize that chemical reactions are reversible (Section 3.3).
b. Describe what is meant by the term *dynamic equilibrium.* **Study Question: 80.**
c. Recognize the difference between reactant-favored and product-favored reactions at equilibrium. **Study Questions: 7, 15.**

Understand the nature of ionic substances dissolved in water
a. Explain the difference between electrolytes and nonelectrolytes and recognize examples of each (Section 3.4 and Figure 3.8). **Study Question: 9.**
b. Predict the solubility of ionic compounds in water (Sections 3.4 and Figure 3.10). **Study Questions: 11, 13, 15, 61, 63, 66, 72, and Go Chemistry Module 5.**
c. Recognize what ions are formed when an ionic compound or acid or base dissolves in water (Sections 3.4–3.6). **Study Questions: 13, 15, 66, 69, 77.**

Recognize common acids and bases, and understand their behavior in aqueous solution
a. Know the names and formulas of common acids and bases (Section 3.6 and Table 3.1). **Study Questions: 21, 27, 30.**
b. Categorize acids and bases as strong or weak.
c. Define and use the Arrhenius concept of acids and bases.
d. Define and use the Brønsted–Lowry concept of acids and bases.
e. Recognize substances that are amphiprotic. **Study Question: 33.**
f. Identify the Brønsted acid and base in a reaction. **Study Questions: 21, 27, 28.**

Recognize the common types of reactions in aqueous solution
a. Recognize the key characteristics of four types of reactions in aqueous solution.

Reaction Type	Key Characteristic
Precipitation	Formation of an insoluble compound
Acid–base	Formation of a salt and water
Gas-forming	Evolution of a water-insoluble gas such as CO_2
Oxidation–reduction	Transfer of electrons (with changes in oxidation numbers)

Study Questions: 51, 52, 55, 56, 82–84.
b. Predict the products of precipitation reactions (Section 3.5), acid–base reactions (Section 3.6), and gas-forming reactions (Section 3.7). These are all examples of exchange reactions, which involve the exchange of anions between the cations involved in the reaction. **Study Questions: 17, 19, 27, 29, 41–43, 83.**

Write chemical equations for the common types of reactions in aqueous solution
a. Write overall balanced equations for precipitation, acid–base, and gas-forming reactions. **Study Questions: 17, 19, 27, 29, 41, 43, 77.**
b. Write net ionic equations (Sections 3.5–3.7). **Study Questions: 17, 19, 35, 37, 39, 69, 73, 74, and Go Chemistry Module 6.**
c. Understand that the net ionic equation for the reaction of a strong acid with a strong base is $H_3O^+(aq) + OH^-(aq) \rightarrow 2\,H_2O(\ell)$ (Section 3.6).

Sign in at **www.cengage.com/owl** to:
- View tutorials and simulations, develop problem-solving skills, and complete online homework assigned by your professor.
- For quick review and exam prep, download Go Chemistry mini lecture modules from OWL (or purchase them at **www.cengagebrain.com**)

Access **How Do I Solve It?** tutorials on how to approach problem solving using concepts in this chapter.

Recognize common oxidizing and reducing agents and identify oxidation–reduction reactions

a. Determine oxidation numbers of elements in a compound and understand that these numbers represent the charge an atom has, or appears to have, when the electrons of the compound are counted according to a set of guidelines (Section 3.8). **Study Questions: 45, 46, 81.**

b. Identify oxidation–reduction reactions (often called *redox reactions*) and identify the oxidizing and reducing agents and substances oxidized and reduced in the reaction (Section 3.8 and Tables 3.3 and 3.4). **Study Questions: 47–50, 67, 85.**

ⓌWL Questions and problems for this chapter are available in OWL. Use the OWL access card that came with this textbook to access assigned questions and problems for this chapter.

4 Stoichiometry: Quantitative Information about Chemical Reactions

A sparkler.

The Chemistry of Pyrotechnics

One of life's pleasures is to watch fireworks, but of course we usually don't think about this as being just an application of chemistry. The sparkler in the photograph, like many forms of fireworks, depends on some very straightforward chemistry. It consists of a mixture of finely powdered metals, such as Al or Fe, other substances such as $KClO_3$, and a binder that holds the mixture onto a wire handle. When ignited, the powdered metal reacts with oxygen in the air (or extracted from $KClO_3$) and the sparks fly!

$$4\ Fe(s) + 3\ O_2(g) \longrightarrow 2\ Fe_2O_3(s)$$

Thermite reaction.

Even more spectacular than the sparkler is the thermite reaction, the reduction of iron(III) oxide with metallic aluminum.

$$Fe_2O_3(s) + 2\ Al(s) \longrightarrow 2\ Fe(\ell) + Al_2O_3(s)$$

In the sparkler, the oxygen that combines with the iron comes from the air. In the thermite reaction, however, the oxygen of iron(III) oxide is transferred to the reducing agent, the aluminum metal.

These reactions generate an enormous amount of energy, sufficient to produce iron in the molten state. Although the thermite reaction in particular was originally developed as a way of producing metals from their oxides (copper, uranium, and others also work well), it was quickly realized that it can be used for welding. For some years it was used to weld the rails of train tracks.

Questions:

1. If a sparkler contains 1.0 g of iron, what is the mass of the product of the reaction of iron with oxygen?
2. What mass of aluminum should be used to completely consume 10.0 g of iron(III) oxide in the thermite reaction?
3. What is the limiting reactant if you combine 20 g of aluminum metal with 10 g of iron(III) oxide?

Answers to these questions are available in Appendix N.

CHAPTER OUTLINE

4.1 Mass Relationships in Chemical Reactions: Stoichiometry [go Chemistry]

4.2 Reactions in Which One Reactant Is Present in Limited Supply [go Chemistry]

4.3 Percent Yield

4.4 Chemical Equations and Chemical Analysis

4.5 Measuring Concentrations of Compounds in Solution

4.6 pH, a Concentration Scale for Acids and Bases [go Chemistry]

4.7 Stoichiometry of Reactions in Aqueous Solution

4.8 Spectrophotometry

CHAPTER GOALS

See Chapter Goals Revisited (page 169) for Study Questions keyed to these goals.

- Perform stoichiometry calculations using balanced chemical equations.

- Understand the meaning of a limiting reactant in a chemical reaction.

- Calculate the theoretical and percent yields of a chemical reaction.

- Use stoichiometry to analyze a mixture of compounds or to determine the formula of a compound.

- Define and use concentrations in solution stoichiometry.

The objective of this chapter is to introduce the quantitative study of chemical reactions. Quantitative studies are needed to determine, for example, how much oxygen is required for the complete combustion of a given quantity of gasoline and what masses of carbon dioxide and water can be obtained. This part of chemistry is fundamental to much of what chemists, chemical engineers, biochemists, molecular biologists, geochemists, and many others do.

4.1 Mass Relationships in Chemical Reactions: Stoichiometry

A balanced chemical equation shows the quantitative relationship between reactants and products in a chemical reaction. On the submicroscopic scale, the stoichiometric coefficients in the equation refer to atoms, molecules, or formula units. On the macroscopic level on which we work in the laboratory, the coefficients refer to the number of moles of each reactant and product. The coefficients in a balanced chemical equation thus allow us to relate the amount (moles) of one substance involved in a chemical reaction to the amount (moles) of another substance involved in that chemical reaction.

Let us apply this concept to the reaction of phosphorus and chlorine in Figure 3.1.

$$P_4(s) + 6\ Cl_2(g) \rightarrow 4\ PCl_3(\ell)$$

1 mol	6 mol	4 mol
124 g	425 g	549 g

The chemical equation for this reaction means that 1 mol (124 g) of solid phosphorus (P_4) can react with 6 mol (425 g) of chlorine gas (Cl_2) to form 4 mol (549 g) of liquid phosphorus trichloride (PCl_3).

Now, suppose you want to use much less P_4 in the reaction, only 1.45 g. What amount of Cl_2 gas would be required and what amount of PCl_3 could be produced? This is an example of a situation common in chemistry and chemical engineering, so let us work carefully through the steps you should follow when solving a stoichiometry problem.

⊖WL

Sign in to OWL at **www.cengage.com/owl** to view tutorials and simulations, develop problem-solving skills, and complete online homework assigned by your professor.

[go Chemistry]

Download mini lecture videos for key concept review and exam prep from OWL or purchase them from **www.cengagebrain.com**

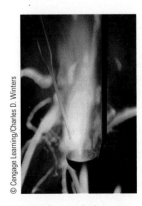

© Cengage Learning/Charles D. Winters

Reaction of P_4 and Cl_2. When white phosphorus, P_4, is placed in a flask of Cl_2, a reaction spontaneously occurs (see also Figure 3.1).

● **Stoichiometry and Reaction Completeness** *Stoichiometry calculations assume reactions are complete.* That is, *all* of at least one of the reactants has been converted to the product or products. Recall from the discussion of chemical equilibrium in Chapter 3 that some reactions do not reach completion, however.

Modules 7: Simple Stoichiometry and 8: Limiting Reactants cover concepts in this section.

● **Stoichiometry and Recipes** In one sense a balanced equation is a "recipe" for a chemical reaction: how much of one chemical is needed to consume another to produce a desired product. Food analogies have often been used in teaching stoichiometry. See, for example, "Learning Stoichiometry with Hamburger Sandwiches," by L. Haim, E. Cortón, S. Kocmur, and L. Galagovsky, *Journal of Chemical Education, Vol. 80,* pp. 1021-1022, 2003.

● **Mass Balance and Moles of Reactants and Products** Mass is always conserved in chemical reactions. Assuming complete conversion of reactants to products, the total mass of the reactants is always the same as that of the products. This does not mean, however, that the total *amount* (moles) of reactants is the same as that of the products. Atoms are rearranged into different "units" (molecules) in the course of a reaction. In the $P_4 + Cl_2$ reaction, 7 mol of reactants gives 4 mol of product.

Part (a): Calculate the mass of Cl_2 required by 1.45 g of P_4

Step 1. *Write the balanced equation* (using correct formulas for reactants and products). This is *always* the first step when dealing with chemical reactions.

$$P_4(s) + 6\ Cl_2(g) \rightarrow 4\ PCl_3(\ell)$$

Step 2. *Calculate amount (moles) from mass (grams).* Recall that chemical equations reflect the relative amounts of reactants and products, not their masses. Therefore, first calculate the amount (moles) of P_4 available.

$$1.45\ \text{g } P_4 \times \frac{1\ \text{mol } P_4}{123.9\ \text{g } P_4} = 0.0117\ \text{mol } P_4$$

$$\uparrow$$

1/molar mass of P_4

Step 3a. *Use a stoichiometric factor.* The next step is to use the balanced equation to relate the amount of P_4 available to the amount of Cl_2 required to completely consume the P_4. This relationship is a **stoichiometric factor**, a mole ratio based on the stoichiometric coefficients for the chemicals in the balanced equation. Here the balanced equation specifies that 6 mol of Cl_2 is required for each mole of P_4, so the stoichiometric factor is (6 mol Cl_2/1 mol P_4).

$$0.0117\ \text{mol } P_4 \times \frac{6\ \text{mol } Cl_2\ \text{required}}{1\ \text{mol } P_4\ \text{available}} = 0.0702\ \text{mol } Cl_2\ \text{required}$$

$$\uparrow$$

stoichiometric factor from balanced equation

This calculation shows that 0.0702 mol of Cl_2 is required to react with all the available phosphorus (1.45 g, 0.0117 mol).

Step 4a. *Calculate mass from amount.* Convert the amount (moles) of Cl_2 calculated in Step 3 to the mass of Cl_2 required.

$$0.0702\ \text{mol } Cl_2 \times \frac{70.91\ \text{g } Cl_2}{1\ \text{mol } Cl_2} = 4.98\ \text{g } Cl_2$$

Part (b) Calculate mass of PCl_3 produced from 1.45 g of P_4 and 4.98 g of Cl_2

What mass of PCl_3 can be produced from the reaction of 1.45 g of phosphorus with 4.98 g of Cl_2? From part (a), we know that these masses are the correct quantities needed for complete reaction. Because mass is conserved, the answer can be obtained by adding the masses of P_4 and Cl_2 used (giving 1.45 g + 4.98 g = 6.43 g of PCl_3 produced). Alternatively, Steps 3 and 4 can be repeated, but with the appropriate stoichiometric factor and molar mass.

Step 3b. *Use a stoichiometric factor.* Convert the amount of available P_4 to the amount of PCl_3 produced. Here the balanced equation specifies that 4 mol of PCl_3 is produced for each mole of P_4 used, so the stoichiometric factor is (4 mol PCl_3/1 mol P_4).

$$0.0117\ \text{mol } P_4 \times \frac{4\ \text{mol } PCl_3\ \text{produced}}{1\ \text{mol } P_4\ \text{available}} = 0.0468\ \text{mol } PCl_3\ \text{produced}$$

$$\uparrow$$

stoichiometric factor from balanced equation

Step 4b. *Calculate the mass of product from its amount.* Convert the amount of PCl_3 produced to its mass in grams.

$$0.0468\ \text{mol } PCl_3 \times \frac{137.3\ \text{g } PCl_3}{1\ \text{mol } PCl_3} = 6.43\ \text{g } PCl_3$$

Stoichiometry Calculations

You are asked to determine what mass of product can be formed from a given mass of reactant. It is not possible to calculate the mass of product in a single step. Instead, you must follow a route such as that illustrated in the strategy map here for the reaction of reactant A to give the product B according to an equation such as $x A \rightarrow y B$.

- The mass (g) of reactant A is converted to the amount (moles) of A using the molar mass of A.
- Next, using the stoichiometric factor, you find the amount (moles) of B.

- Finally, the mass (g) of B is obtained by multiplying amount of B by its molar mass.

When solving a stoichiometry problem, remember that you will always use a stoichiometric factor at some point.

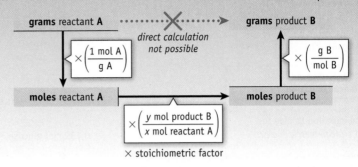

The mole relationships of reactants and products in a reaction can be summarized in an **amounts table**, and you may find these tables helpful in organizing information in stoichiometry problems. The balanced chemical equation is written across the top. The next three lines indicate the following:

- *Initial* amount (moles) of each reactant and product present
- *Change* in amount that occurs during the reaction
- *Final* amount of each reactant and product present after the reaction.

The completed amounts table for the example just worked indicates that the reactants P_4 and Cl_2 were initially present in the correct stoichiometric ratio but no PCl_3 was present. During the course of the reaction, all of the reactants were consumed and formed the product.

Equation	$P_4(s)$	+	$6\ Cl_2(g)$	→	$4\ PCl_3(\ell)$
Initial amount (mol)	0.0117 mol		0.0702		0 mol
	(1.45 g)		(4.98 g)		(0 g)
Change in amount upon reaction (mol)	−0.0117 mol		−0.0702 mol		+0.0468 mol
Amount after complete reaction (mol)	0 mol		0 mol		0.0468 mol
	(0 g)		(0 g)		(6.43 g)

● **Amounts Tables** The mole (and mass) relationships of reactants and products in a reaction can be summarized in an *amounts table*. Such tables will be used extensively when studying chemical equilibria in Chapters 16–18.

☉WL   INTERACTIVE EXAMPLE 4.1   Mass Relations in Chemical Reactions

Problem Glucose, $C_6H_{12}O_6$, reacts with oxygen to give CO_2 and H_2O. What mass of oxygen (in grams) is required for complete reaction of 25.0 g of glucose? What masses of carbon dioxide and water (in grams) are formed?

What Do You Know? This is a stoichiometry problem because you are given the mass of one of the reactants (glucose) and are asked to determine the mass of the other substances involved in the chemical reaction. In addition, you know formulas for the reactants and products and the mass of one of the reactants (glucose). You will need the molar masses of the reactants and products for the calculation.

Strategy Write the balanced chemical equation for this reaction. Then, follow the scheme outlined in Problem-Solving Tip 4.1.

- Find the amount (moles) of glucose available.

- Find the amount (moles) of O_2. Relate the amount of glucose to the amount of O_2 required using the stoichiometric factor based on the coefficients in the balanced equation.

- Find the mass (g) of O_2 required from the amount of O_2.

- Follow the same procedure to calculate the masses of carbon dioxide and water.

Strategy Map 4.1

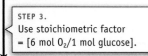

PROBLEM

Calculate mass of O₂ required for combustion of **25.0 g of glucose.**

↓

DATA/INFORMATION

Formulas for reactants and products and the **mass** of one reactant (glucose)

↓

STEP 1. Write the **balanced equation.**

Balanced equation to give the required *stoichiometric factor*

STEP 2. Amount glucose = mass × (1 molar mass).

↓

Amount of **reactant** (glucose)

STEP 3.
Use stoichiometric factor = [6 mol O₂/1 mol glucose].

↓

Amount of **O₂**

STEP 4. Mass of O₂ = mol O₂ × mass/1 mol.

↓

Mass of **O₂**

Solution

Step 1. *Write a balanced equation.*

$$C_6H_{12}O_6(s) + 6\ O_2(g) \rightarrow 6\ CO_2(g) + 6\ H_2O(\ell)$$

Step 2. *Convert the mass of glucose to amount of glucose.*

$$25.0\ \text{g glucose} \times \frac{1\ \text{mol}}{180.2\ \text{g}} = 0.139\ \text{mol glucose}$$

Step 3. *Use the stoichiometric factor.* Here we calculate the amount of O₂ required.

$$0.139\ \text{mol glucose} \times \frac{6\ \text{mol O}_2}{1\ \text{mol glucose}} = 0.832\ \text{mol O}_2$$

Step 4. *Calculate mass from amount.* Convert the required amount of O₂ to a mass in grams.

$$0.832\ \text{mol O}_2 \times \frac{32.00\ \text{g}}{1\ \text{mol O}_2} = 26.6\ \text{g O}_2$$

Repeat Steps 3 and 4 to find the mass of CO₂ produced in the combustion. First, relate the amount (moles) of glucose available to the amount of CO₂ produced using a stoichiometric factor. Then convert the amount of CO₂ to its mass in grams.

$$0.139\ \text{mol glucose} \times \frac{6\ \text{mol CO}_2}{1\ \text{mol glucose}} \times \frac{44.01\ \text{g CO}_2}{1\ \text{mol CO}_2} = 36.6\ \text{g CO}_2$$

Now, how can you find the mass of H₂O produced? You could go through Steps 3 and 4 again. However, recognize that the total mass of reactants

$$25.0\ \text{g C}_6\text{H}_{12}\text{O}_6 + 26.6\ \text{g O}_2 = 51.6\ \text{g reactants}$$

must be the same as the total mass of products. The mass of water that can be produced is therefore

$$\text{Total mass of products} = 51.6\ \text{g} = 36.6\ \text{g CO}_2\ \text{produced} + ?\ \text{g H}_2\text{O}$$

$$\text{Mass of H}_2\text{O produced} = 15.0\ \text{g}$$

Think about Your Answer The results of this calculation can be summarized in an amounts table.

Equation	C₆H₁₂O₆(s)	+	6 O₂(g)	→	6 CO₂(g)	+	6 H₂O(ℓ)
Initial amount (mol)	0.139 mol		6(0.139 mol) = 0.832 mol		0		0
Change in amount upon reaction (mol)	−0.139 mol		−0.832 mol		+0.832 mol		+0.832 mol
Amount after complete reaction (mol)	0		0		0.832 mol		0.832 mol

When you know the mass of all but one of the chemicals in a reaction, you can find the unknown mass using the principle of mass conservation (the total mass of reactants must equal the total mass of products; page 93).

Check Your Understanding

What mass of oxygen, O₂, is required to completely combust 454 g of propane, C₃H₈? What masses of CO₂ and H₂O are produced?

4.2 Reactions in Which One Reactant Is Present in Limited Supply

Reactions are often carried out with an excess of one reactant over that required by stoichiometry. This is usually done to make sure that one of the reactants is consumed completely, even though some of another reactant remains unused.

Suppose you burn a toy sparkler, a wire coated with a mixture of aluminum or iron powder and potassium nitrate or chlorate (Figure 4.1 and page 132). The aluminum or iron burns, consuming oxygen from the air or from the potassium salt and producing a metal oxide.

$$4\ Al(s) + 3\ O_2(g) \rightarrow 2\ Al_2O_3(s)$$

The sparkler burns until the metal powder is consumed completely. What about the oxygen? Four moles of aluminum require three moles of oxygen, but there is much, much more O_2 available in the air than is needed to consume the metal in a sparkler. How much metal oxide is produced? That depends on the quantity of metal powder in the sparkler, not on the quantity of O_2 in the atmosphere. The metal powder in this example is called the **limiting reactant** because its amount determines, or limits, the amount of product formed.

Now let us see how these same principles apply to another example, the reaction of oxygen and carbon monoxide to give carbon dioxide. The balanced equation for the reaction is

$$2\ CO(g) + O_2(g) \rightarrow 2\ CO_2(g)$$

Suppose you have a mixture of four CO molecules and three O_2 molecules. The four CO molecules require only two O_2 molecules (and produce four CO_2 molecules). This means that one O_2 molecule remains after reaction is complete.

Reactants: 4 **CO** and 3 **O_2**   Products: 4 **CO_2** and 1 **O_2**

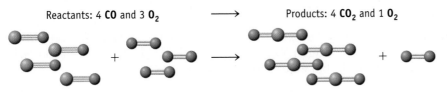

Because more O_2 molecules are available than are required, the number of CO_2 molecules produced is determined by the number of CO molecules available. Carbon monoxide, CO, is therefore the limiting reactant in this case.

A Stoichiometry Calculation with a Limiting Reactant

The first step in the manufacture of nitric acid is the oxidation of ammonia to NO over a platinum-wire gauze (Figure 4.2).

$$4\ NH_3(g) + 5\ O_2(g) \rightarrow 4\ NO(g) + 6\ H_2O(\ell)$$

Suppose equal masses of NH_3 and O_2 are mixed (750. g of each). Are these reactants mixed in the correct stoichiometric ratio or is one of them in short supply? That is, will one of them limit the quantity of NO that can be produced? How much NO can be

FIGURE 4.1 Burning aluminum and iron powder. A toy sparkler contains a metal powder such as Al or Fe and other chemicals such as KNO_3 or $KClO_3$. When ignited, the metal burns with a brilliant white light.

• **Comparing Reactant Ratios** For the CO/O_2 reaction, the stoichiometric ratio of reactants should be (2 mol CO/1 mol O_2). However, the ratio of amounts of reactants available in the text example is (4 mol CO/3 mol O_2) or (1.33 mol CO/1 mol O_2). The fact that the CO/O_2 ratio is not large enough tells us that there is not enough CO to react with all of the available O_2. Carbon monoxide is the limiting reactant, and some O_2 will be left over when all of the CO is consumed.

Module 8: Stoichiometry and Limiting Reactants covers concepts in this section.

© Cengage Learning/Charles D. Winters

FIGURE 4.2
Oxidation of
ammonia.

(a) Burning ammonia on the surface of a platinum wire produces so much heat energy that the wire glows bright red.

(b) Billions of kilograms of HNO_3 are made annually starting with the oxidation of ammonia over a wire gauze containing platinum.

formed if the reaction using this reactant mixture goes to completion? And how much of the excess reactant is left over when the maximum amount of NO has been formed?

Step 1. *Find the amount of each reactant.*

$$750. \text{ g NH}_3 \times \frac{1 \text{ mol NH}_3}{17.03 \text{ g NH}_3} = 44.0 \text{ mol NH}_3 \text{ available}$$

$$750. \text{ g O}_2 \times \frac{1 \text{ mol O}_2}{32.00 \text{ g O}_2} = 23.4 \text{ mol O}_2 \text{ available}$$

Step 2. *What is the limiting reactant?* Examine the ratio of amounts of reactants. Are the reactants present in the correct stoichiometric ratio as given by the balanced equation?

$$\text{Stoichiometric ratio of reactants required by balanced equation} = \frac{5 \text{ mol O}_2}{4 \text{ mol NH}_3} = \frac{1.25 \text{ mol O}_2}{1 \text{ mol NH}_3}$$

$$\text{Ratio of reactants } actually \text{ available} = \frac{23.4 \text{ mol O}_2}{44.0 \text{ mol NH}_3} = \frac{0.532 \text{ mol O}_2}{1 \text{ mol NH}_3}$$

Dividing moles of O_2 available by moles of NH_3 available shows that the ratio of available reactants is much smaller than the 5 mol O_2/4 mol NH_3 ratio required by the balanced equation. Thus, there is not sufficient O_2 available to react with all of the NH_3. In this case, oxygen, O_2, is the limiting reactant. That is, 1 mol of NH_3 requires 1.25 mol of O_2, but we have only 0.532 mol of O_2 available.

Step 3. *Calculate the mass of product.* We can now calculate the expected mass of product, NO, based on the amount of the limiting reactant, O_2.

$$23.4 \text{ mol O}_2 \times \frac{4 \text{ mol NO}}{5 \text{ mol O}_2} \times \frac{30.01 \text{ g NO}}{1 \text{ mol NO}} = 562 \text{ g NO}$$

Step 4. *Calculate the mass of excess reactant.* Ammonia is the "excess reactant" in this NH_3/O_2 reaction because more than enough NH_3 is available to react with 23.4 mol of O_2. Let us calculate the quantity of NH_3 remaining after all the O_2 has been used. To do so, we first need to know the amount of NH_3 required to consume all the limiting reactant, O_2.

$$23.4 \text{ mol O}_2 \text{ available} \times \frac{4 \text{ mol NH}_3 \text{ required}}{5 \text{ mol O}_2} = 18.8 \text{ mol NH}_3 \text{ required}$$

Because 44.0 mol of NH_3 is available, the amount of excess NH_3 can be calculated,

$$\text{Excess } NH_3 = 44.0 \text{ mol } NH_3 \text{ available} - 18.8 \text{ mol } NH_3 \text{ required}$$

$$= 25.2 \text{ mol } NH_3 \text{ remaining}$$

and then converted to a mass.

$$25.2 \text{ mol } NH_3 \times \frac{17.03 \text{ g } NH_3}{1 \text{ mol } NH_3} = 429 \text{ g } NH_3 \text{ in excess of that required}$$

Finally, because 429 g of NH_3 is left over, this means that 321 g of NH_3 has been consumed (= 750. g − 429 g).

When solving limiting reactant problems you may find it helpful to organize your results in an amounts table.

Equation	$4\,NH_3(g)$	+	$5\,O_2(g)$	→	$4\,NO(g)$	+	$6\,H_2O(g)$
Initial amount (mol)	44.0		23.4		0		0
Change in amount upon reaction (mol)	−(4/5)(23.4) = −18.8		−23.4		+(4/5)(23.4) = +18.8		+(6/5)(23.4) = +28.1
Amount after complete reaction (mol)	25.2		0		18.8		28.1

All of the limiting reactant, O_2, is consumed. Of the original 44.0 mol of NH_3, 18.8 mol is consumed and 25.2 mol remains. The balanced equation indicates that the amount of NO produced is equal to the amount of NH_3 consumed, so 18.8 mol of NO is produced from 18.8 mol of NH_3. In addition, 28.1 mol of H_2O is produced.

⬛WL   INTERACTIVE EXAMPLE 4.2   A Reaction with a Limiting Reactant

Problem Methanol, CH_3OH, which is used as a fuel in racing cars and in fuel cells, can be made by the reaction of carbon monoxide and hydrogen.

$$CO(g) + 2\,H_2(g) \rightarrow \underset{\text{methanol}}{CH_3OH(\ell)}$$

Suppose 356 g of CO and 65.0 g of H_2 are mixed and allowed to react.

(a)   What mass of methanol can be produced?

(b)   What mass of the excess reactant remains after the limiting reactant has been consumed?

What Do You Know? You should suspect this problem might involve a limiting reactant because the masses of two reactants are given, and you are asked to determine the mass of the product. The equation for the reaction is given, and you know the masses of CO and H_2 that are available. You will need the molar masses of the two reactants and the product to solve the problem.

Strategy

- Calculate the amounts of each reactant. Compare the ratio of reactant amounts to the stoichiometric ratio defined by the equation, here (2 mol H_2/1 mol CO).

- If (mol H_2 available/mol CO available) is greater than 2/1, then CO is the limiting reactant. If (mol H_2 available/mol CO available) is less than 2/1, then H_2 is the limiting reactant.

- Use the amount of the limiting reactant to calculate the mass of product formed, following the scheme in Problem Solving Tip 4.1.

- The mass of the second reactant required in the reaction is determined from the limiting reactant in a similar calculation. The mass of excess reagent is the difference between its starting mass and what was used in the reaction.

Methanol as a Fuel. Methanol can be used as an automotive fuel when blended with gasoline (in M85, which contains 85% methanol). It is also used as the pure liquid in Indianapolis race cars and in dragsters. Small fuel cells using methanol to produce electricity are also on the market.

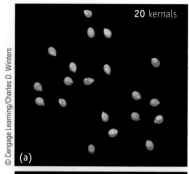

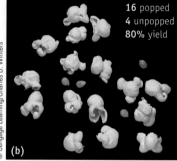

FIGURE 4.3 Percent yield. Although not a chemical reaction, popping corn is a good analogy to the difference between a theoretical yield and an actual yield. Here, we began with 20 popcorn kernels and found that only 16 of them popped. The percent yield from our "reaction" was (16/20) × 100%, or 80%.

FIGURE 4.4 A modern analytical instrument. This nuclear magnetic resonance (NMR) spectrometer is closely related to a magnetic resonance imaging (MRI) instrument found in a hospital. NMR is used to analyze compounds and to decipher their structure. (The instrument is controlled by a computer and console not seen in this photo.)

To provide information to other chemists who might want to carry out a reaction, it is customary to report a percent yield. **Percent yield**, which specifies how much of the theoretical yield was obtained, is defined as

$$\text{Percent yield} = \frac{\text{actual yield}}{\text{theoretical yield}} \times 100\% \qquad (4.1)$$

Suppose you made aspirin in the laboratory by the following reaction:

$$C_7H_6O_3(s) \quad + \quad C_4H_6O_3(\ell) \quad \longrightarrow \quad C_9H_8O_4(s) \quad + \quad CH_3CO_2H(\ell)$$

salicylic acid   acetic anhydride   aspirin   acetic acid

and that you began with 14.4 g of salicylic acid ($C_7H_6O_3$) and an excess of acetic anhydride. That is, salicylic acid is the limiting reactant. If you obtain 6.26 g of aspirin, what is the percent yield of this product? The first step is to find the amount of the limiting reactant, salicylic acid.

$$14.4 \text{ g } C_7H_6O_3 \times \frac{1 \text{ mol } C_7H_6O_3}{138.1 \text{g } C_7H_6O_3} = 0.104 \text{ mol } C_7H_6O_3$$

Next, use the stoichiometric factor from the balanced equation to find the amount of aspirin expected based on the limiting reactant, $C_7H_6O_3$.

$$0.104 \text{ mol } C_7H_6O_3 \times \frac{1 \text{ mol aspirin}}{1 \text{ mol } C_7H_6O_3} = 0.104 \text{ mol aspirin}$$

The maximum amount of aspirin that can be produced—the theoretical yield—is 0.104 mol. Because the quantity you measure in the laboratory is the mass of the product, it is customary to express the theoretical yield as a mass in grams.

$$0.104 \text{ mol aspirin} \times \frac{180.2 \text{ g aspirin}}{1 \text{ mol aspirin}} = 18.8 \text{ g aspirin}$$

Finally, with the actual yield known to be only 6.26 g, the percent yield of aspirin can be calculated.

$$\text{Percent yield} = \frac{6.26 \text{ g aspirin obtained (actual yield)}}{18.8 \text{ g aspirin expected (theoretical yield)}} \times 100\% = 33.3\% \text{ yield}$$

REVIEW & CHECK FOR SECTION 4.3

Aluminum carbide, Al_4C_3, reacts with water to produce methane, CH_4.

$$Al_4C_3(s) + 12 \text{ H}_2O(\ell) \rightarrow 4 \text{ Al(OH)}_3(s) + 3 \text{ CH}_4(g)$$

1. If 125 g of aluminum carbide is decomposed, what is the theoretical yield of methane?

    (a) 4.64 g   (b) 13.9 g   (c) 41.8 g   (d) 154 g

2. If only 13.6 g of methane is obtained, what is the percent yield of this gas?

    (a) 34.1%   (b) 97.8%   (c) 32.5%   (d) 8.83%

4.4 Chemical Equations and Chemical Analysis

Analytical chemists use a variety of approaches to identify substances as well as to measure the quantities of components of mixtures. Analytical chemistry is often done now using instrumental methods (Figure 4.4), but classical chemical reactions and stoichiometry still play a central role.

Quantitative Analysis of a Mixture

Quantitative chemical analysis generally depends on one of the following basic ideas:

- A substance, present in unknown amount, can be allowed to react with a known amount of another substance. If the stoichiometric ratio for their reaction is known, the unknown amount can be determined.
- A material of unknown composition can be converted to one or more substances of known composition. Those substances can be identified, their amounts determined, and these amounts related to the amount of the original, unknown substance.

An example of the first type of analysis is the analysis of a sample of vinegar containing an unknown amount of acetic acid, the ingredient that makes vinegar acidic. The acid reacts readily and completely with sodium hydroxide.

$$CH_3CO_2H(aq) + NaOH(aq) \rightarrow CH_3CO_2Na(aq) + H_2O(\ell)$$
acetic acid

If the exact amount of sodium hydroxide used in the reaction can be measured, the amount of acetic acid present can be calculated. This type of analysis is the subject of a later section in this chapter [▶ Section 4.7].

An example of the second type of analysis is the analysis of a sample of a mineral, thenardite, which is largely sodium sulfate, Na_2SO_4 (Figure 4.5). Sodium sulfate is soluble in water. Therefore, to find the quantity of Na_2SO_4 in an impure mineral sample, we would crush the rock and then wash the powdered sample thoroughly with water to dissolve the sodium sulfate. Next, we would treat this solution of sodium sulfate with barium chloride to precipitate the water-insoluble compound barium sulfate. The barium sulfate is collected on a filter and weighed (Figure 4.6).

$$Na_2SO_4(aq) + BaCl_2(aq) \rightarrow BaSO_4(s) + 2 NaCl(aq)$$

We can then find the amount of sodium sulfate in the mineral sample because it is directly related to the amount of $BaSO_4$.

$$1 \text{ mol } Na_2SO_4(aq) \rightarrow 1 \text{ mol } BaSO_4(s)$$

Example 4.3 illustrates the analysis of another mineral, millerite, in this way.

FIGURE 4.5 Thenardite. The mineral thenardite is sodium sulfate, Na_2SO_4. It is named after the French chemist Louis Thenard (1777–1857). Sodium sulfate is used in making detergents, glass, and paper.

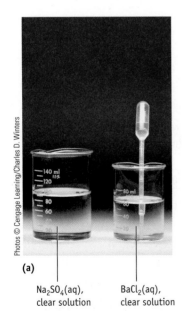

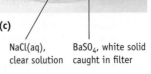

(a)

$Na_2SO_4(aq)$, clear solution

$BaCl_2(aq)$, clear solution

(b)

$BaSO_4$, white solid

$NaCl(aq)$, clear solution

(c)

$NaCl(aq)$, clear solution

$BaSO_4$, white solid caught in filter

(d)

Mass of dry $BaSO_4$ determined

FIGURE 4.6 Analysis for the sulfate content of a sample. The sulfate ions in a solution of Na_2SO_4 react with barium ions (Ba^{2+}) to form insoluble $BaSO_4$. The white precipitate is collected on a filter and weighed. The amount of $BaSO_4$ obtained can be related to the amount of Na_2SO_4 in the sample.

CASE STUDY

Green Chemistry and Atom Economy

Chemists and chemical industries are increasingly following the principles of "green chemistry." Among these principles are that it is better to prevent waste when producing chemicals than to clean it up later and to use and generate compounds that have little or no toxicity to human health and the environment.

Another principle of green chemistry is to try to convert all of the atoms of the reactants into the product; nothing should be wasted. To evaluate how well this occurs chemists usually report the percent yield of the product. As you will come to realize with your own laboratory experience, the percent yield can often be quite low.

Another way to evaluate the efficiency of a reaction is to calculate the "atom economy."

$$\% \text{ atom economy} =$$
$$\frac{\text{molar mass of atoms utilized}}{\text{molar mass of reactants}} \times 100\%$$

The principles of green chemistry helped in making "Green Works" cleaning products from a division of the Clorox Company. Most of the ingredients are naturally occurring substances.

A simple example of the concept is the reaction of methanol and carbon monoxide to produce acetic acid. The atom economy is 100% because all of the atoms of the reactants appear in the product.

$$CH_3OH(\ell) + CO(g) \rightarrow CH_3CO_2H(aq)$$

Ibuprofen is a widely used nonsteroidal anti-inflammatory drug, which is used in the United States in products under tradenames such as Motrin and Advil. A recently developed synthesis of ibuprofen involves the collection of compounds below (which combine in three reaction steps to give ibuprofen and, as a by-product, acetic acid).

Ibruprofen

What is the atom economy for this reaction? The reactants, collectively, have 15 C atoms, 22 H atoms, and 4 O atoms. The "molar mass" of this collection is 266 g/mol. In contrast, ibuprofen has a molar mass of 206 g/mol. Therefore, the atom economy is 77%. This is far superior to a competing commercial process for the synthesis of ibuprofen that has an atom economy of only 40%.

Question:

Methyl methacrylate is the reactant used to prepare plastics you may know as Lucite or some other tradename. The substance was first introduced in 1933, but chemists have long sought better (and less costly) ways to make the compound. In a newly developed process, ethylene (C_2H_4), methanol (CH_3OH), CO, and formaldehyde (CH_2O) combine in two steps to give methyl methacrylate and water. What is the atom economy of this new process? (Not only does the new process have a much better atom economy than the previously used process, it also does not involve HCN, a toxic substance used in another method of making this compound.)

$$H_2C=CH_2 \quad CH_3OH \quad CO \quad H_2CO$$

Methyl methacrylate

The answer to this question is available in Appendix N.

Reference:

M. C. Cann and M. E. Connelly, *Real-World Cases in Green Chemistry*, American Chemical Society, 2000.

EXAMPLE 4.3   Mineral Analysis

Problem Nickel(II) sulfide, NiS, occurs naturally as the relatively rare mineral millerite. One of its occurrences is in meteorites. To analyze a mineral sample for the quantity of NiS, the sample is dissolved in nitric acid to form a solution of $Ni(NO_3)_2$.

$$NiS(s) + 4 HNO_3(aq) \rightarrow Ni(NO_3)_2(aq) + S(s) + 2 NO_2(g) + 2 H_2O(\ell)$$

The aqueous solution of $Ni(NO_3)_2$ is then treated with the organic compound dimethylglyoxime ($C_4H_8N_2O_2$, DMG) to give the red solid $Ni(C_4H_7N_2O_2)_2$.

$$Ni(NO_3)_2(aq) + 2 C_4H_8N_2O_2(aq) \rightarrow Ni(C_4H_7N_2O_2)_2(s) + 2 HNO_3(aq)$$

Suppose a 0.468-g sample containing millerite produces 0.206 g of red, solid $Ni(C_4H_7N_2O_2)_2$. What is the mass percent of NiS in the sample?

What Do You Know? You know the mass of $Ni(C_4H_7N_2O_2)_2$ and the two balanced equations for the reactions leading to its formation. You will need molar masses of $Ni(C_4H_7N_2O_2)_2$ and NiS.

A precipitate of nickel with dimethylglyoxime. Red, insoluble $Ni(C_4H_7N_2O_2)_2$ precipitates when dimethylglyoxime ($C_4H_8N_2O_2$) is added to an aqueous solution of nickel(II) ions.

Strategy The balanced equations for the reactions show the following "road map":

$$1 \text{ mol NiS} \rightarrow 1 \text{ mol Ni(NO}_3)_2 \rightarrow 1 \text{ mol Ni(C}_4\text{H}_7\text{N}_2\text{O}_2)_2$$

From this you see that if you know the amount of $Ni(C_4H_7N_2O_2)_2$, you also know the amount of NiS. Therefore, first calculate the amount of $Ni(C_4H_7N_2O_2)_2$ from its mass. Because 1 mol of NiS was present in the sample for each mole of $Ni(C_4H_7N_2O_2)_2$ isolated, you therefore know the amount of NiS and can then calculate its mass and mass percent in the sample.

Solution The molar mass of $Ni(C_4H_7N_2O_2)_2$ is 288.9 g/mol. The amount of this red solid is

$$0.206 \text{ g Ni(C}_4\text{H}_7\text{N}_2\text{O}_2)_2 \times \frac{1 \text{ mol Ni(C}_4\text{H}_7\text{N}_2\text{O}_2)_2}{288.9 \text{ g Ni(C}_4\text{H}_7\text{N}_2\text{O}_2)_2} = 7.13 \times 10^{-4} \text{ mol Ni(C}_4\text{H}_7\text{N}_2\text{O}_2)_2$$

Because 1 mol of $Ni(C_4H_7N_2O_2)_2$ is ultimately produced from 1 mol of NiS, the amount of NiS in the sample must also have been 7.13×10^{-4} mol.

With the amount of NiS known, we calculate the mass of NiS.

$$7.13 \times 10^{-4} \text{ mol NiS} \times \frac{90.76 \text{ g NiS}}{1 \text{ mol NiS}} = 0.0647 \text{ g NiS}$$

Finally, the mass percent of NiS in the 0.468-g sample is

$$\text{Mass percent NiS} = \frac{0.0647 \text{ g NiS}}{0.468 \text{ g sample}} \times 100\% = 13.8\% \text{ NiS}$$

Think about Your Answer For an analytical procedure to be used, the reactants must be completely converted to products and the product being measured must be isolated without losses in handling. This calls for very careful experimental techniques.

Check Your Understanding

One method for determining the purity of a sample of titanium(IV) oxide, TiO_2, an important industrial chemical, is to react the sample with bromine trifluoride.

$$3 \text{ TiO}_2(s) + 4 \text{ BrF}_3(\ell) \rightarrow 3 \text{ TiF}_4(s) + 2 \text{ Br}_2(\ell) + 3 \text{ O}_2(g)$$

This reaction is known to occur completely and quantitatively. That is, all of the oxygen in TiO_2 is evolved as O_2. Suppose 2.367 g of a TiO_2-containing sample evolves 0.143 g of O_2. What is the mass percent of TiO_2 in the sample?

Determining the Formula of a Compound by Combustion

The empirical formula of a compound can be determined if the percent composition of the compound is known [◀ Section 2.10]. But where do the percent composition data come from? One chemical method that works well for compounds that burn in oxygen is analysis by combustion. In this technique, each element in the compound combines with oxygen to produce the appropriate oxide.

Consider an analysis of the hydrocarbon methane, CH_4. A balanced equation for the combustion of methane shows that every carbon atom in the original compound appears as CO_2 and every hydrogen atom appears in the form of water. In other words, for every mole of CO_2 observed, there must have been one mole of carbon in the unknown compound. Similarly, for every mole of H_2O observed from combustion, there must have been *two* moles of H atoms in the unknown compound.

$$CH_4(g) + 2 O_2(g) \longrightarrow CO_2(g) + 2 H_2O(\ell)$$

In the combustion experiment gaseous carbon dioxide and water are separated (as illustrated in Figure 4.7) and their masses determined. From these masses it is possible to calculate the amounts of C and H in CO_2 and H_2O, respectively and the

FIGURE 4.7 Combustion analysis of a hydrocarbon. If a compound containing C and H is burned in oxygen, CO_2 and H_2O are formed, and the mass of each can be determined. The H_2O is absorbed by magnesium perchlorate, and the CO_2 is absorbed by finely divided NaOH. The mass of each absorbent before and after combustion gives the masses of CO_2 and H_2O. Only a few milligrams of a combustible compound are needed for analysis.

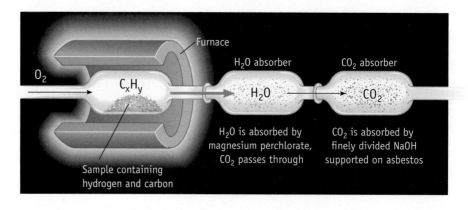

ratio of amounts of C and H in a sample of the original compound can then be found. This ratio gives the empirical formula.

● **Finding an Empirical Formula by Chemical Analysis** Finding the empirical formula of a compound by chemical analysis always uses the following procedure:

1. The unknown but pure compound is converted in a chemical reaction into known products.
2. The reaction products are isolated, and the amount of each is determined.
3. The amount of each product is related to the amount of each element in the original compound.
4. The empirical formula is determined from the relative amounts of elements in the original compound.

⏽WL   **INTERACTIVE EXAMPLE 4.4** Using Combustion Analysis to Determine the Formula of a Hydrocarbon

Problem When 1.125 g of a liquid hydrocarbon, C_xH_y, was burned in an apparatus like that shown in Figure 4.7, 3.447 g of CO_2 and 1.647 g of H_2O were produced. In a separate experiment the molar mass of the compound was found to be 86.2 g/mol. Determine the empirical and molecular formulas for the unknown hydrocarbon, C_xH_y.

What Do You Know? You know the mass of the hydrocarbon, the fact that it contains only C and H, and the molar mass of this compound. You are also provided with the masses of H_2O and CO_2 formed when the hydrocarbon is burned. You will need the molar masses of the three compounds in this calculation.

Strategy The strategy to solve this problem is outlined in the diagram below and the Strategy Map for Example 4.4.

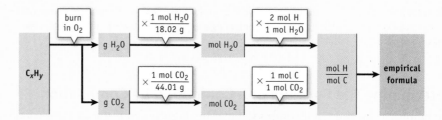

The steps in sequence are as follows:

• Calculate the amount of CO_2 and H_2O and from these values calculate the amounts of C and H.

• Determine the lowest whole-number ratio of amounts of C and H. This gives you the subscripts for C and H in the empirical formula.

• Compare the molar mass of the hydrocarbon with the empirical formula to determine the molecular formula.

Solution The amounts of CO_2 and H_2O isolated from the combustion are

$$3.447 \text{ g } CO_2 \times \frac{1 \text{ mol } CO_2}{44.010 \text{ g } CO_2} = 0.07832 \text{ mol } CO_2$$

$$1.647 \text{ g } H_2O \times \frac{1 \text{ mol } H_2O}{18.015 \text{ g } H_2O} = 0.09142 \text{ mol } H_2O$$

For every mole of CO_2 isolated, 1 mol of C must have been present in the unknown compound.

$$0.07832 \text{ mol } CO_2 \times \frac{1 \text{ mol C in unknown}}{1 \text{ mol } CO_2} = 0.07832 \text{ mol C}$$

For every mole of H_2O isolated, 2 mol of H must have been present in the unknown.

$$0.09142 \text{ mol } H_2O \times \frac{2 \text{ mol H in unknown}}{1 \text{ mol } H_2O} = 0.1828 \text{ mol H}$$

The original 1.125-g sample of compound therefore contained 0.07832 mol of C and 0.1828 mol of H. To determine the empirical formula of the unknown, we find the ratio of moles of H to moles of C [◄ Section 2.10].

$$\frac{0.1828 \text{ mol H}}{0.07832 \text{ mol C}} = \frac{2.335 \text{ mol H}}{1.000 \text{ mol C}}$$

The empirical formula gives the simplest *whole-number* ratio. The translation of this ratio (2.335/1) to a whole-number ratio can usually be done quickly by trial and error. Multiplying the numerator and denominator by 3 gives 7/3. So, we know the ratio is 7 mol H to 3 mol C, which means the empirical formula of the hydrocarbon is C_3H_7.

Comparing the experimental molar mass with the molar mass calculated for the empirical formula,

$$\frac{\text{Experimental molar mass}}{\text{Molar mass of } C_3H_7} = \frac{86.2 \text{ g/mol}}{43.1 \text{ g/mol}} = \frac{2}{1}$$

we find that the molecular formula is twice the empirical formula. That is, the molecular formula is $(C_3H_7)_2$, or C_6H_{14}.

Think about Your Answer As noted in Problem Solving Tip 2.3 (page 82), for problems of this type be sure to use data with enough significant figures to give accurate atom ratios. Finally, note that the determination of the molecular formula does not end the problem for a chemist. In this case, the formula C_6H_{14} is appropriate for several distinctly different compounds. Two of the five compounds having this formula are shown here:

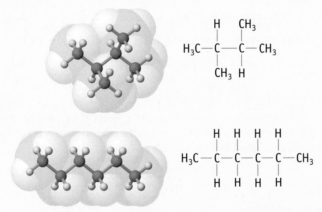

To determine the identity of the unknown compound, more laboratory experiments have to be done. One option is to use an NMR spectrometer, as pictured in Figure 4.4, or to compare the properties of the unknown with values listed in the chemical literature.

Check Your Understanding

A 0.523-g sample of the unknown compound C_xH_y was burned in air to give 1.612 g of CO_2 and 0.7425 g of H_2O. A separate experiment gave a molar mass for C_xH_y of 114 g/mol. Determine the empirical and molecular formulas for the hydrocarbon.

EXAMPLE 4.5   Using Combustion Analysis to Determine an Empirical Formula of a Compound Containing C, H, and O

Problem Suppose you isolate an acid from clover leaves and know that it contains only the elements C, H, and O. Heating 0.513 g of the acid in oxygen produces 0.501 g of CO_2 and 0.103 g of H_2O. What is the empirical formula of the acid $C_xH_yO_z$?

Strategy Map 4.4

PROBLEM

Determine **empirical** and **molecular formulas** of a hydrocarbon.

↓

DATA/INFORMATION

• **Masses** of CO_2 and H_2O from combustion of given mass of C_xH_y
• **Molar mass** of **unknown**

STEP 1. Calculate **amounts** of CO_2 and H_2O.

Amounts of CO_2 and H_2O

STEP 2. Calculate **amount** of **C** and **H** from amounts of CO_2 and H_2O, respectively.

Amounts of **C** and **H** in unknown compound

STEP 3. Find **whole-number ratio** of amounts of **C** and **H**.

C_xH_y = empirical formula

STEP 4. Determine ratio of **experimental molar mass** to **empirical formula mass**.

Molecular formula $[C_xH_y]_n$

What Do You Know? You know the mass of the compound and the fact that it contains only C, H, and O. You are also provided with the masses of H_2O and CO_2 formed when the hydrocarbon is burned.

Strategy In order to determine the empirical formula, you need to determine the amounts (moles) of C, H, and O in the unknown compound. Follow the steps outlined below.

* Determine the amounts of C and H following the procedure in Example 4.4.

* Determine the masses of C and H from the amounts of C and H.

* The mass of O is the mass of the sample minus the masses of C and H.

* From the mass of O determine the amount of O.

* Finally, determine the smallest whole number ratio between the amounts of the three elements. This determines the subscripts for the elements in the empirical formula.

Solution The first step is to determine the amounts of C and H in the sample.

$$0.501 \text{ g } CO_2 \times \frac{1 \text{ mol } CO_2}{44.01 \text{ g } CO_2} \times \frac{1 \text{ mol } C}{1 \text{ mol } CO_2} = 0.0114 \text{ mol } C$$

$$0.103 \text{ g } H_2O \times \frac{1 \text{ mol } H_2O}{18.02 \text{ g } H_2O} \times \frac{2 \text{ mol } H}{1 \text{ mol } H_2O} = 0.0114 \text{ mol } H$$

From these amounts, you can determine the mass of C and the mass of H in the sample.

$$0.0114 \text{ mol } C \times \frac{12.01 \text{ g } C}{1 \text{ mol } C} = 0.137 \text{ g } C$$

$$0.0114 \text{ mol } H \times \frac{1.008 \text{ g } H}{1 \text{ mol } H} = 0.0115 \text{ g } H$$

Using the mass of the original sample and the masses of C and H in the sample, you can now determine the mass of O in the sample. From this, you can find the amount of O in the sample.

$$\text{Mass of sample} = 0.513 \text{ g} = 0.137 \text{ g } C + 0.0115 \text{ g } H + x \text{ g } O$$

$$\text{Mass of O} = 0.365 \text{ g } O$$

$$0.365 \text{ g } O \times \frac{1 \text{ mol } O}{16.00 \text{ g } O} = 0.0228 \text{ mol } O$$

To find the mole ratios of elements, divide the number of moles of each element by the smallest amount present. Because both C and H are present in the same amount in the sample, you know their ratio is 1 C : 1 H. What about O?

$$\frac{0.0228 \text{ mol } O}{0.0114 \text{ mol } C} = \frac{2 \text{ mol } O}{1 \text{ mol } C}$$

The mole ratios show that, for every C atom in the molecule, one H atom and two O atoms occur. The empirical formula of the acid is therefore CHO_2.

Think about Your Answer If the molar mass of the unknown compound were known, you would then be able to derive the molecular formula of the compound.

Check Your Understanding

A 0.1342-g sample of a compound composed of C, H, and O ($C_xH_yO_z$) was burned in oxygen and 0.240 g of CO_2 and 0.0982 g of H_2O was isolated. What is the empirical formula of the compound? If the experimentally determined molar mass is 74.1 g/mol, what is the molecular formula of the compound?

REVIEW & CHECK FOR SECTION 4.4

A 0.509-g sample of an unknown organic compound containing C, H, and O was burned in air to give 1.316 g of CO_2 and 0.269 g of H_2O. What is the empirical formula for this compound?

(a) CHO   (b) C_2H_2O (c) C_4H_4O (d) CHO_4

4.5 Measuring Concentrations of Compounds in Solution

Most chemical studies require quantitative measurements, including experiments involving aqueous solutions. When doing such experiments, we continue to use balanced equations and moles, but we measure volumes of solutions rather than masses of solids, liquids, or gases. Solution concentration, expressed as molarity, relates the volume of solution in liters to the amount of substance in moles.

Solution Concentration: Molarity

The concept of concentration is useful in many contexts. For example, about 5,628,000 people live in Wisconsin (as of July 2008), and the state has a land area of roughly 56,000 square miles; therefore, the average concentration of people is about $(5.6 \times 10^6$ people$/5.6 \times 10^4$ square miles) or 1.0×10^2 people per square mile. In chemistry the amount of solute dissolved in a given volume of solution, the concentration of the solution, can be found in the same way. A useful unit of solute concentration, c, is **molarity**, which is defined as amount of solute per liter of solution.

$$\text{Molarity of } x \ (c_x) = \frac{\text{amount of solute } x \ (\text{mol})}{\text{volume of solution (L)}} \qquad (4.2)$$

For example, if 58.4 g (1.00 mol) of NaCl is dissolved in enough water to give a total solution volume of 1.00 L, the molarity, c, is 1.00 mol/L. This is often abbreviated as 1.00 M, where the capital "M" stands for "moles per liter." Another common notation is to place the formula of the compound in square brackets (for example, [NaCl]); this implies that the concentration of the solute in moles per liter of solution is being specified.

$$c_{NaCl} = [NaCl] = 1.00 \text{ mol/L} = 1.00 \text{ M}$$

It is important to notice that molarity refers to the amount of solute *per liter of solution* and not per liter of solvent. If one liter of water is added to one mole of a solid compound, the final volume will not be exactly one liter, and the final concentration will not be exactly one mol/L (Figure 4.8). When making solutions of a given concentration, it is always the case that we dissolve the solute in a volume of

● **Molar and Molarity** Chemists use "molar" as an adjective to describe a solution. We use "molarity" as a noun. For example, we refer to a 0.1 molar solution or say the solution has a molarity of 0.1 mole per liter.

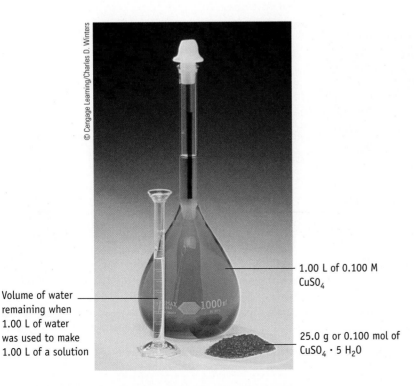

Volume of water remaining when 1.00 L of water was used to make 1.00 L of a solution

1.00 L of 0.100 M $CuSO_4$

25.0 g or 0.100 mol of $CuSO_4 \cdot 5 H_2O$

FIGURE 4.8 Volume of solution versus volume of solvent. To make a 0.100 M solution of $CuSO_4$, 25.0 g or 0.100 mol of $CuSO_4 \cdot 5 H_2O$ (the blue crystalline solid) was placed in a 1.00-L volumetric flask. For this photo, we measured out exactly 1.00 L of water, which was slowly added to the volumetric flask containing 25.0 g of $CuSO_4 \cdot 5 H_2O$. When enough water had been added so that the solution volume was exactly 1.00 L, approximately 8 mL (the quantity in the small graduated cylinder) was left over from the original 1.00 L of water. This emphasizes that molar concentrations are defined as moles of solute per liter of solution and not per liter of water or other solvent.

Ion concentrations for a soluble ionic compound. Here, 1 mol of $CuCl_2$ dissociates to 1 mol of Cu^{2+} ions and 2 mol of Cl^- ions. Therefore, the Cl^- concentration is twice the concentration calculated for $CuCl_2$.

solvent smaller than the desired volume of solution, then add solvent until the final solution volume is reached.

Potassium permanganate, $KMnO_4$, which was used at one time as a germicide in the treatment of burns, is a shiny, purple-black solid that dissolves readily in water to give a deep purple solution. Suppose 0.435 g of $KMnO_4$ has been dissolved in enough water to give 250. mL of solution (Figure 4.9). What is the concentration of $KMnO_4$? The first step is to convert the mass of $KMnO_4$ to an amount (moles) of solute.

$$0.435 \text{ g KMnO}_4 \times \frac{1 \text{ mol KMnO}_4}{158.0 \text{ g KMnO}_4} = 0.00275 \text{ mol KMnO}_4$$

Now that the amount of $KMnO_4$ is known, this information can be combined with the volume of solution—which must be in liters—to give the concentration. Because 250. mL is equivalent to 0.250 L,

$$\text{Concentration of KMnO}_4 = c_{KMnO_4} = [KMnO_4] = \frac{0.00275 \text{ mol KMnO}_4}{0.250 \text{ L}} = 0.0110 \text{ M}$$

The $KMnO_4$ concentration is 0.0110 mol/L, or 0.0110 M. This is useful information, but it is often equally useful to know the concentration of each type of ion in a solution. Like all soluble ionic compounds, $KMnO_4$ dissociates completely into its ions, K^+ and MnO_4^-, when dissolved in water.

$$KMnO_4(aq) \longrightarrow K^+(aq) + MnO_4^-(aq)$$
$$\text{100\% dissociation}$$

One mole of $KMnO_4$ provides 1 mol of K^+ ions and 1 mol of MnO_4^- ions. Accordingly, 0.0110 M $KMnO_4$ gives a concentration of K^+ in the solution of 0.0110 M; similarly, the concentration of MnO_4^- is also 0.0110 M.

Another example of ion concentrations is provided by the dissociation of $CuCl_2$.

$$CuCl_2(aq) \longrightarrow Cu^{2+}(aq) + 2 Cl^-(aq)$$
$$\text{100\% dissociation}$$

If 0.10 mol of $CuCl_2$ is dissolved in enough water to make 1.0 L of solution, the concentration of the copper(II) ion is $[Cu^{2+}] = 0.10$ M. However, the concentration of chloride ions, $[Cl^-]$, is 0.20 M because the compound dissociates in water to provide 2 mol of Cl^- ions for each mole of $CuCl_2$.

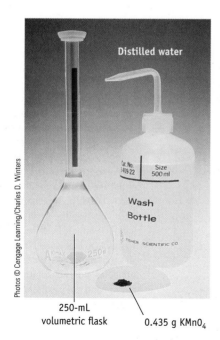

250-mL volumetric flask 0.435 g $KMnO_4$

The $KMnO_4$ is first dissolved in a small amount of water.

Distilled water is added to fill the flask with solution just to the mark on the flask.

A mark on the neck of a volumetric flask indicates a volume of exactly 250. mL at 25 °C.

FIGURE 4.9 Making a solution. A 0.0110 M solution of $KMnO_4$ is made by adding enough water to 0.435 g of $KMnO_4$ to make 0.250 L of solution.

EXAMPLE 4.6   Concentration

Problem If 25.3 g of sodium carbonate, Na_2CO_3, is dissolved in enough water to make 250. mL of solution, what is the concentration of Na_2CO_3? What are the concentrations of the Na^+ and CO_3^{2-} ions?

What Do You Know? You know the mass of the solute, Na_2CO_3, and the volume of the solution. You will need the molar mass of Na_2CO_3 to calculate the amount of this compound.

Strategy The concentration (moles/L) of Na_2CO_3 is the amount of Na_2CO_3 (moles) divided by the volume (in liters). To determine the concentrations of the ions, recognize that one mole of this ionic compound contains two moles of Na^+ and one mole of CO_3^{2-} ions.

$$Na_2CO_3(s) \rightarrow 2\ Na^+(aq) + CO_3^{2-}(aq)$$

Solution Let us first find the amount of Na_2CO_3.

$$25.3\ g\ Na_2CO_3 \times \frac{1\ mol\ Na_2CO_3}{106.0\ g\ Na_2CO_3} = 0.239\ mol\ Na_2CO_3$$

and then the concentration of Na_2CO_3,

$$\text{Concentration of }Na_2CO_3 = \frac{0.239\ mol\ Na_2CO_3}{0.250\ L} = 0.995\ mol/L$$

The ion concentrations follow from the knowledge that each mole of Na_2CO_3 produces 2 mol of Na^+ ions and 1 mol of CO_3^{2-} ions.

$$0.955\ M\ Na_2CO_3(aq) \equiv 2 \times 0.955\ M\ Na^+(aq) + 0.955\ M\ CO_3^{2-}(aq)$$

That is, $[Na^+] = 1.91\ M$ and $[CO_3^{2-}] = 0.955\ M$.

Think about Your Answer Many chemical reactions are carried out in aqueous solution. The amount (moles) of a compound is readily calculated if you know the concentration and volume of solution.

Check Your Understanding

Sodium bicarbonate, $NaHCO_3$, is used in baking powder formulations and in the manufacture of plastics and ceramics, among other things. If 26.3 g of the compound is dissolved in enough water to make 200. mL of solution, what is the concentration of $NaHCO_3$? What are the concentrations of the ions in solution?

Preparing Solutions of Known Concentration

Chemists often have to prepare a given volume of solution of known concentration. There are two generally used ways to do this.

Combining a Weighed Solute with the Solvent

Suppose you wish to prepare 2.00 L of a 1.50 M solution of Na_2CO_3. You have some solid Na_2CO_3 and distilled water. You also have a 2.00-L volumetric flask. To make the solution, you must weigh the required quantity of Na_2CO_3 as accurately as possible considering the number of significant figures desired for the concentration, carefully place all the solid in the volumetric flask, and then add some water to dissolve the solid. After the solid has dissolved completely, more water is added to bring the solution volume to 2.00 L. The solution then has the desired concentration and the volume specified.

But what mass of Na_2CO_3 is required to make 2.00 L of 1.50 M Na_2CO_3? First, calculate the amount of Na_2CO_3 required,

$$2.00\ L \times \frac{1.50\ mol\ Na_2CO_3}{1.00\ L\ solution} = 3.00\ mol\ Na_2CO_3\ required$$

• **Volumetric Flask** A volumetric flask is a special flask with a line marked on its neck (see page 32 and Figures 4.8 and 4.9). If the flask is filled with a solution to this line (at a given temperature), it contains precisely the volume of solution specified.

and then the mass in grams.

$$3.00 \ \text{mol Na}_2\text{CO}_3 \times \frac{106.0 \ \text{g Na}_2\text{CO}_3}{1 \ \text{mol Na}_2\text{CO}_3} = 318 \ \text{g Na}_2\text{CO}_3$$

Thus, to prepare the desired solution, you should dissolve 318 g of Na_2CO_3 in enough water to make 2.00 L of solution.

Diluting a More Concentrated Solution

Another method of making a solution of a given concentration is to begin with a concentrated solution and add more solvent (usually water) until the desired, lower concentration is reached (Figure 4.10). Many of the solutions prepared for your laboratory course are probably made by this dilution method. It is more efficient to store a small volume of a concentrated solution and then, when needed, add water to make a much larger volume of a dilute solution.

Suppose you need 500. mL of 0.0010 M potassium dichromate, $\text{K}_2\text{Cr}_2\text{O}_7$, for use in chemical analysis. You have some 0.100 M $\text{K}_2\text{Cr}_2\text{O}_7$ solution available. To make the required 0.0010 M solution, place a measured volume of the more concentrated $\text{K}_2\text{Cr}_2\text{O}_7$ solution in a flask and then add water until the $\text{K}_2\text{Cr}_2\text{O}_7$ is contained in the appropriate, larger volume of water (Figure 4.10).

What volume of a 0.100 M $\text{K}_2\text{Cr}_2\text{O}_7$ solution must be diluted to make the 0.0010 M solution? If the volume and concentration of a solution are known, the amount of solute is also known. Therefore, the amount of $\text{K}_2\text{Cr}_2\text{O}_7$ that must be in the final dilute solution is

$$\text{Amount of K}_2\text{Cr}_2\text{O}_7 \text{ in dilute solution} = c_{\text{K}_2\text{Cr}_2\text{O}_7} \times V_{\text{K}_2\text{Cr}_2\text{O}_7} = \left(\frac{0.0010 \ \text{mol}}{L} \right) \times (0.500 \ \text{L})$$
$$= 0.00050 \ \text{mol K}_2\text{Cr}_2\text{O}_7$$

The more concentrated solution containing this amount of $\text{K}_2\text{Cr}_2\text{O}_7$ must be placed in a 500.-mL flask and then diluted to the final volume. The volume of 0.100 M $\text{K}_2\text{Cr}_2\text{O}_7$ that must be transferred and diluted is 5.0 mL.

$$0.00050 \ \text{mol K}_2\text{Cr}_2\text{O}_7 \times \frac{1.00 \ \text{L}}{0.100 \ \text{mol K}_2\text{Cr}_2\text{O}_7} = 0.0050 \ \text{L or 5.0 mL}$$

Thus, to prepare 500. mL of 0.0010 M $\text{K}_2\text{Cr}_2\text{O}_7$, place 5.0 mL of 0.100 M $\text{K}_2\text{Cr}_2\text{O}_7$ in a 500.-mL flask and add water until a volume of 500. mL is reached (Figure 4.10).

0.100 M $\text{K}_2\text{Cr}_2\text{O}_7$

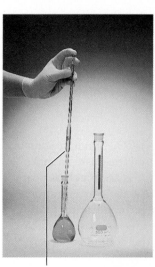

Use a 5.00-mL pipet to withdraw 5.00 mL of 0.100 M $\text{K}_2\text{Cr}_2\text{O}_7$ solution.

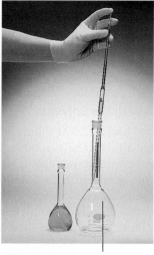

Add the 5.00-mL sample of 0.100 M $\text{K}_2\text{Cr}_2\text{O}_7$ solution to a 500-mL volumetric flask.

Fill the flask to the mark with distilled water to give 0.00100 M $\text{K}_2\text{Cr}_2\text{O}_7$ solution.

FIGURE 4.10 **Making a solution by dilution.** Here, 5.00 mL of a $\text{K}_2\text{Cr}_2\text{O}_7$ solution is diluted to 500. mL. This means the solution is diluted by a factor of 100, from 0.100 M to 0.00100 M.

Preparing a Solution by Dilution

There is a straightforward method to use for problems involving dilution. The central idea is that the amount of solute in the final, dilute solution has to be equal to the amount of solute taken from the more concentrated solution. If c is the concentration (molarity) and V is the volume (and the subscripts d and c identify the dilute and concentrated solutions, respectively), then the amount of solute in either solution (in the case of the $K_2Cr_2O_7$ example in the text) can be calculated as follows:

(a) Amount of $K_2Cr_2O_7$ in the final dilute solution is

$$c_d \times V_d = 0.00050 \text{ mol}$$

(b) Amount of $K_2Cr_2O_7$ taken from the more concentrated solution is

$$c_c \times V_c = 0.00050 \text{ mol}$$

Because both cV products are equal to the same amount of solute, we can use the following equation:

Amount in concentrated solution = Amount in dilute solution

$$c_c \times V_c = c_d \times V_d$$

This equation is valid for all cases in which a more concentrated solution is used to make a more dilute one.

EXAMPLE 4.7   Preparing a Solution by Dilution

Problem What is the concentration of iron(III) ions in a solution prepared by diluting 1.00 mL of a 0.236 M solution of iron(III) nitrate to a volume of 100.0 mL?

What Do I Know? You know the initial concentration and volume and the final volume after dilution.

Strategy First calculate the amount of iron(III) ion in the 1.00-mL sample (amount = concentration × volume). The concentration of the ion in the final, dilute solution is equal to this amount of iron(III) divided by the new volume.

Solution The amount of iron(III) ion in the 1.00 mL sample is

$$\text{Amount of Fe}^{3+} = c_{Fe^{3+}}V_{Fe^{3+}} = \frac{0.236 \text{ mol Fe}^{3+}}{L} \times 1.00 \times 10^{-3} \text{ L} = 2.36 \times 10^{-4} \text{ mol Fe}^{3+}$$

This amount of iron(III) ion is distributed in the new volume of 100.0 mL, so the final concentration of the diluted solution is

$$c_{Fe^{3+}} = [Fe^{3+}] = \frac{2.36 \times 10^{-4} \text{ mol Fe}^{3+}}{0.1000 \text{ L}} = 2.36 \times 10^{-3} \text{ M}$$

Think about Your Answer The sample has been diluted 100-fold, so we would expect the final concentration to be 1/100th of the initial value. See also Problem Solving Tip 4.3, which gives a quick and easy-to-use method for this calculation.

Check Your Understanding

An experiment calls for you to use 250. mL of 1.00 M NaOH, but you are given a large bottle of 2.00 M NaOH. Describe how to make the desired volume of 1.00 M NaOH.

REVIEW & CHECK FOR SECTION 4.5

1. A 1.71-g sample of $Ba(OH)_2(s)$ was dissolved in water to give 250 mL of solution. What is the OH^- concentration in this solution?

    (a) 0.010 M   (b) 0.040 M   (c) 0.080 M   (d) 0.10 M

2. You have a 100.-mL volumetric flask containing 0.050 M HCl. A 2.0-mL sample of this solution was diluted to 10.0 mL. What is the concentration of HCl in the dilute solution?

    (a) 0.10 M   (b) 0.010 M   (c) 1.0×10^{-3} M   (d) 0.25 M

3. Which of the solutions below contains the greatest amount of $OH^-(aq)$?

    (a) 20 mL of 0.10 M NaOH   (c) 10 mL of 0.25 M CsOH

    (b) 15 mL of 0.20 M KOH   (d) 15 mL of 0.15 M LiOH

A CLOSER LOOK

Serial Dilutions

We often find in the laboratory that a solution is too concentrated for the analytical technique we want to use. You might want to analyze a seawater sample for its chloride ion content, for instance. To obtain a solution with a chloride concentration of the proper magnitude for analysis by the Mohr method (Case Study, page 163), for example, you might want to dilute the sample, not once but several times.

Suppose you have 100.0 mL of a seawater sample that has an NaCl concentration of 0.550 mol/L. You transfer 10.00 mL of that sample to a 100.0-mL volumetric flask and fill to the mark with distilled water. You then transfer 5.00 mL of that diluted sample to another 100.0-mL flask and fill to the mark with distilled water. What is the NaCl concentration in the final 100.0-mL sample?

The original solution contains 0.550 mol/L of NaCl. If you remove 10.00 mL, you have removed

$$0.01000 \text{ L} \times 0.550 \text{ mol/L}$$
$$= 5.50 \times 10^{-3} \text{ mol NaCl}$$

and the concentration in 100.0 mL of the diluted solution is

$$c_{NaCl} = 5.50 \times 10^{-3} \text{ mol}/0.1000 \text{ L}$$
$$= 5.50 \times 10^{-2} \text{ M}$$

or 1/10 of the concentration of the original solution (because we diluted the sample by a factor of 10).

Now we take 5.00 mL of the diluted solution and dilute that once again to 100.0 mL. The final concentration is

$$0.00500 \text{ L} \times 5.50 \times 10^{-2} \text{ mol/L}$$
$$= 2.75 \times 10^{-4} \text{ mol NaCl}$$

$$c_{NaCl} = 2.75 \times 10^{-4} \text{ mol}/0.1000 \text{ L}$$
$$= 2.75 \times 10^{-3} \text{ M}$$

This is 1/200 of the concentration of the original solution.

A fair question at this point is why we did not just take 1 mL of the original solution and dilute to 200 mL. The answer is that there is

less error in using larger pipets such as 5.00- or 10.00-mL pipets rather than a 1.00-mL pipet. And then there is a limitation in available glassware. A 200.00-mL volumetric flask is not often available.

Question:
You have a 100.0-mL sample of a blue dye having a concentration of 0.36 M. You dilute a 10.0-mL sample of this to 100.0 mL and then a 2.00-mL sample of that solution to 100.0 mL. What is the final dye concentration?

Answer: 7.2 × 10⁻⁴ M

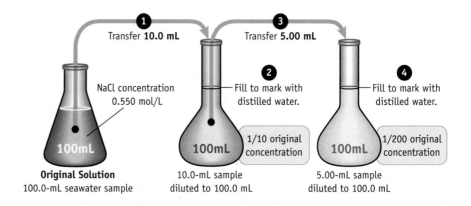

Transfer **10.0 mL**   Transfer **5.00 mL**

NaCl concentration 0.550 mol/L

Fill to mark with distilled water.

Fill to mark with distilled water.

100mL   1/10 original concentration

100mL   1/200 original concentration

100mL

Original Solution
100.0-mL seawater sample

10.0-mL sample diluted to 100.0 mL

5.00-mL sample diluted to 100.0 mL

Module 9: pH covers concepts in this section.

4.6   pH, a Concentration Scale for Acids and Bases

A sample of vinegar, which contains the weak acid acetic acid, has a hydronium ion concentration of 1.6×10^{-3} M and "pure" rainwater has $[H_3O^+] = 2.5 \times 10^{-6}$ M. These small values can be expressed using scientific notation, but a more convenient way to express such numbers is the logarithmic pH scale.

The pH of a solution is the negative of the base-10 logarithm of the hydronium ion concentration.

$$pH = -\log[H_3O^+] \tag{4.3}$$

● **Logarithms** Numbers less than 1 have negative logs. Defining pH as $-\log[H_3O^+]$ produces a positive number. See Appendix A for a discussion of logs.

Taking vinegar, pure water, blood, and ammonia as examples,

pH of vinegar $= -\log (1.6 \times 10^{-3} \text{ M}) = -(-2.80) = 2.80$

pH of pure water (at 25 °C) $= -\log (1.0 \times 10^{-7} \text{ M}) = -(-7.00) = 7.00$

pH of blood $= -\log (4.0 \times 10^{-8} \text{ M}) = -(-7.40) = 7.40$

pH of household ammonia $= -\log (4.3 \times 10^{-12} \text{ M}) = -(-11.37) = 11.37$

you see that something you recognize as acidic has a relatively low pH, whereas ammonia, a common base, has a very low hydronium ion concentration and a high pH.

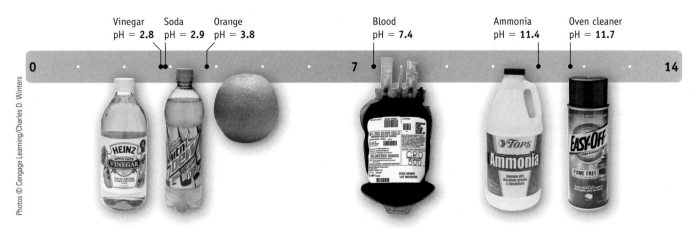

Vinegar pH = **2.8**   Soda pH = **2.9**   Orange pH = **3.8**   Blood pH = **7.4**   Ammonia pH = **11.4**   Oven cleaner pH = **11.7**

Photos © Cengage Learning/Charles D. Winters

FIGURE 4.11 pH values of some common substances. Here, the "bar" is colored red at one end and blue at the other. These are the colors of litmus paper, commonly used in the laboratory to decide whether a solution is acidic (litmus is red) or basic (litmus is blue).

For aqueous solutions at 25 °C, acids have pH values less than 7, bases have values greater than 7, and a pH of 7 represents a neutral solution (Figure 4.11). Blood, which your common sense tells you is likely to be neither acidic nor basic, has a pH slightly greater than 7.

Suppose you know the pH of a solution. To find the hydronium ion concentration you take the antilog of the pH. That is,

$$[H_3O^+] = 10^{-pH} \tag{4.4}$$

For example, the pH of a diet soda is 3.12, and the hydronium ion concentration of the solution is

$$[H_3O^+] = 10^{-3.12} = 7.6 \times 10^{-4} \text{ M}$$

The approximate pH of a solution may be determined using any of a variety of dyes. Litmus paper contains a dye extracted from a type of lichen, but many other dyes are also available (Figure 4.12a). A more accurate measurement of pH is done with a pH meter such as that shown in Figure 4.12b. Here, a pH electrode is immersed in the solution to be tested, and the pH is read from the instrument.

● **Logs and Your Calculator** All scientific calculators have a key marked "log." To find an antilog, use the key marked "10^x" or the inverse log. In determining $[H_3O^+]$ from a pH, when you enter the value of x for 10^x, make sure it has a negative sign.

FIGURE 4.12 Determining pH.

© Cengage Learning/Charles D. Winters

(a) Some household products. Each solution contains a few drops of a universal indicator, a mixture of several acid–base indicators. A color of yellow or red indicates a pH less than 7. A green to purple color indicates a pH greater than 7.

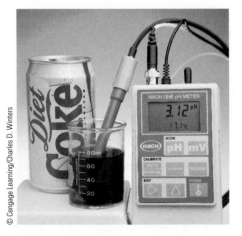

© Cengage Learning/Charles D. Winters

(b) The pH of a soda is measured with a modern pH meter. Soft drinks are often quite acidic, owing to dissolved CO_2 and other ingredients.

● **Weak and Strong Acids and Hydronium Ion Concentration**
Because a weak acid (e.g., acetic acid) does not ionize completely in water, the hydronium ion concentration in an aqueous solution of a weak acid does not equal the concentration of the acid (as is the case for strong acids) but must be calculated using the principles of chemical equilibrium (▶ Chapter 17).

EXAMPLE 4.8   pH of Solutions

Problem

(a)   Lemon juice has $[H_3O^+]$ = 0.0032 M. What is its pH?

(b)   Seawater has a pH of 8.30. What is the hydronium ion concentration of this solution?

(c)   A solution of nitric acid has a concentration of 0.0056 mol/L. What is the pH of this solution?

What Do You Know? In part (a) you are given a concentration and asked to calculate the pH, whereas the opposite is true in (b). For part (c), however, you first must recognize that HNO_3 is a strong acid and is 100% ionized in water.

Strategy Use Equation 4.3 to calculate pH from the H_3O^+ concentration and Equation 4.4 to find $[H_3O^+]$ from the pH.

Solution

(a)   Lemon juice: Because the hydronium ion concentration is known, the pH is found using Equation 4.3.

$$pH = -\log [H_3O^+] = -\log (3.2 \times 10^{-3}) = -(-2.49) = \boxed{2.49}$$

(b)   Seawater: Here pH = 8.30. Therefore,

$$[H_3O^+] = 10^{-pH} = 10^{-8.30} = \boxed{5.0 \times 10^{-9} \text{ M}}$$

(c)   Nitric acid: Nitric acid, a strong acid (Table 3.1, page 111), is completely ionized in aqueous solution. Because the concentration of HNO_3 is 0.0056 mol/L, the ion concentrations are

$$[H_3O^+] = [NO_3^-] = 0.0056 \text{ M}$$

$$pH = -\log [H_3O^+] = -\log (0.0056 \text{ M}) = \boxed{2.25}$$

Think about Your Answer Our answers all agree with the fact that, at 25 °C, solutions with $[H_3O^+] > 10^{-7}$ M have pH < 7, and those with $[H_3O^+] < 10^{-7}$ have pH > 7. A comment on logarithms and significant figures (Appendix A) is useful. The number to the left of the decimal point in a logarithm is called the *characteristic*, and the number to the right is the *mantissa*. The mantissa has as many significant figures as the number whose log was found. For example, the logarithm of 3.2×10^{-3} (two significant figures) is 2.49. (The significant figures are the two numbers to the right of the decimal point.)

Check Your Understanding

(a)   What is the pH of a solution of HCl in which [HCl] = 2.6×10^{-2} M?

(b)   What is the hydronium ion concentration in orange juice with a pH of 3.80?

REVIEW & CHECK FOR SECTION 4.6

1.   Which of the solutions listed below has the lowest pH?

   (a)   0.10 M HCl   (c) 2.5×10^{-5} M HNO_3

   (b)   0.10 M NaOH   (d)   pure H_2O

2.   A 0.365-g sample of HCl is dissolved in enough water to give 2.00×10^2 mL of solution. What is the pH?

   (a)   2.000   (c)   1.301

   (b)   0.0500   (d)   1.000

4.7 Stoichiometry of Reactions in Aqueous Solution

Solution Stoichiometry

Suppose we want to know what mass of $CaCO_3$ is required to react completely with 25 mL of 0.750 M HCl. The first step in finding the answer is to write a balanced equation. In this case, we have a gas-forming exchange reaction involving a metal carbonate and an aqueous acid (Figure 4.13).

$$CaCO_3(s) \quad + \quad 2\ HCl(aq) \rightarrow CaCl_2(aq) + H_2O(\ell) + \quad CO_2(g)$$

metal carbonate + acid → salt + water + carbon dioxide

This problem differs from the previous stoichiometry problems in that the quantity of one reactant (HCl) is given as a volume of a solution of known concentration instead of as a mass in grams. Because our balanced equation is written in terms of amounts (moles), our first step will be to determine the number of moles of HCl present from the known information so that we can relate the amount of HCl available to the amount of $CaCO_3$

$$\text{Amount of HCl} = c_{HCl}V_{HCl} = \frac{0.750\ \text{mol HCl}}{1\ \text{L HCl}} \times 0.025\ \text{L HCl} = 0.019\ \text{mol HCl}$$

This is then related to the amount of $CaCO_3$ required using the stoichiometric factor from the balanced equation.

$$0.019\ \text{mol HCl} \times \frac{1\ \text{mol CaCO}_3}{2\ \text{mol HCl}} = 0.0094\ \text{mol CaCO}_3$$

Finally, the amount of $CaCO_3$ is converted to a mass in grams using the molar mass of $CaCO_3$ as the conversion factor.

$$0.0094\ \text{mol CaCO}_3 \times \frac{100.\ \text{g CaCO}_3}{1\ \text{mol CaCO}_3} = 0.94\ \text{g CaCO}_3$$

If you follow the general scheme outlined in Problem-Solving Tip 4.4 and pay attention to the units on the numbers, you can successfully carry out any kind of stoichiometry calculations involving concentrations.

FIGURE 4.13 A commercial remedy for excess stomach acid. The tablet contains calcium carbonate, which reacts with hydrochloric acid, the acid present in the digestive system. The most obvious product is CO_2 gas, but $CaCl_2(aq)$ is also produced.

© Cengage Learning/Charles D. Winters

PROBLEM SOLVING TIP 4.4

Stoichiometry Calculations Involving Solutions

In Problem-Solving Tip 4.1, you learned about a general approach to stoichiometry problems. We can now modify that scheme for a reaction involving solutions such as $x\ A(aq) + y\ B(aq) \rightarrow$ products.

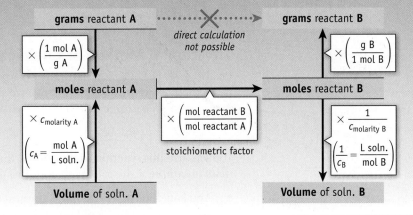

Strategy Map 4.9

PROBLEM

Calculate **volume of HCl solution** required to consume *given mass* of a **reactant (Zn)**

↓

DATA/INFORMATION

- **Mass of Zn**
- **Concentration of HCl**
- Balanced equation

STEP 1. Write balanced equation.

↓

Balanced equation is **given**

STEP 2. Amount reactant (mol) = mass × (1 mol/molar mass).

↓

Amount of Zn

STEP 3. Use **stoichiometric factor** to relate moles Zn to amount (mol) acid required.

↓

Amount of acid required **(mol)**

STEP 4. Volume of acid required = mol HCl required × (1 L/mol HCl).

↓

Volume of acid required **(L)**

UWL   INTERACTIVE EXAMPLE 4.9 **Stoichiometry of a Reaction in Solution**

Problem Metallic zinc reacts with aqueous HCl.

$$Zn(s) + 2\ HCl(aq) \rightarrow ZnCl_2(aq) + H_2(g)$$

What volume of 2.50 M HCl, in milliliters, is required to convert 11.8 g of Zn completely to products?

What Do You Know? The balanced equation for the reaction of Zn and HCl(aq) is provided. You know the mass of zinc and the concentration of HCl(aq).

Strategy

- Calculate the amount of zinc.

- Use a stoichiometric factor (= 2 mol HCl/1 mol Zn) to relate amount of HCl required to amount of Zn available.

- Calculate the volume of HCl solution from the amount of HCl and its concentration.

Solution Begin by calculating the amount of Zn.

$$11.8\ g\ Zn \times \frac{1\ mol\ Zn}{65.38\ g\ Zn} = 0.180\ mol\ Zn$$

Use the stoichiometric factor to calculate the amount of HCl required.

$$0.180\ mol\ Zn \times \frac{2\ mol\ HCl}{1\ mol\ Zn} = 0.360\ mol\ HCl$$

Use the amount of HCl and the solution concentration to calculate the volume.

$$0.360\ mol\ HCl \times \frac{1.00\ L\ solution}{2.50\ mol\ HCl} = 0.144\ L\ HCl$$

The answer is requested in units of milliliters, so we convert the volume to milliliters and find that 144 mL of 2.50 M HCl is required to convert 11.8 g of Zn completely to products.

Think about Your Answer You began with much less than 1 mol of zinc, and the concentration of the HCl solution is 2.50 M. Because the reaction requires 2 mol HCl/1 mol Zn, it makes sense that your answer should be significantly below 1 L of solution needed. Notice also that this is a redox reaction in which zinc is oxidized (oxidation number changes from 0 to +2) and hydrogen, in HCl(aq), is reduced (its oxidation number changes from +1 to 0).

Check Your Understanding

If you combine 75.0 mL of 0.350 M HCl and an excess of Na_2CO_3, what mass of CO_2, in grams, is produced?

$$Na_2CO_3(s) + 2\ HCl(aq) \rightarrow 2\ NaCl(aq) + H_2O(\ell) + CO_2(g)$$

Titration: A Method of Chemical Analysis

Oxalic acid, $H_2C_2O_4$, is a naturally occurring acid. Suppose you are asked to determine the mass of this acid in an impure sample. Because the compound is an acid, it reacts with a base such as sodium hydroxide.

$$H_2C_2O_4(aq) + 2\ NaOH(aq) \rightarrow Na_2C_2O_4(aq) + 2\ H_2O(\ell)$$

You can use this reaction to determine the quantity of oxalic acid present in a given mass of sample if the following conditions are met:

- You can determine when the amount of sodium hydroxide added is just enough to react with all the oxalic acid present in solution.
- You know the concentration of the sodium hydroxide solution and volume that has been added at exactly the point of complete reaction.

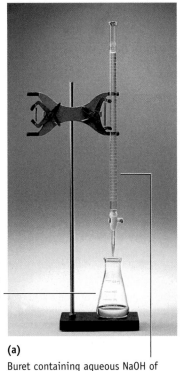

Flask containing aqueous solution of sample being analyzed and an indicator

(a)
Buret containing aqueous NaOH of accurately known concentration.

(b)
A solution of NaOH is added slowly to the sample being analyzed.

(c)
When the amount of NaOH added from the buret equals the amount of H_3O^+ supplied by the acid being analyzed, the dye (indicator) changes color.

Photos © Cengage Learning/Charles D. Winters

FIGURE 4.14 Titration of an acid in aqueous solution with a base. (a) A buret, a volumetric measuring device calibrated in divisions of 0.1 mL (and consequently read to the nearest 0.01 mL), is filled with an aqueous solution of a base of known concentration. **(b)** Base is added slowly from the buret to the solution containing the acid being analyzed and an indicator. **(c)** A change in the color of the indicator signals the equivalence point. (The indicator used here is phenolphthalein.)

These conditions are fulfilled in a titration, a procedure illustrated in Figure 4.14. The solution containing oxalic acid is placed in a flask along with an acid–base indicator, a dye that changes color when the pH of the reaction solution reaches a certain value. Aqueous sodium hydroxide of accurately known concentration is placed in a buret. The sodium hydroxide in the buret is added slowly to the acid solution in the flask. As long as some acid is present in solution, all the base supplied from the buret is consumed, the solution remains acidic, and the indicator color is unchanged. At some point, however, the amount of OH^- added exactly equals the amount of H_3O^+ that can be supplied by the acid. This is called the **equivalence point**. As soon as the slightest excess of base has been added beyond the equivalence point, the solution becomes basic, and the indicator changes color (see Figure 4.14). Example 4.10 shows how to use the equivalence point and the other information to determine the percentage of oxalic acid in a mixture.

H atom lost as H^+

H atom lost as H^+

oxalic acid $H_2C_2O_4$

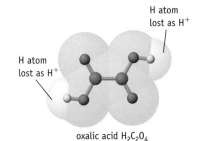

$(-)$

$(-)$

oxalate anion $C_2O_4^{2-}$

Oxalic acid. Oxalic acid has two groups that can supply an H^+ ion to solution. Hence, 1 mol of the acid requires 2 mol of NaOH for complete reaction.

⬙WL **INTERACTIVE EXAMPLE 4.10** Acid–Base Titration

Problem A 1.034-g sample of impure oxalic acid is dissolved in water and an acid–base indicator added. The sample requires 34.47 mL of 0.485 M NaOH to reach the equivalence point. What is the mass of oxalic acid, and what is its mass percent in the sample?

What Do You Know? You know the mass of oxalic acid (the formula is $H_2C_2O_4$) and the volume and concentration of NaOH solution used in the titration.

Strategy Map 4.10

PROBLEM

Calculate the **mass percent of acid** in an impure sample. Determine the **acid content** using an *acid-base titration*.

↓

DATA/INFORMATION

• **Mass** of impure sample containing *acid*.
• **Volume** and **concentration** of *base* used in titration.

> **STEP 1.** Write balanced equation.

Balanced equation for reaction of **acid** (oxalic acid) with **base** (NaOH)

> **STEP 2. Amount** of base (mol) = volume (L) × (mol/L).

Amount of base (mol)

> **STEP 3.** Use a **stoichiometric factor** to relate amount of base (mol) to amount (mol) of acid.

Amount of acid (mol) in impure sample

> **STEP 4. Mass** of acid in sample = mol acid × (g/mol).

Mass of acid in impure sample

> **STEP 5.**
> **Mass %** of acid in sample = (g acid/g sample)100%.

Mass % acid in impure sample

Strategy

• Write a balanced chemical equation for this acid–base reaction.

• Calculate the amount of NaOH used in the titration from its volume and concentration.

• Use the stoichiometric factor defined by the equation to determine the amount of $H_2C_2O_4$.

• Calculate the mass of $H_2C_2O_4$ from the amount and its molar mass.

• Determine the percent by mass of $H_2C_2O_4$ in the sample.

Solution The balanced equation for the reaction of NaOH and $H_2C_2O_4$ is

$$H_2C_2O_4(aq) + 2\ NaOH(aq) \rightarrow Na_2C_2O_4(aq) + 2\ H_2O(\ell)$$

and the amount of NaOH is given by

$$\text{Amount of NaOH} = c_{NaOH} \times V_{NaOH} = \frac{0.485\ \text{mol NaOH}}{\ell} \times 0.03447\ \ell = 0.0167\ \text{mol NaOH}$$

The balanced equation for the reaction shows that 1 mol of oxalic acid requires 2 mol of sodium hydroxide. This is the required stoichiometric factor to obtain the amount of oxalic acid present.

$$0.0167\ \text{mol NaOH} \times \frac{1\ \text{mol } H_2C_2O_4}{2\ \text{mol NaOH}} = 0.00836\ \text{mol } H_2C_2O_4$$

The mass of oxalic acid is found from the amount of the acid and its molar mass.

$$0.00836\ \text{mol } H_2C_2O_4 \times \frac{90.04\ \text{g } H_2C_2O_4}{1\ \text{mol } H_2C_2O_4} = 0.753\ \text{g } H_2C_2O_4$$

This mass of oxalic acid represents 72.8% of the total sample mass.

$$\frac{0.753\ \text{g } H_2C_2O_4}{1.034\ \text{g sample}} \times 100\% = 72.8\%\ H_2C_2O_4$$

Think about Your Answer Problem Solving Tip 4.4 outlines the procedure used to solve this problem.

Check Your Understanding

A 25.0-mL sample of vinegar (which contains the weak acid acetic acid, CH_3CO_2H) requires 28.33 mL of a 0.953 M solution of NaOH for titration to the equivalence point. What is the mass of acetic acid, in grams, in the vinegar sample, and what is the concentration of acetic acid in the vinegar?

$$CH_3CO_2H(aq) + NaOH(aq) \rightarrow NaCH_3CO_2(aq) + H_2O(\ell)$$

Standardizing an Acid or Base

In Example 4.10 the concentration of the base used in the titration was given. In actual practice this usually has to be found by a prior measurement. The procedure by which the concentration of an analytical reagent is determined accurately is called **standardization**, and there are two general approaches.

One approach is to weigh accurately a sample of a pure, solid acid or base (known as a *primary standard*) and then titrate this sample with a solution of the base or acid to be standardized (Example 4.11). An alternative approach to standardizing a solution is to titrate it with another solution that is already standardized (see "Check Your Understanding" in Example 4.11). This is often done using standard solutions purchased from chemical supply companies.

EXAMPLE 4.11   Standardizing an Acid by Titration

Problem Sodium carbonate, Na_2CO_3, is a base, and an accurately weighed sample can be used to standardize an acid. A sample of sodium carbonate (0.263 g) requires 28.35 mL of aqueous HCl for titration to the equivalence point. What is the concentration of the HCl?

What Do You Know? The concentration of the HCl(aq) solution is the unknown in this problem. You know the mass of Na_2CO_3 and the volume of HCl(aq) solution needed to react completely with the Na_2CO_3. You need the molar mass of Na_2CO_3 and a balanced equation for the reaction.

Strategy

- Write a balanced equation for this acid–base reaction.

- Calculate the amount of Na_2CO_3 from the mass and molar mass.

- Use the stoichiometric factor (from the equation) to find the amount of HCl(aq).

- The amount of HCl divided by the volume of solution (in liters) gives its concentration (mol/L).

Solution The balanced equation for the reaction is written first.

$$Na_2CO_3(aq) + 2\ HCl(aq) \rightarrow 2\ NaCl(aq) + H_2O(\ell) + CO_2(g)$$

Calculate the amount of the base, Na_2CO_3, from its mass.

$$0.263\ g\ Na_2CO_3 \times \frac{1\ mol\ Na_2CO_3}{106.0\ g\ Na_2CO_3} = 0.00248\ mol\ Na_2CO_3$$

Next, use the stoichiometric factor to calculate the amount of HCl in 28.35 mL.

$$0.00248\ mol\ Na_2CO_3 \times \frac{2\ mol\ HCl\ required}{1\ mol\ Na_2CO_3\ available} = 0.00496\ mol\ HCl$$

Finally, the 28.35-mL (0.02835-L) sample of aqueous HCl contains 0.00496 mol of HCl, so the concentration of the HCl solution is 0.175 M.

$$[HCl] = \frac{0.00496\ mol\ HCl}{0.02835\ L} = \boxed{0.175\ M\ HCl}$$

Think about Your Answer Sodium carbonate is commonly used as a primary standard. It can be obtained in pure form, can be weighed accurately, and it reacts completely with strong acids.

Check Your Understanding

Hydrochloric acid, HCl, can be purchased from chemical supply houses with a concentration of 0.100 M, and this solution can be used to standardize the solution of a base. If titrating 25.00 mL of a sodium hydroxide solution to the equivalence point requires 29.67 mL of 0.100 M HCl, what is the concentration of the base?

Determining Molar Mass by Titration

In Chapter 2 and in this chapter we used analytical data to determine the empirical formula of a compound. The molecular formula could then be derived if the molar mass were known. If the unknown substance is an acid or a base, it is possible to determine the molar mass by titration.

ⓦWL   INTERACTIVE EXAMPLE 4.12   Determining the Molar Mass of an Acid by Titration

Problem To determine the molar mass of an organic acid, HA, we titrate 1.056 g of HA with standardized NaOH. Calculate the molar mass of HA assuming the acid reacts with 33.78 mL of 0.256 M NaOH according to the equation

$$HA(aq) + NaOH(aq) \rightarrow NaA(aq) + H_2O(\ell)$$

Strategy Map 4.12

PROBLEM
Calculate the **molar mass** of an acid, **HA,** using an *acid-base titration*.

↓

DATA/INFORMATION
• **Mass** of *acid* sample.
• **Volume** and **concentration** of *base* used in titration

↓

> STEP 1. Write balanced equation.

Balanced equation for reaction of acid (HA) with **base** (NaOH)

↓

> STEP 2. **Amount** of base (mol) = volume (L) × (mol/L).

Amount of base (mol)

↓

> STEP 3. Use a **stoichiometric factor** to relate amount of base (mol) to amount (mol) of acid.

Amount of acid **HA** (mol)

↓

> STEP 4. **Molar mass** = mass of acid in sample/amount of acid in sample.

Molar mass of acid **HA**

What Do You Know? You know the mass of the sample of unknown acid, and the volume and concentration of NaOH(aq). From the equation given, you know that the acid and base react in a 1:1 fashion.

Strategy The key to this problem is to recognize that the molar mass of a substance is the ratio of the mass of the sample (g) to its amount (mol). You know the mass, but you need to determine the amount equivalent to that mass. The equation informs you that 1 mol of HA reacts with 1 mol of NaOH, so the amount of HA equals the amount of NaOH used in the titration. And, the amount of NaOH can be calculated from its concentration and volume.

Solution Let us first calculate the amount of NaOH used in the titration.

$$\text{Amount of NaOH} = c_{\text{NaOH}}V_{\text{NaOH}} = \frac{0.256 \text{ mol}}{L} \times 0.03378 \, L = 8.65 \times 10^{-3} \text{ mol NaOH}$$

Next, recognize from the balanced equation that the amount of NaOH used in the titration is the same as the amount of acid titrated. That is,

$$8.65 \times 10^{-3} \text{ mol NaOH} \times \frac{1 \text{ mol HA}}{1 \text{ mol NaOH}} = 8.65 \times 10^{-3} \text{ mol HA}$$

Finally, calculate the molar mass of HA.

$$\text{Molar mass of acid} = \frac{1.056 \text{ g HA}}{8.65 \times 10^{-3} \text{ mol HA}} = 122 \text{ g/mol}$$

Think about Your Answer Note that we *assumed* the acid had the composition HA. That is, it could lose one H^+ ion per molecule. If the acid had the composition H_2A (such as oxalic acid, page 159) or H_3A (for example, citric acid, page 111), the calculated molar mass would not be correct.

Check Your Understanding

An acid reacts with NaOH according to the net ionic equation

$$HA(aq) + OH^-(aq) \longrightarrow A^-(aq) + H_2O(\ell)$$

Calculate the molar mass of HA if 0.856 g of the acid requires 30.08 mL of 0.323 M NaOH.

Titrations Using Oxidation–Reduction Reactions

Analysis by titration is not limited to acid–base chemistry. Many oxidation–reduction reactions go rapidly to completion in aqueous solution, and methods exist to determine their equivalence points.

EXAMPLE 4.13   Using an Oxidation–Reduction Reaction in a Titration

Problem The iron in a sample of an iron ore can be converted quantitatively to the iron(II) ion, Fe^{2+}, in aqueous solution, and this solution can then be titrated with aqueous potassium permanganate, $KMnO_4$. The balanced, net ionic equation for the reaction occurring in the course of this titration is

$$\underset{\text{purple}}{MnO_4^-(aq)} + \underset{\text{colorless}}{5 \text{ Fe}^{2+}(aq)} + 8 \text{ H}_3O^+(aq) \longrightarrow \underset{\text{colorless}}{Mn^{2+}(aq)} + \underset{\text{pale yellow}}{5 \text{ Fe}^{3+}(aq)} + 12 \text{ H}_2O(\ell)$$

A 1.026-g sample of iron-containing ore requires 24.35 mL of 0.0195 M $KMnO_4$ to reach the equivalence point. What is the mass percent of iron in the ore?

What Do You Know? You know the concentration and volume of the $KMnO_4$ solution used to titrate Fe^{2+}(aq) to the equivalence point. The stoichiometric factor relating amounts of $KMnO_4$ and Fe^{2+}(aq) is derived from the balanced equation.

How Much Salt Is There in Seawater?

Saltiness is one of the basic taste sensations, and a taste of seawater quickly reveals it is salty. How did the oceans become salty?

A result of the interaction of atmospheric CO_2 and water is hydronium ions and bicarbonate ions.

$$CO_2(g) + H_2O(\ell) \rightarrow H_2CO_3(aq)$$

$$H_2CO_3(aq) + H_2O(\ell)$$
$$\rightleftharpoons H_3O^+(aq) + HCO_3^-(aq)$$

Indeed, this is the reason rain is normally acidic, and this slightly acidic rainwater can then cause substances such as limestone or corals to dissolve, producing calcium ions and more bicarbonate ions.

$$CaCO_3(s) + H_3O^+(aq) \rightarrow$$
$$Ca^{2+}(aq) + HCO_3^-(aq) + H_2O(\ell)$$

Sodium ions arrive in the oceans by a similar reaction with sodium-bearing minerals such as albite, $NaAlSi_3O_6$. Acidic rain falling on the land extracts sodium ions that are then carried by rivers to the ocean.

The average chloride content of rocks in the earth's crust is only 0.01%, so only a minute proportion of the chloride ion in the oceans can come from the weathering of rocks and minerals. What then is the origin of the chloride ions in seawater? The answer is volcanoes. Hydrogen chloride gas, HCl, is a constituent of volcanic gases. Early in Earth's history, the planet was much hotter, and volcanoes were much more widespread.

The HCl gas emitted from these volcanoes is very soluble in water and quickly dissolves to give a dilute solution of hydrochloric acid. The chloride ions from dissolved HCl gas and sodium ions from weathered rocks are the source of the salt in the sea.

Suppose you are an oceanographer, and you want to determine the concentration of chloride ions in a sample of seawater. How can you do this? And what results might you find?

There are several ways to analyze a solution for its chloride ion content; among them is the classic "Mohr method." Here, a solution containing chloride ions is titrated with standardized silver nitrate. You know that the following reaction should occur:

$$Ag^+(aq) + Cl^-(aq) \rightarrow AgCl(s)$$

and will continue until the chloride ions have been precipitated completely. To detect the equivalence point of the titration of Cl^- with Ag^+, the Mohr method involves the addition of a few drops of a solution of potassium chromate. This "indicator" works because silver chromate is slightly more soluble than AgCl, so the red Ag_2CrO_4 precipitates only after all of the AgCl is precipitated.

$$2 Ag^+(aq) + CrO_4^{2-}(aq) \rightarrow Ag_2CrO_4(s)$$

The appearance of the red color of Ag_2CrO_4 (◄ Figure 3.12d) signals the equivalence point.

John C. Kotz

Question:
Using the following information, calculate the chloride ion concentration in a sample of seawater.

a. Volume of original seawater sample = 100.0 mL.

b. A 10.00-mL sample of the seawater was diluted to 100.0 mL with distilled water.

c. 10.00 mL of the diluted sample was again diluted to 100.0 mL.

d. A Mohr titration was performed on 50.00 mL of the diluted sample (from step c) and required 26.25 mL of 0.100 M $AgNO_3$. What was the chloride ion concentration in the original seawater sample?

The answer to this question is available in Appendix N.

Strategy

- Use the volume and concentration of the $KMnO_4$ solution to calculate the amount of $KMnO_4$ used in the titration.

- Use the stoichiometric factor to determine the amount of Fe^{2+} from the amount of $KMnO_4$.

- Convert the amount of Fe^{2+} to mass of iron using the molar mass of iron.

- Calculate the mass percent of iron in the sample.

Solution First, calculate the amount of $KMnO_4$.

$$\text{Amount of } KMnO_4 = c_{KMnO_4} \times V_{KMnO_4} = \frac{0.0195 \text{ mol } KMnO_4}{\ell} \times 0.02435 \text{ } \ell = 0.000475 \text{ mol}$$

Use the stoichiometric factor to calculate the amount of iron(II) ion.

$$0.000475 \text{ mol } KMnO_4 \times \frac{5 \text{ mol } Fe^{2+}}{1 \text{ mol } KMnO_4} = 0.00237 \text{ mol } Fe^{2+}$$

© Cengage Learning/Charles D. Winters

Using an oxidation–reduction reaction for analysis by titration. Purple, aqueous $KMnO_4$ is added to a solution containing Fe^{2+}. As $KMnO_4$ drops into the solution, colorless Mn^{2+} and pale yellow Fe^{3+} form. Here, an area of the solution containing unreacted $KMnO_4$ is seen. As the solution is mixed, this disappears until the equivalence point is reached.

Next, calculate the mass of iron.

$$0.00237 \text{ mol Fe}^{2+} \times \frac{55.85 \text{ g Fe}^{2+}}{1 \text{ mol Fe}^{2+}} = 0.133 \text{ g Fe}^{2+}$$

Finally, the mass percent can be determined.

$$\frac{0.133 \text{ g Fe}^{2+}}{1.026 \text{ g sample}} \times 100\% = \boxed{12.9\% \text{ iron}}$$

Think about Your Answer The mass of the sample is 1.026 g, so the mass of iron in the sample must be less than this, as confirmed by the answer. The titration of iron(II) ions with $KMnO_4$ is a useful analytical reaction because it is easy to detect when all the iron(II) ion has reacted. The MnO_4^- ion is a deep purple color, but when it reacts with Fe^{2+} the color disappears because the reaction product, Mn^{2+}, is colorless. Therefore, $KMnO_4$ solution is added from a buret until the initially colorless, Fe^{2+}-containing solution just turns a faint purple color (due to unreacted $KMnO_4$), the signal that the equivalence point has been reached.

Check Your Understanding

Vitamin C, ascorbic acid ($C_6H_8O_6$), is a reducing agent. One way to determine the ascorbic acid content of a sample is to mix the acid with an excess of iodine,

$$C_6H_8O_6(aq) + I_2(aq) + 2 H_2O(\ell) \rightarrow C_6H_6O_6(aq) + 2 H_3O^+(aq) + 2 I^-(aq)$$

and then titrate the iodine that did not react with the ascorbic acid with sodium thiosulfate. The balanced, net ionic equation for the reaction occurring in this titration is

$$I_2(aq) + 2 S_2O_3^{2-}(aq) \rightarrow 2 I^-(aq) + S_4O_6^{2-}(aq)$$

Suppose 50.00 mL of 0.0520 M I_2 was added to the sample containing ascorbic acid. After the ascorbic acid/I_2 reaction was complete, the I_2 not used in this reaction required 20.30 mL of 0.196 M $Na_2S_2O_3$ for titration to the equivalence point. Calculate the mass of ascorbic acid in the unknown sample.

REVIEW & CHECK FOR SECTION 4.7

1. A 25-mL sample of 0.40 M HCl is added to 32 mL of 0.30 M NaOH. Is the resulting solution acidic, basic, or neutral?

   (a) acidic   (b) basic   (c) neutral

2. What volume of 0.250 M NaOH is required to react completely with 0.0100 moles of H_2SO_4?

   (a) 80.0 mL   (b) 60.0 mL   (c) 40.0 mL   (d) 125 mL

4.8 Spectrophotometry

Solutions of many compounds are colored, a consequence of the absorption of light (Figure 4.15). It is possible to measure, quantitatively, the extent of light absorption and to relate this to the concentration of the dissolved solute. This kind of experiment, called **spectrophotometry**, is an important analytical method and you may have used it in your laboratory course.

Every substance absorbs or transmits certain wavelengths of radiant energy but not others (Figures 4.15 and 4.16). For example, nickel(II) ions (and chlorophyll) absorb red and blue/violet light while they transmit green light. Your eyes "see" the transmitted or reflected wavelengths, those not absorbed, as the color green. Furthermore, the specific wavelengths absorbed and transmitted are characteristic for a substance, so a spectrum serves as a "fingerprint" of the substance that can help identify an unknown.

CASE STUDY

Forensic Chemistry: Titrations and Food Tampering

The U.S. Food and Drug Administration (FDA) has discovered cases of product tampering involving the addition of bleach to products such as soup, infant formula, and soft drinks. Household bleach is a dilute solution of sodium hypochlorite (NaClO), a compound that is an oxidizing agent and is dangerous if swallowed.

One method of detecting bleach uses starch-iodide paper. The bleach oxidizes the iodide ion to iodine in an acid solution,

$$2 \, I^-(aq) + HClO(aq) + H_3O^+(aq) \rightarrow$$
$$I_2(aq) + 2 \, H_2O(\ell) + Cl^-(aq)$$

and the presence of I_2 is detected by a deep blue color in the presence of starch.

This reaction is also used in the quantitative analysis of solutions containing bleach. Excess iodide ion (in the form of KI) is added to the sample. The bleach in the sample (which forms HClO in acid solution) oxidizes I^- in a ratio of 1 mol HClO to 2 mol I^-. The iodine formed in the reaction is then titrated with sodium thiosulfate, $Na_2S_2O_3$ in another oxidation-reduction reaction (as in "Check Your Understanding" in Example 4.13).

$$I_2(aq) + 2 \, S_2O_3^{2-}(aq) \rightarrow$$
$$2 \, I^-(aq) + S_4O_6^{2-}(aq)$$

The amount of $Na_2S_2O_3$ used in the titration can then be used to determine the amount of NaClO in the original sample.

Question:
Excess KI is added to a 100.0-mL sample of a soft drink that had been contaminated with bleach, NaClO. The iodine (I_2) generated in the solution is then titrated with 0.0425 M $Na_2S_2O_3$ and requires 25.3 mL to reach the equivalence point. What mass of NaClO was contained in the 100.0-mL sample of adulterated soft drink?

The answer to this question is available in Appendix N.

© Cengage Learning/Charles D. Winters

A distinctive blue color is generated when iodine reacts with water-soluble starch.

© Cengage Learning/Charles D. Winters

FIGURE 4.15 Light absorption and color. A beam of white light shines on a solution of nickel(II) ions in water, and the light that emerges is green. The color of a solution is due to the color of the light *not* absorbed by the solution. Here, red and blue/violet light was absorbed, and green light is transmitted.

Now suppose you look at two solutions containing copper(II) ions, one a deeper color than the other. Your common sense tells you that the intensely colored one is the more concentrated (Figure 4.17a). This is true: the intensity of the color is a measure of the concentration of the material in the solution.

In recent years, spectrophotometry has become one of the most frequently used methods of quantitative analysis. It is applicable to many industrial, clinical, and forensic problems involving the quantitative determination of compounds that are colored or that react to form a colored product.

Transmittance, Absorbance, and the Beer–Lambert Law

To understand the exact relationship of light absorption and solution concentration, we need to define several terms. **Transmittance** (*T*) is the ratio of the amount

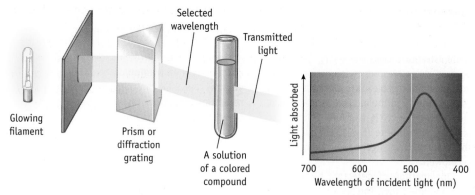

FIGURE 4.16 **An absorption spectrophotometer.** A beam of white light passes through a prism or diffraction grating, which splits the light into its component wavelengths. After passing through the sample, the light reaches a detector. The spectrophotometer "scans" all wavelengths of light and determines the amount of light absorbed at each wavelength. The output is a *spectrum*, a plot of the amount of light absorbed as a function of the wavelength or frequency of the incoming or incident light. Here, the sample absorbs light in the green-blue part of the spectrum and transmits light in the remaining wavelengths. The sample would appear red to orange to your eye.

of light transmitted by or passing through the sample relative to the amount of light that initially fell on the sample (the incident light).

$$\text{Transmittance } (T) = \frac{P}{P_0} = \frac{\text{intensity of transmitted light}}{\text{intensity of incident light}}$$

The **absorbance** of a sample is defined as the negative logarithm of its transmittance. That is, absorbance and transmittance have an inverse relationship. As the absorbance of a solution increases, the transmittance decreases

$$\text{Absorbance} = -\log T = -\log P/P_0$$

Going back to our example of an aqueous solution of copper(II) ions in Figure 4.17a, if you have two solutions of copper(II) sulfate, you may deduce that the bluer solution appears more blue because it absorbs more of the light falling on it

FIGURE 4.17 Light absorption, concentration, and path length.

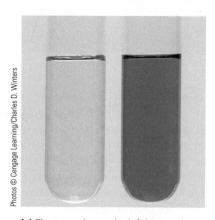

(a) The test tube on the left has a solution of copper(II) sulfate with a concentration of 0.05 M. On the right, the concentration is 1.0 M in copper(II) sulfate. More light is absorbed by the more concentrated sample, and it appears more blue.

(b) The amount of light absorbed by a solution depends on the path length. Here, both solutions have the same concentration, but the distance the light travels is longer in one than the other.

Photos © Cengage Learning/Charles D. Winters

and that this solution has a greater concentration of copper(II) sulfate. That is, the *absorbance, A, of a sample increases as the concentration increases.*

Next, suppose that there are two test tubes, both containing the same solution at the same concentration. The only difference is that one of the test tubes has a smaller diameter than the other (Figure 4.17b). We shine light of the same intensity (P_0) on both test tubes. In the first case, the light has to travel only a short distance through the sample before its strikes your eyes, whereas in the second case it has to pass through more of the sample. In the second case more of the light will be absorbed because the path length is longer. In other words, *absorbance increases as path length increases.*

The two observations described above constitute the **Beer–Lambert law:**

$$\text{Absorbance } (A) \propto \text{path length } (\ell) \times \text{concentration } (c)$$
$$A = \varepsilon \times \ell \times c$$

(4.5)

where

- A, the absorbance of the sample, is a dimensionless number.
- ε, a proportionality constant, is called the *molar absorptivity.* It is a constant for a given substance, provided the temperature and wavelength are constant. It has units of L/mol · cm.
- ℓ and c have the units of length (cm) and concentration (mol/L), respectively.

The important point is that the Beer–Lambert law shows *there is a linear relationship between a sample's absorbance and its concentration for a given path length.*

Spectrophotometric Analysis

There are usually four steps in carrying out a spectrophotometric analysis.

- **Record the absorption spectrum of the substance to be analyzed.** In introductory chemistry laboratories, this is often done using instruments such as the ones shown in Figure 4.18. The result is a spectrum such as that for aqueous permanganate ions (MnO_4^-) in Figure 4.19. The spectrum is a plot of the absorbance of the sample versus the wavelength of incident light. Here, the maximum in absorbance is at about 525 nm.
- **Choose the wavelength for the measurement.** According to the Beer–Lambert law, the absorbance at each wavelength is proportional to concentration. Therefore, in theory we could choose any wavelength for quantitative estimations of concentration. However, the magnitude of the absorbance is important, especially when you are trying to detect very small amounts of material. In the

● **Beer–Lambert Law** The Beer–Lambert law applies strictly to relatively dilute solutions. At higher solute concentrations, the dependence of absorbance on concentration may not be linear.

Spectronic 20 from Spectronic Instruments.

Ocean Optics spectrometer (where the digital data are collected by a computer).

FIGURE 4.18 Spectrophotometers. The instruments illustrated here are often found in introductory chemistry laboratories.

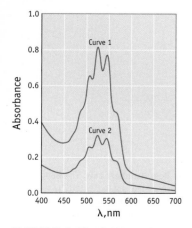

FIGURE 4.19 **The absorption spectrum of solutions of potassium permanganate (KMnO₄) at different concentrations.** The solution for curve 1 has a higher concentration than that for curve 2.

spectra of permanganate ions in Figure 4.19, note that the difference in absorbance between curves 1 and 2 is largest at about 525 nm, and at this wavelength the change in absorbance is greatest for a given change in concentration. That is, the measurement of concentration as a function of concentration is most sensitive at this wavelength. For this reason, *we generally select the wavelength of maximum absorbance for our measurements.*

- **Prepare a calibration plot.** Once we have chosen the wavelength, the next step is to construct a **calibration curve** or **calibration plot** at this wavelength. This consists of a plot of absorbance versus concentration for a series of standard solutions whose concentrations are accurately known. Because of the linear relation between concentration and absorbance (at a given wavelength and path length), this plot is a straight line with a positive slope. (You will prepare a calibration plot in Example 4.14.)

- **Determine the concentration of the species of interest in other solutions.** Once the calibration plot has been made, and the equation for the line is known, you can find the concentration of an unknown sample from its absorbance.

Example 4.14 illustrates the preparation of a calibration curve and its use in determining the concentration of a species in solution.

EXAMPLE 4.14 Using Spectrophotometry in Chemical Analysis

Problem A solution of KMnO₄ has an absorbance of 0.539 when measured at 540 nm in a 1.0-cm cell. What is the concentration of the KMnO₄? Prior to determining the absorbance for the unknown solution, the following calibration data were collected for the spectrophotometer.

Concentration of KMnO₄ (M)	Absorbance
0.0300	0.162
0.0600	0.330
0.0900	0.499
0.120	0.670
0.150	0.840

What Do You Know? The table relates concentration and absorbance for KMnO₄. The absorbance of the unknown sample is given.

Strategy Prepare a calibration plot from the data given above. You can then use this plot to estimate the concentration of the unknown from its absorbance. A more accurate value of the concentration can be obtained if you find the equation for the straight line in the calibration plot and calculate the unknown concentration using this equation.

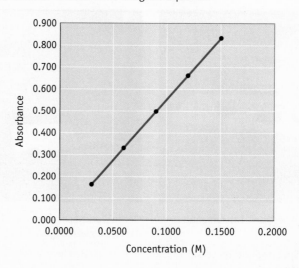

Solution Using Microsoft Excel (or equivalent software) or a calculator, prepare a calibration plot from the experimental data. The equation for the straight line (as determined using Excel) is

$$y = 5.653x - 0.009$$

$$\text{Absorbance} = 5.653 \,(\text{Conc}) - 0.009$$

If we put in the absorbance for the unknown solution,

$$0.539 = 5.653 \,(\text{Conc}) - 0.009$$

$$\text{Unknown concentration} = 0.0969$$

Think about Your Answer The absorbance of the unknown was 0.539. Looking back at our calibration data, we can see that this absorbance falls between the data points for absorbances of 0.499 and 0.670. Our answer determined for the concentration of $KMnO_4$ in the unknown (0.0969 M) falls between the concentrations for these two data points of 0.0900 M and 0.120 M, as it should (see pages 37 and 38 for information on graphing).

Check Your Understanding

Using the following data, calculate the concentration of copper(II) ions in the unknown solution. (The cell path length is 1.00 cm in all cases, and the wavelength used in the determination was 645 nm.)

Calibration Data

Concentration of Cu^{2+} (M)	Absorbance
0.0562	0.720
0.0337	0.434
0.0281	0.332
0.0169	0.219

Absorbance of unknown solution containing Cu^{2+} ions = 0.418

REVIEW & CHECK FOR SECTION 4.8

How is the absorbance (A) of a solution affected if a solution is diluted to twice the initial volume?

(a) A is doubled.   (c) A is not changed.

(b) A is reduced to half the original value.

CHAPTER GOALS REVISITED

Now that you have completed this chapter, you should ask whether you have met the chapter goals. In particular, you should be able to

Perform stoichiometry calculations using balanced chemical equations

a. Understand the principle of the conservation of matter, which forms the basis of chemical stoichiometry. **Study Question: 133.**

b. Calculate the mass of one reactant or product from the mass of another reactant or product by using the balanced chemical equation (Section 4.1). **Study Questions: 2, 3, 5, 77, 81, 93, 95, 97, 100, and Go Chemistry Module 7.**

c. Use amounts tables to organize stoichiometric information. **Study Questions: 7–10.**

 and

Sign in at **www.cengage.com/owl** to:
- View tutorials and simulations, develop problem-solving skills, and complete online homework assigned by your professor.
- For quick review and exam prep, download Go Chemistry mini lecture modules from OWL (or purchase them at **www.cengagebrain.com**)

Access **How Do I Solve It?** tutorials on how to approach problem solving using concepts in this chapter.

Understand the meaning of a limiting reactant in a chemical reaction

a. Determine which of two reactants is the limiting reactant (Section 4.2). **Study Questions: 11–18, 96, 136, and Go Chemistry Module 8.**

b. Determine the yield of a product based on the limiting reactant. **Study Questions: 11–18.**

Calculate the theoretical and percent yields of a chemical reaction

Explain the differences among actual yield, theoretical yield, and percent yield, and calculate percent yield (Section 4.3). **Study Question: 21.**

Use stoichiometry to analyze a mixture of compounds or to determine the formula of a compound

a. Use stoichiometry principles to analyze a mixture (Section 4.4). **Study Questions: 23, 25, 27, 127, 129, 130.**

b. Find the empirical formula of an unknown compound using chemical stoichiometry (Section 4.4). **Study Questions: 29, 31, 33–35.**

Define and use concentrations in solution stoichiometry

a. Calculate the concentration of a solute in a solution in units of moles per liter (molarity), and use concentrations in calculations (Section 4.5). **Study Questions: 37, 39, 41.**

b. Describe how to prepare a solution of a given concentration from the solute and a solvent or by dilution from a more concentrated solution (Section 4.5). **Study Questions: 45, 46, 47, 51.**

c. Calculate the pH of a solution from the concentration of hydronium ion in the solution. Calculate the hydronium ion concentration of a solution from the pH (Section 4.6). **Study Questions: 53–58, and Go Chemistry Module 9.**

d. Solve stoichiometry problems using solution concentrations (Section 4.7). **Study Questions: 59, 61, 62, 63, 65, 106, 107.**

e. Explain how a titration is carried out, explain the procedure of standardization, and calculate concentrations or amounts of reactants from titration data (Section 4.7). **Study Questions: 67, 69, 71, 73.**

f. Understand and use the principles of spectrophotometry to determine the concentration of a species in solution (Section 4.8). **Study Question: 75.**

Key Equations

Equation 4.1 (page 142) Percent yield.

$$\text{Percent yield} = \frac{\text{actual yield}}{\text{theoretical yield}} \times 100\%$$

Equation 4.2 (page 149) Definition of molarity, a measure of the concentration of a solute in a solution.

$$\text{Molarity of } x\ (c_x) = \frac{\text{amount of solute } x\ (\text{mol})}{\text{volume of solution (L)}}$$

A useful form of this equation is

$$\text{Amount of solute } x\ (\text{mol}) = c_x\ (\text{mol/L}) \times \text{volume of solution (L)}$$

Dilution Equation (page 153) This is a shortcut to find, for example, the concentration of a solution (c_d) after diluting some volume (V_c) of a more concentrated solution (c_c) to a new volume (V_d).

$$c_c \times V_c = c_d \times V_d$$

Equation 4.3 (page 154) pH. The pH of a solution is the negative logarithm of the hydronium ion concentration.

$$\text{pH} = -\log[\text{H}_3\text{O}^+]$$

Equation 4.4 (page 155) Calculating $[H_3O^+]$ from pH. The equation for calculating the hydronium ion concentraton of a solution from the pH of the solution.

$$[H_3O^+] = 10^{-pH}$$

Equation 4.5 (page 167) Beer–Lambert law. The absorbance of light (A) by a substance in solution is equal to the product of the molar absorptivity of the substance (ε), the pathlength of the cell (ℓ), and the concentration of the solute (c).

$$\text{Absorbance } (A) \propto \text{path length } (\ell) \times \text{concentration } (c)$$

$$A = \varepsilon \times \ell \times c$$

OWL Questions and problems for this chapter are available in OWL. Use the OWL access card that came with this textbook to access assigned questions and problems for this chapter.

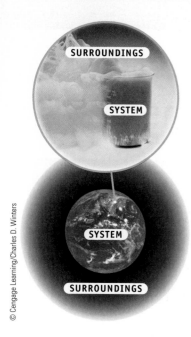

FIGURE 5.1 **Systems and their surroundings.** Earth can be considered a thermodynamic system, with the rest of the universe as its surroundings. A chemical reaction occurring in a laboratory is also a system, with the laboratory its surroundings.

• **Thermal Equilibrium** Although no change is evident at the macroscopic level once thermal equilibrium is reached, on the molecular level transfer of energy between individual molecules will continue. A general feature of systems at equilibrium is that there is no change on a macroscopic level but that processes still occur at the particulate level. (See Section 3.3, page 98.)

Systems and Surroundings

In thermodynamics, the terms *system* and *surroundings* have precise and important scientific meanings. A **system** is defined as an object, or collection of objects, being studied (Figure 5.1). The **surroundings** include everything outside the system that can exchange energy and/or matter with the system. In the discussion that follows, we will need to define systems precisely. If we are studying the energy evolved in a chemical reaction carried out in solution, for example, the system might be defined as the reactants, products, and solvent. The surroundings would be the reaction vessel and the air in the room and anything else in contact with the vessel with which it might exchange energy or matter. At the atomic level, the system could be a single atom or molecule, and the surroundings would be the atoms or molecules in its vicinity. How the system and its surroundings for each situation are defined depends on the information we are trying to obtain or convey.

This concept of a system and its surroundings applies to nonchemical situations as well. If we want to study the energy balance on our planet, we might choose to define Earth as the system and outer space as the surroundings. On a cosmic level, the solar system might be defined as the system being studied, and the rest of the galaxy would be the surroundings.

Directionality and Extent of Transfer of Heat: Thermal Equilibrium

Much of our initial study of thermodynamics will deal with the ways that energy can be transferred between a system and its surroundings or between different parts of the system. One way that energy can be transferred is as heat. Energy is transferred as heat if two objects at different temperatures are brought into contact. In Figure 5.2, for example, the beaker of water and the piece of metal being heated in a Bunsen burner flame have different temperatures. When the hot metal is plunged into the cold water, energy is transferred as heat from the metal to the water. The thermal energy (molecular motion) of the water molecules increases, and the thermal energy of the metal atoms decreases. Eventually, the two objects reach the same temperature, and the system has reached **thermal equilibrium**. The distinguishing feature of thermal equilibrium is that, on the macroscopic scale, no further temperature change occurs; both the metal and water are at the same temperature.

Putting a hot metal bar into a beaker of water and following the temperature change may seem like a rather simple experiment with an obvious outcome. However, the experiment illustrates three principles that are important in our further discussion:

- Energy transfer as heat will occur spontaneously from an object at a higher temperature to an object at a lower temperature.
- Transfer of energy as heat continues until both objects are at the same temperature and thermal equilibrium is achieved.
- After thermal equilibrium is attained, the object whose temperature increased has gained thermal energy, and the object whose temperature decreased has lost thermal energy.

FIGURE 5.2 **Energy transfer.** Energy transfer as heat occurs from the hotter metal cylinder to the cooler water. Eventually, the water and metal reach the same temperature and are said to be in thermal equilibrium.

A CLOSER LOOK

What Is Heat?

Two hundred years ago, scientists characterized heat as a real substance called the caloric fluid. The caloric hypothesis supposed that when a fuel burned and a pot of water was heated, for example, caloric fluid was transferred from the fuel to the water. Burning the fuel released caloric fluid, and the temperature of the water increased as the caloric fluid was absorbed.

Over the next 50 years, however, the caloric hypothesis lost favor, and we now know it is incorrect. Experiments by James Joule (1818–1889) and Benjamin Thompson (1753–1814) that showed the interrelationship between heat and other forms of energy such as mechanical energy provided the key to dispelling this idea. Even so, some of our everyday language retains the influence of this early theory. For example, we often speak of heat flowing as if it were a fluid.

From our discussion so far, we know one thing that "heat" is not—but what is it? Heat

Work and heat. A classic experiment that showed the relationship between work and heat was performed by Benjamin Thompson (also known as Count Rumford) (1753–1814) using an apparatus similar to that shown here. Thompson measured the rise in temperature of water (in the vessel mostly hidden at the back of the apparatus) that resulted from the energy expended to turn the crank.

Richard Howard

is said to be a "process quantity." It is the *process* by which energy is transferred across the boundary of a system owing to a difference in temperature between the two sides of the boundary. In this process, the energy of the object at the lower temperature increases, and the energy of the object at the higher temperature decreases.

Heat is not the only process by which energy can be transferred. Specifically, work is another process that can transfer energy between objects.

The idea of energy transfer by the processes of heat and work is embodied in the definition of thermodynamics: the science of heat and work.

For the specific case where energy is transferred only as heat within an isolated system (that is, a system that cannot transfer either energy or matter with its surroundings), we can also say that the quantity of energy lost as heat by the hotter object and the quantity of energy gained as heat by the cooler object are numerically equal. (This is required by the law of conservation of energy.) When energy is transferred as heat between a system and its surroundings, we describe the directionality of this transfer as **exothermic** or **endothermic** (Figure 5.3).

- In an **exothermic process**, energy is transferred as heat from a system to its surroundings. The energy of the system decreases, and the energy of the surroundings increases.
- An **endothermic process** is the opposite of an exothermic process. Energy is transferred as heat from the surroundings to the system, increasing the energy of the system and decreasing the energy of the surroundings.

Exothermic
$q_{sys} < 0$ SYSTEM

SURROUNDINGS

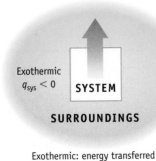

Exothermic: energy transferred
from system to surroundings

SYSTEM

Endothermic
$q_{sys} > 0$

SURROUNDINGS

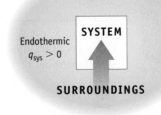

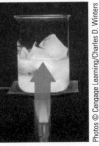

Photos © Cengage Learning/Charles D. Winters

Endothermic: energy transferred
from surroundings to system

FIGURE 5.3 Exothermic and endothermic processes. The symbol q represents the energy transferred as heat, and the subscript sys refers to the system.

REVIEW & CHECK FOR SECTION 5.1

Which of the following processes is endothermic?

(a)   melting of ice at 0 °C

(b)   the reaction of methane and O_2

(c)   condensation of water vapor at 100 °C

(d)   cooling liquid water from 25 °C to 0 °C

5.2   Specific Heat Capacity: Heating and Cooling

When an object is heated or cooled, the quantity of energy transferred depends on three things:

- the quantity of material
- the magnitude of the temperature change
- the identity of the material gaining or losing energy

Specific heat capacity (C) is defined as *the energy transferred as heat that is required to raise the temperature of 1 gram of a substance by one kelvin.* It has units of joules per gram per kelvin (J/g · K). A few specific heat capacities are listed in Figure 5.4, and a longer list of specific heat capacities is given in Appendix D (Table 11).

The energy gained or lost as heat when a given mass of a substance is warmed or cooled can be calculated using Equation 5.1.

$$q = C \times m \times \Delta T \tag{5.1}$$

Here, q is the energy gained or lost as heat by a given mass of substance (m), C is the specific heat capacity, and ΔT is the change in temperature. The change in temperature, ΔT, is calculated as the final temperature minus the initial temperature.

$$\Delta T = T_{final} - T_{initial} \tag{5.2}$$

Calculating a change in temperature using Equation 5.2 will give a result with an algebraic sign that indicates the direction of energy transfer. For example, we can use the specific heat capacity of copper, 0.385 J/g · K, to calculate the energy that

Specific Heat Capacities of Some Elements, Compounds, and Substances

Substances	Specific Heat Capacity (J/g · K)	Molar Heat Capacity (J/mol · K)
Al, aluminum	0.897	24.2
Fe, iron	0.449	25.1
Cu, copper	0.385	24.5
Au, gold	0.129	25.4
Water (liquid)	4.184	75.4
Water (ice)	2.06	37.1
$HOCH_2CH_2OH(\ell)$, ethylene glycol (antifreeze)	2.39	14.8
Wood	1.8	—
Glass	0.8	—

FIGURE 5.4 Specific heat capacity. Metals have different values of specific heat capacity on a per-gram basis. However, their molar heat capacities are all near 25 J/mol · K. Among common substances, liquid water has the highest specific heat capacity on a per-gram or per-mole basis, a fact that plays a significant role in Earth's weather and climate.

must be transferred from the surroundings to a 10.0-g sample of copper if the metal's temperature is raised from 298 K (25 °C) to 598 K (325 °C).

$$q = \left(0.385 \, \frac{J}{g \cdot K}\right)(10.0 \, g)(598 \, K - 298 \, K) = +1160 \, J$$

T_{final} $T_{initial}$
Final temp. Initial temp.

Notice that the answer has a positive sign. This indicates that the energy of the sample of copper has increased by 1160 J, which is in accord with energy being transferred as heat to the copper (the system) from the surroundings.

The relationship between energy, mass, and specific heat capacity has numerous implications. The high specific heat capacity of liquid water, $4.184 \, J/g \cdot K$, is a major reason why large bodies of water have a profound influence on weather. In spring, lakes warm up more slowly than the air. In autumn, the energy given off by a large lake as it cools moderates the drop in air temperature. The relevance of specific heat capacity is also illustrated when bread is wrapped in aluminum foil (specific heat capacity $0.897 \, J/g \cdot K$) and heated in an oven. You can remove the foil with your fingers after taking the bread from the oven. The bread and the aluminum foil are very hot, but the small mass of aluminum foil used and its low specific heat capacity result in only a small quantity of energy being transferred to your fingers (which have a larger mass and a higher specific heat capacity) when you touch the hot foil. This is also the reason why a chain of fast-food restaurants warns you that the filling of an apple pie can be much warmer than the paper wrapper or the pie crust. Although the wrapper, pie crust, and filling are at the same temperature, the quantity of energy transferred to your fingers (or your mouth!) from the filling is greater than that transferred from the wrapper and crust.

• **Molar Heat Capacity** Heat capacities can be expressed on a per-mole basis. The amount of energy that is transferred as heat in raising the temperature of one mole of a substance by one kelvin is the molar heat capacity. For water, the molar heat capacity is $75.4 \, J/mol \cdot K$. The molar heat capacity of metals at room temperature is always near $25 \, J/mol \cdot K$.

A practical example of specific heat capacity. The filling of the apple pie from a fast-food restaurant chain has a higher specific heat (and higher mass) than the pie crust and wrapper. Notice the warning on the wrapper.

○WL **INTERACTIVE EXAMPLE 5.1** Specific Heat Capacity

Problem How much energy must be transferred to raise the temperature of a cup of coffee (250 mL) from 20.5 °C (293.7 K) to 95.6 °C (368.8 K)? Assume that water and coffee have the same density (1.00 g/mL) and specific heat capacity (4.184 J/g · K).

What Do You Know? The energy required to warm a substance is related to its specific heat capacity (C), the mass of the substance, and the temperature change. A value for C is given in the problem.

Strategy You can calculate the mass of coffee from the volume and density (mass = volume × density) and the temperature change from the initial and final temperatures ($\Delta T = T_{final} - T_{initial}$). Use Equation 5.1 to solve for q.

Solution

Mass of coffee = (250 mL)(1.00 g/mL) = 250 g

$\Delta T = T_{final} - T_{initial} = 368.8 \, K - 293.7 \, K = 75.1 \, K$

$q = C \times m \times \Delta T$

$q = (4.184 \, J/g \cdot K)(250 \, g)(75.1 \, K)$

$q = 79,000 \, J \text{ (or 79 kJ)}$

Think about Your Answer The positive sign in the answer indicates that energy has been transferred to the coffee as heat. The thermal energy of the coffee is now higher.

Check Your Understanding

In an experiment, it was determined that 59.8 J was required to raise the temperature of 25.0 g of ethylene glycol (a compound used as antifreeze in automobile engines) by 1.00 K. Calculate the specific heat capacity of ethylene glycol from these data.

Strategy Map 5.1

PROBLEM
Calculate the **energy** required to **raise temperature** of liquid.

↓

DATA/INFORMATION
- Specific heat capacity
- Volume
- Density of liquid
- ΔT

STEP 1. Calculate the **sample mass** and ΔT.

↓

Mass of sample, ΔT, and specific heat capacity

STEP 2. Use **Equation 5.1** to calculate q.

↓

Energy required (q) to raise temperature of liquid by ΔT

Quantitative Aspects of Energy Transferred as Heat

Like melting point, boiling point, and density, specific heat capacity is a characteristic intensive property of a pure substance. The specific heat capacity of a substance can be determined experimentally by accurately measuring temperature changes that occur when energy is transferred as heat from the substance to a known quantity of water (whose specific heat capacity is known).

Suppose a 55.0-g piece of metal is heated in boiling water to 99.8 °C and then dropped into cool water in an insulated beaker (Figure 5.5). Assume the beaker contains 225 g of water and its initial temperature (before the metal was dropped in) was 21.0 °C. The final temperature of the metal and water is 23.1 °C. What is the specific heat capacity of the metal? Here are the important aspects of this experiment.

- Let's define the metal and the water as the system and the beaker and environment as the surroundings and assume that energy is transferred only within the system. (This means that energy is not transferred between the system and the surroundings. This assumption is good, but not perfect; for a more accurate result, we would also want to account for any energy transfer to the surroundings.)
- The water and the metal bar end up at the same temperature. (T_{final} is the same for both.)
- We will also assume energy is transferred *only as heat* within the system.
- The energy transferred as heat from the metal to the water, q_{metal}, has a negative value because the temperature of the metal drops. Conversely, q_{water} has a positive value because its temperature increases.
- The values of q_{water} and q_{metal} are numerically equal but of opposite sign.

Because of the law of conservation of energy, *in an isolated system the sum of the energy changes within the system must be zero.* If energy is transferred only as heat, then

$$q_1 + q_2 + q_3 + \ldots = 0 \qquad (5.3)$$

where the quantities q_1, q_2, and so on represent the energies transferred as heat for the individual parts of the system. For this specific problem, there are thermal energy changes associated with water and metal, q_{water} and q_{metal}, the two components of the system; thus

$$q_{water} + q_{metal} = 0$$

Each of these quantities is related individually to specific heat capacities, masses, and changes of temperature, as defined by Equation 5.1. Thus

$$[C_{water} \times m_{water} \times (T_{final} - T_{initial, water})] + [C_{metal} \times m_{metal} \times (T_{final} - T_{initial, metal})] = 0$$

The specific heat capacity of the metal, C_{metal}, is the unknown in this problem. Using the specific heat capacity of water (4.184 J/g · K) and converting Celsius to kelvin temperatures gives

$$[(4.184 \text{ J/g} \cdot \text{K})(225 \text{ g})(296.3 \text{ K} - 294.2 \text{ K})] + [(C_{metal})(55.0 \text{ g})(296.3 \text{ K} - 373.0 \text{ K})] = 0$$

$$C_{metal} = 0.47 \text{ J/g} \cdot \text{K}$$

FIGURE 5.5 Transfer of energy as heat. When energy is transferred as heat from a hot metal to cool water, the thermal energy of the metal decreases and that of the water increases. The value of q_{metal} is thus negative and that of q_{water} is positive.

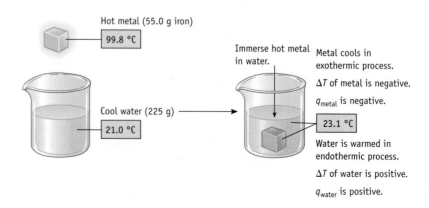

Hot metal (55.0 g iron)
99.8 °C

Immerse hot metal in water.

Cool water (225 g)
21.0 °C

Metal cools in exothermic process.
ΔT of metal is negative.
q_{metal} is negative.

23.1 °C

Water is warmed in endothermic process.
ΔT of water is positive.
q_{water} is positive.

Calculating ΔT

Virtually all calculations that involve temperature in chemistry are expressed in kelvins. In calculating ΔT, however, we can use Celsius temperatures because a kelvin and a Celsius degree are the same size. That is, the difference between two temperatures is the same on both scales. For example, the difference between the boiling and freezing points of water is

ΔT, Celsius = 100 °C − 0 °C = 100 °C

ΔT, kelvin = 373 K − 273 K = 100 K

⦿WL **INTERACTIVE EXAMPLE 5.2** Using Specific Heat Capacity

Problem An 88.5-g piece of iron whose temperature is 78.8 °C (352.0 K) is placed in a beaker containing 244 g of water at 18.8 °C (292.0 K). When thermal equilibrium is reached, what is the final temperature? (Assume no energy is lost to warm the beaker and its surroundings.)

What Do You Know? Iron cools and the water warms until thermal equilibrium is reached. The energies associated with the cooling of iron and the heating of water are determined by the specific heat capacities, masses, and temperature changes for each species. If we define the system as the iron and water, the sum of these two energy quantities will be zero. The final temperature is the unknown in this problem. Masses and initial temperatures are given; the specific heat capacities of iron and water can be found in Appendix D or Figure 5.4.

Strategy The two energy quantities, q_{Fe} and q_{water}, sum to zero ($q_{Fe} + q_{water} = 0$). Each energy quantity is defined using Equation 5.1; the value of ΔT in each is $T_{final} - T_{initial}$. We can use either kelvin or Celsius temperatures (Problem Solving Tip 5.1). Substituting gives the equation

$$C_{water} \times m_{water} \times (T_{final} - T_{initial,water}) + C_{Fe} \times m_{Fe} \times (T_{final} - T_{initial,Fe}) = 0.$$

Substitute in the given information and solve.

Solution

$$[C_{water} \times m_{water} \times (T_{final} - T_{initial, water})] + [C_{Fe} \times m_{Fe} \times (T_{final} - T_{initial, Fe})] = 0$$

$$[(4.184 \text{ J/g} \cdot \text{K})(244 \text{ g})(T_{final} - 292.0 \text{ K})] + [(0.449 \text{ J/g} \cdot \text{K})(88.5 \text{ g})(T_{final} - 352.0 \text{ K})] = 0$$

$$T_{final} = 294 \text{ K (21 °C)}$$

Think about Your Answer Be sure to notice that $T_{initial}$ for the metal and $T_{initial}$ for the water in this problem have different values. Also, the low specific heat capacity and smaller quantity of iron result in the temperature of iron being reduced by about 60 degrees; in contrast, the temperature of the water has been raised by only a few degrees. Finally, as expected, T_{final} (294 K) is between $T_{initial, Fe}$ and $T_{initial, water}$.

Check Your Understanding

A 15.5-g piece of chromium, heated to 100.0 °C, is dropped into 55.5 g of water at 16.5 °C. The final temperature of the metal and the water is 18.9 °C. What is the specific heat capacity of chromium? (Assume no energy is lost to the container or to the surrounding air.)

Strategy Map 5.2

PROBLEM

Calculate **final T** when piece of hot iron is added to **given mass** of cool water.

↓

DATA/INFORMATION

- **Specific heat capacities**
- **masses** of iron and water
- **initial T** for iron and water

↓

Use equation $q_{Fe} + q_{water} = 0$.
Solve for T_{final}.

↓

T_{final} of water and iron

REVIEW & CHECK FOR SECTION 5.2

1. Which of the following processes requires the largest input of energy?

(a) warming 1.0 g of iron by 10 °C (C_{Fe} = 0.449 J/g · K)

(b) warming 2.0 g of copper by 10 °C (C_{Cu} = 0.385 J/g · K)

(c) warming 4.0 g of lead by 10 °C (C_{Pb} = 0.127 J/g · K)

(d) warming 3.0 g of silver by 10 °C (C_{Ag} = 0.236 J/g · K)

2. The samples of water listed below are mixed. After thermal equilibrium is reached, which mixture has the lowest temperature?

    (a) 10 g of water at 20 °C + 10 g of water at 30 °C

    (b) 10 g of water at 20 °C + 20 g of water at 30 °C

    (c) 20 g of water at 20 °C + 10 g of water at 30 °C

    (d) 20 g of water at 20 °C + 20 g of water at 30 °C

5.3   Energy and Changes of State

A *change of state* is a change, for example, between solid and liquid or between liquid and gas. When a solid melts, its atoms, molecules, or ions move about vigorously enough to break free of the attractive forces holding them in rigid positions in the solid lattice. When a liquid boils, the particles move much farther apart from one another, to distances at which attractive forces are minimal. In both cases, energy must be furnished to overcome attractive forces among the particles.

The energy transferred as heat that is required to convert a substance from a solid at its melting point to a liquid is called the **heat of fusion**. The energy transferred as heat to convert a liquid at its boiling point to a vapor is called the **heat of vaporization**. Heats of fusion and vaporization for many substances are provided along with other physical properties in reference books. Values for a few common substances are given in Appendix D (Table 12).

It is important to recognize that *temperature is constant throughout a change of state* (Figure 5.6). During a change of state, the added energy is used to overcome the forces holding one molecule to another, not to increase the temperature (Figures 5.6 and 5.7).

For water, the heat of fusion at 0 °C is 333 J/g, and the heat of vaporization at 100 °C is 2256 J/g. These values can be used to calculate the energy required for a given mass of water to melt or evaporate, respectively. For example, the energy required to convert 500. g of water from the liquid to gaseous state at 100 °C is

$$(2256 \text{ J/g})(500. \text{ g}) = 1.13 \times 10^6 \text{ J } (= 1130 \text{ kJ})$$

In contrast, to melt the same mass of ice to form liquid water at 0 °C requires only 167 kJ.

$$(333 \text{ J/g})(500. \text{ g}) = 1.67 \times 10^5 \text{ J } (= 167 \text{ kJ})$$

Figure 5.6 gives a profile of the energy changes occurring as 500. g of ice at −50 °C is converted to water vapor at 200 °C. This involves a series of steps: (1) warming ice to 0 °C, (2) conversion to liquid water at 0 °C, (3) warming liquid water to 100 °C, (4) evaporation at 100 °C, and (5) warming the water vapor to 200 °C. Each step requires the input of additional energy. The energy transferred as

FIGURE 5.6 Energy transfer as heat and the temperature change for water. This graph shows the energy transferred as heat to 500. g of water and the consequent temperature change as the water warms from −50 °C to 200 °C (at 1 atm).

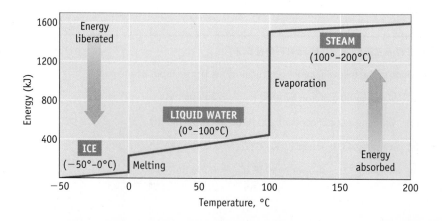

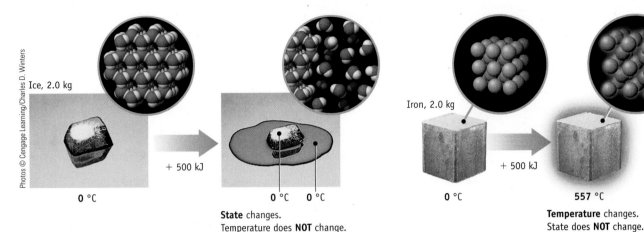

Photos © Cengage Learning/Charles D. Winters

Ice, 2.0 kg

+ 500 kJ

0 °C   0 °C   0 °C

State changes.
Temperature does **NOT** change.

Iron, 2.0 kg

+ 500 kJ

0 °C   557 °C

Temperature changes.
State does **NOT** change.

Transferring 500 kJ of energy as heat to 2.0 kg of ice at 0 °C will cause 1.5 kg of ice to melt to water at 0 °C (and 0.5 kg of ice will remain). No temperature change occurs.

In contrast, transferring 500 kJ of energy as heat to 2.0 kg of iron at 0 °C will cause the temperature to increase to 557 °C (and the metal to expand slightly but not melt).

FIGURE 5.7 Changes of state.

heat to raise the temperature of solid, liquid, and vapor can be calculated with Equation 5.1, using the specific heat capacities of ice, liquid water, and water vapor (which are different), and the energies for the changes of state can be calculated using heats of fusion and vaporization. These calculations are done in Example 5.3.

ⓦWL   INTERACTIVE EXAMPLE 5.3   Energy and Changes of State

Problem Calculate the energy needed to convert 500. g of ice at −50.0 °C to steam at 200.0 °C. (The temperature change occurring in each step is illustrated in Figure 5.6.) The heat of fusion of water is 333 J/g, and the heat of vaporization is 2256 J/g. The specific heat capacities of ice, liquid water, and water vapor are given in Appendix D.

What Do You Know? The overall process of converting ice at −50 °C to steam at 200 °C involves both temperature changes and changes of state; all require input of energy as heat. Given in the problem is the mass of water. Recall that melting occurs at 0 °C and boiling at 100 °C (at 1 atm pressure). You will need the specific heat capacities of ice, liquid water, and steam (available in Appendix D), the heat of fusion of water (333 J/g), and the heat of vaporization (2256 J/g).

Strategy The problem is broken down into a series of steps:

Step 1: Warm the ice from −50 °C to 0 °C.

Step 2: Melt the ice at 0 °C.

Step 3: Raise the temperature of the liquid water from 0 °C to 100 °C.

Step 4: Boil the water at 100 °C.

Step 5: Raise the temperature of the steam from 100 °C to 200 °C.

Use Equation 5.1 and the specific heat capacities of solid, liquid, and gaseous water to calculate the energy transferred as heat associated with the temperature changes. Use the heat of fusion and the heat of vaporization to calculate the energy transferred as heat associated with changes of state. The total energy transferred as heat is the sum of the energies of the individual steps.

Solution

Step 1. (to warm ice from −50.0 °C to 0.0 °C)

$$q_1 = (2.06 \text{ J/g} \cdot \text{K})(500. \text{ g})(273.2 \text{ K} − 223.2 \text{ K}) = 5.15 \times 10^4 \text{ J}$$

Step 2. (to melt ice at 0.0 °C)

$$q_2 = (500. \text{ g})(333 \text{ J/g}) = 1.67 \times 10^5 \text{ J}$$

Strategy Map 5.3

PROBLEM
Calculate **energy required** to heat a mass of water from **−50.0 °C** to steam at **200.0 °C**.

DATA/INFORMATION
- Mass of water
- ΔT
- Heats of fusion and vaporization of water
- Specific heat capacities

Calculate **energy** involved in
STEP 1. warm ice to 0 °C (q_1)
STEP 2. melt ice at 0 °C (q_2)
STEP 3. warm water from 0 °C to 100 °C (q_3)
STEP 4. evaporate water at 100 °C (q_4)
STEP 5. heat steam to 200 °C (q_5)

Values of q_1, q_2, q_3, q_4, and q_5

STEP 6. Sum values of q.

q_{total} for process

Step 3. (to raise temperature of liquid water from 0.0 °C to 100.0 °C)

$$q_3 = (4.184 \text{ J/g} \cdot \text{K})(500. \text{ g})(373.2 \text{ K} - 273.2 \text{ K}) = 2.09 \times 10^5 \text{ J}$$

Step 4. (to evaporate water at 100.0 °C)

$$q_4 = (2256 \text{ J/g})(500. \text{ g}) = 1.13 \times 10^6 \text{ J}$$

Step 5. (to raise temperature of water vapor from 100.0 °C to 200.0 °C)

$$q_5 = (1.86 \text{ J/g} \cdot \text{K})(500. \text{ g}) (473.2 \text{ K} - 373.2 \text{ K}) = 9.30 \times 10^4 \text{ J}$$

The total energy transferred as heat is the sum of the energies of the individual steps.

$$q_{total} = q_1 + q_2 + q_3 + q_4 + q_5$$

$$q_{total} = 1.65 \times 10^6 \text{ J (or 1650 kJ)}$$

Think about Your Answer The conversion of liquid water to steam is the largest increment of energy by a considerable margin. (You may have noticed that it does not take much time to heat water to boiling on a stove, but to boil off the water takes a much longer time.)

Check Your Understanding

Calculate the amount of energy necessary to raise the temperature of 1.00 L of ethanol ($d = 0.7849 \text{ g/cm}^3$) from 25.0 °C to its boiling point (78.3 °C) and then to vaporize the liquid. ($C_{ethanol} = 2.44 \text{ J/g} \cdot \text{K}$; heat of vaporization at 78.3 °C = 38.56 kJ/mol.)

Strategy Map 5.4

PROBLEM

Calculate **mass of ice** needed to cool mass of diet cola from $T_{initial}$ to **0 °C.**

↓

DATA/INFORMATION
- Mass of diet cola
- ΔT
- heat of fusion of water
- specific heat capacity

↓

> **STEP 1.** Calculate q_{cola} from Equation 5.1.

↓

q_{cola} for cooling diet cola

> **STEP 2.** Use $q_{cola} + q_{ice} = 0$ to calculate q_{ice}.

↓

q_{ice} for melting ice

> **STEP 3.** Use heat of fusion and q_{ice} to calculate m_{ice}.

↓

Mass of ice required (m_{ice})

 **INTERACTIVE EXAMPLE 5.4** **Change of State**

Problem What is the minimum mass of ice at 0 °C that must be added to the contents of a can of diet cola (340. mL) to cool the cola from 20.5 °C to 0.0 °C? Assume that the specific heat capacity and density of diet cola are the same as for water.

What Do You Know? The final temperature is 0 °C. Melting ice requires energy as heat, and cooling the cola evolves energy as heat. The sum of the energy changes for the two components in the system is zero; that is, the two energy changes (melting ice, cooling cola) will be the same magnitude but opposite in sign. (We also need to assume there is no transfer of energy between the surroundings and the system.) Use the data given in the question to calculate the energy changes for the two components of this system, q_{cola} and q_{ice}. To solve the problem, we will also need the density and specific heat capacity of water (Appendix D).

Strategy Assuming only energy changes within the system, $q_{cola} + q_{ice} = 0$. The energy evolved by cooling the cola, q_{cola}, is calculated using Equation 5.1. The initial temperature is 20.5 °C and the final temperature is 0 °C. The mass of cola is calculated from the volume and density. The energy as heat required to melt the ice, q_{ice}, is determined from the heat of fusion (333 J/g, from text page 180). The mass of ice is the unknown.

Solution The mass of cola is 340. g [(340. mL)(1.00 g/mL) = 340. g], and its temperature changes from 293.7 K to 273.2 K. The heat of fusion of water is 333 J/g, and the mass of ice is the unknown.

$$q_{cola} + q_{ice} = 0$$

$$C_{cola} \times m_{cola} \times (T_{final} - T_{initial}) + (\text{heat of fusion of water}) (m_{ice}) = 0$$

$$[(4.184 \text{ J/g} \cdot \text{K})(340. \text{ g})(273.2 \text{ K} - 293.7 \text{ K})] + [(333 \text{ J/g}) (m_{ice})] = 0$$

$$m_{ice} = 87.6 \text{ g}$$

Think about Your Answer If more than 87.6 g of ice is added, the final temperature will still be 0 °C when thermal equilibrium is reached, but some ice will remain (see problem below). If less than 87.6 g of ice is added, the final temperature will be greater than 0 °C. In this case, all the ice will melt, and the liquid water formed by melting the ice will absorb additional energy to warm up to the final temperature (an example is given in Study Question 71).

Check Your Understanding

To make a glass of iced tea, you pour 250 mL of tea, whose temperature is 18.2 °C, into a glass containing five ice cubes. Each cube has a mass of 15 g. What quantity of ice will melt, and how much ice will remain floating in the tea? Assume iced tea has a density of 1.0 g/mL and a specific heat capacity of 4.2 J/g · K, that energy is transferred as heat only within the system, ice is at 0.0 °C, and no energy is transferred between system and surroundings.

REVIEW & CHECK FOR SECTION 5.3

1. Which of the following processes requires the largest input of energy as heat?

   (a) raising the temperature of 100 g of water by 1.0 °C

   (b) vaporization of 0.10 g of water at 100 °C

   (c) melting 1.0 g of ice at 0 °C

   (d) warming 1.0 g of ice from -50 °C to 0 °C (specific heat of ice = 2.06 J/g · K)

2. Ice (5.0 g) at 0 °C is added to 25 g of liquid water at 20 °C. Which statement best describes the system when thermal equilibrium is reached?

   (a) Part of the ice has melted; the remaining ice and water are at 0 °C.

   (b) All of the ice melts; liquid water is at a temperature between 0 °C and 10 °C.

   (c) All of the ice melts; liquid water is at a temperature between 10 °C and 20 °C.

   (d) Part of the ice melts; the ice-water mixture is at a temperature between 0 ° C and 10 °C.

5.4 The First Law of Thermodynamics

Recall that *thermodynamics is the science of heat and work.* To this point, we have only considered energy being transferred as heat, but now we need to broaden the discussion to include work.

Energy transferred as heat between a system and its surroundings changes the energy of the system, but work done by a system or on a system will also affect the energy in the system. If a system does work on its surroundings, energy must be expended by the system, and the system's energy will decrease. Conversely, if work is done by the surroundings on a system, the energy of the system will increase.

A system doing work on its surroundings is illustrated in Figure 5.8. A small quantity of dry ice, solid CO_2, is sealed inside a plastic bag, and a weight (a book) is placed on top of the bag. When energy is transferred as heat from the surroundings to the dry ice, the dry ice changes directly from solid to gas at -78 °C in a process called **sublimation**:

$$CO_2(s, -78 \text{ °C}) \rightarrow CO_2(g, -78 \text{ °C})$$

As sublimation proceeds, gaseous CO_2 expands within the plastic bag, lifting the book against the force of gravity. The system (the CO_2 inside the bag) is expending energy to do this work.

Even if the book had not been on top of the plastic bag, work would have been done by the expanding gas because the gas must push back the atmosphere when it expands. Instead of raising a book, the expanding gas moves a part of the atmosphere.

Now let us restate this example in terms of thermodynamics. First, we must identify the system and the surroundings. The system is the CO_2, initially a solid and later a gas. The surroundings consist of objects that exchange energy with the system. This includes the plastic bag, the book, the table-top, and the surrounding air. Sublimation requires energy, which is transferred as heat to the system (the CO_2)

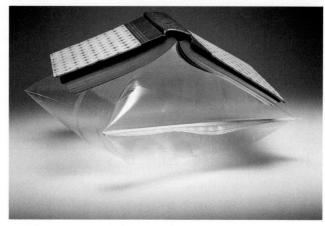

(a) Pieces of dry ice [$CO_2(s)$, $-78\ °C$] are placed in a plastic bag. The dry ice will sublime (change directly from a solid to a gas) upon the input of energy.

(b) Energy is absorbed by $CO_2(s)$ when it sublimes, and the system (the contents of the bag) does work on its surroundings by lifting the book against the force of gravity.

FIGURE 5.8 Energy changes in a physical process.

from the surroundings. At the same time, the system does work on the surroundings by lifting the book. An energy balance for the system will include both quantities, energy transferred as heat and energy transferred as work.

This example can be generalized. For any system, we can identify energy transfers both as heat and as work between the system and surroundings. The change in energy for a system is given explicitly by Equation 5.4,

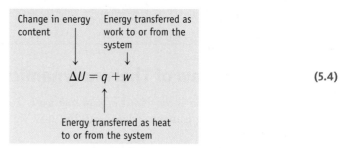

$$\Delta U = q + w \tag{5.4}$$

which is a mathematical statement of the **first law of thermodynamics**: The energy change for a system (ΔU) is the sum of the energy transferred as heat between the system and its surroundings (q) and the energy transferred as work between the system and its surroundings (w).

The equation defining the first law of thermodynamics can be thought of as a version of the general principle of conservation of energy. Because energy is conserved, we must be able to account for any change in the energy of the system. All energy transfers between a system and its surroundings occur by the processes of heat and work. Equation 5.4 thus states that the change in the energy of the system is exactly equal to the sum of all of the energy transfers (heat and/or work) between the system and its surroundings.

The quantity U in Equation 5.4 has a formal name—**internal energy**—and a precise meaning in thermodynamics. The internal energy in a chemical system is the sum of the potential and kinetic energies inside the system, that is, the energies of the atoms, molecules, or ions in the system. The potential energy here is the energy associated with the attractive and repulsive forces between all the nuclei and electrons in the system. It includes the energy associated with bonds in molecules, forces between ions, and forces between molecules. The kinetic energy is the energy of motion of the atoms, ions, and molecules in the system. Actual values of internal energy are rarely determined or needed. Instead, in most instances, we are interested in the *change* in internal energy, a measurable quantity. In fact, Equation 5.4 tells us how to determine ΔU: *Measure the energy transferred as heat and work to or from the system.*

The sign conventions for Equation 5.4 are important. The following table summarizes how the internal energy of a system is affected by energy transferred as heat and work.

Sign Conventions for q and w of the System

Energy transferred as . . .	Sign Convention	Effect on U_{system}
Heat to the system (endothermic)	$q > 0$ (+)	U increases
Heat from the system (exothermic)	$q < 0$ (−)	U decreases
Work done on system	$w > 0$ (+)	U increases
Work done by system	$w < 0$ (−)	U decreases

The work in the example involving the sublimation of CO_2 (Figure 5.8) is of a specific type, called P–V (pressure–volume) work. It is the work (w) associated with a change in volume (ΔV) that occurs against a resisting external pressure (P). For a system in which the external pressure is constant, the value of P–V work can be calculated using Equation 5.5:

Work (at constant pressure)   Change in volume

$$w = -P \times \Delta V \qquad (5.5)$$

Pressure

Under conditions of constant volume, $\Delta V = 0$. If the only type of work possible is P–V work, then the energy transferred as work will also be zero (because $\Delta V = 0$ and so $w = 0$). That is, the energy transferred as heat under conditions of constant volume (q_v) is equal to the change in the internal energy of the system

$$\Delta U = q_v + w_v$$

$$\Delta U = q_v + 0 \text{ when } w_v = 0 \text{ because } \Delta V = 0$$

$$\text{and so } \Delta U = q_v$$

● **Heat and Work** For an in-depth look at heat and work in thermodynamics, see E. A. Gislason and N. C. Craig, *Journal of Chemical Education*, Vol. 64, No. 8, pp. 660–668, 1987.

A CLOSER LOOK

P–V Work

Work is done when an object of some mass is moved against an external resisting force.

To evaluate the work done when a gas is compressed, we can use, for example, a cylinder with a movable piston, as would occur in a bicycle pump (see figure). The drawing on the left shows the initial position of the piston, and the one on the right shows its final position. To depress the piston, we would have to expend some energy (the energy for this process comes from the energy obtained by food metabolism in our bodies). The work required to depress the piston is calculated from a law of physics, $w = F \times d$, or work equals the magnitude of the force (F) applied times the distance (d) over which the force is applied.

Pressure is defined as a force divided by the area over which the force is applied: $P = F/A$. In this example, the force is being applied to a piston with an area A. Substituting $P \times A$ for F in the equation for work gives $w = (P \times A) \times d$. The product of $A \times d$ is equivalent to the change in the volume of the gas in the pump, and, because $\Delta V = V_{final} - V_{initial}$, this change in volume is negative. Finally, because work done on a system is defined as positive, this means that $w = -P\Delta V$.

Pushing down on the piston means we have done work on the system, the gas contained within the cylinder. The gas is now compressed to a smaller volume and has attained a higher energy as a consequence. The additional energy is equal to $-P\Delta V$.

Notice how energy has been converted from one form to another—from chemical energy in food to mechanical energy used to depress the piston, to potential energy stored in a system of a gas at a higher pressure. In each step, energy was conserved, and the total energy of the universe remained constant.

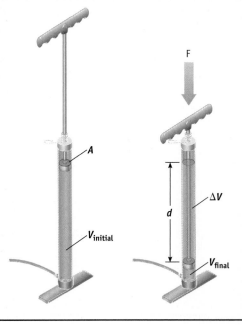

Enthalpy

Most experiments in a chemical laboratory are carried out in beakers or flasks open to the atmosphere, where the external pressure is constant. Similarly, chemical processes that occur in living systems are open to the atmosphere. Because many processes in chemistry and biology are carried out under conditions of constant pressure, it is useful to have a specific measure of the energy transferred as heat under this circumstance.

Under conditions of constant pressure,

$$\Delta U = q_p + w_p$$

where the subscript p indicates conditions of constant pressure. If the only type of work that occurs is $P–V$ work, then

$$\Delta U = q_p - P\Delta V$$

Rearranging this gives

$$q_p = \Delta U + P\Delta V$$

We now introduce a new thermodynamic function called **enthalpy**, H, which is defined as

$$H = U + PV$$

The *change* in enthalpy for a system at constant pressure would be calculated from the following equation:

$$\Delta H = \Delta U + P\Delta V$$

Thus,

$$\Delta H = q_p$$

For a system where the only type of work possible is $P–V$ work, the change in enthalpy, ΔH, is therefore equal to the energy transferred as heat at constant pressure, often symbolized by q_p. The directionality of energy transfer (under conditions of constant pressure) is signified by the sign of ΔH.

- Negative values of ΔH specify that energy is transferred as heat from the system to the surroundings.
- Positive values of ΔH specify that energy is transferred as heat from the surroundings to the system.

Under conditions of constant pressure and where the only type of work possible is $P–V$ work, ΔU ($= q_p - P\Delta V$) and ΔH ($= q_p$) differ by $P\Delta V$ (the energy transferred to or from the system as work). We observe that in many processes—such as the melting of ice—the volume change, ΔV, is small, and hence the amount of energy transferred as work is small. Under these circumstances, ΔU and ΔH have almost the same value. The amount of energy transferred as work will be significant, however, when the volume change is large, as when gases are formed or consumed. In the evaporation or condensation of water, the sublimation of CO_2, and chemical reactions in which the number of moles of gas changes, ΔU and ΔH have significantly different values.

State Functions

Internal energy and enthalpy share a significant characteristic—namely, changes in these quantities depend only on the initial and final states. They do not depend on the path taken on going from the initial state to the final state. No matter how you go from reactants to products in a reaction, for example, the values of ΔH and ΔU for the reaction are always the same. A quantity that has this property is called a **state function**.

● **Energy Transferred as Heat** If the only type of work possible is P–V work, then, under conditions of constant volume, the energy transferred as heat is equal to ΔU

$$\Delta U = q_v$$

Under conditions of constant pressure, the energy transferred as heat is equal to ΔH.

$$\Delta H = q_p$$

● **Enthalpy and Internal Energy Differences** The difference between ΔH and ΔU will be quite small unless a large volume change occurs. For water at 1 atm pressure and 273 K, for example, the difference between ΔH and ΔU ($= P\Delta V$) is 0.142 J/mol for the conversion of ice to liquid water at 273 K, whereas it is 3100 J/mol for the conversion of liquid water to water vapor at 373 K.

Many commonly measured quantities, such as the pressure of a gas, the volume of a gas or liquid, and the temperature of a substance are state functions. For example, if the final temperature of a substance is 75 °C and its inital temperature is 25 °C, the change in temperature, ΔT, is calculated as $T_{final} - T_{initial} = 75\ °C - 25\ °C = 50\ °C$. It does not matter if the substance was heated directly from 25 °C to 75 °C or if the substance was heated from 25 °C to 95 °C and then cooled to 75 °C; the overall change in temperature is still the same, 50 °C.

Not all quantities are state functions; some depend on the pathway taken to get from the initial condition to the final condition. For instance, distance traveled is not a state function (Figure 5.9). The travel distance from New York City to Denver depends on the route taken. Nor is the elapsed time of travel between these two locations a state function. In contrast, the altitude above sea level is a state function; in going from New York City (at sea level) to Denver (1600 m above sea level), there is an altitude change of 1600 m, regardless of the route followed.

Significantly, neither the energy transferred as heat nor the energy transferred as work individually is a state function but their sum, the change in internal energy, ΔU, is. The value of ΔU is fixed by $U_{initial}$ and U_{final}. A transition between the initial and final states can be accomplished by different routes having different values of q and w, but the sum of q and w for each path must always give the same ΔU.

To get a better picture of how it might arise that the sum of two path-dependent functions can result in a state function, consider the size of your bank account. This is a state function. Let us assume that there are two path-dependent ways to transfer money into or out of your account: either as cash or as an electronic payment. You could have arrived at a current bank balance of $25 by having deposited $25 in cash or you could have deposited $100 in cash and then made a $75 electronic payment, or you could have deposited $100 in cash, made an electronic payment of $50 and withdrawn $25 in cash. In each of these cases, the change in your bank balance is determined by taking the final balance ($25) and subtracting the initial balance ($0), giving a value for $\Delta\$$ of +$25, but the pathway to get from $0 to $25 is different. In addition, the quantity of money transferred as cash and the quantity of money transferred by electronic payment are path-dependent with different values in each case, but by taking into account all of the transfers as cash and all of the transfers as electronic payments, we end up with a state function, the overall balance.

As previously stated, enthalpy is a state function. The enthalpy change occurring when 1.0 g of water is heated from 20 °C to 50 °C is independent of how the process is carried out.

FIGURE 5.9 **State functions.** There are many ways to climb a mountain, but the change in altitude from the base of the mountain to its summit is the same. The change in altitude is a state function. The distance traveled to reach the summit is not.

REVIEW & CHECK FOR SECTION 5.4

1. Which of the following processes will lead to a decrease in the internal energy of a system? (1) Energy is transferred as heat to the system; (2) energy is transferred as heat from the system; (3) energy is transferred as work done on the system; or (4) energy is transferred as work done by the system.

   (a) 1 and 3   (b) 2 and 4   (c) 1 and 4   (d) 2 and 3

2. In chemical reactions, a significant amount of energy is transferred as work between systems and surroundings if there is a significant change in volume. This occurs if there is a change in the number of moles of gases. In which of the following reactions is there a significant transfer of energy as work from the system to the surroundings?

   (a) $C(s) + O_2(g) \rightarrow CO_2(g)$

   (b) $CH_4(g) + 2\ O_2(g) \rightarrow CO_2(g) + 2\ H_2O(g)$

   (c) $2\ C(s) + O_2(g) \rightarrow 2\ CO(g)$

   (d) $2\ Mg(s) + O_2(g) \rightarrow 2\ MgO(s)$

5.5 Enthalpy Changes for Chemical Reactions

Enthalpy changes accompany chemical reactions. For example, the **standard reaction enthalpy**, $\Delta_r H°$, for the decomposition of water vapor to hydrogen and oxygen at 25 °C is +241.8 kJ/mol-rxn.

$$H_2O(g) \rightarrow H_2(g) + \tfrac{1}{2} O_2(g) \qquad \Delta_r H° = +241.8 \text{ kJ/mol-rxn}$$

The positive sign of $\Delta_r H°$ in this case indicates that the decomposition is an endothermic process.

There are several important things to know about $\Delta_r H°$.

● **Notation for Thermodynamic Parameters** NIST and IUPAC (International Union of Pure and Applied Chemistry) specify that descriptors of functions such as ΔH should be written as a subscript, between the Δ and the thermodynamic function. Among the subscripts you will see are a lower case r for "reaction," f for "formation," c for "combustion," fus for "fusion," and vap for "vaporization."

- The designation of $\Delta_r H°$ as a "standard enthalpy change" (where the superscript ° indicates standard conditions) means that the pure, unmixed reactants in their standard states have formed pure, unmixed products in their standard states. The **standard state** of an element or a compound is defined as the most stable form of the substance in the physical state that exists at a pressure of 1 bar and at a specified temperature. [Most sources report standard reaction enthalpies at 25 °C (298 K).]
- The "per mol-rxn" designation in the units for $\Delta_r H°$ means this is the enthalpy change for a "mole of reaction" (where *rxn* is an abbreviation for reaction). One mole of reaction is said to have occurred when a chemical reaction occurs exactly in the amounts specified by the coefficients of the balanced chemical equation. For example, for the reaction $H_2O(g) \rightarrow H_2(g) + 1/2\, O_2(g)$, a mole of reaction has occurred when 1 mol of water vapor has been converted completely to 1 mol of H_2 and 1/2 mol of O_2 gas.

● **Moles of Reaction, Mol-rxn** This concept was also described in one of the methods shown for solving limiting reactant problems on page 141.

Now consider the opposite reaction, the combination of hydrogen and oxygen to form 1 mol of water. The magnitude of the enthalpy change for this reaction is the same as that for the decomposition reaction, but the sign of $\Delta_r H°$ is reversed. The exothermic formation of 1 mol of water vapor from 1 mol of H_2 and 1/2 mol of O_2 transfers 241.8 kJ to the surroundings (Figure 5.10).

$$H_2(g) + \tfrac{1}{2} O_2(g) \rightarrow H_2O(g) \qquad \Delta_r H° = -241.8 \text{ kJ/mol-rxn}$$

The value of $\Delta_r H°$ depends on the chemical equation used. Let us write the equation for the formation of water again but without a fractional coefficient for O_2.

$$2\, H_2(g) + O_2(g) \rightarrow 2\, H_2O(g) \qquad \Delta_r H° = -483.6 \text{ kJ/mol-rxn}$$

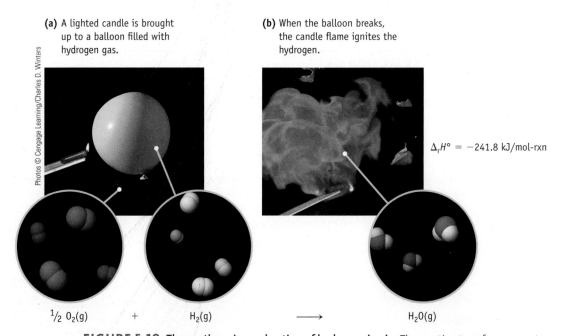

(a) A lighted candle is brought up to a balloon filled with hydrogen gas.

(b) When the balloon breaks, the candle flame ignites the hydrogen.

$\Delta_r H° = -241.8 \text{ kJ/mol-rxn}$

Photos © Cengage Learning/Charles D. Winters

$\tfrac{1}{2} O_2(g) \qquad + \qquad H_2(g) \qquad \longrightarrow \qquad H_2O(g)$

FIGURE 5.10 The exothermic combustion of hydrogen in air. The reaction transfers energy to the surroundings in the form of heat, work, and light.

The value of $\Delta_r H°$ for 1 mol of this reaction, the formation of 2 mol of water, is twice the value for the formation of 1 mol of water.

It is important to identify the states of reactants and products in a reaction because the magnitude of $\Delta_r H°$ depends on whether they are solids, liquids, or gases. For the formation of 1 mol of *liquid water* from the elements, the enthalpy change is -285.8 kJ.

$$H_2(g) + \tfrac{1}{2}\,O_2(g) \rightarrow H_2O(\ell) \qquad \Delta_r H° = -285.8 \text{ kJ/mol-rxn}$$

Notice that this value is not the same as $\Delta_r H°$ for the formation of 1 mol of *water vapor* from hydrogen and oxygen. The difference between the two values is equal to the enthalpy change for the condensation of 1 mol of water vapor to 1 mol of liquid water.

These examples illustrate several general features of enthalpy changes for chemical reactions.

- Enthalpy changes are specific to the reaction being carried out. The identities of reactants and products and their states (s, ℓ, g) are important, as are the amounts of reactants and products.
- The enthalpy change depends on the number of moles of reaction, that is, the number of times the reaction *as written* is carried out.
- $\Delta_r H°$ has a negative value for an exothermic reaction; it has a positive value for an endothermic reaction.
- Values of $\Delta_r H°$ are numerically the same, but opposite in sign, for chemical reactions that are the reverse of each other.

Standard reaction enthalpies can be used to calculate the energy transferred as heat under conditions of constant pressure for any given mass of a reactant or product. Suppose you want to know the energy transferred to the surroundings as heat if 454 g of propane, C_3H_8, is burned (at constant pressure), given the equation for the exothermic combustion and the enthalpy change for the reaction.

$$C_3H_8(g) + 5\,O_2(g) \rightarrow 3\,CO_2(g) + 4\,H_2O(\ell) \qquad \Delta_r H° = -2220 \text{ kJ/mol-rxn}$$

Two steps are needed. First, find the amount of propane present in the sample:

$$454 \text{ g } C_3H_8 \left(\frac{1 \text{ mol } C_3H_8}{44.10 \text{ g } C_3H_8} \right) = 10.3 \text{ mol } C_3H_8$$

Second, use $\Delta_r H°$ to determine $\Delta H°$ for this amount of propane:

$$\Delta_r H° = 10.3 \text{ mol } C_3H_8 \left(\frac{1 \text{ mol-rxn}}{1 \text{ mol } C_3H_8} \right)\left(\frac{-2220 \text{ kJ}}{1 \text{ mol-rxn}} \right) = -22,900 \text{ kJ}$$

⏻WL INTERACTIVE EXAMPLE 5.5   Calculating the Enthalpy Change for a Reaction

Problem Sucrose (table sugar, $C_{12}H_{22}O_{11}$) can be oxidized to CO_2 and H_2O, and the enthalpy change for the reaction can be measured.

$$C_{12}H_{22}O_{11}(s) + 12\,O_2(g) \rightarrow 12\,CO_2(g) + 11\,H_2O(\ell) \qquad \Delta_r H° = -5645 \text{ kJ/mol-rxn}$$

What is the enthalpy change when 5.00 g of sugar is burned under conditions of constant pressure?

What Do You Know? The balanced equation for the combustion and the value of $\Delta_r H°$ are given. Also, the mass of sugar is given.

Strategy First determine the amount (mol) of sucrose in 5.00 g, and then use this with the value given for the enthalpy change for the oxidation of 1 mol of sucrose.

Sidebar notes

- **Fractional Stoichiometric Coefficients** When writing balanced equations to define thermodynamic quantities, chemists often use fractional stoichiometric coefficients. For example, to define $\Delta_r H$ for the decomposition or formation of 1 mol of H_2O, the coefficient for O_2 must be 1/2.

- **Standard Conditions** The superscript ° indicates standard conditions. It is applied to any type of thermodynamic data, such as enthalpy of fusion and vaporization ($\Delta_{fus}H°$ and $\Delta_{vap}H°$) and enthalpy of a reaction ($\Delta_r H°$). Standard conditions refers to reactants and products in their standard states at a pressure of 1 bar. One bar is approximately one atmosphere (1 atm = 1.013 bar; ▶ Appendix B).

- **Burning Sugar and Gummi Bears** A person on a diet might note that a (level) teaspoonful of sugar (about 3.5 g) supplies about 15 Calories (dietary Calories; the conversion is 4.184 kJ = 1 Cal). As diets go, a single spoonful of sugar doesn't have a large caloric content. But will you use just one level teaspoonful? Or just one Gummi Bear? In the Cengage *YouBook*, **click on the bear** for a video of a Gummi bear consumed by an oxidizing agent.

© Cengage Learning/Charles D. Winters

Strategy Map 5.5

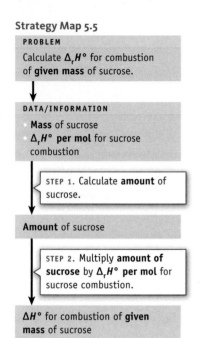

PROBLEM

Calculate $\Delta_r H°$ for combustion of **given mass** of sucrose.

↓

DATA/INFORMATION

- **Mass** of sucrose
- $\Delta_r H°$ **per mol** for sucrose combustion

STEP 1. Calculate **amount** of sucrose.

Amount of sucrose

STEP 2. Multiply **amount of sucrose** by $\Delta_r H°$ **per mol** for sucrose combustion.

$\Delta H°$ for combustion of **given mass** of sucrose

Solution

$$5.00 \text{ g sucrose} \times \frac{1 \text{ mol sucrose}}{342.3 \text{ g sucrose}} = 1.46 \times 10^{-2} \text{ mol sucrose}$$

$$\Delta H° = 1.46 \times 10^{-2} \text{ mol sucrose} \left(\frac{1 \text{ mol-rxn}}{1 \text{ mol sucrose}} \right) \left(\frac{-5645 \text{ kJ}}{1 \text{ mol-rxn}} \right)$$

$$\Delta H° = -82.5 \text{ kJ}$$

Think about Your Answer The calculated value is negative, as expected for a combustion reaction. The magnitude of $\Delta H°$ agrees with the fact that the mass of sucrose used, 5.00 g, is significantly less than that of one mole of sucrose (342.3 g).

Check Your Understanding

The combustion of ethane, C_2H_6, has an enthalpy change of -2857.3 kJ for the reaction as written below. Calculate $\Delta H°$ for the combustion of 15.0 g of C_2H_6.

$$2 \text{ C}_2\text{H}_6(g) + 7 \text{ O}_2(g) \rightarrow 4 \text{ CO}_2(g) + 6 \text{ H}_2\text{O}(g) \qquad \Delta_r H° = -2857.3 \text{ kJ/mol-rxn}$$

REVIEW & CHECK FOR SECTION 5.5

1. For the reaction 2 Hg(ℓ) + O$_2$(g) → 2 HgO(s), $\Delta_r H° = -181.6$ kJ/mol-rxn. What is the enthalpy change to decompose 1.00 mol of HgO(s) to O$_2$(g) and Hg(ℓ)?

    (a) 181.6 kJ   (b) −90.8 kJ   (c) 90.8 kJ   (d) 363.3 kJ

2. For the reaction 2 CO(g) + O$_2$(g) → 2 CO$_2$(g), $\Delta_r H° = -566$ kJ/mol-rxn. What is the enthalpy change for the oxidation of 42.0 g of CO(g)?

    (a) −283 kJ   (b) −425 kJ   (c) −566 kJ   (d) −393.5 kJ

5.6 Calorimetry

The energy evolved or absorbed as heat in a chemical or physical process can be measured by **calorimetry**. The apparatus used in this kind of experiment is a *calorimeter*.

Constant Pressure Calorimetry, Measuring ΔH

A constant pressure calorimeter can be used to measure the amount of energy transferred as heat under constant pressure conditions, that is, the enthalpy change for a chemical reaction.

The constant pressure calorimeter used in general chemistry laboratories is often a "coffee-cup calorimeter." This inexpensive device consists of two nested Styrofoam coffee cups with a loose-fitting lid and a temperature-measuring device such as a thermometer (Figure 5.11) or thermocouple. Styrofoam, a fairly good insulator, minimizes energy transfer as heat between the system and the surroundings. The reaction is carried out in solution in the cup. If the reaction is exothermic, it releases energy as heat to the solution, and the temperature of the solution rises. If the reaction is endothermic, energy is absorbed as heat from the solution, and a decrease in the temperature of the solution will be seen. The change in temperature of the solution is measured. Knowing the mass and specific heat capacity of the solution and the temperature change, the enthalpy change for the reaction can be calculated.

In this type of calorimetry experiment, it will be convenient to define the chemicals and the solution as the system. The surroundings are the cup and everything beyond the cup. As noted above, we assume there is no energy transfer to the cup

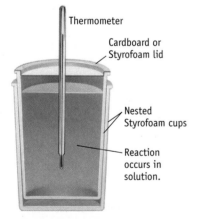

FIGURE 5.11 A coffee-cup calorimeter. A chemical reaction produces a change in temperature of the solution in the calorimeter. The Styrofoam container is fairly effective in preventing the transfer of energy as heat between the solution and its surroundings. Because the cup is open to the atmosphere, this is a constant pressure measurement.

Thermometer

Cardboard or Styrofoam lid

Nested Styrofoam cups

Reaction occurs in solution.

or beyond and that energy is transferred only as heat within the system. Two energy changes occur within the system. One is the change that takes place as the chemical reaction occurs, either releasing the potential energy stored in the reactants or absorbing energy and converting it to potential energy stored in the products. We label this energy as q_r. The other energy change is the energy gained or lost as heat by the solution ($q_{solution}$). Based on the law of conservation of energy,

$$q_r + q_{solution} = 0$$

The value of $q_{solution}$ can be calculated from the specific heat capacity, mass, and change in temperature of the solution. The quantity of energy evolved or absorbed as heat for the reaction (q_r) is the unknown in the equation.

The accuracy of a calorimetry experiment depends on the accuracy of the measured quantities (temperature, mass, specific heat capacity). In addition, it depends on how closely the assumption is followed that there is no energy transfer beyond the solution. A coffee-cup calorimeter is an unsophisticated apparatus, and the results obtained with it are not highly accurate, largely because this assumption is poorly met. In research laboratories, however, calorimeters are used that more effectively limit the energy transfer between system and surroundings. In addition, it is also possible to estimate and correct for the minimal energy transfer that occurs between the system and the surroundings.

ⓌWL   INTERACTIVE EXAMPLE 5.6   Using a Coffee-Cup Calorimeter

Problem Suppose you place 0.0500 g of magnesium chips in a coffee-cup calorimeter and then add 100.0 mL of 1.00 M HCl. The reaction that occurs is

$$Mg(s) + 2 HCl(aq) \rightarrow H_2(g) + MgCl_2(aq)$$

The temperature of the solution increases from 22.21 °C (295.36 K) to 24.46 °C (297.61 K). What is the enthalpy change for the reaction per mole of Mg? Assume that the specific heat capacity of the solution is 4.20 J/g · K and the density of the HCl solution is 1.00 g/mL.

What Do You Know? You know that energy is evolved as heat in this reaction because the temperature of the solution rises. Assuming no energy loss to the surroundings, the sum of the energy evolved as heat in the reaction, q_r, and the energy absorbed as heat by the solution, $q_{solution}$, will be zero, that is, $q_r + q_{solution} = 0$. The value of $q_{solution}$ can be calculated from data given; q_r is the unknown.

Strategy Solving the problem has four steps.

Step 1: Calculate the amount of magnesium.

Step 2: Calculate $q_{solution}$ from the values of the mass, specific heat capacity, and ΔT using Equation 5.1.

Step 3: Calculate q_r, assuming no energy transfer as heat occurs beyond the solution, that is, $q_r + q_{solution} = 0$.

Step 4: Use the value of q_r and the amount of Mg to calculate the enthalpy change per mole of Mg.

Solution

Step 1. Calculate the amount of Mg.

$$0.0500 \text{ g Mg} \times \frac{1 \text{ mol Mg}}{24.31 \text{ g Mg}} = 0.00206 \text{ mol Mg}$$

Step 2. Calculate $q_{solution}$. The mass of the solution is the mass of the 100.0 mL of HCl plus the mass of magnesium.

$$q_{solution} = (100.0 \text{ g HCl } \text{solution} + 0.0500 \text{ g Mg})(4.20 \text{ J/g} \cdot \text{K})(297.61 \text{ K} - 295.36 \text{ K})$$

$$= 945 \text{ J}$$

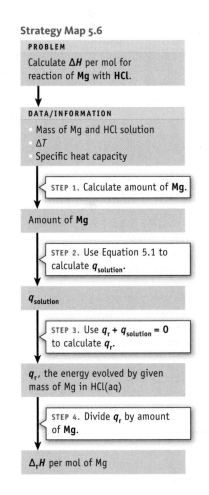

Strategy Map 5.6

PROBLEM

Calculate ΔH per mol for reaction of **Mg** with **HCl**.

DATA/INFORMATION

- Mass of Mg and HCl solution
- ΔT
- Specific heat capacity

STEP 1. Calculate amount of **Mg**.

Amount of **Mg**

STEP 2. Use Equation 5.1 to calculate $q_{solution}$.

$q_{solution}$

STEP 3. Use $q_r + q_{solution} = 0$ to calculate q_r.

q_r, the energy evolved by given mass of Mg in HCl(aq)

STEP 4. Divide q_r by amount of **Mg**.

$\Delta_r H$ per mol of Mg

Step 3. Calculate q_r.

$$q_r + q_{solution} = 0$$

$$q_r + 945 \text{ J} = 0$$

$$q_r = -945 \text{ J}$$

Step 4. Calculate the value of Δ_rH per mole of Mg. The value of q_r found in Step 2 resulted from the reaction of 0.00206 mol of Mg. The enthalpy change per mole of Mg is therefore

$$\Delta_rH = (-945 \text{ J}/0.00206 \text{ mol Mg})$$

$$= -4.59 \times 10^5 \text{ J/mol Mg } (= -459 \text{ kJ/mol Mg})$$

Think about Your Answer The calculation gives the correct sign of q_r and Δ_rH. The negative sign indicates that this is an exothermic reaction. The balanced equation states that one mole of magnesium is involved in one mole of reaction. The enthalpy change per mole of reaction, Δ_rH, is therefore −459 kJ/mol-rxn.

Check Your Understanding

Assume 200. mL of 0.400 M HCl is mixed with 200. mL of 0.400 M NaOH in a coffee-cup calorimeter. The temperature of the solutions before mixing was 25.10 °C; after mixing and allowing the reaction to occur, the temperature is 27.78 °C. What is the enthalpy change when one mole of acid is neutralized? (Assume that the densities of all solutions are 1.00 g/mL and their specific heat capacities are 4.20 J/g · K.)

Constant Volume Calorimetry, Measuring ΔU

Constant volume calorimetry is often used to evaluate the energy released by the combustion of fuels and the caloric value of foods. A weighed sample of a combustible solid or liquid is placed inside a "bomb," often a cylinder about the size of a large fruit juice can with thick steel walls and ends (Figure 5.12). The bomb is placed in a water-filled container with well-insulated walls. After filling the bomb with pure oxygen, the sample is ignited, usually by an electric spark. The heat generated by the combustion reaction warms the bomb and the water around it. The bomb, its

FIGURE 5.12 Constant volume calorimeter. A combustible sample is burned in pure oxygen in a sealed metal container or "bomb." Energy released as heat warms the bomb and the water surrounding it. By measuring the increase in temperature, the energy evolved as heat in the reaction can be determined.

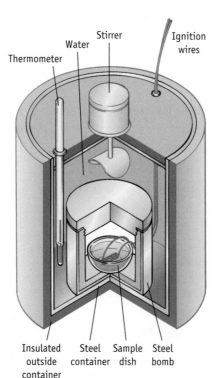

Thermometer   Water   Stirrer   Ignition wires

Insulated outside container   Steel container   Sample dish   Steel bomb

The sample burns in pure oxygen, warming the bomb.   The heat generated warms the water, and ΔT is measured by the thermometer.

contents, and the water are defined as the system. Assessment of energy transfers as heat within the system shows that

$$q_r + q_{bomb} + q_{water} = 0$$

where q_r is the energy released as heat by the reaction, q_{bomb} is the energy involved in heating the calorimeter bomb, and q_{water} is the energy involved in heating the water in the calorimeter. Because the volume does not change in a constant volume calorimeter, energy transfer as work cannot occur (assuming only *P–V* work is possible). Therefore, the energy transferred as heat at constant volume (q_v) is equal to the change in internal energy, ΔU.

● **Calorimetry, ΔU, and ΔH** The two types of calorimetry (constant volume and constant pressure) highlight the differences between enthalpy and internal energy. The energy transferred as heat at constant pressure, q_p, is, by definition, ΔH, whereas the energy transferred as heat at constant volume, q_v, is ΔU.

ᗝWL INTERACTIVE EXAMPLE 5.7 Constant Volume Calorimetry

Problem Octane, C_8H_{18}, a primary constituent of gasoline, burns in air:

$$C_8H_{18}(\ell) + 25/2\ O_2(g) \rightarrow 8\ CO_2(g) + 9\ H_2O(\ell)$$

A 1.00-g sample of octane is burned in a constant volume calorimeter similar to that shown in Figure 5.12. The calorimeter is in an insulated container with 1.20 kg of water. The temperature of the water and the bomb rises from 25.00 °C (298.15 K) to 33.20 °C (306.35 K). The heat required to raise the bomb's temperature (its heat capacity), C_{bomb}, is 837 J/K. (a) What is the heat of combustion per gram of octane? (b) What is the heat of combustion per mole of octane?

What Do You Know? There are energy changes for the three components of this system: the energy evolved in the reaction, q_r; the energy absorbed by the water, q_{water}; and the energy absorbed by the calorimeter, q_{bomb}. You know the following: the molar mass of octane, masses of the sample and the calorimeter water, $T_{initial}$, T_{final}, C_{bomb}, and C_{water}. You can assume no energy loss to the surroundings.

Strategy

- The sum of all the energies transferred as heat in the system = $q_r + q_{bomb} + q_{water} = 0$. The first term, q_r, is the unknown. The second and third terms in the equation can be calculated from the data given: q_{bomb} is calculated from the bomb's heat capacity and ΔT, and q_{water} is determined from the specific heat capacity, mass, and ΔT for water.

- The value of q_r is the energy evolved in the combustion of 1.00 g of octane. Use this and the molar mass of octane (114.2 g/mol) to calculate the energy evolved as heat per mole of octane.

Solution

(a) $q_{water} = C_{water} \times m_{water} \times \Delta T = (4.184\ J/g \cdot K)(1.20 \times 10^3\ g)(306.35\ K - 298.15\ K) = +41.2 \times 10^3\ J$

$q_{bomb} = (C_{bomb})(\Delta T) = (837\ J/K)(306.35\ K - 298.15\ K) = 6.86 \times 10^3\ J$

$q_r + q_{water} + q_{bomb} = 0$

$q_r + 41.2 \times 10^3\ J + 6.86 \times 10^3\ J = 0$

$q_r = -48.1 \times 10^3\ J\ (or\ -48.1\ kJ)$

Heat of combustion per gram = −48.1 kJ

(b) Heat of combustion per mole of octane = $(-48.1\ kJ/g)(114.2\ g/mol) = -5.49 \times 10^3\ kJ/mol$

Think about Your Answer Because the volume does not change, no energy transfer in the form of work occurs. The change of internal energy, $\Delta_r U$, for the combustion of $C_8H_{18}(\ell)$ is -5.49×10^3 kJ/mol. Also note that C_{bomb} has no mass units. It is the heat required to warm the bomb by 1 kelvin.

Strategy Map 5.7

PROBLEM

Calculate **ΔU** per mol for combustion of octane.

↓

DATA/INFORMATION

- Mass of octane
- Mass of water in calorimeter
- C_{water}
- C_{bomb}
- ΔT

STEP 1. Use Equation 5.1 to calculate q_{water}.

↓

q_{water}

STEP 2. Use C_{bomb} and ΔT to calculate q_{bomb}.

↓

q_{bomb}

STEP 3. Use $q_r + q_{water} + q_{bomb} = 0$ to calculate q_r.

↓

q_r, the energy evolved as heat by given mass of octane (kJ)

STEP 4. Multiply q_r by **molar mass** of octane.

↓

ΔU per mol of octane

Check Your Understanding

A 1.00-g sample of ordinary table sugar (sucrose, $C_{12}H_{22}O_{11}$) is burned in a bomb calorimeter. The temperature of 1.50×10^3 g of water in the calorimeter rises from 25.00 °C to 27.32 °C. The heat capacity of the bomb is 837 J/K, and the specific heat capacity of the water is 4.20 J/g · K. Calculate (a) the heat evolved per gram of sucrose and (b) the heat evolved per mole of sucrose.

REVIEW & CHECK FOR SECTION 5.6

1.   A student used a coffee-cup calorimeter to determine the enthalpy of solution for NH_4NO_3. When NH_4NO_3 is added to water, there is a decrease in temperature of the solution. Is the solution process exothermic or endothermic?

    (a)   exothermic   (b)   endothermic

2.   What if, contrary to our assumption of no transfer of energy between system and surroundings in the calorimeter experiment, some energy as heat is transferred from the surroundings to the system. How will this affect the value of $\Delta_{solution}H°$?

    (a)   No effect will be seen.

    (b)   The calculated value of $\Delta_{solution}H°$ will be too small.

    (c)   The calculated value of $\Delta_{solution}H°$ will be too large.

---

go Chemistry

Module 10: Thermochemistry and Hess's Law covers concepts in this section.

5.7   Enthalpy Calculations

Enthalpy changes for an enormous number of chemical and physical processes are available on the World Wide Web and in reference books. These data have been collected by scientists over a number of years from many experiments and are used to calculate enthalpy changes for chemical processes. Now we want to discuss how to use these data.

Hess's Law

The enthalpy change can be measured by calorimetry for many, but not all, chemical processes. Consider, for example, the oxidation of carbon to form carbon monoxide.

$$C(s) + \tfrac{1}{2}\,O_2(g) \rightarrow CO(g)$$

The primary product of the reaction of carbon and oxygen is CO_2, even if a deficiency of oxygen is used. As soon as CO is formed, it reacts with O_2 to form CO_2. Because the reaction cannot be carried out in a way that allows CO to be the sole product, it is not possible to measure the change in enthalpy for this reaction by calorimetry.

    The enthalpy change for the reaction forming $CO(g)$ from $C(s)$ and $O_2(g)$ can be determined indirectly, however, from enthalpy changes for other reactions for which values of $\Delta_r H°$ can be measured. The calculation is based on **Hess's law**, which states that if a reaction is the sum of two or more other reactions, $\Delta_r H°$ for the overall process is the sum of the $\Delta_r H°$ values of those reactions.

    The oxidation of $C(s)$ to $CO_2(g)$ can be viewed as occurring in two steps: first the oxidation of $C(s)$ to $CO(g)$ (Equation 1) and second the oxidation of $CO(g)$ to $CO_2(g)$ (Equation 2). Adding these two equations gives the equation for the oxidation of $C(s)$ to $CO_2(g)$ (Equation 3).

Equation 1:	$C(s) + \tfrac{1}{2}\,O_2(g) \rightarrow CO(g)$	$\Delta_r H_1° = \;?$
Equation 2:	$CO(g) + \tfrac{1}{2}\,O_2(g) \rightarrow CO_2(g)$	$\Delta_r H_2° = -283.0$ kJ/mol-rxn
Equation 3:	$C(s) + O_2(g) \rightarrow CO_2(g)$	$\Delta_r H_3° = -393.5$ kJ/mol-rxn

    Hess's law tells us that the enthalpy change for the overall reaction ($\Delta_r H_3°$) will equal the sum of the enthalpy changes for reactions 1 and 2 ($\Delta_r H_1° + \Delta_r H_2°$). Both $\Delta_r H_2°$ and $\Delta_r H_3°$ can be measured, and these values are then used to calculate the enthalpy change for reaction 1.

$$\Delta_r H_3° = \Delta_r H_1° + \Delta_r H_2°$$

$$-393.5 \text{ kJ/mol-rxn} = \Delta_r H_1° + (-283.0 \text{ kJ/mol-rxn})$$

$$\Delta_r H_1° = -110.5 \text{ kJ/mol-rxn}$$

Hess's law also applies to physical processes. The enthalpy change for the reaction of $H_2(g)$ and $O_2(g)$ to form 1 mol of H_2O vapor is different from the enthalpy change to form 1 mol of liquid H_2O. The difference is the negative of the *enthalpy of vaporization* of water, $\Delta_r H^\circ_2$ ($= -\Delta_{vap}H^\circ$) as shown in the following analysis

Equation 1:	$H_2(g) + \frac{1}{2} O_2(g) \rightarrow H_2O(g)$	$\Delta_r H^\circ_1 = -241.8$ kJ/mol-rxn
Equation 2:	$H_2O(g) \rightarrow H_2O(\ell)$	$\Delta_r H^\circ_2 = -44.0$ kJ/mol-rxn
Equation 3:	$H_2(g) + \frac{1}{2} O_2(g) \rightarrow H_2O(\ell)$	$\Delta_r H^\circ_3 = -285.8$ kJ/mol-rxn

Energy Level Diagrams

When using Hess's law, it is often helpful to represent enthalpy data schematically in an energy level diagram. In such drawings, the various substances being studied—the reactants and products in a chemical reaction, for example—are placed on an arbitrary energy scale. The relative enthalpy of each substance is given by its position on the vertical axis, and numerical differences in enthalpy between them are shown by the vertical arrows. Such diagrams provide a visual perspective on the magnitude and direction of enthalpy changes and show how the enthalpies of the substances are related.

Energy level diagrams that summarize the two examples of Hess's law discussed earlier are shown in Figure 5.13. In Figure 5.13a, the elements $C(s)$ and $O_2(g)$ are at the highest enthalpy. The reaction of carbon and oxygen to form $CO_2(g)$ lowers the enthalpy by 393.5 kJ. This can occur either in a single step, shown on the left in Figure 5.13a, or in two steps via initial formation of $CO(g)$, as shown on the right. Similarly, in Figure 5.13b, the mixture of $H_2(g)$ and $O_2(g)$ is at the highest enthalpy. Both liquid and gaseous water have lower enthalpies, with the difference between the two being the enthalpy of vaporization.

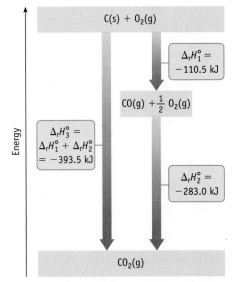

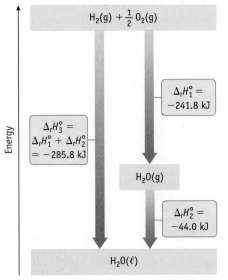

(a) The formation of CO_2 can occur in a single step or in a succession of steps. $\Delta_r H^\circ$ for the overall process is −393.5 kJ, no matter which path is followed.

(b) The formation of $H_2O(\ell)$ can occur in a single step or in a succession of steps. $\Delta_r H^\circ$ for the overall process is −285.8 kJ, no matter which path is followed.

FIGURE 5.13 Energy level diagrams. (a) Relating enthalpy changes in the formation of $CO_2(g)$. **(b)** Relating enthalpy changes in the formation of $H_2O(\ell)$. Enthalpy changes associated with changes between energy levels are given alongside the vertical arrows.

Strategy Map 5.8

PROBLEM

Hess's Law: Calculate $\Delta_r H°$ for **targeted reaction** from $\Delta_r H°$ values for other reactions.

$\downarrow$

DATA/INFORMATION

Three reactions with known $\Delta_r H°$ values

STEP 1. Combine equations with $\Delta_r H°$ **values** to give **targeted net equation.**

Targeted **net equation**

STEP 2. Add $\Delta_r H°$ **values** for equations that sum to targeted equation.

$\Delta_r H°$ for targeted reaction

⊌WL   INTERACTIVE EXAMPLE 5.8   Using Hess's Law

Problem Suppose you want to know the enthalpy change for the formation of methane, CH_4, from solid carbon (as graphite) and hydrogen gas:

$$C(s) + 2\ H_2(g) \rightarrow CH_4(g) \qquad \Delta_r H° = ?$$

The enthalpy change for this reaction cannot be measured in the laboratory because the reaction is very slow. We can, however, measure enthalpy changes for the combustion of carbon, hydrogen, and methane.

Equation 1:	$C(s) + O_2(g) \rightarrow CO_2(g)$	$\Delta_r H_1° = -393.5$ kJ/mol-rxn
Equation 2:	$H_2(g) + \frac{1}{2}\ O_2(g) \rightarrow H_2O(\ell)$	$\Delta_r H_2° = -285.8$ kJ/mol-rxn
Equation 3:	$CH_4(g) + 2\ O_2(g) \rightarrow CO_2(g) + 2\ H_2O(\ell)$	$\Delta_r H_3° = -890.3$ kJ/mol-rxn

Use this information to calculate $\Delta_r H°$ for the formation of methane from its elements.

What Do You Know? This is a Hess's law problem. You need to adjust the three equations so they can be added together to give the targeted equation, $C(s) + 2\ H_2(g) \rightarrow CH_4(g)$. When an adjustment in an equation is made, you need to also adjust the enthalpy change.

Strategy The three reactions (1, 2, and 3), as written, cannot be added together to obtain the equation for the formation of CH_4 from its elements. Methane, CH_4, is a product in the reaction for which we wish to calculate $\Delta_r H°$, but it is a reactant in Equation 3. Water appears in two of these equations although it is not a component of the reaction forming CH_4 from carbon and hydrogen. To use Hess's law to solve this problem, you first have to manipulate the equations and adjust $\Delta_r H°$ values accordingly before adding equations together. Recall, from Section 5.5, that writing an equation in the reverse direction changes the sign of $\Delta_r H°$ and that doubling the amount of reactants and products doubles the value of $\Delta_r H°$. Adjustments to Equations 2 and 3 will produce new equations that, along with Equation 1, can be combined to give the desired net reaction.

Solution To have CH_4 appear as a product in the overall reaction, reverse Equation 3, which changes the sign of its $\Delta_r H°$.

Equation 3′: $CO_2(g) + 2\ H_2O(\ell) \rightarrow CH_4(g) + 2\ O_2(g)$

$$\Delta_r H_3°{}' = -\Delta_r H_3° = +890.3\ \text{kJ/mol-rxn}$$

Next, you see that 2 mol of $H_2(g)$ is on the reactant side in our desired equation. Equation 2 is written for only 1 mol of $H_2(g)$ as a reactant. Therefore, multiply the stoichiometric coefficients in Equation 2 by 2 and multiply the value of its $\Delta_r H°$ by 2.

Equation 2′: $2\ H_2(g) + O_2(g) \rightarrow 2\ H_2O(\ell)$

$$\Delta_r H_2°{}' = 2\ \Delta_r H_2° = 2\ (-285.8\ \text{kJ/mol-rxn}) = -571.6\ \text{kJ/mol-rxn}$$

You now have three equations that, when added together, will give the targeted equation for the formation of methane from carbon and hydrogen. In this summation process, $O_2(g)$, $H_2O(\ell)$, and $CO_2(g)$ all cancel.

Equation 1:	$C(s) + O_2(g) \rightarrow CO_2(g)$	$\Delta_r H_1° = -393.5$ kJ/mol-rxn
Equation 2′:	$2\ H_2(g) + O_2(g) \rightarrow 2\ H_2O(\ell)$	$\Delta_r H_2°{}' = 2\ \Delta_r H_2° = -571.6$ kJ/mol-rxn
Equation 3′:	$CO_2(g) + 2\ H_2O(\ell) \rightarrow CH_4(g) + 2\ O_2\ (g)$	$\Delta_r H_3°{}' = -\Delta_r H_3° = +890.3$ kJ/mol-rxn
Net Equation:	$C(s) + 2\ H_2(g) \rightarrow CH_4(g)$	$\Delta_r H_{net}° = \Delta_r H_1° + 2\ \Delta_r H_2°{}' + (-\Delta_r H_3°{}')$

$$\Delta_r H_{net}° = (-393.5\ \text{kJ/mol-rxn}) + (-571.6\ \text{kJ/mol-rxn}) + (+890.3\ \text{kJ/mol-rxn})$$
$$= -74.8\ \text{kJ/mol-rxn}$$

Thus, for the formation of 1 mol of $CH_4(g)$ from the elements, $\Delta_r H° = -74.8$ kJ/mol-rxn.

Think about Your Answer Notice that the enthalpy change for the formation of the compound from its elements is exothermic, as it is for the great majority of compounds.

Using Hess's Law

How did we know how the three equations should be adjusted in Example 5.8? Here is a general strategy for solving this type of problem.

Step 1. Arrange the given equations to get the reactants and products in the equation whose $\Delta_r H°$ you wish to calculate on the correct sides of the equations. You may need to reverse some of the given equations in order to do this. In Example 5.8, the reactants, C(s) and H_2(g), are reactants

in Equations 1 and 2, but the product, CH_4(g), is a reactant in Equation 3. Equation 3 was reversed to get CH_4 on the product side.

Step 2. Get the correct amounts of the substances on each side. In Example 5.8, only one adjustment was needed. There was 1 mol of H_2 on the left (reactant side) in Equation 2. We needed 2 mol of H_2 in the

overall equation; this required doubling the quantities in Equation 2.

Step 3. Make sure other substances in the equations cancel when the equations are added. In Example 5.8, equal amounts of O_2 and H_2O appeared on the left and right sides in the three equations, and so they canceled when the equations were added together.

Check Your Understanding

Use Hess's law to calculate the enthalpy change for the formation of $CS_2(\ell)$ from C(s) and S(s) [C(s) + 2 S(s) $\rightarrow$ $CS_2(\ell)$] from the following enthalpy values.

$$C(s) + O_2(g) \rightarrow CO_2(g) \qquad \Delta_r H_1° = -393.5 \text{ kJ/mol-rxn}$$

$$S(s) + O_2(g) \rightarrow SO_2(g) \qquad \Delta_r H_2° = -296.8 \text{ kJ/mol-rxn}$$

$$CS_2(\ell) + 3 O_2(g) \rightarrow CO_2(g) + 2 SO_2(g) \qquad \Delta_r H_3° = -1103.9 \text{ kJ/mol-rxn}$$

Standard Enthalpies of Formation

Calorimetry and the application of Hess's law have made available a great many $\Delta_r H°$ values for chemical reactions. The table in Appendix L, for example, lists **standard molar enthalpies of formation, $\Delta_f H°$.** *The standard molar enthalpy of formation is the enthalpy change for the formation of 1 mol of a compound directly from its component elements in their standard states.*

Several examples of standard molar enthalpies of formation will be helpful to illustrate this definition.

$\Delta_f H°$ **for NaCl(s):** At 25 °C and a pressure of 1 bar, Na is a solid, and Cl_2 is a gas. The standard enthalpy of formation of NaCl(s) is defined as the enthalpy change that occurs when 1 mol of NaCl(s) is formed from 1 mol of Na(s) and ½ mol of Cl_2(g).

$$Na(s) + \tfrac{1}{2} Cl_2(g) \rightarrow NaCl(s) \qquad \Delta_f H° = -411.12 \text{ kJ/mol}$$

Notice that a fraction is required as the coefficient for the chlorine gas in this equation because the definition of $\Delta_f H°$ specifies the formation of *one* mole of NaCl(s).

$\Delta_f H°$ **for NaCl(aq):** The enthalpy of formation for an aqueous solution of a compound refers to the enthalpy change for the formation of a 1 mol/L solution of the compound starting with the elements making up the compound. It is thus the enthalpy of formation of the compound plus the enthalpy change that occurs when the substance dissolves in water.

$$Na(s) + \tfrac{1}{2} Cl_2(g) \rightarrow NaCl(aq) \qquad \Delta_f H° = -407.27 \text{ kJ/mol}$$

$\Delta_f H°$ **for $C_2H_5OH(\ell)$:** At 25 °C and 1 bar, the standard states of the elements are C(s, graphite), H_2(g), and O_2(g). The standard enthalpy of formation of $C_2H_5OH(\ell)$ is defined as the enthalpy change that occurs when 1 mol of $C_2H_5OH(\ell)$ is formed from 2 mol of C(s), 3 mol of H_2(g), and 1/2 mol of O_2(g).

$$2 C(s) + 3 H_2(g) + \tfrac{1}{2} O_2(g) \rightarrow C_2H_5OH(\ell) \qquad \Delta_f H° = -277.0 \text{ kJ/mol}$$

• **$\Delta_f H°$ Values** Consult the National Institute for Standards and Technology website (webbook.nist.gov/chemistry) for an extensive compilation of enthalpies of formation.

• **Units for Enthalpy of Formation** The units for values of $\Delta_f H°$ are usually given simply as kJ/mol where the denominator is really mol-rxn. However, because an enthalpy of formation is defined as the change in enthalpy for the formation of 1 mol of compound, it is understood that "per mol" also means "per mol of compound."

Notice that the reaction defining the enthalpy of formation for liquid ethanol is not a reaction a chemist can carry out in the laboratory. This illustrates an important point: *The enthalpy of formation of a compound does not necessarily correspond to a reaction that can be carried out.*

Appendix L lists values of $\Delta_f H°$ for some common substances, and a review of these values leads to some important observations.

- *The standard enthalpy of formation for an element in its standard state is zero.*
- Most $\Delta_f H°$ values are negative, indicating that formation of most compounds from the elements is exothermic. A very few values are positive, and these represent compounds that are unstable with respect to decomposition to the elements. (One example is NO(g) with $\Delta_f H° = +90.29$ kJ/mol.)
- Values of $\Delta_f H°$ can often be used to compare the stabilities of related compounds. Consider the values of $\Delta_f H°$ for the hydrogen halides. Hydrogen fluoride is the most stable of these compounds with respect to decomposition to the elements, whereas HI is the least stable (as indicated by $\Delta_f H°$ of HF being the most negative value and that of HI being the most positive).

● $\Delta_f H°$ Values of Hydrogen Halides

Compound	$\Delta_f H°$ (kJ/mol)
HF(g)	−273.3
HCl(g)	−92.31
HBr(g)	−35.29
HI(g)	+25.36

● Stoichiometric Coefficients In Equation 5.6 a stoichiometric coefficient, *n*, is represented as the number of moles of the substance per mole of reaction.

● Δ = Final − Initial Equation 5.6 is another example of the convention that a change (Δ) is always calculated by subtracting the value for the initial state (the reactants) from the value for the final state (the products).

Enthalpy Change for a Reaction

Using standard molar enthalpies of formation and Equation 5.6, it is possible to calculate the enthalpy change for a reaction under standard conditions.

$$\Delta_r H° = \Sigma n \Delta_f H°(\text{products}) - \Sigma n \Delta_f H°(\text{reactants}) \qquad (5.6)$$

In this equation, the symbol Σ (the Greek capital letter sigma) means "take the sum." To find $\Delta_r H°$, add up the molar enthalpies of formation of the products, each multiplied by its stoichiometric coefficient *n*, and subtract from this the sum of the molar enthalpies of formation of the reactants, each multiplied by its stoichiometric coefficient. This equation is a logical consequence of the definition of $\Delta_f H°$ and Hess's law (see *A Closer Look: Hess's Law and Equation 5.6*, page 200).

Suppose you want to know how much energy is required to decompose 1 mol of calcium carbonate (limestone) to calcium oxide (lime) and carbon dioxide under standard conditions:

$$CaCO_3(s) \rightarrow CaO(s) + CO_2(g) \qquad \Delta_r H° = ?$$

You would use the following enthalpies of formation (from Appendix L):

Compound	$\Delta_f H°$ (kJ/mol)
$CaCO_3(s)$	−1207.6
$CaO(s)$	−635.1
$CO_2(g)$	−393.5

and then use Equation 5.6 to find the standard enthalpy change for the reaction, $\Delta_r H°$.

$$\Delta_r H° = \left[\left(\frac{1 \text{ mol CaO}}{1 \text{ mol-rxn}} \right) \left(\frac{-635.1 \text{ kJ}}{\text{mol CaO}} \right) + \left(\frac{1 \text{ mol CO}_2}{1 \text{ mol-rxn}} \right) \left(\frac{-393.5 \text{ kJ}}{1 \text{ mol CO}_2} \right) \right]$$
$$- \left[\left(\frac{1 \text{ mol CaCO}_3}{1 \text{ mol-rxn}} \right) \left(\frac{-1207.6 \text{ kJ}}{1 \text{ mol CaCO}_3} \right) \right]$$
$$= +179.0 \text{ kJ/mol-rxn}$$

The decomposition of limestone to lime and CO_2 is endothermic. That is, energy (179.0 kJ) must be supplied to decompose 1 mol of $CaCO_3(s)$ to $CaO(s)$ and $CO_2(g)$.

⊙WL **INTERACTIVE EXAMPLE 5.9** Using Enthalpies of Formation

Problem Nitroglycerin is a powerful explosive that forms four different gases when detonated:

$$2\ C_3H_5(NO_3)_3(\ell) \rightarrow 3\ N_2(g) + 1/2\ O_2(g) + 6\ CO_2(g) + 5\ H_2O(g)$$

Calculate the enthalpy change that occurs when 10.0 g of nitroglycerin is detonated. The standard enthalpy of formation of nitroglycerin, $\Delta_f H°$, is −364 kJ/mol. Use Appendix L to find other $\Delta_f H°$ values that are needed.

What Do You Know? From Appendix L, $\Delta_f H°[CO_2(g)] = −393.5$ kJ/mol, $\Delta_f H°[H_2O(g)] = −241.8$ kJ/mol, and $\Delta_f H° = 0$ for $N_2(g)$ and $O_2(g)$. The mass and $\Delta_f H°$ for nitroglycerin are also given.

Strategy Substitute the enthalpy of formation values for products and reactants into Equation 5.6 to determine the enthalpy change for 1 mol of reaction. This represents the enthalpy change for detonation of 2 mol of nitroglycerin. Determine the amount (mol) represented by 10.0 g of nitroglycerin, then use this value with $\Delta_r H°$ and the relationship between moles of nitroglycerin and moles of reaction to obtain the answer.

Solution Using Equation 5.6, we find the enthalpy change for the explosion of 2 mol of nitroglycerin is

$$\Delta_r H° = \left(\frac{6\ \text{mol}\ CO_2}{1\ \text{mol-rxn}}\right)\Delta_f H°[CO_2(g)] + \left(\frac{5\ \text{mol}\ H_2O}{1\ \text{mol-rxn}}\right)\Delta_f H°[H_2O(g)]$$

$$-\left(\frac{2\ \text{mol}\ C_3H_5(NO_3)_3}{1\ \text{mol-rxn}}\right)\Delta_f H°[C_3H_5(NO_3)_3(\ell)]$$

$$\Delta_r H° = \left(\frac{6\ \text{mol}\ CO_2}{1\ \text{mol-rxn}}\right)\left(\frac{-393.5\ kJ}{1\ \text{mol}\ CO_2}\right) + \left(\frac{5\ \text{mol}\ H_2O}{1\ \text{mol-rxn}}\right)\left(\frac{-241.8\ kJ}{1\ \text{mol}\ H_2O}\right)$$

$$-\left(\frac{2\ \text{mol}\ C_3H_5(NO_3)_3}{1\ \text{mol-rxn}}\right)\left(\frac{-364\ kJ}{1\ \text{mol}\ C_3H_5(NO_3)_3}\right) = -2842\ \text{kJ/mol-rxn}$$

The problem asks for the enthalpy change using 10.0 g of nitroglycerin. You next need to determine the amount of nitroglycerin in 10.0 g.

$$10.0\ \text{g nitroglycerin} \left(\frac{1\ \text{mol nitroglycerin}}{227.1\ \text{g nitroglycerin}}\right) = 0.0440\ \text{mol nitroglycerin}$$

The enthalpy change for the detonation of 0.0440 mol of nitroglycerin is

$$\Delta_r H° = 0.0440\ \text{mol nitroglycerin}\left(\frac{1\ \text{mol-rxn}}{2\ \text{mol nitroglycerin}}\right)\left(\frac{-2842\ kJ}{1\ \text{mol-rxn}}\right)$$

$$= -62.6\ kJ$$

Think about Your Answer The large negative value of $\Delta_r H°$ is in accord with the fact that this reaction is highly exothermic.

Check Your Understanding

Calculate the standard enthalpy of combustion for benzene, C_6H_6.

$$C_6H_6(\ell) + 15/2\ O_2(g) \rightarrow 6\ CO_2(g) + 3\ H_2O(\ell) \qquad \Delta_r H° = ?$$

The enthalpy of formation of benzene is known [$\Delta_f H°[C_6H_6(\ell)] = +49.0$ kJ/mol], and other values needed can be found in Appendix L.

Strategy Map 5.9

PROBLEM

Calculate $\Delta_r H°$ for reaction of a **given mass** of compound.

DATA/INFORMATION

- Mass and $\Delta_f H°$ of compound
- $\Delta_f H°$ values for products
- Balanced equation

STEP 1. Calculate $\Delta_r H°$ for reaction of compound using $\Delta_f H°$ **values** for reactants and products.

$\Delta_r H°$ for reaction of compound

STEP 2. Determine **amount** of compound.

Amount of compound

STEP 3. Convert from **mol** of compound to **mol-rxn** and then multiply by $\Delta_r H°$.

$\Delta_r H°$ for reaction of **given mass** of compound

REVIEW & CHECK FOR SECTION 5.7

1. The standard enthalpy of formation for $AlCl_3(s)$ is the enthalpy change for what reaction?

    (a) $Al(s) + 3\ Cl_2(g) \rightarrow AlCl_3(s)$

    (b) $Al(s) + 3\ Cl(g) \rightarrow AlCl_3(s)$

    (c) $3\ HCl(aq) + Al(OH)_3(s) \rightarrow AlCl_3(s) + 3\ H_2O(\ell)$

    (d) $Al(s) + 3/2\ Cl_2(g) \rightarrow AlCl_3(s)$

A CLOSER LOOK

Hess's Law and Equation 5.6

Equation 5.6 is an application of Hess's law. To illustrate this, let us look further at the decomposition of calcium carbonate.

$$CaCO_3(s) \rightarrow CaO(s) + CO_2(g) \qquad \Delta_rH° = ?$$

Because enthalpy is a state function, the change in enthalpy for this reaction is independent of the route from reactants to products. We can imagine an alternate route from reactant to products that involves first converting the reactant ($CaCO_3$) to elements in their standard states, then recombining these elements to give the reaction products. Notice that the enthalpy changes for these processes are the enthalpies of formation of the reactants and products in the equation above:

$$CaCO_3(s) \rightarrow Ca(s) + C(s) + 3/2 \ O_2(g)$$
$$-\Delta_fH°[CaCO_3(s)] = \Delta_rH°_1$$

$$C(s) + O_2(g) \rightarrow CO_2(g)$$
$$\Delta_fH°[CO_2(g)] = \Delta_rH°_2$$

$$Ca(s) + ½ \ O_2(g) \rightarrow CaO(s)$$
$$\Delta_fH°[CaO(s)] = \Delta_rH°_3$$

$$CaCO_3(s) \rightarrow CaO(s) + CO_2(g) \qquad \Delta_rH°_{net}$$

$$\Delta_rH°_{net} = \Delta_rH°_1 + \Delta_rH°_2 + \Delta_rH°_3$$

$$\Delta_rH° = \Delta_fH°[CaO(s)] + \Delta_fH°[CO_2(g)]$$
$$- \Delta_fH°[CaCO_3(s)]$$

That is, the change in enthalpy for the reaction is equal to the enthalpies of formation of products (CO_2 and CaO) minus the enthalpy of formation of the reactant

($CaCO_3$), which is, of course, what one does when using Equation 5.6 for this calculation. The relationship among these enthalpy quantities is illustrated in the energy-level diagram.

Energy level diagram for the decomposition of $CaCO_3(s)$

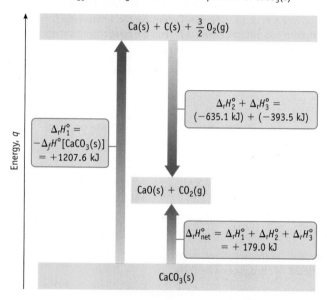

2. Acetic acid is made by the reaction $CH_3OH(\ell) + CO(g) \rightarrow CH_3CO_2H(\ell)$. Determine the enthalpy change for this reaction from the enthalpies of the three reactions below.

$$2 \ CH_3OH(\ell) + 3 \ O_2(g) \rightarrow 2 \ CO_2(g) + 4 \ H_2O(\ell) \qquad \Delta_rH°_1$$

$$CH_3CO_2H(\ell) + 2 \ O_2(g) \rightarrow 2 \ CO_2(g) + 2 \ H_2O(\ell) \qquad \Delta_rH°_2$$

$$2 \ CO(g) + O_2(g) \rightarrow 2 \ CO_2(g) \qquad \Delta_rH°_3$$

(a) $\Delta_rH°_1 + \Delta_rH°_2 + \Delta_rH°_3$

(b) $\Delta_rH°_1 - \Delta_rH°_2 + \Delta_rH°_3$

(c) $1/2 \ \Delta_rH°_1 - \Delta_rH°_2 + 1/2 \ \Delta_rH°_3$

(d) $-1/2 \ \Delta_rH°_1 + \Delta_rH°_2 - 1/2 \ \Delta_rH°_3$

5.8 Product- or Reactant-Favored Reactions and Thermodynamics

At the beginning of this chapter, we noted that thermodynamics would provide answers to four questions:

- How do we measure and calculate the energy changes associated with physical changes and chemical reactions?
- What is the relationship between energy changes, heat, and work?
- How can we determine whether a chemical reaction is product-favored or reactant-favored at equilibrium?
- How can we determine whether a chemical reaction or physical process will occur spontaneously, that is, without outside intervention?

The first two questions were addressed in this chapter, but the other two important questions still remain (for Chapter 19).

In Chapter 3, we learned that chemical reactions proceed toward equilibrium, and spontaneous changes occur in a way that allows a system to approach equilibrium. Reactions in which reactants are largely converted to products when equilibrium is reached are said to be *product-favored at equilibrium*. Reactions in which only small amounts of products are present at equilibrium are called *reactant-favored at equilibrium* (◄ page 100).

CASE STUDY

The Fuel Controversy—Alcohol and Gasoline

It is clear that supplies of fossil fuels are declining and their prices are increasing, just as the nations of the earth have ever greater energy needs. We will have more to say about this in the Interchapter (Energy) that follows. Here, however, let's analyze the debate about replacing gasoline with ethanol (C_2H_5OH).

As Matthew Wald said in the article "Is Ethanol in for the Long Haul?" (*Scientific American*, January 2007), "The U.S. has gone on an ethanol binge." In 2005, the U.S. Congress passed an energy bill stating that ethanol production should be 7.5 billion gallons a year by 2012, up from about 5 billion gallons in 2005. The goal is to at least partially replace gasoline with ethanol.

Is a goal of replacing gasoline completely with ethanol reasonable? This is a lofty goal, given that present gasoline consumption in the U.S. is about 140 billion gallons annually. Again, according to Matthew Wald, "Even if 100 percent of the U.S. corn supply was distilled into ethanol, it would supply only a small fraction of the fuel consumed by the nation's vehicles." Wald's thesis in his article, which is supported by numerous scientific studies, is that if ethanol is to be pursued as an alternative to gasoline, more emphasis would have to be placed on deriving ethanol from sources other than corn, such as cellulose from cornstalks and various grasses.

Beyond this, there are other problems associated with ethanol. One is that it cannot be distributed through a pipeline system as gasoline can. Any water in the pipeline is miscible with ethanol, which causes the fuel value to decline.

Finally, even E85 fuel—a blend of 85% ethanol and 15% gasoline—cannot be used in most current vehicles because relatively few vehicles as yet have engines designed for fuels with a high ethanol content (so-called "flexible fuel" engines). The number of these vehicles would need to be increased in order for E85 to have a significant effect on our gasoline usage.

For more information, see the references in Wald's *Scientific American* article.

Questions:

For the purposes of this analysis, let us use octane (C_8H_{18}) as a substitute for the complex mixture of hydrocarbons in gasoline. Data you will need for this question (in addition to Appendix L) are:

$$\Delta_f H° \ [C_8H_{18}(\ell)] = -250.1 \ \text{kJ/mol}$$

Density of ethanol = 0.785 g/mL

Density of octane = 0.699 g/mL

1. Calculate $\Delta_r H°$ for the combustion of ethanol and octane, and compare the values per mole and per gram. Which provides more energy per mole? Which provides more energy per gram?

2. Compare the energy produced per liter of the two fuels. Which produces more energy for a given volume (something useful to know when filling your gas tank)?

3. What mass of CO_2, a greenhouse gas, is produced per liter of fuel (assuming complete combustion)?

4. Now compare the fuels on an energy-equivalent basis. What volume of ethanol would have to be burned to get the same energy as 1.00 L of octane? When you burn enough ethanol to have the same energy as a liter of octane, which fuel produces more CO_2?

5. On the basis of this analysis and assuming the same price per liter, which fuel will propel your car further? Which will produce less greenhouse gas?

Answers to these questions are available in Appendix N.

© GIPhotoStock Z/Alamy

Ethanol available at a service station. E85 fuel is a blend of 85% ethanol and 15% gasoline. Be aware that you can only use E85 in vehicles designed for the fuel. In an ordinary vehicle, the ethanol leads to deterioration of seals in the engine and fuel system.

FIGURE 5.14 The product-favored oxidation of iron. Iron powder, sprayed into a bunsen burner flame, is rapidly oxidized. The reaction is exothermic and is product-favored.

Let us look back at the many chemical reactions that we have seen. For example, all combustion reactions are exothermic, and the oxidation of iron (Figure 5.14) is clearly exothermic.

$$4 \text{ Fe(s)} + 3 \text{ O}_2(g) \rightarrow 2 \text{ Fe}_2\text{O}_3(s)$$

$$\Delta_r H° = 2 \Delta_f H°[\text{Fe}_2\text{O}_3(s)] = \left(\frac{2 \text{ mol Fe}_2\text{O}_3}{1 \text{ mol-rxn}}\right)\left(\frac{-825.5 \text{ kJ}}{1 \text{ mol Fe}_2\text{O}_3}\right) = -1651.0 \text{ kJ/mol-rxn}$$

The reaction has a negative value for $\Delta_r H°$, and it is also product-favored at equilibrium.

Conversely, the decomposition of calcium carbonate is endothermic.

$$\text{CaCO}_3(s) \rightarrow \text{CaO}(s) + \text{CO}_2(g) \qquad \Delta_r H° = +179.0 \text{ kJ/mol-rxn}$$

The decomposition of CaCO_3 proceeds to an equilibrium that favors the reactants; that is, it is reactant-favored at equilibrium.

Are all exothermic reactions product-favored at equilibrium and all endothermic reactions reactant-favored at equilibrium? From these examples, we might formulate that idea as a hypothesis that can be tested by experiment and by examination of other examples. You would find that *in most cases, product-favored reactions have negative values of $\Delta_r H°$, and reactant-favored reactions have positive values of $\Delta_r H°$*. But this is not *always* true; there are exceptions.

Clearly, a further discussion of thermodynamics must be tied to the concept of equilibrium. This relationship, and the complete discussion of the third and fourth questions, will be presented in Chapter 19.

 and

Sign in at **www.cengage.com/owl** to:
• View tutorials and simulations, develop problem-solving skills, and complete online homework assigned by your professor.
• For quick review and exam prep, download Go Chemistry mini lecture modules from OWL (or purchase them at **www.cengagebrain.com**)

❓ Access **How Do I Solve It?** tutorials on how to approach problem solving using concepts in this chapter.

CHAPTER GOALS REVISITED

Now that you have completed this chapter, you should ask whether you have met the chapter goals. In particular, you should be able to:

Assess the transfer of energy as heat associated with changes in temperature and changes of state
a. Describe the nature of energy transfers as heat (Section 5.1).
b. Recognize and use the language of thermodynamics: the system and its surroundings; exothermic and endothermic reactions (Section 5.1). **Study Questions: 1, 3, 59.**
c. Use specific heat capacity in calculations of energy transfer as heat and of temperature changes (Section 5.2). **Study Questions: 5, 7, 9, 11, 13, 15.**
d. Understand the sign conventions in thermodynamics.
e. Use enthalpy (heat) of fusion and enthalpy (heat) of vaporization to calculate the energy transferred as heat in changes of state (Section 5.3). **Study Questions: 17, 19, 21, 23, 83.**

Understand and apply the first law of thermodynamics
a. Understand the basis of the first law of thermodynamics (Section 5.4).
b. Recognize how energy transferred as heat and work done on or by a system contribute to changes in the internal energy of a system (Section 5.4).

Define and understand state functions (enthalpy, internal energy)
a. Recognize state functions whose values are determined only by the state of the system and not by the pathway by which that state was achieved (Section 5.4).

Describe how energy changes are measured
a. Recognize that when a process is carried out under constant pressure conditions, the energy transferred as heat is the enthalpy change, ΔH (Section 5.5). **Study Questions: 25, 26, 27, 28, 48.**
b. Describe how to measure the quantity of energy transferred as heat in a reaction by calorimetry (Section 5.6). **Study Questions: 29, 30, 31, 32, 34–40.**

Calculate the energy evolved or required for physical changes and chemical reactions using tables of thermodynamic data.

a. Apply Hess's law to find the enthalpy change, $\Delta_r H°$, for a reaction (Section 5.7). **Study Questions: 41–44, 73, 79, and Go Chemistry Module 10.**
b. Know how to draw and interpret energy level diagrams (Section 5.7). **Study Questions: 53, 54, 73, 74, 79, 101, 105.**
c. Use standard molar enthalpies of formation, $\Delta_f H°$, to calculate the enthalpy change for a reaction, $\Delta_r H°$ (Section 5.7). **Study Questions: 47, 49, 51, 55, 56.**

Key Equations

Equation 5.1 (page 176) The energy transferred as heat when the temperature of a substance changes. Calculated from the specific heat capacity (C), mass (m), and change in temperature (ΔT).

$$q(J) = C(J/g \cdot K) \times m(g) \times \Delta T(K)$$

Equation 5.2 (page 176) Temperature changes are always calculated as final temperature minus initial temperature.

$$\Delta T = T_{final} - T_{initial}$$

Equation 5.3 (page 178) If no energy is transferred between a system and its surroundings and if energy is transferred within the system only as heat, the sum of the thermal energy changes within the system equals zero.

$$q_1 + q_2 + q_3 + \ldots = 0$$

Equation 5.4 (page 184) The first law of thermodynamics: The change in internal energy (ΔU) in a system is the sum of the energy transferred as heat (q) and the energy transferred as work (w).

$$\Delta U = q + w$$

Equation 5.5 (page 185) P–V work (w) at constant pressure is the product of pressure (P) and change in volume (ΔV)

$$w = -P \times \Delta V$$

Equation 5.6 (page 198) This equation is used to calculate the standard enthalpy change of a reaction ($\Delta_r H°$) when the enthalpies of formation ($\Delta_f H°$) of all of the reactants and products are known. The parameter n is the stoichiometric coefficient of each product or reactant in the balanced chemical equation.

$$\Delta_r H° = \Sigma n \Delta_f H°(\text{products}) - \Sigma n \Delta_f H°(\text{reactants})$$

OWL Questions and problems for this chapter are available in OWL. Use the OWL access card that came with this textbook to access assigned questions and problems for this chapter.

6 The Structure of Atoms

John C. Kotz

Fireworks The colors of the beautiful displays of fireworks you see on holidays such as July 4 in the United States can light up the night sky. Colors can range from white to red, yellow, orange, and blue. As we shall see in this chapter, the colors arise from the emission of energy by excited atoms.

This photo shows fireworks with blue and yellow light. The next time you attend a fireworks display, watch for the blue fireworks. The color is particularly difficult to produce.

Questions:

This photo shows fireworks with blue and yellow light, but many other colors are possible.

1. Which has the longer wavelength, blue light or yellow light?
2. Which has the greater energy, blue or yellow light?
3. How do the colors of light emitted by excited atoms contribute to our understanding of electronic structure?

Answers to these questions are available in Appendix N.

CHAPTER OUTLINE

6.1 Electromagnetic Radiation

6.2 Quantization: Planck, Einstein, Energy, and Photons

6.3 Atomic Line Spectra and Niels Bohr

6.4 Particle–Wave Duality: Prelude to Quantum Mechanics

6.5 The Modern View of Electronic Structure: Wave or Quantum Mechanics

6.6 The Shapes of Atomic Orbitals

6.7 One More Electron Property: Electron Spin

CHAPTER GOALS

See Chapter Goals Revisited (page 228) for Study Questions keyed to these goals.

- Describe the properties of electromagnetic radiation.

- Understand the origin of light emitted by excited atoms and its relationship to atomic structure.

- Describe the experimental evidence for particle–wave duality.

- Describe the basic ideas of quantum mechanics.

- Define the four quantum numbers (n, ℓ, m_ℓ, and m_s) and recognize their relationship to electronic structure.

OWL

Sign in to OWL at **www.cengage.com/owl** to view tutorials and simulations, develop problem-solving skills, and complete online homework assigned by your professor.

go Chemistry

Download mini lecture videos for key concept review and exam prep from OWL or purchase them from **www.cengagebrain.com**

The picture that most people have of the atom is that of a tiny nucleus, consisting of neutrons and protons, with nearly massless electrons orbiting the nucleus like planets around a star. Unfortunately, this simple picture is only partly correct. The tiny nucleus is indeed composed of neutrons and protons, and electrons are located in the space outside the nucleus, but, as you shall see, a more accurate picture describes the electrons as matter waves, and we can only define regions of space in which it is likely to find an electron.

This chapter lays out the experimental background to our current understanding of atomic structure and describes the modern view of the atom. With this model we can better predict the properties of atoms and better understand the underlying structure of the periodic table.

Much of our understanding of atomic structure comes from a knowledge of how atoms interact with electromagnetic radiation (visible light is one type of electromagnetic radiation) and how excited atoms emit electromagnetic radiation. The first three sections of this chapter describe electromagnetic radiation and its relation to our modern view of the atom.

6.1 Electromagnetic Radiation

In 1864, James Clerk Maxwell (1831–1879) developed a mathematical theory to describe light and other forms of radiation in terms of oscillating, or wave-like, electric and magnetic fields (Figure 6.1). Thus, light, microwaves, television and radio signals, x-rays, and other forms of radiation are now called **electromagnetic radiation**.

Electromagnetic radiation is characterized by its wavelength and frequency.

- **Wavelength**, symbolized by the Greek letter *lambda* (λ), is defined as the distance between successive crests or high points of a wave (or between successive troughs or low points). This distance can be given in meters, nanometers, or whatever unit of length is convenient.

- **Frequency**, symbolized by the Greek letter *nu* (ν), refers to the number of waves that pass a given point in some unit of time, usually per second. The unit for frequency, written either as s^{-1} or 1/s and standing for 1 per second, is now called a **hertz**.

Wavelength and frequency are related to the speed (*c*) at which the wave is propagated (Equation 6.1).

- **Heinrich Hertz** Heinrich Hertz (1857–1894) was the first person to send and receive radio waves. He showed that they could be reflected and refracted the same as light, confirming that different forms of radiation such as radio and light waves are related. Scientists now use "hertz" as the unit of frequency.

$$c \ (\text{m/s}) = \lambda \ (\text{m}) \times \nu \ (1/\text{s}) \qquad (6.1)$$

FIGURE 6.1 Electromagnetic radiation is characterized by its wavelength and frequency. In the 1860s, James Clerk Maxwell developed the currently accepted theory that all forms of radiation are propagated through space as vibrating electric and magnetic fields at right angles to one another. Each field is described by a sine wave (the mathematical function describing the wave). Such oscillating fields emanate from vibrating charges in a source such as a light bulb or radio antenna.

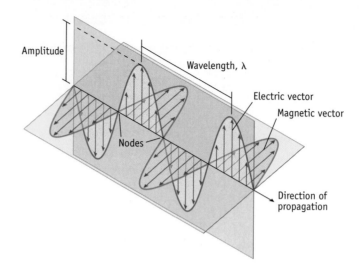

The speed of visible light and all other forms of electromagnetic radiation in a vacuum is a constant: $c = 2.99792458 \times 10^8$ m/s (approximately 186,000 miles/s or 1.079×10^9 km/h). For calculations, we will generally use the value of c with four or fewer significant figures.

As the name suggests, electromagnetic radiation consists of oscillating electric and magnetic disturbances. This is important to recognize because the electric and magnetic fields can interact with charged particles such as electrons in atoms and molecules and with charged atoms in molecules. It is this that allows scientists to probe matter at the atomic and molecular levels using forms of radiation such as x-rays or radio waves (in an MRI or magnetic resonance imaging instrument), for example.

Visible light is only a tiny portion of the total spectrum of electromagnetic radiation (Figure 6.2). Ultraviolet (UV) radiation, the radiation that can lead to sunburn, has wavelengths shorter than those of visible light. X-rays and γ-rays, the latter emitted in the process of radioactive disintegration of some atoms, have even shorter wavelengths. At wavelengths longer than those of visible light, we first encounter infrared radiation (IR). The wavelengths of the radiation used in microwave ovens, in television and radio transmissions, in MRI instruments, and for cell phones are longer still.

EXAMPLE 6.1　　Wavelength-Frequency Conversions

Problem The frequency of the radiation used in cell phones covers a wide range from about 800 MHz to 2 GHz. (MHz stands for "megahertz," where 1 MHz = 10^6 1/s; GHz stands for "gigahertz," where 1 GHz = 10^9 1/s.) What is the wavelength (in meters) of a cell phone signal operating at 1.12 GHz?

What Do You Know? You are given a frequency of radiation, in GHz, and a factor to convert frequency from GHz to Hz (1/s). To calculate the wavelength of this radiation using Equation 6.1, you will need the value of the speed of light, $c = 2.998 \times 10^8$ m/s.

Strategy Rearrange Equation 6.1 to solve for λ. Substitute the values for the speed of light and the frequency (first convert ν to the units 1/s) into the equation and solve.

Solution

$$\lambda = \frac{c}{\nu} = \frac{2.998 \times 10^8 \text{ m/s}}{1.12 \times 10^9 \text{ 1/s}} = 0.268 \text{ m}$$

Think about Your Answer Pay attention to units when solving this problem. Note that by choosing frequency (in 1/s) and the speed of light (in m/s) the calculated wavelength will be in meters (m).

Check Your Understanding

(a) Which color in the visible spectrum has the highest frequency? Which has the lowest frequency?

(b) Is the frequency of the radiation used in a microwave oven (2.45 GHz) higher or lower than that from your favorite FM radio station (for example, 91.7 MHz).

(c) Are the wavelengths of x-rays longer or shorter than those of ultraviolet light?

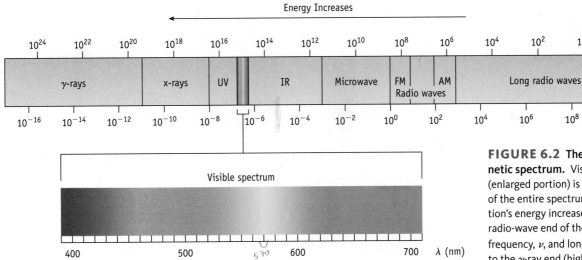

FIGURE 6.2 The electromagnetic spectrum. Visible light (enlarged portion) is a very small part of the entire spectrum. The radiation's energy increases from the radio-wave end of the spectrum (low frequency, ν, and long wavelength, λ) to the γ-ray end (high frequency and short wavelength).

REVIEW & CHECK FOR SECTION 6.1

Which of the following types of electromagnetic radiation has the longest wavelength?

(a) ultraviolet light

(b) x-rays

(c) red light from a laser pointer

(d) microwaves from a microwave oven

6.2 Quantization: Planck, Einstein, Energy, and Photons

Planck's Equation

If a piece of metal is heated to a high temperature, electromagnetic radiation is emitted with wavelengths that depend on temperature. At lower temperatures, the color is a dull red (Figure 6.3). As the temperature increases, the red color brightens, and at even higher temperatures a brilliant white light is emitted.

Your eyes detect the emitted radiation in the visible region of the electromagnetic spectrum. Although not seen, both ultraviolet and infrared radiation are also given off by the hot metal (Figure 6.3). In addition, it is observed that the wavelength of the most intense radiation is related to temperature: As the temperature of the metal is raised, the maximum intensity shifts toward shorter wavelengths, that is, toward the ultraviolet. This corresponds to the change in color observed as the temperature is raised.

At the end of the 19th century, scientists were not able to explain the relationship between the intensity and the wavelength for radiation given off by a heated object (often called *blackbody radiation*). Theories available at the time predicted that the intensity should increase continuously with decreasing wavelength, instead of reaching a maximum and then declining as is actually observed. This perplexing situation became known as the *ultraviolet catastrophe* because predictions failed in the ultraviolet region.

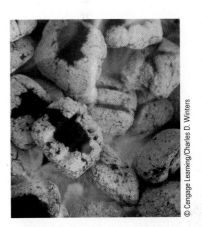

© Cengage Learning/Charles D. Winters

Blackbody radiation. In physics a blackbody is a theoretical concept in which a body absorbs all radiation that falls on it. However, it will emit energy with a temperature-dependent wavelength. The light emanating from the spaces between the burning charcoal briquets in this photo is a close approximation to blackbody radiation. The color of the emitted light depends on the temperature of the briquets.

FIGURE 6.3 The radiation given off by a heated body. When an object is heated, it emits radiation covering a spectrum of wavelengths. For a given temperature, some of the radiation is emitted at long wavelengths and some at short wavelengths. Most, however, is emitted at some intermediate wavelength, the maximum in the curve. As the temperature of the object increases, the maximum moves from the red end of the spectrum to the violet end. At still higher temperatures, intense light is emitted at all wavelengths in the visible region, and the maximum in the curve is in the ultraviolet region. The object is described as "white hot." (Stars are often referred to as "red giants" or "white dwarfs," a reference to their temperatures and relative sizes.)

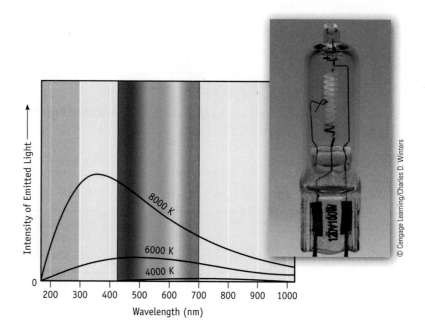

In 1900, a German physicist, Max Planck (1858–1947), offered an explanation for the ultraviolet catastrophe: The electromagnetic radiation emitted originated in vibrating atoms (called *oscillators*) in the heated object. He proposed that each oscillator had a fundamental frequency (ν) of oscillation and that the atoms could only oscillate at either this frequency or whole-number multiples of it ($n\nu$). Because of this, the emitted radiation could have only certain energies, given by the equation

$$E = nh\nu$$

where n must be a positive integer. That is, Planck proposed that the energy is *quantized*. **Quantization** means that only certain energies are allowed. The proportionality constant h in the equation is now called **Planck's constant**, and its experimental value is $6.6260693 \times 10^{-34}$ J · s. The unit of frequency is 1/s, so the energy calculated using this equation is in joules (J). If an oscillator changes from a higher energy to a lower one, energy is emitted as electromagnetic radiation, where the difference in energy between the higher and lower energy states is

$$\Delta E = E_{\text{higher } n} - E_{\text{lower } n} = \Delta n h\nu$$

If the value of Δn is 1, which corresponds to changing from one energy level to the next lower one for that oscillator, then the energy change for the oscillator and the electromagnetic radiation emitted would have an energy equal to

$$E = h\nu \qquad\qquad (6.2)$$

This equation is called **Planck's equation**.

Now, assume as Planck did that there must be a *distribution* of vibrations of atoms in an object—some atoms are vibrating at a high frequency, some are vibrating at a low frequency, but most have some intermediate frequency. The few atoms with high-frequency vibrations are responsible for some of the light, say in the ultraviolet region, and those few with low-frequency vibrations for energies in the infrared region. However, most of the light must come from the majority of the atoms that have intermediate vibrational frequencies. That is, a spectrum of light is emitted with a maximum intensity at some intermediate wavelength, in accord with experiment. The intensity should not become greater and greater on approaching the ultraviolet region. With this realization, the ultraviolet catastrophe was solved.

The key aspect of Planck's work for general chemistry is that Planck introduced the idea of quantized energies based on the equation $E = h\nu$, an equation that was

● **The relationship of energy, wavelength, and frequency**

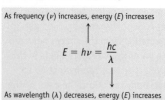

As frequency (ν) increases, energy (E) increases

$$E = h\nu = \frac{hc}{\lambda}$$

As wavelength (λ) decreases, energy (E) increases

to have a profound impact on the work of Albert Einstein in explaining another puzzling phenomenon.

Einstein and the Photoelectric Effect

A few years after Planck's work, Albert Einstein (1879–1955) incorporated Planck's ideas into an explanation of the *photoelectric effect* and in doing so changed the description of electromagnetic radiation.

In the photoelectric effect, electrons are ejected when light strikes the surface of a metal (Figure 6.4), but only if the frequency of the light is high enough. If light with a lower frequency is used, no electrons are ejected, regardless of the light's intensity (its brightness). If the frequency is at or above a minimum, critical frequency, increasing the light intensity causes more electrons to be ejected.

Einstein decided the experimental observations could be explained by combining Planck's equation ($E = h\nu$) with a new idea, that light has particle-like properties. Einstein characterized these massless particles, now called **photons**, as packets of energy, and stated that the energy of each photon is proportional to the frequency of the radiation as defined by Planck's equation. In the photoelectric effect, photons striking atoms on a metal surface will cause electrons to be ejected only if the photons have high enough energy. The greater the number of photons that strike the surface at or above the threshold energy, the greater the number of electrons dislodged. The metal atoms will not lose electrons, however, if no individual photon has enough energy to dislodge an electron from an atom.

● **Einstein and Digital Cameras**
When you take a photograph with a digital camera the detector in the camera (a CCD or "charged-coupled detector") works on a principle much like the photoelectric effect.

Energy and Chemistry: Using Planck's Equation

Compact disc players use lasers that emit red light with a wavelength of 685 nm. What is the energy of one photon of this light? What is the energy of 1 mol of photons of red light? To answer these questions, first convert the wavelength to the frequency of the radiation,

For $\lambda = 685$ nm, $\nu = 4.38 \times 10^{14}$ 1/s (calculated using Equation 6.1)

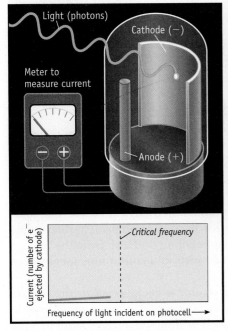

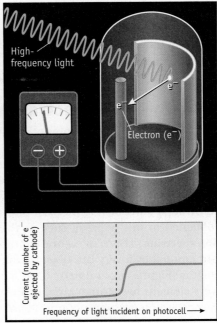

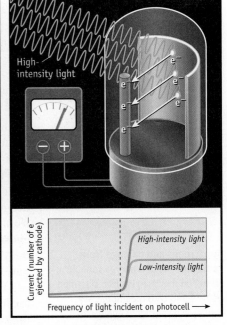

(a) A photocell operates by the photoelectric effect. The main part of the cell is a light-sensitive cathode. This is a material, usually a metal, that ejects electrons if struck by photons of light of sufficient energy. No current is observed until the critical frequency is reached.

(b) When light of higher frequency than the minimum is used, the excess energy of the photon allows the electron to escape the atom with greater velocity. The ejected electrons move to the anode, and a current flows in the cell. Such a device can be used as a switch in electric circuits.

(c) If higher intensity light is used, the only effect is to cause more electrons to be released from the surface. The onset of current is observed at the same frequency as with lower intensity light, but more current flows with more intense light.

FIGURE 6.4 A photoelectric cell.

Strategy Map: Planck's Equation

PROBLEM
Calculate energy of **red light photons**.

↓

KNOWN DATA/INFORMATION
• **Wavelength** of red light (685 nm)

↓

STEP 1. Convert **wavelength** to **frequency**. (Equation 6.1)

↓

Frequency in *cycles per second*

↓

STEP 2. Calculate **energy** using *Planck's equation*. (Equation 6.2)

↓

Obtain energy in **J/photon**.

↓

STEP 3. Use **Avogadro's number** to convert **J/atom** to **J/mol**.

↓

Obtain energy in **J/mol**.

and then use the frequency to calculate the energy per photon.

$$E \text{ per photon} = h\nu = (6.626 \times 10^{-34} \text{ J} \cdot \text{s/photon}) \times (4.38 \times 10^{14} \text{ 1/s})$$
$$= 2.90 \times 10^{-19} \text{ J/photon}$$

Finally, calculate the energy of a mole of photons by multiplying the energy per photon by Avogadro's number:

$$E \text{ per mole} = (2.90 \times 10^{-19} \text{ J/photon}) \times (6.022 \times 10^{23} \text{ photons/mol})$$
$$= 1.75 \times 10^5 \text{ J/mol (or 175 kJ/mol)}$$

The energy of red light photons with a wavelength of 685 nm is 175 kJ/mol, whereas the energy of blue light photons ($\lambda = 400$ nm) is about 300 kJ/mol. The energy of the blue light photons is in the range of the energies necessary to break some chemical bonds in proteins. Higher energy ultraviolet photons are even more likely to cause chemical bond breaking. This is what happens if you spend too much time unprotected in the sun. In contrast, light at the red end of the spectrum and infrared radiation has a lower energy, and, although it is generally not energetic enough to break chemical bonds, it can affect the vibrations of molecules. We sense infrared radiation as heat, such as the heat given off by a glowing burner on an electric stove or a burning log in a fire.

REVIEW & CHECK FOR SECTION 6.2

Various manufacturers have developed mixtures of compounds that protect skin from UVA (400-320 nm) and UVB (320-280 nm) radiation. These sunscreens are given "sun protection factor" (SPF) labels that indicate how long the user can stay in the sun without burning.

1. Calculate the energy of a mole of photons of UVB light with a wavelength of 310 nm.

    (a) 3.9×10^5 J/mol   (b) 1.2×10^5 J/mol   (c) 0.00039 J/mol

2. Which has the greater energy per photon, UVB light at 310 nm or microwave radiation having a frequency of 2.45 GHz (1 GHz = 10^9 s^{-1})?

    (a) UVB   (b) microwave

6.3 Atomic Line Spectra and Niels Bohr

If a high voltage is applied to atoms of an element in the gas phase at low pressure, the atoms absorb energy and are said to be "excited." The excited atoms can then emit light, and a familiar example is the colored light from neon advertising signs.

The light from excited atoms is composed of only a few different wavelengths of light. We can demonstrate this by passing a beam of light from excited neon or hydrogen through a prism (Figure 6.5); only a few colored lines are seen. The spectrum obtained in this manner, such as that for excited H atoms, is called a **line emission spectrum**. This is in contrast to the light falling on Earth from the Sun, or the light emitted by a very hot object, which consists of a continuous spectrum of wavelengths (Figures 6.2 and 6.3).

Every element has a unique emission spectrum, as exemplified by the spectra for hydrogen, mercury, and neon in Figure 6.6. Indeed, the characteristic lines in the emission spectrum of an element can be used in chemical analysis, both to identify the element and to determine how much of it is present.

A goal of scientists in the late 19th century was to explain why excited gaseous atoms emitted light of only certain frequencies. One approach was to look for a mathematical relationship among the observed wavelengths because a regular pattern of information implies a logical explanation. The first steps in this direction were taken by Johann Balmer (1825–1898) and later by Johannes Rydberg

Sunscreens and damage from radiation. Higher-energy sunlight falling on the Earth is often classified as UVA and UVB radiation. UVB radiation has a higher energy than UVA radiation and is largely responsible for sunburns and tanning. Sunscreens containing organic compounds such as 2-ethylhexyl-p-methoxycinnamate and oxybenzone absorb UV radiation, preventing it from reaching your skin.

JGI/Jamie Grill/Blend Images/Jupiter Images

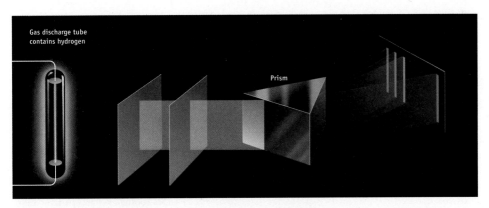

FIGURE 6.5 The visible line emission spectrum of hydrogen. The emitted light is passed through a series of slits to create a narrow beam of light, which is then separated into its component wavelengths by a prism. A photographic plate or photocell can be used to detect the separate wavelengths as individual lines. Hence, the name *line spectrum* for the light emitted by a glowing gas.

(1854–1919). From these studies, an equation—now called the **Balmer equation** (Equation 6.3)—was found that could be used to calculate the wavelength of the red, green, and blue lines in the visible emission spectrum of hydrogen (Figure 6.6).

$$\frac{1}{\lambda} = R\left(\frac{1}{2^2} - \frac{1}{n^2}\right) \quad \text{when } n > 2 \tag{6.3}$$

In this equation n is an integer, and R, now called the **Rydberg constant**, has the value $1.0974 \times 10^7 \text{ m}^{-1}$. If $n = 3$, for example, the wavelength of the red line in the hydrogen spectrum is obtained (6.563×10^{-7} m, or 656.3 nm). If $n = 4$, the wavelength for the green line is calculated. Using $n = 5$ and $n = 6$ in the equation gives the wavelengths of the blue lines. The four visible lines in the spectrum of hydrogen atoms are now known as the **Balmer series**.

The Bohr Model of the Hydrogen Atom

Early in the 20th century, the Danish physicist Niels Bohr (1885–1962) proposed a model for the electronic structure of atoms and with it an explanation for the emission spectra of excited atoms. Bohr proposed a planetary structure for the hydrogen atom in which the electron moved in a circular orbit around the nucleus, similar to a planet revolving about the sun. In proposing this model, however, he had to

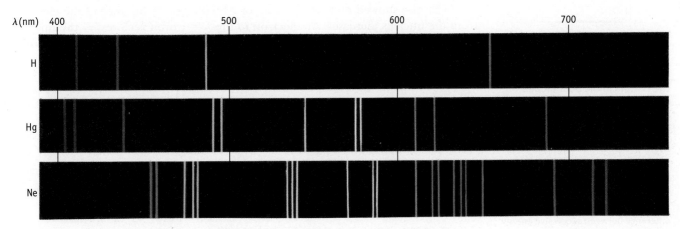

FIGURE 6.6 Line emission spectra of hydrogen, mercury, and neon. Excited gaseous elements produce characteristic spectra that can be used to identify the elements and to determine how much of each element is present in a sample.

contradict the laws of classical physics. According to classical theories, a charged electron moving in the positive electric field of the nucleus should lose energy; a consequence of the loss of energy is that the electron would eventually crash into the nucleus. This is clearly not the case; if it were, matter would eventually self-destruct. To solve this contradiction, Bohr postulated that there are certain orbits corresponding to particular energy levels where this would not occur. As long as an electron is in one of these energy levels, the system is stable. That is, Bohr introduced *quantization* into the description of electronic structure. By combining this quantization postulate with the laws of motion from classical physics, Bohr derived an equation for the total energy possessed by the single electron in the nth orbit (energy level) of the H atom.

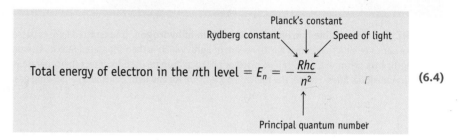

$$\text{Total energy of electron in the } n\text{th level} = E_n = -\frac{Rhc}{n^2} \qquad (6.4)$$

Here, E_n is the energy of the electron (in J/atom); and R, h, and c are constants [the Rydberg constant ($R = 1.097 \times 10^7$ m^{-1}), Planck's constant, and the speed of light, respectively]. The symbol n is a positive, unitless integer called the **principal quantum number**. It can have integral values of 1, 2, 3, and so on.

Equation 6.4 has several important features (illustrated in Figure 6.7) such as the role of n and the significance of the negative sign.

- The quantum number n defines the energies of the allowed orbits in the H atom.
- The energy of an electron in an orbit has a negative value because the electron in the atom has a lower energy than when it is free. The zero of energy occurs when $n = \infty$, that is, when the electron is infinitely separated from the nucleus.
- An atom with its electrons in the lowest possible energy levels is said to be in its **ground state**; for the hydrogen atom, this is the level defined by the quantum number $n = 1$. The energy of this state is $-Rhc/1^2$, meaning that it has an amount of energy equal to Rhc *below* the energy of the infinitely separated electron and nucleus.
- States for the H atom with higher energies (and $n > 1$) are called **excited states**, and, as the value of n increases, these states have less negative energy values.
- Because the energy is dependent on $1/n^2$, the energy levels are progressively closer together as n increases.

Bohr also showed that, as the value of n increases, the distance of the electron from the nucleus increases. An electron in the $n = 1$ orbit is closest to the nucleus and has the lowest (most negative) energy. For higher integer values of n, the electron is further from the nucleus and has a higher (less negative) energy.

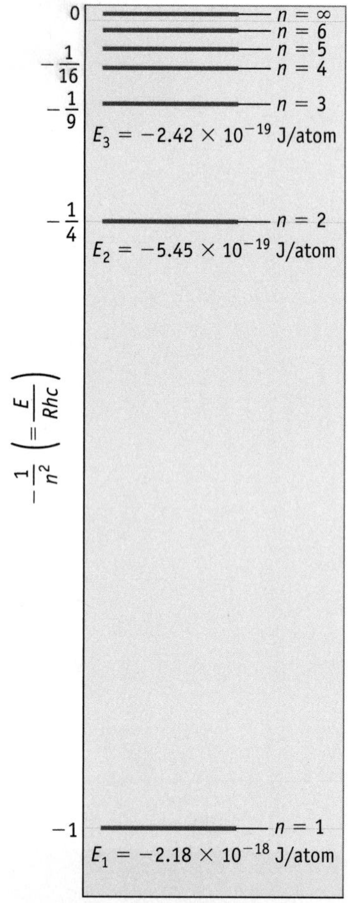

FIGURE 6.7 Energy levels for the H atom in the Bohr model. The energy of the electron in the hydrogen atom depends on the value of the principal quantum number n ($E_n = -Rhc/n^2$). The larger the value of n, the larger the Bohr radius and the less negative the value of the energy. Energies are given in joules per atom (J/atom). Notice that the difference between successive energy levels becomes smaller as n becomes larger.

Energy level diagram (left margin):
- 0 — $n = \infty$
- $-\frac{1}{16}$ — $n = 6$, $n = 5$, $n = 4$
- $-\frac{1}{9}$ — $n = 3$; $E_3 = -2.42 \times 10^{-19}$ J/atom
- $-\frac{1}{4}$ — $n = 2$; $E_2 = -5.45 \times 10^{-19}$ J/atom
- $-\frac{1}{n^2} \left(= \frac{E}{Rhc}\right)$
- -1 — $n = 1$; $E_1 = -2.18 \times 10^{-18}$ J/atom

EXAMPLE 6.2   Energies of the Ground and Excited States of the H Atom

Problem Calculate the energies of the $n = 1$ and $n = 2$ states of the hydrogen atom in joules per atom and in kilojoules per mole. What is the difference in energy of these two states in kJ/mol?

What Do You Know? The $n = 1$ and $n = 2$ states are the first and second states (lowest and next to lowest energy state) in the Bohr description of the hydrogen atom. Use Equation 6.4 to calculate the energy of each state. For the calculations you will need the following constants: R (Rydberg constant) = 1.097×10^7 m^{-1}; h (Planck's constant) = 6.626×10^{-34} J · s; c (speed of light) = 2.998×10^8 m/s; and N (Avogadro's number) = 6.022×10^{23} atoms/mol.

Strategy For each energy level, substituting the appropriate values into Equation 6.4 and solving gives the energy in J/atom. Multiply this value by N to find the energy in J/mol. Subtract the energy for the $n = 1$ level from the energy of the $n = 2$ level to obtain the energy difference.

Solution When $n = 1$, the energy of an electron in a single H atom is

$$E_1 = -Rhc$$
$$E_1 = -(1.097 \times 10^7 \text{ m}^{-1})(6.626 \times 10^{-34} \text{ J} \cdot \text{s})(2.998 \times 10^8 \text{ m/s}) = -2.179 \times 10^{-18} \text{ J/atom}$$

Rhc

In units of kJ/mol,

$$E_1 = \frac{-2.179 \times 10^{-18} \text{ J}}{\text{atom}} \times \frac{6.022 \times 10^{23} \text{ atoms}}{\text{mol}} \times \frac{1 \text{ kJ}}{1000 \text{ J}} = -1312 \text{ kJ/mol}$$

When $n = 2$, the energy is

$$E_2 = -\frac{Rhc}{2^2} = -\frac{E_1}{4} = -\frac{2.179 \times 10^{-18} \text{ J/atom}}{4} = -5.448 \times 10^{-19} \text{ J/atom}$$

In units of kJ/mol,

$$E_2 = \frac{-5.448 \times 10^{-19} \text{ J}}{\text{atom}} \times \frac{6.022 \times 10^{23} \text{ atoms}}{\text{mol}} \times \frac{1 \text{ kJ}}{1000 \text{ J}} = -328.1 \text{ kJ/mol}$$

The difference in energy, ΔE, between the first two energy states of the H atom is

$$\Delta E = E_2 - E_1 = (-328.1 \text{ kJ/mol}) - (-1312 \text{ kJ/mol}) = 984 \text{ kJ/mol}$$

Think about Your Answer The calculated energies are negative, with E_1 more negative than E_2. The $n = 2$ state is higher in energy than the $n = 1$ state by 984 kJ/mol. Also, be sure to notice that 1312 kJ/mol is the value of *Rhc* multiplied by Avogadro's number N (i.e., $NRhc$). This will be useful in future calculations.

Check Your Understanding

Calculate the energy of the $n = 3$ state of the H atom in (a) joules per atom and (b) kilojoules per mole.

The Bohr Theory and the Spectra of Excited Atoms

Bohr's theory describes electrons as having only specific orbits and energies. If an electron moves from one energy level to another, then energy must be absorbed or evolved. This idea allowed Bohr to relate energies of electrons and the emission spectra of hydrogen atoms.

To move an electron from the $n = 1$ state to an excited state, such as the $n = 2$ state, energy must be transferred to the atom. When E_{final} has $n = 2$ and E_{initial} has $n = 1$, then 1.63×10^{-18} J/atom (or 984 kJ/mol) of energy must be transferred (Figure 6.8). This is the difference in energy between final and initial states:

$$\Delta E = E_{\text{final state}} - E_{\text{initial state}} = (-Rhc/2^2) - (-Rhc/1^2) = (0.75)Rhc = 1.63 \times 10^{-18} \text{ J/atom}$$

Moving an electron from the first to the second energy state requires *input* of 1.63×10^{-18} J/atom, no more and no less. If $0.7 Rhc$ or $0.8 Rhc$ is provided, no transition between states is possible. Requiring a specific and precise amount of energy is a consequence of quantization.

Moving an electron from a state of low n to one of higher n always requires that energy be transferred to the atom from the surroundings. The opposite process, in which an electron "falls" from a level of higher n to one of lower n, leads to emission of energy, a transfer of energy, usually as radiation, from the

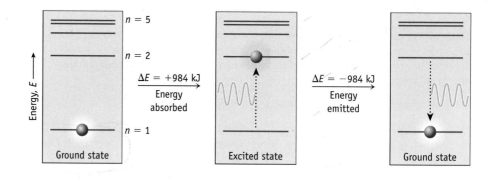

atom to its surroundings. For example, for a transition from the $n = 2$ level to
$n = 1$ level,

$$\Delta E = E_{\text{final state}} - E_{\text{initial state}} = -1.63 \times 10^{-18} \text{ J/atom} \ (= -984 \text{ kJ/mol})$$

The negative sign indicates 1.63×10^{-18} J/atom (or 984 kJ/mol) is *emitted*.

Now we can visualize the mechanism by which the characteristic line emission
spectrum of hydrogen originates according to the Bohr model. Energy is provided
to the atoms from an electric discharge or by heating. Depending on how much
energy is added, some atoms have their electrons excited from the $n = 1$ state to the
$n = 2, 3$, or even higher states. After absorbing energy, these electrons can return to
any lower level (either directly or in a series of steps), releasing energy (Figure 6.9).
We observe this released energy as photons of electromagnetic radiation, and, because only certain energy levels are possible, only photons with particular energies
and wavelengths are emitted.

The energy of any emission line (in J/atom) for excited hydrogen atoms can be
calculated using Equation 6.5.

$$\Delta E = E_{\text{final}} - E_{\text{initial}} = -Rhc \left(\frac{1}{n_{\text{final}}^2} - \frac{1}{n_{\text{initial}}^2} \right) \tag{6.5}$$

[The value of ΔE in J/atom may be related to the wavelength or frequency of radiation using Planck's equation ($\Delta E = h\nu$) or converted to units of kJ/mol if ΔE is
multiplied by $(6.022 \times 10^{23} \text{ atoms/mol})(1 \text{ kJ}/1000 \text{ J})$.]

For hydrogen, a series of emission lines having energies in the ultraviolet region
(called the **Lyman series**; Figure 6.10) arises from electrons moving from states with
$n > 1$ to the $n = 1$ state. The series of lines that have energies in the visible region—
the **Balmer series**—arises from electrons moving from states with $n > 2$ to the
$n = 2$ state. There are also series of lines in the infrared spectral region, arising
from transitions from higher levels to the $n = 3, 4$ or 5 levels.

Bohr's model, introducing quantization into a description of the atom, tied the
unseen (the structure of the atom) to the seen (the observable lines in the hydrogen

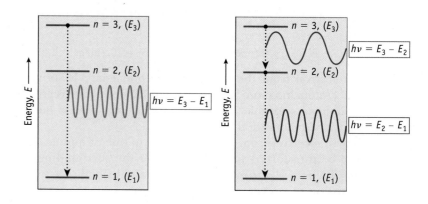

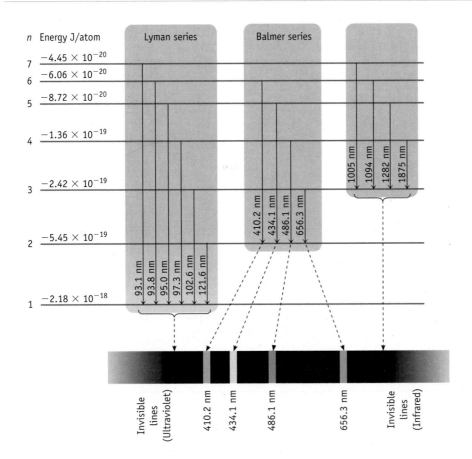

FIGURE 6.10 Some of the electronic transitions that can occur in an excited H atom. The Lyman series of lines in the ultraviolet region results from transitions to the $n = 1$ level. The Balmer series (Figures 6.5 and 6.6) arises from transitions from levels with values of $n > 2$ to $n = 2$. Lines in the infrared region result from transitions from levels with $n > 3$ to the $n = 3$ level. (Transitions from $n = 8$ and higher levels to lower levels occur but are not shown in this figure.)

spectrum). Agreement between theory and experiment is taken as evidence that the theoretical model is valid. It became apparent, however, that Bohr's theory was inadequate. This model of the atom explained only the spectrum of hydrogen atoms and of other systems having one electron (such as He^+), but it failed for all other systems. A better model of electronic structure was needed.

ⓌWL   INTERACTIVE EXAMPLE 6.3   Energies of Emission Lines for Excited Atoms

Problem Calculate the wavelength of the green line in the visible spectrum of excited H atoms.

What Do You Know? The green line in the spectrum of hydrogen arises from the electron transition from the $n = 4$ state ($n_{initial}$) to the $n = 2$ state (n_{final}) (Figure 6.10).

Strategy

- Calculate the difference in energy between the states using Equation 6.5. You can simplify this calculation by using the value of Rhc from Example 6.2 (= 2.179×10^{-18} J/photon).

- Relate the difference in energy to the wavelength of light using the equation $E = hc/\lambda$. (This equation is derived by combining Equations 6.1 and 6.2.)

Solution Calculate ΔE.

$$\Delta E = E_{final} - E_{initial} = \left(-\frac{Rhc}{2^2}\right) - \left(-\frac{Rhc}{4^2}\right)$$

$$\Delta E = -Rhc\left(\frac{1}{4} - \frac{1}{16}\right) = -Rhc(0.1875)$$

$$\Delta E = -(2.179 \times 10^{-18} \text{ J/photon})(0.1875) = -4.086 \times 10^{-19} \text{ J/photon}$$

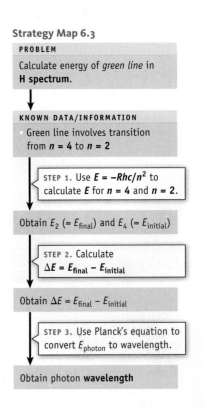

Strategy Map 6.3

PROBLEM

Calculate energy of *green line* in **H spectrum.**

KNOWN DATA/INFORMATION

- Green line involves transition from **$n = 4$** to **$n = 2$**

STEP 1. Use $E = -Rhc/n^2$ to calculate E for $n = 4$ and $n = 2$.

Obtain E_2 (= E_{final}) and E_4 (= $E_{initial}$)

STEP 2. Calculate $\Delta E = E_{final} - E_{initial}$

Obtain $\Delta E = E_{final} - E_{initial}$

STEP 3. Use Planck's equation to convert E_{photon} to wavelength.

Obtain photon **wavelength**

Now apply Planck's equation to calculate the wavelength ($E_{photon} = h\nu = hc/\lambda$, and so $\lambda = hc/E_{photon}$). (Recognize that, while the change in energy has a sign indicating the "direction" of energy transfer, the energy of the photon emitted, E_{photon}, does not have a sign.)

$$\lambda = \frac{hc}{E_{photon}} = \frac{\left(6.626 \times 10^{-34} \dfrac{J \cdot s}{photon}\right)\left(2.998 \times 10^{8}\ m \cdot s^{-1}\right)}{4.086 \times 10^{-19}\ J/photon}$$

$$= 4.862 \times 10^{-7}\ m$$

$$= (4.862 \times 10^{-7}\ m)(1 \times 10^{9}\ nm/m)$$

$$= 486.2\ nm$$

Think about Your Answer You might recall that visible light has wavelengths of 400 to 700 nm. The calculated value is in this region, and your answer has a value appropriate for the green line. The experimentally determined value of 486.1 nm is in excellent agreement with this answer.

Check Your Understanding

The Lyman series of spectral lines for the H atom, in the ultraviolet region, arises from transitions from higher levels to $n = 1$. Calculate the frequency and wavelength of the least energetic line in this series.

REVIEW & CHECK FOR SECTION 6.3

1. Based on Bohr's theory, which of the following transitions for the hydrogen atom will evolve the most energy?

    (a) from $n = 3$ to $n = 2$ (c) from $n = 5$ to $n = 2$

    (b) from $n = 4$ to $n = 2$ (d) from $n = 6$ to $n = 2$

2. Based on Bohr's theory, which of the following transitions for the hydrogen atom will occur with emission of visible light?

    (a) from $n = 3$ to $n = 1$ (c) from $n = 5$ to $n = 3$

    (b) from $n = 4$ to $n = 2$ (d) from $n = 6$ to $n = 4$

6.4 Particle–Wave Duality: Prelude to Quantum Mechanics

The wave nature of electrons.
A beam of electrons was passed through a thin film of MgO. The atoms in the MgO lattice diffracted the electron beam, producing this pattern. Diffraction is best explained by assuming electrons have wave properties.

© R. K. Bohn, Department of Chemistry, University of Connecticut

The photoelectric effect demonstrated that light, usually considered to be a wave, can also have the properties of particles, albeit without mass. But what if matter, which is usually considered to be made of particles, could have wave properties? This was pondered by Louis Victor de Broglie (1892–1987), who, in 1925, proposed that a free electron of mass m moving with a velocity v should have an associated wavelength λ, calculated by the equation

$$\lambda = \frac{h}{mv} \tag{6.6}$$

De Broglie called the wave corresponding to the wavelength calculated from this equation a "matter wave."

This revolutionary idea linked the particle properties of the electron (mass and velocity) with a wave property (wavelength). Experimental proof was soon produced. In 1927, C. J. Davisson (1881–1958) and L. H. Germer (1896–1971) found that diffraction, a property of waves, was observed when a beam of electrons was

CASE STUDY

What Makes the Colors in Fireworks?

Typical fireworks have several important chemical components. For example, there must be an oxidizer. Today, this is usually potassium perchlorate ($KClO_4$), potassium chlorate ($KClO_3$), or potassium nitrate (KNO_3). Potassium salts are used instead of sodium salts because the latter have two important drawbacks. They are hygroscopic—they absorb water from the air—and so do not remain dry on storage. Also, when heated, sodium salts give off an intense, yellow light that is so bright it can mask other colors.

The parts of any fireworks display we remember best are the vivid colors and brilliant flashes. White light can be produced by oxidizing magnesium or aluminum metal at high temperatures. The flashes you see at rock concerts or similar events, for example, are typically $Mg/KClO_4$ mixtures.

Yellow light is easiest to produce because sodium salts give an intense light with a wavelength of 589 nm. Fireworks mixtures

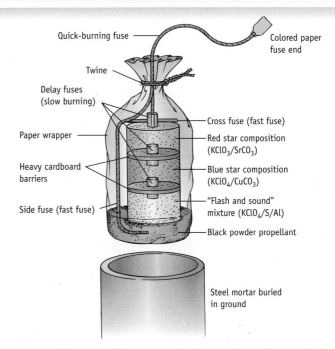

FIGURE B **The design of an aerial rocket for a fireworks display.** When the fuse is ignited, it burns quickly to the delay fuses at the top of the red star mixture as well as to the black powder propellant at the bottom. The propellant ignites, sending the shell into the air. Meanwhile, the delay fuses burn. If the timing is correct, the shell bursts high in the sky into a red star. This is followed by a blue burst and then a flash and sound.

Labels in figure:
Quick-burning fuse
Colored paper fuse end
Twine
Delay fuses (slow burning)
Cross fuse (fast fuse)
Paper wrapper
Red star composition ($KClO_3/SrCO_3$)
Heavy cardboard barriers
Blue star composition ($KClO_4/CuCO_3$)
Side fuse (fast fuse)
"Flash and sound" mixture ($KClO_4/S/Al$)
Black powder propellant
Steel mortar buried in ground

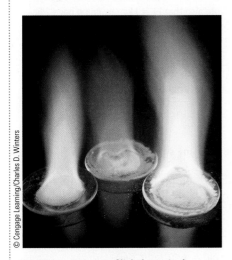

© Cengage Learning/Charles D. Winters

FIGURE A **Emission of light by excited atoms.** Flame tests are often used to identify elements in a chemical sample. Shown here are the colors produced in a flame (burning methanol) by NaCl (yellow), $SrCl_2$ (red), and boric acid (green).

usually contain sodium in the form of non-hygroscopic compounds such as cryolite, Na_3AlF_6. Strontium salts are most often used to produce a red light, and green is produced by barium salts such as $Ba(NO_3)_2$.

The next time you see a fireworks display, watch for the ones that are blue. Blue has always been the most difficult color to produce. Recently, however, fireworks designers have learned that the best way to get a really good "blue" is to decompose copper(I) chloride at low temperatures. To achieve this effect, CuCl is mixed with $KClO_4$, copper powder, and the organic chlorine-containing compound hexachloroethane, C_2Cl_6.

Questions:

1. The main lines in the emission spectrum of sodium are at wavelengths (nm) of 313.5, 589, 590, 818, and 819. Which one or ones are most responsible for the characteristic yellow color of excited sodium atoms?

2. Does the main emission line for $SrCl_2$ (in Figure A) have a longer or shorter wavelength than that of the yellow line from NaCl?

3. Mg is oxidized by $KClO_4$ to make white flashes. One product of the reaction is KCl. Write a balanced equation for the reaction.

Answers to these questions are available in Appendix N.

directed at a thin sheet of metal foil. Furthermore, assuming the electron beam to be a matter wave, de Broglie's relation was followed quantitatively. The experiment was taken as evidence that electrons can be described as having wave properties in certain situations.

De Broglie's equation suggests that, for the wavelength of a matter wave to be measurable, the product of m and v must be very small because h is so small. A 114-g baseball, traveling at 110 mph, for example, has a large mv product (5.6 kg · m/s) and therefore an incredibly small wavelength, 1.2×10^{-34} m! Such a small value cannot be measured with any instrument now available, nor is such a value meaningful. As a consequence, wave properties are never assigned to a baseball or any other massive object. It is possible to observe wave-like properties only for particles of extremely small mass, such as protons, electrons, and neutrons.

Cathode ray tubes, such as were found in television sets before the advent of LCD and plasma TVs, generate a beam of electrons. When the electrons impact the screen, the beam gives rise to tiny flashes of colored light. In contrast to this effect, best explained by assuming electrons are particles, diffraction experiments suggest that electrons are waves. But, how can an electron be both a particle and a wave? In part, we are facing limitations in language; the words *particle* and *wave* accurately describe things encountered on a macroscopic scale. However, they apply less well on the submicroscopic scale associated with subatomic particles. In some experiments, electrons behave like particles. In other experiments, we find that they behave like waves. No single experiment can be done to show the electron behaving *simultaneously* as a wave and a particle. Scientists now accept this **wave–particle duality**—that is, the idea that the electron has the properties of both a wave and a particle. This concept is central to an understanding of the modern model of the atom that we take up in the remainder of the chapter.

EXAMPLE 6.4   Using de Broglie's Equation

Problem Calculate the wavelength associated with an electron of mass $m = 9.109 \times 10^{-28}$ g that travels at 40.0% of the speed of light.

What Do You Know? The equation proposed by de Broglie, $\lambda = h/mv$, relates wavelength to the mass and velocity of a moving particle. Here you are given the mass of the electron and its velocity; you will need Planck's constant, $h = 6.626 \times 10^{-34}$ J · s. Note also that 1 J = 1 kg · m²/s² (◀ page 25). Finally, you will need to manipulate units of m and h to be consistent in the solution.

Strategy
- Express the electron mass in kg and calculate the electron velocity in m · s⁻¹.

- Substitute values of m (in kg), v (in m/s), and h in the de Broglie equation and solve for λ.

Solution
Electron mass = 9.109×10^{-31} kg

Electron speed (40.0% of light speed) = $(0.400)(2.998 \times 10^8$ m · s⁻¹$) = 1.20 \times 10^8$ m · s⁻¹

Substituting these values into de Broglie's equation, we have

$$\lambda = \frac{h}{mv} = \frac{6.626 \times 10^{-34} \text{ (kg} \cdot \text{m}^2/\text{s}^2)(\text{s})}{(9.109 \times 10^{-31} \text{ kg})(1.20 \times 10^8 \text{ m/s})} = 6.07 \times 10^{-12} \text{ m}$$

In nanometers, the wavelength is

$$\lambda = (6.07 \times 10^{-12} \text{ m})(1.00 \times 10^9 \text{ nm/m}) = 6.07 \times 10^{-3} \text{ nm}$$

Think about Your Answer The calculated wavelength is about 1/12 of the diameter of the H atom. Also notice the care taken to monitor units in this problem.

Check Your Understanding
Calculate the wavelength associated with a neutron having a mass of 1.675×10^{-24} g and a kinetic energy of 6.21×10^{-21} J. (Recall that the kinetic energy of a moving particle is $E = \frac{1}{2}mv^2$.)

REVIEW & CHECK FOR SECTION 6.4

The wavelength associated with an electron moving at 40% of the speed of light is 6.07×10^{-3} nm (see Example 6.4). How will the wavelength be changed if the velocity increased to 80% of the speed of light?

(a)   Wavelength will be longer.   (c)   Wavelength will not change.

(b)   Wavelength will be shorter.

6.5 The Modern View of Electronic Structure: Wave or Quantum Mechanics

How does wave–particle duality affect our model of the arrangement of electrons in atoms? Following World War I, German scientists Erwin Schrödinger (1887–1961), Max Born (1882–1970), and Werner Heisenberg (1901–1976) provided the answer.

Erwin Schrödinger received the Nobel Prize in physics in 1933 for a comprehensive theory of the behavior of electrons in atoms. Starting with de Broglie's hypothesis that an electron could be described as a matter wave, Schrödinger developed a model for electrons in atoms that has come to be called **quantum mechanics** or **wave mechanics**. Unlike Bohr's model, Schrödinger's model can be difficult to visualize, and the mathematical approach is complex. Nonetheless, the consequences of the model are important, and understanding its implications is essential to understanding the modern view of the atom.

Let us start by thinking of the behavior of an electron in the atom as a **standing wave**. If you tie down a string at both ends, as you would the string of a guitar, and then pluck it, the string vibrates as a standing wave (Figure 6.11). There are only certain vibrations allowed for the standing wave formed by a plucked guitar string. That is, the vibrations are quantized. Similarly, as Schrödinger showed, only certain matter waves are possible for an electron in an atom.

Following the de Broglie concept, we can adopt the quantum-mechanical view that an electron in an atom behaves as a wave. To describe this wave, physicists introduced the concept of a **wavefunction**, which is designated by the Greek letter ψ (psi). Schrödinger built on the idea that the electron in an atom has the characteristics of a standing wave and wrote an equation defining the energy of an electron in terms of wavefunctions. When this equation was solved for energy—a monumental task in itself—he found the following important outcomes:

- Only certain wavefunctions are found to be acceptable, and each is associated with an allowed energy value. That is, *the energy of the electron in the atom is quantized.*
- The solutions to Schrödinger's equation for an electron in three-dimensional space depend on three integers, n, ℓ, and m_ℓ, which are called **quantum numbers**. Only certain combinations of their values are possible, as we shall outline below.

The next step in understanding the quantum mechanical view is to explore the physical significance of the wavefunction, ψ (psi). Here we owe much to Max Born's interpretation. He said that

(a) the value of the wavefunction ψ at a given point in space is the amplitude (height) of the electron matter wave. This value has both a magnitude and a sign that can be either positive or negative. (Visualize a vibrating string

• Wave Functions and Energy In Bohr's theory, the electron energy for the H atom is given by $E_n = -Rhc/n^2$. Schrödinger's electron wave model gives the same result.

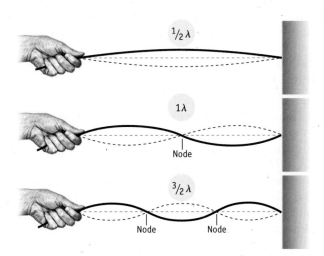

FIGURE 6.11 Standing waves. A two-dimensional standing wave such as a vibrating string must have two or more points of zero amplitude (called **nodes**), and only certain vibrations are possible. These allowed vibrations have wavelengths of $n(\lambda/2)$, where n is an integer ($n = 1, 2, 3, \ldots$). In the first vibration illustrated here, the distance between the ends of the string is half a wavelength, or $\lambda/2$. In the second, the string's length equals one complete wavelength, or $2(\lambda/2)$. In the third vibration, the string's length is $3(\lambda/2)$.

in a guitar or piano, for example (Figure 6.11). Points of positive amplitude are above the axis of the wave, and points of negative amplitude are below it.)

(b) the square of the value of the wavefunction (ψ^2) is related to the *probability* of finding an electron in a tiny region of space. Scientists refer to ψ^2 as a **probability density**. Just as we can calculate the mass of an object from the product of its density and volume, we can calculate the *probability* of finding an electron in a tiny volume from the product of ψ^2 and the volume. The Born interpretation of the wavefunction is that, for a given volume, whenever ψ^2 is large, the probability of finding the electron is larger than when ψ^2 is small.

There is one more important concept to touch on as we try to understand the modern quantum mechanical model. In Bohr's model of the atom, both the energy and location (the orbit) for the electron in the hydrogen atom can be described accurately. However, Werner Heisenberg postulated that, for a tiny object such as an electron in an atom, it is impossible to determine accurately *both* its position and its energy. That is, any attempt to determine accurately either the location or the energy will leave the other uncertain. This is now known as Heisenberg's **uncertainty principle**: *If we choose to know the energy of an electron in an atom with only a small uncertainty, then we must accept a correspondingly large uncertainty in its position.* The importance of this idea is that we can assess only the likelihood, or *probability*, of finding an electron with a given energy within a given region of space. Because electron energy is the key to understanding the chemistry of an atom, chemists accept the notion of knowing only the approximate location of the electron.

Quantum Numbers and Orbitals

The wavefunction for an electron in an atom describes what we call an **atomic orbital**. Following the ideas of Born and Heisenberg, we know the energy of this electron, but we only know the region of space within which it is most probably located. Thus, when an electron has a particular wavefunction, it is said to "occupy" a particular orbital with a given energy, and each orbital is described by three quantum numbers: n, ℓ, and m_ℓ. Let us first describe the quantum numbers and the information they provide and then discuss the connection between quantum numbers and the energies and shapes of atomic orbitals.

n, the Principal Quantum Number ($n = 1, 2, 3, \ldots$)

The principal quantum number n can have any integer value from 1 to infinity. The value of n is the primary factor in determining the *energy* of an orbital. It also defines the *size* of an orbital: for a given atom, the greater the value of n, the greater the size of the orbital.

In atoms having more than one electron, two or more electrons may have the same n value. These electrons are then said to be in the same **electron shell**.

ℓ, the Orbital Angular Momentum Quantum Number ($\ell = 0, 1, 2, 3, \ldots, n - 1$)

Orbitals of a given shell can be grouped into **subshells**, where each subshell is characterized by a different value of the quantum number ℓ. The quantum number ℓ, referred to as the "orbital angular momentum" quantum number, can have any integer value from 0 to a maximum of $n - 1$. This quantum number defines the *characteristic shape of an orbital*; different ℓ values correspond to different orbital shapes.

Because ℓ can be no larger than $n - 1$, the value of n limits the number of subshells possible for each shell. For the shell with $n = 1$, ℓ must equal 0; thus, only one subshell is possible. When $n = 2$, ℓ can be either 0 or 1. Because two values of ℓ are now possible, there are two subshells in the $n = 2$ electron shell.

● **Orbits and Orbitals**
In Bohr's model of the H atom, the electron is confined to a prescribed path around the nucleus, its *orbit*, so we should be able to define its position and energy at a given moment in time. In the modern view, the term *orbital* is used to describe the fact that we have a less definite view of the electron's location. We know its energy but only the region of space within which it is probably located, that is, its orbital.

● **Electron Energy and Quantum Numbers** The electron energy in the H atom depends only on the value of n. In atoms with more electrons, the energy depends on both n and ℓ, as you shall see in Chapter 7.

Subshells are usually identified by letters. For example, an $\ell = 1$ subshell is called a "*p* subshell," and an orbital in that subshell is called a "*p* orbital."

Value of ℓ	Subshell Label
0	*s*
1	*p*
2	*d*
3	*f*

m_ℓ, the Magnetic Quantum Number ($m_\ell = 0, \pm 1, \pm 2, \pm 3, \ldots, \pm \ell$)

The magnetic quantum number, m_ℓ, is related to the *orientation in space of the orbitals within a subshell*. Orbitals in a given subshell differ in their orientation in space, not in their energy.

The value of m_ℓ can range from $+\ell$ to $-\ell$, with 0 included. For example, when $\ell = 2$, m_ℓ can have five values: $+2, +1, 0, -1$, and -2. The number of values of m_ℓ for a given subshell ($= 2\ell + 1$) specifies the number of orbitals in the subshell.

Shells and Subshells

Allowed values of the three quantum numbers are summarized in Table 6.1. By analyzing the sets of quantum numbers in this table, you will discover the following:

- n = the number of subshells in a shell
- $2\ell + 1$ = the number of orbitals in a subshell = the number of values of m_ℓ
- n^2 = the number of orbitals in a shell

The First Electron Shell, $n = 1$

When $n = 1$, the value of ℓ can only be 0, so m_ℓ must also have a value of 0. This means that, in the shell closest to the nucleus, only one subshell exists, and that subshell consists of only a single orbital, the $1s$ orbital.

The Second Electron Shell, $n = 2$

When $n = 2$, ℓ can have two values (0 and 1), so there are two subshells in the second shell. One of these is the $2s$ subshell ($n = 2$ and $\ell = 0$), and the other is the $2p$ subshell ($n = 2$ and $\ell = 1$). Because the values of m_ℓ can be -1, 0, and $+1$ when

● **Orbital Symbols** Early studies of the emission spectra of elements classified lines into four groups on the basis of their appearance. These groups were labeled sharp, principal, diffuse, and fundamental. From these names came the labels we now apply to orbitals: *s*, *p*, *d*, and *f*.

● **Shells, Subshells, and Orbitals—A Summary** Electrons in atoms are arranged in shells. Within each shell, there can be one or more electron subshells, each comprised of one or more orbitals.

	Quantum Number
Shell	n
Subshell	ℓ
Orbital	m_ℓ

Table 6.1 Summary of the Quantum Numbers, Their Interrelationships, and the Orbital Information Conveyed

Principal Quantum Number	Angular Momentum Quantum Number	Magnetic Quantum Number	Number and Type of Orbitals in the Subshell
Symbol = n Values = 1, 2, 3, ...	Symbol = ℓ Values = $0 \ldots n - 1$	Symbol = m_ℓ Values = $+\ell \ldots 0 \ldots -\ell$	n = number of subshells Number of orbitals in shell = n^2 and number of orbitals in subshell = $2\ell + 1$
1	0	0	one $1s$ orbital (one orbital of one type in the $n = 1$ shell)
2	0 1	0 $+1, 0, -1$	one $2s$ orbital three $2p$ orbitals (four orbitals of two types in the $n = 2$ shell)
3	0 1 2	0 $+1, 0, -1$ $+2, +1, 0, -1, -2$	one $3s$ orbital three $3p$ orbitals five $3d$ orbitals (nine orbitals of three types in the $n = 3$ shell)
4	0 1 2 3	0 $+1, 0, -1$ $+2, +1, 0, -1, -2$ $+3, +2, +1, 0, -1, -2, -3$	one $4s$ orbital three $4p$ orbitals five $4d$ orbitals seven $4f$ orbitals (16 orbitals of four types in the $n = 4$ shell)

$\ell = 1$, three 2p orbitals exist. All three orbitals have the same shape. However, because each has a different m$_\ell$ value, the three orbitals differ in their orientation in space.

The Third Electron Shell, $n = 3$

When $n = 3$, three subshells are possible for an electron because ℓ can have the values 0, 1, and 2. The first two subshells within the $n = 3$ shell are the 3s ($\ell = 0$, one orbital) and 3p ($\ell = 1$, three orbitals) subshells. The third subshell is labeled 3d ($n = 3$, $\ell = 2$). Because m$_\ell$ can have five values (-2, -1, 0, $+1$, and $+2$) for $\ell = 2$, there are five d orbitals in this d subshell.

The Fourth Electron Shell, $n = 4$

There are four subshells in the $n = 4$ shell. In addition to 4s, 4p, and 4d subshells, there is the 4f subshell for which $\ell = 3$. Seven such orbitals exist because there are seven values of m$_\ell$ when $\ell = 3$ (-3, -2, -1, 0, $+1$, $+2$, and $+3$).

REVIEW & CHECK FOR SECTION 6.5

1. What label is given to an orbital with quantum numbers $n = 4$ and $\ell = 1$?

   (a) 4s (b) 4p (c) 4d (d) 4f

2. How many orbitals are in the $n = 4$ shell?

   (a) 1   (b) 4   (c) 9   (d) 16

3. Which quantum number, or combination of quantum numbers, is needed to specify a given subshell in an atom?

   (a) n (b) ℓ (c) n and ℓ

6.6 The Shapes of Atomic Orbitals

We often say the electron is assigned to, or "occupies," an orbital. But what does this mean? What is an orbital? What does it look like? To answer these questions, we have to look at the wavefunctions for the orbitals. (To answer the question of *why* the quantum numbers—small, whole numbers—can be related to orbital shape and energy, see *A Closer Look: More About H Atomic Orbital Shapes and Wavefunctions* on page 225.)

s Orbitals

A 1s orbital is associated with the quantum numbers $n = 1$ and $\ell = 0$. If we could photograph a 1s electron at one-second intervals for a few thousand seconds, the composite picture would look like the drawing in Figure 6.12a. This resembles a cloud of dots, and chemists often refer to such representations of electron orbitals as *electron cloud pictures*. In Figure 6.12a, the density of dots is greater close to the nucleus, that is, the electron cloud is denser close to the nucleus. This indicates that the 1s electron is most likely to be found near the nucleus. However, the density of dots declines on moving away from the nucleus and so, therefore, does the probability of finding the electron.

The thinning of the electron cloud at increasing distance is illustrated in a different way in Figure 6.12b. Here we have plotted the square of the wavefunction for the electron in a 1s orbital (ψ^2), times 4π and the distance squared (r^2), as a function of the distance of the electron from the nucleus. This plot represents the probability of finding the electron in a thin spherical shell at a distance r from the nucleus. Chemists refer to the plot of $4\pi r^2 \psi^2$ vs. r as a **surface density plot** or **radial distribution plot**. For the 1s orbital, $4\pi r^2 \psi^2$ is zero at the nucleus—there is no probability the electron will be exactly at the nucleus (where $r = 0$)—but the probability rises rapidly on moving away from the nucleus, reaches a maximum a short distance from the nucleus (at 52.9 pm), and then decreases rapidly as the distance from the nucleus increases. Notice that the probability of finding the electron approaches but never quite reaches zero, even at very large distances.

Figure 6.12a shows that, for the 1s orbital, the probability of finding an electron is the same at a given distance from the nucleus, no matter in which direction you proceed from the nucleus. Consequently, the *1s orbital is spherical in shape.*

Because the probability of finding the electron approaches but never quite reaches zero, there is no sharp boundary beyond which the electron is never found (although the probability can be incalculably small at large distances). Nonetheless, the s orbital (and other types of orbitals as well) is often depicted as having a **boundary surface** (Figure 6.12c), largely because it is easier to draw such pictures. To create Figure 6.12c, we drew a sphere centered on the nucleus in such a way that the probability of finding the electron somewhere inside the sphere is 90%. The choice of 90% is arbitrary—we could have chosen a different value—and if we do, the shape would be the same, but the size of the sphere would be different.

All s orbitals (1s, 2s, 3s ...) are spherical in shape. However, for any atom, the size of s orbitals increases as n increases (Figure 6.13). For a given atom, the 1s orbital is more compact than the 2s orbital, which is in turn more compact than the 3s orbital.

It is important that you avoid some misconceptions about pictures of orbitals.

- First, there is not an impenetrable surface within which the electron is "contained."
- Second, the probability of finding the electron is not the same throughout the volume enclosed by the surface in Figure 6.12c. (An electron in the 1s orbital of a H atom has a greater probability of being 52.9 pm from the nucleus than of being closer or farther away.)
- Third, the terms "electron cloud" and "electron distribution" imply that the electron is a particle, but the basic premise in quantum mechanics is that the electron is treated as a wave, not a particle.

p Orbitals

All atomic orbitals for which $\ell = 1$ (p orbitals) have the same basic shape. If you enclose 90% of the electron density in a p orbital within a surface, the electron cloud is often described as having a shape like a weight lifter's "dumbbell," and chemists

● **Surface Density Plot for 1s** The maximum value of the radial distribution plot for a 1s electron in a hydrogen atom occurs at 52.9 pm. It is interesting to note that this maximum is at exactly the same distance from the nucleus that Niels Bohr calculated for the radius of the orbit occupied by the $n = 1$ electron.

● **Nodal Surfaces** Nodal surfaces that cut through the nucleus occur for all p, d, and f orbitals. These surfaces are usually flat, so they are often referred to as *nodal planes.* In some cases (for example, d_{z^2}), however, the "plane" is not flat and so is better referred to as a *nodal surface.*

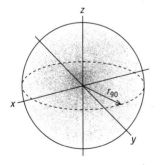

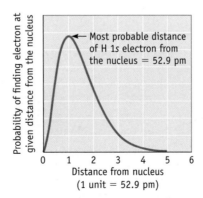

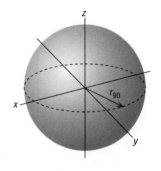

(a) Dot picture of an electron in a 1s orbital. Each dot represents the position of the electron at a different instant in time. Note that the dots cluster closest to the nucleus. r_{90} is the radius of a sphere within which the electron is found 90% of the time.

(b) A plot of the surface density ($4\pi r^2 \psi^2$) as a function of distance for a hydrogen atom 1s orbital. This gives the probability of finding the electron at a given distance from the nucleus.

(c) The surface of the sphere within which the electron is found 90% of the time for a 1s orbital. This surface is often called a "boundary surface." (A 90% surface was chosen arbitrarily. If the choice was the surface within which the electron is found 50% of the time, the sphere would be considerably smaller.)

FIGURE 6.12 Different views of a 1s (n = 1, ℓ = 0) orbital. In panel **(b)** the horizontal axis is marked in units called "Bohr radii," where 1 Bohr radius = 52.9 pm. This is common practice when plotting wave functions.

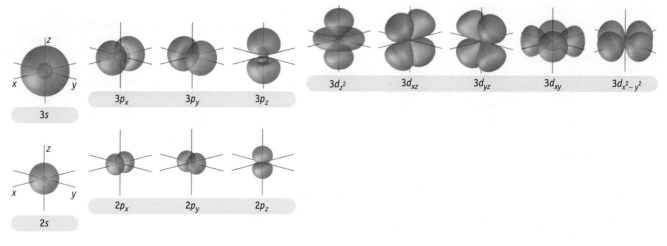

FIGURE 6.13 Atomic orbitals. Boundary surface diagrams for electron densities of 1s, 2s, 2p, 3s, 3p, and 3d orbitals for the hydrogen atom. For the *p* orbitals, the subscript letter indicates the cartesian axis along which the orbital lies. (For more about orbitals, see *A Closer Look: More About H Atom Orbital Shapes and Wavefunctions.*)

- **ℓ and Nodal Surfaces** The number of nodal surfaces passing through the nucleus for an orbital = ℓ.

Orbital	ℓ	Number of Nodal Surfaces through the Nucleus
s	0	0
p	1	1
d	2	2
f	3	3

describe *p* orbitals as having dumbbell shapes (Figures 6.13 and 6.14). A *p* orbital has a **nodal surface**—a surface on which the probability of finding the electron is zero—that passes through the nucleus. (The nodal surface is a consequence of the wavefunction for *p* orbitals, which has a value of zero at the nucleus but which rises rapidly in value on moving way from the nucleus. See *A Closer Look: More About H Atom Orbital Shapes and Wavefunctions.*)

There are three *p* orbitals in a subshell, and all have the same basic shape with one nodal plane through the nucleus. Usually, *p* orbitals are drawn along the *x*-, *y*-, and *z*-axes and labeled according to the axis along which they lie (p_x, p_y, or p_z).

d Orbitals

Orbitals with ℓ = 0, *s* orbitals, have no nodal surfaces through the nucleus, and *p* orbitals, for which ℓ = 1, have one nodal surface through the nucleus. *The value of ℓ is equal to the number of nodal surfaces slicing through the nucleus.* It follows that the five *d* orbitals, for which ℓ = 2, have two nodal surfaces through the nucleus, resulting in four regions of electron density. The d_{xy} orbital, for example, lies in

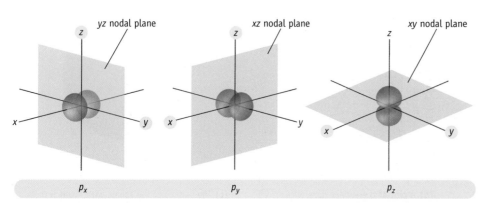

(a) The three *p* orbitals each have one nodal plane (ℓ = 1) that is perpendicular to the axis along which the orbital lies.

FIGURE 6.14 Nodal surfaces of *p* and *d* orbitals. A nodal surface is a surface on which the probability of finding the electron is zero.

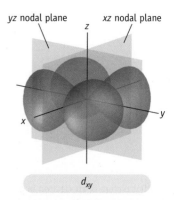

(b) The d_{xy} orbital. All five *d* orbitals have two nodal surfaces (ℓ = 2) passing through the nucleus. Here, the nodal surfaces are the *xz*- and *yz*-planes, so the regions of electron density lie between the *x*- and *y*-axes in the *xy*-plane.

More about H Atom Orbital Shapes and Wavefunctions

There is an important question to answer: How do quantum numbers, which are small, integer numbers, tell us the shape of atomic orbitals? The answer lies in the orbital wavefunctions, which are reasonably simple mathematical equations.

The equations for wavefunctions (ψ) are the product of two functions: the *radial function* and the *angular function*. You need to look at each type to get a picture of an orbital.

Let's first consider the *radial function*, which depends on n and ℓ. This tells us how the value of ψ depends on the distance from the nucleus. The radial functions for the hydrogen atom 1s ($n = 1$ and $\ell = 0$), 2s ($n = 2$ and $\ell = 0$) orbitals, and 2p orbitals ($n = 2$ and $\ell = 1$) are plotted in Figure A. (The horizontal axis has units of a_0, where a_0 is a constant equal to 52.9 pm).

Waves have crests, troughs, and nodes, and plots of the wavefunctions show this. For the 1s orbital of the H atom, the radial wavefunction ψ_{1s} approaches a maximum at the nucleus (Figure A), but the wave's amplitude declines rapidly at points farther removed from the nucleus. The sign of ψ_{1s} is positive at all points in space.

For a 2s orbital, there is a different profile: the sign of ψ_{2s} is positive near the nucleus, drops to zero (there is a node at $r = 2a_0 = 2 \times 52.9$ pm), and then becomes

negative with increasing r before approaching zero at greater distances.

Now to the *angular portion* of the wavefunction: this reflects changes that occur when you travel outward from the nucleus in different directions. It is a function of the quantum numbers ℓ and m_ℓ.

As illustrated in Figure 6.12 the value of ψ_{1s} is the same in every direction. This is a reflection of the fact that, while the radial portion of the wavefunction changes with r, the angular portion for all s orbitals is a constant. As a consequence, all s orbital are spherical.

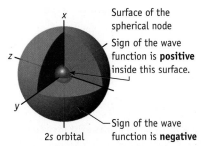

FIGURE B **Wavefunction for a 2s orbital.** A 2s orbital for the H atom showing the spherical node (at $2a_0 = 105.8$ pm) around the nucleus.

For the 2s orbital, you see a node in Figure A at 105.8 pm ($= 2a_0$) when plotting the radial portion of the wavefunction. However, because the angular part of ψ_{2s} has the same value in all directions, this means there is a node—a *spherical nodal surface*—at the same distance in every direction (as illustrated in Figure B). For any orbital, the number of spherical nodes is $n - \ell - 1$.

Now let's look at a p orbital, first the radial portion and then the angular portion. For a p orbital the radial portion of the wavefunction is 0 when $r = 0$. Thus, the value of ψ_{2p} is zero at the nucleus and a nodal surface passes through the nucleus (Figures A and C). This is true for all p orbitals.

But what happens as you move away from the nucleus in one direction, say along the x-axis in the case of the $2p_x$ orbital. Now we see the value of ψ_{2p} rise to a maximum at 105.9 pm ($= 2a_0$) before falling off at greater distances (Figures A and C).

But look at Figure C for the $2p_x$ orbital. Moving away along the $-x$ direction, the value of ψ_{2p} is the same but opposite in sign

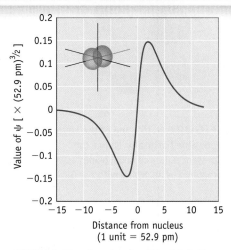

FIGURE C **Wave functions for a 2p orbital.** The sign of ψ for a 2p orbital is positive on one side of the nucleus and negative on the other (but it has a 0 value at the nucleus). A nodal plane separates the two lobes of this "dumbbell-shaped" orbital. (The vertical axis is the value of ψ, and the horizontal axis is the distance from the nucleus, where 1 unit = 52.9 pm.)

to the value in the $+x$ direction. The 2p electron is a wave with a node at the nucleus. (In drawing orbitals, we indicate this with + or − signs or with two different colors as in Figure 6.13.)

Now, what about the *angular portion* of the wavefunction for $2p_x$? The angular portion for all three p orbitals has the same general form: $c(x/r)$ for the p_x orbital, $c(y/r)$ for the p_y orbital, and $c(z/r)$ for the p_z orbital (where c is a constant). Consider a $2p_x$ orbital in Figure D. As long as x has a value, the wavefunction has a value. But when $x = 0$ (in the yz plane), then ψ is zero. This is the nodal plane for the x orbital. Similarly, the angular portion of the wavefunction for the $2p_y$ orbital means its nodal plane is the xz plane.

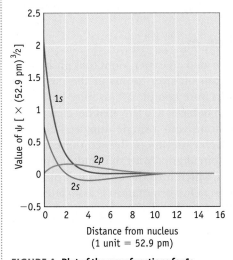

FIGURE A **Plot of the wavefunctions for 1s, 2s, and 2p orbitals versus distance from the nucleus.** As in other plots of wavefunctions, the horizontal axis is marked in units called "Bohr radii," where 1 Bohr radius = 52.9 pm.

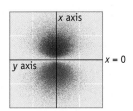

FIGURE D **The $2p_x$ orbital.** The wavefunction has no value when $x = 0$, that is, the yz plane is a nodal plane.

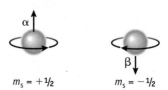

yz nodal plane   z   xz nodal plane
xy nodal plane
y
x
f_{xyz}

FIGURE 6.15 One of the seven possible f orbitals. Notice the presence of three nodal planes as required by an orbital with $\ell = 3$.

the xy-plane, and the two nodal surfaces are the xz- and yz-planes (Figure 6.14). Two other orbitals, d_{xz} and d_{yz}, lie in planes defined by the xz- and yz-axes, respectively; they also have two, mutually perpendicular nodal surfaces (Figure 6.13).

Of the two remaining d orbitals, the $d_{x^2-y^2}$ orbital is easier to visualize. In the $d_{x^2-y^2}$ orbital, the nodal planes bisect the x- and y-axes, so the regions of electron density lie along the x- and y-axes. The d_{z^2} orbital has two main regions of electron density along the z-axis, and a "doughnut" of electron density also occurs in the xy-plane. This orbital has two cone-shaped nodal surfaces.

f Orbitals

Seven f orbitals arise with $\ell = 3$. Three nodal surfaces through the nucleus cause the electron density to lie in up to eight regions of space. One of the f orbitals is illustrated in Figure 6.15.

REVIEW & CHECK FOR SECTION 6.6

1. Which of the following is not a correct representation of an orbital?

    (a) $3s$ (b) $3p$ (c) $3d$ (d) $3f$

2. Which of the following sets of quantum numbers correctly represents a $4p$ orbital?

    (a) $n = 4, \ell = 0, m_\ell = -1$ (c) $n = 4, \ell = 2, m_\ell = 1$

    (b) $n = 4, \ell = 1, m_\ell = 0$ (d) $n = 4, \ell = 1, m_\ell = 2$

3. How many nodal planes exist for a $5d$ orbital?

    (a) 0   (b) 1   (c) 2   (d) 3

6.7 One More Electron Property: Electron Spin

There is one more property of the electron that plays an important role in the arrangement of electrons in atoms and gives rise to properties of elements you observe every day: electron spin.

The Electron Spin Quantum Number, m_s

In 1921 Otto Stern and Walther Gerlach performed an experiment that probed the magnetic behavior of atoms by passing a beam of silver atoms in the gas phase through a magnetic field. Although the results were complex, they were best interpreted by *imagining* the electron has a spin and behaves as a tiny magnet that can be attracted or repelled by another magnet. If atoms with a single unpaired electron are placed in a magnetic field, the Stern-Gerlach experiment showed there are two orientations for the atoms: with the electron spin aligned with the field or opposed to the field. That is, *the electron spin is quantized,* which introduces a fourth quantum number, the **electron spin quantum number**, m_s. One orientation is associated with a value of m_s of $+\frac{1}{2}$ and the other with m_s of $-\frac{1}{2}$.

α

β

$m_s = +\frac{1}{2}$ $m_s = -\frac{1}{2}$

When it was recognized that electron spin is quantized, scientists realized that a complete description of an electron in any atom requires four quantum numbers, n, ℓ, m_ℓ, and m_s. The important consequences of this fact are explored in Chapter 7.

A CLOSER LOOK

Paramagnetism and Ferromagnetism

Magnetic materials are relatively common, and many are important in our economy. For example, a large magnet is at the heart of the magnetic resonance imaging (MRI) used in medicine, and tiny magnets are found in stereo speakers and in telephone handsets. Magnetic oxides are used in recording tapes and computer disks.

The magnetic materials we use are **ferromagnetic**. The magnetic effect of ferromagnetic materials is much larger than that of paramagnetic ones. Ferromagnetism occurs when the spins of unpaired electrons in a cluster of atoms (called a *domain*) in the solid align themselves in the same direction. Only the metals of the iron, cobalt, and nickel subgroups, as well as a few other metals such as neodymium, exhibit this prop-

erty. They are also unique in that, once the domains are aligned in a magnetic field, the metal is permanently magnetized.

Many alloys exhibit greater ferromagnetism than do the pure metals themselves. One example of such a material is Alnico, which is composed of aluminum, nickel, and cobalt as well as copper and iron. The strongest permanent magnet is an alloy of neodymium, iron, and boron ($Nd_2Fe_{14}B$).

Magnets. Many common consumer products such as loudspeakers contain permanent magnets.

© Cengage Learning/Charles D. Winters

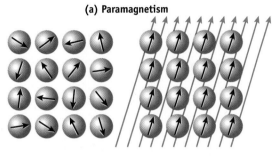

(a) Paramagnetism

No Magnetic Field   External Magnetic Field

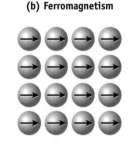

(b) Ferromagnetism

The spins of unpaired electrons align themselves in the same direction

Magnetism. **(a)** Paramagnetism: In the absence of an external magnetic field, the unpaired electrons in the atoms or ions of the substance are randomly oriented. If a magnetic field is imposed, however, these spins will tend to become aligned with the field. **(b)** Ferromagnetism: The spins of the unpaired electrons in a cluster of atoms or ions align themselves in the same direction even in the absence of a magnetic field.

Diamagnetism and Paramagnetism

A hydrogen atom has a single electron. If a hydrogen atom is placed in a magnetic field, the magnetic field of the single electron will tend to align with the external field like the needle of a compass aligns with the magnetic lines of force on the Earth.

In contrast, helium atoms, each with two electrons, are not attracted to a magnet. In fact, they are slightly repelled by the magnet. To account for this observation, we assume the two electrons of helium have *opposite* spin orientations. We say their spins are *paired*, and the result is that the magnetic field of one electron can be canceled out by the magnetic field of the second electron with opposite spin. To account for this, the two electrons are assigned different values of m_s.

It is important to understand the relationship between electron spin and magnetism. *Elements and compounds that have unpaired electrons are attracted to a magnet.* Such species are said to be **paramagnetic**. The effect can be quite weak, but, by placing a sample of an element or compound in a magnetic field, it can be observed (Figure 6.16). For example, the oxygen you breathe is paramagnetic. You can observe this experimentally because liquid oxygen sticks to a magnet of the kind you may have in the speakers of a music player (Figure 6.16b).

Substances in which all electrons are paired (with the two electrons of each pair having opposite spins) experience a slight repulsion when subjected to a magnetic field; they are called **diamagnetic**. Therefore, by determining the magnetic behavior of a substance we can gain information about the electronic structure.

In summary, *substances in which the constituent ions or atoms contain unpaired electrons are paramagnetic and are attracted to a magnetic field.* Substances in which all electrons are paired with partners of opposite spin are diamagnetic. This explanation opens the way to understanding the arrangement of electrons in atoms with more than one electron as you shall learn in the next chapter.

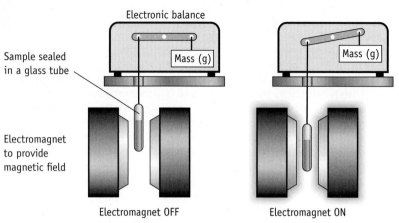

Electronic balance

Sample sealed in a glass tube

Mass (g)

Mass (g)

Electromagnet to provide magnetic field

Electromagnet OFF   Electromagnet ON

(a) A magnetic balance is used to measure the magnetism of a sample. The sample is first weighed with the electromagnet turned off. The magnet is then turned on and the sample reweighed. If the substance is paramagnetic, the sample is drawn into the magnetic field, and the apparent weight increases.

(b) Liquid oxygen (boiling point 90.2 K) clings to a strong magnet. Elemental oxygen is paramagnetic because it has unpaired electrons. (See Chapter 9.)

FIGURE 6.16 Observing and measuring paramagnetism.

   and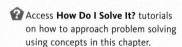

Sign in at **www.cengage.com/owl** to:
- View tutorials and simulations, develop problem-solving skills, and complete online homework assigned by your professor.
- For quick review and exam prep, download Go Chemistry mini lecture modules from OWL (or purchase them at **www.cengagebrain.com**)

? Access **How Do I Solve It?** tutorials on how to approach problem solving using concepts in this chapter.

CHAPTER GOALS REVISITED

Now that you have studied this chapter, you should ask whether you have met the chapter goals. In particular, you should be able to:

Describe the properties of electromagnetic radiation
a. Use the terms wavelength, frequency, amplitude, and node (Section 6.1). **Study Question: 3.**
b. Use Equation 6.1 ($c = \lambda v$), relating wavelength (λ) and frequency (v) of electromagnetic radiation and the speed of light (c). **Study Question: 3.**
c. Recognize the relative wavelength (or frequency) of the various types of electromagnetic radiation (Figure 6.2). **Study Question: 1.**
d. Understand that the energy of a photon, a massless particle of radiation, is proportional to its frequency (Planck's equation, Equation 6.2) (Section 6.2). **Study Questions: 5, 7, 8, 9, 12, 14, 56, 57, 58, 63, 64, 66, 72, 73, 83.**

Understand the origin of light emitted by excited atoms and its relationship to atomic structure
a. Describe the Bohr model of the atom, its ability to account for the emission line spectra of excited hydrogen atoms, and the limitations of the model (Section 6.3). **Study Question: 74.**
b. Understand that, in the Bohr model of the H atom, the electron can occupy only certain energy states, each with an energy proportional to $1/n^2$ ($E_n = -Rhc/n^2$), where n is the principal quantum number (Equation 6.4, Section 6.3). If an electron moves from one energy state to another, the amount of energy absorbed or emitted in the process is equal to the difference in energy between the two states (Equation 6.5, Section 6.3). **Study Questions: 16, 18, 19, 21, 22, 60.**

Describe the experimental evidence for particle–wave duality
a. Understand that in the modern view of the atom, electrons can be described either as particles or as waves (Section 6.4). The wavelength of an electron or any subatomic particle is given by de Broglie's equation (Equation 6.6). **Study Questions: 23–26, 82.**

A CLOSER LOOK

Quantized Spins and MRI

Just as electrons can be thought of as having a spin, so do atomic nuclei. In the hydrogen atom, the single proton can also be thought of having a spin. For most heavier atoms, the atomic nucleus includes both protons and neutrons, and the entire entity has a spin. This property is important, because nuclear spin allows scientists to detect these atoms in molecules and to learn something about their chemical environments.

The technique used to detect the spins of atomic nuclei is *nuclear magnetic resonance* (NMR). It is one of the most powerful methods currently available to determine molecular structures. About 20 years ago, it was adapted as a diagnostic technique in medicine, where it is known as *magnetic resonance imaging* (MRI).

Just as electron spin is quantized, so too is nuclear spin. The H atom nucleus (often referred to simply as a *proton*) can spin in either of two directions. If the H atom is placed in a strong, external magnetic field, however, the spinning nuclear magnet can align itself with or against the external field. If a sample of ethanol (CH_3CH_2OH), for example, is placed in a strong magnetic field, a slight excess of the H atom nuclei (and ^{13}C atom nuclei) is aligned with the lines of force of the field.

The nuclei aligned with the field have a slightly lower energy than when aligned against the field. The NMR and MRI technologies depend on the fact that energy in the radio-frequency region can be absorbed by the sample and can cause the nuclear spins to switch alignments—that is, to move

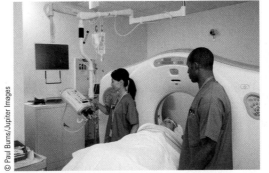
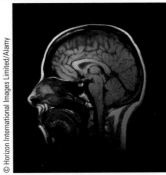

Magnetic resonance imaging. (a) MRI instrument. The patient is placed inside a large magnet, and the tissues to be examined are irradiated with radio-frequency radiation. **(b)** An MRI image of the human brain.

to a higher energy state. This reemission of energy is detected by the instrument.

The most important aspect of the magnetic resonance technique is that the difference in energy between two different spin states depends on the electronic environment of atoms in the molecule. In the case of ethanol, the three CH_3 protons are different from the two CH_2 protons, and both sets are different from the OH proton. These three different sets of H atoms absorb radiation of slightly different energies. The instrument measures the frequencies absorbed, and a scientist familiar with the technique can quickly distinguish the three different environments in the molecule.

The MRI technique closely resembles the NMR method. Hydrogen is abundant in the human body as water and in numerous organic molecules. In the MRI device,

the patient is placed in a strong magnetic field, and the tissues being examined are irradiated with pulses of radio-frequency radiation.

The MRI image is produced by detecting how fast the excited nuclei "relax"; that is, how fast they return to the lower energy state from the higher energy state. The "relaxation time" depends on the type of tissue. When the tissue is scanned, the H atoms in different regions of the body show different relaxation times, and an accurate "image" is built up.

MRI gives information on soft tissue—muscle, cartilage, and internal organs—which is unavailable from x-ray scans. This technology is also noninvasive, and the magnetic fields and radio-frequency radiation used are not harmful to the body.

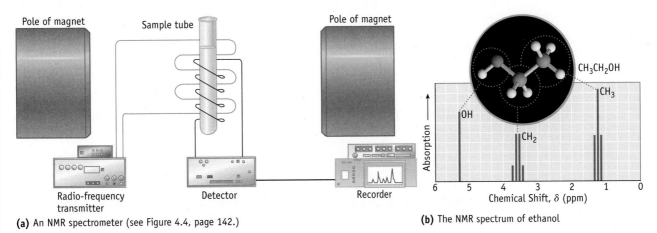

(a) An NMR spectrometer (see Figure 4.4, page 142.)

(b) The NMR spectrum of ethanol

Nuclear magnetic resonance. (a) A schematic diagram of an NMR spectrometer. **(b)** The NMR spectrum of ethanol showing that the three different types of protons appear in distinctly different regions of the spectrum. The pattern observed for the CH_2 and CH_3 protons is characteristic of these groups of atoms and signals the chemist that they are present in the molecule.

Describe the basic ideas of quantum mechanics
a. Recognize the significance of quantum mechanics in describing atomic structure (Section 6.5).
b. Understand that an orbital for an electron in an atom corresponds to the allowed energy of that electron.
c. Understand that the position of the electron is not known with certainty; only the probability of the electron being at a given point of space can be calculated. This is a consequence of the Heisenberg uncertainty principle.

Define the four quantum numbers (n, ℓ, m_ℓ, and m_s), and recognize their relationship to electronic structure
a. Describe the allowed energy states of the orbitals in an atom using three quantum numbers n, ℓ, and m_ℓ (Section 6.5). **Study Questions: 27–34, 36, 37, 38–41, 43, 44, 80, 83.**
b. Describe shapes of the orbitals (Section 6.6). **Study Questions: 46, 53, 67f, 68e.**
c. Recognize the spin quantum number, m_s, which has values of $\pm\frac{1}{2}$. Classify substances as paramagnetic (attracted to a magnetic field; characterized by unpaired electron spins) or diamagnetic (repelled by a magnetic field, all electrons paired) (Section 6.7). **Study Questions: 68j, 69.**

Key Equations

Equation 6.1 (page 205) The product of the wavelength (λ) and frequency (ν) of electromagnetic radiation is equal to the speed of light (c).

$$c = \lambda \times \nu$$

Equation 6.2 (page 208) Planck's equation: The energy of a photon, a massless particle of radiation, is proportional to its frequency (ν). The proportionality constant, h, is called Planck's constant (6.626×10^{-34} J · s).

$$E = h\nu$$

Equation 6.3 (page 211) The Rydberg equation can be used to calculate the wavelengths of the lines in the Balmer series of the hydrogen spectrum. The Rydberg constant, R, is 1.0974×10^7 m^{-1}, and n is 3 or larger.

$$\frac{1}{\lambda} = R\left(\frac{1}{2^2} - \frac{1}{n^2}\right) \text{ when } n > 2$$

Equation 6.4 (page 212) In Bohr's theory, the energy of the electron, E_n, in the nth quantum level of the H atom is proportional to $1/n^2$, where n is a positive integer (the principal quantum number) and $Rhc = 2.179 = 10^{-18}$ J/atom or $NRhc = 1312$ kJ/mol.

$$E_n = -\frac{Rhc}{n^2}$$

Equation 6.5 (page 214) The energy change for an electron moving between two quantum levels (n_{final} and n_{initial}) in the H atom.

$$\Delta E = E_{\text{final}} - E_{\text{initial}} = -Rhc\left(\frac{1}{n_{\text{final}}^2} - \frac{1}{n_{\text{initial}}^2}\right)$$

Equation 6.6 (page 216) De Broglie's equation: The wavelength of a particle (λ) is related to its mass (m) and speed (v) and to Planck's constant (h).

$$\lambda = \frac{h}{mv}$$

OWL Questions and problems for this chapter are available in OWL. Use the OWL access card that came with this textbook to access assigned questions and problems for this chapter.

7 The Structure of Atoms and Periodic Trends

The Bridgeman Art Library International

Deep blue sapphire. Deep blue sapphire is aluminum oxide with traces of iron(II) and titanium(IV) ions.

Rubies and Sapphires—Pretty Stones

People have been fascinated by, and have coveted, precious stones such as sapphires and rubies for hundreds of years.

Sapphires and rubies are just aluminum oxide with traces of impurities. But it is the impurities that make simple aluminum oxide into a beautiful gemstone.

Rubies come in a variety of colors. Most are red, but they can have tints of brown, purple, and pink. The price of a ruby usually depends on its color, the most valuable being deep red, said to resemble "pigeon blood."

The most valuable rubies are clear, but others often have inclusions of tiny needles of titanium dioxide (rutile), which

Synthetic rubies. Rubies are aluminum oxide with a trace of chromium(III) in place of Al^{3+} ions.

© Gabbro/Alamy

makes them a little cloudy. In fact, if a ruby is not slightly cloudy, it may be a synthetic stone with much less value.

Most rubies come from southeast and central Asia and east Africa, and there are a few deposits in the United States. Interestingly, as glaciers have receded in Greenland, large ruby deposits have been discovered.

Synthetic rubies were first made in the 19th century, the best known process being that developed by Auguste Verneuil in 1902. In this process, aluminum oxide and barium fluoride are fused with a chromium(III) compound. The synthetic materials produced in this way are now used in commercial lasers such as those in players of CDs and DVDs.

Sapphires are also based on aluminum oxide with other impurities that lead to their usual blue color. Like rubies, they can be synthesized, and synthetic sapphires are used in jewelry as well.

The red color of rubies comes from the substitution of a few of the Al^{3+} ions in Al_2O_3 by Cr^{3+} ions because their radii are similar. In contrast, the blue color of sapphires comes from a trace of elements such as iron and titanium. When iron and titanium are present as Fe^{2+} and Ti^{4+}, the result is the deep blue color for which sapphires are famous.

Questions:

1. What is the electron configuration for the Cr atom and for the Cr^{3+} ion?
2. Is chromium in any of its ions (Cr^{2+}, Cr^{3+}, CrO_4^{2-}) paramagnetic?
3. The radius of the Cr^{3+} ion is 64 pm. How does this compare with the radius of the Al^{3+} ion?
4. What is the electron configuration for the Fe^{2+} and Ti^{4+} ions?

Answers to these questions are available in Appendix N.

CHAPTER OUTLINE

7.1 The Pauli Exclusion Principle

7.2 Atomic Subshell Energies and Electron Assignments

7.3 Electron Configurations of Atoms

7.4 Electron Configurations of Ions

7.5 Atomic Properties and Periodic Trends **go Chemistry**

7.6 Periodic Trends and Chemical Properties

CHAPTER GOALS

See Chapter Goals Revisited (page 258) for Study Questions keyed to these goals.

- Recognize the relationship of the four quantum numbers (n, ℓ, m_ℓ, and m_s) to atomic structure.

- Write the electron configuration for atoms and monatomic ions.

- Rationalize trends in atom and ion sizes, ionization energy, and electron attachment enthalpy.

The wave mechanical model of the atom developed in Chapter 6 describes electrons as matter waves. The electrons are assigned to orbitals, regions of space in which the probability of finding an electron is high. The orbitals are arranged in subshells that are in turn part of electron shells. One objective of this chapter is to apply this model to the electronic structure of all of the elements.

A second objective is to explore some of the properties of elements, among them the energy changes that occur when atoms lose or gain electrons to form ions and the sizes of atoms and ions. These properties are directly related to the arrangement of electrons in atoms and thus to the chemistry of the elements and their compounds.

7.1 The Pauli Exclusion Principle

To make the quantum theory consistent with experiment, the Austrian physicist Wolfgang Pauli (1900–1958) stated in 1925 his **exclusion principle**: *No more than two electrons can occupy the same orbital, and, if there are two electrons in the same orbital, they must have opposite spins.* This leads to the general statement that no two electrons in an atom can have the same set of four quantum numbers (n, ℓ, m_ℓ, and m_s).

An electron assigned to the 1s orbital of the H atom may have the set of quantum numbers $n = 1$, $\ell = 0$, $m_\ell = 0$, and $m_s = +\frac{1}{2}$. Let us represent an orbital by a box and the electron spin by an arrow ($\uparrow$ or $\downarrow$). A representation of the hydrogen atom is then:

Electrons in 1s orbital: $\boxed{\uparrow}$ Quantum number set
1s $n = 1$, $\ell = 0$, $m_\ell = 0$, $m_s = +\frac{1}{2}$

The choice of m_s (either $+\frac{1}{2}$ or $-\frac{1}{2}$) and the direction of the electron spin arrow are arbitrary; that is, we could choose either value, and the arrow may point in either direction. Diagrams such as these are called **orbital box diagrams**.

A helium atom has two electrons, both assigned to the 1s orbital. The Pauli exclusion principle requires that each electron must have a different set of quantum numbers, so the orbital box diagram now is:

1s

Two electrons in 1s orbital: $\boxed{\uparrow\downarrow}$ ◄——— This electron has $n = 1$, $\ell = 0$, $m_\ell = 0$, $m_s = -\frac{1}{2}$

——— This electron has $n = 1$, $\ell = 0$, $m_\ell = 0$, $m_s = +\frac{1}{2}$

OWL

Sign in to OWL at **www.cengage.com/owl** to view tutorials and simulations, develop problem-solving skills, and complete online homework assigned by your professor.

Download mini lecture videos for key concept review and exam prep from OWL or purchase them from **www.cengagebrain.com**

- **Orbitals Are Not Boxes** Orbitals are not boxes in which electrons are placed. Thus, it is not conceptually correct to talk about electrons being in orbitals or occupying orbitals, although this is commonly done for the sake of simplicity.

By having opposite or "paired" spins, the two electrons in the $1s$ orbital of a He atom have different sets of the four quantum numbers. Our understanding of orbitals and the knowledge that an orbital can accommodate no more than two electrons tell us the maximum number of electrons that can occupy each electron shell or subshell. For example, because two electrons can be assigned to each of the three orbitals in a p subshell, p subshells can hold a maximum of six electrons. By the same reasoning, the five orbitals of a d subshell can accommodate a total of 10 electrons, and the seven f orbitals can accommodate 14 electrons. Recall that there are n subshells in the nth shell, and that there are n^2 orbitals in that shell (◄ Table 6.1, page 221). Thus, *the maximum number of electrons in any shell is $2n^2$.* The relationship among the quantum numbers and the numbers of electrons is shown in Table 7.1.

REVIEW & CHECK FOR SECTION 7.1

How many electrons can be accommodated in the $n = 7$ shell?

(a) 18   (b) 36   (c) 72   (d) 98

Table 7.1 **Number of Electrons Accommodated in Electron Shells and Subshells with $n = 1$ to 6**

Electron Shell (n)	Subshells Available	Orbitals Available $(2\ell + 1)$	Number of Electrons Possible in Subshell $[2(2\ell + 1)]$	Maximum Electrons Possible for nth Shell $(2n^2)$
1	s	1	2	2
2	s	1	2	8
	p	3	6	
3	s	1	2	18
	p	3	6	
	d	5	10	
4	s	1	2	32
	p	3	6	
	d	5	10	
	f	7	14	
5	s	1	2	50
	p	3	6	
	d	5	10	
	f	7	14	
	g^*	9	18	
6	s	1	2	72
	p	3	6	
	d	5	10	
	f^*	7	14	
	g^*	9	18	
	h^*	11	22	

*These orbitals are not occupied in the ground state of any known element.

7.2 Atomic Subshell Energies and Electron Assignments

Our goal in this section is to understand and predict the distribution of electrons in atoms with many electrons. The procedure by which electrons are assigned to orbitals is known as the *Aufbau principle* (the German word *Aufbau* means "building up"). Electrons in an atom are assigned to shells (defined by the quantum number n) and subshells (defined by the quantum numbers n and ℓ) in order of increasingly higher energy. In this way, the total energy of the atom is as low as possible.

Order of Subshell Energies and Assignments

Quantum theory and the Bohr model state that the energy of the H atom, with a single electron, depends only on the value of n (Equation 6.4, $E_n = -Rhc/n^2$). For atoms with more than one electron, however, the situation is more complex. The order of subshell energies for $n = 1, 2$, and 3 in Figure 7.1a shows that subshell energies in multielectron atoms depend on *both n and ℓ*.

Based on theoretical and experimental studies of electron distributions in atoms, chemists have found there are two general rules that help predict these arrangements:

1. Electrons are assigned to subshells in order of increasing "$n + \ell$" value.
2. For two subshells with the same value of "$n + \ell$," electrons are assigned first to the subshell of lower n.

The following are examples of these rules:

- Electrons are assigned to the 2s subshell $(n + \ell = 2 + 0 = 2)$ before the 2p subshell $(n + \ell = 2 + 1 = 3)$.

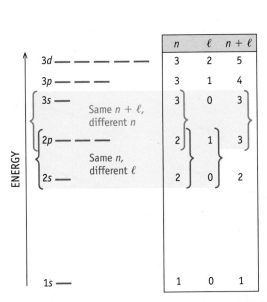

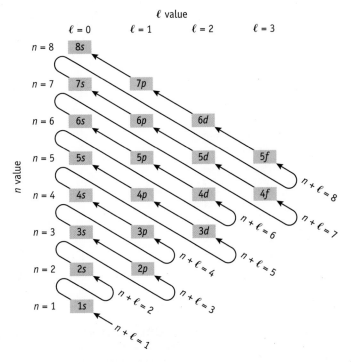

(a) Order of subshell energies in a multielectron atom. Energies of electron shells increase with increasing n, and, within a shell, subshell energies increase with increasing ℓ. (The energy axis is not to scale.) The energy gaps between subshells of a given shell become smaller as n increases.

(b) Subshell filling order. Subshells in atoms are filled in order of increasing $n + \ell$. When two subshells have the same $n + \ell$ value, the subshell of lower n is filled first. To use the diagram, begin at 1s and follow the arrows of increasing $n + \ell$. (Thus, the order of filling is 1s ⟹ 2s ⟹ 2p ⟹ 3s ⟹ 3p ⟹ 4s ⟹ 3d and so on.)

FIGURE 7.1 Subshell energies and filling order in a multielectron atom.

- Electrons are assigned to $2p$ orbitals ($n + \ell = 2 + 1 = 3$) before the $3s$ subshell ($n + \ell = 3 + 0 = 3$). (n for the $2p$ electrons is less than n for the $3s$ electrons.)
- Electrons are assigned to $4s$ orbitals ($n + \ell = 4 + 0 = 4$) before the $3d$ subshell ($n + \ell = 3 + 2 = 5$). ($n + \ell$ is less for $4s$ than for $3d$.)

Figure 7.1b summarizes the assignment of electrons according to increasing $n + \ell$ values. The discussion that follows explores the underlying causes and their consequences and connects atomic electron configurations to the periodic table.

Effective Nuclear Charge, *Z**

The order in which electrons are assigned to subshells in an atom, and many atomic properties, can be rationalized by introducing the concept of **effective nuclear charge (*Z**)**. This is the *net charge experienced by a particular electron in a multielectron atom resulting from a balance of the attractive force of the nucleus and the repulsive forces of other electrons.*

The surface density plot ($4\pi r^2 \psi^2$) for a $2s$ electron for lithium is plotted in Figure 7.2. (Lithium has three protons in the nucleus, two $1s$ electrons in the first shell, and a $2s$ electron in the second shell.) The probability of finding the $2s$ electron (recorded on the vertical axis) changes as one moves away from the nucleus (horizontal axis). The region in which the two $1s$ electrons have their highest probability is shaded in this figure. Observe that the $2s$ electron wave occurs partly within the region of space occupied by $1s$ electrons. Chemists say that the $2s$ orbital *penetrates* the region defining the $1s$ orbital.

At a large distance from the nucleus, the lithium $2s$ electron will experience a $+1$ charge, the net effect of the two $1s$ electrons (total charge $= -2$) and the nucleus ($+3$ charge.) The $1s$ electrons are said to *screen* the $2s$ electron from experiencing the full nuclear charge. However, this screening of the nuclear charge varies with the distance of the $2s$ electron from the nucleus. As the $2s$ electron wave penetrates the $1s$ electron region, it experiences an increasingly higher net positive

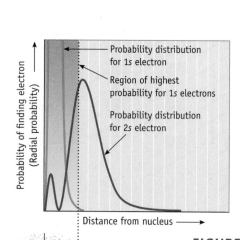

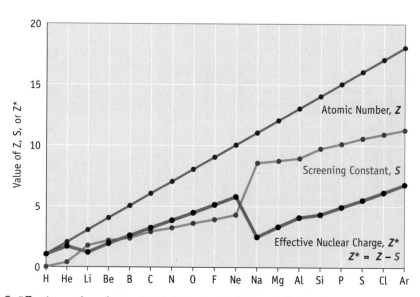

FIGURE 7.2 Effective nuclear charge, *Z**. *(left)* The two $1s$ electrons of lithium have their highest probability in the shaded region, but this region is penetrated by the $2s$ electron (whose approximate surface density plot is shown here). As the $2s$ electron penetrates the $1s$ region, however, the $2s$ electron experiences a larger and larger positive charge, to a maximum of $+3$. On average, the $2s$ electron experiences a charge, called the *effective nuclear charge (Z*)*, which is smaller than $+3$ but greater than $+1$. *(above)* The value of Z^* *for the highest occupied orbital* is given by $(Z - S)$, where Z is the atomic number and S is the screening constant. (S reflects how much the inner electrons shield or screen the outermost electron from the nucleus.) Notice that S increases greatly on going from Ne in the second period to Na in the third period.

charge. Very near the nucleus, the 1s electrons do not effectively screen the electron from the nucleus, and the 2s electron experiences a charge close to +3. Figure 7.2 shows that a 2s electron has some probability of being inside and outside the region occupied by the 1s electrons. Thus, on average, a 2s electron experiences a positive charge greater than +1 but much smaller than +3. The *average* charge experienced by the electron is called the *effective nuclear charge* (Z^*). In the case of the 2s electron in Li, the value of Z^* is 1.28 (Table 7.2).

In the hydrogen atom, with only one electron, the 2s and 2p subshells have the same energy. However, in atoms with two or more electrons, the energies of the 2s and 2p subshells are different. Why should this be true? It is observed that the relative extent to which an outer electron penetrates inner orbitals occurs in the order $s > p > d > f$. Thus, the effective nuclear charge experienced by electrons in a multielectron atom is in the order $ns > np > nd > nf$. The values of Z^* for s and p electrons for the second- and third-period elements (Table 7.2) illustrate this. In each case, Z^* is greater for s electrons than for p electrons. In a given shell, s electrons always have a lower energy than p electrons; p electrons have a lower energy than d electrons, and d electrons have a lower energy than f electrons. A consequence of this is that subshells within an electron shell are filled in the order ns before np before nd before nf. (See *A Closer Look: Orbital Energies, Z*, and Electron Configurations*, page 244.)

With this background on the order of shell and subshell energies and filling order, we turn to the periodic table as a guide to electron arrangements in atoms.

Table 7.2 Effective Nuclear Charges, Z*, for n = 2 Elements

Atom	$Z^*(2s)$	$Z^*(2p)$
Li	1.28	
B	2.58	2.42
C	3.22	3.14
N	3.85	3.83
O	4.49	4.45
F	5.13	5.10

REVIEW & CHECK FOR SECTION 7.2

To which of the subshells should an electron be assigned first in each of the following pairs?

(a) 4s or 4p   (b) 5d or 6s   (c) 4f or 5s

7.3 Electron Configurations of Atoms

Arrangements of electrons in the elements up to 109—their **electron configurations**—are given in Table 7.3. Specifically, these are the ground state electron configurations, where electrons are found in the shells, subshells, and orbitals that result in the lowest energy for the atom. In general, electrons are assigned to orbitals in order of increasing $n + \ell$. The emphasis here, however, will be to connect the configurations of the elements with their positions in the periodic table (Figure 7.3).

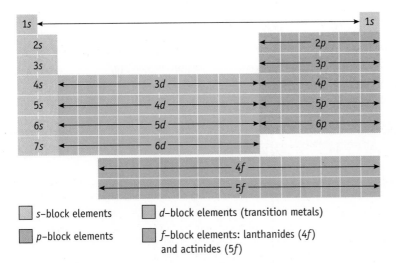

s–block elements
p–block elements
d–block elements (transition metals)
f–block elements: lanthanides (4f) and actinides (5f)

FIGURE 7.3 Electron configurations and the periodic table. The periodic table can serve as a guide in determining the order of filling of atomic orbitals. As one moves from left to right in a period, electrons are assigned to the indicated orbitals. (See Table 7.3.)

Table 7.3 Ground State Electron Configurations

Z	Element	Configuration	Z	Element	Configuration	Z	Element	Configuration
1	H	$1s^1$	37	Rb	$[Kr]5s^1$	74	W	$[Xe]4f^{14}5d^46s^2$
2	He	$1s^2$	38	Sr	$[Kr]5s^2$	75	Re	$[Xe]4f^{14}5d^56s^2$
3	Li	$[He]2s^1$	39	Y	$[Kr]4d^15s^2$	76	Os	$[Xe]4f^{14}5d^66s^2$
4	Be	$[He]2s^2$	40	Zr	$[Kr]4d^25s^2$	77	Ir	$[Xe]4f^{14}5d^76s^2$
5	B	$[He]2s^22p^1$	41	Nb	$[Kr]4d^45s^1$	78	Pt	$[Xe]4f^{14}5d^96s^1$
6	C	$[He]2s^22p^2$	42	Mo	$[Kr]4d^55s^1$	79	Au	$[Xe]4f^{14}5d^{10}6s^1$
7	N	$[He]2s^22p^3$	43	Tc	$[Kr]4d^55s^2$	80	Hg	$[Xe]4f^{14}5d^{10}6s^2$
8	O	$[He]2s^22p^4$	44	Ru	$[Kr]4d^75s^1$	81	Tl	$[Xe]4f^{14}5d^{10}6s^26p^1$
9	F	$[He]2s^22p^5$	45	Rh	$[Kr]4d^85s^1$	82	Pb	$[Xe]4f^{14}5d^{10}6s^26p^2$
10	Ne	$[He]2s^22p^6$	46	Pd	$[Kr]4d^{10}$	83	Bi	$[Xe]4f^{14}5d^{10}6s^26p^3$
11	Na	$[Ne]3s^1$	47	Ag	$[Kr]4d^{10}5s^1$	84	Po	$[Xe]4f^{14}5d^{10}6s^26p^4$
12	Mg	$[Ne]3s^2$	48	Cd	$[Kr]4d^{10}5s^2$	85	At	$[Xe]4f^{14}5d^{10}6s^26p^5$
13	Al	$[Ne]3s^23p^1$	49	In	$[Kr]4d^{10}5s^25p^1$	86	Rn	$[Xe]4f^{14}5d^{10}6s^26p^6$
14	Si	$[Ne]3s^23p^2$	50	Sn	$[Kr]4d^{10}5s^25p^2$	87	Fr	$[Rn]7s^1$
15	P	$[Ne]3s^23p^3$	51	Sb	$[Kr]4d^{10}5s^25p^3$	88	Ra	$[Rn]7s^2$
16	S	$[Ne]3s^23p^4$	52	Te	$[Kr]4d^{10}5s^25p^4$	89	Ac	$[Rn]6d^17s^2$
17	Cl	$[Ne]3s^23p^5$	53	I	$[Kr]4d^{10}5s^25p^5$	90	Th	$[Rn]6d^27s^2$
18	Ar	$[Ne]3s^23p^6$	54	Xe	$[Kr]4d^{10}5s^25p^6$	91	Pa	$[Rn]5f^26d^17s^2$
19	K	$[Ar]4s^1$	55	Cs	$[Xe]6s^1$	92	U	$[Rn]5f^36d^17s^2$
20	Ca	$[Ar]4s^2$	56	Ba	$[Xe]6s^2$	93	Np	$[Rn]5f^46d^17s^2$
21	Sc	$[Ar]3d^14s^2$	57	La	$[Xe]5d^16s^2$	94	Pu	$[Rn]5f^67s^2$
22	Ti	$[Ar]3d^24s^2$	58	Ce	$[Xe]4f^15d^16s^2$	95	Am	$[Rn]5f^77s^2$
23	V	$[Ar]3d^34s^2$	59	Pr	$[Xe]4f^36s^2$	96	Cm	$[Rn]5f^76d^17s^2$
24	Cr	$[Ar]3d^54s^1$	60	Nd	$[Xe]4f^46s^2$	97	Bk	$[Rn]5f^97s^2$
25	Mn	$[Ar]3d^54s^2$	61	Pm	$[Xe]4f^56s^2$	98	Cf	$[Rn]5f^{10}7s^2$
26	Fe	$[Ar]3d^64s^2$	62	Sm	$[Xe]4f^66s^2$	99	Es	$[Rn]5f^{11}7s^2$
27	Co	$[Ar]3d^74s^2$	63	Eu	$[Xe]4f^76s^2$	100	Fm	$[Rn]5f^{12}7s^2$
28	Ni	$[Ar]3d^84s^2$	64	Gd	$[Xe]4f^75d^16s^2$	101	Md	$[Rn]5f^{13}7s^2$
29	Cu	$[Ar]3d^{10}4s^1$	65	Tb	$[Xe]4f^96s^2$	102	No	$[Rn]5f^{14}7s^2$
30	Zn	$[Ar]3d^{10}4s^2$	66	Dy	$[Xe]4f^{10}6s^2$	103	Lr	$[Rn]5f^{14}6d^17s^2$
31	Ga	$[Ar]3d^{10}4s^24p^1$	67	Ho	$[Xe]4f^{11}6s^2$	104	Rf	$[Rn]5f^{14}6d^27s^2$
32	Ge	$[Ar]3d^{10}4s^24p^2$	68	Er	$[Xe]4f^{12}6s^2$	105	Db	$[Rn]5f^{14}6d^37s^2$
33	As	$[Ar]3d^{10}4s^24p^3$	69	Tm	$[Xe]4f^{13}6s^2$	106	Sg	$[Rn]5f^{14}6d^47s^2$
34	Se	$[Ar]3d^{10}4s^24p^4$	70	Yb	$[Xe]4f^{14}6s^2$	107	Bh	$[Rn]5f^{14}6d^57s^2$
35	Br	$[Ar]3d^{10}4s^24p^5$	71	Lu	$[Xe]4f^{14}5d^16s^2$	108	Hs	$[Rn]5f^{14}6d^67s^2$
36	Kr	$[Ar]3d^{10}4s^24p^6$	72	Hf	$[Xe]4f^{14}5d^26s^2$	109	Mt	$[Rn]5f^{14}6d^77s^2$
			73	Ta	$[Xe]4f^{14}5d^36s^2$			

*This table follows the general convention of writing the orbitals in order of increasing n when writing electron configurations. For a given n, the subshells are listed in order of increasing ℓ.

Electron Configurations of the Main Group Elements

Hydrogen, the first element in the periodic table, has one electron in a $1s$ orbital. One way to depict its electron configuration is with the orbital box diagram used earlier, but an alternative and more frequently used method is the *spdf* notation. Using this method, the electron configuration of H is $1s^1$, read "one *s* one." This indicates that there is one electron (indicated by the superscript) in the $1s$ orbital.

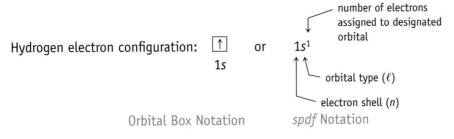

Hydrogen electron configuration:

number of electrons assigned to designated orbital

$1s^1$

orbital type (ℓ)

electron shell (n)

Orbital Box Notation   *spdf* Notation

Lithium. Lithium-ion batteries are increasingly used in consumer products. All Group 1A elements such as lithium share an outer electron configuration of ns^1.

Lithium (Li) and Other Elements of Group 1A

Lithium, with three electrons, is the first element in the second period of the periodic table. The first two electrons are in the $1s$ subshell, and the third electron must be in the $2s$ subshell of the $n = 2$ shell. The *spdf* notation, $1s^2 2s^1$, is read "one *s* two, two *s* one."

Lithium: *spdf* notation $1s^2 2s^1$

Box notation $1s$ $2s$ $2p$

Electron configurations are often written in abbreviated form by writing in brackets the symbol for the noble gas preceding the element (called the **noble gas notation**) and then indicating any electrons beyond those in the noble gas by using *spdf* or orbital box notation. In lithium, the arrangement preceding the $2s$ electron is the electron configuration of the noble gas helium. Instead of writing out $1s^2 2s^1$ for lithium's configuration, it can be written as $[\text{He}]2s^1$.

The electrons included in the noble gas notation are often referred to as the **core electrons** of the atom. The core electrons can generally be ignored when considering the chemistry of an element. The electrons beyond the core electrons—the $2s^1$ electron in the case of lithium—are called **valence electrons**; these are the electrons that determine the chemical properties of an element.

All the elements of Group 1A have one electron assigned to an s orbital of the nth shell, for which n is the number of the period in which the element is found (Figure 7.3). For example, potassium is the first element in the $n = 4$ row (the fourth period), so potassium has the electron configuration of the element preceding it in the table (Ar) plus a final electron assigned to the $4s$ orbital: $[\text{Ar}]4s^1$.

Beryllium (Be) and Other Elements of Group 2A

All elements of Group 2A have electron configurations of [electrons of preceding noble gas] ns^2, where n is the period in which the element is found in the periodic table. Beryllium, for example, has two electrons in the $1s$ orbital plus two additional electrons.

Beryllium: *spdf* notation $1s^2 2s^2$ or $[\text{He}]2s^2$

Box notation $1s$ $2s$ $2p$

Beryllium. Beryllium-copper alloy is used in the oil and gas industry and in electronics. All Group 2A elements like beryllium share an outer electron configuration of ns^2.

Because all the elements of Group 1A have the valence electron configuration ns^1 and those in Group 2A have ns^2, these elements are called **s-block elements**.

Boron. Boron carbide, B_4C, is used in body armor. All Group 3A elements share an outer electron configuration of ns^2np^1.

Boron (B) and Other Elements of Group 3A

Boron (Group 3A) is the first element in the block of elements on the right side of the periodic table. Because the $1s$ and $2s$ orbitals are filled in a boron atom, the fifth electron must be assigned to a $2p$ orbital.

Boron:	*spdf* notation	$1s^2 2s^2 2p^1$	or	$[He]2s^2 2p^1$

Box notation ⊞ ⊞ ⊡ □ □
$1s$ $2s$ $2p$

Elements from Group 3A through Group 8A are often called the **p-block elements**. All have the outer shell configuration ns^2np^x, where x varies from 1 to 6. The elements in Group 3A, for example, have two s electrons and one p electron (ns^2np^1) in their outer shells.

Carbon (C) and Other Elements of Group 4A

Carbon (Group 4A) is the second element in the p block, with two electrons assigned to the $2p$ orbitals. You can write the electron configuration of carbon by referring to the periodic table: Starting at H and moving from left to right across the successive periods, you write $1s^2$ to reach the end of period 1 and then $2s^2$ and finally $2p^2$ to bring the electron count to six. For carbon to be in its lowest energy (ground) state, these electrons must be assigned to different p orbitals, and both will have the same spin direction.

Carbon:	*spdf* notation	$1s^2 2s^2 2p^2$	or	$[He]2s^2 2p^2$

Box notation ⊞ ⊞ ⊡ ⊡ □
$1s$ $2s$ $2p$

● **Hund's Rule and Exchange Energy**
The reason a configuration with the maximum number of unpaired electrons with parallel spins is more stable is due to the *exchange energy*, a concept that comes out of quantum mechanics. Due to the exchange interaction, each pair of unpaired electrons with parallel spins leads to a lowering of the overall energy of an atom. The first of the configurations below is more stable than the second by an amount of energy called the *exchange energy*.

When assigning electrons to p, d, or f orbitals, each successive electron is assigned to a different orbital of the subshell, and each electron has the same spin as the previous one, until the subshell is half full. Additional electrons must then be assigned to half-filled orbitals. This procedure follows **Hund's rule**, which states that *the most stable arrangement of electrons is that with the maximum number of unpaired electrons, all with the same spin direction.*

All elements in Group 4A have similar outer shell configurations, ns^2np^2, where n is the period in which the element is located in the periodic table.

Nitrogen (N) and Oxygen (O) and Elements of Groups 5A and 6A

Nitrogen (Group 5A) has five valence electrons. Besides the two $2s$ electrons, it has three electrons, all with the same spin, in three different $2p$ orbitals.

Nitrogen:	*spdf* notation	$1s^2 2s^2 2p^3$	or	$[He]2s^2 2p^3$

Box notation ⊞ ⊞ ⊡ ⊡ ⊡
$1s$ $2s$ $2p$

Oxygen (Group 6A) has six valence electrons. Two of these six electrons are assigned to the $2s$ orbital, and the other four electrons are assigned to $2p$ orbitals.

Oxygen:	*spdf* notation	$1s^2 2s^2 2p^4$	or	$[He]2s^2 2p^4$

Box notation ⊞ ⊞ ⊞ ⊡ ⊡
$1s$ $2s$ $2p$

The fourth $2p$ electron must pair up with one already present. It makes no difference to which orbital this electron is assigned (the $2p$ orbitals all have the same energy), but it must have a spin opposite to the other electron already assigned to that orbital so that each electron has a different set of quantum numbers.

All elements in Group 5A have an outer shell configuration of ns^2np^3, and all elements in Group 6A have an outer shell configuration of ns^2np^4, where n is the period in which the element is located in the periodic table.

Fluorine (F) and Neon (Ne) and Elements of Groups 7A and 8A

Fluorine (Group 7A) has seven electrons in the $n = 2$ shell. Two of these electrons occupy the $2s$ subshell, and the remaining five electrons occupy the $2p$ subshell.

Fluorine: *spdf* notation $1s^22s^22p^5$ or $[He]2s^22p^5$

Box notation   [⇅] [⇅] [⇅][⇅][↑]
                1s   2s   2p

All halogen atoms have similar outer shell configurations, ns^2np^5, where n is the period in which the element is located.

Like all the elements in Group 8A, neon is a noble gas. The Group 8A elements (except helium) have eight electrons in the shell of highest n value, so all have the outer shell configuration ns^2np^6, where n is the period in which the element is found. That is, all the noble gases have filled ns and np subshells. The nearly complete chemical inertness of the noble gases is associated with this electron configuration.

Neon: *spdf* notation $1s^22s^22p^6$ or $[He]2s^22p^6$

Box notation   [⇅] [⇅] [⇅][⇅][⇅]
                1s   2s   2p

Elements of Period 3

The elements of the third period have valence electron configurations similar to those of the second period, except that the preceding noble gas is neon and the valence shell is the third energy level. For example, silicon has four electrons and a neon core. Because it is the second element in the p block, it has two electrons in $3p$ orbitals. Thus, its electron configuration is

Silicon: *spdf* notation $1s^22s^22p^63s^23p^2$ or $[Ne]3s^23p^2$

Box notation   [⇅] [⇅] [⇅][⇅][⇅] [⇅] [↑][↑][ ]
                1s   2s   2p   3s   3p

EXAMPLE 7.1   Electron Configurations

Problem Give the electron configuration of sulfur, using the *spdf*, noble gas, and orbital box notations.

What Do You Know? From the periodic table: sulfur, atomic number 16, is the sixth element in the third period ($n = 3$) and is in the p-block. You need to know the order of filling (Figure 7.1).

Strategy

• For the *spdf* and orbital box notations: Place the 16 electrons for sulfur into orbitals, based on the order of filling.

• For the noble gas notation: The first 10 electrons are identified by the symbol of the previous noble gas, Ne. The remaining 6 electrons are assigned to the $3s$ and $3p$ subshells. Make sure Hund's rule is followed in the box notation.

Fluorine. Fluorine is found as the fluoride ion in the mineral fluorite (CaF_2), sometimes called *fluorspar*. The mineral, which is the state mineral of Illinois, comes in many colors. All Group 7A elements such as fluorine share an outer electron configuration of ns^2np^5.

Silicon. Si is the fourth element in the third period. The earth's crust is largely composed of silicon-containing minerals. This photo shows some elemental silicon and a thin wafer of silicon on which are etched electronic circuits.

Sulfur. Sulfur is widely distributed on earth. Like all Group 6A elements, it has the outer electron configuration ns^2np^4.

Solution Sulfur, atomic number 16, is the sixth element in the third period ($n = 3$) and is in the p block. The last six electrons assigned to the atom, therefore, have the configuration $3s^23p^4$. These are preceded by the completed shells $n = 1$ and $n = 2$, the electron arrangement for Ne.

Complete *spdf* notation:	$1s^22s^22p^63s^23p^4$
spdf with noble gas notation:	$[Ne]3s^23p^4$
Orbital box notation:	$[Ne]$ ⊞ ⊞⊞⊞
	3s 3p

Think about Your Answer The noble gas and box notations convey slightly different information. The noble gas notation separates out the core electrons (which are not involved in chemical reactions). The box notation allows a quick identification of the number of unpaired electron.

Check Your Understanding

(a)   What element has the configuration $1s^22s^22p^63s^23p^5$?

(b)   Using *spdf* notation and a box diagram, show the electron configuration of phosphorus.

EXAMPLE 7.2   Electron Configurations and Quantum Numbers

Problem Write the electron configuration for Al using the noble gas notation, and give a set of quantum numbers for each of the electrons with $n = 3$ (the valence electrons).

What Do You Know? The periodic table informs you that Al (atomic number 13) is the third element after the noble gas neon.

Strategy Aluminum is the third element in the third period. It therefore has three electrons with $n = 3$. Two of the electrons are assigned to $3s$, and the remaining electron is assigned to $3p$.

Solution The element is preceded by the noble gas neon, so the electron configuration is $[Ne]3s^23p^1$. Using the box notation, the configuration is

Aluminum configuration: $[Ne]$ ⊞ ⊞⊞⊞
                              3s   3p

The possible sets of quantum numbers for the two $3s$ electrons are

	n	ℓ	m_ℓ	m_s
For ↑	3	0	0	$+\frac{1}{2}$
For ↓	3	0	0	$-\frac{1}{2}$

For the single $3p$ electron, one of six possible sets is $n = 3$, $\ell = 1$, $m_\ell = +1$, and $m_s = +\frac{1}{2}$.

Think about Your Answer It is only a convention that assigns $+\frac{1}{2}$ to an "up" arrow for electron spin. Two electrons in the same orbital have opposite electron spins. Which is $+\frac{1}{2}$ and which is $-\frac{1}{2}$ is not important.

Check Your Understanding

Write one possible set of quantum numbers for the valence electrons of calcium.

Electron Configurations of the Transition Elements

The elements of the fourth through the seventh periods use d and f subshells, in addition to s and p subshells, to accommodate electrons (see Figure 7.3 and Tables 7.3 and 7.4). Elements whose atoms are filling d subshells are called

transition elements. Those elements for which atoms are filling f subshells are sometimes called the **inner transition elements** or, more usually, the **lanthanides** (filling $4f$ orbitals, Ce through Lu) and **actinides** (filling $5f$ orbitals, Th through Lr).

In a given period in the periodic table, the transition elements are always preceded by two s-block elements. After filling the ns orbital in the period, we begin filling the $(n-1)d$ orbitals. Scandium, the first transition element, has the configuration $[Ar]3d^14s^2$, and titanium follows with $[Ar]3d^24s^2$ (Table 7.4).

The general procedure for assigning electrons would suggest that the configuration of a chromium atom would be $[Ar]3d^44s^2$. The actual configuration, however, has one electron assigned to each of the six available $3d$ and $4s$ orbitals: $[Ar]3d^54s^1$. This configuration has a lower total energy (due to maximizing the number of unpaired electron spins) than the alternative configuration.

Following chromium, atoms of manganese, iron, and nickel have the configurations that would be expected from the order of orbital filling in Figure 7.1. Copper ($[Ar]3d^{10}4s^1$) is the second "exception" in this series; it has a single electron in the $4s$ orbital, and the remaining 10 electrons beyond the argon core are assigned to the $3d$ orbitals. Zinc, with the configuration $[Ar]3d^{10}4s^2$, ends the first transition series.

The fifth period transition elements follow the pattern of the fourth period but have more exceptions to the general rules of orbital filling.

Lanthanides and Actinides

The sixth period includes the lanthanide series. As the first element in the d-block, lanthanum has the configuration $[Xe]5d^16s^2$. The next element, cerium (Ce), is set out in a separate row at the bottom of the periodic table, and it is with the elements in this row (Ce through Lu) that electrons are first assigned to f orbitals. Thus, the configuration of cerium is $[Xe]4f^15d^16s^2$. Moving across the lanthanide series, the pattern continues (although with occasional variations in

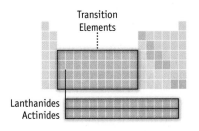

Transition Elements

Lanthanides
Actinides

Lanthanides. These elements have many uses. This is a sample of lanthanum metal.

● **Writing Configurations for Transition Metals** We follow the convention of writing configurations with shells listed in order of increasing n and, within a given shell, writing subshells in order of increasing ℓ. As an alternate representation, some chemists write them as, for example, $[Ar]4s^23d^2$ to reflect the order of orbital filling.

Table 7.4 Orbital Box Diagrams for the Elements Ca Through Zn

		3d	4s
Ca	$[Ar]4s^2$	☐☐☐☐☐	↑↓
Sc	$[Ar]3d^14s^2$	↑☐☐☐☐	↑↓
Ti	$[Ar]3d^24s^2$	↑↑☐☐☐	↑↓
V	$[Ar]3d^34s^2$	↑↑↑☐☐	↑↓
Cr*	$[Ar]3d^54s^1$	↑↑↑↑↑	↑
Mn	$[Ar]3d^54s^2$	↑↑↑↑↑	↑↓
Fe	$[Ar]3d^64s^2$	↑↓ ↑↑↑↑	↑↓
Co	$[Ar]3d^74s^2$	↑↓ ↑↓ ↑↑↑	↑↓
Ni	$[Ar]3d^84s^2$	↑↓ ↑↓ ↑↓ ↑↑	↑↓
Cu*	$[Ar]3d^{10}4s^1$	↑↓ ↑↓ ↑↓ ↑↓ ↑↓	↑
Zn	$[Ar]3d^{10}4s^2$	↑↓ ↑↓ ↑↓ ↑↓ ↑↓	↑↓

*See *A Closer Look: Questions About Transition Element Electron Configurations*, page 246.

Orbital Energies, Z*, and Electron Configurations

The graph below shows how the energies of different atomic orbitals change as we go from one element to the next in the periodic table. Here are some significant observations drawn from the graph.

(1) Generally, as one moves from left to right in any period, the energies of the atomic orbitals decrease. This trend correlates with the changes in Z*, the effective charge experienced by an electron in a given orbital (Figure 7.2). As noted, Z* is a parameter related to the nuclear charge and the shielding of all the other electrons. When moving from one element to the next across each period, the attractive effect of an additional proton in the nucleus outweighs the repulsive effect of an additional electron.

(2) Electrons are classified as either valence electrons or core electrons. We can now see from this graph the obvious distinction in these groups. Valence electrons are found in orbitals that have less negative energies. In contrast, the core electrons are in orbitals that have very low energy, because they experience a much higher Z*. Furthermore, as one moves from one period to the next period (for example, from Ar in period 3 to K in period 4) the energies of core orbitals plummet. The very low energy of electrons in these orbitals rules out the ability of these electrons to participate in chemical reactions.

(3) Notice also the orbital energies of the first transition series. Each step going across the transition series involves adding a proton to the nucleus and a 3d electron to an inner shell. The 4s energy level decreases only slightly in this series of elements, reflecting the fact that the additional 3d electrons effectively shield a 4s electron from the higher charge resulting from addition of a proton. The additional nuclear charge has a greater effect on the 3d orbitals, and their energies decrease to a greater extent across the transition series. When the end of the transition series is reached, the 3d electrons become part of the core and their energies plummet.

(4) Finally, note the exceptional behavior of Cr and Cu in this sequence. These elements have unexpected electron configurations (page 246), arising because another factor, exchange energy (page 240), is a contributor to overall energy in the system.

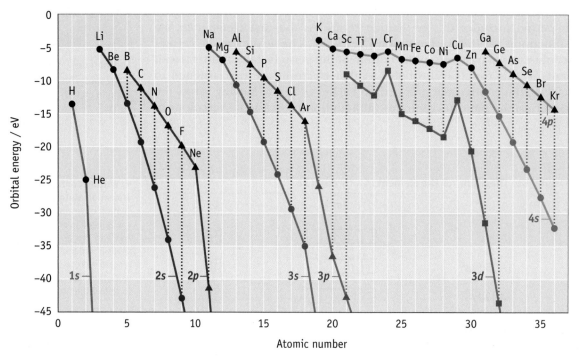

Variation in orbital energies with atomic number. Note that the energies are in electron-volts (eV), where 1 eV = 1.602 × 10⁻¹⁹ J). (Figure redrawn from *Chemical Structure and Reactivity*, J. Keeler and P. Wothers, Oxford, 2008; used with permission.)

occupancy of the 5d and 4f orbitals). The lanthanide series ends with 14 electrons being assigned to the seven 4f orbitals in the last element, lutetium (Lu, [Xe]4f¹⁴5d¹6s²) (see Table 7.3).

The seventh period also includes an extended series of elements utilizing f orbitals, the actinides. Actinium has the configuration [Rn]6d¹7s². The next element is thorium (Th), which is followed by protactinium (Pa) and uranium (U). The electron configuration of uranium is [Rn]5f³6d¹7s².

EXAMPLE 7.3   Electron Configurations of the Transition Elements

Problem Using the *spdf* and noble gas notations, give electron configurations for (a) technetium, Tc, and (b) osmium, Os.

What Do You Know? Base your answer on the positions of the elements in the periodic table. Technetium is the seventh element in period 5 and osmium is the 22nd element in the 6th period.

Strategy For each element, find the preceding noble gas, and then note the number of *s*, *p*, *d*, and *f* electrons that lead from the noble gas to the element.

Solution

(a) Technetium, Tc: The noble gas that precedes Tc is krypton, Kr, at the end of the $n = 4$ row. After 36 electrons are assigned to the core as [Kr], seven electrons remain. Two of these electrons are in the 5*s* orbital, and the remaining five are in 4*d* orbitals. Therefore, the technetium configuration is $[Kr]4d^55s^2$.

(b) Osmium, Os: Osmium is a sixth-period element and the 22nd element following the noble gas xenon. Of the 22 electrons to be added after the Xe core, two are assigned to the 6*s* orbital and 14 to 4*f* orbitals. The remaining six are assigned to 5*d* orbitals. Thus, the osmium configuration is $[Xe]4f^{14}5d^66s^2$.

Think about Your Answer In general, it is best to use the periodic table as a guide to determining electron configurations. Keep in mind, however, that some exceptions exist, particularly in the heavier transition elements and the lanthanides and actinides.

Check Your Understanding

Using the periodic table and without looking at Table 7.3, write electron configurations for the following elements:

(a) P   (c) Zr   (e) Pb

(b) Zn   (d) In   (f) U

Use the *spdf* and noble gas notations. When you have finished, check your answers with Table 7.3.

REVIEW & CHECK FOR SECTION 7.3

1. Find Se in the periodic table. What is the electron configuration beyond its noble gas core?

   (a) $4p^4$ (b) $4s^24p^4$ (c) $3d^{10}4s^24p^4$ (d) [Ar]

2. Based on electron configurations, which of the following atoms has the smallest number of unpaired electrons?

   (a) Cr   (b) Fe   (c) Ni   (d) Mn

7.4 Electron Configurations of Ions

The chemistry of the elements is often the chemistry of their anions and cations, so we want to determine their electron configurations. To form a cation from a neutral atom, one or more of the valence electrons is removed. Electrons are always removed first from the electron shell of highest *n*. If several subshells are present within the *n*th shell, the electron or electrons of maximum ℓ are removed. Thus, a sodium ion is formed by removing the $3s^1$ electron from the Na atom,

$$Na: [1s^22s^22p^63s^1] \rightarrow Na^+: [1s^22s^22p^6] + e^-$$

and Ge^{2+} is formed by removing two 4*p* electrons from a germanium atom,

$$Ge: [Ar]3d^{10}4s^24p^2 \rightarrow Ge^{2+}: [Ar]3d^{10}4s^2 + 2 e^-$$

Questions about Transition Element Electron Configurations

Why don't all of the $n = 3$ subshells fill before beginning to fill the $n = 4$ subshells? Why is scandium's configuration $[Ar]3d^1 4s^2$ and not $[Ar]3d^3$? Why is chromium's configuration $[Ar]3d^5 4s^1$ and not $[Ar]3d^4 4s^2$? As we search for explanations keep in mind that the most stable configuration will be that with the lowest *total* energy.

From scandium to zinc the energies of the $3d$ orbitals are always lower than the energy of the $4s$ orbital (see *A Closer Look: Orbital Energies, Z*, and Electron Configurations*, page 244), so for scandium the configuration $[Ar]3d^3$ would seem to be preferred. One way to understand why it is not is to consider the effect of electron–electron repulsion in $3d$ and $4s$ orbitals. The most stable configuration will be the one that most effectively minimizes electron–electron repulsions.

Plots of $3d$ and $4s$ orbitals (as was done for $1s$ and $2s$ in Figure 7.2) show the most probable distance of a $3d$ electron from the nucleus is less than that for a $4s$ electron. Being closer

to the nucleus, the $3d$ orbitals are more compact than the $4s$ orbital. This means the $3d$ electrons are closer together, so two $3d$ electrons would repel each other more strongly than two $4s$ electrons, for example. A consequence is that placing electrons in the slightly higher energy $4s$ orbital lessens the effect of electron–electron repulsions and lowers the overall energy of the atom.

For chromium the $4s$ and $3d$ orbital energies (see *A Closer Look: Orbital Energies, Z*, and Electron Configurations*, page 244) are closer together than other 4th period transition elements. Again, thinking in terms of total energy, it is the effect of exchange energy (page 240) that tips the balance and causes the configuration $[Ar]3d^5 4s^1$, in which all electron spins are the same, to be most stable.

For more on these questions see the following papers in the *Journal of Chemical Education* and Chapter 9 of the book *The Periodic Table* by E. Scerri.

- E. Scerri, *The Periodic Table*, Oxford, 2007.
- F. L. Pilar, "4s is Always Above 3d," *Journal of Chemical Education*, Vol. 55, pp. 1–6, 1978.
- E. R. Scerri, "Transition Metal Configurations and Limitations of the Orbital Approximation," *Journal of Chemical Education*, Vol. 66, pp. 481–483, 1989.
- L. G. Vanquickenborne, K. Pierloot, and D. Devoghel, "Transition Metals and the Aufbau Principle," *Journal of Chemical Education*, Vol. 71, pp. 469–471, 1994.
- M. P. Melrose and E. R. Scerri, "Why the 4s Orbital is Occupied before the 3d," *Journal of Chemical Education*, Vol. 73, pp. 498–503, 1996.
- W. H. E. Schwarz, "The Full Story of the Electron Configurations of the Transition Elements," *Journal of Chemical Education*, Vol. 87, pp. 444–448, 2010.

The same general rule applies to transition metal atoms. This means, for example, that the titanium(II) cation has the configuration $[Ar]3d^2$:

$$Ti: [Ar]3d^2 4s^2 \rightarrow Ti^{2+}: [Ar]3d^2 + 2\ e^-$$

Iron(II) and iron(III) cations have the configurations $[Ar]3d^6$ and $[Ar]3d^5$, respectively:

$$Fe: [Ar]3d^6 4s^2 \rightarrow Fe^{2+}: [Ar]3d^6 + 2\ e^-$$

$$Fe^{2+}: [Ar]3d^6 \rightarrow Fe^{3+}: [Ar]3d^5 + e^-$$

Note that in the ionization of transition metals the ns electrons are always lost before $(n - 1)d$ electrons. All transition metals lose their ns electrons first, and the cations formed all have electron configurations of the general type [noble gas core] $(n - 1)d^x$. This point is important to remember because the magnetic properties of transition metal cations are determined by the number of unpaired electrons in d orbitals. For example, the Fe^{3+} ion is paramagnetic to the extent of five unpaired electrons (See Figures 6.16, 7.4, and 7.5 and *A Closer Look: Paramagnetism and Ferromagnetism*, page 227). If three $3d$ electrons had been removed instead of two $4s$ electrons and one $3d$ electron, the Fe^{3+} ion would still be paramagnetic but only to the extent of three unpaired electrons.

FIGURE 7.4 Formation of iron(III) chloride. Iron reacts with chlorine (Cl_2) to produce $FeCl_3$. The paramagnetic Fe^{3+} ion has the configuration $[Ar]3d^5$.

EXAMPLE 7.4   Configurations of Transition Metal Ions

Problem Give the electron configurations for Cu, Cu^+ and Cu^{2+}. Is either of the ions paramagnetic? How many unpaired electrons does each have?

What Do You Know? When forming a transition metal ion, the outermost ns electrons are lost first, followed by the $(n - 1)d$ electrons.

Strategy Start with the electron configuration of copper (Table 7.4). It is always observed that the *ns* electron or electrons are ionized first, followed by one or more (*n* − 1) *d* electrons.

Solution Copper has only one electron in the 4*s* orbital and 10 electrons in 3*d* orbitals:

Cu: [Ar]3*d*^{10}4*s*^{1} $\boxed{\uparrow\downarrow}\boxed{\uparrow\downarrow}\boxed{\uparrow\downarrow}\boxed{\uparrow\downarrow}\boxed{\uparrow\downarrow}$ $\boxed{\uparrow}$
                                                    3*d*   4*s*

When copper is oxidized to Cu$^+$, the 4s electron is lost.

Cu$^+$:   [Ar]3*d*^{10} $\boxed{\uparrow\downarrow}\boxed{\uparrow\downarrow}\boxed{\uparrow\downarrow}\boxed{\uparrow\downarrow}\boxed{\uparrow\downarrow}$ $\boxed{}$
                                            3*d*   4*s*

The copper(II) ion is formed from copper(I) by removal of one of the 3*d* electrons.

Cu^{2+}:   [Ar]3*d*^{9} $\boxed{\uparrow\downarrow}\boxed{\uparrow\downarrow}\boxed{\uparrow\downarrow}\boxed{\uparrow\downarrow}\boxed{\uparrow}$ $\boxed{}$
                                            3*d*   4*s*

A copper(II) ion (Cu^{2+}) has one unpaired electron, so it is paramagnetic. In contrast, Cu$^+$, with no unpaired electrons, is diamagnetic.

Think about Your Answer Note that the electron configuration for Cu is [Ar]3*d*^{10}4*s*^{1} and not [Ar]3*d*^{9}4*s*^{2}. Copper is one of two exceptions in the first transition series in the order of filling (Cr is the other) (See Table 7.4).

Check Your Understanding

Depict the electron configurations for V^{2+}, V^{3+}, and Co^{3+}. Use orbital box diagrams and the noble gas notation. Are any of the ions paramagnetic? If so, give the number of unpaired electrons.

REVIEW & CHECK FOR SECTION 7.4

1. Three of the four ions shown below have the same electron configuration but the fourth does not. Identify the one that doesn't match.

    (a) Fe^{3+} (b) Cr$^+$ (c) Mn^{2+} (d) Fe^{2+}

2. Which of the following ions has the largest number of unpaired electrons?

    (a) Cr^{3+} (b) Cu^{2+} (c) Co^{3+} (d) Ni^{2+}

7.5 Atomic Properties and Periodic Trends

Once electron configurations were understood, chemists realized that *similarities in properties of the elements are the result of similar valence shell electron configurations.* An objective of this section is to describe how atomic electron configurations are related to some of the physical and chemical properties of the elements and why those properties, such as atomic size, change in a predictable manner when moving down groups and across periods.

Atomic Size

The sizes of atoms (Figure 7.6) influence many aspects of their properties and reactivity. Size can determine the number of atoms that may surround and be bound to a central atom and can be a determining factor in the shape of a molecule. As you will see in the next chapter in particular, the shapes of molecules are important in how they function.

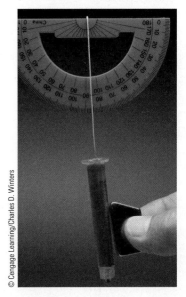

(a) A sample of iron(III) oxide is packed into a plastic tube and suspended from a thin nylon filament.

(b) When a powerful magnet is brought near, the paramagnetic iron(III) ions in Fe$_2$O$_3$ cause the sample to be attracted to the magnet.

FIGURE 7.5 Paramagnetism of transition metals and their compounds. The magnet is made of neodymium, iron, and boron [Nd$_2$Fe$_{14}$B]. These powerful magnets are used in acoustic speakers.

Module 11: Periodic Trends covers concepts in this section.

FIGURE 7.6 Atomic radii in picometers for main group elements. 1 pm = 1×10^{-12} m = 1×10^{-3} nm. (*Data taken from J. Emsley: The Elements,* Clarendon Press, Oxford, *1998, 3rd ed.*)

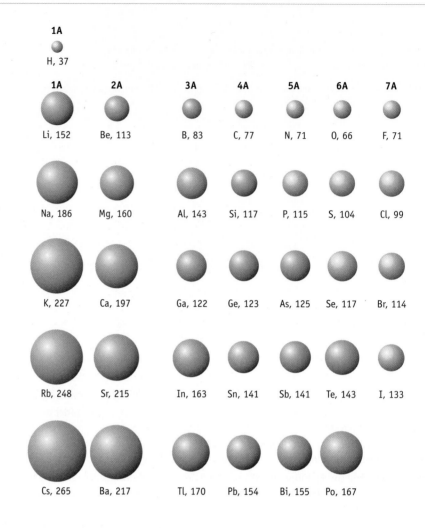

1A
H, 37

1A	2A	3A	4A	5A	6A	7A
Li, 152	Be, 113	B, 83	C, 77	N, 71	O, 66	F, 71
Na, 186	Mg, 160	Al, 143	Si, 117	P, 115	S, 104	Cl, 99
K, 227	Ca, 197	Ga, 122	Ge, 123	As, 125	Se, 117	Br, 114
Rb, 248	Sr, 215	In, 163	Sn, 141	Sb, 141	Te, 143	I, 133
Cs, 265	Ba, 217	Tl, 170	Pb, 154	Bi, 155	Po, 167	

☐ Main Group Metals
☐ Transition Metals
☐ Metalloids
☐ Nonmetals

© Ambient Images Inc./Alamy

Atom size is also geochemically and technologically important. For example, one atom may take the place of a similarly sized atom in an alloy, a solid solution of various elements in a metallic matrix. Our technology-based industrial society depends on these materials. Substituting one element for another of similar size can often lead to alloys with different properties. But this is nothing new: humans made implements from iron meteorites (Figure 7.7) over 5000 years ago. Iron meteorites are an alloy composed of about 90% iron and 10% nickel, two elements of very similar size.

We know an orbital has no sharp boundary (◄ Figure 6.12a), so how can we define the size of an atom? There are actually several ways, and they can give slightly different results.

One of the simplest and most useful ways to define atomic size is to relate it to the distance between atoms in a sample of the element. Let us consider a diatomic molecule such as Cl_2 (Figure 7.8a). The radius of a Cl atom is assumed to be half the experimentally determined distance between the centers of

FIGURE 7.7 The Willamette iron meteorite. This 14150-kg meteorite is composed of 92.38% Fe and 7.62% Ni (with traces of Ga, Ge, and Ir). Iron and nickel atoms have nearly identical radii (124 pm and 125 pm, respectively), so one can substitute for the other in the solid. (The meteorite was discovered in Oregon in 1902, and it is now on display at the American Museum of Natural History in Washington, DC.)

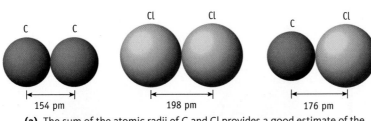

FIGURE 7.8
Determining atomic radii.

(a) The sum of the atomic radii of C and Cl provides a good estimate of the C–Cl distance in a molecule having such a bond.

A distance equivalent to 4 times the radius of an aluminum atom

(b) Pictured here is a tiny piece of an aluminum crystal. Each sphere represents an aluminum atom. Measuring the distance shown allows a scientist to estimate the radius of an aluminum atom.

the two atoms (198 pm), so the radius of one Cl atom is 99 pm. Similarly, the C—C distance in diamond is 154 pm, so a radius of 77 pm can be assigned to carbon. To test these estimates, we can add them together to estimate the C—Cl distance in CCl_4. The predicted distance of 176 pm agrees with the experimentally measured C—Cl distance of 176 pm. (Radii determined this way are often called *covalent radii*.)

This approach to determining atomic radii applies only if molecular compounds of the element exist (and so it is largely limited to nonmetals and metalloids). For metals, atomic radii are sometimes estimated from measurements of the atom-to-atom distance in a crystal of the element (Figure 7.8b).

Some interesting periodic trends are seen immediately on looking at the table of radii (Figure 7.6). *For the main group elements, atomic radii generally increase going down a group in the periodic table and decrease going across a period.* These trends reflect two important effects:

- The size of an atom is determined by the outermost electrons. In going from the top to the bottom of a group in the periodic table, the outermost electrons are assigned to orbitals with increasingly higher values of the principal quantum number, n. Because the underlying electrons require some space, these higher energy electrons are necessarily further from the nucleus.

- For main group elements of a given period, the principal quantum number, n, of the valence electron orbitals is the same. In going from one element to the next across a period, the effective nuclear charge (Z^*) increases (Figure 7.2). This results in an increased attraction between the nucleus and the valence electrons, and the atomic radius decreases.

The periodic trend in the atomic radii of transition metal atoms (Figure 7.9) across a period is somewhat different from that for main group elements. Going from left to right in a given period, the radii initially decrease. However, the sizes of the elements in the middle of a transition series change very little, and a small increase in size occurs at the end of the series. The size of transition metal atoms is determined largely by electrons in the outermost shell—that is, by the electrons of the ns subshell—but electrons are being added to the $(n-1)d$ orbitals across the series. The increased nuclear charge on the atoms as one moves from left to right should cause the radius to decrease. This effect, however, is mostly canceled out by increased electron–electron repulsion. On reaching the Group 1B and 2B elements at the end of the series, the size increases slightly because the d subshell is filled and electron–electron repulsions dominate.

Ionization Energy

Ionization energy *(IE)* is the energy required to remove an electron from an atom in the gas phase (Table 7.5).

$$\text{Atom in ground state(g)} \rightarrow \text{Atom}^+(g) + e^-$$

$$\Delta U \equiv \text{ionization energy, } IE$$

To separate an electron from an atom, energy must be supplied to overcome the attraction of the nuclear charge. Because energy must be supplied, ionization energies always have positive values.

● **Atomic Radii—Caution** Numerous tabulations of atomic and covalent radii exist, and the values quoted in them often differ slightly. The variation comes about because several methods are used to determine the radii of atoms.

● **Valence and Core Electrons** Removal of core electrons requires much more energy than removal of a valence electron. Core electrons are not lost in chemical reactions. (See the plot of orbital energies on page 244.)

FIGURE 7.9 Trends in atomic radii for the transition elements. Atomic radii of the Group 1A and 2A metals and the transition metals of the fourth, fifth, and sixth periods.

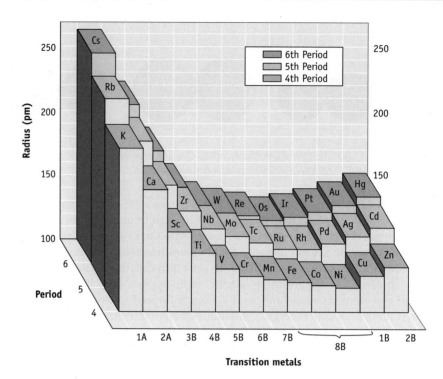

● **Measuring Ionization Energy** Ionization energy values can be measured accurately as compared with the estimations that must be made when measuring atomic radii.

Atoms other than hydrogen have a series of ionization energies as electrons are removed sequentially. For example, the first three ionization energies of magnesium are

First ionization energy, $IE_1 = 738$ kJ/mol

$$Mg(g) \quad \rightarrow \quad Mg^+(g) + e^-$$
$$1s^2 2s^2 2p^6 3s^2 \qquad \quad 1s^2 2s^2 2p^6 3s^1$$

Second ionization energy, $IE_2 = 1451$ kJ/mol

$$Mg^+(g) \quad \rightarrow \quad Mg^{2+}(g) + e^-$$
$$1s^2 2s^2 2p^6 3s^1 \qquad \quad 1s^2 2s^2 2p^6$$

Third ionization energy, $IE_3 = 7732$ kJ/mol

$$Mg^{2+}(g) \quad \rightarrow \quad Mg^{3+}(g) + e^-$$
$$1s^2 2s^2 2p^6 \qquad \quad 1s^2 2s^2 2p^5$$

Table 7.5 First, Second, and Third Ionization Energies for the Main Group Elements in Periods 2–4 (kJ/mol)

2nd Period	Li	Be	B	C	N	O	F	Ne
1st	513	899	801	1086	1402	1314	1681	2080
2nd	7298	1757	2427	2352	2856	3388	3374	3952
3rd	11815	14848	3660	4620	4578	5300	6050	6122
3rd Period	**Na**	**Mg**	**Al**	**Si**	**P**	**S**	**Cl**	**Ar**
1st	496	738	577	787	1012	1000	1251	1520
2nd	4562	1451	1817	1577	1903	2251	2297	2665
3rd	6912	7732	2745	3231	2912	3361	3826	3928
4th Period	**K**	**Ca**	**Ga**	**Ge**	**As**	**Se**	**Br**	**Kr**
1st	419	590	579	762	947	941	1140	1351
2nd	3051	1145	1979	1537	1798	2044	2104	2350
3rd	4411	4910	2963	3302	2735	2974	3500	3565

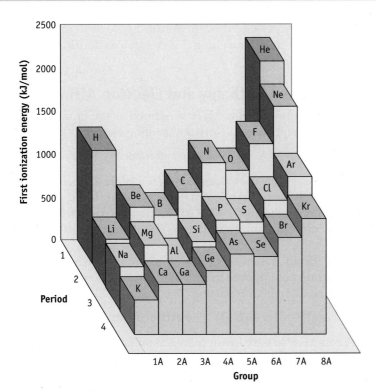

FIGURE 7.10 First ionization energies of the main group elements in the first four periods. (For specific values of the ionization energies for these elements see Table 7.5 and Appendix F.)

Removing each subsequent electron requires more energy because the electron is being removed from an increasingly positive ion (Table 7.5), but there is a particularly large increase in ionization energy for removing the third electron to give Mg^{3+}. The first two ionization steps are for the removal of electrons from the outermost or valence shell of electrons. The third electron, however, must come from the $2p$ subshell, which has a much lower energy than the $3s$ subshell (Figure 7.1). *This large increase is experimental evidence for the electron shell structure of atoms.*

For main group (*s*- and *p*-block) elements, *first ionization energies generally increase across a period and decrease down a group* (Figure 7.10 and Table 7.5, and Appendix F). The trend across a period corresponds to the increase in effective nuclear charge, Z^*, with increasing atomic number (Figure 7.2). As Z^* increases across a period, the energy required to remove an electron increases. Going down a group, the ionization energy decreases. The electron removed is increasingly farther from the nucleus and thus is held less strongly.

Notice that the trends in atomic radius and ionization energy for a given period are both linked to Z^* although they are inversely related. *Owing to an increase in effective nuclear charge across a period, the atomic radius generally decreases and the ionization energy increases.*

A closer look at ionization energies reveals variations to the general trend in a period. One variation occurs on going from *s*-block to *p*-block elements—from beryllium to boron, for example. The $2p$ electrons are slightly higher in energy than the $2s$ electrons so the ionization energy for boron is slightly less than that for beryllium. Another dip to lower ionization energy occurs on going from nitrogen to oxygen. No change occurs in either n or ℓ, but electron–electron repulsions increase for the following reason. In Groups 3A through 5A, electrons are assigned to separate p orbitals (p_x, p_y, and p_z). Beginning in Group 6A, however, two electrons must be assigned to the same p orbital. The fourth p electron shares an orbital with another electron and thus experiences greater repulsion than it would if it had been assigned to an orbital of its own.

$$O \text{ (oxygen atom)} \xrightarrow{+1314 \text{ kJ/mol}} O^+ \text{ (oxygen cation)} + e^-$$

[Ne] ↿⇂ ↿⇂ ↑ ↑   [Ne] ↿⇂ ↑ ↑ ↑
  2s   2p   2s   2p

The greater repulsion experienced by the fourth $2p$ electron makes it easier to remove. The usual trend resumes on going from oxygen to fluorine to neon, however, reflecting the increase in Z^*.

Electron Attachment Enthalpy and Electron Affinity

The **electron attachment enthalpy**, $\Delta_{EA}H$, is defined as the enthalpy change occurring when a gaseous atom adds an electron, forming a gaseous anion.

$$A(g) + e^-(g) \rightarrow A^-(g) \qquad \text{Electron attachment enthalpy} = \Delta_{EA}H$$

As illustrated in Figure 7.11, the value of $\Delta_{EA}H$ for many elements is negative, indicating that this process is exothermic and that energy is evolved. For example, $\Delta_{EA}H$ for fluorine is quite exothermic (-328 kJ/mol), whereas boron has a much less negative value of -26.7 kJ/mol. Values of $\Delta_{EA}H$ for a number of elements are given in Appendix F.

The **electron affinity**, EA, of an atom is closely related to $\Delta_{EA}H$. Electron affinity is equal in magnitude but opposite in sign to the internal energy change associated with a gaseous atom adding an electron.

$$A(g) + e^-(g) \rightarrow A^-(g) \qquad \text{electron affinity, } EA = -\Delta U$$

We expect EA and $\Delta_{EA}H$ to have nearly identical numerical values. However, current convention gives the two values opposite signs.

Because electron attachment enthalpy and ionization energy represent the energy involved in the gain or loss of an electron by an atom, respectively, it makes sense that periodic trends in these properties are also related. The increase in effective nuclear charge of atoms across a period (Figure 7.2) makes it more difficult to ionize the atom, and it also increases the attraction of the atom for an additional electron. Thus, an element with a high ionization energy generally has a more negative value for its electron attachment enthalpy.

As seen in Figure 7.11, the values of $\Delta_{EA}H$ generally become more negative on moving across a period, but the trend is not smooth. The elements in Group 2A and 5A appear as variations to the general trend, corresponding to cases where the added electron would start a p subshell or would be paired with another electron in the p subshell, respectively.

● **Electron Affinity, Electron Attachment Enthalpy, and Sign Conventions** Changes in sign conventions for electron affinities over the years have caused confusion. For a useful discussion of electron affinity, see J. C. Wheeler: "Electron affinities of the alkaline earth metals and the sign convention for electron affinity," *Journal of Chemical Education*, Vol. 74, pp. 123–127, 1997. In this paper the electron affinity is taken as equivalent to the ionization energy of the anion (which is in fact the way electron affinities can be measured experimentally).

FIGURE 7.11 Electron attachment enthalpy. The larger the affinity of an atom for an electron, the more negative the value for $\Delta_{EA}H$. For numerical values, see Appendix F. (Data were taken from H. Hotop and W. C. Lineberger: "Binding energies of atomic negative ions," *Journal of Physical Chemistry*, Reference Data, Vol. 14, p. 731, 1985.)

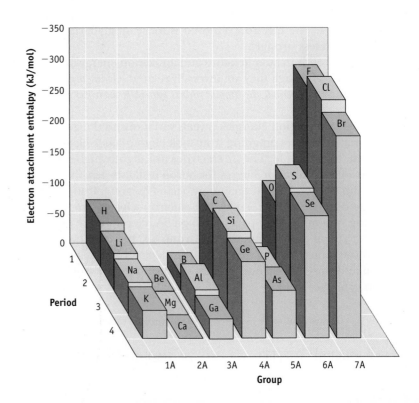

The value of electron attachment enthalpy usually becomes less negative on descending a group of the periodic table. Electrons are added increasingly farther from the nucleus, so the attractive force between the nucleus and electrons decreases. This general trend does not apply to second-period elements, however. For example, the value of the electron attachment enthalpy of fluorine is higher (less negative) than the value for chlorine. The same phenomenon is observed in Groups 3A through 6A. One explanation is that significant electron–electron repulsions occur in small anions such as F^-. That is, adding an electron to the seven electrons already present in the $n = 2$ shell of the small F atom leads to considerable repulsion between electrons. Chlorine has a larger atomic volume than fluorine, so adding an electron leads to a lesser degree of electron–electron repulsions.

A few elements, such as nitrogen and the Group 2A elements, have no affinity for electrons and are listed as having a $\Delta_{EA}H$ value of zero. The noble gases are generally not listed in tables of $\Delta_{EA}H$ values. They have no affinity for electrons, because any additional electron must be added to the next higher electron shell with a considerably higher energy.

No atom has a negative $\Delta_{EA}H$ value for a second electron. So what accounts for the existence of ions such as O^{2-} that occur in many compounds? The answer is that doubly charged anions can be stabilized in crystalline environments by electrostatic attraction to neighboring positive ions (▶ Chapters 8 and 13).

● **Electron Attachment Enthalpy for the Halogens**

Element	$\Delta_{EA}H$ (kJ/mol)	Atomic Radius (pm)
F	−328.0	77
Cl	−349.0	99
Br	−324.7	114
I	−295.2	133

EXAMPLE 7.5   Periodic Trends

Problem Compare the three elements C, O, and Si.

(a) Place them in order of increasing atomic radius.

(b) Which has the largest ionization energy?

(c) Which has the more negative electron attachment enthalpy, O or C?

What Do You Know? Carbon and silicon are the first and second elements in Group 4A, and oxygen is the first element in Group 6A.

Strategy Use the trends in atomic properties in Figures 7.6, 7.9–7.11, Table 7.5, and Appendix F.

Solution

(a) *Atomic size:* Atomic radii decrease on moving across a period, so oxygen has a smaller radius than carbon. However, the radius increases on moving down a periodic group. Because C and Si are in the same group (Group 4A), silicon must be larger than carbon. The trend is O < C < Si.

(b) *Ionization energy:* Ionization energies generally increase across a period and decrease down a group. Thus, the trend in ionization energies is Si (787 kJ/mol) < C (1086 kJ/mol) < O (1314 kJ/mol).

(c) *Electron attachment enthalpy:* Values generally become less negative down a group (except for the second period elements) and more negative across a period. Therefore, O (−141.0 kJ/mol) has a more negative $\Delta_{EA}H$ than C (−121.9 kJ/mol).

Think about Your Answer Notice that the trends in atom size and ionization energy are in the opposite order, as expected based on the values of Z*.

Check Your Understanding

Compare the three elements B, Al, and C.

(a) Place the three elements in order of increasing atomic radius.

(b) Rank the elements in order of increasing ionization energy. (Try to do this without looking at Table 7.5, then compare your prediction with the table.)

(c) Which element, B or C, is expected to have the more negative electron attachment enthalpy value?

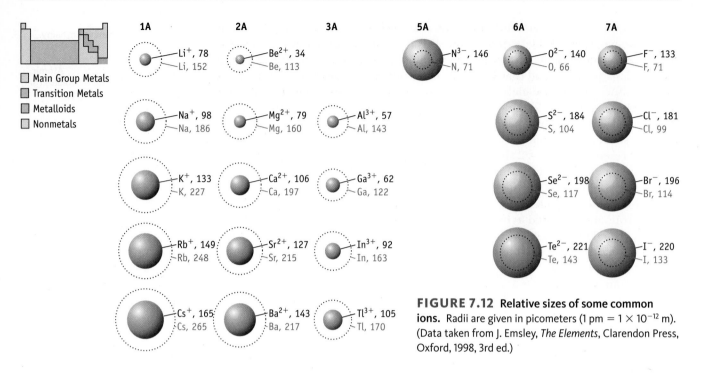

FIGURE 7.12 **Relative sizes of some common ions.** Radii are given in picometers (1 pm = 1 × 10⁻¹² m). (Data taken from J. Emsley, *The Elements*, Clarendon Press, Oxford, 1998, 3rd ed.)

Trends in Ion Sizes

The trend in the sizes of ions down a periodic group is the same as that for neutral atoms: Positive and negative ions increase in size when descending the group (Figure 7.12). Pause for a moment, however, and compare the ionic radii with the atomic radii, as also illustrated in Figure 7.12. When an electron is removed from an atom to form a cation, the size shrinks considerably. The radius of a cation is *always* smaller than that of the atom from which it is derived. For example, the radius of Li is 152 pm, whereas the radius of Li⁺ is only 78 pm. When an electron is removed from a Li atom, the attractive force of three protons is now exerted on only two electrons, so the remaining electrons are drawn closer to the nucleus. The decrease in ion size is especially great when the last electron of a particular shell is removed, as is the case for Li. The loss of the 2s electron from Li leaves Li⁺ with no electrons in the n = 2 shell.

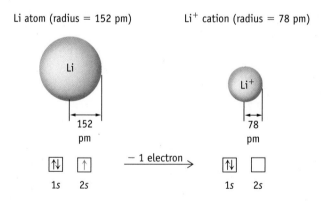

A large decrease in size is also expected if two or more electrons are removed from an atom. For example, an aluminum ion, Al³⁺, has a radius of 57 pm, whereas the radius of an aluminum atom is 143 pm.

Al atom (radius = 143 pm)   Al³⁺ cation (radius = 57 pm)

[Ne]⥮ ↑ □ □ $\xrightarrow{\text{−3 electrons}}$ [Ne]□ □ □ □

    3s   3p   3s   3p

You can also see in Figure 7.12 that anions are *always* larger than the atoms from which they are derived. Here, the argument is the opposite of that used to explain positive ion radii. The F atom, for example, has nine protons and nine electrons. On forming the anion, the nuclear charge is still +9, but the anion has 10 electrons. The F^- ion is larger than the F atom because of increased electron–electron repulsions.

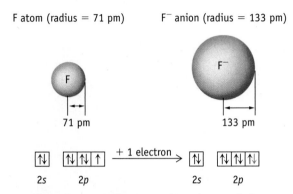

F atom (radius = 71 pm)   F$^-$ anion (radius = 133 pm)

71 pm   133 pm

Finally, it is useful to compare the sizes of isoelectronic ions across the periodic table. **Isoelectronic** ions have the same number of electrons (but a different number of protons). One such series of ions is N^{3-}, O^{2-}, F^-, Na^+, and Mg^{2+}:

Ion	N^{3-}	O^{2-}	F^-	Na^+	Mg^{2+}
Number of electrons	10	10	10	10	10
Number of protons	7	8	9	11	12
Ionic radius (pm)	146	140	133	98	79

All these ions have 10 electrons, but they differ in the number of protons. As the number of protons increases in a series of isoelectronic ions, the balance between electron-proton attraction and electron–electron repulsion shifts in favor of attraction, and the radius decreases.

REVIEW & CHECK FOR SECTION 7.5

1. What is the trend in sizes of the ions K^+, S^{2-}, and Cl^-?

   (a) $K^+ > S^{2-} > Cl^-$ (b) $K^+ < S^{2-} < Cl^-$ (c) $S^{2-} < Cl^- < K^+$ (d) $S^{2-} > Cl^- > K^+$

2. Locate the elements C, N, Si, and P in the periodic table. Based on general trends, identify the elements in this group with the highest and lowest ionization energy.

   (a) C highest, P lowest   (c) N highest, Si lowest

   (b) C highest, Si lowest   (d) C lowest, N highest

7.6 Periodic Trends and Chemical Properties

Atomic and ionic radii, ionization energies, and electron attachment enthalpies are properties associated with atoms and their ions. It is reasonable to expect that knowledge of these properties will be useful as we explore the chemistry involving formation of ionic compounds.

The periodic table was created by grouping together elements having similar chemical properties (Figure 7.13). Alkali metals, for example, characteristically form compounds containing a 1+ ion, such as Li^+, Na^+, or K^+. Thus, the reaction between sodium and chlorine gives the ionic compound, NaCl (composed of Na^+ and Cl^- ions) [Figure 1.2, page 3], and potassium and water react to form an aqueous solution of KOH, a solution containing the hydrated ions $K^+(aq)$ and $OH^-(aq)$.

$$2\ Na(s) + Cl_2(g) \rightarrow 2\ NaCl(s)$$

$$2\ K(s) + 2\ H_2O(\ell) \rightarrow 2\ K^+(aq) + 2\ OH^-(aq) + H_2(g)$$

The facile formation of Na^+ and K^+ ions in chemical reactions agrees with the fact that alkali metals have low ionization energies.

Ionization energies also account for the fact that these reactions of sodium and potassium do not produce compounds such as $NaCl_2$ or $K(OH)_2$. The formation of an Na^{2+} or K^{2+} ion would be a very unfavorable process. Removing a second electron from these metals requires a great deal of energy because a core electron would

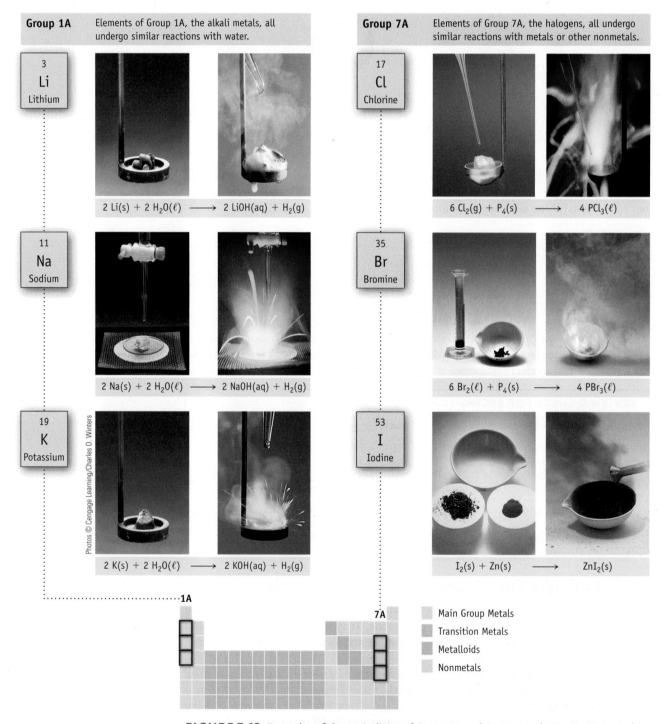

Group 1A	Elements of Group 1A, the alkali metals, all undergo similar reactions with water.

3 Li Lithium

$$2\ Li(s) + 2\ H_2O(\ell) \longrightarrow 2\ LiOH(aq) + H_2(g)$$

11 Na Sodium

$$2\ Na(s) + 2\ H_2O(\ell) \longrightarrow 2\ NaOH(aq) + H_2(g)$$

19 K Potassium

$$2\ K(s) + 2\ H_2O(\ell) \longrightarrow 2\ KOH(aq) + H_2(g)$$

Group 7A	Elements of Group 7A, the halogens, all undergo similar reactions with metals or other nonmetals.

17 Cl Chlorine

$$6\ Cl_2(g) + P_4(s) \longrightarrow 4\ PCl_3(\ell)$$

35 Br Bromine

$$6\ Br_2(\ell) + P_4(s) \longrightarrow 4\ PBr_3(\ell)$$

53 I Iodine

$$I_2(s) + Zn(s) \longrightarrow ZnI_2(s)$$

Photos © Cengage Learning/Charles D. Winters

1A

7A

Main Group Metals
Transition Metals
Metalloids
Nonmetals

FIGURE 7.13 Examples of the periodicity of Group 1A and Group 7A elements. Dmitri Mendeleev developed the first periodic table by listing elements in order of increasing atomic weight. Every so often, an element had properties similar to those of a lighter element, and these were placed in vertical columns or groups. We now recognize that the elements should be listed in order of increasing atomic number and that the periodic occurrence of similar properties is related to the electron configurations of the elements.

CASE STUDY

Metals in Biochemistry and Medicine

Many main group and transition metals play an important role in biochemistry and in medicine. Your body has low levels of the following metals in the form of various compounds: Ca, 1.5%; Na, 0.1%; Mg, 0.05%, as well as iron, cobalt, zinc, and copper, all less than about 0.05%. (Levels are percentages by mass.)

Much of the 3–4 g of iron in your body is found in hemoglobin, the substance responsible for carrying oxygen to cells. Iron deficiency is marked by fatigue, infections, and mouth inflammations.

Iron in your diet can come from eggs and brewer's yeast, which has a very high iron content. In addition, foods such as some breakfast cereals are "fortified" with metallic iron. (In an interesting experiment you can do at home, you can remove the iron by stirring the cereal with a strong magnet.) Vitamin pills often contain iron(II) compounds with anions such as sulfate and succinate ($C_4H_4O_4^{2-}$).

The average person has about 75 mg of copper, about one third of which is found in the muscles. Copper is involved in many biological functions, and a deficiency shows up in many ways: anemia, degeneration of the nervous system, and impaired immunity. Wilson's disease, a genetic disorder, leads to the over-accumulation of copper in the body and results in hepatic and neurological damage.

Like silver ions (page 126), copper ions can also act as a bacteriocide. Scientists from Britain and India recently investigated a long-held belief among people in India that storing water in brass pitchers can ward off illness. (Brass is an alloy of copper and zinc.) They filled brass pitchers

Filling a brass water jug for drinking water in India. Copper ions released in tiny amounts from the brass kill bacteria in contaminated water.

Jeremy Horner/Corbis

with previously sterile water to which they had added *E. coli* (a bacterium that lives in the lower intestine of many warm-blooded animals and consequently is found in their feces). Other brass pitchers were filled with contaminated river water from India. In both cases, they found fecal bacteria counts dropped from as high as 1,000,000 bacteria per milliliter to zero in two days. In contrast, bacteria levels stayed high in plastic or earthenware pots. Apparently, just enough copper ions are released by the brass to kill the bacteria but not enough to affect humans.

Questions:

1. Give the electron configurations for iron and the iron(II) and iron(III) ions.

2. In hemoglobin, iron can be in the iron(II) or iron(III) state. Is either of these iron ions paramagnetic?

3. Give the electron configurations for copper and the copper(I) and copper(II) ions. Is copper in any of these forms paramagnetic?

4. Why are copper atoms (radius = 128 pm) slightly larger than iron atoms (radius = 124 pm)?

5. In hemoglobin, the iron is enclosed by the porphyrin group (shown below), a flat grouping of carbon, hydrogen, and nitrogen atoms. (This is in turn encased in a protein.) When iron is in the form of the Fe^{3+} ion, it just fits into the space within the four N atoms, and the arrangement is flat. Speculate on what occurs to the structure when iron is reduced to the Fe^{2+} ion.

Answers to these questions are available in Appendix N.

---

have to be removed. The energetic barrier to this process is the underlying reason that *main group metals generally form cations with an electron configuration equivalent to that of the preceding noble gas.*

Why isn't Na_2Cl another possible product from the sodium and chlorine reaction? This formula would imply that the compound contains Na^+ and Cl^{2-} ions. Adding two electrons per atom to Cl means that the second electron must enter the next higher shell at much higher energy. Thus, anions such as Cl^{2-} are not known. This example leads us to a general statement: *Nonmetals generally acquire enough electrons to form an anion with the electron configuration of the next noble gas.*

We can use similar logic to rationalize other observations. Ionization energies increase on going from left to right across a period. We have seen that elements from Groups 1A and 2A form ionic compounds, an observation directly related to the low ionization energies for these elements. Ionization energies for elements toward the middle and right side of a period, however, are sufficiently large that cation formation is unfavorable. On the right side of the second period, oxygen and fluorine much prefer taking on electrons to giving them up; these elements have high ionization energies and relatively large, negative electron attachment enthalpies. Thus, oxygen and fluorine form anions and not cations when they react.

Finally, let us think for a moment about carbon, the basis of thousands of chemical compounds. Its ionization energy is not favorable for cation formation, and it also does not generally form anions. Thus, we do not find many binary ionic compounds containing carbon; instead, we find carbon sharing electrons with other elements in compounds such as CO_2 and CCl_4, and we will take up those kinds of compounds in the next two chapters.

REVIEW & CHECK FOR SECTION 7.6

1. Which of the following is an incorrect formula?

    (a) MgCl   (b) $CaCl_2$ (c) $AlCl_3$ (d) RbCl

2. Which of the following is not the formula for a common ion?

    (a) Cl^- (b) K^+ (c) O^{2-} (d) S^{3-}

---

 and

Sign in at **www.cengage.com/owl** to:
- View tutorials and simulations, develop problem-solving skills, and complete online homework assigned by your professor.
- For quick review and exam prep, download Go Chemistry mini lecture modules from OWL (or purchase them at **www.cengagebrain.com**)

Access **How Do I Solve It?** tutorials on how to approach problem solving using concepts in this chapter.

CHAPTER GOALS REVISITED

Now that you have studied this chapter, you should ask whether you have met the chapter goals. In particular, you should be able to:

Recognize the relationship of the four quantum numbers (n, ℓ, m_ℓ, and m_s) to atomic structure

a. Recognize that each electron in an atom has a different set of the four quantum numbers, n, ℓ, m_ℓ, and m_s (Sections 6.5–6.7, 7.1–7.3). **Study Questions: 11, 13, 35, 37, 52.**

b. Understand that the Pauli exclusion principle leads to the conclusion that no atomic orbital can be assigned more than two electrons and that the two electrons in an orbital must have opposite spins (different values of m_s) (Section 7.1).

Write the electron configuration for atoms and monatomic ions

a. Recognize that electrons are assigned to the subshells of an atom in order of increasing energy (Aufbau principle, Section 7.2). In the H atom, the energies increase with increasing n, but, in a many-electron atom, the energies depend on both n and ℓ (Figure 7.1).

b. Understand effective nuclear charge, Z^*, and its ability to explain why different subshells in the same shell of multielectron atoms have different energies. Also, understand the role of Z^* in determining the properties of atoms (Sections 7.2 and 7.5).

c. Using the periodic table as a guide, depict electron configurations of neutral atoms (Section 7.3) and monatomic ions (Section 7.4) using the orbital box or *spdf* notation. In both cases, configurations can be abbreviated with the noble gas notation. **Study Questions: 1, 3, 5–10, 13, 15, 18, 20, 21, 33–36, 39, 71.**

d. When assigning electrons to atomic orbitals, apply the Pauli exclusion principle and Hund's rule (Sections 7.3 and 7.4). **Study Question: 59.**

e. Understand the role magnetism plays in revealing atomic structure (Section 7.4). **Study Questions: 19–21, 33, 39, 52.**

Rationalize trends in atom and ion sizes, ionization energy, and electron attachment enthalpy

a. Predict how properties of atoms—size, ionization energy *(IE)*, and electron attachment enthalpy ($\Delta_{EA}H$)—change on moving down a group or across a period of the periodic table (Section 7.5). The general periodic trends for these properties are as follows: **Study Questions: 24, 26, 28, 30, 32, 40–43, 46–49, 57, 58, 64, and Go Chemisty Module 11.**

  (i) Atomic size decreases across a period and increases down a group.

  (ii) *IE* increases across a period and decreases down a group.

  (iii) The value of $\Delta_{EA}H$ becomes more negative across a period and becomes less negative down a group.

b. Recognize the role that ionization energy and electron attachment enthalpy play in forming ionic compounds (Section 7.6). **Study Questions: 45, 72.**

OWL Questions and problems for this chapter are available in OWL. Use the OWL access card that came with this textbook to access assigned questions and problems for this chapter.

8 Bonding and Molecular Structure

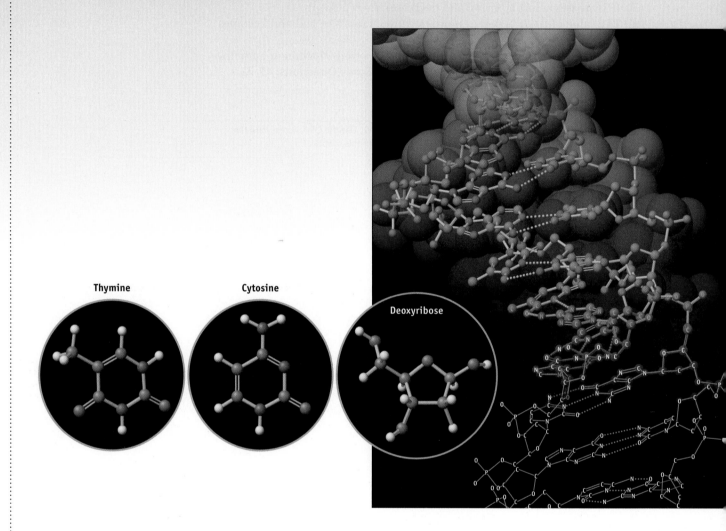

Thymine

Cytosine

Deoxyribose

Chemical Bonding in DNA

The theme of this chapter and the next is molecular bonding and structure, and the subject is well illustrated by the structure of DNA. This molecule is a helical coil of two chains of tetrahedral phosphate groups and deoxyribose groups. Organic bases (such as thymine and cytosine) on one chain interact with complementary bases on the other chain.

Questions:

Among the many questions you can answer from studying this chapter are the following:

1. Why are there four bonds to carbon and phosphorus?
2. Why are the C atoms and P atoms in the backbone and the C atoms in deoxyribose surrounded by other atoms at an angle of 109°?
3. What are the angles in the six-member rings of the bases thymine and cytosine? Why are the six-member rings flat?
4. Are thymine and cytosine polar molecules?

Answers to these questions are available in Appendix N.

CHAPTER OUTLINE

8.1   Chemical Bond Formation

8.2   Covalent Bonding and Lewis Structures go Chemistry

8.3   Atom Formal Charges in Covalent Molecules and Ions

8.4   Resonance

8.5   Exceptions to the Octet Rule

8.6   Molecular Shapes

8.7   Bond Polarity and Electronegativity

8.8   Bond and Molecular Polarity go Chemistry

8.9   Bond Properties: Order, Length, and Energy

8.10 DNA, Revisited

CHAPTER GOALS

See Chapter Goals Revisited (page 305) for Study Questions keyed to these goals.

- Understand the difference between ionic and covalent bonds.

- Draw Lewis electron dot structures for small molecules and ions.

- Use the valence shell electron-pair repulsion (VSEPR) theory to predict the shapes of simple molecules and ions and to understand the structures of more complex molecules.

- Use electronegativity and formal charge to predict the charge distribution in molecules and ions, to define the polarity of bonds, and to predict the polarity of molecules.

- Understand the properties of covalent bonds and their influence on molecular structure.

⊎WL

Sign in to OWL at **www.cengage.com/owl** to view tutorials and simulations, develop problem-solving skills, and complete online homework assigned by your professor.

Scientists have long known that the key to interpreting the properties of a chemical substance is first to recognize and understand its structure and bonding. **Structure** refers to the way atoms are arranged in space, and **bonding** describes the forces that hold adjacent atoms together.

Our discussion of structure and bonding begins with small molecules and then progresses to larger molecules. From compound to compound, atoms of the same element participate in bonding and structure in a predictable way. This consistency allows us to develop principles that apply to many different chemical compounds, including such complex molecules as DNA.

8.1 Chemical Bond Formation

When a chemical reaction occurs between two atoms, their valence electrons are reorganized so that a net attractive force—a **chemical bond**—occurs between the atoms. There are two general types of bonds, ionic and covalent, and their formation can be depicted using Lewis symbols.

An **ionic bond** forms when *one or more valence electrons is transferred from one atom to another*, creating positive and negative ions. When sodium and chlorine react (Figure 8.1a), an electron is transferred from a sodium atom to a chlorine atom to form Na^+ and Cl^-.

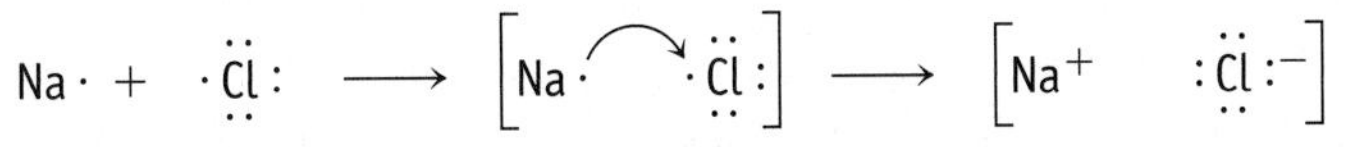

go Chemistry

Download mini lecture videos for key concept review and exam prep from OWL or purchase them from **www.cengagebrain.com**

| metal atom | nonmetal atom | electron transfer from reducing agent to oxidizing agent | ionic compound. Ions have noble gas electron configurations. |

The "bond" is the attractive force between the oppositely charged positive and negative ions.

- **Valence Electron Configurations and Ionic Compound Formation**

For the formation of NaCl, Na $[1s^22s^22p^63s^1]$ gives up an electron to form Na^+ $[1s^22s^22p^6]$, which is equivalent to the Ne configuration. Cl $\{[Ne]3s^23p^5\}$ gains an electron to form Cl^- $\{[Ne]3s^23p^6\}$, which is equivalent to the Ar configuration.

FIGURE 8.1 Formation of ionic compounds. Both reactions shown here are quite exothermic, as reflected by the very negative molar enthalpies of formation for the reaction products.

(a) The reaction of elemental sodium and chlorine to give sodium chloride. $\Delta_f H°$ [NaCl(s)] = −411.12 kJ/mol

(b) The reaction of elemental calcium and oxygen to give calcium oxide. $\Delta_f H°$ [CaO(s)] = −635.09 kJ/mol

Covalent bonding, in contrast, *involves sharing of valence electrons between atoms.* Two chlorine atoms, for example, share a pair of electrons, one electron from each atom, to form a covalent bond.

$$:\ddot{C}l\cdot + \cdot\ddot{C}l: \longrightarrow :\ddot{C}l:\ddot{C}l:$$

As bonding is described in greater detail, you will discover that the two types of bonding—complete electron transfer and the equal sharing of electrons—are extreme cases. In most chemical compounds, electrons are shared unequally, with the extent of sharing varying widely from very little sharing (largely ionic) to considerable sharing (largely covalent).

Ionic bonding will be described in more detail in Chapter 13, whereas the present chapter focuses on bonding in covalent compounds.

REVIEW & CHECK FOR SECTION 8.1

Which of the following compounds is not ionic?

(a) NaBr   (b) $CaBr_2$ (c) $SiBr_4$ (d) $FeBr_2$

---

go Chemistry

Module 12: Drawing Lewis Electron Dot Structures covers concepts in this section.

8.2 Covalent Bonding and Lewis Structures

There are many examples of compounds having covalent bonds, including the gases in our atmosphere (O_2, N_2, H_2O, and CO_2), common fuels (CH_4), and most of the compounds in your body. Covalent bonding is also responsible for the atom-to-atom connections in polyatomic ions such as CO_3^{2-}, CN^-, NH_4^+, NO_3^-, and PO_4^{3-}. We will develop the basic principles of structure and bonding using these and other small molecules and ions, but the same principles apply to larger molecules from aspirin to proteins and DNA with thousands of atoms.

The molecules and ions just mentioned are composed entirely of *nonmetal* atoms. A point that needs special emphasis is that, in molecules or ions made up *only* of nonmetal atoms, the atoms are attached by covalent bonds. Conversely, the presence of a metal in a formula is often a signal that the compound is likely to be ionic.

Valence Electrons and Lewis Symbols for Atoms

The electrons in an atom are of two types: **valence electrons** and **core electrons**. Chemical reactions result in the loss, gain, or rearrangement of valence electrons. The core electrons are not involved in bonding or in chemical reactions.

Table 8.1 Core and Valence Electrons for Several Common Elements

Element	Periodic Group	Core Electrons	Valence Electrons	Total
Main Group Elements				
Na	1A	$1s^2 2s^2 2p^6 = [\text{Ne}]$	$3s^1$	$[\text{Ne}]3s^1$
Si	4A	$1s^2 2s^2 2p^6 = [\text{Ne}]$	$3s^2 3p^2$	$[\text{Ne}]3s^2 3p^2$
As	5A	$1s^2 2s^2 2p^6 3s^2 3p^6 3d^{10} = [\text{Ar}]3d^{10}$	$4s^2 4p^3$	$[\text{Ar}]3d^{10}4s^2 4p^3$
Transition Elements				
Ti	4B	$1s^2 2s^2 2p^6 3s^2 3p^6 = [\text{Ar}]$	$3d^2 4s^2$	$[\text{Ar}]3d^2 4s^2$
Co	8B	$[\text{Ar}]$	$3d^7 4s^2$	$[\text{Ar}]3d^7 4s^2$
Mo	6B	$[\text{Kr}]$	$4d^5 5s^1$	$[\text{Kr}]4d^5 5s^1$

For main group elements (elements of the A groups in the periodic table), the valence electrons are the *s* and *p* electrons in the outermost shell (Table 8.1). All electrons in inner shells are core electrons. A useful guideline for *main group elements* is that *the number of valence electrons is equal to the group number*. The fact that all elements in a periodic group have the same number of valence electrons accounts for the similarity of chemical properties among members of the group.

Valence electrons for transition elements include the electrons in the *ns* and $(n-1)d$ orbitals (Table 8.1). The remaining electrons are core electrons. As with main group elements, the valence electrons for transition metals determine the chemical properties of these elements.

The American chemist Gilbert Newton Lewis (1875–1946) introduced a useful way to represent electrons in the valence shell of an atom. The element's symbol represents the atomic nucleus together with the core electrons. Up to four valence electrons, represented by dots, are placed one at a time around the symbol; then, if any valence electrons remain, they are paired with ones already there. Chemists now refer to these pictures as **Lewis electron dot symbols**. Lewis dot symbols for main group elements of the second and third periods are shown in Table 8.2.

Arranging the valence electrons of a main group element around an atom in four groups suggests that the valence shell can accommodate four pairs of electrons. Because this represents eight electrons in all, this is referred to as an **octet** of electrons. An octet of electrons surrounding an atom is regarded as a stable configuration. The noble gases, with the exception of helium, have eight valence electrons and demonstrate a notable lack of reactivity. (Helium and neon do not undergo any chemical reactions, and the other noble gases have very limited chemical reactivity.) Because chemical reactions involve changes in the valence electron shell, the limited reactivity of the noble gases is taken as evidence of the stability of their noble gas $(ns^2 np^6)$ electron configuration. Hydrogen, which in its compounds has two electrons in its valence shell, obeys the spirit of this rule by matching the electron configuration of He.

● **Gilbert Newton Lewis (1875–1946)** Lewis introduced the theory of shared electron-pair chemical bonds in a paper published in the *Journal of the American Chemical Society* in 1916. Lewis also made major contributions in acid–base chemistry, thermodynamics, and the interaction of light with substances.

Lewis was born in Massachusetts but raised in Nebraska. After earning his B.A. and Ph.D. at Harvard, he began his career in 1912 at the University of California at Berkeley. He was not only a productive researcher but was also an influential teacher. Among his ideas was the use of problem sets in teaching, a practice still in use today.

Gilbert Newton Lewis

Table 8.2 Lewis Electron Dot Symbols for Main Group Atoms

| 1A | 2A | 3A | 4A | 5A | 6A | 7A | 8A |
ns^1	ns^2	$ns^2 np^1$	$ns^2 np^2$	$ns^2 np^3$	$ns^2 np^4$	$ns^2 np^5$	$ns^2 np^6$
Li·	·Be·	·Ḃ·	·Ċ·	·N̈·	:Ö·	:F̈·	:N̈e:
Na·	·Mg·	·Äl·	·Si·	·P̈·	:S̈·	:C̈l·	:Är:

Lewis Electron Dot Structures and the Octet Rule

In a simple description of covalent bonding, a bond results when one or more electron pairs are shared between two atoms. The electron pair bond between the two atoms of an H_2 molecule is represented by a pair of dots or, alternatively, a line.

The representation of a molecule in this fashion is called a **Lewis electron dot structure** or just a **Lewis structure**.

Simple Lewis structures, such as that for F_2, can be drawn starting with Lewis dot symbols for atoms and arranging the valence electrons to form bonds. Fluorine, an element in Group 7A, has seven valence electrons. The Lewis symbol shows that an F atom has a single unpaired electron along with three electron pairs. In F_2, the single electrons, one on each F atom, pair up in the covalent bond.

In the Lewis structure for F_2 the pair of electrons in the F—F bond is the bonding pair, or **bond pair**. The other six pairs reside on single atoms and are called **lone pairs**. Because they are not involved in bonding, they are also called **nonbonding electrons**.

Carbon dioxide, CO_2, and dinitrogen, N_2, are examples of molecules in which two atoms share more than one electron pair.

In carbon dioxide, the carbon atom shares two pairs of electrons with each oxygen and so is linked to each O atom by a double bond. The valence shell of each oxygen atom in CO_2 has two bonding pairs and two lone pairs. In dinitrogen, the two nitrogen atoms share three pairs of electrons, so they are linked by a triple bond. In addition, each N atom has a single lone pair.

An important observation can be made about the molecules you have seen so far: Each atom (except H) has a share in four pairs of electrons, so each has achieved a *noble gas configuration. Each atom is surrounded by an octet of eight electrons.* (Hydrogen typically forms a bond to only one other atom, resulting in two electrons in its valence shell.) *The tendency of molecules and polyatomic ions to have structures in which eight electrons surround each atom* is known as the **octet rule**. As an example, a triple bond is necessary in dinitrogen in order to have an octet around each nitrogen atom. The carbon atom and both oxygen atoms in CO_2 achieve the octet configuration by forming double bonds.

The octet rule is extremely useful, but keep in mind that it is more a *guideline* than a rule. Particularly for the second period elements C, N, O, and F, a Lewis structure in which each atom achieves an octet is likely to be correct. Although there are a few exceptions, if an atom such as C, N, O, or F in a Lewis structure does not follow the octet rule, you should question the structure's validity. If a structure obeying the octet rule cannot be written, then it is possible an incorrect formula has been assigned to the compound or the atoms have been assembled in an incorrect way.

● **Importance of Lone Pairs** Lone pairs can be important in a structure. Since they are in the same valence electron shell as the bonding electrons, they can influence molecular shape. (See Section 8.6.)

● **Exceptions to the Octet Rule** Although the octet rule is widely applicable, there are exceptions. Fortunately, many will be obvious, such as when there are more than four bonds to an element or when an odd number of electrons occurs. (See Section 8.5.)

Drawing Lewis Electron Dot Structures

There is a systematic approach to constructing Lewis structures of molecules and ions. Let us take formaldehyde, CH_2O, as an example.

1. *Determine the arrangement of atoms within a molecule.* The central atom is *usually* the one with the lowest affinity for electrons (least negative electron attachment enthalpy, page 252).

   > In CH_2O, the central atom is C. You will come to recognize that certain elements often appear as the central atom, among them C, N, P, and S. Halogens are often terminal atoms forming a single bond to one other atom, but they can be the central atom when combined with O in oxoacids (such as in $HClO_4$). Oxygen is the central atom in water, but in conjunction with nitrogen, phosphorus, and the halogens it is usually a terminal atom. Hydrogen is a terminal atom because it typically bonds to only one other atom.

2. *Determine the total number of valence electrons in the molecule or ion.* In a neutral molecule, this number will be the sum of the valence electrons for each atom.

   - For an *anion*, *add* the number of electrons equal to the negative charge.
   - For a *cation*, *subtract* the number of electrons equal to the positive charge.

   The number of valence electron *pairs* will be half the total number of valence electrons. For CH_2O,

   $$\text{Valence electrons} = 12 \text{ electrons (or 6 electron pairs)}$$
   $$= 4 \text{ (for C)} + 2 \times 1 \text{ (for two H atoms)} + 6 \text{ (for O)}$$

3. *Place one pair of electrons between each pair of bonded atoms to form a single bond.*

   Here, three electron pairs are used to make three single bonds (which are represented by single lines). Three pairs of electrons remain to be used.

4. *Use any remaining pairs as lone pairs around each terminal atom (except H) so that each terminal atom is surrounded by eight electrons.* If, after this is done, there are electrons left over, assign them to the central atom. (If the central atom is an element in the third or higher period, it can have more than eight electrons. See page 277.)

   Here, all six pairs have been assigned, but notice that the C atom has a share in only three pairs.

5. *If the central atom has fewer than eight electrons at this point, change one or more of the lone pairs on the terminal atoms into a bonding pair between the central and terminal atom to form a multiple bond.*

   As a general rule, double or triple bonds are most often encountered when *both* atoms are from the following list: C, N, or O. That is, bonds such as C=C, C=N, and C=O will be encountered frequently.

● **Choosing the Central Atom**

1. The electronegativities of atoms can also be used to choose the central atom. Electronegativity is discussed in Section 8.7.

2. For simple compounds, the first atom in a formula is often the central atom (e.g., SO_2, NH_4^+, NO_3^-). This is not always a reliable predictor, however. Notable exceptions include water (H_2O) and most common acids (HNO_3, H_2SO_4), in which the acidic hydrogen is usually written first but where another atom (such as N or S) is the central atom.

Strategy Map 8.1

PROBLEM

Draw the **Lewis electron dot structure** for ClO_3^-.

↓

DATA/INFORMATION

• **Formula** of the ion

↓

> **STEP 1.** Decide on **central atom**

Cl is central atom, surrounded by three **O** atoms.

> **STEP 2.** Calculate number of **valence electrons**

ClO_3^- has **26** valence electrons or **13 pairs**.

> **STEP 3.** Form **single bonds** between *central atom* and *terminal atoms*

Single bonds

O
|
O—Cl—O

> **STEP 4.** Place **electron pairs** on *terminal atoms* so each has an **octet** of electrons.

Electron pairs ⟶ :Ö:
|
:Ö—Cl—Ö:

> **STEP 5.** Place **remaining electron pairs** on the *central atom* to achieve octet.

Complete

$$\left[\begin{array}{c} :\overset{..}{O}: \\ | \\ :\overset{..}{O}-Cl-\overset{..}{O}: \end{array} \right]^-$$

⊙WL INTERACTIVE EXAMPLE 8.1 Drawing Lewis Electron Dot Structures

Problem Draw Lewis structures for the chlorate ion (ClO_3^-) and the nitronium ion (NO_2^+).

What Do You Know? The formula of each ion is given, and you can expect that each atom will achieve an octet configuration.

Strategy Follow the five steps outlined for CH_2O in the preceding text.

Solution for chlorate ion

1. Cl is the central atom, and the O atoms are terminal atoms.

2. Valence electrons = 26 (13 pairs)
   = 7 (for Cl) + 18 (six for each O) + 1 (for the negative charge)

3. Three electron pairs form single bonds from Cl to the O terminal atoms.

$$\begin{array}{c} O \\ | \\ O-Cl-O \end{array}$$

4. Distribute three lone pairs on each of the terminal O atoms to complete the octet of electrons around each of these atoms.

$$\left[\begin{array}{c} :\overset{..}{O}: \\ | \\ :\overset{..}{O}-Cl-\overset{..}{O}: \end{array} \right]^-$$

5. One pair of electrons remains, and it is placed on the central Cl atom to complete its octet.

$$\left[\begin{array}{c} :\overset{..}{O}: \\ | \\ :\overset{..}{O}-Cl-\overset{..}{O}: \end{array} \right]^-$$

Each atom now has a share in four pairs of electrons, and the Lewis structure is complete.

Solution for nitronium ion

1. Nitrogen is the central atom. The electron affinity of nitrogen is lower than that of oxygen. (The electron attachment enthalpy of N is less negative than that of O.)

2. Valence electrons = 16 (8 pairs)
   = 5 (for N) + 12 (six for each O) − 1 (for the positive charge)

3. Two electron pairs form single bonds from the nitrogen to each oxygen:

$$O-N-O$$

4. Distribute the remaining six pairs of electrons on the terminal O atoms:

$$\left[:\overset{..}{\underset{..}{O}}-N-\overset{..}{\underset{..}{O}}: \right]^+$$

5. The central nitrogen atom is two electron pairs short of an octet. Thus, a lone pair of electrons on each oxygen atom is converted to a bonding electron pair to give two N=O double bonds. Each atom in the ion now has four electron pairs. Nitrogen has four bonding pairs, and each oxygen atom has two lone pairs and shares two bond pairs.

Move lone pairs to create double bonds and satisfy the octet for N.

$$\left[:\overset{..}{\underset{..}{O}}-N-\overset{..}{\underset{..}{O}}: \right]^+ \longrightarrow \left[\overset{..}{\underset{..}{O}}=N=\overset{..}{\underset{..}{O}} \right]^+$$

Think about Your Answer Why don't we take two lone pairs from one side and none from the other (to give a NO triple bond and a NO single bond, respectively)? We shall discuss that after describing charge distribution in molecules and ions (page 289).

Check Your Understanding

Draw Lewis structures for NH_4^+, CO, NO^+, and SO_4^{2-}.

Useful Ideas to Consider When Drawing Lewis Electron Dot Structures

- The octet rule is a useful guideline when drawing Lewis structures.
- Carbon forms four bonds (four single bonds; two single bonds and one double bond; two double bonds; or one single bond and one triple bond). In uncharged species, nitrogen forms three bonds and oxygen forms two bonds. Hydrogen typically forms only one bond to another atom.

- When multiple bonds are formed, both atoms involved are usually one of the following: C, N, and O. Oxygen has the ability to form multiple bonds with a variety of elements. Carbon forms many compounds having multiple bonds to another carbon or to N or O.
- Nonmetals may form single, double, and triple bonds but never quadruple bonds.

- Always account for single bonds and lone pairs before determining whether multiple bonds are present.
- Be alert for the possibility the molecule or ion you are working on is isoelectronic (page 269) with a species you have seen before. If so, it may have a similar Lewis structure.

Predicting Lewis Structures

Chemists find Lewis structures useful to gain a perspective on the structure and chemistry of a molecule or ion. The guidelines for drawing Lewis structures are helpful, but you will find you can also rely on patterns of bonding in related molecules.

Hydrogen Compounds

Some common compounds and ions formed from second-period nonmetal elements and hydrogen are shown in Table 8.3. Their Lewis structures illustrate the fact that the Lewis symbol for an element is a useful guide in determining the number of bonds formed by the element. For example, if there is no charge, nitrogen has five valence electrons. Two electrons occur as a lone pair; the other three occur as unpaired electrons. To reach an octet, it is necessary to pair each of the unpaired electrons with an electron from another atom. Thus, N is predicted to form three bonds in uncharged molecules, and this is indeed the case. Similarly, carbon is expected to form four bonds, oxygen two, and fluorine one.

Table 8.3 Lewis Structures of Common Hydrogen-Containing Molecules and Ions of Second-Period Elements

EXAMPLE 8.2 Predicting Lewis Electron Dot Structures

Problem Draw Lewis electron dot structures for CCl_4 and NF_3.

What Do You Know? Carbon tetrachloride (CCl_4) and NF_3 have stoichiometries similar to CH_4 and NH_3, so you might reasonably expect similar Lewis structures.

Strategy Recall that carbon is expected to form four bonds and nitrogen three bonds to achieve an octet of electrons. In addition, halogen atoms have seven valence electrons, so both Cl and F can attain an octet by forming one covalent bond, just as hydrogen does.

Solution

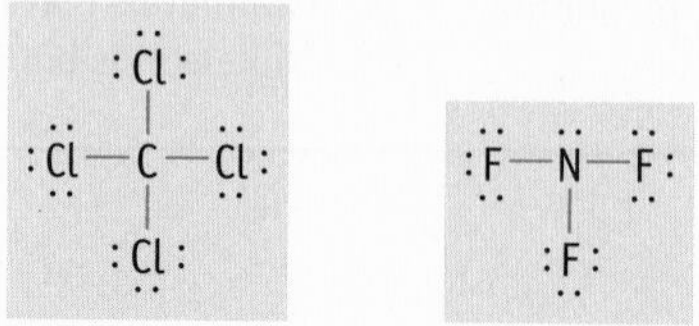

carbon tetrachloride   nitrogen trifluoride

Think about Your Answer As a check, count the number of valence electrons for each molecule, and verify that all are shown in the Lewis structure.

CCl_4: Valence electrons = 4 (for C) + 4 × 7 (for Cl) = 32 electrons (16 pairs)

The structure shows eight electrons in single bonds and 24 electrons as lone pair electrons, for a total of 32 electrons. The structure is correct.

NF_3: Valence electrons = 5 (for N) + 3 × 7 (for F) = 26 electrons (13 pairs)

The structure shows six electrons in single bonds and 20 electrons as lone pair electrons, for a total of 26 electrons. The structure is correct.

Check Your Understanding

Predict Lewis structures for methanol, CH_3OH, and hydroxylamine, H_2NOH. (In each molecule the central C or N atom is surrounded by H atoms and one OH group.)

Oxoacids and Their Anions

Lewis structures of common acids and their anions are shown in Table 8.4. In the absence of water, these acids are covalently bonded molecular compounds, a conclusion that we should draw because all elements in the formula are nonmetals. (Nitric acid, for example, has properties that we associate with a covalent compound: It is a colorless liquid with a boiling point of 83 °C.) In aqueous solution, however, HNO_3, H_2SO_4, and $HClO_4$ ionize to give a hydronium ion and the appropriate anion. A Lewis structure for the nitrate ion, for example, can be created using the guidelines on page 265, and the result is a structure with two N—O single bonds and one N=O double bond. To form nitric acid from the nitrate ion, a hydrogen ion is attached utilizing a lone pair on one of the O atoms.

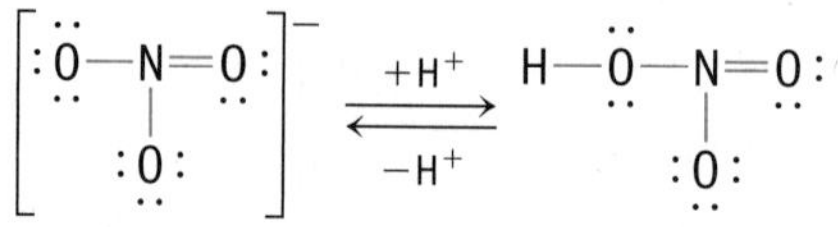

nitrate ion   nitric acid

A characteristic property of acids in aqueous solution is their ability to donate a hydrogen ion (H^+) to water to give the hydronium ion. The NO_3^- anion is formed when the acid, HNO_3, loses a hydrogen ion. The H^+ ion separates from the acid by breaking the H—O bond, the electrons of the bond staying with the O atom. As a result, HNO_3 and NO_3^- have the same number of electrons, 24, and their structures are closely related.

Table 8.4 Lewis Structures of Common Oxoacids and Their Anions

Isoelectronic Species

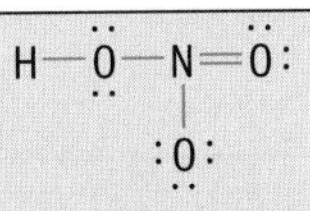

The species NO^+, N_2, CO, and CN^- are similar in that they each have two atoms and the same total number of valence electrons, 10, which leads to a similar Lewis structure for each molecule or ion. The two atoms in each are linked with a triple bond. With three bonding pairs and one lone pair, each atom thus has an octet of electrons.

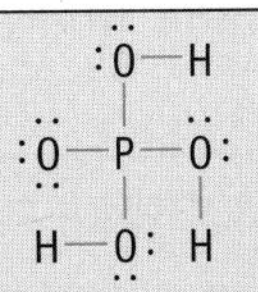

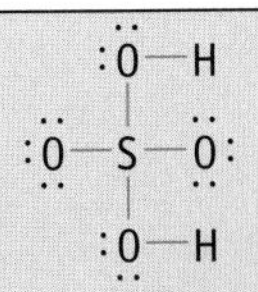

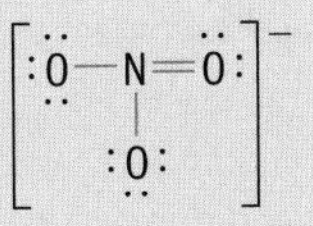

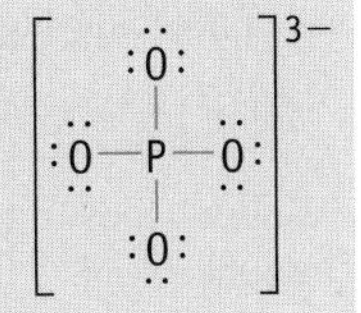

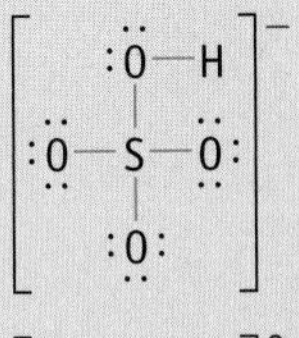

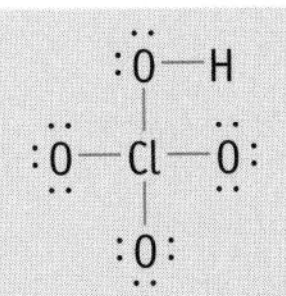

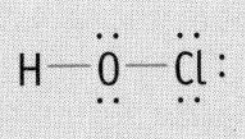

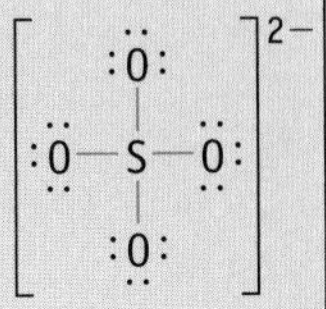

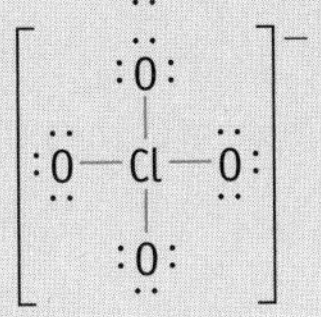

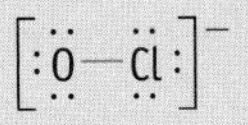

$$[:N{\equiv}O:]^+ \qquad :N{\equiv}N: \qquad :C{\equiv}O: \qquad [:C{\equiv}N:]^-$$

Molecules and ions having the *same number of valence electrons and similar Lewis structures* are said to be **isoelectronic** (Table 8.5). You will find it helpful to recognize isoelectronic molecules and ions because these species have similar electronic (Lewis) structures.

There are similarities and important differences in chemical properties of isoelectronic species. For example, both carbon monoxide, CO, and cyanide ion, CN^-, are very toxic, which results from the fact that they can bind to the iron of hemoglobin in blood and block the uptake of oxygen. They are different, though, in their

● **Isoelectronic and Isostructural** The term *isostructural* is often used in conjunction with isoelectronic species. Species that are isostructural have the same structure. For example, the PO_4^{3-}, SO_4^{2-}, and ClO_4^- ions in Table 8.4 all have four oxygens bonded to the central atom. In addition, they are isoelectronic in that all have 32 valence electrons.

Table 8.5 Some Common Isoelectronic Molecules and Ions

Formulas	Representative Lewis Structure	Formulas	Representative Lewis Structure
BH_4^-, CH_4, NH_4^+	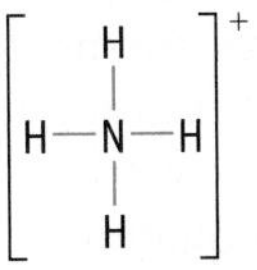	CO_3^{2-}, NO_3^-	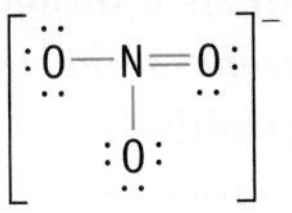
NH_3, H_3O^+	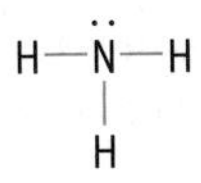	PO_4^{3-}, SO_4^{2-}, ClO_4^-	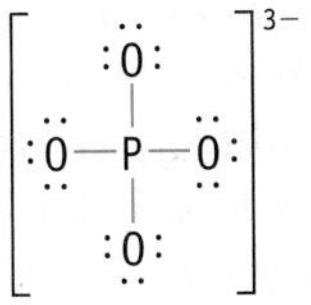
CO_2, OCN^-, SCN^-, N_2O NO_2^+, OCS, CS_2	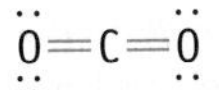		

acid–base chemistry. In aqueous solution, cyanide ion readily adds H^+ to form hydrogen cyanide, whereas CO is not protonated in water.

REVIEW & CHECK FOR SECTION 8.2

1. Which of the following describes the sulfur atom in the Lewis structure of the sulfite ion, SO_3^{2-}?

   (a) Sulfur has three single bonds to oxygen and a lone pair.

   (b) Sulfur bonds to two oxygen atoms with single bonds and one oxygen atom with a double bond.

   (c) Sulfur forms double bonds to three oxygen atoms.

   (d) Sulfur forms three single bonds to oxygen and has no lone pairs.

2. Which one of the species in the list below is NOT isoelectronic with N_3^-?

   (a) N_2O (b) NO_2^+ (c) O_3 (d) CO_2

3. Identify the species in the list below that is NOT isoelectronic with the others.

   (a) O_3 (b) OCN^- (c) NO_2^- (d) SO_2

8.3 Atom Formal Charges in Covalent Molecules and Ions

You have seen that Lewis structures show how electron pairs are placed in a covalently bonded species, whether it is a neutral molecule or a polyatomic ion. Now we turn to one of the consequences of the placement of electron pairs in this way: Individual atoms can be negatively or positively charged or have no electric charge. The location of a positive or negative charge in a molecule or ion will influence, among other things, the atom at which a reaction occurs. For example, does a positive H^+ ion attach itself to the Cl or the O of ClO^-? Is the product HClO or HOCl? It is reasonable to expect H^+ to attach to the more negatively charged atom, and we can predict this by evaluating atom formal charges in molecules and ions.

The **formal charge** is the charge that would reside on an atom in a molecule or polyatomic ion if we assume that all bonding electrons are shared equally. The formal charge for an atom in a molecule or ion is calculated based on the Lewis structure of the molecule or ion, using Equation 8.1,

$$\text{Formal charge of an atom in a molecule or ion} = \text{NVE} - [\text{LPE} + \tfrac{1}{2}(\text{BE})] \qquad (8.1)$$

where NVE = number of valence electrons in the uncombined atom (and equal to its group number in the periodic table), LPE = number of lone pair electrons on an atom, and BE = number of bonding electrons around an atom. The term in square brackets is the number of electrons assigned by the Lewis structure to an atom in a molecule or ion. *The difference between this term and the number of valence electrons on the uncombined atom is the formal charge.* An atom in a molecule or ion will be positive if it "contributes" more electrons to bonding than it "gets back." The atom's formal charge will be negative if the reverse is true.

There are two important assumptions in Equation 8.1. First, lone pairs are assumed to "belong" to the atom on which they reside in the Lewis structure. Second, bond pairs are assumed to be divided equally between the bonded atoms. (The factor of $\frac{1}{2}$ divides the bonding electrons equally between the atoms linked by the bond.)

The sum of the formal charges on the atoms in a molecule or ion always equals the net charge on the molecule or ion. Consider the hypochlorite ion, ClO^-. Oxygen is in Group 6A and so has six valence electrons. However, the oxygen atom in OCl^- can lay claim to seven electrons (six lone pair electrons and one bonding electron), so the atom has a formal charge of -1. The O atom has formally gained an electron by bonding to chlorine.

$$\text{Formal charge} = -1 = 6 - [6 + \tfrac{1}{2}(2)]$$

$$\left[\; :\!\ddot{C}l - \ddot{\underset{\cdot\cdot}{O}}\!: \;\right]^{-} \qquad \Big\} \; \textit{Sum of formal charges} = -1$$

$$\text{Formal charge} = 0 = 7 - [6 + \tfrac{1}{2}(2)]$$

Assume a covalent bond, so bonding electrons are divided equally between Cl and O.

● **HClO$_x$ Acids and Formal Charge** Both ClO^- and ClO_3^- ions attract a proton to give the corresponding acid, HClO and HClO$_3$. In all of the HClO$_x$ acids, the H$^+$ ion is attached to an O atom, owing to the negative formal charge on that atom. (See Table 8.4.)

The formal charge on the Cl atom in ClO^- is zero. So we have -1 for oxygen and 0 for chlorine, and the sum of these equals the net charge of -1 for the ion. An important conclusion we can draw from the formal charges in ClO^- is that, if an H$^+$ ion approaches the ion, it should attach itself to the negatively charged O atom to give hypochlorous acid, HOCl.

EXAMPLE 8.3   Calculating Formal Charges

Problem Calculate formal charges for the atoms of the ClO_3^- ion.

What Do You Know? The Lewis structure of ClO_3^- is required and was developed in Example 8.1.

Strategy The first step is always to write the Lewis structure for the molecule or ion. Then Equation 8.1 can be used to calculate the formal charges.

Solution

$$\text{Formal charge} = -1 = 6 - [6 + \tfrac{1}{2}(2)]$$

$$\left[\; :\!\ddot{\underset{\cdot\cdot}{O}}\! - \overset{\displaystyle :\!\ddot{O}\!:}{\underset{\displaystyle}{Cl}} - \ddot{\underset{\cdot\cdot}{O}}\!: \;\right]^{-}$$

$$\text{Formal charge} = +2 = 7 - [2 + \tfrac{1}{2}(6)]$$

The formal charge on each O atom is -1, whereas for the Cl atom it is $+2$.

A CLOSER LOOK

Comparing Oxidation Number and Formal Charge

In Chapter 3 you learned to calculate the oxidation number of an atom as a way to tell if a reaction involves oxidation and reduction. Are an atom's oxidation number and its formal charge related? To answer this question, let us look at the hydroxide ion. The formal charges are -1 on the O atom and 0 on the H atom. Recall that these formal charges are calculated assuming the O—H bond electrons are shared equally in an O—H covalent bond.

$$\text{Formal charge} = -1 = 6 - [6 + \tfrac{1}{2}(2)]$$

$$\left[\; :\!\ddot{O} - H \;\right]^{-} \qquad \textit{Sum of formal charges} = -1$$

$$\text{Formal charge} = 0 = 1 - [0 + \tfrac{1}{2}(2)]$$

In contrast, in Chapter 3 (page 122), you learned that O has an oxidation number of -2 and H has a number of $+1$. *Oxidation numbers are determined by assuming that the bond between a pair of atoms is ionic*, not covalent. For OH$^-$ this means the pair of electrons between O and H is located *fully* on O. Thus, the O atom now has eight valence electrons instead of six and a charge of -2. The H atom now has no valence electrons and a charge of $+1$.

$$\text{Oxidation number} = -2$$

$$\left[\; :\!\ddot{O}: \; H \;\right]^{-} \qquad \begin{array}{l}\textit{Sum of oxidation}\\ \textit{numbers} = -1\end{array}$$

Assume an ionic bond   Oxidation number $= +1$

Formal charges and oxidation numbers are calculated using different assumptions. Both are useful, but for different purposes. Oxidation numbers allow us to follow changes in redox reactions. Formal charges provide insight into the distribution of charges in molecules and polyatomic ions.

Think about Your Answer Notice that the sum of the formal charges on all the atoms equals the charge on the ion.

Check Your Understanding

Calculate formal charges on each atom in **(a)** CN^- and **(b)** SO_3^{2-}.

REVIEW & CHECK FOR SECTION 8.3

What is the formal charge of the P atom in the anion $H_2PO_4^-$? (The dihydrogen phosphate ion is the anion derived from H_3PO_4 by loss of H^+.)

(a)   +3   (b)   0   (c)   +1   (d)   +5

8.4 Resonance

Ozone, O_3, an unstable, blue, diamagnetic gas with a characteristic pungent odor, protects the earth and its inhabitants from intense ultraviolet radiation from the sun. An important feature of its structure is that the two oxygen–oxygen bonds are the same length, which suggests that the two bonds are equivalent (as described more fully in Section 8.9). That is, equal O—O bond lengths imply an equal number of bond pairs in each O—O bond. Using the guidelines for drawing Lewis structures, however, you might come to a different conclusion. There are two possible ways of writing the Lewis structure for the molecule:

Alternative Ways of Drawing the Ozone Structure

Double bond on the left: $\overset{..}{O}=\overset{..}{O}-\overset{..}{\underset{..}{O}}:$

Double bond on the right: $:\overset{..}{\underset{..}{O}}-\overset{..}{O}=\overset{..}{O}$

These structures are equivalent in that each has a double bond on one side of the central oxygen atom and a single bond on the other side. If either were the actual structure of ozone, one bond (O=O) should be shorter than the other (O—O). The actual structure of ozone shows this is not the case. The inescapable conclusion is that these Lewis structures do not correctly represent the bonding in ozone.

Linus Pauling (1901–1994) proposed the theory of **resonance** to solve the problem. *Resonance structures are used to represent bonding in a molecule or ion when a single Lewis structure fails to describe accurately the actual electronic structure.* The alternative structures shown for ozone are called **resonance structures**. They have identical patterns of bonding and equal energy. The actual structure of this molecule is a *composite,* or **hybrid,** of the equivalent resonance structures. In this resonance hybrid, the bonds between the oxygens are between a single bond and a double bond in length, in this case corresponding to one and a half bonds. This is a reasonable conclusion because we see that the O—O bonds both have a length of 127.8 pm, intermediate between the average length of an O=O double bond (121 pm) and an O—O single bond (132 pm). Because we cannot accurately draw fractions of a bond, chemists draw the resonance structures and connect them with double-headed arrows (↔) to indicate that the true structure is a composite of these extreme structures.

$$:\overset{..}{\underset{..}{O}}-\overset{..}{O}=\overset{..}{\underset{..}{O}}: \longleftrightarrow :\overset{..}{\underset{..}{O}}=\overset{..}{O}-\overset{..}{\underset{..}{O}}:$$

Benzene, C_6H_6, is the classic example of the use of resonance to represent a structure. The benzene molecule is a six-member ring of carbon atoms with six

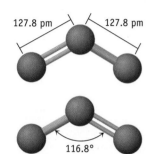

127.8 pm   127.8 pm

116.8°

Ozone, O_3, is a bent molecule with oxygen–oxygen bonds of the same length.

● **Depicting Resonance Structures** The use of an arrow (↔) as a symbol to link resonance structures and the term *resonance* are somewhat unfortunate. An arrow might seem to imply that a change is occurring, and the term *resonance* has the connotation of vibrating or alternating back and forth between different forms. Neither view is correct. Electron pairs are not actually moving from one atom to another.

A CLOSER LOOK

Resonance

- Resonance is a means of representing the bonding in a molecule or polyatomic ion when a single Lewis structure fails to give an accurate picture.
- The atoms must have the same arrangement in each resonance structure. Attaching the atoms in a different fashion creates a different compound.

- Resonance structures differ only in the assignment of electron-pair positions, never atom positions.
- Resonance structures differ in the number of bond pairs between a given pair of atoms.

- Resonance is not meant to indicate the motion of electrons.
- The actual structure of a molecule is a composite or hybrid of the resonance structures.
- There will always be at least one multiple bond (double or triple) in each resonance structure.

equivalent carbon–carbon bonds (and a hydrogen atom attached to each carbon atom). The carbon–carbon bonds are 139 pm long, intermediate between the average length of a C=C double bond (134 pm) and a C—C single bond (154 pm). Two resonance structures that differ only in double bond placement can be written for the molecule. A hybrid of these two structures, however, will lead to a molecule with six equivalent carbon–carbon bonds.

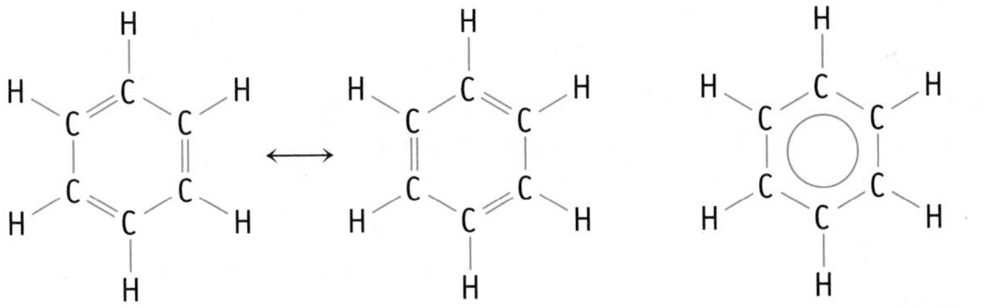

resonance structures of benzene, C_6H_6 abbreviated representation of resonance structures

Let us apply the concepts of resonance to describe bonding in the carbonate ion, CO_3^{2-}, an anion with 24 valence electrons (12 pairs).

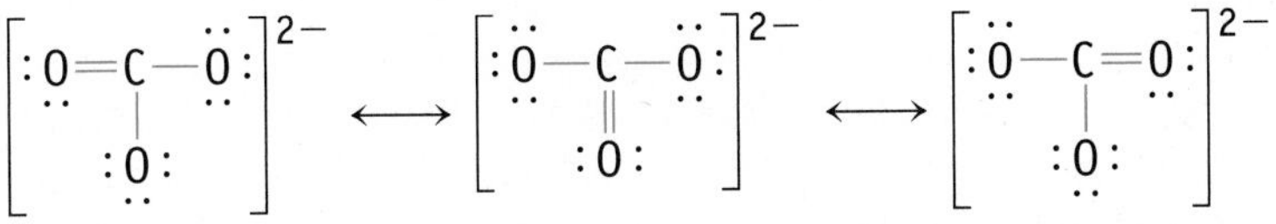

Three equivalent structures can be drawn for this ion, differing only in the location of the C=O double bond, but no single structure correctly describes this ion. Instead, the actual structure is a composite of the three structures, in good agreement with experimental results. In the CO_3^{2-} ion, all three carbon–oxygen bond distances are 129 pm, intermediate between C—O single bond (143 pm) and C=O double bond (122 pm) distances.

Formal charges can be calculated for each atom in the resonance structure for a molecule or ion. For example, using one of the resonance structures for the nitrate ion, we find that the central N atom has a formal charge of +1, and the singly bonded O atoms are both −1. The doubly bonded O atom has no charge. The net charge for the ion is thus −1.

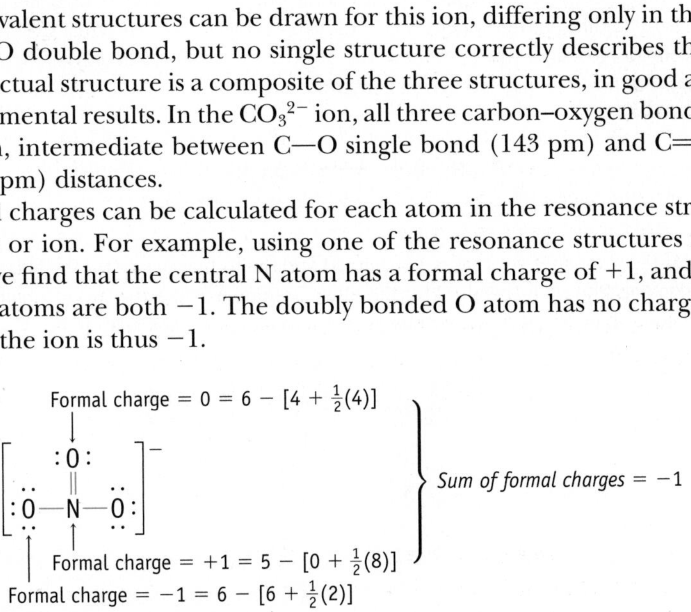

Formal charge $= 0 = 6 - [4 + \frac{1}{2}(4)]$

Sum of formal charges $= -1$

Formal charge $= +1 = 5 - [0 + \frac{1}{2}(8)]$

Formal charge $= -1 = 6 - [6 + \frac{1}{2}(2)]$

Is this a reasonable charge distribution for the nitrate ion? The answer is no. The actual structure of the nitrate ion is a resonance hybrid of three equivalent resonance structures. Because the three oxygen atoms in NO_3^- are equivalent, the charge on one oxygen atom should not be different from the other two. This can be resolved, however, if the formal charges are averaged to give a formal charge of $-\frac{2}{3}$ on the oxygen atoms. Summing the charges on the three oxygen atoms and the $+1$ charge on the nitrogen atom then gives -1, the charge on the ion.

In the resonance structures for O_3, CO_3^{2-}, and NO_3^- all the possible resonance structures are equally likely; they are "equivalent" structures. The molecule or ion therefore has a symmetrical distribution of electrons over all the atoms involved—that is, its electronic structure consists of an equal "mixture," or "hybrid," of the resonance structures.

EXAMPLE 8.4   Drawing Resonance Structures

Problem Draw resonance structures for the nitrite ion, NO_2^-. Are the N—O bonds single, double, or intermediate in value? What are the formal charges on the N and O atoms?

What Do You Know? The N atom is the central atom in the Lewis structure, bonded to two terminal oxygen atoms. You need to determine whether two or more equivalent Lewis structures can be written for this ion.

Strategy

*   Using the guidelines for drawing Lewis structures, determine whether there is more than one way to achieve an octet of electrons around each atom.

*   Use one of the Lewis structures to determine the formal charges on the atoms.

*   If the formal charges on the two oxygen atoms are different, then the formal charge on oxygen will be an average of the two values.

Solution Nitrogen is the central atom in the nitrite ion, which has a total of 18 valence electrons.

Valence electrons = 5 (for the N atom) + 12 (6 for each O atom) + 1 (for negative charge)

After forming N—O single bonds and distributing lone pairs on the terminal O atoms, one electron pair remains, which is placed on the central N atom.

$$\left[:\ddot{O}-\ddot{N}-\ddot{O}: \right]^-$$

To complete the octet of electrons about the N atom, form a N=O double bond. Because there are two ways to do this, two equivalent structures can be drawn, and the actual structure must be a composite or resonance hybrid of these two structures. The nitrogen–oxygen bonds are neither single nor double bonds but have an intermediate value.

$$\left[:\ddot{O}=\ddot{N}-\ddot{O}: \right]^- \longleftrightarrow \left[:\ddot{O}-\ddot{N}=\ddot{O}: \right]^-$$

Taking one of the resonance structures, you should find the formal charge for the N atom is 0. The charge on one O atom is 0 and -1 for the other O atom. Because the two resonance structures are of equal importance, however, the net formal charge on each O atom is $-\frac{1}{2}$.

Formal charge =   Formal charge =
$0 = 6 - [4 + \frac{1}{2}(4)]$ $-1 = 6 - [6 + \frac{1}{2}(2)]$

$$\left[\ddot{O}=\ddot{N}-\ddot{O}: \right]^-$$

Formal charge $= 0 = 5 - [2 + \frac{1}{2}(6)]$

A CLOSER LOOK

Now that you have learned about Lewis structures, resonance, and formal charge, we want to discuss a genuine scientific controversy: the disagreement among chemists about drawing Lewis structures for such simple species as sulfate ion and phosphate ion. For example, two different Lewis structures have been proposed for the sulfate ion:

A Scientific Controversy—Are There Double Bonds in Sulfate and Phosphate Ions?

When drawing Lewis structures, the most important step is first to satisfy the octet rule for each atom in a molecule or ion. If a single Lewis structure does not accurately represent a species, resonance can be introduced. [If two resonance structures are different (as with ions such as OCN^- or a compound like N_2O), a decision can be made among the possible choices by giving preference to structures that minimize formal charges and place negative charges on the most electronegative atoms (page 289).] In the large majority of cases, Lewis structures created by this approach can then be used to determine molecular structure and provide a qualitative picture of bonding.

In this book, we choose to represent the sulfate ion with Lewis structure 1 in which all of the atoms achieve an octet of electrons. The S—O bond order is 1, the sulfur atom has a formal charge of +2, and each oxygen has a formal charge of −1.

The bonding in SO_4^{2-} would involve single bonds between S and O. (As you shall see in Chapter 9, these bonds use the $3s$ and $3p$ orbitals for S and sp^3 hybridization.) Some chemists, however, choose to represent sulfate by a different Lewis structure (structure 2, one of six equivalent structures). Structure 2 is formally a resonance structure of 1. Here, the average sulfur–oxygen bond order is 1.5. This structure is attractive because it minimizes formal charges: sulfur has a formal charge of 0, two of the oxygen atoms have a formal charge of zero, and the other two have formal charges of −1. However, to describe the bonding here, six sulfur orbitals must be used in forming the six bonds. This means that $3d$ orbitals on sulfur must also be involved when describing bonding, in addition to the $3s$ and $3p$ orbitals.

A similar situation is encountered with the phosphate ion and a number of other related species. Several resonance structures can be drawn with or without double bonds.

The question: Which is the better representation of bonding in these species, 1 or 2? Here we find a controversy. A paper in the *Journal of Chemical Education*[1] presents arguments supporting the description of sulfate by structure 2. Among data offered to support multiple bonding between sulfur and oxygen is the observation that the sulfate ion has shorter SO bonds (149 pm) than the known S—O single bond length (about 170 pm) but longer than the double bond in sulfur monoxide (128 pm).

Support for structure 1 comes from a theoretical study also reported in the *Journal of Chemical Education*.[2] The authors of this paper conclude that structure 1 for sulfate is more reasonable and that structure 2 "should be given no weight at all." The basic argument is that the energy of the $3d$ orbitals is too high to allow any significant involvement of these orbitals in bonding. In keeping with this the authors determined that ". . . the calculated d-orbital occupancies . . . are quite small (less than 0.19 e in the entire shell) and inconsistent with any significant valence shell expansion." (In structure 2 the d orbital occupancy would be 2.0 electrons.)

We give more credence to the theoretical study in reference 2 and believe that structure 1 better represents the sulfate ion (and related species) than 2. Others, however, believe structure 2 is the better structure. In any case, there is an important message here: *Lewis structures represent only an approximation when representing structure and bonding.*

References:

1. R. F. See, *J. Chem. Ed.*, **2009**, *86*, 1241–1247.
2. L. Suidan, J. K. Badenhoop, E. D. Glendening, and F. Weinhold, *J. Chem. Ed.*, **1995**, *72*, 583–586.

Additional comments on this matter are found in:

F. Weinhold and C. R. Landis, *Science*, **2007**, *316*, 61.

G. Frenking, R. Tanner, F. Weinhold, and C. R. Landis, *Science*, **2007**, *318*, 746a.

Think about Your Answer The NO bonds in NO_2^- are equivalent, each being intermediate in length between a single N—O bond and a double N=O bond.

Check Your Understanding

Draw resonance structures for the bicarbonate ion, HCO_3^-.

(a) Does HCO_3^- have the same number of resonance structures as the CO_3^{2-} ion?

(b) What are the formal charges on the O and C atoms in HCO_3^-? What is the average formal charge on the O atoms? Compare this with the O atoms in CO_3^{2-}.

(c) Protonation of HCO_3^- gives H_2CO_3. How do formal charges predict where the H^+ ion will be attached?

REVIEW & CHECK FOR SECTION 8.4

1. For which of the following species, SO_4^{2-}, SO_3^{2-}, SO_3, SO_2, is the bonding described by two or more resonance structures?

    (a) SO_3^{2-} and SO_4^{2-} (b) SO_2 and SO_3 (c) SO_2 only   (d) SO_3^{2-}

2. What are the formal charges on nitrogen and oxygen in N_2O_4?

    (a) Formal charges on N and O are 0.

    (b) Formal charge on each N is +1, formal charge on each oxygen is $-\frac{1}{2}$.

    (c) Formal charge on each N is +2, formal charge on each oxygen is -1.

    (d) Formal charge on each N is $+\frac{1}{2}$, formal charge on each oxygen is $-\frac{1}{4}$.

8.5 Exceptions to the Octet Rule

Although the vast majority of molecular compounds and ions obey the octet rule, there are exceptions. These include molecules and ions that have fewer than four pairs of electrons on a central atom, those that have more than four pairs, and those that have an odd number of electrons.

Compounds in Which an Atom Has Fewer Than Eight Valence Electrons

Boron, a metalloid in Group 3A, has three valence electrons and so is expected to form three covalent bonds with other nonmetallic elements. This results in a valence shell for boron in its compounds with only six electrons, two short of an octet. Many boron compounds of this type are known, including such common compounds as boric acid ($B(OH)_3$), borax ($Na_2B_4O_5(OH)_4 \cdot 8\ H_2O$) (Figure 8.2), and the boron trihalides (BF_3, BCl_3, BBr_3, and BI_3).

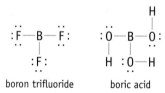

boron trifluoride   boric acid

Boron compounds such as BF_3 that are two electrons short of an octet are quite reactive. The boron atom can accommodate a fourth electron pair when that pair is provided by another atom, and molecules or ions with lone pairs can fulfill this role. Ammonia, for example, reacts with BF_3 to form $H_3N{\rightarrow}BF_3$.

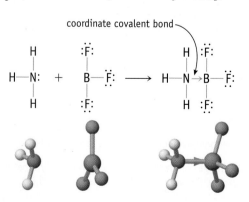

coordinate covalent bond

If a bonding pair of electrons originates on one of the bonded atoms, the bond is called a **coordinate covalent bond**. In Lewis structures, a coordinate covalent

B atom surrounded by 4 electron pairs

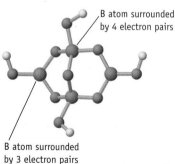

B atom surrounded by 3 electron pairs

FIGURE 8.2 The anion in Borax. Borax is a common mineral that is used in soaps and contains an interesting anion, $B_4O_5(OH)_4^{2-}$. The anion has two B atoms surrounded by four electron pairs, and two B atoms surrounded by only three pairs.

bond is often designated by an arrow that points away from the atom donating the electron pair.

Compounds in Which an Atom Has More Than Eight Valence Electrons

Elements in the third or higher periods form compounds and ions in which the central element can be surrounded by more than four valence electron pairs (Table 8.6). With most compounds and ions in this category, the central atom is bonded to fluorine, chlorine, or oxygen.

It is often obvious from the formula of a compound that an octet around an atom has been exceeded. As an example, consider sulfur hexafluoride, SF_6, a gas formed by the reaction of sulfur and excess fluorine. Sulfur is the central atom in this compound, and fluorine typically bonds to only one other atom with a single electron pair bond (as in HF and CF_4). Six S—F bonds are required in SF_6, meaning there will be six electron pairs in the valence shell of the sulfur atom.

If there are more than four groups bonded to a central atom, this is a reliable signal that there are more than eight electrons around a central atom. But be careful—the central atom octet can also be exceeded with four or fewer atoms bonded to the central atom. Consider three examples from Table 8.6: The central atom in SF_4, ClF_3, and XeF_2 has five electron pairs in its valence shell.

A useful observation is that *only elements of the third and higher periods in the periodic table form compounds and ions in which an octet is exceeded.* Second-period elements (B, C, N, O, F) are restricted to a maximum of eight electrons in their compounds. For example, nitrogen forms compounds and ions such as NH_3, NH_4^+, and NF_3, but NF_5 is unknown. Phosphorus, the third-period element just below nitrogen in the periodic table, forms many compounds similar to nitrogen (PH_3, PH_4^+, PF_3), but it also readily accommodates five or six valence electron pairs in compounds such as PF_5 or in ions such as PF_6^-. Arsenic, antimony, and bismuth, the elements below phosphorus in Group 5A, resemble phosphorus in their behavior.

The usual explanation for the contrasting behavior of second- and third-period elements centers on the number of orbitals in the valence shell of an atom. Second-period elements have four valence orbitals (one 2s and three 2p orbitals). Two electrons per orbital result in a total of eight electrons being accommodated around an atom. For elements in the third and higher periods, the *d* orbitals in the outer shell are traditionally included among valence orbitals for the elements. Thus, for phosphorus, the 3d orbitals are included with the 3s and 3p orbitals as valence orbitals. The extra orbitals provide the element with an opportunity to accommodate up to 12 electrons.

● **Xenon Compounds** Compounds of xenon are among the more interesting entries in Table 8.6 because noble gas compounds were not discovered until the early 1960s. One of the more intriguing compounds is XeF_2, in part because of the simplicity of its synthesis. Xenon difluoride can be made by placing a flask containing xenon gas and fluorine gas in the sunlight. After several weeks, crystals of colorless XeF_2 are found in the flask (▶ page 308).

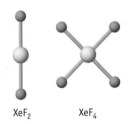

● **A Note on *d* Orbital Use** Chemists often invoke the use of *d* orbitals to explain bonding in molecules or ions if a central atom is surrounded by 5 or 6 single bonds or lone pairs. Some also assume *d* orbitals are involved in multiple bonds in ions such as SO_4^{2-} and PO_4^{3-}. This is not necessary, however, because bonding in these and similar ions is adequately explained using s and p orbitals, and there is little theoretical evidence for multiple bonds. See page 275.

Table 8.6 Lewis Structures in Which the Central Atom Exceeds an Octet

Group 4A	Group 5A	Group 6A	Group 7A	Group 8
SiF_5^-	PF_5	SF_4	ClF_3	XeF_2
SiF_6^{2-}	PF_6^-	SF_6	BrF_5	XeF_4

EXAMPLE 8.5 **Lewis Structures in Which the Central Atom Has More Than Eight Electrons**

Problem Sketch the Lewis structure of the $[ClF_4]^-$ ion.

What Do You Know? Chlorine is the central atom, bonded to four fluorine atoms. The ion has 36 valence electrons (18 pairs) [7 (for Cl) + 4 × 7 (for F) + 1 (for ion charge) = 36].

Strategy Use the guidelines on page 265 to complete the structure.

Solution Draw the ion with four single covalent Cl-F bonds.

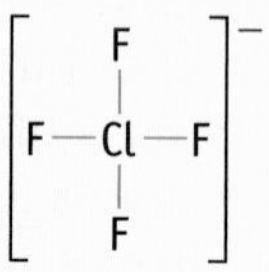

Place lone pairs on the terminal atoms. Because two electron pairs remain after placing lone pairs on the four F atoms and because we know that Cl can accommodate more than four pairs, these two pairs are placed on the central Cl atom.

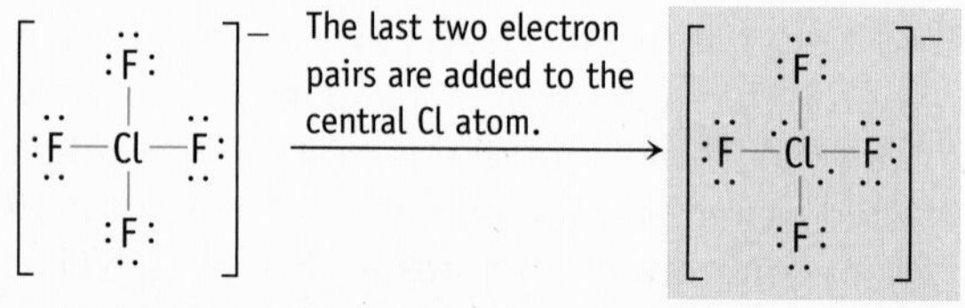

Think about Your Answer The central atom, chlorine, has more than 8 electrons (4 bond pairs and 2 lone pairs.) Because chlorine is in the third period of the periodic table, this is reasonable.

Check Your Understanding

Sketch the Lewis structures for $[ClF_2]^+$ and $[ClF_2]^-$. How many lone pairs and bond pairs surround the Cl atom in each ion?

Molecules with an Odd Number of Electrons

Two nitrogen oxides—NO, with 11 valence electrons, and NO_2, with 17 valence electrons—are among a very small group of stable molecules with an odd number of electrons. Because they have an odd number of electrons, it is impossible to draw a structure obeying the octet rule; at least one electron must be unpaired.

Even though NO_2 does not obey the octet rule, an electron dot structure can be written that approximates the bonding in the molecule. This Lewis structure places the unpaired electron on nitrogen. Two resonance structures show that the nitrogen-oxygen bonds are expected to be equivalent.

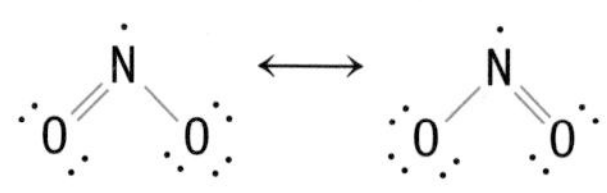

Experimental evidence for NO indicates that the bonding between N and O is intermediate between a double and a triple bond. It is not possible to write a Lewis structure for NO that is in accord with the properties of this substance, so a different theory is needed to understand bonding in this molecule. We shall return to compounds of this type when molecular orbital theory is introduced in Section 9.3.

The two nitrogen oxides, NO and NO_2, are members of a class of chemical substances called free radicals. A **free radical** is a chemical species with an unpaired electron. Free radicals are generally quite reactive. Atoms such as H and Cl, for example, are free radicals and readily combine with each other to give molecules such as H_2, Cl_2, and HCl.

CASE STUDY

Hydroxyl Radicals, Atmospheric Chemistry, and Hair Dyes

Free radical is a term chemists use for molecules containing an unpaired electron in their valence shell. Such molecules are often highly reactive and are crucially important in environmental chemistry and biochemistry. One example is the hydroxyl radical, ·OH. A paper in *Science* magazine recently called hydroxyl radicals "the 'detergent' of the atmosphere. . . . Hydroxyl radicals are the single most important oxidant in the atmosphere because they are the agent primarily responsible for removing the majority of gases emitted into the atmosphere by natural and anthropogenic activity."

The *Science* paper describes what happens when NO_2 molecules, free radicals and a common air pollutant, react with water in the presence of sunlight. One product is the hydroxyl radical, ·OH.

$$NO_2(g) + H_2O(g) \rightarrow$$
$$\cdot OH(g) + HONO(g)$$

The other product, HONO, decomposes with sunlight, to also give a hydroxyl radical and another free radical, NO.

$$HONO(g) \rightarrow HO\cdot(g) + \cdot NO(g)$$

© Cengage Learning/Charles D. Winters

The hydroxyl radicals then go on to react with trace gases (such as SO_2, NH_3, CO, and hydrocarbons) in the atmosphere.

Hydroxyl radicals also play a role in dyeing your hair. Most hair dyes are a mixture of a dye, ammonia, and hydrogen peroxide (H_2O_2). These species react to produce the perhydroxyl anion (HOO^-) and the ·OH radical. The HOO^- ion is a base, and the basic solution causes the hair to swell. This helps in allowing the perhydroxyl anion to react with the melamine pigment in the hair and lighten it. The problem with this approach is that the ·OH radical damages hair, so chemists

came up with a new product. This is a mixture of ammonium carbonate, $(NH_4)_2CO_3$, and H_2O_2 in the presence of sodium glycinate, $Na[H_2NCH_2CO_2]$. The mixture is less basic and produces the peroxymonocarbonate ion (HCO_4^-) and a carbonate radical anion ($CO_3\cdot^-$). The HCO_4^- ion functions as the bleach, and the glycinate ion scavenges the carbonate radicals that could otherwise damage the hair.

Questions:

1. Draw a Lewis structure for HONO. Are there any resonance structures? What is the formal charge on each atom?

2. Draw a Lewis structure for the peroxymonocarbonate ion. (One feature of this ion is a C-O-O-H grouping.) Give the formal charge on each atom and draw any resonance structures.

Answers to these questions are available in Appendix N.

References:
1. S. Li, J. Matthews, and A. Sinha, *Science*, **2008**, *319*, 1657.
2. L. Jarvis, *Chemical & Engineering News*, February 11, 2008, 32.

Free radicals are involved in many reactions in the environment. For example, small amounts of NO are released from vehicle exhausts. The NO rapidly forms NO_2, which is even more harmful to human health and to plants. Exposure to NO_2 at concentrations of 50–100 parts per million can lead to significant inflammation of lung tissue. Nitrogen dioxide is also generated by natural processes. For example, when hay, which has a high level of nitrates, is stored in silos on farms, NO_2 can be generated as the hay ferments, and there have been reports of farm workers dying from exposure to this gas in the silo (because NO_2 reacts with water in the lungs to produce nitric acid).

The two nitrogen oxides, NO and NO_2, are unique in that they can be isolated, and neither has the extreme reactivity of most free radicals. When cooled, however, two NO_2 molecules join or "dimerize" to form colorless N_2O_4; the unpaired electrons combine to form an N—N bond in N_2O_4 (Figure 8.3).

REVIEW & CHECK FOR SECTION 8.5

1. In which of the following sulfur species, SF_2, SF_3^+, SF_4, SF_5^+, and SF_6, does S exceed the octet configuration?

    (a) SF_5^+ and SF_6

    (b) SF_4, SF_5^+, and SF_6

    (c) SF_6 only

    (d) all five species

2. How many lone pairs of electrons are on the central atom Xe in XeF_2?

    (a) 0   (b) 1   (c) 2   (d) 3

FIGURE 8.3 **Free radical chemistry.** When cooled, the concentration of the brown gas NO_2 drops dramatically. Nitrogen dioxide is a free radical, and two NO_2 molecules couple to form colorless N_2O_4, a molecule with an N—N single bond. (N_2O_4 has a boiling point of about 21 °C, so it is liquid at 0 °C. This photo shows some NO_2 remains at this temperature.)

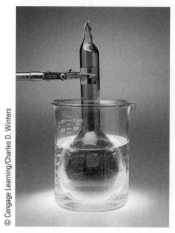

A flask of brown NO_2 gas in warm water

When cooled, NO_2 free radicals couple to form N_2O_4 molecules.
————————————→
N_2O_4 gas is colorless.

A flask of NO_2 gas in ice water

8.6 Molecular Shapes

Lewis structures show the atom connectivity in molecules and polyatomic ions, but they do not show three-dimensional geometry. However, the three-dimensional geometry of molecules and ions is often crucial to their function, so we want to take the next step: Use the Lewis structures to predict the three-dimensional structure.

The **valence shell electron-pair repulsion (VSEPR)** model is a method for predicting the shapes of covalent molecules and polyatomic ions. This model is based on the idea that *bond and lone electron pairs in the valence shell of an element repel each other and seek to be as far apart as possible.* The positions assumed by the bond and lone electron pairs thus define the angles between bonds to surrounding atoms. VSEPR is remarkably successful in predicting structures of molecules and polyatomic ions of main group elements, although it is less effective (and seldom used) in predicting structures of compounds containing transition metals.

To have a sense of how valence shell electron pairs determine structure, blow up several balloons to a similar size. Imagine that each balloon represents an electron cloud. When two, three, four, five, or six balloons are tied together at a central point (representing the nucleus and core electrons of a central atom), the balloons naturally form the shapes shown in Figure 8.4. These geometric arrangements minimize interactions between the balloons.

Central Atoms Surrounded Only by Single-Bond Pairs

The simplest application of VSEPR theory is with molecules and ions in which all the electron pairs around the central atom are involved in single covalent bonds. Figure 8.5 illustrates the geometries predicted for molecules or ions with the general formulas AX_n, where A is the central atom and n is the number of X groups bonded to it.

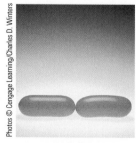

Linear

Trigonal planar

Tetrahedral

Trigonal bipyramidal

Octahedral

FIGURE 8.4 **Balloon models of electron-pair geometries for two to six electron pairs.** If two to six balloons of similar size and shape are tied together, they will naturally assume the arrangements shown. These pictures illustrate the predictions of VSEPR.

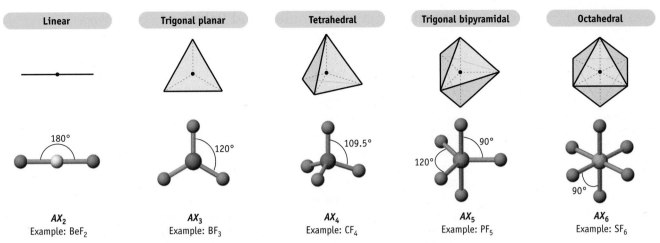

FIGURE 8.5 Various geometries predicted by VSEPR. Geometries predicted by VSEPR for molecules that contain only single covalent bonds around the central atom.

The linear geometry for two single bond pairs and the trigonal-planar geometry for three single bond pairs involve a central atom that does not have an octet of electrons (see Section 8.5). In contrast, the central atom in a tetrahedral molecule obeys the octet rule with four bond pairs. The central atoms in trigonal-bipyramidal and octahedral molecules have five and six bonding pairs, respectively, and are expected only when the central atom is an element in Period 3 or higher in the periodic table (▶ page 284).

EXAMPLE 8.6   Predicting Molecular Shapes

Problem Predict the shape of silicon tetrachloride, $SiCl_4$.

What Do You Know? You know the compound's formula and the number of valence electrons. In $SiCl_4$ you have 32 valence electrons or 16 pairs. You also know Si is the central atom. This information enables you to draw the electron dot structure.

Strategy The first step is to draw the Lewis structure. However, you do not need to draw it in any particular way because its purpose is only to describe the number of bonds around an atom and to determine if there are any lone pairs. The shape of the molecule is dictated by the positions of the four electron pairs used in bonding to the four chlorine atoms.

Solution The Lewis structure of $SiCl_4$ has four electron pairs, all of them bond pairs, around the central Si atom. Therefore, a tetrahedral structure is predicted for the $SiCl_4$ molecule, with Cl—Si—Cl bond angles of 109.5°. This agrees with the actual structure for $SiCl_4$.

Lewis structure   Molecular geometry

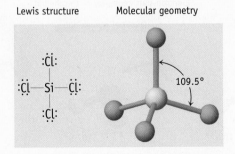

Think about Your Answer Be sure to recognize that all four positions in a tetrahedral geometry are equivalent.

Check Your Understanding

What is the shape of the dichloromethane (CH_2Cl_2) molecule? Predict the Cl—C—Cl bond angle.

Central Atoms with Single-Bond Pairs and Lone Pairs

To see how lone pairs affect the geometry of a molecule or polyatomic ion, return to the balloon models in Figure 8.4. If you assume the balloons represent *all* the electron pairs in the valence shell, the model predicts the "electron-pair geometry" rather than the "molecular geometry." The **electron-pair geometry** is the geometry assumed by *all* the valence electron pairs around a central atom (both bond and lone pairs), whereas the **molecular geometry** describes the arrangement in space of the central atom and the atoms directly attached to it. It is important to recognize that *lone pairs of electrons on the central atom occupy spatial positions, even though their location is not included in the verbal description of the shape of the molecule or ion.*

Let us use the VSEPR model to predict the molecular geometry and bond angles in the NH_3 molecule. On drawing the Lewis structure, we see there are four pairs of electrons in the nitrogen valence shell: three bond pairs and one lone pair. Thus, the predicted *electron-pair geometry* is tetrahedral. The *molecular geometry,* however, is said to be *trigonal pyramidal* because that describes the location of the atoms. The nitrogen atom is at the apex of the pyramid, and the three hydrogen atoms form the trigonal base.

H—N—H
|
H

Lewis structure

→

electron-pair
geometry, tetrahedral

→

molecular geometry

→

Actual H–N–H
angle = 107.5°

molecular geometry,
trigonal pyramidal

Effect of Lone Pairs on Bond Angles

Because the electron-pair geometry in NH_3 is tetrahedral, we would expect the H—N—H bond angle to be 109.5°. However, the experimentally determined bond angles in NH_3 are 107.5°, and the H—O—H angle in water is smaller still (104.5°) (Figure 8.6). These angles are close to the tetrahedral angle but not exactly that value. This highlights the fact that VSEPR theory does not give exact bond angles; it can only predict approximate geometry. Small variations in geometry (e.g., bond angles a few degrees different from predicted) are quite common and often arise

FIGURE 8.6 The molecular geometries of methane, ammonia, and water. All have four electron pairs around the central atom and have a tetrahedral electron-pair geometry. The decrease in bond angles in the series can be explained by the fact that the lone pairs have a larger spatial requirement than the bond pairs.

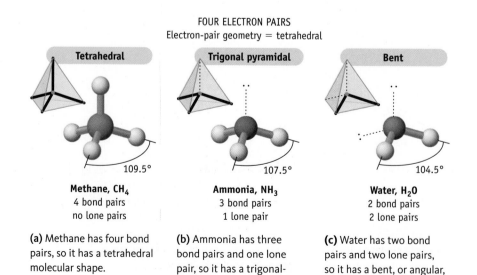

FOUR ELECTRON PAIRS
Electron-pair geometry = tetrahedral

Tetrahedral   Trigonal pyramidal   Bent

109.5°   107.5°   104.5°

Methane, CH_4 Ammonia, NH_3 Water, H_2O
4 bond pairs   3 bond pairs   2 bond pairs
no lone pairs   1 lone pair   2 lone pairs

(a) Methane has four bond pairs, so it has a tetrahedral molecular shape.

(b) Ammonia has three bond pairs and one lone pair, so it has a trigonal-pyramidal molecular shape.

(c) Water has two bond pairs and two lone pairs, so it has a bent, or angular, molecular shape.

because there is a difference between the spatial requirements of lone pairs and bond pairs. Lone pairs of electrons seem to occupy a larger volume than bonding pairs, and the increased volume of lone pairs causes bond pairs to squeeze closer together. In general, the relative strengths of repulsions are in the order

Lone pair–lone pair > lone pair–bond pair > bond pair–bond pair

The different spatial requirements of lone pairs and bond pairs can be used to predict variations in the bond angles in a series of molecules. For example, the bond angles decrease in the series CH_4, NH_3, and H_2O as the number of lone pairs on the central atom increases (Figure 8.6).

ⓦWL   INTERACTIVE EXAMPLE 8.7   Finding the Shapes of Molecules and Polyatomic Ions

Problem What are the shapes of the ions H_3O^+ and ClF_2^+?

What Do You Know? To determine electron-pair and molecular geometries you must first draw the Lewis structure.

Strategy

- Draw the Lewis structures for each ion.

- Count the number of lone and bond pairs around the central atom.

- Use Figure 8.5 to decide on the electron-pair geometry. The location of the atoms in the ion gives the geometry of the ion.

Solution

(a) The Lewis structure of the hydronium ion, H_3O^+, shows that the oxygen atom is surrounded by four electron pairs, so the electron-pair geometry is tetrahedral. Because three of the four pairs are used to bond terminal atoms, the central O atom and the three H atoms form a trigonal-pyramidal molecular shape like that of NH_3.

Lewis structure	electron-pair geometry, tetrahedral	molecular geometry	molecular geometry, trigonal pyramidal

(b) Chlorine is the central atom in ClF_2^+. It is surrounded by four electron pairs, so the electron-pair geometry around chlorine is tetrahedral. Because only two of the four pairs are bonding pairs, the ion has a bent geometry.

Lewis structure	electron-pair geometry, tetrahedral	molecular geometry	molecular geometry, bent or angular

Think about Your Answer In each of these ions the occurrence of lone pairs on the central atom influences the molecular geometry. In both ions the angle (H–O–H or F–Cl–F) is likely to be slightly less than 109.5°.

Check Your Understanding

Give the electron-pair geometry and molecular shape for BF_3 and BF_4^-. What is the effect on the molecular geometry of adding an F^- ion to BF_3 to give BF_4^-?

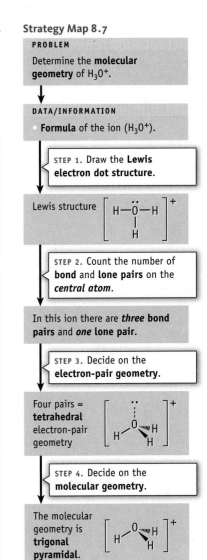

Strategy Map 8.7

PROBLEM
Determine the **molecular geometry** of H_3O^+.

DATA/INFORMATION
- **Formula** of the ion (H_3O^+).

STEP 1. Draw the **Lewis electron dot structure**.

Lewis structure

STEP 2. Count the number of **bond** and **lone pairs** on the **central atom**.

In this ion there are **three bond pairs** and **one lone pair**.

STEP 3. Decide on the **electron-pair geometry**.

Four pairs = **tetrahedral** electron-pair geometry

STEP 4. Decide on the **molecular geometry**.

The molecular geometry is **trigonal pyramidal**.

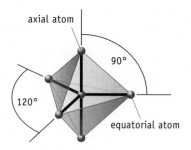

FIGURE 8.7 The trigonal bipyramid showing the axial and equatorial atoms. The angles between atoms in the equator are 120°. The angles between equatorial and axial atoms are 90°. If there are lone pairs they are generally found in the equatorial positions.

Central Atoms with More Than Four Valence Electron Pairs

The situation becomes more complicated if the central atom has five or six electron pairs, some of which are lone pairs. A trigonal-bipyramidal structure (Figures 8.5 and 8.7) has two sets of positions that are not equivalent. The positions in the trigonal plane lie in the equator of an imaginary sphere around the central atom and are called *equatorial* positions. The north and south poles in this representation are called *axial* positions. Each equatorial atom has two neighboring groups (the axial atoms) at 90°, and each axial atom has three groups (the equatorial atoms) at 90°. If lone pairs are present in the valence shell, they require more space than bonding pairs and so prefer to occupy equatorial positions rather than axial positions.

The entries in the top row of Figure 8.8 show species having a total of five valence electron pairs, with zero, one, two, and three lone pairs. In SF_4, with one lone pair, the molecule assumes a *"seesaw"* shape with the lone pair in one of the equatorial positions. The ClF_3 molecule has three bond pairs and two lone pairs. The two lone pairs in ClF_3 are in equatorial positions; two bond pairs are axial, and the third is in the equatorial plane, so the molecular geometry is *T-shaped*. The third molecule shown is XeF_2. Here, all three equatorial positions are occupied by lone pairs, so the molecular geometry is *linear*.

The geometry assumed by six electron pairs is *octahedral* (Figure 8.8), and all the angles at adjacent positions are 90°. Unlike the trigonal bipyramid, the octahedron has no distinct axial and equatorial positions; all positions are the same. Therefore, if the molecule has one lone pair, as in BrF_5, it makes no difference which position it occupies. The lone pair is often drawn in the top or bottom position to make it easier to visualize the molecular geometry, which in this case is *square pyramidal*. If two pairs of electrons in an octahedral arrangement are lone pairs, they seek to be as far apart as possible. The result is a *square planar* molecule, as illustrated by XeF_4.

EXAMPLE 8.8   Predicting Molecular Shape

Problem What is the shape of the ICl_4^- ion?

What Do You Know? You know the number of valence electrons, so you can draw the Lewis electron dot structure.

Strategy Draw the Lewis structure, and then decide on the electron-pair geometry. The position of the atoms gives the molecular geometry of the ion. (See Example 8.7 and Figure 8.8.)

Solution A Lewis structure for the ICl_4^- ion shows that the central iodine atom has six electron pairs in its valence shell. Two of these are lone pairs. Placing the lone pairs on opposite sides leaves the four chlorine atoms in a square-planar geometry.

electron-pair geometry, octahedral $\longrightarrow$ molecular geometry, square planar

Think about Your Answer Square-planar geometry allows the two lone pairs on the central atom to be as far apart as possible.

Check Your Understanding

Draw the Lewis structure for ICl_2^-, and then decide on the geometry of the ion.

Multiple Bonds and Molecular Geometry

Double and triple bonds involve more electron pairs than single bonds, but this has little effect on the overall molecular shape. All of the electron pairs in a multiple bond are shared between the same two nuclei and therefore occupy the

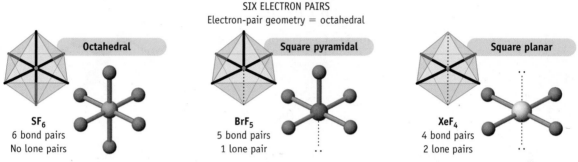

FIVE ELECTRON PAIRS
Electron-pair geometry = trigonal bipyramidal

Trigonal bipyramidal

PF₅
5 bond pairs
No lone pairs

Seesaw

SF₄
4 bond pairs
1 lone pair

T-shaped

ClF₃
3 bond pairs
2 lone pairs

Linear

XeF₂
2 bond pairs
3 lone pairs

SIX ELECTRON PAIRS
Electron-pair geometry = octahedral

Octahedral

SF₆
6 bond pairs
No lone pairs

Square pyramidal

BrF₅
5 bond pairs
1 lone pair

Square planar

XeF₄
4 bond pairs
2 lone pairs

FIGURE 8.8 Electron-pair and molecular geometries for molecules and ions with five or six electron pairs around the central atom.

same region of space. Therefore, all electron pairs in a multiple bond count as one bonding group and affect the molecular geometry the same as a single bond does. For example, the carbon atom in CO_2 has no lone pairs and participates in two double bonds. Each double bond is a region of electron density and effectively counts as one bond for the purpose of predicting geometry; the structure of CO_2 is linear.

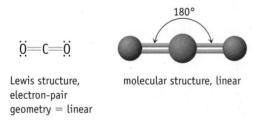

180°

Lewis structure,
electron-pair
geometry = linear

molecular structure, linear

When resonance structures are possible, the geometry can be predicted from any of the Lewis resonance structures or from the resonance hybrid structure. For example, the geometry of the CO_3^{2-} ion is predicted to be trigonal planar because the carbon atom has three sets of bonds and no lone pairs.

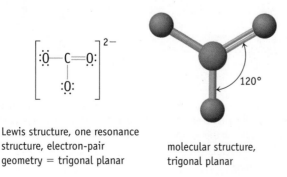

120°

Lewis structure, one resonance
structure, electron-pair
geometry = trigonal planar

molecular structure,
trigonal planar

The NO_2^- ion also has a trigonal-planar electron-pair geometry. Because there is a lone pair on the central nitrogen atom, and bonds in the other two positions, the geometry of the ion is angular or bent.

$$\left[:\ddot{O}-\ddot{N}=O: \right]^{-}$$

Lewis structure, one
resonance structure,
electron-pair geometry
= trigonal planar

115°

molecular structure,
angular or bent

The techniques just outlined can be used to find the geometries around the atoms in more complicated molecules. Consider, for example, cysteine, one of the natural amino acids.

$$H-\ddot{S}-\underset{|}{\overset{H}{\underset{H}{C_3}}}-\underset{|}{\overset{H}{\underset{:N-H}{C_2}}}-\underset{}{\overset{:O:}{\overset{\parallel}{C_1}}}-\ddot{O}-H$$

Cysteine, $HSCH_2CH(NH_2)CO_2H$

Four pairs of electrons occur around the S, N, C_2, and C_3 atoms, and around the oxygen attached to the hydrogen, so the electron-pair geometry around each is tetrahedral. Thus, the H—S—C, S—C—H, H—N—H, and C—O—H angles are predicted to be approximately 109°. The angle made by O—C_1—O is 120° because the electron-pair geometry around C_1 is trigonal planar.

EXAMPLE 8.9 Finding the Shapes of Molecules and Ions

Problem What are the shapes of the nitrate ion, NO_3^-, and $XeOF_4$?

What Do You Know? You know the number of valence electrons in each species. Also, note in the discussion above that multiple bonds are treated as one bonding group in determining geometry.

Strategy Questions on molecular structure are answered using the following roadmap:
Formula → Lewis structure → electron-pair geometry → molecular geometry.

Solution

(a) The NO_3^- and CO_3^{2-} ions are isoelectronic. Thus, like the carbonate ion described in the text above, the electron-pair geometry and molecular shape of NO_3^- are trigonal planar.

(b) The $XeOF_4$ molecule has a Lewis structure with a total of six electron pairs about the central Xe atom, one of which is a lone pair. It has a square-pyramidal molecular structure.

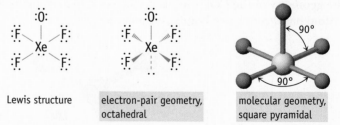

Lewis structure

electron-pair geometry,
octahedral

molecular geometry,
square pyramidal

Think about Your Answer Two structures are possible for $XeOF_4$ based on the position occupied by the oxygen, but there is no way to predict which is correct. The actual structure is the one shown, with the oxygen in the apex of the square pyramid.

Check Your Understanding

Use Lewis structures and the VSEPR model to determine the electron-pair and molecular geometries for (a) the phosphate ion, PO_4^{3-}; (b) the sulfite ion, SO_3^{2-}; and (c) IF_5.

8.7 Bond Polarity and Electronegativity

The models used to represent covalent and ionic bonding are the extreme situations in bonding. Pure covalent bonding, in which atoms share an electron pair equally, occurs *only* when two identical atoms are bonded. When two dissimilar atoms form a covalent bond, the electron pair will be unequally shared. The result is a **polar covalent bond**, a bond in which the two atoms have partial charges (Figure 8.9).

Bonds are polar because not all atoms hold onto their valence electrons with the same force or take on additional electrons with equal ease. Recall from the discussion of atom properties that different elements have different values of ionization energy and electron affinity (electron attachment enthalpy) (Section 7.5). These differences in behavior for free atoms carry over to atoms in molecules.

If a bond pair is not equally shared between atoms, the bonding electrons are on average nearer to one of the atoms. The atom toward which the pair is displaced has a larger share of the electron pair and thus acquires a partial negative charge. At the same time, the atom at the other end of the bond is depleted in electrons and acquires a partial positive charge. The bond between the two atoms has a positive end and a negative end; that is, it has negative and positive poles. The bond is called a **polar bond**.

In ionic compounds, displacement of the bonding pair to one of the two atoms is essentially complete, and + and − symbols are written alongside the atom symbols in the Lewis drawings. For a polar covalent bond, the polarity is indicated by writing the symbols δ^+ and δ^- alongside the atom symbols, where δ (the Greek lowercase letter "delta") stands for a *partial* charge. With so many atoms to use in covalent bond formation, it is not surprising that bonds between atoms can fall anywhere in a continuum from pure covalent to pure ionic (Figure 8.10). There is no sharp dividing line between an ionic bond and a covalent bond.

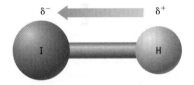

FIGURE 8.9 A polar covalent bond. Iodine has a larger share of the bonding electrons, and hydrogen has a smaller share. The result is that I has a partial negative charge (δ^-), and H has a partial positive charge (δ^+). Sometimes bond polarity is shown by an arrow pointing from + to − (as shown here).

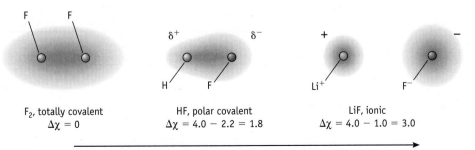

F_2, totally covalent
$\Delta\chi = 0$

HF, polar covalent
$\Delta\chi = 4.0 - 2.2 = 1.8$

LiF, ionic
$\Delta\chi = 4.0 - 1.0 = 3.0$

Increasing ionic character

FIGURE 8.10 Covalent to ionic bonding. As the electronegativity difference increases between the atoms of a bond, the bond becomes increasingly ionic.

Linus Pauling (1901–1994). Linus Pauling was born in Portland, Oregon, earned a B.Sc. degree in chemical engineering from Oregon State College in 1922, and completed his Ph.D. in chemistry at the California Institute of Technology in 1925. In chemistry, he is well known for his book *The Nature of the Chemical Bond.* He also studied protein structures and, in the words of Francis Crick, was "one of the founders of molecular biology." It was this work and his study of chemical bonding that were cited in the award of the Nobel Prize in Chemistry in 1954. Although chemistry was the focus of his life, at the urging of his wife, Ava Helen, he was also involved in nuclear disarmament issues, and he received the Nobel Peace Prize in 1962 for the role he played in advocating for the nuclear test ban treaty.

In the 1930s, Linus Pauling proposed a parameter called *atom electronegativity* that allows us to decide if a bond is polar, which atom of the bond is partially negative and which is partially positive, and if one bond is more polar than another. The **electronegativity**, χ, of an atom is defined as a measure of *the ability of an atom in a molecule to attract electrons to itself.*

Looking at the values of electronegativity in Figure 8.11 you will notice several important features:

- The element with the largest electronegativity is fluorine; it is assigned a value of $\chi = 4.0$. The element with the smallest value is the alkali metal cesium.
- Electronegativities generally increase from left to right across a period and decrease down a group. This is the opposite of the trend observed for metallic character.
- Metals typically have low values of electronegativity, ranging from slightly less than 1 to about 2.
- Electronegativity values for the metalloids are around 2, whereas nonmetals have values greater than 2.

Notice that there is a large *difference* in electronegativity for atoms from the left- and right-hand sides of the periodic table. For cesium fluoride, for example, the difference in electronegativity values, $\Delta\chi$, is 3.2 [= 4.0 (for F) − 0.8 (for Cs)]. The bond is decidedly ionic in CsF, with Cs the cation (Cs^+) and F the anion (F^-). In contrast, the electronegativity difference between H and F in HF is only 1.8 [= 4.0 (for F) − 2.2 (for H)]. We conclude that bonding in HF must be more covalent, as expected for a compound formed from two nonmetals. Because the electronegativities of hydrogen and fluorine are different, the H—F bond is polar. In a polar bond, the more electronegative atom takes on the partial negative charge, and the less electronegative atom takes on the partial positive charge, thus the hydrogen is the positive end of this molecule and fluorine is the negative end ($H^{\delta+}$—$F^{\delta-}$).

HF

1A	2A											H 2.2		3A	4A	5A	6A	7A
Li 1.0	Be 1.6													B 2.0	C 2.5	N 3.0	O 3.5	F 4.0
Na 0.9	Mg 1.3	3B	4B	5B	6B	7B		8B		1B	2B			Al 1.6	Si 1.9	P 2.2	S 2.6	Cl 3.2
K 0.8	Ca 1.0	Sc 1.4	Ti 1.5	V 1.6	Cr 1.7	Mn 1.5	Fe 1.8	Co 1.9	Ni 1.9	Cu 1.9	Zn 1.6	Ga 1.8	Ge 2.0	As 2.2	Se 2.6	Br 3.0
Rb 0.8	Sr 1.0	Y 1.2	Zr 1.3	Nb 1.6	Mo 2.2	Tc 1.9	Ru 2.2	Rh 2.3	Pd 2.2	Ag 1.9	Cd 1.7	In 1.8	Sn 2.0	Sb 1.9	Te 2.1	I 2.7
Cs 0.8	Ba 0.9	La 1.1	Hf 1.3	Ta 1.5	W 2.4	Re 1.9	Os 2.2	Ir 2.2	Pt 2.3	Au 2.5	Hg 2.0	Tl 1.6	Pb 2.3	Bi 2.0	Po 2.0	At 2.2

■ <1.0 □ 1.5–1.9 ▨ 2.5–2.9
■ 1.0–1.4 ▨ 2.0–2.4 ■ 3.0–4.0

FIGURE 8.11 Pauling electronegativity values for the elements. Nonmetals have high values of electronegativity, the metalloids have intermediate values, and the metals have low values. (Values for these elements as well as for the noble gases and for the lanthanides and actinides are widely available; the values in this table are from J. Emsley, *The Elements*, 3rd ed., Clarendon Press, Oxford, 1998.)

EXAMPLE 8.10   Estimating Bond Polarities

Problem For each of the following bond pairs, decide which bond is more polar and indicate the negative and positive poles.

(a)   B—F and B—Cl

(b)   Si—O and P—P

What Do You Know? The more polar bond is the one for which there is the larger difference in electronegativity of the two atoms.

Strategy Locate the elements in the periodic table. Recall that electronegativity generally increases across a period and up a group.

Solution

(a)   B and F lie relatively far apart in the periodic table. B is a metalloid, and F is a nonmetal. Here, χ for B = 2.0, and χ for F = 4.0. Similarly, B and Cl are relatively far apart in the periodic table, but Cl is below F in the periodic table (χ for Cl = 3.2) and is therefore less electronegative than F. The difference in electronegativity for B—F is 2.0, and for B—Cl it is 1.2. Both bonds are expected to be polar, with B positive and the halide atom negative, but a B—F bond is more polar than a B—Cl bond.

(b)   Because the bond is between two atoms of the same kind, the P—P bond is nonpolar. Silicon is in Group 4A and the third period, whereas O is in Group 6A and the second period. Consequently, O has a greater electronegativity (3.5) than Si (1.9), and the bond is highly polar ($\Delta\chi = 1.6$), with O the more negative atom.

Think about Your Answer Polar bonds arise because one atom attracts the pair of electrons in a chemical bond more strongly than the atom to which it is bonded.

Check Your Understanding

For each of the following pairs of bonds, decide which is more polar. For each polar bond, indicate the positive and negative poles. First, make your prediction from the relative atom positions in the periodic table, then check your prediction by calculating $\Delta\chi$.

(a)   H—F and H—I   (b)   B—C and B—F   (c)   C—Si and C—S

Charge Distribution: Combining Formal Charge and Electronegativity

The way electrons are distributed in a molecule or ion is called its **charge distribution**. Charge distribution can profoundly affect the properties of a molecule. Examples include physical properties such as melting and boiling points, and chemical properties such as susceptibility to attack by an anion or cation or whether the molecule is an acid or a base.

We saw earlier (see page 270) that formal charge calculations can often locate the site of a charge in a molecule or an ion. However, this can sometimes lead to results that are incorrect because formal charge calculations assume that there is equal sharing of electrons in all bonds. The ion BF_4^- illustrates this point. Boron has a formal charge of -1 in this ion, and the formal charge calculated for the fluorine atoms is 0. Based on the electronegativity difference between fluorine and boron ($\Delta\chi = 2.0$), the B—F bonds are expected to be polar and fluorine is the negative end of the bond ($B^{\delta+}$—$F^{\delta-}$). So, in this instance, predictions based on electronegativity and formal charge work in opposite directions. The formal charge calculation places the negative charge on boron, but the electronegativity difference leads us to say the negative charge is distributed onto the fluorine atoms, effectively spreading it out over the molecule.

Linus Pauling pointed out two basic guidelines to use when describing charge distributions in molecules and ions. The first is the **electroneutrality principle**: Electrons will be distributed in such a way that the charges on all atoms are as close to zero as possible. Second, he noted that if a negative charge is present, it should reside on the most electronegative atoms. Similarly, positive charges are expected on the least electronegative atoms. The effect of these principles is clearly seen in the

Formal charge =
$0 = 7 - [6 + \frac{1}{2}(2)]$

Formal charge =
$-1 = 3 - [0 + \frac{1}{2}(8)]$

Formal charges for the B and F atoms of the BF_4^- anion.

case of BF_4^-, where it turns out that the negative charge is indeed distributed over the four fluorine atoms, as predicted by the atoms' electronegativities, rather than residing on boron, as would be predicted by formal charge calculations.

Considering the concepts of electronegativity and formal charge together can help to decide which of several resonance structures is the more important. For example, Lewis structure A for CO_2 is the logical one to draw. But what is wrong with B, in which each atom also has an octet of electrons?

| Formal charges | 0 0 0 | +1 0 −1 |

Resonance structures

$$\overset{\cdot\cdot}{O}=C=\overset{\cdot\cdot}{O} \qquad :\overset{}{O}\equiv C-\overset{\cdot\cdot}{O}:$$

A   B

For structure A, each atom has a formal charge of 0, a favorable situation. In B, however, one oxygen atom has a formal charge of +1, and the other has −1. This is contrary to the principle of electroneutrality. In addition, B places a positive charge on an O atom, even though oxygen has a greater electronegativity than carbon. Thus, we can conclude that structure B is a much less satisfactory structure than A.

Now use the same logic to decide which of the three possible resonance structures for the OCN^- ion is the most reasonable. Formal charges for each atom are given above each element's symbol.

| Formal charges | −1 0 0 | 0 0 −1 | +1 0 −2 |

Resonance structures

$$\left[:\overset{\cdot\cdot}{O}-C\equiv N:\right]^- \longleftrightarrow \left[:\overset{\cdot\cdot}{O}=C=\overset{\cdot\cdot}{N}\right]^- \longleftrightarrow \left[:O\equiv C-\overset{\cdot\cdot}{N}:\right]^-$$

A   B   C

Structure C is not expected to contribute significantly to the overall electronic structure of the ion. It has a −2 formal charge on the N atom and a +1 formal charge on the O atom. Not only is the charge on the N atom high, but O is more electronegative than N and would be expected to take on a more negative charge than N. Structure A is more significant than structure B because the negative charge in A is placed on the most electronegative atom (O). We predict, therefore, that structure A is the best representation for this ion and that the carbon–nitrogen bond will most resemble a triple bond. The result for OCN^- also allows us to predict that protonation of the ion will lead to HOCN and not HNCO. That is, an H^+ ion will add to the more negative oxygen atom.

● **Formal Charges in OCN⁻**

Example of formal charge calculation: For resonance structure C for OCN⁻, we have

O = 6 − [2 + (½)(6)] = +1

C = 4 − [0 + (½)(8)] = 0

N = 5 − [6 + (½)(2)] = −2

Sum of formal charges = −1 = charge on the ion.

EXAMPLE 8.11   Calculating Formal Charges

Problem Boron-containing compounds often have a boron atom with only three bonds (and no lone pairs). Why not form a double bond with a terminal atom to complete the boron octet? To answer this, consider possible resonance structures of BF_3, and calculate the atoms' formal charges. Are the bonds polar in BF_3? If so, which is the more negative atom?

What Do You Know? Boron trifluoride is one of the examples of a compound that does not obey the octet rule. The most common Lewis structure given for BF_3 shows the boron atom with three pairs of electrons.

Strategy

- Draw the commonly used Lewis structure for BF_3. Then draw a resonance structure (one of three possible) in which boron achieves an octet structure, using one of the electron pairs for fluorine to create a double bond between boron and fluorine.

- Determine the formal charges on each atom in the two structures.

- Decide the importance of the different resonance structures based on the positions of charges in the molecule.

Solution The two possible structures for $\overset{\cdot\cdot}{B}F_3$ are illustrated here with the calculated formal charges on the B and F atoms.

Formal charge = 0
= 7 − [6 + $\frac{1}{2}$(2)]

Formal charge = +1
= 7 − [4 + $\frac{1}{2}$(4)]

$$:\overset{\cdot\cdot}{F}:$$
$$|$$
$$:\overset{\cdot\cdot}{F}—\overset{\cdot\cdot}{B}—\overset{\cdot\cdot}{F}:$$

Formal charge = 0
= 3 − [0 + $\frac{1}{2}$(6)]

$$:\overset{\cdot\cdot}{F}:$$
$$\|$$
$$:\overset{\cdot\cdot}{F}—\overset{\cdot\cdot}{B}—\overset{\cdot\cdot}{F}:$$

Formal charge = −1
= 3 − [0 + $\frac{1}{2}$(8)]

The structure on the left is strongly preferred because all atoms have a zero formal charge and the very electronegative F atom does not have a charge of +1.

F ($\chi = 4.0$) is more electronegative than B ($\chi = 2.0$), so the B—F bond is highly polar, the F atom being partially negative and the B atom partially positive.

Think about Your Answer When there are two resonance forms that are dissimilar, you need to decide which better represents the bonding in the compound. When discussing structures, you can say that one resonance form is more important than the other.

Check Your Understanding

Draw the resonance structures for SCN^-. What are the formal charges on each atom in each resonance structure? What are the bond polarities? Do they agree with the formal charges?

REVIEW & CHECK FOR SECTION 8.7

1. Which bond among those listed below is the most polar?

   (a) C—Cl   (b) Si—Cl   (c) B—Cl   (d) Al—Cl

2. The compound ICl is a red crystalline compound that melts to a red liquid at 27° C. Which of the following best describes the bonding in this compound?

   (a) The I—Cl bond is not polar.

   (b) The I—Cl bond is polar with I being the negative end.

   (c) The I—Cl bond is polar with I being the positive end.

3. Three resonance forms can be drawn for the molecule N_2O. Which resonance form is likely to more closely resemble the structure of this molecule?

$$:N≡N—\overset{\cdot\cdot}{\underset{\cdot\cdot}{O}}:\qquad \overset{\cdot\cdot}{\underset{\cdot\cdot}{N}}=N=\overset{\cdot\cdot}{O}\qquad :\overset{\cdot\cdot}{N}—N≡O:$$

   (a)   (b)   (c)

8.8 Bond and Molecular Polarity

The term "polar" was used in Section 8.7 to describe a bond in which one atom has a partial positive charge and the other a partial negative charge. Because most molecules have polar bonds, molecules as a whole can also be polar. In a polar molecule, electron density accumulates toward one side of the molecule, giving that side a partial negative charge, δ^-, and leaving the other side with a partial positive charge, δ^+ (Figure 8.12a).

Before describing the factors that determine whether a molecule is polar, let us look at the experimental measurement of the polarity of a molecule. When placed in an electric field created by a pair of oppositely charged plates, polar molecules

Module 13: Molecular Polarity covers concepts in this section.

FIGURE 8.12 Polar molecules in an electric field.

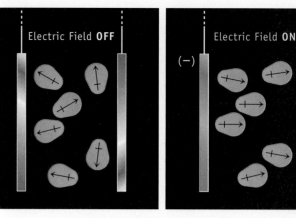

(a) A representation of a polar molecule. To indicate the direction of molecular polarity, an arrow is drawn with the head pointing to the negative side and a plus sign placed at the positive end.

(b) When placed in an electric field (between charged plates), polar molecules experience a force that tends to align them with the field. The negative end of the molecules is attracted to the positive plate, and vice versa. The orientation of the polar molecule affects the electrical capacitance of the plates (their ability to hold a charge), and this provides a way to measure experimentally the magnitude of the dipole moment.

● **Peter Debye and Dipoles** The commonly used unit of dipole moments is named in honor of Peter Debye (1884–1966). He was born in The Netherlands, but attended university in Germany and later studied for his Ph.D. in physics in Munich. He developed a theory on the diffraction of x-rays by solids, a new concept for magnetic cooling, and (with E. Hückel) a model for interionic attractions in aqueous solution. As his interests turned more to chemistry, he worked on methods of determining the shapes of polar molecules. Debye received the Nobel Prize in Chemistry in 1936 and was later a professor at Cornell University.

experience a force that tends to align them with the field (Figure 8.12). The positive end of each molecule is attracted to the negative plate, and the negative end is attracted to the positive plate (Figure 8.12b). The extent to which the molecules line up with the field depends on their **dipole moment**, μ. In a molecule with partial charges equal to $+q$ and $-q$ and in which these charges are separated by a distance, d, in the molecule, the magnitude of the dipole moment is given by the equation $\mu = q \times d$. The SI unit of the dipole moment is the coulomb-meter, but dipole moments have traditionally been given using a derived unit called the *debye* (D; 1 D = 3.34 $\times$ 10^{-30} C · m). Experimental values of some dipole moments are listed in Table 8.7.

To predict if a molecule is polar, we need to consider if the molecule has polar bonds and how these bonds are positioned relative to one another. Diatomic molecules composed of two atoms with different electronegativities are always polar (Table 8.7); there is one bond, and the molecule has a positive and a negative end. But what happens with a molecule composed of three or more atoms, in which there are two or more polar bonds?

Table 8.7 **Dipole Moments of Selected Molecules**

Molecule (AX)	Moment (μ, D)	Geometry	Molecule (AX₂)	Moment (μ, D)	Geometry
HF	1.78	Linear	H_2O	1.85	Bent
HCl	1.07	Linear	H_2S	0.95	Bent
HBr	0.79	Linear	SO_2	1.62	Bent
HI	0.38	Linear	CO_2	0	Linear
H_2	0	Linear			
Molecule (AX₃)	**Moment (μ, D)**	**Geometry**	**Molecule (AX₄)**	**Moment (μ, D)**	**Geometry**
NH_3	1.47	Trigonal pyramidal	CH_4	0	Tetrahedral
NF_3	0.23	Trigonal pyramidal	CH_3Cl	1.92	Tetrahedral
BF_3	0	Trigonal planar	CH_2Cl_2	1.60	Tetrahedral
			$CHCl_3$	1.04	Tetrahedral
			CCl_4	0	Tetrahedral

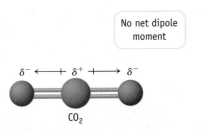

No net dipole moment

CO₂

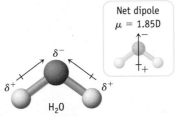

Net dipole $\mu = 1.85D$

H₂O

FIGURE 8.13 Polarity of triatomic molecules, *AX₂*.

(a) For CO₂, the CO bonds are polar, but the electron density is distributed symmetrically over the molecule, and the charges of δ⁻ lie 180° apart. (Charges calculated using advanced molecular modeling software: O = −0.38 and C = +0.77.) The molecule has no net dipole.

(b) In a water molecule, the O atom is negative, and the H atoms are positive. (Calculated charges: H = +0.19 and O = −0.38.) However, the positively charged H atoms lie on one side of the molecule, and the negatively charged O atom is on the other side. The molecule is polar.

Consider first a linear triatomic molecule such as carbon dioxide, CO₂ (Figure 8.13). Here, each C=O bond is polar, with the oxygen atom the negative end of the bond dipole. The terminal atoms are at the same distance from the C atom; they both have the same δ⁻ charge, and they are symmetrically arranged around the central C atom. Therefore, CO₂ has no molecular dipole, even though each bond is polar. This is analogous to a tug-of-war in which the people at opposite ends of the rope are pulling with equal force.

In contrast, water is a bent triatomic molecule. Because O has a larger electronegativity ($\chi = 3.5$) than H ($\chi = 2.2$), each of the O—H bonds is polar, with the H atoms having the same δ⁺ charge and oxygen having a negative charge (δ⁻) (Figure 8.13). Electron density accumulates on the O side of the molecule, making the molecule electrically "lopsided" and therefore polar ($\mu = 1.85$ D).

In trigonal-planar BF₃, the B—F bonds are highly polar because F is much more electronegative than B (χ of B = 2.0 and χ of F = 4.0) (Figure 8.14). The molecule is nonpolar, however, because the three terminal F atoms have the same δ⁻ charge, are the same distance from the boron atom, and are arranged symmetrically around the central boron atom. In contrast, the trigonal-planar molecule phosgene is polar (Cl₂CO, $\mu = 1.17$ D) (Figure 8.14). Here, the angles are all about 120°, so the O and Cl atoms are symmetrically arranged around the C atom. The electronegativities of

● **Partial Charges, δ**
The water molecule has a small positive charge on each H atom and a small negative charge on the O atom. When the atoms of molecular models are marked δ±, this is meant to indicate a small positive or negative charge but with no indication of the actual magnitude of the charge. (In the case of water, the actual charge on each hydrogen must be 1/2 the value of the charge on the oxygen to maintain electrical neutrality, but this is not indicated in the drawing shown below.)

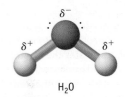

H₂O

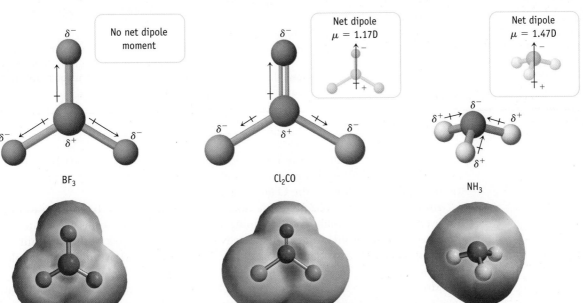

No net dipole moment

BF₃

Net dipole $\mu = 1.17D$

Cl₂CO

Net dipole $\mu = 1.47D$

NH₃

FIGURE 8.14 Polar and nonpolar molecules of the type AX₃. In BF₃, the negative charge on the F atoms is distributed symmetrically, so the molecular dipole is zero. In contrast, in Cl₂CO and NH₃, the negative charge in the molecules is shifted to one side and the positive charge to the other side. These charge distributions are reflected in the electrostatic potential surfaces. (See *A Closer Look: Visualizing Charge Distributions and Molecular Polarity—Electrostatic Potential Surfaces and Partial Charge*, page 294.)

A CLOSER LOOK

Visualizing Charge Distributions and Molecular Polarity— Electrostatic Potential Surfaces and Partial Charge

In Chapter 6, you learned about atomic orbitals, regions of space within which an electron is most probably found. The boundary surface of these orbitals was created in such a way that the electron wave amplitude at all points of the surface was the same value. When we use advanced molecular modeling software, we can generate the same type of pictures for molecules, and in Figure A you see a surface defining the electron density in the HF molecule. The electron density surface, calculated using software from Wavefunction, is made up of all of the points in space around the HF molecule where the electron density is 0.002 e⁻/Å³ (where 1 Å = 0.1 nm). You can see that the surface bulges toward the F end of the molecule, an indication of the larger size of the F atom. The larger size of the F atom here is mainly related to the fact that it has more valence electrons than H, and to a lesser extent to the fact that H—F bond is polar and electron density in that bond is shifted toward the F atom.

We can add another layer of information to our picture. The electron density surface can be colored according to the *electrostatic potential*. (Hence, this figure is called an *electrostatic potential surface*.) The computer program calculates the electrostatic potential that would be observed by a proton (H^+) on the surface. This is the sum of the attractive and repulsive forces on that proton due to the nuclei and the electrons in the molecule. Regions of the molecule in which there is an attractive potential are colored red. That is, this is a region of negative charge on the molecule. Repulsive potentials occur in regions where the molecule is positively charged; these regions are colored blue. As

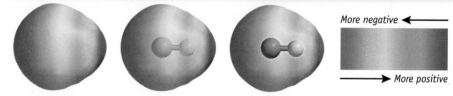

FIGURE A **Three views of the electrostatic potential surface for HF.**
(*left*) The electron density surface around HF. The F atom is at the left. The surface is made up of all of the points in space around the HF molecule where the electron density is 0.002 e⁻/Å³ (where 1 Å = 0.1 nm). (*middle*) The surface is made more transparent, so you can see the HF atom nuclei inside the surface. (*right*) The front of the electron density surface has been "peeled away" for a view of the HF molecule inside.
Color scheme: The colors on the electron density surface reflect the charge in the different regions of the molecule. Colors toward the blue end of the spectrum indicate a positive charge, whereas colors to the red end of the spectrum indicate a negative charge.

might be expected, the net electrostatic potential will change continuously as one moves from a negative portion of a molecule to a positive portion, and this is indicated by a progression of colors from red to blue (from negative to positive).

The electrostatic potential surface for HF shows the H atom is positive (the H atom end of the molecule is blue), and the F atom is negative (the F atom end is red). This is, of course, what we would predict based on electronegativity.

The program also calculated that the F atom has a charge of −0.3 and H has a charge of +0.3. Finally, the calculated dipole moment for the molecule is about 1.7 D, in good agreement with the experimental value in Table 8.7.

Other examples of electrostatic potential surfaces illustrate the polarity of water and methylamine, CH_3NH_2.

The surface for water shows the O atom of the molecule bears a partial negative

charge and the H atoms are positive. The surface for the amine shows the molecule is polar and that the region around the N atom is negative. Indeed, we know from experiment that an H^+ ion will attack the N atom to give the cation $CH_3NH_3^+$.

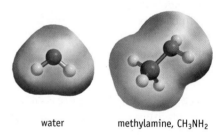

water   methylamine, CH_3NH_2

FIGURE B **Visualizing charge distribution**

You will see other examples of useful information chemists can learn from these surfaces in this book and as you move on to other courses in chemistry,

the three types of atoms in the molecule differ, however: $\chi(O) > \chi(Cl) > \chi(C)$. There is therefore a net displacement of electron density away from the center of the molecule, more toward the O atom than the Cl atoms.

Ammonia, like BF_3, has AX_3 stoichiometry and polar bonds. In contrast to BF_3, however, NH_3 is a trigonal-pyramidal molecule. The positive H atoms are located in the base of the pyramid, and the negative N atom is on the apex of the pyramid. As a consequence, NH_3 is polar (Figure 8.14). Indeed, trigonal-pyramidal molecules are generally polar.

Molecules like carbon tetrachloride, CCl_4, and methane, CH_4, are nonpolar, owing to their symmetrical, tetrahedral structures. The four atoms bonded to C have the same partial charge and are the same distance from the C atom. Tetrahedral molecules with both Cl and H atoms ($CHCl_3$, CH_2Cl_2, and CH_3Cl) are polar, however (Figure 8.15). The electronegativity for H atoms (2.2) is less than that of Cl atoms (3.2), and the carbon–hydrogen distance is different from the carbon–

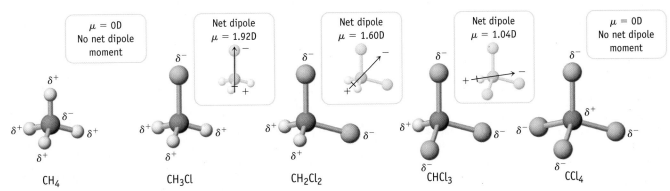

FIGURE 8.15 Polarity of tetrahedral molecules. The electronegativities of the atoms involved are in the order Cl (3.2) > C (2.5) > H (2.2). This means the C—H and C—Cl bonds are polar with a net displacement of electron density away from the H atoms and toward the Cl atoms [H $^{\delta+}$—C $^{\delta-}$ and C $^{\delta+}$—Cl $^{\delta-}$]. Although the electron-pair geometry around the C atom in each molecule is tetrahedral, only in CH_4 and CCl_4 are the polar bonds totally symmetrical in their arrangement. Therefore, CH_3Cl, CH_2Cl_2, and $CHCl_3$ are polar molecules, with the negative end toward the Cl atoms and the positive end toward the H atoms.

chlorine distances. Because Cl is more electronegative than H, the Cl atoms are on the more negative side of the molecule. This means the positive end of the molecular dipole is toward the H atom.

To summarize this discussion of molecular polarity, look again at Figure 8.5 (page 281). These are sketches of molecules of the type AX_n where A is the central atom and X is a terminal atom. You can predict that a molecule AX_n will not be polar, regardless of whether the A—X bonds are polar, if

- all the terminal atoms (or groups), X, are identical, and
- all the X atoms (or groups) are arranged symmetrically around the central atom, A.

On the other hand, if one of the X atoms (or groups) is different in the structures in Figure 8.5 (as in Figures 8.14 and 8.15), or if one of the X positions is occupied by a lone pair, the molecule will be polar.

◉WL INTERACTIVE EXAMPLE 8.12 Molecular Polarity

Problem Are sulfur tetrafluoride (SF_4) and nitrogen trifluoride (NF_3) polar or nonpolar? If polar, indicate the negative and positive sides of the molecule.

What Do You Know? Based on the discussion in this chapter, you know how to derive the molecular geometry of the molecules and you know how to decide which bonds are polar. These are the two features you use to decide if the molecule is polar.

Strategy Draw the Lewis structure, decide on the electron-pair geometry, and determine the molecular geometry. If the molecular geometry is one of the highly symmetrical geometries in Figure 8.5, the molecule is not polar. If it does not fit one of these categories, it will be polar if the bonds are polar.

Solution

(a) The S—F bonds in sulfur tetrafluoride, SF_4, are polar, the bond dipole having F as the negative end (χ for S is 2.6 and χ for F is 4.0). The molecule has a trigonal bipyramidal electron-pair geometry (Figure 8.8). Because the lone pair occupies one of the positions, the S—F bonds are not arranged symmetrically. The axial S—F bond dipoles cancel each other because they point in opposite directions. The equatorial S—F bonds, however, both point to one side of the molecule, so SF_4 is polar.

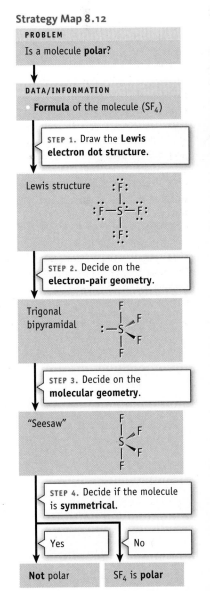

Strategy Map 8.12

PROBLEM

Is a molecule **polar?**

↓

DATA/INFORMATION

- **Formula** of the molecule (SF_4)

STEP 1. Draw the **Lewis electron dot structure.**

Lewis structure

STEP 2. Decide on the **electron-pair geometry.**

Trigonal bipyramidal

STEP 3. Decide on the **molecular geometry.**

"Seesaw"

STEP 4. Decide if the molecule is **symmetrical.**

Yes — **Not** polar

No — SF_4 **is polar**

(b) NF$_3$ has the same trigonal-pyramidal structure as NH$_3$. Because F is more electronegative than N, each bond is polar, the more negative end being the F atom. Because this molecule contains polar bonds and because the geometry is not symmetrical but instead has three positions of the tetrahedron occupied by bonding groups and one by a lone pair, the NF$_3$ molecule as a whole is expected to be polar.

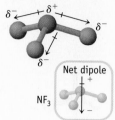

Think about Your Answer Notice that the dipole moment for NF$_3$ is quite small (0.23 D, Table 8.7), much smaller than the dipole moment of NH$_3$. This illustrates the effect of lone pairs on molecular polarity. For NH$_3$, the bond polarity is N$^{\delta-}$—H$^{\delta+}$, and the dipole created by these bonds is reinforced by the negatively charged lone pair on the N atom. In contrast, the polarity of the bonds in NF$_3$ is N$^{\delta+}$—F$^{\delta-}$. This is counterbalanced by the lone pair, and the overall dipole is diminished. These effects are illustrated by the electrostatic potential surfaces for these two molecules.

NH$_3$ NF$_3$

Ammonia, μ = 1.47 D   Nitrogen trifluoride, μ = 0.23 D

Check Your Understanding

For each of the following molecules, decide whether the molecule is polar and which side is positive and which negative: BFCl$_2$, NH$_2$Cl, and SCl$_2$.

EXAMPLE 8.13   Molecular Polarity

Problem 1,2-Dichloroethylene can exist in two forms. Is either of these planar molecules polar?

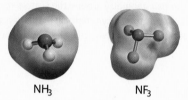

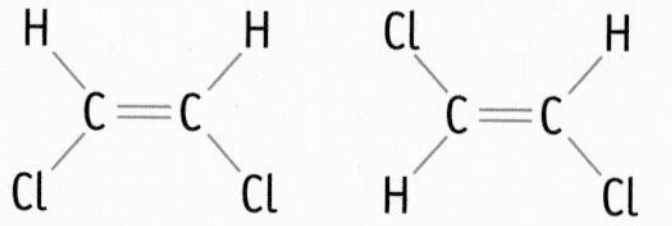

What Do You Know? The structures for the two molecules are given. Note that these are planar molecules.

Strategy

•   Use electronegativity values to decide on the bond polarity.

•   Decide if the electron density in the bonds is distributed symmetrically or if it is shifted to one side of the molecule.

Solution Here, the H and Cl atoms are arranged around the C=C double bonds with all bond angles 120° (and all the atoms lie in one plane). The electronegativities of the atoms involved are in the order Cl (3.2) > C (2.5) > H (2.2). This means the C—H and C—Cl bonds are polar with a net displacement of electron density away from the H atoms and toward the Cl atoms [H$^{\delta+}$—C$^{\delta-}$ and C$^{\delta+}$—Cl$^{\delta-}$]. In structure A, the Cl atoms are located on one side of the molecule, so electrons in the H—C and C—Cl bonds are displaced toward the side of the molecule with Cl atoms and away from the side with the H atoms. Molecule A is polar. In molecule B, the displacement of electron density toward the Cl atom on one end of the molecule is counterbalanced by an opposing displacement on the other end. Molecule B is not polar.

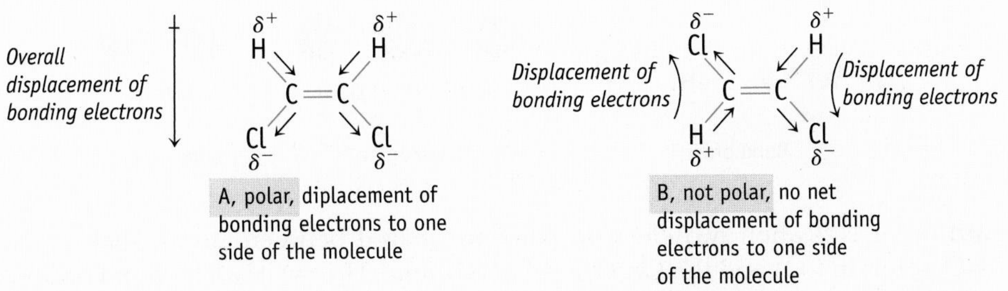

Overall displacement of bonding electrons

Displacement of bonding electrons

Displacement of bonding electrons

A, polar, diplacement of bonding electrons to one side of the molecule

B, not polar, no net displacement of bonding electrons to one side of the molecule

Think about Your Answer The electrostatic potential surfaces reflect the fact that molecule A is polar because the electron density is shifted to one side of the molecule. Molecule B is not polar because the electron density is distributed symmetrically.

A = *cis*-1,2-dichloroethylene

B = *trans*-1,2-dichloroethylene

Check Your Understanding

The electrostatic potential surface for $OSCl_2$ is pictured here.

(a) Draw a Lewis electron dot picture for the molecule, and give the formal charge of each atom.

(b) What is the molecular geometry of $OSCl_2$? Is it polar?

REVIEW & CHECK FOR SECTION 8.8

1. Which of the hydrogen halides is the most polar?

    (a) HF   (b) HCl   (c) HBr   (d) HI

2. Which of the molecules listed below is nonpolar?

    (a) BCl_3 (b) PCl_3 (c) $POCl_3$ (d) SO_2Cl_2

8.9 Bond Properties: Order, Length, and Energy

Bond Order

The **order of a bond** is the number of bonding electron pairs shared by two atoms in a molecule (Figure 8.16). You will encounter bond orders of 1, 2, and 3, as well as fractional bond orders.

When the bond order is 1, there is only a single covalent bond between a pair of atoms. Examples are the bonds in molecules such as H_2, NH_3, and CH_4. The

FIGURE 8.16 Bond order. The four C—H bonds in methane each have a bond order of 1. The two C=O bonds of CO_2 each have a bond order of two, whereas the nitrogen–nitrogen bond in N_2 has an order of 3.

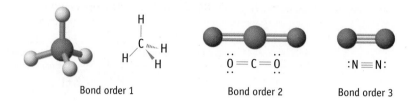

Bond order 1   Bond order 2   Bond order 3

bond order is 2 when two electron pairs are shared between atoms, such as the C=O bonds in CO_2 and the C=C bond in ethylene, $H_2C=CH_2$. The bond order is 3 when two atoms are connected by three bonds. Examples include the carbon–oxygen bond in carbon monoxide, CO, and the nitrogen–nitrogen bond in N_2.

Fractional bond orders occur in molecules and ions having resonance structures. For example, what is the bond order for each oxygen–oxygen bond in O_3? Each resonance structure of O_3 has one O—O single bond and one O=O double bond, for a total of three shared bonding pairs accounting for two oxygen–oxygen links.

We can define the bond order between any bonded pair of atoms X and Y as

$$\text{Bond order} = \frac{\text{number of shared pairs in all X—Y bonds}}{\text{number of X—Y links in the molecule or ion}} \qquad (8.2)$$

For ozone, there are three bond pairs involved in two oxygen–oxygen links, so the bond order for each oxygen–oxygen bond is $\frac{3}{2}$, or 1.5.

Bond Length

Bond length, the distance between the nuclei of two bonded atoms, is clearly related to the sizes of the atoms (Section 7.5). In addition, for a given pair of atoms, the order of the bond plays a role.

Table 8.8 lists average bond lengths for a number of common chemical bonds. It is important to recognize that these are *average* values. Neighboring parts of a molecule can affect the length of a particular bond, so there can be a range of values for a particular bond type. For example, Table 8.8 gives the average C—H bond as 110 pm. However, in methane, CH_4, the measured bond length is 109.4 pm, whereas the C—H bond is only 105.9 pm long in acetylene, H—C≡C—H. Variations as great as 10% from the average values listed in Table 8.8 are possible.

Because atom sizes vary in a regular fashion with the position of the element in the periodic table (Figure 7.6), we can predict trends in bond lengths. For example, the H—X distance in the hydrogen halides increases in the order predicted by the relative sizes of the halogens: H—F < H—Cl < H—Br < H—I. Likewise, bonds between carbon and another element in a given period decrease going from left to right, in a predictable fashion; for example, C—C > C—N > C—O. Trends for multiple bonds are similar. A C=O bond is shorter than a C=S bond, and a C=N bond is shorter than a C=C bond.

The effect of bond order is evident when bonds between the same two atoms are compared. For example, the bonds become shorter as the bond order increases in the series C—O, C=O, and C≡O:

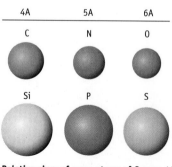

Relative sizes of some atoms of Groups 4A, 5A, and 6A.

Bond lengths are related to atom sizes.

C—H	N—H	O—H
110	98	94 pm
Si—H	**P—H**	**S—H**
145	138	132 pm

Bond	C—O	C=O	C≡O
Bond order	1	2	3
Average bond length (pm)	143	122	113

Table 8.8 Some Average Single- and Multiple-Bond Lengths in Picometers (pm)*

					Single Bond Lengths					
					Group					
1A	4A	5A	6A	7A	4A	5A	6A	7A	7A	7A
H	C	N	O	F	Si	P	S	Cl	Br	I
H 74	110	98	94	92	145	138	132	127	142	161
C	154	147	143	141	194	187	181	176	191	210
N		140	136	134	187	180	174	169	184	203
O			132	130	183	176	170	165	180	199
F				128	181	174	168	163	178	197
Si					234	227	221	216	231	250
P						220	214	209	224	243
S							208	203	218	237
Cl								200	213	232
Br									228	247
I										266

Multiple Bond Lengths			
C=C	134	C≡C	121
C=N	127	C≡N	115
C=O	122	C≡O	113
N=O	115	N≡O	108

*1 pm = 10^{-12} m.

The carbonate ion, CO_3^{2-}, has three equivalent resonance structures. Each CO bond has a bond order of 1.33 (or $\frac{4}{3}$) because four electron pairs are used to form three carbon–oxygen links. The CO bond distance (129 pm) is intermediate between a C—O single bond (143 pm) and a C=O double bond (122 pm).

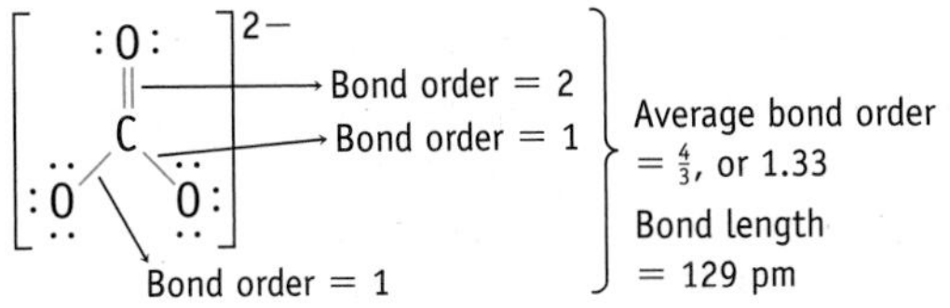

Bond Dissociation Enthalpy

The **bond dissociation enthalpy** is the enthalpy change for breaking a bond in a molecule with the reactants and products *in the gas phase. The process of breaking bonds in a molecule is always endothermic, so* $\Delta_r H$ *for bond breaking is always positive.*

$$\text{Molecule (g)} \underset{\text{Energy released} = \Delta H < 0}{\overset{\text{Energy supplied} = \Delta H > 0}{\rightleftarrows}} \text{Molecular fragments (g)}$$

Suppose you wish to break the carbon–carbon bonds in ethane (H_3C—CH_3), ethylene (H_2C=CH_2), and acetylene (HC≡CH). The carbon–carbon bond orders in these molecules are 1, 2, and 3, respectively, and these bond orders are reflected in the bond dissociation enthalpies. Breaking the single C—C bond in ethane requires the least energy in this group, whereas breaking the C—C triple bond in acetylene requires the most energy.

$$H_3C\text{—}CH_3(g) \rightarrow H_3C(g) + CH_3(g) \qquad \Delta_rH = +368 \text{ kJ/mol-rxn}$$

$$H_2C\text{=}CH_2(g) \rightarrow H_2C(g) + CH_2(g) \qquad \Delta_rH = +682 \text{ kJ/mol-rxn}$$

$$HC\text{≡}CH(g) \rightarrow HC(g) + CH(g) \qquad \Delta_rH = +962 \text{ kJ/mol-rxn}$$

The energy supplied to break carbon–carbon bonds must be the same as the energy released when the same bonds form. *The formation of bonds from atoms or radicals in the gas phase is always exothermic.* This means, for example, that Δ_rH for the formation of $H_3C\text{—}CH_3$ from two $CH_3(g)$ radicals is -368 kJ/mol-rxn.

$$H_3C\cdot(g) + \cdot CH_3(g) \rightarrow H_3C\text{—}CH_3(g) \qquad \Delta_rH = -368 \text{ kJ/mol-rxn}$$

Generally, the bond energy for a given type of bond (a C—C bond, for example) varies somewhat, depending on the compound, just as bond lengths vary from one molecule to another. Thus, data provided in tables must be viewed as *average bond dissociation enthalpies* (Table 8.9). The values in such tables may be used to *estimate* the enthalpy change for a reaction, as described below.

In reactions between molecules, bonds in reactants are broken, and new bonds are formed as products form. If the total energy released when new bonds form exceeds the energy required to break the original bonds, the overall

• Variability in Bond Dissociation Enthalpies The values of Δ_rH for ethane, ethylene, and acetylene in the text are for those molecules in particular. The bond dissociation enthalpies in Table 8.9 are average values for a range of molecules containing the indicated bond.

Table 8.9 Some Average Bond Dissociation Enthalpies (kJ/mol)*

	H	C	N	O	F	Si	P	S	Cl	Br	I
Single Bonds											
H	436	413	391	463	565	328	322	347	432	366	299
C		346	305	358	485	—	—	272	339	285	213
N			163	201	283	—	—	—	192	—	—
O				146	—	452	335	—	218	201	201
F					155	565	490	284	253	249	278
Si						222	—	293	381	310	234
P							201	—	326	—	184
S								226	255	—	—
Cl									242	216	208
Br										193	175
I											151

Multiple Bonds

N=N	418	C=C	610
N≡N	945	C≡C	835
C=N	615	C=O	745
C≡N	887	C≡O	1046
O=O (in O₂)	498		

*Sources of dissociation enthalpies: I. Klotz and R. M. Rosenberg, *Chemical Thermodynamics,* 4th Ed., p. 55, New York, John Wiley, 1994; and J. E. Huheey, E. A. Keiter, and R. L. Keiter, *Inorganic Chemistry* 4th Ed., Table E. 1, New York, HarperCollins, 1993. See also Lange's *Handbook of Chemistry,* J. A. Dean (ed.), McGraw-Hill Inc., New York.

reaction is exothermic. If the opposite is true, then the overall reaction is endothermic.

Let us use bond dissociation enthalpies to estimate the enthalpy change for the hydrogenation of propene to propane:

$$H-\underset{\underset{H}{|}}{\overset{\overset{H}{|}}{C}}-\overset{H}{C}=\overset{H}{C}-H(g) + H-H(g) \longrightarrow H-\underset{\underset{H}{|}}{\overset{\overset{H}{|}}{C}}-\underset{\underset{H}{|}}{\overset{\overset{H}{|}}{C}}-\underset{\underset{H}{|}}{\overset{\overset{H}{|}}{C}}-H(g)$$

$$\text{propene} \qquad\qquad\qquad\qquad \text{propane}$$

The first step is to identify what bonds are broken and what bonds are formed. In this case, the C=C bond in propene and the H—H bond in hydrogen are broken. A C—C bond and two C—H bonds are formed.

Bonds broken: 1 mol of C=C bonds and 1 mol of H—H bonds

$$H-\underset{\underset{H}{|}}{\overset{\overset{H}{|}}{C}}-\overset{H}{C}=\overset{H}{C}-H(g) + H-H(g)$$

Energy required = 610 kJ for C=C bonds + 436 kJ for H—H bonds = 1046 kJ/mol-rxn

---

Ibuprofen, A Study in Green Chemistry

Ibuprofen, $C_{13}H_{18}O_2$, is now one of the world's most common, over-the-counter drugs. It is an effective anti-inflammatory drug and is used to treat arthritis and similar conditions. Unlike aspirin it does not decompose in solution, so ibuprofen can be applied to the skin as a topical gel, thus avoiding the gastrointestinal problems sometimes associated with aspirin.

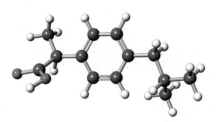

FIGURE A   Ibuprofen, $C_{13}H_{18}O_2$

Ibuprofen was first synthesized in the 1960s by the Boots Company in England, and it soon became an important product. Millions of pounds are sold every year.

Chemists realized, however, that the method of making ibuprofen took six steps, wasted chemicals, and produced by-products that required disposal. Many of the atoms of the reagents used to make the drug did not end up in the drug itself, that is, the method had a very poor atom economy (page 144). Recognizing this, chemists looked for new approaches for the synthesis of ibuprofen. A method was soon found that uses only three steps, severely reduces waste, and uses reagents that can be recovered and reused. A plant in Texas now makes more than 7 million pounds of ibuprofen annually by the new method, enough to make over 6 billion tablets.

Questions:
1. The third and final step in the synthesis involves the transformation of an OH group with CO to a carboxylic acid group ($-CO_2H$) (Figure B). Using bond enthalpy data, is this an endothermic or exothermic step?
2. Do any of the atoms in an ibuprofen molecule have a formal charge of zero?
3. What is the most polar bond in the molecule?
4. Is the molecule polar?
5. What is the shortest bond in the molecule?

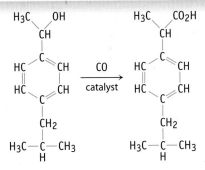

FIGURE B The final step in the synthesis of ibuprofen.

6. What bond or bonds have the highest bond order?
7. Are there any 120° bond angles in ibuprofen? Any 180° angles?
8. If you were to titrate 200. mg of ibuprofen to the equivalence point with 0.0259 M NaOH, what volume of NaOH would be required?

Answers to these questions are available in Appendix N.

Bonds formed: 1 mol of C—C bonds and 2 mol of C—H bonds

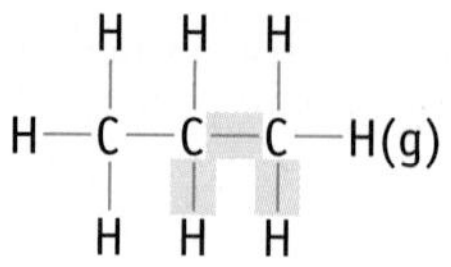

$$
\begin{array}{c}
\quad\; H \;\; H \;\; H \\
\quad\; | \;\;\;\; | \;\;\;\; | \\
H - C - C - C - H(g) \\
\quad\; | \;\;\;\; | \;\;\;\; | \\
\quad\; H \;\; H \;\; H
\end{array}
$$

Energy evolved = 346 kJ for C—C bonds + 2 mol × 413 kJ/mol for C—H bonds =
1172 kJ/mol-rxn

By combining the enthalpy changes for breaking bonds and for making bonds, we can estimate $\Delta_r H$ for the hydrogenation of propene and predict that the reaction is exothermic.

$$\Delta_r H = 1046 \text{ kJ/mol-rxn} - 1172 \text{ kJ/mol-rxn} = -126 \text{ kJ/mol-rxn}$$

In general, the enthalpy change for any reaction can be estimated using the equation

$$\Delta_r H = \Sigma\Delta H(\text{bonds broken}) - \Sigma\Delta H(\text{bonds formed}) \tag{8.3}$$

Such calculations can give acceptable results in many cases.

● **$\Delta_r H$ from Enthalpies of Formation**
Using $\Delta_f H°$ values for propane and propene, we calculate $\Delta_r H° = -125.1$ kJ/mol–rxn. The bond dissociation enthalpy calculation is in excellent agreement with that from enthalpies of formation in this case.

EXAMPLE 8.14   Using Bond Dissociation Enthalpies

Problem Acetone, a common industrial solvent, can be converted to 2-propanol, rubbing alcohol, by hydrogenation. Calculate the enthalpy change for this reaction using bond enthalpies.

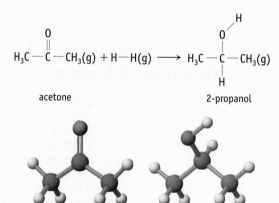

acetone   2-propanol

What Do You Know? You know the molecular structures of the reactants and the products.

Strategy Determine which bonds are broken and which are formed. Add up the enthalpy changes for breaking bonds in the reactants and for forming bonds in the product. The difference in the sums of bond dissociation enthalpies can be used as an estimate of the enthalpy change of the reaction (Equation 8.3).

Solution

Bonds broken: 1 mol of C=O bonds and 1 mol of H—H bonds

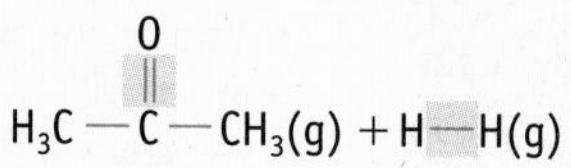

$$
\begin{array}{c}
\quad\quad O \\
\quad\quad \| \\
H_3C - C - CH_3(g) + H - H(g)
\end{array}
$$

$\Sigma\Delta H$(bonds broken) = 745 kJ for C=O bonds + 436 kJ for H—H bonds = 1181 kJ/mol-rxn

Bonds formed: 1 mol of C—H bonds, 1 mol of C—O bonds, and 1 mol of O—H bonds

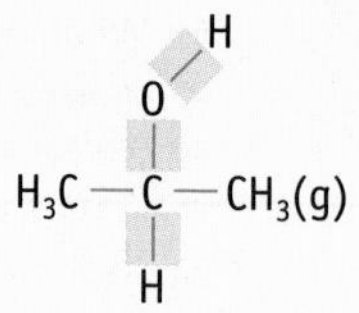

$$
\begin{array}{c}
\quad\quad\quad H \\
\quad\quad\; O \;\; / \\
\quad\quad\; | \\
H_3C - C - CH_3(g) \\
\quad\quad\; |\\
\quad\quad\; H
\end{array}
$$

$\Sigma \Delta H$(bonds formed) = 413 kJ for C—H + 358 kJ for C—O + 463 kJ for O—H = 1234 kJ/mol-rxn

$\Delta_r H = \Sigma \Delta H$(bonds broken) $- \Sigma \Delta H$(bonds formed)

$\Delta_r H$ = 1181 kJ $-$ 1234 kJ = -53 kJ/mol-rxn

Think about Your Answer The overall reaction is predicted to be exothermic by 53 kJ per mol of product formed. This is in good agreement with the value calculated from $\Delta_f H°$ values (-55.8 kJ/mol-rxn).

Check Your Understanding

Using the bond dissociation enthalpies in Table 8.9, estimate the enthalpy of combustion of gaseous methane, CH_4 to give water vapor and carbon dioxide gas.

REVIEW & CHECK FOR SECTION 8.9

1. Which of the following species has the largest N—O bond?

    (a) NO_3^- (b) NO^+ (c) NO_2^- (d) H_2NOH

2. Which of the following species has the largest C—N bond order?

    (a) CN^- (b) OCN^- (c) CH_3NH_2 (d) $N(CH_3)_3$

A CLOSER LOOK

DNA—Watson, Crick, and Franklin

DNA is the substance in every plant and animal that carries the exact blueprint of that plant or animal. The structure of this molecule, the cornerstone of life, was uncovered in 1953, and James D. Watson, Francis Crick, and Maurice Wilkins shared the 1962 Nobel Prize in Medicine and Physiology for the work. It was one of the most important scientific discoveries of the 20th century. The story of this discovery has been told by Watson in his book *The Double Helix*.

When Watson was a graduate student at Indiana University, he had an interest in the gene and said he hoped that its biological role might be solved "without my learning any chemistry." Later, however, he and Crick found out just how useful chemistry can be when they began to unravel the structure of DNA.

Solving important problems requires teamwork among scientists of many kinds, so Watson went to Cambridge in 1951. There he met Crick, who, Watson said, talked louder and faster than anyone else. Crick shared Watson's belief in the fundamental importance of DNA, and the pair soon learned that Maurice Wilkins and Rosalind Franklin at King's College in London were using a technique called *x-ray crystallography* to learn more about DNA's structure. Watson and Crick believed that understanding this structure was crucial to understanding genetics. To solve the structural problem, however, they needed experimental data of the type that could come from the experiments at King's College.

The King's College group was initially reluctant to share their data, and, what is more, they did not seem to share Watson and Crick's sense of urgency. There was also an ethical dilemma: Could Watson and Crick work on a problem that others had claimed as theirs? "The English sense of fair play

Rosalind Franklin of King's College, London. She died in 1958 at the age of 37. Because Nobel Prizes are never awarded posthumously, she did not share in this honor with Watson, Crick, and Wilkins. For more on Rosalind Franklin, read *Rosalind Franklin: The Dark Lady of DNA*, by Brenda Maddox.

would not allow Francis to move in on Maurice's problem," said Watson.

Watson and Crick approached the problem through a technique chemists now use frequently—model building. They built models of the pieces of the DNA chain, and they tried various chemically reasonable ways of fitting them together. Finally, they discovered that one arrangement was "too pretty not to be true." Ultimately, the experimental evidence of Wilkins and Franklin confirmed the "pretty structure" to be the real DNA structure.

James D. Watson and Francis Crick. In a photo taken in 1953, Watson (*left*) and Crick (*right*) stand by their model of the DNA double helix (at The University of Cambridge, England). Together with Maurice Wilkins, Watson and Crick received the Nobel Prize in Medicine and Physiology in 1962.

3. Use bond dissociation enthalpies to estimate the enthalpy change for the decomposition of hydrogen peroxide.

$$2\,H_2O_2(g) \rightarrow 2\,H_2O(g) + O_2(g)$$

(a) +352 kJ/mol-rxn

(c) +1037 kJ/mol-rxn

(b) −206 kJ/mol-rxn

(d) −2204 kJ/mol-rxn

8.10 DNA, Revisited

This chapter opened with some questions about the structure of DNA, one of the key molecules in all biological organisms. Now that you have studied this chapter, we can say more about the structure of this important molecule and why it looks the way it does.

As shown in Figure 8.17, each strand of the double-stranded DNA molecule consists of three repeating parts: a phosphate, a deoxyribose molecule (a sugar molecule with a five-member ring), and a nitrogen-containing base. (The bases in DNA can be one of four molecules: adenine, guanine, cytosine, and thymine; in Figure 8.17, the base is adenine.) Two units of the backbone (without the adenine on the deoxyribose ring) are also illustrated in Figure 8.17a.

One important point here is that the repeating unit in the backbone of DNA consists of the atoms O—P—O—C—C—C. Each atom has a tetrahedral electron-pair geometry. Therefore, the chain cannot be linear. In fact, the chain twists as one moves along the backbone. This twisting gives DNA its helical shape.

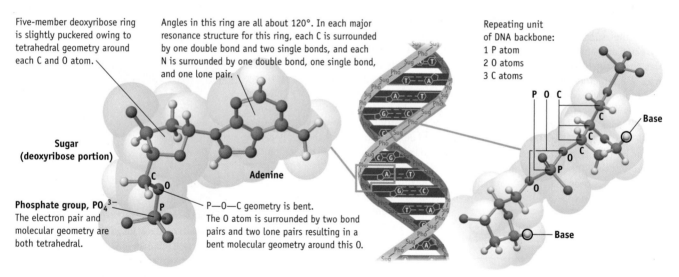

Five-member deoxyribose ring is slightly puckered owing to tetrahedral geometry around each C and O atom.

Angles in this ring are all about 120°. In each major resonance structure for this ring, each C is surrounded by one double bond and two single bonds, and each N is surrounded by one double bond, one single bond, and one lone pair.

Repeating unit of DNA backbone:
1 P atom
2 O atoms
3 C atoms

Sugar (deoxyribose portion)

Adenine

Phosphate group, PO_4^{3-} The electron pair and molecular geometry are both tetrahedral.

P—O—C geometry is bent. The O atom is surrounded by two bond pairs and two lone pairs resulting in a bent molecular geometry around this O.

(a) A repeating unit consists of a phosphate portion, a deoxyribose portion (a sugar molecule with a five-member ring), and a nitrogen-containing base (here adenine) attached to the deoxyribose ring.

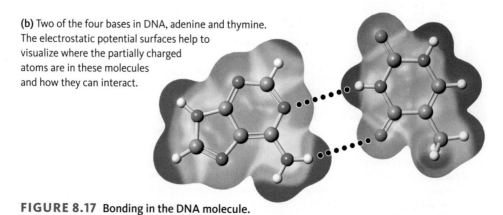

(b) Two of the four bases in DNA, adenine and thymine. The electrostatic potential surfaces help to visualize where the partially charged atoms are in these molecules and how they can interact.

FIGURE 8.17 Bonding in the DNA molecule.

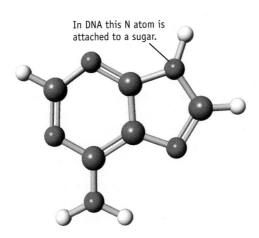

In DNA this N atom is attached to a sugar.

FIGURE 8.18 Adenine, one of the four bases in the DNA molecule. Notice that all of the C and N atoms of the five- and six-member rings have trigonal-planar electron-pair geometry. The N atom in the five-member ring is attached to a sugar in DNA.

Why are there two strands in DNA with the O—P—O—C—C—C backbone on the outside and the nitrogen-containing bases on the inside? This structure arises from the polarity of the bonds in the base molecules attached to the backbone. For example, the N–H bonds in the adenine molecule are very polar, which leads to a special type of intermolecular force—hydrogen bonding—binding the adenine to thymine in the neighboring chain (Figure 8.17b). You will learn more about this in Chapter 12 when we explore intermolecular forces.

The rings in the nitrogen-containing bases are all flat with trigonal-planar electron-pair geometry around each atom in the rings. Let us examine the electron-pair geometries in one of the bases, adenine (Figure 8.18). There are two major resonance structures for this ring system. In these, each carbon atom is surrounded by one double bond and two single bonds, leading to a trigonal-planar electron-pair geometry. Each nitrogen atom in the rings, except the one attached to the sugar, is surrounded by a double bond, a single bond, and a lone pair, likewise leading to trigonal-planar electron-pair geometries around these atoms. The nitrogen attached to the sugar, however, is different from what we would normally predict. It is surrounded by three single bonds and one lone pair. We would normally expect it to have a tetrahedral electron-pair geometry (and trigonal-pyramidal molecular geometry), but it does not. Instead, the bonding groups assume a trigonal-planar geometry, and the lone pair is in a plane perpendicular to the bonds. After studying Chapter 9, you will understand how this allows the electrons in this lone pair to interact with the electrons in the rings' double bonds in a favorable way.

CHAPTER GOALS REVISITED

Now that you have studied this chapter, you should ask whether you have met the chapter goals. In particular, you should be able to:

Understand the difference between ionic and covalent bonds

a. Describe the basic forms of chemical bonding—ionic and covalent—and the differences between them, and predict from the formula whether a compound has ionic or covalent bonding, based on whether a metal is part of the formula (Section 8.1).

b. Write Lewis symbols for atoms (Section 8.2).

Draw Lewis electron dot structures for small molecules and ions

a. Draw Lewis structures for molecular compounds and ions (Section 8.2). **Study Questions: 5–12.**

b. Understand and apply the octet rule; recognize exceptions to the octet rule (Sections 8.2–8.5). **Study Questions: 5–12, 60.**

c. Write resonance structures, understand what resonance means, and know how and when to use this means of representing bonding (Section 8.4). **Study Questions: 9, 10.**

Use the valence shell electron-pair repulsion (VSEPR) theory to predict the shapes of simple molecules and ions and to understand the structures of more complex molecules.

a. Predict the shape or geometry of molecules and ions of main group elements using VSEPR theory (Section 8.6). Table 8.10 shows a summary of the relation between valence electron pairs, electron-pair and molecular geometry, and molecular polarity. **Study Questions: 17–24, and Go Chemistry Module 12.**

Use electronegativity and formal charge to predict the charge distribution in molecules and ions, to define the polarity of bonds, and to predict the polarity of molecules.

a. Calculate formal charges for atoms in a molecule based on the Lewis structure (Section 8.3). **Study Questions: 13–16, 38.**

b. Define electronegativity and understand how it is used to describe the unequal sharing of electrons between atoms in a bond (Section 8.7).

c. Combine formal charge and electronegativity to gain a perspective on the charge distribution in covalent molecules and ions (Section 8.7). **Study Questions: 27–38, 79.**

d. Understand why some molecules are polar whereas others are nonpolar (Section 8.8). (See Table 8.7.)

e. Predict the polarity of a molecule (Section 8.8). **Study Questions: 41, 42, 82, 83, 85, 90, and Go Chemistry Module 13.**

Understand the properties of covalent bonds and their influence on molecular structure

a. Define and predict trends in bond order, bond length, and bond dissociation enthalpy (Section 8.9). **Study Questions: 43–50, 62, 85.**

b. Use bond dissociation enthalpies in calculations (Section 8.9 and Example 8.14). **Study Questions: 51–56, 73.**

Table 8.10 Summary of Molecular Shapes and Molecular Polarity

Pairs of Valence Electrons	Electron-Pair Geometry	Number of Bond Pairs	Number of Lone Pairs	Molecular Geometry	Molecular Dipole?*	Examples
2	Linear	2	0	Linear	No	$BeCl_2$
3	Trigonal planar	3	0	Trigonal planar	No	BF_3, BCl_3
		2	1	Bent	Yes	$SnCl_2(g)$
4	Tetrahedral	4	0	Tetrahedral	No	CH_4, BF_4^-
		3	1	Trigonal pyramidal	Yes	NH_3, PF_3
		2	2	Bent	Yes	H_2O, SCl_2
5	Trigonal bipyramidal	5	0	Trigonal bipyramidal	No	PF_5
		4	1	Seesaw	Yes	SF_4
		3	2	T-shaped	Yes	ClF_3
		2	3	Linear	No	XeF_2, I_3^-
6	Octahedral	6	0	Octahedral	No	SF_6, PF_6^-
		5	1	Square pyramidal	Yes	ClF_5
		4	2	Square planar	No	XeF_4

*For molecules of AX_n, where the X atoms are identical.

Key Equations

Equation 8.1 (page 270) Used to calculate the formal charge on an atom in a molecule.

Formal charge of an atom in a molecule or ion = NVE − [LPE + ½(BE)]

Equation 8.2 (page 298) Used to calculate bond order.

$$\text{Bond order} = \frac{\text{number of shared pairs in all X—Y bonds}}{\text{number of X—Y links in the molecule or ion}}$$

Equation 8.3 (page 302) Used to estimate the enthalpy change for a reaction using bond dissociation enthalpies.

$$\Delta_r H = \Sigma \Delta H(\text{bonds broken}) - \Sigma \Delta H(\text{bonds formed})$$

OWL Questions and problems for this chapter are available in OWL. Use the OWL access card that came with this textbook to access assigned questions and problems for this chapter.

9 Bonding and Molecular Structure: Orbital Hybridization and Molecular Orbitals

© Gary J. Schrobilgen

The Noble Gases: Not So Inert

Generations of chemistry professors once taught that the noble gases were chemically inert. However, in 1962 chemists learned this was not so. Xenon at the very least was found to form compounds! The first was an ionic compound, now known to be $XeF^+Pt_2F_{11}^-$. However, this was followed shortly thereafter with the discovery of a large number of covalently bonded compounds, including XeF_4, XeF_6, $XeOF_4$, and XeO_3.

Since 1962, the field of noble gas chemistry has expanded with the discovery of such interesting molecules as FXeOXeF and, at low temperatures, species such as HArF, HXeH, HXeCl, and HKrF.

Initially, xenon compounds were thought to form only under the most severe conditions. Therefore, it was again a surprise when it was learned that irradiating a mixture of xenon and fluorine gases at room temperature with sunlight gave crystals of XeF_2 (as seen in the photo).

Questions:

1. What is the most reasonable structure of XeF_2? Does knowing that the molecule has no dipole moment confirm your structural choice? Why or why not?
2. Describe the bonding in XeF_2 using valence bond theory.
3. Predict a structure for FXeOXeF.

Answers to these questions are available in Appendix N.

CHAPTER OUTLINE

9.1 Orbitals and Theories of Chemical Bonding

9.2 Valence Bond Theory **go Chemistry**

9.3 Molecular Orbital Theory

CHAPTER GOALS

See Chapter Goals Revisited (page 336) for Study Questions keyed to these goals.

- Understand the differences between valence bond theory and molecular orbital theory.

- Identify the hybridization of an atom in a molecule or ion.

- Understand the differences between bonding and antibonding molecular orbitals and be able to write the molecular orbital configurations for simple diatomic molecules.

OWL

Sign in to OWL at **www.cengage.com/owl** to view tutorials and simulations, develop problem-solving skills, and complete online homework assigned by your professor.

Just how are molecules held together? Millions of organic compounds are based on carbon atoms surrounded tetrahedrally by other atoms. How can a carbon atom be bonded to atoms at the corners of a tetrahedron when its atomic orbitals don't point in those directions? How can a dye molecule absorb light? Why is oxygen paramagnetic, and how is this property connected with bonding in the molecule? These are just a few of the fundamental and interesting questions that are raised in this chapter and that require us to take a more advanced look at bonding.

9.1 Orbitals and Theories of Chemical Bonding

From Chapter 6, you know that the location of the valence electrons in an atom is described by an orbital model. It seems reasonable that an orbital model can also be used to describe electrons in molecules.

Two common approaches to rationalizing chemical bonding based on orbitals are **valence bond (VB) theory** and **molecular orbital (MO) theory**. The former was developed largely by Linus Pauling and the latter by another American scientist, Robert S. Mulliken (1896–1986). The valence bond approach is closely tied to Lewis's idea of bonding electron pairs between atoms and lone pairs of electrons localized on a particular atom. In contrast, Mulliken's approach was to derive molecular orbitals that are "spread out," or *delocalized*, over the molecule. One way to do this is to combine atomic orbitals to form a set of orbitals that are the property of the molecule and then to distribute the electrons of the molecule within those orbitals.

Why are two theories used? Is one more correct than the other? Actually, both give good descriptions of the bonding in molecules and polyatomic ions, but they are most useful in different situations. Valence bond theory is generally the method of choice to provide a qualitative, visual picture of molecular structure and bonding. This theory is particularly useful for molecules made up of many atoms. In contrast, molecular orbital theory is used when a more quantitative picture of bonding is needed. Valence bond theory provides a good description of bonding for molecules in their ground or lowest energy state. Molecular orbital theory is essential if we want to describe molecules in higher-energy, excited states. Among other things, this is important in explaining the colors of compounds. Finally, for a few molecules such as NO and O_2, MO theory is the only one of the two theories that can describe their bonding accurately.

go Chemistry

Download mini lecture videos for key concept review and exam prep from OWL or purchase them from **www.cengagebrain.com**

- **Bonds Are a "Figment of Our Own Imagination"** C. A. Coulson, a prominent theoretical chemist at the University of Oxford, England, has said that "Sometimes it seems to me that a bond between atoms has become so real, so tangible, so friendly, that I can almost see it. Then I awake with a little shock, for a chemical bond is not a real thing. It does not exist. No one has ever seen one. No one ever can. It is a figment of our own imagination" (*Chemical and Engineering News,* January 29, 2007, page 37). Nonetheless, bonds are a useful figment, and this chapter presents some of these useful ideas.

Module 14: Hybrid Atomic Orbitals covers concepts in this section.

9.2 Valence Bond Theory
The Orbital Overlap Model of Bonding

What happens if two atoms at an infinite distance apart are brought together to form a bond? This process is often illustrated with H_2 because, with two electrons and two nuclei, this is the simplest neutral molecule known (Figure 9.1). Initially, when two hydrogen atoms are widely separated, they do not interact. If the atoms move closer together, however, the electron on one atom begins to experience an attraction to the positive charge of the nucleus of the other atom. Because of the attractive forces, the electron clouds on the atoms distort as the electron of one atom is drawn toward the nucleus of the second atom, and the potential energy of the system is lowered.

Calculations show that when the distance between the H atoms is 74 pm, the potential energy reaches a minimum and the H_2 molecule is most stable. Significantly, 74 pm corresponds to the experimentally measured bond distance in the H_2 molecule.

When the distance between the H atoms is less than 74 pm, the potential energy rises rapidly, and the H_2 molecule is less stable. At separations less than 74 pm there is a significant repulsive force between the nuclei of the two atoms and between the electrons of the atoms. In Figure 9.1 for the H_2 molecule, you see that 74 pm is the distance at which there is a minimum in the potential energy diagram.

In the H_2 molecule two electrons, one from each atom, pair up to form the bond. There is a net stabilization, representing the extent to which the energies of the two electrons are lowered from their value in the free atoms. The net stabilization (the extent to which the potential energy is lowered) can be calculated, and the calculated value closely matches the experimentally determined bond energy. Agreement between theory and experiment on both bond distance and energy is evidence that this theoretical approach has merit.

Bond formation is depicted in Figures 9.1 and 9.2 as occurring when the electron clouds on the two atoms interpenetrate or overlap. This **orbital overlap** increases the probability of finding the bonding electrons in the region of space between the two nuclei. *The idea that bonds are formed by overlap of atomic orbitals is the basis for valence bond theory.*

When the single covalent bond is formed in H_2, the $1s$ electron cloud of each atom is distorted in a way that gives the electrons a higher probability of being in the region between the two hydrogen atoms (Figure 9.2a). This makes sense because this distortion results in the electrons being situated so that they are attracted

FIGURE 9.1 Potential energy change during H—H bond formation from isolated hydrogen atoms. There is little or no overlap of 1s orbitals when the atoms are 400 pm apart. When the separation is about 200 pm there is some overlap, but the attractive forces are weak. The lowest potential energy is reached at an H—H separation of 74 pm. Here there is a balance of the attractive and repulsive forces. At H—H distances less than 74 pm, repulsions between the nuclei and between the electrons of the two atoms increase rapidly, and the curve rises steeply. (The red color reflects the increase in electron density between the H atoms as the distance decreases.)

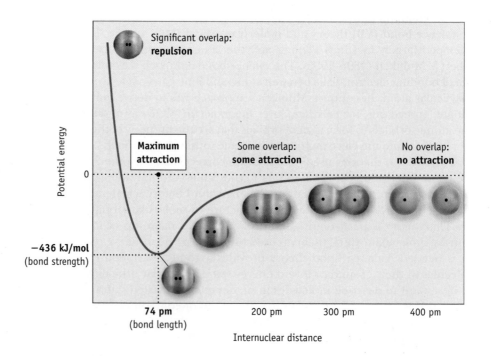

equally to the two positively charged nuclei. Placing the electrons between the nuclei also matches the Lewis electron dot model.

The covalent bond that arises from the overlap of two *s* orbitals, one from each of two atoms as in H_2, is called a **sigma (σ) bond**. *A sigma bond is a bond in which electron density is greatest along the axis of the bond.*

In summary, the main points of the valence bond approach to bonding are as follows:

- Orbitals overlap to form a bond between two atoms.
- Two electrons, *of opposite spin*, can be accommodated in the overlapping orbitals. Usually, one electron is supplied by each of the two bonded atoms.
- Because of orbital overlap, the bonding electrons have a higher probability of being found within a region of space influenced by both nuclei. Both electrons are thus simultaneously attracted to both nuclei.

What happens for elements beyond hydrogen? In the Lewis structure of HF, for example, a bonding electron pair is placed between H and F, and three lone pairs of electrons are depicted as localized on the F atom (Figure 9.2b). To use an orbital approach, look at the valence shell electrons and orbitals for each atom that can overlap. The hydrogen atom will use its 1*s* orbital in bond formation. The electron configuration of fluorine is $1s^2 2s^2 2p^5$, and the unpaired electron for this atom is assigned to one of the 2*p* orbitals. A sigma bond results from overlap of the hydrogen 1*s* and this fluorine 2*p* orbital.

Formation of the H—F bond is similar to formation of an H—H bond. A hydrogen atom approaches a fluorine atom along the axis containing the 2*p* orbital with a single electron. The orbitals (1*s* on H and 2*p* on F) distort as each atomic nucleus influences the electron and orbital of the other atom. Still closer together, the 1*s* and 2*p* orbitals overlap, and the two electrons, with opposite spins, pair up to give a σ bond (Figure 9.2b). There is an optimum distance (92 pm) at which the energy is lowest, and this corresponds to the bond distance in HF. The net stabilization achieved in this process is the energy for the H—F bond.

The remaining electrons on the fluorine atom (a pair of electrons in the 2*s* orbital and two pairs of electrons in the other two 2*p* orbitals) are not involved in bonding. They are nonbonding electrons, the lone pairs associated with this element in the Lewis structure.

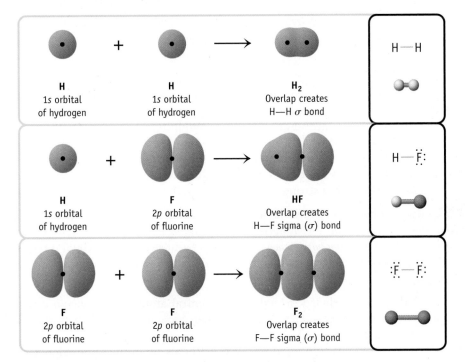

(a) Overlap of hydrogen 1s orbitals to form the H—H sigma (σ) bond.

(b) Overlap of hydrogen 1s and fluorine 2p orbitals to form the sigma (σ) bond in HF.

(c) Overlap of 2p orbitals on two fluorine atoms forming the sigma (σ) bond in F_2.

H
1s orbital
of hydrogen

H
1s orbital
of hydrogen

H_2
Overlap creates
H—H σ bond

H—H

H
1s orbital
of hydrogen

F
2p orbital
of fluorine

HF
Overlap creates
H—F sigma (σ) bond

H—F̈:

F
2p orbital
of fluorine

F
2p orbital
of fluorine

F_2
Overlap creates
F—F sigma (σ) bond

:F̈—F̈:

FIGURE 9.2 Covalent bond formation in H_2, HF, and F_2.

Extension of this model gives a description of bonding in F_2. The $2p$ orbitals on the two atoms overlap, and the single electron from each atom is paired in the resulting σ bond (Figure 9.2c). The $2s$ and the $2p$ electrons not involved in the bond are the lone pairs on each atom.

Hybridization of Atomic Orbitals

The simple picture using orbital overlap to describe bonding in H_2, HF, and F_2 works well, but we run into difficulty when molecules with more atoms are considered. For example, a Lewis dot structure of methane, CH_4, shows four C—H covalent bonds. VSEPR theory predicts, and experiments confirm, that the electron-pair geometry of the C atom in CH_4 is tetrahedral, with an angle of 109.5° between the bond pairs. The hydrogens are identical in this structure. This means that four equivalent bonding electron pairs occur around the C atom. An orbital picture of the bonds should convey both the geometry and the fact that all C—H bonds are the same.

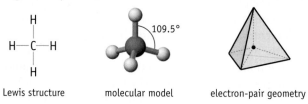

| Lewis structure | molecular model | electron-pair geometry |

If we apply the orbital overlap model used for H_2 and F_2 without modification to describe the bonding in CH_4, a problem arises. The three p-orbitals for the valence electrons of carbon ($2p_x$, $2p_y$, $2p_z$) are at right angles, 90°, and do not match the tetrahedral angle of 109.5°.

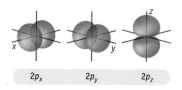

The spherical $2s$ orbital could bond in any direction. Furthermore, a carbon atom in its ground state ($1s^2 2s^2 2p^2$) has only two unpaired electrons (in the $2p$ orbitals), not the four that are needed to allow formation of four bonds.

To describe the bonding in methane and other molecules, Linus Pauling proposed the theory of **orbital hybridization**. He suggested that a new set of orbitals, called **hybrid orbitals**, could be created by mixing the s, p, and (when required) d atomic orbitals on an atom (Figure 9.3). There are three important principles that govern the outcome.

- The number of hybrid orbitals is always equal to the number of atomic orbitals that are mixed to create the hybrid orbital set.
- Hybrid orbital sets are always built by combining an s orbital with as many p orbitals (and d orbitals if necessary) to have enough hybrid orbitals to accommodate the bond and lone pairs on the central atom.
- The hybrid orbitals are directed toward the terminal atoms, leading to better orbital overlap and a stronger bond between the central and terminal atoms.

The sets of hybrid orbitals that arise from mixing s, p, and d atomic orbitals are illustrated in Figure 9.3. *The hybrid orbitals required by an atom in a molecule or ion are chosen to match the electron-pair geometry of the atom* because a hybrid orbital is required for each sigma bond electron pair and each lone pair. The following types of hybridization are important:

- *sp*: If the valence shell s orbital on the central atom in a molecule or ion is mixed with a valence shell p orbital on that same atom, two sp hybrid orbitals are created. They are separated by 180°.
- *sp²*: If an s orbital is combined with two p orbitals, all in the same valence shell, three sp^2 hybrid orbitals are created. They are in the same plane and are separated by 120°.

• **Hybridization and Geometry** Hybridization reconciles the electron-pair geometry with the orbital overlap criterion of bonding. A statement such as "the atom is tetrahedral because it is sp^3 hybridized" is backward. That the electron-pair geometry around the atom is tetrahedral is a fact. Hybridization is one way to rationalize that strong bonds can occur in this geometry.

- **sp³**: When the s orbital in a valence shell is combined with three p orbitals, the result is four hybrid orbitals, each labeled sp³. The hybrid orbitals are separated by 109.5°, the tetrahedral angle.
- **sp³d and sp³d²**: If one or two d orbitals are combined with s and p orbitals in the same valence shell, two other hybrid orbital sets are created. These are utilized by the central atom of a molecule or ion with a trigonal-bipyramidal or octahedral electron-pair geometry, respectively.

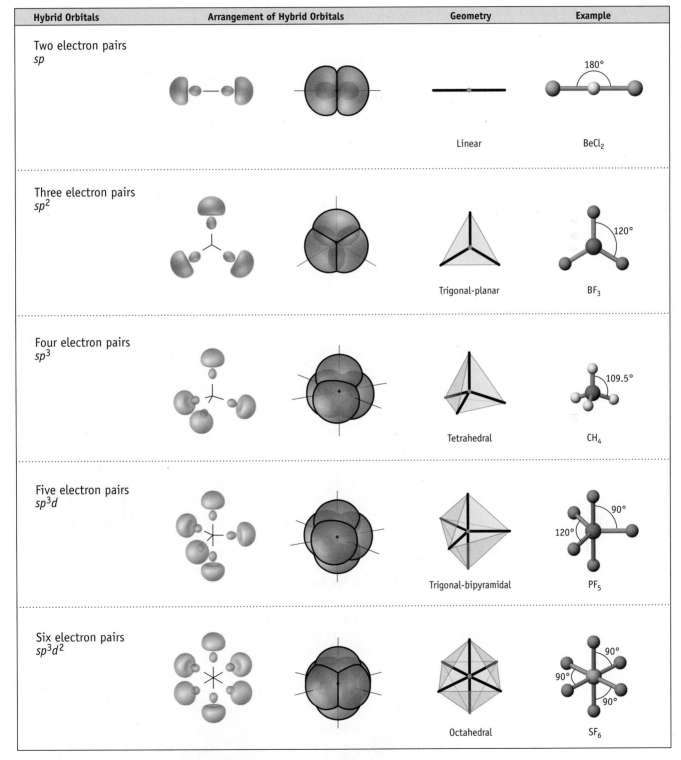

FIGURE 9.3 Hybrid orbitals for two to six electron pairs. The geometry of the hybrid orbital sets for two to six valence shell electron pairs is given in the right column. In forming a hybrid orbital set, the s orbital is always used, plus as many p orbitals (and d orbitals) as are required to give the necessary number of σ-bonding and lone-pair orbitals.

FIGURE 9.4 Bonding in the methane (CH_4) molecule.

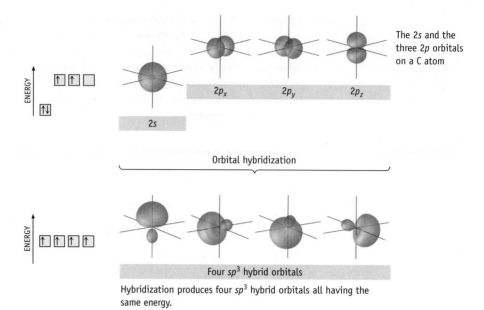

Orbital hybridization

The 2s and the three 2p orbitals on a C atom

$2p_x$ $2p_y$ $2p_z$

2s

Four sp^3 hybrid orbitals

Hybridization produces four sp^3 hybrid orbitals all having the same energy.

Valence Bond Theory for Methane, CH_4

In methane, four orbitals directed to the corners of a tetrahedron are needed to match the electron-pair geometry on the central carbon atom. By combining the four valence shell orbitals, the 2s and all three of the 2p orbitals on carbon, a new set of four hybrid orbitals is created that has tetrahedral geometry (Figures 9.3 and 9.4). Each of the four hybrid orbitals is labeled sp^3 to indicate the atomic orbital combination (an s orbital and three p orbitals) from which it is derived. All four sp^3 orbitals have an identical shape, and the angle between them is 109.5°, the tetrahedral angle. Because the orbitals have the same energy, one electron can be assigned to each according to Hund's rule (◀ Section 7.3, page 240). Then, each C—H bond is formed by overlap of one of the carbon sp^3 hybrid orbitals with the 1s orbital from a hydrogen atom; one electron from the C atom is paired with an electron from an H atom.

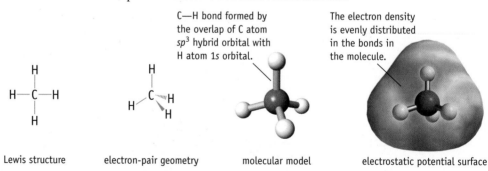

C—H bond formed by the overlap of C atom sp^3 hybrid orbital with H atom 1s orbital.

The electron density is evenly distributed in the bonds in the molecule.

Lewis structure   electron-pair geometry   molecular model   electrostatic potential surface

Valence Bond Theory for Ammonia, NH_3

The Lewis structure for ammonia shows there are four electron pairs in the valence shell of nitrogen: three bond pairs and a lone pair. VSEPR theory predicts a tetrahedral electron-pair geometry and a trigonal-pyramidal molecular geometry. The actual structure is a close match to the predicted structure; the H—N—H bond angles are 107.5° in this molecule.

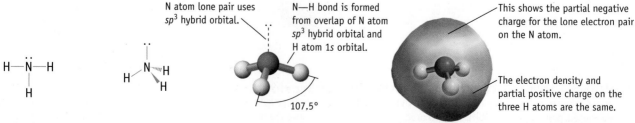

N atom lone pair uses sp^3 hybrid orbital.

N—H bond is formed from overlap of N atom sp^3 hybrid orbital and H atom 1s orbital.

This shows the partial negative charge for the lone electron pair on the N atom.

107.5°

The electron density and partial positive charge on the three H atoms are the same.

Lewis structure   electron-pair geometry   molecular model   electrostatic potential surface

Based on the electron-pair geometry of NH_3, we predict sp^3 hybridization to accommodate the four electron pairs on the N atom. The lone pair is assigned to one of the hybrid orbitals, and each of the other three hybrid orbitals is occupied by a single electron. Overlap of each of the singly occupied, sp^3 hybrid orbitals with a $1s$ orbital from a hydrogen atom, and pairing of the electrons in these orbitals, create the N—H bonds.

Valence Bond Theory for Water, H₂O

The oxygen atom of water has two bonding pairs and two lone pairs in its valence shell, and the H—O—H angle is 104.5°. Four sp^3 hybrid orbitals are created from the $2s$ and $2p$ atomic orbitals of oxygen. Two of these sp^3 orbitals are occupied by unpaired electrons and are used to form O—H bonds. Lone pairs occupy the other two hybrid orbitals.

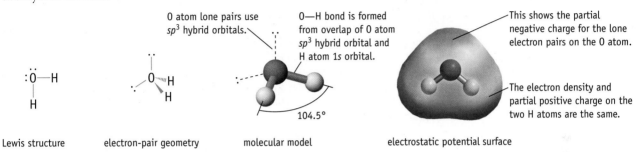

O atom lone pairs use sp^3 hybrid orbitals.

O—H bond is formed from overlap of O atom sp^3 hybrid orbital and H atom $1s$ orbital.

This shows the partial negative charge for the lone electron pairs on the O atom.

The electron density and partial positive charge on the two H atoms are the same.

104.5°

Lewis structure   electron-pair geometry   molecular model   electrostatic potential surface

ⓌWL INTERACTIVE EXAMPLE 9.1 Valence Bond Description of Bonding in Ethane

Problem Describe the bonding in ethane, C_2H_6, using valence bond theory.

What Do You Know? The formula is known, so the number of valence electrons can be calculated. From this you can derive the electron-pair geometry and the bonding model.

Strategy

- Draw the Lewis structure and determine the electron-pair geometry at both carbon atoms.

- Assign hybridization to the carbon atoms based upon the electron-pair geometry.

- Describe covalent bonds based on orbital overlap, and place electron pairs in their proper locations.

Solution Each carbon atom has an octet configuration, sharing electron pairs with three hydrogen atoms and with the other carbon atom. The electron pairs around carbon have tetrahedral geometry, so carbon is assigned sp^3 hybridization. The C—C bond is formed by overlap of sp^3 orbitals on each C atom, and each of the C—H bonds is formed by overlap of an sp^3 orbital on carbon with a hydrogen $1s$ orbital.

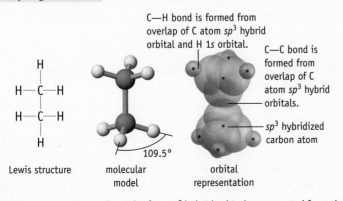

C—H bond is formed from overlap of C atom sp^3 hybrid orbital and H $1s$ orbital.

C—C bond is formed from overlap of C atom sp^3 hybrid orbitals.

sp^3 hybridized carbon atom

109.5°

Lewis structure   molecular model   orbital representation

Think about Your Answer For carbon, the four sp^3 hybrid orbitals are created from the four valence orbitals (one $2s$ and three $2p$ atomic orbitals). All the valence orbitals on carbon and hydrogen are utilized in forming bonds.

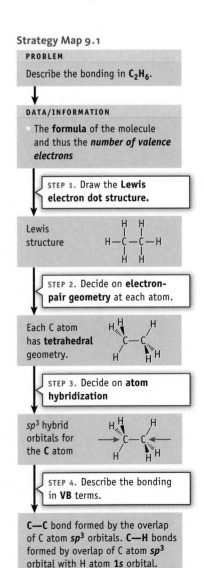

Strategy Map 9.1

PROBLEM

Describe the bonding in C_2H_6.

DATA/INFORMATION

- The **formula** of the molecule and thus the **number of valence electrons**

STEP 1. Draw the **Lewis electron dot structure.**

Lewis structure

STEP 2. Decide on **electron-pair geometry** at each atom.

Each C atom has **tetrahedral** geometry.

STEP 3. Decide on **atom hybridization**

sp^3 hybrid orbitals for the **C** atom

STEP 4. Describe the bonding in **VB** terms.

C—C bond formed by the overlap of C atom sp^3 orbitals. **C—H** bonds formed by overlap of C atom sp^3 orbital with H atom **1s** orbital.

Check Your Understanding

Use valence bond theory to describe the bonding in $CHCl_3$.

EXAMPLE 9.2   Valence Bond Description of Bonding in Methanol

Problem Describe the bonding in the methanol molecule, CH_3OH, using valence bond theory.

What Do You Know? The formula, CH_3OH, helps to define how atoms are linked together. Three hydrogen atoms are linked to carbon. The fourth bond from carbon is to oxygen, and oxygen is attached to the remaining hydrogen.

Strategy First, construct the Lewis structure for the molecule. The electron-pair geometry around each atom determines the hybrid orbital set used by that atom.

Solution The electron-pair geometry around both the C and O atoms in CH_3OH is tetrahedral. Thus, we may assign sp^3 hybridization to each atom, and the C—O bond is formed by overlap of sp^3 orbitals on these atoms. Each C—H bond is formed by overlap of a carbon sp^3 orbital with a hydrogen $1s$ orbital, and the O—H bond is formed by overlap of an oxygen sp^3 orbital with the hydrogen $1s$ orbital. Two lone pairs on oxygen occupy the remaining sp^3 orbitals on the atom.

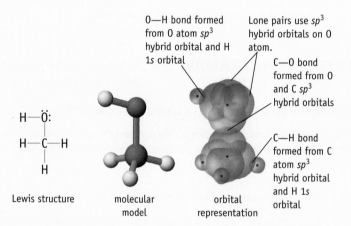

O—H bond formed from O atom sp^3 hybrid orbital and H $1s$ orbital

Lone pairs use sp^3 hybrid orbitals on O atom.

C—O bond formed from O and C sp^3 hybrid orbitals

C—H bond formed from C atom sp^3 hybrid orbital and H $1s$ orbital

Lewis structure   molecular model   orbital representation

Think about Your Answer Notice that one end of the CH_3OH molecule (the CH_3 or methyl group) is just like the CH_3 group in the methane molecule, and the OH group resembles the OH group in water. This example also shows how to predict the structure and bonding in a complicated molecule by looking at each part separately. This is an important idea when dealing with molecules made up of many atoms.

Check Your Understanding

Use valence bond theory to describe the bonding in methylamine, CH_3NH_2.

methylamine, CH_3NH_2

Hybrid Orbitals for Molecules and Ions with Trigonal-Planar Electron-Pair Geometries

The central atoms in species such as BF_3, O_3, NO_3^-, and CO_3^{2-} all have a trigonal-planar electron-pair geometry, which requires a central atom with three hybrid orbitals in a plane, 120° apart. Three hybrid orbitals mean three atomic orbitals must be combined, and the combination of an s orbital with two p orbitals is appropriate (Figure 9.5). If p_x and p_y orbitals are used in hybrid orbital formation, the three hybrid sp^2 orbitals will lie in the xy-plane. The p_z orbital not used to form these hybrid orbitals is perpendicular to the plane containing the three sp^2 orbitals.

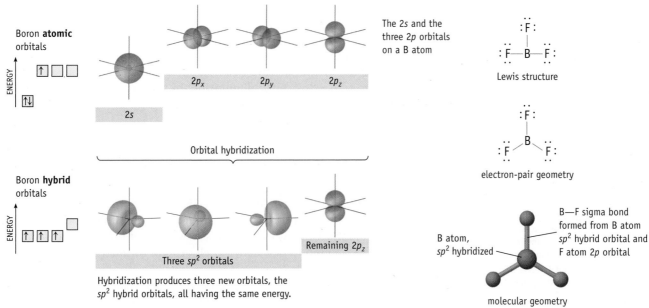

FIGURE 9.5 Bonding in a trigonal-planar molecule.

Boron trifluoride has a trigonal-planar electron-pair and molecular geometry. Each boron–fluorine bond in this compound results from overlap of an sp^2 orbital on boron with a p orbital on fluorine. Notice that the p_z orbital on boron, which is not used to form the sp^2 hybrid orbitals, is not occupied by electrons.

Hybrid Orbitals for Molecules and Ions with Linear Electron-Pair Geometries

For molecules in which the central atom has a linear electron-pair geometry, two hybrid orbitals, 180° apart, are required. One s and one p orbital can be hybridized to form two sp hybrid orbitals (Figure 9.6). If the p_z orbital is used, then the sp orbitals are oriented along the z-axis. The p_x and p_y orbitals are perpendicular to this axis.

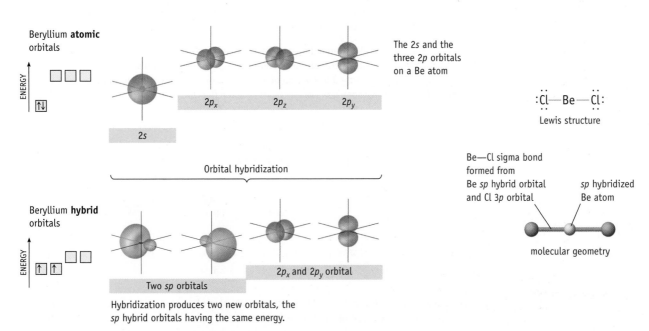

FIGURE 9.6 Bonding in a linear molecule. Because only one p orbital is incorporated in the hybrid orbital, two p orbitals remain unhybridized. These orbitals are perpendicular to each other and to the axis along which the two sp hybrid orbitals lie.

Beryllium dichloride, $BeCl_2$, is a solid under ordinary conditions. When it is heated to over 520 °C, however, it vaporizes to give $BeCl_2$ vapor. In the gas phase, $BeCl_2$ is a linear molecule, so sp hybridization is appropriate for the beryllium atom in this species. Combining beryllium's $2s$ and $2p_z$ orbitals gives the two sp hybrid orbitals that lie along the z-axis. Each Be—Cl bond arises by overlap of an sp hybrid orbital on beryllium with a $3p$ orbital on chlorine. In this molecule, there are only two electron pairs around the beryllium atom, so the p_x and p_y orbitals are not occupied (Figure 9.6).

Hybrid Orbitals for Molecules and Ions with Trigonal-Bipyramidal or Octahedral Electron-Pair Geometries: *d*-Orbital Participation

• **Do *nd* orbitals participate in bonding?** Because the *nd* orbitals are at a relatively high energy their involvement in bonding is believed to be minimal. Using molecular orbital theory, it is possible to describe the bonding in compounds with expanded octets without using *d* orbitals to form hybrid orbital sets. (See page 335.)

Using valence bond theory to explain the bonding in compounds having five or six electron pairs on a central atom (such as PF_5 or SF_6) requires the atom to have five or six hybrid orbitals, which must be created from five or six atomic orbitals. This requires the use of atomic orbitals from the *d* subshell in hybrid orbital formation. The *d* orbitals are considered to be valence shell orbitals for main group elements of the third period and beyond.

Five coordination and trigonal-bipyramidal electron-pair geometries are matched to sp^3d hybridization. One *s*, three *p*, and one *d* orbital combine to produce five sp^3d hybrid orbitals. To accommodate six electron pairs in the valence shell of an atom, six sp^3d^2 hybrid orbitals can be created from one *s*, three *p*, and two *d* orbitals. The six sp^3d^2 hybrid orbitals are directed to the corners of an octahedron (Figure 9.3) and can accommodate the valence electron pairs for a compound that has an octahedral electron-pair geometry.

EXAMPLE 9.3   Hybridization Involving *d* Orbitals

Problem Describe the bonding in PF_5 using valence bond theory.

What Do You Know? Phosphorus is the central atom in this compound, and it is bonded to five fluorine atoms.

Strategy The first step in this problem is to identify the hybridization of the central atom. A roadmap to follow is Formula → Lewis structure → electron-pair geometry → hybridization. Bonding can then be described as the overlap of the P atom hybrid orbitals with appropriate orbitals on the fluorine atoms (the half-filled $2p$ orbital on each fluorine).

Solution The P atom is surrounded by five electron pairs, so PF_5 has a trigonal-bipyramidal electron-pair and molecular geometry (see Figure 9.3). The hybridization scheme is therefore sp^3d. Each P—F bond is formed by overlap of an sp^3d orbital on phosphorus with a $2p$ orbital on fluorine. There are three lone pairs of electrons on each fluorine atom.

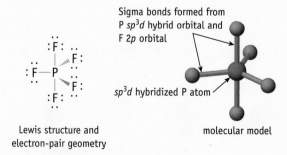

Sigma bonds formed from P sp^3d hybrid orbital and F $2p$ orbital

sp^3d hybridized P atom

Lewis structure and electron-pair geometry

molecular model

Think about Your Answer To decide on the hybridization of an atom in a molecule you need to know the electron-pair geometry around that atom.

Check Your Understanding

Describe the bonding in XeF_2 (page 308) using valence bond theory.

EXAMPLE 9.4   Recognizing Hybridization

Problem Identify the hybridization of the central atom in the following compounds and ions:

(a) SF_3^+ (b) SO_4^{2-} (c) SF_4 (d) I_3^-

What Do You Know? To determine hybridization, you need to know the electron-pair geometry.

Strategy The roadmap here is the same as in Example 9.3: Formula → Lewis structure → electron-pair geometry → hybridization.

Solution The structures for SF_3^+, and SO_4^{2-} are written as follows:

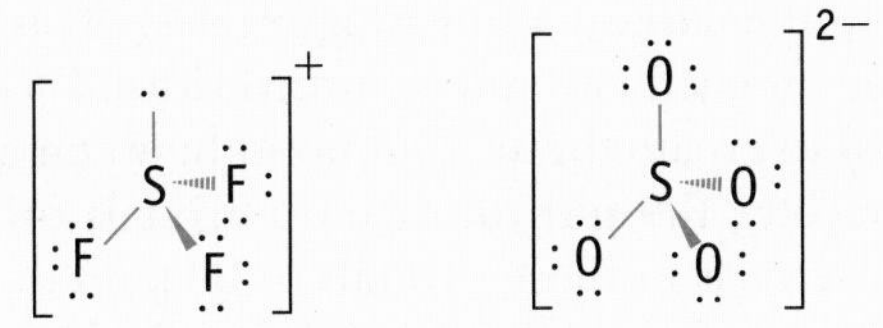

Four electron pairs surround the central atom in each of these ions, and the electron-pair geometry for these atoms is tetrahedral. Thus, *sp³* hybridization for the central atom is used to describe the bonding.

For SF_4 and I_3^-, five pairs of electrons are in the valence shell of the central atom and both have trigonal-bipyramidal electron-pair geometry. For both species *sp³d* hybridization is appropriate for the central atom.

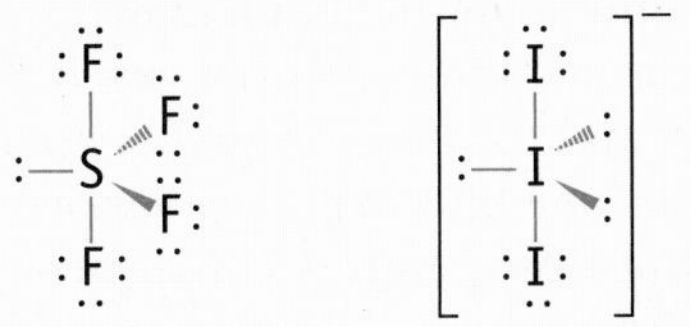

Think about Your Answer Notice that in both SF_4 and I_3^- the central atom lone pairs are in equatorial positions, a feature that leads to minimum repulsion of the valence electron pairs.

Check Your Understanding

Identify the hybridization of the underlined central atom in the following compounds and ions:

(a) $\underline{B}H_4^-$ (c) $O\underline{S}F_4$ (e) $\underline{B}Cl_3$

(b) $\underline{S}F_5^-$ (d) $\underline{Cl}F_3$ (f) $\underline{Xe}O_6^{4-}$

Multiple Bonds

According to valence bond theory, bond formation requires that two orbitals on adjacent atoms overlap. Many molecules have double or triple bonds, that is, there are two or three bonds, respectively, between pairs of atoms. Therefore, according to valence bond theory, a double bond requires *two* sets of overlapping orbitals and *two* electron pairs. For a triple bond, *three* sets of atomic orbitals are required, each set accommodating a pair of electrons.

Double Bonds

Consider ethylene, $H_2C{=}CH_2$, a common molecule with a double bond. The molecular structure of ethylene places all six atoms in a plane, with H—C—H and H—C—C angles of approximately 120°. Each carbon atom has trigonal-planar geometry, so *sp²* hybridization is assumed for these atoms.

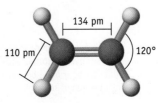

ethylene, C_2H_4

A description of bonding in ethylene starts with each carbon atom having three sp^2 hybrid orbitals in the molecular plane and an unhybridized p orbital perpendicular to that plane. Because each carbon atom is involved in four bonds, a single unpaired electron is placed in each of these orbitals.

[↑] Unhybridized p orbital. Used for π bonding in C_2H_4.

[↑|↑|↑] Three sp^2 hybrid orbitals. Used for C—H and C—C σ bonding in C_2H_4.

The C—H bonds of C_2H_4 arise from overlap of sp^2 orbitals on carbon with hydrogen $1s$ orbitals. After accounting for these bonds, one sp^2 orbital on each carbon atom remains. These hybrid orbitals point toward each other and overlap to form one of the bonds linking the carbon atoms (Figure 9.7). This leaves only one other orbital unaccounted for on each carbon, an unhybridized p orbital, and it is these orbitals that can be used to create the second bond between carbon atoms in C_2H_4. If they are aligned correctly, the unhybridized p orbitals on the two carbons can overlap, allowing the electrons in these orbitals to be paired. The overlap does not occur directly along the C—C axis, however. Instead, the arrangement compels these orbitals to overlap sideways, and the electron pair occupies an orbital with electron density above and below the plane containing the six atoms.

This description results in two types of bonds in C_2H_4. One type is the C—H and C—C bonds that arise from the overlap of atomic orbitals so that the bonding electrons that lie along the bond axes form sigma (σ) bonds. The other is the bond formed by sideways overlap of p atomic orbitals, called a **pi (π) bond**. In a π bond, the overlap region is above and below the internuclear axis, and the electron density of the π bond is above and below the bond axis.

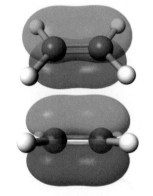

Computer-generated representation of the π bond in ethylene, C_2H_4

Be sure to notice that a π bond can form *only* if (a) there are unhybridized p orbitals on adjacent atoms and (b) the p orbitals are perpendicular to the plane of the molecule and parallel to one another. This happens only if the sp^2 orbitals of both carbon atoms are in the same plane. A consequence of this is that both atoms involved in the π bond have trigonal-planar geometry, and the six atoms in and around the π bond (the two atoms involved in the π bond and the four atoms attached to the π-bonded atoms) lie in one plane.

Double bonds between carbon and oxygen, sulfur, or nitrogen are quite common. Consider formaldehyde, CH_2O, in which a carbon–oxygen π bond occurs (Figure 9.8). A trigonal-planar electron-pair geometry indicates sp^2 hybridization for the C atom. The σ bonds from carbon to the O atom and the two H atoms form by overlap of sp^2 hybrid orbitals with half-filled orbitals from the oxygen and two hydrogen atoms. An unhybridized p orbital on carbon is oriented perpendicular to

(a) Lewis structure and bonding of ethylene, C_2H_4.

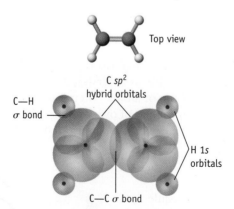

Top view

C sp^2 hybrid orbitals

C—H σ bond

H $1s$ orbitals

C—C σ bond

(b) The C—H σ bonds are formed by overlap of C atom sp^2 hybrid orbitals with H atom $1s$ orbitals. The σ bond between C atoms arises from overlap of sp^2 orbitals.

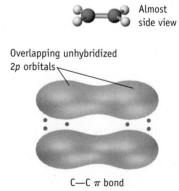

Almost side view

Overlapping unhybridized $2p$ orbitals

C—C π bond

(c) The carbon–carbon π bond is formed by overlap of an unhybridized $2p$ orbital on each atom. Note the lack of electron density along the C—C bond axis from this bond.

FIGURE 9.7 The valence bond model of bonding in ethylene, C_2H_4. Each C atom is assumed to be sp^2 hybridized.

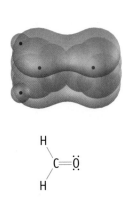

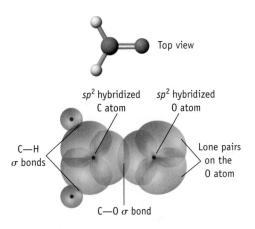

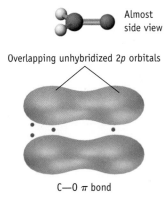

(a) Lewis structure and bonding of formaldehyde, CH_2O.

(b) The C—H σ bonds are formed by overlap of C atom sp^2 hybrid orbitals with H atom $1s$ orbitals. The σ bond between C and O atoms arises from overlap of sp^2 orbitals.

(c) The C—O π bond comes from the side-by-side overlap of unhybridized p orbitals on the two atoms.

FIGURE 9.8 Valence bond description of bonding in formaldehyde, CH_2O.

the molecular plane (just as for the carbon atoms of C_2H_4). This p orbital is available for π bonding, this time with an oxygen orbital.

What orbitals on oxygen are used in this model? The approach in Figure 9.8 assumes sp^2 hybridization for oxygen. This uses one O atom sp^2 orbital in σ bond formation, leaving two sp^2 orbitals to accommodate lone pairs. The remaining p orbital on the O atom participates in the π bond.

EXAMPLE 9.5   Bonding in Acetic Acid

Problem Using valence bond theory, describe the bonding in acetic acid, CH_3CO_2H, the important ingredient in vinegar.

What Do You Know? You know the formula for the compound of interest. You may recognize from the earlier discussion of acids that this compound contains the CO_2H group in which both oxygen atoms are attached to the carbon.

Strategy Write a Lewis electron dot structure, and determine the geometry around each atom using VSEPR theory. Use this geometry to decide on the hybrid orbitals used in σ bonding. If unhybridized p orbitals are available on adjacent C and O atoms, then C—O π bonding can occur.

Solution The carbon atom of the CH_3 group has tetrahedral electron-pair geometry, which means it is sp^3 hybridized. Three sp^3 orbitals are used to form the C—H bonds. The fourth sp^3 orbital is used to bond to the adjacent carbon atom. This carbon atom has a trigonal-planar electron-pair geometry; it must be sp^2 hybridized. The C—C bond is formed using one of these hybrid orbitals, and the other two sp^2 orbitals are used to form the σ bonds to the two oxygens. The oxygen of the O—H group has four electron pairs; it must be tetrahedral and sp^3 hybridized. Thus, this O atom uses two sp^3 orbitals to bond to the adjacent carbon and the hydrogen, and two sp^3 orbitals accommodate the two lone pairs.

Finally, the carbon–oxygen double bond can be described by assuming the C and O atoms are both sp^2 hybridized (like the C—O π bond in formaldehyde, Figure 9.8). The unhybridized p orbital remaining on each atom is used to form the carbon–oxygen π bond, and the lone pairs on the O atom are accommodated in sp^2 hybrid orbitals.

Lewis dot structure   molecular model

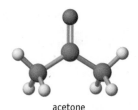

acetone

Think about Your Answer Notice that the hybridization around each central atom in a complex structure is determined by considering its electron-pair geometry.

Check Your Understanding

Use valence bond theory to describe the bonding in acetone, CH_3COCH_3.

Triple Bonds

Acetylene, H—C≡C—H, is an example of a molecule with a triple bond. VSEPR theory predicts that the four atoms lie in a straight line with H—C—C angles of 180°. This implies that the carbon atom is *sp* hybridized (Figure 9.9). For each carbon atom, there are two *sp* orbitals, one directed toward hydrogen and used to create the C—H σ bond, and the second directed toward the other carbon and used to create a σ bond between the two carbon atoms. Two unhybridized *p* orbitals remain on each carbon, and they are oriented so that it is possible to form *two* π bonds in HC≡CH.

⇅↑↑   Two unhybridized *p* orbitals. Used for π bonding in C_2H_2.

⇅↑↑   Two *sp* hybrid orbitals. Used for C—H and C—C σ bonding in C_2H_2.

These π bonds are perpendicular to the molecular axis and perpendicular to each other. Three electrons on each carbon atom are paired to form the triple bond consisting of a σ bond and two π bonds (Figure 9.9).

Now that we have examined several cases involving double and triple bonds, let us summarize several important points:

- In valence bond theory a double bond *always* consists of a σ bond and a π bond. Similarly, a triple bond *always* consists of a σ bond and *two* π bonds.
- A π bond may form only if unhybridized *p* orbitals remain on the bonded atoms.
- If a Lewis structure shows multiple bonds, the atoms involved must be either sp^2 or *sp* hybridized. Only in this manner will unhybridized *p* orbitals be available to form a π bond.

Cis-Trans Isomerism: A Consequence of π Bonding

Ethylene, C_2H_4, is a planar molecule, a geometry that allows the unhybridized *p* orbitals on the two carbon atoms to line up and form a π bond (Figure 9.7). Let us speculate on what would happen if one end of the ethylene molecule were twisted relative to the other end (Figure 9.10). This action would distort the molecule away from planarity, and the *p* orbitals would rotate out of alignment. Rotation would diminish the extent of overlap of these orbitals, and, if a twist of 90° were achieved, the two *p* orbitals would no longer overlap at all; the π bond would be broken. However, so much energy is required to break this bond (about 260 kJ/mol) that rotation around a C=C bond is not expected to occur at room temperature.

H—C≡C—H

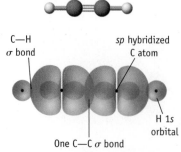

C—H σ bond   *sp* hybridized C atom

One C—C σ bond   H 1s orbital

Two C—C π bonds

C—C π bond 1

C—C π bond 2

FIGURE 9.9 Bonding in acetylene.

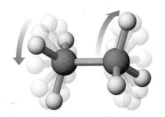

FIGURE 9.10 Rotation around bonds.

(a) In ethane nearly free rotation can occur around the single (σ) bond.

(b) Rotation around the C=C bond in ethylene is severely restricted because doing so would mean breaking the π bond, a process requiring considerable energy.

A consequence of restricted rotation is that isomers occur for many compounds containing a C=C bond. **Isomers** are compounds that have the same formula but different structures. In this case, the two isomeric compounds differ with respect to the orientation of the groups attached to the carbons of the double bond. Two isomers of $C_2H_2Cl_2$ are *cis*- and *trans*-1,2-dichloroethylene. Their structures resemble ethylene, except that two hydrogen atoms have been replaced by chlorine atoms. Because a large amount of energy is required to break the π bond, the *cis* compound cannot rearrange to the *trans* compound under ordinary conditions. Each compound can be obtained separately, and each has its own identity. *Cis*-1,2-dichloroethylene boils at 60.3 °C, whereas *trans*-1,2-dichloroethylene boils at 47.5 °C.

cis-1,2-dichloroethylene   *trans*-1,2-dichloroethylene

Although *cis* and *trans* isomers do not interconvert at ordinary temperatures, they will do so at higher temperatures. If the temperature is sufficiently high, the molecular motions can become sufficiently energetic that rotation around the C=C bond can occur. This may also occur under other special conditions, such as when the molecule absorbs light energy.

Benzene: A Special Case of π Bonding

Benzene, C_6H_6, is the simplest member of a large group of substances known as *aromatic* compounds, a historical reference to their odor. It occupies a pivotal place in the history and practice of chemistry.

To 19th-century chemists, benzene was a perplexing substance with an unknown structure. Based on its chemical reactions, however, August Kekulé (1829–1896) suggested that the molecule has a planar, symmetrical ring structure. We know now he was correct. The ring is flat, and all the carbon–carbon bonds are the same length, 139 pm, a distance intermediate between the average single bond (154 pm) and double bond (134 pm) lengths. Assuming the molecule has two resonance structures with alternating double bonds, the observed structure is rationalized. The C—C bond order in C_6H_6(1.5) is the average of a single and a double bond.

resonance structures   resonance hybrid

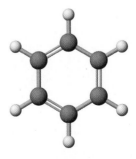

benzene, C_6H_6

FIGURE 9.11 **Bonding in benzene, C_6H_6.** *(left)* The C atoms of the ring are bonded to each other through σ bonds using C atom sp^2 hybrid orbitals. The C—H bonds also use C atom sp^2 hybrid orbitals. The π framework of the molecule arises from overlap of C atom p orbitals not used in hybrid orbital formation. Because these orbitals are perpendicular to the ring, π electron density is above and below the plane of the ring. *(right)* A composite of σ and π bonding in benzene.

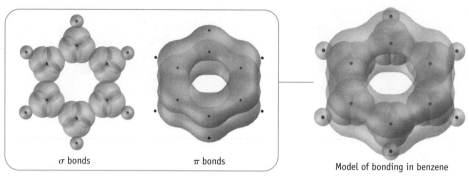

σ bonds $\qquad$ π bonds

σ and π bonding in benzene

Model of bonding in benzene

Understanding the bonding in benzene (Figure 9.11) is important because the benzene ring structure occurs in an enormous number of chemical compounds. We assume that the trigonal-planar carbon atoms have sp^2 hybridization. Each C—H bond is formed by overlap of an sp^2 orbital of a carbon atom with a $1s$ orbital of hydrogen, and the C—C σ bonds arise by overlap of sp^2 orbitals on adjacent carbon atoms. After accounting for the σ bonding, an unhybridized p orbital remains on each C atom, and each is occupied by a single electron. These six orbitals and six electrons form π bonds. Because all carbon–carbon bond lengths are the same, each p orbital overlaps equally well with the p orbitals of both adjacent carbons, and the π interaction is unbroken around the six-member ring.

REVIEW & CHECK FOR SECTION 9.2

1. What is the hybridization of S in SF_4?

   (a) sp^3 $\qquad$ (b) sp^3d $\qquad$ (c) sp^3d^2 $\qquad$ (d) sp^2

2. What orbitals overlap to form the B—F σ bonds in BF_4^-?

   (a) Boron sp^3 + fluorine $2p$ $\qquad$ (c) Boron sp^3 + fluorine $2s$

   (b) Boron sp^2 + fluorine $2p$ $\qquad$ (d) Boron sp^2 + fluorine $2s$

3. Which of the following is incorrect?

   (a) The hybridization of N in NH_4^+ is sp^3. $\qquad$ (c) The hybridization of I in ICl_4^- is sp^3.

   (b) The hybridization of S in SO_4^{2-} is sp^3. $\qquad$ (d) The hybridization of O in H_3O^+ is sp^3.

4. π bonds between two atoms are formed by overlap of what orbitals?

   (a) sp^3 on atom A with sp^3 on atom B $\qquad$ (c) sp on atom A with sp on atom B

   (b) sp^2 on atom A with sp^2 on atom B $\qquad$ (d) p on atom A with p on atom B

5. Which description of the bond between nitrogen atoms in N_2 is correct?

   (a) The nitrogens are attached by a single σ bond.

   (b) The nitrogens are attached by one σ and two π bonds.

   (c) The nitrogen atoms are sp^3 hybridized.

   (d) The bonds in N_2 are formed by overlap of sp orbitals.

9.3 Molecular Orbital Theory

Molecular orbital (MO) theory is an alternative way to view orbitals in molecules. In contrast to the localized bond and lone pair electrons of valence bond theory, MO theory assumes that pure atomic orbitals of the atoms in the molecule combine to produce orbitals that are spread out, or delocalized, over several atoms or even over an entire molecule. These orbitals are called **molecular orbitals**.

One reason for learning about the MO concept is that it correctly predicts the electronic structures of molecules such as O_2 that do not follow the electron-pairing assumptions of the Lewis approach. The rules of Section 8.2 would guide you to draw the electron dot structure of O_2 with all the electrons paired, which fails to explain its paramagnetism (Figure 9.12). The molecular orbital approach can account for this property, but valence bond theory cannot. To see how MO theory can be used to describe the bonding in O_2 and other diatomic molecules, let us first describe four principles of the theory.

Principles of Molecular Orbital Theory

In MO theory, we begin with a given arrangement of atoms in the molecule at the known bond distances and then determine the *sets* of molecular orbitals. One way to do this is to combine available valence orbitals on all the constituent atoms. These molecular orbitals more or less encompass all the atoms of the molecule, and the valence electrons for all the atoms in the molecule occupy the molecular orbitals. Just as with orbitals in atoms, electrons are assigned in order of increasing orbital energy and according to the Pauli principle and Hund's rule (Sections 7.1 and 7.3).

The **first principle of molecular orbital theory** is that *the total number of molecular orbitals is always equal to the total number of atomic orbitals contributed by the atoms that have combined.* To illustrate this orbital conservation principle, let us consider the H_2 molecule.

Molecular Orbitals for H_2

Molecular orbital theory specifies that when the $1s$ orbitals of two hydrogen atoms overlap, *two* molecular orbitals result. One molecular orbital results from the *addition* of the $1s$ atomic orbital wave functions, leading to an increased probability

• **Orbitals and Electron Waves**
Orbitals are characterized as electron waves; therefore, a way to view molecular orbital formation is to assume that two electron waves, one from each atom, interfere with each other. The interference can be constructive, giving a bonding MO, or destructive, giving an antibonding MO. (See Figure 9.13.)

(a) Oxygen gas can be liquified by passing it through a coil immersed in liquid nitrogen (at −196 °C). Here you see the liquid dripping into an insulated flask.

(b) Liquid oxygen (boiling point, −183 °C) is a pale blue liquid.

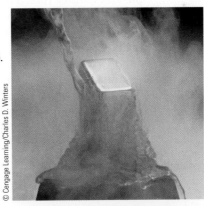

(c) Because O_2 molecules have two unpaired electrons, oxygen in the liquid state is paramagnetic and clings to a relatively strong neodymium magnet.

(d) In contrast, liquid N_2 is diamagnetic and does not stick to the magnet. It just splashes on the surface when poured onto the magnet.

FIGURE 9.12 The paramagnetism of liquid oxygen.

FIGURE 9.13 Bonding molecular orbital in H_2. The matter waves for the 1s electrons overlap along the bond axis. There is an enhancement of the wave amplitude between the nuclei, the result of which is that there is an increased probability of finding the electrons between the nuclei.

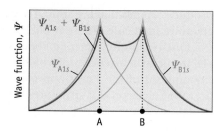

that electrons will reside in the bond region between the two nuclei (Figures 9.13 and 9.14). This is called a **bonding molecular orbital**. It is also a σ orbital because the region of electron probability lies directly along the bond axis. This molecular orbital is labeled σ_{1s}, the subscript 1s indicating that 1s atomic orbitals were used to create the molecular orbital.

The other molecular orbital is constructed by *subtracting* one atomic orbital wave function from the other (Figure 9.14). When this happens, the probability of finding an electron between the nuclei in the molecular orbital is reduced, and the probability of finding the electron in other regions is higher. Without significant electron density between them, the nuclei repel one another. This type of orbital is called an **antibonding molecular orbital**. Because it is also a σ orbital, derived from 1s atomic orbitals, it is labeled σ^*_{1s}. The asterisk signifies that it is antibonding. *Antibonding orbitals have no counterpart in valence bond theory.*

A **second principle of molecular orbital theory** is that *the bonding molecular orbital is lower in energy than the parent orbitals, and the antibonding orbital is higher in energy* (Figure 9.14). This means that the energy of a group of atoms is lower than the energy of the separated atoms when electrons occupy bonding molecular orbitals. Chemists say the system is "stabilized" by chemical bond formation. Conversely, the system is "destabilized" when electrons occupy antibonding orbitals because the energy of the system is higher than that of the atoms themselves.

A **third principle of molecular orbital theory** is that the *electrons of the molecule are assigned to orbitals of successively higher energy* according to the Pauli exclusion principle and Hund's rule. This is analogous to the procedure for building up electronic structures of atoms. Thus, electrons occupy the lowest energy orbitals available, and when two electrons are assigned to an orbital, their spins must be paired. Because the energy of the electrons in the bonding orbital of H_2 is lower than the energy of either parent 1s electron (Figure 9.14b), the H_2 molecule is more stable than two separate H atoms. We write the electron configuration of H_2 as $(\sigma_{1s})^2$.

What would happen if we tried to combine two helium atoms to form dihelium, He_2? Both He atoms have a 1s valence orbital that can produce the same kind of molecular orbitals as in H_2. Unlike H_2, however, four electrons need to be assigned

FIGURE 9.14 Molecular orbitals for H_2.

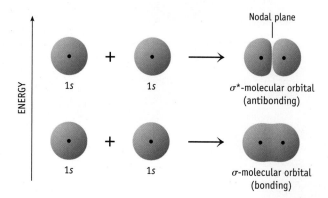

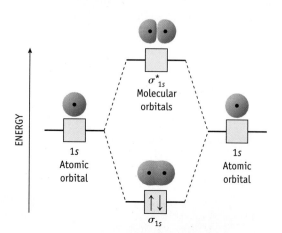

(a) Bonding and antibonding σ-molecular orbitals are formed from two 1s atomic orbitals on adjacent atoms. Notice the presence of a node between the nuclei in the antibonding orbital. (The node is a plane on which there is zero probability of finding an electron.)

(b) A molecular orbital diagram for H_2. The two electrons are placed in the σ_{1s} orbital. This molecular orbital is lower in energy.

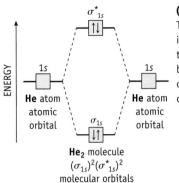

(a) Dihelium molecule, He$_2$.
This diagram provides a rationalization for the nonexistence of the molecule. In He$_2$, both the bonding (σ_{1s}) and antibonding orbitals (σ^*_{1s}) would be fully occupied.

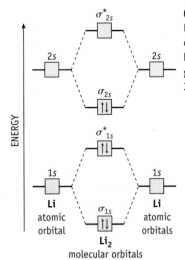

(b) Dilithium molecule, Li$_2$.
Notice that the molecular orbitals are created by combining orbitals of similar energies. Dilithium will utilize 1s and 2s atomic orbitals.

FIGURE 9.15 Molecular orbital diagrams for atoms with 1s and 2s orbitals.

to these orbitals (Figure 9.15a). The pair of electrons in the σ_{1s} orbital stabilizes He$_2$. The two electrons in σ^*_{1s}, however, destabilize the He$_2$ molecule. The energy decrease from the electrons in the σ_{1s}-bonding molecular orbital is fully offset by the energy increase due to the electrons in the σ^*_{1s}-antibonding molecular orbital. Thus, molecular orbital theory predicts that He$_2$ has no net stability; that is, two He atoms should have no tendency to combine. This confirms what we already know, that elemental helium exists in the form of single atoms and not as a diatomic molecule.

Bond Order

Bond order was defined in Section 8.9 as the net number of bonding electron pairs linking a pair of atoms. This same concept can be applied directly to molecular orbital theory, but now bond order is defined as

$$\text{Bond order} = 1/2 \text{ (number of electrons in bonding MOs} - \text{number of electrons in antibonding MOs)} \qquad (9.1)$$

In the H$_2$ molecule, there are two electrons in a bonding orbital and none in an antibonding orbital, so H$_2$ has a bond order of 1. In contrast, in the hypothetical molecule He$_2$ the stabilizing effect of the σ_{1s} pair would be canceled by the destabilizing effect of the σ^*_{1s} pair, and so the bond order would be 0.

Fractional bond orders are possible. Consider the ion He$_2^+$. Its molecular orbital electron configuration is $(\sigma_{1s})^2(\sigma^*_{1s})^1$. In this ion, there are two electrons in a bonding molecular orbital, but only one in an antibonding orbital. MO theory predicts that He$_2^+$ should have a bond order of 0.5; that is, a weak bond should exist between helium atoms in such a species. Interestingly, this ion has been identified in the gas phase using special experimental techniques.

EXAMPLE 9.6   Molecular Orbitals and Bond Order

Problem Write the electron configuration of the H$_2^-$ ion in molecular orbital terms. What is the bond order of the ion?

What Do You Know? Combining the 1s orbitals on the two hydrogen atoms will give two molecular orbitals (Figure 9.14). One is a bonding orbital (σ_{1s}) and the second is an antibonding orbital (σ^*_{1s}).

Strategy Count the number of valence electrons in the ion, and then place those electrons in the MO diagram for the H$_2$ molecule. Find the bond order from Equation 9.1.

Solution This ion has three electrons (one each from the H atoms plus one for the negative charge). Therefore, its electronic configuration is $(\sigma_{1s})^2(\sigma^*_{1s})^1$, identical with the configuration for He$_2^+$. This means H$_2^-$ has a net bond order of 0.5.

Think about Your Answer There is a weak bond in this ion, so it is predicted to exist only under special circumstances.

Check Your Understanding

What is the electron configuration of the H_2^+ ion? Compare the bond order of this ion with He_2^+ and H_2^-. Do you expect H_2^+ to exist?

Molecular Orbitals of Li_2 and Be_2

A **fourth principle of molecular orbital theory** is that *atomic orbitals combine to form molecular orbitals most effectively when the atomic orbitals are of similar energy.* This principle becomes important when we move past He_2 to Li_2, dilithium, and heavier molecules such as O_2 and N_2.

A lithium atom has electrons in two orbitals of the s type ($1s$ and $2s$), so a $1s \pm 2s$ combination is theoretically possible. Because the $1s$ and $2s$ orbitals are quite different in energy, however, this interaction can be disregarded. Thus, the molecular orbitals come only from $1s \pm 1s$ and $2s \pm 2s$ combinations (Figure 9.15b). This means the molecular orbital electron configuration of dilithium, Li_2, is

Li_2 MO Configuration: $(\sigma_{1s})^2(\sigma^*_{1s})^2(\sigma_{2s})^2$

The bonding effect of the σ_{1s} electrons is canceled by the antibonding effect of the σ^*_{1s} electrons, so these pairs make no net contribution to bonding in Li_2. Bonding in Li_2 is due to the electron pair assigned to the σ_{2s} orbital, and the bond order is 1.

The fact that the σ_{1s} and σ^*_{1s} electron pairs of Li_2 make no net contribution to bonding is exactly what you observed in drawing electron dot structures in Section 8.2: *Core electrons are ignored.* In molecular orbital terms, core electrons are assigned to bonding and antibonding molecular orbitals that offset one another.

A diberyllium molecule, Be_2, is not expected to exist. Its electron configuration would be

Be_2 MO Configuration: [core electrons]$(\sigma_{2s})^2(\sigma^*_{2s})^2$

The effects of σ_{2s} and σ^*_{2s} electrons cancel, and there is no net bonding. The bond order is 0, so the molecule does not exist.

• **Diatomic Molecules** Molecules such as H_2, Li_2, and N_2, in which two identical atoms are bonded, are examples of *homonuclear* diatomic molecules.

EXAMPLE 9.7   Molecular Orbitals in Homonuclear Diatomic Molecules

Problem Should the Be_2^+ ion exist? Describe its electron configuration in molecular orbital terms, and give the net bond order.

What Do You Know? The MO diagram in Figure 9.15b applies to this question because we are dealing only with $1s$ and $2s$ orbitals in the original atoms.

Strategy Count the number of electrons in the ion, and place them in the MO diagram in Figure 9.15b. Write the electron configuration, and calculate the bond order from Equation 9.1.

Solution The Be_2^+ ion has only seven electrons (in contrast to eight for Be_2), of which four are core electrons. (The four core electrons are assigned to σ_{1s} and σ^*_{1s} molecular orbitals.) The remaining three electrons are assigned to the σ_{2s} and σ^*_{2s} molecular orbitals, so the MO electron configuration is [core electrons]$(\sigma_{2s})^2(\sigma^*_{2s})^1$. This means the net bond order is 0.5.

Think about Your Answer Be_2^+ is predicted to exist under special circumstances. Scientists might search for such species in gas discharge experiments (gaseous atoms subjected to a strong electrical potential) and use spectroscopy to confirm their existence.

Check Your Understanding

Could the anion Li_2^- exist? What is the ion's bond order?

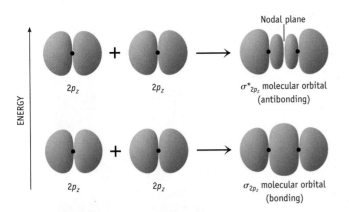

(a) Sigma (σ) molecular orbitals from p atomic orbitals.
Sigma-bonding (σ_{2p}) and antibonding (σ^*_{2p}) molecular orbitals arise from overlap of $2p$ orbitals. Each orbital can accommodate two electrons. The p orbitals in electron shells of higher n give molecular orbitals of the same basic shape.

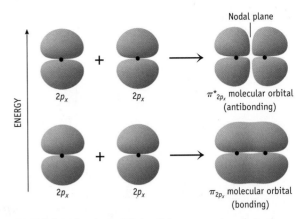

(b) Pi (π) molecular orbitals. Sideways overlap of atomic $2p$ orbitals that lie in the same direction in space gives rise to pi-bonding (π_{2p}) and pi-antibonding (π^*_{2p}) molecular orbitals. The p orbitals in shells of higher n give molecular orbitals of the same basic shape.

FIGURE 9.16 Combinations of p atomic orbitals to yield σ and π molecular orbitals.

Molecular Orbitals from Atomic p Orbitals

With the principles of molecular orbital theory in place, we are ready to account for bonding in such important homonuclear diatomic molecules as N_2, O_2, and F_2. To describe the bonding in these molecules, we will have to use both s and p valence orbitals in forming molecular orbitals.

For p-block elements, sigma-bonding and antibonding molecular orbitals are formed by their s orbitals interacting as in Figure 9.14. Similarly, it is possible for a p orbital on one atom to interact with a p orbital on the other atom to produce a pair of σ-bonding and σ^*-antibonding molecular orbitals (Figure 9.16a).

In addition, each p-block atom has *two* p orbitals in planes perpendicular to the σ bond connecting the two atoms. These p orbitals can interact in a side-by-side fashion to give *two* π-bonding molecular orbitals (π_p) and *two* π-antibonding molecular orbitals (π^*_p) (Figure 9.16b).

Electron Configurations for Homonuclear Molecules for Boron Through Fluorine

Orbital interactions in a second-period, homonuclear, diatomic molecule lead to the energy level diagram in Figure 9.17. Electron assignments can be made using this diagram, and the results for the diatomic molecules B_2 through F_2 are assembled in Table 9.1, which has two noteworthy features.

First, notice the correlation between the electron configurations and the bond orders, bond lengths, and bond energies at the bottom of Table 9.1. As the bond order between a pair of atoms increases, the energy required to break the bond increases, and the bond distance decreases. Dinitrogen, N_2, with a bond order of 3, has the largest bond energy and shortest bond distance.

Second, notice the configuration for dioxygen, O_2. Dioxygen has 12 valence electrons (six from each atom), so it has the molecular orbital configuration

O_2 MO Configuration: [core electrons]$(\sigma_{2s})^2(\sigma^*_{2s})^2(\sigma_{2p})^2(\pi_{2p})^4(\pi^*_{2p})^2$

This configuration leads to a bond order of 2 in agreement with experiment, and it specifies two unpaired electrons (in π^*_{2p} molecular orbitals). Thus, molecular orbital theory succeeds where valence bond theory fails. MO theory explains both the observed bond order and, as illustrated in Figure 9.12, the paramagnetic behavior of O_2.

• **Phases of Atomic Orbitals and Molecular Orbitals** Recall from page 225 in Chapter 6 that electron orbitals describe electron waves and as such have positive and negative phases. For this reason, the atomic orbitals in Figure 9.16 are drawn with two different colors. Looking at the p orbitals in Figure 9.16, you see that a bonding MO is formed when p orbitals with the same wave function sign overlap (+ with +). An antibonding orbital arises if they overlap out of phase (+ with −). (Wave function phases are signified by different colors.)

• **HOMO and LUMO** Chemists often refer to the highest energy MO that contains electrons as the HOMO (for "highest occupied molecular orbital"). For O_2, this is the π^*_{2p} orbital. Chemists also use the term LUMO for the "lowest unoccupied molecular orbital." For O_2, this would be σ^*_{2p}.

FIGURE 9.17 Molecular orbitals for homonuclear diatomic molecules of second period elements

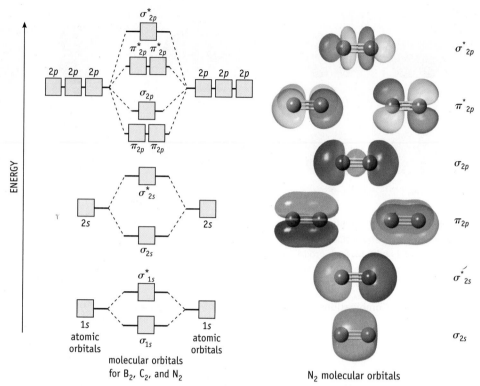

(a) Energy level diagram for B₂, C₂, and N₂. Although the diagram leads to the correct conclusions regarding bond order and magnetic behavior for O_2 and F_2, the energy ordering of the MOs in this figure is correct only for B_2, C_2, and N_2. For O_2 and F_2, the σ_{2p} MO is lower in energy than the π_{2p} MOs. See *A Closer Look*, page 331.

(b) Calculated molecular orbitals for N₂. (Color scheme: occupied MOs are blue/green. Unoccupied MOs are red/yellow. The different colors in a given orbital reflect the different phases [positive or negative signs] of the wave functions.)

EXAMPLE 9.8   Electron Configuration for a Homonuclear Diatomic Ion

Problem Potassium superoxide, KO_2, is one of the products from the reaction of K and O_2. This is an ionic compound, and the anion is the superoxide ion, O_2^-. Write the molecular orbital electron configuration for the ion. Predict its bond order and magnetic behavior.

What Do You Know? The 2s and 2p orbitals on the two oxygen atoms can be combined to give the series of molecular orbitals shown in Table 9.1. The O_2^- ion has 13 valence electrons.

Strategy Use the energy level diagram for O_2 in Table 9.1 to generate the electron configuration of this ion, and use Equation 9.1 to determine the bond order. The magnetism is determined by whether there are unpaired electrons.

Solution The MO electron configuration for O_2^- is

O_2^- *MO Configuration:*   [core electrons]$(\sigma_{2s})^2(\sigma^*_{2s})^2(\sigma_{2p})^2(\pi_{2p})^4(\pi^*_{2p})^3$

The ion is predicted to be paramagnetic to the extent of one unpaired electron, a prediction confirmed by experiment. The bond order is 1.5, because there are eight bonding electrons and five antibonding electrons. The bond order for O_2^- is lower than O_2 so we predict the O—O bond in O_2^- should be longer than the oxygen–oxygen bond in O_2. The superoxide ion in fact has an O—O bond length of 134 pm, whereas the bond length in O_2 is 121 pm.

Think about Your Answer The superoxide ion (O_2^-), contains an odd number of electrons. This is another diatomic species (in addition to NO and O_2) for which it is not possible to write a Lewis structure that accurately represents the bonding.

Check Your Understanding

The cations O_2^+ and N_2^+ are important components of Earth's upper atmosphere. Write the electron configuration for O_2^+. Predict its bond order and magnetic behavior.

A CLOSER LOOK

Molecular Orbitals for Molecules Formed from *p*-Block Elements

Several features of the molecular orbital energy level diagram in Figure 9.17 should be described in more detail.

(a) The bonding and antibonding σ orbitals from 2s interactions are lower in energy than the σ and π MOs from 2p interactions. The reason is that 2s orbitals have a lower energy than 2p orbitals in the separated atoms.

(b) The energy separation of the bonding and antibonding orbitals is greater for σ_{2p} than for π_{2p}. This happens because p orbitals overlap to a greater extent when they are oriented head to head (to give σ_{2p} MOs) than when they are side by side (to give π_{2p} MOs). The greater the orbital overlap, the greater the stabilization of the bonding MO and the greater the destabilization of the antibonding MO.

You may have been surprised that Figure 9.17 shows the π_{2p} orbitals lower in energy than the σ_{2p} orbital. Why would these orbitals be lower? A more sophisticated approach takes into account the "mixing" of s and p atomic orbitals, which have similar energies. This causes the σ_{2s} and σ^*_{2s} molecular orbitals to be lower in energy than otherwise expected, and the σ_{2p} and σ^*_{2p} orbitals to be higher in energy.

The mixing of s and p orbitals is important for B₂, C₂, and N₂, so Figure 9.17 applies strictly only to these molecules. For O₂ and F₂, σ_{2p} is lower in energy than π_{2p}, and Table 9.1 takes this into account.

Table 9.1 Molecular Orbital Occupations and Physical Data for Homonuclear Diatomic Molecules of Second-Period Elements

	B₂	C₂	N₂		O₂	F₂
σ^*_{2p}	☐	☐	☐	σ^*_{2p}	☐	☐
π^*_{2p}	☐☐	☐☐	☐☐	π^*_{2p}	↑ ↑	↑↓ ↑↓
σ_{2p}	☐	☐	↑↓	π_{2p}	↑↓ ↑↓	↑↓ ↑↓
π_{2p}	↑ ↑	↑↓ ↑↓	↑↓ ↑↓	σ_{2p}	↑↓	↑↓
σ^*_{2s}	↑↓	↑↓	↑↓	σ^*_{2s}	↑↓	↑↓
σ_{2s}	↑↓	↑↓	↑↓	σ_{2s}	↑↓	↑↓
Bond order	One	Two	Three		Two	One
Bond-dissociation energy (kJ/mol)	290	620	945		498	155
Bond distance (pm)	159	131	110		121	143
Observed magnetic behavior (paramagnetic or diamagnetic)	Para	Dia	Dia		Para	Dia

*The π_{2p} and σ_{2p} orbitals are inverted in energy on going to elements later in the second period. See *A Closer Look* at the top of this page.

Electron Configurations for Heteronuclear Diatomic Molecules

The molecules CO, NO, and ClF, containing two different elements, are examples of *heteronuclear* diatomic molecules. Molecular orbital descriptions for heteronuclear diatomic molecules resemble those for homonuclear diatomic molecules, but there are significant differences.

Carbon monoxide is an important molecule, so it is worth looking at its bonding. First, in its MO diagram (Figure 9.18) notice that similar atomic orbitals for the individual atoms have different energies. The 2s and 2p orbitals of O are at a lower relative energy than the 2s and 2p orbitals of C (◀ page 244). Second, these can nonetheless be combined in the same way as in homonuclear diatomic molecules to give an ordering of molecular orbitals that is similar to homonuclear diatomics.

The 10 valence electrons of CO are added to the available molecular orbitals from lowest to highest energy as we have done before, and the molecular electron configuration for CO is

CO MO Configuration: [core electrons]$(\sigma_{2s})^2(\sigma^*_{2s})^2(\pi_{2p})^4(\sigma_{2p})^2$

This clearly shows that CO has a bond order of 3, as expected from an electron dot structure.

FIGURE 9.18 **The molecular orbitals of carbon monoxide and the HOMO of the molecule.** The highest occupied molecular orbital (the HOMO) for CO, a sigma orbital, contains an electron pair that chemists identify as the lone pair on the C atom. The CO dipole is small (0.12 D) because the lone pair on the carbon atom slightly balances the greater electronegativity of O relative to C. (Compare with NF_3 on page 296.)

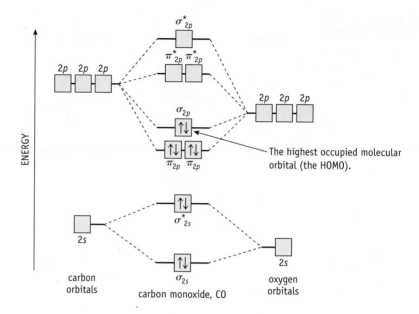

The highest occupied molecular orbital (the HOMO).

Resonance and MO Theory

Ozone, O_3, is a simple triatomic molecule with equal oxygen–oxygen bond lengths. Equal X—O bond lengths are also observed in other molecules and ions, such as SO_2, NO_2^-, and HCO_2^-. Valence bond theory introduced resonance to rationalize the equivalent bonding to the oxygen atoms in these structures, but MO theory provides a useful view of this problem.

To understand the bonding in ozone, let's begin by looking at the valence bond picture. First, assume that all three O atoms are sp^2 hybridized. The central atom uses its sp^2 hybrid orbitals to form two σ bonds and to accommodate a lone pair. The terminal atoms use their sp^2 hybrid orbitals to form one σ bond and to accommodate two lone pairs. Seven of the nine valence electron pairs in O_3 are either lone pairs or bonding pairs in the σ framework of O_3 (Figure 9.19a). The π bond in ozone arises from the two remaining pairs (Figure 9.19b). Because we have assumed that each oxygen atom in O_3 is sp^2 hybridized, an unhybridized p orbital perpendicular to the O_3 plane remains on each of the three oxygen atoms. These orbitals are in the correct orientation to form π bonds.

To simplify matters, let's now apply MO theory only to these three p orbitals that can be involved in π bonding in ozone (rather than on all of the orbitals in the molecule). A principle of MO theory is that the number of molecular orbitals must equal the number of atomic orbitals. Thus, the three $2p$ atomic orbitals must be combined in a way that forms three molecular orbitals.

One π_p MO for ozone is a bonding orbital because the three p orbitals are "in phase" across the molecule (Figure 9.19b). Another π_p MO is an antibonding orbital because the atomic orbital on the central atom is "out of phase" with the terminal atom p orbitals. The third π_p MO is a *nonbonding* orbital because the middle p orbital does not participate in the MO. (As the name implies, electrons in this molecular orbital neither help nor hinder the bonding in the molecule.)

One of the two pairs of π_p electrons of O_3 occupies the lowest energy or π_p-bonding MO, which is delocalized, or "spread over," the molecule (just as the resonance hybrid of valence bond theory implies). The π_p-nonbonding orbital is also occupied, but the electrons in this orbital are concentrated near the two

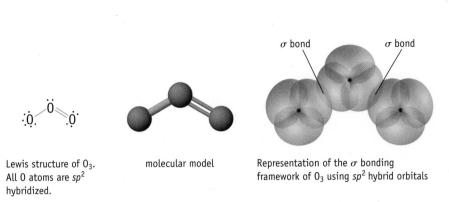

Lewis structure of O_3. All O atoms are sp^2 hybridized.

molecular model

Representation of the σ bonding framework of O_3 using sp^2 hybrid orbitals

(a) Ozone, O_3, has equal-length oxygen–oxygen bonds, which, in valence bond theory, would be rationalized by resonance structures. The σ bonding framework of O_3 is formed by utilizing sp^2 hybrid orbitals on each O atom. These accommodate two σ bonds and five lone pairs.

FIGURE 9.19. A bonding model for ozone, O_3.

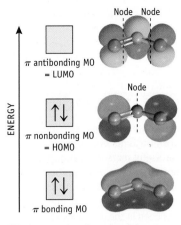

(b) The π molecular orbital diagram for ozone. Notice that, as in the other MO diagrams illustrated (especially Figure 9.20), the energy of the molecular orbitals increases as the number of nodes increases.

terminal oxygens. Thus, there is a net of only one pair of π_p-bonding electrons for two O—O bonds, giving a π bond order for O_3 of 0.5. Because the σ bond order is 1.0 and the π bond order is 0.5, the net oxygen–oxygen bond order is 1.5—the same value given by valence bond theory.

The observation that two of the π molecular orbitals for ozone extend over three atoms illustrates an important point regarding molecular orbital theory: *Orbitals can extend beyond two atoms.* In valence bond theory, in contrast, all representations for bonding are based on being able to localize pairs of electrons in bonds between two atoms. To further illustrate the MO approach, look again at benzene (Figure 9.20). On page 324, we noted that the π electrons in this molecule were spread out over all six carbon atoms. We can now see how the same case can be made with MO theory. Six p orbitals contribute to the π system. Based on the

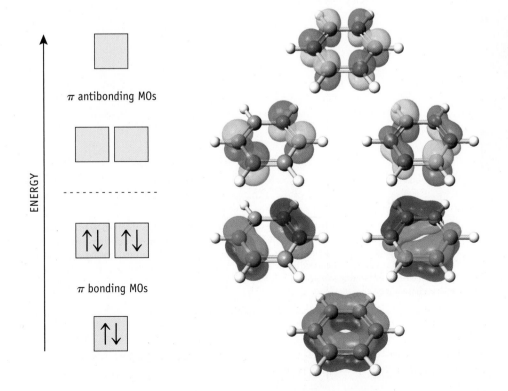

FIGURE 9.20 **Molecular orbital energy level diagram for benzene.** Because there are six unhybridized p orbitals (one on each C atom), six π molecular orbitals can be formed—three bonding and three antibonding. The three bonding molecular orbitals accommodate the six π electrons.

premise that the number of molecular orbitals must equal the number of atomic orbitals, there must be six π molecular orbitals in benzene. An energy level diagram for benzene shows that the six π electrons reside in the three lowest-energy (bonding) molecular orbitals.

CASE STUDY

Green Chemistry, Safe Dyes, and Molecular Orbitals

An important obligation of chemists is not only to search for useful materials but also to ensure they are safe to use and to produce them in the safest manner possible. This is the essence of "green chemistry," and one recent success has been in the dye industry.

Dyes and pigments have been used for centuries because humans want colorful clothing and objects that have color. Dyes were extracted from plants such as onion skins, carrots, turmeric, butternuts, walnuts, tea, red cabbage, and fruits and berries. Some came from animals, and one, Tyrian purple (Figure A), was especially prized. This dye, which was known to the ancient Greeks, is extracted from a marine gastropod, the *Murex brandaris*. The problem is that it takes over 12,000 of these organisms to produce 1.4 g of the dye, which is only enough to dye a portion of one garment. This is the reason the dye is also called *royal purple* because only royalty could afford to acquire it.

The first synthetic dye was produced quite by accident in 1856. William Perkin, then 18, was searching for an antimalarial drug, but the route he chose to make it led instead to the dye mauveine. Because dyes were so highly prized, his discovery led to the synthesis of many more dyes. Indeed, some say Perkin's discovery led to the modern chemical industry.

Chemists have developed large numbers of dyes, among them azo dyes, so named because they contain a N=N double bond (an azo group). One example is "butter yellow" (Figure B). It was given this name because it was originally used to give butter a more appealing yellow color. The problem is that the molecule is carcinogenic.

The fact that some dyes were unsafe led chemists to look for ways to achieve the desired color and to ensure they were not toxic. In the case of butter yellow, the solution was to add an $-NO_2$ group to the molecule to give a new yellow dye that is easy to synthesize and safe to use.

The structure of Tyrian purple is typical of many dyes. Notice that it has alternating double bonds (C=C and C=O) extending across the molecule. Azo dyes similarly have extended frameworks of alternating double bonds. Indeed, all organic dye molecules have extended π bonding networks, and it is this that leads to their color. The extended π bonds give rise to low-energy π antibonding molecular orbitals. When visible light strikes such a molecule, an electron can be excited from a bonding or nonbonding molecular orbital to an antibonding molecular orbital, and light is absorbed (Figure C). Your eye sees the wavelengths of light not absorbed by the molecule, so it has the color of the remaining wavelengths of light. For example, on page 166 you see a case where a substance absorbs wavelengths of light in the blue region of the spectrum, so you see red light.

Questions:

1. What is the empirical formula of Tyrian purple?
2. Butter yellow absorbs light with a wavelength of 408 nm, whereas the nitrated form absorbs at 478 nm. Which absorbs the higher energy light?
3. How many alternating double bonds are there in Tyrian purple? In nitrated butter yellow?

Answers to these questions are available in Appendix N.

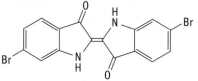

FIGURE A   The structure of Tyrian purple or 6,6′-dibromoindigo. Its formula is $C_{16}H_8N_2O_2Br_2$.

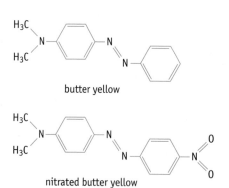

butter yellow

nitrated butter yellow

FIGURE B   The structure of "butter yellow" and "nitrated butter yellow." The first is a synthetic dye originally used to color butter. It is in the class of azo dyes, all characterized by an N=N double bond. The nitrated version is safe for human consumption.

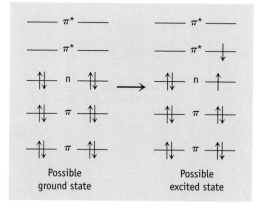

Possible ground state   Possible excited state

FIGURE C Light is absorbed by an azo dye by the promotion of an electron from a nonbonding molecular orbital to an antibonding π molecular orbital. (Light absorption by all molecules occurs by promotion of an electron from a low-lying to a higher-lying molecular orbital.)

Three-Center Bonds and Hybrid Orbitals with *d* Orbitals

Although a large majority of known compounds can be described by Lewis electron dot pictures, nature (and chemistry) has a surprise now and then. You have already seen several examples (most notably O_2 and NO) where it is not possible to draw simple dot structures. In these instances we turned to MO theory to understand the bonding.

An interesting chemical species is the ion HF_2^-, formed in a product-favored reaction between F^- and HF. This ion has a linear structure (F—H—F) with the hydrogen midway between the two fluorine atoms. Of the 16 valence electrons in the ion, there are 4 that can be utilized in bonding (the other 12 are lone pairs on F).

But how can we account for bonding in the ion? The best way is to use MO theory. We begin with three atomic orbitals, the $2p_z$ orbitals on each fluorine atom and the 1s orbital on hydrogen.

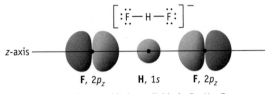

$$\left[:\ddot{F} - H - \ddot{F}: \right]^-$$

z-axis ————

F, $2p_z$ **H, 1s**   **F, $2p_z$**

valence orbitals available in F—H—F⁻

The bonding MO of the FHF⁻ ion extends across the ion. The three-center bond has a bond order of ½.

As in the model to account for π bonding in ozone (Figure 9.19), the 1s and $2p_z$ orbitals combine to form three molecular orbitals, one bonding, one nonbonding, and one antibonding. One pair of valence electrons is placed in the bonding MO and one in the nonbonding MO, to give a net of one bond. This result, which nicely accounts for the bonding, is often referred to as a *three-center/four-electron bond* model.

Another example for which an acceptable Lewis dot structure cannot be drawn is for diborane, B_2H_6, the simplest member of a large group of compounds. As the structure below shows, each boron atom has distorted tetrahedral geometry.

![diborane structures]

(a)   **(b)**

The molecule has two terminal hydrogen atoms bonded to each boron atom, and two hydrogen atoms bridge the two boron atoms. Chemists refer to diborane as "electron deficient" because two B atoms and six H atoms do not contribute enough electrons for eight, two-electron bonds.

We can again account for the bonding in this molecule using MO theory with a three-center bond involving an H atom. Let's begin by assuming that each B atom is sp^3 hybridized. The four terminal H atoms bond to the B atoms with two-electron bonds using 8 of the 12 electrons and two of the sp^3 hybrid orbitals on each B atom. The 4 remaining electrons account for the B—H—B bridges. Three molecular orbitals encompassing the B—H—B atoms in each bridge can be constructed from the remaining two sp^3 hybrid orbitals on each B atom and a H atom 1s orbital. Again, one MO is bonding, one is nonbonding, and one is antibonding. Two electrons are assigned to the bonding orbital, giving us a *three-center/two-electron* bond. Not surprisingly, this kind of bond is weaker than a typical two-electron/two-center bond, and diborane dissociates to two BH_3 molecules at fairly low temperatures.

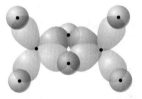

An overlap of sp^3 orbitals, one from each B atom, and an H 1s creates two, three-center/two-electron bonds.

And finally we return to an issue introduced earlier: *whether d orbitals are involved in bonding in some main group element compounds.* Theory suggests that *d* orbitals are not significantly involved in bonding in ions such as SO_4^{2-} and PO_4^{3-} (page 275). If we accept the premise that *d* orbitals are too high in energy to be used in bonding in main group element compounds, do we need to describe PF_5 and SF_6 bonding, for example, using sp^3d and sp^3d^2 hybrid orbitals? Another bonding model is needed and molecular orbital theory provides the alternative without the use of *d* orbitals.

Consider SF_6, for instance. The molecule has a total of 48 valence electrons. Of these, 36 are involved as lone pairs on the F atoms, so 12 electrons remain to account for 6 S-F bonds. Let's look specifically at the bonding between sulfur and two fluorine atoms across from each other along each of the three axes.

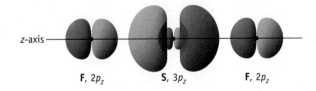

z-axis ————

F, $2p_z$ **S, $3p_z$**   **F, $2p_z$**

Along the z-axis, for example, bonding, nonbonding, and antibonding orbitals can be constructed from combinations of the $3p_z$ orbital on sulfur and the two $2p_z$ orbitals on fluorine. Four electrons are placed in the bonding and nonbonding orbitals. This description is repeated on the x- and y-axis, which means that two electron pairs are involved in the F—S—F group along each axis. That is, each of the three F—S—F "groups" is bonded with a three-center/four-electron bond, thus accounting nicely for the bonding in SF_6 without using *d* orbitals.

REVIEW & CHECK FOR SECTION 9.3

1. What is the N—O bond order in nitrogen monoxide, NO?

   (a) 2   (b) 2.5   (c) 3.0   (d) 1

2. What is the HOMO in the peroxide ion, O_2^{2-}?

   (a) π_{2p} (b) π^*_{2p} (c) σ^*_{2p} (d) σ^*_{2s}

3. Which of the following species is diamagnetic?

   (a) O_2 (b) B_2 (c) C_2 (d) N_2^+

4. Among the known dioxygen species (O_2^+, O_2, O_2^-, and O_2^{2-}), which is expected to have the shortest bond length?

   (a) O_2^+ (b) O_2 (c) O_2^- (d) O_2^{2-}

 and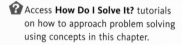

Sign in at **www.cengage.com/owl** to:
- View tutorials and simulations, develop problem-solving skills, and complete online homework assigned by your professor.
- For quick review and exam prep, download Go Chemistry mini lecture modules from OWL (or purchase them at **www.cengagebrain.com**)

❓ Access **How Do I Solve It?** tutorials on how to approach problem solving using concepts in this chapter.

CHAPTER GOALS REVISITED

Now that you have studied this chapter, you should ask whether you have met the chapter goals. In particular, you should be able to:

Understand the differences between valence bond theory and molecular orbital theory

a. Describe the main features of valence bond theory and molecular orbital theory, the two commonly used theories for covalent bonding (Section 9.1).

b. Recognize that the premise for valence bond theory is that bonding results from the overlap of atomic orbitals. By virtue of the overlap of orbitals, electrons are concentrated (or localized) between two atoms (Section 9.2).

c. Distinguish how sigma (σ) and pi (π) bonds arise. For σ bonding, orbitals overlap in a head-to-head fashion, concentrating electrons along the bond axis. Sideways overlap of p atomic orbitals results in π bond formation, with electron density above and below the molecular plane (Section 9.2).

d. Understand how molecules having double bonds can have isomeric forms. **Study Questions: 19, 20.**

Identify the hybridization of an atom in a molecule or ion

a. Use the concept of hybridization to rationalize molecular structure (Section 9.2). **Study Questions: 1–18, 31–33, 37, 42, 45, 46, 48, 54, 55, 57–62, and Go Chemistry Module 14.**

Hybrid Orbitals	Atomic Orbitals Used	Number of Hybrid Orbitals	Electron-Pair Geometry
sp	$s + p$	2	Linear
sp^2	$s + p + p$	3	Trigonal-planar
sp^3	$s + p + p + p$	4	Tetrahedral
sp^3d	$s + p + p + p + d$	5	Trigonal-bipyramidal
sp^3d^2	$s + p + p + p + d + d$	6	Octahedral

Understand the differences between bonding and antibonding molecular orbitals and be able to write the molecular orbital configurations for simple diatomic molecules.

a. Understand molecular orbital theory (Section 9.3), in which atomic orbitals are combined to form bonding orbitals, nonbonding orbitals, or antibonding orbitals that are delocalized over several atoms. In this description, the electrons of the molecule or ion are assigned to the orbitals beginning with the one at lowest energy, according to the Pauli exclusion principle and Hund's rule.

b. Use molecular orbital theory to explain the properties of O_2 and other diatomic molecules. **Study Questions: 21–30, 49–53.**

Key Equation

Equation 9.1 (page 327) Used to calculate the order of a bond from the molecular orbital electron configuration.

Bond order = 1/2 (number of electrons in bonding MOs
  − number of electrons in antibonding MOs)

OWL Questions and problems for this chapter are available in OWL. Use the OWL access card that came with this textbook to access assigned questions and problems for this chapter.

10 Carbon: Not Just Another Element

John C. Katz

Cacao pods on a tree in an island in the Caribbean and the ultimate product.

Theobromine, $C_7H_8N_4O_2$, one of many molecules in chocolate. It is a smooth muscle stimulator.

© Cengage Learning/Charles D. Winters

Caffeine, $C_8H_{10}N_4O_2$, like theobromine is a member of a class of compounds called *xanthines*.

The Food of the Gods

There is a tree that grows in the tropics called *Theobroma cacao,* a name given to it in 1735 by the great biologist Linneas. The root of the name comes from the Greek, meaning "food of the gods."

What could be the "food of the gods"? Chocolate, of course.

The history of chocolate can be traced back to Central America. When the Spanish came to the New World, they discovered cocoa beans were highly prized by the natives—100 beans would purchase a slave in 1500 AD—and that a drink made from the beans was quite delicious. The Spanish explorers first took cocoa to Spain in the late 1500s, and by the 17th century chocolate was popular with the Spanish court.

The appeal of chocolate soon spread to the rest of Europe, and, in the late 1800s two Swiss, Henry Nestlé and Daniel Peter, added dried milk to produce the first milk chocolate.

This chapter is all about organic chemistry, the branch of chemistry that studies molecules based on carbon. Carbon compounds are the basis of all living species on this planet. Chocolate certainly fits that definition, as there are almost 400 organic compounds in chocolate.

The main ingredient that gives chocolate its character is the organic compound theobromine. In spite of the name, there is no bromine in the molecule; it is named for the tree where it is found. In medicine, it can be used as a vasodilator, a diuretic, and a heart stimulator. It has even been reputed to be an aphrodisiac! But it can also cause restlessness, tremors, sleeplessness, and anxiety.

Chocolate products usually contain only a few milligrams of theobromine per gram and are quite safe to consume. However, chocolate can be toxic to dogs, and they should not be given chocolate even if they have a sweet tooth.

Theobromine is one of a class of molecules called *xanthines*, a class that includes caffeine. In fact, caffeine is metabolized in the body to theobromine, among other xanthine derivatives.

Questions:

1. How do theobromine and caffeine differ structurally?
2. A 5.00-g sample of Hershey's cocoa contains 2.16% theobromine. What is the mass of the compound in the sample?

Answers to these questions are available in Appendix N.

Reference:

G. Tannenbaum, *Journal of Chemical Education,* Vol. 81, p. 1131, 2004.

CHAPTER OUTLINE

10.1 Why Carbon?

10.2 Hydrocarbons

10.3 Alcohols, Ethers, and Amines

10.4 Compounds with a Carbonyl Group

10.5 Polymers

CHAPTER GOALS

See Chapter Goals Revisited (page 382) for Study Questions keyed to these goals.

- Classify organic compounds based on formula and structure.

- Recognize and draw structures of structural isomers and stereoisomers for carbon compounds.

- Name and draw structures of common organic compounds.

- Know the common reactions of organic functional groups.

- Relate properties to molecular structure.

- Identify common polymers.

The vast majority of the millions of chemical compounds currently known are organic; that is, they are compounds built on a carbon framework. Organic compounds vary greatly in size and complexity, from the simplest hydrocarbon, methane, to molecules made up of many thousands of atoms. As you read this chapter, you will see why the range of possible materials is huge and why they are so interesting and very often useful.

⊌WL

Sign in to OWL at **www.cengage.com/owl** to view tutorials and simulations, develop problem-solving skills, and complete online homework assigned by your professor.

10.1 Why Carbon?

We begin this discussion of organic chemistry with a question: What features of carbon lead to both the abundance and the complexity of organic compounds? The answers to this question revolve around two main issues: structural diversity and stability.

Structural Diversity

With four electrons in its outer shell, carbon will form four bonds to reach an octet configuration. In contrast, the elements boron and nitrogen generally form three bonds in molecular compounds; oxygen forms two bonds; and hydrogen and the halogens form one bond. With a larger number of bonds comes the opportunity to create more complex structures. This will become increasingly evident in this brief tour of organic chemistry.

A carbon atom can reach an octet of electrons in various ways (Figure 10.1):

- *By forming four single bonds.* A carbon atom can bond to four other atoms, which can be either atoms of other elements (often H, N, or O) or other carbon atoms.
- *By forming a double bond and two single bonds.* The carbon atoms in ethylene, $H_2C=CH_2$, are linked in this way.
- *By forming two double bonds,* as in carbon dioxide ($O=C=O$).
- *By forming a triple bond and a single bond,* an arrangement seen in acetylene, $HC\equiv CH$.

Recognize, with each of these arrangements, the various possible geometries around carbon: tetrahedral, trigonal-planar, and linear. Carbon's tetrahedral

go Chemistry

Download mini lecture videos for key concept review and exam prep from OWL or purchase them from **www.cengagebrain.com**

ethylene, $H_2C=CH_2$

acetylene, $HC\equiv CH$

Ethylene and acetylene. These two-carbon hydrocarbons can be the building blocks of more complex molecules. These are their common names; their systematic names are ethene and ethyne, respectively.

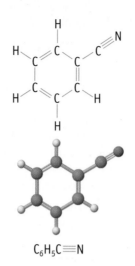

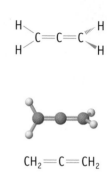

CH₃COH

CH_3COH

(a) Acetic acid. One carbon atom in this compound is attached to four other atoms by single bonds and has tetrahedral geometry. The second carbon atom, connected by a double bond to one oxygen and by single bonds to the other oxygen and to the first carbon, has trigonal-planar geometry.

$C_6H_5C{\equiv}N$

(b) Benzonitrile. Six trigonal-planar carbon atoms make up the benzene ring. The seventh C atom, bonded by a single bond to carbon and a triple bond to nitrogen, has a linear geometry.

$CH_2{=}C{=}CH_2$

(c) Carbon is linked by double bonds to two other carbon atoms in C_3H_4, a linear molecule commonly called allene.

FIGURE 10.1 Ways that carbon atoms bond.

geometry is of special significance because it leads to three-dimensional chains and rings of carbon atoms, as in propane and cyclopentane.

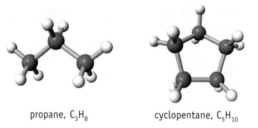

propane, C_3H_8 cyclopentane, C_5H_{10}

The ability to form multiple bonds leads to families of compounds with double and triple bonds.

Isomers

A hallmark of carbon chemistry is the remarkable array of isomers that can exist. **Isomers** are compounds that have identical composition but different structures. Two broad categories exist: structural isomers and stereoisomers.

Structural isomers are compounds having the same elemental composition but in which the atoms are linked together in different ways. Ethanol and dimethyl ether are structural isomers, as are 1-butene and 2-methylpropene.

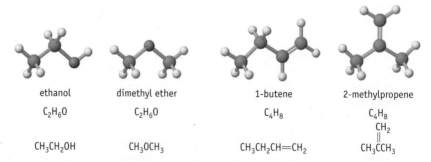

ethanol	dimethyl ether	1-butene	2-methylpropene
C_2H_6O	C_2H_6O	C_4H_8	C_4H_8
			CH_2
			$\parallel$
CH_3CH_2OH	CH_3OCH_3	$CH_3CH_2CH{=}CH_2$	CH_3CCH_3

Stereoisomers are compounds with the same formula and the same attachment of atoms. However, the atoms have different orientations in space. Two types of stereoisomers exist: *geometric isomers* and *optical isomers*.

A CLOSER LOOK

Writing Formulas and Drawing Structures

In Chapter 2, you learned that there are various ways of presenting structures (page 59). It is appropriate to return to this topic as we look at organic compounds. Consider methane and ethane, for example. We can represent these molecules in several ways:

1. *Molecular formula:* CH_4 or C_2H_6. This type of formula gives information on composition only.

2. *Condensed formula:* For ethane, this would typically be written CH_3CH_3. This method of writing the formula gives some information on the way atoms are connected.

3. *Structural formula:* You will recognize this formula as the Lewis structure. An elaboration on the condensed formula in (2), this representation defines more clearly how the atoms are connected,

but it fails to describe the shapes of molecules.

$$H-\underset{\underset{H}{|}}{\overset{\overset{H}{|}}{C}}-H \qquad H-\underset{\underset{H}{|}}{\overset{\overset{H}{|}}{C}}-\underset{\underset{H}{|}}{\overset{\overset{H}{|}}{C}}-H$$

methane, CH_4 ethane, C_2H_6

4. *Perspective drawings:* These drawings are used to convey the three-dimensional nature of molecules. Bonds extending out of the plane of the paper are drawn with wedges, and bonds behind the plane of the paper are represented as dashed wedges (page 60). Using these guidelines, the structures of methane and ethane could be drawn as follows:

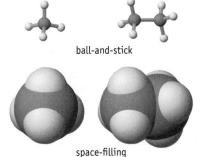

5. *Computer-drawn ball-and-stick and space-filling models.*

ball-and-stick

space-filling

Cis- and *trans*-2-butene are **geometric isomers**. Geometric isomerism in these compounds occurs as a result of the C=C double bond. *Recall that the carbon atom and the attached groups cannot rotate around a double bond* (page 323). Thus, the geometry around the C=C double bond is fixed in space. *Cis-trans* isomerism occurs if each carbon atom involved in the double bond has two different groups attached. If, on the adjacent carbons, it turns out that identical groups are on the same side of the double bond, then it is a *cis* isomer. If those groups appear on opposite sides, a *trans* isomer is produced.

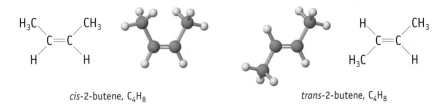

cis-2-butene, C_4H_8 *trans*-2-butene, C_4H_8

Optical isomerism is a second type of stereoisomerism. Optical isomers are molecules that have nonsuperimposable mirror images (Figure 10.2). Molecules (and other objects) that have nonsuperimposable mirror images are termed **chiral**. Pairs of nonsuperimposable, mirror-image molecules are called **enantiomers**.

Pure samples of enantiomers have the same physical properties, such as melting point, boiling point, density, and solubility in common solvents. They differ in one significant way, however: When a beam of plane-polarized light passes through a solution of a pure enantiomer, the plane of polarization rotates. The two enantiomers rotate polarized light to an equal extent, but in opposite directions (Figure 10.3). The term *optical isomerism* is used because this effect involves light.

The most common examples of chiral compounds are those in which four different atoms (or groups of atoms) are attached to a tetrahedral carbon atom. Lactic acid, found in milk and a product of normal human metabolism, is an example of a chiral compound (Figure 10.2). Optical isomerism is particularly important in the amino acids and other biologically important molecules. Among the many

FIGURE 10.2 Optical isomers.
Lactic acid, CH₃CH(OH)CO₂H, is
produced when milk is fermented to
make cheese. It is also found in other
sour foods such as sauerkraut and is a
preservative in pickled foods such as
onions and olives. In our bodies, it is
produced by muscle activity and
normal metabolism.

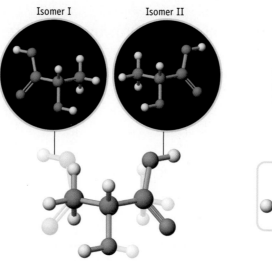

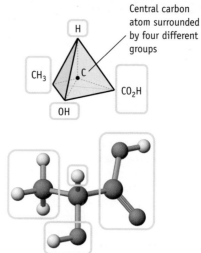

(a) Optical isomerism occurs if a molecule and its mirror image cannot be superimposed. The situation is seen if four different groups are attached to carbon.

(b) Lactic acid is a chiral molecule because four different groups (H, OH, CH₃, and CO₂H) are attached to the central carbon atom.

interesting examples is a compound, frontalin, produced naturally by male elephants (see *A Closer Look: Chirality and Elephants* on the next page).

Stability of Carbon Compounds

Carbon compounds are notable for their resistance to chemical change. This resistance is a result of two things: strong bonds and slow reactions.

Strong bonds are needed for molecules to survive in their environment. Energetic collisions between molecules in gases, liquids, and solutions can provide enough energy to break some bonds, and bonds can be broken if the energy associated with photons of visible and ultraviolet light exceeds the bond energy. Carbon–carbon bonds are relatively strong, however, as are the bonds between carbon and most other atoms. The average C—C bond energy is 346 kJ/mol; the C—H bond energy is 413 kJ/mol; and carbon–carbon double and triple bond energies are even higher (◄ Section 8.9). Contrast these values with bond energies for the Si—H bond (328 kJ/mol) and the Si—Si bond (222 kJ/mol). The consequence of high bond energies for bonds to carbon is that, for the most part, organic compounds do not decompose thermally under normal conditions.

Oxidation of most organic compounds is strongly product-favored, but most organic compounds survive continual and prolonged contact with O_2. The reason is that reactions of most organic compounds with oxygen are very slow. Typically, organic compounds burn only if their combustion is initiated by heat or by a spark. As

FIGURE 10.3 Rotation of plane-polarized light by an optical isomer.
Monochromatic light (light of only one wavelength) is produced by a sodium lamp. After it passes through a polarizing filter, the light vibrates in only one direction—it is polarized. A solution of an optical isomer placed between the first and second polarizing filters causes rotation of the plane of polarized light. The second filter is rotated to a point where a maximum of light is transmitted and the angle of rotation is calculated. The magnitude and direction of rotation are unique physical properties of the optical isomer being tested.

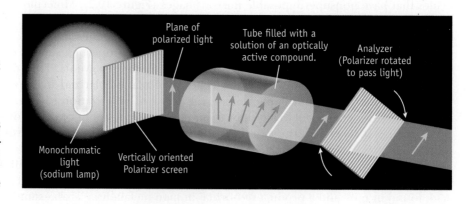

Chirality and Elephants

During a period known as musth, male elephants undergo a time of heightened sexual activity. They can become more aggressive and can work themselves into a frenzy. At the same time, the males also produce a chemical signal. A secretion containing the enantiomers of frontalin ($C_8H_{14}O_2$) is emitted from a gland between the eye and the ear. Young males produce mixtures containing more of one enantiomer than the other, whereas older elephants produce a more balanced and more concentrated mixture. When that occurs in older elephants, other males are repelled, but ovulating female elephants are more highly attracted.

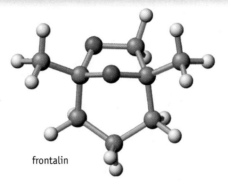

frontalin

One enantiomer of frontalin.

John C. Kotz

An African elephant in musth. Fluid containing the enantiomers of frontalin flows from a gland between the elephant's eye and ear.

a consequence, oxidative degradation is not a barrier to the existence of organic compounds.

REVIEW & CHECK FOR SECTION 10.1

(a)   (b)   (c)

1. Which pair of the three compounds pictured above are stereoisomers?

    (a) a and b   (b) a and c   (c) b and c

2. Which pair of compounds are structural isomers?

    (a) a and b   (b) b and c

3. How many chiral carbon atoms are there in frontalin? (See *A Closer Look: Chirality and Elephants*, above)

    (a) 0   (b) 1   (c) 2   (d) 3

10.2 Hydrocarbons

Hydrocarbons, compounds made only of carbon and hydrogen, are divided into several subgroups: alkanes, cycloalkanes, alkenes, alkynes, and aromatic compounds (Table 10.1). We begin our discussion by considering alkanes and cycloalkanes, compounds in which each carbon atom is linked with four single bonds to other carbon atoms or to hydrogen.

Alkanes

Alkanes have the general formula C_nH_{2n+2}, with n having integer values (Table 10.2). Formulas of specific compounds can be generated from this general formula, the first four of which are CH_4 (methane), C_2H_6 (ethane), C_3H_8 (propane), and C_4H_{10} (butane) (Figure 10.4). Methane has four hydrogen atoms arranged tetrahedrally around a single carbon atom. Replacing a hydrogen atom in methane by a —CH_3 group gives ethane. If an H atom of ethane is replaced by yet another —CH_3 group, propane results. Butane is derived from propane by replacing an H atom of

Table 10.1 Some Types of Hydrocarbons

Type of Hydrocarbon	Characteristic Features	General Formula	Example
Alkanes	C—C single bonds and all C atoms have four single bonds	C_nH_{2n+2}	CH_4, methane C_2H_6, ethane
Cycloalkanes	C—C single bonds, C atoms arranged in a ring	C_nH_{2n}	C_6H_{12}, cyclohexane
Alkenes	C=C double bond	C_nH_{2n}	$H_2C=CH_2$, ethylene
Alkynes	C≡C triple bond	C_nH_{2n-2}	HC≡CH, acetylene
Aromatics	Rings with π bonding extending over several C atoms	—	Benzene, C_6H_6

one of the chain-ending carbon atoms with a —CH_3 group. In all of these compounds, each C atom is attached to four other atoms, either C or H, so alkanes are often called **saturated compounds**.

Structural Isomers

Structural isomers are possible for all alkanes larger than propane. For example, there are two structural isomers for C_4H_{10} and three for C_5H_{12}. As the number of carbon atoms in an alkane increases, the number of possible structural isomers greatly increases; there are five isomers possible for C_6H_{14}, nine isomers for C_7H_{16}, 18 for C_8H_{18}, 75 for $C_{10}H_{22}$, and 1858 for $C_{14}H_{30}$.

To recognize the isomers corresponding to a given formula, keep in mind the following points:

- Each alkane has a framework of tetrahedral carbon atoms, and each carbon has four single bonds.
- An effective approach to deriving isomer structures is to create a framework of carbon atoms and then fill the remaining positions around carbon with H atoms so that each C atom has four bonds.

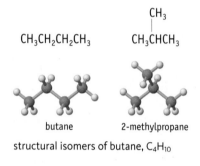

$CH_3CH_2CH_2CH_3$ CH_3CHCH_3 (with CH_3 above)

butane   2-methylpropane

structural isomers of butane, C_4H_{10}

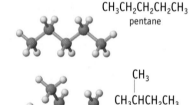

$CH_3CH_2CH_2CH_2CH_3$
pentane

$CH_3CHCH_2CH_3$ (with CH_3 above)
2-methylbutane

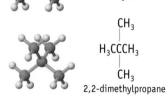

H_3CCCH_3 (with CH_3 above and CH_3 below)
2,2-dimethylpropane

structural isomers of pentane, C_5H_{12}

Table 10.2 Selected Hydrocarbons of the Alkane Family, C_nH_{2n+2}*

Name	Molecular Formula	State at Room Temperature
Methane	CH_4	
Ethane	C_2H_6	
Propane	C_3H_8	Gas
Butane	C_4H_{10}	
Pentane	C_5H_{12} (pent- = 5)	
Hexane	C_6H_{14} (hex- = 6)	
Heptane	C_7H_{16} (hept- = 7)	
Octane	C_8H_{18} (oct- = 8)	Liquid
Nonane	C_9H_{20} (non- = 9)	
Decane	$C_{10}H_{22}$ (dec- = 10)	
Octadecane	$C_{18}H_{38}$ (octadec- = 18)	
Eicosane	$C_{20}H_{42}$ (eicos- = 20)	Solid

*This table lists only selected alkanes. At 25 °C straight-chain alkanes with 11–17 C atoms are liquid. Those with more than 17 C atoms are solid.

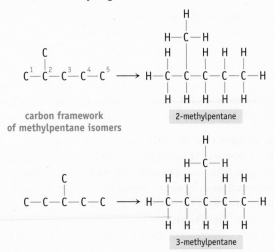

FIGURE 10.4 Alkanes. The lowest–molar-mass alkanes, all gases under normal conditions, are methane, ethane, propane, and butane.

- Nearly free rotation occurs around carbon–carbon single bonds. Therefore, when atoms are assembled to form the skeleton of an alkane, the emphasis is on how carbon atoms are attached to one another and not on how they might lie relative to one another in the plane of the paper.

⬤WL INTERACTIVE EXAMPLE 10.1 Drawing Structural Isomers of Alkanes

Problem Draw structures of the five isomers of C_6H_{14}. Are any of these isomers chiral?

What Do You Know? Each structure must have six carbon and 14 hydrogen atoms. There must be four single bonds to each carbon atom, and each hydrogen will form one bond.

Strategy Focus first on the different frameworks that can be built from six carbon atoms. Having created a carbon framework, fill hydrogen atoms into the structure so that each carbon has four bonds.

Solution

Step 1. Placing six carbon atoms in a chain gives the framework for the first isomer. Now fill in hydrogen atoms: three on the carbons on the ends of the chain and two on each of the carbons in the middle. You have created the first isomer, hexane.

carbon framework of hexane

hexane

Step 2. Draw a chain of five carbon atoms, then add the sixth carbon atom to one of the carbons in the middle of this chain. (Adding it to a carbon at the end of the chain gives a six-carbon chain, the same framework drawn in Step 1.) Two different carbon frameworks can be built from the five-carbon chain, depending on whether the sixth carbon is linked to the 2 or 3 position. For each of these frameworks, fill in the hydrogens.

carbon framework of methylpentane isomers

2-methylpentane

3-methylpentane

Strategy Map 10.1

PROBLEM
Draw *structural isomers* of an **alkane**.

DATA/INFORMATION
- The **formula** is known, so the *longest possible* **C atom chain** is known.

STEP 1. Draw the *longest* **C atom chain** possible.

Simplest **straight-chain alkane**

STEP 2. Draw C atom chain *one atom shorter* and place a C atom at other places on the chain. Fill in **H** atoms.

Alkane with **one less C atom** but with **substituent group** with **one C atom**

STEP 3. Draw *shorter chains,* and place remaining C atoms at various positions on the chain.

Remaining isomers

● **Chirality in Alkanes** To be chiral, *a compound must have at least one C atom attached to four different groups.* Thus, the C_7H_{16} isomer here is chiral.

$$CH_3$$
$$H—C^*—CH_2CH_3$$
$$CH_2CH_2CH_3$$

The center of chirality is often indicated with an asterisk.

● **Naming Guidelines** For more details on naming organic compounds, see Appendix E.

one possible isomer of an alkane with the formula C_7H_{16}

Module 15: Naming Organic Compounds covers concepts in this section.

Step 3. Draw a chain of four carbon atoms. Add in the two remaining carbons, again being careful not to extend the chain length. Two different structures are possible: one with the remaining carbon atoms in the 2 and 3 positions, and another with both extra carbon atoms attached at the 2 position. Fill in the 14 hydrogens. You have now drawn the fourth and fifth isomers.

carbon atom frameworks for dimethylbutane isomers

2,3-dimethylbutane

2,2-dimethylbutane

None of the isomers of C_6H_{14} is chiral. *To be chiral, a compound must have at least one C atom with four different groups attached. This condition is not met in any of these isomers.*

Think about Your Answer Should we look for structures in which the longest chain is three carbon atoms? Try it, but you will see that it is not possible to add the three remaining carbons to a three-carbon chain without creating one of the carbon chains already drawn in a previous step. Thus, we have completed the analysis, with five isomers of this compound being identified.

Names have been given to each of these compounds. See the text that follows this Example, and see Appendix E for guidelines on nomenclature.

Check Your Understanding

(a) Draw the nine isomers having the formula C_7H_{16}. (*Hint:* There is one structure with a seven-carbon chain, two structures with six-carbon chains, five structures with a five-carbon chain [one is illustrated in the margin], and one structure with a four-carbon chain.)

(b) Identify the isomers of C_7H_{16} that are chiral.

Naming Alkanes

With so many possible isomers for a given alkane, chemists need a systematic way of naming them. The guidelines for naming alkanes and their derivatives follow:

- The names of alkanes end in "-ane."
- The names of alkanes with chains of one to 10 carbon atoms are given in Table 10.2. After the first four compounds, the names are derived from Greek and Latin numbers—pentane, hexane, heptane, octane, nonane, decane—and this regular naming continues for higher alkanes.
- When naming a specific alkane, the root of the name corresponds to the longest carbon chain in the compound. One isomer of C_5H_{12} has a three-carbon chain

Drawing Structural Formulas

$$
\begin{array}{ccc}
& C & \\
& | & \\
C-C-C-C-C & \\
1 \;\; 2 \;\; 3 \;\; 4 \;\; 5 &
\end{array}
\qquad
\begin{array}{c}
C \\
| \\
C-C-C-C \\
|2 \;\; 3 \;\; 4 \;\; 5 \\
C \\
1
\end{array}
\qquad
\begin{array}{c}
C \\
| \\
C-C-C-C-C \\
5 \;\; 4 \;\; 3 \;\; 2 \;\; 1
\end{array}
$$

Remember that Lewis structures do not indicate the geometry of molecules.

with two —CH$_3$ groups on the second C atom of the chain. Thus, its name is based on propane.

$$
\begin{array}{c}
CH_3 \\
| \\
H_3C - C - CH_3 \\
| \\
CH_3
\end{array}
$$

2,2-dimethylpropane

- Substituent groups on a hydrocarbon chain are identified by a name and the position of substitution in the carbon chain; this information precedes the root of the name. The position is indicated by a number that refers to the carbon atom to which the substituent is attached. Numbering of the carbon atoms in a chain should begin at the end of the carbon chain that allows the first substituent encountered to have the lowest possible number. (If there is no distinction at this point, then number so to give the second substituent encountered the lowest number.) Both —CH$_3$ groups in 2,2-dimethylpropane are located at the 2 position.
- Names of hydrocarbon substituents, called **alkyl groups**, are derived from the name of the hydrocarbon. The group —CH$_3$, derived by taking a hydrogen from methane, is called the methyl group; the —C$_2$H$_5$ group is the ethyl group.
- If two or more of the same substituent groups are present in the molecule, the prefixes di-, tri-, and tetra- are added. When different substituent groups are present, they are generally listed in alphabetical order.

● **Systematic and Common Names** The IUPAC (International Union of Pure and Applied Chemistry) has formulated rules for systematic names, which are generally used in this book. (See Appendix E.) However, many organic compounds are known by common names. For example, 2,2-dimethylpropane is also called neopentane.

EXAMPLE 10.2   Naming Alkanes

Problem Give the systematic name for

$$
\begin{array}{ccc}
CH_3 & & CH_2CH_3 \\
| & & | \\
CH_3CHCH_2CH_2CHCH_2CH_3
\end{array}
$$

What Do You Know? You know the condensed formula and can recognize that the compound is an alkane. You also know the rules for naming alkanes.

Strategy Identify the longest carbon chain and base the name of the compound on that alkane. Identify the substituent groups on the chain and their locations. When there are two or more substituents (the groups attached to the chain), number the parent chain from the end that gives the lower number to the substituent encountered first. If the substituents are different, list them in alphabetical order. (For more on naming compounds, see Appendix E.)

Solution Here, the longest chain has seven C atoms, so the root of the name is *heptane*. There is a methyl group (—CH$_3$) on C-2 and an ethyl group (—C$_2$H$_5$) on C-5. Giving the substituents in alphabetical order and numbering the chain from the end having the methyl group, the systematic name is 5-ethyl-2-methylheptane.

FIGURE 10.5 Paraffin wax and mineral oil. These common consumer products are mixtures of alkanes.

Think about Your Answer Notice that the carbon atoms in the longest chain are numbered so that the lower number is given to the substituent encountered first.

Check Your Understanding

Name the nine isomers of C_7H_{16} in "Check Your Understanding" in Example 10.1.

Properties of Alkanes

Methane, ethane, propane, and butane are gases at room temperature and pressure, whereas the higher-molar-mass compounds are liquids or solids (Table 10.2). An increase in melting point and boiling point with molar mass in a series of similar compounds is a general phenomenon (▶ Sections 12.3 and 12.4).

You already know about alkanes in a nonscientific context because several are common fuels. Natural gas, gasoline, kerosene, fuel oils, and lubricating oils are all mixtures of various alkanes. White mineral oil is also a mixture of alkanes, as is paraffin wax (Figure 10.5).

Pure alkanes are colorless. The gases and liquids have noticeable but not unpleasant odors. All of these substances are insoluble in water, a property typical of compounds that are nonpolar or nearly so. Low polarity is expected for alkanes because the electronegativities of carbon ($\chi = 2.5$) and hydrogen ($\chi = 2.2$) are not greatly different (◀ Section 8.8).

All alkanes burn readily in air to give CO_2 and H_2O in very exothermic reactions. This is, of course, the reason they are widely used as fuels.

$$CH_4(g) + 2\ O_2(g) \rightarrow CO_2(g) + 2\ H_2O(\ell) \qquad \Delta_rH° = -890.3\ \text{kJ/mol-rxn}$$

Other than in combustion reactions, alkanes exhibit relatively low chemical reactivity. One reaction that does occur, however, is the replacement of the hydrogen atoms of an alkane by chlorine atoms on reaction with Cl_2. It is formally an oxidation because Cl_2, like O_2, is a strong oxidizing agent. These reactions, which can be initiated by ultraviolet radiation, are free radical reactions. Highly reactive Cl atoms are formed from Cl_2 under ultraviolet (UV) radiation. Reaction of methane with Cl_2 under these conditions proceeds in a series of steps, eventually yielding CCl_4, commonly known as carbon tetrachloride. (HCl is the other product of these reactions.)

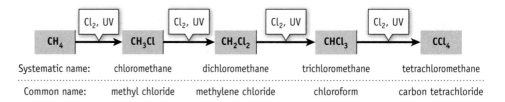

Systematic name:	chloromethane	dichloromethane	trichloromethane	tetrachloromethane
Common name:	methyl chloride	methylene chloride	chloroform	carbon tetrachloride

The last three compounds are used as solvents, albeit less frequently today because of their toxicity.

Cycloalkanes, C_nH_{2n}

Cycloalkanes are constructed with tetrahedral carbon atoms joined together to form a ring. Cyclopropane and cyclobutane are the simplest cycloalkanes, although the bond angles in these species are much less than 109.5°. As a result, chemists say they are **strained hydrocarbons**, so named because an unfavorable geometry is imposed around carbon. One of the features of strained hydrocarbons is that the C—C bonds are weaker and the molecules readily undergo ring-opening reactions that relieve the bond angle strain.

The most common cycloalkane is cyclohexane, C_6H_{12}, which has a nonplanar ring with six —CH_2 groups. If the carbon atoms were in the form of a regular

cyclopropane, C_3H_6 cyclobutane, C_4H_8

Cyclopropane and cyclobutane. Cyclopropane was at one time used as a general anesthetic in surgery. However, its explosive nature when mixed with oxygen soon eliminated this application.

hexagon with all carbon atoms in one plane, the C—C—C bond angles would be 120°. To have tetrahedral bond angles of 109.5° around each C atom, the ring has to pucker. The C_6 ring is flexible and exists in two interconverting forms (see below for *A Closer Look: Flexible Molecules*).

Alkenes and Alkynes

The diversity seen for alkanes is repeated with **alkenes**, hydrocarbons with one or more C=C double bonds. The presence of a double bond adds two features missing in alkanes: the possibility of geometric isomerism and increased reactivity.

The general formula for alkenes with a single double bond is C_nH_{2n}. The first two members of the series of alkenes are ethene, C_2H_4 (common name, ethylene), and propene, C_3H_6 (common name, propylene). Only a single structure can be drawn for these compounds. As with alkanes, the occurrence of isomers begins with species containing four carbon atoms. Four alkene isomers have the formula C_4H_8, and each has distinct chemical and physical properties (Table 10.3). There are three structural isomers, one of which ($CH_3CH=CHCH_3$) exists as two stereoisomers.

C_2H_4
Systematic name: ethene
Common name: ethylene

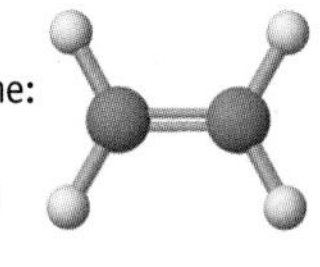

C_3H_6
Systematic name: propene
Common name: propylene

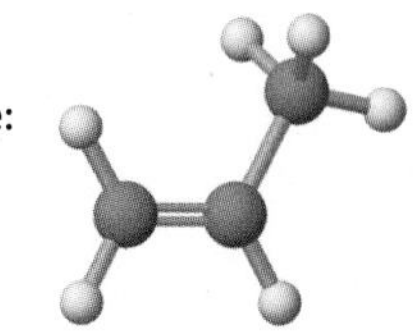

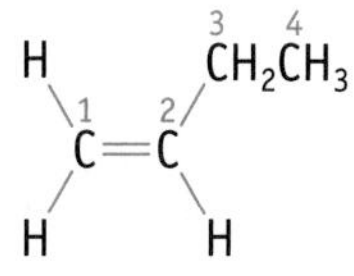

1-butene 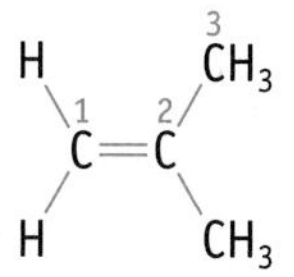 2-methylpropene 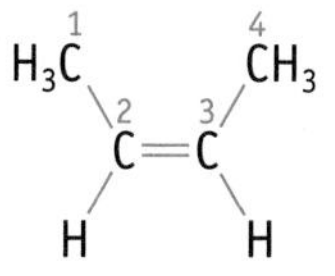 *cis*-2-butene 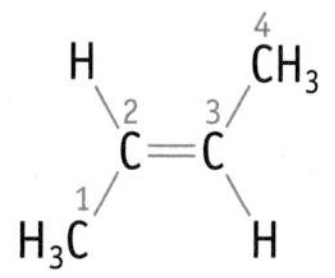 *trans*-2-butene

Alkene names end in "-ene." As with alkanes, the root for the name of an alkene is determined by the longest carbon chain that contains the double bond. The position of the double bond is indicated with a number, and, when appropriate, the prefix *cis* or *trans* is added.

Three of the C_4H_8 isomers have four-carbon chains and so are butenes. One has a three-carbon chain and is a propene. Notice that the carbon chain is numbered from the end that gives the double bond the lowest number. In the first isomer at the left, the double bond is between C atoms 1 and 2, so the name is 1-butene and not 3-butene.

A CLOSER LOOK

Flexible Molecules

Most organic molecules are flexible; that is, they can twist and bend in various ways. Few molecules better illustrate this behavior than cyclohexane. Two structures are possible: "chair" and "boat" forms. These forms can interconvert by partial rotation of several bonds.

The more stable structure is the chair form, which allows the hydrogen atoms to remain as far apart as possible. A side view of this form of cyclohexane reveals two sets of hydrogen atoms in this molecule. Six hydrogen atoms, called the *equatorial hydrogens*, lie in a plane around the carbon ring. The other six hydrogens are positioned above and below the plane and are called *axial hydrogens*. Flexing the ring (a rotation around the C—C single bonds) moves the hydrogen atoms between axial and equatorial environments.

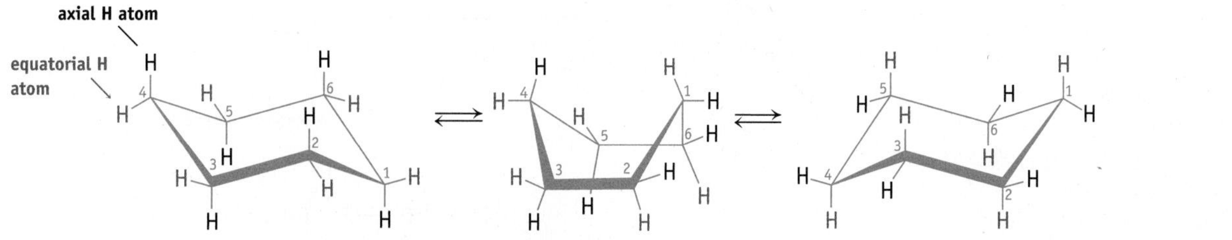

chair form boat form chair form

Table 10.3 Properties of Butene Isomers

Name	Boiling Point	Melting Point	Dipole Moment (D)	$\Delta_f H°$ (gas) (kJ/mol)
1-Butene	−6.26 °C	−185.4 °C	—	−0.63
2-Methylpropene	−6.95 °C	−140.4 °C	0.503	−17.9
Cis-2-butene	3.71 °C	−138.9 °C	0.253	−7.7
Trans-2-butene	0.88 °C	−105.5 °C	0	−10.8

EXAMPLE 10.3 Determining Isomers of Alkenes from a Formula

Problem Draw structures for the six possible alkene isomers with the formula C_5H_{10}. Give the systematic name of each.

What Do You Know? When linking the five carbon atoms together, two will be joined with a double bond. Each carbon must have four bonds, and hydrogen atoms will fill into the remaining positions.

Strategy A procedure that involved drawing the carbon skeleton and then adding hydrogen atoms served well when drawing structures of alkanes (Example 10.1), and a similar approach can be used here. It will be necessary to put one double bond into the framework and to be alert for *cis-trans* isomerism.

Solution

Step 1. A five-carbon chain with one double bond can be constructed in two ways. *Cis-trans* isomers are possible for 2-pentene.

1-pentene

cis-2-pentene

trans-2-pentene

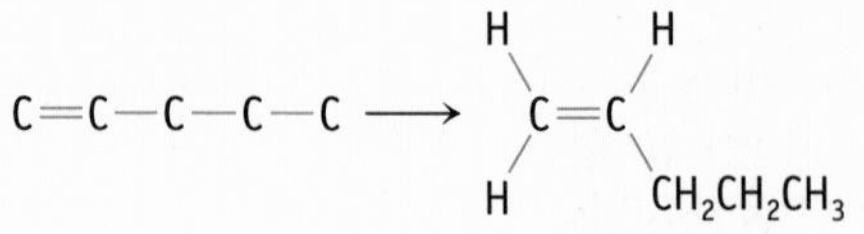

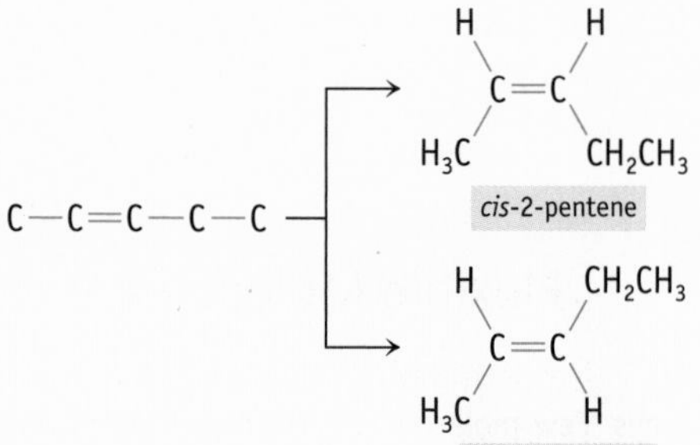

Step 2. Draw the possible four-carbon chains containing a double bond. Add the fifth carbon atom to either the 2 or 3 position. When all the possible combinations are found, fill in the hydrogen atoms. This results in three more structures:

2-methyl-1-butene

3-methyl-1-butene

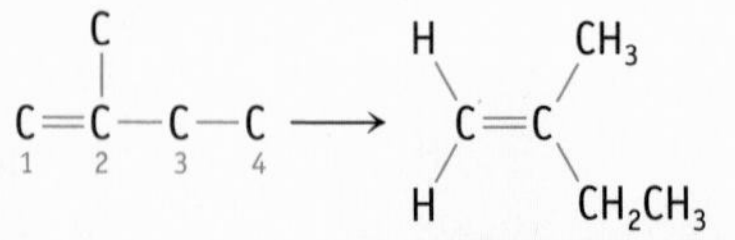

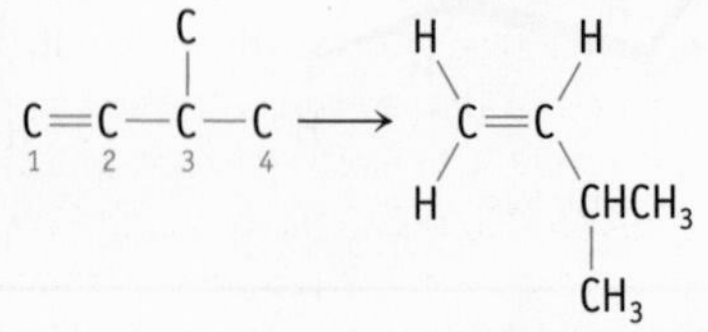

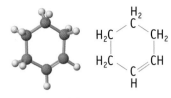

cyclohexene, C_6H_{10}

$H_2C=CHCH=CH_2$

1,3-butadiene, C_4H_6

Cycloalkenes and dienes. Cyclohexene, C_6H_{10} (*top*), and 1,3-butadiene (C_4H_6) (*bottom*).

$$C-C=C-C \longrightarrow \text{2-methyl-2-butene}$$

2-methyl-2-butene

Think about Your Answer With questions like this, it is important to be very organized in your approach. When you complete your answer, you should look carefully to see that each structure is unique, that is, that no two are the same.

Check Your Understanding

There are 17 possible alkene isomers with the formula C_6H_{12}. Draw structures of the five isomers in which the longest chain has six carbon atoms, and give the name of each. Which of these isomers is chiral? (There are also eight isomers in which the longest chain has five carbon atoms, and four isomers in which the longest chain has four carbon atoms. How many can you find?)

Hydrocarbons exist that have two or more double bonds. Butadiene, for example, has two double bonds and is known as a *diene*. Many natural products have numerous double bonds (Figure 10.6). There are also cyclic hydrocarbons, such as cyclohexene, with double bonds.

Alkynes, compounds with a carbon–carbon triple bond, have the general formula (C_nH_{2n-2}). Table 10.4 lists alkynes that have four or fewer carbon atoms. The first member of this family is ethyne (common name, acetylene), a gas used as a fuel in metal cutting torches.

Properties of Alkenes and Alkynes

Like alkanes, alkenes and alkynes are colorless. Low–molar-mass compounds are gases, whereas compounds with higher molar masses are liquids or solids. Alkenes and alkynes are also oxidized by O_2 to give CO_2 and H_2O.

Alkenes and alkynes have an elaborate chemistry. We get some insight into their chemical behavior by recognizing that they are called **unsaturated compounds**. Carbon atoms are capable of bonding to a maximum of four other atoms, and they do so in alkanes and cycloalkanes. In alkenes, however, each carbon atom linked by a double bond is bonded to a total of only three atoms. In alkynes, each carbon atom linked by a triple bond is bonded to a total of only two atoms. It is possible to

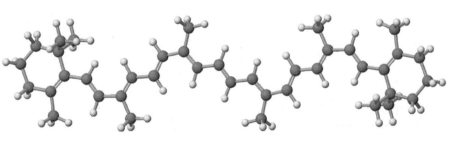

FIGURE 10.6 Carotene, a naturally occurring compound with 11 C=C bonds. The π electrons can be excited by visible light in the blue-violet region of the spectrum. As a result, carotene appears orange-yellow to the observer. Carotene or carotene-like molecules are partnered with chlorophyll in nature in the role of assisting in the harvesting of sunlight. Green leaves have a high concentration of carotene. In autumn, green chlorophyll molecules are destroyed, and the yellows and reds of carotene and related molecules are seen. The red color of tomatoes, for example, comes from a molecule very closely related to carotene. As a tomato ripens, its chlorophyll disintegrates, and the green color is replaced by the red of the carotene-like molecule.

Table 10.4 **Some Simple Alkynes C$_n$H$_{2n-2}$**

Structure	Systematic Name	Common Name	BP (°C)
HC≡CH	ethyne	acetylene	−85
CH$_3$C≡CH	propyne	methylacetylene	−23
CH$_3$CH$_2$C≡CH	1-butyne	ethylacetylene	9
CH$_3$C≡CCH$_3$	2-butyne	dimethylacetylene	27

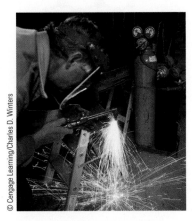

An oxy-acetylene torch. The reaction of ethyne (acetylene) with oxygen produces a very high temperature. Oxy-acetylene torches, used in welding, take advantage of this fact.

increase the number of groups attached to a carbon atom in an alkene or alkyne by **addition reactions**, in which molecules with the general formula X—Y (such as hydrogen, halogens, hydrogen halides, and water) add across the carbon–carbon double or triple bond (Figure 10.7). For an alkene the result is a compound with four groups bonded to each carbon.

$$\underset{H}{\overset{H}{}}C=C\underset{H}{\overset{H}{}} + X—Y \longrightarrow H—\underset{H}{\overset{X}{C}}—\underset{H}{\overset{Y}{C}}—H$$

X—Y = H$_2$, Cl$_2$, Br$_2$; H—Cl, H—Br, H—OH, HO—Cl

The products of some addition reactions are substituted alkanes. For example, the addition of bromine to ethene (ethylene) forms 1,2-dibromoethane.

$$\underset{H}{\overset{H}{}}C=C\underset{H}{\overset{H}{}} + Br_2 \longrightarrow H—\underset{H}{\overset{Br}{C}}—\underset{H}{\overset{Br}{C}}—H$$

1,2-dibromoethane

The addition of 2 mol of chlorine to ethyne (acetylene) gives 1,1,2,2-tetrachloroethane.

$$HC≡CH + 2\ Cl_2 \longrightarrow Cl—\underset{H}{\overset{Cl}{C}}—\underset{H}{\overset{Cl}{C}}—Cl$$

1,1,2,2-tetrachloroethane

● **Nomenclature of Substituted Alkanes** The substituent groups in substituted alkanes are identified by the name and position of the substituent on the alkane chain.

During the 1860s, the Russian chemist Vladimir Markovnikov examined a large number of alkene addition reactions. In cases in which two isomeric products were possible, he found that one was more likely to predominate. Based on these results, Markovnikov formulated a rule (now called *Markovnikov's rule*) stating that, when a

FIGURE 10.7 **Bacon fat and addition reactions.** The fat in bacon is partially unsaturated. Like other unsaturated compounds, bacon fat reacts with Br$_2$ in an addition reaction. Here, you see the color of Br$_2$ vapor fade when a strip of bacon is introduced.

a few minutes

reagent HX adds to an unsymmetrical alkene, the hydrogen atom in the reagent becomes attached to the carbon that already has the largest number of hydrogens. An example of Markovnikov's rule is the reaction of 2-methylpropene with HCl that results in formation of 2-chloro-2-methylpropane rather than 1-chloro-2-methylpropane.

$$H_3C \ \ \ \ \ \ \ \ \ \ \ \ \ \ \ \ \ \ Cl \ \ \ \ \ \ \ \ \ \ \ \ \ \ \ \ \ \ H$$

$$\diagdown$$

$$C{=}CH_2 + HCl \longrightarrow H_3C{-}C{-}CH_3 \ \ + \ \ H_3C{-}C{-}CH_2Cl$$

$$\diagup \ \ \ \ \ \ \ \ \ \ \ \ \ \ \ \ \ \ \ | \ \ \ \ \ \ \ \ \ \ \ \ \ \ \ \ \ \ \ |$$

$$H_3C \ \ \ \ \ \ \ \ \ \ \ \ \ \ \ \ \ \ CH_3 \ \ \ \ \ \ \ \ \ \ \ \ \ \ \ CH_3$$

2-methylpropene   2-chloro-2-methylpropane   1-chloro-2-methylpropane
                         sole product   *not* formed

If the reagent added to a double bond is hydrogen (X—Y = H_2), the reaction is called **hydrogenation**. Hydrogenation is usually a very slow reaction, but it can be speeded up by adding a catalyst, often a specially prepared form of a metal, such as platinum, palladium, and rhodium. You may have heard the term *hydrogenation* because certain foods contain "hydrogenated" or "partially hydrogenated" ingredients. One brand of crackers has a label that says, "Made with 100% pure vegetable shortening . . . (partially hydrogenated soybean oil with hydrogenated cottonseed oil)." One reason for hydrogenating an oil is to make it less susceptible to spoilage; another is to convert it from a liquid to a solid.

> **● Catalysts** A catalyst is a substance that causes a reaction to occur at a faster rate without itself being permanently changed in the reaction. We will describe catalysts in more detail in Chapter 15.

EXAMPLE 10.4   Reaction of an Alkene

Problem Draw the structure of the compound obtained from the reaction of Br_2 with propene, and name the compound.

What Do You Know? Propene is the three-carbon alkene. Addition reactions are among the most common reactions of alkenes, and it is known that bromine is one of several common reagents that adds to double bonds.

Strategy Bromine adds across the C=C double bond. The name of the product is based on the name of the carbon chain and indicates the positions of the Br atoms.

Solution

$$H \ \ \ \ \ \ H \ Br \ \ Br$$

$$\diagdown \ \ \diagup \ | \ \ \ \ |$$

$$C{=}C \ \ + Br_2 \longrightarrow H{-}C{-}C{-}CH_3$$

$$\diagup \ \ \diagdown \ | \ \ \ \ |$$

$$H \ \ \ \ \ \ CH_3 \ \ \ \ \ \ \ \ \ \ \ \ \ \ \ \ \ H \ \ \ H$$

propene   1,2-dibromopropane

Think about Your Answer This reaction converts an unsaturated hydrocarbon to a substituted alkane. It is named as an alkane (propane) with substituents (Br atoms) identified by name and position on the three-carbon chain.

Check Your Understanding

(a)   Draw the structure of the compound obtained from the reaction of HBr with ethylene, and name the compound.

(b)   Draw the structure of the product of the reaction of Br_2 with *cis*-2-butene, and name this compound.

Aromatic Compounds

Benzene, C_6H_6, is a key molecule in chemistry. It is the simplest **aromatic compound**, a class of compounds so named because they have significant, and usually not unpleasant, odors. Other members of this class, which are all based on benzene, include toluene and naphthalene. A source of many aromatic compounds is coal.

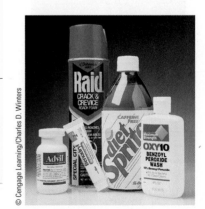

Some products containing compounds based on benzene.
Examples include sodium benzoate in soft drinks, ibuprofen in Advil, and benzoyl peroxide in Oxy-10.

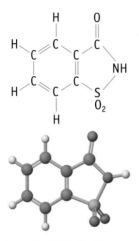

Saccharin ($C_7H_5NO_3S$). This compound, an artificial sweetener, contains an aromatic ring.

These compounds, along with other volatile substances, are released when coal is heated to a high temperature in the absence of air (Table 10.5).

benzene   toluene   naphthalene

Benzene occupies a pivotal place in the history and practice of chemistry. Michael Faraday discovered this compound in 1825 as a by-product of illuminating gas, a fuel produced by heating coal. Today, benzene is an important industrial chemical, usually ranking among the top 25 chemicals in production annually in the United States. It is used as a solvent and is also the starting point for making thousands of different compounds by replacing the H atoms of the ring.

Toluene was originally obtained from tolu balsam, the pleasant-smelling gum of a South American tree, *Toluifera balsamum*. This balsam has been used in cough syrups and perfumes. Naphthalene is an ingredient in "moth balls," although 1,4-dichlorobenzene is now more commonly used. Aspartame and another artificial sweetener, saccharin, are also benzene derivatives.

The Structure of Benzene

The formula of benzene suggested to 19th-century chemists that this compound should be unsaturated, but, if viewed this way, its chemistry was perplexing. Whereas alkenes readily undergo addition reactions, benzene does not do so under similar conditions. We now recognize that benzene's different reactivity relates to its structure and bonding, both of which are quite different from the structure and bonding in alkenes. Benzene has equivalent carbon–carbon bonds, 139 pm in length, intermediate between a C—C single bond (154 pm) and a C=C double bond (134 pm). The π bonds are formed by the continuous overlap of the p orbitals on the six carbon atoms (page 324). Using valence bond terminology, the structure is represented by two resonance structures.

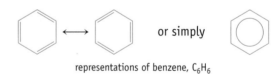

representations of benzene, C_6H_6

● **August Kekulé and the Structure of Benzene** The structural question was solved by August Kekulé (1829–1896). Kekulé, one of the most prominent organic chemists in Europe in the late 19th century, argued for the ring structure with alternating double bonds based on the number of isomers possible for the structure. The legend in chemistry is that Kekulé proposed the ring structure after dreaming of a snake biting its tail.

Table 10.5 Some Aromatic Compounds from Coal Tar

Common Name	Formula	Boiling Point (°C)	Melting Point (°C)
Benzene	C_6H_6	80	+6
Toluene	$C_6H_5CH_3$	111	−95
o-Xylene	$1,2\text{-}C_6H_4(CH_3)_2$	144	−25
m-Xylene	$1,3\text{-}C_6H_4(CH_3)_2$	139	−48
p-Xylene	$1,4\text{-}C_6H_4(CH_3)_2$	138	+13
Naphthalene	$C_{10}H_8$	218	+80

Benzene Derivatives

Toluene, chlorobenzene, benzoic acid, aniline, styrene, and phenol are common examples of benzene derivatives.

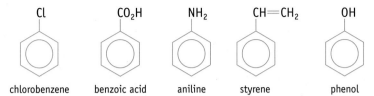

| chlorobenzene | benzoic acid | aniline | styrene | phenol |

The systematic nomenclature for benzene derivatives with two or more substituent groups involves naming these groups and identifying their positions on the ring by numbering the six carbon atoms (▶ Appendix E). Some common names, which are based on an older naming scheme, are also used. This scheme identified isomers of disubstituted benzenes with the prefixes *ortho* (*o-*, substituent groups on adjacent carbons in the benzene ring), *meta* (*m-*, substituents separated by one carbon atom), and *para* (*p-*, substituent groups on carbons on opposite sides of the ring).

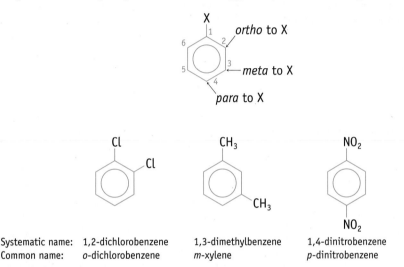

| Systematic name: | 1,2-dichlorobenzene | 1,3-dimethylbenzene | 1,4-dinitrobenzene |
| Common name: | *o*-dichlorobenzene | *m*-xylene | *p*-dinitrobenzene |

● **Drawing Aromatic Rings** When drawing benzene rings chemists often allow the vertices of the hexagon to represent the carbon atoms and do not show the H atoms attached to those carbon atoms.

EXAMPLE 10.5   Isomers of Substituted Benzenes

Problem Draw and name the isomers of $C_6H_3Cl_3$.

What Do You Know? From the formula you can infer that $C_6H_3Cl_3$ is a substituted benzene with three hydrogen atoms replaced by chlorine atoms.

Strategy Begin by drawing the carbon framework of benzene, and attach a chlorine atom to one of the carbon atoms. Place a second Cl atom on the ring in the *ortho, meta,* and *para* positions. Add the third Cl in one of the remaining positions, being careful not to repeat a structure already drawn.

Solution The three isomers of $C_6H_3Cl_3$ are shown here. They are named as derivatives of benzene by specifying the number of substituent groups with the prefix "tri-," the name of the substituent, and the positions of the three groups around the six-member ring.

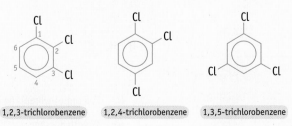

| 1,2,3-trichlorobenzene | 1,2,4-trichlorobenzene | 1,3,5-trichlorobenzene |

Think about Your Answer Are there other possibilities? Try moving the chlorine atoms around in each isomer. In every case, you will find that moving one Cl atom to a different position generates one of these three isomers. For example, in the first structure, moving the Cl atom at position 1 to either position 5 or 6 leads to structures identical to the second structure.

Check Your Understanding

Aniline, $C_6H_5NH_2$, is the common name for aminobenzene. Draw a structure for *p*-diaminobenzene, a compound used in dye manufacture. What is the systematic name for *p*-diaminobenzene?

Properties of Aromatic Compounds

Benzene is a colorless liquid, and simple substituted benzenes are liquids or solids under normal conditions. The properties of aromatic hydrocarbons are typical of hydrocarbons in general: They are insoluble in water, soluble in nonpolar solvents, and oxidized by O_2 to form CO_2 and H_2O.

One of the most important properties of benzene and other aromatic compounds is an unusual stability that is associated with the unique π bonding in this molecule. Because the π bonding in benzene is typically described using resonance structures, the extra stability is termed **resonance stabilization**. The extent of resonance stabilization in benzene is evaluated by comparing the energy evolved in the hydrogenation of benzene to form cyclohexane

$$C_6H_6(\ell) + 3\ H_2(g) \xrightarrow{\text{catalyst}} C_6H_{12}(\ell) \qquad \Delta_rH° = -206.7 \text{ kJ/mol-rxn}$$

with the energy evolved in hydrogenation of three isolated double bonds.

$$3\ H_2C\!=\!CH_2(g) + 3\ H_2(g) \rightarrow 3\ C_2H_6(g) \qquad \Delta_rH° = -410.8 \text{ kJ/mol-rxn}$$

The hydrogenation of benzene is about 200 kJ/mol less exothermic than the hydrogenation of three moles of ethene. The difference is attributable to the added stability associated with π bonding in benzene.

Although aromatic compounds are unsaturated hydrocarbons, they do not undergo the addition reactions typical of alkenes and alkynes. Instead, *substitution reactions* occur, in which one or more hydrogen atoms are replaced by other groups. Such reactions require higher temperatures and a strong Brønsted acid such as H_2SO_4 or a Lewis acid such as $AlCl_3$ or $FeBr_3$.

● **Lewis Acids and Bases** G. N. Lewis defined an acid as an electron-pair acceptor (such as $AlCl_3$) and a base as an electron-pair donor (such as NH_3). (See Chapter 17, Section 11.)

Nitration: $\quad C_6H_6(\ell)\ +\ HNO_3(\ell) \xrightarrow{H_2SO_4} C_6H_5NO_2(\ell)\ +\ H_2O(\ell)$

Alkylation: $\quad C_6H_6(\ell)\ +\ CH_3Cl(\ell) \xrightarrow{AlCl_3} C_6H_5CH_3(\ell)\ +\ HCl(g)$

Halogenation: $\ C_6H_6(\ell)\ +\ Br_2(\ell) \xrightarrow{FeBr_3} C_6H_5Br(\ell)\ +\ HBr(g)$

REVIEW & CHECK FOR SECTION 10.2

1. What is the systematic name for this alkane?

(a) nonane

(b) 2-ethyl-5-methylhexane

(c) 2, 5-dimethylheptane

(d) dimethyloctane

2. Which statement below correctly describes the following compound?

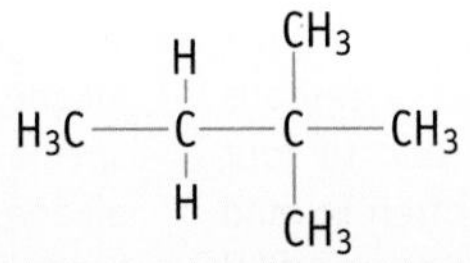

$$H_3C-\underset{\underset{H}{|}}{\overset{\overset{H}{|}}{C}}-\underset{\underset{CH_3}{|}}{\overset{\overset{CH_3}{|}}{C}}-CH_3$$

(a) The compound is an isomer of pentane, is chiral, and is named 2,3-dimethylbutane.

(b) The compound is an isomer of octane, is not chiral, and is named 2,2-dimethylbutane.

(c) The compound is an isomer of hexane, is not chiral, and is named 2,2-dimethylbutane.

(d) The compound is an isomer of hexane, is not chiral, and is named 3,3-dimethylbutane.

3. Consider the following list of compounds:

1. C_2H_4 2. C_5H_{10} 3. $C_{14}H_{30}$ 4. C_7H_8

(i) Which compound or compounds in the list can be an alkane?

(a) only 1   (b) 2 and 3   (c) only 3   (d) 3 and 4

(ii) Which compound or compounds in the list can be an alkene?

(a) only 1   (b) only 2   (c) 1 and 2   (d) 3 and 4

4. What is the product of the following reaction?

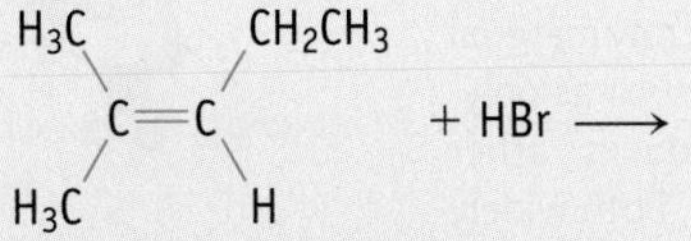

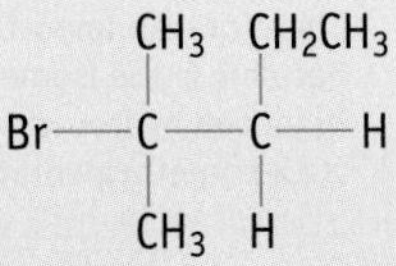

$$\underset{H_3C}{\overset{H_3C}{>}}C=C\underset{H}{\overset{CH_2CH_3}{<}} + HBr \longrightarrow$$

$$Br-\underset{\underset{CH_3}{|}}{\overset{\overset{CH_3}{|}}{C}}-\underset{\underset{H}{|}}{\overset{\overset{CH_2CH_3}{|}}{C}}-H \qquad H-\underset{\underset{CH_3}{|}}{\overset{\overset{CH_3}{|}}{C}}-\underset{\underset{H}{|}}{\overset{\overset{CH_2CH_3}{|}}{C}}-Br$$

        (a) 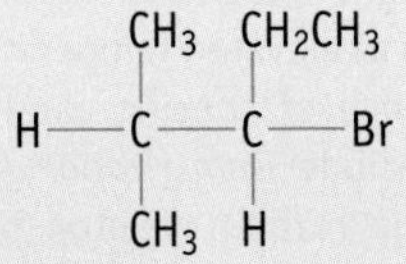   **(b)**

5. How many isomers are possible for $C_6H_4(CH_3)Cl$, a benzene derivative?

(a) 1   (b) 2   (c) 3

10.3 Alcohols, Ethers, and Amines

Organic compounds often contain other elements in addition to carbon and hydrogen. Two elements in particular, oxygen and nitrogen, add a rich dimension to carbon chemistry.

Organic chemistry organizes compounds containing elements other than carbon and hydrogen as derivatives of hydrocarbons. Formulas (and structures) are represented by substituting one or more hydrogens in a hydrocarbon molecule by a **functional group**. A functional group is an atom or group of atoms attached to a carbon atom in the hydrocarbon. Formulas of hydrocarbon derivatives are then written as R—X, in which R is a hydrocarbon lacking a hydrogen atom, and X is the functional group (such as $-OH$, $-NH_2$, a halogen atom, or $-CO_2H$) that has replaced the hydrogen. The chemical and physical properties of the hydrocarbon derivatives are a blend of the properties associated with hydrocarbons and the group that has been substituted for hydrogen.

Table 10.6 identifies some common functional groups and the families of organic compounds resulting from their attachment to a hydrocarbon.

Petroleum Chemistry

Much of the world's current technology relies on petroleum. Burning fuels derived from petroleum provides by far the largest amount of energy in the industrial world. Petroleum and natural gas are also the chemical raw materials used in the manufacture of many plastics, pharmaceuticals, and a vast array of other compounds.

The petroleum that is pumped out of the ground is a complex mixture whose composition varies greatly, depending on its source. The primary components of petroleum are always alkanes, but, to varying degrees, nitrogen- and sulfur-containing compounds are also present. Aromatic compounds are present as well, but alkenes and alkynes are not.

An early step in the petroleum refining process is distillation, in which the crude mixture is separated into a series of fractions based on boiling point: first a gaseous fraction (mostly alkanes with one to four carbon atoms; this fraction is often burned off), and then gasoline, kerosene, and fuel oils. After distillation, considerable material, in the form of a semi-solid, tar-like residue, remains.

The petrochemical industry seeks to maximize the amounts of the higher-valued fractions of petroleum produced and to make specific compounds for which a particular need exists. This means carrying out chemical reactions involving the raw materials on a huge scale. One process to which petroleum is subjected is known as *cracking*. At very high temperatures, bond breaking or "cracking" can occur, and longer-chain hydrocarbons will fragment into smaller molecular units. These reactions are carried out in the presence of a wide array of catalysts, materials that speed up reactions and direct them toward specific products. Among the important products of cracking is ethylene, which serves as the raw material for the formation of materials such as polyethylene. Cracking also produces other alkenes and gaseous hydrogen, both widely used raw materials in the chemical industry.

Other important reactions involving petroleum are run at elevated temperatures and in the presence of specific catalysts. Such reactions include *isomerization* reactions, in which the carbon skeleton of an alkane rearranges to form a new isomeric species, and *reformation* processes, in which alkanes become cycloalkanes or aromatic hydrocarbons. Each process is directed toward achieving a specific goal, such as increasing the proportion of branched-chain hydrocarbons in gasoline to obtain higher octane ratings. A great amount of chemical research has gone into developing and understanding these highly specialized processes.

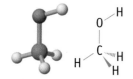

A modern petrochemical plant.

© iStockphoto.com/Olivier Lantzendörffer

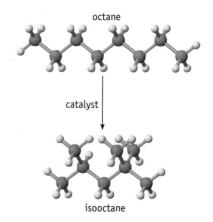

Producing gasoline. Branched hydrocarbons have a higher octane rating in gasoline. Therefore, an important process in producing gasoline is the isomerization of octane to a branched hydrocarbon such as isooctane, 2,2,4-trimethylpentane.

Alcohols and Ethers

Methanol, CH₃OH, the simplest alcohol. Methanol is often called *wood alcohol* because it was originally produced by heating wood in the absence of air.

If one of the hydrogen atoms of an alkane is replaced by a hydroxyl (—OH) group, the result is an **alcohol**, ROH. Methanol, CH_3OH, and ethanol, CH_3CH_2OH, are the most important alcohols, but others are also commercially important (Table 10.7). Notice that several have more than one OH functional group.

More than 5×10^8 kg of methanol is produced in the United States annually. Most of this production is used to make formaldehyde (CH_2O) and acetic acid (CH_3CO_2H), both important chemicals in their own right. Methanol is also used as a solvent, as a de-icer in gasoline, and as a fuel in high-powered racing cars. It is found in low concentration in new wine, where it contributes to the odor, or "bouquet." Like ethanol, methanol causes intoxication, but methanol differs in being more poisonous, largely because the human body converts it to formic acid (HCO_2H) and formaldehyde (CH_2O). These compounds attack the cells of the retina in the eye, leading to permanent blindness.

Ethanol is the "alcohol" of alcoholic beverages, in which it is formed by the anaerobic (without air) fermentation of sugar. On a much larger scale, ethanol for use

Table 10.6 Common Functional Groups and Derivatives of Alkanes

Functional Group*	General Formula*	Class of Compound	Examples
F, Cl, Br, I	RF, RCl, RBr, RI	Haloalkane	CH_3CH_2Cl, chloroethane
OH	ROH	Alcohol	CH_3CH_2OH, ethanol
OR′	ROR′	Ether	$(CH_3CH_2)_2O$, diethyl ether
NH_2†	RNH_2	(Primary) Amine	$CH_3CH_2NH_2$, ethylamine
$\overset{\displaystyle O}{\overset{\displaystyle \|}{-CH}}$	RCHO	Aldehyde	CH_3CHO, ethanal (acetaldehyde)
$\overset{\displaystyle O}{\overset{\displaystyle \|}{-C-R'}}$	RCOR′	Ketone	CH_3COCH_3, propanone (acetone)
$\overset{\displaystyle O}{\overset{\displaystyle \|}{-C-OH}}$	RCO_2H	Carboxylic acid	CH_3CO_2H, ethanoic acid (acetic acid)
$\overset{\displaystyle O}{\overset{\displaystyle \|}{-C-OR'}}$	RCO_2R'	Ester	$CH_3CO_2CH_3$, methyl acetate
$\overset{\displaystyle O}{\overset{\displaystyle \|}{-C-NH_2}}$	$RCONH_2$	Amide	CH_3CONH_2, acetamide

* R and R′ can be the same or different hydrocarbon groups.
† Secondary amines (R_2NH) and tertiary amines (R_3N) are also possible, see discussion in the text.

as a fuel is made by fermentation of corn and other plant materials. Some ethanol (about 5%) is made from petroleum, by the reaction of ethylene and water.

$$\underset{\text{ethylene}}{\overset{H}{\underset{H}{C}}=\overset{H}{\underset{H}{C}}}\text{(g)} + H_2O\text{(g)} \xrightarrow{\text{catalyst}} \underset{\text{ethanol}}{H-\overset{H}{\underset{H}{C}}-\overset{H}{\underset{H}{C}}-OH(\ell)}$$

Beginning with three-carbon alcohols, structural isomers are possible. For example, 1-propanol and 2-propanol (common names, propyl alcohol and isopropyl alcohol) are different compounds (Table 10.7).

Ethylene glycol and glycerol are common alcohols having two and three —OH groups, respectively. Ethylene glycol is used as antifreeze in automobiles. Glycerol's

Table 10.7 Some Important Alcohols

Condensed Formula	BP (°C)	Systematic Name	Common Name	Use
CH_3OH	65.0	Methanol	Methyl alcohol	Fuel, gasoline additive, making formaldehyde
CH_3CH_2OH	78.5	Ethanol	Ethyl alcohol	Beverages, gasoline additive, solvent
$CH_3CH_2CH_2OH$	97.4	1-Propanol	Propyl alcohol	Industrial solvent
$CH_3CH(OH)CH_3$	82.4	2-Propanol	Isopropyl alcohol	Rubbing alcohol
$HOCH_2CH_2OH$	198	1,2-Ethanediol	Ethylene glycol	Antifreeze
$HOCH_2CH(OH)CH_2OH$	290	1,2,3-Propanetriol	Glycerol (glycerin)	Moisturizer in consumer products

Rubbing alcohol. Common rubbing alcohol is 2-propanol, also called *isopropyl alcohol.*

FIGURE 10.8 Nitroglycerin.
(a) Concentrated nitric acid and glycerin react to form an oily, highly unstable compound called *nitroglycerin,* $C_3H_5(ONO_2)_3$. **(b)** Nitroglycerin is more stable if absorbed onto an inert solid, a combination called *dynamite.* **(c)** The fortune of Alfred Nobel (1833–1896), built on the manufacture of dynamite, now funds the Nobel Prizes.

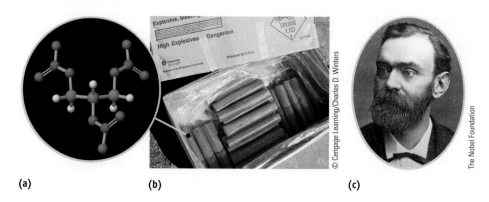

(a)   (b)   (c)

most common use is as a softener in soaps and lotions. It is also a raw material for the preparation of nitroglycerin (Figure 10.8).

Systematic name:	1,2-ethanediol	1,2,3-propanetriol
Common name:	ethylene glycol	glycerol or glycerin

Ethers have the general formula ROR′. The best-known ether is diethyl ether, $CH_3CH_2OCH_2CH_3$. Lacking an —OH group, the properties of ethers are in sharp contrast to those of alcohols. Diethyl ether, for example, has a lower boiling point (34.5 °C) than ethanol, CH_3CH_2OH (78.3 °C), and is only slightly soluble in water.

EXAMPLE 10.6   Structural Isomers of Alcohols

Problem How many different alcohols are derivatives of pentane? Draw their structures, and name each alcohol.

What Do You Know? The formula for pentane is C_5H_{12}. In an alcohol an —OH group will replace one H atom.

Strategy Pentane, C_5H_{12}, has a five-carbon chain. An —OH group can replace a hydrogen atom on one of the carbon atoms. Alcohols are named as derivatives of the alkane (pentane) by replacing the "-e" at the end with "-ol" and indicating the position of the —OH group by a numerical prefix (Appendix E).

Solution Three different alcohols are possible, depending on whether the —OH group is placed on the first, second, or third carbon atom in the chain. (The fourth and fifth positions are identical to the second and first positions in the chain, respectively.)

1-pentanol

2-pentanol

3-pentanol

Think about Your Answer Additional structural isomers with the formula $C_5H_{11}OH$ are possible in which the longest carbon chain has three C atoms (one isomer) or four C atoms (four isomers).

Check Your Understanding

Draw the structure of 1-butanol and alcohols that are structural isomers of the compound.

Properties of Alcohols

Methane, CH_4, is a gas (boiling point, $-161\ ^\circ C$) with low solubility in water. Methanol, CH_3OH, by contrast, is a liquid that is *miscible* with water in all proportions. The boiling point of methanol, $65\ ^\circ C$, is $226\ ^\circ C$ higher than the boiling point of methane. What a difference the addition of a single atom into the structure can make in the properties of simple molecules!

Alcohols are related to water, with one of the H atoms of H_2O being replaced by an organic group. If a methyl group is substituted for one of the hydrogens of water, methanol results. Ethanol has a $-C_2H_5$ (ethyl) group, and propanol has a $-C_3H_7$ (propyl) group in place of one of the hydrogens of water. Viewing alcohols as related to water also helps in understanding their properties.

The two parts of methanol, the $-CH_3$ group and the $-OH$ group, contribute to its properties. For example, methanol will burn, a property associated with hydrocarbons. On the other hand, its boiling point is more like that of water. The temperature at which a substance boils is related to the forces of attraction between molecules, called *intermolecular forces:* The stronger the attractive, intermolecular forces in a sample, the higher the boiling point (▶ Section 12.4). These forces are particularly strong in water, a result of the polarity of the $-OH$ group in this molecule (◀ Section 8.8). Methanol is also a polar molecule, and it is the polar $-OH$ group that leads to a high boiling point. In contrast, methane is nonpolar and its low boiling point is the result of weak intermolecular forces.

It is also possible to explain the differences in the solubility of methane, methanol, and other alcohols in water (Figure 10.9). The solubility of methanol and ethylene glycol is conferred by the polar $-OH$ portion of the molecule. Methane, which is nonpolar, has low water-solubility.

Methanol is often added to automobile gasoline tanks in the winter to prevent water in the fuel lines from freezing. It is soluble in water and lowers the water's freezing point.

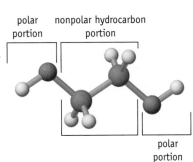

polar portion   nonpolar hydrocarbon portion

polar portion

Ethylene glycol is used in automobile radiators. It is soluble in water, and lowers the freezing point and raises the boiling point of the water in the cooling system. (▶ Section 14.4.)

Ethylene glycol, a major component of automobile antifreeze, is completely miscible with water.

FIGURE 10.9 Properties and uses of two alcohols, methanol and ethylene glycol.

As the size of the alkyl group in an alcohol increases, the alcohol boiling point rises, a general trend seen in families of similar compounds and related to molar mass (see Table 10.7). The solubility in water in this series decreases. Methanol and ethanol are completely miscible in water, whereas 1-propanol is moderately water-soluble; 1-butanol is less soluble than 1-propanol. With an increase in the size of the hydrocarbon group, the organic group (the nonpolar part of the molecule) has become a larger fraction of the molecule, and properties associated with nonpolarity begin to dominate. Space-filling models show that in methanol, the polar and nonpolar parts of the molecule are approximately similar in size, but in 1-butanol the —OH group is less than 20% of the molecule. The molecule is less like water and more "organic." Electrostatic potential surfaces amplify this point.

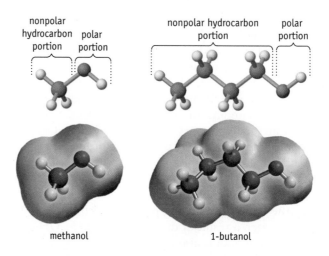

methanol   1-butanol

Amines

It is often convenient to think about water and ammonia as being similar molecules: They are the simplest hydrogen compounds of adjacent second-period elements. Both are polar and exhibit some similar chemistry, such as protonation (to give H_3O^+ and NH_4^+) and deprotonation (to give OH^- and NH_2^-).

This comparison of water and ammonia can be extended to alcohols and amines. Alcohols have formulas related to water in which one hydrogen in H_2O is replaced with an organic group (R—OH). In organic **amines**, one or more hydrogen atoms of NH_3 are replaced with an organic group. Amine structures are similar to ammonia's structure; that is, the geometry about the N atom is trigonal pyramidal.

Amines are categorized based on the number of organic substituents as primary (one organic group), secondary (two organic groups), or tertiary (three organic groups). As examples, consider the three amines with methyl groups: CH_3NH_2, $(CH_3)_2NH$, and $(CH_3)_3N$.

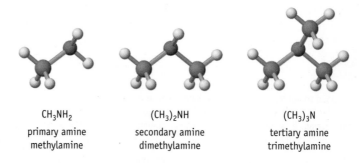

CH_3NH_2

primary amine
methylamine

$(CH_3)_2NH$

secondary amine
dimethylamine

$(CH_3)_3N$

tertiary amine
trimethylamine

Properties of Amines

Amines usually have offensive odors. You know what the odor is if you have ever smelled decaying fish. Two appropriately named amines, putrescine and cadaverine, add to the odor of urine, rotten meat, and bad breath.

$H_2NCH_2CH_2CH_2CH_2NH_2$
putrescine
1,4-butanediamine

$H_2NCH_2CH_2CH_2CH_2CH_2NH_2$
cadaverine
1,5-pentanediamine

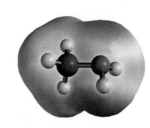

Electrostatic potential surface for methylamine. The surface for methylamine shows that this water-soluble amine is polar with the partial negative charge on the N atom.

The smallest amines are water-soluble, but most amines are not. All amines are bases, however, and they react with acids to give salts, many of which are water-soluble. As with ammonia, the reactions involve adding H^+ to the lone pair of electrons on the N atom. This is illustrated by the reaction of aniline (aminobenzene) with H_2SO_4 to give anilinium hydrogen sulfate.

$$C_6H_5NH_2(aq) + H_2SO_4(aq) \longrightarrow C_6H_5NH_3^+(aq) + HSO_4^-(aq)$$

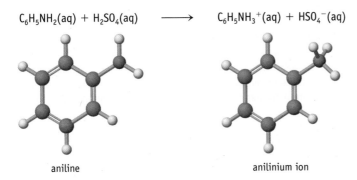

aniline   anilinium ion

The facts that an amine can be protonated and that the proton can be removed again by treating the compound with a base have practical and physiological importance. Nicotine in cigarettes is normally found in the protonated form. (This water-soluble form is often also used in insecticides.) Adding a base such as ammonia removes the H^+ ion to leave nicotine in its "free-base" form.

$$NicH_2^{2+}(aq) + 2\ NH_3(aq) \rightarrow Nic(aq) + 2\ NH_4^+(aq)$$

In this form, nicotine is much more readily absorbed by the skin and mucous membranes, so the compound is a much more potent poison.

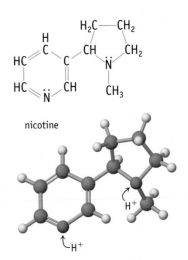

nicotine

Nicotine, an amine. Two nitrogen atoms in the nicotine molecule can be protonated, which is the form in which nicotine is normally found. The protons can be removed, by treating it with a base. This "free-base" form is much more poisonous and addictive.

REVIEW & CHECK FOR SECTION 10.3

1. How many different compounds (alcohols and ethers) exist with the molecular formula $C_4H_{10}O$?

   (a) 2　　　　　　(b) 3　　　　　　(c) 4　　　　　　(d) more than 4

2. Which of the following compounds is not chiral, that is, which does not possess a carbon atom attached to four different groups?

   (a) 2-propanol　　　　　　　　　(c) 2-methyl-3-pentanol

   (b) 2-butanol　　　　　　　　　　(d) 1,2-propanediol

3. What is the hybridization of nitrogen in dimethylamine?

   (a) sp^3　　　　　　　　　　　　(c) sp

   (b) sp^2　　　　　　　　　　　　(d) nitrogen is not hybridized

4. What chemical reagent will react with the ethylammonium ion $[CH_3CH_2NH_3]^+$ to form ethylamine?

   (a) O_2　　　　　　(b) N_2　　　　　　(c) H_2SO_4　　　　　　(d) NaOH

10.4 Compounds with a Carbonyl Group

Formaldehyde, acetic acid, and acetone are among the organic compounds referred to in previous examples. These compounds have a common structural feature: Each contains a trigonal-planar carbon atom doubly bonded to an oxygen. The $C=O$ group is called the **carbonyl group**, and these compounds are members of a large class of compounds called **carbonyl compounds**.

An Awakening with L-DOPA

From about 1917 to 1928, millions of people worldwide were affected by a condition known as *encephalitis lethargica*, or a form of sleeping sickness. Those who suffered from the condition were in a state of semi-consciousness that lasted for decades. In his book, *Awakenings*, Oliver Sacks wrote about treating a patient with the compound L-DOPA, which "was started in early March 1969 and raised by degrees to 5.0 g a day. Little effect was seen for two weeks, and then a sudden 'conversion' took place. . . . Mr. L enjoyed a mobility, a health, and a happiness which he had not known in thirty years. Everything about him filled with delight: he was like a man who had awoken from a nightmare or a serious illness . . . ."

Robert DeNiro as Leonard Lowe and Robin Williams as Malcolm Sayer, a fictionalized portrayal of Oliver Sacks, in the movie version of *Awakenings*.

If you have read the book or have seen the movie of the same name, you know that Mr. L eventually could not tolerate the treatment, but that Sacks treated many others who benefitted from it. L-DOPA is now widely used in the treatment of another condition, Parkinson's disease, a degenerative disorder of the central nervous system.

L-DOPA or L-dopamine (L-3,4-dihydroxy–phenylalanine) is chiral. The symbol L stands for "levo," which means a solution of the compound rotates polarized light to the left.

The compound is also a derivative of phenylalanine, one of many naturally occurring alpha-amino acids that play such an important role in protein formation and other natural processes.

L-DOPA also illustrates why chiral molecules are so interesting to chemists: Only the "levo" enantiomer is physiologically active. The enantiomer that rotates polarized light in the opposite direction has no biological function.

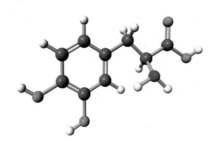

L-DOPA, $C_9H_{11}NO_4$, a treatment for Parkinson's disease

When L-DOPA is ingested, it is metabolized to dopamine in a process that removes the carboxylic acid group, $-CO_2H$, and it is dopamine that is physiologically active. Dopamine is a neurotransmitter that occurs in a wide variety of animals.

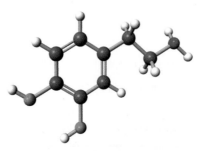

dopamine, $C_8H_{11}NO_2$, a neurotransmitter

Interestingly, both L-DOPA and dopamine are closely related to another amine, epinephrine. This is sometimes referred to as adrenaline, the hormone that is released from the adrenal glands when there is an emergency or danger threatens.

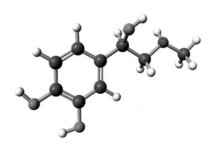

epinephrine or adrenaline, $C_9H_{13}NO_3$

Questions:
1. L-DOPA is chiral. What is the center of chirality in the molecule?
2. Is either dopamine or epinephrine chiral? If so, what is the center of chirality?
3. If you are treated with 5.0 g of L-DOPA, what amount (in moles) is this?

Answers to these questions are available in Appendix N.

References:
1. Oliver Sacks, *Awakenings*, Vintage Books, New York, 1999.
2. N. Angier, "A Molecule of Motivation, Dopamine Excels at its Task," *New York Times*, October 27, 2009.

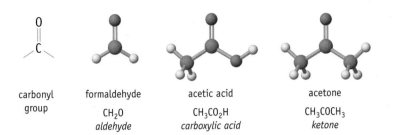

| carbonyl group | formaldehyde CH_2O *aldehyde* | acetic acid CH_3CO_2H *carboxylic acid* | acetone CH_3COCH_3 *ketone* |

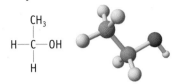

Primary alcohol: ethanol

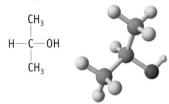

Secondary alcohol: 2-propanol

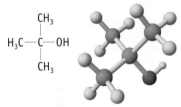

Tertiary alcohol: 2-methyl-2-propanol

In this section, we will examine five groups of carbonyl compounds (Table 10.6, page 359):

- *Aldehydes* (RCHO) have an organic group (—R) and an H atom attached to a carbonyl group.
- *Ketones* (RCOR′) have two —R groups attached to the carbonyl carbon; they may be the same groups, as in acetone, or different groups.
- *Carboxylic acids* (RCO₂H) have an —R group and an —OH group attached to the carbonyl carbon.
- *Esters* (RCO₂R′) have —R and —OR′ groups attached to the carbonyl carbon.
- *Amides* (RCONR₂′, RCONHR′, and RCONH₂) have an —R group and an amino group (—NH₂, —NHR, —NR₂) bonded to the carbonyl carbon.

Aldehydes, ketones, and carboxylic acids are oxidation products of alcohols and, indeed, are commonly made by this route. The product obtained through oxidation of an alcohol depends on the alcohol's structure, which is classified according to the number of carbon atoms bonded to the C atom bearing the —OH group. *Primary alcohols* have one carbon and two hydrogen atoms attached, whereas *secondary alcohols* have two carbon atoms and one hydrogen atom attached. *Tertiary alcohols* have three carbon atoms attached to the C atom bearing the —OH group.

A *primary alcohol* is oxidized in two steps, first to an aldehyde and then to a carboxylic acid:

$$R-CH_2-OH \xrightarrow[\text{agent}]{\text{oxidizing}} R-\overset{O}{\underset{}{C}}-H \xrightarrow[\text{agent}]{\text{oxidizing}} R-\overset{O}{\underset{}{C}}-OH$$

primary   aldehyde   carboxylic acid
alcohol

For example, the air oxidation of ethanol in wine produces wine vinegar, the most important ingredient of which is acetic acid.

$$H-\overset{H}{\underset{H}{C}}-\overset{H}{\underset{H}{C}}-OH(\ell) \xrightarrow{\text{oxidizing agent}} H-\overset{H}{\underset{H}{C}}-\overset{O}{\underset{}{C}}-OH(\ell)$$

ethanol   acetic acid

Acids have a sour taste. The word "vinegar" (from the French *vin aigre*) means sour wine. A device to test one's breath for alcohol relies on the oxidation of ethanol (Figure 10.10).

Oxidation of a *secondary alcohol* produces a ketone:

$$R-\overset{OH}{\underset{H}{C}}-R' \xrightarrow[\text{agent}]{\text{oxidizing}} R-\overset{O}{\underset{}{C}}-R'$$

secondary alcohol   ketone
(—R and —R′ are organic groups. They may be the same or different.)

Common oxidizing agents used for these reactions are reagents such as $KMnO_4$ and $K_2Cr_2O_7$ (◄ Table 3.3).

© Cengage Learning/Charles D. Winters

FIGURE 10.10 Alcohol tester.
This device for testing a person's breath for the presence of ethanol relies on the oxidation of the alcohol. If present, ethanol is oxidized by potassium dichromate, $K_2Cr_2O_7$, to acetaldehyde, and then to acetic acid. The yellow-orange dichromate ion is reduced to green Cr^{3+}(aq), the color change indicating that ethanol was present.

Finally, tertiary alcohols do *not* react with the usual oxidizing agents.

$$(CH_3)_3COH \xrightarrow{\text{oxidizing agent}} \text{no reaction}$$

Aldehydes and Ketones

Aldehydes and **ketones** can have pleasant odors and are often used in fragrances. Benzaldehyde is responsible for the odor of almonds and cherries; cinnamaldehyde is found in the bark of the cinnamon tree; and the ketone 4-(*p*-hydroxyphenyl)-2-butanone is responsible for the odor of ripe raspberries (a favorite of the authors of this book). Table 10.8 lists several simple aldehydes and ketones.

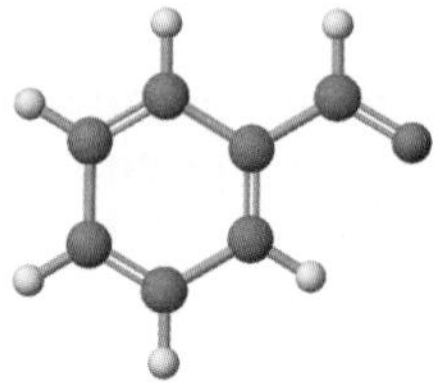

benzaldehyde, C_6H_5CHO 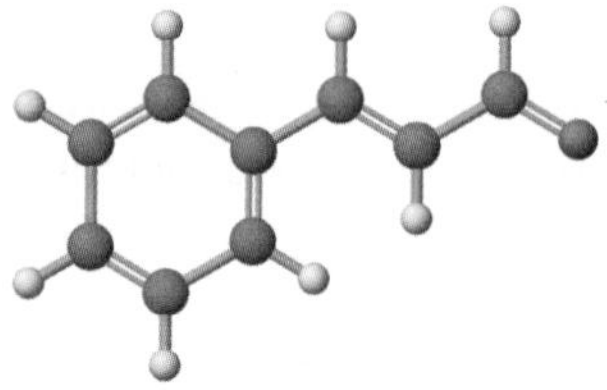   *trans*-cinnamaldehyde, $C_6H_5CH{=}CHCHO$

Aldehydes and ketones are the oxidation products of primary and secondary alcohols, respectively. The reverse reactions—reduction of aldehydes to primary alcohols and reduction of ketones to secondary alcohols—are also known. Commonly used reagents for such reductions are $NaBH_4$ and $LiAlH_4$, although H_2 is used on an industrial scale.

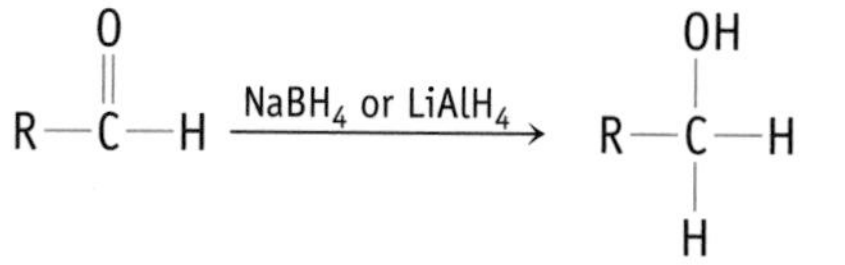

aldehyde → primary alcohol

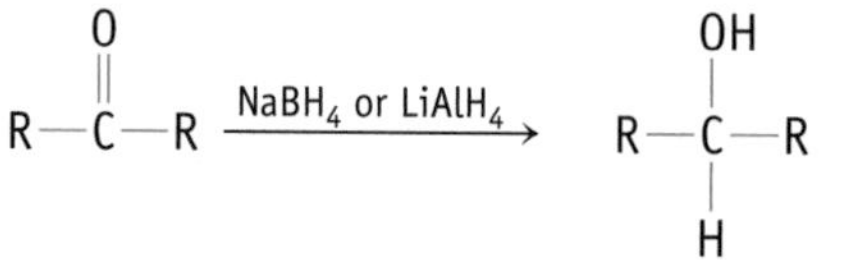

ketone → secondary alcohol

Table 10.8 Simple Aldehydes and Ketones

Structure	Common Name	Systematic Name	BP (°C)
HCH (O)	Formaldehyde	Methanal	−19
CH₃CH (O)	Acetaldehyde	Ethanal	20
CH₃CCH₃ (O)	Acetone	Propanone	56
CH₃CCH₂CH₃ (O)	Methyl ethyl ketone	Butanone	80
CH₃CH₂CCH₂CH₃ (O)	Diethyl ketone	3-Pentanone	102

Aldehydes and odors. The odors of almonds and cinnamon are due to aldehydes, whereas the odor of fresh raspberries comes from a ketone.

© Cengage Learning/Charles D. Winters

Carboxylic Acids

Acetic acid is the most common and most important **carboxylic acid**. For many years, acetic acid was made by oxidizing ethanol produced by fermentation. Now, however, acetic acid is generally made by combining carbon monoxide and methanol in the presence of a catalyst:

$$CH_3OH(\ell) + CO(g) \xrightarrow{\text{catalyst}} CH_3CO_2H(\ell)$$

methanol   acetic acid

About 1 billion kilograms of acetic acid are produced annually in the United States for use in plastics, synthetic fibers, and fungicides.

Many organic acids are found naturally (Table 10.9). Acids are recognizable by their sour taste (Figure 10.11) and are found in common foods: Citric acid in fruits, acetic acid in vinegar, and tartaric acid in grapes are just three examples.

Some carboxylic acids have common names derived from the source of the acid (Table 10.9). Because formic acid is found in ants, its name comes from the Latin word for ant (*formica*). Butyric acid gives rancid butter its unpleasant odor, and the name is related to the Latin word for butter (*butyrum*). The systematic names of acids (Table 10.10) are formed by dropping the "-e" on the name of the corresponding alkane and adding "-oic" (and the word "acid").

A CLOSER LOOK

Glucose and Other Sugars

Glucose, the most common, naturally occurring carbohydrate, has the alcohol and carbonyl functional groups.

As their name implies, formulas of most carbohydrates can be written as though they are a combination of carbon and water, $C_x(H_2O)_y$. Thus, the formula of glucose, $C_6H_{12}O_6$, is equivalent to $C_6(H_2O)_6$. This compound is a sugar, or, more accurately, a **monosaccharide**.

Carbohydrates are polyhydroxy aldehydes or ketones. Glucose is an interesting molecule that exists in three different isomeric forms. Two of the isomers contain six-member rings; the third isomer features a chain structure. In solution, the three forms rapidly interconvert.

Notice that glucose is a chiral molecule. In the chain structure, four of the carbon atoms are bonded to four different groups. In nature, glucose occurs in just one of its enantiomeric forms; thus, a solution of glucose rotates polarized light.

α-D-glucose   open-chain form   β-D-glucose

Knowing glucose's structure allows one to predict some of its properties. With five polar —OH groups in the molecule, glucose is, not surprisingly, soluble in water.

The aldehyde group is susceptible to chemical oxidation to form a carboxylic acid, and detection of glucose (in urine or blood) takes advantage of this fact. Diagnostic tests for glucose involve oxidation with subsequent detection of the products.

Glucose is in a class of sugar molecules called hexoses, monosaccharides having six carbon atoms. 2-Deoxyribose, the sugar in the backbone of the DNA molecule, is a pentose, a molecule with five carbon atoms.

Glucose and other monosaccharides serve as the building blocks for larger carbohydrates. Sucrose, or "table sugar," is a disaccharide and is formed from a molecule of glucose and a molecule of fructose, another monosaccharide. Starch is a polymer composed of many monosaccharide units.

α-D-glucose

fructose

The structure of sucrose. Sucrose is formed from α-D-glucose and fructose. An ether linkage is formed by loss of H_2O from two —OH groups.

deoxyribose, a pentose, part of the DNA backbone

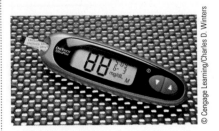

Home test for glucose.

© Cengage Learning/Charles D. Winters

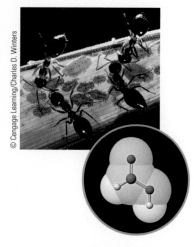

Formic acid, HCO₂H. This acid puts the sting in ant bites.

Table 10.9 Some Naturally Occurring Carboxylic Acids

Name	Structure	Natural Source
Benzoic acid	⬡—CO_2H	Berries
Citric acid	$HO_2C-CH_2-\overset{\overset{OH}{\mid}}{\underset{\underset{CO_2H}{\mid}}{C}}-CH_2-CO_2H$	Citrus fruits
Lactic acid	$H_3C-\overset{}{\underset{\underset{OH}{\mid}}{CH}}-CO_2H$	Sour milk
Malic acid	$HO_2C-CH_2-\overset{}{\underset{\underset{OH}{\mid}}{CH}}-CO_2H$	Apples
Oleic acid	$CH_3(CH_2)_7-CH=CH-(CH_2)_7-CO_2H$	Vegetable oils
Oxalic acid	HO_2C-CO_2H	Rhubarb, spinach, cabbage, tomatoes
Stearic acid	$CH_3(CH_2)_{16}-CO_2H$	Animal fats
Tartaric acid	$HO_2C-\overset{}{\underset{\underset{OH}{\mid}}{CH}}-\overset{}{\underset{\underset{OH}{\mid}}{CH}}-CO_2H$	Grape juice, wine

Because of the substantial electronegativity of oxygen, the two O atoms of the carboxylic acid group are slightly negatively charged, and the H atom of the —OH group is positively charged. This charge distribution has several important implications:

- The polar acetic acid molecule dissolves readily in water, which you already know because vinegar is an aqueous solution of acetic acid. (Acids with larger organic groups are less soluble, however.)
- The hydrogen of the —OH group is the acidic hydrogen. As noted in Chapter 3, acetic acid is a weak acid in water, as are most other organic acids.

Carboxylic acids undergo a number of reactions. Among these is the reduction of the acid (with reagents such as $LiAlH_4$ or $NaBH_4$) first to an aldehyde and then

FIGURE 10.11 Acetic acid in bread. Acetic acid is produced in bread leavened with the yeast *Saccharomyces exigus*. Another group of bacteria, *Lactobacillus sanfrancisco*, contributes to the flavor of sourdough bread. These bacteria metabolize the sugar maltose, excreting acetic acid and lactic acid, $CH_3CH(OH)CO_2H$, thereby giving the bread its unique sour taste.

Table 10.10 Some Simple Carboxylic Acids

Structure	Common Name	Systematic Name	BP (°C)
$H\overset{\overset{O}{\|}}{C}OH$	Formic acid	Methanoic acid	101
$CH_3\overset{\overset{O}{\|}}{C}OH$	Acetic acid	Ethanoic acid	118
$CH_3CH_2\overset{\overset{O}{\|}}{C}OH$	Propionic acid	Propanoic acid	141
$CH_3(CH_2)_2\overset{\overset{O}{\|}}{C}OH$	Butyric acid	Butanoic acid	163
$CH_3(CH_2)_3\overset{\overset{O}{\|}}{C}OH$	Valeric acid	Pentanoic acid	187

to an alcohol. For example, acetic acid is reduced first to acetaldehyde and then to ethanol.

$$CH_3CO_2H \xrightarrow{LiAlH_4} CH_3CHO \xrightarrow{LiAlH_4} CH_3CH_2OH$$

acetic acid   acetaldehyde   ethanol

Yet another important aspect of carboxylic acid chemistry is the reaction with bases to give carboxylate anions. For example, acetic acid reacts with hydroxide ions to give acetate ions and water.

$$CH_3CO_2H(aq) + OH^-(aq) \rightarrow CH_3CO_2^-(aq) + H_2O(\ell)$$

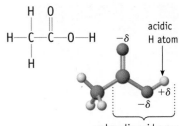

Acetic acid. The H atom of the carboxylic acid group ($-CO_2H$) is the acidic proton of this and other carboxylic acids.

Esters

Carboxylic acids (RCO_2H) react with alcohols ($R'OH$) to form esters (RCO_2R') in an **esterification** reaction. (These reactions are generally carried out in the presence of strong acids because acids speed up the reaction.)

carboxylic acid   alcohol   ester

$$CH_3COH + CH_3CH_2OH \xrightarrow{H_3O^+} CH_3COCH_2CH_3 + H_2O$$

acetic acid   ethanol   ethyl acetate

When a carboxylic acid and an alcohol react to form an ester, the OR' group of the alcohol ends up as part of the ester (as shown above). This fact is known because of isotope labeling experiments. If the reaction is run using an alcohol in which the alcohol oxygen is ^{18}O, all of the ^{18}O ends up in the ester molecule.

Table 10.11 lists a few common esters and the acid and alcohol from which they are formed. The two-part name of an ester is given by (1) the name of the hydrocarbon group from the alcohol and (2) the name of the carboxylate group derived from the acid name by replacing "-ic" with "-ate." For example, ethanol (commonly called *ethyl alcohol*) and acetic acid combine to give the ester ethyl acetate.

An important reaction of esters is their **hydrolysis** (literally, reaction with water), a reaction that is the reverse of the formation of the ester. The reaction, generally done in the presence of a base such as NaOH, produces the alcohol and a sodium salt of the carboxylic acid:

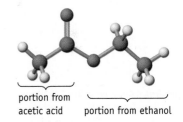

portion from acetic acid   portion from ethanol

ethyl acetate, an ester
$CH_3CO_2CH_2CH_3$

$$RCOR' + NaOH \xrightarrow[\text{in water}]{\text{heat}} RCO^-Na^+ + R'OH$$

ester   carboxylate salt   alcohol

$$CH_3COCH_2CH_3 + NaOH \xrightarrow[\text{in water}]{\text{heat}} CH_3CO^-Na^+ + CH_3CH_2OH$$

ethyl acetate   sodium acetate   ethanol

The carboxylic acid can be recovered if the sodium salt is treated with a strong acid such as HCl:

$$CH_3CO^-Na^+(aq) + HCl(aq) \longrightarrow CH_3COH(aq) + NaCl(aq)$$

sodium acetate   acetic acid

Esters. Many fruits such as bananas and strawberries as well as consumer products (here, perfume and oil of wintergreen) contain esters.

Table 10.11 Some Acids, Alcohols, and Their Esters

Acid	Alcohol	Ester	Odor of Ester
CH_3CO_2H acetic acid	CH_3 | $CH_3CHCH_2CH_2OH$ 3-methyl-1-butanol	O CH_3 || | $CH_3COCH_2CH_2CHCH_3$ 3-methylbutyl acetate	Banana
$CH_3CH_2CH_2CO_2H$ butanoic acid	$CH_3CH_2CH_2CH_2OH$ 1-butanol	O || $CH_3CH_2CH_2COCH_2CH_2CH_2CH_3$ butyl butanoate	Pineapple
$CH_3CH_2CH_2CO_2H$ butanoic acid	⬡—CH_2OH benzyl alcohol	O || $CH_3CH_2CH_2COCH_2$—⬡ benzyl butanoate	Rose

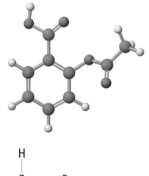

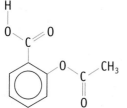

Aspirin, a commonly used analgesic. It is based on benzoic acid with an acetate group, —O_2CCH_3, in the *ortho* position. Aspirin has both carboxylic acid and ester functional groups.

Unlike the acids from which they are derived, esters often have pleasant odors (Table 10.11). Typical examples are methyl salicylate, or "oil of wintergreen," and benzyl acetate. Methyl salicylate is derived from salicylic acid, the parent compound of aspirin.

$$\underset{\substack{\text{salicylic acid}}}{\overset{\substack{O \\ \|\|}}{\text{⬡—COH}}} + \underset{\substack{\text{methanol}}}{CH_3OH} \longrightarrow \underset{\substack{\text{methyl salicylate,} \\ \text{oil of wintergreen}}}{\overset{\substack{O \\ \|\|}}{\text{⬡—COCH}_3}} + H_2O$$

Benzyl acetate, the active component of "oil of jasmine," is formed from benzyl alcohol ($C_6H_5CH_2OH$) and acetic acid. The chemicals are inexpensive, so synthetic jasmine is a common fragrance in less expensive perfumes and toiletries.

$$\underset{\substack{\text{acetic acid}}}{\overset{\substack{O \\ \|\|}}{CH_3COH}} + \underset{\substack{\text{benzyl alcohol}}}{\text{⬡—CH}_2OH} \longrightarrow \underset{\substack{\text{benzyl acetate} \\ \text{oil of jasmine}}}{\overset{\substack{O \\ \|\|}}{CH_3COCH_2\text{—⬡}}} + H_2O$$

Amides

An acid and an alcohol react by loss of water to form an ester. In a similar manner, another class of organic compounds—amides—form when an acid reacts with an amine, again with loss of water.

$$\underset{\substack{\text{carboxylic acid}}}{\overset{\substack{O \\ \|\|}}{R-C-(OH)}} + \underset{\substack{\text{amine}}}{\overset{\substack{R' \\ \|}}{(H)-N-R'}} \longrightarrow \underset{\substack{\text{amide}}}{\overset{\substack{O \quad R' \\ \|\| \quad \|}}{R-C-N-R'}} + H_2O$$

Amides have an organic group and an amino group (—NH_2, —NHR', or —$NR'R$) attached to the carbonyl group.

The C atom involved in the amide bond has three bonded groups and no lone pairs around it. We would predict it should be sp^2 hybridized with trigonal-planar geometry and bond angles of approximately 120°—and this is what is found. However, the structure of the amide group offers a surprise. The N atom is also observed to have trigonal-planar geometry with bonds to three attached atoms at 120°. Because the amide nitrogen is surrounded by four pairs of electrons, we would have predicted the N atom would have sp^3 hybridization and bond angles of about 109°.

Based on the observed geometry of the amide N atom, the atom is assigned sp^2 hybridization. To rationalize the observed angle and to rationalize sp^2 hybridization, we can introduce a second resonance form of the amide.

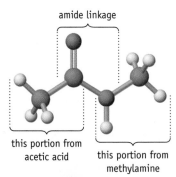

amide linkage

this portion from acetic acid this portion from methylamine

(A) ⟷ (B)

Form B contains a C=N double bond, and the O and N atoms have negative and positive charges, respectively. The N atom can be assigned sp^2 hybridization, and the π bond in B arises from overlap of p orbitals on C and N.

The second resonance structure for an amide link also explains why the carbon–nitrogen bond is relatively short, about 132 pm, a value between that of a C—N single bond (149 pm) and a C=N double bond (127 pm). In addition, restricted rotation occurs around the C=N bond, making it possible for isomeric species to exist if the two groups bonded to N are different.

The amide grouping is particularly important in some synthetic polymers (Section 10.5) and in proteins, where it is referred to as a *peptide* link. The compound *N*-acetyl-*p*-aminophenol, an analgesic known by the generic name acetaminophen, is another amide. Use of this compound as an analgesic was apparently discovered by accident when a common organic compound called acetanilide (like acetaminophen but without the —OH group) was mistakenly put into a prescription for a patient. Acetanilide acts as an analgesic, but it can be toxic. An —OH group *para* to the amide group makes the compound nontoxic, an interesting example of how a seemingly small structural difference affects chemical function.

An amide, *N*-methylacetamide. The *N*-methyl portion of the name derives from the amine portion of the molecule, where the *N* indicates that the methyl group is attached to the nitrogen atom. The "-acet" portion of the name indicates the acid on which the amide is based. The electrostatic potential surface shows the polarity and planarity of the amide linkage.

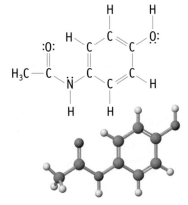

Acetaminophen, *N*-acetyl-*p*-aminophenol. This analgesic is an amide. It is used in over-the-counter painkillers such as Tylenol.

EXAMPLE 10.7 Functional Group Chemistry

Problem

(a) Draw the structure of the product of the reaction between propanoic acid and 1-propanol. What is the systematic name of the reaction product, and what functional group does it contain?

(b) What is the result of reacting 2-butanol with an oxidizing agent? Give the name, and draw the structure of the reaction product.

What Do You Know? From the material covered in this chapter, you should know the names, structures, and common chemical reactions of organic compounds mentioned in this question.

Strategy Determine the products of these reactions, based on the discussion in the text. Propanoic acid is a carboxylic acid (page 368), and 1-propanol and 2-butanol are both alcohols. Consult the discussion regarding their chemistry.

Solution

(a) Carboxylic acids such as propanoic acid react with alcohols to give esters.

$$CH_3CH_2\overset{\overset{\displaystyle O}{\|}}{C}OH + CH_3CH_2CH_2OH \longrightarrow CH_3CH_2\overset{\overset{\displaystyle O}{\|}}{C}OCH_2CH_2CH_3 + H_2O$$

propanoic acid   1-propanol   propyl propanoate, an ester

(b) 2-Butanol is a secondary alcohol. Such alcohols are oxidized to ketones.

$$CH_3\overset{\overset{\displaystyle OH}{|}}{C}HCH_2CH_3 \xrightarrow{\text{oxidizing agent}} CH_3\overset{\overset{\displaystyle O}{\|}}{C}CH_2CH_3$$

2-butanol   butanone, a ketone

Think about Your Answer Students sometimes find themselves overwhelmed by the large amount of information presented in organic chemistry. Your study of this material will be more successful if you carefully organize information based on the type of compound.

Check Your Understanding

(a) Name each of the following compounds and its functional group.

$$(1)\ CH_3CH_2CH_2OH \quad (2)\ CH_3\overset{\overset{\displaystyle O}{\|}}{C}OH \quad (3)\ CH_3CH_2NH_2$$

(b) Name the product from the reaction of compounds 1 and 2 above.

(c) What is the name and structure of the product from the oxidation of 1 with an excess of oxidizing agent?

(d) Give the name and structure of the compound that results from combining 2 and 3.

(e) What is the result of adding an acid (say HCl) to compound 3?

REVIEW & CHECK FOR SECTION 10.4

1. How many aldehydes and ketones are possible that have the formula $C_5H_{10}O$ and that have a five-carbon chain?

   (a) 2 aldehydes and 1 ketone

   (b) 1 aldehyde and 3 ketones

   (c) 1 aldehyde and 1 ketone

   (d) 1 aldehyde and 2 ketones

2. Addition of water to 2-butene gives a single product. Oxidation of this product with $K_2Cr_2O_7$ gives a single compound. What is its name?

   (a) butanal

   (b) 2-butanone

   (c) 2-butanol

   (d) butane

3. What is the bond angle of the O—C—O group of atoms in benzoic acid and what is the hybridization of the carbonyl carbon atom?

   (a) 90°, not hybridized

   (b) 109.5°, sp^3 hybridized

   (c) 180°, sp hybridized

   (d) 120°, sp^2 hybridized

4. A sample of ethanol is divided into two portions. One portion is oxidized with excess oxidizing agent to give an acid. The acid and the remaining alcohol react to give an ester. What is the name of the ester?

   (a) ethyl propanoate

   (b) ethanoic acid

   (c) ethyl ethanoate

   (d) methyl propanoate

10.5 Polymers

We turn now to the very large molecules known as *polymers*. These can be either synthetic materials or naturally occurring substances such as proteins or nucleic acids. Although many different types of polymers are known and they have widely different compositions and structures, their properties are understandable, based on the principles developed for small molecules.

Classifying Polymers

The word *polymer* means "many parts" (from the Greek, *poly* and *meros*). **Polymers** are giant molecules made by chemically joining together many small molecules called **monomers**. Polymer molar masses range from thousands to millions.

Extensive use of synthetic polymers is a fairly recent development. A few synthetic polymers (Bakelite, rayon, and celluloid) were made early in the 20th century, but most of the products with which you are likely to be familiar originated in the last 75 years. By 1976, synthetic polymers outstripped steel as the most widely used materials in the United States. The average production of synthetic polymers in the United States is now 150 kg or more per person annually.

The polymer industry classifies polymers in several different ways. One is their response to heating. **Thermoplastics** (such as polyethylene) soften and flow when they are heated and harden when they are cooled. **Thermosetting plastics** (such as Formica) are initially soft but set to a solid when heated and cannot be resoftened. Another classification scheme depends on the end use of the polymer—for example, plastics, fibers, elastomers, coatings, and adhesives.

A more chemically oriented approach to polymer classification is based on the method of synthesis. **Addition polymers** are made by directly adding monomer units together. **Condensation polymers** are made by combining monomer units and splitting out a small molecule, often water.

Addition Polymers

Polyethylene, polystyrene, and polyvinyl chloride (PVC) are common addition polymers (Figure 10.12). They are built by "adding together" simple alkenes such as ethylene ($CH_2{=}CH_2$), styrene ($C_6H_5CH{=}CH_2$), and vinyl chloride ($CH_2{=}CHCl$). These and other addition polymers (Table 10.12), all derived from alkenes, have widely varying properties and uses.

Polyethylene and Other Polyolefins

Polyethylene is by far the leader in amount of polymer produced. Ethylene (C_2H_4), the monomer from which polyethylene is made, is a product of petroleum refining and one of the top five chemicals produced in the United States. When ethylene is

(a) High-density polyethylene.

(b) Polystyrene.

(c) Polyvinyl chloride.

FIGURE 10.12 Common polymer-based consumer products. Recycling information is provided on most plastics (often molded into the bottom of bottles). High-density polyethylene is designated with a "2" inside a triangular symbol and the letters "HDPE." Polystyrene is designated by "6" with the symbol PS, and polyvinyl chloride, PVC, is designated with a "3" inside a triangular symbol with the symbol "V" or "PVC" below.

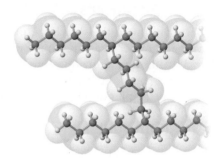

(a) The linear form, high-density polyethylene (HDPE).

(b) Branched chains occur in low-density polyethylene (LDPE).

(c) Cross-linked polyethylene (CLPE).

FIGURE 10.13 Polyethylene.

heated to between 100 and 250 °C at a pressure of 1000 to 3000 atm in the presence of a catalyst, polymers with molar masses up to several million are formed. The reaction can be expressed as the chemical equation:

$$n\ H_2C{=}CH_2 \longrightarrow \left(\begin{array}{cc} H & H \\ | & | \\ C-C \\ | & | \\ H & H \end{array}\right)_n$$

ethylene   polyethylene

The abbreviated formula of the reaction product, $(\!{-}CH_2CH_2{-}\!)_n$, shows that polyethylene is a chain of carbon atoms, each bearing two hydrogens. The chain length for polyethylene can be very long. A polymer with a molar mass of 1 million would contain almost 36,000 ethylene molecules linked together.

Samples of polyethylene formed under various pressures and catalytic conditions have different properties, as a result of different molecular structures. For example, when chromium(III) oxide is used as a catalyst, the product is almost exclusively a linear chain (Figure 10.13a). If ethylene is heated to 230 °C at high pressure, however, irregular branching occurs. Still other conditions lead to cross-linked polyethylene, in which different chains are linked together (Figures 10.13b and c).

The high–molar-mass chains of linear polyethylene pack closely together and result in a material with a density of 0.97 g/cm³. This material, referred to as high-density polyethylene (HDPE), is hard and tough, which makes it suitable for items such as milk bottles. If the polyethylene chain contains branches, however, the chains cannot pack as closely together, and a lower-density material (0.92 g/cm³) known as low-density polyethylene (LDPE) results. This material is softer and more flexible than HDPE. It is used in plastic wrap and sandwich bags, among other things. Linking up the polymer chains in cross-linked polyethylene (CLPE) causes the material to be even more rigid and inflexible. Plastic bottle caps are often made of CLPE.

Polymers formed from substituted ethylenes ($CH_2{=}CHX$) have a range of properties and uses (Table 10.12). Sometimes, the properties are predictable based on the molecule's structure. Polymers without polar substituent groups, such as polystyrene, often dissolve in organic solvents, a property useful for some types of fabrication (Figure 10.14).

Christopher Springmann, Springmann Productions

Polyethylene film. The polymer film is produced by extruding the molten plastic through a ring-like gap and inflating the film like a balloon.

polymers based on substituted ethylenes, $H_2C{=}CHX$

$$\left(\!CH_2CH\!\right)_n \quad \left(\!CH_2CH\!\right)_n \quad \left(\!CH_2CH\!\right)_n$$
$$\qquad\ \ |\qquad\qquad\quad |\qquad\qquad\quad |$$
$$\qquad\ \ OH\qquad\qquad OCCH_3\qquad\qquad \bigcirc$$
$$\qquad\qquad\qquad\qquad\quad \|$$
$$\qquad\qquad\qquad\qquad\quad O$$

polyvinyl alcohol   polyvinyl acetate   polystyrene

Table 10.12 Ethylene Derivatives That Undergo Addition Polymerization

Formula	Monomer Common Name	Polymer Name (Trade Names)	Uses
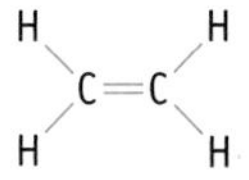 H₂C=CH₂ (Ethylene formula)	Ethylene	Polyethylene (polythene)	Squeeze bottles, bags, films, toys and molded objects, electric insulation
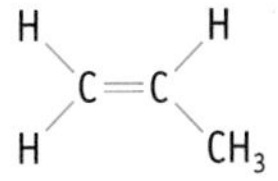 CH₂=CHCH₃ (Propylene formula)	Propylene	Polypropylene (Vectra, Herculon)	Bottles, films, indoor-outdoor carpets
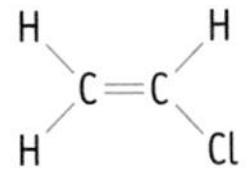 CH₂=CHCl (Vinyl chloride formula)	Vinyl chloride	Polyvinyl chloride (PVC)	Floor tile, raincoats, pipe
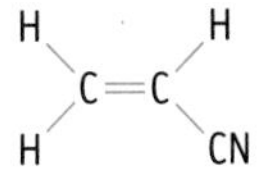 CH₂=CHCN (Acrylonitrile formula)	Acrylonitrile	Polyacrylonitrile (Orlan, Acrilan)	Rugs, fabrics
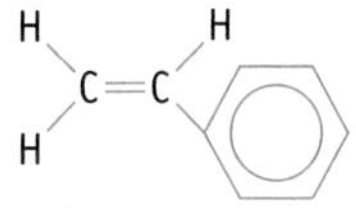 CH₂=CH–C₆H₅ (Styrene formula)	Styrene	Polystyrene (Styrofoam, Styron)	Food and drink coolers, building material insulation
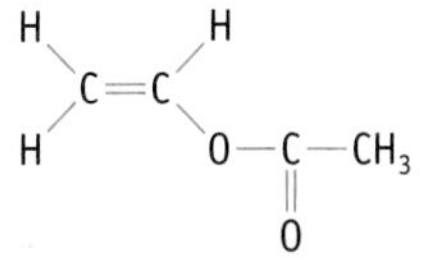 CH₂=CH–O–C(=O)CH₃ (Vinyl acetate formula)	Vinyl acetate	Polyvinyl acetate (PVA)	Latex paint, adhesives, textile coatings
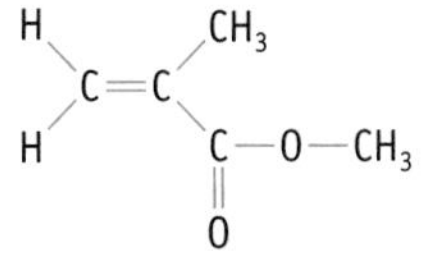 CH₂=C(CH₃)–C(=O)–O–CH₃ (Methyl methacrylate formula)	Methyl methacrylate	Polymethyl methacrylate (Plexiglas, Lucite)	High-quality transparent objects, latex paints, contact lenses
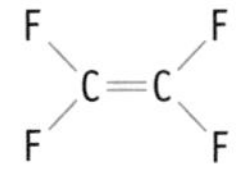 CF₂=CF₂ (Tetrafluoroethylene formula)	Tetrafluoroethylene	Polytetrafluoroethylene (Teflon)	Gaskets, insulation, bearings, pan coatings

Polyvinyl alcohol is a polymer with little affinity for nonpolar solvents but an affinity for water, which is not surprising, based on the large number of polar OH groups (Figure 10.15). Vinyl alcohol itself is not a stable compound (it isomerizes to acetaldehyde CH_3CHO), so polyvinyl alcohol cannot be made from this compound. Instead, it is made by hydrolyzing the ester groups in polyvinyl acetate.

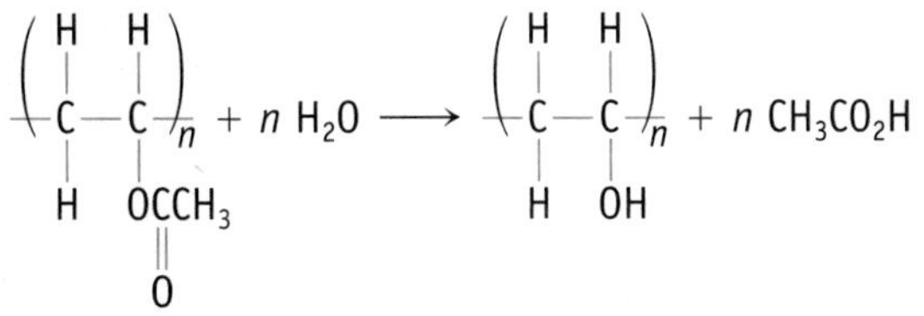

$$\left(\begin{array}{c} H\ H \\ -C-C- \\ H\ OCCH_3 \\ \quad \| \\ \quad O \end{array}\right)_n + n\,H_2O \longrightarrow \left(\begin{array}{c} H\ H \\ -C-C- \\ H\ OH \end{array}\right)_n + n\,CH_3CO_2H$$

Solubility in water or organic solvents can be a liability for polymers. The many uses of polytetrafluoroethylene [Teflon, $-(CF_2CF_2)_n-$] stem from the fact that it does not interact with water or organic solvents.

Polystyrene, with $n = 5700$, is a clear, hard, colorless solid that can be molded easily at 250 °C. You are probably more familiar with the very light, foam-like material known as Styrofoam that is used widely for food and beverage containers and for home insulation (Figure 10.14). Styrofoam is produced by a process called "expansion molding." Polystyrene beads containing 4% to 7% of a low-boiling liquid like pentane are placed in a mold and heated with steam or hot air. Heat causes the

War II. All nylon was diverted to making parachutes and other military gear. It was not until about 1952 that nylon reappeared in the consumer marketplace.

Figure 10.19 illustrates why nylon makes such a good fiber. To have good tensile strength (the ability to resist tearing), the polymer chains should be able to attract one another, albeit not so strongly that the plastic cannot be drawn into fibers. Ordinary covalent bonds between the chains (cross-linking) would be too strong. Instead, cross-linking occurs by a somewhat weaker intermolecular force called *hydrogen bonding* (▶ Section 12.3) between the hydrogens of N—H groups on one chain and the carbonyl oxygens on another chain. The polarities of the $N^{\delta-}$—$H^{\delta+}$ group and the $C^{\delta+}$=$O^{\delta-}$ group lead to attractive forces between the polymer chains of the desired magnitude.

EXAMPLE 10.8 Condensation Polymers

Problem What is the repeating unit of the condensation polymer obtained by combining $HO_2CCH_2CH_2CO_2H$ (succinic acid) and $H_2NCH_2CH_2NH_2$ (1,2-ethylenediamine)?

What Do You Know? Carboxylic acids and amines react to form amides, splitting out water. Here we have a diacid and diamine that will react. The repeating unit will be the shortest sequence that when repeated gives a long polymer chain.

Strategy Recognize that the polymer will link the two monomer units through the amide linkage. The smallest repeating unit of the chain will contain two parts, one from the diacid and the other from the diamine.

Solution The repeating unit of this polyamide is

$$\left(-\underset{\underset{O}{\|}}{C}CH_2CH_2\underset{\underset{O}{\|}}{C} - \underset{\underset{H}{|}}{N}CH_2CH_2\underset{\underset{H}{|}}{N} - \right)_n$$

amide linkage

Think about Your Answer Alternating fragments of the diacid and diamine appear in the polymer chain. The fragments are linked by amide bonds making this a polyamide.

Check Your Understanding

Kevlar is a polymer that is now well known because it is used to make sports equipment and bulletproof vests. This polymer has the formula shown below. Is this a condensation polymer or an addition polymer? What chemicals could be used to make this polymer? Write a balanced equation for the formation of Kevlar.

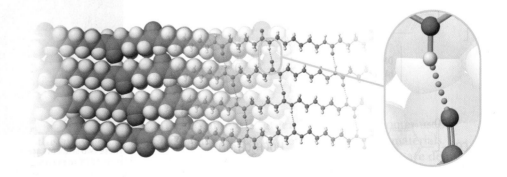

amide group

FIGURE 10.19 Hydrogen bonding between polyamide chains. Carbonyl oxygen atoms with a partial negative charge on one chain interact with an amine hydrogen with a partial positive charge on a neighboring chain. (This form of bonding is described in more detail in Section 12.3.)

CASE STUDY

Green Adhesives

Chemist Kaichang Li was trained in the chemistry of wood and is now doing research at Oregon State University. Oregon has a beautiful and rugged coast, and Li went there in search of mussels to make a special dish. As the waves pounded onshore, he was struck by the fact that the mussels could cling stubbornly to the rocks in spite of the force of the waves and tides. What glue enabled them to do this?

Back in his lab Li found that the strands of glue were largely protein-based. Proteins are simply polymers of amino acids with an

amide link between units (page 371 and Figure A). Li realized that such polymers could have enormous application in the wood industry.

Adhesives, or glues in common terminology, have been known and used for thousands of years. Early glues were based on animal or plant products. Now, however, adhesives are largely synthetic, among them condensation polymers based on the combination of phenol or urea with formaldehyde. These have been used for well over a half-century in the manufacture of plywood and particle board, and your home or dormitory likely contains a significant amount of these building materials. Unfortunately, they have a disadvantage. In their manufacture and use, formaldehyde, a suspected carcinogen, can be released into the air.

Li's work with the mussels eventually led to a new, safer adhesive that could be used in these same wood products. His first problem was how to make a protein-based adhesive in the laboratory. The idea came to him one day at lunch when he was eating tofu, a soy-based food very high in protein. Why not modify soy protein to make a new adhesive? Using mussels as his model, Li did exactly that, and, as he said, "We turned soy proteins into mussel adhesive proteins."

Scientists at Hercules Chemical Company provided expertise to cure (or

Courtesy of Oregon State University

Professor K. Li, a discoverer of "green" adhesives.

harden) the new "green" adhesive, and the Columbia Forest Products Company adopted the environmentally friendly adhesive for use in plywood and particle board.

In 2007 Li and his coworkers, as well as Columbia Forest Products and Hercules, shared a Presidential Green Chemistry Award.

Questions:

1. Draw structures of phenol, urea, and formaldehyde.
2. Describe the bonding in formaldehyde.
3. It has been said that nylon is similar to a protein. Compare and contrast the structures of nylon 6,6 and a protein.

Answers to these questions are available in Appendix N.

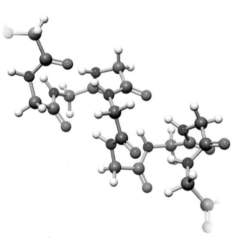

FIGURE A A portion of a protein chain made of repeating glycine molecules ($H_2NCH_2CO_2H$).

REVIEW & CHECK FOR SECTION 10.5

Polyacrylic acid, shown below, is made from which of the following monomers? (The sodium salt of this polymer, sodium polyacrylate, and cellulose are the important ingredients in disposable baby diapers.)

$$\left[-CH_2-\underset{\underset{\underset{OH}{|}}{\underset{\|}{C}}=O}{\overset{\overset{H}{|}}{C}} -CH_2-\underset{\underset{\underset{OH}{|}}{\underset{\|}{C}}=O}{\overset{\overset{H}{|}}{C}} -CH_2-\underset{\underset{\underset{OH}{|}}{\underset{\|}{C}}=O}{\overset{\overset{H}{|}}{C}} - \right]_n$$

(a) $CH_2{=}CH_2$ (b) $CH_2{=}\underset{\underset{CN}{|}}{CH}$ (c) $CH_2{=}\underset{\underset{CO_2H}{|}}{CH}$ (d) $CH_2{=}\underset{\underset{OH}{|}}{CH}$

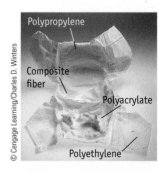

© Cengage Learning/Charles D. Winters

Polypropylene
Composite fiber
Polyacrylate
Polyethylene

Polymers in a disposable baby diaper. At least three polymeric materials are used: sodium polyacrylate, polypropylene, and polyethylene.

and

CHAPTER GOALS REVISITED

Now that you have studied this chapter, you should ask whether you have met the chapter goals. In particular, you should be able to:

Classify organic compounds based on formula and structure

a. Understand the factors that contribute to the large numbers of organic compounds and the wide array of structures (Section 10.1). **Study Question: 103.**

Recognize and draw structures of structural isomers and stereoisomers for carbon compounds

a. Recognize and draw structures of geometric isomers and optical isomers (Section 10.1). **Study Questions: 11, 12, 15, 19, 69.**

Name and draw structures of common organic compounds

a. Draw structural formulas, and name simple hydrocarbons, including alkanes, alkenes, alkynes, and aromatic compounds (Section 10.2). **Study Questions: 1–16, 69, 70, 77, and Go Chemistry Module 15.**

b. Identify possible isomers for a given formula (Section 10.2). **Study Questions: 6, 8, 11, 15, 16, 19–22, 28.**

c. Name and draw structures of alcohols and amines (Section 10.3). **Study Questions: 37–42.**

d. Name and draw structures of carbonyl compounds—aldehydes, ketones, acids, esters, and amides (Section 10.4). **Study Questions: 47–50.**

Know the common reactions of organic functional groups

a. Predict the products of the reactions of alkenes, aromatic compounds, alcohols, amines, aldehydes and ketones, and carboxylic acids. **Study Questions: 23–26, 29, 30, 33–36, 43–46, 51–56, 59, 62, 71–74, 89, 91, 93, 94.**

Relate properties to molecular structure

a. Describe the physical and chemical properties of the various classes of hydrocarbon compounds (Section 10.2). **Study Question: 17.**

b. Recognize the connection between the structures and the properties of alcohols (Section 10.3). **Study Questions: 45, 46, 72.**

c. Know the structures and properties of some natural products, including carbohydrates (Section 10.4). **Study Questions: 57, 58, 83, 84, 87.**

Identify common polymers

a. Write equations for the formation of addition polymers and condensation polymers, and describe their structures (Section 10.5). **Study Questions: 63–66.**

b. Relate properties of polymers to their structures (Section 10.5). **Study Question: 105.**

11 Gases and Their Properties

©Davis Barber/PhotoEdit

The Atmosphere and Altitude Sickness

Some of you may have dreamed of climbing to the summits of the world's tallest mountains, or you may be an avid skier and visit high-mountain ski areas. In either case, "acute mountain sickness" (AMS) is a possibility. AMS is common at higher altitudes and is characterized by a headache, nausea, insomnia, dizziness, lassitude, and fatigue. It can be prevented by a slow ascent, and its symptoms can be relieved by a mild pain reliever.

AMS and more serious forms of high-altitude sickness are generally due to oxygen deprivation, also called *hypoxia*. The oxygen concentration in Earth's atmosphere is 21%. As you go higher into the atmosphere, the concentration remains 21%, but the atmospheric pressure drops. When you reach 3000 m (the altitude of some ski resorts), the barometric pressure is about 70% of that at sea level. At 5000 m, barometric pressure is only 50% of sea level, and on the summit of Mt. Everest, it is only 29% of the sea level pressure. At sea level, your blood is nearly saturated with oxygen, but as the partial pressure of oxygen [$P(O_2)$] drops, the percent saturation drops as well. At $P(O_2)$ of 50 mm Hg, hemoglobin in the red blood cells is about 80% saturated. Other saturation levels are given in the table (for a pH of 7.4).

$P(O_2)$(mm Hg)	Approximate Percent Saturation
90	95%
80	92%
70	90%
60	85%
50	80%
40	72%

For more information on the atmosphere, see page 404.

Questions:

1. Assume a sea level pressure of 1 atm (760 mm Hg). What are the O_2 partial pressures at a 3000-m ski resort and on top of Mt. Everest?
2. What are the approximate blood saturation levels under these conditions?

Answers to these questions are available in Appendix N.

CHAPTER OUTLINE

11.1 Gas Pressure

11.2 Gas Laws: The Experimental Basis

11.3 The Ideal Gas Law

11.4 Gas Laws and Chemical Reactions

11.5 Gas Mixtures and Partial Pressures

11.6 The Kinetic-Molecular Theory of Gases **go Chemistry**

11.7 Diffusion and Effusion

11.8 Nonideal Behavior of Gases

CHAPTER GOALS

See Chapter Goals Revisited (page 413) for Study Questions keyed to these goals.

- Understand the basis of the gas laws and know how to use those laws (Boyle's law, Charles's law, Avogadro's hypothesis, Dalton's law).

- Use the ideal gas law.

- Apply the gas laws to stoichiometric calculations.

- Understand kinetic-molecular theory as it is applied to gases, especially the distribution of molecular speeds (energies).

- Recognize why gases do not behave like ideal gases under some conditions.

Hot air balloons, SCUBA diving tanks, and automobile air bags (Figure 11.1) depend on the properties of gases. Aside from understanding how these work, there are other reasons for studying gases. First, some common elements and compounds (such as oxygen, nitrogen, and methane) exist in the gaseous state under normal conditions of pressure and temperature. Second, atmospheric gases provide one means of transferring energy and material throughout the globe, and they are a source of life-sustaining chemicals. In addition, the question of "global warming" is one of the major debates of our time, and gases such as carbon dioxide and methane play a role in this apparent problem.

But there is yet another reason for studying gases. Of the three states of matter, the behavior of gases is reasonably simple when viewed at the molecular level, and, as a result, it is well understood. It is possible to describe the properties of gases *qualitatively* in terms of the behavior of the molecules that make up the gas. Even more impressive, it is possible to describe the properties of gases *quantitatively* using simple mathematical models. One objective of scientists is to develop precise mathematical and conceptual models of natural phenomena, and a study of gas behavior

OWL

Sign in to OWL at **www.cengage.com/owl** to view tutorials and simulations, develop problem-solving skills, and complete online homework assigned by your professor.

go Chemistry

Download mini lecture videos for key concept review and exam prep from OWL or purchase them from **www.cengagebrain.com**

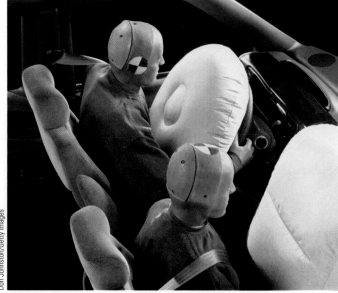

Don Johnston/Getty Images

FIGURE 11.1 Automobile air bags. Most automobiles are now equipped with air bags to protect the driver and passengers in the event of a head-on or side crash. Such bags are inflated with nitrogen gas, which is generated by the explosive decomposition of sodium azide (in the presence of KNO_3 and silica):

$$2 \ NaN_3(s) \rightarrow 2 \ Na(s) + 3 \ N_2(g)$$

The air bag is fully inflated in about 0.050 s. This is important because the typical automobile collision lasts about 0.125 s.

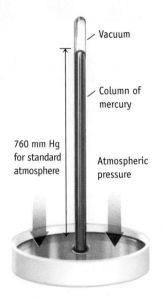

FIGURE 11.2 A barometer. The pressure of the atmosphere on the surface of the mercury in the dish is balanced by the downward pressure exerted by the column of mercury. The barometer was invented in 1643 by Evangelista Torricelli (1608–1647). A unit of pressure called the *torr* in his honor is equivalent to 1 mm Hg.

will introduce you to this approach. A remarkably accurate description of most gases requires knowing only four quantities: the pressure (*P*), volume (*V*), and temperature (*T*, kelvins) of the gas, and its amount (*n*, mol).

11.1 Gas Pressure

Pressure is the force exerted on an object divided by the area over which it is exerted, and a barometer is a device used to measure atmospheric pressure. A barometer can be made by filling a tube with a liquid, often mercury, and inverting the tube in a dish containing the same liquid (Figure 11.2). If the air has been removed completely from the vertical tube, the liquid in the tube assumes a level such that the pressure exerted by the mass of the column of liquid in the tube is balanced by the pressure of the atmosphere pressing down on the surface of the liquid in the dish.

Pressure is often reported in units of **millimeters of mercury (mm Hg)**, the height (in mm) of the mercury column in a mercury barometer above the surface of the mercury in the dish. At sea level, this height is about 760 mm. Pressures are also reported as **standard atmospheres (atm)**, a unit defined as follows:

$$1 \text{ standard atmosphere (1 atm)} = 760 \text{ mm Hg (exactly)}$$

The SI unit of pressure is the **pascal (Pa)**.

$$1 \text{ pascal (Pa)} = 1 \text{ newton/meter}^2$$

(The newton is the SI unit of force.) Because the pascal is a very small unit compared with ordinary pressures, the unit kilopascal (kPa) is more often used. Another unit used for gas pressures is the **bar**, where 1 bar = 100,000 Pa. To summarize, the units used in science for pressure are

$$1 \text{ atm} = 760 \text{ mm Hg (exactly)} = 101.325 \text{ kilopascals (kPa)} = 1.01325 \text{ bar}$$

or

$$1 \text{ bar} = 1 \times 10^5 \text{ Pa (exactly)} = 1 \times 10^2 \text{ kPa} = 0.9872 \text{ atm}$$

EXAMPLE 11.1　Pressure Unit Conversions

Problem Convert a pressure of 635 mm Hg into its corresponding value in units of atmospheres (atm), bars, and kilopascals (kPa).

What Do You Know? You will need the following conversion factors:

$$1 \text{ atm} = 760 \text{ mm Hg} = 1.013 \text{ bar} \qquad 760 \text{ mm Hg} = 101.3 \text{ kPa}$$

Strategy Use the relationships between millimeters of Hg, atmospheres, bars, and pascals described in the text. (The use of dimensional analysis, page 34, is highly recommended.)

Solution

(a) Convert pressure in units of mm Hg to units of atm.

$$635 \text{ mm Hg} \times \frac{1 \text{ atm}}{760 \text{ mm Hg}} = 0.836 \text{ atm}$$

(b) Convert pressure in units of atm to units of bar.

$$0.836 \text{ atm} \times \frac{1.013 \text{ bar}}{1 \text{ atm}} = 0.846 \text{ bar}$$

(c) Convert pressure in units of mm Hg to units of kilopascals.

$$635 \text{ mm Hg} \times \frac{101.3 \text{ kPa}}{760 \text{ mm Hg}} = 84.6 \text{ kPa}$$

A CLOSER LOOK

Measuring Gas Pressure

Pressure is the force exerted on an object divided by the area over which the force is exerted:

$$Pressure = force/area$$

This book, for example, weighs more than 4 lb and has an area of 85 in^2, so it exerts a pressure of about 0.05 lb/in^2 when it lies flat on a surface. (In metric units, the pressure is about 320 Pa.)

Now consider the pressure that the column of mercury exerts on the mercury in the dish in the barometer shown in Figure 11.2. This pressure exactly balances the pressure of the atmosphere. Thus, the pressure of the atmosphere (or of any other gas) can be measured by relating it to the height of the column of mercury (or any other liquid) the gas can support.

Mercury is the liquid of choice for barometers because of its high density. A barometer filled with water would be over 10 m in height. [The water column is about 13.6 times as high as a column of mercury because mercury's density (13.53 g/cm^3) is

13.6 times that of water (density = 0.997 g/cm^3, at 25 °C).]

In the laboratory, we often use a U-tube manometer, which is a mercury-filled, U-shaped glass tube. The closed side of the tube has been evacuated so that no gas remains to exert pressure on the mercury on that side. The other side is open to the gas whose pressure we want to measure. When the gas presses on the mercury in the open side, the gas pressure is read directly (in mm Hg) as the difference in mercury levels on the closed and open sides.

You may have used a tire gauge to check the pressure in your car or bike tires. In the United States, such gauges usually indicate the pressure in pounds per square inch (psi) where 1 atm = 14.7 psi. Some newer gauges give the pressure in kilopascals as well. Be sure to recognize that the reading on the scale refers to the pressure *in excess of atmospheric pressure*. (A flat tire is not a vacuum; it contains air at atmospheric pressure.) For example, if the gauge reads 35 psi (2.4 atm), the pressure in the tire is actually about 50 psi or 3.4 atm.

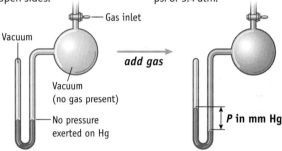

Think about Your Answer The original pressure, 635 mm Hg, is less than 1 atm, so the pressure is also less than 1 bar and less than 100 kPa.

Check Your Understanding

Rank the following pressures in decreasing order of magnitude (from largest to smallest): 75 kPa 250 mm Hg, 0.83 bar, and 0.63 atm.

REVIEW & CHECK FOR SECTION 11.1

1. In a lab experiment you measure a gas pressure of 120 mm Hg. This pressure in units of kPa is approximately

   (a) 12 kPa   (b) 16 kPa   (c) 900 kPa

2. Suppose you construct a barometer filled with an oil ($d = 1.25$ g/cm^3) instead of mercury ($d = 13.5$ g/cm^3) in the vertical tube. If the atmosphere has a pressure of 0.95 atm, what is the approximate height of the oil in the tube?

   (a) 7.2 m   (b) 7.8 m   (c) 8.3 m

11.2 Gas Laws: The Experimental Basis

Boyle's Law: The Compressibility of Gases

When you pump up the tires of your bicycle, the pump squeezes the air into a smaller volume. This shows that a gas is compressed into a smaller volume when the pressure is increased. While studying the compressibility of gases, the Englishman Robert Boyle (1627–1691) (see page 394) observed that the volume of a fixed amount of gas at a given temperature is inversely proportional to the pressure exerted by the gas. All gases behave in this manner, and we now refer to this relationship as **Boyle's law**.

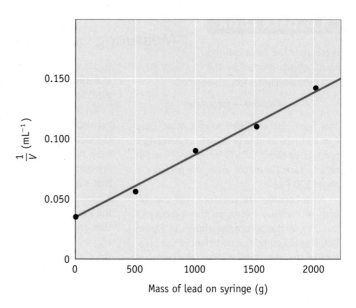

A syringe filled with air was sealed. Pressure was applied by adding lead shot to the beaker on top of the syringe. As the mass of lead increased, the pressure on the air in the sealed syringe increased, and the gas was compressed.

A plot of $1/V$ (volume of air in the syringe) versus P (as measured by the mass of lead) is a straight line.

FIGURE 11.3 An experiment to demonstrate Boyle's law.

Boyle's law can be demonstrated in many ways. In Figure 11.3, a hypodermic syringe is filled with air (n moles) and sealed. When pressure is applied to the movable plunger of the syringe, the air inside is compressed. As the pressure (P) increases on the syringe, the gas volume in the syringe (V) decreases. When $1/V$ of the gas in the syringe is plotted as a function of P (as measured by mass of lead), a straight line results. This type of plot demonstrates that the pressure and volume of the gas are inversely proportional; that is, they change in opposite directions.

Mathematically, we can write Boyle's law as:

$$P \propto \frac{1}{V} \text{ when } n \ (= \text{amount of gas}) \text{ and } T \text{ are constant}$$

where the symbol $\propto$ means "proportional to."

When two quantities are proportional to each other, they can be equated if a *proportionality constant*, here called C_B, is introduced.

$$P = C_B \times \frac{1}{V} \quad \text{or} \quad PV = C_B \text{ at constant } n \text{ and } T$$

This form of Boyle's law expresses the fact that *the product of the pressure and volume of a gas sample is a constant at a given temperature*, where the constant C_B is determined by the amount of gas (in moles) and its temperature (in kelvins). Because the PV product is always equal to C_B [assuming no change in the number of moles of gas (n) and the temperature (T)], it follows that, if PV is known for a gas sample under one set of conditions (P_1 and V_1), then it is known for another set of conditions (P_2 and V_2).

$$P_1V_1 = P_2V_2 \text{ at constant } n \text{ and } T \qquad (11.1)$$

This form of **Boyle's law** is useful when we want to know, for example, what happens to the volume of a given amount of gas when the pressure changes at a constant temperature.

ᴖWL   INTERACTIVE EXAMPLE 11.2   Boyle's Law

Problem A bicycle pump has a volume of 1400 cm³. If a sample of air in the pump has a pressure of 730 mm Hg, what is the pressure when the volume is reduced to 170 cm³?

What Do You Know? This is a Boyle's law problem because you are dealing with changes in pressure and/or volume of a given amount of gas at constant temperature. Here the original pressure and volume of gas (P_1 and V_1) and the new volume (V_2) are known, but the final pressure (P_2) is not known. Both volumes are in the same unit so no unit changes are necessary.

Initial Conditions	**Final Conditions**
$P_1 = 730$ mm Hg	$P_2 = ?$
$V_1 = 1400$ cm³	$V_2 = 170$ cm³

Strategy Use Boyle's law, Equation 11.1.

Solution You can solve this problem by substituting data into a rearranged version of Boyle's law.

$$P_2 = P_1 (V_1/V_2)$$

$$P_2 = (730 \text{ mm Hg}) \times \frac{1400 \text{ cm}^3}{170 \text{ cm}^3} = 6.0 \times 10^3 \text{ mm Hg}$$

Think about Your Answer You know P and V change in opposite directions. In this case, the volume has decreased, so the new pressure (P_2) must be greater than the original pressure (P_1). The answer of about 6000 mm Hg is reasonable because the volume decreased by a factor of more than 8.

Check Your Understanding

A large balloon contains 65.0 L of helium gas at 25 °C and a pressure of 745 mm Hg. The balloon ascends to 3000 m, at which the external pressure has decreased by 30.%. What would be the volume of the balloon, assuming it expands so that the internal and external pressures are equal. (Assume the temperature is still 25 °C.)

Strategy Map 11.2

PROBLEM

Calculate the **pressure** when the *volume of a gas* is **reduced**.

↓

DATA/INFORMATION

- *Initial* pressure
- *Initial* volume
- *Final* volume

Rearrange **Boyle's law** (Equation 11.1) to calculate **final pressure** (P_2).

Pressure (P_2) after reducing the volume

The Effect of Temperature on Gas Volume: Charles's Law

In 1787, the French scientist Jacques Charles (1746–1823) discovered that the volume of a fixed quantity of gas at constant pressure decreases with decreasing temperature. The demonstration in Figure 11.4 shows a dramatic decrease in volume with temperature, but Figure 11.5 illustrates more quantitatively how the volumes of two different gas samples change with temperature (at a constant pressure). When

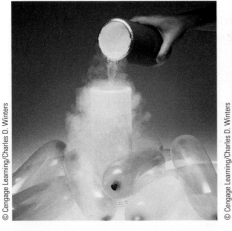

(a) Air-filled balloons are placed in liquid nitrogen (77 K). The volume of the gas in the balloons is dramatically reduced at this temperature.

(b) Here all of the balloons have been placed in the flask of liquid nitrogen.

(c) When the balloons are removed they warm to room temperature and reinflate to their original volume.

FIGURE 11.4 A dramatic illustration of Charles's law.

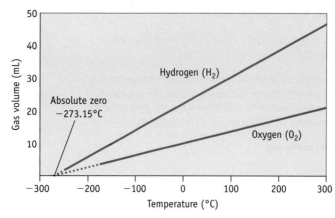

T (°C)	T (K)	Vol. H_2 (mL)	Vol. O_2 (mL)
300	573	47.0	21.1
200	473	38.8	17.5
100	373	30.6	13.8
0	273	22.4	10.1
−100	173	14.2	6.39
−200	73	6.00	—

FIGURE 11.5 Charles's law. The solid lines represent the volumes of samples of hydrogen (0.00100 mol) and oxygen (0.000450 mol) at a pressure of 1.00 atm but at different temperatures. The volumes decrease as the temperature is lowered (at constant pressure). These lines, if extended, intersect the temperature axis at approximately −273 °C.

the plots of volume versus temperature are extrapolated to lower temperatures, they all reach zero volume at the same temperature, −273.15 °C. (Of course, gases will not actually reach zero volume; they liquefy above that temperature.) This temperature is significant, however. William Thomson (1824–1907), also known as Lord Kelvin, proposed a temperature scale—now known as the Kelvin scale—for which the zero point is −273.15 °C (◀ page 22).

When Kelvin temperatures are used with volume measurements, the volume–temperature relationship is

$$V = C_c \times T$$

where C_c is a proportionality constant (which depends on the amount of gas and its pressure). This is **Charles's law**, which states that if a given amount of gas is held at a constant pressure, its volume is directly proportional to the Kelvin temperature.

Writing Charles's law another way, we have $V/T = C_c$; that is, the volume of a gas divided by the temperature of the gas (in kelvins) is constant for a given sample of gas at a specified pressure. Therefore, if we know the volume and temperature of a given quantity of gas (V_1 and T_1), we can find the volume, V_2, at some other temperature, T_2, using the equation

$$\frac{V_1}{T_1} = \frac{V_2}{T_2} \quad \text{at constant } n \text{ and } P \tag{11.2}$$

A calculation using Charles's law is illustrated by the following example. Be sure to notice that the temperature T *must always be expressed in kelvins.*

EXAMPLE 11.3   Charles's Law

Problem A sample of CO_2 in a gas-tight syringe (as in Figure 11.3) has a volume of 25.0 mL at room temperature (20.0 °C). What is the final volume of the gas if you hold the syringe in your hand to raise its temperature to 37 °C?

What Do You Know? This is a Charles's law problem: There is a change in volume of a gas sample with temperature at constant pressure. You know the original volume and temperature, and you want to know the volume at a new temperature. Be sure to recall that, to use Charles's law, the temperature must be expressed in kelvins.

Initial Conditions	Final Conditions
V_1 = 25.0 mL	V_2 = ?
T_1 = 20.0 + 273.2 = 293.2 K	T_2 = 37 + 273 = 310. K

Strategy Use Charles's Law, Equation 11.2.

Solution Rearrange Equation 11.2 to solve for V_2:

$$V_2 = V_1 \times \frac{T_2}{T_1} = 25.0 \text{ mL} \times \frac{310. \text{ K}}{293.2 \text{ K}} = 26.4 \text{ mL}$$

Think about Your Answer While the temperature increase in degrees Celsius seemed large, the change in volume is small. Mathematically, this is because the change depends on the *ratio* of Kelvin temperatures, which is only slightly greater than 1.

Check Your Understanding

A balloon is inflated with helium to a volume of 45 L at room temperature (25 °C). If the balloon is cooled to −10 °C, what is the new volume of the balloon? Assume that the pressure does not change.

Combining Boyle's and Charles's Laws: The General Gas Law

You now know that the volume of a given amount of gas is inversely proportional to its pressure at constant temperature (Boyle's law) and directly proportional to the Kelvin temperature at constant pressure (Charles's law). But what if you need to know what happens to the gas when two of the three parameters (*P*, *V*, and *T*) change? For example, what would happen to the pressure of a sample of nitrogen in an automobile air bag if the same amount of gas were placed in a smaller bag and heated to a higher temperature? You can deal with this situation by combining the two equations that express Boyle's and Charles's laws.

$$\frac{P_1 V_1}{T_1} = \frac{P_2 V_2}{T_2} \quad \text{for a given amount of gas, } n \qquad (11.3)$$

This equation is sometimes called the **general gas law** or **combined gas law**. It applies specifically to situations in which the *amount of gas does not change*.

ⓌWL INTERACTIVE EXAMPLE 11.4 General Gas Law

Problem Helium-filled balloons are used to carry scientific instruments high into the atmosphere. Suppose a balloon is launched when the temperature is 22.5 °C and the barometric pressure is 754 mm Hg. If the balloon's volume is 4.19×10^3 L (and no helium escapes from the balloon), what will the volume be at a height of 20 miles, where the pressure is 76.0 mm Hg and the temperature is −33.0 °C?

What Do You Know? Here you know the initial volume, temperature, and pressure of the gas. You want to know the volume of the same amount of gas at a new pressure and temperature.

Initial Conditions	Final Conditions
$V_1 = 4.19 \times 10^3$ L	$V_2 = ?$ L
$P_1 = 754$ mm Hg	$P_2 = 76.0$ mm Hg
$T_1 = 22.5$ °C (295.7 K)	$T_2 = -33.0$ °C (240.2 K)

Strategy It is most convenient to use Equation 11.3, the general gas law.

Solution You can rearrange the general gas law to calculate the new volume V_2:

$$V_2 = \left(\frac{T_2}{P_2}\right) \times \left(\frac{P_1 V_1}{T_1}\right) = V_1 \times \frac{P_1}{P_2} \times \frac{T_2}{T_1}$$

$$= 4.19 \times 10^3 \text{ L} \times \left(\frac{754 \text{ mm Hg}}{76.0 \text{ mm Hg}}\right) \times \left(\frac{240.2 \text{ K}}{295.7 \text{ K}}\right) = 3.38 \times 10^4 \text{ L}$$

Strategy Map 11.4

PROBLEM

Calculate the **volume** of a gas sample after it undergoes a change in **T** and **P**.

DATA/INFORMATION

• *Initial* pressure, volume, and temperature
• *Final* pressure and temperature

Rearrange the **general gas law** (Equation 11.3) to calculate **final volume** (V_2).

Volume (V_2) after change in *T* and *P*

A weather balloon is filled with helium. As it ascends into the troposphere, does the volume increase or decrease?

Think about Your Answer The pressure decreased by almost a factor of 10, which should lead to about a ten-fold volume increase. This increase is partly offset by a drop in temperature that leads to a volume decrease. On balance, the volume increases because the pressure has dropped so substantially.

Check Your Understanding

You have a 22-L cylinder of helium at a pressure of 150 atm (above atmospheric pressure) and at 31 °C. How many balloons can you fill, each with a volume of 5.0 L, on a day when the atmospheric pressure is 755 mm Hg and the temperature is 22 °C?

The general gas law leads to other, useful predictions of gas behavior. For example, if a given amount of gas is held in a closed container, the pressure of the gas will increase with increasing temperature.

$$\frac{P_1}{T_1} = \frac{P_2}{T_2} \text{ when } V_1 = V_2, \text{ so } P_2 = P_1 \times \frac{T_2}{T_1}$$

That is, when T_2 is greater than T_1, P_2 will be greater than P_1. In fact, this is the reason tire manufacturers recommend checking tire pressures when the tires are cold. After driving for some distance, friction warms a tire and increases the internal pressure. Filling a warm tire to the recommended pressure may lead to an underinflated tire.

Avogadro's Hypothesis

Front and side air bags are now common in automobiles. In the event of an accident, a bag is rapidly inflated with nitrogen gas generated by a chemical reaction. The air bag unit has a sensor that is sensitive to sudden deceleration of the vehicle and will send an electrical signal that will trigger the reaction (Figures 11.1 and 11.6). In many types of air bags, the explosion of sodium azide generates nitrogen gas.

$$2 \text{ NaN}_3(s) \rightarrow 2 \text{ Na}(s) + 3 \text{ N}_2(g)$$

(The sodium produced in the reaction is converted to harmless salts by adding KNO_3 and silica to the mixture.) Driver-side air bags inflate to a volume of about 35–70 L, and passenger air bags inflate to about 60–160 L. The final volume of the bag will depend on the amount of nitrogen gas generated.

The relationship between volume and amount of gas was first noted by Amedeo Avogadro. In 1811, he used work on gases by the chemist (and early experimenter with hot air balloons) Joseph Gay-Lussac (1778–1850) to propose that *equal volumes of gases under the same conditions of temperature and pressure have equal numbers of particles* (either molecules or atoms, depending on the composition of

When a car decelerates in a collision, an electrical contact is made in the sensor unit. The propellant (green solid) detonates, releasing nitrogen gas, and the folded nylon bag explodes out of the plastic housing.

Driver-side air bags inflate with 35–70 L of N_2 gas, whereas passenger air bags hold about 60–160 L.

FIGURE 11.6 Automobile air bags.

The bag deflates within 0.2 second, the gas escaping through holes in the bottom of the bag.

the gas). This idea is now called **Avogadro's hypothesis**. Stated another way, the volume of a gas at a given temperature and pressure is directly proportional to the amount of gas in moles:

$$V \propto n \quad \text{at constant } T \text{ and } P$$

EXAMPLE 11.5   Avogadro's Hypothesis

Problem Ammonia can be made directly from the elements:

$$N_2(g) + 3 H_2(g) \rightarrow 2 NH_3(g)$$

If you begin with 15.0 L of $H_2(g)$, what volume of $N_2(g)$ is required for complete reaction (both gases being at the same T and P)? What is the theoretical yield of NH_3, in liters, under the same conditions?

What Do You Know? From Avogadro's hypothesis, you know that gas volume is proportional to the amount of gas. You know the volume of hydrogen, and, from the balanced equation, you know the stoichiometric factors that relate the known amount of H_2 to the unknown amounts of N_2 and NH_3.

Strategy This is a stoichiometry problem where you can substitute gas volumes for moles. That is, you can calculate the volumes of N_2 required and NH_3 produced (here in liters) by multiplying the volume of H_2 available by a stoichiometric factor obtained from the chemical equation.

Solution

$$V \text{ (N}_2 \text{ required)} = (15.0 \text{ L } H_2 \text{ available}) \left(\frac{1 \text{ L } N_2 \text{ required}}{3 \text{ L } H_2 \text{ available}} \right) = 5.00 \text{ L } N_2 \text{ required}$$

$$V \text{ (NH}_3 \text{ produced)} = (15.0 \text{ L } H_2 \text{ available}) \left(\frac{2 \text{ L } NH_3 \text{ produced}}{3 \text{ L } H_2 \text{ available}} \right) = 10.0 \text{ L } NH_3 \text{ produced}$$

Think about Your Answer The balanced equation informs you that the amount of N_2 required is only 1/3 of the amount of H_2 present, and the amount of NH_3 produced is 2/3 of the amount of H_2.

Check Your Understanding

Methane burns in oxygen to give CO_2 and H_2O, according to the balanced equation

$$CH_4(g) + 2 O_2(g) \rightarrow CO_2(g) + 2 H_2O(g)$$

If 22.4 L of gaseous CH_4 is burned, what volume of O_2 is required for complete combustion? What volumes of CO_2 and H_2O are produced? Assume all gases have the same temperature and pressure.

REVIEW & CHECK FOR SECTION 11.2

1. A sample of a gas has a volume of 222 mL at 695 mm Hg and 0 °C. What is the volume of this same sample of gas if it is measured at 333 mm Hg and 0 °C?

   (a) 894 mL   (b) 463 mL   (c) 657 mL   (d) 359 mL

2. The volume of a gas sample is 235 mL at a temperature of 25 °C. At what temperature would that same gas sample have a volume of 310. mL, if the pressure of the gas sample is held constant?

   (a) −47.0 °C   (b) 69.4 °C   (c) 33.1 °C   (d) 120. °C

3. A gas at 25 °C in a 10.0-L tank has a pressure of 1.00 atm. The gas is transferred to a 20.0-L tank at 0 °C. Which statement is true?

   (a) The pressure of the gas is halved.   (c) The gas pressure is 1.00 atm

   (b) The pressure in the tank is doubled.   (d) None of the above

Studies on Gases—Robert Boyle and Jacques Charles

Robert Boyle (1627–1691) was born in Ireland as the 14th and last child of the first Earl of Cork. In his book *Uncle Tungsten*, Oliver Sacks tells us that "Chemistry as a true science made its first emergence with the work of Robert Boyle in the middle of the seventeenth century. Twenty years [Isaac] Newton's senior, Boyle was born at a time when the practice of alchemy still held sway, and he maintained a variety of alchemical beliefs and practices, side by side with his scientific ones. He believed gold could be created, and that he had succeeded in creating it (Newton, also an alchemist, advised him to keep silent about this)."

Boyle examined crystals, explored color, devised an acid–base indicator from the syrup of violets, and provided the first modern definition of an element. He was also a physiologist and was the first to show that the healthy human body has a constant temperature. Today, Boyle is best known for his studies of gases, which were described in his book *The Sceptical Chymist*, published in 1680.

The French chemist and inventor Jacques Alexandre César Charles began his career as a clerk in the finance ministry, but his real interest was science. He developed several inventions and was best known in his lifetime for inventing the hydrogen balloon. In August 1783, Charles exploited his recent studies on hydrogen gas by inflating a balloon with this gas. Because hydrogen would escape easily from a paper bag, he made a silk bag coated with rubber. Inflating the bag took several days and required nearly 225 kg of sulfuric acid and 450 kg of iron to produce the H_2 gas. The balloon stayed aloft for almost 45 minutes and traveled about 15 miles. When it landed in a village, however, the people were so terrified they tore it to shreds. Several months later, Charles and a passenger flew a new hydrogen-filled balloon some distance across the French countryside and ascended to the then-incredible altitude of 2 miles.

See K. R. Williams, "Robert Boyle: Founder of Modern Chemistry," *Journal of Chemical Education*, Vol. 86, page 148, 2009.

Robert Boyle (1627–1691).

Oesper Collection in the History of Chemistry, University of Cincinnati

Jacques Alexandre César Charles (1746–1823).

Image Courtesy of the Library of Congress

National Air & Space Museum, Smithsonian Institution (SI 2004-29394)

Jacques Charles and Aivé Roberts ascended over Paris on December 1, 1783, in a hydrogen-filled balloon.

11.3 The Ideal Gas Law

Four interrelated quantities can be used to describe a gas: pressure, volume, temperature, and amount (moles). We know from experiments that the following gas laws can be used to describe the relationship of these properties (Section 11.2).

Boyle's Law	Charles's Law	Avogadro's Hypothesis
$V \propto (1/P)$	$V \propto T$	$V \propto n$
(constant T, n)	(constant P, n)	(constant T, P)

If all three laws are combined, the result is

$$V \propto \frac{nT}{P}$$

This can be made into a mathematical equation by introducing a proportionality constant, now labeled **R**. This constant, called the **gas constant**, is a *universal constant*, a number you can use to interrelate the properties of any gas:

$$V = R\left(\frac{nT}{P}\right)$$

or

$$PV = nRT \qquad (11.4)$$

densities
fore, gase
into the
(methan

Calcula

When a
done is t
method
erted by

WL

Problem
made to
mula is (
to deter
70.5 mm
its mole

What Do
tempera

$m =$

$V =$

Strategy
to the m
late the a
amount.

Solution

Use this
$(d = PM)$

With this
mole of g

E

Therefore

In the alt

You now

Think ab
formula o

The equation $PV = nRT$ is called the **ideal gas law**. It describes the behavior of a so-called ideal gas. As you will learn in Section 11.8, however, there is no such thing as an "ideal" gas. Nonetheless, real gases at pressures around one atmosphere or less and temperatures around room temperature usually behave close enough to the ideal that $PV = nRT$ adequately describes their behavior.

To use the equation $PV = nRT$, we need a value for R. This is readily determined experimentally. By carefully measuring P, V, n, and T for a sample of gas, we can calculate the value of R from these values using the ideal gas law equation. For example, under conditions of **standard temperature and pressure (STP)** (a gas temperature of 0 °C or 273.15 K and a pressure of 1 atm), 1 mol of an ideal gas occupies 22.414 L, a quantity called the **standard molar volume**. Substituting these values into the ideal gas law gives a value for R:

$$R = \frac{PV}{nT} = \frac{(1.0000 \text{ atm})(22.414 \text{ L})}{(1.0000 \text{ mol})(273.15 \text{ K})} = 0.082057 \frac{\text{L} \cdot \text{atm}}{\text{K} \cdot \text{mol}}$$

With a value for R, we can now use the ideal gas law in calculations.

WL INTERACTIVE EXMPLE 11.6   Ideal Gas Law

Problem The nitrogen gas in an automobile air bag, with a volume of 65 L, exerts a pressure of 829 mm Hg at 25 °C. What amount of N_2 gas (in moles) is in the air bag?

What Do You Know? You are given P, V, and T for a gas sample and want to calculate the amount of gas (n). $P = 829$ mm Hg, $V = 65$ L, $T = 25$ °C, $n = ?$ You also know $R = 0.082057$ L · atm/K · mol.

Strategy Use the ideal gas law, Equation 11.4.

Solution To use the ideal gas law with R having units of L · atm/K · mol, the pressure must be expressed in atmospheres and the temperature in kelvins. Therefore, you should first convert the pressure and temperature to values with these units.

$$P = 829 \text{ mm Hg} \left(\frac{1 \text{ atm}}{760 \text{ mm Hg}} \right) = 1.09 \text{ atm}$$

$$T = 25 + 273 = 298 \text{ K}$$

Now substitute the values of P, V, T, and R into the ideal gas law, and solve for the amount of gas, n:

$$n = \frac{PV}{RT} = \frac{(1.09 \text{ atm})(65 \text{ L})}{(0.082057 \text{ L} \cdot \text{atm/K} \cdot \text{mol})(298 \text{ K})} = 2.9 \text{ mol}$$

Notice that units of atmospheres, liters, and kelvins cancel to leave the answer in units of moles.

Think about Your Answer You know that 1 mol of an ideal gas at STP occupies about 22.4 L, so it is reasonable to guess that 65 L of gas (under only slightly different conditions) would be about 3 mol.

Check Your Understanding

The balloon used by Jacques Charles in his historic balloon flight in 1783 (see page 394) was filled with about 1300 mol of H_2. If the temperature of the gas was 23 °C and the gas pressure was 750 mm Hg, what was the volume of the balloon?

The Density of Gases

The density of a gas at a given temperature and pressure is a useful quantity (Figure 11.7). Because the amount (n, mol) of any compound is given by its mass (m) divided by its molar mass (M), we can substitute m/M for n in the ideal gas equation.

$$PV = \left(\frac{m}{M} \right) RT$$

• **Properties of an Ideal Gas** For ideal gases, it is assumed that there are no forces of attraction between molecules and that the molecules themselves occupy no volume.

• **STP—What Is It?** A gas is at STP, or standard temperature and pressure, when its temperature is 0 °C or 273.15 K and its pressure is 1 atm. Under these conditions, exactly 1 mol of an ideal gas occupies 22.414 L.

Strategy Map 11.6

PROBLEM
Calculate the amount of a gas using the **ideal gas law.**

↓

DATA/INFORMATION
- **Pressure** of the gas
- **Volume** of the gas
- **Temperature** of the gas

↓

Rearrange the **ideal gas law** (Equation 11.4) to calculate the **amount** (n).

↓

Amount of gas, n

FIGURE 11.
(a) The balloon
equal amounts
temperature an
yellow balloon
low-density gas
STP). The other
a higher density
(b) A hot-air ba
the heated air i
lower density th
air.

© Cengage Learning/Charles D. Winters

FIGURE 11.8
Because carbon d
extinguishers is de
settles on top of a
it. (When CO_2 gas
the tank, it expand
cantly. The white c
moisture from the

Check Your Understanding

A 0.105-g sample of a gaseous compound has a pressure of 561 mm Hg in a volume of 125 mL at 23.0 °C. What is its molar mass?

REVIEW & CHECK FOR SECTION 11.3

1. Which gas has the greatest density at 25 °C and 1.00 atm pressure?

    (a) O_2 (b) N_2 (c) H_2 (d) CO_2 (e) Xe

2. Gas A is in a 2-L flask with a pressure of 1 atm at 298 K. Gas B is in a 4-L flask with a pressure of 0.5 atm at 273 K. Which gas is present in greater amount, A or B?

    (a) A   (b) B

3. Measured at a pressure of 1.0 atm and 298 K, gas C has a density of 2.0 g/L. The density of gas D, measured at a pressure of 0.5 atm and 298 K, is 1.3 g/L. Which gas, C or D, has the larger molar mass?

    (a) C   (b) D

11.4 Gas Laws and Chemical Reactions

Many industrially important reactions involve gases. Two examples are the combination of nitrogen and hydrogen to produce ammonia,

$$N_2(g) + 3\ H_2(g) \rightarrow 2\ NH_3(g)$$

and the electrolysis of aqueous NaCl to produce hydrogen and chlorine,

$$2\ NaCl(aq) + 2\ H_2O(\ell) \rightarrow 2\ NaOH(aq) + H_2(g) + Cl_2(g)$$

If we want to understand the quantitative aspects of such reactions, we need to carry out stoichiometry calculations. The scheme in Figure 11.9 connects these calculations for gas reactions with the stoichiometry calculations in Chapter 4.

OWL   INTERACTIVE EXAMPLE 11.9   Gas Laws and Stoichiometry

Problem You are asked to design an air bag for a car. You know that the bag should be filled with gas with a pressure higher than atmospheric pressure, say 829 mm Hg, at a temperature of 22.0 °C. The bag has a volume of 45.5 L. What quantity of sodium azide, NaN_3, should be used to generate the required quantity of gas? The gas-producing reaction is

$$2\ NaN_3(s) \rightarrow 2\ Na(s) + 3\ N_2(g)$$

What Do You Know? You know the pressure, volume, and temperature of the N_2 gas to be produced, and you know the balanced equation that connects the reactant, NaN_3, to the product, N_2.

$$P = 829 \text{ mm Hg (1 atm/760 mm Hg)} = 1.09 \text{ atm}$$

$$V = 45.5 \text{ L}$$

$$T = 22.0 \text{ °C, or } 295.2 \text{ K}$$

You want to know the mass of NaN_3 required to produce a given amount of N_2. This will require the molar mass of NaN_3.

Strategy The general logic to be used here follows a pathway in Figure 11.9 and in the Strategy Map in the margin. First, use $PV = nRT$ with gas data to calculate the amount of N_2 required. Then use that amount with a stoichiometric factor to calculate the amount of NaN_3 required. Finally, use the molar mass of NaN_3 to calculate the mass of NaN_3 required.

Strategy Map 11.9

PROBLEM

Calculate the **mass of reactant** needed to produce a gas at a known **P, V**, and **T**.

↓

DATA/INFORMATION

- **Pressure** of the gas
- **Volume** of the gas
- **Temperature** of the gas
- Balanced equation

↓

STEP 1. Calculate the *amount of gas (n)* using the **ideal gas law.**

↓

Amount of gas, *n*

↓

STEP 2. Use **stoichiometric factor** to relate *amount of gas (n)* to *amount of reactant* required.

↓

Amount of reactant

↓

STEP 3. Calculate **mass of reactant** from its amount.

↓

Mass of reactant

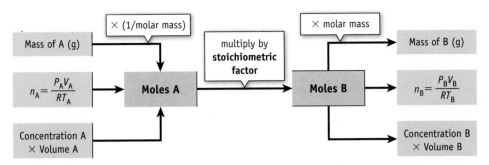

FIGURE 11.9 A scheme for stoichiometry calculations. Here, A and B may be either reactants or products. The amount of A (mol) can be calculated from its mass in grams and its molar mass, from the concentration and volume of a solution, or from P, V, and T data by using the ideal gas law. Once the amount of B is determined, this value can be converted to a mass or solution concentration or volume, or to a property of a gas using the ideal gas law.

Solution

Step 1: Find the amount (mol) of gas required.

$$n = N_2 \text{ required (mol)} = \frac{PV}{RT}$$

$$n = \frac{(1.09 \text{ atm})(45.5 \text{ L})}{(0.082057 \text{ L} \cdot \text{atm/K} \cdot \text{mol})(295.2 \text{ K})} = 2.05 \text{ mol N}_2$$

Step 2: Calculate the quantity of sodium azide that will produce 2.05 mol of N$_2$ gas.

$$\text{Mass of NaN}_3 = 2.05 \text{ mol N}_2 \left(\frac{2 \text{ mol NaN}_3}{3 \text{ mol N}_2} \right) \left(\frac{65.01 \text{ g}}{1 \text{ mol NaN}_3} \right) = \boxed{88.8 \text{ g NaN}_3}$$

Think about Your Answer You know that at STP, 1 mol of an ideal gas has a volume of 22.4 L. While not at STP exactly, the conditions are close to it, so a volume of 45.5 L should correspond to about 2 mol of N$_2$. Based on the stoichiometry of the reaction, you will need 2/3 of this amount of NaN$_3$, or about 1.4 mol. Based on the molar mass of NaN$_3$ (65 g/mol) this will be about 90 g. The amount of N$_2$ produced is substantial, so it is reasonable that around 90 g of sodium azide is required.

Check Your Understanding

If you need 25.0 L of N$_2$ gas, at $P = 1.10$ atm and 25.0 °C, what mass of Na will be produced?

EXAMPLE 11.10 Gas Laws and Stoichiometry

Problem You wish to prepare some deuterium gas, D$_2$, for use in an experiment. One way to do this is to react heavy water, D$_2$O, with an active metal such as lithium.

$$2 \text{ Li(s)} + 2 \text{ D}_2\text{O}(\ell) \rightarrow 2 \text{ LiOD(aq)} + \text{D}_2(\text{g})$$

What amount of D$_2$ (in moles) can be prepared from 0.125 g of Li metal combined with 15.0 mL of D$_2$O ($d = 1.11$ g/mL). If dry D$_2$ gas is captured in a 1450-mL flask at 22.0 °C, what is the pressure of the gas in mm Hg? (Deuterium, D, has an atomic weight of 2.0147 g/mol.)

What Do You Know? You know the quantities of each of the two reactants, which you can use to calculate the amount of D$_2$ gas that can be produced. Based on the amount (moles) of D$_2$ gas, its temperature, and the volume of the flask, you can calculate the pressure.

Strategy You are combining two reactants with no guarantee they are in the correct stoichiometric ratio. Therefore, you should approach this as a limiting reactant problem. Once the limiting reactant is known, the amount of D$_2$ produced can be calculated. From this value, and the values of V and T, the pressure is calculated using the ideal gas law, $PV = nRT$.

© Cengage Learning/Charles D. Winters

Lithium metal (in the spoon) reacts with drops of water, H$_2$O, to produce LiOH and hydrogen gas, H$_2$. If heavy water, D$_2$O, is used, deuterium gas, D$_2$, can be produced.

Solution

Step 1. Calculate the amounts (mol) of Li and of D_2O:

$$0.125 \text{ g Li} \left(\frac{1 \text{ mol Li}}{6.941 \text{ g Li}} \right) = 0.0180 \text{ mol Li}$$

$$15.0 \text{ mL } D_2O \left(\frac{1.11 \text{ g } D_2O}{1 \text{ mL } D_2O} \right) \left(\frac{1 \text{ mol } D_2O}{20.03 \text{ g } D_2O} \right) = 0.831 \text{ mol } D_2O$$

Step 2. Decide which reactant is the limiting reactant:

$$\text{Ratio of moles of reactants available} = \frac{0.831 \text{ mol } D_2O}{0.0180 \text{ mol Li}} = \frac{46.2 \text{ mol } D_2O}{1 \text{ mol Li}}$$

The balanced equation shows that the ratio should be 1 mol of D_2O to 1 mol of Li. From the calculated values, we see that D_2O is in large excess, and so Li is the limiting reactant. Therefore, further calculations are based on the amount of Li available.

Step 3. Use the limiting reactant to calculate the amount of D_2 produced:

$$0.0180 \text{ mol Li} \left(\frac{1 \text{ mol } D_2 \text{ produced}}{2 \text{ mol Li}} \right) = 0.00900 \text{ mol } D_2 \text{ produced}$$

Step 4. Calculate the pressure of D_2:

$P = ?$ $T = 22.0 \text{ °C, or } 295.2 \text{ K}$

$V = 1450 \text{ mL, or } 1.45 \text{ L}$ $n = 0.00900 \text{ mol } D_2$

$$P = \frac{nRT}{V} = \frac{(0.00900 \text{ mol})(0.082057 \text{ L} \cdot \text{atm/K} \cdot \text{mol})(295.2 \text{ K})}{1.45 \text{ L}} = \boxed{0.150 \text{ atm}}$$

Think about Your Answer Be sure to recognize that this is a limiting reactant problem. The two reactants were not present in the correct stoichiometric ratio.

Check Your Understanding

Gaseous ammonia is synthesized by the reaction

$$N_2(g) + 3 H_2(g) \rightarrow 2 NH_3(g)$$

Assume that 355 L of H_2 gas at 25.0 °C and 542 mm Hg is combined with excess N_2 gas. What amount of NH_3 gas, in moles, can be produced? If this amount of NH_3 gas is stored in a 125-L tank at 25.0 °C, what is the pressure of the gas?

REVIEW & CHECK FOR SECTION 11.4

Diborane reacts with O_2 to give boric oxide and water vapor.

$$B_2H_6(g) + 3 O_2(g) \rightarrow B_2O_3(s) + 3 H_2O(g)$$

1.   What mass of O_2 gas is required to react completely with 1.5 L of B_2H_6 at 25 °C and 0.15 atm?

   (a) 0.29 g   (b) 0.59 g   (c) 0.44 g   (d) 0.88 g

2.   If you mix 1.5 L of B_2H_6 with 4.0 L of O_2, each at 0 °C and a pressure of 1.0 atm, what amount of water is produced?

   (a) 0.18 mol   (b) 0.067 ml   (c) 0.060 mol   (d) 0.25 mol

11.5 Gas Mixtures and Partial Pressures

The air you breathe is a mixture of nitrogen, oxygen, argon, carbon dioxide, water vapor, and small amounts of other gases (Table 11.1). Each of these gases exerts its own pressure, and atmospheric pressure is the sum of the pressures

Table 11.1 Major Components of Atmospheric Dry Air

Constituent	Molar Mass*	Mole Percent	Partial Pressure at STP (atm)
N_2	28.01	78.08	0.7808
O_2	32.00	20.95	0.2095
Ar	39.95	0.934	0.00934
CO_2	44.01	0.0385	0.00039

*The average molar mass of dry air = 28.960 g/mol.

exerted by each gas. The pressure of each gas in the mixture is called its **partial pressure**.

John Dalton (1766–1844) was the first to observe that the pressure of a mixture of ideal gases is the sum of the partial pressures of the different gases in the mixture. This observation is now known as **Dalton's law of partial pressures** (Figure 11.10). Mathematically, we can write Dalton's law of partial pressures as

$$P_{total} = P_1 + P_2 + P_3 \ldots \tag{11.6}$$

where P_1, P_2, and P_3 are the pressures of the different gases in a mixture, and P_{total} is the total pressure.

In a mixture of gases, each gas behaves independently of all others in the mixture. Therefore, we can consider the behavior of each gas in a mixture separately. As an example, take a mixture of three ideal gases, labeled A, B, and C. There are n_A moles of A, n_B moles of B, and n_C moles of C. Assume that the mixture ($n_{total} = n_A + n_B + n_C$) is contained in a given volume (V) at a given temperature (T). We can calculate the pressure exerted by each gas from the ideal gas law equation:

$$P_A V = n_A RT \qquad P_B V = n_B RT \qquad P_C V = n_C RT$$

where each gas (A, B, and C) is in the same volume V and is at the same temperature T. According to Dalton's law, the total pressure exerted by the mixture is the sum of the pressures exerted by each component:

$$P_{total} = P_A + P_B + P_C = n_A\left(\frac{RT}{V}\right) + n_B\left(\frac{RT}{V}\right) + n_C\left(\frac{RT}{V}\right)$$

$$P_{total} = (n_A + n_B + n_C)\left(\frac{RT}{V}\right)$$

$$P_{total} = (n_{total})\left(\frac{RT}{V}\right) \tag{11.7}$$

FIGURE 11.10 Dalton's law.

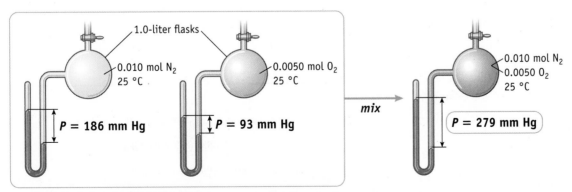

In a 1.0-L flask at 25 °C, 0.010 mol of N_2 exerts a pressure of 186 mm Hg, and 0.0050 mol of O_2 in a 1.0-L flask at 25 °C exerts a pressure of 93 mm Hg.

The N_2 and O_2 samples are both placed in a 1.0-L flask at 25 °C. The total pressure, 279 mm Hg, is the sum of the pressures that each gas alone exerts in the flask.

For mixtures of gases, it is convenient to introduce a quantity called the **mole fraction**, **X**, which is defined as the number of moles of a particular substance in a mixture divided by the total number of moles of all substances present. Mathematically, the mole fraction of a substance A in a mixture with B and C is expressed as

$$X_A = \frac{n_A}{n_A + n_B + n_C} = \frac{n_A}{n_{total}}$$

Now we can combine this equation (written as $n_{total} = n_A / X_A$) with the equations for P_A and P_{total}, and derive the equation

$$P_A = X_A P_{total} \tag{11.8}$$

This equation tells us that *the pressure of a gas in a mixture of gases is the product of its mole fraction and the total pressure of the mixture.* In other words, the partial pressure of a gas is directly related to the fraction of particles of that gas in the mixture. For example, the mole fraction of N_2 in air is 0.78, so, at STP, its partial pressure is 0.78 atm or 590 mm Hg.

Strategy Map 11.11

PROBLEM

Calculate the **partial pressure** of two gases in a mixture.

↓

DATA/INFORMATION

- **Mass** of each gas
- **Total pressure**

STEP 1. Calculate *amount* (*n*) of each gas from its **mass.**

Amount of each gas, *n*

STEP 2. Calculate **mole fraction** of each gas, **X.**

Mole fraction of each gas, **X**

STEP 3. Calculate **partial pressure** of gases using Equation 11.8.

Partial pressure of each gas

⊘WL INTERACTIVE EXAMPLE 11.11 Partial Pressures of Gases

Problem Halothane, $C_2HBrClF_3$, is a nonflammable, nonexplosive, and nonirritating gas that is commonly used as an inhalation anesthetic. The total pressure of a mixture of 15.0 g of halothane vapor and 23.5 g of oxygen gas is 855 mm Hg. What is the partial pressure of each gas?

What Do You Know? You know the identity and mass of each gas, so you can calculate the amount of each. You also know the total pressure of the gas mixture.

Strategy The partial pressure of a gas is given by the total pressure of the mixture multiplied by the mole fraction of the gas (Equation 11.8). Because you can calculate the amount of each gas, you can determine the total amount (moles) of gas and thus the mole fraction of each.

Solution

Step 1. Calculate mole fractions.

$$\text{Amount of } C_2HBrClF_3 = 15.0 \text{ g}\left(\frac{1 \text{ mol}}{197.4 \text{ g}}\right) = 0.0760 \text{ mol}$$

$$\text{Amount of } O_2 = 23.5 \text{ g}\left(\frac{1 \text{ mol}}{32.00 \text{ g}}\right) = 0.734 \text{ mol}$$

$$\text{Total amount of gas} = 0.0760 \text{ mol } C_2HBrClF_3 + 0.734 \text{ mol } O_2 = 0.810 \text{ mol}$$

$$\text{Mole fraction of } C_2HBrClF_3 = \frac{0.0760 \text{ mol } C_2HBrClF_3}{0.810 \text{ total moles}} = 0.0938$$

Because the sum of the mole fraction of halothane and of O_2 must equal 1.0000, this means that the mole fraction of oxygen is 0.906.

$$X_{halothane} + X_{oxygen} = 1.0000$$

$$0.0938 + X_{oxygen} = 1.0000$$

$$X_{oxygen} = 0.906$$

Step 2. Calculate partial pressures.

$$\text{Partial pressure of halothane} = P_{halothane} = X_{halothane} \cdot P_{total}$$

$$P_{halothane} = 0.0938 \cdot P_{total} = 0.0938 \,(855 \text{ mm Hg})$$

$$P_{halothane} = 80.2 \text{ mm Hg}$$

The total pressure of the mixture is the sum of the partial pressures of the gases in the mixture.

$$P_{halothane} + P_{oxygen} = 855 \text{ mm Hg}$$

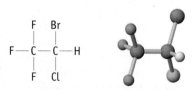

1,1,1-trifluorobromochloroethane, halothane

and so

$$P_{oxygen} = 855 \text{ mm Hg} - P_{halothane}$$

$$P_{oxygen} = 855 \text{ mm Hg} - 80.2 \text{ mm Hg} = \boxed{775 \text{ mm Hg}}$$

Think about Your Answer The amount of halothane is about 1/10th of the amount of oxygen so we would expect the ratio of partial pressures of the two gases to be in a similar ratio.

Check Your Understanding

The halothane–oxygen mixture described in this Example is placed in a 5.00-L tank at 25.0 °C. What is the total pressure (in mm Hg) of the gas mixture in the tank? What are the partial pressures (in mm Hg) of the gases?

REVIEW & CHECK FOR SECTION 11.5

Acetylene reacts with O_2 to give carbon dioxide and water vapor.

$$2 \text{ C}_2\text{H}_2(g) + 5 \text{ O}_2(g) \rightarrow 4 \text{ CO}_2(g) + 2 \text{ H}_2\text{O}(g)$$

If you mix C_2H_2 and O_2 in the correct stoichiometric ratio, and if the total pressure of the mixture is 140. mm Hg, what are the partial pressures of the gases before the reaction occurs?

(a)   100. mm Hg C_2H_2 and 40. mm Hg O_2

(b)   140. mm Hg C_2H_2 and 700. mm Hg O_2

(c)   40. mm Hg C_2H_2 and 100. mm Hg O_2

(d)   140. mm Hg C_2H_2 and 56.0 mm Hg O_2

11.6 The Kinetic-Molecular Theory of Gases

So far, we have discussed the macroscopic properties of gases, properties such as pressure and volume that result from the behavior of a system with a large number of particles. Now we turn to the kinetic-molecular theory (◄ page 7) for a description of the behavior of matter at the molecular or atomic level. Hundreds of experimental observations have led to the following postulates regarding the behavior of gases.

- Gases consist of particles (molecules or atoms) whose separation is much greater than the size of the particles themselves (Figure 11.11).
- The particles of a gas are in continual, random, and rapid motion. As they move, they collide with one another and with the walls of their container, but they do so in a way in which the total energy is unchanged.
- The average kinetic energy of gas particles is proportional to the gas temperature. *All gases, regardless of their molecular mass, have the same average kinetic energy at the same temperature.*

Let us discuss the behavior of gases from this point of view.

Molecular Speed and Kinetic Energy

If your friend walks into your room carrying a pizza, how do you know it? In scientific terms, we know that the odor-causing molecules of food enter the gas phase and drift through space until they reach the cells of your body that react to odors. The same thing happens in the laboratory when bottles of aqueous ammonia (NH_3) and hydrochloric acid (HCl) sit side by side (Figure 11.12). Molecules of the two compounds enter the gas phase and drift along until they encounter one another, at which time they react and form a cloud of tiny particles of solid ammonium chloride (NH_4Cl).

If you change the temperature of the environment of the containers in Figure 11.12 and measure the time needed for the cloud of ammonium chloride to form, you would find the time would be longer at lower temperatures. The rea-

FIGURE 11.11 A molecular view of gases and liquids. The fact that a large volume of N_2 gas can be condensed to a small volume of liquid indicates that the distance between molecules in the gas phase is very large as compared with the distances between molecules in liquids.

go Chemistry

Module 16: Gas Laws and the Kinetic Molecular Theory covers concepts in this section.

FIGURE 11.12 The movement of gas molecules. Open dishes of aqueous ammonia and hydrochloric acid are placed side by side. When molecules of NH_3 and HCl escape from solution to the atmosphere and encounter one another, a cloud of solid ammonium chloride, NH_4Cl, is observed.

The Earth's Atmosphere

Earth's atmosphere is a fascinating mixture of gases in more or less distinct layers with widely differing temperatures.

Up through the troposphere, there is a gradual decline in temperature (and pressure) with altitude. The temperature climbs again in the stratosphere due to the absorption of energy from the sun by stratospheric ozone, O_3.

Above the stratosphere, the pressure declines because there are fewer molecules present. At still higher altitudes, we observe a dramatic increase in temperature in the thermosphere. This is an illustration of the difference between *temperature* and *thermal energy*. The temperature of a gas reflects the average kinetic energy of the molecules of the gas, whereas the thermal energy present in an object is the *total* kinetic energy of the molecules. In the thermosphere, the few molecules present have a very high temperature, but the thermal energy is exceedingly small because there are so few molecules.

Gases within the troposphere are well mixed by convection. Pollutants that are evolved on Earth's surface can rise up to the stratosphere, but it is said that the stratosphere acts as a "thermal lid" on the troposphere and prevents significant mixing of polluting gases into the stratosphere and beyond.

The pressure of the atmosphere declines with altitude, and so the partial pressure of O_2 declines. The figure shows why climbers have a hard time breathing on Mt. Everest, where the altitude is 29,028 ft (8848 m) and the O_2 partial pressure is only 29% of the sea level partial pressure. With proper training, a climber could reach the summit without supplemental oxygen. However, this same feat would not be possible if Everest were farther north. Earth's atmosphere thins toward the poles, and so the O_2 partial pressure would be even less if Everest's summit were in the northern part of North America, for example.

See G. N. Eby, *Environmental Geochemistry*, Cengage Learning/Brooks/Cole, 2004.

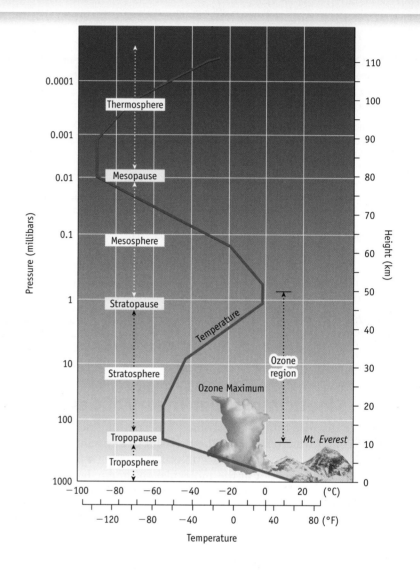

Average Composition of Earth's Atmosphere to a Height of 25 km

Gas	Volume %	Source	Gas	Volume %	Source
N_2	78.08	Biologic	He	0.0005	Radioactivity
O_2	20.95	Biologic	H_2O	0 to 4	Evaporation
Ar	0.93	Radioactivity	CH_4	0.00017	Biologic
CO_2	0.0385	Biologic, industrial	N_2O	0.00003	Biologic, industrial
Ne	0.0018	Earth's interior	O_3	0.000004	Photochemical

son for this is that the speed at which molecules move depends on the temperature. Let us expand on this idea.

The molecules in a gas sample do not all move at the same speed. Rather, as illustrated in Figure 11.13 for O_2 molecules, there is a distribution of speeds. Figure 11.13 shows the number of particles in a gas sample that are moving at certain speeds at a given temperature, and there are two important observations we can make. First, at a given temperature some molecules in a sample have high speeds,

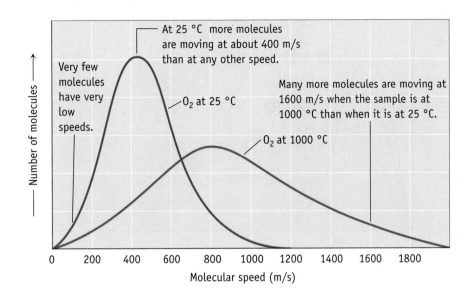

FIGURE 11.13 The distribution of molecular speeds. A graph of the number of molecules with a given speed versus that speed shows the distribution of molecular speeds. The red curve shows the effect of increased temperature. Even though the curve for the higher temperature is "flatter" and broader than the one at a lower temperature, the areas under the curves are the same because the number of molecules in the sample is fixed.

and others have low speeds. Most of the molecules, however, have some intermediate speed, and their most probable speed corresponds to the maximum in the curve. For oxygen gas at 25 °C, for example, most molecules have speeds in the range from 200 m/s to 700 m/s, and their most probable speed is about 400 m/s. (These are very high speeds, indeed. A speed of 400 m/s corresponds to about 900 miles per hour!)

A second observation regarding the distribution of speeds is that as the temperature increases the most probable speed increases, and the number of molecules traveling at very high speeds increases greatly.

The kinetic energy of a single molecule of mass m in a gas sample is given by the equation

$$KE = \frac{1}{2}(mass)(speed)^2 = \frac{1}{2} mu^2$$

where u is the speed of that molecule. We can calculate the kinetic energy of a single gas molecule from this equation but not of a collection of molecules because not all of the molecules in a gas sample are moving at the same speed. However, we can calculate the average kinetic energy of a collection of molecules by relating it to other averaged quantities of the system. In particular, the average kinetic energy of a mole of molecules in a gas sample is related to the average speed:

$$\overline{KE} = \frac{1}{2}Nm\overline{u^2}$$

where N is Avogadro's number. (The horizontal bar over the symbols $\overline{KE}$ and $\overline{u}$ indicate an average value.) The product of the mass per molecule and Avogadro's constant is the molar mass, so we can write

$$\overline{KE} = \frac{1}{2}M\overline{u^2}$$

where the molar mass *(M)* has units of kg/mol (and KE is in joules). This equation states that the average kinetic energy of the molecules in a gas sample, $\overline{KE}$, is related to $\overline{u^2}$, the average of the squares of their speeds (called the "mean square speed").

Experiments also show that the average kinetic energy, $\overline{KE}$, of a mole of gas molecules is directly proportional to temperature with a proportionality constant of $\frac{3}{2}R$,

$$\overline{KE} = \frac{3}{2}RT$$

where R is the gas constant expressed in SI units (8.314472 J/K · mol).

• **Maxwell-Boltzmann Curves** Plots such as those in Figures 11.13 and 11.14 are often referred to as Maxwell-Boltzmann curves. They are named for two scientists who studied the physical properties of gases: James Clerk Maxwell (1831–1879) and Ludwig Boltzmann (1844–1906).

FIGURE 11.14 The effect of molecular mass on the distribution of speeds. At a given temperature, molecules with higher masses have lower speeds.

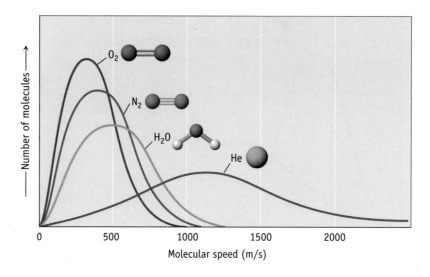

The two kinetic energy equations can be combined to yield an equation that relates mass, average speed, and temperature (Equation 11.9).

$$\sqrt{\overline{u^2}} = \sqrt{\frac{3RT}{M}} \qquad (11.9)$$

Here, the square root of the mean square speed ($\sqrt{\overline{u^2}}$, called the **root-mean-square speed**, or **rms speed**), the temperature (T, in kelvins), and the molar mass (M) are related. This equation shows that the speeds of gas molecules are indeed related to the temperature (Figure 11.13). The rms speed is a useful quantity because of its relationship to the average kinetic energy and because it is very close to the true average speed for a sample. (The average speed is 92% of the rms speed.)

All gases have the same average kinetic energy at the same temperature. However, if you compare a sample of one gas with another, say compare O_2 and N_2, this does not mean the molecules have the same rms speed (Figure 11.14). Instead, Equation 11.9 shows that the smaller the molar mass of the gas the greater the rms speed.

EXAMPLE 11.12 Molecular Speed

Problem Calculate the rms speed of oxygen molecules at 25 °C.

What Do You Know? You know the molar mass of O_2 and the temperature, the main determinants of speed.

Strategy Use Equation 11.9 with M in units of kg/mol. The reason for this is that R is in units of J/K · mol, and 1 J = 1 kg · m²/s².

Solution The molar mass of O_2 is 32.0×10^{-3} kg/mol.

$$\sqrt{\overline{u^2}} = \sqrt{\frac{3(8.3145 \text{ J/K} \cdot \text{mol})(298 \text{ K})}{32.0 \times 10^{-3} \text{ kg/mol}}} = \sqrt{2.32 \times 10^5 \text{ J/kg}}$$

To obtain the answer in meters per second, use the relation 1 J = 1 kg · m²/s². This means you have

$$\sqrt{\overline{u^2}} = \sqrt{2.32 \times 10^5 \text{ kg} \cdot \text{m}^2/(\text{kg} \cdot \text{s}^2)} = \sqrt{2.32 \times 10^5 \text{ m}^2/\text{s}^2} = 482 \text{ m/s}$$

Think about Your Answer The calculated speed is equivalent to about 1100 miles per hour! This is somewhat greater than the speed of sound in air, 343 m/s.

Check Your Understanding

Calculate the rms speeds of helium atoms and N_2 molecules at 25 °C.

Kinetic-Molecular Theory and the Gas Laws

The gas laws, which come from experiment, can be explained by the kinetic-molecular theory. The starting place is to describe how pressure arises from collisions of gas molecules with the walls of the container holding the gas (Figure 11.15). Remember that pressure is related to the force of the collisions (Section 11.1).

$$\text{Gas pressure} = \frac{\text{force of collisions}}{\text{area}}$$

The force exerted by the collisions depends on the number of collisions and the average force per collision. When the temperature of a gas is increased, we know the average kinetic energy of the molecules increases. This causes the average force of the collisions with the walls to increase as well. (This is much like the difference in the force exerted by a car traveling at high speed versus one moving at only a few kilometers per hour.) Also, because the speed of gas molecules increases with temperature, more collisions occur per second. Thus, the collective force per square centimeter is greater, and the pressure increases. Mathematically, this is related to the direct proportionality between P and T when n and V are fixed, that is, $P = (nR/V)T$.

Increasing the number of molecules of a gas at a fixed temperature and volume does not change the average collision force, but it does increase the number of collisions occurring per second. Thus, the pressure increases, and we can say that P is proportional to n when V and T are constant, that is, $P = n(RT/V)$.

If the pressure is to remain constant when either the number of molecules of gas or the temperature is increased, then the volume of the container (and the area over which the collisions can take place) must increase. This is expressed by stating that V is proportional to nT when P is constant $[V = nT(R/P)]$, a statement that is a *combination of Avogadro's hypothesis and Charles's law.*

Finally, if the temperature is constant, the average impact force of molecules of a given mass with the container walls must be constant. If n is kept constant while the volume of the container is made smaller, the number of collisions with the container walls per second must increase. This means the pressure increases, and so P is proportional to $1/V$ when n and T are constant, as stated by *Boyle's law,* that is, $P = (1/V)(nRT)$.

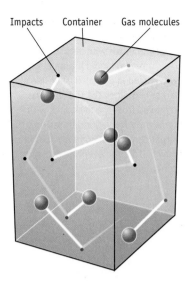

Impacts   Container   Gas molecules

FIGURE 11.15 Gas pressure. According to the kinetic-molecular theory, gas pressure is caused by gas molecules bombarding the container walls.

REVIEW & CHECK FOR SECTION 11.6

The species identified with each curve in the Maxwell-Boltzmann plot below are: (Assume all gases are at the same temperature.)

(a)   A = Xe, B = O_2, C = Ne, and D = He

(b)   A = Xe, B = He, C = Ne, and D = O_2

(c)   A = He, B = Ne, C = O_2, and D = Xe

(d)   A = He, B = O_2, C = Ne, and D = Xe

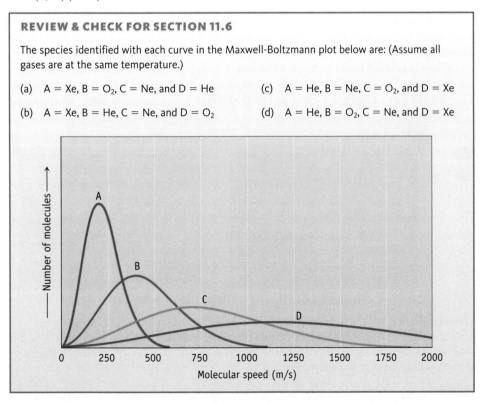

FIGURE 11.16 Gaseous diffusion. Here bromine diffuses out of a flask and mixes with air in the bottle over time. **(a)** Liquid bromine, Br_2, was placed in a small flask inside a larger container. **(b)** The cork was removed from the flask, and, with time, bromine vapor diffused into the larger container. Bromine vapor is now distributed evenly in the containers.

time →

(a)   (b)

11.7 Diffusion and Effusion

When a warm pizza is brought into a room, the volatile aroma-causing molecules vaporize into the atmosphere, where they mix with the oxygen, nitrogen, carbon dioxide, water vapor, and other gases present. Even if there were no movement of the air in the room caused by fans or people moving about, the odor would eventually reach everywhere in the room. This mixing of molecules of two or more gases due to their random molecular motions is the result of **diffusion**. Given time, the molecules of one component in a gas mixture will thoroughly and completely mix with all other components of the mixture (Figure 11.16).

Diffusion is also illustrated by the experiment in Figure 11.17. Here, we have placed cotton moistened with hydrochloric acid at one end of a U-tube and cotton moistened with aqueous ammonia at the other end. Molecules of HCl and NH_3 diffuse into the tube, and, when they meet, they produce white, solid NH_4Cl (just as in Figure 11.12).

$$HCl(g) + NH_3(g) \rightarrow NH_4Cl(s)$$

We find that the gases do not meet in the middle. Rather, because the heavier HCl molecules diffuse less rapidly than the lighter NH_3 molecules, the molecules meet closer to the HCl end of the U-tube.

Closely related to diffusion is **effusion**, which is the movement of gas through a tiny opening in a container into another container where the pressure is very low (Figure 11.17). Thomas Graham (1805–1869), a Scottish chemist, studied the

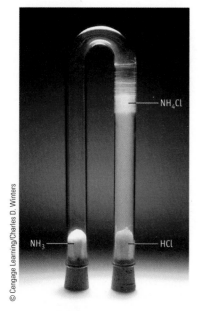

Gaseous diffusion. Here, HCl gas (from hydrochloric acid) and ammonia gas (from aqueous ammonia) diffuse from opposite ends of a glass U-tube. When they meet, they produce white, solid NH_4Cl. It is clear that the NH_4Cl is formed closer to the end from which the HCl gas begins because HCl molecules move slower, on average, than NH_3 molecules.

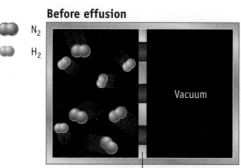

Before effusion   **During effusion**

N_2

H_2

Vacuum

Porous barrier

Gaseous effusion. H_2 and N_2 gas molecules effuse through the pores of a porous barrier. Lighter molecules (H_2) with higher average speeds strike the barrier more often and pass more often through it than heavier, slower molecules (N_2) at the same temperature. According to Graham's law, H_2 molecules effuse 3.72 times faster than N_2 molecules.

FIGURE 11.17 Gaseous diffusion and effusion.

effusion of gases and found that the rate of effusion of a gas—the amount of gas moving from one place to another in a given amount of time—is inversely proportional to the square root of its molar mass. Based on these experimental results, the rates of effusion of two gases can be compared:

$$\frac{\text{Rate of effusion of gas 1}}{\text{Rate of effusion of gas 2}} = \sqrt{\frac{\text{molar mass of gas 2}}{\text{molar mass of gas 1}}} \qquad \textbf{(11.10)}$$

The relationship in Equation 11.10—now known as **Graham's law**—is readily derived from Equation 11.9 by recognizing that the rate of effusion depends on the speed of the molecules. The ratio of the rms speeds is the same as the ratio of the effusion rates:

$$\frac{\text{Rate of effusion of gas 1}}{\text{Rate of effusion of gas 2}} = \frac{\sqrt{\overline{u^2} \text{ of gas 1}}}{\sqrt{\overline{u^2} \text{ of gas 2}}} = \frac{\sqrt{3RT/(M \text{ of gas 1})}}{\sqrt{3RT/(M \text{ of gas 2})}}$$

Canceling out like terms gives the expression in Equation 11.10.

EXAMPLE 11.13   Using Graham's Law of Effusion to Calculate a Molar Mass

Problem Tetrafluoroethylene, C_2F_4, effuses through a barrier at a rate of 4.6×10^{-6} mol/h. An unknown gas, consisting only of boron and hydrogen, effuses at the rate of 5.8×10^{-6} mol/h under the same conditions. What is the molar mass of the unknown gas?

What Do You Know? You have two gases, one with a known molar mass (C_2F_4, 100.0 g/mol) and the other unknown. The rate of effusion for both gases is known.

Strategy Substitute the experimental data into Graham's law equation (Equation 11.10).

Solution

$$\frac{5.8 \times 10^{-6} \text{ mol/h}}{4.6 \times 10^{-6} \text{ mol/h}} = 1.3 = \sqrt{\frac{100.0 \text{ g/mol}}{M \text{ of unknown}}}$$

To solve for the unknown molar mass, square both sides of the equation and rearrange to find M for the unknown.

$$1.6 = \frac{100.0 \text{ g/mol}}{M \text{ of unknown}}$$

$$M = \boxed{63 \text{ g/mol}}$$

Think about Your Answer From Graham's law, we know that a light molecule will effuse more rapidly than a heavier one. Because the unknown gas effuses more rapidly than C_2F_4 ($M = 100.0$ g/mol), the unknown must have a molar mass less than 100 g/mol. A boron–hydrogen compound corresponding to this molar mass is B_5H_9, called *pentaborane*.

Check Your Understanding

A sample of pure methane, CH_4, is found to effuse through a porous barrier in 1.50 minutes. Under the same conditions, an equal number of molecules of an unknown gas effuses through the barrier in 4.73 minutes. What is the molar mass of the unknown gas?

REVIEW & CHECK FOR SECTION 11.7

In Figure 11.17, ammonia gas and hydrogen chloride are introduced from opposite ends of a glass U-tube. The gases react to produce white, solid NH_4Cl. What are the relative root mean square speeds of HCl and NH_3?

(a) rms for HCl/rms for NH_3 = 2.2

(b) rms for HCl/rms for NH_3 = 1.5

(c) rms for HCl/rms for NH_3 = 0.68

(d) rms for HCl/rms for NH_3 = 0.46

SCUBA Diving—An Applicaton of the Gas Laws

Diving with a self-contained underwater breathing apparatus (SCUBA) is exciting. If you want to dive much beyond about 60 ft (18 m), however, you need to take special precautions.

When you breathe air from a SCUBA tank, the pressure of the gas in your lungs is equal to the pressure exerted on your body. When you are at the surface, atmospheric pressure is about 1 atm, and, because air has an oxygen concentration of 21%, the partial pressure of O_2 is about 0.21 atm. If you are at a depth of about 33 ft, the water pressure is 2 atm. This means the oxygen partial pressure is double the surface partial pressure, or about 0.4 atm. Similarly, the partial pressure of N_2, which is about 0.8 atm at the surface, doubles to about 1.6 atm at a depth of 33 ft. The solubility of gases in water (and in blood) is directly proportional to pressure. Therefore, more oxygen and nitrogen dissolve in blood under these conditions, and this can lead to several problems.

Nitrogen narcosis, also called *rapture of the deep* or the *martini effect*, results from the toxic effect on nerve conduction of N_2 dissolved in blood. Its effect is comparable to drinking a martini on an empty stomach or taking laughing gas (nitrous oxide, N_2O) at the dentist; it makes you slightly giddy. In severe cases, it can impair a diver's judgment and even cause a diver to take

Image copyright © Jon Milnes. Used under license from Shutterstock.com

the regulator out of his or her mouth and hand it to a fish! Some people can go as deep as 130 ft with no problem, but others experience nitrogen narcosis at 80 ft.

Another problem with breathing air at depths beyond 100 ft or so is oxygen toxicity. Our bodies are regulated for a partial pressure of O_2 of 0.21 atm. At a depth of 130 ft, the

SCUBA diving. Ordinary recreational dives can be made with compressed air to depths of about 60 feet. With a gas mixture called *Nitrox* (which has up to 36% O_2), one can stay at such depths for a longer period. To go even deeper, however, divers must breathe special gas mixtures such as Trimix. This is a breathing mixture consisting of oxygen, helium, and nitrogen.

partial pressure of O_2 is comparable to breathing 100% oxygen at sea level. These higher partial pressures can harm the lungs and cause central nervous system damage. Oxygen toxicity is the reason deep dives are done not with compressed air but with gas mixtures with a much lower percentage of O_2, say about 10%.

Because of the risk of nitrogen narcosis, divers going beyond about 130 ft, such as those who work for offshore oil drilling companies, use a mixture of oxygen and helium. This solves the nitrogen narcosis problem, but it introduces another. If the diver has a voice link to the surface, the diver's speech sounds like Donald Duck! Speech is altered because the velocity of sound in helium is different from that in air.

11.8 Nonideal Behavior of Gases

If you are working with a gas at approximately room temperature and a pressure of 1 atm or less, the ideal gas law is remarkably successful in relating the amount of gas and its pressure, volume, and temperature. At higher pressures or lower temperatures, however, deviations from the ideal gas law occur. The origin of these deviations is explained by the breakdown of the assumptions used when describing ideal gases, specifically the assumptions that the particles have no size and that there are no forces between them.

At standard temperature and pressure (STP), the volume occupied by a single molecule is *very* small relative to its share of the total gas volume. Relatively speaking, a helium atom with a radius of 31 pm has about the same space to move about as a pea has inside a basketball. Now suppose the pressure is increased significantly, to 1000 atm. The volume available to each molecule is a sphere with a radius of only about 200 pm, which means the situation is now like that of a pea inside a sphere a bit larger than a Ping-Pong ball.

The kinetic-molecular theory and the ideal gas law are concerned with the volume available to the molecules to move about, not the total volume of the container. For example, suppose you have a flask marked with a volume of 500 mL. This does not mean the space available to molecules is 500 mL. Rather, the available volume is less than 500 mL because the molecules themselves occupy some of the volume. At low pressures, the volume occupied by the gas molecules in the container is so small in comparison to the space available that neglecting to subtract it

• **How Much Space Does a Molecule Take Up?** The volume of a methane (CH_4) molecule is 3.31×10^{-23} cm^3. Suppose you have 1.00 mol of CH_4 gas in a 22.4-L flask at 273 K. The pressure would be high, 22.4 atm, and 19.9 cm^3 of the space in the flask would be occupied by CH_4 molecules or about 0.09% of the flask volume.

from the total volume does not introduce significant error. The problem is that the volume occupied by gas molecules is not negligible at higher pressures.

Another assumption of the kinetic-molecular theory is that the atoms or molecules of the gas never stick to one another by some type of intermolecular force. This is clearly not true as well. All gases can be liquefied—although some gases require a very low temperature (Figure 11.11)—and the only way this can happen is if there are forces between the molecules. When a molecule is about to strike the wall of its container, other molecules in its vicinity exert a slight pull on the molecule and pull it away from the wall. The effect of the intermolecular forces is that molecules strike the wall with less force than in the absence of intermolecular attractive forces. Thus, because collisions between molecules in a real gas and the wall are softer, the observed gas pressure is less than that predicted by the ideal gas law. This effect can be particularly pronounced when the temperature is low (near the condensation temperature).

The Dutch physicist Johannes van der Waals (1837–1923) studied the breakdown of the ideal gas law equation and developed an equation to correct for the errors arising from nonideality. This equation is known as the **van der Waals equation**:

Table 11.2 van der Waals Constants

Gas	a Values (atm·L²/mol²)	b Values (L/mol)
He	0.034	0.0237
Ar	1.34	0.0322
H_2	0.244	0.0266
N_2	1.39	0.0391
O_2	1.36	0.0318
CO_2	3.59	0.0427
Cl_2	6.49	0.0562
H_2O	5.46	0.0305

observed pressure   container V

$$\left(P + a\left[\frac{n}{V}\right]^2\right)(V - bn) = nRT \qquad (11.11)$$

correction for intermolecular forces   correction for molecular volume

where a and b are experimentally determined constants (Table 11.2). Although Equation 11.11 might seem complicated at first glance, the terms in parentheses are those of the ideal gas law, each corrected for the effects discussed previously. The pressure correction term, $a(n/V)^2$, accounts for intermolecular forces. Owing to intermolecular forces, the observed gas pressure is lower than the ideal pressure ($P_{observed} < P_{ideal}$ where P_{ideal} is calculated using the equation $PV = nRT$). Therefore, the term $a(n/V)^2$ is *added* to the observed pressure. The constant a typically has values in the range 0.01 to 10 atm · L²/mol².

The actual volume available to the molecules is smaller than the volume of the container because the molecules themselves take up space. Therefore, an amount is *subtracted* from the container volume ($= bn$) to take this into account. Here, n is the number of moles of gas, and b is an experimental quantity that corrects for the molecular volume. Typical values of b range from 0.01 to 0.1 L/mol, roughly increasing with increasing molecular size.

As an example of the importance of these corrections, consider a sample of 4.00 mol of chlorine gas, Cl_2, in a 4.00-L tank at 100.0 °C. The ideal gas law would lead you to expect a pressure of 30.6 atm. A better estimate of the pressure, obtained from the van der Waals equation, is 26.0 atm, about 4.6 atm less than the ideal pressure!

© Cengage Learning/Charles D. Winters

A flask of helium gas for filling party balloons. The tank holds about 15 cubic feet (about 425 L) of helium. (*Note:* These tanks are not designed to be reused! Dispose of them properly. Also, use helium sparingly. The supply of the gas on the planet is not limitless.)

REVIEW & CHECK FOR SECTION 11.8

You can purchase tanks of helium at a party store to fill balloons. Such a tank may hold 343 mol of He in a volume of 473 L. Use van der Waals's equation to calculate the pressure in the tank at 25 °C.

(a) 18.0 atm   (b) 17.7 atm   (c) 17.3 atm

What to Do with All of That CO₂? More on Green Chemistry

Most scientists, including the authors of this book, believe that global warming is occurring and that it is a problem. They also believe this increase is related to the increase in the percentage of greenhouse gases, especially CO_2, in the atmosphere. The percentage of CO_2 in the atmosphere has been on the rise since the beginning of the Industrial Revolution and is projected to continue to rise in the future. Are there ways to stabilize the concentration of CO_2 at the current level, or at least limit its future increase? This in turn focuses attention on the obvious source of CO_2 in the atmosphere, the combustion of fossil fuels.

What is wrong with CO_2 in the atmosphere? After all, it is used by plants as their carbon source, and the oceans can absorb the gas, much of which ends up incorporated in corals. But there is a danger these "sinks" could be overwhelmed.

The United States obtains over half its electricity from coal-fired power plants. So, one question is why not capture and dispose of the CO_2 from power plants. In fact, this is being considered. One way of removing CO_2 generated by burning fossil fuels involves geological sequestration, a process in which CO_2 is pumped into rock formations thou-sands of feet underground. There, it would presumably remain trapped for thousands of years.

An experiment is underway at the Mountaineer Power Plant in New Haven, West Virginia, to do just that. For this process to succeed, however, the CO_2 to be sequestered needs to be pure, free of impurities. This is accomplished in two steps. The flue gas from the coal-based electricity generating plant is first routed through a scrubber, to remove most of the SO_2, a typical contaminant produced in coal combustion. The gas stream is then cooled to about 35 °F and introduced under pressure into a cooled slurry of ammonium carbonate, $(NH_4)_2CO_3$. As set up in this pilot program, a portion of the CO_2 reacts with the slurry to form ammonium hydrogen carbonate, $NH_4(HCO_3)$:

$$(NH_4)_2CO_3 + CO_2 + H_2O \rightleftharpoons 2\,NH_4(HCO_3)$$

The remaining flue gases, mainly N_2 and unreacted CO_2, exit the system. The slurry is then separated and heated, reversing the chemical process, giving back pure CO_2 and regenerating $(NH_4)_2CO_3$ which is recycled back into the system. The pure CO_2 is then pumped underground.

The pilot project is currently set up to remove about 100,000 tons of CO_2 annually. This is only about 1.5% of CO_2 from the exhaust gases, but it is believed that the process could be scaled up to remove as much as 90% of the CO_2 in the future if the method proves economically feasible.

Questions:

1. A large coal-fired plant can burn 10,000 tons of coal a day. We will make the assumption for this problem that the coal is pure carbon, so this is about 9.1×10^9 g of carbon. What mass of CO_2 will be generated?

2. What mass of ammonium carbonate is needed to remove 1.0 million grams of CO_2 per day?

Answers to these questions are available in Appendix N.

Reference:

New York Times, p. A1, September 22, 2009.

Captured, Then Buried

A coal-fired electricity plant in West Virginia plans to capture some of the carbon dioxide it emits and inject it underground. Some environmentalists are wary.

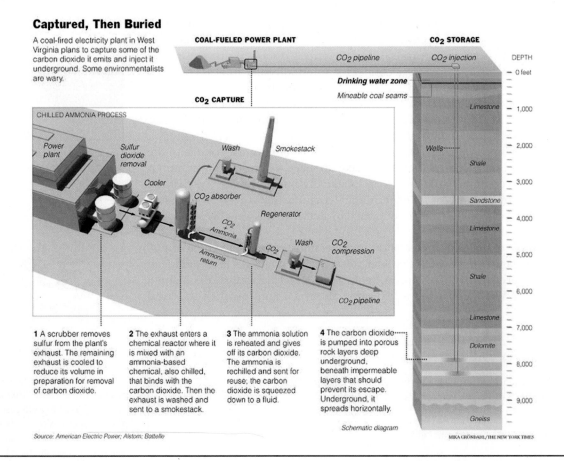

1 A scrubber removes sulfur from the plant's exhaust. The remaining exhaust is cooled to reduce its volume in preparation for removal of carbon dioxide.

2 The exhaust enters a chemical reactor where it is mixed with an ammonia-based chemical, also chilled, that binds with the carbon dioxide. Then the exhaust is washed and sent to a smokestack.

3 The ammonia solution is reheated and gives off its carbon dioxide. The ammonia is rechilled and sent for reuse; the carbon dioxide is squeezed down to a fluid.

4 The carbon dioxide is pumped into porous rock layers deep underground, beneath impermeable layers that should prevent its escape. Underground, it spreads horizontally.

Schematic diagram

Source: American Electric Power; Alstom; Battelle

MIKA GRÖNDAHL/THE NEW YORK TIMES

CHAPTER GOALS REVISITED

Now that you have studied this chapter, you should ask whether you have met the chapter goals. In particular, you should be able to:

Understand the basis of the gas laws and how to use those laws (Boyle's law, Charles's law, Avogadro's hypothesis, Dalton's law)

a. Describe how pressure measurements are made and the units of pressure, especially atmospheres (atm) and millimeters of mercury (mm Hg) (Section 11.1). **Study Questions: 1, 3.**
b. Understand the basis of the gas laws and how to apply them (Section 11.2). **Study Questions: 6, 8, 10, 12, 14–16.**

Use the ideal gas law

a. Understand the origin of the ideal gas law and how to use the equation (Section 11.3). **Study Questions: 17–24, 59, 63, 77, 85, 102.**
b. Calculate the molar mass of a compound from a knowledge of the pressure of a known quantity of a gas in a given volume at a known temperature (Section 11.3). **Study Questions: 25–30, 66, 89, 90.**

Apply the gas laws to stoichiometric calculations

a. Apply the gas laws to a study of the stoichiometry of reactions (Section 11.4). **Study Questions: 31–36, 65, 79, 82.**
b. Use Dalton's law of partial pressures (Section 11.5). **Study Questions: 39, 40, 70, 80, 87.**

Understand kinetic-molecular theory as it is applied to gases, especially the distribution of molecular speeds (energies)

a. Apply the kinetic-molecular theory of gas behavior at the molecular level (Section 11.6). **Study Questions: 41–46, 107, and Go Chemistry Module 16.**
b. Understand the phenomena of diffusion and effusion and how to use Graham's law (Section 11.7). **Study Questions: 47–50, 75, 76.**

Recognize why gases do not behave like ideal gases under some conditions

a. Appreciate the fact that gases usually do not behave as ideal gases. Deviations from ideal behavior are largest at high pressure and low temperature (Section 11.8). **Study Questions: 51, 52.**

 and

Sign in at **www.cengage.com/owl** to:
- View tutorials and simulations, develop problem-solving skills, and complete online homework assigned by your professor.
- For quick review and exam prep, download Go Chemistry mini lecture modules from OWL (or purchase them at **www.cengagebrain.com**)

Access **How Do I Solve It?** tutorials on how to approach problem solving using concepts in this chapter.

Key Equations

Equation 11.1 (page 388) Boyle's law (where P is the pressure and V is the volume).

$$P_1V_1 = P_2V_2 \text{ at constant } n \text{ and } T$$

Equation 11.2 (page 390) Charles's law (where T is the Kelvin temperature).

$$\frac{V_1}{T_1} = \frac{V_2}{T_2} \text{ at constant } n \text{ and } P$$

Equation 11.3 (page 391) General gas law (combined gas law).

$$\frac{P_1V_1}{T_1} = \frac{P_2V_2}{T_2} \text{ for a given amount of gas, } n$$

Equation 11.4 (page 394) Ideal gas law (where n is the amount of gas in moles and R is the universal gas constant, $0.082057 \text{ L} \cdot \text{atm/K} \cdot \text{mol}$).

$$PV = nRT$$

Equation 11.5 (page 396) Density of gases (where d is the gas density in g/L and M is the molar mass of the gas).

$$d = \frac{m}{V} = \frac{PM}{RT}$$

Equation 11.6 (page 401) Dalton's law of partial pressures. The total pressure of a gas mixture is the sum of the partial pressures of the component gases (P_n).

$$P_{\text{total}} = P_1 + P_2 + P_3 + \ldots$$

Equation 11.7 (page 401) The total pressure of a gas mixture is equal to the total number of moles of gases multiplied by (RT/V).

$$P_{\text{total}} = (n_{\text{total}})\left(\frac{RT}{V}\right)$$

Equation 11.8 (page 402) The pressure of a gas (A) in a mixture is the product of its mole fraction (X_A) and the total pressure of the mixture.

$$P_A = X_A P_{\text{total}}$$

Equation 11.9 (page 406) The rms speed $(\sqrt{u^2}\,)$ depends on the molar mass of a gas (M) and its temperature (T).

$$\sqrt{u^2} = \sqrt{\frac{3RT}{M}}$$

Equation 11.10 (page 409) Graham's law. The rate of effusion of a gas is inversely proportional to the square root of its molar mass.

$$\frac{\text{Rate of effusion of gas 1}}{\text{Rate of effusion of gas 2}} = \sqrt{\frac{\text{molar mass of gas 2}}{\text{molar mass of gas 1}}}$$

Equation 11.11 (page 411) The van der Waals equation, which relates pressure, volume, temperature, and amount of gas for a nonideal gas.

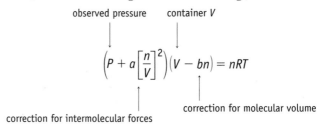

$$\left(P + a\left[\frac{n}{V}\right]^2\right)(V - bn) = nRT$$

observed pressure · container V · correction for intermolecular forces · correction for molecular volume

---

ⓦWL Questions and problems for this chapter are available in OWL. Use the OWL access card that came with this textbook to access assigned questions and problems for this chapter.

12 Intermolecular Forces and Liquids

Geckos Can Climb Up der Waals

Just like the comic-book hero Spiderman, a little lizard, the gecko, can climb walls and hang from the ceiling.

This fact intrigued Kellar Autumn, a professor of biology at Lewis and Clark College in Portland, Oregon. An interdisciplinary team of Autumn and his students, along with scientists and engineers from Stanford University, and the Universities of California at Berkeley and Santa Barbara, realized that gecko toes are covered with an array of stiff hairlike setae. Each of these is about as long as the thickness of a human hair or about 0.1 mm long. But each seta is further divided into about 1000 even more minute pads called *spatulae*. And these are only about 200 nm wide, a distance smaller than the wavelength of visible light!

The design of gecko feet is the secret to their wall-crawling abilities. Autumn said, "We discovered that the seta is 10 times more adhesive than predicted from prior measurements on whole animals. In fact, the adhesive is so strong that a single seta can lift the weight of an ant. A million setae, which could easily fit onto the area of a dime, could lift a 45-pound child. Our discovery explains why the gecko can support its entire body weight with only a single finger."

"When the gecko attaches itself to a surface, it uncurls its toes like a party favor that uncurls when you blow into it," Autumn says. "But," he adds, "getting yourself to stick isn't really that difficult. It's getting off the surface that is the major problem. When a gecko runs, it has to attach and detach its feet 15 times a second."

But what is the "adhesive effect" that allows a gecko to climb a wall? It is an intermolecular force, called a *van der Waals force*. This is ordinarily a weak force that operates only over a very short distance. However, each spatula of the millions in each toe experiences an attractive van der Waals force with the surface. It has been calculated that the force between each spatula and the surface is about 0.4 μN (micronewtons). But given that there are millions of spatula in each toe, the total force can easily be 10 N or more.

One of the topics in this chapter is the importance of van der Waals forces. As you learn about this, think about the possibilities of using the design of gecko feet to make even more useful adhesives.

References:

1. www.kellarautumn.com/photography/Welcome.html
2. geckolab.lclark.edu/dept/AutumnLab/ Welcome.html

A Tokay gecko. This is a nocturnal arboreal gecko found from northeast India to southeast Asia.

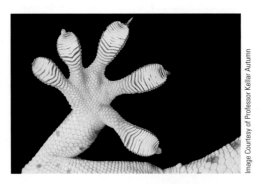

A close-up of a gecko foot. Notice the adhesive lamellae at the end of each toe. The lamellae consist of arrays of setae.

Rows of setae. Each is about 75 μm long. Each seta is composed of millions of spatulae, the basis of the gecko's stickiness.

Image Courtesy of Professor Kellar Autumn

CHAPTER OUTLINE

12.1 States of Matter and Intermolecular Forces

12.2 Interactions between Ions and Molecules with a Permanent Dipole

12.3 Interactions between Molecules with a Dipole

12.4 Intermolecular Forces Involving Nonpolar Molecules

12.5 A Summary of van der Waals Intermolecular Forces

12.6 Properties of Liquids

CHAPTER GOALS

See Chapter Goals Revisited (page 442) for Study Questions keyed to these goals.

- Describe intermolecular forces and their effects.
- Understand the importance of hydrogen bonding.
- Understand the properties of liquids.

⌚WL

Sign in to OWL at **www.cengage.com/owl** to view tutorials and simulations, develop problem-solving skills, and complete online homework assigned by your professor.

go Chemistry

Download mini lecture videos for key concept review and exam prep from OWL or purchase them from **www.cengagebrain.com**

In the last chapter, we studied gases, one of the states of matter. Most of the time we were able to assume that the attractions between the molecules, intermolecular forces, were negligible and so we could ignore them. At the end of that chapter, however, we saw that at high pressures and/or low temperatures this assumption did not work as well; we had to consider intermolecular forces when dealing with the nonideal behavior of gases. In the other two states of matter, the liquid and solid states, intermolecular forces play a major role. In order to describe these states of matter, we will first need to explore the different types and relative strengths of intermolecular forces.

The primary objectives of this chapter are to examine the intermolecular forces that allow molecules to interact and then to look at liquids, a state of matter that results from such interactions. You will find this a useful chapter because it explains, among other things, why your body is cooled when you sweat and why ice can float on liquid water, a property shared by very few other substances.

12.1 States of Matter and Intermolecular Forces

The kinetic-molecular theory of gases (◄ Section 11.6) assumes that gas molecules or atoms are widely separated and that these particles can be considered to be independent of one another. Consequently, we can relate the properties of gases under most conditions by the ideal gas law equation, $PV = nRT$ (◄ Equation 11.4). In real gases, however, there are intermolecular forces between molecules. If these forces are strong enough, the gas can condense to a liquid and eventually to a solid. The existence of intermolecular forces in liquids makes the picture more complex, and it is not possible to create a simple "ideal liquid equation."

How different are the states of matter at the particulate level? We can get a sense of this by comparing the volumes occupied by equal numbers of molecules of a material in different states. Figure 12.1a shows a flask containing about 300 mL of liquid nitrogen. If all of the liquid were allowed to evaporate, the gaseous nitrogen, at 1 atm and room temperature, would fill a large balloon (more than 200 L). A great amount of space exists between molecules in a gas, whereas in liquids the molecules are much closer together.

The increase in volume when converting liquids to gases is strikingly large. In contrast, no dramatic change in volume occurs when a solid is converted to a liquid. Figure 12.1b shows the same amount of liquid and solid benzene, C_6H_6, side by side.

Nitrogen gas

Liquid nitrogen

(a) When a 300-mL sample of liquid nitrogen evaporates, it will produce more than 200 L of gas at 25 °C and 1.0 atm. In the liquid phase, the molecules of N_2 are close together; in the gas phase, they are far apart.

© Cengage Learning/Charles D. Winters

(b) The same volume of liquid benzene, C_6H_6, is placed in two test tubes, and one tube *(right)* is cooled, freezing the liquid. The solid and liquid states have almost the same volume, showing that the molecules are packed together almost as tightly in the liquid state as they are in the solid state.

© Cengage Learning/Charles D. Winters

Liquid benzene   Solid benzene

FIGURE 12.1 Contrasting gases, liquids, and solids.

They are not appreciably different in volume. This means the molecules in the liquid are packed together about as tightly as the molecules in the solid phase.

Intermolecular forces influence chemistry in many ways:

- They are directly related to properties such as melting point, boiling point, and the energy needed to convert a solid to a liquid or a liquid to a vapor.
- They are important in determining the solubility of gases, liquids, and solids in various solvents.
- They are crucial in determining the structures of biologically important molecules such as DNA and proteins.

Bonding in ionic compounds depends on the electrostatic forces of attraction between oppositely charged ions (▶ Section 13.4). Similarly, the intermolecular forces attracting one molecule to another are electrostatic. The energy associated with the attractive forces between the ions in ionic compounds are usually in the range of 700 to 1100 kJ/mol, and most covalent bond energies are in the range of 100 to 400 kJ/mol (◀ Table 8.9). As a rough guideline, intermolecular forces are generally less than about 15% of the values of covalent bond energies. Nonetheless, these interactions can have a profound effect on molecular properties.

Collectively, forces between molecules are called **van der Waals forces**, and these include the attractive and repulsive forces between

- molecules with permanent dipoles (dipole-dipole forces, Section 12.3)
- polar molecules and nonpolar ones (dipole-induced dipole forces, Section 12.4)
- nonpolar molecules (induced dipole-induced dipole forces, also known as *London forces*, Section 12.4)

12.2 Interactions between Ions and Molecules with a Permanent Dipole

Module 17: Intermolecular Forces covers concepts in this section.

Many molecules are polar, a result of the polarity of individual bonds and a nonsymmetrical geometry (Section 8.7). Conceptually, we can view polar molecules as having positive and negative ends. What happens when a polar molecule encounters an ionic compound? The negative end of the dipole is attracted to a positive cation (Figure 12.2). Similarly, the positive end of the dipole is attracted to a negative anion. Forces of attraction between a positive or negative ion and polar molecules—ion–dipole forces—are less than those for ion–ion attractions, but they are greater than other types of forces between molecules, whether polar or nonpolar.

Ion–dipole attractions can be evaluated using Coulomb's law (◀ Equation 2.3), which informs us that the force of attraction between two charged objects depends

Table 12.1 **Radii and Enthalpies of Hydration of Alkali Metal Ions**

Cation	Ion Radius (pm)	Enthalpy of Hydration (kJ/mol)
Li^+	78	−515
Na^+	98	−405
K^+	133	−321
Rb^+	149	−296
Cs^+	165	−263

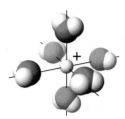

water surrounding
a cation

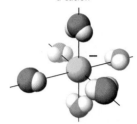

water surrounding
an anion

FIGURE 12.2 Ion–dipole interactions. When an ionic compound such as NaCl is placed in water, the polar water molecules surround the cations and anions.

on the product of their charges divided by the square of the distance between them (Section 2.7). Therefore, when a polar molecule encounters an ion, the attractive forces depend on three factors:

- The distance between the ion and the dipole. The closer they are, the stronger the attraction.
- The charge on the ion. The higher the ion charge, the stronger the attraction.
- The magnitude of the dipole. The greater the magnitude of the dipole, the stronger the attraction.

The formation of hydrated ions in aqueous solution is one of the most important examples of the interaction between an ion and a polar molecule (Figure 12.2). The enthalpy change associated with the hydration of ions—which is generally called the *enthalpy of solvation* or, for ions in water, the *enthalpy of hydration*—is substantial. The solvation enthalpy for an individual ion cannot be measured directly, but values can be estimated. For example, the hydration of sodium ions is described by the following reaction:

$$Na^+(g) + water \rightarrow Na^+(aq) \qquad \Delta_{hydration}H° = -405 \text{ kJ/mol}$$

The enthalpy of hydration depends on the charge of the ion and on $1/d$, where d is the distance between the center of the ion and the oppositely charged "pole" of the dipole.

The effect of ion radius is illustrated by the enthalpies of hydration of the alkali metal cations in Table 12.1, and the effects of radius *and* charge by $\Delta_{hydration}H°$ for Mg^{2+}, Li^+, and K^+ (Figure 12.3). It is interesting to compare these values with the enthalpy of hydration of the H^+ ion, estimated to be −1090 kJ/mol. This extraordinarily large value is due to the tiny size of the H^+ ion.

● **Coulomb's Law** The force of attraction between oppositely charged particles depends directly on the product of their charges and inversely on the square of the distance (d) between the ions ($1/d^2$) (◄ Equation 2.3, page 68). The energy of the attraction is also proportional to the charge product, but it is inversely proportional to the distance between them ($1/d$).

EXAMPLE 12.1   Hydration Energy

Problem Explain why the enthalpy of hydration of Na^+ (−405 kJ/mol) is more exothermic than that of Cs^+ (−263 kJ/mol), whereas that of Mg^{2+} is much more exothermic (−1922 kJ/mol) than that of either Na^+ or Cs^+.

What Do You Know? You know the ion charges and their sizes (◄ Figure 7.12): $Na^+ = 98$ pm, $Cs^+ = 165$ pm, and $Mg^{2+} = 79$ pm.

Strategy The energy associated with ion–dipole attractions depends directly on the size of the ion charge and the magnitude of the dipole, and inversely on the distance between them. Here the water dipole is a constant factor, so the answer depends on ion size and charge.

Solution From the ion sizes, we can predict that the distances between the center of the positive charge of the ion and the water dipole will vary in this order: $Mg^{2+} < Na^+ < Cs^+$. The hydration energy varies in the reverse order (with the hydration energy of Mg^{2+} being the most negative value). Notice also that Mg^{2+} has a 2+ charge, whereas the other ions are 1+. The greater charge on Mg^{2+} leads to a much greater force of ion–dipole attraction than for the other two ions, which have only a 1+ charge. As a result, the hydration energy for Mg^{2+} is much more negative than for the other two ions.

FIGURE 12.3 Enthalpy of hydration. The energy evolved when an ion is hydrated depends on the dipole moment of water, the ion charge, and the distance d between centers of the ion and the polar water molecule. The distance d increases as ion size increases.

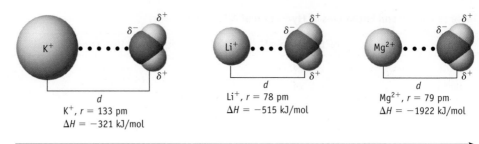

K^+, $r = 133$ pm
$\Delta H = -321$ kJ/mol

Li^+, $r = 78$ pm
$\Delta H = -515$ kJ/mol

Mg^{2+}, $r = 79$ pm
$\Delta H = -1922$ kJ/mol

Increasing force of attraction; more exothermic enthalpy of hydration

Thinking about Your Answer The charge difference between Mg^{2+} and the other ions has a much greater effect than the size difference. The values for hydration energies for the three ions are $Mg^{2+}(aq) = -1922$ kJ/mol (from Table 12.1), $Na^+(aq) = -405$ kJ/mol, and $Cs^+(aq) = -263$ kJ/mol.

Check Your Understanding

Which should have the more negative hydration energy, F^- or Cl^-? Explain briefly.

REVIEW & CHECK FOR SECTION 12.2

Which should have the more negative enthalpy of hydration?

(a) $MgCl_2$ (b) $AlCl_3$

12.3 Interactions between Molecules with a Dipole
Dipole–Dipole Forces

When a polar molecule encounters another polar molecule, of the same or a different kind, the positive end of one molecule is attracted to the negative end of the other polar molecule. This is called a **dipole-dipole interaction**.

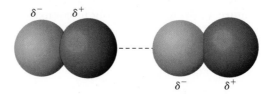

For polar molecules, dipole–dipole attractions influence, among other things, the evaporation of a liquid and the condensation of a gas (Figure 12.4). An energy change occurs in both processes. Evaporation requires the input of energy, specifically the enthalpy of vaporization ($\Delta_{vap}H°$) [see Sections 5.3 and 12.6]. The value for the enthalpy of vaporization has a positive sign, indicating that evaporation is an endothermic process. The enthalpy change for the condensation process—the reverse of evaporation—has a negative value.

The greater the forces of attraction between molecules in a liquid, the greater the energy that must be supplied to separate them. Thus, we expect polar compounds to have a higher value for their enthalpy of vaporization than nonpolar compounds with similar molar masses. For example, notice in Table 12.2 that $\Delta_{vap}H°$ for polar molecules is greater than for nonpolar molecules of approximately the same size and mass.

The boiling point of a liquid also depends on intermolecular forces of attraction. As the temperature of a substance is raised, its molecules gain kinetic energy. Eventually, when the boiling point is reached, the molecules have sufficient kinetic

A CLOSER LOOK

Hydrated Salts

Solid salts with waters of hydration are common. The formulas of these compounds are given by appending a specific number of water molecules to the end of the formula, as in $BaCl_2 \cdot 2\ H_2O$. Sometimes, the water molecules simply fill in empty spaces in a crystalline lattice, but often the cation in these salts is directly bound to water molecules. For example, the compound $CrCl_3 \cdot 6\ H_2O$ is better written as $[Cr(H_2O)_4Cl_2]Cl \cdot 2\ H_2O$. Four of the six water molecules are bound to the Cr^{3+} ion by ion–dipole attractive forces; the remaining two water molecules are in the lattice. Common examples of hydrated salts are listed in the table.

© Cengage Learning/Charles D. Winters

Compound	Common Name	Uses
$Na_2CO_3 \cdot 10\ H_2O$	Washing soda	Water softener
$Na_2S_2O_3 \cdot 5\ H_2O$	Hypo	Photography
$MgSO_4 \cdot 7\ H_2O$	Epsom salt	Cathartic, dyeing and tanning
$CaSO_4 \cdot 2\ H_2O$	Gypsum	Wallboard
$CuSO_4 \cdot 5\ H_2O$	Blue vitriol	Biocide

Hydrated cobalt(II) chloride, $CoCl_2 \cdot 6\ H_2O$. In the solid state, the compound is best described by the formula $[Co(H_2O)_4Cl_2] \cdot 2\ H_2O$. The cobalt(II) ion is surrounded by four water molecules and two chloride ions in an octahedral arrangement. In water, the ion is completely hydrated, now being surrounded by six water molecules. Cobalt(II) ions and water molecules interact by ion–dipole forces. This is an example of a coordination compound, a class of compounds discussed in detail in Chapter 22.

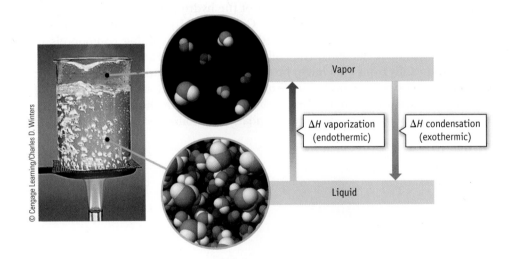

© Cengage Learning/Charles D. Winters

Vapor

ΔH vaporization (endothermic)

ΔH condensation (exothermic)

Liquid

FIGURE 12.4 Evaporation at the molecular level. Energy must be supplied to separate molecules in the liquid state against intermolecular forces of attraction.

energy to escape the forces of attraction of their neighbors. For molecules of similar molar mass, the greater the polarity, the higher the temperature required for the liquid to boil. In Table 12.2, you see that the boiling point for polar ICl is greater than that for nonpolar Br_2, for example.

Intermolecular forces also influence solubility. A qualitative observation on solubility is that "like dissolves like." In other words, polar molecules are likely to dissolve in a polar solvent, and nonpolar molecules are likely to dissolve in a nonpolar solvent (Figure 12.5) (◀ Chapter 8). The converse is also true; that is, it is unlikely that polar molecules will dissolve in nonpolar solvents or that nonpolar molecules will dissolve in polar solvents.

For example, water and ethanol (C_2H_5OH) can be mixed in any ratio to give a homogeneous mixture. In contrast, water does not dissolve in gasoline to an appreciable extent. The difference in these two situations is that ethanol and water are

Table 12.2 **Molar Masses, Boiling Points, and** $\Delta_{vap}H°$ **of Nonpolar and Polar Substances**

	Nonpolar				Polar		
	M (g/mol)	BP (°C)	$\Delta_{vap}H°$ (kJ/mol)		M (g/mol)	BP (°C)	$\Delta_{vap}H°$ (kJ/mol)
N_2	28	−196	5.57	CO	28	−192	6.04
SiH_4	32	−112	12.10	PH_3	34	−88	14.06
GeH_4	77	−90	14.06	AsH_3	78	−62	16.69
Br_2	160	59	29.96	ICl	162	97	—

polar molecules, whereas the hydrocarbon molecules in gasoline (e.g., octane, C_8H_{18}) are nonpolar. The water–ethanol interactions are strong enough that the energy expended in pushing water molecules apart to make room for ethanol molecules is compensated for by the energy of attraction between the two kinds of polar molecules. In contrast, water–hydrocarbon attractions are weak. The hydrocarbon molecules cannot disrupt the stronger water–water attractions.

Hydrogen Bonding

Hydrogen fluoride, water, ammonia, and many other compounds with O—H and N—H bonds have exceptional properties, and we can see that by looking at the boiling points for hydrogen compounds of elements in Groups 4A through 7A in Figure 12.6. Generally, the boiling points of related compounds increase with molar mass, and this trend is seen in the boiling points of the hydrogen compounds of Group 4A elements, for example ($CH_4 < SiH_4 < GeH_4 < SnH_4$). The same effect is also operating for the heavier molecules of the hydrogen-containing compounds of Group 5A, 6A, and 7A elements. The boiling points of NH_3, H_2O, and HF, however, deviate significantly from what might be expected based on molar mass alone. If we extrapolate the curve for the boiling points of H_2Te, H_2Se, and H_2S, the boiling point of water is predicted to be around −90 °C. However, the boiling point of water is almost 200 °C higher than this value! Similarly, the boiling points of NH_3 and HF

(a) Ethylene glycol ($HOCH_2CH_2OH$), a polar compound used as antifreeze in automobiles, dissolves in water.

(b) Nonpolar motor oil (a hydrocarbon) dissolves in nonpolar solvents such as gasoline or CCl_4. It will not dissolve in a polar solvent such as water, however. Commercial spot removers use nonpolar solvents to dissolve oil and grease from fabrics.

FIGURE 12.5 "Like dissolves like."

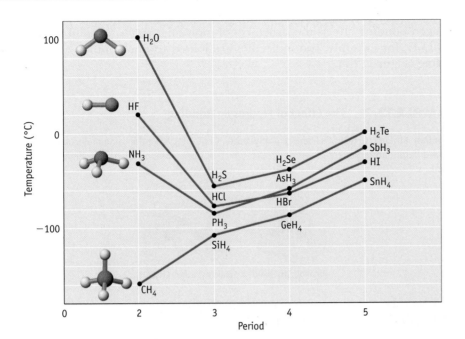

FIGURE 12.6 The boiling points of some simple hydrogen compounds. The effect of hydrogen bonding is apparent in the unusually high boiling points of H_2O, HF, and NH_3. (Also, notice that the boiling point of HCl is somewhat higher than expected based on the data for HBr and HI. It is apparent that some degree of hydrogen bonding also occurs in liquid HCl.)

are much higher than would be expected based on molar mass. Because the temperature at which a substance boils depends on the attractive forces between molecules, the extraordinarily high boiling points of H_2O, HF, and NH_3 indicate strong intermolecular attractions.

Why should H_2O, NH_3, and HF have such strong intermolecular forces? The answer starts with the electronegativities of N (3.0), O (3.5), and F (4.0), which are among the highest of all the elements, whereas the electronegativity of hydrogen is much lower (2.2). This large difference in electronegativity means that N—H, O—H, and F—H bonds are very polar. In bonds between H and N, O, or F, the more electronegative element takes on a significant negative charge (◄ Figure 8.11), and the hydrogen atom acquires a significant positive charge.

There is an unusually strong attraction between an electronegative atom with a lone pair of electrons (most often, an N, O, or F atom in another molecule or even in the same molecule) and the hydrogen atom of the N—H, O—H, or F—H bond. This type of interaction is known as a **hydrogen bond**. Hydrogen bonds are an extreme form of dipole–dipole interaction where one atom involved is always H and the other atom is highly electronegative, most often O, N, or F. A hydrogen bond can be represented as

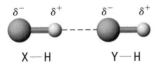

The hydrogen atom becomes a bridge between the two electronegative atoms X and Y, and the dashed line represents the hydrogen bond. The most pronounced effects of hydrogen bonding occur where both X and Y are N, O, or F. Energies associated with most hydrogen bonds involving these elements are in the range of 5 to 30 kJ/mol.

Types of Hydrogen Bonds [X—H - - - :Y]

N—H - - - :N—	O—H - - - :N—	F—H - - - :N—
N—H - - - :O—	O—H - - - :O—	F—H - - - :O—
N—H - - - :F—	O—H - - - :F—	F—H - - - :F—

Hydrogen bonding has important implications for any property of a compound that is influenced by intermolecular forces of attraction. For example,

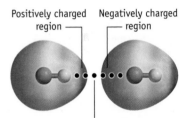

Positively charged region   Negatively charged region

Hydrogen bond

Hydrogen bonding between HF molecules. The partially negative F atom of one HF molecule interacts through hydrogen bonding with a neighboring HF molecule. (Red regions of the molecule are negatively charged, whereas blue regions are positively charged. For more on electrostatic potential surfaces, ◄ page 294.)

● **The Importance of H Atom Charge Density in Hydrogen Bonding** Not only are the bonds between H and O, N, and F significantly polar, but the H atom has an extraordinarily small radius. This means that the partial charge on the H atom is concentrated in a small volume; that is, it has a high charge density. The result is that it is strongly attractive to the negative charge on a neighboring molecule.

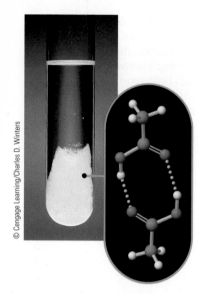

FIGURE 12.7 Hydrogen bonding. Two acetic acid molecules can interact through hydrogen bonds. This photo shows partly solid glacial acetic acid. Notice that the solid is denser than the liquid, a property shared by virtually all substances, the notable exception being water.

hydrogen bonding affects the structures of molecular solids. In solid acetic acid, CH_3CO_2H, for example, two molecules are joined to one another by hydrogen bonding (Figure 12.7).

EXAMPLE 12.2 The Effect of Hydrogen Bonding

Problem Ethanol, CH_3CH_2OH, and dimethyl ether, CH_3OCH_3, have the same molecular formula but a different arrangement of atoms. Predict which of these compounds has the higher boiling point.

ethanol, CH_3CH_2OH dimethyl ether, CH_3OCH_3

What Do You Know? You know the molecular structures for the two compounds and that the compound with the stronger intermolecular forces will have the higher boiling point.

Strategy Inspect the structure of each molecule to decide whether each is polar and, if polar, whether hydrogen bonding is possible.

Solution Although these two compounds have identical masses, they have different structures. Ethanol possesses a polar O—H group and so can participate in hydrogen bonding. Dimethyl ether is a polar molecule, and the O atom has a partial negative charge. However, no H atom is attached to the O atom. Thus, there is no opportunity for hydrogen bonding in dimethyl ether. We can predict, therefore, that intermolecular forces will be larger in ethanol than in dimethyl ether and that ethanol will have the higher boiling point.

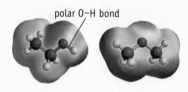

polar O–H bond

$$CH_3CH_2-\overset{..}{\underset{|}{O}}:\cdots H-\overset{..}{\underset{|}{O}}:$$
$$\qquad\qquad H \qquad\qquad CH_2CH_3$$

hydrogen bonding in ethanol, CH_3CH_2OH

Think about Your Answer According to the literature, ethanol boils at 78.3 °C, whereas dimethyl ether has a boiling point of −24.8 °C, more than 100 °C lower. Under similar conditions (room temperature and 1 atm pressure), dimethyl ether is a gas, whereas ethanol is a liquid.

Check Your Understanding

Using structural formulas, describe the hydrogen bonding between methanol (CH_3OH) molecules. What physical properties of methanol are likely to be affected by hydrogen bonding?

Hydrogen Bonding and the Unusual Properties of Water

One of the most striking differences between our planet and others in our solar system is the presence of large amounts of water on Earth. Three fourths of the planet is covered by oceans; the polar regions are vast ice fields; and even soil and rocks hold large amounts of water. Although we tend to take water for granted, almost no other substance behaves in a similar manner. Water's unique features are a consequence of the ability of H_2O molecules to cling tenaciously to one another by hydrogen bonding.

The unusually high intermolecular forces of attraction between water molecules are a result of the fact that each water molecule can participate in *four* hydrogen bonds. An individual water molecule has two polar O—H bonds and two lone pairs. Both hydrogen atoms are available to hydrogen bond to oxygen atoms in adjacent water molecules. In addition, the oxygen lone pairs can participate in hydrogen bonding to the hydrogen atoms in two other water molecules (Figure 12.8a).

Earth from space. Three quarters of the earth is covered by water, and vast amounts of water are locked in the polar ice caps. Many of the unique properties of water depend on the presence of hydrogen bonding in water.

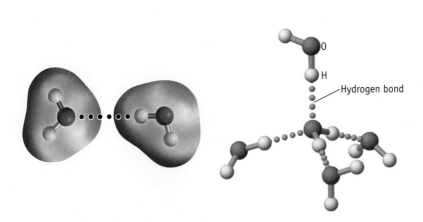

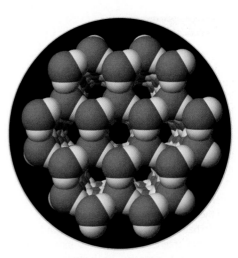

(a) Electrostatic potential surfaces for two water molecules shows the hydrogen bond involving the negatively charged O atom of one molecule and the positively charged H atom of a neighboring molecule.

(b) The oxygen atom of a water molecule attaches itself to two other water molecules by hydrogen bonds. Notice that the four groups that surround an oxygen atom are arranged as a distorted tetrahedron. Each oxygen atom is covalently bonded to two hydrogen atoms and hydrogen bonded to hydrogen atoms from two other molecules. The hydrogen bonds are longer than the covalent bonds.

(c) In ice, the structural unit shown in part (b) is repeated in the crystalline lattice. This computer-generated structure shows a small portion of the extensive lattice. Notice the six-member, hexagonal rings. The vertices of each hexagon are O atoms, and each side is composed of two oxygen atoms with a hydrogen atom in between. One of the oxygen atoms is covalently bonded to the hydrogen atom, and the other is attracted to it by a hydrogen bond.

FIGURE 12.8 Hydrogen bonding in water and the structure of ice.

The result, seen particularly in ice, is a tetrahedral arrangement for the hydrogen atoms around each oxygen, involving two covalently bonded hydrogen atoms and two hydrogen-bonded hydrogen atoms.

As a consequence of the regular arrangement of water molecules linked by hydrogen bonding, ice has an open-cage structure with lots of empty space (Figure 12.8c). The result is that ice has a density about 10% less than that of liquid water, which explains why ice floats. (In contrast, virtually all other solids sink in their liquid phase.) We can also see in this structure that the oxygen atoms are arranged at the corners of puckered, hexagonal rings. Snowflakes are always based on six-sided figures (◄ page 60), a reflection of this internal molecular structure of ice.

When ice melts at 0 °C, the regular structure imposed on the solid state by hydrogen bonding breaks down, and a relatively large increase in density occurs (Figure 12.9). Another surprising thing occurs when the temperature of liquid water is raised from 0 °C to 4 °C: The density of water increases. For almost every other substance known, density decreases as the temperature is raised. Once again, hydrogen bonding is the reason for water's seemingly odd behavior. At a temperature just above the melting point, some of the water molecules continue to cluster in ice-like arrangements, which require extra space. As the temperature is raised from 0 °C to 4 °C, the final vestiges of the ice structure disappear, and the volume contracts further, giving rise to the increase in density. Water's density reaches a maximum at about 4 °C. From this point, the density declines with increasing temperature in the normal fashion.

Because of the way that water's density changes as the temperature approaches the freezing point, lakes do not freeze from the bottom up in the winter. When lake water cools with the approach of winter, its density increases, the cooler water sinks, and the warmer water rises. This "turn over" process continues until all the water reaches 4 °C, the maximum density. (This is the way oxygen-rich water moves to the lake bottom to restore the oxygen used during the summer and nutrients are brought to the top layers of the lake.) As the temperature decreases further, the

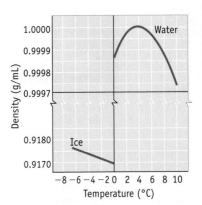

FIGURE 12.9 The temperature dependence of the densities of ice and water.

Hydrogen Bonding & Methane Hydrates: Opportunities and Problems

Cold water under pressure begins to form complex solid structures at temperatures above 273 K. These are unstable networks of hydrogen-bonded water molecules with large, open cavities. If a small "guest" molecule of the right size is present, however, it can be trapped in the cavities, and the network does not collapse to form the usual ice structure. This phenomenon is most often observed when cold water is saturated with methane, and the result is methane hydrate (Figure A).

Methane hydrates have been known for years, but interest in them has increased because vast deposits of hydrate were recently discovered deep within sediments on the floor of the oceans. It is estimated that global methane hydrate deposits contain approximately 10^{13} tons of carbon, or about twice the combined amount in all known reserves of coal, oil, and natural gas. Methane hydrate is also an efficient energy storehouse; a cubic meter of methane hydrate releases about 160 cubic meters of methane gas.

If methane is to be captured from hydrates and used as a fuel, there are problems to be solved. One significant problem is how to bring commercially useful quantities to the surface from deep in the ocean. Yet another is the possibility of a large, uncontrolled release of methane. Methane is a very effective greenhouse gas, so the release of a significant quantity into the atmosphere could damage the earth's climate. (Indeed, some believe a massive release of methane from hydrates about 55 million years ago led to significant global warming.)

In May 2010 methane hydrates may have led to an environmental catastrophe. The Deepwater Horizon rig used for drilling for oil in very deep water in the Gulf of Mexico was destroyed by a fire triggered by a methane explosion. There is a strong suggestion that the ultimate cause may have been the explosive release of the gas from methane hydrates under the sea bed. The methane shot to the surface through the drilling pipe and was ignited. This then led to an uncontrolled leak of oil from the well, followed by attempts to shut it down. The first attempt to cap the well was to lower a 100-ton steel and concrete dome over the well, which would contain the oil and allow it to be pumped to the surface (Figure B). This was a failure because methane coming from the rupture formed a slush of methane hydrate. The "crystals" of hydrate were trapped in the dome, increased its buoyancy, and prevented it from being anchored to the sea floor.

FIGURE B A containment dome used to control the oil leak in the Gulf of Mexico, May, 2010. The first containment dome that was tried did not work because methane hydrate prevented it from being anchored to the sea floor.

The realization that methane hydrates may be the cause of a significant accident will surely lead to reassessment of oil exploration in very deep water. But it may also renew interest in using more accessible methane hydrates as an energy source. For example, in Alaska's North Slope there are recoverable deposits of 2400 billion cubic meters of methane, roughly the equivalent of four years of U.S. natural gas production And there are extensive deposits in other places in the world. Indeed, the deepest oceans may be a better source of methane than of oil.

Questions:

1. In a methane hydrate the methane molecule is trapped in a cage of water molecules. Describe the structure: (a) how many water molecules make up the cage, (b) how many hydrogen bonds are involved, and (c) how many faces does the cage have?

2. One cubic meter of methane hydrate has 164 m^3 of CH_4 (at STP). If you burn the methane in 1.00 m^3 of hydrate [to give $CO_2(g)$ and $H_2O(g)$], how much energy can be obtained?

Answers to these questions are available in Appendix N

Reference:

A good article on methane hydrates is: E. Suess, G. Bohrmann, J. Greinert, and E. Lausch, *Scientific American*, November 1999, pp. 76–83.

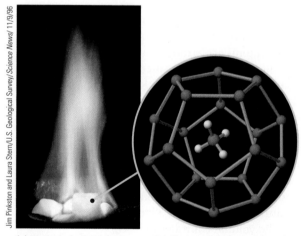

(a) Methane hydrate burns as methane gas escapes from the solid hydrate.

(b) Methane hydrate consists of a lattice of water molecules with methane molecules trapped in the cavity.

FIGURE A Methane hydrate. When a sample is brought to the surface from the depths of the ocean, the methane oozes out of the solid, and the gas readily burns. The structure of the solid methane hydrate consists of methane molecules trapped in a lattice of water molecules. The lattice shown here is a common structural unit of a more complex structure. Each point of the lattice shown here is an O atom of an H_2O molecule. The edges consist of an O—H—O series of atoms connected by a hydrogen bond and covalent bond. (Other, more complex structural units are known.) Such structures are often called "clathrates."

colder water stays on the top of the lake, because water cooler than 4 °C is less dense than water at 4 °C. With further heat loss, ice can then begin to form on the surface, floating there and protecting the underlying water and aquatic life from further heat loss.

Extensive hydrogen bonding is also the origin of the extraordinarily high heat capacity of water. Although liquid water does not have the regular structure of ice, hydrogen bonding still occurs. When the temperature is raised, there must be a significant input of energy to disrupt the intermolecular forces and to raise the temperature even a small amount.

The high specific heat capacity of water is, in large part, why oceans and lakes have such an enormous effect on weather. In autumn, when the temperature of the air is lower than the temperature of the ocean or lake, water transfers energy as heat to the atmosphere, moderating the drop in air temperature. Furthermore, so much energy is available to be transferred for each degree drop in temperature that the decline in water temperature is gradual. For this reason, the temperature of the ocean or of a large lake is generally higher than the average air temperature until late in the autumn.

REVIEW & CHECK FOR SECTION 12.3

1. In which of the following substances would hydrogen bonding be expected?

    (a) CH_4 (b) CH_3OH (c) H_3COCH_3 (d) $H_2C=CH_2$

2. Which compound has the strongest intermolecular forces?

    (a) H_2O (b) H_2S (c) H_2Se (d) H_2Te

12.4 Intermolecular Forces Involving Nonpolar Molecules

Many important molecules such as O_2, N_2, and the halogens are not polar. Why, then, does O_2 dissolve in polar water? Why can the N_2 of the atmosphere be liquefied? Some intermolecular forces must be acting between O_2 and water and between N_2 molecules, but what is their nature?

Dipole-Induced Dipole Forces

Polar molecules such as water can induce, or create, a dipole in molecules that do not have a permanent dipole. To see how this can happen, picture a polar water molecule approaching a nonpolar molecule such as O_2 (Figure 12.10). The electron cloud of an isolated (gaseous) O_2 molecule is symmetrically distributed between the two oxygen atoms. As the negative end of the polar H_2O molecule approaches, however, the O_2 electron cloud becomes distorted. In this process, the O_2 molecule itself becomes polar; that is, a dipole is *induced* in the otherwise nonpolar O_2 molecule. The result is that H_2O and O_2 molecules are now attracted to one another, albeit only weakly. Oxygen can dissolve in water because a force of attraction exists between water's permanent dipole and the induced dipole in O_2. Chemists refer to such interactions as **dipole-induced dipole interactions**.

The process of inducing a dipole is called **polarization**, and the degree to which the electron cloud of an atom or a molecule can be distorted depends on the **polarizability** of that atom or molecule. The electron cloud of an atom or molecule with a large, extended electron cloud such as I_2 can be polarized more readily than the electron cloud in a much smaller atom or molecule, such as He or H_2, in which the valence electrons are close to the nucleus and more tightly held. In general, for an analogous series of substances, say the halogens or alkanes (such as CH_4, C_2H_6, C_3H_8, and so on), the higher the molar mass, the greater the polarizability of the molecule.

(a) A polar molecule such as water can induce a dipole in nonpolar O_2 by distorting the molecule's electron cloud.

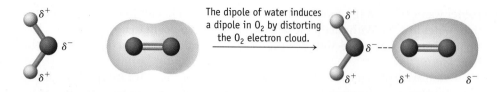

The dipole of water induces a dipole in O_2 by distorting the O_2 electron cloud.

(b) Nonpolar I_2 dissolves in polar ethanol (C_2H_5OH). The intermolecular force involved is a dipole-induced dipole force.

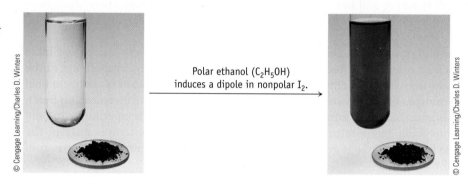

Polar ethanol (C_2H_5OH) induces a dipole in nonpolar I_2.

© Cengage Learning/Charles D. Winters

© Cengage Learning/Charles D. Winters

FIGURE 12.10 Dipole-induced dipole interaction.

The solubilities of common gases in water illustrate the effect of interactions between a dipole and an induced dipole. In Table 12.3, you see a trend to higher solubility with increasing mass of the nonpolar gas. As the molar mass of the gas increases, the polarizability of the electron cloud increases and the strength of the dipole-induced dipole interaction increases.

London Dispersion Forces: Induced Dipole-Induced Dipole Forces

Iodine, I_2, is a solid and not a gas around room temperature and pressure, illustrating that nonpolar molecules must also experience intermolecular forces. An indication of the magnitude of these forces is provided by the enthalpy of vaporization of the substance at its boiling point. The data in Table 12.4 suggest that these forces can range from very weak (N_2, O_2, and CH_4 with low enthalpies of vaporization and very low boiling points) to more substantial (I_2 and benzene).

To understand how two nonpolar molecules can attract each other, recall that the electrons in atoms or molecules are in a state of constant motion. When two atoms or nonpolar molecules approach each other, attractions or repulsions between their electrons and nuclei can lead to distortions in their electron clouds (Figure 12.11).

Table 12.3 The Solubility of Some Gases in Water*

	Molar Mass (g/mol)	Solubility at 20 °C (g gas/100 g water)†
H_2	2.01	0.000160
N_2	28.0	0.00190
O_2	32.0	0.00434

*Data taken from J. Dean: *Lange's Handbook of Chemistry.* 14th Ed., pp. 5.3–5.8, New York, McGraw-Hill, 1992.
†Measured under conditions where pressure of gas + pressure of water vapor = 760 mm Hg.

Table 12.4 Enthalpies of Vaporization and Boiling Points of Some Nonpolar Substances

	$\Delta_{vap}H°$ (kJ/mol)	Element/Compound BP (°C)
N_2	5.57	−196
O_2	6.82	−183
CH_4 (methane)	8.2	−161.5
Br_2	29.96	+58.8
C_6H_6 (benzene)	30.7	+80.1
I_2	41.95	+185

Hydrogen Bonding in Biochemistry

It is arguable that our world is what it is because of hydrogen bonding in water and in biochemical systems. Perhaps the most important occurrence is in DNA and RNA where the organic bases adenine, cytosine, guanine, and thymine (in DNA) or uracil (in RNA) are attached to sugar-phosphate chains (**Figure A**). The chains in DNA are joined by the pairing of bases, adenine with thymine and guanine with cytosine.

Figure B illustrates the hydrogen bonding between adenine and thymine. These models show that the molecules naturally fit together to form a six-sided ring, where two of the six sides involve hydrogen bonds. One side consists of a N · · · H—N grouping, and the other side is N—H · · · O. Here, the electrostatic potential surfaces show that the N atoms of adenine and the O atoms of thymine bear partial negative charges, and the H atoms of the N—H groups bear a positive charge. These charges and the geometry of the bases lead to these very specific interactions.

The fact that base pairing through hydrogen bonding leads to the joining of the sugar-phosphate chains of DNA, and to the double helical form of DNA, was first recognized by James Watson and Francis Crick on the basis of experimental work by Rosalind Franklin and Maurice Wilkins in the 1950s. Determination of the DNA structure was a key development in the molecular biology revolution in the last part of the 20th century. (See page 303 for more on these scientists.)

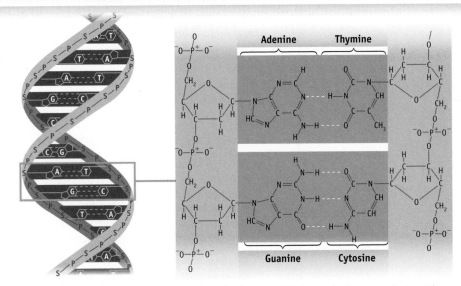

FIGURE A **Hydrogen bonding in DNA.** With the four bases in DNA, the usual pairings are adenine with thymine and guanine with cytosine. This pairing is promoted by hydrogen bonding.

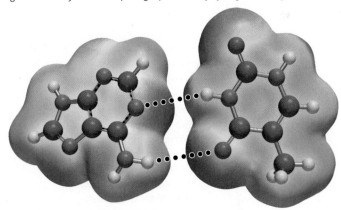

FIGURE B **Hydrogen bonding between adenine and thymine.** Electrostatic potential surfaces depict the hydrogen bonding interactions between adenine and thymine. The polar N—H bond on one molecule can hydrogen bond to an electronegative N or O atom in a neighboring molecule.

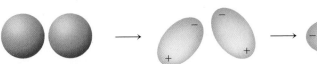

| Two nonpolar atoms or molecules (depicted as having an electron cloud that has a time-averaged spherical shape). | Momentary attractions and repulsions between nuclei and electrons in neighboring molecules lead to induced dipoles. | Correlation of the electron motions between the two atoms or molecules (which are now polar) leads to a lower energy and stabilizes the system. |

FIGURE 12.11 **Induced dipole–induced dipole interactions or London dispersion forces.** Momentary attractions and repulsions between nuclei and electrons create induced dipoles and lead to a net stabilization due to attractive forces. Nonpolar Br_2 and I_2 both exemplify such forces. They are a liquid and a solid, respectively, indicating that there are forces between the molecules sufficient to cause them to be in a condensed phase.

Br_2 I_2

That is, dipoles can be induced momentarily in neighboring atoms or molecules, and these *induced* dipoles lead to intermolecular attractions. The intermolecular force of attraction in liquids and solids composed of nonpolar molecules is an **induced dipole-induced dipole force**. Chemists often call them **London dispersion forces**. *Dispersion forces actually arise between all molecules, both nonpolar and polar, but dispersion forces are the only intermolecular forces between nonpolar molecules.*

EXAMPLE 12.3   Intermolecular Forces

Problem Suppose you have a mixture of solid iodine, I_2, and the liquids water and carbon tetrachloride (CCl_4). What intermolecular forces exist between each possible pair of compounds?

What Do You Know? Iodine, I_2, is a nonpolar molecule composed of large iodine atoms. It has an extensive electron cloud and is polarizable. CCl_4 is a symmetrical tetrahedral molecule and is not polar. Polar H_2O could be involved in hydrogen bonding with other water molecules or with other molecules with highly polar groups.

Strategy You know whether each substance is polar or nonpolar so you need only to determine the types of intermolecular forces that can exist between the different pairs.

Solution Nonpolar iodine, I_2, is easily polarized, and iodine can interact with polar water molecules by dipole-induced dipole forces. Nonpolar carbon tetrachloride can interact with nonpolar iodine only by dispersion forces. Water and CCl_4 could interact by dipole-induced dipole forces.

Think about Your Answer The photos here show the result of mixing these three compounds. Iodine does dissolve to a small extent in water to give a brown solution. When this brown solution is added to a test tube containing CCl_4, the liquid layers do not mix. (Polar water does

Nonpolar I_2
Polar H_2O

Nonpolar CCl_4

Shake the test tube

Polar H_2O

Nonpolar CCl_4 and I_2

© Cengage Learning/ Charles D. Winters

© Cengage Learning/ Charles D. Winters

not dissolve in nonpolar CCl_4.) (Notice the more dense CCl_4 layer [$d = 1.58$ g/mL] is underneath the less dense water layer.) When the test tube is shaken, nonpolar I_2 is extracted into nonpolar CCl_4, as evidenced by the disappearance of the color of I_2 in the water layer (top) and the appearance of the purple I_2 color in the CCl_4 layer (bottom).

Check Your Understanding

You mix water, CCl_4, and hexane ($CH_3CH_2CH_2CH_2CH_2CH_3$). What type of intermolecular forces can exist between each pair of these compounds?

REVIEW & CHECK FOR SECTION 12.4

1. In which of the following solids are the component particles attracted to each other only by induced dipole-induced dipole forces?

   (a) ice   (c) NaCl

   (b) solid NH_3 at low temperature   (d) I_2

2. Which of the following compounds has the largest intermolecular forces?

   (a) pentane, C_5H_{12} (c) heptane, C_7H_{16}

   (b) hexane, C_6H_{14} (d) octane, C_8H_{18}

Table 12.5 **Summary of Intermolecular Forces (in descending order of strength)**

Type of Interaction	Factors Responsible for Interaction	Example
Ion–dipole	Ion charge, magnitude of dipole	$Na^+ \ldots H_2O$
Hydrogen bonding, X—H . . . :Y	Very polar X—H bond and atom Y with lone pair of electrons (where X and Y = F, N, O). Typical energy = 20 kJ/mol.	$H_2O \ldots H_2O$
Dipole–dipole . . . $(CH_3)_2O$	Dipole moment (depends on electronegativities and molecular structure). Typical energy = 5–20 kJ/mol.	$(CH_3)_2O \ldots (CH_3)_2O$
Dipole-induced dipole	Dipole moment of polar molecule and polarizability of nonpolar molecule. Typical energy <2 kJ/mol.	$H_2O \ldots I_2$
Induced dipole-induced dipole (London dispersion forces)	Polarizability	$I_2 \ldots I_2$

12.5 A Summary of van der Waals Intermolecular Forces

Van der Waals intermolecular forces involve molecules that are polar or those in which polarity can be induced (Table 12.5). It is important to recognize that

- several types of intermolecular forces contribute to the overall ability of a molecule (Figure 12.12) to interact with another of the same or different kind.
- induced dipole-induced dipole forces can be quite substantial, even in polar molecules (Figure 12.12).

EXAMPLE 12.4   Intermolecular Forces

Problem Decide which are the most important intermolecular forces between each of the following and compare the strength of the interactions: (a) between molecules in liquid methane, CH_4; (b) between molecules of water and methanol (CH_3OH) in a mixture of these species; and (c) between molecules of bromine and water in a mixture of these compounds.

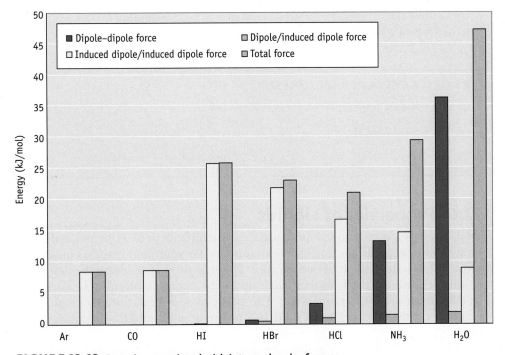

FIGURE 12.12 Energies associated with intermolecular forces.

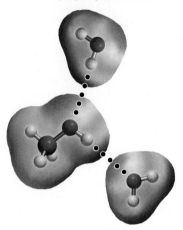

Hydrogen bonding involving methanol (CH_3OH) and water.

What Do You Know? Methane, CH_4, is a symmetrical, tetrahedral molecule and so is not polar. Both water and methanol are polar, and both have $-OH$ groups that can be involved in hydrogen bonding. Bromine, Br_2, is nonpolar, but it is a large molecule and has an extensive, polarizable electron cloud.

Strategy Determine the type of interaction possible between pairs of molecules based on the structure and characteristics of each species. To evaluate their relative importance, consult Table 12.5 on relative strengths of interactions.

Solution

(a) Methane is not polar. Therefore, the only way methane molecules can interact with one another is through induced dipole-induced dipole forces (dispersion forces).

(b) Both water and methanol are polar, and both have an O—H bond. They therefore interact through the special dipole–dipole force called hydrogen bonding as well as dispersion forces.

$$\delta^+H \diagdown_{\delta^+H} \overset{\delta^-}{O} \cdots \overset{\delta^+}{H} - \overset{\delta^-}{O} \diagdown_{CH_3} \quad \text{and} \quad \delta^+H \diagdown_{H_3C} \overset{\delta^-}{O} \cdots \overset{\delta^+}{H} - \overset{\delta^-}{O} \diagdown_{H^{\delta^+}}$$

(c) Nonpolar molecules of bromine, Br_2, and polar water molecules interact through dipole-induced dipole forces (and dispersion forces). (This is similar to the I_2–ethanol interaction in Figure 12.10.)

In order of increasing strength, the likely order of interactions is

liquid CH_4 < H_2O and Br_2 < H_2O and CH_3OH

Think about Your Answer The fact that CH_4 is a liquid only at very low temperatures suggests it has the weakest attractive forces. We would expect the highest forces of attraction to be those involving hydrogen bonding. Thus, the strongest interaction is expected to be between H_2O and CH_3OH, whereas the weakest interaction occurs between two nonpolar molecules of CH_4.

Check Your Understanding

Decide which type of intermolecular force is involved in (a) liquid O_2; (b) liquid CH_3OH; and (c) N_2 dissolved in H_2O. Place the interactions in order of increasing strength.

REVIEW & CHECK FOR SECTION 12.5

Rank the following molecules in order of increasing intermolecular forces: SO_2, CH_3OH, and He.

(a)   He < SO_2 < CH_3OH

(b)   He < CH_3OH < SO_2

(c) CH_3OH < He < SO_2

(d) SO_2 < CH_3OH < He

12.6 Properties of Liquids

Of the three states of matter, liquids are the most difficult to describe precisely. The molecules in a gas under normal conditions are far apart and may be considered more or less independent of one another. The structures of solids can be described readily because the particles that make up solids—atoms, molecules, or ions—are usually in an orderly arrangement. The particles of a liquid interact with their neighbors, like the particles in a solid, but, unlike in solids, there is little long-range order.

In spite of the difficulty in describing liquids, we can still consider the behavior of liquids at the molecular level. Here we want to look further at the process of va-

A Pet Food Catastrophe

In early 2007 pet owners all over the United States reported pets becoming seriously ill or dying from eating apparently adulterated food. Dogs and cats developed symptoms of kidney failure, which included loss of appetite, vomiting, extreme thirst, and lethargy. Many ultimately died. Round, greenish-brown kidney stones were also found to be clogging the kidneys of stricken pets. This was mysterious at first, but, within two months, chemists and toxicologists traced the ailments to two compounds in the pet food: melamine and cyanuric acid.

Melamine   Cyanuric acid

Melamine is used to make plastics and fertilizers. It is not approved for food use. Cyanuric acid is used to stabilize chlorine in swimming pools and to sanitize food processing equipment. The acid is sometimes formed as a by-product in the manufacture of melamine, so melamine samples can be contaminated with cyanuric acid.

But why was melamine in pet food in the first place? A U.S. manufacturer of pet food, which also supplied other companies that in turn distributed the food under their brand names, had purchased wheat gluten from a supplier in China. Wheat gluten is a concentrated vegetable protein used as a thickener and binder in pet food. The speculation is that the Chinese supplier had added melamine to the wheat gluten to raise the apparent nitrogen content. When products are tested for protein content, any nitrogen found is assumed to come from protein. Adding melamine is an inexpensive way to deceive

the buyer into thinking that the product contains a higher percentage of protein.

Melamine and cyanuric acid by themselves are not toxic, but when they are present together they are. The reason is hydrogen bonding! Mixing melamine and cyanuric acid in water produces insoluble crystals of melamine cyanurate.

The structure above shows only the hydrogen bonding between two molecules. However, additional hydrogen bonding to other molecules is possible, and the result is further aggregation of cyanuric acid and melamine to form an insoluble material.

The behavior of the hydrogen-bonded complex depends on pH. In the stomach, a very acidic medium with a low pH, hydrogen bonding does not occur between these compounds, and the compounds are soluble. The body seeks to get rid of these chemicals through the kidneys, however, and there the compounds encounter a near neutral pH. Hydrogen bonding can then occur, and the molecules aggregate to form solid melamine cyanurate.

The adulteration of food with melamine was not limited to the pet food industry. In 2008, a year after the pet food adulteration scandal, melamine was found in milk products including infant formula. Several thousand babies in China became ill, having suffered acute kidney failure, and several fatalities resulted. These babies had been fed formula milk contaminated with melamine.

Questions:
1. Calculate the weight percent of nitrogen in melamine and cyanuric acid and compare this value to the average percent of nitrogen in protein (14%).
2. An infant formula was found to have 0.14 ppm melamine. (ppm stands for parts per million. This means, for example, there is 1 g of a substance per million grams of the product.) What mass of melamine is there in a package of infant formula with a mass of 454 g (one pound)?

Answers to these questions are available in Appendix N.

---

porization, at the vapor pressure of liquids, at their boiling points and critical properties, and at surface tension, capillary action, and viscosity.

Vaporization and Condensation

Vaporization or evaporation is the process in which a substance in the liquid state becomes a gas. In this process, molecules escape from the liquid surface and enter the gaseous state.

To understand evaporation, we have to look at molecular energies. Molecules in a liquid have a range of energies (Figure 12.13) that resembles the distribution of energies for molecules of a gas (◄ Figure 11.14). As with gases, the average energy for molecules in a liquid depends only on temperature: The higher the

FIGURE 12.13 The distribution of energy among molecules in a liquid sample. T_2 is a higher temperature than T_1, and at the higher temperature, there are more molecules with an energy greater than E.

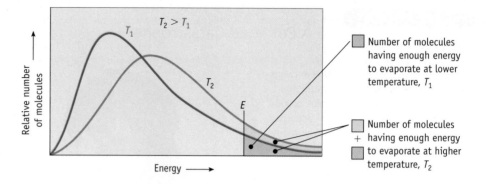

■ Number of molecules having enough energy to evaporate at lower temperature, T_1

■ Number of molecules
+ having enough energy to evaporate at higher
■ temperature, T_2

temperature, the higher the average energy and the greater the relative number of molecules with high kinetic energy. In a sample of a liquid, at least a few molecules have more kinetic energy than the potential energy of the intermolecular attractive forces holding the liquid molecules to one another. If these high-energy molecules are at the surface of the liquid and if they are moving in the right direction, they can break free of their neighbors and enter the gas phase (Figure 12.14).

Vaporization is an endothermic process because energy is required to overcome the intermolecular forces of attraction holding the molecules together. The energy required to vaporize a sample is often given as the standard **molar enthalpy of vaporization**, $\Delta_{vap}H°$ (in units of kilojoules per mole; see Tables 12.4 and 12.6 and Figure 12.4).

$$liquid \xrightarrow[\substack{energy\ absorbed \\ by\ liquid}]{vaporization} vapor \qquad \Delta_{vap}H° = molar\ enthalpy\ of\ vaporization$$

A molecule in the gas phase can transfer some of its kinetic energy by colliding with slower gaseous molecules and solid objects. If this molecule loses sufficient energy and comes in contact with the surface of the liquid, it can reenter the liquid phase in the process called **condensation**.

$$vapor \xrightarrow[\substack{energy\ released \\ by\ vapor}]{condensation} liquid \qquad -\Delta_{vap}H° = molar\ enthalpy\ of\ condensation$$

Condensation is the reverse of vaporization and so is an exothermic process. Energy is transferred to the surroundings. *The enthalpy change for condensation is equal but opposite in sign to the enthalpy of vaporization.* For example, the enthalpy change for the vaporization of 1.00 mol of water at 100 °C is +40.7 kJ. On condensing 1.00 mol of water vapor to liquid water at 100 °C, the enthalpy change is −40.7 kJ.

In the discussion of intermolecular forces, we pointed out the relationship between the $\Delta_{vap}H°$ values for various substances and the temperatures at which they

FIGURE 12.14 Evaporation. Some molecules at the surface of a liquid have enough energy to escape the attractions of their neighbors and enter the gaseous state. At the same time, some molecules in the gaseous state can reenter the liquid.

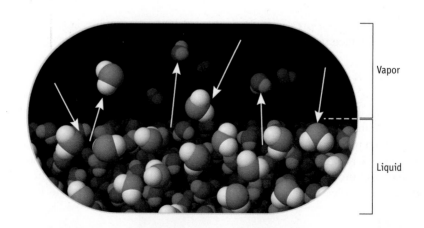

Vapor

Liquid

Table 12.6 Molar Enthalpies of Vaporization and Boiling Points for Common Substances*

Compound	Molar Mass (g/mol)	$\Delta_{vap}H°$ (kJ/mol)†	Boiling Point (°C) (Vapor Pressure = 760 mm Hg)
Polar Compounds			
HF	20.0	25.2	19.7
HCl	36.5	16.2	−84.8
HBr	80.9	19.3	−66.4
HI	127.9	19.8	−35.6
NH_3	17.0	23.3	−33.3
H_2O	18.0	40.7	100.0
SO_2	64.1	24.9	−10.0
Nonpolar Compounds			
CH_4 (methane)	16.0	8.2	−161.5
C_2H_6 (ethane)	30.1	14.7	−88.6
C_3H_8 (propane)	44.1	19.0	−42.1
C_4H_{10} (butane)	58.1	22.4	−0.5
Monatomic Elements			
He	4.0	0.08	−268.9
Ne	20.2	1.7	−246.1
Ar	39.9	6.4	−185.9
Xe	131.3	12.6	−108.0
Diatomic Elements			
H_2	2.0	0.90	−252.9
N_2	28.0	5.6	−195.8
O_2	32.0	6.8	−183.0
F_2	38.0	6.6	−188.1
Cl_2	70.9	20.4	−34.0
Br_2	159.8	30.0	58.8

*Data taken from D. R. Lide: *Basic Laboratory and Industrial Chemicals,* Boca Raton, FL, CRC Press, 1993.
†$\Delta_{vap}H°$ is measured at the normal boiling point of the liquid.

boil (Table 12.6). Both properties reflect the attractive forces between particles in the liquid. The boiling points of nonpolar liquids (e.g., the hydrocarbons, atmospheric gases, and the halogens) increase with increasing atomic or molecular mass, a reflection of increased intermolecular dispersion forces. The alkanes (such as methane) listed in Table 12.6 show this trend clearly. Similarly, the boiling points and enthalpies of vaporization of the heavier hydrogen halides (HX, where X = Cl, Br, and I) increase with increasing molecular mass. For these molecules, dispersion forces and ordinary dipole–dipole forces account for their intermolecular attractions (Figure 12.12). Because dispersion forces become increasingly important with increasing mass, the enthalpies of vaporization and the boiling points are in the order HCl < HBr < HI. Among the hydrogen halides HF is the exception; the high enthalpy of vaporization and boiling point are a direct result of extensive hydrogen bonding.

⬤WL   INTERACTIVE EXAMPLE 12.5   Enthalpy of Vaporization

Problem You put 925 mL of water (about 4 cupsful) in a pan at 100 °C, and the water slowly evaporates. How much energy is transferred as heat to vaporize all the water?

What Do You Know? You know the volume of water and wish to know the energy required for evaporation. Three pieces of information are needed to solve this problem:

1. $\Delta_{vap}H°$ for water = +40.7 kJ/mol at 100 °C.

2. Density of water at 100 °C = 0.958 g/cm³.

3. Molar mass of water = 18.02 g/mol.

Strategy Map 12.5

PROBLEM
Calculate the **energy** required to **evaporate** water sample.

↓

DATA/INFORMATION
- Enthalpy of vaporization at boiling point
- Volume of water

STEP 1. Use **density** and **molar mass** to convert *volume* of water to *amount* (mol).

↓

Amount of sample

STEP 2. Multiply **amount** of sample by **molar enthalpy**.

↓

Energy required (kJ) to evaporate sample at boiling point

FIGURE 12.15 Rainstorms release an enormous quantity of energy. When water vapor condenses, energy is transferred to the surroundings. The enthalpy of condensation of water is large, so a large quantity of energy is released in a rainstorm.

Strategy $\Delta_{vap}H°$ has units of kilojoules per mole, so you first must find the mass of water and then the amount. Finally, use the enthalpy of vaporization (in kJ/mol) to calculate the energy required as heat.

Solution Using the density of water at this temperature, we find that 925 mL of water is equivalent to 886 g, and this mass is in turn equivalent to 49.2 mol of water.

$$925 \text{ mL} \left(\frac{0.958 \text{ g}}{1 \text{ mL}} \right) \left(\frac{1 \text{ mol}}{18.02 \text{ g}} \right) = 49.2 \text{ mol H}_2\text{O}$$

Therefore, the amount of energy required is

$$49.2 \text{ mol H}_2\text{O} \left(\frac{40.7 \text{ kJ}}{\text{mol}} \right) = 2.00 \times 10^3 \text{ kJ}$$

Think about Your Answer 2000 kJ is equivalent to the energy obtained by burning about 60 g of carbon.

Check Your Understanding

The molar enthalpy of vaporization of methanol, CH_3OH, is 35.2 kJ/mol at 64.6 °C. How much energy is required to evaporate 1.00 kg of methanol at 64.6 °C?

Liquid water is an exceptional substance; an enormous amount of energy as heat is required to convert liquid water to water vapor. This fact is important to your own physical well-being. When you exercise vigorously, your body responds by sweating. Energy from your body is transferred to sweat in the process of evaporation, and your body is cooled.

Enthalpies of vaporization and condensation of water also play a role in weather (Figure 12.15). For example, if enough water condenses from the air to fall as an inch of rain on an acre of ground, the energy released exceeds 2.0×10^8 kJ! This is equivalent to about 50 tons of exploded dynamite, the energy released by a small bomb.

Vapor Pressure

If you put some water in an open beaker, it will eventually evaporate completely. Air movement and gas diffusion remove the water vapor from the vicinity of the liquid surface, so many water molecules are not able to return to the liquid.

If you put water in a sealed flask (Figure 12.16), however, the water vapor cannot escape, and some will recondense to form liquid water. Eventually, the masses of liquid and vapor in the flask remain constant. This is another example of a **dynamic equilibrium** (◀ page 98).

$$\text{liquid} \rightleftharpoons \text{vapor}$$

FIGURE 12.16 Vapor pressure. A volatile liquid is placed in an evacuated flask *(left)*. At the beginning, no molecules of the liquid are in the vapor phase. After a short time, however, some of the liquid evaporates, and the molecules now in the vapor phase exert a pressure. The pressure of the vapor measured when the liquid and the vapor are in equilibrium is called the *equilibrium vapor pressure (right)*.

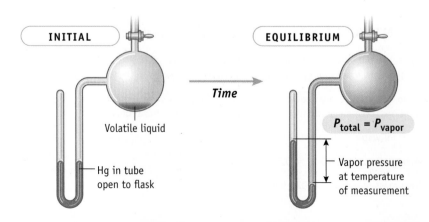

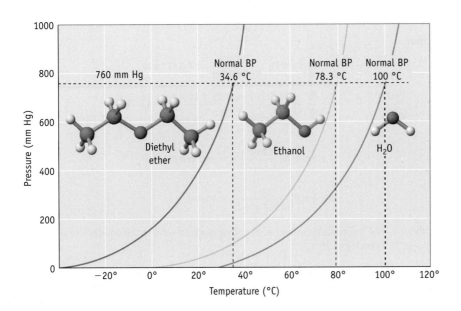

FIGURE 12.17 Vapor pressure curves for diethyl ether [(C₂H₅)₂O], ethanol (C₂H₅OH), and water. Each curve represents conditions of *T* and *P* at which the two phases, liquid and vapor, are in equilibrium. These compounds exist as liquids for temperatures and pressures to the left of the curve and as gases under conditions to the right of the curve.

Molecules still move continuously from the liquid phase to the vapor phase and from the vapor phase back to the liquid phase. At equilibrium, the rate at which molecules move from liquid to vapor is the same as the rate at which they move from vapor to liquid; thus, there is no net change in the masses of the two phases.

When a liquid–vapor equilibrium has been established, the equilibrium vapor pressure (often just called the vapor pressure) can be measured. The **equilibrium vapor pressure** of a substance is the pressure exerted by the vapor in equilibrium with the liquid phase. Conceptually, the vapor pressure of a liquid is a measure of the tendency of its molecules to escape from the liquid phase and enter the vapor phase at a given temperature. This tendency is referred to qualitatively as the **volatility** of the compound. The higher the equilibrium vapor pressure at a given temperature, the more volatile the substance.

As described previously (Figure 12.13), the distribution of molecular energies in the liquid phase is a function of temperature. At a higher temperature, more molecules have sufficient energy to escape the surface of the liquid. The equilibrium vapor pressure must, therefore, increase with temperature.

It is useful to represent vapor pressure as a function of temperature. Figure 12.17 shows the vapor pressure curves for several liquids as a function of temperature. *All points along the vapor pressure versus temperature curves represent conditions of pressure and temperature at which liquid and vapor are in equilibrium.* For example, at 60 °C the vapor pressure of water is 149 mm Hg. (Vapor pressures of water at various temperatures are given in Appendix G.) If water is placed in an evacuated flask that is maintained at 60 °C, liquid water will evaporate until the pressure exerted by the water vapor is 149 mm Hg (assuming enough water is in the flask so that some liquid remains when equilibrium is reached).

⍅WL   INTERACTIVE EXAMPLE 12.6   Using Vapor Pressure

Problem You place 2.00 L of water in an open container in your dormitory room; the room has a amount of 4.25×10^4 L. You seal the room and wait for the water to evaporate. Will all of the water evaporate at 25 °C? (At 25 °C the density of water is 0.997 g/mL, and its vapor pressure is 23.8 mm Hg.)

What Do You Know? You know the volume of water, the density of the water, the volume of the room, and the vapor pressure of the water at 25 °C.

Strategy One approach to solving this problem is to use the ideal gas law to calculate the amount of liquid water that must evaporate for the vapor produced to exert a pressure of 23.8 mm Hg in a volume of 4.25×10^4 L at 25 °C. Next, determine the volume of liquid water from the amount and compare the answer to the 2.00 L of water available.

Strategy Map 12.6

PROBLEM

Will a sample of water **evaporate completely** in a given volume at known T?

DATA/INFORMATION

- Volume of water
- Density of water
- Vapor pressure at T

STEP 1. Use **ideal gas law** to calculate *amount of water vapor* needed to have a P equal to vapor pressure in *given volume.*

Amount of water (mol) in the vapor phase to achieve P equal to the vapor pressure at given T.

STEP 2. Convert **amount of water** to a **volume of liquid water** using **molar mass** and **density**.

Volume of water that evaporated to achieve the **vapor pressure** at given T.

STEP 3. Compare *volume of water evaporated* to *volume available before evaporation.*

Volume of water that evaporated is *less* than that available.
Some liquid water remains.

Solution

$$P = 23.8 \text{ mm Hg}\left(\frac{1 \text{ atm}}{760 \text{ mm Hg}}\right) = 0.0313 \text{ atm}$$

$$n = \frac{PV}{RT} = \frac{(0.0313 \text{ atm})(4.25 \times 10^4 \text{ L})}{\left(0.082057 \dfrac{\text{L} \cdot \text{atm}}{\text{K} \cdot \text{mol}}\right)(298 \text{ K})} = 54.4 \text{ mol}$$

$$54.4 \text{ mol H}_2\text{O}\left(\frac{18.02 \text{ g}}{1 \text{ mol H}_2\text{O}}\right) = 980. \text{ g H}_2\text{O}$$

$$980. \text{ g H}_2\text{O}\left(\frac{1 \text{ mL}}{0.997 \text{ g H}_2\text{O}}\right) = 983 \text{ mL}$$

Only about half of the available water needs to evaporate to achieve the equilibrium water vapor pressure of 23.8 mm Hg at 25 °C.

Think about Your Answer Another approach to the problem is to calculate the pressure exerted if all of the water had evaporated. The answer would be 48.4 mm Hg. This is about twice the equilibrium vapor pressure at 25 °C, so only about half of the water needs to evaporate, as you found in the other approach.

Check Your Understanding

Examine the vapor pressure curve for ethanol in Figure 12.17.

(a) What is the approximate vapor pressure of ethanol at 40 °C?

(b) Are liquid and vapor in equilibrium when the temperature is 60 °C and the pressure is 600 mm Hg? If not, does liquid evaporate to form more vapor, or does vapor condense to form more liquid?

Vapor Pressure, Enthalpy of Vaporization, and the Clausius–Clapeyron Equation

Plotting the vapor pressure for a liquid at a series of temperatures results in a curved line (Figure 12.17). However, the German physicist R. Clausius (1822–1888) and the Frenchman B. P. E. Clapeyron (1799–1864) showed that, for a pure liquid, a linear relationship exists between the reciprocal of the Kelvin temperature ($1/T$) and the logarithm of the vapor pressure ($\ln P$) (Figure 12.18).

$$\ln P = -(\Delta_{\text{vap}}H^\circ/RT) + C \tag{12.1}$$

Here, $\Delta_{\text{vap}}H^\circ$ is the enthalpy of vaporization of the liquid, R is the ideal gas constant ($8.314472 \text{ J/K} \cdot \text{mol}$), and C is a constant characteristic of the liquid in question. This equation, now called the **Clausius–Clapeyron equation**, provides a method of obtaining values for $\Delta_{\text{vap}}H^\circ$. The equilibrium vapor pressure of a liquid can be measured at several different temperatures, and the logarithm of these pressures is plotted versus $1/T$. The result is a straight line with a slope of $-\Delta_{\text{vap}}H^\circ/R$. For example, plotting data for water (Figure 12.18), we find the slope of the line is -4.90×10^3 K^{-1}, which gives $\Delta_{\text{vap}}H^\circ = 40.7 \text{ kJ/mol}$.

As an alternative to plotting $\ln P$ versus $1/T$, we can write the following equation that allows us to calculate $\Delta_{\text{vap}}H^\circ$ if we know the vapor pressure of a liquid at two different temperatures.

$$\ln P_2 - \ln P_1 = \left[\frac{-\Delta_{\text{vap}}H^\circ}{RT_2} + C\right] - \left[\frac{-\Delta_{\text{vap}}H^\circ}{RT_1} + C\right]$$

This can be simplified to

$$\ln\frac{P_2}{P_1} = -\frac{\Delta_{\text{vap}}H^\circ}{R}\left[\frac{1}{T_2} - \frac{1}{T_1}\right] \tag{12.2}$$

Ethylene glycol has a vapor pressure of 14.9 mm Hg (P_1) at 373 K (T_1), and a vapor pressure of 49.1 mm Hg (P_2) at 398 K (T_2). Using Equation 12.2, we can calculate the enthalpy of vaporization, 59.0 kJ/mol.

$$\ln\left(\frac{49.1 \text{ mm Hg}}{14.9 \text{ mm Hg}}\right) = -\frac{\Delta_{vap}H°}{0.0083145 \text{ kJ/K} \cdot \text{mol}}\left[\frac{1}{398 \text{ K}} - \frac{1}{373 \text{ K}}\right]$$

$$1.192 = -\frac{\Delta_{vap}H°}{0.0083145 \text{ kJ/K} \cdot \text{mol}}\left(-\frac{0.000168}{K}\right)$$

$$\Delta_{vap}H° = 59.0 \text{ kJ/mol}$$

Boiling Point

If you have a beaker of water open to the atmosphere, the atmosphere presses down on the surface. If enough energy is added, a temperature is eventually reached when the vapor pressure of the liquid equals the atmospheric pressure. At this temperature, bubbles of the liquid's vapor will not be crushed by the atmospheric pressure. Instead, bubbles can rise to the surface, and the liquid boils (Figure 12.19).

The **boiling point** of a liquid is the temperature at which its vapor pressure is equal to the external pressure. If the external pressure is 760 mm Hg, this temperature is called the **normal boiling point**. This point is highlighted on the vapor pressure curves for the substances in Figure 12.17.

The normal boiling point of water is 100 °C, and in a great many places in the United States, water boils at or near this temperature. If you live at higher altitudes, however, such as in Salt Lake City, Utah, where the barometric pressure is about 650 mm Hg, water will boil at a noticeably lower temperature. The curve in Figure 12.17 shows that a pressure of 650 mm Hg corresponds to a boiling temperature of about 95 °C. Food, therefore, has to be cooked a little longer in Salt Lake City to achieve the same result as in New York City at sea level.

Critical Temperature and Pressure

On first thought, it might seem that vapor pressure–temperature curves (such as shown in Figure 12.17) should continue upward without limit, but this is not so. Instead, when a high enough temperature and pressure are reached, the interface between the liquid and the vapor disappears at the **critical point**. The temperature at which this occurs is the **critical temperature, T_c,** and the corresponding pressure is the **critical pressure, P_c** (Figure 12.20). The substance that exists under these conditions is called a **supercritical fluid**. It is like a gas under such a high pressure that its density resembles that of a liquid, while its viscosity (resistance to flow) remains close to that of a gas.

Consider what the substance looks like at the molecular level under these conditions. The molecules have been forced almost as close together as they are in the liquid state, but, unlike the situation in liquids, each molecule in the supercritical fluid has enough kinetic energy to exceed the forces holding molecules together.

For most substances, the critical point is at a very high temperature and pressure (Table 12.7). Water, for instance, has a critical temperature of 374 °C and a critical pressure of 217.7 atm.

Supercritical fluids can have beneficial uses, such as the ability to extract caffeine from coffee beans. As described in *A Closer Look: Supercritical CO$_2$ and Green Chemistry* (page 442), this property can also be useful in our movement to "green chemistry."

Surface Tension, Capillary Action, and Viscosity

Molecules in the interior of a liquid interact with molecules all around them (Figure 12.21). In contrast, molecules on the surface of a liquid are affected only by those molecules located at or below the surface layer. This leads to a net inward force of attraction on the surface molecules, contracting the surface area and making the liquid behave as though it had a skin. The toughness of this skin is measured

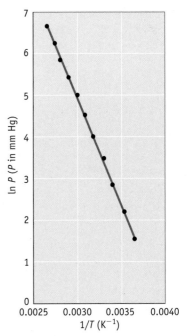

FIGURE 12.18 Clausius–Clapeyron equation. When the natural logarithm of the vapor pressure (ln P) of water at various temperatures (T) is plotted against $1/T$, a straight line is obtained. The slope of the line equals $-\Delta_{vap}H°/R$. Values of T and P for water are from Appendix G.

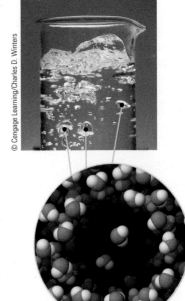

FIGURE 12.19 Vapor pressure and boiling. When the vapor pressure of the liquid equals the atmospheric pressure, bubbles of vapor begin to form within the body of liquid, and the liquid boils.

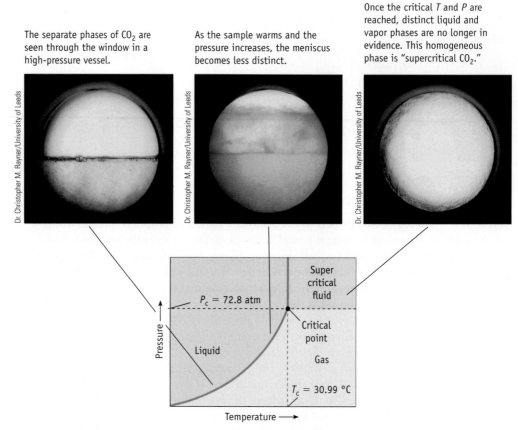

The separate phases of CO_2 are seen through the window in a high-pressure vessel.

As the sample warms and the pressure increases, the meniscus becomes less distinct.

Once the critical T and P are reached, distinct liquid and vapor phases are no longer in evidence. This homogeneous phase is "supercritical CO_2."

FIGURE 12.20 Critical temperature and pressure for CO_2. The pressure versus temperature curve representing equilibrium conditions for liquid and gaseous carbon dioxide ends at the critical point; above that temperature and pressure, water becomes a supercritical fluid.

TABLE 12.7 Critical Temperatures and Pressures for Common Compounds*

Compound	T_c (°C)	P_c (atm)
CH_4 (methane)	−82.6	45.4
C_2H_6 (ethane)	32.3	49.1
C_3H_8 (propane)	96.7	41.9
C_4H_{10} (butane)	152.0	37.3
CCl_2F_2 (CFC-12)	111.8	40.9
NH_3	132.4	112.0
H_2O	374.0	217.7
CO_2	30.99	72.8
SO_2	157.7	77.8

*Data taken from D. R. Lide: *Basic Laboratory and Industrial Chemicals,* Boca Raton, FL, CRC Press, 1993.

by its **surface tension**—the energy required to break through the surface or to disrupt a liquid drop and spread the material out as a film. Surface tension causes water drops to be spheres and not little cubes, for example (Figure 12.22a), because a sphere has a smaller surface area than any other shape of the same volume.

Capillary action is closely related to surface tension. When a small-diameter glass tube is placed in water, the water rises in the tube, just as water rises in a piece of paper in water (Figures 12.22b and 12.22c). Because polar Si—O bonds are present on the surface of glass, polar water molecules are attracted to the surface by **adhesive forces**. These forces are strong enough that they can compete with the **cohesive forces** between the water molecules themselves. Thus, some water molecules can adhere to the walls; other water molecules are attracted to them by cohesive forces and build a "bridge" back into the liquid. The adhesive forces between the water and the glass are great enough that the water level rises in the tube. The rise will continue until the attractive forces—adhesion between water and glass, cohesion between water molecules—are balanced by the force of gravity pulling down on the water column. These forces lead to the characteristic concave, or downward-curving, meniscus seen with water in a test tube (Figure 12.22c).

In some liquids, cohesive forces (high surface tension) are much greater than adhesive forces with glass. Mercury is one example. Mercury does not climb the walls of a glass capillary. In fact, when it is in a glass tube, mercury will form a convex, or upward-curving, meniscus (Figure 12.22c).

One other important property of liquids in which intermolecular forces play a role is **viscosity**, the resistance of liquids to flow. When you turn over a glassful of water, it empties quickly. In contrast, it takes much more time to empty a glassful of olive oil or honey. Olive oil consists of molecules with long chains of carbon atoms,

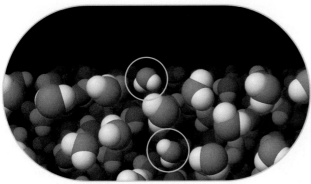

Water molecules on the surface are not completely surrounded by other water molecules.

Water molecules under the surface are completely surrounded by other water molecules.

FIGURE 1
forces in a l'
molecule at
different th
molecule in the ..

and it is about 70 times more viscous than ethanol, a small molecule with only two carbons and one oxygen. Longer chains have greater intermolecular forces because there are more atoms to attract one another, with each atom contributing to the total force. Honey (a concentrated aqueous solution of sugar molecules), however, is also a viscous liquid, even though the size of the molecules is fairly small. In this case, the sugar molecules have numerous —OH groups. These lead to greater forces of attraction due to hydrogen bonding.

REVIEW & CHECK FOR SECTION 12.6

1. *Vapor pressure:* Suppose 0.50 g of pure water is sealed in an evacuated 5.0-L flask and the whole assembly is heated to 60 °C. Will any liquid water be left in the flask or does all of the water evaporate?

   (a)   liquid is left   (b)   all the water evaporates

2. *Using the Clausius-Clapeyron equation:* Calculate the enthalpy of vaporization of diethyl ether, $(C_2H_5)_2O$ (Figure 12.17). This compound has vapor pressures of 57.0 mm Hg and 534 mm Hg at −22.8 °C and 25.0 °C, respectively.

   (a)   29.0 kJ/mol   (b)   122 kJ/mol   (c)   2.44 kJ/mol

FIGURE 12.22 Adhesive and cohesive forces.

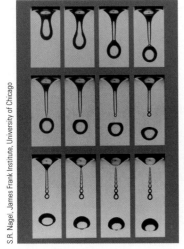

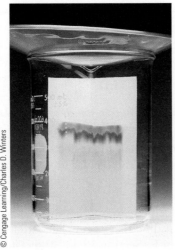

(a) A series of photographs showing the different stages when a water drop falls. The drop was illuminated by a strobe light of 5-ms duration. (The total time for this sequence was 0.05 second.) Water droplets take a spherical shape because of surface tension.

(b) Capillary action. Polar water molecules are attracted to the — OH bonds in paper fibers, and water rises in the paper. If a line of ink is placed in the path of the rising water, the different components of the ink are attracted differently to the water and paper and are separated in a process called paper chromatography.

(c) Water (top layer) forms a concave meniscus, while mercury (bottom layer) forms a convex meniscus. The different shapes are determined by the adhesive forces of the molecules of the liquid with the walls of the tube and the cohesive forces between molecules of the liquid.

Supercritical CO$_2$ and Green Chemistry

uch has been written recently about CO$_2$, particularly its role as a greenhouse gas. But there are useful sides to this molecule: one is as a solvent when its physical state is a supercritical fluid.

Carbon dioxide is widely available, essentially nontoxic, nonflammable, and inexpensive. Importantly, it is relatively easy to reach its critical temperature of 30.99 °C and critical pressure of 72.8 atm. In the supercritical state, the molecules are in close proximity, so the substance behaves as a liquid. However, because the kinetic energy of CO$_2$ molecules is greater than the energy associated with the forces of attraction between molecules, supercritical CO$_2$ has some properties of a gas. As a result, the supercritical fluid has the density of a liquid but the low viscosity of a gas.

One use of supercritical CO$_2$ is to extract caffeine from coffee. Coffee beans are treated with steam to bring the caffeine to the surface. The beans are then immersed in supercritical CO$_2$, which selectively dissolves the caffeine but leaves intact the compounds that give flavor to coffee. (Decaffeinated coffee contains less than 3% of the original caffeine.) The solution of caffeine in supercritical CO$_2$ is poured off, and the CO$_2$ is evaporated, trapped, and reused.

Supercritical CO$_2$ is also being used more and more as a replacement for organic solvents used to clean parts during the manufacture of electronic devices and for

Image courtesy of SiOx Machines AB

Liquid CO$_2$ inside high-pressure chamber

dry cleaning clothing. This is an important advance because the organic solvents used for these purposes are often atmospheric pollutants.

Solvents such as alcohols and toluene (C$_6$H$_5$CH$_3$) are classified as VOCs (volatile organic compounds). To give you an idea of the scale of their use, it has been estimated that, between the United States and Canada, 26 million tons of VOCs are released into the atmosphere each year.

In addition to VOCs, there are halogenated compounds such as perchloroethylene (PERC) used in dry cleaning. Replacing VOCs and halogen-bearing solvents with supercritical CO$_2$ would help our environment greatly. Simply replacing VOCs, PERC, and related compounds with liquid or supercritical CO$_2$ won't work, however, because grease and dirt do not dis-

solve in CO$_2$. Fortunately, the problem was solved by Professor J. DeSimone of the University of North Carolina. His solution was to add a surfactant, just as you find in detergents in the kitchen. A surfactant is generally a long chain organic molecule with one end that can bind to an organic particle such as grease and another end that is soluble in the bulk solvent, in this case CO$_2$.

"CO$_2$-phobic" hydrocarbon head

$$\left[H_2C - CH \right]_n$$
$$C = O$$
$$O$$
$$CH_2CF_2CF_2CF_2CF_2CF_2CF_2CF_3$$

"CO$_2$-philic" fluorocarbon tail

The "CO$_2$-phobic" ends of these molecules surround the grease or dirt to be removed while the "CO$_2$-philic" ends extend into the supercritical fluid. The dirt is taken into the solvent and washed away, leaving no toxic chemicals behind. The CO$_2$ is evaporated and recovered, and the dirt is collected.

Reference:

Real World Cases in Green Chemistry, M. C. Cann and M. E. Connelly, American Chemical Society, Washington, DC, 2000.

OWL and go Chemistry

Sign in at **www.cengage.com/owl** to:
- View tutorials and simulations, develop problem-solving skills, and complete online homework assigned by your professor.
- For quick review and exam prep, download Go Chemistry mini lecture modules from OWL (or purchase them at **www.cengagebrain.com**)

Access **How Do I Solve It?** tutorials on how to approach problem solving using concepts in this chapter.

CHAPTER GOALS REVISITED

Now that you have studied this chapter, you should ask whether you have met the chapter goals. In particular, you should be able to:

Describe intermolecular forces and their effects

a. Describe the various intermolecular forces found in liquids and solids (Sections 12.3 and 12.4). **Study Questions: 1–10, 29–32, 34, 36, 43, and Go Chemistry Module 17.**

b. Tell when two molecules can interact through a dipole–dipole attraction and when hydrogen bonding may occur. The latter occurs most strongly when H is attached to O, N, or F (Section 12.3). **Study Questions: 1–8.**

c. Identify instances in which molecules interact by induced dipoles (dispersion forces) (Section 12.4). **Study Questions: 1–8.**

Understand the importance of hydrogen bonding

a. Explain how hydrogen bonding affects the properties of water (Section 12.3). **Study Questions: 1–8.**

Understand the properties of liquids

a. Explain the processes of evaporation and condensation, and use the enthalpy of vaporization in calculations (Section 12.6). **Study Questions: 11, 12, 18, 35, 57.**

b. Define the equilibrium vapor pressure of a liquid, and explain the relation-
    ship between the vapor pressure and boiling point of a liquid (Section 12.6).
    Study Questions: 14, 15, 17, 19, 20, 33, 42, 54.
c. Describe the phenomena of the critical temperature, T_c, and critical pressure,
    P_c, of a substance (Section 12.6). Study Questions: 23, 24, 53.
d. Describe how intermolecular interactions affect the cohesive forces between
    identical liquid molecules, the energy necessary to break through the surface
    of a liquid (surface tension), capillary action, and the resistance to flow, or
    viscosity, of liquids (Section 12.6). Study Questions: 25, 26, 27, 28, 45, 46.
e. Use the Clausius–Clapeyron equation, which connects temperature, vapor
    pressure, and enthalpy of vaporization for liquids (Section 12.6). Study
    Questions: 21, 22, 38, 39.

Key Equations

Equation 12.1 (page 438) The Clausius–Clapeyron equation relates the equilibrium
vapor pressure, P, of a volatile liquid to the molar enthalpy of vaporization ($\Delta_{vap}H°$)
at a given temperature, T. (R is the universal constant, 8.314472 J/K · mol.)

$$\ln P = -(\Delta_{vap}H°/RT) + C$$

Equation 12.2 (page 438) This modification of the Clausius–Clapeyron equation
allows you to calculate $\Delta_{vap}H°$ if you know the vapor pressures at two different
temperatures.

$$\ln \frac{P_2}{P_1} = -\frac{\Delta_{vap}H°}{R}\left[\frac{1}{T_2} - \frac{1}{T_1}\right]$$

OWL Questions and problems for this chapter are available in OWL. Use the OWL access card that
came with this textbook to access assigned questions and problems for this chapter.

13 The Chemistry of Solids

Lithium and "Green Cars"

The focus of much of this chapter is on metals and metal-containing compounds, the vast majority of which are solids.

You are well aware of the use of metals in our world: copper in electric wiring; aluminum in airplanes and soft drink cans; lead and zinc in batteries; and iron in cars, buses, and bridges. You may be less aware of metals such as the platinum group metals (platinum, palladium, rhodium, iridium, and ruthenium) in the catalytic converter in your car or in the chemicals industry. Approximately 2700 metric tons of these metals is used in cars and another 1000 metric tons in industry annually. Fortunately, many of the platinum group metals are recycled, but scientists are concerned about the depletion of metals such as copper, zinc, gallium, indium, and hafnium. They won't become "extinct" because their atoms are immutable, but they may be increasingly difficult to recover.

One metal that seems, fortunately, to be quite abundant on the earth is lithium. We say "fortunately" because lithium compounds are used in ceramics and glass, in reagents for the production of pharmaceuticals, and in lubricating greases. But their most important and largest use may be in batteries for appliances and most recently in cars. Your laptop computer may have a "lithium ion" battery, and several brands of hybrid cars now use lithium-based batteries.

Much of the world's lithium comes from northern Chile and southern Bolivia. There, groundwater with high concentrations of Group 1A salts is pumped into ponds. The solution slowly evaporates in the intense sun of the high Andes plateau,

Lithium and lithium salts. The high plateau in Bolivia where ground water contains high concentrations of Group 1A salts, among them lithium carbonate. The water is left in the sun to deposit the salts by evaporation.

A sample of lithium metal and a view of the solid state structure of lithium.

and, after about a year, the highly concentrated solution is finally taken to a chemical plant where it is evaporated to give white, powdered lithium carbonate. It is estimated that Bolivia alone has a reserve of about 73 million metric tons of Li_2CO_3. To produce lithium metal, the carbonate is converted to LiCl, which is then electrolyzed to produce the metal.

In 2008 *Forbes* magazine said that "the green-car revolution could make lithium one of the planet's most strategic commodities." This remains to be seen, however. Some automobile companies are planning to continue using nickel-metal hydride batteries well into the future.

Questions:

1. What mass of lithium can be obtained from 73 million metric tons of lithium carbonate? (1 metric ton = 1000 kg.)
2. Describe the unit cell of lithium (see Figure).
3. The lithium unit cell is a cube with sides of 351 pm. Use this information, and a knowledge of unit cells, to calculate the density of lithium metal.
4. In the process of making lithium metal, Li_2CO_3 is converted to LiCl. Suggest a way to do this.

Answers to these questions are available in Appendix N.

CHAPTER OUTLINE

13.1 Crystal Lattices and Unit Cells

13.2 Structures and Formulas of Ionic Solids **go Chemistry**

13.3 Bonding in Metals and Semiconductors

13.4 Bonding in Ionic Compounds: Lattice Energy

13.5 The Solid State: Other Types of Solid Materials

13.6 Phase Changes Involving Solids

13.7 Phase Diagrams

CHAPTER GOALS

See Chapter Goals Revisited (page 467) for Study Questions keyed to these goals.

- Understand cubic unit cells.
- Relate unit cells for ionic compounds to formulas.
- Describe bonding in ionic and metallic solids and the nature of semiconductors.
- Describe the properties of solids.
- Interpret phase diagrams.

⊎WL

Many kinds of solids exist in the world around us (Figure 13.1 and Table 13.1). As the description of lithium and lithium-based batteries shows, solid-state chemistry is one of the booming areas of science, especially because it relates to the development of interesting new materials. As we describe various kinds of solids, we hope to provide a glimpse of the reasons this area is exciting.

13.1 Crystal Lattices and Unit Cells

In both gases and liquids, molecules move continually and randomly, and they rotate and vibrate as well. Because of this movement, an orderly arrangement of molecules in the gaseous or liquid state is not possible. In solids, however, the molecules, atoms, or ions cannot change their relative positions (although they vibrate and

Download mini lecture videos for key concept review and exam prep from OWL or purchase them from **www.cengagebrain.com**

Table 13.1 Structures and Properties of Various Types of Solid Substances

Type	Examples	Structural Units	Forces Holding Units Together	Typical Properties
Ionic	NaCl, K_2SO_4, $CaCl_2$, $(NH_4)_3PO_4$	Positive and negative ions; no discrete molecules	Ionic; attractions among charges on positive and negative ions	Hard; brittle; high melting point; poor electric conductivity as solid, good as liquid; often water-soluble
Metallic	Iron, silver, copper, other metals and alloys	Metal atoms (positive metal ions with delocalized electrons)	Metallic; electrostatic attraction among metal ions and electrons	Malleable; ductile; good electric conductivity in solid and liquid; good heat conductivity; wide range of hardness and melting points
Molecular	H_2, O_2, I_2, H_2O, CO_2, CH_4, CH_3OH, CH_3CO_2H	Molecules	Dispersion forces, dipole–dipole forces, hydrogen bonds	Low to moderate melting points and boiling points; soft; poor electric conductivity in solid and liquid
Network	Graphite, diamond, quartz, feldspars, mica	Atoms held in an infinite two- or three-dimensional network	Covalent; directional electron-pair bonds	Wide range of hardness and melting points (three-dimensional bonding > two-dimensional bonding); poor electric conductivity, with some exceptions
Amorphous	Glass, polyethylene, nylon	Covalently bonded networks with no long-range regularity	Covalent; directional electron-pair bonds	Noncrystalline; wide temperature range for melting; poor electric conductivity, with some exceptions

FIGURE 13.1
Some common
solids.

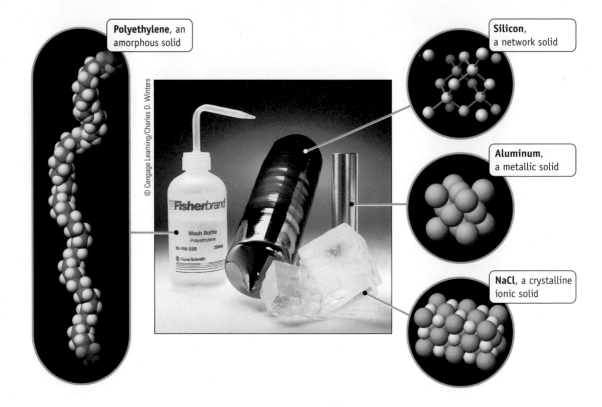

occasionally rotate). Thus, a regular, repeating pattern of atoms or molecules within the structure—a long-range order—is a characteristic of most solids. The beautiful, external (macroscopic) regularity of many crystalline compounds (such as those of salt, Figure 13.1) is a consequence of this internal order.

Structures of solids can be described as three-dimensional lattices of atoms, ions, or molecules. For a crystalline solid, we can identify the **unit cell**, the smallest repeating unit that has all of the symmetry characteristic of the way the atoms, ions, or molecules are arranged in the solid.

To understand unit cells, consider first a two-dimensional lattice model, the repeating pattern of circles shown in Figure 13.2. The yellow square at the left is a unit cell because the overall pattern can be created from a group of these cells by joining them edge to edge. It is also a requirement that unit cells reflect the stoichiometry of the solid. Here, the square unit cell at the left contains one smaller sphere and one fourth of each of the four larger circles, giving a total of one small and one large circle per two-dimensional unit cell.

You may recognize that it is possible to draw other unit cells for this two-dimensional lattice. One option is the square in the middle of Figure 13.2 that fully encloses a single large circle and parts of small circles that add up to one net small circle. Yet another possible unit cell is the parallelogram at the right. Other unit cells can be drawn, but it is conventional to draw unit cells in which atoms or ions are placed at the **lattice points**; that is, at the corners of the cube or other geometric object that constitutes the unit cell.

FIGURE 13.2 Unit cells for a flat, two-dimensional solid made from circular "atoms." A lattice can be represented as being built from repeating unit cells. This two-dimensional lattice can be built by translating the unit cells throughout the plane of the figure. Each unit cell must move by the length of one side of the cell. In this figure, all unit cells contain a net of one large circle and one small circle. Be sure to notice that several unit cells are possible, with two of the most obvious being squares.

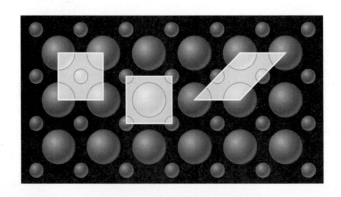

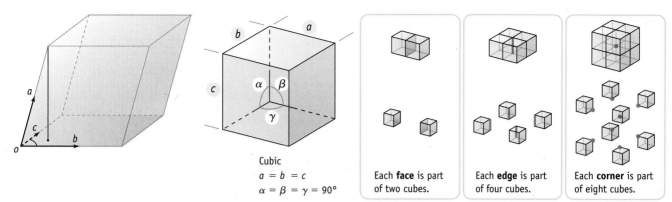

(a) Types of cells. Of the possible unit cells, all are parallelepipeds (except for the hexagonal cell), figures in which opposite sides are parallel. In a cube, all angles (*a-o-c*, *a-o-b*, and *c-o-b*; where *o* is the origin) are 90°, and all sides are equal. In other cells, the angles and sides may be the same or different. For example, in a tetragonal cell, the angles are 90°, but $a = b \neq c$. In a triclinic cell, the sides have different lengths, the angles are different, and none equals 90°.

(b) Cubic cell. The cube is one of the seven basic unit cells that describe crystal systems. In a cube, all sides are of equal length, and all angles are 90°.

(c) Building a lattice. Stacking cubes to build a crystal lattice. Each crystal face is part of two cubes; each edge is part of four cubes; and each corner is part of eight cubes.

FIGURE 13.3 Unit cells.

The three-dimensional lattices of solids can be built by assembling three-dimensional unit cells much like building blocks (Figure 13.3). The assemblage of these three-dimensional unit cells defines the **crystal lattice**.

To construct crystal lattices, nature can use any of seven three-dimensional unit cells. They differ from one another in that their sides have different relative lengths and their edges meet at different angles. The simplest of the seven crystal lattices is the **cubic unit cell**, a cell with edges of equal length that meet at 90° angles (Figure 13.3). We shall look in detail at only this structure because cubic unit cells are commonly encountered.

Within the cubic class, three cell symmetries occur: **primitive cubic (pc)**, **body-centered cubic (bcc)**, and **face-centered cubic (fcc)** (Figure 13.4). All three have identical atoms, molecules, or ions at the corners of the cubic unit cell. The bcc and

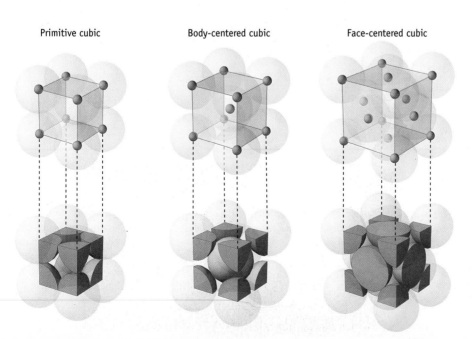

Primitive cubic   Body-centered cubic   Face-centered cubic

FIGURE 13.4 The three cubic unit cells. The top row shows the lattice points of the three cells, and the bottom row shows the same cells using space-filling spheres. The spheres in each figure represent identical atoms (or ions) centered on the lattice points. Because eight unit cells share a corner atom, only ⅛ of each corner atom lies within a given unit cell; the remaining ⅞ lies in seven other unit cells. Because each face of an fcc unit cell is shared with another unit cell, one half of each atom in the face of a face-centered cube lies in a given unit cell, and the other half lies in the adjoining cell.

FIGURE 13.5 Metals use four different unit cells. Three are based on the cube, and the fourth is the hexagonal unit cell (see the next page). (Many metals can crystallize in more than one structure. Mn, for example, can be bcc or fcc, and La exists in all but the primitive unit cell.)

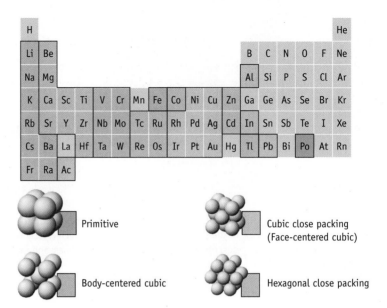

Primitive

Cubic close packing (Face-centered cubic)

Body-centered cubic

Hexagonal close packing

fcc arrangements, however, differ from the primitive cube in that they have additional particles at other locations. The bcc structure is called *body-centered* because it has an additional particle, of the same type as those at the corners, at the center of the cube. The fcc arrangement is called *face-centered* because it has a particle, of the same type as the corner atoms, in the center of each of the six faces of the cube. Examples of each structure are found among the crystal lattices of the metals (Figure 13.5). The alkali metals, for example, are body-centered cubic, whereas nickel, copper, and aluminum are face-centered cubic. Notice that only one metal, polonium, has a primitive cubic lattice.

When the cubes pack together to make a three-dimensional crystal of a metal, the atom at each corner is shared among eight cubes (Figures 13.3, 13.4, and 13.6a). Because of this, only one eighth of each corner atom is actually within a given unit cell. Furthermore, because a cube has eight corners, and because one eighth of the atom at each corner "belongs to" a particular unit cell, the corner atoms contribute a net of one atom to a given unit cell. Thus, *the primitive cubic arrangement has one net atom within the unit cell.*

(8 corners of a cube)($\frac{1}{8}$ of each corner atom within a unit cell) =

1 net atom per unit cell for the primitive cubic unit cell

In contrast to the primitive cubic lattice, a body-centered cube has an additional atom wholly within the unit cell at the cube's center. The center particle is present

FIGURE 13.6 Atom sharing at cube corners and faces.

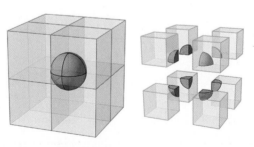

(a) In any cubic lattice, each corner particle is shared equally among eight cubes, so one eighth of the particle is within a particular cubic unit cell.

(b) In a face-centered lattice, each particle on a cube face is shared equally between two unit cells. One half of each particle of this type is within a given unit cell.

A CLOSER LOOK

Packing Oranges and Marbles

It is a "rule" that nature does things as efficiently as possible. You know this if you have ever tried to stack some oranges into a pile that doesn't fall over and that takes up as little space as possible. How did you do it? Clearly, the pyramid arrangement below on the right works, whereas the cubic one on the left does not.

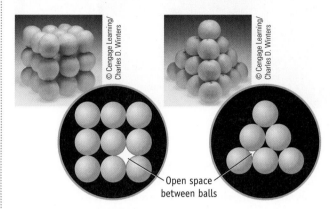

If you could look inside the pile, you would find that less open space is left in the pyramid stacking than in the cube stacking. Only 52% of the space is filled in the cubic packing arrangement. (If you could stack oranges as a body-centered cube, that would be slightly better; 68% of the space is used.) However, the best method is the pyramid stack, which is really a face-centered cubic arrangement. Oranges, atoms, or ions packed this way occupy 74% of the available space.

To fill three-dimensional space, the most efficient way to pack oranges or atoms is to begin with a hexagonal arrangement of spheres, as in this arrangement of marbles.

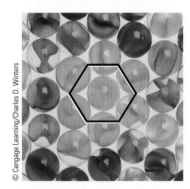

Succeeding layers of atoms or ions are then stacked one on top of the other in two different ways. Depending on the stacking pattern (Figure A), you will get either a **cubic close-packed (ccp)** or **hexagonal close-packed (hcp)** arrangement.

In the hcp arrangement, additional layers of particles are placed above and below a given layer, fitting into the same depressions on either side of the middle layer. In a three-dimensional crystal, the layers repeat their pattern in the manner ABABAB.... Atoms in each A layer are directly above the ones in another A layer; the same holds true for the B layers.

In the ccp arrangement, the atoms of the "top" layer (A) rest in depressions in the middle layer (B), and those of the "bottom" layer (C) are oriented opposite to those in the top layer. In a crystal, the pattern is repeated ABCABCABC.... By turning the whole crystal, you can see that the ccp arrangement is the face-centered cubic structure (Figure B).

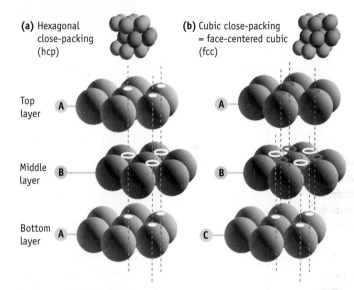

FIGURE A Efficient packing. The most efficient ways to pack atoms or ions in crystalline materials are hexagonal close-packing (hcp) and cubic close packing (ccp).

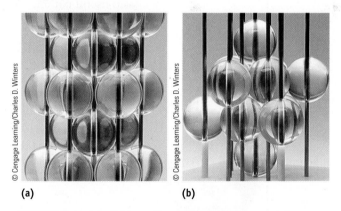

FIGURE B Models of close packing. (a) A model of hexagonal close-packing, where the layers repeat in the order ABABAB.... **(b)** A face-centered unit cell (cubic close-packing), where the layers repeat in the order ABCABC.... (A kit from which these models can be built is available from the Institute for Chemical Education at the University of Wisconsin at Madison.)

FIGURE 13.7 X-ray crystallography. In the x-ray diffraction experiment, a beam of x-rays is directed at a crystalline solid. The photons of the x-ray beam are scattered by the atoms of the solid. The scattered x-rays are detected by a photographic film or an electronic detector, and the pattern of scattered x-rays is related to the locations of the atoms or ions in the crystal.

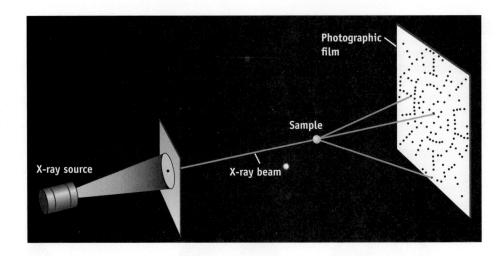

in addition to those at the cube corners, so *the body-centered cubic arrangement has a net of two atoms within the unit cell.*

In a face-centered cubic arrangement, there is an atom on each of the six faces of the cube in addition to those at the cube corners. One half of each atom on a face belongs to a given unit cell (Figure 13.6b). Three net particles are therefore contributed by the particles on the faces of the cube:

(6 faces of a cube)(½ of an atom within a unit cell) =

3 net face-centered atoms within a face-centered cubic unit cell

Thus, *the face-centered cubic arrangement has a net of four atoms within the unit cell,* one contributed by the corner atoms and another three contributed by the atoms centered in the six faces.

An experimental technique, x-ray crystallography, can be used to determine the structure of a crystalline substance (Figure 13.7). Once the structure is known, the information can be combined with other experimental information to calculate such useful parameters as the radius of an atom (Example 13.1).

ⓦWL   INTERACTIVE EXAMPLE 13.1   Determining an Atom Radius from Lattice Dimensions

Problem Aluminum has a density of 2.699 g/cm³, and the atoms are packed in a face-centered cubic crystal lattice. What is the radius of an aluminum atom?

What Do You Know? You are trying to find the radius of an aluminum atom and begin with a knowledge of its unit cell geometry. It is important to recognize the atoms in a face-centered cube do not touch along the cell edges but do so in the faces (as illustrated just below).

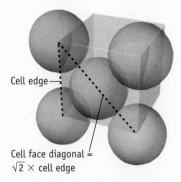

Cell edge

Cell face diagonal = $\sqrt{2}$ × cell edge

One face of a face-centered cubic unit cell. This shows the cell face diagonal, $\sqrt{2}$ × edge, is equal to four times the radius of the atoms in the lattice.

Aluminum metal. The metal has a face-centered cubic unit cell with a net of four Al atoms in each unit cell.

Strategy Map 13.1

DATA/INFORMATION
- Al unit cell is **FCC** with $d = 2.699$ g/cm^3
- Need Al **atomic weight**
- Need **Avogadro's number**

STEP 1. Calculate **mass** of unit cell ($= 4$ Al atoms).

Mass of Al atom, **mass** of unit cell

STEP 2. Calculate **volume** of unit cell from mass and density.

Volume of unit cell

STEP 3. Calculate **length** of unit cell **edge** from cell volume.

Length of unit cell edge

STEP 4. Relate **atom radius** to **unit cell edge length**.

Radius of atom

Strategy If you know the cell edge length, you can calculate the diagonal distance across a face, and the illustration above shows the atom radius is one fourth of this distance. So, the problem comes down to finding the length of the cell edge. This dimension is the cube root of the cell volume. But where do you find the cell volume? From the density and mass of the unit cell. The density is given, and the mass of the cell is 4 times the mass of an atom. As described in Chapter 2, the mass of an Al atom can be found from its atomic mass and Avogadro's number. Therefore, you should proceed as follows:

1. Find the mass of a unit cell from the knowledge that it is face-centered cubic.

2. Combine the density of aluminum with the mass of the unit cell to find the cell volume.

3. Find the length of a side of the unit cell from its volume.

4. Calculate the atom radius from the edge dimension.

Solution

1. *Calculate the mass of the unit cell.*

$$\text{Mass of 1 Al atom} = \left(\frac{26.98 \text{ g}}{1 \text{ mol}}\right)\left(\frac{1 \text{ mol}}{6.022 \times 10^{23} \text{ atoms}}\right) = 4.480 \times 10^{-23} \text{ g/atom}$$

$$\text{Mass of unit cell} = \left(\frac{4.480 \times 10^{-23} \text{ g}}{1 \text{ Al atom}}\right)\left(\frac{4 \text{ Al atoms}}{1 \text{ unit cell}}\right) = 1.792 \times 10^{-22} \text{ g/unit cell}$$

2. *Calculate the volume of the unit cell from the unit cell mass and density.*

$$\text{Volume of unit cell} = \left(\frac{1.792 \times 10^{-22} \text{ g}}{\text{unit cell}}\right)\left(\frac{1 \text{ cm}^3}{2.699 \text{ g}}\right) = 6.640 \times 10^{-23} \text{ cm}^3/\text{unit cell}$$

3. *Calculate the length of a unit cell edge.* The length of the unit cell edge is the cube root of the cell volume.

$$\text{Length of unit cell edge} = \sqrt[3]{6.640 \times 10^{-23} \text{ cm}^3} = 4.049 \times 10^{-8} \text{ cm}$$

4. *Calculate the atom radius.* As illustrated in the model above, the diagonal distance across the face of the cell is equal to four times the Al atom radius.

$$\text{Cell face diagonal} = 4 \times (\text{Al atom radius})$$

The cell diagonal is the hypotenuse of a right isosceles triangle, so, using the Pythagorean theorem,

$$(\text{Diagonal distance})^2 = 2 \times (\text{edge})^2$$

Taking the square root of both sides, we have

$$\text{Diagonal distance} = \sqrt{2} \times (\text{cell edge})$$
$$= \sqrt{2} \times (4.049 \times 10^{-8} \text{ cm}) = 5.727 \times 10^{-8} \text{ cm}$$

We divide the diagonal distance by 4 to obtain the Al atom radius in cm.

$$\text{Al atom radius} = \frac{5.727 \times 10^{-8} \text{ cm}}{4} = 1.432 \times 10^{-8} \text{ cm}$$

Atomic dimensions are often expressed in picometers, so we convert the radius to that unit.

$$1.432 \times 10^{-8} \text{ cm}\left(\frac{1 \text{ m}}{100 \text{ cm}}\right)\left(\frac{1 \text{ pm}}{1 \times 10^{-12} \text{ m}}\right) = 143.2 \text{ pm}$$

Think about Your Answer The radius in Figure 7.8 was calculated from experimental data as in this Example.

Check Your Understanding

(a) *Determining an Atom Radius from Lattice Dimensions:* Gold has a face-centered unit cell, and its density is 19.32 g/cm^3. Calculate the radius of a gold atom.

(b) *The Structure of Solid Iron:* Iron has a density of 7.8740 g/cm^3, and the radius of an iron atom is 126 pm. Verify that solid iron has a body-centered cubic unit cell. (Be sure to note that the atoms in a body-centered cubic unit cell touch along the diagonal across the cell. They do not touch along the edges of the cell.) (*Hint:* The diagonal distance across the unit cell = edge $\times \sqrt{3}$.)

A portion of the solid state structure of potassium.

REVIEW & CHECK FOR SECTION 13.1

1. A portion of the crystalline lattice for potassium is illustrated in the margin. In what type of unit cell are the K atoms arranged?

   (a) primitive cubic   (c) face-centered cubic

   (b) body-centered cubic

2. If one edge of the potassium unit cell is 533 pm, what is the density of potassium?

   (a) 0.858 g/cm³   (b) 1.17 g/cm³   (c) 0.428 g/cm³

13.2 Structures and Formulas of Ionic Solids

The lattices of many ionic compounds are built by taking a primitive cubic or face-centered cubic lattice of ions of one type and placing ions of opposite charge in the holes within the lattice. This produces a three-dimensional lattice of regularly placed ions. The smallest repeating unit in these structures is, by definition, the unit cell for the ionic compound.

The choice of the lattice and the number and location of the holes in the lattice that are filled are the keys to understanding the relationship between the lattice structure and the formula of a salt. Consider, for example, the ionic compound cesium chloride, CsCl (Figure 13.8). The structure of CsCl has a primitive cubic unit cell of chloride ions. The cesium ion fits into a hole in the center of the cube. (An equivalent unit cell has a primitive cubic unit cell of Cs⁺ ions with a Cl⁻ ion in the center of the cube.)

Next, consider the structure for NaCl. An extended view of the lattice and one unit cell are illustrated in Figures 13.9a and 13.9b, respectively. The Cl⁻ ions are arranged in a face-centered cubic unit cell, and the Na⁺ ions are arranged in a regular manner between these ions. Notice that each Na⁺ ion is surrounded by six Cl⁻ ions in an octahedral geometry. Thus, the Na⁺ ions are said to be in **octahedral holes** (Figure 13.9c).

The formula of an ionic compound must always be reflected in the composition of its unit cell; therefore, the formula can always be derived from the unit cell structure. The formula for NaCl can be related to this structure by counting the number of cations and anions contained in one unit cell. A face-centered cubic lattice of Cl⁻ ions has a net of four Cl⁻ ions within the unit cell. There is one Na⁺ ion in the center of the unit cell, contained totally within the unit cell. In addition, there are 12 Na⁺ ions along the edges of the unit cell. Each of these Na⁺ ions is shared among four unit cells, so each contributes one fourth of an Na⁺ ion to the unit cell, giving three additional Na⁺ ions within the unit cell.

(1 Na⁺ ion in the center of the unit cell) + (¼ of a Na⁺ ion in each edge × 12 edges)

= net of 4 Na⁺ ions in NaCl unit cell

This accounts for all of the ions contained in the unit cell: four Cl⁻ and four Na⁺ ions. Thus, a unit cell of NaCl has a 1:1 ratio of Na⁺ and Cl⁻ ions, as the formula requires.

FIGURE 13.8 Cesium chloride (CsCl) unit cell. The unit cell of CsCl may be viewed in two ways. The only requirement is that the unit cell must have a net of one Cs⁺ ion and one Cl⁻ ion. Either way, it is a primitive cubic unit cell of ions of one type (Cl⁻ on the left or Cs⁺ on the right). Generally, ionic lattices are assembled by placing the larger ions (here Cl⁻) at the lattice points and placing the smaller ions (here Cs⁺) in the lattice holes.

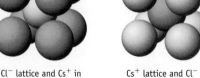
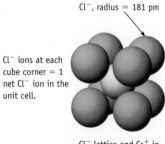

Cl⁻, radius = 181 pm

Cs⁺, radius = 165 pm

Cl⁻ ions at each cube corner = 1 net Cl⁻ ion in the unit cell.

One Cs⁺ ion at each cube corner = 1 net Cs⁺ ion in the unit cell.

Cl⁻ lattice and Cs⁺ in lattice hole

Cs⁺ lattice and Cl⁻ in lattice hole

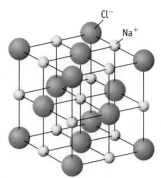

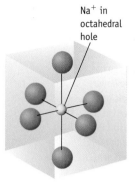

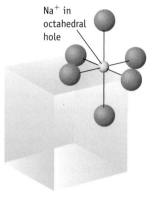

NaCl unit cell (expanded)

1 hole of this kind
in the center of the
unit cell

12 holes of this kind in the 12
edges of the unit cell
(a net of 3 holes)

(a) Cubic NaCl. The solid is based on a face-centered cubic unit cell of Na^+ and Cl^- ions.

FIGURE 13.9 Sodium chloride.

(b) Expanded view of the NaCl lattice. (The lines are not bonds; they are there to help visualize the lattice.) The smaller Na^+ ions (silver) are packed into a face-centered cubic lattice of larger Cl^- ions (yellow).

(c) Octahedral holes. A view showing the octahedral holes in the lattice.

Another common unit cell has a different placement of ions. Again, there are ions of one type in a face-centered cubic unit cell. Ions of the other type are located in **tetrahedral holes**, wherein each ion is surrounded by four oppositely charged ions. As illustrated in Figure 13.10, there are eight tetrahedral holes in a face-centered cubic unit cell. In ZnS (zinc blende), the sulfide ions (S^{2-}) form a face-centered cubic unit cell. The zinc ions (Zn^{2+}) then occupy one half of the tetrahedral holes, and each Zn^{2+} ion is surrounded by four S^{2-} ions. The unit cell has four S^{2-} ions (making up the fcc lattice) and four Zn^{2+} ions contained wholly within the unit cell (in tetrahedral holes). This 1 : 1 ratio of ions matches the ratio in the formula.

In summary, compounds with the formula MX commonly form one of three possible crystal structures:

1. M^{n+} ions occupying the cubic hole in a primitive cubic X^{n-} lattice. Example: CsCl.
2. M^{n+} ions in all the octahedral holes in a face-centered cubic X^{n-} lattice. Example: NaCl.
3. M^{n+} ions occupying half of the tetrahedral holes in a face-centered cube lattice of X^{n-} ions. Example: ZnS.

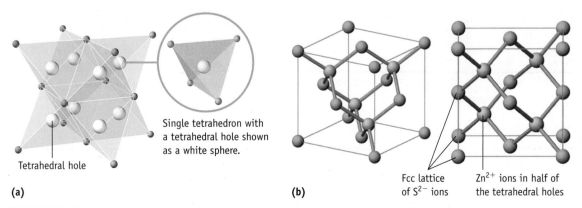

Tetrahedral hole

Single tetrahedron with
a tetrahedral hole shown
as a white sphere.

Fcc lattice
of S^{2-} ions

Zn^{2+} ions in half of
the tetrahedral holes

(a)

(b)

FIGURE 13.10 Tetrahedral holes and two views of the ZnS (zinc blende) unit cell. **(a)** The tetrahedral holes in a face-centered cubic lattice. **(b)** This unit cell is an example of a face-centered cubic lattice of ions of one type with ions of the opposite type in one half of the tetrahedral holes.

CASE STUDY

High-Strength Steel and Unit Cells

A few decades ago if you rolled a car over or suffered a front end or side collision, you were likely to be seriously injured or killed. In recent years, though, the use of seat belts and air bags has greatly increased survivability. However, there is an additional factor—the steels used in cars have improved greatly and have led to both stronger frames and energy-absorbing front ends. At the same time cars are lighter and, partly as a result, more fuel-efficient. Much of this is due to the "design" of better steel.

Iron and steel are still made the same way they have always been: First, iron ore is reduced with carbon to produce raw iron. This is then heated in a "basic oxygen furnace" to remove impurities (▶ Chapter 22). Stronger steels were traditionally produced by adding very small amounts of other metals at this point to produce alloys that have very specific properties, but this can be expensive. So, the steel industry has developed new methods for producing steels that are less expensive.

When steel is produced, it is not a single, large crystal of iron atoms neatly arranged in a lattice. Rather, it consists of grains or crystallites of iron in various forms as well as carbon. The characteristics of this material can be changed by rolling a slab at room temperature, by annealing, or by quenching. Quenching refers to a process in which the steel slab is rolled at room temperature, heated to a high temperature, and then rapidly cooled or "quenched" in cold water.

FIGURE A Modern high-strength steels can save the lives of passengers in serious accidents such as rollovers.

Ferrite has a unit cell consisting only of iron atoms (Figure C); carbon atoms are not part of in the lattice. However, when the steel is heated, the lattice can rearrange, and carbon atoms become part of the unit cell. This new form with an embedded carbon atom is called the *austinite form*. When the steel is then quenched, another rearrangement occurs, this time to the martensite form, still with an embedded carbon atom.

Steel that has gone through this process will consist of crystallites of ferrite, austenite, and martensite. Depending on the precise conditions used, the proportion of these can vary, and so can the strength of the steel. For example, steel with a high proportion of martensite is very strong and is used in car bumpers.

Some steels may have added boron, which gives an ultra-high-strength steel that can be used in door pillars and beams. The lives of passengers can be saved in the event of a rollover.

Questions:

1. What is the unit cell of ferrite?
2. Describe the unit cell of austenite.
3. What change has occurred on going from the austenite form to the martensite form?

Answers to these questions are available in Appendix N.

Reference:

New York Times, September 25, 2009, page D1.

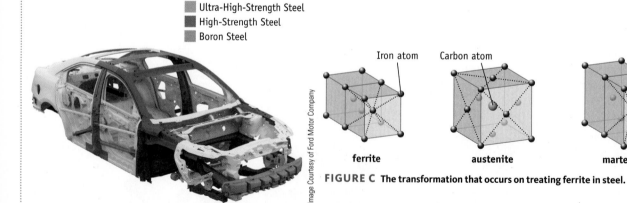

Ultra-High-Strength Steel
High-Strength Steel
Boron Steel

FIGURE B Steels used in a recent automobile.

Iron atom   Carbon atom

ferrite   **austenite**   **martensite**

FIGURE C The transformation that occurs on treating ferrite in steel.

Chemists and geologists in particular have observed that the sodium chloride or "rock salt" structure is adopted by many ionic compounds, most especially by all the alkali metal halides (except CsCl, CsBr, and CsI), all the oxides and sulfides of the alkaline earth metals, and all the oxides of formula MO of the transition metals of the fourth period.

EXAMPLE 13.2 Ionic Structure and Formula

Problem One unit cell of the common mineral perovskite is illustrated in the margin. This compound is composed of calcium and titanium cations and oxide anions. Based on the unit cell, what is the formula of perovskite?

What Do You Know? The cell is composed of the ions Ca^{2+}, Ti^{4+}, and O^{2-}. The Ca^{2+} ion is wholly within the cell, the Ti^{4+} ions are at the corners of the cell, and the O^{2-} ions are in the cell edges. From this you can decide how many net ions are within the unit cell and thus know the formula.

Strategy Based on the locations of the ions, decide on the net number of ions of each kind in the cell.

Solution

> Number of Ti^{4+} ions:
>
> > (8 Ti^{4+} ions at cube corners) $\times$ ($\frac{1}{8}$ of each ion inside unit cell) = 1 net Ti^{4+} ion
>
> Number of Ca^{2+} ions:
>
> > One ion is in the cube center = 1 net Ca^{2+} ion
>
> Number of O^{2-} ions:
>
> > (12 O^{2-} ions in cube edges) $\times$ ($\frac{1}{4}$ of each ion inside unit cell) = 3 net O^{2-} ions
>
> Thus, the formula of perovskite is $CaTiO_3$.

Think about Your Answer $CaTiO_3$ is a reasonable formula. One Ca^{2+} ion and three O^{2-} ions would require a titanium ion with a 4+ charge, a reasonable value because titanium is in Group 4B of the periodic table.

Check Your Understanding

If an ionic solid has an fcc lattice of anions (X) and all of the tetrahedral holes are occupied by metal cations (M), is the formula of the compound MX, MX_2, or M_2X?

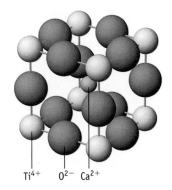

Ti^{4+} O^{2-} Ca^{2+}

The unit cell of the mineral perovskite. See Example 13.2.

EXAMPLE 13.3 The Relation of the Density of an Ionic Compound and Its Unit Cell Dimensions

Problem Magnesium oxide has a face-centered cubic unit cell of oxide ions with magnesium ions in octahedral holes. If the radius of Mg^{2+} is 79 pm and the density of MgO is 3.56 g/cm³, what is the radius of an oxide ion?

What Do You Know? This is similar to Example 13.1 in that you wish to know the size of one ion or atom in the cell. You are given the density of the compound and the fact that the unit cell is face-centered cubic. In addition, you know the radius of a Mg^{2+} ion.

Strategy You can use the atomic masses of Mg and O and Avogadro's number to calculate the mass of one formula unit of MgO. You know that the unit cell is face-centered cubic. Because the lattice points are O^{2-} ions, this tells you that there are 4 O^{2-} ions within the cell. There are Mg^{2+} ions in the octahedral holes, so you know there are also 4 Mg^{2+} ions within the cell. The unit cell thus contains 4 MgO units. This, along with the mass per formula unit, can be used to calculate the mass per unit cell. The density of the cell relates the mass of the cell and the cell volume, so you can now calculate the volume of the unit cell and from this the length of one edge (Length = volume$^{1/3}$). As illustrated in the margin, the edge of the unit cell is twice the radius of a Mg^{2+} ion (2 $\times$ 79 pm) plus twice the radius of an O^{2-} ion (the unknown).

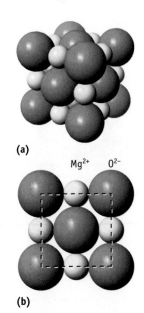

(a)

Mg^{2+} O^{2-}

(b)

Magnesium oxide. (a) A unit cell showing oxide ions in a face-centered cubic lattice with magnesium ions in the octahedral holes. **(b)** One face of the cell.

Solution

1. *Calculate the mass of the unit cell.* An ionic compound of formula MX (here MgO) and based on a face-centered cubic lattice of X^- ions with M^+ ions in the octahedral holes has 4 MX units per unit cell.

$$\text{Unit cell mass} = \left(\frac{40.304\text{ g}}{1\text{ mol MgO}}\right)\left(\frac{1\text{ mol MgO}}{6.022 \times 10^{23}\text{ units of MgO}}\right)\left(\frac{4\text{ MgO units}}{1\text{ unit cell}}\right)$$

$$= 2.677 \times 10^{-22}\text{ g/unit cell}$$

2. *Calculate the volume of the unit cell from the unit cell mass and density.*

$$\text{Unit cell volume} = \left(\frac{2.667 \times 10^{-22}\text{ g}}{\text{unit cell}}\right)\left(\frac{1\text{ cm}^3}{3.56\text{ g}}\right) = 7.52 \times 10^{-23}\text{ cm}^3/\text{unit cell}$$

3. *Calculate the edge dimension of the unit cell in pm.*

$$\text{Unit cell edge} = (7.52 \times 10^{-23}\text{ cm}^3)^{1/3} = 4.22 \times 10^{-8}\text{ cm}$$

$$\text{Unit cell edge} = 4.22 \times 10^{-8}\text{ cm}\left(\frac{1\text{ m}}{100\text{ cm}}\right)\left(\frac{1 \times 10^{12}\text{ pm}}{1\text{ m}}\right) = 422\text{ pm}$$

4. *Calculate the oxide ion radius.*

One face of the MgO unit cell is shown on the previous page. The O^{2-} ions define the lattice, and the Mg^{2+} and O^{2-} ions along the cell edge just touch one another. This means that one edge of the cell is equal to one O^{2-} radius (x) plus twice the Mg^{2+} radius plus one more O^{2-} radius.

$$\text{MgO unit cell edge} = x\text{ pm} + 2(79\text{ pm}) + x\text{ pm} = 422\text{ pm}$$

$$x = \boxed{\text{oxide ion radius} = 132\text{ pm}}$$

Think about Your Answer Chemists often check the chemical literature to judge the reasonableness of an answer. The calculated result here is very close to the value in Table 7.12 (140 pm).

Check Your Understanding

Potassium chloride has the same unit cell as NaCl. Using the ion sizes in Figure 7.12, calculate the density of KCl.

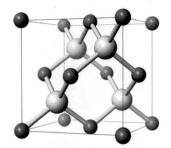

Unit cell of silicon carbide, SiC.

REVIEW & CHECK FOR SECTION 13.2

1. The unit cell of silicon carbide, SiC, is illustrated in the margin. In what type of unit cell are the (dark gray) C atoms arranged?

    (a) primitive cubic   (c) face-centered cubic

    (b) body-centered cubic

2. If one edge of the silicon carbide unit cell is 436.0 pm, what is the calculated density of this compound?

    (a) 0.803 g/cm³   (b) 0.311 g/cm³   (c) 3.21 g/cm³

13.3 Bonding in Metals and Semiconductors

Molecular orbital (MO) theory was introduced in Chapter 9 to rationalize covalent bonding in molecules, but it can also be used to describe metallic bonding. A metal is a kind of "supermolecule," and to describe the bonding in a metal we have to look at all the atoms in a given sample.

Even a tiny piece of metal contains a very large number of atoms and an even larger number of valence orbitals. In 1 mol of lithium atoms, for example, there are 6×10^{23} atoms. Considering only the $2s$ valence orbitals of lithium, there are 6×10^{23} atomic orbitals, from which 6×10^{23} molecular orbitals can be created

FIGURE 13.11 **Bands of molecular orbitals in a metal crystal.** Here, the 2s valence orbitals of Li atoms are combined to form molecular orbitals. As more and more atoms with the same valence orbitals are added, the number of molecular orbitals grows until the orbitals are so close in energy that they merge into a band of molecular orbitals. If 1 mol of Li atoms, each with its 2s valence orbital, is combined, 6×10^{23} molecular orbitals are formed. However, only 1 mol of electrons, or 3×10^{23} electron pairs, is available, so only half of these molecular orbitals are filled. (See Section 9.3 for the discussion of molecular orbital theory.)

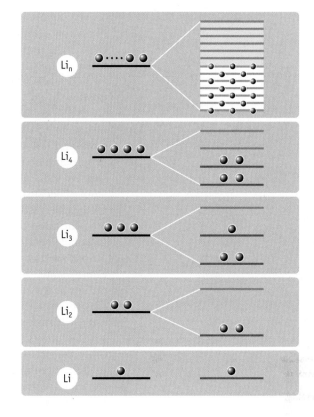

(◄ Section 9.3). The molecular orbitals that we envision in lithium will span all the atoms in the crystalline solid. A mole of lithium has 1 mol of valence electrons, and these electrons occupy the lower-energy bonding orbitals. The bonding is described as *delocalized* because the electrons are associated with all the atoms in the crystal and not with a specific bond between two atoms.

This theory of metallic bonding is called **band theory.** An energy-level diagram would show the bonding and antibonding molecular orbitals blending together into a band of molecular orbitals (Figure 13.11), with the individual MOs being so close together in energy that they are not distinguishable. The band is composed of as many molecular orbitals as there are contributing atomic orbitals, and each molecular orbital can accommodate two electrons of opposite spin.

In metals, there are not enough electrons to fill all of the molecular orbitals. In 1 mol of Li atoms, for example, 6×10^{23} electrons, or 3×10^{23} electron pairs, are sufficient to fill only half of the 6×10^{23} molecular orbitals. The lowest energy for a system occurs with all electrons in orbitals with the lowest possible energy, but this is reached only at 0 K. The highest filled level at 0 K is called the **Fermi level** (Figure 13.12).

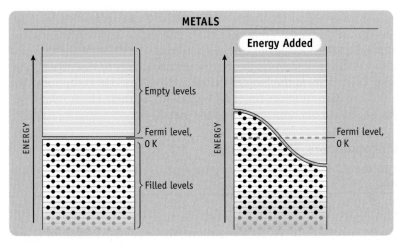

Metals. The highest filled level at 0 K is referred to as the *Fermi level*.

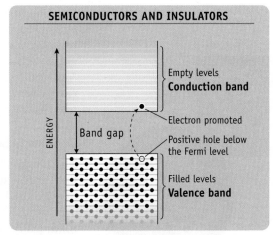

Semiconductors. In contrast to metals, the band of filled levels (the valence band) is separated from the band of empty levels (the conduction band) by a band gap. In insulators, the energy of the band gap is large.

FIGURE 13.12 **Band theory applied to metals, semiconductors, and insulators.** The bonding in metals and semiconductors can be described using molecular orbital theory. Molecular orbitals are constructed from the valence orbitals on each atom and are delocalized over all the atoms.

In metals at temperatures above 0 K, thermal energy will cause some electrons to occupy orbitals above the Fermi level. Even a small input of energy (for example, raising the temperature a few degrees above 0 K) will cause electrons to move from filled orbitals to higher-energy orbitals. For each electron promoted, two singly occupied levels result: a negative electron in an orbital above the Fermi level and a positive "hole"—from the absence of an electron—below the Fermi level.

The positive holes and negative electrons in a piece of metal account for its electrical conductivity. Electrical conductivity arises from the movement of electrons and holes in singly occupied states in the presence of an applied electric field. When an electric field is applied to the metal, negative electrons move toward the positive side, and the positive "holes" move to the negative side. (Positive holes "move" because an electron from an adjacent atom can move into the hole, thereby creating a fresh "hole.")

The band of energy levels in a metal is essentially continuous, that is, the energy gaps between levels are extremely small. A consequence of this is that a metal can absorb energy of nearly any wavelength, causing an electron to move to a higher energy state. The now-excited system can immediately emit a photon of the same energy as the electron returns to the original energy level. This rapid and efficient absorption *and* reemission of light make polished metal surfaces be reflective and appear lustrous (shiny).

The molecular orbital picture for metallic bonding provides an interpretation for other physical characteristics of metals. For example, most metals are malleable and ductile, meaning they can be rolled into sheets and drawn into wires. In these processes, the metal atoms must be able to move fairly freely with respect to their nearest neighbors. This is possible because metallic bonding is delocalized—that is, nondirectional. The layers of atoms can slip past one another relatively easily (as if the delocalized electrons were ball bearings that facilitate this motion), while at the same time the layers are bonded through coulombic attractions between the nuclei and the electrons.

In contrast to metals, rigid network solids such as diamond, silicon, and silica (SiO_2) have localized bonding, which anchors the component atoms or ions in fixed positions. Movement of atoms in these structures relative to their neighbors requires breaking covalent bonds. As a result, such substances are typically hard and brittle. They will not deform under stress as metals do, but instead tend to cleave along crystal planes.

Semiconductors

Semiconducting materials are at the heart of all solid-state electronic devices, including such well-known devices as computer chips and diode lasers. Semiconductors do not conduct electricity easily but can be encouraged to do so by the input of energy. This property allows devices made from semiconductors to essentially have "on" and "off" states, which form the basis of the binary logic used in computers. We can understand how semiconductors function by looking at their electronic structure, following the band theory approach used for metals.

Bonding in Semiconductors: The Band Gap

The Group 4A elements carbon (in the diamond form), silicon, and germanium have similar structures. Each atom is surrounded by four other atoms at the corners of a tetrahedron (Figure 13.13). Using the band model of bonding, the orbitals of each atom are combined to form molecular orbitals that are delocalized over the solid. Unlike metals, however, the result for carbon, silicon, and germanium is two bands, a lower-energy valence band and a higher-energy conduction band. In metals, there is only a small energy barrier for an electron to go from the filled molecular orbitals to empty molecular orbitals, and electricity can flow easily. However, in electrical insulators, such as diamond, and in semiconductors, such as silicon and germanium, the valence and conduction bands are separated from each other resulting in a **band gap**, a barrier to the promotion of electrons to higher energy levels (Figure 13.12). In the Group 4A elements, the orbitals of the valence band are completely filled, but the conduction band is empty.

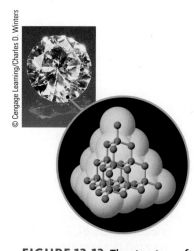

FIGURE 13.13 The structure of diamond, an insulator. The structures of silicon and germanium, semiconductors, are similar in that each atom is bound tetrahedrally to four others.

Semiconductors can conduct a current because thermal energy is sufficient to promote a few electrons from the valence band, across the band gap, to the conduction band (Figure 13.14). Conduction then occurs when the electrons in the conduction band migrate in one direction and the positive holes in the valence band migrate in the opposite direction.

The band gap in diamond is 580 kJ/mol—so large that electrons are trapped in the filled valence band and cannot make the transition to the conduction band, even at elevated temperatures. Thus, it is not possible to create positive "holes," and so diamond is an insulator, a nonconductor. Semiconductors, in contrast, have a smaller band gap. For common semiconducting materials, this band gap is usually in the range of 10 to 240 kJ/mol. (The band gap is 106 kJ/mol in silicon, whereas it is 68 kJ/mol in germanium.) The magnitude of the band gap in semiconductors is such that these substances are able to conduct small quantities of current under ambient conditions, but, as their name implies, they are much poorer conductors than metals.

Pure silicon and germanium are called **intrinsic semiconductors**, with the name referring to the fact that this is an intrinsic or naturally occurring property of the pure material. In intrinsic semiconductors, the number of electrons in the conduction band is determined by the temperature and the magnitude of the band gap. The smaller the band gap, the smaller the energy required to promote a significant number of electrons. As the temperature increases, more electrons are promoted into the conduction band, and a higher conductivity results.

There are also **extrinsic semiconductors**. The conductivity of these materials is controlled by adding small numbers of different atoms (typically 1 in 10^6 to 1 in 10^8) called *dopants*. That is, the characteristics of semiconductors can be changed by altering their chemical makeup.

Suppose a few silicon atoms in the silicon lattice are replaced by aluminum atoms (or atoms of some other Group 3A element). Aluminum has only three valence electrons, whereas silicon has four. Four Si-Al bonds are created per aluminum atom in the lattice, but these bonds must be deficient in electrons. According to band theory, the Si-Al bonds form a discrete but empty band at an energy level higher than the valence band but lower than the conduction band. This level is referred to as an *acceptor level* because it can accept electrons from the valence band. The gap between the valence band and the acceptor level is usually quite small, so electrons can be promoted readily to the acceptor level. The positive holes created in the valence band are able to move under the influence of an electric potential, so current results from the hole mobility. Because *positive* holes are created in an aluminum-doped semiconductor, this is called a **p-type semiconductor** (Figure 13.14b, left).

● **Intrinsic and Extrinsic** The term *intrinsic* means something is "essential" or "belongs naturally." The converse, *extrinsic*, means that a property is not part of the essential nature of an object.

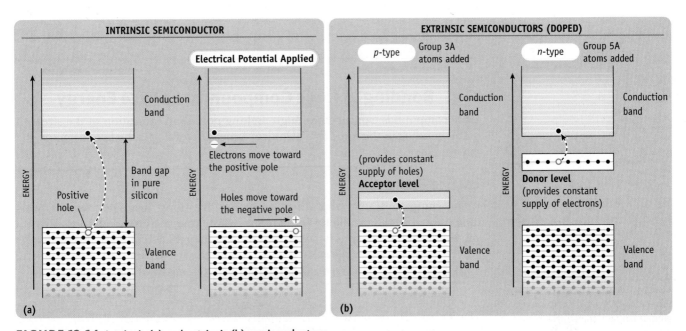

FIGURE 13.14 Intrinsic (a) and extrinsic (b) semiconductors.

Now suppose phosphorus atoms (or atoms of some other Group 5A element such as arsenic) are incorporated into the silicon lattice instead of aluminum atoms. This material is also a semiconductor, but it now has extra electrons because each phosphorus atom has one more valence electron than the silicon atom it replaces in the lattice. Semiconductors doped in this manner have a discrete, partially filled donor level that resides just below the conduction band. Electrons can be promoted to the conduction band from this donor band, and electrons in the conduction band carry the charge. Such a material, consisting of *negative* charge carriers, is called an **n-type semiconductor** (Figure 13.14b, right).

One group of materials that have desirable semiconducting properties is the III-V semiconductors, so called because they are formed by combining elements from Group 3A (such as Ga and In) with elements from Group 5A (such as As or Sb).

GaAs is a common semiconducting material that has electrical conductivity properties that are sometimes preferable to those of pure silicon or germanium. The crystal structure of GaAs is similar to that of diamond and silicon; each Ga atom is tetrahedrally coordinated to four As atoms, and vice versa.

It is also possible for Group 2B and 6A elements to form semiconducting compounds, such as CdS. The farther apart the elements are found in the periodic table, however, the more ionic the bonding becomes. As the ionic character of the bonding increases, the band gap will increase, and the material will become an insulator rather than a semiconductor. For example, the band gap in GaAs is 140 kJ/mol, whereas it is 232 kJ/mol in CdS.

These materials can also be modified by substituting other atoms into the structure. For example, in one widely used semiconductor, aluminum atoms are substituted for gallium atoms in GaAs, giving materials with a range of compositions ($Ga_{1-x}Al_xAs$). The importance of this modification is that the band gap depends on the relative proportions of the elements, so it is possible to control the size of the band gap by adjusting the stoichiometry. As Al atoms are substituted for Ga atoms, for example, the band gap energy increases. This consideration is important for the specific uses of these materials in devices such as LEDs (light-emitting diodes).

Gallium arsenide, GaAs. The unit cell of GaAs has Ga atoms in an fcc lattice with As atoms in tetrahedral holes. It is an important semiconductor used in infrared emitting diodes, laser diodes, and solar cells.

REVIEW & CHECK FOR SECTION 13.3

Germanium was one of the first materials used as a semiconductor. Given that it has a band gap of 68 kJ/mol, what is the maximum wavelength of light that can excite an electron transition across the band gap of Ge? To what region of the electromagnetic spectrum does this correspond?

(a) 1800 nm, infrared

(b) 29 nm, ultraviolet

(c) 15×10^6 nm, microwave

13.4 Bonding in Ionic Compounds: Lattice Energy

Ionic compounds typically have high melting points, an indication of the strength of the bonding in the ionic crystal lattice. A measure of that is the lattice energy, the main topic of this section.

Lattice Energy

Ionic compounds have positive and negative ions arranged in a three-dimensional lattice in which there are extensive attractions between ions of opposite charge and repulsions between ions of like charge. Each of these interactions is governed by an equation related to Coulomb's law (◄ page 68). For example, $U_{\text{ion pair}}$, the energy of attractive interactions between a pair of ions, is given by

$$U_{\text{ion pair}} = C \times \frac{(n^+e)(n^-e)}{d}$$

The symbol C represents a constant, d is the distance between the ion centers, and e is the charge on an electron. The term n^+e represents the charge on the cation in the ion pair, and the term n^-e is the charge on the negative ion. Because n^-e is negative, the energy will have a negative value. Be sure to notice that *the energy depends directly on the charges on the ions and inversely on the distance between them.*

In an extended ionic lattice, there are multiple cation–anion interactions. Taking NaCl as an example (Figure 13.9), focus on an Na^+ ion in the center of the unit cell. We see it is surrounded by, and attracted to, six Cl^- ions. Just a bit farther away from this Na^+ ion, however, there are 12 Na^+ ions along the edges of the cubes, and there is a force of repulsion between the center Na^+ and these ions. And if we still focus on the Na^+ ion in the "center" of the unit cell, we see there are eight more Cl^- ions at the corners of the cube, and these are attracted to the center Na^+ ion. Taking into account *all* of the interactions between the ions in a lattice, we can calculate the **lattice energy,** $\Delta_{lattice}U$, the energy of formation of one mole of a solid crystalline ionic compound when ions in the gas phase combine (Table 13.2). For sodium chloride, this reaction is

$$Na^+(g) + Cl^-(g) \rightarrow NaCl(s)$$

When dealing with ionic compounds chemists often use **lattice enthalpy,** $\Delta_{lattice}H$. The same trends are seen in both lattice energy and enthalpy, and, because we are dealing with a condensed phase, the numerical values are nearly identical.

We shall focus here on the dependence of lattice enthalpy on ion charges and sizes. As given by Coulomb's law, the higher the ion charges, the greater the attraction between oppositely charged ions, so $\Delta_{lattice}H$ has a larger negative value for more highly charged ions. This is illustrated by the lattice enthalpies of MgO and NaF. The value of $\Delta_{lattice}H$ for MgO (-4050 kJ/mol) is about four times more negative than the value for NaF (-926 kJ/mol) because the charges on the Mg^{2+} and O^{2-} ions $[(2+) \times (2-)]$ are twice as large as those on Na^+ and F^- ions.

Because the attraction between ions is inversely proportional to the distance between them, the effect of ion size on lattice enthalpy is also predictable: A lattice built from smaller ions generally leads to a more negative value for the lattice enthalpy (Table 13.2 and Figure 13.15). For alkali metal halides, for example, the lattice enthalpy for lithium compounds is generally more negative than that for potassium compounds because the Li^+ ion is much smaller than the K^+ cation. Similarly, lattice enthalpies of fluorides are more negative than those for iodides with the same cation.

Calculating a Lattice Enthalpy from Thermodynamic Data

Lattice enthalpies can be calculated using a thermodynamic approach known as a **Born–Haber cycle,** an application of Hess's law (◀ page 194). The relationship of various enthalpy changes is illustrated in Figure 13.16 for solid sodium chloride.

Table 13.2 Lattice Energies of Some Ionic Compounds

Compound	$\Delta_{lattice}U$ (kJ/mol)
LiF	−1037
LiCl	−852
LiBr	−815
LiI	−761
NaF	−926
NaCl	−786
NaBr	−752
NaI	−702
KF	−821
KCl	−717
KBr	−689
KI	−649

Source: D. Cubicciotti: Lattice energies of the alkali halides and electron affinities of the halogens. *Journal of Chemical Physics*, Vol. 31, p. 1646, 1959.

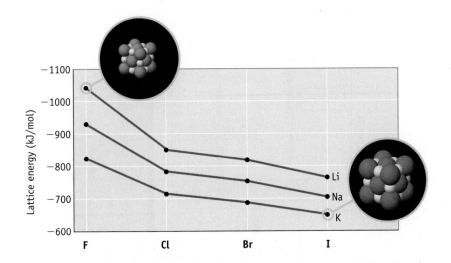

FIGURE 13.15 Lattice energy. $\Delta_{lattice}U$ is illustrated for the formation of the alkali metal halides, MX(s), from the ions $M^+(g) + X^-(g)$.

Steps 1 and 2 in Figure 13.16 involve formation of $Na^+(g)$ and $Cl^-(g)$ ions from the elements; the enthalpy change for each of these steps is known (Appendices F and L). Step 3 in Figure 13.16 gives the lattice enthalpy, $\Delta_{lattice}H$. $\Delta_f H°$ is the standard molar enthalpy of formation of NaCl(s) (Appendix L). The enthalpy values for each step are related by the following equation:

$$\Delta_f H° \, [NaCl(s)] = \Delta H_{Step\,1a} + \Delta H_{Step\,1b} + \Delta H_{Step\,2a} + \Delta H_{Step\,2b} + \Delta H_{Step\,3}$$

Because the values for all of these quantities are known except for $\Delta H_{Step\,3}$ ($\Delta_{lattice}H$), the value for this step can be calculated.

Step 1a. Enthalpy of formation of Cl(g) $= +121.3$ kJ/mol (Appendix L)
Step 1b. $\Delta H°$ for $Cl(g) + e^- \rightarrow Cl^-(g)$ $= -349$ kJ/mol (Appendix F)
Step 2a. Enthalpy of formation of Na(g) $= +107.3$ kJ/mol (Appendix L)
Step 2b. $\Delta H°$ for $Na(g) \rightarrow Na^+(g) + e^-$ $= +496$ kJ/mol (Appendix F)

The standard enthalpy of formation of NaCl(s), $\Delta_f H°$, is -411.12 kJ/mol. Combining this with the known values of Steps 1 and 2, we can calculate $\Delta H_{step\,3}$, which is the lattice enthalpy, $\Delta_{lattice}H$.

Step 3. Formation of NaCl(s) from the ions in the gas phase $= \Delta H_{step\,3}$
$$\Delta H_{step\,3} = \Delta_f H° \, [NaCl(s)] - \Delta H_{Step\,1a} - \Delta H_{Step\,1b} - \Delta H_{Step\,2a} - \Delta H_{Step\,2b}$$
$$= -411.12 \text{ kJ/mol} - 121.3 \text{ kJ/mol} - (-349 \text{ kJ/mol})$$
$$- 107.3 \text{ kJ/mol} - 496 \text{ kJ/mol}$$
$$= -787 \text{ kJ/mol}$$

REVIEW & CHECK FOR SECTION 13.4

1. Which sodium halide has the most negative lattice enthalpy?

   (a) NaI   (b) NaBr   (c) NaCl   (d) NaF

2. The lattice enthalpy for NaCl is defined as the enthalpy change for which of the following reactions?

   (a) $Na(s) + Cl_2(g) \rightarrow NaCl(s)$ (c) $Na^+(g) + Cl^-(g) \rightarrow NaCl(s)$

   (b) $Na(g) + Cl(g) \rightarrow NaCl(s)$ (d) $NaOH(s) + HCl(g) \rightarrow NaCl(s) + H_2O(\ell)$

3. Calculate the molar enthalpy of formation, $\Delta_f H°$, of solid sodium iodide using the approach outlined in Figure 13.16. The required data can be found in Appendices F and L and in Table 13.2.

   (a) $+1710$ kJ/mol   (c) -287 kJ/mol

   (b) -702 kJ/mol   (d) $+1120$ kJ/mol

FIGURE 13.16 Born–Haber cycle for the formation of NaCl(s) from the elements. The calculation in the text uses enthalpy values, and the value obtained is the lattice enthalpy, $\Delta_{lattice}H$. The difference between $\Delta_{lattice}U$ and $\Delta_{lattice}H$ is generally small and can be corrected for, if desired. (Note that the energy diagram is not to scale.)

Calculation of lattice energies by this procedure (a Born–Haber cycle) is named for Max Born (1882–1970) and Fritz Haber (1868–1934), German scientists who played prominent roles in thermodynamic research.

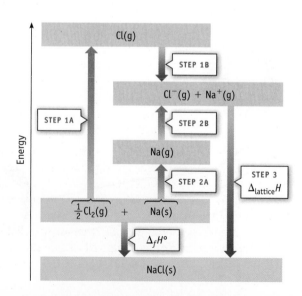

13.5   The Solid State: Other Types of Solid Materials

So far, we have described the structures of metals and simple ionic solids. Now we turn briefly to the other categories of solids: molecular solids, network solids, and amorphous solids (Table 13.1).

Molecular Solids

Compounds such as H_2O and CO_2 exist as solids under appropriate conditions. In these cases, it is molecules, rather than atoms or ions, that pack in a regular fashion in a three-dimensional lattice. You have already seen an example of a molecular solid: ice (◄ Figure 12.8).

The way molecules are arranged in a crystalline lattice depends on the shape of the molecules and the types of intermolecular forces. Molecules tend to pack in the most efficient manner and to align in ways that maximize intermolecular forces of attraction. Thus, the water structure was established to gain the maximum intermolecular attraction through hydrogen bonding.

It is from structural studies on molecular solids that most of the information on molecular geometries, bond lengths, and bond angles discussed in Chapter 8 was assembled.

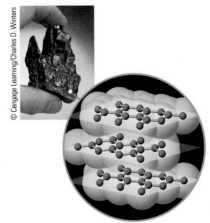

Graphite, a network solid.

Network Solids

Network solids are composed entirely of a three-dimensional array of covalently bonded atoms. Common examples include two allotropes of carbon: graphite and diamond. Elemental silicon is also a network solid with a diamond-like structure.

Graphite consists of carbon atoms bonded together in flat sheets that cling only weakly to one another. Within the layers, each carbon atom is surrounded by three other carbon atoms in a trigonal-planar arrangement. The layers can slip easily over another, which explains why graphite is soft, a good lubricant, and used in pencil lead. (Pencil "lead" is not the element lead but rather a composite of clay and graphite.)

Diamonds, which have a relatively low density ($d = 3.51$ g/cm^3), are the hardest material and the best conductor of heat known. They are transparent to visible light, as well as to infrared and ultraviolet radiation. In addition to their use in jewelry, diamonds are used as abrasives and in diamond-coated cutting tools. In the structure of diamond (Figure 13.13), each carbon atom is bonded to four other carbon atoms at the corners of a tetrahedron, and this pattern extends throughout the solid.

Silicates, compounds composed of silicon and oxygen, are also network solids and represent an enormous class of chemical compounds. You know them in the form of sand, quartz, talc, and mica, or as a major constituent of rocks such as granite. The structure of quartz is illustrated in Figure 13.17. It consists of tetrahedral silicon atoms covalently bonded to oxygen atoms in a giant three-dimensional lattice.

Most network solids are hard and rigid and are characterized by high melting and boiling points. These characteristics reflect the fact that a great deal of energy must be provided to break the covalent bonds in the lattice. For example, silicon dioxide melts at temperatures higher than 1600 °C.

Amorphous Solids

A characteristic property of pure crystalline solids—whether metals, ionic solids, or molecular solids—is that they melt at a specific temperature. For example, water in the form of ice melts at 0 °C, aspirin at 135 °C, lead at 327.5 °C, and NaCl at 801 °C. Because they are specific and reproducible values, melting points are often used as a means of identifying chemical compounds.

Another property of crystalline solids is that they form well-defined crystals, with smooth, flat faces. When a sharp force is applied to a crystal, it will most often cleave to give smooth, flat faces. The resulting solid particles are smaller versions of the original crystal (Figure 13.18a).

FIGURE 13.17 Silicon dioxide, a network solid. Common quartz, SiO_2, is a network solid consisting of silicon and oxygen atoms.

FIGURE 13.18 Crystalline and amorphous solids.

(a) A salt crystal, a crystalline solid, can be cleaved cleanly into smaller and smaller crystals that are duplicates of the larger crystal.

(b) Glass is an amorphous solid composed of silicon and oxygen atoms. It has, however, no long-range order as in crystalline quartz.

Many common solids, including ones that we encounter every day, do not have these properties, however. These are *amorphous solids*, solids that do not have a regular structure. Glass is a good example. When glass is heated, it softens over a wide temperature range, a property useful for artisans and craftsmen who can create beautiful and functional products for our enjoyment and use. Glass also possesses a property that we would rather it not have: When glass breaks, it leaves randomly shaped pieces. Other amorphous solids that behave similarly include common polymers such as polyethylene, nylon, and other plastics.

REVIEW & CHECK FOR SECTION 13.5

1. Which of the following allotropes of carbon is not a network solid?

    (a) graphite   (b) diamond   (c) Buckyballs (C_{60})

2. A soft, white waxy solid melts over a temperature range from 120 °C to 130 °C. It doesn't dissolve in water and it doesn't conduct electricity. These properties are consistent with its identity as

    (a) a network solid   (c) an ionic solid

    (b) an amorphous solid   (d) a metallic solid

13.6 Phase Changes Involving Solids

The shape of a crystalline solid is a reflection of its internal structure. But what about physical properties of solids, such as the temperatures at which they melt? This and many other physical properties of solids are of interest to chemists, geologists, and engineers, among others.

Melting: Conversion of Solid into Liquid

The melting point of a solid is the temperature at which the lattice collapses and the solid is converted into a liquid. Like any phase change, melting requires energy, called the *enthalpy of fusion* (given in kilojoules per mole) (◄ Chapter 5).

Energy absorbed as heat on melting = enthalpy of fusion = $\Delta_{fusion}H$ (kJ/mol)

Energy evolved as heat on freezing = enthalpy of crystallization = $-\Delta_{fusion}H$ (kJ/mol)

Enthalpies of fusion can range from just a few thousand joules per mole to many thousands of joules per mole (Table 13.3). A low melting temperature will certainly mean a low value for the enthalpy of fusion, whereas high melting points

Graphene—The Hottest New Network Solid

One of the most interesting developments in chemistry in the last 20 years has been the discovery of new forms of carbon. First, there were buckyballs and then single-wall and multiwall carbon nanotubes.

Common graphite, from which your pencil lead is made, consists of six-member rings of carbon atoms connected in sheets, and the sheets stack one on top of another like cards in a deck. But if carbon compounds are heated under the right conditions, the carbon atoms assemble into sheets, and the sheets close on themselves to form tubes. These are called **nanotubes** because the tubes are only a nanometer or so in diameter. Sometimes they are single tubes, and other times there are tubes within tubes. Carbon nanotubes are at least 100 times stronger than steel but only one sixth as dense, and they conduct heat and electricity far better than copper. Thus, there has been enormous interest in their commercial applications, but it has also been difficult to make them with consistent properties.

Now there is **graphene**, a single sheet of six-member carbon atoms. Researchers in England, Andre Geim and Kostya Novoselov, discovered graphene in a simple way: Put a flake of graphite on Scotch tape, fold the tape over, and then pull it apart. The graphite layers come apart, and, if you do it enough times, only one layer—one C atom thick!—is left on the tape. Now graphene can be made in much

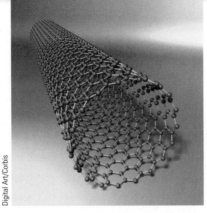

Digital Art/Corbis

Carbon nanotube. These tubes are assembled from six-member rings of carbon atoms and are typically about 1 nm in diameter.

Image courtesy: Dmitriy Dikin (Northwestern University)

Graphene "paper." Researchers at Northwestern University have reassembled graphene sheets to create a tough, flexible, paper-like material.

larger quantities by other methods, and more than 1000 scientific papers have been published on this substance. Chemists and physicists have found that the material is extremely strong and stiff and that it is a better conductor of electrons than any other material at room temperature. It was no surprise when Geim and Novoselov received the 2010 Nobel Prize in physics for their discovery.

Chemical and Engineering News (March 2009) said that although it was "considered a hypothetical curiosity just a few years ago, one-atom-thick graphene has stormed the world of scientific research." Research is

proceeding worldwide on its use in electronics and the dream is of making flexible displays. Watch for graphene-based devices to make a difference in your life in the next few years!

Questions:

1. Based on a C–C distance of 139 pm, what is the side-to-side dimension of a planar, C_6 ring?
2. If a graphene sheet has a width of 1.0 micrometer, how many C_6 rings are joined across the sheet?
3. Estimate the thickness of a sheet of graphene (in pm). How did you determine this value?

Answers to these questions are available in Appendix N.

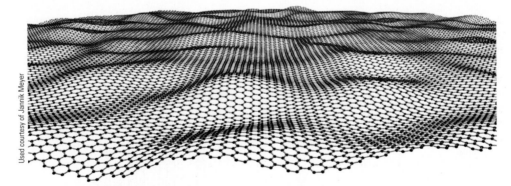

Used courtesy of Jannik Meyer

Graphene is a single sheet of six-member carbon rings. This latest material in the world of carbon chemistry is a faster conductor of electrons than any other material.

are associated with high enthalpies of fusion. Figure 13.19 shows the enthalpies of fusion for the metals of the fourth through the sixth periods. Here we see that transition metals have high enthalpies of fusion, with many of those in the sixth period being extraordinarily high. This trend parallels the trend seen with the melting points for these elements. Tungsten, which has the highest melting point of all the known elements except for carbon, also has the highest enthalpy of fusion among

Table 13.3 **Melting Points and Enthalpies of Fusion of Some Elements and Compounds**

Element or Compound	Melting Point (°C)	Enthalpy of Fusion (kJ/mol)	Type of Interparticle Forces
Metals			
Hg	−39	2.29	Metal bonding.
Na	98	2.60	
Al	660	10.7	
Ti	1668	20.9	
W	3422	35.2	
Molecular Solids: Nonpolar Molecules			
O_2	−219	0.440	Dispersion forces only.
F_2	−220	0.510	
Cl_2	−102	6.41	
Br_2	−7.2	10.8	
Molecular Solids: Polar Molecules			
HCl	−114	1.99	All three HX molecules have dipole–dipole forces. Dispersion forces increase with size and molar mass.
HBr	−87	2.41	
HI	−51	2.87	
H_2O	0	6.01	Hydrogen bonding, dispersion forces.
Ionic Solids			
NaF	996	33.4	All ionic solids have extended ion–ion interactions. Note the general trend is the same as for lattice energies (see Section 13.4 and Figure 13.15).
NaCl	801	28.2	
NaBr	747	26.1	
NaI	660	23.6	

FIGURE 13.19
Enthalpy of fusion of fourth-, fifth-, and sixth-period metals. Enthalpies of fusion range from 2–5 kJ/mol for Group 1A elements to 35.2 kJ/mol for tungsten. Notice that enthalpies of fusion generally increase for group 4B–8B metals on descending the periodic table.

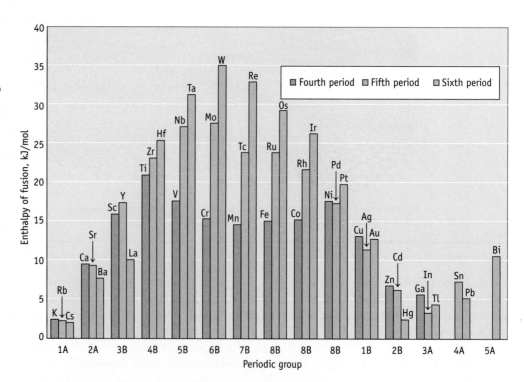

the transition metals. This is the reason tungsten is used for the filaments in incandescent lightbulbs; no other material has been found to work better since the invention of the lightbulb in 1908.

Table 13.3 presents some data for several basic types of substances: metals, polar and nonpolar molecules, and ionic solids. In general, nonpolar substances that form molecular solids have low melting points. Melting points increase within a series of related molecules, however, as the size and molar mass increase. This happens because London dispersion forces are generally larger when the molar mass is larger. Thus, increasing amounts of energy are required to break down the intermolecular forces in the solid, a principle that is reflected in an increasing enthalpy of fusion.

The ionic compounds in Table 13.3 have higher melting points and higher enthalpies of fusion than the molecular solids. This property is due to the strong ion–ion forces present in ionic solids, forces that are reflected in high lattice energies. Because ion–ion forces depend on ion size (as well as ion charge), there is a good correlation between lattice energy and the position of the metal or halogen in the periodic table. For example, the data in Table 13.3 show a decrease in melting point and enthalpy of fusion for sodium salts as the halide ion increases in size. This parallels the decrease in lattice energy seen with increasing ion size.

LED lights. Incandescent lights, with a tungsten filament, may soon be a thing of the past. They can be replaced by LED (light-emitting diode) lamps with a much lower energy consumption. The LED element is a semiconductor that releases light energy when electrons recombine with holes.

Sublimation: Conversion of Solid into Vapor

Molecules can escape directly from the solid to the gas phase by sublimation (Figure 13.20).

$$\text{Solid} \rightarrow \text{gas} \qquad \text{Energy required as heat} = \Delta_{\text{sublimation}}H$$

Sublimation, like fusion and evaporation, is an endothermic process. The energy required as heat is called the **enthalpy of sublimation**. Water, which has a molar enthalpy of sublimation of 51 kJ/mol, can be converted from solid ice to water vapor quite readily. A good example of this phenomenon is the sublimation of frost from grass and trees as night turns to day on a cold morning in the winter.

REVIEW & CHECK FOR SECTION 13.6

1. Which of the following Group 1A halides is expected to have the highest enthalpy of fusion?

   (a) NaCl   (b) KCl   (c) NaBr   (d) KBr

2. Suppose you wanted to cool 100. g of water from 20 °C to 0 °C using dry ice, $CO_2(s)$. The enthalpy of sublimation of $CO_2(s)$ is 25.2 kJ/mol. What mass of dry ice should you need?

   (a) 0.33 g   (b) 15 g   (c) 3.5 g   (d) 150 g

Iodine sublimes
when heated.

FIGURE 13.20 Sublimation. Sublimation is the conversion of a solid directly to its vapor. Here, iodine (I_2) sublimes when warmed. If an ice-filled test tube is inserted into the flask, the vapor deposits on the cold surface.

13.7 Phase Diagrams

Depending on the conditions of temperature and pressure, a substance can exist as a gas, a liquid, or a solid. In addition, under certain specific conditions, two (or even three) states can coexist in equilibrium. It is possible to summarize this information in the form of a graph called a **phase diagram**. Phase diagrams are used to illustrate the relationship between phases of matter and the pressure and temperature.

Water

Figure 13.21 illustrates the phase diagram for water. The lines in a phase diagram identify the conditions under which two phases exist at equilibrium. Conversely, all points that do not fall on the lines in the figure represent conditions under which there is only one state that is stable. Line *A–B* represents conditions for solid-vapor equilibrium, and line *A–C* for liquid–solid equilibrium. The line from point *A* to point *D*, representing the temperature and pressure combination at which the liquid and vapor phases are in equilibrium, is the same curve plotted for water vapor pressure in Figure 12.17. Recall that the normal boiling point, 100 °C in the case of water, is the temperature at which the equilibrium vapor pressure is 760 mm Hg.

Point *A*, appropriately called the **triple point**, indicates the conditions under which all three phases coexist in equilibrium. For water, the triple point is at $P = 4.6$ mm Hg and $T = 0.01$ °C.

The line *A–C* shows the conditions of pressure and temperature at which solid–liquid equilibrium exists. (Because no vapor pressure is involved here, the pressure referred to is the external pressure on the liquid.) For water, this line has a negative slope; the change for water is approximately -0.01 °C for each one-atmosphere increase in pressure. That is, the higher the external pressure, the lower the melting point.

The negative slope of the water solid–liquid equilibrium line can be explained from our knowledge of the structure of water and ice. When the pressure on an object increases, common sense tells us that the volume of the object will become smaller, giving the substance a higher density. Because ice is less dense than liquid water (due to the open lattice structure of ice, Figure 12.8), ice and water in equilibrium respond to increased pressure (at constant *T*) by melting ice to form more water because the same mass of liquid water requires less volume.

Phase Diagrams and Thermodynamics

Let us explore the water phase diagram further by correlating phase changes with thermodynamic data. Suppose we begin with ice at -10 °C and under a pressure of 500 mm Hg (point *a* on Figure 13.21). As ice is heated (at constant *P*), it absorbs about 2.1 J/g · K in warming from point *a* to point *b* at a temperature between 0 °C and 0.01 °C. At this point, the solid is in equilibrium with liquid water. Solid-liquid equilibrium is maintained until 333 J/g has been transferred to the sample and it has become liquid water at this temperature. If the liquid, still under a pressure of 500 mm Hg, absorbs 4.184 J/g · K, it warms to point *c*. The temperature at point *c* is about 89 °C, and equilibrium is established between liquid water and water vapor. The equilibrium vapor pressure of the liquid water is 500 mm Hg. If about 2300 J/g is transferred to the liquid–vapor sample, the equilibrium vapor pressure remains 500 mm Hg until the liquid is completely converted to vapor at 89 °C.

Carbon Dioxide

The features of the phase diagram for CO_2 (Figure 13.22) are generally the same as those for water but with some important differences.

In contrast to water, the CO_2 solid–liquid equilibrium line has a positive slope. Once again, increasing pressure on the solid in equilibrium with the liquid will shift the equilibrium to the more dense phase, but for CO_2 this will be the solid. Because solid CO_2 is denser than the liquid, the newly formed solid CO_2 sinks to the bottom in a container of liquid CO_2.

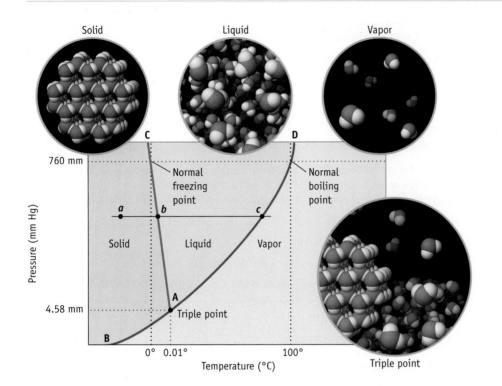

Another feature of the CO_2 phase diagram is the triple point that occurs at a pressure of 5.19 atm (3940 mm Hg) and 216.6 K (-56.6 °C). Carbon dioxide cannot be a liquid at pressures lower than this.

At pressures around normal atmospheric pressure, CO_2 will be either a solid or a gas, depending on the temperature. [At a pressure of 1 atm, solid CO_2 is in equilibrium with the gas at a temperature of 197.5 K (-78.7 °C).] As a result, as solid CO_2 warms above this temperature, it sublimes rather than melts. Carbon dioxide is called *dry ice* for this reason; it looks like water ice, but it does not melt.

From the CO_2 phase diagram, we can also learn that CO_2 gas can be converted to a liquid at room temperature (20–25 °C) by exerting a moderate pressure on the gas. In fact, CO_2 is regularly shipped in tanks as a liquid to laboratories and industrial companies.

Finally, the critical pressure and temperature for CO_2 are 73 atm and 31 °C, respectively. Because the critical temperature and pressure are easily attained in the laboratory, it is possible to observe the transformation to supercritical CO_2 (Figure 12.20).

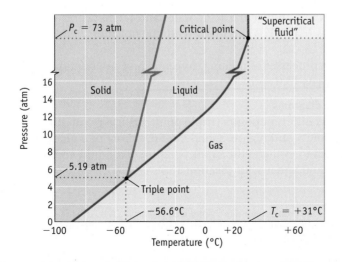

FIGURE 13.22 The phase diagram of CO₂. Notice in particular the positive slope of the solid–liquid equilibrium line. (For more on the critical point, see page 439.)

REVIEW & CHECK FOR SECTION 13.7

1. Refer to the phase diagram for CO_2 in the text. Which of the following statements is *not* true?

    (a) Liquid and gaseous CO_2 exist in equilibrium at approximately 0 °C and 12 atm.

    (b) Liquid CO_2 can only exist at a pressure greater than 5.19 atm.

    (c) Increasing the pressure leads to an increase in the melting point of $CO_2(s)$.

    (d) Supercritical CO_2 can be obtained at 25 °C if a high enough pressure is applied.

2. Nitrogen has the following properties: normal melting point, 63.3 K; normal boiling point, 77.4 K; triple point, 0.123 atm and 63.2 K. Sketch a phase diagram based on these data. Determine from your drawing which of the following statements is *not* correct.

    (a) At 75 K and 0.20 atm N_2 is a gas.

    (b) At 66 K and 0.50 atm N_2 is a gas.

    (c) At 60 K N_2 will be either a solid or a gas, depending on the pressure.

    (d) Solid and liquid coexist at equilibrium at 63.3 K and 1.0 atm.

 and

Sign in at **www.cengage.com/owl** to:
- View tutorials and simulations, develop problem-solving skills, and complete online homework assigned by your professor.
- For quick review and exam prep, download Go Chemistry mini lecture modules from OWL (or purchase them at **www.cengagebrain.com**)

? Access **How Do I Solve It?** tutorials on how to approach problem solving using concepts in this chapter.

CHAPTER GOALS REVISITED

Now that you have studied this chapter, you should ask whether you have met the chapter goals. In particular, you should be able to:

Understand cubic unit cells
a. Describe the three types of cubic unit cells: primitive cubic (pc), body-centered cubic (bcc), and face-centered cubic (fcc) (Section 13.1).
b. Relate atom size and unit cell dimensions. **Study Questions: 7, 8, 10, 38, 39, 41, 44, 46, 48, 61, and Go Chemistry Module 18.**

Relate unit cells for ionic compounds to formulas
a. Understand the relation of unit cell structure and formula for ionic compounds. (Section 13.2). **Study Questions: 3, 4, 5, 6, 8, 40, and Go Chemistry Module 18.**

Describe bonding in ionic and metallic solids and the nature of semiconductors.
a. Understand the band theory of bonding in metals and the electrical conductivity of metals. (Section 13.3). **Study Questions: 11, 13.**
b. Understand the nature of semiconductors. (Section 13.3). **Study Questions: 15–18, 51, 53.**

Describe the properties of solids
a. Understand lattice energy and how it is calculated (Section 13.4). **Study Questions: 19, 21, 22, 24, 50.**
b. Characterize different types of solids: metallic (e.g., copper), ionic (e.g., NaCl and CaF_2), molecular (e.g., water and I_2), network (e.g., diamond), and amorphous (e.g., glass and many synthetic polymers) (Table 13.1). **Study Questions: 25–30.**
c. Define the processes of melting, freezing, and sublimation and their enthalpies (Sections 13.6 and 13.7). **Study Questions: 31, 32.**

Interpret phase diagrams
a. Identify the different points (triple point, critical point, normal boiling point, freezing point) and regions (solid, liquid, vapor) of a phase diagram, and use the diagram to evaluate the vapor pressure of a liquid and the relative densities of a liquid and a solid (Section 13.7). **Study Questions: 33–36.**

OWL Questions and problems for this chapter are available in OWL. Use the OWL access card that came with this textbook to access assigned questions and problems for this chapter.

14 Solutions and Their Behavior

John C. Kotz

The Seax, an ocean-going yacht, in the Fiji Islands.

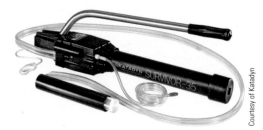

Courtesy of Katadyn

Portable reverse osmosis water purifier carried by deep water sailors.

Survival at Sea In the late 1980s, Bill and Simone Butler set off from Florida to sail their 38-foot boat around the world. Everything went well as they sailed through the Panama Canal and into the Pacific Ocean. Somewhere west of Costa Rica, however, on June 15, 1989, their boat was attacked by a school of whales in the middle of the night. The boat was holed and was going to sink. They had only enough time to launch their small life raft, grab some fishing line and a few other supplies, and one other important device: a portable water purifier that worked by reverse osmosis.

The Butlers drifted toward the east in their little raft for 66 days. They subsisted on the raw fish they could catch, but most importantly they had fresh water to drink from the reverse osmosis device. The device was able to produce 2–3 L of pure water a day, which saved their lives. Although they lost some weight, they were in good health when they were rescued by Costa Rican fishermen on August 19, 1989.

Question:

What is the osmotic pressure of seawater at 25 °C with a total dissolved ion concentration of 1.15 M?

The answer to this question is available in Appendix N.

CHAPTER OUTLINE

14.1 Units of Concentration

14.2 The Solution Process

14.3 Factors Affecting Solubility: Pressure and Temperature

14.4 Colligative Properties

14.5 Colloids

CHAPTER GOALS

See Chapter Goals Revisited (page 502) for Study Questions keyed to these goals.

- Calculate and use the solution concentration units molality, mole fraction, and weight percent.

- Understand the solution process.

- Understand and use the colligative properties of solutions.

- Recognize the properties and importance of colloids

We come into contact with solutions every day: aqueous solutions of ionic salts, gasoline with additives to improve its properties, and household cleaners such as ammonia in water. We purposely make solutions. Adding sugar, flavoring, and sometimes CO_2 to water produces a palatable soft drink. Athletes drink commercial beverages with dissolved salts to match salt concentrations in body fluids precisely, thus allowing the fluid to be taken into the body more rapidly. In medicine, saline solutions (aqueous solutions containing NaCl and other soluble salts) are infused into the body to replace lost fluids.

A **solution** is a homogeneous mixture of two or more substances in a single phase. By convention, the component present in largest amount is identified as the **solvent** and the other component(s) as the **solute**(s) (Figure 14.1). Although other types of solutions exist (such as alloys, solid solutions of metals), the objective in this chapter is to develop an understanding of gases, liquids, and solids dissolved in liquid solvents.

Experience tells you that adding a solute to a pure liquid will change the properties of the liquid. Indeed, that is the reason some solutions are made. For instance, adding antifreeze to the water in your car's radiator prevents the coolant from boiling in the summer and freezing in the winter. The changes that occur in the freezing and boiling points when a substance is dissolved in a pure liquid are two observations we want to examine in detail. These properties, as well as the osmotic pressure of a solution and changes in solvent vapor pressure, are examples of colligative properties. **Colligative properties** are properties of solutions that depend only on the number of solute particles per solvent molecule and not on the identity of the solute.

OWL

Sign in to OWL at **www.cengage.com/ owl** to view tutorials and simulations, develop problem-solving skills, and complete online homework assigned by your professor.

go Chemistry

Download mini lecture videos for key concept review and exam prep from OWL or purchase them from **www.cengagebrain.com**

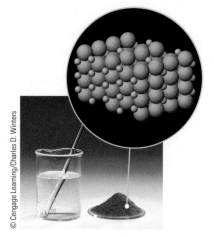

(a) Copper(II) chloride, the solute, is added to water, the solvent.

(b) Interactions between water molecules and Cu^{2+} and Cl^- ions allow the solid to dissolve. The ions are now sheathed with water molecules.

FIGURE 14.1 Making a solution of copper(II) chloride (the solute) in water (the solvent). When ionic compounds dissolve in water, each ion is surrounded by water molecules. The number of water molecules depends on ion size and charge.

FIGURE 14.2 Preparing 0.100 molal and 0.100 molar solutions. In the flask on the right, 0.100 mol (19.4 g) of K_2CrO_4 was mixed with enough water to make 1.000 L of solution. (The volumetric flask was filled to the mark on its neck, indicating that the volume is 1.000 L. Slightly less than 1.00 L of water was added.) If 1.00 kg of water was added to 0.100 mol of K_2CrO_4 in the flask on the left, the volume of solution is greater than 1.000 L. (The small pile of yellow solid in front of the flasks is 0.100 mol of K_2CrO_4.)

Notice the difference in water levels in the two flasks

© Cengage Learning/Charles D. Winters

0.100 molal solution
Water added to flask = 1.00 kg
Volume of solution > 1.00 L

0.100 molar solution
Water added to flask < 1.00 L
Volume of solution = 1.00 L

14.1 Units of Concentration

To analyze the colligative properties of solutions, we need ways of measuring solute concentrations that reflect the number of molecules or ions of solute per molecule of solvent.

Molarity, a concentration unit useful in stoichiometry calculations, is not useful when dealing with most colligative properties. Recall that molarity (M) is defined as the number of moles of solute per liter of solution (◄ page 149), so using molarity does not allow us to identify the exact amount of solvent used to make the solution. This fact is illustrated in Figure 14.2. The flask on the right contains a 0.100 M aqueous solution of potassium chromate. It was made by adding enough water to 0.100 mol of K_2CrO_4 to make 1.000 L of solution. The amount of solvent (water) that was actually added is not readily known. If 1.00 kg of water had been added to 0.100 mol of K_2CrO_4, as illustrated with the flask on the left in Figure 14.2, the volume of solution would be greater than 1.000 L.

Three concentration units are described here that reflect the number of molecules or ions of solute per solvent molecule: molality, mole fraction, and weight percent.

The **molality**, m, of a solution is defined as the amount of solute (mol) per kilogram of solvent.

$$\text{Concentration } (c, \text{ mol/kg}) = \text{molality of solute} = \frac{\text{amount of solute (mol)}}{\text{mass of solvent (kg)}} \quad (14.1)$$

The molality of K_2CrO_4 in the flask on the left side of Figure 14.2 is 0.100 mol/kg. It was prepared from 0.100 mol (19.4 g) of K_2CrO_4 and 1.00 kg (1.000 L × 1.00 kg/L) of water.

Notice that different quantities of water were used to make the 0.100 M (0.100 molar) and 0.100 m (0.100 molal) solutions of K_2CrO_4. This means the *molarity and the molality of a given solution cannot be the same*, although the difference may be negligibly small when the solution is quite dilute.

The **mole fraction**, X, of a solution component is defined as the amount of that component (n_A) divided by the total amount of all of the components of the mixture ($n_A + n_B + n_C + \ldots$). Mathematically it is represented as

• **Mole Fraction and Gases** The mole fraction of a gas in a mixture of gases was defined on page 402.

$$\text{Mole fraction of A } (X_A) = \frac{n_A}{n_A + n_B + n_C + \ldots} \quad (14.2)$$

Consider a solution that contains 1.00 mol (46.1 g) of ethanol, C_2H_5OH, in 9.00 mol (162 g) of water. The mole fraction of alcohol is 0.100, and that of water is 0.900.

$$X_{ethanol} = \frac{1.00 \text{ mol ethanol}}{1.00 \text{ mol ethanol} + 9.00 \text{ mol water}} = 0.100$$

$$X_{water} = \frac{9.00 \text{ mol water}}{1.00 \text{ mol ethanol} + 9.00 \text{ mol water}} = 0.900$$

Notice that the sum of the mole fractions of the components in the solution equals 1.000, a relationship that must be true, based on how mole fraction is defined.

Weight percent is the mass of one component divided by the total mass of the mixture, multiplied by 100%:

$$\text{Weight \% A} = \frac{\text{mass of A}}{\text{mass of A} + \text{mass of B} + \text{mass of C} + \dots} \times 100\% \qquad \textbf{(14.3)}$$

The alcohol–water mixture has 46.1 g of ethanol and 162 g of water, so the total mass of solution is 208 g, and the weight % of alcohol is

$$\text{Weight \% ethanol} = \frac{46.1 \text{ g ethanol}}{46.1 \text{ g ethanol} + 162 \text{ g water}} \times 100\% = 22.2\%$$

Weight percent is a common unit in consumer products (Figure 14.3). Vinegar, for example, is an aqueous solution containing approximately 5% acetic acid and 95% water. The label on a common household bleach lists its active ingredient as 6.00% sodium hypochlorite (NaOCl) and 94.00% inert ingredients.

Naturally occurring solutions are often very dilute. Environmental chemists, biologists, geologists, oceanographers, and others frequently use **parts per million (ppm)** to express their concentrations. The unit ppm refers to relative quantities by mass; 1.0 ppm represents 1.0 g of a substance in a sample with a total mass of 1.0 million g or 1 mg in 1000 g.

FIGURE 14.3 Weight percent. The composition of many common products is often given in terms of weight percent. Here, the label on the cat spray indicates it contains 1.15% active ingredients.

EXAMPLE 14.1 Calculating Mole Fractions, Molality, and Weight Percent

Problem Assume you add 1.2 kg of ethylene glycol, $HOCH_2CH_2OH$, as an antifreeze to 4.0 kg of water in the radiator of your car. What are the mole fraction, molality, and weight percent of the ethylene glycol?

What Do You Know? You know the identity and masses of solute and solvent.

Strategy The amount of solute and solvent can be calculated using the mass and molar mass of each material. The masses and amounts of solute and solvent can then be combined to calculate the concentration in each of the desired concentration units using Equations 14.1–14.3.

Solution The 1.2 kg of ethylene glycol (molar mass = 62.1 g/mol) is equivalent to 19 mol, and 4.0 kg of water represents 220 mol.

Mole fraction:

$$X_{glycol} = \frac{19 \text{ mol ethylene glycol}}{19 \text{ mol ethylene glycol} + 220 \text{ mol water}} = 0.080$$

Molality:

$$c_{glycol} = \frac{19 \text{ mol ethylene glycol}}{4.0 \text{ kg water}} = 4.8 \text{ mol/kg} = 4.8 \text{ } m$$

Weight percent:

$$\text{Weight \%} = \frac{1.2 \times 10^3 \text{ g ethylene glycol}}{1.2 \times 10^3 \text{ g ethylene glycol} + 4.0 \times 10^3 \text{ g water}} \times 100\% = 23\%$$

Commercial antifreeze. This solution contains ethylene glycol, $HOCH_2CH_2OH$, an organic alcohol that is readily soluble in water. Regulations specify that the weight percent of ethylene glycol in ethylene glycol–based antifreeze must be at least 75%. (The remainder of the solution can be other glycols and water.)

Think about Your Answer Although the numerical values are very different, the information contained in these values is similar; each relates to the relative numbers of solvent and solute particles.

Check Your Understanding

(a) Mole fraction, molality, and weight percent: If you dissolve 10.0 g (about one heaping tea-spoonful) of sugar (sucrose, $C_{12}H_{22}O_{11}$), in a cup of water (250. g), what are the mole fraction, molality, and weight percent of sugar?

(b) Parts per million: Seawater has a sodium ion concentration of 1.08×10^4 ppm. If the sodium is present in the form of dissolved sodium chloride, what mass of NaCl is in each liter of sea-water? Seawater is denser than pure water because of dissolved salts. Its density is 1.05 g/mL.

REVIEW & CHECK FOR SECTION 14.1

1. You dissolve 1.0 mol of urea (H_2NCONH_2) in 270 g of water. The mole fraction of urea is

   (a) 3.7×10^{-3} (b) 0.063   (c) 16   (d) 1.0

   and its molality is

   (a) 1.0   (b) 3.7   (c) 0.063   (d) 0.27

2. An aqueous solution has 1.0 g of CH_3OH in 50.0 g of water. Which concentration has the largest numerical value for the methanol content?

   (a) weight percent

   (b) mole fraction

   (c) molality

14.2 The Solution Process

• **Unsaturated** The term unsaturated is used when referring to solutions with solute concentrations that are less than that of a saturated solution.

If solid $CuCl_2$ is added to a beaker of water, the salt will begin to dissolve (Figure 14.1). The amount of solid diminishes, and the concentrations of $Cu^{2+}(aq)$ and $Cl^-(aq)$ in the solution increase. If we continue to add $CuCl_2$, however, we will eventually reach a point when no additional $CuCl_2$ seems to dissolve. The concentrations of $Cu^{2+}(aq)$ and $Cl^-(aq)$ will not increase further, and any additional solid $CuCl_2$ added after this point will remain as a solid at the bottom of the beaker. We say that such a solution is **saturated**.

Although no change is observed on the macroscopic level, it is a different matter on the particulate level. The process of dissolving continues, with Cu^{2+} and Cl^- ions leaving the solid state and entering solution. However, at the same time solid $CuCl_2(s)$ is being formed from $Cu^{2+}(aq)$ and $Cl^-(aq)$. The rates at which $CuCl_2$ is dissolving and reprecipitating are equal in a saturated solution, so that no net change in the concentration of ions is observed on the macroscopic level.

This process is another example of a dynamic equilibrium (◀ page 100), and we can describe the situation in terms of an equation with substances linked by a set of double arrows ($\rightleftharpoons$):

$$CuCl_2(s) \rightleftharpoons Cu^{2+}(aq) + 2\ Cl^-(aq)$$

A saturated solution gives us a way to define precisely the solubility of a solid in a liquid. **Solubility** is the concentration of solute in equilibrium with undissolved solute in a saturated solution. The solubility of $CuCl_2$, for example, is 70.6 g in 100 mL of water at 0 °C. If we add 100.0 g of $CuCl_2$ to 100 mL of water at 0 °C, we can expect 70.6 g to dissolve, and 29.4 g of solid to remain.

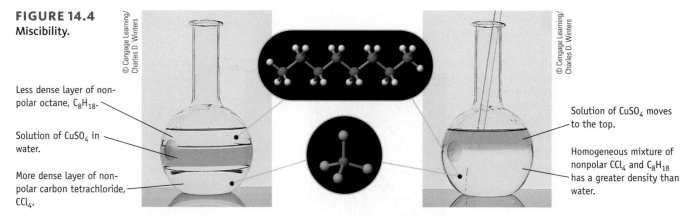

FIGURE 14.4
Miscibility.

Less dense layer of non-polar octane, C_8H_{18}.

Solution of $CuSO_4$ in water.

More dense layer of non-polar carbon tetrachloride, CCl_4.

Solution of $CuSO_4$ moves to the top.

Homogeneous mixture of nonpolar CCl_4 and C_8H_{18} has a greater density than water.

(a) Before mixing. The colorless, denser bottom layer is nonpolar carbon tetrachloride, CCl_4. The blue middle layer is a solution of $CuSO_4$ in water, and the colorless, less dense top layer is nonpolar octane, C_8H_{18}. This mixture was prepared by carefully layering one liquid on top of another, without mixing.

(b) After mixing. After stirring the mixture, the two nonpolar liquids form a homogeneous mixture. This layer of mixed liquids is under the water layer because the mixture of CCl_4 and C_8H_{18} has a greater density than water.

Liquids Dissolving in Liquids

If two liquids mix to an appreciable extent to form a solution, they are said to be **miscible**. In contrast, **immiscible** liquids do not mix to form a solution; they exist in contact with each other as separate layers (Figure 14.4).

The polar compounds ethanol (C_2H_5OH) and water are miscible in all proportions, as are the nonpolar liquids octane (C_8H_{18}) and carbon tetrachloride (CCl_4). On the other hand, neither C_8H_{18} nor CCl_4 is miscible with water. Observations like these have led to a familiar rule of thumb: *Like dissolves like.* That is, two or more nonpolar liquids frequently are miscible, just as are two or more polar liquids.

What is the molecular basis for the "like dissolves like" guideline? In pure water and pure ethanol, the major force between molecules is hydrogen bonding involving —OH groups. When the two liquids are mixed, hydrogen bonding between ethanol and water molecules also occurs and assists in the solution process. In contrast, molecules of pure octane or pure CCl_4, both of which are nonpolar, are held together in the liquid phase by dispersion forces (◄ Section 12.4). The energy associated with these forces of attraction is similar in value to the energy due to the forces of attraction between octane and CCl_4 molecules when these nonpolar liquids are mixed. Thus, little or no energy change occurs when octane–octane and CCl_4–CCl_4 attractive forces are replaced with octane–CCl_4 forces. The solution process is expected to be nearly energy neutral. So, why do the liquids mix? The answer lies deeper in thermodynamics. As you shall see in Chapter 19, spontaneous changes, such as the mixing of liquids, are accompanied by an increase in entropy, a thermodynamic function that is a measure of the wider dispersal of the energy of the particles in the mixture relative to the pure liquids (Figure 14.5).

In contrast, polar and nonpolar liquids usually do not mix to an appreciable degree; when placed together in a container, they separate into two distinct layers

● **Entropy and the Solution Process** Although the energetics of solution formation are important, it is generally accepted that entropy is a more important contributor to the solution process. (See Chapter 19 and T. P. Silverstein: "The real reason why oil and water don't mix." *Journal of Chemical Education*, Vol. 75, pp. 116–118, 1998.)

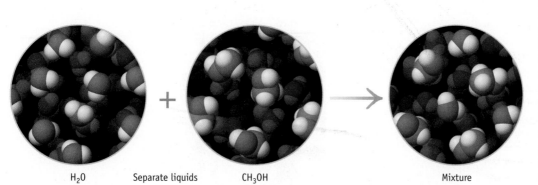

H_2O Separate liquids CH_3OH Mixture

FIGURE 14.5 Driving the solution process—entropy. When two similar liquids—here water and methanol—are mixed, the molecules intermingle, and the energy of the system is more dispersed than in the two, separate pure liquids. A measure of this energy dispersal is entropy, a thermodynamic function described in more detail in Chapter 19.

A CLOSER LOOK

Supersaturated Solutions

Although at first glance it may seem a contradiction, it is possible for a solution to hold more dissolved solute than the amount in a saturated solution. Such solutions are referred to as **supersaturated** solutions. Supersaturated solutions are unstable, and the excess solid eventually crystallizes from the solution until the equilibrium concentration of the solute is reached.

The solubility of substances often decreases if the temperature is lowered. Supersaturated solutions are usually made by preparing a saturated solution at a given temperature and then carefully cooling it. If the rate of crystallization is slow, the solid may not precipitate when the solubility is exceeded. The result is a solution that has more solute than the amount defined by equilibrium conditions; it is supersaturated.

When disturbed in some manner, a supersaturated solution moves toward equilibrium by precipitating solute. This change can occur rapidly, often with the evolution of thermal energy. In fact, supersaturated solutions are used in "heat packs" to apply

Supersaturated solutions. When a supersaturated solution is disturbed, the dissolved salt (here sodium acetate, $NaCH_3CO_2$) rapidly crystallizes.

heat to injured muscles. When crystallization of sodium acetate ($NaCH_3CO_2$) from a supersaturated solution in a heat pack is

Heat of crystallization. A heat pack relies on the heat evolved by the crystallization of sodium acetate.

initiated, the temperature of the heat pack rises to about 50 °C and crystals of solid sodium acetate are detected inside the bag.

(Figure 14.4). The explanation is complex and involves the interplay of the enthalpy of mixing and entropy. The enthalpy of mixing is zero or nearly so, but mixing dissimilar liquids leads to a decrease in entropy. As explained in Chapter 19, this means that mixing dissimilar liquids is not thermodynamically favorable.

Solids Dissolving in Water

The "like dissolves like" guideline also holds for molecular solids dissolving in liquids. Nonpolar solids such as naphthalene, $C_{10}H_8$, dissolve readily in nonpolar solvents such as benzene, C_6H_6, and hexane, C_6H_{14}. Iodine, I_2, a nonpolar inorganic solid, dissolves in water to some extent, but, given a choice, it dissolves to a larger extent in a nonpolar liquid such as CCl_4 (Figure 14.6). Sucrose (sugar), a polar molecular solid, is not very soluble in nonpolar solvents but is readily soluble in water, a fact that we know well because of its use to sweeten beverages. The presence of O—H groups in the structure of sugar and other substances such as glucose

FIGURE 14.6 Solubility of nonpolar iodine in polar water and nonpolar carbon tetrachloride. When a solution of nonpolar I_2 in water (the brown layer on top in the left test tube) is shaken with nonpolar CCl_4 (the colorless bottom layer in the left test tube), the I_2 transfers preferentially to the nonpolar solvent. Evidence for this is the purple color of the bottom CCl_4 layer in the test tube on the right.

Nonpolar I_2
Polar H_2O

Nonpolar CCl_4

Shake the test tube

Polar H_2O

Nonpolar CCl_4 and I_2

Photos © Cengage Learning/Charles D. Winters

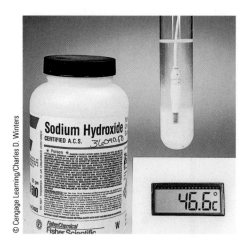

(a) Dissolving NaOH in water is a strongly exothermic process.

FIGURE 14.7 **Dissolving ionic solids and enthalpy of solution.**

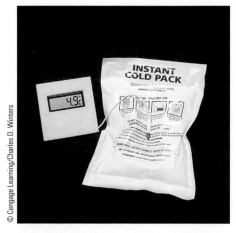

(b) A "cold pack" contains solid ammonium nitrate, NH_4NO_3, and a package of water. When the water and NH_4NO_3 are mixed and the salt dissolves, the temperature of the system drops, owing to the endothermic enthalpy of solution of ammonium nitrate ($\Delta_{soln}H° = +25.7$ kJ/mol).

allows these molecules to interact with polar water molecules through hydrogen bonding.

"Like dissolves like" is a somewhat less effective but still useful guideline when considering the solubility of ionic solids. Thus, we can reasonably predict that ionic compounds, which can be considered extreme examples of polar compounds, will not dissolve in nonpolar solvents. This fact is amply borne out by observation. Sodium chloride, for example, will not dissolve in liquids such as hexane or CCl_4, but it does have a significant solubility in water. Many ionic compounds are soluble in water, but, according to the solubility guidelines on page 104, there are many other ionic solids that are not.

Predicting the solubility of ionic compounds in water is complicated. As mentioned earlier, two factors—enthalpy and entropy—together determine the extent to which one substance dissolves in another. For ionic compounds dissolving in water, entropy usually (but not always) favors solution. A favorable enthalpy factor (negative ΔH) generally leads to a compound being soluble. For example, when sodium hydroxide dissolves in water, the solution warms up (Figure 14.7a), and sodium hydroxide dissolves readily in water. An unfavorable enthalpy factor, however, does not guarantee that an ionic compound will not dissolve. When ammonium nitrate dissolves in water, the solution becomes colder (Figure 14.7b), but ammonium nitrate is still very soluble in water.

Network solids, including graphite, diamond, and quartz sand (SiO_2), do not dissolve in water. (Where would all the beaches be if sand dissolved in water?) The covalent chemical bonding in network solids is simply too strong to be broken; the lattice remains intact when in contact with water.

Enthalpy of Solution

To understand the energetics of the solution process, let us view it at the molecular level. We will use the process of dissolving potassium fluoride, KF, in water to illustrate what occurs, and the energy-level diagram in Figure 14.8 will help us to follow the changes.

Solid potassium fluoride has an ionic crystal lattice with alternating K^+ and F^- ions held in place by attractive forces due to their opposite charges. In water, these ions are separated from each other and *hydrated*; that is, they are surrounded by water molecules (Figure 14.1). Ion–dipole forces of attraction bind water molecules strongly to each ion. The energy change occurring on going from the

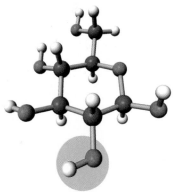

OH group

Like dissolves like. Glucose has five —OH groups on each molecule, groups that allow it to form hydrogen bonds with water molecules. As a result, glucose dissolves readily in water.

FIGURE 14.8 Model for energy changes on dissolving KF. An estimate of the magnitude of the energy change on dissolving an ionic compound in water is achieved by imagining it as occurring in two steps at the particulate level. Here, KF is first separated into cations and anions in the gas phase with an expenditure of 821 kJ per mol of KF. These ions are then hydrated, with $\Delta_{\text{hydration}}H$ estimated to be −837 kJ. Thus, the net energy change is −16 kJ, a slightly exothermic enthalpy of solution.

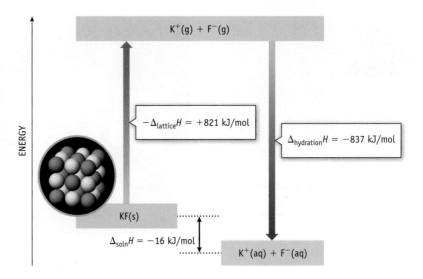

reactant, KF(s), to the products, K^+(aq) and F^-(aq), is the sum of the energies of two individual steps:

1. Energy must be supplied to separate the ions in the lattice against their attractive forces. This is the reverse of the process defining the lattice enthalpy of an ionic compound (◀ page 460), and its value will be equal to $-\Delta_{\text{lattice}}H$. Separating the ions from one another is highly endothermic because the attractive forces between ions are strong.
2. Energy is evolved when the individual ions dissolve in water, where each ion becomes surrounded by water molecules. Again, strong forces of attraction (ion–dipole forces) are involved. This process, referred to as **hydration** when water is the solvent, is strongly exothermic.

We can therefore represent the process of dissolving KF in terms of these chemical equations:

Step 1: $KF(s) \longrightarrow K^+(g) + F^-(g)$ $-\Delta_{\text{lattice}}H$
Step 2: $K^+(g) + F^-(g) \longrightarrow K^+(aq) + F^-(aq)$ $\Delta_{\text{hydration}}H$

The overall reaction is the sum of these two steps. The enthalpy of the overall reaction, called the **enthalpy of solution** ($\Delta_{\text{soln}}H$), is the sum of the two enthalpies.

Overall: $KF(s) \longrightarrow K^+(aq) + F^-(aq)$ $\Delta_{\text{soln}}H = -\Delta_{\text{lattice}}H + \Delta_{\text{hydration}}H$

We can use this equation to estimate the value of $\Delta_{\text{hydration}}H$. For example, the lattice energy for KF is −821 kJ/mol [calculated using a Born-Haber cycle calculation (◀ page 462)], and the value of $\Delta_{\text{soln}}H$ determined by a calorimetry experiment is −16.4 kJ/mol. From these two values, we can determine $\Delta_{\text{hydration}}H$ to be −837 kJ/mol.

As a general rule, to be soluble, an ionic compound will have an enthalpy of solution that is exothermic or only slightly endothermic (Figure 14.9). In the latter instance, it is assumed that the enthalpy-disfavored solution process will be balanced by a favorable entropy of solution. If the enthalpy of solution is very endothermic—because of a low hydration energy, for example—then the compound is unlikely to be soluble. We can reasonably speculate that nonpolar solvents would not solvate ions strongly, and that solution formation would thus be energetically unfavorable. We therefore predict that an ionic compound, such as copper(II) sulfate, is not very soluble in nonpolar solvents such as carbon tetrachloride and octane (Figure 14.4).

It is also important to recognize that the enthalpy of solution is the difference between two very large numbers. Small variations in either lattice energy or hydration enthalpies can determine whether a salt dissolves endothermically or exothermically.

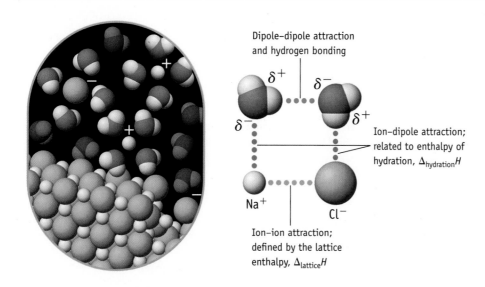

Dipole–dipole attraction and hydrogen bonding

$\delta+$ $\delta-$

$\delta-$ $\delta+$

Ion–dipole attraction; related to enthalpy of hydration, $\Delta_{hydration}H$

Na$^+$ Cl$^-$

Ion–ion attraction; defined by the lattice enthalpy, $\Delta_{lattice}H$

FIGURE 14.9 Dissolving an ionic solid in water. This process is a balance of forces. There are inter-molecular forces between water molecules, and ion–ion forces are at work in the ionic crystal lattice. To dissolve, the ion–dipole forces between water and the ions (as measured by $\Delta_{hydration}H$) must overcome the ion–ion forces (as measured by $\Delta_{lattice}H$) and the intermolecular forces in water.

Finally, notice that the two energy quantities, $\Delta_{lattice}H$ and $\Delta_{hydration}H$, are both affected by ion sizes and ion charges (◄ pages 419 and 460). A salt composed of smaller ions is expected to have a greater (more negative) lattice enthalpy because the ions can be closer together and experience higher attractive forces. However, the small size will also allow a closer approach of solvent molecules and a greater solvation enthalpy. The net result is that correlations of solubility with structure (ionic radii) or thermodynamic parameters ($\Delta_{lattice}H$) are generally not successful.

Enthalpy of Solution: Thermodynamic Data

As mentioned earlier, the enthalpy of solution for a salt can be measured using a calorimeter. This is usually done in an open system such as the coffee-cup calorimeter described in Section 5.6. For an experiment run under standard conditions, the resulting measurement produces a value for the standard enthalpy of solution, $\Delta_{soln}H°$, where standard conditions refer to a concentration of 1 molal.

Tables of thermodynamic values often include values for the enthalpies of formation of aqueous solutions of salts. For example, a value of $\Delta_fH°$ for NaCl(aq) of -407.3 kJ/mol is listed in Table 14.1 and Appendix L. This value refers to the formation of a 1 m solution of NaCl from the elements. It may be considered to involve the enthalpies of two steps: (1) the formation of NaCl(s) from the elements Na(s) and Cl$_2$(g) in their standard states, and (2) the formation of a 1 m solution by dissolving solid NaCl in water:

Formation of NaCl(s):	Na(s) + ½ Cl$_2$(g) ⟶ NaCl(s)	$\Delta_fH° = -411.1$ kJ/mol
Dissolving NaCl:	NaCl(s) ⟶ NaCl(aq, 1 m)	$\Delta_{soln}H° = +3.8$ kJ/mol
Net process:	Na(s) + ½ Cl$_2$(g) ⟶ NaCl(aq, 1 m)	$\Delta_fH° = -407.3$ kJ/mol

Table 14.1
Data for Calculating Enthalpy of Solution

Compound	$\Delta_fH°$ (s) (kJ/mol)	$\Delta_fH°$ (aq, 1 m) (kJ/mol)
LiF	−616.9	−611.1
NaF	−573.6	−572.8
KF	−568.6	−585.0
RbF	−557.7	−583.8
LiCl	−408.7	−445.6
NaCl	−411.1	−407.3
KCl	−436.7	−419.5
RbCl	−435.4	−418.3
NaOH	−425.9	−469.2
NH$_4$NO$_3$	−365.6	−339.9

EXAMPLE 14.2   Calculating an Enthalpy of Solution

Problem Determine the enthalpy of solution for NH$_4$NO$_3$, the compound used in cold packs.

What Do You Know? The solution process for NH$_4$NO$_3$ is represented by the equation

$$NH_4NO_3(s) \longrightarrow NH_4NO_3(aq)$$

The enthalpies of formation of NH$_4$NO$_3$ in the solid state (-365.6 kJ/mol) and in solution (-339.9 kJ/mol) are given in Table 14.1.

Strategy Equation 5.6 (page 198) states that the enthalpy change for a process is the difference between the enthalpy of the final state (here the compound in solution) and the initial state (the solid).

Solution The enthalpy change for this process is calculated using Equation 5.6 as follows:

$$\Delta_{soln}H° = \sum[n\Delta_fH° \text{ (product)}] - \sum[n\Delta_fH° \text{ (reactant)}]$$
$$= \Delta_fH° \left[NH_4NO_3(aq)\right] - \Delta_fH° \left[NH_4NO_3(s)\right] \text{ (where } n = 1 \text{ for both)}$$
$$= -339.9 \text{ kJ} - (-365.6 \text{ kJ}) = +25.7 \text{ kJ}$$

Think about Your Answer The process is endothermic, as indicated by the fact that $\Delta_{soln}H°$ has a positive value and as verified by the experiment in Figure 14.7b.

Check Your Understanding

Use the data in Table 14.1 to calculate the enthalpy of solution for NaOH.

REVIEW & CHECK FOR SECTION 14.2

Given the enthalpy of formation data below, calculate $\Delta_{soln}H°$ for LiOH.

$$\Delta_fH° \text{ [LiOH(s)]} = -484.93 \text{ kJ/mol} \qquad \Delta_fH° \text{ [LiOH(aq)]} = -508.48 \text{ kJ/mol}$$

(a) +23.55 kJ/mol   (c) −23.55 kJ/mol

(b) +993.41 kJ/mol   (d) −993.41 kJ/mol

14.3 Factors Affecting Solubility: Pressure and Temperature

Table 14.2
Henry's Law Constants (25 °C)*

Gas	k_H (mol/kg · bar)
N_2	6.0×10^{-4}
O_2	1.3×10^{-3}
CO_2	0.034

*From http://webbook.nist.gov/
chemistry/. *Note:* 1 bar = 0.9869 atm.

Pressure and temperature are two external factors that influence solubility. Both affect the solubility of gases in liquids, whereas only temperature is a factor in the solubility of solids in liquids.

Dissolving Gases in Liquids: Henry's Law

The solubility of a gas in a liquid is directly proportional to the gas pressure. This is a statement of **Henry's law,**

$$S_g = k_H P_g \qquad \qquad (14.4)$$

where S_g is the gas solubility (in mol/kg), P_g is the partial pressure of the gaseous solute, and k_H is Henry's law constant (Table 14.2), a constant characteristic of the solute and solvent.

Carbonated soft drinks illustrate how Henry's law works. These beverages are packed under pressure in a chamber filled with carbon dioxide gas, some of which dissolves in the beverage. When the can or bottle is opened, the partial pressure of CO_2 above the solution drops, which causes the solubility of CO_2 to drop. Gas bubbles out of the solution (Figure 14.10).

Henry's law has important consequences in SCUBA diving. When you dive, the pressure of the air you breathe must be balanced against the external pressure of the water. In deeper dives, the pressure of the gases in the SCUBA gear must be several atmospheres and, as a result, more gas dissolves in the blood. This can lead to a problem. If you ascend too rapidly, you can experience a painful and potentially lethal condition referred to as "the bends," in which nitrogen gas bubbles form in the blood as the solubility of nitrogen decreases with decreasing pressure. In an effort to prevent the bends, divers may use a helium–oxygen mixture (rather than nitrogen–oxygen) because helium is not as soluble in blood as nitrogen.

We can better understand the effect of pressure on solubility by examining the system at the particulate level. The solubility of a gas is defined as the concentration of the dissolved gas in equilibrium with the substance in the gaseous state. At equilibrium, the rate at which solute gas molecules escape the solution and enter the gaseous state equals the rate at which gas molecules reenter the solution. An increase in pressure

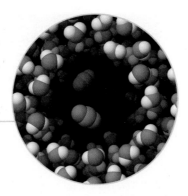

FIGURE 14.10 Gas solubility and pressure. Carbonated beverages are bottled under CO_2 pressure. When the bottle is opened, the pressure is released, and bubbles of CO_2 form within the liquid and rise to the surface. After some time, an equilibrium between dissolved CO_2 and atmospheric CO_2 is reached. The beverage tastes flat when most of its dissolved CO_2 is lost.

results in more molecules of gas striking the surface of the liquid and entering solution in a given time. The solution eventually reaches a new equilibrium when the concentration of gas dissolved in the solvent is high enough that the rate of gas molecules escaping the solution again equals the rate of gas molecules entering the solution.

• **Limitations of Henry's Law** Henry's law holds quantitatively only for gases that do not interact chemically with the solvent. It does not accurately predict the solubility of NH_3 in water, for example, because this compound gives small concentrations of NH_4^+ and OH^- in water.

EXAMPLE 14.3 Using Henry's Law

Problem What is the concentration of O_2 in a freshwater stream in equilibrium with air at 25 °C and 1.0 bar? Express the answer in grams of O_2 per kg of solvent.

What Do You Know? You know the total air pressure and temperature, and you can look up the Henry's law constant at the indicated temperature (1.3×10^{-3} mol/kg · bar). (Note that 1 bar is approximately 1 atm; page 386.)

Strategy Because you are trying to find the solubility of O_2, you first must find the partial pressure of O_2 from its mole fraction in air (0.21) and the specified atmospheric pressure (1.0 bar). Then use Henry's Law to calculate its molar solubility.

Solution

(a) The mole fraction of O_2 in air is 0.21, and, assuming the total pressure is 1.0 bar, the partial pressure of O_2 is 0.21 bar.

(b) Using the O_2 partial pressure for P_g in Henry's law, we have:

$$\text{Solubility of } O_2 = k_H P_g = \left(\frac{1.3 \times 10^{-3} \text{ mol}}{\text{kg} \cdot \text{bar}} \right)(0.21 \text{ bar}) = 2.7 \times 10^{-4} \text{ mol/kg}$$

(c) Calculate the concentration in g/kg from the concentration in mol/kg and the molar mass of O_2:

$$\text{Solubility of } O_2 = \left(\frac{2.7 \times 10^{-4} \text{ mol}}{\text{kg}} \right)\left(\frac{32.0 \text{ g}}{\text{mol}} \right) = 0.0087 \text{ g/kg}$$

Think about Your Answer The concentration of O_2 is 8.7 ppm (8.7 mg/1000 g). This concentration is quite low, but it is sufficient to provide the oxygen required by aquatic life.

Check Your Understanding

What is the concentration of CO_2 in water at 25 °C when the partial pressure is 0.33 bar? (Although CO_2 reacts with water to give traces of H_3O^+ and HCO_3^-, the reaction occurs to such a small extent that Henry's law is obeyed at low CO_2 partial pressures.)

Temperature Effects on Solubility: Le Chatelier's Principle

The solubility of all gases in water decreases with increasing temperature. You may realize this from everyday observations such as the appearance of bubbles of air as water is heated below the boiling point.

To understand the effect of temperature on the solubility of gases, let us reexamine the enthalpy of solution. Gases that dissolve to an appreciable extent in water usually do so in an exothermic process

$$\text{Gas} + \text{liquid solvent} \xrightleftharpoons{\Delta_{soln}H < 0} \text{saturated solution} + \text{energy}$$

FIGURE 14.11
The temperature dependence of the solubility of some ionic compounds in water. Most compounds, such as NH_4Cl, decrease in solubility with decreasing temperature.

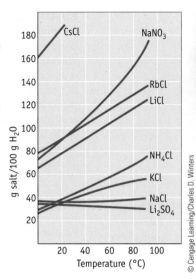

(a) Temperature dependence of the solubility of some ionic compounds.

(b) NH_4Cl dissolved in water.

(c) NH_4Cl precipitates when the solution is cooled in ice.

The reverse process, loss of dissolved gas molecules from a solution, requires energy as heat. At equilibrium, the rates of the two processes are the same.

To understand how temperature affects solubility, we turn to **Le Chatelier's principle**, which states that a change in any of the factors determining an equilibrium causes the system to adjust by shifting in the direction that reduces or counteracts the effect of the change. If a solution of a gas in a liquid is heated, for example, the equilibrium will shift to absorb some of the added energy. That is, the reaction

$$\text{Gas + liquid solvent} \underset{\text{Add energy. Equilibrium shifts left.}}{\overset{\substack{\text{Exothermic process}\\ \Delta_{soln}H \text{ is negative.}}}{\rightleftharpoons}} \text{saturated solution + energy}$$

shifts to the left if the temperature is raised because energy is absorbed in the process that produces free gas molecules and pure solvent. This shift corresponds to less gas dissolved and a lower solubility at higher temperature—the observed result.

The solubility of solids in water is also affected by temperature, but, unlike the situation involving solutions of gases, no general pattern of behavior is observed. In Figure 14.11, the solubilities of several salts are plotted versus temperature. The solubility of many salts increases with increasing temperature, but there are notable exceptions. Predictions based on whether the enthalpy of solution is positive or negative work most of the time, but exceptions do occur.

Chemists take advantage of the variation of solubility with temperature to purify compounds. An impure sample of a compound that is more soluble at higher temperatures is dissolved by heating the solution. The solution is then cooled to decrease the solubility (Figure 14.11c). When the limit of solubility is reached at the lower temperature, crystals of the pure compound form. If the process is done slowly and carefully, it is sometimes possible to obtain very large crystals (Figure 14.12).

FIGURE 14.12 Giant crystals of potassium dihydrogen phosphate. The crystal being measured by this researcher at Lawrence Livermore Laboratory in California weighs 318 kg and measures 66 × 53 × 58 cm. The crystals were grown by suspending a thumbnail-sized seed crystal in a 6-foot tank of saturated KH_2PO_4. The temperature of the solution was gradually reduced from 65 °C over a period of about 50 days. The crystals are sliced into thin plates, which are used to convert light from a giant laser from infrared to ultraviolet.

REVIEW & CHECK FOR SECTION 14.3

Soft drinks are carbonated by adding CO_2 under pressure. Assume that CO_2 is added to a bottle of soft drink containing 710 g of diet soda. The pressure of CO_2 is 4.0 bar. (CO_2 obeys Henry's law at pressures up to about 5 bar.) What amount of CO_2 is dissolved in 710 g of diet soda?

(a) 0.14 mol   (b) 0.097 mol   (c) 0.034 mol   (d) 0.13 mol

Exploding Lakes and Diet Cokes

On Thursday, August 21, 1986, people and animals around Lake Nyos in Cameroon, a small nation on the west coast of Africa, suddenly collapsed and died. More than 1700 people and hundreds of animals were dead, but there was no apparent cause—no fire, no earthquake, no storm. What had brought on this disaster?

Some weeks later, the mystery was solved. Lake Nyos and nearby Lake Monoun are crater lakes, which formed when cooled volcanic craters filled with water. Importantly, Lake Nyos contains an enormous amount of dissolved carbon dioxide, which was generated as a result of volcanic activity deep in the earth. Under the high pressure at the bottom of the lake, a very large amount of CO_2 dissolved in the water.

But on that evening in 1986, something disturbed the lake. The CO_2-saturated water at the bottom of the lake was carried to the surface, where, under lower pressure, the gas was much less soluble. About one cubic kilometer of carbon dioxide was released into the atmosphere, much like the explosive release of CO_2 from a can of carbonated beverage that has been shaken. The CO_2 shot up about 260 feet; then, because this gas is more dense than air, it hugged the ground and began to move with the prevailing breeze at about 45 miles per hour. When it reached the villages 12 miles away, vital oxygen was displaced. The result was that both people and animals were asphyxiated.

In most lakes, this situation would not occur because lake water "turns over" as the seasons change. In the autumn, the top layer of water in a lake cools; its density increases; and the water sinks. This process continues, with warmer water coming to the surface and cooler water sinking. Dissolved CO_2 at the bottom of a lake would normally be expelled in this turnover process, but geologists found that the lakes in Cameroon are different. The chemocline, the boundary between deep water, rich in gas and minerals, and the upper layer, full of freshwater, stayed intact. As carbon dioxide continued to enter the lake through vents in the bottom of the lake, the water became saturated with this gas. It is presumed that a minor disturbance—perhaps a small earthquake, a strong wind, or an underwater landslide—caused the lake water to turn over and led to the explosive release of CO_2.

The explosive release of CO_2 that occurred in Lake Nyos is much like what occurs when you drop a candy like a Mentos into a bottle of Diet Coke. Carbonated sodas are bottled under a high pressure of CO_2. Some of the gas dissolves in the soda, but some also remains in the small space above the liquid (called the *headspace*). The pressure of the CO_2 in the headspace is between 2 and 4 atm. When the bottle cap is removed, the CO_2 in the headspace escapes rapidly. Some of the dissolved CO_2 also comes out of solution, and you see bubbles of gas rising to the surface. If the bottle remains open, this continues until equilibrium is established with CO_2 in the atmosphere (where the partial pressure of CO_2 is 3.75×10^{-4} atm),

$$CO_2(\text{solution}) \rightleftharpoons CO_2(g)$$

and the soda goes "flat." If the newly opened soda bottle is undisturbed, however, the loss of CO_2 from solution is rather slow because bubble formation is not rapid, and your soda keeps its fizz.

To understand how CO_2 can be released explosively, you have to understand that impurities in the water or the rough surfaces of an ice cube or bottle or drinking glass can serve as "nucleation sites" on which a bubble can form. The more nucleation sites available, the more rapid the bubble formation. The surface of a Mentos apparently has many such sites and pro-

A Mentos was dropped into large bottles of Diet Coke. For more information, see J. E. Baur and M. B. Baur, *Journal of Chemical Education,* Vol. 83, pages 577–580, 2006.

motes very rapid bubble formation. In the case of the African lake, perhaps an earthquake or underwater landslide opened up nucleation sites, and the CO_2 was released explosively.

Questions:

1. If the headspace of a soda is 25 mL and the pressure of CO_2 in the space is 4.0 atm (≈ 4.0 bar) at 25 °C, what amount of CO_2 is contained in the headspace?

2. If the CO_2 in the headspace escapes into the atmosphere where the partial pressure of CO_2 is 3.7×10^{-4} atm, what volume would the CO_2 occupy (at 25 °C)? By what amount did the CO_2 expand when it was released?

3. What is the solubility of CO_2 in water at 25 °C when the pressure of the gas is 3.7×10^{-4} bar?

4. After opening a 1.0 L soda, what mass of the dissolved CO_2 is released to the atmosphere in order to reach equilibrium? Assume that the CO_2 dissolved in the soda in the sealed bottle was in equilibrium with the CO_2 in the headspace at 4.0 bar pressure.

Answers to these questions are available in Appendix N.

Lake Nyos in Cameroon (western Africa), the site of a natural disaster. In 1986, a huge bubble of CO_2 escaped from the lake and asphyxiated more than 1700 people.

Module 19: Colligative Properties
covers concepts in this section.

14.4 Colligative Properties

If you dissolve some salt in water, the vapor pressure of the water over the solution will decrease. In addition, the solution will freeze below 0 °C and boil above 100 °C. These observations refer to the **colligative properties** of the solution, properties that depend on the relative numbers of solute and solvent particles in a solution and not on their identity.

Changes in Vapor Pressure: Raoult's Law

The equilibrium vapor pressure at a particular temperature is the pressure of the vapor when the liquid and the vapor are in equilibrium (◄ page 436). When the vapor pressure of the solvent over a solution is measured at a given temperature, it is experimentally observed that

- The vapor pressure of the solvent over the solution is lower than the vapor pressure of the pure solvent.
- The vapor pressure of the solvent, $P_{solvent}$, is proportional to the relative number of solvent molecules in the solution; that is, the solvent vapor pressure is proportional to the solvent mole fraction, $P_{solvent} \propto X_{solvent}$.

Because solvent vapor pressure is proportional to the relative number of solvent molecules, we can write the following equation for the equilibrium vapor pressure of the solvent over a solution:

$$P_{solvent} = X_{solvent}\, P°_{solvent} \tag{14.5}$$

This equation, called **Raoult's law**, tells us that the vapor pressure of solvent over a solution ($P_{solvent}$) is some fraction of the pure solvent equilibrium vapor pressure ($P°_{solvent}$). For example, if 95% of the molecules in a solution are solvent molecules ($X_{solvent} = 0.95$), then the vapor pressure of the solvent ($P_{solvent}$) is 95% of $P°_{solvent}$.

An ideal solution is one that obeys Raoult's law. However, just as no gas is truly ideal, no solution is ideal. Although Raoult's law describes a simplified model of a solution, it is a good approximation of solution behavior in many instances, especially at low solute concentration.

For Raoult's law to hold, the forces of attraction between solute and solvent molecules must be the same as those between solvent molecules in the pure solvent. This is frequently the case when molecules with similar structures are involved. Solutions of one hydrocarbon in another (hexane, C_6H_{14}, dissolved in octane, C_8H_{18}, for example) follow Raoult's law quite closely. If solvent–solute interactions are stronger than solvent–solvent interactions, the actual vapor pressure will be lower than calculated by Raoult's law. If the solvent–solute interactions are weaker than solvent–solvent interactions, the vapor pressure will be higher.

EXAMPLE 14.4   Using Raoult's Law

Problem You dissolve 651 g of ethylene glycol, $HOCH_2CH_2OH$, in 1.50 kg of water. What is the vapor pressure of the water over the solution at 90 °C? Assume ideal behavior for the solution.

What Do You Know? This is a Raoult's law problem where you want to know $P_{solvent}$. To calculate $P_{solvent}$ you need the mole fraction of solvent and the vapor pressure of the pure solvent. You can calculate the mole fraction of solute from the masses of solute and solvent and the molar masses of these species. The vapor pressure of pure water at 90 °C (= 525.8 mm Hg) is found in Appendix G.

Strategy First calculate the mole fraction of the solvent (water) and then combine that with the vapor pressure of pure solvent at the specified temperature using Raoult's law (Equation 14.5).

Solution

(a) Calculate the amounts of water and ethylene glycol and, from these, the mole fraction of water.

$$\text{Amount of water} = 1.50 \times 10^3 \text{ g} \left(\frac{1 \text{ mol}}{18.02 \text{ g}} \right) = 83.2 \text{ mol water}$$

$$\text{Amount of ethylene glycol} = 651 \text{ g} \left(\frac{1 \text{ mol}}{62.07 \text{ g}} \right) = 10.5 \text{ mol glycol}$$

$$X_{water} = \frac{83.2 \text{ mol water}}{83.2 \text{ mol water} + 10.5 \text{ mol glycol}} = 0.888$$

(b) Next apply Raoult's law.

$$P_{water} = X_{water} P^\circ_{water} = (0.888)(525.8 \text{ mm Hg}) = \boxed{467 \text{ mm Hg}}$$

Think about Your Answer Although a substantial mass of ethylene glycol was added to the water, it led to a decrease in the vapor pressure of the solvent of only 59 mm Hg, or about 11%:

$$\Delta P_{water} = P_{water} - P^\circ_{water} = 467 \text{ mm Hg} - 525.8 \text{ mm Hg} = -59 \text{ mm Hg}$$

Ethylene glycol is ideal for use as antifreeze. It dissolves easily in water, is noncorrosive, and is relatively inexpensive. Because of its high boiling point, it will not evaporate readily. It is, however, toxic to animals, so it is being replaced by less toxic propylene glycol.

Check Your Understanding

Assume you dissolve 10.0 g of sucrose ($C_{12}H_{22}O_{11}$) in 225 mL (225 g) of water and warm the water to 60 °C. What is the vapor pressure of the water over this solution? (Appendix G lists $P^\circ(H_2O)$ at various temperatures.)

Adding a nonvolatile solute to a solvent lowers the vapor pressure of the solvent (Example 14.4). Raoult's law can be modified to calculate directly the lowering of the vapor pressure, $\Delta P_{solvent}$, as a function of the mole fraction of the solute.

$$\Delta P_{solvent} = P_{solvent} - P^\circ_{solvent}$$

Substituting Raoult's law for $P_{solvent}$, we have

$$\Delta P_{solvent} = (X_{solvent} P^\circ_{solvent}) - P^\circ_{solvent} = -(1 - X_{solvent})P^\circ_{solvent}$$

In a solution that has only the volatile solvent and one nonvolatile solute, the sum of the mole fraction of solvent and solute must be 1:

$$X_{solvent} + X_{solute} = 1$$

Therefore, $1 - X_{solvent} = X_{solute}$, and the equation for $\Delta P_{solvent}$ can be rewritten as

$$\Delta P_{solvent} = -X_{solute} P^\circ_{solvent} \qquad (14.6)$$

Thus, the *decrease* in the vapor pressure of the solvent is proportional to the mole fraction (the relative number of particles) of solute.

Boiling Point Elevation

Suppose you have a solution of a nonvolatile solute in the volatile solvent benzene. If the solute concentration is 0.200 mol in 100. g of benzene (C_6H_6) (= 2.00 mol/kg), this means that $X_{benzene} = 0.865$. Using $X_{benzene}$ and applying Raoult's law, we can calculate that the vapor pressure of the solvent at 60 °C will drop from 400. mm Hg for the pure solvent to 346 mm Hg for the solution:

$$P_{benzene} = X_{benzene} P^\circ_{benzene} = (0.865)(400. \text{ mm Hg}) = 346 \text{ mm Hg}$$

FIGURE 14.13 Lowering the vapor pressure of benzene by addition of a nonvolatile solute. The curve drawn in red represents the vapor pressure of pure benzene, and the curve in blue represents the vapor pressure of a solution containing 0.200 mol of a solute dissolved in 0.100 kg of solvent (2.00 *m*). This graph was created by doing a series of calculations such as those shown in the text. As an alternative, the graph could be created by measuring various vapor pressures for the solution in a laboratory experiment.

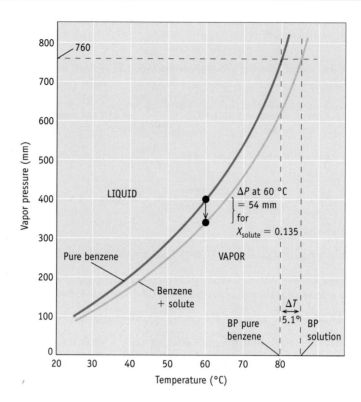

This point is marked on the vapor pressure graph in Figure 14.13. Now, what is the vapor pressure when the temperature of the solution is raised another 10 °C? The vapor pressure of pure benzene, $P°_{benzene}$, becomes larger with increasing temperature, so $P_{benzene}$ for the solution must also become larger. This new point, and additional ones calculated in the same way for other temperatures, define the vapor pressure curve for the solution (the lower curve in Figure 14.13).

An important observation we can make in Figure 14.13 is that the vapor pressure lowering caused by the nonvolatile solute leads to an increase in the boiling point. The normal boiling point of a liquid is the temperature at which its vapor pressure is equal to 1 atm or 760 mm Hg (◄ page 439). In Figure 14.13, we see that the normal boiling point of pure benzene (at 760 mm Hg) is about 80 °C. Tracing the vapor pressure curve for the solution, we also see that the vapor pressure reaches 760 mm Hg at a temperature about 5 °C higher than this value.

An important question is how the boiling point of the solution varies with solute concentration. In fact, a simple relationship exists: the boiling point elevation, ΔT_{bp}, is directly proportional to the molality of the solute.

$$\text{Elevation in boiling point} = \Delta T_{bp} = K_{bp}m_{solute} \tag{14.7}$$

In this equation, K_{bp} is a proportionality constant called the **molal boiling point elevation constant**. It has the units of degrees/molal (°C/*m*). Values for K_{bp} are determined experimentally, and different solvents have different values (Table 14.3). Formally, the value of K_{bp} corresponds to the elevation in boiling point for a 1 *m* solution.

Eugenol, $C_{10}H_{12}O_2$, is an important component in oil of cloves, a commonly used spice.

ⓌWL INTERACTIVE EXAMPLE 14.5 Boiling Point Elevation

Problem Eugenol, a compound found in nutmeg and cloves, has the formula $C_{10}H_{12}O_2$. What is the boiling point of a solution containing 0.144 g of this compound dissolved in 10.0 g of benzene?

What Do You Know? You know the identify and mass of both the solute and the solvent.

Strategy

- Calculate the solution concentration (molality, m, in mol/kg) from the amount of eugenol and the mass of solvent (kg).

- Calculate the change in boiling point using Equation 14.7 (with a value of K_{bp} from Table 14.3).

- Add the change to the boiling temperature of pure benzene to obtain the new boiling point.

Solution

(a) Solution concentration

$$0.144 \text{ g eugenol}\left(\frac{1 \text{ mol eugenol}}{164.2 \text{ g}}\right) = 8.77 \times 10^{-4} \text{ mol eugenol}$$

$$c_{eugenol} = \frac{8.77 \times 10^{-4} \text{ mol eugenol}}{0.0100 \text{ kg benzene}} = 8.77 \times 10^{-2} \text{ m}$$

(b) Boiling point elevation

$$\Delta T_{bp} = (2.53 \text{ °C/}m)(0.0877 \text{ m}) = 0.222 \text{ °C}$$

Because the boiling point rises relative to that of the pure solvent, the boiling point of the solution is

$$80.10 \text{ °C} + 0.222 \text{ °C} = \boxed{80.32 \text{ °C}}$$

Think about Your Answer Keep in mind that the boiling point elevation is proportional to the solute concentration. At high concentrations, sizable increases in boiling point are possible.

Check Your Understanding

What quantity of ethylene glycol, $HOCH_2CH_2OH$, must be added to 125 g of water to raise the boiling point by 1.0 °C? Express the answer in grams.

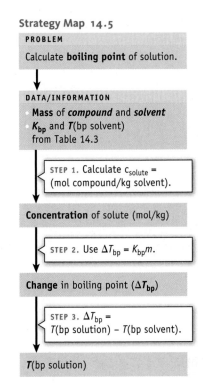

Strategy Map 14.5

PROBLEM

Calculate **boiling point** of solution.

DATA/INFORMATION

- Mass of *compound* and *solvent*
- K_{bp} and T(bp solvent) from Table 14.3

STEP 1. Calculate $c_{solute} = $ (mol compound/kg solvent).

Concentration of solute (mol/kg)

STEP 2. Use $\Delta T_{bp} = K_{bp}m$.

Change in boiling point (ΔT_{bp})

STEP 3. $\Delta T_{bp} = T$(bp solution) – T(bp solvent).

T(bp solution)

The elevation of the boiling point of a solvent on adding a solute has many practical consequences. One of them is the summer protection your car's engine receives from "all-season" antifreeze. The main ingredient of commercial antifreeze is ethylene glycol, $HOCH_2CH_2OH$. The car's radiator and cooling system are sealed to keep the coolant under pressure, ensuring that it will not vaporize at normal engine temperatures. When the air temperature is high in the summer, however, the radiator could "boil over" if it were not protected with "antifreeze." By adding this nonvolatile liquid, the solution in the radiator has a higher boiling point than that of pure water.

Table 14.3 Some Boiling Point Elevation and Freezing Point Depression Constants

Solvent	Normal Boiling Point (°C) Pure Solvent	K_{bp} (°C/m)	Normal Freezing Point (°C) Pure Solvent	K_{fp} (°C/m)
Water	100.00	+0.5121	0.0	−1.86
Benzene	80.10	+2.53	5.50	−5.12
Camphor	207.4	+5.611	179.75	−39.7
Chloroform ($CHCl_3$)	61.70	+3.63	—	—

FIGURE 14.14 Freezing a solution.

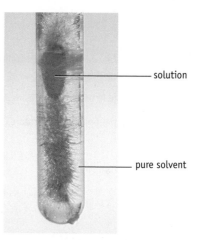

solution

pure solvent

(a) Adding antifreeze to water prevents the water from freezing. Here, a jar of pure water *(left)* and a jar of water to which automobile antifreeze had been added *(right)* were kept overnight in the freezing compartment of a home refrigerator.

(b) When a solution freezes, it is pure solvent that solidifies. To take this photo, a purple dye was dissolved in water, and the solution was frozen slowly. Pure ice formed along the walls of the tube, and the dye stayed in solution. The concentration of the solute increased as more and more solvent was frozen out, and the resulting solution had a lower and lower freezing point. Eventually, the system contains pure, colorless ice that formed along the walls of the tube and a concentrated solution of dye in the center of the tube.

Freezing Point Depression

● **Why Is the Boiling Point of a Solution Elevated and Its Freezing Point Depressed?** The answer to this question is related to entropy, a thermodynamic function discussed in Chapter 19. You can consult a website that has an *extensive* discussion of entropy (entropysite.oxy.edu) or the specific site at which colligative properties are discussed: http://entropysite.oxy .edu/entropy_is_simple/index.html.

Another consequence of dissolving a solute in a solvent is that the freezing point of the solution is lower than that of the pure solvent (Figure 14.14). For an ideal solution, the depression of the freezing point is given by an equation similar to that for the elevation of the boiling point:

$$\text{Freezing point depression} = \Delta T_{fp} = K_{fp}m_{solute} \tag{14.8}$$

where K_{fp} is the **freezing point depression constant** in degrees Celsius per molal $(°C/m)$. Values of K_{fp} for a few common solvents are given in Table 14.3. The values are negative quantities, so the result of the calculation is a negative value for ΔT_{fp}, signifying a decrease in temperature.

The practical aspects of freezing point changes from pure solvent to solution are similar to those for boiling point elevation. The very name of the liquid you add to the radiator in your car, antifreeze, indicates its purpose (Figure 14.14a). The label on the container of antifreeze tells you, for example, to add 6 qt (5.7 L) of antifreeze to a 12-qt (11.4-L) cooling system to lower the freezing point to −34 °C and to raise the boiling point to +109 °C.

EXAMPLE 14.6   Freezing Point Depression

Problem What mass of ethylene glycol, HOCH₂CH₂OH, must be added to 5.50 kg of water to lower the freezing point of the water from 0.0 °C to −10.0 °C?

What Do You Know? In some ways this is the reverse of Example 14.5. Here you know the change in freezing point, but wish to know how much solute is needed to produce that value of ΔT.

Strategy The solution concentration (molality, m) can be calculated from ΔT_{fp} and K_{fp} (Table 14.3) using Equation 14.8. Combine the concentration with the mass of solvent to obtain the amount of solute and then its mass.

Solution

(a) Calculate the solute concentration in a solution with a freezing point depression of −10.0 °C.

$$\text{Solute concentration } (m) = \frac{\Delta T_{fp}}{K_{fp}} = \frac{-10.0\ °C}{-1.86\ °C/m} = 5.38\ m$$

(b) Calculate the amount of solute from the concentration and solvent mass.

$$\left(\frac{5.38\ \text{mol glycol}}{1.00\ \text{kg water}}\right)(5.50\ \text{kg water}) = \boxed{29.6\ \text{mol glycol}}$$

(c) Calculate the mass of the solute, ethylene glycol.

$$29.6\ \text{mol glycol}\left(\frac{62.07\ g}{1\ \text{mol}}\right) = \boxed{1840\ g\ \text{glycol}}$$

Think about Your Answer The value of K_{fp} tells us that the freezing point of water goes down by 1.86 °C for a 1 molal solution. In this particular problem, the freezing point depression was −10.0 °C, so a molality around 5 m is reasonable. The density of ethylene glycol is 1.11 kg/L, so the volume of glycol to be added is: (1.84 kg)(1 L/1.11 kg) = 1.66 L.

Check Your Understanding

In the northern United States, summer cottages are usually closed up for the winter. When doing so, the owners "winterize" the plumbing by putting antifreeze in the toilet tanks, for example. Will adding 525 g of $HOCH_2CH_2OH$ to 3.00 kg of water ensure that the water will not freeze at −25 °C?

Osmotic Pressure

Osmosis is the movement of solvent molecules through a semipermeable membrane from a region of lower solute concentration to a region of higher solute concentration. This movement can be demonstrated with a simple experiment. The beaker in Figure 14.15 contains pure water, and the bag and tube hold a concentrated sugar

FIGURE 14.15 The process of osmosis.

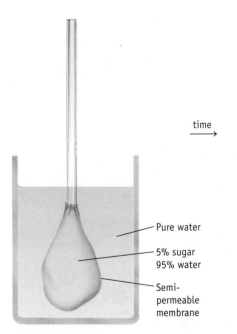
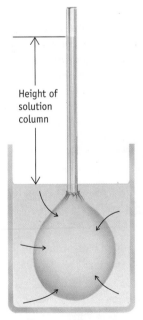

(a) The bag attached to the tube contains a solution that is 5% sugar and 95% water. The beaker contains pure water. The bag is made of a material that is semipermeable, meaning that it allows water, but not sugar molecules, to pass through.

(b) Over time, water flows from the region of low solute concentration (pure water) to the region of higher solute concentration (the sugar solution). Flow continues until the pressure exerted by the column of solution in the tube above the water level in the beaker is great enough to result in equal rates of passage of water molecules in both directions. The height of the column of solution is a measure of the osmotic pressure.

solution. The liquids are separated by a semipermeable membrane, a thin sheet of material (such as a vegetable tissue or cellophane) through which only certain types of molecules can pass. Here, water molecules can pass through the membrane, but larger sugar molecules (or hydrated ions) cannot (Figure 14.16). When the experiment is begun, the liquid levels in the beaker and the tube are the same. Over time, however, the level of the sugar solution inside the tube rises, the level of pure water in the beaker falls, and the sugar solution becomes more dilute. Eventually, no further net change occurs; equilibrium is reached.

From a molecular point of view, the semipermeable membrane does not present a barrier to the movement of water molecules, so they move through the membrane in both directions. Over time, more water molecules pass through the membrane from the pure water side to the solution side than in the opposite direction. The net effect is that water molecules move from regions of low solute concentration to regions of high solute concentration. The same is true for any solvent, as long as the membrane allows solvent molecules but not solute molecules or ions to pass through.

Why does the system eventually reach equilibrium? Clearly, the solution in the tube in Figure 14.15 can never reach zero sugar or salt concentration, which would be required to equalize the number of water molecules moving through the membrane in each direction in a given time. The answer lies in the fact that the solution moves higher and higher in the tube as osmosis continues and water moves into the sugar solution. Eventually, the pressure exerted by this column of solution counterbalances the pressure exerted by the water moving through the membrane from the pure water side, and no further net movement of water occurs. An equilibrium of forces is achieved. The pressure created by the column of solution for the system at equilibrium is called the **osmotic pressure**, Π. A measure of this pressure is the difference between the height of the solution in the tube and the level of pure water in the beaker.

From experimental measurements on dilute solutions, it is known that osmotic pressure and concentration (c) are related by the equation

$$\Pi = cRT \tag{14.9}$$

In this equation, c is the molar concentration (in moles per liter); R is the gas constant; and T is the absolute temperature (in kelvins). Using a value for the gas law constant of 0.082057 L · atm/K · mol allows calculation of the osmotic pressure Π in atmospheres. This equation is analogous to the ideal gas law ($PV = nRT$), with Π taking the place of P and c being equivalent to n/V.

FIGURE 14.16 Osmosis at the particulate level. Osmotic flow through a membrane that is selectively permeable (semipermeable) to water. Dissolved substances such as hydrated ions or large sugar molecules cannot diffuse through the membrane.

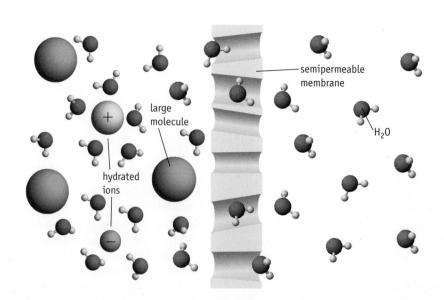

Reverse Osmosis for Pure Water

Finding sources of freshwater for humans and agriculture has been a constant battle for centuries, and, if we continue using Earth's water at the present rate, those problems may increase. Although the earth has abundant water, 97% of it is too salty to drink or to use on crops. A large portion of the remaining 3% is locked in the form of ice in the polar regions and is not easily obtained.

One of the oldest ways to obtain freshwater from seawater is by evaporation. This

A reverse osmosis plant.

is, however, very energy-intensive, and the salt and other materials left behind may not be useful.

Reverse osmosis is another method of obtaining freshwater from seawater or groundwater. In this technique a pressure greater than the osmotic pressure of the impure water is applied to force water through a semipermeable membrane from a region of high solute concentration to one of lower solute concentration, that is, in the reverse direction that the water would move by osmosis.

Although reverse osmosis has been known for over 200 years, only in the last few decades has it been exploited. Now some municipalities obtain drinking water that way, and pharmaceutical companies use it to obtain highly purified water. More than 15,000 reverse osmosis plants are in operation or in the planning stage worldwide.

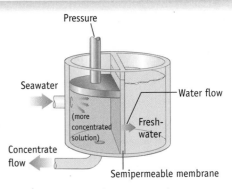

Reverse osmosis. Drinking water can be produced from seawater by reverse osmosis. The osmotic pressure of seawater is approximately 22 atm. To obtain freshwater at a reasonable rate, reverse osmosis requires a pressure of about 50 atm. For comparison, bicycle tires usually have an air pressure of 2–3 atm. (See the portable reverse osmosis device on page 472.)

Even solutions with low solute concentrations have a significant osmotic pressure. For example, the osmotic pressure of a 1.00×10^{-3} M solution at 298K is 18.5 mm Hg. Such pressures are easily and accurately measured. Because of this, the concentrations of very dilute solutions can be determined using this technique. Compare this to the effect of a solute on freezing point, for example. A dilute aqueous solution of similar concentration, 1.00×10^{-3} mol/kg solvent, would be expected to lower the freezing point by about 0.002 °C, too small to measure accurately. For this reason osmosis is a particularly useful technique when dealing with solutions of compounds having a very high molar mass such as polymers and large biomolecules.

Other examples of osmosis are shown in Figure 14.17. In this case, the egg's membrane serves as the semipermeable membrane. Osmosis occurs in one direction if the concentration of solute is greater inside the egg than in the exterior

(a) A fresh egg is placed in dilute acetic acid. The acid reacts with the $CaCO_3$ of the shell but leaves the egg membrane intact.

(b) If the egg, with its shell removed, is placed in pure water, the egg swells.

(c) If the egg, with its shell removed, is placed in a concentrated sugar solution, the egg shrivels.

FIGURE 14.17 An experiment to observe osmosis. You can try this experiment in your kitchen. In the first step, use vinegar as a source of acetic acid.

solution and occurs in the other direction if the concentration solution is less inside the egg than it is in the exterior solution. In both cases, solvent flows from the region of low solute concentration to the region of high solute concentration.

EXAMPLE 14.7 Determining the Osmotic Pressure of a Solution of a Polymer

Problem Polyvinyl alcohol is a water soluble polymer with an average molar mass of 28,000. You dissolve 1.844 g of this polymer in water to give 150. mL of solution. What is the osmotic pressure, measured at 27 °C?

What Do You Know? Osmotic pressure is calculated by Equation 14.9, $\Pi = cRT$. You are given the mass and molar mass of the polymer and the temperature and will need the gas law constant.

Strategy Calculate the concentration of the polymer, c, in mol/L. The temperature must be expressed in kelvins. Substitute these values into Equation 14.9 to solve for the osmotic pressure.

Solution

$$c = 1.844 \text{ g}(1 \text{ mol}/28,000 \text{ g})/(0.150 \text{ L}) = 4.4 \times 10^{-4} \text{ mol/L}$$

$$\Pi = cRT$$

$$\Pi = (4.4 \times 10^{-4} \text{ mol/L})(0.08206 \text{ L atm/mol K})(300 \text{ K})$$

$$\Pi = 0.011 \text{ atm } (= 8.2 \text{ mm Hg})$$

Think about Your Answer If the osmotic pressure is measured in a device similar to that shown in Figure 14.15, the height of the column of aqueous solution supported would be about 110 mm (8.2 mm $\times$ 13.5 mm H_2O/mm Hg = 110 mm). This would be easily measurable in the laboratory.

Check Your Understanding

Bradykinin is a small peptide (9 amino acids; 1060 g/mol) that lowers blood pressure by causing blood vessels to dilate. What is the osmotic pressure of a solution of this protein at 20 °C if 0.033 g of the peptide is dissolved in water to give 50.0 mL of solution?

Colligative Properties and Molar Mass Determination

Earlier in this book you learned how to calculate a molecular formula from an empirical formula when given the molar mass. But how do you know the molar mass of an unknown compound? An experiment must be carried out to find this crucial piece of information, and one way to do so is to use a colligative property of a solution of the compound. The strategy map used with Example 14.8 represents the basic approach for each of the colligative properties studied.

⏻WL INTERACTIVE EXAMPLE 14.8 Determining Molar Mass from Boiling Point Elevation

Problem A solution prepared from 1.25 g of oil of wintergreen (methyl salicylate) in 99.0 g of benzene has a boiling point of 80.31 °C. Determine the molar mass of methyl salicylate.

What Do You Know? The molar mass of a compound is the quotient of the mass of a sample (g) and the amount (mol) represented by that sample. Here you know the mass of the sample (1.25 g), so you need to find the amount that corresponds to this mass.

Strategy

- Determine ΔT_{bp} from the given boiling point of the solution. (The boiling point of pure benzene and K_{bp} for benzene are given in Table 14.3.) Use the equation $\Delta T_{bp} = K_{bp} \cdot m$ to calculate the solution concentration in units of mol/kg.

- Knowing the mass of solvent, calculate the amount of solute (mol).

- Combine the mass and amount of solute to give the molar mass [molar mass = mass (g)/amount (mol)].

Solution

(a) Use the boiling point elevation to calculate the solution concentration:

Boiling point elevation $(\Delta T_{bp}) = 80.31\ °C - 80.10\ °C = 0.21\ °C$

$$c_{solute} = \frac{\Delta T_{bp}}{K_{bp}} = \frac{0.21\ °C}{2.53\ °C/m} = 0.083\ m$$

(b) Calculate the amount of solute in the solution from the solution concentration:

$$\text{Amount of solute} = \left(\frac{0.083\ mol}{1.00\ kg}\right)(0.099\ kg\ solvent) = 0.0082\ mol\ solute$$

(c) Combine the amount of solute with its mass to obtain its molar mass.

$$\frac{1.25\ g}{0.0082\ mol} = 150\ g/mol$$

Think about Your Answer Methyl salicylate has the formula $C_8H_8O_3$ and a molar mass of 152.14 g/mol. Given that ΔT has only two significant figures, the calculated value is acceptably close to the actual value.

Check Your Understanding

An aluminum-containing compound has the empirical formula $(C_2H_5)_2AlF$. Find the molecular formula if 0.448 g of the compound dissolved in 23.46 g of benzene has a freezing point of 5.265 °C.

Strategy Map 14.8

PROBLEM

Calculate **molar mass** of unknown.

↓

DATA/INFORMATION
- **Mass** of unknown and solvent
- **Boiling point** of solution (80.31 °C)

↓

STEP 1. Use $\Delta T_{bp} = K_{bp}m$.

Concentration of solute (mol/kg)

STEP 2. Mol solute = (mol/kg)(kg solvent).

Amount of solute (mol)

STEP 3. Molar mass = g solute/mol solute.

Molar mass of unknown **(g/mol)**

EXAMPLE 14.9 Osmotic Pressure and Molar Mass

Problem Beta-carotene is the most important of the A vitamins. Calculate the molar mass of β-carotene if 10.0 mL of a solution in chloroform containing 7.68 mg of β-carotene has an osmotic pressure of 26.57 mm Hg at 25.0 °C.

What Do You Know? As in Example 14.8, the molar mass of a compound is the quotient of the mass of a sample and the amount represented by that sample. Here you know the mass of the sample (7.68 mg), so you need to find the amount equivalent to this mass.

Strategy

- Use Equation 14.9 to calculate the solution concentration from the osmotic pressure.

- Use the volume and concentration of the solution to calculate the amount of solute.

- Find the molar mass of the solute from its mass and amount.

Solution

(a) Calculate the concentration of β-carotene from Π, R, and T.

$$\text{Concentration (mol/L)} = \frac{\Pi}{RT} = \frac{(26.57\ mm\ Hg)\left(\dfrac{1\ atm}{760\ mm\ Hg}\right)}{(0.082057\ L \cdot atm/K \cdot mol)(298.2\ K)}$$

$$= 1.429 \times 10^{-3}\ mol/L$$

(b) Calculate the amount of β-carotene dissolved in 10.0 mL of solvent.

$$(1.429 \times 10^{-3}\ mol/L)(0.0100\ L) = 1.43 \times 10^{-5}\ mol$$

(c) Combine the amount of solute with its mass to calculate its molar mass:

$$\frac{7.68 \times 10^{-3}\ g}{1.43 \times 10^{-5}\ mol} = 538\ g/mol$$

Think about Your Answer Beta-carotene is a hydrocarbon with the formula $C_{40}H_{56}$ (molar mass = 536.9 g/mol).

Osmosis and Medicine

Osmosis is of practical significance for people in the health professions. Patients who become dehydrated through illness often need to be given water and nutrients intravenously. Water cannot simply be dripped into a patient's vein, however. Rather, the intravenous solution must have the same overall solute concentration as the patient's blood: the solution must be isoosmotic or **isotonic** (Figure B, middle). If pure water was used, the inside of a blood cell would have a higher solute concentration (lower water concentration), and water would flow into the cell. This hypotonic situation would cause the red blood cells to burst (lyse) (Figure B, right). The opposite situation, hypertonicity, occurs if the intravenous solution is more concentrated than the contents of the blood cell (Figure B, left). In this case, the cell would lose water and shrivel up (crenate). To combat this, a dehydrated patient is rehydrated in the hospital with a sterile saline solution that is 0.16 M NaCl, a solution that is isotonic with the cells of the body.

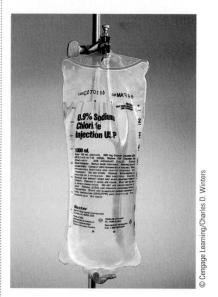

FIGURE A An isotonic saline solution. This solution has the same molality as body fluids.

© Cengage Learning/Charles D. Winters

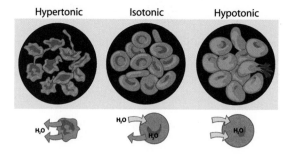

FIGURE B Osmosis and living cells. *(middle)* A cell placed in an isotonic solution. The net movement of water into and out of the cell is zero because the concentration of solutes inside and outside the cell is the same. *(left)* In a hypertonic solution, the concentration of solutes outside the cell is greater than that inside. There is a net flow of water out of the cell, causing the cell to dehydrate, shrink, and perhaps die. *(right)* In a hypotonic solution, the concentration of solutes outside the cell is less than that inside. There is a net flow of water into the cell, causing the cell to swell and perhaps to burst (or lyse).

Check Your Understanding

A 1.40-g sample of polyethylene, a common plastic, is dissolved in enough organic solvent to give 100.0 mL of solution. What is the average molar mass of the polymer if the measured osmotic pressure of the solution is 1.86 mm Hg at 25 °C?

Colligative Properties of Solutions Containing Ions

In the northern United States, it is common practice to scatter salt on snowy or icy roads or sidewalks. When the sun shines on the snow or patch of ice, a small amount melts, and some salt dissolves in the water. As a result of the dissolved solute, the freezing point of the solution is lower than 0 °C. The solution "eats" its way through the ice, breaking it up, and the icy patch is no longer dangerous for drivers or for people walking.

Salt (NaCl) is the most common substance used on roads because it is inexpensive and dissolves readily in water. Its relatively low molar mass means that the effect per gram is large. In addition, salt is especially effective because it is an electrolyte. That is, it dissolves to give ions in solution:

$$NaCl(s) \rightarrow Na^+(aq) + Cl^-(aq)$$

Recall that colligative properties depend not on what is dissolved but *only on the number of particles of solute per solvent particle*. When 1 mol of NaCl dissolves, 2 mol of ions form, which means that the effect on the freezing point of water should be twice as large as that expected for a mole of sugar. A 0.100 *m* solution of

Table 14.4 Freezing Point Depressions of Some Ionic Solutions

Mass %	m (mol/kg)	ΔT_{fp} (measured, °C)	ΔT_{fp} (calculated, °C)	$\dfrac{\Delta T_{fp},\ \text{measured}}{\Delta T_{fp},\ \text{calculated}}$
NaCl				
0.00700	0.0120	−0.0433	−0.0223	1.94
0.500	0.0860	−0.299	−0.160	1.87
1.00	0.173	−0.593	−0.322	1.84
2.00	0.349	−1.186	−0.649	1.83
Na₂SO₄				
0.00700	0.00493	−0.0257	−0.00917	2.80
0.500	0.0354	−0.165	−0.0658	2.51
1.00	0.0711	−0.320	−0.132	2.42
2.00	0.144	−0.606	−0.268	2.26

NaCl really contains two solutes, $0.100\ m$ Na^+ and $0.100\ m$ Cl^-. What we should use to estimate the freezing point depression is the *total* molality of solute particles:

$$m_{total} = m(Na^+) + m(Cl^-) = (0.100 + 0.100)\ \text{mol/kg} = 0.200\ \text{mol/kg}$$

$$\Delta T_{fp} = (-1.86\ °C/m)(0.200\ m) = -0.372\ °C$$

To estimate the freezing point depression for an ionic compound, first find the molality of solute from the mass and molar mass of the compound and the mass of the solvent. Then, multiply the molality by the number of ions in the formula: two for NaCl, three for Na_2SO_4, four for $LaCl_3$, five for $Al_2(SO_4)_3$, and so on.

Table 14.4 shows that as the concentration of NaCl decreases, ΔT_{fp} for NaCl approaches but does not quite reach a value that is two times larger than the value determined assuming no dissociation. Likewise, ΔT_{fp} for Na_2SO_4 approaches but does not reach a value that is three times larger. The ratio of the experimentally observed value of ΔT_{fp} to the value calculated, assuming no dissociation, is called the **van't Hoff factor** after Jacobus Henrikus van't Hoff (1852–1911), who studied this phenomenon. The van't Hoff factor is represented by i.

$$i = \frac{\Delta T_{fp},\ \text{measured}}{\Delta T_{fp},\ \text{calculated}} = \frac{\Delta T_{fp},\ \text{measured}}{K_{fp}\ m}$$

or

$$\Delta T_{fp}\ \text{measured} = K_{fp} \times m \times i \tag{14.10}$$

The numbers in the last column of Table 14.4 are van't Hoff factors. These values can be used in calculations of any colligative property. Vapor pressure lowering, boiling point elevation, freezing point depression, and osmotic pressure are all larger for electrolytes than for nonelectrolytes of the same molality.

The van't Hoff factor approaches a whole number (2, 3, and so on) only with very dilute solutions. In more concentrated solutions, the experimental freezing point depressions indicate that there are fewer ions in solution than expected. This behavior, which is typical of all ionic compounds, is a consequence of the strong attractions between ions. The result is as if some of the positive and negative ions are paired, decreasing the total molality of particles. Indeed, in more concentrated solutions, and especially in solvents less polar than water, ions are extensively associated in ion pairs and in even larger clusters.

EXAMPLE 14.10 Freezing Point and Ionic Solutions

Problem A 0.0200 m aqueous solution of an ionic compound, $Co(NH_3)_4Cl_3$, freezes at -0.0640 °C. How many moles of ions does 1 mol of the salt produce on being dissolved in water?

What Do You Know? You know the freezing point depression (-0.0640 °C), the solution concentration, and K_{fp}.

Strategy The van't Hoff factor, i, is the ratio of the measured ΔT_{fp} to the calculated freezing point depression. First, calculate ΔT_{fp} expected for a solution in which no ions are produced. Compare this value with the actual value of ΔT_{fp}. The ratio ($= i$) will reflect the number of ions produced.

Solution

(a) Calculate the freezing-point depression expected for a 0.0200 m solution assuming that the salt does not dissociate into ions.

$$\Delta T_{fp} \text{ calculated} = K_{fp}m = (-1.86 \text{ °C})(0.0200 \text{ } m) = -0.0372 \text{ °C}$$

(b) Compare the calculated freezing point depression with the measured depression. This gives the van't Hoff factor:

$$i = \frac{\Delta T_{fp}, \text{ measured}}{\Delta T_{fp}, \text{ calculated}} = \frac{-0.0640 \text{ °C}}{-0.0372 \text{ °C}} = 1.72$$

The i value is much greater that 1 and is approaching 2. Therefore, we assume each molecule affords 2 ions in solution.

Think about Your Answer We find i is approaching 2, meaning that the complex dissociates into two ions: $[Co(NH_3)_4Cl_2]^+$ and Cl^-. As you will see in Chapter 22, the cation is a Co^{3+} ion surrounded octahedrally by 4 NH_3 molecules and two Cl^- ions.

Check Your Understanding

Calculate the freezing point of 525 g of water that contains 25.0 g of NaCl. Assume i, the van't Hoff factor, is 1.85 for NaCl.

REVIEW & CHECK FOR SECTION 14.4

1. *Vapor pressure:* Arrange the following aqueous solutions in order of increasing vapor pressure at 25 °C: 0.35 m $C_2H_4(OH)_2$ (ethylene glycol, nonvolatile solute); 0.50 m sugar; 0.20 m KBr; and 0.20 m Na_2SO_4.

    (a) $C_2H_4(OH)_2 < $ sugar $ < $ KBr $ < Na_2SO_4$ (c) sugar $ < C_2H_4(OH)_2 < $ KBr $ < Na_2SO_4$

    (b) $Na_2SO_4 < $ sugar $ < $ KBr $ < C_2H_4(OH)_2$ (d) KBr $ < $ sugar $ < Na_2SO_4 < C_2H_4(OH)_2$

2. One gram (1.00 g) of each of the listed substances is dissolved in 50 g of benzene. Which of the solutions has the lowest freezing point?

    (a) buckyballs, C_{60} (c) trinitrotoluene, $C_7H_5(NO_2)_3$

    (b) acetaminophen, $C_8H_9NO_2$ (d) naphthalene, $C_{10}H_8$

3. Samples of each of the substances listed below are dissolved in 125 g of water. Which of the solutions has the highest boiling point?

    (a) 3.0 g sucrose, $C_{12}H_{22}O_{11}$ (c) 1.0 g propylene glycol, $C_3H_6(OH)_2$

    (b) 1.0 g glycerol, $C_3H_5(OH)_3$ (d) 2.0 g glucose, $C_6H_{12}O_6$

4. *Molar mass:* Erythritol is a compound that occurs naturally in algae and fungi. It is about twice as sweet as sucrose. A solution of 2.50 g of erythritol in 50.0 g of water freezes at -0.762 °C. What is the molar mass of the compound?

    (a) 26.9 g/mol   (c) 122 g/mol

    (b) 35.5 g/mol   (d) 224 g/mol

14.5 Colloids

Earlier in this chapter, we defined a solution broadly as a homogeneous mixture of two or more substances in a single phase. To this definition we should add that, in a true solution, no settling of the solute should be observed and the solute particles should be in the form of ions or relatively small molecules. Thus, NaCl and sugar form true solutions in water. You are also familiar with suspensions, which result, for example, if a handful of fine sand is added to water and shaken vigorously. Sand particles are still visible and gradually settle to the bottom of the beaker or bottle. **Colloidal dispersions**, also called **colloids**, represent a state intermediate between a solution and a suspension (Figure 14.18). Colloids include many of the foods you eat and the materials around you; among them are JELL-O®, milk, fog, and porcelain (Table 14.5).

Around 1860, the British chemist Thomas Graham (1805–1869) found that substances such as starch, gelatin, glue, and albumin from eggs diffused only very slowly when placed in water, compared with sugar or salt. In addition, the former substances differ significantly in their ability to diffuse through a thin membrane: Sugar molecules can diffuse through many membranes, but the very large molecules that make up starch, gelatin, glue, and albumin do not. Moreover, Graham found that he could not crystallize these substances, whereas he could crystallize sugar, salt, and other materials that form true solutions. Graham coined the word "colloid" (from the Greek, meaning "glue") to describe this class of substances that are distinctly different from true solutions and suspensions.

We now know that it is possible to crystallize some colloidal substances, albeit with difficulty, so there really is no sharp dividing line between these classes based on this property. Colloids do, however, have two distinguishing characteristics. First, colloids generally have high molar masses; this is true of proteins such as hemoglobin that have molar masses in the thousands. Second, the particles of a colloid are relatively large (say, 1000 nm in diameter). As a consequence, they exhibit the **Tyndall effect**; they scatter visible light when dispersed in a solvent, making the mixture appear cloudy (Figure 14.19). Third, even though colloidal particles are large, they are not so large that they settle out.

Graham also gave us the words **sol** for a colloidal dispersion of a solid substance in a fluid medium and **gel** for a colloidal dispersion that has a structure that prevents it from being mobile. JELL-O® is a sol when the solid is first mixed with boiling water, but it becomes a gel when cooled. Other examples of gels are the gelatinous precipitates of $Al(OH)_3$, $Fe(OH)_3$, and $Cu(OH)_2$ (Figure 14.20).

Colloidal dispersions consist of finely divided particles that, as a result, have a very high surface area. For example, if you have one millionth of a mole of colloidal particles, each assumed to be a sphere with a diameter of 200 nm, the total surface

FIGURE 14.18 Gold colloid. A water-soluble salt of $[AuCl_4]^-$ is reduced to give colloidal gold metal. The colloidal gold gives the dispersion its red color (when the particles have a length or diameter of less than 100 nm). (Similarly, colloidal gold is used to give a beautiful red color to glass.) Since the days of alchemy, some have claimed that drinking a colloidal gold solution "cleared the mind, increased intelligence and willpower, and balanced the emotions."

Dust in the air scatters the light coming through the trees in a forest along the Oregon coast.

A narrow beam of light from a laser is passed through an NaCl solution *(left)* and then a colloidal mixture of gelatin and water *(right)*.

FIGURE 14.19 The Tyndall effect. Colloidal dispersions scatter light, a phenomenon known as the Tyndall effect.

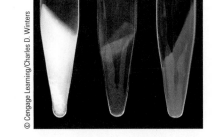

FIGURE 14.20 Gelatinous precipitates. *(left)* $Al(OH)_3$, *(center)* $Fe(OH)_3$, and *(right)* $Cu(OH)_2$.

Table 14.5 **Types of Colloids**

Type	Dispersing Medium	Dispersed Phase	Examples
Aerosol	Gas	Liquid	Fog, clouds, aerosol sprays
Aerosol	Gas	Solid	Smoke, airborne viruses, automobile exhaust
Foam	Liquid	Gas	Shaving cream, whipped cream
Foam	Solid	Gas	Styrofoam, marshmallow
Emulsion	Liquid	Liquid	Mayonnaise, milk, face cream
Gel	Solid	Liquid	Jelly, JELL-O®, cheese, butter
Sol	Liquid	Solid	Gold in water, milk of magnesia, mud
Solid sol	Solid	Solid	Milkglass

area of the particles would be on the order of 200 million cm^2, or the size of several football fields. It is not surprising, therefore, that many of the properties of colloids depend on the properties of surfaces.

Types of Colloids

Colloids are classified according to the state of the dispersed phase and the dispersing medium. Table 14.5 lists several types of colloids and gives examples of each.

Colloids with water as the dispersing medium can be classified as **hydrophobic** (from the Greek, meaning "water-fearing") or **hydrophilic** ("water-loving"). A hydrophobic colloid is one in which only weak attractive forces exist between the water and the surfaces of the colloidal particles. Examples include dispersions of metals (Figure 14.18) and of nearly insoluble salts in water. When compounds like AgCl precipitate, the result is often a colloidal dispersion. The precipitation reaction occurs too rapidly for ions to gather from long distances and make large crystals, so the ions aggregate to form small particles that remain suspended in the liquid.

Why don't the particles come together (coagulate) and form larger particles? The answer is that the colloidal particles carry electric charges. To see how this happens, suppose AgCl ion pairs come together to form a tiny particle. If Ag$^+$ ions are still present in substantial concentration in the solution, these positive ions could be attracted to negative Cl$^-$ ions on the surface of the particle. Thus the original clump of AgCl ion pairs becomes positively charged, allowing it to attract a secondary layer of anions. The particles, now surrounded by layers of ions, repel one another and are prevented from coming together to form a precipitate (Figure 14.21).

FIGURE 14.21 Hydrophobic colloids. A hydrophobic colloid is stabilized by positive ions absorbed onto each particle and a secondary layer of negative ions. Because the particles bear similar charges, they repel one another, and precipitation is prevented.

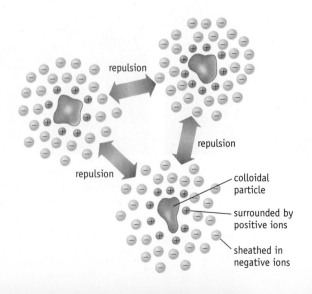

Soil particles are often carried by water in rivers and streams as hydrophobic colloids. When river water carrying large amounts of colloidal particles meets seawater with its high concentration of salts, the particles coagulate to form the silt seen at the mouth of the river (Figure 14.22). Municipal water treatment plants often add salts such as $Al_2(SO_4)_3$ to clarify water. In aqueous solution, aluminum ions exist as $[Al(H_2O)_6]^{3+}$ cations, which neutralize the charge on the hydrophobic colloidal soil particles, causing these particles to aggregate and settle out.

Hydrophilic colloids are strongly attracted to water molecules. They often have groups such as —OH and —NH₂ on their surfaces. These groups form strong hydrogen bonds to water, thereby stabilizing the colloid. Proteins and starch are important examples of hydrophilic colloids, and homogenized milk is the most familiar example.

Emulsions are colloidal dispersions of one liquid in another, such as oil or fat in water. Familiar examples include salad dressing, mayonnaise, and milk. If vegetable oil and vinegar are mixed to make a salad dressing, the mixture quickly separates into two layers because the nonpolar oil molecules do not interact with the polar water and acetic acid (CH_3CO_2H) molecules. So why are milk and mayonnaise apparently homogeneous mixtures that do not separate into layers? The answer is that they contain an **emulsifying agent** such as soap or a protein. Lecithin is a phospholipid found in egg yolks, so mixing egg yolks with oil and vinegar stabilizes the colloidal dispersion known as mayonnaise. To understand this process further, let us look into the functioning of soaps and detergents, substances known as surfactants.

Surfactants

Soaps and detergents are emulsifying agents. Soap is made by heating a fat with sodium or potassium hydroxide, which produces the anions of long chain carboxylic acids, sometimes referred to as fatty acids. An example is sodium stearate.

$$\underbrace{H_3C(CH_2)_{16}}_{\substack{\text{hydrocarbon tail} \\ \text{soluble in oil}}} - \overset{\displaystyle \overset{O}{\|}}{\underbrace{C - O^-}_{\substack{\text{polar head} \\ \text{soluble in water}}}} Na^+$$

sodium stearate, a soap

The fatty acid anion has a split personality: It has a nonpolar, hydrophobic hydrocarbon tail that is soluble in other similar hydrocarbons and a polar, hydrophilic head that is soluble in water.

Oil cannot be readily washed away from dishes or clothing with water because oil is nonpolar and thus insoluble in water. Instead, we add soap to the water to clean away the oil. The nonpolar molecules of the oil interact with the nonpolar hydrocarbon tails of the soap molecules, leaving the polar heads of the soap to interact with surrounding water molecules. The oil and water then mix (Figure 14.23).

FIGURE 14.22 Formation of silt. Silt forms at a river delta as colloidal soil particles come in contact with saltwater in the ocean. Here, the Ashley and Cooper Rivers empty into the Atlantic Ocean at Charleston, South Carolina. The high concentration of ions in seawater causes the colloidal soil particles to coagulate.

NASA/Peter Arnold, Inc.

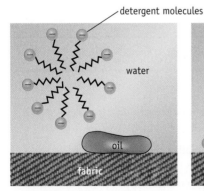
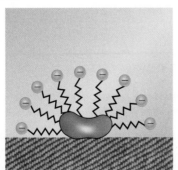
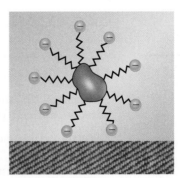
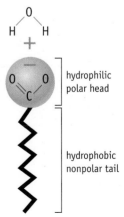

FIGURE 14.23 The cleaning action of soap. Soap molecules interact with water through the charged, hydrophilic end of the molecule. The long, hydrocarbon end of the molecule is hydrophobic, but it can bind through dispersion forces with hydrocarbons and other nonpolar substances.

FIGURE 14.24 Effect of a detergent on the surface tension of water. Sulfur (density = 2.1 g/cm³) is carefully placed on the surface of water (density, 1.0 g/cm³) *(left)*. The surface tension of the water keeps the denser sulfur afloat. Several drops of detergent are then placed on the surface of the water *(right)*. The surface tension of the water is reduced, and the sulfur sinks to the bottom of the beaker.

add surfactant →

● **Soaps and Surfactants** A sodium soap is a solid at room temperature, whereas potassium soaps are usually liquids. About 30 million tons of household and toilet soap, and synthetic and soap-based laundry detergents, are produced annually worldwide.

If the oily material on a piece of clothing or a dish also contains some dirt particles, that dirt can now be washed away.

Substances such as soaps that affect the properties of surfaces, and therefore affect the interaction between two phases, are called surface-active agents, or **surfactants**, for short. A surfactant used for cleaning is called a **detergent**. One function of a surfactant is to lower the surface tension of water, which enhances the cleansing action of the detergent (Figure 14.24).

Many detergents used in the home and industry are synthetic. One example is sodium laurylbenzenesulfonate, a biodegradable compound.

$$CH_3CH_2CH_2CH_2CH_2CH_2CH_2CH_2CH_2CH_2CH_2CH_2-\bigcirc-SO_3^- \; Na^+$$

sodium laurylbenzenesulfonate

In general, synthetic detergents use the sulfonate group, $—SO_3^-$, as the polar head instead of the carboxylate group, $—CO_2^-$. The carboxylate anions form an insoluble precipitate with any Ca^{2+} or Mg^{2+} ions present in water. Because hard water is characterized by high concentrations of these ions, using soaps containing carboxylates produces bathtub rings and telltale gray clothing. The synthetic sulfonate detergents have the advantage that they do not form such precipitates because their calcium salts are more soluble in water.

REVIEW & CHECK FOR SECTION 14.5

Figure 14.18 illustrates a gold colloid. Assume the colloidal particles are spherical clumps of gold atoms and are 100 nm in diameter. What is the approximate number of gold atoms that could form such a particle? (The radius of a gold atom is 144 pm.)

(a) 300 atoms   (b) 40 million atoms   (c) 2 billion atoms

CHAPTER GOALS REVISITED

Now that you have studied this chapter, you should ask whether you have met the chapter goals. In particular, you should be able to:

Calculate and use the solution concentration units molality, mole fraction, and weight percent

a. Define the terms solution, solvent, solute, and colligative properties.

b. Use the following concentration units: molality, mole fraction, and weight percent (Section 14.1) **Study Questions: 1–12.**

c. Understand the distinctions between saturated, unsaturated, and supersaturated solutions (Section 14.2). **Study Questions: 17, 18, 94.**

d. Define and illustrate the terms *miscible* and *immiscible* (Section 14.2). **Study Questions: 13, 96.**

Understand the solution process

a. Describe the process of dissolving a solute in a solvent, including the energy changes that may occur (Section 14.2). **Study Questions: 14–16, 71.**

b. Understand the relationship of lattice enthalpy and enthalpy of hydration to the enthalpy of solution for an ionic solute (Section 14.2). **Study Question: 91.**

c. Describe the effect of pressure and temperature on the solubility of a solute (Section 14.3). **Study Questions: 17, 18.**

d. Use Henry's law to calculate the solubility of a gas in a solvent (Section 14.3). **Study Questions: 19–22, 67, 81, 82.**

e. Apply Le Chatelier's principle to the change in solubility of gases with temperature changes (Section 14.3).

Understand and use the colligative properties of solutions

a. Calculate the mole fraction of a solvent ($X_{solvent}$) and the effect of a solute on solvent vapor pressure ($P_{solvent}$) using Raoult's law (Section 14.4). **Study Questions: 23–26, 73, 74, 98.**

b. Calculate the boiling point elevation or freezing point depression caused by a solute in a solvent (Section 14.4). **Study Questions: 27–34, 45, 53, 60, and Go Chemistry Module 19.**

c. Calculate the osmotic pressure (Π) for solutions (Section 14.4). **Study Questions: 35–38, 78, 88.**

d. Use colligative properties to determine the molar mass of a solute (Section 14.4). **Study Questions: 39–44, 64, 89, 90.**

e. Characterize the effect of ionic solutes on colligative properties (Section 14.4). **Study Questions: 47, 48, 52, 53, 60, 61, 93.**

f. Use the van't Hoff factor, *i*, in calculations involving colligative properties (Section 14.4). **Study Questions: 45, 46, 62, 63, 75, 77, 79, 80, 102.**

Recognize the properties and importance of colloids (Section 14.5).

Key Equations

Equation 14.1 (page 474) Molality is defined as the amount of solute per kilogram of solvent.

$$\text{Concentration } (c, \text{ mol/kg}) = \text{molality of solute} = \frac{\text{amount of solute (mol)}}{\text{mass of solvent (kg)}}$$

Equation 14.2 (page 474) The mole fraction, X, of a solution component is defined as the number of moles of a given component of a mixture (n_A, mol) divided by the total number of moles of all of the components of the mixture.

$$\text{Mole fraction of A } (X_A) = \frac{n_A}{n_A + n_B + n_C + \ldots}$$

Equation 14.3 (page 475) Weight percent is the mass of one component divided by the total mass of the mixture (multiplied by 100%).

$$\text{Weight \% A} = \frac{\text{mass of A}}{\text{mass of A} + \text{mass of B} + \text{mass of C} + \ldots} \times 100\%$$

Equation 14.4 (page 482) Henry's law: The solubility of a gas, S_g, is equal to the product of the partial pressure of the gaseous solute (P_g) and a constant (k_H) characteristic of the solute and solvent.

$$S_g = k_H P_g$$

Equation 14.5 (page 486) Raoult's law: The equilibrium vapor pressure of a solvent over a solution at a given temperature, $P_{solvent}$, is the product of the mole fraction of the solvent ($X_{solvent}$) and the vapor pressure of the pure solvent ($P°_{solvent}$).

$$P_{solvent} = X_{solvent}\, P°_{solvent}$$

Equation 14.6 (page 487) The decrease in the vapor pressure of the solvent over a solution, $\Delta P_{solvent}$, depends on the mole fraction of the solute (X_{solute}) and the vapor pressure of the pure solvent ($P°_{solvent}$).

$$\Delta P_{solvent} = -X_{solute}\, P°_{solvent}$$

Equation 14.7 (page 488) The elevation in boiling point of the solvent in a solution, ΔT_{bp}, is the product of the molality of the solute, m_{solute}, and a constant characteristic of the solvent, K_{bp}.

$$\text{Elevation in boiling point} = \Delta T_{bp} = K_{bp} m_{solute}$$

Equation 14.8 (page 490) The depression of the freezing point of the solvent in a solution, ΔT_{fp}, is the product of the molality of the solute, m_{solute}, and a constant characteristic of the solvent, K_{fp}.

$$\text{Freezing point depression} = \Delta T_{fp} = K_{fp} m_{solute}$$

Equation 14.9 (page 492) The osmotic pressure, Π, is the product of the solute concentration c (in mol/L), the universal gas constant R (0.082057 L · atm/ K · mol), and the temperature T (in kelvins).

$$\Pi = cRT$$

Equation 14.10 (page 497) This modified equation for freezing point depression accounts for the possible dissociation of a solute. The van't Hoff factor, i, the ratio of the measured freezing point depression and the freezing point depression calculated assuming no solute dissociation, is related to the relative number of particles produced by a dissolved solute.

$$\Delta T_{fp} \text{ measured} = K_{fp} \times m \times i$$

☉WL Questions and problems for this chapter are available in OWL. Use the OWL access card that came with this textbook to access assigned questions and problems for this chapter.

15 Chemical Kinetics: The Rates of Chemical Reactions

Fading of the color of phenolphthalein with time

(elapsed time about 3 minutes)

Where Did the Indicator Go?

The indicator phenolphthalein is often used for the titration of a weak acid using a strong base. A change from colorless to pale pink indicates that the equivalence point in the reaction has been reached (◄ page 159). If more base is added to the solution, the color of the indicator intensifies to a bright red color.

If the solution containing phenolphthalein has a pH higher than about 12, another phenomenon is observed. Slowly, the red color fades, and the solution becomes colorless. This is due to a chemical reaction of the anion of phenolphthalein with hydroxide ion, as shown in the equation. The reaction is slow, and it is easy to measure the rate of this reaction by monitoring the intensity of color of the solution.

This chapter is about a fundamental area of chemistry: *chemical kinetics*. This encompasses the study of reaction rates and their interpretation based on reaction mechanisms. In Study Question 75, you will see some data that will allow you to determine how the rate of the phenolphthalein reaction depends on the hydroxide ion concentration, and you will derive an equation that will allow you to predict the results under other conditions.

The reaction of phenolphthalein and hydroxide ion.

CHAPTER OUTLINE

15.1 Rates of Chemical Reactions

15.2 Reaction Conditions and Rate

15.3 Effect of Concentration on Reaction Rate

15.4 Concentration–Time Relationships: Integrated Rate Laws

15.5 A Microscopic View of Reaction Rates

15.6 Reaction Mechanisms

CHAPTER GOALS

See Chapter Goals Revisited (page 544) for Study Questions keyed to these goals.

- Understand rates of reaction and the conditions affecting rates.

- Derive a rate equation, rate constant, and reaction order from experimental data.

- Use integrated rate laws.

- Understand the collision theory of reaction rates and the role of activation energy.

- Relate reaction mechanisms and rate laws.

When carrying out a chemical reaction, chemists are concerned with two issues: the *rate* at which the reaction proceeds and the *extent* to which the reaction is product-favored at equilibrium. Chapter 5 began to address the second question, and Chapters 16 and 19 will develop that topic further. In this chapter, we turn to the first-mentioned question, **chemical kinetics**, a study of the rates of chemical reactions.

The study of kinetics is divided into two parts. The first part is at the *macroscopic level*, which addresses rates of reactions: what reaction rate means, how to determine a reaction rate experimentally, and how factors such as temperature and the concentrations of reactants influence rates. The second part of this subject considers chemical reactions at the *particulate level*. Here, the concern is with the **reaction mechanism**, the detailed pathway taken by atoms and molecules as a reaction proceeds. The goal is to reconcile data in the macroscopic world of chemistry with an understanding of how and why chemical reactions occur at the particulate level—and then to apply this information to control important reactions.

15.1 Rates of Chemical Reactions

You encounter the concept of rates in many nonchemical circumstances. Common examples are the speed of an automobile, given in terms of the distance traveled per unit time (for example, kilometers per hour) and the rate of flow of water from a faucet, given as volume per unit time (perhaps liters per minute). In each case, a change is measured over an interval of time. Similarly, the rate of a chemical reaction refers to the change in concentration of a reactant or product per unit of time.

$$\text{Rate of reaction} = \frac{\text{change in concentration}}{\text{change in time}}$$

Two measurements are made to determine the average speed of an automobile: distance traveled and time elapsed. Average speed is the distance traveled divided by the time elapsed, or $\Delta(\text{distance})/\Delta(\text{time})$. If an automobile travels 3.9 km in 4.5 min (0.075 h), its average speed is (3.9 km/0.075 h), or 52 kph (or 32 mph).

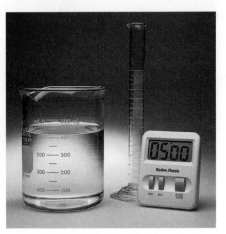
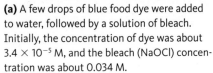

(a) A few drops of blue food dye were added to water, followed by a solution of bleach. Initially, the concentration of dye was about 3.4×10^{-5} M, and the bleach (NaOCl) concentration was about 0.034 M.

(b) The color fades as the dye reacts with the bleach. The absorbance of the solution can be measured at various times using a spectrophotometer (page 164), and these values can be used to determine the concentration of the dye.

FIGURE 15.1 An experiment to measure rate of reaction.

Average rates of chemical reactions can be determined similarly. Two quantities, concentration and time, are measured. Concentrations can be determined in a variety of ways, such as by using a pH meter to determine the concentration of hydronium ion in a solution or by measuring the absorbance of light by a solution, a property that can be related to the concentration of a species in solution (Figure 15.1). The average rate of the reaction is the change in the concentration per unit time—that is, $\Delta(\text{concentration})/\Delta(\text{time})$.

Let's consider the decomposition of N_2O_5 in a solvent. This reaction occurs according to the following equation:

$$N_2O_5 \rightarrow 2\ NO_2 + \tfrac{1}{2}\ O_2$$

Concentrations and time elapsed for a typical experiment done at 30.0 °C are presented as a graph in Figure 15.2.

FIGURE 15.2 A plot of reactant concentration versus time for the decomposition of N_2O_5. The average rate for a 15-minute interval from 40 to 55 minutes is 0.0080 mol/L · min. The instantaneous rate calculated when $[N_2O_5] = 0.34$ M is 0.0014 mol/L · min.

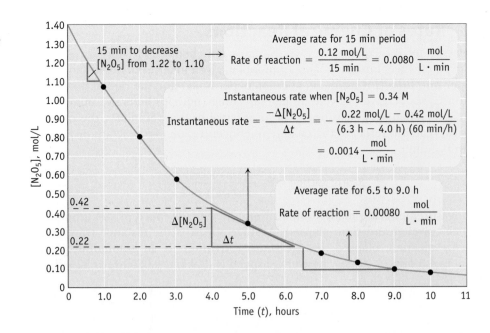

The rate of this reaction for any interval of time can be expressed as the change in concentration of N_2O_5 divided by the change in time:

$$\text{Rate of reaction} = -\frac{\text{change in } \left[N_2O_5\right]}{\text{change in time}} = -\frac{\Delta\left[N_2O_5\right]}{\Delta t}$$

The minus sign is required because the concentration of N_2O_5 decreases with time (that is, $\Delta[N_2O_5] = [N_2O_5](\text{final}) - [N_2O_5](\text{initial})$ is negative), and rate is always expressed as a positive quantity. Using data from Figure 15.2, the rate of disappearance of N_2O_5 between 40 and 55 minutes is given by

$$\text{Rate} = -\frac{\Delta\left[N_2O_5\right]}{\Delta t} = -\frac{(1.10 \text{ mol/L}) - (1.22 \text{ mol/L})}{55 \text{ min} - 40 \text{ min}} = +\frac{0.12 \text{ mol/L}}{15 \text{ min}}$$

$$\text{Rate} = 0.0080 \frac{\text{mol } N_2O_5 \text{ consumed}}{L \cdot \text{min}}$$

Note the units for reaction rates; if concentration is expressed in mol/L, the units for rate will be mol/L · time.

During a chemical reaction, amounts of reactants decrease with time, and amounts of products increase. For the decomposition of N_2O_5, we could also express the rate either as $\Delta[NO_2]/\Delta t$ or as $\Delta[O_2]/\Delta t$. Rates based on changes in concentrations of products will have a positive sign because the concentration is increasing. Furthermore, the numerical values of rates defined in these ways will be different from the value of $\Delta[N_2O_5]/\Delta t$. Note that the rate of decomposition of N_2O_5 is one half the rate of formation of NO_2 and twice the rate of formation of O_2. A rate expression relating all of these rates can be determined by dividing each rate expression by the stoichiometric coefficient of the substance in the balanced chemical equation.

$$\text{Rate of reaction} = -\frac{\Delta\left[N_2O_5\right]}{\Delta t} = +\frac{1}{2}\frac{\Delta\left[NO_2\right]}{\Delta t} = +2\frac{\Delta\left[O_2\right]}{\Delta t}$$

For the 15-minute interval between 40 and 55 minutes, the average rates for the formation of NO_2 and O_2 are

$$\text{Rate} = \frac{\Delta\left[NO_2\right]}{\Delta t} = \frac{0.0080 \text{ mol } N_2O_5 \text{ consumed}}{L \cdot \text{min}} \times \frac{2 \text{ mol } NO_2 \text{ formed}}{1 \text{ mol } N_2O_5 \text{ consumed}}$$

$$= 0.016 \frac{\text{mol } NO_2 \text{ formed}}{L \cdot \text{min}}$$

$$\text{Rate} = \frac{\Delta\left[O_2\right]}{\Delta t} = \frac{0.0080 \text{ mol } N_2O_5 \text{ consumed}}{L \cdot \text{min}} \times \frac{\frac{1}{2} \text{ mol } O_2 \text{ formed}}{1 \text{ mol } N_2O_5 \text{ consumed}}$$

$$= 0.0040 \frac{\text{mol } O_2 \text{ formed}}{L \cdot \text{min}}$$

The graph of $[N_2O_5]$ versus time in Figure 15.2 does not give a straight line because, as described in Section 15.3, the rate of the reaction changes during the course of the reaction. The concentration of N_2O_5 decreases rapidly at the beginning of the reaction but more slowly near the end. We can verify this by comparing the rate of disappearance of N_2O_5 calculated previously (the concentration decreased by 0.12 mol/L in 15 minutes) to the average rate of reaction calculated for the time interval from 6.5 to 9.0 hours (when the concentration drops by 0.12 mol/L in 150 minutes). The average rate in this later stage of this reaction is only one tenth of the previous value.

$$-\frac{\Delta\left[N_2O_5\right]}{\Delta t} = -\frac{(0.10 \text{ mol/L}) - (0.22 \text{ mol/L})}{540 \text{ min} - 390 \text{ min}} = +\frac{0.12 \text{ mol/L}}{150 \text{ min}}$$

$$= 0.00080 \frac{\text{mol}}{L \cdot \text{min}}$$

● **Representing Concentrations**
Recall that square brackets around a formula indicate its concentrations in mol/L (Section 4.5).

● **Calculating Changes** When calculating a change in a quantity, we always do so by subtracting the initial quantity from the final quantity: $\Delta c = c_{\text{final}} - c_{\text{initial}}$.

● **Reaction Rates and Stoichiometry**
For the general reaction $aA + bB \rightarrow cC + dD$, the international convention defines the reaction rate as

$$\text{Reaction rate} = -\frac{1}{a}\frac{\Delta\left[A\right]}{\Delta t} = -\frac{1}{b}\frac{\Delta\left[B\right]}{\Delta t}$$
$$= +\frac{1}{c}\frac{\Delta\left[C\right]}{\Delta t} = +\frac{1}{d}\frac{\Delta\left[D\right]}{\Delta t}$$

Note that this gives us the reaction rate on a "per mole of reaction" basis. For more on this topic, see K. T. Quisenberry and J. Tellinghuisen, *Journal of Chemical Education*, Vol. 83, pp. 510–512, 2006.

The procedure we have used to calculate the reaction rate gives the average rate over the chosen time interval.

We might also ask what the instantaneous rate is at a single point in time. In an automobile, the instantaneous rate can be read from the speedometer. For a chemical reaction, we can extract the instantaneous rate from the concentration–time graph by drawing a line tangent to the concentration–time curve at a particular time (Figure 15.2). The instantaneous rate is obtained from the slope of this line. For example, when $[N_2O_5] = 0.34$ mol/L and $t = 5.0$ hours, the rate is

$$\text{Rate when } \left[N_2O_5\right] \text{ is 0.34 M} = -\frac{\Delta\left[N_2O_5\right]}{\Delta t} = +\frac{0.20 \text{ mol/L}}{140 \text{ min}}$$

$$= 0.0014\frac{\text{mol}}{\text{L} \cdot \text{min}}$$

At that particular moment in time, ($t = 5.0$ hours), N_2O_5 is being consumed at a rate of 0.0014 mol/L $\cdot$ min.

● **The Slope of a Line** The instantaneous rate in Figure 15.2 can be determined from an analysis of the slope of the line. See pages 37–38 for more on finding the slope of a line.

EXAMPLE 15.1   Relative Rates and Stoichiometry

Problem Relate the rates for the disappearance of reactants and formation of products for the following reaction:

$$4 \text{ PH}_3(g) \rightarrow \text{P}_4(g) + 6 \text{ H}_2(g)$$

What Do You Know? The balanced chemical equation for the reaction is given. The stoichiometric coefficients in this equation will be used to evaluate the relative rates for the disappearance of the starting material and formation of the products.

Strategy In this reaction, PH_3 disappears, and P_4 and H_2 are formed. Consequently, the value of $\Delta[PH_3]/\Delta t$ will be negative, whereas $\Delta[P_4]/\Delta t$ and $\Delta[H_2]/\Delta t$ will be positive. To relate the rates to each other, we divide $\Delta[\text{reagent}]/\Delta t$ by its stoichiometric coefficient in the balanced equation.

Solution Because four moles of PH_3 disappear for every one mole of P_4 formed, the numerical value of the rate of formation of P_4 is one fourth of the rate of disappearance of PH_3. Similarly, P_4 is formed at only one sixth of the rate that H_2 is formed.

$$\text{Reaction rate} = -\frac{1}{4}\left(\frac{\Delta\left[PH_3\right]}{\Delta t}\right) = +\frac{\Delta\left[P_4\right]}{\Delta t} = +\frac{1}{6}\left(\frac{\Delta\left[H_2\right]}{\Delta t}\right)$$

Think about Your Answer A general equation to determine the relative rates of disappearance and formation from a balanced chemical equation is given in the margin on page 509.

Check Your Understanding

What are the relative rates of appearance or disappearance of each product and reactant, respectively, in the decomposition of nitrosyl chloride, NOCl?

$$2 \text{ NOCl}(g) \rightarrow 2 \text{ NO}(g) + \text{Cl}_2(g)$$

EXAMPLE 15.2   Rate of Reaction

Problem Data collected on the concentration of dye as a function of time (Figure 15.1) are given in the graph below. What is the average rate of change of the dye concentration over the first 2 minutes? What is the average rate of change during the fifth minute (from $t = 4.0$ minutes to $t = 5.0$ minutes)? Estimate the instantaneous rate at 4.0 minutes.

What Do You Know? The concentration of dye as a function of time is presented as a graph. From this curve you can identify the concentration of dye at specific times.

Strategy To find the average rate, calculate the difference in concentration at the beginning and end of a time period ($\Delta c = c_{final} - c_{initial}$) and divide by the elapsed time. To find the instantaneous rate at 4 minutes, draw a line tangent to the graph at the specified time. The negative of the slope of the line is the instantaneous rate.

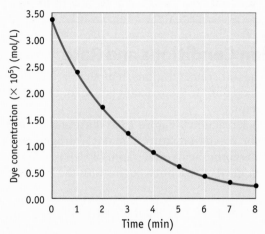

Solution The concentration of dye decreases from 3.4×10^{-5} mol/L at $t = 0$ minutes to 1.7×10^{-5} mol/L at $t = 2.0$ minutes. The average rate of the reaction in this interval of time is

$$\text{Average rate} = -\frac{\Delta[\text{Dye}]}{\Delta t} = -\frac{(1.7 \times 10^{-5}\ \text{mol/L}) - (3.4 \times 10^{-5}\ \text{mol/L})}{2.0\ \text{min}}$$

$$\text{Average rate} = \frac{8.5 \times 10^{-6}\ \text{mol}}{\text{L} \cdot \text{min}}$$

The concentration of dye decreases from 0.90×10^{-5} mol/L at $t = 4.0$ minutes to 0.60×10^{-5} mol/L at $t = 5.0$ minutes. The average rate of the reaction in this interval of time is

$$\text{Average rate} = -\frac{\Delta[\text{Dye}]}{\Delta t} = -\frac{(0.60 \times 10^{-5}\ \text{mol/L}) - (0.90 \times 10^{-5}\ \text{mol/L})}{1.0\ \text{min}}$$

$$\text{Average rate} = \frac{3.0 \times 10^{-6}\ \text{mol}}{\text{L} \cdot \text{min}}$$

From the slope of the line tangent to the curve, the instantaneous rate at 4 minutes is found to be $+3.5 \times 10^{-6}$ mol/L · min.

Think about Your Answer Notice that the rate of reaction decreases as the concentration of dye decreases. This tells us that the rate of the reaction is related to the concentration of dye. We can also deduce this qualitatively from the fact that graphing concentration vs. time gave a curved line.

Check Your Understanding

Sucrose decomposes to fructose and glucose in acid solution. A plot of the concentration of sucrose as a function of time is given in the margin. What is the rate of change of the sucrose concentration over the first 2 hours? What is the rate of change over the last 2 hours? Estimate the instantaneous rate at 4 hours.

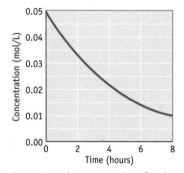

Concentration versus time for the decomposition of sucrose.
("Check Your Understanding" in Example 15.2.)

REVIEW & CHECK FOR SECTION 15.1

1. Compare the rates of disappearance of NO(g) and O_2(g) for the reaction
    2 NO(g) + O_2(g) → 2 NO_2(g).

    (a) The rates are equal.

    (b) The rate of disappearance of NO(g) is twice the rate of disappearance of O_2(g).

    (c) The rate of disappearance of NO(g) is one half the rate of disappearance of O_2(g).

    (d) The rates are not related.

2. Use the graph provided in Example 15.2 to estimate the average rate of disappearance of dye for the period 1-3 minutes.

(a) 0.6 mol/L · min

(c) 0.6×10^{-5} mol/L · min

(b) 1.2×10^{-5} mol/L · min

(d) 1.2 mol/L · min

15.2 Reaction Conditions and Rate

Several factors—reactant concentrations, temperature, and presence of catalysts—affect the rate of a reaction. If the reactant is a solid, the surface area available for reaction is also a factor.

The "iodine clock reaction" (Figure 15.3) illustrates the **effect of concentration and temperature**. The reaction mixture contains hydrogen peroxide (H_2O_2), iodide ion (I^-), vitamin C (ascorbic acid), and starch (which is an indicator of the presence of iodine, I_2). A sequence of reactions begins with the slow oxidation of iodide ion to I_2 by H_2O_2.

$$H_2O_2(aq) + 2\ I^-(aq) + 2\ H_3O^+(aq) \rightarrow 4\ H_2O(\ell) + I_2(aq)$$

As soon as I_2 is formed in the solution, vitamin C rapidly reduces it to I^-.

$$2\ H_2O(\ell) + I_2(aq) + C_6H_8O_6(aq) \rightarrow C_6H_6O_6(aq) + 2\ H_3O^+(aq) + 2\ I^-(aq)$$

When all of the vitamin C has been consumed, I_2 remains in solution and forms a blue-black complex with starch. The time measured represents how long it has taken for the given amount of vitamin C to react. For the first experiment (A in Figure 15.3) the time required is 51 seconds. When the concentration of iodide ion is smaller (B), the time required for the vitamin C to be consumed is longer, 1 minute 33 seconds. Finally, when the concentrations are again the same as in experiment B but the reaction mixture is heated, the reaction occurs more rapidly (56 seconds). This experiment illustrates two features that are true of most reactions:

● **Effect of Temperature on Reaction Rate** Cooking involves chemical reactions, and a higher temperature results in foods cooking faster. In the laboratory, reaction mixtures are often heated to make reactions occur faster.

- If the concentration of a reactant is increased, the reaction rate will often increase as well.
- Chemical reactions occur more rapidly at higher temperatures.

(a) Initial experiment.
The blue color of the starch–iodine complex develops in 51 seconds.

(b) Change concentration.
The blue color of the starch–iodine complex develops in 1 minute 33 seconds when the solution is less concentrated than A.

(c) Change temperature.
The blue color of the starch–iodine complex develops in 56 seconds when the solution is the same concentration as in B but at a higher temperature.

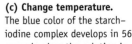
Hot bath

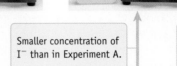
Smaller concentration of I^- than in Experiment A.

Same concentrations as in Experiment B, but at a higher temperature.

Solutions containing vitamin C, H_2O_2, I^-, and starch are mixed.

Photos © Cengage Learning/Charles D. Winters

FIGURE 15.3 The iodine clock reaction. This reaction illustrates the effects of concentration and temperature on reaction rate. (You can do these experiments yourself with reagents available in the supermarket. For details, see S. W. Wright: "The vitamin C clock reaction," *Journal of Chemical Education*, Vol. 79, p. 41, 2002.)

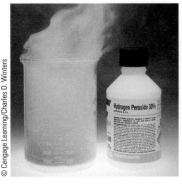

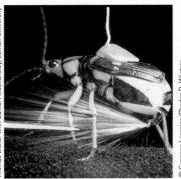

(a) The rate of decomposition of hydrogen peroxide is increased by the catalyst MnO_2. Here, H_2O_2 (as a 30% aqueous solution) is poured onto the black solid MnO_2 and rapidly decomposes to O_2 and H_2O. Steam forms because of the high heat of reaction.

(b) A bombardier beetle uses the catalyzed decomposition of H_2O_2 as a defense mechanism. The energy involved in the reaction lets the insect eject hot water and other irritating chemicals with explosive force.

(c) Enzyme catalysis. Here, the enzyme catalase, found in potatoes, is used to catalyze H_2O_2 decomposition, and bubbles of O_2 gas are seen rising in the solution.

FIGURE 15.4 Catalyzed decomposition of H_2O_2.

Catalysts are substances that accelerate chemical reactions but are not themselves consumed. Consider the effect of a catalyst on the decomposition of hydrogen peroxide, H_2O_2, to form water and oxygen.

$$2\ H_2O_2(aq) \rightarrow O_2(g) + 2\ H_2O(\ell)$$

This decomposition is very slow; a solution of H_2O_2 can be stored for many months with only minimal change in concentration. Adding manganese(IV) oxide, an iodide-containing salt, or a compound called catalase—an example of a biological catalyst or *enzyme*—causes this reaction to occur rapidly, as shown by vigorous bubbling as gaseous oxygen escapes from the solution (Figure 15.4).

The *surface area* of a solid reactant can also affect the reaction rate. Only molecules at the surface of a solid can come in contact with other reactants. The smaller the particles of a solid, the more molecules are found on the solid's surface. With very small particles, the effect of surface area on rate can be quite dramatic (Figure 15.5). Farmers know that explosions of fine dust particles (suspended in the air in an enclosed silo or at a feed mill) represent a major hazard.

(a)

REVIEW & CHECK FOR SECTION 15.2

Which of the factors listed below affect the rates of chemical reactions?

(a) catalysts   (c) concentration of reactants   (e) all of the above

(b) temperature   (d) surface area of solid

15.3 Effect of Concentration on Reaction Rate

One important goal in studying the kinetics of a reaction is to determine its mechanism; that is, how the reaction occurs at the molecular level. The place to begin is to learn how concentrations of reactants affect the reaction rate.

The effect of concentration can be determined by evaluating how the rate is affected when the concentrations of the reactants are varied (with the temperature held constant). Consider, for example, the decomposition of N_2O_5 to NO_2 and O_2. Figure 15.2 presented data on the concentration of N_2O_5 as a function of time. We previously calculated that, when $[N_2O_5] = 0.34$ mol/L, the instantaneous rate of disappearance of N_2O_5 is 0.0014 mol/L · min. An evaluation of the instantaneous rate of the reaction when $[N_2O_5] = 0.68$ mol/L reveals a rate of 0.0028 mol/L · min.

(b)

FIGURE 15.5 The combustion of lycopodium powder. (a) The spores of this common fern burn only with difficulty when piled in a dish. **(b)** If the spores are ground to a fine powder and sprayed into a flame, combustion is rapid.

That is, doubling the concentration of N_2O_5 doubles the reaction rate. A similar exercise shows that if $[N_2O_5]$ is 0.17 mol/L (half of 0.34 mol/L), the reaction rate is also halved. From these results, we know that the rate for this reaction must be directly proportional to the N_2O_5 concentration:

$$N_2O_5 \rightarrow 2\ NO_2 + \tfrac{1}{2}\ O_2$$

$$\text{Rate of reaction} \propto [N_2O_5]$$

where the symbol $\propto$ means "proportional to."

Different relationships between reaction rate and reactant concentration are encountered in other reactions. For example, the reaction rate could be independent of concentration, or it may depend on the reactant concentration raised to some power n (that is, $[\text{reactant}]^n$). If the reaction involves several reactants, the reaction rate may depend on the concentrations of each of them or on only one of them. Finally, the concentration of a catalyst or even the concentrations of the products may affect the rate.

Rate Equations

The relationship between reactant concentration and reaction rate is expressed by an equation called a **rate equation**, or **rate law**. For the decomposition of N_2O_5 the rate equation is

$$N_2O_5(g) \rightarrow 2\ NO_2(g) + \tfrac{1}{2}\ O_2$$

$$\text{Rate of reaction} = -\Delta[N_2O_5]/\Delta t = +(1/2)\Delta[NO_2]/\Delta t = +2\Delta[O_2]/\Delta t = k[N_2O_5]$$

where the proportionality constant, k, is called the **rate constant**. This rate equation tells us that this reaction rate is proportional to the concentration of the reactant. Based on this equation, we can determine that when $[N_2O_5]$ is doubled, the reaction rate doubles (and we know $[N_2O_5]$ decreases twice as fast as $[O_2]$ increases).

Generally, for a reaction such as

$$a\ A + b\ B \rightarrow x\ X$$

the rate equation has the form

$$\text{Rate of reaction} = k[A]^m[B]^n$$

● **Exponents on Reactant Concentrations and Reaction Stoichiometry** It is important to recognize that the exponents m and n are not necessarily the stoichiometric coefficients (a and b) for the balanced chemical equation.

The rate equation expresses the fact that the rate of reaction is proportional to the reactant concentrations, each concentration being raised to some power. The exponents in this equation are often positive whole numbers, but they can also be negative numbers, fractions, or zero, and they are determined by experiment.

If a homogeneous catalyst is present, its concentration might also be included in the rate equation, even though the catalytic species in not a product or reactant in the equation for the reaction. Consider, for example, the decomposition of hydrogen peroxide in the presence of a catalyst such as iodide ion.

$$H_2O_2(aq) \xrightarrow{\ I^-(aq)\ } H_2O(\ell) + \tfrac{1}{2}\ O_2(g)$$

Experiments show that this reaction has the following rate equation:

$$\text{Reaction rate} = -\Delta[H_2O_2]/\Delta t = +\Delta[H_2O]/\Delta t = +2\Delta[O_2]/\Delta t = k[H_2O_2][I^-]$$

Here, the concentration of I^- appears in the rate law, even though it is not involved in the balanced equation.

● **The Nature of Catalysts** A catalyst does not appear as a reactant in the balanced, overall equation for the reaction, but it may appear in the rate expression. A common practice is to identify catalysts by name or symbol above the reaction arrow, as shown in the example. A homogeneous catalyst is one in the same phase as the reactants. For example, both H_2O_2 and I^- are dissolved in water.

The Order of a Reaction

The **order** of a reaction with respect to a particular reactant is the exponent of its concentration term in the rate expression, and the **overall reaction order** is the sum of the exponents on all concentration terms. Consider, for example, the reaction of NO and Cl_2:

$$2\ NO(g) + Cl_2(g) \rightarrow 2\ NOCl(g)$$

The experimentally determined rate equation for this reaction is

$$\text{Reaction rate} = -(1/2)\Delta[\text{NO}]/\Delta t = -\Delta[\text{Cl}_2]/\Delta t = +(1/2)\Delta[\text{NOCl}]/\Delta t = k[\text{NO}]^2[\text{Cl}_2]$$

This reaction is second order in NO, first order in Cl_2, and third order overall. How is this related to the following experimental data for the rate of disappearance of NO?

Experiment	[NO] mol/L	[Cl₂] mol/L	Rate mol/L · s
1	0.250	0.250	1.43×10^{-6}
	$\downarrow \times 2$	$\downarrow$ No change	$\downarrow \times 4$
2	0.500	0.250	5.72×10^{-6}
3	0.250	0.500	2.86×10^{-6}
4	0.500	0.500	11.4×10^{-6}

- *Compare Experiments 1 and 2:* Here, $[\text{Cl}_2]$ is held constant, and $[\text{NO}]$ is doubled. The change in $[\text{NO}]$ leads to a reaction rate increase by a factor of 4; that is, the rate is proportional to the *square* of the NO concentration.
- *Compare Experiments 1 and 3:* In experiments 1 and 3, $[\text{NO}]$ is held constant, and $[\text{Cl}_2]$ is doubled, causing the rate to double. That is, the rate is proportional to $[\text{Cl}_2]$.
- *Compare Experiments 1 and 4:* Both $[\text{NO}]$ and $[\text{Cl}_2]$ are doubled from 0.250 M to 0.500 M. From previous experiments, we know that doubling $[\text{NO}]$ should cause a four-fold increase, and doubling $[\text{Cl}_2]$ causes a two-fold increase. Therefore, doubling both concentrations should cause an eight-fold increase, as is observed $[(1.43 \times 10^{-6}\ \text{mol/L} \cdot \text{s}) \times 8 = 11.4 \times 10^{-6}\ \text{mol/L} \cdot \text{s}]$.

The decomposition of ammonia on a platinum surface at 856 °C is a zero-order reaction.

$$\text{NH}_3(g) \rightarrow \tfrac{1}{2}\,\text{N}_2(g) + \tfrac{3}{2}\,\text{H}_2(g)$$

This means that the reaction rate is independent of NH_3 concentration.

$$\text{Reaction rate} = -\Delta[\text{NH}_3]/\Delta t = +2\Delta[\text{N}_2]/\Delta t = +(2/3)\Delta[\text{H}_2]/\Delta t = k[\text{NH}_3]^0 = k$$

Reaction order is important because it gives some insight into the most interesting question of all—how the reaction occurs. This is described further in Section 15.6.

The Rate Constant, *k*

The rate constant, *k*, is a proportionality constant that relates rate and concentration at a given temperature. It is an important quantity because it enables you to find the reaction rate for a new set of concentrations. To see how to use *k*, consider the substitution of Cl^- ion by water in the cancer chemotherapy agent cisplatin, $\text{Pt}(\text{NH}_3)_2\text{Cl}_2$.

$$\text{Pt}(\text{NH}_3)_2\text{Cl}_2(aq) + \text{H}_2\text{O}(\ell) \longrightarrow [\text{Pt}(\text{NH}_3)_2(\text{H}_2\text{O})\text{Cl}]^+(aq) + \text{Cl}^-(aq)$$

The rate law for this reaction is

$$\text{Reaction rate} = -\Delta[\text{Pt}(\text{NH}_3)_2\text{Cl}_2]\Delta t = k[\text{Pt}(\text{NH}_3)_2\text{Cl}_2]$$

and the rate constant, *k*, is 0.27/h at 25 °C. Knowing *k* allows you to calculate the rate at a particular reactant concentration—for example, when $[\text{Pt}(\text{NH}_3)_2\text{Cl}_2] = 0.018\ \text{mol/L}$:

$$\text{Reaction rate} = (0.27/\text{h})(0.018\ \text{mol/L}) = 0.0049\ \text{mol/L} \cdot \text{h}$$

● **Time and Rate Constants** The time in a rate constant can be seconds, minutes, hours, days, years, or whatever time unit is appropriate. The fraction 1/time can also be written as time^{-1}. For example, $1/y$ is equivalent to y^{-1}, and 1/s is equivalent to s^{-1}.

● **Some Rate Constants**

First Order	*k* (1/s)
$2\ \text{N}_2\text{O}_5(g)$ $\rightarrow 4\ \text{NO}_2(g) + \text{O}_2(g)$	3.38×10^{-5} at 25 °C
$\text{C}_2\text{H}_6(g) \rightarrow 2\ \text{CH}_3(g)$	5.36×10^{-4} at 700 °C
Sucrose(aq, H_3O^+) $\rightarrow$ fructose(aq) + glucose(aq)	6.0×10^{-5} at 25 °C

Second Order	*k* (L/mol · s)
$2\ \text{NOBr}(g)$ $\rightarrow 2\ \text{NO}(g) + \text{Br}_2(g)$	0.80 at 10 °C
$\text{H}_2(g) + \text{I}_2(g)$ $\rightarrow 2\ \text{HI}(g)$	0.0242 at 400 °C

As noted earlier, reaction rates have units of mol/L · time when concentrations are given as moles per liter. Rate constants must have units consistent with the units for the other terms in the rate equation.

- First-order reactions: The units of k are 1/time.
- Second-order reactions: The units of k are L/mol · time.
- Zero-order reaction: The units of k are mol/L · time.

Determining a Rate Equation

One way to determine a rate equation is by using the "method of initial rates." The initial rate is the instantaneous reaction rate at the start of the reaction (the rate at $t = 0$). An approximate value of the initial rate can be obtained by mixing the reactants and determining the reaction rate after 1% to 2% of the limiting reactant has been consumed. Measuring the rate during the initial stage of a reaction is convenient because initial concentrations are known.

As an example of the determination of a reaction rate by the method of initial rates, let us consider the reaction of sodium hydroxide with methyl acetate to produce acetate ion and methanol.

$$CH_3CO_2CH_3(aq) \quad + \quad OH^-(aq) \quad \longrightarrow \quad CH_3CO_2^-(aq) \quad + \quad CH_3OH(aq)$$

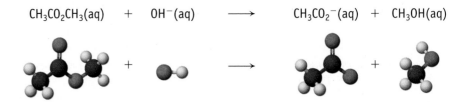

Reactant concentrations and initial rates for this reaction for several experiments at 25 °C are collected in the table below.

	Initial Concentrations (mol/L)		Initial Reaction Rate
Experiment	[CH₃CO₂CH₃]	[OH⁻]	(mol/L · s) at 25 °C
1	0.050	0.050	0.00034
	↓ no change	↓ × 2	↓ × 2
2	0.050	0.10	0.00069
	↓ × 2	↓ no change	↓ × 2
3	0.10	0.10	0.00137

As you can see from the data in the table, when the initial concentration of one reactant (either $CH_3CO_2CH_3$ or OH^-) is doubled while the concentration of the other reactant is held constant, the initial reaction rate doubles. This rate doubling shows that the reaction rate is directly proportional to the concentrations of both $CH_3CO_2CH_3$ and OH^-; thus, the reaction is first order in each of these reactants and second order overall. The rate law that reflects these experimental observations is

$$\text{Reaction rate} = -\Delta[CH_3CO_2H]/\Delta t = k[CH_3CO_2CH_3][OH^-]$$

Using this equation, we can predict that doubling both concentrations at the same time should cause the rate to go up by a factor of 4. What happens, however, if one concentration is doubled and the other is halved? The rate equation tells us the rate should not change!

If the rate equation is known, the value of k, the rate constant, can be found by substituting values for the rate and concentration into the rate equation. Using the data for the methyl acetate/hydroxide ion reaction from the first experiment, we have

$$\text{Reaction rate} = 0.00034 \text{ mol/L} \cdot s = k(0.050 \text{ mol/L})(0.050 \text{ mol/L})$$

$$k = \frac{0.00034 \text{ mol/L} \cdot s}{(0.050 \text{ mol/L})(0.050 \text{ mol/L})} = 0.14 \text{ L/mol} \cdot s$$

OWL   INTERACTIVE EXAMPLE 15.3   Determining a Rate Equation

Problem The rate of the reaction between CO and NO_2 at 540 K

$$CO(g) + NO_2(g) \rightarrow CO_2(g) + NO(g)$$

was measured starting with various concentrations of CO and NO_2. Determine the rate equation and the value of the rate constant.

Experiment	Initial Concentrations		Initial Rate
	[CO], mol/L	[NO₂], mol/L	(mol/L · h)
1	5.10×10^{-4}	0.350×10^{-4}	3.4×10^{-8}
2	5.10×10^{-4}	0.700×10^{-4}	6.8×10^{-8}
3	5.10×10^{-4}	0.175×10^{-4}	1.7×10^{-8}
4	1.02×10^{-3}	0.350×10^{-4}	6.8×10^{-8}
5	1.53×10^{-3}	0.350×10^{-4}	10.2×10^{-8}

What Do You Know? The table contains concentrations of the two reactants and initial rates for five experiments.

Strategy For a reaction involving several reactants, the general approach is to keep the concentration of one reactant constant and then decide how the rate of reaction changes as the concentration of the other reagent is varied. Because the rate is proportional to the concentration of a reactant, R, raised to some power n (the reaction order)

$$\text{Rate} \propto [R]^n$$

we can write the general equation for the ratio of rates in two experiments

$$\frac{\text{Rate in experiment 2}}{\text{Rate in experiment 1}} = \frac{[R_2]^n}{[R_1]^n} = \left(\frac{[R_2]}{[R_1]}\right)^n$$

If [R] is doubled and the rate doubles from experiment 1 to experiment 2, then $n = 1$. If [R], doubles and the rate goes up by 4, then $n = 2$.

Solution In the first three experiments, the concentration of CO is held constant. In the second experiment, the NO_2 concentration has been doubled relative to experiment 1, leading to a two-fold increase in the rate. Thus, $n = 1$ and the reaction is first order in NO_2.

$$\frac{\text{Rate in experiment 2}}{\text{Rate in experiment 1}} = \frac{6.8 \times 10^{-8} \text{ mol/L} \cdot \text{h}}{3.4 \times 10^{-8} \text{ mol/L} \cdot \text{h}} = \left(\frac{0.700 \times 10^{-4}}{0.350 \times 10^{-4}}\right)^n$$

$$2 = (2)^n$$

and so $n = 1$.

This finding is confirmed by experiment 3. Decreasing $[NO_2]$ in experiment 3 to half its original value causes the rate to decrease by half.

The data in experiments 1 and 4 (with constant $[NO_2]$) show that doubling [CO] doubles the rate, and the data from experiments 1 and 5 show that tripling the concentration of CO triples the rate. These results mean that the reaction is first order in [CO]. Thus, we now know the rate equation is

$$\text{Reaction rate} = -\Delta[CO]/\Delta t = -\Delta[NO_2]/\Delta t = +\Delta[CO_2]/\Delta t = +\Delta[NO]/\Delta t = k[CO][NO_2]$$

The rate constant, k, can be found by inserting data for one of the experiments into the rate equation. Using data from experiment 1, for example,

$$\text{Rate} = 3.4 \times 10^{-8} \text{ mol/L} \cdot \text{h} = k(5.10 \times 10^{-4} \text{ mol/L})(0.350 \times 10^{-4} \text{ mol/L})$$

$$k = 1.9 \text{ L/mol} \cdot \text{h}$$

Think about Your Answer This reaction is first order in each reactant (and second order overall), and you know the value of the rate constant k. The same rate law would apply if the reaction were carried out at a different temperature, but the rate constant k would have a different value.

Strategy Map 15.3

PROBLEM
Derive the **rate equation** and **value of k** for a given reaction.

DATA/INFORMATION
- **5 experiments** measuring *initial rate* as a function of **reactant concentration**

STEP 1. **Compare rates** for two experiments where the concentration of reactant **A** is *constant* and reactant **B** is *varied*.

Gives the **dependence of rate** on one reactant (**B**)

STEP 2. **Compare rates** for two experiments where the concentration of reactant **B** is *constant* and reactant **A** is *varied*.

Gives the **dependence of rate** on the other reactant (**A**)

STEP 3. Use the *rate dependence* on **A** and **B** to write the **rate equation**.

Rate = k [A]^n[B]^m

STEP 4. Substitute the **rate data** for one experiment into the **rate equation** to calculate k.

Value of rate constant k

Check Your Understanding

The initial rate of the reaction of nitrogen monoxide and oxygen

$$2 \text{ NO(g)} + O_2(g) \rightarrow 2 \text{ NO}_2(g)$$

was measured for various initial concentrations of NO and O_2. Determine the rate equation from these data. What is the value of the rate constant, k, and what are its units?

| Experiment | Initial Concentrations (mol/L) | | Initial Rate for [NO] |
	[NO]	[O₂]	(mol NO/L · s)
1	0.020	0.010	0.028
2	0.020	0.020	0.057
3	0.020	0.040	0.114
4	0.040	0.020	0.227
5	0.010	0.020	0.014

Be sure to note that the reaction rate here is defined as

$$\text{Reaction rate} = -(1/2)\Delta[\text{NO}]/\Delta t = -\Delta[O_2]/\Delta t = +(1/2)\Delta[\text{NO}_2]/\Delta t = k[\text{NO}]^m[O_2]^n$$

EXAMPLE 15.4   Using a Rate Equation to Determine Rates

Problem Using the rate equation and rate constant determined for the reaction of CO and NO_2 at 540 K in Example 15.3, determine the initial rate of the reaction when $[CO] = 3.8 \times 10^{-4}$ mol/L and $[NO_2] = 0.650 \times 10^{-4}$ mol/L.

What Do You Know? The rate law and the value for the rate constant (1.9 L/mol · h) are both known. The concentrations of the two reactants are given.

Strategy A rate equation consists of three parts: a rate, a rate constant (k), and the concentration terms. if two of these parts are known (here k and the concentrations), the third can be calculated.

Solution Substitute k (= 1.9 L/mol · h) and the concentration of each reactant into the rate law determined in Example 15.3.

$$\text{Reaction rate} = k[\text{CO}][\text{NO}_2] = (1.9 \text{ L/mol} \cdot \text{h})(3.8 \times 10^{-4} \text{ mol/L})(0.650 \times 10^{-4} \text{ m}$$

$$\text{Reaction rate} = 4.7 \times 10^{-8} \text{ mol/L} \cdot \text{h}$$

Think about Your Answer As a check on the calculated result, it is sometimes useful to make an educated guess at the answer before carrying out the mathematical solution. We know that the reaction here is first order in both reactants. Comparing the concentration values given in this problem with the concentration values in experiment 1 in Example 15.3, we notice that [CO] is about three fourths of the concentration value, whereas [NO₂] is almost twice the value. The effects do not precisely offset each other, but we might predict that the difference in rates between this experiment and experiment 1 will be fairly small, with the rate of this experiment being just a little greater. The calculated value bears this out.

Check Your Understanding

The rate constant, k, at 25 °C is 0.27/h for the reaction

$$\text{Pt(NH}_3)_2\text{Cl}_2(\text{aq}) + H_2O(\ell) \rightarrow [\text{Pt(NH}_3)_2(H_2O)\text{Cl}]^+(\text{aq}) + Cl^-(\text{aq})$$

and the rate equation is

$$\text{Reaction rate} = k[\text{Pt(NH}_3)_2\text{Cl}_2]$$

Calculate the rate of reaction when the concentration of $\text{Pt(NH}_3)_2\text{Cl}_2$ is 0.020 M.

15.4 Concentration–Time Relationships: Integrated Rate Laws

It is often important for a chemist to know how long a reaction must proceed to reach a predetermined concentration of some reactant or product, or what the re-actant and product concentrations will be after some time has elapsed. For this reason it would be useful to have a mathematical equation that relates time and concentration—an equation that describes concentration-time curves like the one shown in Figure 15.2—and that can be used to determine this information. With such an equation, we could calculate the concentration at any given time or the length of time needed for a given amount of reactant to react.

go Chemistry

Module 20: Chemical Kinetics covers concepts in this section.

First-Order Reactions

Suppose the reaction "R → products" is first order. This means the reaction rate is directly proportional to the concentration of R raised to the first power, or, mathematically,

$$-\frac{\Delta[R]}{\Delta t} = k[R]$$

Using calculus, this relationship can be transformed into a very useful equation called an **integrated rate equation** (because integral calculus is used in its derivation).

$$\ln \frac{[R]_t}{[R]_0} = -kt \tag{15.1}$$

Here, $[R]_0$ and $[R]_t$ are concentrations of the reactant at time $t = 0$ and at a later time, t, respectively. The ratio of concentrations, $[R]_t/[R]_0$, is the fraction of reactant that remains after a given time has elapsed. Notice the negative sign in the equation. The ratio $[R]_t/[R]_0$ is less than 1 because $[R]_t$ is always less than $[R]_0$; the reactant R is consumed during the reaction. This means the logarithm of $[R]_t/[R]_0$ is negative, so the other side of the equation must also bear a negative sign.

  Equation 15.1 can be used to carry out the following calculations:

● **Initial and Final Time, t** The time $t = 0$ does not need to correspond to the actual beginning of the experiment. It can be the time when instrument readings were started, for example, even though the reaction may have already begun.

- If $[R]_t/[R]_0$ is measured in the laboratory after some amount of time has elapsed, then k can be calculated.
- If $[R]_0$ and k are known, then the concentration of material remaining after a given amount of time ($[R]_t$) can be calculated.
- If k is known, then the time elapsed until a specific fraction ($[R]_t/[R]_0$) remains can be calculated.

  Finally, notice that k for a first-order reaction is independent of concentration; k has units of time^{-1} (y^{-1} or s^{-1}, for example). This means we can choose any conve-nient unit for $[R]_t$ and $[R]_0$: moles per liter, moles, grams, number of atoms, num-ber of molecules, or gas pressure.

EXAMPLE 15.5   The First-Order Rate Equation

Problem In the past, cyclopropane, C_3H_6, was used in a mixture with oxygen as an anesthetic. (This practice has almost ceased today because the compound is flammable.) When heated, cyclopropane rearranges to propene in a first-order process.

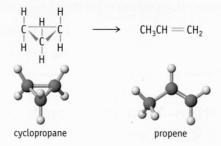

cyclopropane   propene

$$\text{Rate} = k[\text{cyclopropane}] \qquad k = 2.42 \text{ h}^{-1} \text{ at } 500 \text{ °C}$$

If the initial concentration of cyclopropane is 0.050 mol/L, how much time (in hours) must elapse for its concentration to drop to 0.010 mol/L?

What Do You Know? The reaction is first order in cyclopropane. You know the rate constant, k, and the initial and final concentrations of this reactant, $[R]$ and $[R]_0$.

Strategy Use Equation 15.1 to calculate the time (t) elapsed to reach a concentration of 0.010 mol/L.

Solution Values for $[\text{cyclopropane}]_t$, $[\text{cyclopropane}]_0$, and k are substituted into Equation 15.1; t (time) is the unknown:

$$\ln\frac{[0.010]}{[0.050]} = -(2.42 \text{ h}^{-1})t$$

$$t = \frac{-\ln(0.20)}{2.42 \text{ h}^{-1}} = \frac{-(-1.61)}{2.42 \text{ h}^{-1}} = \boxed{0.665 \text{ h}}$$

Think about Your Answer Cycloalkanes with fewer than five carbon atoms are strained because the C—C—C bond angles cannot match the preferred 109.5°. Because of ring strain, the cyclopropane ring opens readily to form propene.

Check Your Understanding

Sucrose, a sugar, decomposes in acid solution to give glucose and fructose. The reaction is first order in sucrose, and the rate constant at 25 °C is $k = 0.21 \text{ h}^{-1}$. If the initial concentration of sucrose is 0.010 mol/L, what is its concentration after 5.0 h?

EXAMPLE 15.6   Using the First-Order Rate Equation

Problem Hydrogen peroxide decomposes in a dilute sodium hydroxide solution at 20 °C in a first-order reaction:

$$H_2O_2(aq) \rightarrow H_2O(\ell) + \tfrac{1}{2} O_2(g)$$
$$\text{Rate} = k[H_2O_2] \text{ with } k = 1.06 \times 10^{-3} \text{ min}^{-1}$$

What is the fraction remaining after 100. min? What is the concentration of H_2O_2 after 100. minutes if the initial concentration of H_2O_2 is 0.020 mol/L??

What Do You Know? This is a first-order reaction. The rate constant k, initial concentration of H_2O_2, and elapsed time are given.

Strategy Because the reaction is first order in H_2O_2, we use Equation 15.1. Here, $[H_2O_2]_0$, k, and t are known, and we are asked to find the value of the fraction remaining. Recall that

$$\frac{[R]_t}{[R]_0} = \text{fraction remaining}$$

Once this value is known, and knowing $[H_2O_2]_0$, we can calculate $[H_2O_2]_t$.

Solution Substitute the known values into Equation 15.1.

$$\ln \frac{[H_2O_2]_t}{[H_2O_2]_0} = -kt = -(1.06 \times 10^{-3} \text{ min}^{-1})(100. \text{ min})$$

$$\ln \frac{[H_2O_2]_t}{[H_2O_2]_0} = -0.106$$

Taking the antilogarithm of -0.106 [i.e., the inverse of the logarithm of -0.106 or $e^{-0.106}$], we find the fraction remaining to be 0.90.

$$\text{Fraction remaining} = \frac{[H_2O_2]_t}{[H_2O_2]_0} = 0.90$$

The calculated fraction remaining is 0.90, thus the concentration of H_2O_2 remaining is 90% of the initial concentration.

$$[H_2O_2]_t = 0.90 \, [H_2O_2]_0 = 0.90 \, (0.020 \text{ mol/L}) = 0.018 \text{ mol/L}$$

Think about Your Answer Although H_2O_2 is unstable, its rate of decomposition is very slow, particularly in a dilute solution. However, sodium hydroxide catalyzes the decomposition. The rate of the reaction can be studied by measuring the volume of O_2 gas evolved as a function of time.

Check Your Understanding

Gaseous azomethane ($CH_3N_2CH_3$) decomposes to ethane and nitrogen when heated:

$$CH_3N_2CH_3(g) \rightarrow CH_3CH_3(g) + N_2(g)$$

The disappearance of azomethane is a first-order reaction with $k = 3.6 \times 10^{-4} \text{ s}^{-1}$ at 600 K.

(a) A sample of gaseous $CH_3N_2CH_3$ is placed in a flask and heated at 600 K for 150 seconds. What fraction of the initial sample remains after this time?

(b) How long must a sample be heated so that 99% of the sample has decomposed?

Second-Order Reactions

Suppose the reaction "R → products" is second order. The rate equation is

$$-\frac{\Delta[R]}{\Delta t} = k[R]^2$$

Using calculus, this relationship can be transformed into the following equation that relates reactant concentration and time:

$$\frac{1}{[R]_t} - \frac{1}{[R]_0} = kt \qquad \qquad \textbf{(15.2)}$$

The same symbolism used with first-order reactions applies: $[R]_0$ is the concentration of reactant at the time $t = 0$; $[R]_t$ is the concentration at a later time; and k is the second-order rate constant, which has the units of L/mol · time.

EXAMPLE 15.7   Using the Second-Order Integrated Rate Equation

Problem The gas-phase decomposition of HI

$$HI(g) \rightarrow \tfrac{1}{2} H_2(g) + \tfrac{1}{2} I_2(g)$$

has the rate equation

$$-\frac{\Delta[HI]}{\Delta t} = k[HI]^2$$

where $k = 30.$ L/mol · min at 443 °C. How much time does it take for the concentration of HI to drop from 0.010 mol/L to 0.0050 mol/L at 443 °C?

What Do You Know? Equation 15.2 is used for a second-order reaction. The rate constant, k, and the initial and final concentrations of HI are given; the elapsed time is the unknown.

Strategy Substitute the values of $[HI]_0$, $[HI]_t$, and k into Equation 15.2, and solve for the unknown, t.

Solution Here, $[HI]_0 = 0.010$ mol/L and $[HI]_t = 0.0050$ mol/L. Using Equation 15.2, we have

$$\frac{1}{0.0050 \text{ mol/L}} - \frac{1}{0.010 \text{ mol/L}} = (30. \text{ L/mol} \cdot \text{min})t$$

$$(2.0 \times 10^2 \text{ L/mol}) - (1.0 \times 10^2 \text{ L/mol}) = (30. \text{ L/mol} \cdot \text{min})t$$

$$t = 3.3 \text{ minutes}$$

Think about Your Answer In the solution of this problem, we kept track of the units for each quantity. This leads to an answer for time elapsed with the unit minutes.

Check Your Understanding

Using the rate constant for HI decomposition given in this Example, calculate the concentration of HI after 12 minutes if $[HI]_0 = 0.010$ mol/L.

Zero-Order Reactions

If a reaction (R → products) is zero order, the rate equation is

$$-\frac{\Delta[R]}{\Delta t} = k[R]^0$$

This equation leads to the integrated rate equation

$$[R]_0 - [R]_t = kt \tag{15.3}$$

where the units of k are mol/L · s.

Graphical Methods for Determining Reaction Order and the Rate Constant

● **Finding the Slope of a Line** See pages 37–38 for a description of methods for finding the slope of a line.

Chemists find it is convenient to determine the order of a reaction and its rate constant using graphical methods. Equations 15.1, 15.2, and 15.3, if rearranged slightly, have the form $y = mx + b$. This is the equation for a straight line, where m is the slope of the line and b is the y-intercept. In these equations, $x = t$ in each case.

Zero order	First order	Second order
$\underset{\substack{\downarrow \quad\;\; \downarrow \;\; \downarrow \\ y \quad\;\; mx \;\; b}}{[R]_t = -kt + [R]_0}$	$\underset{\substack{\downarrow \quad\;\;\; \downarrow \;\;\; \downarrow \\ y \quad\;\;\; mx \;\;\; b}}{\ln [R]_t = -kt + \ln [R]_0}$	$\underset{\substack{\downarrow \quad\;\; \downarrow \;\; \downarrow \\ y \quad\;\; mx \;\; b}}{\frac{1}{[R]_t} = +kt + \frac{1}{[R]_0}}$

As an example of the graphical method for determining reaction order, consider the decomposition of azomethane.

$$CH_3N_2CH_3(g) \rightarrow CH_3CH_3(g) + N_2(g)$$

The decomposition of azomethane was followed at 600 K by observing the decrease in its partial pressure with time (Figure 15.6). (Recall from Chapter 11 that pressure is proportional to concentration at a given temperature and volume.) The third column of the data table lists values of $\ln P(CH_3N_2CH_3)$. As shown in Figure 15.6, a graph of $\ln P(CH_3N_2CH_3)$ versus time produces a straight line, which shows that the reaction is first order in $CH_3N_2CH_3$. The slope of the line can be measured, and the negative of the slope equals the rate constant for the reaction, $3.6 \times 10^{-4} \text{ s}^{-1}$.

t(s)	$P \times 10^2$ atm	ln P
0	8.20	−2.50
1000	5.72	−2.86
2000	3.99	−3.22
3000	2.78	−3.58
4000	1.94	−3.94

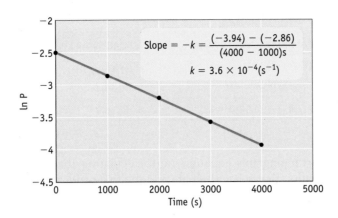

$$\text{Slope} = -k = \frac{(-3.94) - (-2.86)}{(4000 - 1000)\text{s}}$$

$$k = 3.6 \times 10^{-4}(\text{s}^{-1})$$

FIGURE 15.6 The decomposition of azomethane, $CH_3N_2CH_3$. If data for the decomposition of azomethane,

$$CH_3N_2CH_3(g) \rightarrow CH_3CH_3(g) + N_2(g)$$

are plotted as the natural logarithm of the $CH_3N_2CH_3$ pressure versus time, the result is a straight line with a negative slope. This indicates a first-order reaction. The rate constant $k = -$slope.

The decomposition of NO_2 is a second-order process.

$$NO_2(g) \rightarrow NO(g) + \tfrac{1}{2} O_2(g)$$
$$\text{Rate} = k[NO_2]^2$$

This fact can be verified by showing that a plot of $1/[NO_2]$ versus time is a straight line (Figure 15.7). Here, the slope of the line is equal to k.

For a zero-order reaction (Figure 15.8 on page 524), a plot of concentration versus time gives a straight line with a slope equal to the negative of the rate constant.

Table 15.1 (on page 524) summarizes the relationships between concentration and time for first-, second-, and zero-order processes.

Half-Life and First-Order Reactions

The **half-life**, $t_{1/2}$, of a reaction is the time required for the concentration of a reactant to decrease to one half its initial value. Half-life is a convenient way to describe the rate at which a reactant is consumed in a chemical reaction: *The longer the half-life, the slower the reaction.* Half-life is used primarily when dealing with first-order processes.

The half-life, $t_{1/2}$, is the time when the fraction of the reactant R remaining is equal to $1/2$.

$$[R]_t = \tfrac{1}{2}[R]_0 \quad \text{or} \quad \frac{[R]_t}{[R]_0} = \tfrac{1}{2}$$

Here, $[R]_0$ is the initial concentration, and $[R]_t$ is the concentration after the reaction is half completed. To evaluate $t_{1/2}$ for a first-order reaction, we substitute $[R]_t/[R]_0 = \tfrac{1}{2}$ and $t = t_{1/2}$ into the integrated first-order rate equation (Equation 15.1),

$$\ln (\tfrac{1}{2}) = -kt_{1/2} \quad \text{or} \quad \ln 2 = kt_{1/2}$$

Time (min)	$[NO_2]$ (mol/L)	$1/[NO_2]$ (L/mol)
0	0.020	50
0.50	0.015	67
1.0	0.012	83
1.5	0.010	100
2.0	0.0087	115

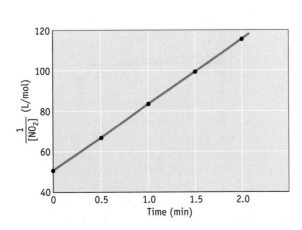

FIGURE 15.7 A second-order reaction. A plot of $1/[NO_2]$ versus time for the decomposition of NO_2,

$$NO_2(g) \rightarrow NO(g) + \tfrac{1}{2} O_2(g)$$

results in a straight line. This confirms this is a second-order reaction. The slope of the line equals the rate constant for this reaction.

Table 15.1 Characteristic Properties of Reactions of the Type "R ⟶ Products"

Order	Rate Equation	Integrated Rate Equation	Straight-Line Plot	Slope	k Units
0	$-\Delta[R]/\Delta t = k[R]^0$	$[R]_0 - [R]_t = kt$	$[R]_t$ vs. t	$-k$	mol/L · time
1	$-\Delta[R]/\Delta t = k[R]^1$	$\ln([R]_t/[R]_0) = -kt$	$\ln[R]_t$ vs. t	$-k$	1/time
2	$-\Delta[R]/\Delta t = k[R]^2$	$(1/[R]_t) - (1/[R]_0) = kt$	$1/[R]_t$ vs. t	k	L/mol · time

● **Half-Life and Radioactivity**
Half-life is a term often encountered when dealing with radioactive elements. Radioactive decay is a first-order process, and half-life is commonly used to describe how rapidly a radioactive element decays. See Chapter 23 and Example 15.9.

Rearranging this equation (and calculating that $\ln 2 = 0.693$) provides a useful equation that relates half-life and the first-order rate constant:

$$t_{1/2} = \frac{0.693}{k} \qquad (15.4)$$

This equation identifies an important feature of first-order reactions: $t_{1/2}$ *is independent of concentration.*

To illustrate the concept of half-life, consider again the first-order decomposition of azomethane, $CH_3N_2CH_3$.

$$CH_3N_2CH_3(g) \rightarrow CH_3CH_3(g) + N_2(g)$$

$$\text{Rate} = k[CH_3N_2CH_3] \text{ with } k = 3.6 \times 10^{-4} \text{ s}^{-1} \text{ at } 600 \text{ K}$$

● **Half-Life Equations for Other Reaction Orders**
For a zero-order reaction, R → products

$$t_{1/2} = \frac{[R]_0}{2k}$$

For a second-order reaction, R → products

$$t_{1/2} = \frac{1}{k[R]_0}$$

Note that in both cases the half-life depends on the initial concentration.

Given a rate constant of $3.6 \times 10^{-4} \text{ s}^{-1}$, we calculate a half-life of 1.9×10^3 s or 32 minutes.

$$t_{1/2} = \frac{0.693}{3.6 \times 10^{-4} \text{ s}^{-1}} = 1.9 \times 10^3 \text{ s (or 32 min)}$$

The partial pressure of azomethane has been plotted as a function of time in Figure 15.9, and this graph shows that P(azomethane) decreases by half every 32 minutes. The initial pressure was 820 mm Hg, but it dropped to 410 mm Hg in 32 minutes, and then dropped to 205 mm Hg in another 32 minutes. That is, after two half-lives (64 minutes), the pressure is $(\frac{1}{2}) \times (\frac{1}{2}) = (\frac{1}{2})^2 = \frac{1}{4}$ or 25% of the initial pressure. After three half-lives, the pressure has dropped further to 102 mm Hg or 12.5% of the initial value and is equal to $(\frac{1}{2}) \times (\frac{1}{2}) \times (\frac{1}{2}) = (\frac{1}{2})^3 = \frac{1}{8}$ of the initial value.

It can be hard to visualize whether a reaction is fast or slow from the rate constant value. Can you tell from the rate constant, $k = 3.6 \times 10^{-4} \text{ s}^{-1}$, whether the

FIGURE 15.8 Plot of a zero-order reaction. A graph of the concentration of ammonia, $[NH_3]_t$, against time for the decomposition of NH_3.

$$NH_3(g) \rightarrow \tfrac{1}{2} N_2(g) + \tfrac{3}{2} H_2(g)$$

on a metal surface at 856 °C is a straight line, indicating that this is a zero-order reaction. The rate constant, k, for this reaction is found from the slope of the line; $k = -$slope. (The points chosen to calculate the slope are given in red.)

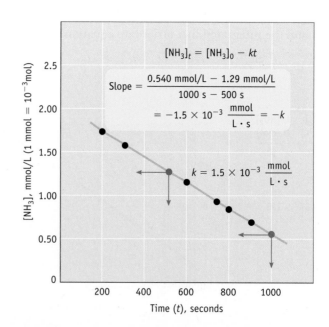

$$[NH_3]_t = [NH_3]_0 - kt$$

$$\text{Slope} = \frac{0.540 \text{ mmol/L} - 1.29 \text{ mmol/L}}{1000 \text{ s} - 500 \text{ s}}$$

$$= -1.5 \times 10^{-3} \frac{\text{mmol}}{\text{L} \cdot \text{s}} = -k$$

$$k = 1.5 \times 10^{-3} \frac{\text{mmol}}{\text{L} \cdot \text{s}}$$

$[NH_3]$, mmol/L (1 mmol = 10^{-3}mol)

Time (t), seconds

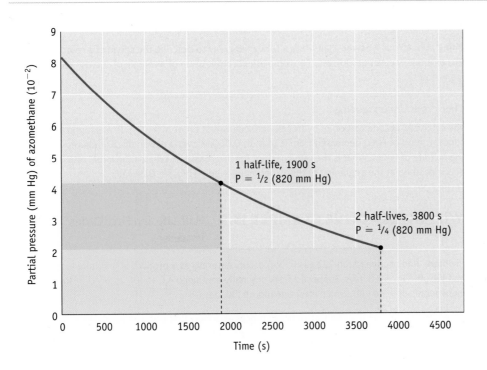

FIGURE 15.9 Half-life of a first-order reaction. The curve is a plot of the pressure of $CH_3N_2CH_3$ as a function of time. (The compound decomposes to CH_3CH_3 and N_2 with $k = 3.6 \times 10^{-4}\ s^{-1}$). The pressure of $CH_3N_2CH_3$ is halved every 1900 seconds (32 minutes). (This plot of pressure versus time is similar in shape to plots of concentration versus time for all other first-order reactions.)

In-figure labels:

1 half-life, 1900 s
P = ½ (820 mm Hg)

2 half-lives, 3800 s
P = ¼ (820 mm Hg)

y-axis: Partial pressure (mm Hg) of azomethane (10^{-2})
x-axis: Time (s)

azomethane decomposition will take seconds, hours, or days to reach completion? Probably not, but this can be assessed from the value of the half-life for the reaction (32 minutes). Now we know that we would only have to wait a few hours for the reactant to be essentially consumed.

EXAMPLE 15.8   Half-Life and a First-Order Process

Problem Sucrose, $C_{12}H_{22}O_{11}$, decomposes to fructose and glucose in acid solution with the rate law

$$\text{Rate} = k[C_{12}H_{22}O_{11}] \qquad k = 0.216\ h^{-1} \text{ at } 25\ °C$$

(a)   What is the half-life of $C_{12}H_{22}O_{11}$ at this temperature?

(b)   What amount of time is required for 87.5% of the initial concentration of $C_{12}H_{22}O_{11}$ to decompose?

What Do You Know? The decomposition of sucrose is a first-order reaction. The rate constant for the reaction (at 25 °C) is given.

Strategy (a) Use Equation 15.4 to calculate the half-life from the rate constant. (b) After 87.5% of the $C_{12}H_{22}O_{11}$ has decomposed, 12.5% (or one eighth of the sample) remains. To reach this point, three half-lives are required.

Half-Life	Fraction Remaining
1	0.5
2	0.25
3	0.125

Therefore, we multiply the half-life calculated in part (a) by 3.

Solution

(a)   The half-life for the reaction is

$$t_{1/2} = 0.693/k = 0.693/(0.216\ h^{-1}) = 3.21 \text{ hours}$$

(b)   Three half-lives must elapse before the fraction remaining is 0.125, so

$$\text{Time elapsed} = 3 \times 3.21\ h = \boxed{9.63 \text{ hours}}$$

Think about Your Answer Half-life is a convenient way to describe the speed of a reaction. In this example, we can quickly see that complete decomposition of a sample of sucrose requires many hours.

Check Your Understanding

The catalyzed decomposition of hydrogen peroxide is first order in $[H_2O_2]$. It was found that the concentration of H_2O_2 decreased from 0.24 M to 0.060 M over a period of 282 minutes. What is the half-life of H_2O_2? What is the rate constant for this reaction? What is the initial rate of decomposition at the beginning of this experiment (when $[H_2O_2] = 0.24$ M)?

Strategy Map 15.9

PROBLEM

Find the **final number of atoms** of a radioactive isotope after a **given time period.**

↓

DATA/INFORMATION

- Half-life for decay
- Initial number of atoms
- Time elapsed

STEP 1. Use Equation 15.4 to calculate k from $t_{1/2}$.

Value of k

STEP 2. Use Equation 15.1 where k, $[R]_0$, and t are known to calculate $[R]_t$.

Value of $[R]_t$, the **final number of atoms** of radioactive isotope after time t.

⊙WL   INTERACTIVE EXAMPLE 15.9   Half-Life and First-Order Processes

Problem Radioactive radon-222 gas (^{222}Rn) occurs naturally as a product of uranium decay. The half-life of ^{222}Rn is 3.8 days. Suppose a flask originally contained 4.0×10^{13} atoms of ^{222}Rn. How many atoms of ^{222}Rn will remain after one month (30. days)?

What Do You Know? All radioactive decay processes follow first-order kinetics. The half-life of ^{222}Rn and the number of atoms initially present are known.

Strategy First, the rate constant, k, must be found from the half-life using Equation 15.4. Then, using Equation 15.1, and knowing the number of atoms at the beginning ($[R]_0$), the elapsed time (30. days), and the rate constant, we can calculate the number of atoms remaining ($[R]_t$).

Solution The rate constant, k, is

$$k = \frac{0.693}{t_{1/2}} = \frac{0.693}{3.8\ \text{d}} = 0.18\ \text{d}^{-1}$$

Now use Equation 15.1 to calculate the number of atoms remaining after 30. days.

$$\ln \frac{[Rn]_t}{4.0 \times 10^{13}\ \text{atoms}} = -(0.18\ \text{d}^{-1})(30.\ \text{d}) = -5.5$$

$$\frac{[Rn]_t}{4.0 \times 10^{13}\ \text{atoms}} = e^{-5.5} = 0.0042$$

$$[Rn]_t = 2 \times 10^{11}\ \text{atoms}$$

Think about Your Answer Thirty days is approximately 8 half-lives for this element. This means that the number of atoms present at the end of the month is approximately $(1/2)^8$ or $1/256$th of the original number.

Check Your Understanding

Americium is used in smoke detectors and in medicine for the treatment of certain malignancies. One isotope of americium, ^{241}Am, has a rate constant, k, for radioactive decay of 0.0016 y^{-1}. In contrast, radioactive iodine-125, which is used for studies of thyroid functioning, has a rate constant for decay of 0.011 d^{-1}.

(a)   What are the half-lives of these isotopes?

(b)   Which isotope decays faster?

(c)   If you are given a dose of iodine-125 containing 1.6×10^{15} atoms, how many atoms remain after 2.0 days?

REVIEW & CHECK FOR SECTION 15.4

1.   The decomposition of N_2O_5 is a first-order process. How many half-lives would be required to decompose 99% of the sample?

(a)   7   (b)   between 6 and 7   (c)   5   (d)   between 5 and 6

2. Which of the following will confirm that the decomposition of SO_2Cl_2 (to form SO_2 and Cl_2) is a first-order process?

   (a) A graph of $[SO_2Cl_2]$ vs. time gives a curved line.

   (b) A graph of $\ln[SO_2Cl_2]$ vs. time gives a curved line.

   (c) A graph of $1/[SO_2Cl_2]$ vs time gives a straight line.

   (d) A graph of $\ln[SO_2Cl_2]$ vs. time gives a straight line.

3. The equation for the decomposition of $NO_2(g)$ at 573 K is $2\,NO_2(g) \rightarrow 2\,NO(g) + O_2(g)$. Using the concentration-time data below, determine the order of the reaction with respect to $[NO_2]$.

$[NO_2]$, M	time, min
0.20	0
0.095	5
0.063	10
0.047	15

   (a) first order   (b) second order   (c) zero order

15.5 A Microscopic View of Reaction Rates

Throughout this book, we have turned to the particulate level of chemistry to understand chemical phenomena. Rates of reaction are no exception. Looking at the way reactions occur at the atomic and molecular levels can give us some insight into the various influences on rates of reactions.

Let us review the macroscopic observations we have made so far concerning reaction rates. We know that there are wide differences in rates of reactions—from very fast reactions like the explosion that occurs when hydrogen and oxygen are exposed to a spark or flame (◀ Figure 1.16), to slow reactions like the formation of rust that occurs over days, weeks, or years. For a specific reaction, factors that influence reaction rate include the concentrations of the reactants, the temperature of the reaction system, and the presence of catalysts. In this section we want to look at each of these influences in more depth in the context of the **collision theory of reaction rates**, which states that three conditions must be met for a reaction to occur:

1. The reacting molecules must collide with one another.
2. The reacting molecules must collide with sufficient energy to initiate the process of breaking and forming bonds.
3. The molecules must collide in an orientation that can lead to rearrangement of the atoms and the formation of products.

Collision Theory: Concentration and Reaction Rate

Consider the gas-phase reaction of nitric oxide and ozone, an environmentally important reaction:

$$NO(g) + O_3(g) \rightarrow NO_2(g) + O_2(g)$$

The rate law for this product-favored reaction is first order in each reactant: Rate $= k[NO][O_3]$. How can this reaction have this rate law?

Let us consider the reaction at the particulate level and imagine a flask containing a mixture of NO and O_3 molecules in the gas phase. Both kinds of molecules are in rapid and random motion within the flask. They strike the walls of the vessel, but they also collide with other molecules. Thus, it is reasonable to propose that the rate of their reaction should be related to the number of collisions, which is in turn related to their concentrations (Figure 15.10). Doubling the concentration of one reagent in the NO + O_3 reaction, say NO, will lead to twice the number of

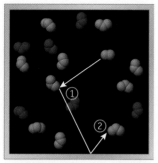

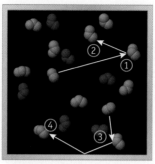

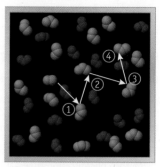

(a) **1 NO : 16 O₃** — *2 hits/second.*
A single NO molecule, moving among sixteen O₃ molecules, is shown colliding with two of them per second.

(b) **2 NO : 16 O₃** — *4 hits/second.*
If two NO molecules move among 16 O₃ molecules, we would predict that four NO—O₃ collisions would occur per second.

(c) **1 NO : 32 O₃** — *4 hits/second.*
If the number of O₃ molecules is doubled (to 32), the frequency of NO—O₃ collisions is also doubled, to four per second.

FIGURE 15.10 The effect of concentration on the frequency of molecular collisions.

molecular collisions. Figure 15.10a shows a single molecule of one of the reactants (NO) moving randomly among sixteen O_3 molecules. In a given time period, it might collide with two O_3 molecules. The number of NO—O_3 collisions will double, however, if the concentration of NO molecules is doubled (to 2, as shown in Figure 15.10b) or if the number of O_3 molecules is doubled (to 32, as in Figure 15.10c). Thus, we can explain the dependence of reaction rate on concentration: The number of collisions between the two reactant molecules is directly proportional to the concentration of each reactant, and the rate of the reaction shows a first-order dependence on each reactant.

Collision Theory: Temperature and Reaction Rate

In a laboratory or in the chemical industry, a chemical reaction is often carried out at elevated temperatures because this allows the reaction to occur more rapidly. Conversely, it is sometimes desirable to lower the temperature to slow down a chemical reaction (to avoid an uncontrollable reaction or a potentially dangerous explosion). Chemists are very aware of the effect of temperature on the rate of a reaction.

A discussion of the effect of temperature on reaction rate begins with reference to distribution of energies for molecules in a sample of a gas or liquid. Recall from studying gases and liquids that the molecules in a sample have a wide range of energies, described earlier as a Boltzmann distribution of energies (◄ Figure 11.14). In any sample of a gas or liquid, some molecules have very low energies; others have very high energies; but most have some intermediate energy. As the temperature increases, the average energy of the molecules in the sample increases, as does the fraction having higher energies (Figure 15.11).

Collision Theory: Activation Energy

Molecules require some minimum energy to react. Chemists visualize this as an energy barrier that must be surmounted by the reactants for a reaction to occur (Figure 15.12). The energy required to surmount the barrier is called the **activation energy**, E_a. If the barrier is low, the energy required is low, and a high proportion of the molecules in a sample may have sufficient energy to react. In such a case, the reaction will be fast. If the barrier is high, the activation energy is high, and only a few reactant molecules in a sample may have sufficient energy to react. In this case, the reaction will be slow.

To illustrate an activation energy barrier, consider the high temperature conversion of NO_2 and CO to NO and CO_2 or the reverse reaction (Figure 15.13). At the molecular level, we imagine that the reaction involves the transfer of an O atom

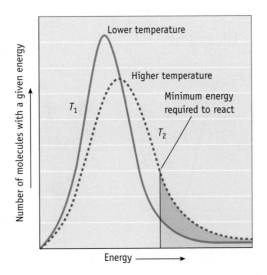

FIGURE 15.11 Energy distribution curve. The vertical axis gives the relative number of molecules possessing the energy indicated on the horizontal axis. The graph indicates the minimum energy required for an arbitrary reaction. At a higher temperature, a larger fraction of the molecules have sufficient energy to react. (Recall Figure 11.14, the Boltzmann distribution function, for a collection of gas molecules.)

from an NO_2 molecule to a CO molecule (or, in the reverse reaction, the transfer of an O atom from CO_2 to NO).

$$NO_2(g) + CO(g) \rightleftharpoons NO(g) + CO_2(g)$$

We can describe this process by using an energy diagram or **reaction coordinate** diagram. The horizontal axis describes the reaction progress as the reaction proceeds, and the vertical axis represents the potential energy of the system during the reaction. When NO_2 and CO approach and O atom transfer begins, an N—O bond is being broken, and a C=O bond is forming. Energy input (the activation energy) is required for this to occur. The energy of the system reaches a maximum at the **transition state**. At the transition state, sufficient energy has been concentrated in the appropriate bonds; bonds in the reactants can now break, and new bonds can form to give products. The system is poised to go on to products. Alternatively, it can return to the reactants. Because the transition state is at a maximum in potential energy, it cannot be isolated. Using computer molecular modeling techniques, however, chemists can describe what the transition state must look like.

In the NO_2 + CO reaction, 132 kJ/mol is required to reach the transition state; that is, the top of the energy barrier. As the reaction continues toward the products—as the N—O bond is finally broken and a C=O bond forms—the reaction evolves energy, 358 kJ/mol. The net energy change involved in this exothermic reaction is −226 kJ/mol.

$$\Delta U = +132 \text{ kJ/mol} + (-358 \text{ kJ/mol}) = -226 \text{ kJ/mol}$$

FIGURE 15.12 An analogy to chemical activation energy. For the volleyball to go over the net, the player must give it sufficient energy.

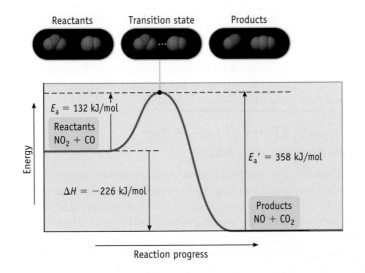

FIGURE 15.13 Activation Energy. The reaction of NO_2 and CO (to give NO and CO_2) has an activation energy barrier of 132 kJ/mol. The reverse reaction (NO + CO_2 → NO_2 + CO) has an activation energy barrier of 358 kJ/mol. The net energy change for the reaction of NO_2 and CO is −226 kJ/mol.

Reaction Coordinate Diagrams

Reaction coordinate diagrams (Figure 15.13) convey a great deal of information. A reaction that would have an energy diagram like that in Figure 15.13 is the substitution of a halogen atom of CH_3Cl by an ion such as F^-. Here, the F^- ion attacks the molecule from the side opposite the Cl substituent. As F^- begins to form a bond to carbon, the C—Cl bond weakens, and the CH_3 portion of the molecule changes shape. As time progresses, the products CH_3F and Cl^- are formed.

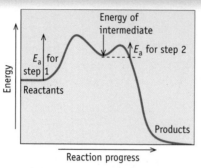

FIGURE A **A reaction coordinate diagram for a two-step reaction, a process involving an intermediate.**

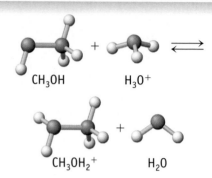

The diagram in Figure A describes a reaction that occurs in two steps. An example of such a reaction is the substitution of the —OH group on methanol by a halide ion in the presence of acid. In the first step, an H^+ ion attaches to the O of the C—O—H group in a rapid, reversible reaction. The energy of this protonated species, $CH_3OH_2^+$, a reaction intermediate, is higher than the energies of the reactants and is represented by the dip in the curve shown in Figure A. In the second step, a halide ion, say Br^-, attacks the intermediate to produce methyl bromide, CH_3Br, and water. There is an activation energy barrier in both the first step and second step.

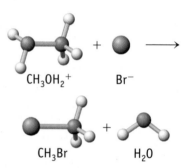

Notice in Figure A, as in Figure 15.13, that the energy of the products is lower than the energy of the reactants. The reaction is exothermic.

What happens if NO and CO_2 are mixed to form NO_2 and CO? Now the reaction requires 358 kJ/mol-rxn to reach the transition state, and 132 kJ/mol-rxn is evolved on proceeding to the product, NO_2 and CO. The reaction in this direction is endothermic, requiring a net input of +226 kJ/mol-rxn.

Collision Theory: Activation Energy and Temperature

The conversion of NO_2 and CO to products at room temperature is slow because only a small fraction of the molecules have enough energy to reach the transition state. The rate can be increased by heating the sample. Raising the temperature increases the reaction rate by increasing the fraction of molecules with enough energy to surmount the activation energy barrier (Figure 15.11).

Collision Theory: Effect of Molecular Orientation on Reaction Rate

Having a sufficiently high energy is necessary, but it is not sufficient to ensure that a given collision of reactants will form products. The reactant molecules must also come together in the correct orientation. For the reaction of NO_2 and CO, we can imagine that the transition state structure has one of the O atoms of NO_2 beginning to bind to the C atom of CO in preparation for O atom transfer (Figure 15.13). The lower the probability of achieving the proper alignment, the smaller the value of k, and the slower the reaction.

Imagine what happens when two or more complicated molecules collide. In only a small fraction of the collisions will the molecules come together in exactly the right orientation. Thus, only a tiny fraction of the collisions can be effective. No wonder some reactions are slow. Conversely, it is amazing that so many are fast!

The Arrhenius Equation

The observation that reaction rates depend on the energy and frequency of collisions between reacting molecules, on the temperature, and on whether the collisions have the correct geometry is summarized by the **Arrhenius equation**:

$$k = \text{rate constant} = A e^{-E_a/RT}$$

Frequency factor ⟋   ⟍ Fraction of molecules with minimum
                               energy for reaction   **(15.5)**

In this equation, k is the rate constant, R is the gas constant with a value of 8.314510×10^{-3} kJ/K·mol, and T is the kelvin temperature. The parameter A is called the **frequency factor**. It is related to the number of collisions and to the fraction of collisions that have the correct geometry; A is specific to each reaction and is temperature dependent. The factor $e^{-E_a/RT}$ represents *the fraction of molecules having the minimum energy required for reaction*; its value is always less than 1. As the table in the margin shows, this fraction changes significantly with temperature.

The Arrhenius equation is significant because

- it can be used to calculate E_a from the temperature dependence of the rate constant.
- it can be used to calculate the rate constant for a given temperature, if E_a and A are known.

If rate constants of a given reaction are measured at several temperatures then one can apply graphical techniques to determine the activation energy of a reaction. Taking the natural logarithm of each side of Equation 15.5, we have

$$\ln k = \ln A + \left(-\frac{E_a}{RT}\right)$$

Rearranging this expression slightly shows that $\ln k$ and $1/T$ are related linearly.

$$\ln k = -\frac{E_a}{R}\left(\frac{1}{T}\right) + \ln A \quad \leftarrow \text{Arrhenius equation} \qquad \textbf{(15.6)}$$

$$\downarrow \qquad\quad \downarrow \qquad\quad \downarrow$$
$$y \;\; = \;\; mx \;\;\; + b \quad \leftarrow \text{Equation for straight line}$$

This means that, if the natural logarithm of k ($\ln k$) is plotted versus $1/T$, the result is a downward-sloping straight line with a slope of $(-E_a/R)$. The activation energy, E_a, can be obtained from the slope of this line ($E_a = -R \times \text{slope}$).

> ● E_a, **Reaction Rates, and Temperature** An often-used rule of thumb is that reaction rates double for every 10 °C rise in temperature in the vicinity of room temperature.

> ● **Interpreting the Arrhenius Equation** (a) The exponential term. This gives the fraction of molecules having sufficient energy for reaction and is a function of T.

Temperature (K)	Value of $e^{-E_a/RT}$ for $E_a =$ 40 kJ/mol-rxn
298	9.7×10^{-8}
400	5.9×10^{-6}
600	3.3×10^{-4}

> (b) Significance of A. Although a complete understanding of A goes beyond the level of this text, it can be noted that A becomes smaller as the reactants become larger. It reflects the fact that larger molecules have a smaller probability of coming together in the appropriate geometry.

EXAMPLE 15.10   Determination of E_a from the Arrhenius Equation

Problem Using the experimental data shown in the table, calculate the activation energy E_a for the reaction

$$2\,N_2O(g) \rightarrow 2\,N_2(g) + O_2(g)$$

Experiment	Temperature (K)	k (L/mol · s)
1	1125	11.59
2	1053	1.67
3	1001	0.380
4	838	0.0011

What Do You Know? The rate constants are given at several temperatures.

Strategy To solve this problem graphically, we first need to calculate ln k and $1/T$ for each data point. These data are then plotted, and E_a is calculated from the resulting straight line (slope $= -E_a/R$).

Solution First, calculate $1/T$ and ln k.

Experiment	$1/T$ (K^{-1})	ln k
1	8.889×10^{-4}	2.4501
2	9.497×10^{-4}	0.513
3	9.990×10^{-4}	-0.968
4	11.9×10^{-4}	-6.81

Plotting these data gives the graph shown in Figure 15.14. Choosing the large blue points on the graph, the slope is found to be

$$\text{Slope} = \frac{\Delta \ln k}{\Delta(1/T)} = \frac{2.0 - (-5.6)}{(9.0 - 11.5)(10^{-4})K^{-1}} = -3.0 \times 10^4 \text{ K}$$

The activation energy is evaluated from the slope.

$$\text{Slope} = -\frac{E_a}{R} = -\frac{E_a}{8.31 \times 10^{-3} \text{ kJ/K} \cdot \text{mol}} = -3.0 \times 10^4 \text{ K}$$

$$E_a = 250 \text{ kJ/mol}$$

Think about Your Answer Notice the units of the answer. Using the value for $R = 8.31 \times 10^{-3}$ kJ/mol will give an answer with the units kJ/mol.

Check Your Understanding

Rate constants were determined for the decomposition of acetaldehyde (CH$_3$CHO) in the temperature range 700 to 1000 K. Use these data to determine E_a for the reaction using a graphical method.

T (K)	700	760	840	1000
k (L/mol · s)	0.011	0.105	2.17	145

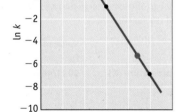

FIGURE 15.14 Arrhenius plot. A plot of ln k versus $1/T$ for the reaction 2 N$_2$O(g) → 2 N$_2$(g) + O$_2$(g). The slope of the line can be used to calculate E_a. See Example 15.10.

The activation energy, E_a, for a reaction can be obtained algebraically if k is known at two different temperatures. We can write an equation for each set of these conditions:

$$\ln k_1 = -\left(\frac{E_a}{RT_1}\right) + \ln A \quad \text{and} \quad \ln k_2 = -\left(\frac{E_a}{RT_2}\right) + \ln A$$

If one of these equations is subtracted from the other, we have

$$\ln k_2 - \ln k_1 = \ln \frac{k_2}{k_1} = -\frac{E_a}{R}\left[\frac{1}{T_2} - \frac{1}{T_1}\right] \tag{15.7}$$

Example 15.11 demonstrates the use of this equation.

EXAMPLE 15.11   Calculating E$_a$ Numerically

Problem Use values of k determined at two different temperatures to calculate the value of E_a for the decomposition of HI:

$$2 \text{ HI(g)} \rightarrow \text{H}_2\text{(g)} + \text{I}_2\text{(g)}$$

$$k_1 = 2.15 \times 10^{-8} \text{ L/(mol} \cdot \text{s) at } 6.50 \times 10^2 \text{ K } (T_1)$$

$$k_2 = 2.39 \times 10^{-7} \text{ L/(mol} \cdot \text{s) at } 7.00 \times 10^2 \text{ K } (T_2)$$

What Do You Know? Values for the rate constant at two temperatures are given.

Strategy Substitute values of k_1, k_2, T_1, and T_2 into Equation 15.7 and solve for E_a.

Solution

$$\ln \frac{2.39 \times 10^{-7} \text{ L/(mol·s)}}{2.15 \times 10^{-8} \text{ L/(mol·s)}} = -\frac{E_a}{8.315 \times 10^{-3} \text{ kJ/K·mol}} \times \left[\frac{1}{7.00 \times 10^2 \text{ K}} - \frac{1}{6.50 \times 10^2 \text{ K}} \right]$$

Solving this equation gives $E_a = 180$ kJ/mol.

Think about Your Answer Another way to write the difference in fractions in brackets is

$$\left[\frac{1}{T_2} - \frac{1}{T_1} \right] = \frac{T_1 - T_2}{T_1 T_2}$$

This expression is sometimes easier to use.

Check Your Understanding

The colorless gas N_2O_4 decomposes to the brown gas NO_2 in a first-order reaction.

$$N_2O_4(g) \rightarrow 2 \ NO_2(g)$$

The rate constant $k = 4.5 \times 10^3$ s^{-1} at 274 K and $k = 1.00 \times 10^4$ s^{-1} at 283 K. What is the activation energy, E_a?

Effect of Catalysts on Reaction Rate

Catalysts are substances that speed up the rate of a chemical reaction. We have seen several examples of catalysts in earlier discussions in this chapter: MnO_2, iodide ion, the enzyme catalase in a potato, and hydroxide ion all catalyze the decomposition of hydrogen peroxide (Figure 15.4).

Catalysts are not consumed in a chemical reaction. They are, however, intimately involved in the details of the reaction at the particulate level. Their function is to provide a different pathway with a lower activation energy for the reaction. To illustrate how a catalyst participates in a reaction, let us consider the isomerization of *cis*-2-butene, to the slightly more stable isomer, *trans*-2-butene.

● **Enzymes: Biological Catalysts** Catalase is an enzyme whose function is to speed up the decomposition of hydrogen peroxide. This enzyme ensures that hydrogen peroxide, which is highly toxic, does not build up in the body.

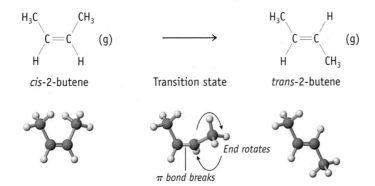

cis-2-butene   Transition state   *trans*-2-butene

End rotates

π bond breaks

● **Butene Isomerization** Isomerization of *cis*-2-butene is a first-order process with the rate law Rate = k[*cis*-2-butene]. It is suggested to occur by rotation around the carbon–carbon double bond. The rate at which a molecule will isomerize is related to the fraction of molecules that have a high enough energy.

The activation energy for the uncatalyzed conversion is relatively large— 264 kJ/mol—because the π bond must be broken to allow one end of the molecule to rotate into a new position. Because of the high activation energy, this is a slow reaction, and rather high temperatures are required for it to occur at a reasonable rate.

The *cis*- to *trans*-2-butene reaction is greatly accelerated by a catalyst, iodine. In the presence of iodine, this reaction can be carried out at a temperature several hundred degrees lower than for the uncatalyzed reaction. Iodine is not consumed (nor is it a product), and it does not appear in the overall balanced equation. It does appear in the reaction rate law, however; the rate of the reaction depends on the square root of the iodine concentration:

$$\text{Rate} = k[\textit{cis}\text{-2-butene}][I_2]^{1/2}$$

The presence of I_2 changes the way the reaction occurs; that is, it changes the mechanism of the reaction (Figure 15.15). The best hypothesis is that iodine molecules first dissociate to form iodine atoms (Step 1). An iodine atom then adds to one of the C atoms of the C=C double bond (Step 2). This converts the double bond between the carbon atoms to a single bond (the π bond is broken) and allows the ends of the molecule to twist freely relative to each other (Step 3). If the iodine atom then dissociates from the intermediate, the double bond can re-form in the *trans* configuration (Step 4). The iodine atom catalyzing the rotation is now free to add to another molecule of *cis*-2-butene. The result is a kind of chain reaction, as one molecule of *cis*-2-butene after another is converted to the *trans* isomer. The chain is broken if the iodine atom recombines with another iodine atom to re-form molecular iodine.

An energy profile for the catalyzed reaction (Figure 15.16) shows that the overall energy barrier is much lower than for the uncatalyzed reaction. Five separate steps are identified for the mechanism in the energy profile. This proposed mechanism also includes a series of chemical species called **reaction intermediates**, species formed in one step of the reaction and consumed in a later step. Iodine atoms are intermediates, as are the free radical species formed when an iodine atom adds to *cis*-2-butene.

Five important points are associated with this mechanism:

- Iodine molecules, I_2, dissociate to atoms and then re-form. On the macroscopic level, the concentration of I_2 is unchanged. Iodine does not appear in the balanced, stoichiometric equation even though it appears in the rate equation. This is generally true of catalysts.
- Both the catalyst I_2 and the reactant *cis*-2-butene are in the gas phase. If a catalyst is present in the same phase as the reacting substance, it is called a *homogeneous catalyst*.
- Iodine atoms and the radical species formed by addition of an iodine atom to a 2-butene molecule are intermediates.

FIGURE 15.15 The mechanism of the iodine-catalyzed isomerization of *cis*-2-butene. *Cis*-2-butene is converted to *trans*-2-butene in the presence of a catalytic amount of iodine. Catalyzed reactions are often pictured in such diagrams to emphasize what chemists refer to as a "catalytic cycle."

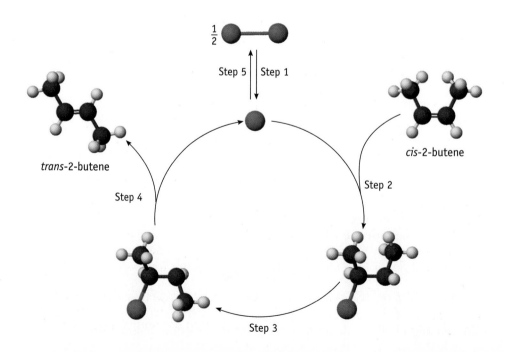

trans-2-butene

cis-2-butene

Step 5 | Step 1

Step 2

Step 4

Step 3

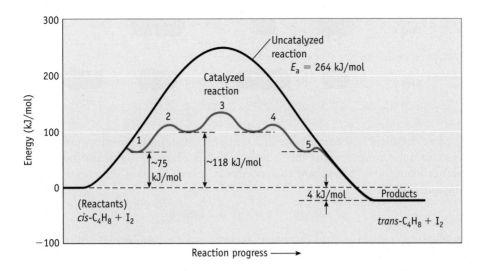

FIGURE 15.16 Energy profile for the iodine-catalyzed reaction of *cis*-2-butene. A catalyst accelerates a reaction by altering the mechanism so that the activation energy is lowered. With a smaller barrier to overcome, more reacting molecules have sufficient energy to surmount the barrier, and the reaction occurs more rapidly. The energy profile for the uncatalyzed conversion of *cis*-2-butene to *trans*-2-butene is shown by the black curve, and that for the iodine-catalyzed reaction is represented by the red curve. Notice that the shape of the barrier has changed because the mechanism has changed.

- The activation energy barrier to reaction is significantly lower because the mechanism changed. Dropping the activation energy from 264 kJ/mol for the uncatalyzed reaction to about 150 kJ/mol for the catalyzed process makes the catalyzed reaction 10^{15} times faster!
- The diagram of energy-versus-reaction progress has five energy barriers (five humps appear in the curve). This feature in the diagram means that the reaction occurs in a series of five steps.

What we have described here are two different reaction mechanisms for the isomerization of *cis*-2-butene. The uncatalyzed isomerization reaction is a one-step reaction mechanism, whereas the catalyzed mechanism involves a series of steps. We shall discuss reaction mechanisms in more detail in the next section.

REVIEW & CHECK FOR SECTION 15.5

1. Which of the following graphs will produce a straight line?

    (a) k vs. T (b) $\ln E_a$ vs. T (c) $\ln k$ vs. $1/T$ (d) E_a vs. $1/T$

2. What is the primary reason that increasing the temperature causes an increase in reaction rate?

    (a) An increase in T lowers the activation energy.

    (b) A higher proportion of reactant molecules exceeds the activation energy.

    (c) In a closed vessel, increasing the temperature results in a higher pressure.

    (d) Increasing the temperature provides other reaction pathways.

15.6 Reaction Mechanisms

Rate laws help us understand **reaction mechanisms**, the sequence of bond-making and bond-breaking steps that occurs during the conversion of reactants to products. We want to analyze the changes that molecules undergo when they react. We then want to relate this description back to the macroscopic world, to the experimental observations of reaction rates.

Based on the rate equation for a reaction, and by applying chemical intuition, chemists can often make an educated guess about the mechanism for a reaction. In some reactions, the conversion of reactants to products in a single step is envisioned as the logical mechanism. For example, the uncatalyzed isomerization of *cis*-2-butene to *trans*-2-butene is best described as a single-step reaction (Figure 15.16).

FIGURE 15.17 A reaction mechanism. A representation of the proposed two-step mechanism by which NO and Br_2 are converted to NOBr.

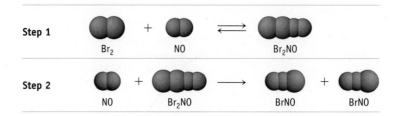

Most chemical reactions occur in a sequence of steps, however. A multiple-step mechanism was proposed for the iodine-catalyzed 2-butene isomerization reaction. Another example of a reaction that occurs in several steps is the reaction of bromine and NO:

$$Br_2(g) + 2\ NO(g) \rightarrow 2\ BrNO(g)$$

A single-step reaction would require that three reactant molecules collide simultaneously in just the right orientation. The probability of this occurring is small; thus, it would be reasonable to look for a mechanism that occurs in a series of steps, with each step involving only one or two molecules. In one possible mechanism, Br_2 and NO might combine in an initial step to produce an intermediate species, Br_2NO (Figure 15.17). This intermediate would then react with another NO molecule to give the reaction products. The equation for the overall reaction is obtained by adding the equations for these two steps:

Step 1: $\qquad Br_2(g) + NO(g) \rightleftharpoons Br_2NO(g)$

Step 2: $\qquad \underline{Br_2NO(g) + NO(g) \rightarrow 2\ BrNO(g)}$

Overall Reaction: $\quad Br_2(g) + 2\ NO(g) \rightarrow 2\ BrNO(g)$

Each step in a multistep reaction sequence is an **elementary step**, defined by a chemical equation that describes a single molecular event such as the formation or rupture of a chemical bond resulting from a molecular collision. Each step has its own activation energy, E_a, and rate constant, k. Adding the equations for each step must give the balanced equation for the overall reaction, and the time required to complete all of the steps defines the overall reaction rate. The series of steps constitutes a possible reaction mechanism.

Molecularity of Elementary Steps

Elementary steps are classified by the number of reactant molecules (or ions, atoms, or free radicals) that come together. This whole, positive number is called the **molecularity** of the elementary step. When one molecule is the only reactant in an elementary step, the reaction is a **unimolecular** process. A **bimolecular** elementary process involves two molecules, which may be identical (A + A → products) or different (A + B → products). The mechanism proposed for the decomposition of ozone in the stratosphere illustrates the use of these terms.

Step 1: $\qquad$ Unimolecular $\qquad O_3(g) \rightarrow O_2(g) + O(g)$

Step 2: $\qquad$ Bimolecular $\qquad \underline{O_3(g) + O(g) \rightarrow 2\ O_2(g)}$

Overall Reaction: $\qquad\qquad\qquad 2\ O_3(g) \rightarrow 3\ O_2(g)$

A **termolecular** elementary step involves three molecules, which could be the same or different (3 A → products; 2 A + B → products; or A + B + C → products). Be aware, however, the simultaneous collision of three molecules has a low probability, unless one of the molecules involved is in high concentration, such as a solvent molecule. In fact, most termolecular processes involve the collision of two reactant molecules and a third, inert molecule. The function of the inert molecule is to absorb the excess energy produced when a new chemical bond is formed by the first two molecules. For example, N_2 is unchanged in a termolecular reaction between oxygen molecules and oxygen atoms that produces ozone in the upper atmosphere:

$$O(g) + O_2(g) + N_2(g) \rightarrow O_3(g) + \text{energetic } N_2(g)$$

● **Rate Laws and Mechanisms** Rate laws are derived by experiment; they are macroscopic observations. Mechanisms are schemes we propose that speculate on how reactions occur at the particulate level.

The probability that four or more molecules will simultaneously collide with sufficient kinetic energy and proper orientation to react is so small that reaction molecularities greater than three are never proposed.

Rate Equations for Elementary Steps

The experimentally determined rate equation for a reaction cannot be predicted from its overall stoichiometry. In contrast, *the rate equation for any elementary step is defined by the reaction stoichiometry.* The rate equation of an elementary step is given by the product of the rate constant for that step and the concentrations of the reactants in that step. We can therefore write the rate equation for any elementary step, as shown by examples in the following table:

Elementary Step	Molecularity	Rate Equation
A → product	Unimolecular	Rate = $k[A]$
A + B → product	Bimolecular	Rate = $k[A][B]$
A + A → product	Bimolecular	Rate = $k[A]^2$
2 A + B → product	Termolecular	Rate = $k[A]^2[B]$

For example, the rate laws for each of the two elementary steps in the decomposition of ozone are

$$\text{Rate for (unimolecular) Step 1} = k[O_3]$$

$$\text{Rate for (bimolecular) Step 2} = k'[O_3][O]$$

When a reaction mechanism consists of two elementary steps, the two steps will likely occur at different rates. The two rate constants (k and k' in this example) are not expected to have the same value (nor the same units, if the two steps have different molecularities).

Molecularity and Reaction Order

A unimolecular elementary step must be first order; a bimolecular elementary step must be second order; and a termolecular elementary step must be third order. Such a direct relation between molecularity and order is emphatically not true for a multistep reaction. If you learn from an experiment that a reaction is first order, you cannot conclude that it occurs in a single, unimolecular elementary step. Similarly, a second-order rate equation does not imply that the reaction occurs in a single, bimolecular elementary step. An illustration of this is the decomposition of N_2O_5:

$$2\ N_2O_5(g) \rightarrow 4\ NO_2(g) + O_2(g)$$

Here, the rate law is "Rate = $k[N_2O_5]$," but chemists are fairly certain the mechanism involves a series of unimolecular and bimolecular steps.

To see how the experimentally observed rate equation for the overall reaction is connected with a possible mechanism or sequence of elementary steps requires some chemical intuition. We will provide only a glimpse of the subject in the next section.

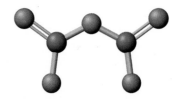

N_2O_5 Decomposition. The first step in the decomposition of N_2O_5 is thought to be the cleavage of one of the N—O bonds in the N—O—N link to give the odd-electron molecules NO_2 and NO_3. These react further to give the final products.

EXAMPLE 15.12   Elementary Steps

Problem The hypochlorite ion undergoes self-oxidation–reduction to give chlorate, ClO_3^-, and chloride ions.

$$3\ ClO^-(aq) \rightarrow ClO_3^-(aq) + 2\ Cl^-(aq)$$

This reaction is thought to occur in two steps:

Step 1: $ClO^-(aq) + ClO^-(aq) \rightarrow ClO_2^-(aq) + Cl^-(aq)$

Step 2: $ClO_2^-(aq) + ClO^-(aq) \rightarrow ClO_3^-(aq) + Cl^-(aq)$

What is the molecularity of each step? Write the rate equation for each reaction step. Show that the sum of these reactions gives the equation for the net reaction.

What Do You Know? A two-step mechanism for the reaction of OCl^- to form Cl^- and ClO_3^- is proposed.

Strategy The molecularity is the number of ions or molecules involved in a reaction step. The rate equation involves the concentration of each ion or molecule in an elementary step, raised to the power of its stoichiometric coefficient.

Solution Because two ions are involved in each elementary step, each step is bimolecular. The rate equation for any elementary step involves the product of the concentrations of the reactants. Thus, in this case, the rate equations are

Step 1:   Rate $= k[ClO^-]^2$

Step 2:   Rate $= k'[ClO_2^-][ClO^-]$

From the equations for the two elementary steps, we see that the ClO_2^- ion is an intermediate, a product of the first step and a reactant in the second step. It therefore cancels out, and we are left with the stoichiometric equation for the overall reaction:

Step 1: $ClO^-(aq) + ClO^-(aq) \rightarrow ClO_2^-(aq) + Cl^-(aq)$

Step 2: $ClO_2^-(aq) + ClO^-(aq) \rightarrow ClO_3^-(aq) + Cl^-(aq)$

Sum of steps: $3\ ClO^-(aq) \rightarrow ClO_3^-(aq) + 2\ Cl^-(aq)$

Think about Your Answer Other mechanisms are possible. The next question to ask is "What is the evidence that will let you decide between several different mechanisms?"

Check Your Understanding

Nitrogen monoxide is reduced by hydrogen to give nitrogen and water:

$$2\ NO(g) + 2\ H_2(g) \rightarrow N_2(g) + 2\ H_2O(g)$$

One possible mechanism for this reaction is

$$2\ NO(g) \rightarrow N_2O_2(g)$$

$$N_2O_2(g) + H_2(g) \rightarrow N_2O(g) + H_2O(g)$$

$$N_2O(g) + H_2(g) \rightarrow N_2(g) + H_2O(g)$$

What is the molecularity of each of the three steps? What is the rate equation for the third step? Identify the intermediates in this reaction; how many different intermediates are there? Show that the sum of these elementary steps gives the equation for the overall reaction.

Reaction Mechanisms and Rate Equations

The dependence of rate on concentration is an experimental fact. Mechanisms, by contrast, are constructs of our imagination, intuition, and good "chemical sense." To describe a mechanism, we need to make a guess (a good guess, we hope) about how the reaction occurs at the particulate level. Several mechanisms can always be proposed that correspond to the observed rate equation, and a postulated mechanism can be wrong. A good mechanism is a worthy goal because it allows us to understand the chemistry better. A practical consequence of a good mechanism is that it allows us to predict, for example, how to control a reaction better and how to design new experiments.

One of the important guidelines of kinetics is that *products of a reaction can never be produced at a rate faster than the rate of the slowest step*. If one step in a multistep reaction is slower than the others, then *the rate of the overall reaction is limited by the combined rates of all elementary steps up through the slowest step in the mechanism*. Often the overall reaction rate and the rate of the slow step are nearly the same. If the slow step determines the rate of the reaction, it is called the **rate-determining step**, or rate-limiting step.

Imagine that a reaction takes place with a mechanism involving two sequential steps, and assume that we know the rates of both steps. The first step is slow and the second is fast:

Elementary Step 1: $\quad\quad\quad\quad$ A + B $\xrightarrow[\text{Slow, } E_a \text{ large}]{k_1}$ X + M

Elementary Step 2: $\quad\quad\quad\quad$ M + A $\xrightarrow[\text{Fast, } E_a \text{ small}]{k_2}$ Y

Overall Reaction: $\quad\quad\quad\quad$ 2A + B $\xrightarrow{\quad\quad}$ X + Y

In the first step, A and B come together and slowly react to form one of the products (X) plus another reactive species, M. Almost as soon as M is formed, however, it is rapidly consumed by reacting with another molecule of A to form the second product Y. The rate-determining elementary step in this example is the first step. That is, the rate of the first step is equal to the rate of the overall reaction. This step is bimolecular and so it has the rate equation:

$$\text{Rate} = k_1[A][B]$$

where k_1 is the rate constant for that step. The overall reaction is expected to have this same second-order rate equation.

Let us apply these ideas to the mechanism of a real reaction: the second-order reaction of nitrogen dioxide with fluorine.

Overall Reaction: $\quad\quad\quad\quad$ $2\,NO_2(g) + F_2(g) \rightarrow 2\,FNO_2(g)$

$$\text{Rate} = k[NO_2][F_2]$$

The rate equation immediately rules out the possibility that the reaction occurs in a single step. If the equation for the reaction represented an elementary step, the rate law would have a second-order dependence on $[NO_2]$. Because a single-step reaction is ruled out, the mechanism must include at least two steps. We can also conclude from the rate law that the rate-determining elementary step must involve NO_2 and F_2 in a 1:1 ratio. One possible mechanism proposes that molecules of NO_2 and F_2 first react to produce one molecule of the product (FNO_2) plus one F atom. In a second step, the fluorine atom produced in the first step reacts with additional NO_2 to give a second molecule of product. If the first, bimolecular step is rate determining, the rate equation would be "Rate $= k_1[NO_2][F_2]$," the same as the experimentally observed rate equation. The experimental rate constant would be the same as k_1.

● **Can You Derive a Mechanism?** At this introductory level, you cannot be expected to derive reaction mechanisms. Given a mechanism, however, you can decide whether it agrees with experimental rate laws.

Elementary Step 1:   Slow $\quad NO_2(g) + F_2(g) \xrightarrow{k_1} FNO_2(g) + F(g)$

Elementary Step 2:   Fast $\quad NO_2(g) + F(g) \xrightarrow{k_2} FNO_2(g)$

Overall Reaction: $\quad\quad\quad$ $2\,NO_2(g) + F_2(g) \xrightarrow{\quad} 2\,FNO_2(g)$

The fluorine atom formed in the first step of the NO_2/F_2 reaction is a reaction intermediate. It does not appear in the equation describing the overall reaction. Reaction intermediates usually have only a fleeting existence, but occasionally they have long enough lifetimes to be observed. The detection and identification of an intermediate are strong evidence for the proposed mechanism.

Enzymes—Nature's Catalysts

Within any living organism, there are untold numbers of chemical reactions occurring, many of them extremely rapidly. In many cases, enzymes, biological catalysts, speed up reactions that would normally move at a snail's pace from reactants to products. Typically, enzyme-catalyzed reactions are 10^7 to 10^{14} times faster than uncatalyzed reactions.

Enzymes are typically large proteins, often containing metal ions such as Zn^{2+}. They are thought to function by bringing the reactants together in just the right orientation in a site where specific bonds can be broken and/or made.

Carbonic anhydrase is one of many enzymes important in biological processes (Figure A). Carbon dioxide dissolves in water to a small extent to produce carbonic acid, which ionizes to give H_3O^+ and HCO_3^- ions.

$$CO_2(g) \rightleftharpoons CO_2(aq) \quad (1)$$

$$CO_2(aq) + H_2O(\ell) \rightleftharpoons H_2CO_3(aq) \quad (2)$$

$$H_2CO_3(aq) + H_2O(\ell)$$
$$\rightleftharpoons H_3O^+(aq) + HCO_3^-(aq) \quad (3)$$

Carbonic anhydrase speeds up reactions 1 and 2. Many of the H_3O^+ ions produced by ionization of H_2CO_3 (reaction 3) are picked up by hemoglobin in the blood as hemoglobin loses O_2. The resulting HCO_3^- ions are transported back to the lungs. When hemoglobin again takes on O_2, it releases H_3O^+ ions. These ions and HCO_3^- re-form H_2CO_3, from which CO_2 is liberated and exhaled.

You can do an experiment that illustrates the effect of carbonic anhydrase. First, add a small amount of NaOH to a cold, aqueous solution of CO_2. The solution becomes basic immediately because there is not enough H_2CO_3 in the solution to use up the NaOH. After some seconds, however, dissolved CO_2 slowly produces more H_2CO_3, which consumes NaOH, and the solution is again acidic.

Now try the experiment again, this time adding a few drops of blood to the solution (Figure A). Carbonic anhydrase in blood speeds up reactions 1 and 2 by a factor of about 10^7, as evidenced by the more rapid reaction under these conditions.

In 1913, Leonor Michaelis and Maud L. Menten proposed a general theory of enzyme action based on kinetic observations. They assumed that the substrate, S (the reactant), and the enzyme, E, form a complex, ES. This complex then breaks down, releasing the enzyme and the product, P.

$$E + S \rightleftharpoons ES$$

$$ES \rightarrow P + E$$

When the substrate concentration is low, the rate of the reaction is first order in S (Figure B). As [S] increases, however, the active sites in the enzyme become saturated with substrate, and the rate reaches its maximum value. Now the kinetics are zero order in substrate.

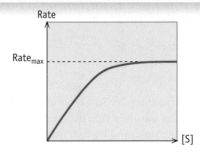

FIGURE B Rate of enzyme-catalyzed reaction. This plot of substrate concentration [S] versus reaction velocity is typical of reactions catalyzed by enzymes that follow the Michaelis–Menten model.

Questions:

1. Catalase can decompose hydrogen peroxide to O_2 and water about 10^7 times faster than the uncatalyzed reaction. If the latter requires one year, how much time is required by the enzyme-catalyzed reaction?

2. According to the Michaelis–Menten model, if 1/Rate is plotted versus 1/[S], the intercept of the plot (when 1/[S] = 0) is $1/Rate_{max}$. Find $Rate_{max}$ for a reaction involving carbonic anhydrase.

[S], mol/L	Rate (millimoles/min)
2.500	0.588
1.00	0.500
0.714	0.417
0.526	0.370
0.250	0.256

Answers to these questions are available in Appendix N.

Photos © Cengage Learning/Charles D. Winters

(a) $t = 0$ **(b)** $t = 3$ sec   **(c)** $t = 15$ sec   **(d)** $t = 17$ sec   **(e)** $t = 21$ sec

FIGURE A CO_2 in water. **(a)** A few drops of blood are added to a cold solution of CO_2 in water. **(b)** A few drops of a dye (bromthymol blue) are added to the solution, the yellow color indicating an acidic solution. **(c, d)** A less-than-stoichiometric amount of sodium hydroxide is added, converting the H_2CO_3 to HCO_3^- (and CO_3^{2-}). The blue color of the dye indicates a basic solution. **(e)** The blue color begins to fade after some seconds as CO_2 forms more H_2CO_3. The amount of H_2CO_3 formed is finally sufficient to consume the added NaOH, and the solution is again acidic. Blood is a source of the enzyme *carbonic anhydrase,* so the formation of H_2CO_3 is noticeably more rapid than the reaction in the absence of blood.

EXAMPLE 15.13 Elementary Steps and Reaction Mechanisms

Problem Oxygen atom transfer from NO_2 to CO produces nitrogen monoxide and carbon dioxide (Figure 15.13):

$$NO_2(g) + CO(g) \rightarrow NO(g) + CO_2(g)$$

The rate equation for this reaction at temperatures less than 500 K is Rate $= k[NO_2]^2$. Can this reaction occur in one bimolecular step?

What Do You Know? The reaction equation and the rate law for the reaction are given.

Strategy Write the rate law based on the equation for the NO_2 + CO reaction occurring as if it were an elementary step. If this rate law corresponds to the observed rate law, then a one-step mechanism is possible.

Solution If the reaction occurs by the collision of one NO_2 molecule with one CO molecule, the rate equation would be

$$Rate = k[NO_2][CO]$$

This does not agree with experiment, so the mechanism must involve more than a single step. In one possible mechanism, the reaction occurs in two bimolecular steps, the first one slow and the second one fast:

Elementary Step 1:	Slow, rate-determining	$2 NO_2(g) \longrightarrow NO_3(g) + NO(g)$

Elementary Step 2:	Fast	$NO_3(g) + CO(g) \longrightarrow NO_2(g) + CO_2(g)$

Overall Reaction:		$NO_2(g) + CO(g) \longrightarrow NO(g) + CO_2(g)$

The first (rate-determining) step has a rate equation that agrees with experiment, so this is a possible mechanism.

Think about Your Answer Because the rate equation is second order in $[NO_2]$, the rate-determining step in this multistep reaction must involve the collision of two NO_2 molecules.

Check Your Understanding

The Raschig reaction produces hydrazine, N_2H_4, an industrially important reducing agent, from NH_3 and OCl^- in basic, aqueous solution. A proposed mechanism is

Step 1: Fast $NH_3(aq) + OCl^-(aq) \rightarrow NH_2Cl(aq) + OH^-(aq)$

Step 2: Slow $NH_2Cl(aq) + NH_3(aq) \rightarrow N_2H_5^+(aq) + Cl^-(aq)$

Step 3: Fast $N_2H_5^+(aq) + OH^-(aq) \rightarrow N_2H_4(aq) + H_2O(\ell)$

(a) What is the overall equation?

(b) Which step of the three is rate determining?

(c) Write the rate equation for the rate-determining elementary step.

(d) What reaction intermediates are involved?

Reaction Mechanisms Involving an Equilibrium Step

A common two-step reaction mechanism involves an initial fast reaction that produces an intermediate, followed by a slower second step in which the intermediate is converted to the final product. The rate of the reaction is determined by the second step, for which a rate law can be written. The rate of that step, however, depends on the concentration of the intermediate. An intermediate, whose concentration will probably not be measurable, cannot appear as a term in the overall rate equation. We must find a way to replace the expression for the intermediate with another expression in terms of quantities measurable in the lab.

The reaction of nitrogen monoxide and oxygen is an example of a two-step reaction where the first step is fast and the second step is rate determining.

$$2 \, NO(g) + O_2(g) \rightarrow 2 \, NO_2(g)$$

$$Rate = k[NO]^2[O_2]$$

The experimentally determined rate law shows second-order dependence on NO and first-order dependence on O_2. Although this rate law would be correct for a termolecular reaction, experimental evidence indicates that an intermediate is formed in this reaction. A possible two-step mechanism that proceeds through an intermediate is

Elementary Step 1:   Fast, equilibrium

$$NO(g) + O_2(g) \underset{k_{-1}}{\overset{k_1}{\rightleftharpoons}} OONO(g)$$
intermediate

Elementary Step 2:   Slow, rate-determining

$$NO(g) + OONO(g) \overset{k_2}{\rightarrow} 2 \, NO_2(g)$$

Overall Reaction:

$$2 \, NO(g) + O_2(g) \rightarrow 2 \, NO_2(g)$$

The second step of this reaction is the slow step, and the overall rate depends on it. We can write a rate law for the second step:

$$Rate = k_2[NO][OONO]$$

This rate law cannot be compared directly with the experimental rate law because it contains the concentration of an intermediate, OONO. To eliminate the concentration of intermediate from this rate expression, we look at the rapid first step in this reaction sequence that involves an equilibrium between the intermediate species and the reactants.

At the beginning of the reaction, NO and O_2 react rapidly and produce the intermediate OONO. The rate of formation can be defined by a rate law with a rate constant k_1:

$$Rate \ of \ production \ of \ OONO = k_1[NO][O_2]$$

Because the intermediate is consumed only very slowly in the second step, it is possible for the OONO to revert to NO and O_2 before it reacts further:

$$Rate \ of \ reverse \ reaction \ (OONO \rightarrow NO + O_2) = k_{-1}[OONO]$$

As NO and O_2 form OONO, their concentrations drop, so the rate of the forward reaction decreases. At the same time, the concentration of OONO builds up, so the rate of the reverse reaction increases. At equilibrium, the rates of the forward and reverse reactions become the same.

Rate of forward reaction = rate of reverse reaction

$$k_1[NO][O_2] = k_{-1}[OONO]$$

Rearranging this equation, we find

$$\frac{k_1}{k_{-1}} = \frac{[OONO]}{[NO][O_2]} = K$$

Both k_1 and k_{-1} are constants (they will change only if the temperature changes). We can define a new constant K equal to the ratio of these two constants and called the **equilibrium constant**, which is equal to the quotient $[OONO]/[NO][O_2]$. From this, we can derive an expression for the concentration of OONO:

$$[OONO] = K[NO][O_2]$$

● **Mechanisms with an Initial Equilibrium** In this mechanism, the forward and reverse reactions in the first elementary step are so much faster than the second elementary step that equilibrium is established before any significant amount of OONO is consumed by NO to give NO_2. The state of equilibrium for the first step remains throughout the lifetime of the overall reaction.

Relating Rate Equations and Reaction Mechanisms

The connection between an experimental rate equation and the proposed reaction mechanism is important in chemistry.

1. Experiments must first be performed to determine the experimental rate equation.
2. A mechanism for the reaction is proposed on the basis of the experimental rate equation, the principles of stoichi-

ometry, molecular structure and bonding, general chemical experience, and intuition.

3. The proposed reaction mechanism is used to derive a rate equation. This rate equation can contain only those species present in the overall chemical reaction. If the derived and experimental rate equations are the same, the postulated

mechanism *may* be a reasonable hypothesis of the reaction sequence.

4. If more than one mechanism can be proposed and they all predict derived rate equations in agreement with experiment, then more experiments must be done.

If $K[NO][O_2]$ is substituted for [OONO] in the rate law for the rate-determining elementary step, we have

$$\text{Rate} = k_2[NO][OONO] = k_2[NO]\{K[NO][O_2]\}$$

$$= k_2K[NO]^2[O_2]$$

Because both k_2 and K are constants, their product is another constant k', and we have

$$\text{Rate} = k'[NO]^2[O_2]$$

This is exactly the rate law derived from experiment. Thus, the sequence of reactions on which the rate law is based may be a reasonable mechanism for this reaction. It is not the only possible mechanism, however. This rate equation is also consistent with the reaction occurring in a single termolecular step. Another possible mechanism is illustrated in Example 15.14.

● **Equilibrium Constant** The important concept of chemical equilibrium was introduced in Chapter 3 and will be described in more detail in Chapters 16–19.

EXAMPLE 15.14   Reaction Mechanism Involving an Equilibrium Step

Problem The NO + O_2 reaction described in the text could also occur by the following mechanism:

Elementary Step 1: Fast, equilibrium

$$NO(g) + NO(g) \underset{k_{-1}}{\overset{k_1}{\rightleftharpoons}} N_2O_2(g) \text{ intermediate}$$

Elementary Step 2: Slow, rate-determining

$$N_2O_2(g) + O_2(g) \xrightarrow{k_2} 2\ NO_2(g)$$

Overall Reaction: 2 NO(g) + O_2(g) → 2 NO_2(g)

Show that this mechanism leads to the following experimental rate law: Rate = $k[NO]^2[O_2]$.

What Do You Know? A possible mechanism for the reaction of NO and O_2 is given.

Strategy The rate law for the rate-determining elementary step is

$$\text{Rate} = k_2[N_2O_2][O_2]$$

The intermediate N_2O_2 cannot appear in the final derived rate law. However, we can find an expression relating the concentration of the unknown to the concentrations of the reactants, using the equilibrium constant expression for the first step.

Solution [N_2O_2] and [NO] are related by the equilibrium constant.

$$\frac{k_1}{k_{-1}} = \frac{[N_2O_2]}{[NO]^2} = K$$

Solving this equation for [N₂O₂] gives $[N_2O_2] = K[NO]^2$. When this is substituted into the derived rate law

$$\text{Rate} = k_2\{K[NO]^2\}[O_2]$$

the resulting equation is identical with the experimental rate law where $k_2K = k$.

Think about Your Answer Three mechanisms have been proposed for the NO + O₂ reaction. The challenge for chemists is to decide which is correct. In this case, further experimentation detected the species OONO as a short-lived intermediate, thus providing evidence for the mechanism involving this intermediate.

Check Your Understanding

One possible mechanism for the decomposition of nitryl chloride, NO₂Cl, is

Elementary Step 1: Fast, equilibrium $\quad NO_2Cl(g) \underset{k_{-1}}{\overset{k_1}{\rightleftharpoons}} NO_2(g) + Cl(g)$

Elementary Step 2: Slow $\quad NO_2Cl(g) + Cl(g) \xrightarrow{k_2} NO_2(g) + Cl_2(g)$

What is the overall reaction? What rate law would be derived from this mechanism? What effect does increasing the concentration of the product NO₂ have on the reaction rate?

REVIEW & CHECK FOR SECTION 15.6

1. The rate equation for a reaction A + B → C was determined by experiment to be Rate = k[A][B]. From this we can conclude

    (a) the reaction occurs in a single elementary step

    (b) this reaction might occur in a single elementary step

    (c) this reaction must involve several elementary steps

2. A reaction is believed to occur by the following mechanism:

    Step 1: 2 A ⇌ I   (Fast, equilibrium)

    Step 2: I + B → C   (Slow)

    Overall: 2 A + B → C

    What experimentally determined rate law would lead to this mechanism?

    (a) Rate = k[A][B]   (c) Rate = k[A]²[B]

    (b) Rate = k[A]²   (d) Rate = k[I][B]

OWL and **go Chemistry**

Sign in at **www.cengage.com/owl** to:
• View tutorials and simulations, develop problem-solving skills, and complete online homework assigned by your professor.
• For quick review and exam prep, download Go Chemistry mini lecture modules from OWL (or purchase them at **www.cengagebrain.com**)

Access **How Do I Solve It?** tutorials on how to approach problem solving using concepts in this chapter.

CHAPTER GOALS REVISITED

Now that you have studied this chapter, you should ask whether you have met the chapter goals. In particular, you should be able to:

Understand rates of reaction and the conditions affecting rates

a. Explain the concept of reaction rate (Section 15.1).

b. Derive the average and instantaneous rates of a reaction from concentration-time data (Section 15.1). **Study Question: 5.**

c. Describe factors that affect reaction rate (i.e., reactant concentrations, temperature, presence of a catalyst, and the state of the reactants) (Section 15.2). **Study Questions: 8, 9, 10, 66, 80, 81, 85.**

Derive a rate equation, rate constant, and reaction order from experimental data

a. Define the various parts of a rate equation (the rate constant and order of reaction), and understand their significance (Section 15.3). **Study Questions: 12–14.**

b. Derive a rate equation from experimental information (Section 15.3). **Study Questions: 11–14, 52, 60.**

Use integrated rate laws

a. Describe and use the relationships between reactant concentration and time for zero-order, first-order, and second-order reactions (Section 15.4 and Table 15.1). **Study Questions: 15–24, and Go Chemistry Module 20.**

b. Apply graphical methods for determining reaction order and the rate constant from experimental data (Section 15.4 and Table 15.1). **Study Questions: 31–36, 54, 55, 57.**

c. Use the concept of half-life ($t_{1/2}$), especially for first-order reactions (Section 15.4). **Study Questions: 25–30, 64, 75.**

Understand the collision theory of reaction rates and the role of activation energy

a. Describe the collision theory of reaction rates (Section 15.5).

b. Relate activation energy (E_a) to the rate of a reaction (Section 15.5). **Study Question: 40.**

c. Use collision theory to describe the effect of reactant concentration on reaction rate (Section 15.5). **Study Question: 84.**

d. Understand the effect of molecular orientation on reaction rate (Section 15.5).

e. Describe the effect of temperature on reaction rate using the collision theory of reaction rates and the Arrhenius equation (Section 15.5 and Equations 15.5–15.7). **Study Question: 84.**

f. Use Equations 15.5, 15.6, and 15.7 to calculate the activation energy from rate constants at different temperatures (Section 15.5). **Study Questions: 37, 39, 40, 62, 67, 68, 76c.**

g. Understand reaction coordinate diagrams (Section 15.5). **Study Questions: 41, 42, 47, 87.**

Relate reaction mechanisms and rate laws

a. Describe the functioning of a catalyst and its effect on the activation energy and mechanism of a reaction (Section 15.5). **Study Questions: 82, 88.**

b. Understand the concept of a reaction mechanism (a proposed sequence of bond-making and bond-breaking steps that occurs during the conversion of reactants to products) and the relation of the mechanism to the overall, stoichiometric equation for a reaction (Section 15.6).

c. Describe the elementary steps of a mechanism, and give their molecularity (Section 15.6). **Study Questions: 43–48, 78.**

d. Define the rate-determining step in a mechanism, and identify any reaction intermediates (Section 15.6). **Study Questions: 45, 48, 59, 65, 74, 82, 83.**

Key Equations

Equation 15.1 (page 519) Integrated rate equation for a first-order reaction (in which $-\Delta[R]/\Delta t = k[R]$).

$$\ln \frac{[R]_t}{[R]_0} = -kt$$

Here, $[R]_0$ and $[R]_t$ are concentrations of the reactant at time $t = 0$ and at a later time, t. The ratio of concentrations, $[R]_t/[R]_0$, is the fraction of reactant that remains after a given time has elapsed.

Equation 15.2 (page 521) Integrated rate equation for a second-order reaction (in which $-\Delta[R]/\Delta t = k[R]^2$).

$$\frac{1}{[R]_t} - \frac{1}{[R]_0} = kt$$

Equation 15.3 (page 522) Integrated rate equation for a zero-order reaction (in which $-\Delta[R]/\Delta t = k[R]^0$).

$$[R]_0 - [R]_t = kt$$

Equation 15.4 (page 524) The relation between the half-life ($t_{1/2}$) and the rate constant (k) for a first-order reaction.

$$t_{1/2} = \frac{0.693}{k}$$

Equation 15.5 (page 531) Arrhenius equation in exponential form.

$$k = \text{rate constant} = A e^{-E_a/RT}$$

Frequency factor

Fraction of molecules with minimum energy for reaction

A is the frequency factor; E_a is the activation energy; T is the temperature (in kelvins); and R is the gas constant ($= 8.314510 \times 10^{-3}$ kJ/K · mol).

Equation 15.6 (page 531) Expanded Arrhenius equation in logarithmic form.

$$\ln k = -\frac{E_a}{R}\left(\frac{1}{T}\right) + \ln A \longleftarrow \text{Arrhenius equation}$$

$$y \quad = \quad mx \quad + b \longleftarrow \text{Equation for straight line}$$

Equation 15.7 (page 532) A version of the Arrhenius equation used to calculate the activation energy for a reaction when you know the values of the rate constant at two temperatures (in kelvins).

$$\ln k_2 - \ln k_1 = \ln \frac{k_2}{k_1} = -\frac{E_a}{R}\left[\frac{1}{T_2} - \frac{1}{T_1}\right]$$

OWL Questions and problems for this chapter are available in OWL. Use the OWL access card that came with this textbook to access assigned questions and problems for this chapter.

(a) Combining solutions of $NaHCO_3$ and $CaCl_2$ produces solid $CaCO_3$ and CO_2 gas.

(b) Dry ice (the white solid) is added to the slurry of $CaCO_3$ precipitated in (a).

(c) The calcium carbonate dissolves on adding sufficient dry ice (CO_2) to give $Ca^{2+}(aq)$ and $HCO_3^-(aq)$.

FIGURE 16.1 Equilibrium in the $CO_2/Ca^{2+}/H_2O$ system. See Figure 3.5, which shows the same reaction and illustrates the chemistry of the system.

These experiments illustrate an important feature of chemical reactions: *All chemical reactions are reversible, at least in principle.* This was a key point in our earlier discussion of equilibrium (◄ Chapter 3).

Our next step will be to move from a qualitative to a quantitative description of equilibrium systems. Among other things, this will lead us to the subject of product- and reactant-favored reactions (◄ page 100). Recall that a reaction that has a greater concentration of products than reactants once it has reached equilibrium is said to be product-favored. Similarly, a reaction that has a greater concentration of reactants than products at equilibrium is said to be reactant-favored.

REVIEW & CHECK FOR SECTION 16.1

What would happen in the experiment in Figure 16.1c if you added some additional Ca^{2+} ions (in the form of $CaCl_2$)?

(a) More $CaCO_3$ would precipitate.   (c) Nothing would happen.

(b) Some $CaCO_3$ would dissolve.

16.2 The Equilibrium Constant and Reaction Quotient

Chemical equilibria can also be described in a quantitative fashion. The concentrations of reactants and products when a reaction has reached equilibrium are related. For the reaction of hydrogen and iodine to produce hydrogen iodide, for example, a very large number of experiments have shown that at equilibrium the ratio of the square of the HI concentration to the product of the H_2 and I_2 concentrations is a constant.

$$H_2(g) + I_2(g) \rightleftharpoons 2 HI(g)$$

$$\frac{[HI]^2}{[H_2][I_2]} = \text{constant } (K) \text{ at equilibrium}$$

This constant is always the same within experimental error for all experiments done *at a given temperature*. Suppose, for example, the concentrations of H_2 and I_2 in a flask are each initially 0.0175 mol/L at 425 °C and no HI is present. Over time, the concentrations of H_2 and I_2 will decrease, and the concentration of HI will increase until a state of equilibrium is reached (Figure 16.2). If the gases in

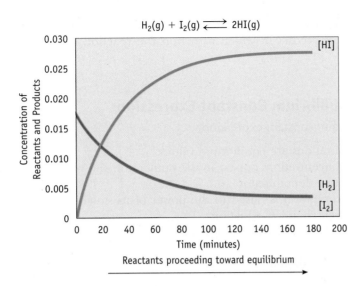

$$H_2(g) + I_2(g) \rightleftharpoons 2HI(g)$$

FIGURE 16.2 The reaction of H_2 and I_2 reaches equilibrium. The final concentrations of H_2, I_2, and HI depend on the initial concentrations of H_2 and I_2. If one begins with a different set of initial concentrations, the equilibrium concentrations will be different, but the quotient $[HI]^2/[H_2][I_2]$ will always be the same at a given temperature.

the flask are then analyzed, the observed concentrations would be $[H_2] = [I_2] = 0.0037$ mol/L and $[HI] = 0.0276$ mol/L. The following table—which we call an **ICE table for initial, change, and equilibrium concentrations**—summarizes these results:

Equation	$H_2(g)$	+	$I_2(g)$	$\rightleftharpoons$	2 HI(g)
I = Initial concentration (M)	0.0175		0.0175		0
C = Change in concentration as reaction proceeds to equilibrium (M)	−0.0138		−0.0138		+0.0276
E = Equilibrium concentration (M)	0.0037		0.0037		0.0276

● **ICE Table: Initial, Change, and Equilibrium** Throughout our discussions of chemical equilibria, we shall express the quantitative information for reactions in an amounts table or ICE table (◀ Chapter 4, page 135). These tables show what the initial (*I*) concentrations are, how those concentrations change (*C*) on proceeding to equilibrium, and what the concentrations are at equilibrium (*E*). Some students call these **RICE tables**, where R stands for "reaction."

The second line in the table gives the change in concentration of reactants and products on proceeding to equilibrium. Changes are always equal to the difference between the equilibrium and initial concentrations.

Change in concentration = equilibrium concentration − initial concentration

Putting the equilibrium concentration values from the ICE table into the expression for the constant (K) gives a value of 56 (to two significant figures).

$$\frac{[HI]^2}{[H_2][I_2]} = \frac{(0.0276)^2}{(0.0037)(0.0037)} = 56$$

Other experiments can be done on the H_2/I_2 reaction with different concentrations of reactants, or done using mixtures of reactants and products. Regardless of the initial amounts, when equilibrium is achieved, the ratio $[HI]^2/[H_2][I_2]$ is always the same, 56, at this temperature.

The observation that the product and reactant concentrations for the H_2 and I_2 reaction are always in the same ratio can be generalized to other reactions. For the general chemical reaction

$$aA + bB \rightleftharpoons cC + dD$$

we can define the equilibrium constant, K.

When the reaction is at equilibrium

Equilibrium constant = $K = \dfrac{[C]^c[D]^d}{[A]^a[B]^b}$ **(16.1)**

Equation 16.1 is called the **equilibrium constant expression**. If the ratio of products to reactants as defined by Equation 16.1 matches the equilibrium constant value, the system is known to be at equilibrium. Conversely, if the ratio has a

FIGURE 16.3 **Burning sulfur.** Elemental sulfur burns in oxygen with a beautiful blue flame to give SO_2 gas.

different value, the system is not at equilibrium, and, as you will see later in this section, we can predict in which direction the reaction will proceed to reach equilibrium.

Writing Equilibrium Constant Expressions

In an equilibrium constant expression,

- all concentrations are equilibrium values.
- product concentrations appear in the numerator, and reactant concentrations appear in the denominator.
- each concentration is raised to the power of its stoichiometric coefficient in the balanced chemical equation.
- the value of the constant K depends on the particular reaction and on the temperature.
- values of K are dimensionless.

Reactions Involving Solids

The oxidation of solid, yellow sulfur produces colorless sulfur dioxide gas in a product-favored reaction (Figure 16.3).

$$S(s) + O_2(g) \rightleftharpoons SO_2(g)$$

The general principle when writing an equilibrium constant expression is to place product concentrations in the numerator and reactant concentrations in the denominator. In reactions involving solids, however, experiments show that the equilibrium concentrations of other reactants or products—here, O_2 and SO_2—do not depend on the amount of solid present (as long as some solid is present at equilibrium). Therefore, the concentration of a solid such as sulfur is not included in the equilibrium constant expression.

$$K = \frac{[SO_2]}{[O_2]}$$

In general, *the concentrations of any solid reactants and products are not included in the equilibrium constant expression.*

Reactions in Solution

There are also special considerations for reactions occurring in solution, particularly aqueous solutions in which water is either a reactant or a product. Consider ammonia, which is a weak base owing to its incomplete reaction with water.

$$NH_3(aq) + H_2O(\ell) \rightleftharpoons NH_4^+(aq) + OH^-(aq)$$

Because the water concentration is very high in a dilute ammonia solution, the concentration of water is essentially unchanged by the reaction. The general rule *for reactions in aqueous solution is that the molar concentration of water is not included in the equilibrium constant expression.* Thus, for aqueous ammonia we write

$$K = \frac{[NH_4^+][OH^-]}{[NH_3]}$$

Reactions Involving Gases: K_c and K_p

Concentration data can be used to calculate equilibrium constants for both aqueous and gaseous systems. In these cases, the symbol K is sometimes given the subscript "c" for "concentration," as in K_c. For gases, however, equilibrium constant expressions can be written in another way—in terms of partial pressures of reactants and products. If you rearrange the ideal gas law, $[PV = nRT]$, and recognize that the "gas concentration," (n/V), is equivalent to P/RT, you see that the partial pressure of a gas is proportional to its concentration $[P = (n/V)RT]$. If reactant and

A CLOSER LOOK

Activities and Units of K

In the text we state that "Values of K are dimensionless." After all our care in using units, this might seem sloppy. However, advanced thermodynamics informs us that equilibrium constants should really be calculated from the "activities" of reactants and products and not from their concentrations or partial pressures. Activities can be thought of as "effective" concentrations or partial pressures.

The activity of a substance in solution is obtained by calculating the ratio of its concentration, [X], relative to a standard concentration (1 M) and then multiplying this ratio by a correction factor called an *activity coefficient*. Because it involves a ratio of concentrations, the activity is dimensionless. (Likewise, the activity of a gas is obtained from the ratio of its partial pressure, P_X, relative to a standard pressure (1 atm or 1 bar) and then multiplying by an activity coefficient.) In general chemistry, we assume all the activity coefficients are equal to 1 and so the activity of a solute or a gas is numerically the same as its concentration or partial pressure, respectively. This assumption is best met for solutes in very dilute solutions or gases at low pressures. Regardless of the values of activity coefficients, values of K are properly calculated using dimensionless quantities and so they have no units.

Another consequence of using activities is that the "concentration" of solids does not appear in the K expression. This is because the activity of a solid is 1. Similarly, water or other solvent is not included because its activity is 1.

product quantities are given in partial pressures (in atmospheres or in bars), then K is given the subscript "p," as in K_p.

$$H_2(g) + I_2(g) \rightleftharpoons 2\ HI(g)$$

$$K_p = \frac{P_{HI}^2}{P_{H_2}P_{I_2}}$$

Notice that the basic form of the equilibrium constant expression is the same as for K_c. In some cases, the numerical values of K_c and K_p are the same, but they are different when the numbers of moles of gaseous reactants and products are different. *A Closer Look: Equilibrium Constant Expressions for Gases—K_c and K_p* shows how K_c and K_p are related and how to convert from one to the other.

EXAMPLE 16.1 Writing Equilibrium Constant Expressions

Problem Write the equilibrium constant expressions for the following reactions.

(a) $N_2(g) + 3\ H_2(g) \rightleftharpoons 2\ NH_3(g)$

(b) $H_2CO_3(aq) + H_2O(\ell) \rightleftharpoons HCO_3^-(aq) + H_3O^+(aq)$

What Do You Know? You have balanced chemical equations, from which you can write the equilibrium constant expressions.

Strategy Product concentrations always appear in the numerator and reactant concentrations appear in the denominator. Each concentration should be raised to a power equal to the stoichiometric coefficient in the balanced equation. In reaction (b), the water concentration does not appear in the equilibrium constant expression.

Solution

(a) $\quad K = \dfrac{[NH_3]^2}{[N_2][H_2]^3}$

(b) $\quad K = \dfrac{[HCO_3^-][H_3O^+]}{[H_2CO_3]}$

Think about Your Answer Always check to make sure you have the products in the numerator and reactants in the denominator. Confusing these is a common source of student error.

Check Your Understanding

Write the equilibrium constant expression for each of the following reactions in terms of concentrations.

(a) $CO_2(g) + C(s) \rightleftharpoons 2\ CO(g)$

(b) $[Cu(NH_3)_4]^{2+}(aq) \rightleftharpoons Cu^{2+}(aq) + 4\ NH_3(aq)$

(c) $CH_3CO_2H(aq) + H_2O(\ell) \rightleftharpoons CH_3CO_2^-(aq) + H_3O^+(aq)$

A CLOSER LOOK

Equilibrium Constant Expressions for Gases—K_c and K_p

Many metal carbonates, such as limestone, decompose on heating to give the metal oxide and CO_2 gas.

$$CaCO_3(s) \rightleftharpoons CaO(s) + CO_2(g)$$

The equilibrium constant for this reaction can be expressed either in terms of the number of moles per liter of CO_2, $K_c = [CO_2]$, or in terms of the pressure of CO_2, $K_p = P_{CO_2}$. From the ideal gas law, you know that

$$P = (n/V)RT =$$
$$\text{(concentration in mol/L)} \times RT$$

For this reaction, we can therefore say that $P_{CO_2} = [CO_2]RT = K_p$. Because $K_c = [CO_2]$, we find that $K_p = K_c(RT)$. That is, the values of K_p and K_c are not the same; for the decomposition of calcium carbonate, K_p is the product of K_c and the factor RT.

Consider the equilibrium constant for the reaction of N_2 and H_2 to produce ammonia in terms of partial pressures, K_p.

$$N_2(g) + 3 H_2(g) \rightleftharpoons 2 NH_3(g)$$

$$K_p = \frac{(P_{NH_3})^2}{(P_{N_2})(P_{H_2})^3}$$

Does K_c, the equilibrium constant in terms of concentrations, have the same value as, or a different value than, K_p? We can answer this question by substituting for each pressure in K_p the equivalent expression $[C](RT)$. That is,

$$K_p = \frac{\{[NH_3](RT)\}^2}{\{[N_2](RT)\}\{[H_2](RT)\}^3} =$$
$$\frac{[NH_3]^2}{[N_2][H_2]^3} \times \frac{1}{(RT)^2} = \frac{K_c}{(RT)^2}$$

or $K_p = K_c(RT)^{-2}$

Once again, you see that K_p and K_c are not the same but are related by some function of RT.

Looking carefully at this example and others, we find that

$$K_p = K_c(RT)^{\Delta n}$$

where Δn is the change in the number of moles of gas on going from reactants to products.

Δn = total moles of gaseous products − total moles of gaseous reactants

For the decomposition of $CaCO_3$,

$$\Delta n = 1 - 0 = 1$$

whereas the value of Δn for the ammonia synthesis is

$$\Delta n = 2 - 4 = -2$$

The Meaning of the Equilibrium Constant, K

Table 16.1 lists a few equilibrium constants for different kinds of reactions. A large value of K means that the concentrations of the products are higher than the concentrations of the reactants at equilibrium. That is, the products are favored over the reactants at equilibrium.

$K > 1$: Reaction is product-favored at equilibrium. The concentrations of products are greater than the concentrations of the reactants at equilibrium.

Table 16.1 Selected Equilibrium Constant Values

Reaction	Equilibrium Constant, K (at 25 °C)	Product- or Reactant-Favored at Equilibrium
Combination Reaction of Nonmetals		
$S(s) + O_2(g) \rightleftharpoons SO_2(g)$	4.2×10^{52}	$K > 1$; product-favored
$2 H_2(g) + O_2(g) \rightleftharpoons 2 H_2O(g)$	3.2×10^{81}	$K > 1$; product-favored
$N_2(g) + 3 H_2(g) \rightleftharpoons 2 NH_3(g)$	3.5×10^8	$K > 1$; product-favored
Ionization of Weak Acids and Bases		
$HCO_2H(aq) + H_2O(\ell) \rightleftharpoons HCO_2^-(aq) + H_3O^+(aq)$ formic acid	1.8×10^{-4}	$K < 1$; reactant-favored
$CH_3CO_2H(aq) + H_2O(\ell) \rightleftharpoons CH_3CO_2^-(aq) + H_3O^+(aq)$ acetic acid	1.8×10^{-5}	$K < 1$; reactant-favored
$H_2CO_3(aq) + H_2O(\ell) \rightleftharpoons HCO_3^-(aq) + H_3O^+(aq)$ carbonic acid	4.2×10^{-7}	$K < 1$; reactant-favored
$NH_3(aq) + H_2O(\ell) \rightleftharpoons NH_4^+(aq) + OH^-(aq)$ ammonia	1.8×10^{-5}	$K < 1$; reactant-favored
Dissolution of "Insoluble" Solids		
$CaCO_3(s) \rightleftharpoons Ca^{2+}(aq) + CO_3^{2-}(aq)$	3.8×10^{-9}	$K < 1$; reactant-favored
$AgCl(s) \rightleftharpoons Ag^+(aq) + Cl^-(aq)$	1.8×10^{-10}	$K < 1$; reactant-favored

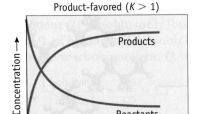

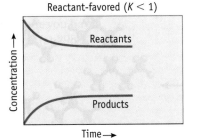

(a) The concentrations of products exceeds the concentrations of reactants at equilibrium, and the reaction is product-favored.

(b) The reactant concentrations exceed the product concentrations at equilibrium, and the reaction is reactant-favored.

FIGURE 16.4 The difference between reactions that are product-favored or reactant-favored at equilibrium.

An example is the reaction of nitrogen monoxide and ozone.

$$NO(g) + O_3(g) \rightleftharpoons NO_2(g) + O_2(g)$$

$$K = \frac{[NO_2][O_2]}{[NO][O_3]} = 6 \times 10^{34} \text{ at 25 °C}$$

The very large value of K indicates that, at equilibrium, $[NO_2][O_2] \gg [NO][O_3]$. If stoichiometric amounts of NO and O_3 are mixed and allowed to come to equilibrium, virtually none of the reactants will be found (Figure 16.4a). Essentially, all will have been converted to NO_2 and O_2. A chemist would say that "the reaction has gone to completion."

Conversely, a small value of K means that very little of the products exists when equilibrium has been achieved (Figure 16.4b). That is, the reactants are favored over the products at equilibrium.

$K < 1$: Reaction is reactant-favored at equilibrium. Concentrations of reactants are greater than concentrations of products at equilibrium.

This is true for the formation of ozone from oxygen.

$$3/2\ O_2(g) \rightleftharpoons O_3(g)$$

$$K = \frac{[O_3]}{[O_2]^{3/2}} = 2.5 \times 10^{-29} \text{ at 25 °C}$$

The very small value of K indicates that, at equilibrium, $[O_3] \ll [O_2]^{3/2}$. If O_2 is placed in a flask, very little O_2 will have been converted to O_3 when equilibrium has been achieved.

When K is close to 1, it may not be immediately clear whether the reactant concentrations are larger than the product concentrations, or vice versa. It will depend on the form of K and thus on the reaction stoichiometry. Calculations of the concentrations will have to be done.

The Reaction Quotient, Q

The equilibrium constant, K, for a reaction has a particular numerical value when the reactants and products are at equilibrium. When the reactants and products in a reaction are not at equilibrium, however, it is convenient to calculate the reaction quotient, Q. For the general reaction of A and B to give C and D,

$$aA + bB \rightleftharpoons cC + dD$$

the reaction quotient is defined as

$$\text{Reaction quotient} = Q = \frac{[C]^c[D]^d}{[A]^a[B]^b} \tag{16.2}$$

(a) Not at e

Here, four isc
butane mole
proceed to c
reach equilib

> ### REVIEW & CHECK FOR SECTION 16.2
>
> 1. Which of the following is the correct form of the equilibrium constant expression for the reaction below?
>
> $$2\ SO_3(g) \rightleftharpoons 2\ SO_2(g) + O_2(g)$$
>
>   (a) $K = [SO_2][O_2]/[SO_3]$ (c) $K = [SO_3]^2/[SO_2]^2[O_2]$
>
>   (b) $K = [SO_2]^2[O_2]/[SO_3]^2$ (d) $K = [SO_2][O_2]^2$
>
> 2. At 2000 K the equilibrium constant for the formation of NO(g) is 4.0×10^{-4}.
>
> $$N_2(g) + O_2(g) \rightleftharpoons 2\ NO(g)$$
>
> You have a flask in which, at 2000 K, the concentration of N_2 is 0.50 mol/L, that of O_2 is 0.25 mol/L, and that of NO is 4.2×10^{-3} mol/L. Is the system at equilibrium?
>
>   (a) Yes, it is at equilibrium.
>
>   (b) No, it is not at equilibrium, and the reaction proceeds further to the right.
>
>   (c) No, it is not at equilibrium, and the reaction proceeds to the left, turning products into reactants.

16.3 Determining an Equilibrium Constant

When the experimental values of the concentrations of all of the reactants and products are known at equilibrium, an equilibrium constant can be calculated by substituting the data into the equilibrium constant expression. Consider this concept as it applies to the oxidation of sulfur dioxide.

$$2\ SO_2(g) + O_2(g) \rightleftharpoons 2\ SO_3(g)$$

In an experiment done at 852 K, the equilibrium concentrations are found to be $[SO_2] = 3.61 \times 10^{-3}$ mol/L, $[O_2] = 6.11 \times 10^{-4}$ mol/L, and $[SO_3] = 1.01 \times 10^{-2}$ mol/L. Substituting these data into the equilibrium constant expression, we can calculate the value of K.

$$K = \frac{[SO_3]^2}{[SO_2]^2[O_2]} = \frac{(1.01 \times 10^{-2})^2}{(3.61 \times 10^{-3})^2(6.11 \times 10^{-4})} = 1.28 \times 10^4 \text{ at } 852 \text{ K}$$

(Notice that K has a large value; at equilibrium, the oxidation of sulfur dioxide is product-favored at 852 K.)

More commonly, an experiment will provide information on the initial quantities of reactants and the concentration at equilibrium of only one of the reactants or of one of the products. The equilibrium concentrations of the rest of the reactants and products must then be inferred from the reaction stoichiometry. As an example, again consider the oxidation of sulfur dioxide to sulfur trioxide. Suppose that 1.00 mol of SO_2 and 1.00 mol of O_2 are placed in a 1.00-L flask, this time at 1000 K. When equilibrium has been achieved, 0.925 mol of SO_3 has been formed. Let us use this information to calculate the equilibrium constant for the reaction. After writing the equilibrium constant expression in terms of concentrations, we set up an ICE table (page 551) showing the initial concentrations, the changes in those concentrations on proceeding to equilibrium, and the concentrations at equilibrium.

● **Comparing Q ar**

Relative
Magnitude   **Di**

$Q < K$	Re
$Q = K$	Re
$Q > K$	Re

Equation	2 SO₂(g)	+	O₂(g)	⇌	2 SO₃(g)
Initial (M)	1.00		1.00		0
Change (M)	−2x		−x		+2x
Equilibrium (M)	1.00 − 2x		1.00 − x		2x
	1.00 − 0.925 = 0.075		1.00 − 0.925/2 = 0.54		= 0.925

The quantities in the ICE table result from the following analysis:

- Line 1: Original concentrations.
- Line 2: The amount of O_2 consumed is designated as $-x$ mol/L. We give it a minus sign because O_2 is consumed. It then follows from the reaction stoichiometry that the amount of SO_2 consumed is $-2x$, and the amount of SO_3 produced is $+2x$.
- Lines 3 and 4:
  - (a) We know from experiment that $[SO_3]$ at equilibrium is 0.925 M. Therefore, $2x = 0.925$ M.
  - (b) The equilibrium concentration of SO_2 is equal to the initial concentration minus what was consumed. Therefore, $[SO_2]$ is $(1.00 - 2x)$M or 0.075 M.
  - (c) The amount of O_2 consumed is half of the amount of SO_3 produced or x ($= 0.463$ mol/L). Therefore, the equilibrium concentration of O_2 is 0.54 M.

With the equilibrium concentrations now known, it is possible to calculate K.

$$K = \frac{[SO_3]^2}{[SO_2]^2[O_2]} = \frac{(0.925)^2}{(0.075)^2(0.54)} = 2.8 \times 10^2 \text{ at 1000 K}$$

⬛**WL** **INTERACTIVE EXAMPLE 16.3** **Calculating an Equilibrium Constant (K_c) Using Concentrations**

Problem In aqueous solution iron(III) ions react with iodide ions to give iron(II) ions and triiodide ions, I_3^-. Suppose the initial concentration of Fe^{3+} ions is 0.200 M, the initial I^- ion concentration is 0.300 M, and the equilibrium concentration of I_3^- ions is 0.0866 M. What is the value of K_c?

$$2\ Fe^{3+}(aq) + 3\ I^-(aq) \rightleftharpoons 2\ Fe^{2+}(aq) + I_3^-(aq)$$

What Do You Know? You are given the balanced equation (from which the equilibrium constant expression can be written), initial concentrations of the reactants, and the concentration of one product (I_3^-) after equilibrium has been reached.

Strategy

- Set up an ICE table and enter the initial concentrations.

- Decide how each concentration changes. Let us begin by saying that x mol/L of I_3^- ions are produced on proceeding to equilibrium. Based on the reaction stoichiometry, this means that $2x$ mol/L of Fe^{2+} must also be produced, $2x$ mol/L of Fe^{3+} ions are consumed, and $3x$ mol/L of I^- ions are consumed.

- The equilibrium concentration of I_3^- is known (0.0866 M), so this is the quantity x.

- Knowing that $x = 0.0866$ M, calculate the equilibrium concentrations for each species.

- Enter the equilibrium concentrations of reactants and products into the equilibrium constant expression and solve for K_c.

Solution The strategy outlined leads to the ICE table below.

Equation	2 Fe³⁺	+	3 I⁻	⇌	2 Fe²⁺	+	I₃⁻
Initial (M)	0.200		0.300		0		0
Change (M)	$-2x$		$-3x$		$+2x$		$+x$
Equilibrium (M)	$0.200 - 2(0.0866)$		$0.300 - 3(0.0866)$		$2(0.0866)$		0.0866
	$= 0.027$		$= 0.040$		$= 0.173$		

The concentration of each substance at equilibrium is now known, and K can be calculated.

$$K_c = \frac{[Fe^{2+}]^2[I_3^-]}{[Fe^{3+}]^2[I^-]^3} = \frac{(0.173)^2(0.0866)}{(0.027)^2(0.040)^3} = 5.6 \times 10^4$$

Strategy Map 16.3

PROBLEM

Calculate K_c for reaction of Fe^{3+} with I^- to give Fe^{2+} and I_3^-.

DATA/INFORMATION KNOWN

- Balanced equation
- **Initial concentrations** of reactants
- **Equilibrium concentration** of one product

STEP 1. Organize information.

Write K_c expression, set up **ICE table**, and enter **known concentrations** in the table.

STEP 2. Enter **concentration changes** in ICE table and derive **equilibrium concentrations**.

Complete **ICE table** with equilibrium concentrations known.

STEP 3. Enter **equilibrium concentrations** in K_c expression and solve for K_c.

Value of K_c

Think about Your Answer The key to this problem is that the concentration of one product, I_3^-, was known at equilibrium. The concentrations of the other product and of the reactants could be derived from this based on the reaction stoichiometry. Also notice that the calculated K is much greater than one, so the reaction is product-favored at equilibrium. This is consistent with the equilibrium concentrations calculated in the ICE table.

Check Your Understanding

A solution is prepared by dissolving 0.050 mol of diiodocyclohexane, $C_6H_{10}I_2$, in the solvent CCl_4. The total solution volume is 1.00 L. When the reaction

$$C_6H_{10}I_2 \rightleftharpoons C_6H_{10} + I_2$$

has come to equilibrium at 35 °C, the concentration of I_2 is 0.035 mol/L.

(a)   What are the concentrations of $C_6H_{10}I_2$ and C_6H_{10} at equilibrium?

(b)   Calculate K_c, the equilibrium constant.

EXAMPLE 16.4 Calculating an Equilibrium Constant (K_p) Using Partial Pressures

Problem Suppose a tank initially contains H_2S at a pressure of 10.00 atm and a temperature of 800 K. When the reaction has come to equilibrium, the partial pressure of S_2 vapor is 0.020 atm. Calculate K_p.

$$2\ H_2S(g) \rightleftharpoons 2\ H_2(g) + S_2(g)$$

What Do You Know? You are given the balanced equation (from which the equilibrium constant expression for K_p can be written), the initial pressure of the reactant (H_2S), and the pressure of one product (S_2) after equilibrium has been reached.

Strategy

• Set up an ICE table and enter the initial pressures of H_2S, H_2, and S_2.

• Define the change in S_2 concentration as $+x$.

• From the reaction stoichiometry you know that the change in the partial pressure of H_2 is twice that of S_2 ($= +2x$ atm) and that the change in H_2S partial pressure is $-2x$.

• The equilibrium partial pressure of S_2 is given, so you know that $x = 0.020$ atm. From this calculate the equilibrium partial pressures for H_2S and H_2.

• Enter the equilibrium pressures of reactants and products into the equilibrium constant expression and solve for K_p.

Solution Following the strategy outlined gives the ICE table below.

Equation	2 H₂S(g)	⇌	2 H₂(g)	+	S₂(g)
Initial (atm)	10.00		0		0
Change (atm)	$-2x$		$+2x$		$+x$
Equilibrium (atm)	$10.00 - 2x$		$2(0.020)$		0.020
	$10.00 - 0.040 = 9.96$		$= 0.040$		

Now that the partial pressures of all of the reactants and products are known, K_p can be calculated.

$$K_p = \frac{(P_{H_2})^2 P_{S_2}}{(P_{H_2S})^2} = \frac{(0.040)^2(0.020)}{(9.96)^2} = 3.2 \times 10^{-7}$$

Think about Your Answer The value of K_p calculated is much less than 1, therefore this reaction is reactant-favored at equilibrium, as also indicated by the concentrations calculated in the ICE table. The value of K_p is the same as the value of K_c only when the number of moles of gaseous reactants is the same as the number of moles of gaseous products. This is not true here, so $K_p \neq K_c$. (See *A Closer Look*, page 554).

Check Your Understanding

Pure PCl_5 gas is placed in a 2.00-L flask. After heating to 250 °C the pressure of PCl_5 is initially 2.000 atm. However, the gas slowly but only partially decomposes to gaseous PCl_3 and Cl_2. When equilibrium is reached, the partial pressure of Cl_2 is 0.814 atm. Calculate K_p for the decomposition.

REVIEW & CHECK FOR SECTION 16.3

You place 0.010 mol of $N_2O_4(g)$ in a 2.0-L flask at 200 °C. After reaching equilibrium, $[N_2O_4] = 0.0042$ M. Calculate K_c for the following reaction: $N_2O_4(g) \rightleftharpoons 2\ NO_2(g)$

(a) 1640

(c) 3.1×10^{-4}

(b) 6.1×10^{-4}

(d) 8.8×10^{-6}

16.4 Using Equilibrium Constants in Calculations

In many cases, the value of K and the initial concentrations of reactants are known, and you want to know the concentrations present at equilibrium. As we look at several examples of this situation, we will again use ICE tables that summarize the initial conditions, the changes occurring on proceeding to equilibrium, and the final conditions.

OWL   INTERACTIVE EXAMPLE 16.5   Calculating Equilibrium Concentrations

Problem The equilibrium constant K_c (= 55.64) for

$$H_2(g) + I_2(g) \rightleftharpoons 2\ HI(g)$$

has been determined at 425 °C. If 1.00 mol each of H_2 and I_2 is placed in a 0.500-L flask at 425 °C, what are the concentrations of H_2, I_2, and HI when equilibrium has been achieved?

What Do You Know? You are given the balanced equation (from which the equilibrium constant expression can be written), the value of K, and the initial amounts of the reactants and the volume of the container (from which initial concentrations of the reactants can be calculated).

Strategy

*   Write the equilibrium constant expression and set up an ICE table.

*   Enter the initial concentrations of H_2 and I_2 on the first line (I).

*   Assign the variable x to represent changes in concentration. Based on the reaction stoichiometry the change in $[H_2]$ and $[I_2]$ is $-x$ and the change in [HI] is $+2x$. Enter these values into the change (C) line of the table.

*   Enter the expressions for the final equilibrium concentrations of all three species on the third line (E) of the ICE table, and then transfer these expressions to the equilibrium constant expression and solve for x.

*   Use the calculated value of x to solve for the final concentration of each species.

Strategy Map 16.5

PROBLEM

What are the **equilibrium concentrations** of H_2, I_2, and **HI** in formation of HI from H_2 and I_2?

DATA/INFORMATION KNOWN

*   Balanced equation
*   Value of K_c
*   **Initial concentrations** of reactants

STEP 1. Organize information.

Write K_c expression, set up **ICE table**, and enter **known concentrations** in the table.

STEP 2. Enter *concentration changes* in ICE table and derive **equilibrium concentrations** in terms of unknown quantity x.

Equilibrium concentrations of H_2, I_2, and **HI** are defined in terms of the unknown quantity x.

STEP 3. Enter **equilibrium concentrations** in K_c expression and solve for x.

Value of x

STEP 4. Use value of x to derive **equilibrium concentrations**.

Values of equilibrium concentrations

Module 21: Chemical Equilibrium
covers concepts in this section.

Solution Write the equilibrium constant expression

$$K = \frac{[HI]^2}{[H_2][I_2]} = 55.64$$

and then set up the ICE table as outlined in the Strategy.

Equation	$H_2(g)$	+	$I_2(g)$	$\rightleftharpoons$	2 HI(g)
Initial (M)	1.00 mol/0.500 L = 2.00 M		1.00 mol/0.500 L = 2.00 M		0
Change (M)	$-x$		$-x$		$+2x$
Equilibrium (M)	$2.00 - x$		$2.00 - x$		$2x$

Now the expressions for the equilibrium concentrations can be substituted into the equilibrium constant expression.

$$55.64 = \frac{(2x)^2}{(2.00 - x)(2.00 - x)} = \frac{(2x)^2}{(2.00 - x)^2}$$

In this case, you can solve for the unknown quantity x by taking the square root of both sides of the equation,

$$\sqrt{K} = 7.459 = \frac{2x}{2.00 - x}$$

$$7.459 (2.00 - x) = 14.9 - 7.459x = 2x$$

$$14.9 = 9.459x$$

$$x = 1.58$$

With x known, you can solve for the equilibrium concentrations of the reactants and products.

$$[H_2] = [I_2] = 2.00 - x = 0.42 \text{ M}$$

$$[HI] = 2x = 3.16 \text{ M}$$

Think about Your Answer It is always wise to verify the answer by substituting the value of each concentration back into the equilibrium expression to see if your calculated K agrees with the one given in the problem. In this case, $(3.16)^2/(0.42)^2 = 57$. The slight discrepancy with the given value, $K = 55.64$, occurs because we know $[H_2]$ and $[I_2]$ to only two significant figures.

Check Your Understanding

At some temperature, $K_c = 33$ for the reaction

$$H_2(g) + I_2(g) \rightleftharpoons 2 \text{ HI}(g)$$

Assume the initial concentrations of both H_2 and I_2 are 6.00×10^{-3} mol/L. Find the concentration of each reactant and product at equilibrium.

Calculations Where the Solution Involves a Quadratic Expression

Suppose you are studying the decomposition of PCl_5 to form PCl_3 and Cl_2. You know that $K = 1.20$ at a given temperature.

$$PCl_5(g) \rightleftharpoons PCl_3(g) + Cl_2(g)$$

If the initial concentration of PCl_5 is 1.60 M, what will be the concentrations of reactant and products when the system reaches equilibrium? Following the procedures outlined in Example 16.5, you would set up an ICE table to define the equilibrium concentrations of reactants and products.

Reaction	$PCl_5(g)$	$\rightleftharpoons$	$PCl_3(g)$	+	$Cl_2(g)$
Initial (M)	1.60		0		0
Change (M)	$-x$		$+x$		$+x$
Equilibrium (M)	$1.60 - x$		x		x

Substituting into the equilibrium constant expression, we have

$$K = 1.20 = \frac{[PCl_3][Cl_2]}{[PCl_5]} = \frac{(x)(x)}{1.60 - x}$$

Expanding the algebraic expression results in a quadratic equation,

$$x^2 + 1.20x - 1.92 = 0$$

Using the quadratic formula (Appendix A; $a = 1$, $b = 1.20$, and $c = -1.92$), we find two roots to the equation: $x = 0.91$ and -2.11. Because a negative value of x (which represents a negative concentration) is not chemically meaningful, the answer is $x = 0.91$. Therefore, at equilibrium we have

$$[PCl_5] = 1.60 - 0.91 = 0.69 \text{ M}$$

$$[PCl_3] = [Cl_2] = 0.91 \text{ M}$$

Although a solution to a quadratic equation can always be obtained using the quadratic formula, in many instances an acceptable answer can be obtained by using a realistic approximation to simplify the equation. To illustrate this, consider another equilibrium, the dissociation of I_2 molecules to form I atoms, for which $K = 5.6 \times 10^{-12}$ at 500 K.

$$I_2(g) \rightleftharpoons 2\,I(g)$$

$$K = \frac{[I]^2}{[I_2]} = 5.6 \times 10^{-12}$$

Assuming the initial I_2 concentration is 0.45 M and setting up the ICE table in the usual manner, we have

Reaction	$I_2(g)$	$\rightleftharpoons$	2 I(g)
Initial (M)	0.45		0
Change (M)	$-x$		$+2x$
Equilibrium (M)	$0.45 - x$		$2x$

For the equilibrium constant expression, we again arrive at a quadratic equation.

$$K = 5.6 \times 10^{-12} = \frac{(2x)^2}{(0.45 - x)}$$

Although we could solve this equation using the quadratic formula, there is a simpler way to reach an answer. Notice that the value of K is very small, indicating that the amount of I_2 that will be dissociated ($= x$) is very small. In fact, K is so small that subtracting x from the original reactant concentration (0.45 mol/L) in the denominator of the equilibrium constant expression will leave the denominator essentially unchanged. That is, $(0.45 - x)$ is essentially equal to 0.45. Thus, we drop x in the denominator and have a simpler equation to solve.

$$K = 5.6 \times 10^{-12} = \frac{(2x)^2}{(0.45)}$$

The solution to this equation gives $x = 7.9 \times 10^{-7}$. From this value, we can determine that $[I_2] = 0.45 - x = 0.45$ mol/L and $[I] = 2x = 1.6 \times 10^{-6}$ mol/L. Notice that the answer to the I_2 dissociation problem confirms the assumption the dissociation of I_2 is so small that $[I_2]$ at equilibrium is essentially equal to the initial concentration.

When is it possible to simplify a quadratic equation? The decision depends on both the value of the initial concentration of the reactant and the value of x, which is in turn related to the value of K. Consider the general reaction

$$A \rightleftharpoons B + C$$

● **Solving Quadratic Equations**
Quadratic equations are usually solved using the quadratic formula (Appendix A). An alternative is the *method of successive approximations*, also outlined in Appendix A. Most equilibrium expressions can be solved quickly by this method, and you are urged to try it. This will remove the uncertainty of whether K expressions need to be solved exactly. (There are, however, rare cases in which this does not work.)

PROBLEM SOLVING TIP 16.1

When Do You Need to Use the Quadratic Equation?

In most equilibrium calculations, the quantity x may be neglected in the denominator of the equation $K = x^2/([A]_0 - x)$ if x is less than 10% of the quantity of reactant initially present. The guideline presented in the text for making the approximation that $[A]_0 - x = [A]_0$ when $100 \times K < [A]_0$ reflects this fact.

In general, when K is about 1 or greater, the approximation cannot be made. If K is much less than 1 and $100 \times K < [A]_0$ (you will see many such cases in Chapter 17), the approximate expression ($K = x^2/[A]_0$) gives an acceptable answer. If you are not certain, then first make the assumption that the unknown (x) is small, and solve the approximate expression ($K = (x)^2/[A]_0$). Next, compare the "approximate" value of x with $[A]_0$. If x has a value equal to or less than 10% of $[A]_0$, then there is no need to solve the full equation using the quadratic formula.

where $K = [B][C]/[A]$. Assume you know K and the initial concentration of A ($= [A]_0$) and wish to find the equilibrium concentrations of B and C ($= x$). The equilibrium constant expression now is

$$K = \frac{[B][C]}{[A]} = \frac{(x)(x)}{[A]_0 - x}$$

When K is very small, the value of x will be much less than $[A]_0$, so $[A]_0 - x \cong [A]_0$. Therefore, we can write the following expression.

$$K = \frac{[B][C]}{[A]} \approx \frac{(x)(x)}{[A]_0} \tag{16.3}$$

If $100 \times K < [A]_0$, the approximate expression will give acceptable values of equilibrium concentrations (to two significant figures). For more about this useful guideline, see *Problem Solving Tip 16.1*.

Strategy Map 16.6

PROBLEM

What are the **equilibrium concentrations** of N$_2$, O$_2$, and NO in formation of NO from N$_2$ and O$_2$?

↓

DATA/INFORMATION KNOWN

- Balanced equation
- Value of K_c
- **Initial concentrations** of reactants

STEP 1. Enter *concentration changes* in ICE table and derive **equilibrium concentrations** in terms of unknown quantity x.

Equilibrium concentrations of N$_2$, O$_2$, and **NO** are defined in terms of the unknown quantity x.

STEP 2. Enter **equilibrium concentrations** in K_c **expression** and solve for x.

Value of x

STEP 3. Use value of x to derive **equilibrium concentrations**.

Values of equilibrium concentrations

⦿WL   INTERACTIVE EXAMPLE 16.6   Calculating Equilibrium Concentrations Using an Equilibrium Constant

Problem The reaction

$$N_2(g) + O_2(g) \rightleftharpoons 2\, NO(g)$$

contributes to air pollution whenever a fuel is burned in air at a high temperature, as in a gasoline engine. At 1500 K, $K = 1.0 \times 10^{-5}$. Suppose a sample of air has $[N_2] = 0.80$ mol/L and $[O_2] = 0.20$ mol/L before any reaction occurs. Calculate the equilibrium concentrations of reactants and products after the mixture has been heated to 1500 K.

What Do You Know? As in Example 16.5, you know the value of K and can write the equilibrium expression from the balanced equation. You also know the initial concentrations of the reactants and can define the equilibrium concentrations in terms of the amounts of N$_2$ and O$_2$ consumed ($= x$).

Strategy Set up an ICE table, and then substitute the equilibrium concentrations into the equilibrium constant expression. The result will be a quadratic equation. This expression can be solved using the methods outlined in Appendix A or by using the guideline in the text to simplify the calculation.

Solution We first set up an ICE table where the amounts of N$_2$ and O$_2$ consumed are designated as x.

Equation	N$_2$(g)	+	O$_2$(g)	⇌	2 NO(g)
Initial (M)	0.80		0.20		0
Change (M)	$-x$		$-x$		$+2x$
Equilibrium (M)	$0.80 - x$		$0.20 - x$		$2x$

Next, the equilibrium concentrations are substituted into the equilibrium constant expression.

$$K = 1.0 \times 10^{-5} = \frac{[NO]^2}{[N_2][O_2]} = \frac{[2x]^2}{(0.80 - x)(0.20 - x)}$$

We refer to the guideline (Equation 16.3) to decide whether an approximate solution is possible. Here, $100 \times K \,(= 1.0 \times 10^{-3})$ is smaller than either of the initial reactant concentrations (0.80 and 0.20). This means we can use the approximate expression

$$K = 1.0 \times 10^{-5} = \frac{[NO]^2}{[N_2][O_2]} = \frac{(2x)^2}{(0.80)(0.20)}$$

Solving this expression, we find

$$1.6 \times 10^{-6} = 4x^2$$

$$x = 6.3 \times 10^{-4}$$

Therefore, the reactant and product concentrations at equilibrium are

$$[N_2] = 0.80 - 6.3 \times 10^{-4} \approx 0.80 \text{ M}$$

$$[O_2] = 0.20 - 6.3 \times 10^{-4} \approx 0.20 \text{ M}$$

$$[NO] = 2x = 1.3 \times 10^{-3} \text{ M}$$

Think about Your Answer The value of x obtained using the approximation is the same as that obtained from the quadratic formula. If the full equilibrium constant expression is expanded, we have

$$(1.0 \times 10^{-5})(0.80 - x)(0.20 - x) = 4x^2$$

$$(1.0 \times 10^{-5})(0.16 - 1.00x + x^2) = 4x^2$$

$$\underset{ax^2}{(4 - 1.0 \times 10^{-5})x^2} + \underset{bx}{(1.0 \times 10^{-5})x} - \underset{c}{0.16 \times 10^{-5}} = 0$$

The two roots to this equation are:

$$x = 6.3 \times 10^{-4} \text{ or } x = -6.3 \times 10^{-4}$$

The only meaningful root is identical to the approximate answer obtained above. The approximation is indeed valid in this case.

Check Your Understanding

The decomposition of $PCl_5(g)$ to form $PCl_3(g)$ and $Cl_2(g)$ has $K_c = 33.3$ at a high temperature. If the initial concentration of PCl_5 is 0.1000 M, what are the equilibrium concentrations of the reactants and products?

REVIEW & CHECK FOR SECTION 16.4

Graphite and carbon dioxide are kept at constant volume at 1000 K until the reaction

$$C(graphite) + CO_2(g) \rightleftharpoons 2\,CO(g)$$

has come to equilibrium. At this temperature, $K = 0.021$. The initial concentration of CO_2 is 0.012 mol/L. Calculate the equilibrium concentration of CO.

(a) 0.012 M

(c) 0.0057 M

(b) 0.011 M

16.5 More about Balanced Equations and Equilibrium Constants

Chemical equations can be balanced using different sets of stoichiometric coefficients. For example, the equation for the oxidation of carbon to give carbon monoxide can be written

$$C(s) + \tfrac{1}{2} O_2(g) \rightleftharpoons CO(g)$$

In this case, the equilibrium constant expression would be

$$K_1 = \frac{[CO]}{[O_2]^{1/2}} = 4.6 \times 10^{23} \text{ at } 25\ °C$$

You can write the chemical equation equally well, however, as

$$2\,C(s) + O_2(g) \rightleftharpoons 2\,CO(g)$$

and the equilibrium constant expression would now be

$$K_2 = \frac{[CO]^2}{[O_2]} = 2.1 \times 10^{47} \text{ at } 25\ °C$$

When you compare the two equilibrium constant expressions you find that $K_2 = (K_1)^2$; that is,

$$K_2 = \frac{[CO]^2}{[O_2]} = \left\{ \frac{[CO]}{[O_2]^{1/2}} \right\}^2 = K_1^{\,2}$$

When the stoichiometric coefficients of a balanced equation are multiplied by some factor, the equilibrium constant for the new equation (K_{new}) is the old equilibrium constant (K_{old}) raised to the power of the multiplication factor.

In the case of the oxidation of carbon, the second equation was obtained by multiplying the first equation by two. Therefore, K_2 is the *square* of K_1 ($K_2 = K_1^{\,2}$).

Let us consider what happens if a chemical equation is reversed. Here, we will compare the value of K for formic acid transferring an H^+ ion to water

$$HCO_2H(aq) + H_2O(\ell) \rightleftharpoons HCO_2^-(aq) + H_3O^+(aq)$$

$$K_1 = \frac{[HCO_2^-][H_3O^+]}{[HCO_2H]} = 1.8 \times 10^{-4} \text{ at } 25\ °C$$

with the opposite reaction, the gain of an H^+ ion by the formate ion, HCO_2^-.

$$HCO_2^-(aq) + H_3O^+(aq) \rightleftharpoons HCO_2H(aq) + H_2O(\ell)$$

$$K_2 = \frac{[HCO_2H]}{[HCO_2^-][H_3O^+]} = 5.6 \times 10^3 \text{ at } 25\ °C$$

Here, $K_2 = 1/K_1$.

The equilibrium constants for a reaction and its reverse are the reciprocals of one another.

It is often useful to add two equations to obtain the equation for a net process. As an example, consider the reactions that take place when silver chloride dissolves in water (to a *very* small extent) and ammonia is added to the solution. Ammonia reacts with silver ions to form a water-soluble compound, $Ag(NH_3)_2Cl$ (Figure 16.6). Adding the equation for dissolving solid AgCl to the equation for the reaction of Ag^+ ions with ammonia gives the equation for the net reaction, dissolving solid AgCl in aqueous ammonia. (All equilibrium constants are given at 25 °C.)

$$AgCl(s) \rightleftharpoons Ag^+(aq) + Cl^-(aq) \qquad K_1 = [Ag^+][Cl^-] = 1.8 \times 10^{-10}$$

$$Ag^+(aq) + 2\,NH_3(aq) \rightleftharpoons [Ag(NH_3)_2]^+(aq) \qquad K_2 = \frac{[Ag(NH_3)_2^+]}{[Ag^+][NH_3]^2} = 1.1 \times 10^7$$

Net equation:

$$AgCl(s) + 2\,NH_3(aq) \rightleftharpoons [Ag(NH_3)_2]^+(aq) + Cl^-(aq)$$

Balanced Equations and Equilibrium Constants

You should now know

1. how to write an equilibrium constant expression from the balanced equation, recognizing that the concentrations of solids and liquids used as solvents do not appear in the expression.

2. that when the stoichiometric coefficients in a balanced equation are changed by a factor of n, $K_{new} = (K_{old})^n$.

3. that when a balanced equation is reversed, $K_{new} = 1/K_{old}$.

4. that when several balanced equations (each with its own equilibrium constant, K_1, K_2, etc.) are added to obtain a net, balanced equation, $K_{net} = K_1 \times K_2 \times K_3 \times \dots$.

AgCl(s) in water   After adding NH₃(aq)

FIGURE 16.6 **Dissolving silver chloride in aqueous ammonia.** *(left)* A precipitate of AgCl(s) is suspended in water. *(right)* When aqueous ammonia is added, the ammonia reacts with the trace of silver ion in solution, the equilibrium shifts, and the silver chloride dissolves.

To obtain the equilibrium constant for the net reaction, K_{net}, we *multiply* the equilibrium constants for the two reactions, K_1 by K_2.

$$K_{net} = K_1 \times K_2 = [Ag^+][Cl^-] \times \frac{[Ag(NH_3)_2{}^+]}{[Ag^+][NH_3]^2} = \frac{[Ag(NH_3)_2{}^+][Cl^-]}{[NH_3]^2}$$

$$K_{net} = K_1 \times K_2 = 2.0 \times 10^{-3}$$

When two or more chemical equations are added to produce a net equation, the equilibrium constant for the net equation is the product of the equilibrium constants for the added equations.

EXAMPLE 16.7   Balanced Equations and Equilibrium Constants

Problem A mixture of nitrogen, hydrogen, and ammonia is brought to equilibrium. When the equation is written using whole-number coefficients, as follows, the value of K is 3.5×10^8 at 25 °C.

Equation 1: $N_2(g) + 3 H_2(g) \rightleftharpoons 2 NH_3(g)$ $K_1 = 3.5 \times 10^8$

However, the equation can also be written as given in Equation 2. What is the value of K_2?

Equation 2: $\frac{1}{2} N_2(g) + \frac{3}{2} H_2(g) \rightleftharpoons NH_3(g)$ $K_2 = ?$

The decomposition of ammonia to the elements (Equation 3) is the reverse of its formation (Equation 1). What is the value of K_3?

Equation 3: $2 NH_3(g) \rightleftharpoons N_2(g) + 3 H_2(g)$ $K_3 = ?$

What Do You Know? You know the value of K for a given balanced equation. You want to know how the value of K changes as the stoichiometric coefficients change or when the equation is reversed.

Strategy Determine how the desired reaction is related to the given reaction(s). (Was a given chemical reaction multiplied by a factor? Was a reaction reversed? Were two or more reactions added?) Use the relationships discussed to determine the effect(s) of these transformations on the given K value(s). See also *Problem Solving Tip 16.2*.

Solution Equation 2 can be obtained by multiplying Equation 1 by 1/2. Thus K_2 is equal to K_1 raised to the one-half power, $K_1^{1/2}$. To confirm this relationship between K_1 and K_2, write the equilibrium constant expressions for these two balanced equations.

$$K_1 = \frac{[NH_3]^2}{[N_2][H_2]^3} \qquad K_2 = \frac{[NH_3]}{[N_2]^{1/2}[H_2]^{3/2}}$$

Writing these expressions makes it clear that K_2 is the square root of K_1.

$$K_2 = (K_1)^{1/2} = \sqrt{K_1} = \sqrt{3.5 \times 10^8} = 1.9 \times 10^4$$

Equation 3 is the reverse of Equation 1, and its equilibrium constant expression is

$$K_3 = \frac{[N_2][H_2]^3}{[NH_3]^2}$$

In this case, K_3 is the reciprocal of K_1. That is, $K_3 = 1/K_1$.

$$K_3 = \frac{1}{K_1} = \frac{1}{3.5 \times 10^8} = 2.9 \times 10^{-9}$$

Think about Your Answer

Notice that the production of ammonia from the elements has a large equilibrium constant and is product-favored (see Section 16.2). As expected, the reverse reaction, the decomposition of ammonia to its elements, has a small equilibrium constant and is reactant-favored.

Check Your Understanding

The conversion of oxygen to ozone has a very small equilibrium constant.

$$3/2\ O_2(g) \rightleftharpoons O_3(g) \qquad K = 2.5 \times 10^{-29}$$

(a) What is the value of K when the equation is written using whole-number coefficients?

$$3\ O_2(g) \rightleftharpoons 2\ O_3(g)$$

(b) What is the value of K for the conversion of ozone to oxygen?

$$2\ O_3(g) \rightleftharpoons 3\ O_2(g)$$

REVIEW & CHECK FOR SECTION 16.5

The following equilibrium constants are given at 500 K:

$H_2(g) + Br_2(g) \rightleftharpoons 2\ HBr(g)$	$K_p = 7.9 \times 10^{11}$
$H_2(g) \rightleftharpoons 2\ H(g)$	$K_p = 4.8 \times 10^{-41}$
$Br_2(g) \rightleftharpoons 2\ Br$	$K_p = 2.2 \times 10^{-15}$

Calculate K_p for the reaction $H + Br \rightarrow HBr$.

(a) $K_p = 8.3 \times 10^{-44}$ (b) $K_p = 7.5 \times 10^{66}$ (c) $K_p = 2.7 \times 10^{33}$

16.6 Disturbing a Chemical Equilibrium

The equilibrium between reactants and products may be disturbed in three ways: (1) by changing the temperature, (2) by changing the concentration of a reactant or product, or (3) by changing the volume (for systems involving gases) (Table 16.2).

Table 16.2 Effects of Disturbances on Equilibrium Composition

Disturbance	Change as Mixture Returns to Equilibrium	Effect on Equilibrium	Effect on K
Reactions Involving Solids, Liquids, or Gases			
Rise in temperature	Energy is consumed by system	Shift in endothermic direction	Change
Drop in temperature	Energy is generated by system	Shift in exothermic direction	Change
Addition of reactant*	Some of added reactant is consumed	Product concentration increases	No change
Addition of product*	Some of added product is consumed	Reactant concentration increases	No change
Reactions Involving Gases			
Decrease in volume, increase in pressure	Pressure decreases	Composition changes to reduce total number of gas molecules	No change
Increase in volume, decrease in pressure	Pressure increases	Composition changes to increase total number of gas molecules	No change

*Does not apply when an insoluble solid reactant or product is added. Recall that their "concentrations" do not appear in the reaction quotient.

A change in any of the factors that determine the equilibrium conditions of a system will cause the system to change in such a manner as to reduce or counteract the effect of the change. This statement is often referred to as *Le Chatelier's principle* (◄ page 483). It is a shorthand way of saying how the quantities of reactants and products are adjusted so that equilibrium is restored, that is, so that the reaction quotient is once again equal to the equilibrium constant.

Effect of the Addition or Removal of a Reactant or Product

If the concentration of a reactant or product is changed from its equilibrium value *at a given temperature*, equilibrium will be reestablished eventually. The new equilibrium concentrations of reactants and products will be different, but the value of the equilibrium constant expression will still equal K (Table 16.2). To illustrate this, let us return to the butane/isobutane equilibrium (with $K = 2.5$).

$$\underset{\text{butane}}{CH_3CH_2CH_2CH_3} \rightleftharpoons \underset{\text{isobutane}}{CH_3\overset{\overset{\textstyle CH_3}{|}}{C}HCH_3} \qquad K = 2.5$$

Suppose the equilibrium mixture consists of two molecules of butane and five molecules of isobutane (Figure 16.7). The reaction quotient, Q, is 5/2 (or 2.5/1), the

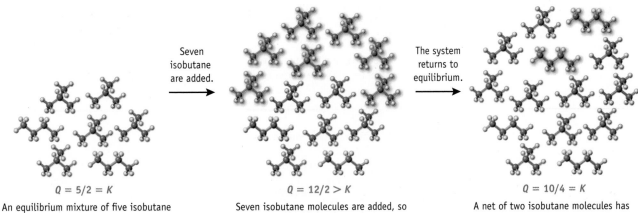

Seven isobutane are added.

The system returns to equilibrium.

$Q = 5/2 = K$

An equilibrium mixture of five isobutane molecules and two butane molecules.

$Q = 12/2 > K$

Seven isobutane molecules are added, so the system is no longer at equilibrium.

$Q = 10/4 = K$

A net of two isobutane molecules has changed to butane molecules, to once again give an equilibrium mixture where the ratio of isobutane to butane is 5:2 (or 2.5:1).

FIGURE 16.7 Addition of more reactant or product to an equilibrium system.

value of the equilibrium constant for the reaction. Now we add seven more molecules of isobutane to the mixture to give a ratio of 12 isobutane molecules to two butane molecules. The reaction quotient is now 6/1. This means Q is greater than K, and the system will change to reestablish equilibrium. To do so, some molecules of isobutane must be changed into butane molecules, a process that continues until the ratio [isobutane]/[butane] is once again 2.5/1. In this particular case, if two of the 12 isobutane molecules change to butane, the ratio of isobutane to butane is again equal to K ($= 10/4 = 2.5/1$), and equilibrium is reestablished.

Strategy Map 16.8

PROBLEM

What are the **equilibrium concentrations** of reactant and product after adding **excess reactant?**

↓

DATA/INFORMATION KNOWN

- Balanced equation
- Value of K_c
- **Initial concentrations** of **reactant** and **product** and amount of **excess reactant** added

STEP 1. Enter **concentration changes** in ICE table and derive **equilibrium concentrations** in terms of unknown quantity **x**.

Equilibrium concentrations of **reactant** and **product** are defined in terms of the unknown quantity **x**.

STEP 2. Enter **equilibrium concentrations** in K_c **expression** and solve for **x**.

Value of **x**

STEP 3. Use value of **x** to derive **equilibrium concentrations**.

Values of equilibrium concentrations

⏻WL   INTERACTIVE EXAMPLE 16.8   Effect of Concentration Changes on Equilibrium

Problem Assume equilibrium has been established in a 1.00-L flask with [butane] = 0.500 mol/L and [isobutane] = 1.25 mol/L.

$$\text{Butane} \rightleftharpoons \text{Isobutane} \qquad K = 2.50$$

Then 1.50 mol of butane is added. What are the concentrations of butane and isobutane when equilibrium is reestablished?

What Do You Know? Here you know the value of K, the balanced equation, the original equilibrium concentrations of reactant and product, and the amount of reactant added to the system at equilibrium.

Strategy After adding excess butane, $Q < K$. To reestablish equilibrium, the concentration of butane must decrease by an amount x and that of isobutane must increase, also by an amount x. Use an ICE table to track the changes.

Solution First organize the information in a modified ICE table.

Equation	Butane	⇌	Isobutane
Initial (M)	0.500		1.25
Concentration immediately on adding butane (M)	0.500 + 1.50		1.25
Change in concentration to reestablish equilibrium (M)	$-x$		$+x$
Equilibrium (M)	0.500 + 1.50 − x		1.25 + x

The entries in this table were arrived at as follows:

(a) The concentration of butane when equilibrium is reestablished will be the original equilibrium concentration plus what was added (1.50 mol/L) minus the concentration of butane that is converted to isobutane to reestablish equilibrium. The quantity of butane converted to isobutane is unknown and so is designated as x.

(b) The concentration of isobutane when equilibrium is reestablished is the concentration that was already present (1.25 mol/L) plus the concentration formed (x mol/L) on reestablishing equilibrium.

Having defined [butane] and [isobutane] when equilibrium is reestablished and remembering that K is a constant ($= 2.50$), we can write

$$K = 2.50 = \frac{[\text{isobutane}]}{[\text{butane}]}$$

We now calculate the new equilibrium composition:

$$2.50 = \frac{1.25 + x}{0.500 + 1.50 - x} = \frac{1.25 + x}{2.00 - x}$$

$$2.50\,(2.00 - x) = 1.25 + x$$

$$x = 1.07 \text{ mol/L}$$

[butane] = 0.500 + 1.50 − x = 0.93 M and [isobutane] = 1.25 + x = 2.32 M

Answer check: new ratio [isobutane]/[butane] = 2.32/0.93 = 2.5

Think about Your Answer It is useful to see a graphical solution to this problem because it emphasizes the point that, after adding excess butane, some butane is consumed on proceeding to the new equilibrium mixture and additional isobutane is formed. In the diagram here any mixture of butane and isobutane lying on the equilibrium line represents an equilibrium mixture. Mixtures in the blue portion of the diagram have $Q > K$, whereas those in the tan portion have $Q < K$. In this case, adding butane produced a mixture in the $Q < K$ region of the diagram. As predicted by Le Chatelier's principle, the added "stress" on the system is relieved by converting some butane to isobutane to achieve a new equilibrium mixture when $Q = K$.

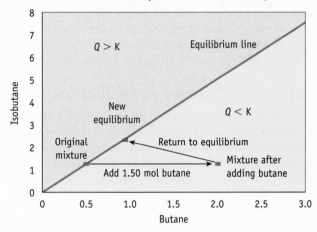

A graphical representation of the butane-isobutane problem.

Check Your Understanding

Equilibrium exists between butane and isobutane when [butane] = 0.20 M and [isobutane] = 0.50 M. An additional 2.00 mol/L of isobutane is added to the mixture. What are the concentrations of butane and isobutane after equilibrium has again been attained?

Effect of Volume Changes on Gas-Phase Equilibria

For a reaction that involves gases, what happens to equilibrium concentrations or pressures if the size of the container is changed? (Such a change occurs, for example, when fuel and air are compressed in an automobile engine.) To answer this question, recall that concentrations are in moles per liter. If the volume of a gas changes, its concentration therefore must also change, and the equilibrium composition can change. As an example, consider the following equilibrium:

$$2 \ NO_2(g) \rightleftharpoons N_2O_4(g)$$

brown gas   colorless gas

$$K_c = \frac{[N_2O_4]}{[NO_2]^2} = 170 \text{ at } 298 \text{ K}$$

What happens to this equilibrium if the volume of the flask holding the gases is suddenly halved? The immediate result is that the concentrations of both gases will double. For example, assume equilibrium is established when $[N_2O_4]$ is 0.0280 mol/L and $[NO_2]$ is 0.0128 mol/L. When the volume is halved, $[N_2O_4]$ becomes 0.0560 mol/L, and $[NO_2]$ is 0.0256 mol/L. The reaction quotient, Q, under these circumstances is $(0.0560)/(0.0256)^2 = 85.4$, a value less than K. Because Q is less than K, the quantity of product must increase at the expense of the reactants to return to equilibrium, and the new equilibrium composition will have a higher concentration of N_2O_4 than immediately after the volume change.

$$2 \ NO_2(g) \rightleftharpoons N_2O_4(g)$$

decrease volume of container
NO_2 is converted to N_2O_4
until equilibrium is attained

The concentration of NO_2 decreases twice as much as the concentration of N_2O_4 increases because one molecule of N_2O_4 is formed by consuming two molecules of

NO_2. This occurs until the reaction quotient, $Q = [N_2O_4]/[NO_2]^2$, is once again equal to K. The net effect of the volume decrease is to decrease the number of molecules in the gas phase.

The conclusions for the NO_2/N_2O_4 equilibrium can be generalized:

- For reactions involving gases, the stress of a volume decrease (a pressure increase) will be counterbalanced by a change in the equilibrium composition to one having a smaller number of gas molecules.
- For a volume increase (a pressure decrease), the equilibrium composition will favor the side of the reaction with the larger number of gas molecules.
- For a reaction in which there is no change in the number of gas molecules in proceeding from reactants to products, such as in the reaction of H_2 and I_2 to produce HI [$H_2(g) + I_2(g) \rightleftharpoons 2\ HI(g)$], a volume change will have no effect.

Effect of Temperature Changes on Equilibrium Composition

The value of the equilibrium constant for a given reaction varies with temperature. Changing the temperature of a system at equilibrium is therefore different in some ways from the other means we have studied of disturbing a chemical equilibrium because the equilibrium constant itself will be different at the new temperature from what it was at the previous temperature. Predicting the exact changes in equilibrium compositions with temperature is beyond the scope of this text, but you can make a qualitative prediction about the effect if you know whether the reaction is exothermic or endothermic. As an example, consider the endothermic reaction of N_2 with O_2 to give NO.

$$N_2(g) + O_2(g) \rightleftharpoons 2\ NO(g) \qquad \Delta_rH° = +180.6\ \text{kJ/mol-rxn}$$

$$K = \frac{[NO]^2}{[N_2][O_2]}$$

Le Chatelier's principle allows us to predict how the value of K will vary with temperature. The formation of NO from N_2 and O_2 is endothermic; that is, energy must be provided as heat for the reaction to occur. We might imagine that heat is a "reactant." If the system is at equilibrium and the temperature then increases, the system will adjust to alleviate this "stress." The way to counteract the energy input is to use up some of the energy added as heat by consuming N_2 and O_2 and producing more NO as the system returns to equilibrium. This raises the value of the numerator ($[NO]^2$) and lowers the value of the denominator ($[N_2][O_2]$) in the reaction quotient, Q, resulting in a higher value of K.

This prediction is borne out. The following table lists the equilibrium constant for this reaction at various temperatures. As predicted, the equilibrium constant and thus the proportion of NO in the equilibrium mixture increases with temperature.

Equilibrium Constant, K	Temperature (K)
4.5×10^{-31}	298
6.7×10^{-10}	900
1.7×10^{-3}	2300

As another example, consider the combination of molecules of the brown gas NO_2 to form colorless N_2O_4. An equilibrium between these compounds is readily achieved in a closed system (Figure 16.8).

$$2\ NO_2(g) \rightleftharpoons N_2O_4(g) \qquad \Delta_rH° = -57.1\ \text{kJ/mol-rxn}$$

$$K = \frac{[N_2O_4]}{[NO_2]^2}$$

Equilibrium Constant, K	Temperature (K)
1300	273
170	298

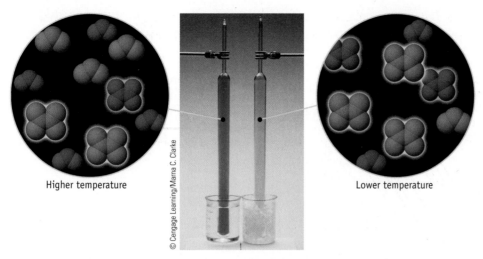

Higher temperature   Lower temperature

© Cengage Learning/Marna C. Clarke

FIGURE 16.8 Effect of temperature on an equilibrium. The tubes in the photograph both contain gaseous NO_2 (brown) and N_2O_4 (colorless) at equilibrium. K is larger at the lower temperature because the equilibrium favors colorless N_2O_4. This is clearly seen in the tube at the right, where the gas in the ice bath at 0 °C is only slightly brown, which indicates a smaller concentration of the brown gas NO_2. At 50 °C (the tube at the left), the equilibrium is shifted toward NO_2, as indicated by the darker brown color.

Here, the reaction is exothermic, so we might imagine heat as being a reaction "product." By lowering the temperature of the system, as in Figure 16.8, some energy is removed as heat. The removal of energy can be counteracted if the reaction produces energy as heat by the combination of NO_2 molecules to give more N_2O_4. Thus, the equilibrium concentration of NO_2 decreases; the concentration of N_2O_4 increases; and the value of K is larger at lower temperatures.

In summary,

- when the temperature of a system at equilibrium increases, the equilibrium will shift in the direction that absorbs energy as heat (Table 16.2)—that is, in the endothermic direction.
- if the temperature decreases, the equilibrium will shift in the direction that releases energy as heat—that is, in the exothermic direction.
- changing the temperature changes the value of K.

REVIEW & CHECK FOR SECTION 16.6

The formation of ammonia from its elements is an important industrial process.

$$3 H_2(g) + N_2(g) \rightleftharpoons 2 NH_3(g)$$

1. Does the reaction shift to the right or to the left, or does it remain unchanged, when extra H_2 is added?

    (a) shift left   (c) unchanged

    (b) shift right

2. Does the reaction shift to the right or to the left, or does it remain unchanged, when the volume of the system is increased?

    (a) shift left   (c) unchanged

    (b) shift right

3. For the equilibrium $2 SO_2(g) + O_2(g) \rightleftharpoons 2 SO_3(g)$, K has the following values: 4.0×10^{24} at 300 K, 2.5×10^{10} at 500 K, and 3.0×10^4 at 700 K. Is this reaction exothermic or endothermic?

    (a) exothermic   (b) endothermic

CASE STUDY

Applying Equilibrium Concepts—The Haber-Bosch Ammonia Process

Nitrogen-containing substances are used around the world to stimulate the growth of field crops. Farmers from Portugal to Tibet have used animal waste for centuries as a "natural" fertilizer. In the 19th century, industrialized countries imported nitrogen-rich marine bird manure from Peru, Bolivia, and Chile, but the supply of this material was clearly limited. In 1898, William Ramsay (the discoverer of the noble gases) pointed out that the amount of "fixed nitrogen" in the world was being depleted, and Ramsay predicted that world food shortages would occur by the mid-20th century as a result. That Ramsay's prediction failed to materialize was due in part to the work of Fritz Haber (1868–1934). In about 1908, Haber developed a method to make ammonia directly from the elements,

$$N_2(g) + 3\ H_2(g) \rightleftharpoons 2\ NH_3(g)$$

and, a few years later, Carl Bosch (1874–1940) perfected the industrial scale synthesis. Ammonia is now made for pennies per kilogram and is consistently ranked in the top five chemicals produced in the United States, with 15–20 billion kilograms produced annually. Not only is ammonia used directly as a fertilizer, but it is also a starting material for making nitric acid and ammonium nitrate, among other things.

The manufacture of ammonia (Figure A) is a good example of the role that kinetics and chemical equilibria play in practical chemistry.

The $N_2 + H_2$ reaction is exothermic and product-favored ($K > 1$ at 25 °C).

At 25 °C, K (calc'd value) = 3.5×10^8
and $\Delta_r H° = -92.2$ kJ/mol-rxn

Unfortunately, the reaction at 25 °C is slow, so it is carried out at a higher temperature to increase the reaction rate. The problem with this, however, is that the equilibrium constant declines with temperature, as predicted by Le Chatelier's principle.

At 450 °C, K (experimental value) = 0.16
and $\Delta_r H° = -111.3$ kJ/mol-rxn

Thus, the yield declines with increasing temperature.

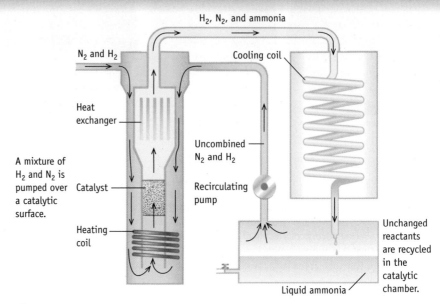

Figure A **The Haber-Bosch process for ammonia synthesis.** A mixture of H_2 and N_2 is pumped over a catalytic surface. The NH_3 is collected as a liquid (at −33 °C), and unchanged reactants are recycled in the catalytic chamber.

There are two things that can be done. The first is to raise the pressure. This does not change the value of K, but an increase in pressure can be compensated by converting 4 mol of reactants to 2 mol of product.

In an industrial ammonia plant, it is necessary to balance reaction rate (improved at higher temperature) with product yield (K is smaller at higher temperatures). Additionally, a catalyst is often used to accelerate the reaction. An effective catalyst for the Haber-Bosch process is Fe_3O_4 mixed with KOH, SiO_2, and Al_2O_3 (all inexpensive chemicals). Because the catalyst is not effective below 400 °C, the process is carried out at 450–500 °C and 250 atm pressure.

Questions:

1. Anhydrous ammonia is used directly as a fertilizer, but much of it is also converted to other fertilizers, ammonium nitrate and urea.
   (a) How is NH_3 converted to ammonium nitrate?

   (b) Urea is formed in the reaction of ammonia and CO_2.

   $$2\ NH_3(g) + CO_2(g) \rightleftharpoons (NH_2)_2CO(s) + H_2O(g)$$

   Which would favor urea production, high temperature or high pressure? ($\Delta_f H°$ for solid urea = −333.1 kJ/mol-rxn)

2. One important aspect of the Haber-Bosch process is the source of the hydrogen. This is made from natural gas in a process called *steam reforming*.

   $$CH_4(g) + H_2O(g) \rightarrow CO(g) + 3\ H_2(g)$$
   $$CO(g) + H_2O(g) \rightarrow CO_2(g) + H_2(g)$$

   (a) Are the two reactions above endo- or exothermic?
   (b) To obtain the H_2 necessary to manufacture 15 billion kilograms of NH_3, how much CH_4 is required, and what mass of CO_2 is produced as a by-product?

Answers to these questions are available in Appendix N.

CHAPTER GOALS REVISITED

Now that you have studied this chapter, you should ask whether you have met the chapter goals. In particular, you should be able to:

Understand the nature and characteristics of chemical equilibria

a.   Chemical reactions are reversible and equilibria are dynamic (Section 16.1).

Understand the significance of the equilibrium constant, _K_, and reaction quotient, _Q_.

a. Write the reaction quotient, _Q_, for a chemical reaction (Section 16.2). When the system is at equilibrium, the reaction quotient is called the _equilibrium constant expression_ and has a constant value called the _equilibrium constant_, which is symbolized by _K_ (Equation 16.1). **Study Questions: 1–4.**

b. Recognize that the concentrations of solids, pure liquids, and solvents (e.g., water) are not included in the equilibrium constant expression (Section 16.2).

c. Recognize that a large value of _K_ (_K_ > 1) means the reaction is product-favored, and the product concentrations are greater than the reactant concentrations at equilibrium. A small value of _K_ (_K_ < 1) indicates a reactant-favored reaction in which the product concentrations are smaller than the reactant concentrations at equilibrium (Section 16.2). **Study Questions: 66, 68.**

d. Appreciate the fact that equilibrium concentrations may be expressed in terms of reactant and product concentrations (in moles per liter) and that _K_ is then sometimes designated as _K_c. Alternatively, concentrations of gases may be represented by partial pressures, and _K_ for such cases is designated _K_p (Section 16.2). **Study Questions: 53, 60.**

Understand how to use _K_ in quantitative studies of chemical equilibria

a. Use the reaction quotient (_Q_) to decide whether a reaction is at equilibrium (_Q_ = _K_), or if there will be a net conversion of reactants to products (_Q_ < _K_) or products to reactants (_Q_ > _K_) to attain equilibrium (Section 16.2). **Study Questions: 3–6, 35, 36.**

b. Calculate an equilibrium constant given the reactant and product concentrations at equilibrium (Section 16.3). **Study Questions: 7–11, 29, 33, 34, 44, 61a.**

c. Use equilibrium constants to calculate the concentration (or pressure) of a reactant or a product at equilibrium (Section 16.4). **Study Questions: 16, 17, 32, 36, 42, 46, 47, 50–54, 56, 59–61, and Go Chemistry Module 21.**

d. Know how _K_ changes as different stoichiometric coefficients are used in a balanced equation, if the equation is reversed, or if several equations are added to give a new net equation (Section 16.5). **Study Questions: 19–24, 31, 37.**

Predict how a system at equilibrium will respond if the reaction conditions are changed.

a. Know how to predict, using Le Chatelier's principle, the effect of a disturbance on a chemical equilibrium—a change in temperature, a change in concentrations, or a change in volume or pressure for a reaction involving gases (Section 16.6 and Table 16.2). **Study Questions: 25–28, 39, 41, 54, 62.**

Key Equations

Equation 16.1 (page 551) The equilibrium constant expression. At equilibrium, the ratio of products to reactants (each raised to the power corresponding to its stoichiometric coefficient) has a constant value, _K_ (at a particular temperature). For the general reaction $aA + bB \rightleftharpoons cC + dD$,

$$\text{Equilibrium constant} = K = \frac{[C]^c[D]^d}{[A]^a[B]^b}$$

Equation 16.2 (page 555) For the general reaction $aA + bB \rightleftharpoons cC + dD$, the ratio of product to reactant concentrations at any point in the reaction is the reaction quotient, _Q_.

$$\text{Reaction quotient} = Q = \frac{[C]^c[D]^d}{[A]^a[B]^b}$$

Equation 16.3 (page 564) This approximation is used to solve for the equilibrium concentrations of _B_ and _C_ (= _x_) in the general reaction $A \rightleftharpoons B + C$ when the value of $100 \times K$ is less than the original concentration of $A (= [A]_0)$.

$$K = \frac{[B][C]}{[A]} \approx \frac{(x)(x)}{[A]_0}$$

17 Principles of Chemical Reactivity: The Chemistry of Acids and Bases

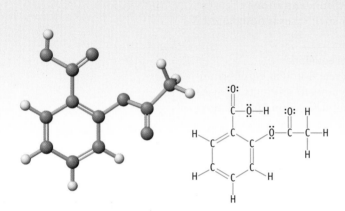

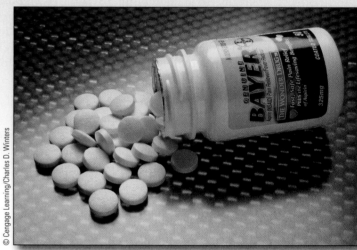

© Cengage Learning/Charles D. Winters

Aspirin Is Over 100 Years Old!

Aspirin is one of the most successful nonprescription drugs ever made. Americans swallow more than 50 million aspirin tablets a day, mostly for the pain-relieving (analgesic) effects of the drug. Aspirin also wards off heart disease and thrombosis (blood clots), and it has even been suggested as a possible treatment for certain cancers and for senile dementia.

Hippocrates (460–370 BC), the ancient Greek physician, recommended an infusion of willow bark to ease the pain of childbirth. It was not until the 19th century that an Italian chemist, Raffaele Piria, isolated salicylic acid, the active compound in the bark. Soon thereafter, it was found that the acid could be extracted from a wild flower, *Spiraea ulmaria*. It is from the name of this plant that the name "aspirin" (a + spiraea) is derived.

Hippocrates's willow bark extract, salicylic acid, is an analgesic, but it is also very irritating to the stomach lining. It was therefore an important advance when chemists at Bayer Chemicals in Germany found, in 1897, that a derivative of salicylic acid, acetylsalicylic acid, was also a useful drug and had fewer side effects. This is the compound we now call "aspirin."

Acetylsalicylic acid slowly reverts to salicylic acid, and acetic acid in the presence of moisture; therefore, if you smell the characteristic odor of acetic acid in an old bottle of aspirin, the tablets are too old and should be discarded.

Aspirin (acetylsalicylic acid) is a component of various over-the-counter medicines, such as Anacin, Ecotrin, Excedrin, and Alka-Seltzer. The latter is a combination of aspirin with citric acid and sodium bicarbonate. Sodium bicarbonate is a base and reacts with the acid to produce the sodium salt of acetylsalicylic acid, a form of aspirin that is water-soluble and quicker acting.

Questions:

1. Aspirin has a pK_a of 3.49, and that of acetic acid is 4.74. Which is the stronger acid?
2. Identify the acidic H atom in aspirin.
3. Write an equation for the ionization of aspirin.

Answers to these questions are available in Appendix N.

CHAPTER OUTLINE

17.1 Acids and Bases: A Review

17.2 The Brønsted–Lowry Concept of Acids and Bases Extended

17.3 Water and the pH Scale

17.4 Equilibrium Constants for Acids and Bases

17.5 Acid–Base Properties of Salts

17.6 Predicting the Direction of Acid–Base Reactions

17.7 Types of Acid–Base Reactions

17.8 Calculations with Equilibrium Constants **go Chemistry**

17.9 Polyprotic Acids and Bases

17.10 Molecular Structure, Bonding, and Acid–Base Behavior

17.11 The Lewis Concept of Acids and Bases

CHAPTER GOALS

See Chapter Goals Revisited (page 616) for Study Questions keyed to these goals.

- Use the Brønsted–Lowry and Lewis theories of acids and bases.

- Apply the principles of chemical equilibrium to acids and bases in aqueous solution.

- Predict the outcome of reactions of acids and bases.

- Understand the influence of structure and bonding on acid–base properties.

OWL

Sign in to OWL at **www.cengage.com/ owl** to view tutorials and simulations, develop problem-solving skills, and complete online homework assigned by your professor.

Acids and bases are among the most common substances in nature. Amino acids are the building blocks of proteins. The repository of genetic information in your cells is DNA, deoxyribonucleic acid. The pH of the lakes, rivers, and oceans is affected by dissolved acids and bases, and many bodily functions depend on acids and bases. You were introduced to acids and bases and to some of their chemistry in Chapter 3, but this chapter and the next take up the chemistry of this important class of substances in more detail.

17.1 Acids and Bases: A Review

In Chapter 3, you were introduced to two definitions of acids and bases: the Arrhenius definition and the Brønsted–Lowry definition. According to the Arrhenius definition, an acid is any substance that, when dissolved in water, increases the concentration of hydrogen ions, H^+ (◄ page 110). An Arrhenius base is any substance that increases the concentration of hydroxide ions, OH^-, when dissolved in water. Based on the Arrhenius definition, hydrochloric acid is therefore classified as an acid, and sodium hydroxide is classified as a base.

$$HCl(aq) \rightarrow H^+(aq) + Cl^-(aq)$$

$$NaOH(aq) \rightarrow Na^+(aq) + OH^-(aq)$$

Using this definition, reactions between acids and bases involve the combination of H^+ and OH^- ions to form water (and a salt).

$$NaOH(aq) + HCl(aq) \rightarrow H_2O(\ell) + NaCl(aq)$$

The Brønsted–Lowry definition of acids and bases is more general and views acid–base behavior in terms of proton transfer from one substance to another. *A Brønsted–Lowry acid is a proton (H^+) donor, and a Brønsted–Lowry base is a proton acceptor*, a definition that extends the list of acids and bases and the scope of acid–base reactions. In the following reaction, HCl acts as a Brønsted–Lowry acid, and water acts

go Chemistry

Download mini lecture videos for key concept review and exam prep from OWL or purchase them from **www.cengagebrain.com**

as a Brønsted–Lowry base because HCl transfers a H^+ ion to H_2O to form the hydronium ion, H_3O^+.

$$HCl(aq) + H_2O(\ell) \rightleftharpoons H_3O^+(aq) + Cl^-(aq)$$

We will begin this chapter by looking at Brønsted–Lowry acid–base chemistry in more detail.

REVIEW & CHECK FOR SECTION 17.1

Which of the following is a list of Brønsted–Lowry acids?

(a) CH_3CO_2H, $Al(OH)_3$, H_3PO_4

(b) NH_4^+, NH_3, HCl

(c) H_2CO_3, CH_3CO_2H, H_3PO_4

17.2 The Brønsted–Lowry Concept of Acids and Bases Extended

A wide variety of Brønsted–Lowry acids is known. These include some molecular compounds such as nitric acid,

$$HNO_3(aq) \quad + \quad H_2O(\ell) \rightleftharpoons NO_3^-(aq) + H_3O^+(aq)$$
$$\text{acid}$$

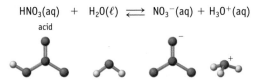

cations such as NH_4^+,

$$NH_4^+(aq) \quad + \quad H_2O(\ell) \rightleftharpoons NH_3(aq) + H_3O^+(aq)$$
$$\text{acid}$$

and anions.

$$H_2PO_4^-(aq) \quad + \quad H_2O(\ell) \quad \rightleftharpoons \quad HPO_4^{2-}(aq) \quad + \quad H_3O^+(aq)$$

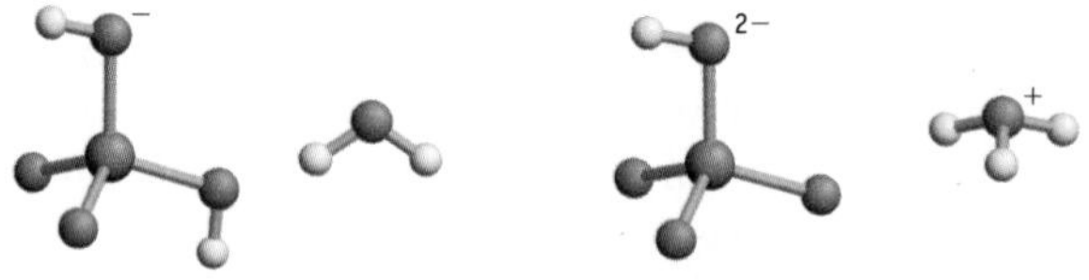

Similarly, many different types of species can act as Brønsted–Lowry bases in their reactions with water. These include some molecular compounds,

$$NH_3(aq) \quad + \quad H_2O(\ell) \rightleftharpoons NH_4^+(aq) + OH^-(aq)$$
$$\text{base}$$

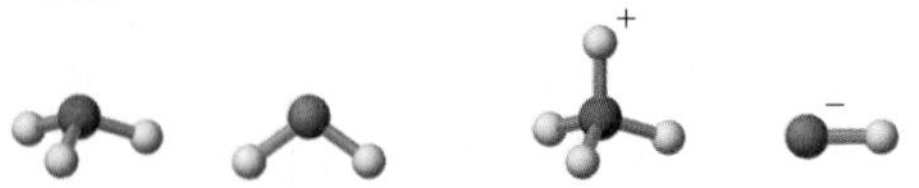

and anions.

$$CO_3^{2-}(aq) \quad + \quad H_2O(\ell) \quad \rightleftharpoons \quad HCO_3^-(aq) \quad + \quad OH^-(aq)$$

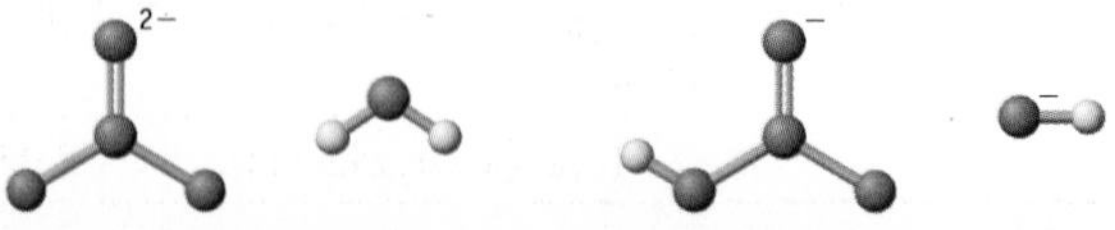

Hydrated metal cations can also function as acids and bases. Some of these ions act as Brønsted–Lowry acids:

$$[Fe(H_2O)_6]^{3+}(aq) + H_2O(\ell) \rightleftharpoons [Fe(H_2O)_5(OH)]^{2+}(aq) + H_3O^+(aq)$$

while others act as Brønsted–Lowry bases.

$$[Al(H_2O)_5(OH)]^{2+}(aq) + H_2O(\ell) \rightleftharpoons [Al(H_2O)_6]^{3+}(aq) + OH^-(aq)$$

Acids such as HF, HCl, HNO₃, and CH₃CO₂H (acetic acid) are all capable of donating one proton and so are called **monoprotic acids.** Other acids, called **polyprotic acids** (Table 17.1), are capable of donating two or more protons. Sulfuric acid is a familiar example of a polyprotic acid.

$$H_2SO_4(aq) \quad + \quad H_2O(\ell) \rightleftharpoons HSO_4^-(aq) \quad + \quad H_3O^+(aq)$$

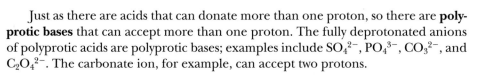

$$HSO_4^-(aq) \quad + \quad H_2O(\ell) \rightleftharpoons SO_4^{2-}(aq) \quad + \quad H_3O^+(aq)$$

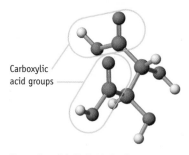

Tartaric acid, $H_2C_4H_4O_6$, is a naturally occurring diprotic acid. Tartaric acid and its potassium salt are found in grapes and other fruits. The acidic protons are the H atoms of the $-CO_2H$ or carboxylic acid groups.

Just as there are acids that can donate more than one proton, so there are **polyprotic bases** that can accept more than one proton. The fully deprotonated anions of polyprotic acids are polyprotic bases; examples include SO_4^{2-}, PO_4^{3-}, CO_3^{2-}, and $C_2O_4^{2-}$. The carbonate ion, for example, can accept two protons.

$$CO_3^{2-}(aq) + H_2O(\ell) \rightleftharpoons HCO_3^-(aq) + OH^-(aq)$$
base

$$HCO_3^-(aq) + H_2O(\ell) \rightleftharpoons H_2CO_3(aq) + OH^-(aq)$$
base

Some molecules (such as water) and ions can behave either as Brønsted acids or bases and are referred to as being **amphiprotic** (◀ page 113). An example of an amphiprotic anion that is particularly important in biochemical systems is the dihydrogen phosphate anion (Table 17.1).

$$H_2PO_4^-(aq) + H_2O(\ell) \rightleftharpoons H_3O^+(aq) + HPO_4^{2-}(aq)$$
acid

$$H_2PO_4^-(aq) + H_2O(\ell) \rightleftharpoons H_3PO_4(aq) + OH^-(aq)$$
base

Table 17.1 Polyprotic Acids and Bases

Acid Form	Amphiprotic Form	Base Form
H₂S (hydrosulfuric acid or hydrogen sulfide)	HS⁻ (hydrogen sulfide ion)	S²⁻ (sulfide ion)
H₃PO₄ (phosphoric acid)	H₂PO₄⁻ (dihydrogen phosphate ion) HPO₄²⁻ (hydrogen phosphate ion)	PO₄³⁻ (phosphate ion)
H₂CO₃ (carbonic acid)	HCO₃⁻ (hydrogen carbonate ion or bicarbonate ion)	CO₃²⁻ (carbonate ion)
H₂C₂O₄ (oxalic acid)	HC₂O₄⁻ (hydrogen oxalate ion)	C₂O₄²⁻ (oxalate ion)

Conjugate Acid–Base Pairs

The reaction of the hydrogen corbonate ion and water exemplifies a feature of Brønsted acid–base chemistry: a Brønsted acid and base produce a new acid and base.

conjugate pair 1

conjugate pair 2

$$\underset{\text{base}}{HCO_3^-(aq)} + \underset{\text{acid}}{H_2O(\ell)} \rightleftharpoons \underset{\text{base}}{OH^-(aq)} + \underset{\text{acid}}{H_2CO_3(aq)}$$

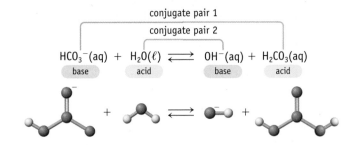

In the forward direction, HCO_3^- is the Brønsted base because it captures H^+ from the Brønsted acid, H_2O. The products are a new Brønsted acid and base. In the reverse direction, the H_2CO_3 is the acid, and OH^- is the base.

A **conjugate acid–base pair** consists of two species that differ from each other by the presence of one hydrogen ion. Thus, H_2CO_3 and HCO_3^- comprise a conjugate acid–base pair. In this pair, HCO_3^- is the conjugate base of the acid H_2CO_3, and H_2CO_3 is the conjugate acid of the base HCO_3^-. There is a second conjugate acid–base pair in this reaction: H_2O and OH^-. In fact, *every reaction between a Brønsted acid and a Brønsted base involves two conjugate acid–base pairs* (Table 17.2).

REVIEW & CHECK FOR SECTION 17.2

1. H_3PO_4, phosphoric acid, can donate two protons to water to form the monohydrogen phosphate ion, HPO_4^{2-}. Is the monohydrogen phosphate ion an acid, a base, or amphiprotic?

   (a) acid (b) base (c) amphiprotic

2. The cyanide ion, CN^-, accepts a proton from water to form HCN. Is CN^- a Brønsted acid or base or is it amphiprotic?

   (a) acid (b) base (c) amphiprotic

3. In the following reaction, identify the acid on the left and its conjugate base on the right.

   $$HNO_3(aq) + NH_3(aq) \rightleftharpoons NH_4^+(aq) + NO_3^-(aq)$$

   (a) acid = NH_3 and conjugate base = NH_4^+

   (b) acid = HNO_3 and conjugate base = NO_3^-

   Similarly, identify the base on the left and its conjugate acid on the right.

   (a) base = NH_3 and conjugate acid = NH_4^+

   (b) base = HNO_3 and conjugate acid = NO_3^-

4. Identify the conjugate acid/base pairs in the reaction of HF and acetic acid.

   $$HF(aq) + CH_3CO_2^-(aq) \rightleftharpoons F^-(aq) + CH_3CO_2H(aq)$$

   (a) $HF/CH_3CO_2^-$ and F^-/CH_3CO_2H

   (b) HF/CH_3CO_2H and $F^-/CH_3CO_2^-$

   (c) HF/F^- and $CH_3CO_2H/CH_3CO_2^-$

17.3 Water and the pH Scale

Because we generally use aqueous solutions of acids and bases and because the acid–base reactions in your body occur in your aqueous interior, we want to consider the behavior of water in terms of chemical equilibria.

Table 17.2 Acid–Base Reactions and Conjugate Acid–Base Pairs*

Name	Acid 1		Base 2		Base 1		Acid 2
Hydrochloric acid	HCl	+	H_2O	$\rightleftharpoons$	Cl^-	+	H_3O^+
Nitric acid	HNO_3	+	H_2O	$\rightleftharpoons$	NO_3^-	+	H_3O^+
Carbonic acid	H_2CO_3	+	H_2O	$\rightleftharpoons$	HCO_3^-	+	H_3O^+
Acetic acid	CH_3CO_2H	+	H_2O	$\rightleftharpoons$	$CH_3CO_2^-$	+	H_3O^+
Hydrocyanic acid	HCN	+	H_2O	$\rightleftharpoons$	CN^-	+	H_3O^+
Hydrogen sulfide	H_2S	+	H_2O	$\rightleftharpoons$	HS^-	+	H_3O^+
Ammonia	H_2O	+	NH_3	$\rightleftharpoons$	OH^-	+	NH_4^+
Carbonate ion	H_2O	+	CO_3^{2-}	$\rightleftharpoons$	OH^-	+	HCO_3^-
Water	H_2O	+	H_2O	$\rightleftharpoons$	OH^-	+	H_3O^+

*Acid 1 and base 1 are a conjugate pair, as are base 2 and acid 2.

Water Autoionization and the Water Ionization Constant, K_w

An acid such as HCl does not need to be present for the hydronium ion to exist in water. In fact, two water molecules can interact with each other to produce a hydronium ion and a hydroxide ion by proton transfer from one water molecule to the other.

$$2\ H_2O(\ell) \rightleftharpoons H_3O^+(aq) + OH^-(aq)$$

This **autoionization** reaction of water was demonstrated a century ago by Friedrich Kohlrausch (1840–1910). He found that, even after water is painstakingly purified, it still conducts electricity to a very small extent. We now know this is because autoionization produces very low concentrations of H_3O^+ and OH^- ions. *Water autoionization is the cornerstone of our concepts of aqueous acid–base behavior.*

The water autoionization equilibrium lies far to the left side. In fact, in pure water at 25 °C, only about two out of a billion (10^9) water molecules are ionized at any instant. To express this idea more quantitatively, we can write the equilibrium constant expression for autoionization.

$$K_w = [H_3O^+][OH^-] = 1.0 \times 10^{-14} \text{ at 25 °C} \tag{17.1}$$

There are several important aspects of this equation.

- Based on the rules for writing equilibrium constants, we do not include the concentration of water.
- The equilibrium constant is given a special symbol, K_w, and is known as the **autoionization constant for water**.
- Because autoionization is the only source of hydronium and hydroxide ions in pure water, we know that $[H_3O^+]$ must be equal to $[OH^-]$. Electrical conductivity measurements of pure water show that $[H_3O^+] = [OH^-] = 1.0 \times 10^{-7}$ M at 25 °C, so K_w has a value of 1.0×10^{-14} at 25 °C.

In pure water, the hydronium ion and hydroxide ion concentrations are equal, and the water is said to be neutral. If some acid or base is added to pure water, however, the equilibrium

$$2\ H_2O(\ell) \rightleftharpoons H_3O^+(aq) + OH^-(aq)$$

● K_w **and Temperature** The equation $K_w = [H_3O^+][OH^-]$ is valid for pure water and for any aqueous solution. The numerical value for K_w, however, is temperature dependent. Because the autoionization reaction is endothermic, K_w increases with temperature.

T (°C)	K_w
10	0.29×10^{-14}
15	0.45×10^{-14}
20	0.68×10^{-14}
25	1.01×10^{-14}
30	1.47×10^{-14}
50	5.48×10^{-14}

● **The Magnitude of $[H_3O^+]$ in Pure Water** The ionization of two water molecules out of a billion produces a H_3O^+ concentration of 1×10^{-7} M. To have a sense of this tiny amount, this would be like identifying just 14 people out of the current population of the earth, about 7 billion.

is disturbed. Adding acid raises the concentration of the H_3O^+ ions, so the solution is acidic. To oppose this increase, Le Chatelier's principle (◀ Section 16.6) predicts that a small fraction of the H_3O^+ ions will react with OH^- ions from water autoionization to form water. This lowers $[OH^-]$ until the product of $[H_3O^+]$ and $[OH^-]$ is again equal to 1.0×10^{-14} at 25 °C. Similarly, adding a base to pure water gives a basic solution because the OH^- ion concentration has increased. Le Chatelier's principle predicts that some of the added OH^- ions will react with H_3O^+ ions present in the solution from water autoionization, thereby lowering $[H_3O^+]$ until the value of the product of $[H_3O^+]$ and $[OH^-]$ equals 1.0×10^{-14} at 25 °C.

Thus, for aqueous solutions at 25 °C, we can say that

- In a neutral solution, $[H_3O^+] = [OH^-]$.
  Both are equal to 1.0×10^{-7} M.
- In an acidic solution, $[H_3O^+] > [OH^-]$.
  $[H_3O^+] > 1.0 \times 10^{-7}$ M and $[OH^-] < 1.0 \times 10^{-7}$ M.
- In a basic solution, $[H_3O^+] < [OH^-]$.
  $[H_3O^+] < 1.0 \times 10^{-7}$ M and $[OH^-] > 1.0 \times 10^{-7}$ M.

EXAMPLE 17.1   Hydronium and Hydroxide Ion Concentrations in a Solution of a Strong Base

Problem What are the hydroxide and hydronium ion concentrations in a 0.0012 M solution of NaOH at 25 °C?

What Do You Know? You know the concentration of NaOH and that it is a strong base, 100% dissociated into ions in water.

Strategy Because NaOH is a strong base we assume that the OH^- ion concentration is the same the NaOH concentration. The H_3O^+ ion concentration can then be calculated using Equation 17.1.

Solution The initial concentration of OH^- is 0.0012 M.

$$0.0012 \text{ mol NaOH per liter} \rightarrow 0.0012 \text{ M Na}^+(aq) + 0.0012 \text{ M OH}^-(aq)$$

Substituting the OH^- concentration into Equation 17.1, we have

$$K_w = 1.0 \times 10^{-14} = [H_3O^+][OH^-] = [H_3O^+](0.0012)$$

and so

$$[H_3O^+] = \frac{1.0 \times 10^{-14}}{0.0012} = 8.3 \times 10^{-12} \text{ M}$$

Think about Your Answer Why didn't we take into account the ions produced by water autoionization when we calculated the concentration of hydroxide ions? It should add OH^- and H_3O^+ ions to the solution. If *x is equal to the concentration of OH^-* ions generated by the autoionization of water, then, when equilibrium is achieved,

$$[OH^-] = (0.0012 \text{ M} + OH^- \text{ from water autoionization}) = (0.0012 \text{ M} + x)$$

In pure water, the concentration of OH^- ion generated by autoionization is 1.0×10^{-7} M. Le Chatelier's principle (◀ Section 16.6) suggests that the concentration should be even smaller when OH^- ions are already present in solution from NaOH; that is, x should be much less than 1.0×10^{-7} M. This means x in the term $(0.0012 + x)$ is insignificant compared with 0.0012. (Following the rules for significant figures, the sum of 0.0012 and a number even smaller than 1.0×10^{-7} is 0.0012.) Thus, the equilibrium concentration of OH^- is equivalent to the concentration of NaOH in the solution.

Check Your Understanding

A solution of the strong acid HCl has $[HCl] = 4.0 \times 10^{-3}$ M. What are the concentrations of H_3O^+ and OH^- in this solution at 25 °C? (Recall that because HCl is a strong acid, it is 100% ionized in water.)

The pH Scale

The **pH** of a solution is defined as the negative of the base-10 logarithm (log) of the hydronium ion concentration (◄ Section 4.6, page 154).

$$pH = -\log[H_3O^+] \qquad \text{(4.3 and 17.2)}$$

In a similar way, we can define the pOH of a solution as the negative of the base-10 logarithm of the hydroxide ion concentration.

$$pOH = -\log[OH^-] \qquad \text{(17.3)}$$

• **Working with Logarithms** See Appendix A for more on using logarithms.

In pure water, the hydronium and hydroxide ion concentrations are both 1.0×10^{-7} M. Therefore, for pure water at 25 °C

$$pH = -\log(1.0 \times 10^{-7}) = 7.00$$

In the same way, you can show that the pOH of pure water is also 7.00 at 25 °C.

If we take the negative logarithms of both sides of the expression $K_w = [H_3O^+][OH^-]$, we obtain another useful equation.

$$K_w = 1.0 \times 10^{-14} = [H_3O^+][OH^-]$$
$$-\log K_w = -\log(1.0 \times 10^{-14}) = -\log([H_3O^+][OH^-])$$
$$pK_w = 14.00 = -\log([H_3O^+]) + (-\log[OH^-])$$

$$pK_w = 14.00 = pH + pOH \qquad \text{(17.4)}$$

The sum of the pH and pOH of a solution must be equal to 14.00 at 25 °C.

As illustrated in Figures 4.11 and 17.1, solutions with pH less than 7.00 (at 25 °C) are acidic, whereas solutions with pH greater than 7.00 are basic. Solutions with pH = 7.00 at 25 °C are neutral.

Calculating pH

The calculation of pH from the hydronium ion concentration, or the concentration of hydronium ion from pH, was introduced in Chapter 4 (◄ page 154). The *Review & Check* questions below review those calculations.

REVIEW & CHECK FOR SECTION 17.3

1. What is the pH of a 0.0012 M NaOH solution at 25 °C?

   (a) 2.92   (b) 11.08   (c) 8.67

2. The pH of a diet soda is 4.32 at 25 °C. What is the hydronium ion concentration in the soda?

   (a) 4.8×10^{-5} M   (b) 2.1×10^{-10} M   (c) 2.1×10^{-4} M

3. If the pH of a solution containing the strong base $Sr(OH)_2$ is 10.46 at 25 °C, what is the concentration of $Sr(OH)_2$?

   (a) 3.5×10^{-11} M   (b) 2.9×10^{-4} M   (c) 6.9×10^{-11} M   (d) 1.4×10^{-4} M

---

	pH	$[H_3O^+]$	$[OH^-]$	pOH
Basic	14.00	1.0×10^{-14}	1.0×10^0	0.00
	10.00	1.0×10^{-10}	1.0×10^{-4}	4.00
Neutral	7.00	1.0×10^{-7}	1.0×10^{-7}	7.00
	4.00	1.0×10^{-4}	1.0×10^{-10}	10.00
Acidic	0.00	1.0×10^0	1.0×10^{-14}	14.00

FIGURE 17.1 pH and pOH. This figure illustrates the relationship of hydronium ion and hydroxide ion concentrations and of pH and pOH.

17.4 Equilibrium Constants for Acids and Bases

In Chapter 3, it was stated that acids and bases can be divided roughly into those that are strong electrolytes (such as HCl, HNO$_3$, and NaOH) and those that are weak electrolytes (such as CH$_3$CO$_2$H and NH$_3$) (Figure 17.2) (◄ Table 3.1, Common Acids and Bases, page 111). Hydrochloric acid is a strong acid, so 100% of the acid ionizes to produce hydronium and chloride ions. In contrast, acetic acid is a weak electrolyte because it ionizes only to a very small extent in water.

$$CH_3CO_2H(aq) + H_2O(\ell) \rightleftharpoons H_3O^+(aq) + CH_3CO_2^-(aq)$$

The acid, its anion, and the hydronium ion are all present at equilibrium in solution, but the ions are present in very low concentration relative to the acid concentration. For example, in a 0.100 M solution of acetic acid, [H$_3$O$^+$] and [CH$_3$CO$_2^-$] are each about 0.0013 M whereas that of un-ionized acetic acid, [CH$_3$CO$_2$H], is 0.099 M.

Similarly, ammonia is a weak base.

$$NH_3(aq) + H_2O(\ell) \rightleftharpoons NH_4^+(aq) + OH^-(aq)$$

Only about 1% of ammonia molecules in a 0.100 M solution react with water to produce the ammonium and hydroxide ions.

One way to define the relative strengths of a series of acids is to measure the pH of solutions of acids of equal concentration: the lower the pH, the greater the concentration of hydronium ion, the stronger the acid. Similarly, for a series of weak bases, [OH$^-$] will increase, and the pH will increase as the bases become stronger.

- For a strong monoprotic acid, [H$_3$O$^+$] in solution is equal to the original acid concentration. Similarly, for a strong monoprotic base, [OH$^-$] will be equal to the original base concentration.
- For a weak acid, [H$_3$O$^+$] will be much less than the original acid concentration. That is, [H$_3$O$^+$] will be smaller than if the acid were a strong acid of the same concentration. Similarly, a weak base will give a smaller [OH$^-$] than if the base were a strong base of the same concentration.
- For a series of weak monoprotic acids (of the type HA) of the same concentration, [H$_3$O$^+$] will increase (and the pH will decline) as the acids become stronger. Similarly, for a series of weak bases, [OH$^-$] will increase (and the pH will increase) as the bases become stronger.

Strong Acid

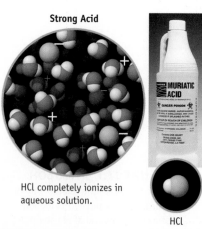

HCl completely ionizes in aqueous solution.

HCl

(a) Hydrochloric acid, a strong acid, is sold for household use as "muriatic acid." The acid completely ionizes in water.

Weak Acid

Acetic acid, CH$_3$CO$_2$H, ionizes only slightly in water.

CH$_3$CO$_2$H

(b) Vinegar is a solution of acetic acid, a weak acid that ionizes only to a small extent in water.

Weak Base

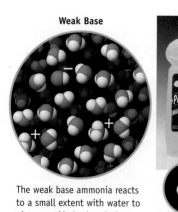

The weak base ammonia reacts to a small extent with water to give a weakly basic solution.

NH$_3$

(c) Ammonia is a weak base, ionizing to a small extent in water.

Photos © Cengage Learning/Charles D. Winters

FIGURE 17.2 Strong and weak acids and bases.

Strong or Weak?

How can you tell whether an acid or a base is weak? The easiest way is to remember those few that are strong. All others are probably weak.

Common strong acids include the following:

Hydrohalic acids: HCl, HBr, and HI (but not HF)

Nitric acid: HNO_3

Sulfuric acid: H_2SO_4 (for loss of first H^+ only)

Perchloric acid: $HClO_4$

Some common strong bases include the following:

All Group 1A hydroxides: LiOH, NaOH, KOH, RbOH, CsOH

Group 2A hydroxides: $Sr(OH)_2$ and $Ba(OH)_2$. [$Mg(OH)_2$ and $Ca(OH)_2$ are not considered strong bases because they do not dissolve appreciably in water.]

The relative strength of an acid or base can also be expressed quantitatively with an equilibrium constant, often called an **ionization constant**. For the general acid HA, we can write

$$HA(aq) + H_2O(\ell) \rightleftharpoons H_3O^+(aq) + A^-(aq)$$

$$K_a = \frac{[H_3O^+][A^-]}{[HA]} \qquad (17.5)$$

where the equilibrium constant, K, has a subscript "a" to indicate that it is an equilibrium constant for an acid in water. For weak acids, the value of K_a is less than 1 because the product $[H_3O^+][A^-]$ is less than the equilibrium concentration of the weak acid $[HA]$. For a series of acids, the acid strength increases as the value of K_a increases.

Similarly, we can write the equilibrium expression for a weak base B in water. Here, we label K with a subscript "b." Its value is less than 1 for weak bases.

$$B(aq) + H_2O(\ell) \rightleftharpoons BH^+(aq) + OH^-(aq)$$

$$K_b = \frac{[BH^+][OH^-]}{[B]} \qquad (17.6)$$

Some acids and bases are listed in Table 17.3, each with its value of K_a or K_b. The following are important concepts concerning this table.

- Acids are listed in the table at the left, and their conjugate bases are listed on the right.
- A large value of K indicates that ionization products are strongly favored, whereas a small value of K indicates that reactants are favored.
- The strongest acids are at the upper left. They have the largest K_a values. K_a values become smaller on descending the chart as acid strength declines.
- The strongest bases are at the lower right. They have the largest K_b values. K_b values become larger on descending the chart as base strength increases.
- *The weaker the acid, the stronger its conjugate base.* That is, the smaller the value of K_a, the larger the value of K_b.
- Some acids or bases are listed as having K_a or K_b values that are large or very small. Aqueous acids that are stronger than H_3O^+ are completely ionized (HNO_3, for example), so their K_a values are "large." Their conjugate bases (such as NO_3^-) do not produce meaningful concentrations of OH^- ions, so their K_b values are "very small." Similar arguments follow for strong bases and their conjugate acids.

Table 17.3 **Ionization Constants for Some Acids and Their Conjugate Bases at 25 °C**

Acid Name	Acid	K_a	Base	K_b	Base Name
Perchloric acid	$HClO_4$	Large	ClO_4^-	Very small	Perchlorate ion
Sulfuric acid	H_2SO_4	Large	HSO_4^-	Very small	Hydrogen sulfate ion
Hydrochloric acid	HCl	Large	Cl^-	Very small	Chloride ion
Nitric acid	HNO_3	Large	NO_3^-	Very small	Nitrate ion
Hydronium ion	H_3O^+	1.0	H_2O	1.0×10^{-14}	Water
Sulfurous acid	H_2SO_3	1.2×10^{-2}	HSO_3^-	8.3×10^{-13}	Hydrogen sulfite ion
Hydrogen sulfate ion	HSO_4^-	1.2×10^{-2}	SO_4^{2-}	8.3×10^{-13}	Sulfate ion
Phosphoric acid	H_3PO_4	7.5×10^{-3}	$H_2PO_4^-$	1.3×10^{-12}	Dihydrogen phosphate ion
Hexaaquairon(III) ion	$[Fe(H_2O)_6]^{3+}$	6.3×10^{-3}	$[Fe(H_2O)_5OH]^{2+}$	1.6×10^{-12}	Pentaaquahydroxoiron(III) ion
Hydrofluoric acid	HF	7.2×10^{-4}	F^-	1.4×10^{-11}	Fluoride ion
Nitrous acid	HNO_2	4.5×10^{-4}	NO_2^-	2.2×10^{-11}	Nitrite ion
Formic acid	HCO_2H	1.8×10^{-4}	HCO_2^-	5.6×10^{-11}	Formate ion
Benzoic acid	$C_6H_5CO_2H$	6.3×10^{-5}	$C_6H_5CO_2^-$	1.6×10^{-10}	Benzoate ion
Acetic acid	CH_3CO_2H	1.8×10^{-5}	$CH_3CO_2^-$	5.6×10^{-10}	Acetate ion
Propanoic acid	$CH_3CH_2CO_2H$	1.3×10^{-5}	$CH_3CH_2CO_2^-$	7.7×10^{-10}	Propanoate ion
Hexaaquaaluminum ion	$[Al(H_2O)_6]^{3+}$	7.9×10^{-6}	$[Al(H_2O)_5OH]^{2+}$	1.3×10^{-9}	Pentaaquahydroxoaluminum ion
Carbonic acid	H_2CO_3	4.2×10^{-7}	HCO_3^-	2.4×10^{-8}	Hydrogen carbonate ion
Hexaaquacopper(II) ion	$[Cu(H_2O)_6]^{2+}$	1.6×10^{-7}	$[Cu(H_2O)_5OH]^+$	6.3×10^{-8}	Pentaaquahydroxocopper(II) ion
Hydrogen sulfide	H_2S	1×10^{-7}	HS^-	1×10^{-7}	Hydrogen sulfide ion
Dihydrogen phosphate ion	$H_2PO_4^-$	6.2×10^{-8}	HPO_4^{2-}	1.6×10^{-7}	Hydrogen phosphate ion
Hydrogen sulfite ion	HSO_3^-	6.2×10^{-8}	SO_3^{2-}	1.6×10^{-7}	Sulfite ion
Hypochlorous acid	$HClO$	3.5×10^{-8}	ClO^-	2.9×10^{-7}	Hypochlorite ion
Hexaaqualead(II) ion	$[Pb(H_2O)_6]^{2+}$	1.5×10^{-8}	$[Pb(H_2O)_5OH]^+$	6.7×10^{-7}	Pentaaquahydroxolead(II) ion
Hexaaquacobalt(II) ion	$[Co(H_2O)_6]^{2+}$	1.3×10^{-9}	$[Co(H_2O)_5OH]^+$	7.7×10^{-6}	Pentaaquahydroxocobalt(II) ion
Boric acid	$B(OH)_3(H_2O)$	7.3×10^{-10}	$B(OH)_4^-$	1.4×10^{-5}	Tetrahydroxoborate ion
Ammonium ion	NH_4^+	5.6×10^{-10}	NH_3	1.8×10^{-5}	Ammonia
Hydrocyanic acid	HCN	4.0×10^{-10}	CN^-	2.5×10^{-5}	Cyanide ion
Hexaaquairon(II) ion	$[Fe(H_2O)_6]^{2+}$	3.2×10^{-10}	$[Fe(H_2O)_5OH]^+$	3.1×10^{-5}	Pentaaquahydroxoiron(II) ion
Hydrogen carbonate ion	HCO_3^-	4.8×10^{-11}	CO_3^{2-}	2.1×10^{-4}	Carbonate ion
Hexaaquanickel(II) ion	$[Ni(H_2O)_6]^{2+}$	2.5×10^{-11}	$[Ni(H_2O)_5OH]^+$	4.0×10^{-4}	Pentaaquahydroxonickel(II) ion
Hydrogen phosphate ion	HPO_4^{2-}	3.6×10^{-13}	PO_4^{3-}	2.8×10^{-2}	Phosphate ion
Water	H_2O	1.0×10^{-14}	OH^-	1.0	Hydroxide ion
Hydrogen sulfide ion*	HS^-	1×10^{-19}	S^{2-}	1×10^5	Sulfide ion
Ethanol	C_2H_5OH	Very small	$C_2H_5O^-$	Large	Ethoxide ion
Ammonia	NH_3	Very small	NH_2^-	Large	Amide ion
Hydrogen	H_2	Very small	H^-	Large	Hydride ion

*The values of K_a for HS^- and K_b for S^{2-} are estimates.

Increasing Acid Strength

Increasing Base Strength

To illustrate some of these ideas, let us compare some common acids and bases. For example, HF is a stronger acid than HClO, which is in turn stronger than HCO_3^-,

$$\text{Increasing acid strength} \longrightarrow$$

HCO_3^-	HClO	HF
$K_a = 4.8 \times 10^{-11}$	$K_a = 3.5 \times 10^{-8}$	$K_a = 7.2 \times 10^{-4}$

and their conjugate bases become stronger from F^- to ClO^- to CO_3^{2-}.

$$\longleftarrow \text{Increasing base strength}$$

CO_3^{2-}	ClO^-	F^-
$K_b = 2.1 \times 10^{-4}$	$K_b = 2.9 \times 10^{-7}$	$K_a = 1.4 \times 10^{-11}$

Nature abounds in acids and bases (Figure 17.3). Many naturally occurring acids contain the carboxyl group ($-CO_2H$) (see page 579), and a few are illustrated here. Notice that the organic portion of the molecule has an effect on its relative strength (as described further in Section 17.10).

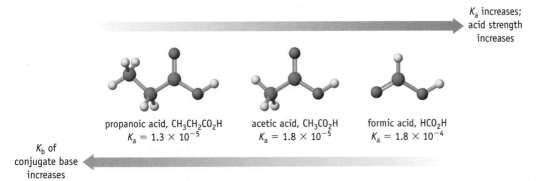

K_a increases; acid strength increases

K_b of conjugate base increases

propanoic acid, $CH_3CH_2CO_2H$
$K_a = 1.3 \times 10^{-5}$

acetic acid, CH_3CO_2H
$K_a = 1.8 \times 10^{-5}$

formic acid, HCO_2H
$K_a = 1.8 \times 10^{-4}$

There are many naturally occurring weak bases as well (Figure 17.3). Ammonia and its conjugate acid, the ammonium ion, are part of the nitrogen cycle in the environment. Biological systems reduce nitrate ion to NH_3 and NH_4^+ and incorporate nitrogen into amino acids and proteins. Many bases are derived from NH_3 by replacement of the H atoms with organic groups.

ammonia
$K_b = 1.8 \times 10^{-5}$

methylamine
$K_b = 5.0 \times 10^{-4}$

aniline
$K_b = 4.0 \times 10^{-10}$

Ammonia is a weaker base than methylamine (K_b for $NH_3 < K_b$ for CH_3NH_2). This means that the conjugate acid of ammonia, NH_4^+ ($K_a = 5.6 \times 10^{-10}$), is stronger than the methylammonium ion, the conjugate acid of methylamine, $CH_3NH_3^+$ ($K_a = 2.0 \times 10^{-11}$).

K_a Values for Polyprotic Acids

Like all polyprotic acids, phosphoric acid ionizes in a series of steps, three in this case.

First ionization step: $K_{a1} = 7.5 \times 10^{-3}$

$$H_3PO_4(aq) + H_2O(\ell) \rightleftharpoons H_2PO_4^-(aq) + H_3O^+(aq)$$

Second ionization step: $K_{a2} = 6.2 \times 10^{-8}$

$$H_2PO_4^-(aq) + H_2O(\ell) \rightleftharpoons HPO_4^{2-}(aq) + H_3O^+(aq)$$

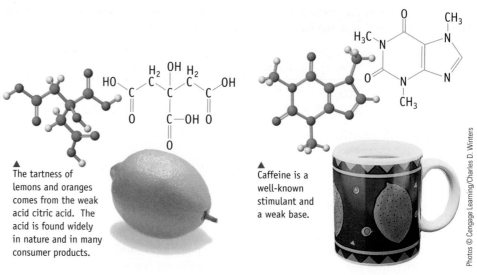

The tartness of lemons and oranges comes from the weak acid citric acid. The acid is found widely in nature and in many consumer products.

Caffeine is a well-known stimulant and a weak base.

FIGURE 17.3 Natural acids and bases. Hundreds of acids and bases occur in nature. Our foods contain a wide variety, and biochemically important molecules are often acids and bases.

Third ionization step: $K_{a3} = 3.6 \times 10^{-13}$

$$HPO_4{}^{2-}(aq) + H_2O(\ell) \rightleftharpoons PO_4{}^{3-}(aq) + H_3O^+(aq)$$

We observe that the K_a value for each successive step becomes smaller because it is more difficult to remove H^+ from a negatively charged ion, such as $H_2PO_4{}^-$, than from a neutral molecule, such as H_3PO_4. Similarly, the larger the negative charge of the anionic acid, the more difficult it is to remove H^+. Finally, for many inorganic polyprotic acids, K_a values become smaller by about 10^5 for each proton removed.

Logarithmic Scale of Relative Acid Strength, pK_a

Many chemists and biochemists use a logarithmic scale to report and compare relative acid strengths.

$$pK_a = -\log K_a \tag{17.7}$$

The pK_a of an acid is the negative log of the K_a value (just as pH is the negative log of the hydronium ion concentration). For example, acetic acid has a pK_a value of 4.74.

$$pK_a = -\log (1.8 \times 10^{-5}) = 4.74$$

The pK_a value becomes smaller as the acid strength increases.

—— Acid strength increases ⟶

Propanoic acid	Acetic acid	Formic acid
$CH_3CH_2CO_2H$	CH_3CO_2H	HCO_2H
$K_a = 1.3 \times 10^{-5}$	$K_a = 1.8 \times 10^{-5}$	$K_a = 1.8 \times 10^{-4}$
p$K_a = 4.89$	p$K_a = 4.74$	p$K_a = 3.74$

—— pK_a decreases ⟶

Relating the Ionization Constants for an Acid and Its Conjugate Base

Let us look again at Table 17.3. From the top of the table to the bottom, the strengths of the acids decline (K_a becomes smaller), and the strengths of their conjugate bases increase (the values of K_b increase). Examining a few cases shows that the product of K_a for an acid and K_b for its conjugate base is equal to a constant, specifically K_w.

• **A Relation Among pK Values** A useful relationship for an acid–conjugate base pair follows from Equation 17.8.
$$pK_w = pK_a + pK_b$$

$$K_a \times K_b = K_w \tag{17.8}$$

Photos © Cengage Learning/Charles D. Winters

Consider the specific case of the ionization of a weak acid, say HCN, and the interaction of its conjugate base, CN^-, with H_2O.

Weak acid: $HCN(aq) + H_2O(\ell) \rightleftharpoons H_3O^+(aq) + CN^-(aq)$ $K_a = 4.0 \times 10^{-10}$
Conjugate base: $CN^-(aq) + H_2O(\ell) \rightleftharpoons HCN(aq) + OH^-(aq)$ $K_b = 2.5 \times 10^{-5}$
$\overline{2 H_2O(\ell) \rightleftharpoons H_3O^+(aq) + OH^-(aq)}$ $K_w = 1.0 \times 10^{-14}$

Adding the equations gives the chemical equation for the autoionization of water, and the numerical value is indeed 1.0×10^{-14}. That is,

$$K_a \times K_b = \left(\frac{[H_3O^+][CN^-]}{[HCN]}\right)\left(\frac{[HCN][OH^-]}{[CN^-]}\right) = [H_3O^+][OH^-] = K_w$$

Equation 17.8 is useful because K_b can be calculated from K_a. The value of K_b for the cyanide ion, for example, is

$$K_b \text{ for } CN^- = \frac{K_w}{K_a \text{ for HCN}} = \frac{1.0 \times 10^{-14}}{4.0 \times 10^{-10}} = 2.5 \times 10^{-5}$$

REVIEW & CHECK FOR SECTION 17.4

Use Table 17.3 to answer the following questions.

1. Which of the following is the strongest acid?
   (a) HF   (b) H_2S (c) HOCl   (d) CH_3CO_2H

2. Which acid has the strongest conjugate base?
   (a) HNO_2 (b) $C_6H_5CO_2H$ (c) HCN   (d) HCl

3. Which of the following has a pK_a value of 4.20?
   (a) $C_6H_5CO_2H$ (b) CH_3CO_2H (c) HCO_2H (d) HF

4. What is the pK_a for the conjugate acid of ammonia?
   (a) 4.74   (b) 9.25   (c) 5.60   (d) 7.00

5. K_a for lactic acid, $CH_3CHOHCO_2H$, is 1.4×10^{-4}. What is pK_b for the conjugate base of this acid, $CH_3CHOHCO_2^-$?
   (a) 3.85   (b) 7.00   (c) 10.15   (d) 12.60

17.5 Acid–Base Properties of Salts

A number of the acids and bases listed in Table 17.3 are cations or anions. As described earlier, anions can act as Brønsted bases because they can accept a proton from an acid to form the conjugate acid of the ion.

$$CO_3^{2-}(aq) + H_2O(\ell) \rightleftharpoons HCO_3^-(aq) + OH^-(aq)$$
$$K_b = 2.1 \times 10^{-4}$$

You should also notice that many metal cations in water are Brønsted acids.

$$[Al(H_2O)_6]^{3+}(aq) + H_2O(\ell) \rightleftharpoons [Al(H_2O)_5(OH)]^{2+}(aq) + H_3O^+(aq)$$
$$K_a = 7.9 \times 10^{-6}$$

Table 17.4 summarizes the acid–base properties of some common cations and anions. As you look over this table, notice the following points:

- Anions that are conjugate bases of strong acids (for example, Cl^- and NO_3^-) are such weak bases that they have no effect on solution pH.
- There are numerous basic anions (such as CO_3^{2-}). All are the conjugate bases of weak acids.
- The acid–base behavior of anions of polyprotic acids depends on the extent of deprotonation. For example, a fully deprotonated anion (such as CO_3^{2-}) will be basic. A partially deprotonated anion (such as HCO_3^-) is amphiprotic. Its behavior will depend on the other species in the reaction.

● **Hydrolysis Reactions and Hydrolysis Constants** Chemists often say that, when ions interact with water to produce acidic or basic solutions, the ions "hydrolyze" in water, or they undergo "hydrolysis." Thus, some books refer to the K_a and K_b values of ions as "hydrolysis constants," K_h.

Many aqueous metal cations are Brønsted acids. A pH measurement of a dilute solution of copper(II) sulfate shows that the solution is clearly acidic. Among the common cations, Al^{3+} and transition metal ions form acidic solutions in water.

Table 17.4 Acid and Base Properties of Some Ions in Aqueous Solution

Neutral		Basic			Acidic
Anions	Cl^- NO_3^- Br^- ClO_4^- I^-	$CH_3CO_2^-$ CN^- HCO_2^- PO_4^{3-} CO_3^{2-} HCO_3^- S^{2-} HS^- F^- NO_2^-		SO_4^{2-} HPO_4^{2-} SO_3^{2-} OCl^-	HSO_4^- $H_2PO_4^-$ HSO_3^-
Cations	Li^+ Na^+ Ca^{2+} K^+ Ba^{2+}	$[Al(H_2O)_5(OH)]^{2+}$ (for example)			$[Al(H_2O)_6]^{3+}$ and hydrated transition metal cations (such as $[Fe(H_2O)_6]^{3+}$) NH_4^+

- Alkali metal and alkaline earth cations have no measurable effect on solution pH.
- Basic cations are conjugate bases of acidic cations such as $[Al(H_2O)_6]^{3+}$.
- Acidic cations fall into two categories: (a) metal cations with 2+ and 3+ charges and (b) ammonium ions (and its organic derivatives). All metal cations are hydrated in water, forming ions such as $[M(H_2O)_6]^{n+}$. However, only when M is a 2+ or 3+ ion, particularly a transition metal ion, does the ion act as an acid.

⬤WL   INTERACTIVE EXAMPLE 17.2   Acid–Base Properties of Salts

Problem Decide whether each of the following will give rise to an acidic, basic, or neutral solution in water.

(a) $NaNO_3$

(b) K_3PO_4

(c) $FeCl_2$

(d) $NaHCO_3$

(e) NH_4F

What Do You Know? You know the salt formulas and, from Tables 17.3 and 17.4, you know the effect of the constituent ions on pH.

Strategy First, identify the cation and anion in each salt. Next, use Tables 17.3 and 17.4 to describe the acid–base properties of each ion. Finally decide on the overall acid–base properties of the salt.

Solution

(a) $NaNO_3$: This salt gives a neutral, aqueous solution (pH = 7). Neither the sodium ion, Na^+, nor the nitrate ion, NO_3^- (the very weak conjugate base of a strong acid), affects the solution pH.

(b) K_3PO_4: An aqueous solution of K_3PO_4 should be basic (pH > 7) because PO_4^{3-} is the conjugate base of the weak acid HPO_4^{2-}. The K^+ ion, like the Na^+ ion, does not affect the solution pH.

(c) $FeCl_2$: An aqueous solution of $FeCl_2$ should be weakly acidic (pH < 7). The Fe^{2+} ion in water, $[Fe(H_2O)_6]^{2+}$, is a Brønsted acid. In contrast, Cl^- is the very weak conjugate base of the strong acid HCl, so it does not contribute excess OH^- ions to the solution.

(d) $NaHCO_3$: Some additional information is needed concerning salts of amphiprotic anions such as HCO_3^-. Because they have an ionizable hydrogen, they can act as acids,

$$HCO_3^-(aq) + H_2O(\ell) \rightleftharpoons CO_3^{2-}(aq) + H_3O^+(aq) \qquad K_a = 4.8 \times 10^{-11}$$

but because they are anions, they can also act as bases and accept an H^+ ion from water.

$$HCO_3^-(aq) + H_2O(\ell) \rightleftharpoons H_2CO_3(aq) + OH^-(aq) \qquad K_b = 2.4 \times 10^{-8}$$

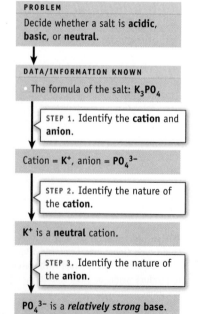

Strategy Map 17.2

PROBLEM

Decide whether a salt is **acidic, basic,** or **neutral.**

⬇

DATA/INFORMATION KNOWN

- The formula of the salt: K_3PO_4

STEP 1. Identify the **cation** and **anion.**

Cation = K^+, anion = PO_4^{3-}

STEP 2. Identify the nature of the **cation.**

K^+ is a **neutral** cation.

STEP 3. Identify the nature of the **anion.**

PO_4^{3-} is a *relatively strong* base.

STEP 4. Decide on the acid–base properties of the salt.

K_3PO_4 is **basic.**

Aqueous Solutions of Salts

Because aqueous solutions of salts are found in our bodies and throughout our economy and environment, it is important to know how to predict their acid and base properties. Information on the pH of an aqueous solution of a salt is summarized in Table 17.4. Consider also the following examples:

Cation	Anion	pH of the Solution
From strong base (Na^+)	From strong acid (Cl^-)	= 7 (neutral)
From strong base (K^+)	From weak acid ($CH_3CO_2^-$)	> 7 (basic)
From weak base (NH_4^+)	From strong acid (Cl^-)	< 7 (acidic)
From any weak base (BH^+)	From any weak acid (A^-)	Depends on relative acid strength of BH^+ and base strength of A^-

Whether the solution is acidic or basic will depend on the relative magnitude of K_a and K_b. In the case of the hydrogen carbonate anion, K_b is larger than K_a, so $[OH^-]$ is larger than $[H_3O^+]$, and an aqueous solution of $NaHCO_3$ will be slightly basic.

(e) NH_4F: What happens if you have a salt based on an acidic cation and a basic anion? One example is ammonium fluoride. Here, the ammonium ion would decrease the pH, and the fluoride ion would increase the pH.

$$NH_4^+(aq) + H_2O(\ell) \rightleftharpoons H_3O^+(aq) + NH_3(aq) \qquad K_a(NH_4^+) = 5.6 \times 10^{-10}$$

$$F^-(aq) + H_2O(\ell) \rightleftharpoons HF(aq) + OH^-(aq) \qquad K_b(F^-) = 1.4 \times 10^{-11}$$

Because $K_a(NH_4^+) > K_b(F^-)$, the ammonium ion is a stronger acid than the fluoride ion is a base. The resulting solution should be slightly acidic.

Think about Your Answer There are several important points here:

- Anions that are conjugate bases of strong acids—such as Cl^- and NO_3^-—have no effect on solution pH.

- To determine whether a salt is acidic, basic, or neutral, we must take into account both the cation and the anion. When a salt has an acidic cation and a basic anion, the pH of the solution will be determined by the ion that is the stronger acid or base.

Check Your Understanding

For each of the following salts in water, predict whether the pH will be greater than, less than, or equal to 7.

(a) KBr   (b) NH_4NO_3 (c) $AlCl_3$ (d) Na_2HPO_4

REVIEW & CHECK FOR SECTION 17.5

1. Adding NaH_2PO_4 to water will cause the pH to

   (a) increase   (b) decrease   (c) stay the same

2. Adding KCN to water will cause the pH to

   (a) increase   (b) decrease   (c) stay the same

17.6 Predicting the Direction of Acid–Base Reactions

According to the Brønsted–Lowry theory, all acid–base reactions can be written as equilibria involving the acid and base and their conjugates.

Acid + base $\rightleftharpoons$ conjugate base of the acid + conjugate acid of the base

In Sections 17.4 and 17.5, we used equilibrium constants to provide information about the relative strengths of acids and bases. Now we want to show how these constants can be used to decide whether a particular acid–base reaction is product- or reactant–favored at equilibrium.

Hydrochloric acid is a strong acid. Its equilibrium constant for reaction with water is very large, with the equilibrium lying completely to the right.

$$HCl(aq) + H_2O(\ell) \rightleftharpoons H_3O^+(aq) + Cl^-(aq)$$

Strong acid ($\approx$ 100% ionized), $K \gg 1$

$[H_3O^+] \approx$ initial concentration of the acid

Of the two acids here, HCl is stronger than H_3O^+. Of the two bases, H_2O and Cl^-, water is the stronger base and wins out in the competition for the proton. Thus, the equilibrium lies to the side of the chemical equation having the weaker acid and base.

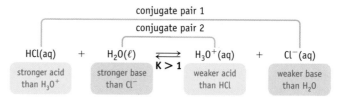

In contrast to HCl and other strong acids, acetic acid, a *weak* acid, ionizes to only a very small extent (Table 17.3).

$$CH_3CO_2H(aq) + H_2O(\ell) \rightleftharpoons H_3O^+(aq) + CH_3CO_2^-(aq)$$

Weak acid ($<$ 100% ionized), $K = 1.8 \times 10^{-5}$

$[H_3O^+] \ll$ initial concentration of the acid

When equilibrium is achieved in a 0.1 M aqueous solution of CH_3CO_2H, the concentrations of $H_3O^+(aq)$ and $CH_3CO_2^-(aq)$ are each only about 0.001 M. Approximately 99% of the acetic acid is not ionized.

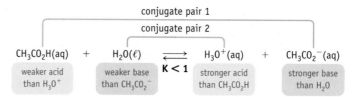

● **General Principle of Reactivity**
A general principle in chemistry is that more reactive substances react to form less reactive substances. Here you see the result of this for acid–base reactions: The stronger acid and stronger base always react to form the weaker acid and weaker base.

Again, the equilibrium lies toward the side of the reaction having the weaker acid and base.

These two examples of the relative extent of acid–base reactions illustrate a general principle:

> All proton transfer reactions proceed from the stronger acid and base to the weaker acid and base.

Using this principle and Table 17.3, you can predict which reactions are product-favored and which are reactant-favored at equilibrium. Consider the possible reaction of phosphoric acid and acetate ion to give acetic acid and the dihydrogen phosphate ion. Table 17.3 informs us that H_3PO_4 is a stronger acid ($K_a = 7.5 \times 10^{-3}$) than acetic acid ($K_a = 1.8 \times 10^{-5}$), and the acetate ion ($K_b = 5.6 \times 10^{-10}$) is a stronger base than the dihydrogen phosphate ion ($K_b = 1.3 \times 10^{-12}$).

Brønsted acids			
Brønsted bases			
$H_3PO_4(aq)$ +	$CH_3CO_2^-(aq)$ $\rightleftharpoons$	$H_2PO_4^-(aq)$ +	$CH_3CO_2H(aq)$
stronger acid than CH_3CO_2H	stronger base than $H_2PO_4^-$	weaker base than $CH_3CO_2^-$	weaker acid than H_3PO_4

Thus, mixing phosphoric acid with sodium acetate will produce a significant amount of dihydrogen phosphate ion and acetic acid. That is, the equilibrium is predicted to lie to the right because the reaction proceeds from the stronger acid–base combination to the weaker acid–base combination.

EXAMPLE 17.3   Reactions of Acids and Bases

Problem Write a balanced, net ionic equation for the reaction that occurs between acetic acid and sodium bicarbonate. Decide whether the equilibrium lies predominantly to the left or to the right.

What Do You Know? You know the identity of the acid, and you know that the other reactant, $NaHCO_3$, is a water-soluble salt that forms Na^+ and HCO_3^- ions in water. You also know that the HCO_3^- ion is amphiprotic, but it must act as a base toward acetic acid in this reaction. You need to know the ionization constants for acetic acid and the HCO_3^- ion and for their conjugates.

Strategy

- Identify the products of the acid–base reaction (which arise by H^+ transfer from acetic acid to the base, HCO_3^-). That is, identify the conjugate base of acetic acid and the conjugate acid of HCO_3^-.

- Write a balanced chemical equation for the equilibrium between the acid and base and their conjugate base and acid.

- Use Table 17.3 to decide which is the weaker of the two acids (or the weaker of the two bases). The equilibrium will favor the weaker acid and base.

Solution The conjugate base of acetic acid is the acetate ion, $CH_3CO_2^-$, and the conjugate acid of the base HCO_3^- is H_2CO_3. Hydrogen ion transfer from the acid to the base occurs by the following net ionic equation:

$$CH_3CO_2H(aq) \quad + \quad HCO_3^-(aq) \quad \rightleftarrows \quad CH_3CO_2^-(aq) \quad + \quad H_2CO_3(aq)$$
$$K_a = 1.8 \times 10^{-5} \qquad K_b = 2.4 \times 10^{-8} \qquad K_b = 5.6 \times 10^{-10} \qquad K_a = 4.2 \times 10^{-7}$$

The value of K_a or K_b (from Table 17.3) is given for each reactant and product. The equilibrium constants show that H_2CO_3 is a weaker acid than CH_3CO_2H, and $CH_3CO_2^-$ is a weaker base than HCO_3^-. The reaction favors the side having the weaker acid and base—that is, the right side.

Think about Your Answer One product of the reaction of acetic acid and $NaHCO_3$ is H_2CO_3, which dissociates into CO_2 and H_2O. As you can see in the photograph in the margin, CO_2 bubbles out of the solution. The equilibrium for the dissociation of H_2CO_3 to form CO_2 and H_2O lies far to the right.

$$H_2CO_3(aq) \rightleftarrows CO_2(g) + H_2O(\ell)$$

The loss of a product, CO_2, causes the equilibrium to shift further to the right. See the discussion of gas-forming reactions in Chapter 3 and of Le Chatelier's principle in Section 16.6.

Check Your Understanding

(a) Which is the stronger Brønsted acid, HCO_3^- or NH_4^+? Which has the stronger conjugate base?

(b) Is a reaction between HCO_3^- ions and NH_3 product- or reactant-favored at equilibrium?

$$HCO_3^-(aq) + NH_3(aq) \rightleftarrows CO_3^{2-}(aq) + NH_4^+(aq)$$

(c) You mix solutions of sodium hydrogen phosphate and ammonia. The net ionic equation for a possible reaction is

$$HPO_4^{2-}(aq) + NH_3(aq) \rightleftarrows PO_4^{3-}(aq) + NH_4^+(aq)$$

Does the equilibrium lie to the left or to the right for this reaction?

Reaction of vinegar and baking soda. This reaction involves the weak acid acetic acid and the weak base HCO_3^- from sodium hydrogen carbonate. Based on the values of the equilibrium constants, the reaction is predicted to proceed to the right to produce acetate ion, CO_2, and water.

REVIEW & CHECK FOR SECTION 17.6

1. Is the reaction of NaCN and HCl in water product- or reactant-favored at equilibrium?

    (a) reactant-favored (b) product-favored

2. In the following reaction, does the equilibrium lie predominantly to the left or to the right?

    $$HS^-(aq) + H_3PO_4(aq) \rightleftharpoons H_2S(aq) + H_2PO_4^-(aq)$$

    (a) left   (b) right

17.7 Types of Acid–Base Reactions

The reaction of hydrochloric acid and sodium hydroxide is the classic example of a strong acid–strong base reaction, whereas the reaction of citric acid and bicarbonate ion represents the reaction of a weak acid and weak base (Figure 17.4). These are two of four general types of acid–base reactions (Table 17.5). Because acid–base reactions are among the most important classes of chemical reactions, it is useful for you to know the outcome of the various types of these reactions.

The Reaction of a Strong Acid with a Strong Base

Strong acids and bases are effectively 100% ionized in solution. Therefore, the complete ionic equation for the reaction of HCl (strong acid) and NaOH (strong base) is

$$H_3O^+(aq) + Cl^-(aq) + Na^+(aq) + OH^-(aq) \rightleftharpoons 2\,H_2O(\ell) + Na^+(aq) + Cl^-(aq)$$

which leads to the following net ionic equation

$$H_3O^+(aq) + OH^-(aq) \rightleftharpoons 2\,H_2O(\ell) \qquad K = 1/K_w = 1.0 \times 10^{14}$$

The net ionic equation for the reaction of any strong acid with any strong base is always simply the reaction of hydronium ion and hydroxide ion to give water (◀ Section 3.6). Because this reaction is the reverse of the autoionization of water, it has an equilibrium constant of $1/K_w$. This very large value of K shows that, for all practical purposes, the reactants are completely consumed to form products. Thus, if equal numbers of moles of NaOH and HCl are mixed, the result is just a solution of NaCl in water. The constituents of NaCl, Na^+ and Cl^- ions, which arise from a strong base and a strong acid, respectively, produce a neutral aqueous solution. For this reason, reactions of strong acids and bases are often called "neutralizations."

> Mixing equal amounts (moles) of a strong base and a strong acid produces a neutral solution (pH = 7.00 at 25 °C).

The Reaction of a Weak Acid with a Strong Base

Consider the reaction of the naturally occurring weak acid formic acid, HCO_2H, with sodium hydroxide. The net ionic equation is

$$HCO_2H(aq) + OH^-(aq) \rightleftharpoons H_2O(\ell) + HCO_2^-(aq)$$

© Cengage Learning/Charles D. Winters

FIGURE 17.4 Reaction of a weak acid with a weak base. The bubbles coming from the tablet are carbon dioxide. This arises from the reaction of a weak Brønsted acid (citric acid) with a weak Brønsted base (HCO_3^-). The reaction is driven to completion by gas evolution.

● **Formic Acid + NaOH** The equilibrium constant for the reaction of formic acid and sodium hydroxide is 1.8×10^{10}. Can you confirm this? (See Study Question 17.97.)

Table 17.5 **Characteristics of Acid–Base Reactions**

Type	Example	Net Ionic Equation	Species Present after Equal Amounts Are Mixed; pH
Strong acid + strong base	HCl + NaOH	$H_3O^+(aq) + OH^-(aq) \rightleftharpoons 2\,H_2O(\ell)$	Cl^-, Na^+, pH = 7
Strong acid + weak base	HCl + NH$_3$	$H_3O^+(aq) + NH_3(aq) \rightleftharpoons NH_4^+(aq) + H_2O(\ell)$	Cl^-, NH_4^+, pH < 7
Weak acid + strong base	HCO$_2$H + NaOH	$HCO_2H(aq) + OH^-(aq) \rightleftharpoons HCO_2^-(aq) + H_2O(\ell)$	HCO_2^-, Na^+, pH > 7
Weak acid + weak base	HCO$_2$H + NH$_3$	$HCO_2H(aq) + NH_3(aq) \rightleftharpoons HCO_2^-(aq) + NH_4^+(aq)$	HCO_2^-, NH_4^+, pH dependent on K_a and K_b of conjugate acid and base

In the reaction of formic acid with NaOH, OH^- is a much stronger base than HCO_2^- ($K_b = 5.6 \times 10^{-11}$), and the reaction is predicted to proceed to the right. If equal amounts of weak acid and base are mixed, the final solution will contain sodium formate ($NaHCO_2$), a salt that is 100% dissociated in water. The Na^+ ion is a Group IA cation and so gives a neutral solution. The formate ion, however, is the conjugate base of a weak acid (Table 17.3), so the solution is basic. This example leads to a useful general conclusion:

> Mixing equal amounts (moles) of a strong base and a weak acid produces a salt whose anion is the conjugate base of the weak acid. The solution is basic, with the pH depending on K_b for the anion.

The Reaction of a Strong Acid with a Weak Base

The net ionic equation for the reaction of the strong acid HCl and the weak base NH_3 is

$$H_3O^+(aq) + NH_3(aq) \rightleftharpoons H_2O(\ell) + NH_4^+(aq)$$

The hydronium ion, H_3O^+, is a much stronger acid than NH_4^+ ($K_a = 5.6 \times 10^{-10}$), and NH_3 is a stronger base ($K_b = 1.8 \times 10^{-5}$) than H_2O. Therefore, reaction is predicted to proceed to the right and essentially to completion. Thus, after mixing equal amounts of HCl and NH_3, the solution contains the salt ammonium chloride, NH_4Cl. The Cl^- ion has no effect on the solution pH (Tables 17.3 and 17.4). However, the NH_4^+ ion is the conjugate acid of the weak base NH_3, so the solution at the conclusion of the reaction is acidic. In general, we can conclude that

● **Ammonia + HCl** The equilibrium constant for the reaction of a strong acid with aqueous ammonia is 1.8×10^9. Can you confirm this? (See Study Question 17.96.)

> Mixing equal amounts (moles) of a strong acid and a weak base produces a salt whose cation is the conjugate acid of the weak base. The solution is acidic, with the pH depending on K_a for the cation.

The Reaction of a Weak Acid with a Weak Base

If acetic acid, a weak acid, is mixed with ammonia, a weak base, the following reaction occurs.

$$CH_3CO_2H(aq) + NH_3(aq) \rightleftharpoons NH_4^+(aq) + CH_3CO_2^-(aq)$$

You know that the reaction is product-favored because CH_3CO_2H is a stronger acid than NH_4^+ and NH_3 is a stronger base than $CH_3CO_2^-$ (Table 17.3). Thus, if equal amounts of the acid and base are mixed, the resulting solution contains ammonium acetate, $NH_4CH_3CO_2$. Is this solution acidic or basic? On page 591 you learned that this depends on the relative values of K_a for the conjugate acid (here, NH_4^+; $K_a = 5.6 \times 10^{-10}$) and K_b for the conjugate base (here, $CH_3CO_2^-$; $K_b = 5.6 \times 10^{-10}$). In this case, the values of K_a and K_b are the same, so the solution is predicted to be neutral.

● **K for Reaction of Weak Acid and Weak Base** The equilibrium constant for the reaction between a weak acid and a weak base is $K_{net} = K_w/(K_a \cdot K_b)$. Can you confirm this? (See Study Question 17.121.)

> Mixing equal amounts (moles) of a weak acid and a weak base produces a salt whose cation is the conjugate acid of the weak base and whose anion is the conjugate base of the weak acid. The solution pH depends on the relative K_a and K_b values.

© Cengage Learning/Charles D. Winters

A weak acid reacting with a weak base. Baking powder contains the weak acid calcium dihydrogen phosphate, $Ca(H_2PO_4)_2$. This can react with the basic HCO_3^- ion in baking soda to give HPO_4^{2-}, CO_2 gas, and water.

REVIEW & CHECK FOR SECTION 17.7

1. Equal amounts (moles) of HCl(aq) and NaCN(aq) are mixed. The resulting solution is

   (a) acidic   (b) basic   (c) neutral

2. Equal amounts (moles) of acetic acid(aq) and sodium sulfite, Na_2SO_3(aq), are mixed. The resulting solution is

   (a) acidic   (b) basic   (c) neutral

3. Equal amounts (moles) of NaOH(aq) and NaH_2PO_4(aq) are mixed. The resulting solution is

   (a) acidic   (b) basic   (c) neutral

17.8 Calculations with Equilibrium Constants

Determining K from Initial Concentrations and Measured pH

The K_a and K_b values found in Table 17.3 and in the more extensive tables in Appendices H and I were all determined by experiment. There are several experimental methods available, but one approach, illustrated by the following example, is to determine the pH of the solution.

Strategy Map 17.4

PROBLEM
Calculate K_a for the ionization of **lactic acid**, a weak acid.

↓

DATA/INFORMATION KNOWN
• Acid concentration
• Solution pH

STEP 1. Write **balanced equation** and K_a **expression**, and set up **ICE table**.

K_a expression and ICE table

STEP 2. Enter **equilibrium concentrations** in ICE table.

At equilibrium:
$[H_3O^+]$ = [Conjugate base] = x
[Lactic acid] = original conc. − x

STEP 3. Convert **pH** to **[H₃O⁺]**, and solve for x

Equilibrium concentrations of **lactic acid**, **hydronium ion**, and **lactate ion**

STEP 4. Enter **equilibrium concentrations** into K_a **expression** and solve.

Value of K_a

⬮**WL** **INTERACTIVE EXAMPLE 17.4** Calculating a K_a Value from a Measured pH

Problem A 0.10 M aqueous solution of lactic acid, $CH_3CHOHCO_2H$, has a pH of 2.43. What is the value of K_a for lactic acid?

What Do You Know? To calculate K_a, you need to know the equilibrium concentration of each species. The pH of the solution allows you to calculate the equilibrium concentration of H_3O^+, and you can derive the other equilibrium concentrations from this.

Strategy

• Write the equilibrium constant expression, set up an ICE table, and convert pH to $[H_3O^+]$, the hydronium ion equilibrium concentration.

• Enter the initial concentration of $CH_3CHOHCO_2H$ on the initial (I) line of the ICE table and the equilibrium concentration of H_3O^+ on the equilibrium (E) line.

• Assign the variable x to represent changes in concentration. Based on the reaction stoichiometry the change in $[CH_3CHOHCO_2H]$ is $-x$ and the change in $[CH_3CHOHCO_2^-]$ is $+x$. Enter these values in the change (C) line of the table.

• Recognize that the hydronium ion equilibrium concentration, $[H_3O^+]$, is equal to x.

• Enter the expressions for the final equilibrium concentrations of all three species on the E line of the ICE table. Using this, determine the values of the other equilibrium concentrations.

• Use the equilibrium concentrations to solve for K_a.

Solution The equation for the reaction of lactic acid with water and its equilibrium constant expression are

$$CH_3CHOHCO_2H(aq) + H_2O(\ell) \rightleftharpoons CH_3CHOHCO_2^-(aq) + H_3O^+(aq)$$
$$\text{lactic acid} \qquad\qquad\qquad \text{lactate ion}$$

$$K_a \text{ (lactic acid)} = \frac{[H_3O^+][CH_3CHOHCO_2^-]}{[CH_3CHOHCO_2H]}$$

Begin by converting the pH to $[H_3O^+]$:

$$[H_3O^+] = 10^{-pH} = 10^{-2.43} = 3.7 \times 10^{-3} \text{ M}$$

Enter in the ICE table the concentrations before equilibrium is established, the change that occurs as the reaction proceeds to equilibrium, and the concentrations when equilibrium has been achieved. (See Examples 16.2–16.5.)

Equilibrium	$CH_3CHOHCO_2H$	+	H_2O	$\rightleftharpoons$	$CH_3CHOHCO_2^-$	+	H_3O^+
Initial (M)	0.10				0		0
Change (M)	$-x$				$+x$		$+x$
Equilibrium (M)	$(0.10 - x)$				x		x

The following points can be made concerning the ICE table.

• The quantity x represents the equilibrium concentrations of hydronium ion and lactate ion. That is, at equilibrium $x = [H_3O^+] = [CH_3CHOHCO_2^-] = 3.7 \times 10^{-3}$ M.

• By stoichiometry, x is also the concentration of acid that ionized on proceeding to equilibrium.

With these points in mind, we can calculate K_a for lactic acid.

$$K_a \text{ (lactic acid)} = \frac{[H_3O^+][CH_3CHOHCO_2^-]}{[CH_3CHOHCO_2H]}$$

$$= \frac{(3.7 \times 10^{-3})(3.7 \times 10^{-3})}{0.10 - 0.0037} = 1.4 \times 10^{-4}$$

Comparing this value of K_a with others in Table 17.3, we see lactic acid is similar to formic acid in its strength.

Think about Your Answer Hydronium ion, H_3O^+, is present in solution from lactic acid ionization and from water autoionization. Le Chatelier's principle informs us that the H_3O^+ from lactic acid will suppress the H_3O^+ coming from the water autoionization. Because $[H_3O^+]$ from water must be less than 10^{-7} M, the pH is almost completely a reflection of H_3O^+ from lactic acid. (See Example 17.1.)

Check Your Understanding

A solution prepared from 0.055 mol of butanoic acid dissolved in sufficient water to give 1.0 L of solution has a pH of 2.72. Determine K_a for butanoic acid. The acid ionizes according to the balanced equation

$$CH_3CH_2CH_2CO_2H(aq) + H_2O(\ell) \rightleftharpoons H_3O^+(aq) + CH_3CH_2CH_2CO_2^-(aq)$$

lactic acid, $CH_3CHOHCO_2H$

Lactic acid, $CH_3CHOHCO_2H$. Lactic acid is a weak monoprotic acid that occurs naturally in sour milk and arises from metabolism in the human body.

Module 22: Equilibrium: pH of a Weak Acid covers concepts in this section.

There is an important point to notice in Example 17.4. The lactic acid concentration at equilibrium was given by $(0.10 - x)$ where x was found from the solution pH to be 3.7×10^{-3} M. By the usual rules governing significant figures $(0.10 - 0.0037)$ is equal to 0.10. The acid is weak, so very little of it ionizes (approximately 4%), and the equilibrium concentration of lactic acid is essentially equal to the initial acid concentration. Neglecting to subtract 0.0037 from 0.10 has no effect on the answer, which is known only to two significant figures.

Like lactic acid, most weak acids (HA) are so weak that the equilibrium concentration of the acid, [HA], is effectively its initial concentration ($= [HA]_0$). This leads to the useful conclusion that the denominator in the equilibrium constant expression for dilute solutions of most weak acids is simply $[HA]_0$, the original or initial concentration of the weak acid.

$$HA(aq) + H_2O(\ell) \rightleftharpoons H_3O^+(aq) + A^-(aq)$$

$$K_a = \frac{[H_3O^+][A^-]}{[HA]_0 - [H_3O^+]} \approx \frac{[H_3O^+][A^-]}{[HA]_0}$$

Analysis shows that

The approximation that

$$[HA]_{\text{equilibrium}} = [HA]_0 - [H_3O^+] \approx [HA]_0$$

is valid whenever $[HA]_0$ is greater than or equal to $100 \times K_a$.

This is the same approximation we used in Chapter 16 when deciding whether we needed to solve quadratic equations exactly (◄ Problem Solving Tip 16.1).

What Is the pH of an Aqueous Solution of a Weak Acid or Base?

Knowing values of the equilibrium constants for weak acids and bases enables us to calculate the pH of a solution of a weak acid or base.

Strategy Map 17.5

PROBLEM
Calculate the **pH** for a weak acid solution knowing the value of K_a.

$\downarrow$

DATA/INFORMATION KNOWN
• Acid concentration
• Value of K_a

STEP 1. Write **balanced equation** and K_a **expression**, and set up **ICE table**.

K_a expression and ICE table

STEP 2. Enter **equilibrium concentrations** in ICE table.

At equilibrium:
$[H_3O^+]$ = [conjugate base] = x
[acid] = original conc. − x

STEP 3. Enter **equilibrium concentrations** in K_a.

K_a **expression** with equilibrium concentrations in terms of x

STEP 4. Solve K_a **expression** for x.

x = value of $[H_3O^+]$

STEP 5. Convert $[H_3O^+]$ to **pH**.

pH of solution

OWL **INTERACTIVE EXAMPLE 17.5** **Calculating Equilibrium Concentrations and pH from K_a**

Problem Calculate the pH of a 0.020 M solution of benzoic acid ($C_6H_5CO_2H$) given that $K_a = 6.3 \times 10^{-5}$ for the acid.

$$C_6H_5CO_2H(aq) + H_2O(\ell) \rightleftharpoons H_3O^+(aq) + C_6H_5CO_2^-(aq)$$

What Do You Know? You know the value of K_a for the acid and its initial concentration. You need to find the equilibrium concentration of H_3O^+ in order to calculate the pH.

Strategy This is similar to Examples 16.5 and 16.6, where you wanted to find the concentration of a reaction product. The strategy is the same:

• Write the equilibrium constant expression and set up an ICE table.

• Enter the initial concentration of $C_6H_5CO_2H$ on the initial (I) line of the ICE table.

• The variable x represents changes in concentration, so the change in $[C_6H_5CO_2H]$ is $-x$ and the change in product concentrations is $+x$. Enter these values in the change (C) line of the table.

• Enter the expressions for the final equilibrium concentrations of all three species on the E line of the ICE table, and then transfer these expressions to the equilibrium constant expression and solve for x.

• Convert $[H_3O^+]$ (= x) to pH.

Solution Organize the information in an ICE table.

Equilibrium	$C_6H_5CO_2H$	+	H_2O	$\rightleftharpoons$	$C_6H_5CO_2^-$	+	H_3O^+
Initial (M)	0.020				0		0
Change (M)	$-x$				$+x$		$+x$
Equilibrium (M)	$(0.020 - x)$				x		x

According to reaction stoichiometry,

$$[H_3O^+] = [C_6H_5CO_2^-] = x \text{ at equilibrium}$$

the concentration of acid ionized is x. Thus, the benzoic acid concentration at equilibrium is

$$[C_6H_5CO_2H] = \text{initial acid concentration} - \text{concentration of acid that ionized}$$

$$[C_6H_5CO_2H] = 0.020 - x$$

Substituting these equilibrium concentrations into the K_a expression, we have

$$K_a = \frac{[H_3O^+][C_6H_5CO_2^-]}{[C_6H_5CO_2H]}$$

$$6.3 \times 10^{-5} = \frac{(x)(x)}{0.020 - x}$$

The value of x is small compared with 0.020 (because $[HA]_0 > 100 \times K_a$; 0.020 M $> 6.3 \times 10^{-3}$). Therefore, you can use the approximate expression.

$$K_a = 6.3 \times 10^{-5} = \frac{x^2}{0.020}$$

Solving for x, we have

$$x = \sqrt{K_a \times (0.020)} = 0.0011 \text{ M}$$

and we find that

$$[H_3O^+] = [C_6H_5CO_2^-] = 0.0011 \text{ M}$$

and

$$[C_6H_5CO_2H] = (0.020 - x) = 0.019 \text{ M}$$

Finally, the pH of the solution is found to be

$$pH = -\log(1.1 \times 10^{-3}) = 2.96$$

Problem What is the pH

What Do You Know? You
conjugate base of a weak
(You also know that the s
you know the value of K_b
find the equilibrium conce

Strategy This is similar t
a reaction product.

- Write the balanced e

- Enter the initial con

- The variable x repre
  the change in produ
  table.

- Enter the expressio
  of the ICE table, an
  for x.

- The value of $x = [C$
  $[H_3O^+][OH^-]$, then

Solution The value of K

C

Set up an ICE table to s
solution.

Equilibrium	CH₃
Initial (M)	0.0
Change (M)	$-x$
Equilibrium (M)	(0.0

Next, substitute the val

The acetate ion, a weak
concentration of hydro
we use the approximat

$$K_b = $$

$$x = $$

To calculate the pH of
tions, it is always true

Therefore,

The acetate ion gives

Think about Your Answer We made the approximation that $(0.020 - x) \approx 0.020$. If we do not make the approximation and instead solve the exact expression, $x = [H_3O^+] = 0.0011$ M. This is the same answer to two significant figures that we obtained from the "approximate" expression. Finally, notice that we again ignored any H_3O^+ that arises from water ionization.

Check Your Understanding

What are the equilibrium concentrations of acetic acid, the acetate ion, and H_3O^+ for a 0.10 M solution of acetic acid ($K_a = 1.8 \times 10^{-5}$)? What is the pH of the solution?

EXAMPLE 17.6 Calculating Equilibrium Concentrations and pH from K_a and Using the Method of Successive Approximations

Problem What is the pH of a 0.0010 M solution of formic acid? What is the concentration of formic acid at equilibrium? The acid is moderately weak, with $K_a = 1.8 \times 10^{-4}$.

$$HCO_2H(aq) + H_2O(\ell) \rightleftharpoons HCO_2^-(aq) + H_3O^+(aq)$$

What Do You Know? You know the value of K_a for the acid and its initial concentration. You need to find the equilibrium concentration of H_3O^+ in order to calculate the pH.

Strategy This is similar to Example 17.5, where you wanted to find the concentration of a reaction product, except that an approximate solution will not be possible. The strategy is the same:

- Write the equilibrium constant expression and set up an ICE table.

- Enter the initial concentration of HCO_2H on the initial (I) line of the ICE table.

- The variable x represents changes in concentration, so the change in $[HCO_2H]$ is $-x$ and the change in product concentrations is $+x$. Enter these values in the change (C) line of the table.

- Enter the expressions for the final equilibrium concentrations of all three species on the E line of the ICE table, and then transfer these expressions to the equilibrium constant expression and solve for x.

- Convert $[H_3O^+]$ ($= x$) to pH.

Solution The ICE table is as follows.

Equilibrium	HCO₂H + H₂O ⇌	HCO₂⁻ +	H₃O⁺
Initial (M)	0.0010	0	0
Change (M)	$-x$	$+x$	$+x$
Equilibrium (M)	$(0.0010 - x)$	x	x

Substituting the values in the table into the K_a expression we have

$$K_a = \frac{[H_3O^+][HCO_2^-]}{[HCO_2H]} = 1.8 \times 10^{-4} = \frac{(x)(x)}{0.0010 - x}$$

In this example, $[HA]_0$ ($= 0.0010$ M) is *not* greater than $100 \times K_a$ ($= 1.8 \times 10^{-2}$), so the usual approximation is not reasonable. Thus, we have to find the equilibrium concentrations by solving the "exact" expression. This can be solved with the quadratic formula or by successive approximations (Appendix A). Let us use the successive approximation method here.

To use the **successive approximations approach**, begin by solving the approximate expression for x.

$$1.8 \times 10^{-4} = \frac{(x)(x)}{0.0010}$$

Solving this, we find $x = 4.2 \times 10^{-4}$. Put this value into the expression for x in the denominator of the exact expression.

$$1.8 \times 10^{-4} = \frac{(x)(x)}{0.0010 - x} = \frac{(x)(x)}{0.0010 - 4.2 \times 10^{-4}}$$

Think about Your Answer The hydroxide ion concentration (x) is indeed quite small relative to the initial acetate ion concentration so the approximate expression is appropriate. (Note that $100 \times K_b$ is less than the initial base concentration.)

Check Your Understanding

The weak base, ClO^- (hypochlorite ion), is used in the form of NaClO as a disinfectant in swimming pools and water treatment plants. What are the concentrations of HClO and OH^- and the pH of a 0.015 M solution of NaClO?

Strategy Map 17.8

PROBLEM

Calculate **pH** of solution after the reaction of a **weak base** and a **strong acid**.

↓

DATA/INFORMATION KNOWN

- Reactant concentrations
- Reactant volumes

STEP 1. Write **balanced equation** and solve **stoichiometry problem**.

Know **amounts of reactants** and derive **amount of product**.

STEP 2. Decide if product is **weak acid** or **weak base**.

Product is a **weak acid** (NH_4^+)

STEP 3. Calculate concentration of NH_4^+.

Concentration of weak acid whose K_a is known.

STEP 4. Solve for $[H_3O^+]$ as in Examples 17.5 - 17.7.

Value of $[H_3O^+]$

STEP 5. Convert $[H_3O^+]$ to **pH**.

pH of solution

⬤WL INTERACTIVE EXAMPLE 17.8 Calculating the pH after the Reaction of a Weak Base with a Strong Acid

Problem What is the pH of the solution that results from mixing 25 mL of 0.016 M NH_3 and 25 mL of 0.016 M HCl?

What Do You Know? You know this is an acid–base reaction (HCl + NH_3), and you know the amount of each reactant (calculated from the volume and concentration of each). Because equal volumes and concentrations are involved, neither the acid nor the base is in excess, and none will remain after the reaction. To find the pH of the solution after reaction, you need to know the amount of product, whether it is a weak acid or weak base, and its concentration. The final step then involves a weak acid or weak base equilibrium calculation (Examples 17.5–17.7).

Strategy This question involves three problems in one:

(a) *Writing a balanced equation:* First write a balanced equation for the reaction that occurs and then decide whether the reaction products are acids or bases. Here, the weak acid NH_4^+ is the product of interest.

(b) *Stoichiometry problem:* Finding the "initial" NH_4^+ concentration is a stoichiometry problem: What amount of NH_4^+ (in moles) is produced in the HCl + NH_3 reaction, and in what volume of solution is the NH_4^+ ion found?

(c) *Equilibrium problem:* Calculating the pH involves first solving an equilibrium problem. The crucial piece of information needed for this calculation is the "initial" concentration of NH_4^+ from part (b).

Solution If equal amounts (moles) of base (NH_3) and acid (HCl) are mixed, the result should be an acidic solution because the significant species remaining in solution upon completion of the reaction is NH_4^+, the conjugate acid of the weak base ammonia (see Tables 17.3 and 17.5).

(a) *Writing balanced equations*

The equation for the product-favored reaction of HCl (the supplier of hydronium ion) with NH_3 to give NH_4^+:

$$NH_3(aq) + H_3O^+(aq) \rightarrow NH_4^+(aq) + H_2O(\ell)$$

The equation for the reactant-favored reaction of NH_4^+, the product, with water:

$$NH_4^+(aq) + H_2O(\ell) \rightleftharpoons H_3O^+(aq) + NH_3(aq)$$

(b) *Stoichiometry problem*

Amount of HCl and NH_3 consumed:

$$(0.025 \text{ L HCl})(0.016 \text{ mol/L}) = 4.0 \times 10^{-4} \text{ mol HCl}$$

$$(0.025 \text{ L } NH_3)(0.016 \text{ mol/L}) = 4.0 \times 10^{-4} \text{ mol } NH_3$$

Amount of NH_4^+ produced upon completion of the reaction:

$$4.0 \times 10^{-4} \text{ mol } NH_3 \left(\frac{1 \text{ mol } NH_4^+}{1 \text{ mol } NH_3} \right) = 4.0 \times 10^{-4} \text{ mol } NH_4^+$$

What is the pH After Mixing Equal Numbers of Moles of an Acid and a Base

Table 17.5 summarizes the outcome of mixing various types of acids and bases. But how do you calculate a numerical value for the pH, particularly in the case of mixing a weak acid with a strong base or a weak base with a strong acid? The strategy (Example 17.8) is to recognize that this involves two calculations: a stoichiometry calculation and an equilibrium calculation. The key to this is that you need to know the concentration of the weak acid or weak base produced when the acid and base are mixed. Answering the following questions will guide you to an answer:

(a) What amounts of acid and base are used (in moles)? (This is a stoichiometry problem.)

(b) What is the total volume of the solution after mixing the acid and base solutions?

(c) What is the concentration of the weak acid or base produced on mixing the acid and base solutions?

(d) Using the concentration found in Step (c), what is the hydronium ion concentration in the solution? (This is an equilibrium problem.)

(e) Calculate the pH of the solution from $[H_3O^+]$.

Concentration of NH_4^+: Combining 25 mL each of HCl and NH_3 gives a total solution volume of 50. mL. Therefore, the concentration of NH_4^+ is

$$[NH_4^+] = \frac{4.0 \times 10^{-4}\ \text{mol}}{0.050\ \text{L}} = 8.0 \times 10^{-3}\ \text{M}$$

(c) *Acid–base equilibrium problem*

With the initial concentration of ammonium ion known, set up an ICE table to find the equilibrium concentration of hydronium ion.

Equilibrium	NH_4^+	$+$	H_2O	$\rightleftharpoons$	NH_3	$+$	H_3O^+
Initial (M)	0.0080				0		0
Change (M)	$-x$				$+x$		$+x$
Equilibrium (M)	$(0.0080 - x)$				x		x

Next, substitute the values in the table into the K_a expression for the ammonium ion. Thus, we have

$$K_a = 5.6 \times 10^{-10} = \frac{[H_3O^+][NH_3]}{[NH_4^+]} = \frac{(x)(x)}{0.0080 - x}$$

The ammonium ion is a very weak acid, as reflected by the very small value of K_a. Therefore, x, the concentration of hydronium ion generated by reaction of ammonium ion with water, is assumed to be very small, and the approximate expression is used to solve for x. (Here $100 \times K_a$ is much less than the original acid concentration.)

$$K_a = 5.6 \times 10^{-10} \approx \frac{x^2}{0.0080}$$

$$x = \sqrt{(5.6 \times 10^{-10})(0.0080)}$$

$$= [H_3O^+] = [NH_3] = 2.1 \times 10^{-6}\ \text{M}$$

$$pH = -\log(2.1 \times 10^{-6}) = 5.67$$

Think about Your Answer As predicted (Table 17.5), the solution after mixing equal amounts of a strong acid and weak base is weakly acidic.

Check Your Understanding

Calculate the pH after mixing 15 mL of 0.12 M acetic acid with 15 mL of 0.12 M NaOH. What are the major species in solution at equilibrium (besides water), and what are their concentrations?

CASE STUDY

Would You Like Some Belladonna Juice in Your Drink?

The belladonna plant produces black, cherry-sized berries that look good enough to eat. But you should think twice because the plant is also called *deadly nightshade*. The berries contain atropine, a base, which is toxic and has been used by women and men for centuries to eliminate old lovers and political enemies.

The name for the plant, *belladonna*, comes from the Italian meaning "beautiful woman." Courtesans and actresses in Venice in the 16th century would put a drop or two of the juice in their eyes. It would dilate the pupils and give them a "fashionable, doe-like" appearance that could last several days.

Atropine was used for some years by physicians to dilate the eye, and it is used to treat such conditions as bradycardia, an extremely low heart rate. It is also an antidote for sarin, a nerve gas. In fact, victims of a sarin gas attack in Japan in the 1990s were treated with atropine, and soldiers in the first Gulf War in 1990–1991 carried atropine as an antidote in the event of a nerve gas attack.

© WILDLIFE GmbH/Alamy

Berries of the belladonna plant, a source of atropine

But atropine is mostly remembered in history as a poison. A fatal dose is around 100 mg. One legend is that Cleopatra at first wanted to commit suicide by taking atropine. She had a slave try it first, but the slave's death was not pleasant, so she opted for the bite of an asp, a deadly snake.

At the height of the Roman Empire, poisoning with atropine and other agents was so often used that a law was passed in 82 BC to "suppress domestic poisoning." It apparently had little effect.

The symptoms of atropine poisoning are well known. Physicians say the patient is "red as a beet" because the blood vessels dilate. Or you can be "blind as a bat" because the pupils are so dilated that vision is blurred. "Dry as a bone" is another symptom arising from the suppression of the salivary glands. In addition, you can be "hot as a hare" because your body temperature goes up and stays up for some hours. Finally, you can be "mad as a hatter" (like the hatter in *Alice in Wonderland*) because you act like a drunk person.

Atropine is not very soluble in water. However, because atropine is a base—like so many other naturally occurring substances such as caffeine, nicotine, morphine, heroin, and amphetamine—it reacts with acids. For example, reaction with sulfuric acid forms the sulfate salt. This salt is so soluble that 1 mL of a saturated solution of this compound contains many times the amount needed for a fatal dose.

If you are interested in the history of atropine as a murder weapon be sure to read about the attempted murder of Alexandra Agutter by her husband in Scotland in the 1990s. Just like an episode of CSI!

Atropine, $C_{17}H_{23}NO_3$.

Questions:

1. If a fatal dose of atropine is 100. mg, what amount of compound (moles) is this?
2. When atropine is added to sulfuric acid, a proton attaches to the molecule. What is the site of attachment?
3. The pK_a of the conjugate acid of atropine is 4.35. How does this compare with the pK_a values for the conjugate acids of ammonia, methylamine, and aniline (page 587)?
4. Atropine is chiral. (See Chapter 10.) Which C atom is the site of chirality? (Of the two enantiomers, only one is physiologically active.)

Answers to these questions are available in Appendix N.

Reference:

J. Emsley, *Molecules of Murder*, Royal Society of Chemistry, London, 2008.

REVIEW & CHECK FOR SECTION 17.8

1. What is $[H_3O^+]$ in a 0.10 M solution of HCN at 25 °C? (K_a for HCN $= 4.0 \times 10^{-10}$)

   (a) 1.6×10^{-9} M

   (b) 6.3×10^{-6} M

   (c) 2.0×10^{-5} M

   (d) 4.0×10^{-11} M

2. A 0.040 M solution of an acid, HA, has a pH of 3.02 at 25 °C. What is K_a for this acid?

   (a) 2.3×10^{-5}

   (b) 5.7×10^{-4}

   (c) 2.4×10^{-2}

   (d) 4.3×10^{-10}

3. What are the pH and ion concentrations in a solution of 0.10 M sodium formate, $NaCHO_2$? K_b for the formate ion, HCO_2^- is 5.6×10^{-11}.

	pH	$[Na^+]$	$[CHO_2^-]$	$[OH^-]$
(a)	5.63	0.10	0.10	2.4×10^{-6}
(b)	8.37	0.10	0.10	2.4×10^{-6}
(c)	8.22	0.050	0.050	1.7×10^{-6}
(d)	5.63	0.10	0.10	4.2×10^{-9}

4. You mix 0.40 g of NaOH with 100 mL of 0.10 M acetic acid. What is the pH of the resulting solution?

    (a) less than 7   (b) equal to 7   (c) greater than 7

17.9 Polyprotic Acids and Bases

Because polyprotic acids are capable of donating more than one proton (Table 17.1), they present us with additional challenges when predicting the pH of their solutions. For many inorganic polyprotic acids, such as phosphoric acid, carbonic acid, and hydrosulfuric acid (H_2S), the ionization constant for each successive loss of a proton is about 10^4 to 10^6 smaller than the previous ionization step. This means that the first ionization step of a polyprotic acid produces up to about a million times more H_3O^+ ions than the second step. For this reason, *the pH of many inorganic polyprotic acids depends primarily on the hydronium ion generated in the first ionization step; the hydronium ion produced in the second step can be neglected.* The same principle applies to the fully deprotonated conjugate bases of polyprotic acids. This is illustrated by the calculation of the pH of a solution of carbonate ion, an important base in our environment (Example 17.9).

EXAMPLE 17.9   Calculating the pH of the Solution of a Polyprotic Base

Problem The carbonate ion, CO_3^{2-}, is a base in water, forming the hydrogen carbonate ion, which in turn can form carbonic acid.

$$CO_3^{2-}(aq) + H_2O(\ell) \rightleftharpoons HCO_3^-(aq) + OH^-(aq) \qquad K_{b1} = 2.1 \times 10^{-4}$$

$$HCO_3^-(aq) + H_2O(\ell) \rightleftharpoons H_2CO_3(aq) + OH^-(aq) \qquad K_{b2} = 2.4 \times 10^{-8}$$

What is the pH of a 0.10 M solution of Na_2CO_3?

What Do You Know? You know the balanced equations and the values of K_b for the ions as well as the concentration of the carbonate ion.

Strategy The first ionization constant, K_{b1}, is much larger than the second, K_{b2}, so the hydroxide ion concentration in the solution results almost entirely from the first step. Therefore, you can calculate the OH^- concentration produced in the first ionization step, but we will test the conclusion that OH^- produced in the second step is negligible.

Solution Set up an ICE table for the reaction of the carbonate ion (Equilibrium Table 1).

Equilibrium Table 1—Reaction of CO_3^{2-} Ion

Equilibrium	CO_3^{2-}	+	H_2O	$\rightleftharpoons$	HCO_3^-	+	OH^-
Initial (M)	0.10				0		0
Change	$-x$				$+x$		$+x$
Equilibrium (M)	$(0.10 - x)$				x		x

Based on this table, the equilibrium concentration of OH^- ($= x$) can then be calculated.

$$K_{b1} = 2.1 \times 10^{-4} = \frac{[HCO_3^-][OH^-]}{[CO_3^{2-}]} = \frac{x^2}{0.10 - x}$$

A polyprotic acid. Malic acid $C_4H_4O_4$ is a diprotic acid occurring in apples. It is also classified as an alpha-hydroxy acid because it has an OH group on the C atom next to the CO_2H (in the alpha position). It is one of a large group of natural alpha-hydroxy acids such as lactic acid, citric acid, and ascorbic acid. Alpha-hydroxy acids have been touted as an ingredient in "anti-aging" skin creams. They work by accelerating the natural process by which skin replaces the outer layer of cells with new cells.

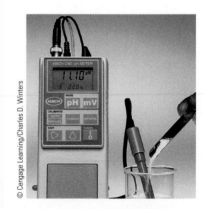

Sodium carbonate, a polyprotic base. This common substance is a base in aqueous solution. Its primary use is in the glass industry. Although it used to be manufactured, it is now mined as the mineral trona, $Na_2CO_3 \cdot NaHCO_3 \cdot 2\ H_2O$.

Because K_{b1} is relatively small, it is reasonable to make the approximation that $(0.10 - x) \approx 0.10$. Therefore,

$$x = [HCO_3^-] = [OH^-] = \sqrt{(2.1 \times 10^{-4})(0.10)} = 4.6 \times 10^{-3} \text{ M}$$

Using this value of $[OH^-]$, you can calculate the pOH of the solution,

$$pOH = -\log (4.6 \times 10^{-3}) = 2.34$$

and then use the relationship $pH + pOH = 14$ to calculate the pH.

$$pH = 14 - pOH = \boxed{11.66}$$

Finally, you can conclude that the concentration of the carbonate ion is, to a good approximation, 0.10 M.

$$[CO_3^{2-}] = 0.10 - 0.0046 \approx 0.10 \text{ M}$$

The HCO_3^- ion produced in the first step could acquire another proton to give H_2CO_3 and this could affect the pH. But does this occur to a meaningful extent? To test this, set up a second ICE Table.

Equilibrium Table 2—Reaction of HCO_3^- Ion

Equilibrium	HCO_3^-	+	H_2O	$\rightleftharpoons$	H_2CO_3	+	OH^-
Initial (M)	4.6×10^{-3}				0		4.6×10^{-3}
Change (M)	$-y$				$+y$		$+y$
Equilibrium (M)	$(4.6 \times 10^{-3} - y)$				y		$(4.6 \times 10^{-3} + y)$

Because K_{b2} is so small, the second step occurs to a much smaller extent than the first step. This means the amount of H_2CO_3 and OH^- produced in the second step ($= y$) is much smaller than 10^{-3} M. Therefore, it is reasonable that both $[HCO_3^-]$ and $[OH^-]$ are very close to 4.6×10^{-3} M.

$$K_{b2} = 2.4 \times 10^{-8} = \frac{[H_2CO_3][OH^-]}{[HCO_3^-]} = \frac{(y)(4.6 \times 10^{-3})}{4.6 \times 10^{-3}}$$

Because $[HCO_3^-]$ and $[OH^-]$ have nearly identical values, they cancel from the expression, and we find that $[H_2CO_3]$ is simply equal to K_{b2}.

$$y = [H_2CO_3] = K_{b2} = 2.4 \times 10^{-8} \text{ M}$$

The amount of OH^- produced in this reaction is negligible. The hydroxide ion is essentially all produced in the first equilibrium process.

Think about Your Answer It is almost always the case that the pH of a solution of an inorganic polyprotic acid is due to the hydronium ion generated in the first ionization step. Similarly, the pH of a polyprotic base is due to the OH^- ion produced in the first ionization step.

Check Your Understanding

What is the pH of a 0.10 M solution of oxalic acid, $H_2C_2O_4$? What are the concentrations of H_3O^+, $HC_2O_4^-$, and the oxalate ion, $C_2O_4^{2-}$? (See Appendix H for K_a values.)

REVIEW & CHECK FOR SECTION 17.9

Hydrazine (N_2H_4) is like CO_3^{2-} in that it is a polyprotic base ($K_{b1} = 8.5 \times 10^{-7}$ and $K_{b2} = 8.9 \times 10^{-16}$). The two conjugate acids are $N_2H_5^+$ and $N_2H_6^{2+}$. What is the expected pH of a 0.025 M solution of N_2H_4?

(a) 3.83   (b) 8.32   (c) 10.16

17.10 Molecular Structure, Bonding, and Acid–Base Behavior

One of the most interesting aspects of chemistry is the correlation between a molecule's structure and bonding and its chemical properties. Because so many acids and bases play such a key role in chemistry, it is especially useful to see if there are some general principles governing acid–base behavior.

Acid Strength of the Hydrogen Halides, HX

Aqueous HF is a weak Brønsted acid in water, whereas the other hydrohalic acids—aqueous HCl, HBr, and HI—are all strong acids. Experiments show that the acid strength increases in the order HF << HCl < HBr < HI. A detailed analysis of the factors that lead to these differences in acid strength in this group is interesting (see *A Closer Look, Acid Strengths and Molecular Structure,* page 608), and predictions about relative acid strength can be made based on the sum of two energy quantities, the energy required to break the H—X bond (the bond dissociation enthalpy, page 299) and the electron attachment enthalpy of the halogen (page 252). That is, when a hydrohalic acid HX ionizes in water,

$$HX(aq) + H_2O(\ell) \rightarrow H_3O^+(aq) + X^-(aq)$$

the sum of the two energy terms

Change in energy $\propto$ HX bond dissociation enthalpy + electron attachment enthalpy of X

should correlate with acid strength. Stronger acids should result when the H—X bond is more easily broken (as signaled by a smaller, positive value of ΔH for bond dissociation) and a more negative value for the electron attachment enthalpy of X.

The bond dissociation enthalpy and electron attachment enthalpy effects can work together (a weak H—X bond and a large, negative electron attachment enthalpy of the X group) to produce a strong acid, but they can also work in opposite directions. The balance of the two effects is important. Let us examine some data for the Group 7A binary acids, HX.

| | ——— Increasing acid strength ——→ | | | |
	HF	**HCl**	**HBr**	**HI**
pK_a	+3.14	−7	−9	−10
H—X bond strength (kJ/mol)	565	432	366	299
Electron attachment enthalpy of X (kJ/mol)	−328	−349	−325	−295
Sum (kJ/mol)	237	83	41	4

In this series of acids, the bond enthalpy factor dominates: the weakest acid, HF, has the strongest H—X bond, and the strongest acid, HI, has the weakest H—X bond. However, the electron attachment enthalpy of X is most negative with Cl, less so with F and Br, and much less so with I. A less negative electron attachment enthalpy should lead to a weaker acid, but it is the *sum* of the two effects that leads to the observation that HI is the strongest acid. In Figure 17.6 you see there is a good correlation between an acid's pK_a and the *sum* of the enthalpies for bond dissociation and electron attachment enthalpy.

Comparing Oxoacids: HNO₂ and HNO₃

Nitrous acid (HNO₂) and nitric acid (HNO₃) are representative of several series of **oxoacids**. Oxoacids contain an atom (usually a nonmetal atom) bonded to one or more oxygen atoms, some with hydrogen atoms attached. Besides those based on N, you are familiar with the sulfur- and chlorine-based oxoacids (Table 17.6). In all these series of related compounds, the acid strength increases as the number of oxy-

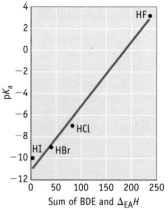

FIGURE 17.6 The effect of H—X bond enthalpy and X electron attachment enthalpy on acid strength. Stronger acids have weaker H—X bonds and X atoms with more negative electron attachment enthalpies. (BDE is the bond dissociation enthalpy of the H—X bond, and $\Delta_{EA}H$ is the electron attachment enthalpy of the halogen atom, see page 253.) See M. Moran, *Journal of Chemical Education,* Vol. 83, pp. 800–803, 2006.

Table 17.6 Oxoacids

Acid	pK_a
Cl-Based Oxoacids	
HOCl	7.46
HOClO (HClO₂)	~ 2
HOClO₂ (HClO₃)	~ −3
HOClO₃ (HClO₄)	~ −8
S-Based Oxoacids	
(HO)₂SO [H₂SO₃]	1.92, 7.21
(HO)₂SO₂ [H₂SO₄]	~ −3, 1.92

Linus Pauling stated the general relationship that, for oxoacids with the general formula (HO)ₙE(O)ₘ, the value of pK_a is about 8–5m. When $n > 1$, the pK_a increases by about 5 for each successive loss of a proton.

A CLOSER LOOK

Acid Strengths and Molecular Structure

Although predictions about acid strength in aqueous solution are fairly simple to make, a complete explanation can be complicated.

Acid strength is sometimes correlated with the strength and the polarity of the H—X bond, readily identifiable characteristics derived from the structure of the acid, the reactant in the ionization process. We need to point out, however, that when one is assessing any chemical reaction, it is necessary to consider both reactants and products. Looking only at the reactant when dealing with acid ionization only takes you halfway.

When evaluating the strength of an acid HX(aq), we are looking at the following reaction:

$$HX(aq) + H_2O(\ell) \rightleftharpoons H_3O^+(aq) + X^-(aq)$$

To fully explain the extent of ionization, we must consider characteristics of both the acid and the anion. The ability of the anion to spread out the negative charge across the ion, for example, and the solvation of the anion by the solvent are among the issues that must have some relevance in an explanation of the strength of the acid.

How enthalpy changes contribute to acid strength can be assessed using a thermochemical cycle (such as the one used to evaluate a lattice enthalpy, page 461). Consider the relative acid strengths of the hydrogen halides. The enthalpy change for the ionization of an acid in water can be related to other enthalpy changes as shown in the diagram. The solvation of H$^+$ (Step 5) and the ionization energy of H(g) (Step 3) are common to all of the hydrogen halides and do not contribute to the differences among hydrogen halides, but the four remaining terms are different. For the hydrogen halides, the bond dissociation

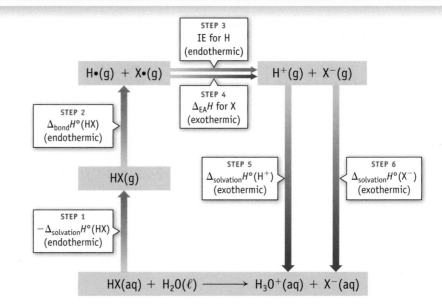

enthalpy (Step 2) of HF is much larger than the dissociation enthalpies of the other hydrogen halides. However, it is compensated for significantly by the enthalpy of solvation of the anion (Step 6), which for the fluoride ion is much more exothermic than the solvation energies of the other halide ions. The electron attachment enthalpy is also a contributor to the differences in overall enthalpy changes. Electron attachment enthalpy values vary among the halogens, but to a smaller extent than the variation in bond energy and the enthalpy of solvation of the halide ion. Differences in solvation energy for the molecular species (Step 1) are minimal.

A complete analysis of strengths of acids in aqueous solution will include consideration of both enthalpies and entropies. Entropy has yet to be discussed in this text

in detail (▶ Chapter 19) but we noted earlier (◀ page 478) that entropy plays an important role in solution chemistry, specifically in determining solubilities. It is not surprising that entropy has a role in determining acid strength, too. Indeed, differences in entropy changes are significant in accounting for the differences in acid strength of the hydrogen halides.

Although all of these terms contribute to acid strength, acid strength can often be correlated with a subset of this information, as can be seen by the examples presented in this section. We point out, however, that correlations, while highly useful to a chemist because they can be used to make important predictions, are at best only partial explanations.

gen atoms bonded to the central element increases. Thus, nitric acid (HNO$_3$) is a stronger acid than nitrous acid (HNO$_2$).

HNO$_3$, strong acid,
pK_a = −1.4

HNO$_2$, weak acid,
pK_a = +3.35

Let's apply the bond enthalpy/electron affinity analysis to HNO$_3$ and HNO$_2$.

	Increasing acid strength ⟶	
	HNO₂	**HNO₃**
pK_a	+3.35	−1.4
H—O bond strength (kJ/mol)	328	423
Electron attachment enthalpy of X (kJ/mol)	−219	−377
Sum (kJ/mol)	109	46

nitrous acid, HNO_2 nitric acid, HNO_3

FIGURE 17.7 Electrostatic potential surfaces for the nitrogen oxoacids. Both surfaces show the O—H bond is quite polar. More importantly, calculations show that the OH bond becomes more polar as more O atoms are added to N.

Partial Charges

Molecule	H Atom	O Atom of OH	N Atom
HNO₂	+0.39	−0.35	+0.14
HNO₃	+0.41	−0.47	+0.76

As in the case of the Group 7A acids, we again see that acid strength correlates with the sum of the enthalpies for bond breaking and X group (NO₂ or NO₃) electron attachment enthalpy.

Our previous analysis of the Group 7A acids showed that, in that series, the H—X bond strength was the more important factor; as the H—X bond became stronger, the acid became weaker. However, the data for HNO₃ and HNO₂ show this is not true here. The O—H bond is stronger in the stronger acid HNO₃. Instead, the electron attachment term is the more important term in this correlation. These same effects are observed for other oxyacids such as the chlorine-based oxoacids HOCl < HOClO < HOClO₂ < HOClO₃ and the S-based oxoacids (Table 17.6). How is this to be interpreted?

In HNO₃, there are two other oxygen atoms bonded to the central nitrogen atom, whereas in HNO₂ only one other oxygen is bonded to the nitrogen atom. By attaching more electronegative O atoms to nitrogen, we are increasing the electron affinity of the group attached to hydrogen, and anything that increases the affinity of the X group for an electron should also make HX a stronger acid and X⁻ a weaker conjugate base. This is another way of saying that if X⁻ has a way to accommodate and stabilize a negative charge, it will be a weaker conjugate base. In the case of oxoacids, additional oxygen atoms have the effect of stabilizing the anion because the negative charge on the anion can be dispersed over more atoms. In the nitrate ion, for example, the negative charge is shared equally over the three oxygen atoms. This is represented symbolically by the three resonance structures for this ion.

$$\left[\begin{array}{c} :O: \\ \| \\ N \\ :O \quad O: \end{array}\right]^- \longleftrightarrow \left[\begin{array}{c} :O: \\ | \\ N \\ O \quad O: \end{array}\right]^- \longleftrightarrow \left[\begin{array}{c} :O: \\ | \\ N \\ :O \quad O \end{array}\right]^-$$

In nitrite ion, only two atoms share the negative charge. Therefore, NO₂⁻ is a stronger conjugate base than NO₃⁻. In general, greater stabilization of the products formed by ionizing the acid contributes to increased acidity.

Table 17.7 and Figure 17.7 show that another empirical correlation can be made between the structure of an acid and its acidity: In a series of related acids, the larger the formal charge on the central atom, the stronger the acid (Table 17.7). For example, the N atom formal charge in the weak acid HNO₂ is 0, whereas it is +1 in the strong acid HNO₃, and these are reflected by the results of theoretical calculations cited in Figure 17.7.

In summary, molecules such as the oxoacids can behave as stronger Brønsted acids when the anion created by loss of H⁺ is stable and able to accommodate the negative charge. These conditions are promoted by

- the presence of electronegative atoms attached to the central atom.
- the possibility of resonance structures for the anion, which lead to delocalization of the negative charge over the anion and thus to a more stable ion.
- greater positive formal charge of the central atom.

Why Are Carboxylic Acids Brønsted Acids?

There is a large class of organic acids, typified by acetic acid (CH₃CO₂H) (Figure 17.8), called *carboxylic acids* because all have the carboxylic acid group, —CO₂H (◀ page 367). The arguments used to explain the acidity of oxoacids can

Table 17.7 Correlation of Atom Formal Charge and pK_a

	Central Atom Formal Charge	pK_a
:O—H | :O—S—O: | :O—H	+2	−3
H :O—H | | :O—S—O:	+1	1.92
:O—Cl—O—H	+1	2
:Cl—O—H	0	7.46

FIGURE 17.8 Electrostatic potential surface and partial charges for acetic acid. The H atoms of the molecule are all positively charged, but the H atom of the OH is much more highly charged. As expected, both of the electronegative O atoms have a partial negative charge. The table below gives some of the computer-calculated partial charges on the acid.

Atom or Group	Calc'd Partial Charge
H of OH	+0.43
O of OH	−0.62
H of CH$_3$	+0.23

also be applied to carboxylic acids. The O—H bond in these compounds is polar, a prerequisite for ionization.

polar O—H bond broken by interaction of positively charged H atom with hydrogen-bonded H$_2$O

In addition, carboxylate anions are stabilized by delocalizing the negative charge over the two oxygen atoms.

The simple carboxylic acids, RCO$_2$H in which R is a hydrocarbon group (◄ Section 10.4), do not differ markedly in acid strength (compare acetic acid, pK_a = 4.74, and propanoic acid, pK_a = 4.89, Table 17.3). The acidity of carboxylic acids is enhanced, however, if electronegative substituents replace the hydrogens in the alkyl (–CH$_3$ or –C$_2$H$_5$) group. Compare, for example, the pK_a values of a series of acetic acids in which hydrogen is replaced sequentially by the more electronegative element chlorine.

Acid		pK_a Value	
CH$_3$CO$_2$H	Acetic acid	4.74	
ClCH$_2$CO$_2$H	Chloroacetic acid	2.85	Increasing acid
Cl$_2$CHCO$_2$H	Dichloroacetic acid	1.49	strength
Cl$_3$CCO$_2$H	Trichloroacetic acid	0.7	

As in the nitrogen oxoacids, increasingly electronegative substituents lead to an increase in acid strength. One argument is that electronegative substituents stabilize the negative charge of the anion. That is, the Cl$_3$CCO$_2$$^-$ anion is more stable than the H$_3$CCO$_2$$^-$ anion, so the formation of Cl$_3$CCO$_2$$^-$ is more favored than the formation of CH$_3$CO$_2$$^-$. Thus, the Cl$_3CCO_2$$^-$ anion is a weaker base than the CH$_3$CO$_2$$^-$ anion.

Finally, why are the C—H hydrogens of carboxylic acids not dissociated as H$^+$ instead of (or in addition to) the O—H hydrogen atom? The calculated partial positive charges listed in Figure 17.8 show that the H atoms of the CH$_3$ group have a much smaller positive partial charge than the O—H hydrogen atom. Furthermore, in carboxylic acids, the C atom of the CH$_3$ group is not sufficiently electronegative to accommodate the negative charge left if the bond breaks as C—H → C:$^-$ + H$^+$, and the product anion is not well stabilized.

Why Are Hydrated Metal Cations Brønsted Acids?

When a coordinate covalent bond is formed between a metal cation and a water molecule, the positive charge of the metal ion and its small size means that the electrons of the H$_2$O→M^{n+} bond are very strongly attracted to the metal (Section 17.11). As a result, the O—H bonds of the bound water molecules are polarized, just as in oxoacids and carboxylic acids. The net effect is that a H atom of a coordinated water molecule is removed as H$^+$ more readily than in an uncoordinated water molecule. Thus, a hydrated metal cation functions as a Brønsted acid or proton donor.

$$[Cu(H_2O)_6]^{2+} + H_2O(\ell) \rightleftharpoons [Cu(H_2O)_5(OH)]^+(aq) + H_3O^+(aq)$$

The acidity of the hydrated metal ion increases with increasing charge. Consulting Table 17.3, you see that the K_a of $+3$ ions (for example Al^{3+} and Fe^{3+}) is greater than for $+2$ cations (Cu^{2+}, Pb^{2+}, Co^{2+}, Fe^{2+}, Ni^{2+}). Ions with a single positive charge such as Na^+ and K^+ are not acidic. (This is similar to the effect of central atom formal charge in a series of related acids. See Table 17.7.)

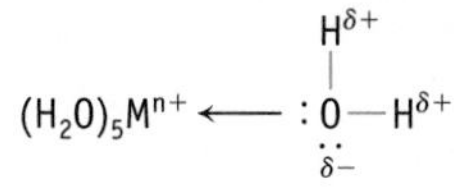

Why Are Anions Brønsted Bases?

Anions, particularly oxoanions such as PO_4^{3-}, are Brønsted bases. The negatively charged anion interacts with the positively charged H atom of a polar water molecule, and an H^+ ion is transferred to the anion.

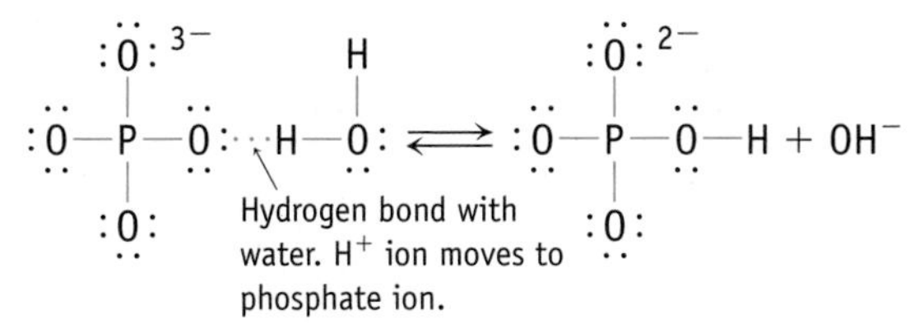

The data in Table 17.8 show that, in a series of related anions, the basicity of an anionic base increases as the negative charge of the anion increases.

REVIEW & CHECK FOR SECTION 17.10

1. Which of the following is the stronger acid?

    (a) H_2SeO_4 (b) H_2SeO_3

2. Which of the following should be the stronger acid?

    (a) $[Fe(H_2O)_6]^{2+}$ (b) $[Fe(H_2O)_6]^{3+}$

3. Which of the following should be the stronger acid?

    (a) $HOCl$ (b) $HOBr$

● **Polarization of O—H Bonds** Water molecules attached to a metal cation have strongly polarized O—H bonds.

Table 17.8 Basic Oxoanions

Anion	pK_b
PO_4^{3-}	1.55
HPO_4^{2-}	6.80
$H_2PO_4^-$	11.89
CO_3^{2-}	3.68
HCO_3^-	7.62
SO_3^{2-}	6.80
HSO_3^-	12.08

17.11 The Lewis Concept of Acids and Bases

The concept of acid–base behavior advanced by Brønsted and Lowry in the 1920s works well for reactions involving proton transfer. However, a more general acid–base concept was developed by Gilbert N. Lewis in the 1930s. This concept is based on the sharing of electrons pairs between an acid and a base.

> A **Lewis acid** is a substance that can accept a pair of electrons from another atom to form a new bond, and a **Lewis base** is a substance that can donate a pair of electrons to another atom to form a new bond.

This means that an acid–base reaction in the Lewis sense occurs when a molecule (or ion) donates a pair of electrons to another molecule (or ion).

$$\begin{array}{ccc} A & + & B: & \rightarrow & B{\rightarrow}A \\ acid & & base & & adduct \end{array}$$

The product is often called an **acid–base adduct.** In Section 8.5, this type of chemical bond was called a *coordinate covalent bond.*

Formation of a hydronium ion from H^+ and water, and the formation of the ammonium ion from H^+ and ammonia, are good examples of a Lewis acid–base reaction (Figure 17.9). The H^+ ion has no electrons in its valence ($1s$) shell, and the

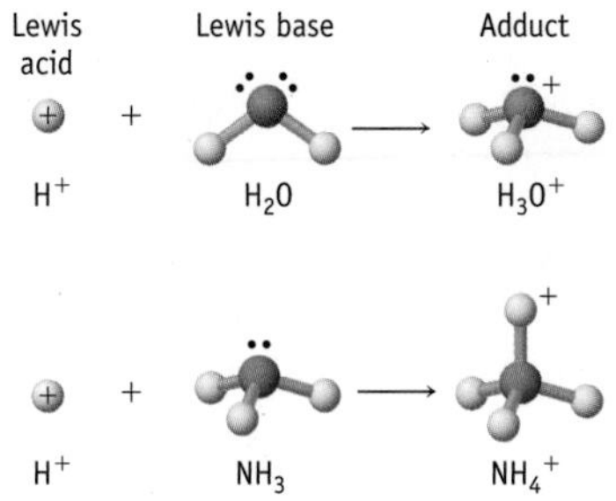

FIGURE 17.9 Protonation of water and ammonia are examples of Lewis acid–base reactions.

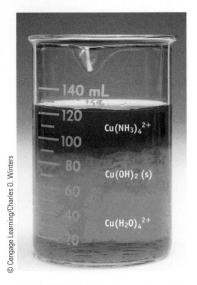

FIGURE 17.10 The Lewis acid–base complex ion [Cu(NH₃)₄]²⁺. Here, aqueous ammonia was added to aqueous CuSO₄ (the light blue solution at the bottom of the beaker). The small concentration of OH⁻ in NH₃(aq) first led to the formation of insoluble blue-white Cu(OH)₂ (the solid in the middle of the beaker). With additional NH₃, however, the deep blue, soluble complex ion formed (the solution at the top of the beaker). The model in the text shows the copper(II)–ammonia complex ion.

water molecule has two unshared pairs of electrons (located in sp^3 hybrid orbitals). One of the lone pairs of a water molecule oxygen atom can be shared with an H^+ ion, thus forming an O—H bond in an H_3O^+ ion. A similar interaction occurs between H^+ and the nitrogen lone pair on ammonia to form the ammonium ion.

Lewis acid–base reactions are very common. In general, they involve Lewis acids that are cations or neutral molecules with an available, empty valence orbital and bases that are anions or neutral molecules with a lone electron pair.

Cationic Lewis Acids

Just as H^+ and water form a Lewis acid–base adduct, metal cations interact with water molecules to form hydrated cations (Figure 17.10 and page 419). In these species, coordinate covalent bonds form between the metal cation and a lone pair of electrons on the O atom of each water. For example, an iron(II) ion, Fe^{2+}, forms six coordinate covalent bonds to water.

$$Fe^{2+}(aq) + 6\ H_2O(\ell) \rightarrow [Fe(H_2O)_6]^{2+}(aq)$$

Solutions of transition metal cations are generally very colorful (Figures 17.10 and 17.11 and Section 22.3). Chemists call the species formed by coordination of water to a metal ion **complex ions** or, because of the presence of coordinate covalent bonds, **coordination complexes**. Several are listed in Table 17.3 as acids, and their behavior is described further in Chapter 22.

Ammonia is an excellent Lewis base and, like water, it also combines with metal cations to give adducts (complex ions), which are often very colorful. For example, copper(II) ions, light blue in aqueous solution (Figure 17.10), react with ammonia to give a deep blue adduct with four ammonia molecules surrounding each Cu^{2+} ion.

$$Cu^{2+}(aq) + 4\ NH_3(aq) \longrightarrow [Cu(NH_3)_4]^{2+}(aq)$$

light blue   deep blue

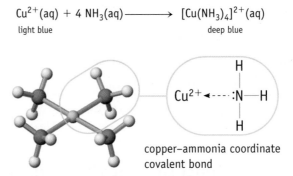

copper–ammonia coordinate covalent bond

Hydroxide ion, OH^-, is an excellent Lewis base and binds readily to metal cations to give metal hydroxides. An important feature of the chemistry of some metal

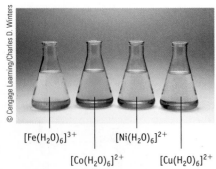

(a) Solutions of the nitrate salts of iron(III), cobalt(II), nickel(II), and copper(II) all have characteristic colors.

[Fe(H₂O)₆]³⁺   [Ni(H₂O)₆]²⁺
[Co(H₂O)₆]²⁺   [Cu(H₂O)₆]²⁺

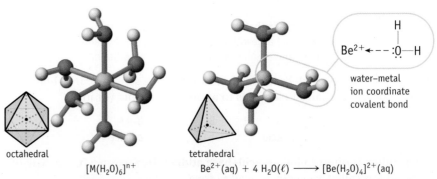

octahedral
[M(H₂O)₆]ⁿ⁺

tetrahedral
$$Be^{2+}(aq) + 4\ H_2O(\ell) \longrightarrow [Be(H_2O)_4]^{2+}(aq)$$

water–metal ion coordinate covalent bond

(b) Models of complex ions (Lewis acid–base adducts) formed between a metal cation and water molecules. Such complexes often have six or four water molecules arranged octahedrally or tetrahedrally around the metal cation.

FIGURE 17.11 Metal cations in water.

Table 17.9 Some Common Amphoteric Metal Hydroxides*

Hydroxide	Reaction as a Base	Reaction as an Acid
$Al(OH)_3$	$Al(OH)_3(s) + 3\ H_3O^+(aq) \rightleftharpoons Al^{3+}(aq) + 6\ H_2O(\ell)$	$Al(OH)_3(s) + OH^-(aq) \rightleftharpoons [Al(OH)_4]^-(aq)$
$Zn(OH)_2$	$Zn(OH)_2(s) + 2\ H_3O^+(aq) \rightleftharpoons Zn^{2+}(aq) + 4\ H_2O(\ell)$	$Zn(OH)_2(s) + 2\ OH^-(aq) \rightleftharpoons [Zn(OH)_4]^{2-}(aq)$
$Sn(OH)_4$	$Sn(OH)_4(s) + 4\ H_3O^+(aq) \rightleftharpoons Sn^{4+}(aq) + 8\ H_2O(\ell)$	$Sn(OH)_4(s) + 2\ OH^-(aq) \rightleftharpoons [Sn(OH)_6]^{2-}(aq)$
$Cr(OH)_3$	$Cr(OH)_3(s) + 3\ H_3O^+(aq) \rightleftharpoons Cr^{3+}(aq) + 6\ H_2O(\ell)$	$Cr(OH)_3(s) + OH^-(aq) \rightleftharpoons [Cr(OH)_4]^-(aq)$

* The aqueous metal cations are best described as $[M(H_2O)_6]^{n+}$.

hydroxides is that they are **amphoteric**. An amphoteric metal hydroxide can behave as an acid or a base (Table 17.9). The chemistry of aluminum hydroxide, $Al(OH)_3$, is one of the best examples of this behavior (Figure 17.12). Adding OH^- to a precipitate of $Al(OH)_3$ produces the water-soluble complex ion $[Al(OH)_4]^-$. In this reaction, $Al(OH)_3$ is acting as a Lewis acid.

$$Al(OH)_3(s) + \underset{\text{base}}{OH^-(aq)} \rightarrow [Al(OH)_4]^-(aq)$$
$$\text{acid}$$

If acid is added to the $Al(OH)_3$ precipitate, it again dissolves. This time, however, aluminum hydroxide is acting as a base.

$$\underset{\text{base}}{Al(OH)_3(s)} + \underset{\text{acid}}{3\ H_3O^+(aq)} \rightarrow Al^{3+}(aq) + 6\ H_2O(\ell)$$

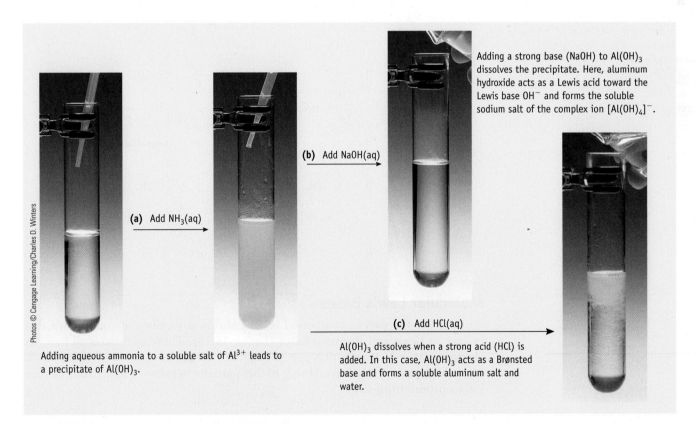

(a) Add $NH_3(aq)$

(b) Add NaOH(aq)

Adding a strong base (NaOH) to $Al(OH)_3$ dissolves the precipitate. Here, aluminum hydroxide acts as a Lewis acid toward the Lewis base OH^- and forms the soluble sodium salt of the complex ion $[Al(OH)_4]^-$.

(c) Add HCl(aq)

Adding aqueous ammonia to a soluble salt of Al^{3+} leads to a precipitate of $Al(OH)_3$.

$Al(OH)_3$ dissolves when a strong acid (HCl) is added. In this case, $Al(OH)_3$ acts as a Brønsted base and forms a soluble aluminum salt and water.

FIGURE 17.12 The amphoteric nature of $Al(OH)_3$. Aluminum hydroxide is formed by the reaction of aqueous Al^{3+} and ammonia.

$$Al^{3+}(aq) + 3\ NH_3(aq) + 3\ H_2O(\ell) \rightleftharpoons Al(OH)_3(s) + 3\ NH_4^+(aq)$$

Reactions of solid $Al(OH)_3$ with aqueous NaOH and HCl demonstrate that aluminum hydroxide is amphoteric.

Molecular Lewis Acids

Lewis's acid–base concept also accounts for the fact that oxides of nonmetals such as CO_2 and SO_2 behave as acids (◄ Section 3.6). Because oxygen is more electronegative than C, the C—O bonding electrons in CO_2 are polarized away from carbon and toward oxygen. This causes the carbon atom to be slightly positive, and it is this atom that the negatively charged Lewis base OH^- can attack to give, ultimately, the bicarbonate ion.

● **CO₂ in Basic Solution** This reaction of CO_2 with OH^- is the first step in the precipitation of $CaCO_3$ when CO_2 is bubbled into a solution of $Ca(OH)_2$ (Figure 3.5, page 99).

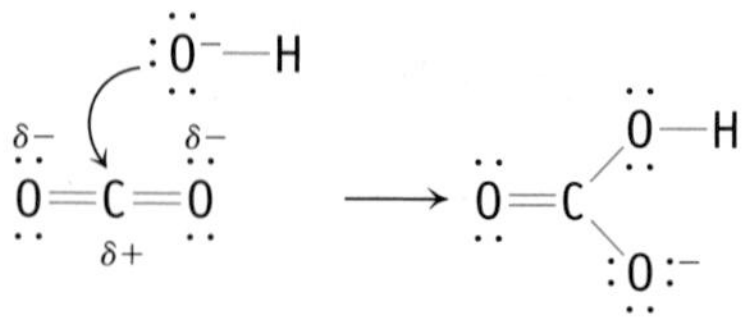

Similarly, SO_2 reacts with aqueous OH^- to form the HSO_3^- ion.

Compounds based on the Group 3A elements boron and aluminum are among the most-studied Lewis acids. One example is a reaction in organic chemistry that is catalyzed by the Lewis acid $AlCl_3$. The mechanism of this important reaction—called the *Friedel-Crafts reaction*—is illustrated here. In the first step, a Lewis base, the Cl^- ion, transfers from the reactant, here CH_3COCl, to the Lewis acid to give $[AlCl_4]^-$ and an organic cation. The organic cation attacks a benzene molecule to give a cationic intermediate, and this then interacts with $[AlCl_4]^-$ to produce HCl and the final organic product.

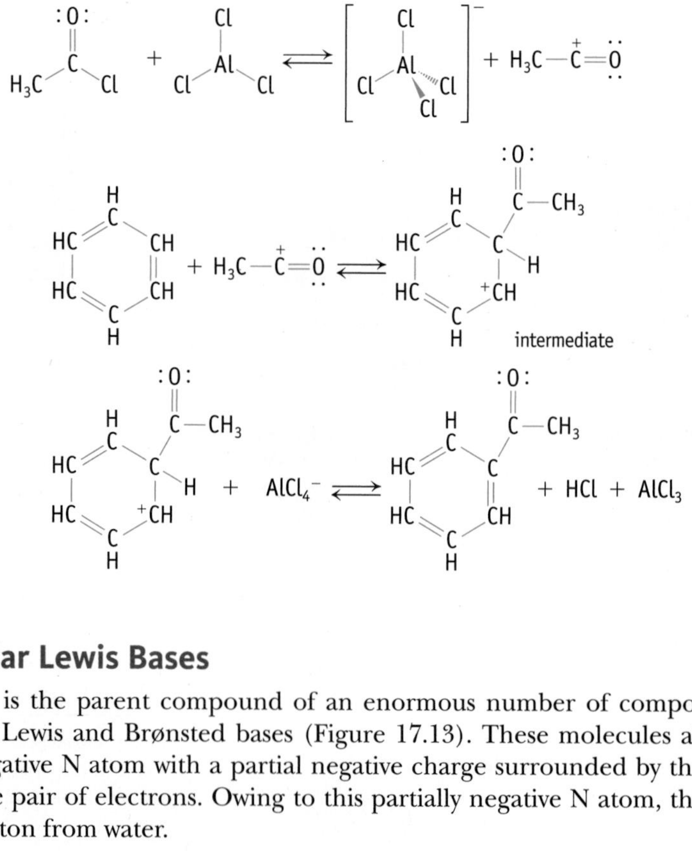

Molecular Lewis Bases

Ammonia is the parent compound of an enormous number of compounds that behave as Lewis and Brønsted bases (Figure 17.13). These molecules all have an electronegative N atom with a partial negative charge surrounded by three bonds and a lone pair of electrons. Owing to this partially negative N atom, they can extract a proton from water.

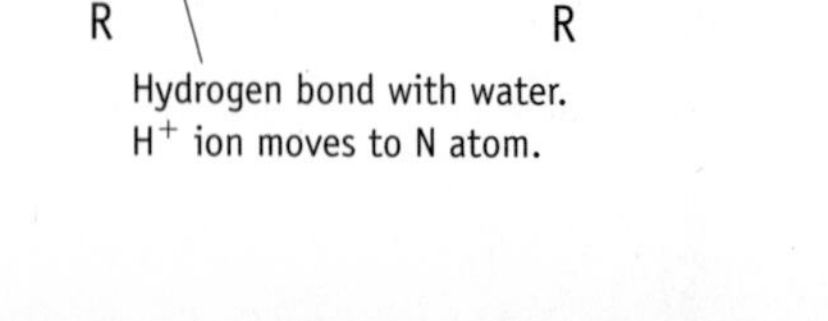

Hydrogen bond with water.
H^+ ion moves to N atom.

trimethylamine   pyridine   nicotine   glycine, an amino acid

FIGURE 17.13 Nitrogen-based Lewis and Brønsted bases. All have an N atom surrounded by three bonds and a lone pair of electrons.

In addition, the lone pair can be used to form a coordinate covalent bond to Lewis acids such as a metal cation (◄ page 612).

Ammonia is widely distributed in nature and is involved as a Lewis base in numerous reactions. One example where this is important is in the conversion of ammonia to urea (NH_2CONH_2) in natural systems. The process begins with the reaction of bicarbonate ion with ATP (adenosine triphosphate), and a subsequent step in the mechanism is the following:

intermediate

Here, the Lewis base ammonia attacks a carbon atom with a partial positive charge. The dihydrogen phosphate ion is then released, yielding the $NH_2CO_2^-$ ion, which eventually forms urea in another step in this reaction mechanism.

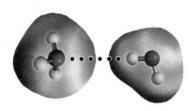

Electrostatic potential surfaces for NH_3 and H_2O. The red or negative region of these surfaces corresponds to the lone pair of electrons on N or the two pairs on O. The N atom of NH_3 has a calculated partial charge of -1.0, and the H atoms have a charge of $+0.33$. The N atom lone pair of NH_3 is involved in forming a hydrogen bond with the polar O—H bond of water. The NH_3 is both a Lewis and a Brønsted base and can remove the proton from water to form NH_4^+ and OH^-.

REVIEW & CHECK FOR SECTION 17.11

1. Which of the following can act as a Lewis acid? (*Hint:* In each case, draw the Lewis electron dot structure of the molecule or ion. Are there lone pairs of electrons on the central atom? If so, it can be a Lewis base. Does the central atom lack an electron pair? If so, it can behave as a Lewis acid.)

   (a) PH_3 (c) H_2S

   (b) BCl_3 (d) HS^-

2. The molecule whose structure is illustrated here is amphetamine, a stimulant. Which description best fits this molecule? (*Note:* There may be more than one answer.)

   (a) Brønsted acid   (b) Lewis acid   (c) Brønsted base   (d) Lewis base

and

CHAPTER GOALS REVISITED

Now that you have studied this chapter, you should ask whether you have met the chapter goals. In particular, you should be able to:

Use the Brønsted–Lowry and Lewis theories of acids and bases
a. Define and use the Brønsted concept of acids and bases (Sections 17.1 and 17.2).
b. Recognize common monoprotic and polyprotic acids and bases, and write balanced equations for their ionization in water (Section 17.2).
c. Appreciate when a substance can be amphiprotic (Section 17.2). **Study Questions: 5, 6.**
d. Recognize the Brønsted acid and base in a reaction, and identify the conjugate partner of each (Section 17.2). **Study Questions: 1–4, 7, 8.**
e. Use the water ionization constant, K_w (Section 17.3).
f. Use the pH concept (Section 17.3). **Study Questions: 9–14.**
g. Identify common strong acids and bases (Tables 3.1 and 17.3).
h. Recognize some common weak acids and understand that they can be neutral molecules (such as acetic acid), cations (NH_4^+ or hydrated metal ions such as $[Fe(H_2O)_6]^{2+}$, or anions (such as HCO_3^-) (Table 17.3).

Apply the principles of chemical equilibrium to acids and bases in aqueous solution
a. Write equilibrium constant expressions for weak acids and bases (Section 17.4).
b. Calculate pK_a from K_a (or K_a from pK_a), and understand how pK_a is correlated with acid strength (Section 17.4). **Study Questions: 15, 16, 23–28, 104, 106.**
c. Understand the relationship between K_a for a weak acid and K_b for its conjugate base (Section 17.4). **Study Questions: 17, 18, 29–32.**
d. Write equations for acid–base reactions, and decide whether they are product- or reactant-favored at equilibrium (Section 17.6 and Table 17.5). **Study Questions: 33–40.**
e. Calculate the equilibrium constant for a weak acid (K_a) or a weak base (K_b) from experimental information (such as pH, $[H_3O^+]$, or $[OH^-]$) (Section 17.8 and Example 17.4). **Study Questions: 41–44, 85, 107.**
f. Use the equilibrium constant and other information to calculate the pH of a solution of a weak acid or weak base (Section 17.8 and Examples 17.5 and 17.6). **Study Questions: 47–56, 65, 66, 89, 91, 99, and Go Chemistry Module 22.**
g. Describe the acid–base properties of salts, and calculate the pH of a solution of a salt of a weak acid or of a weak base (Section 17.5 and Example 17.7). **Study Questions: 57–60, 84, 92, 93, 103, 121.**

Predict the outcome of reactions of acids and bases
a. Recognize the type of acid–base reaction, and describe its result (Section 17.7).
b. Calculate the pH after an acid–base reaction (Section 17.7 and Example 17.8). **Study Questions: 61–64, 98, 102.**

Understand the influence of structure and bonding on acid–base properties
a. Appreciate the connection between the structure of a compound and its acidity or basicity (Section 17.10). **Study Questions: 69–72, 74, 114.**
b. Characterize a compound as a Lewis base (an electron-pair donor) or a Lewis acid (an electron-pair acceptor) (Section 17.11). **Study Questions: 73–76, 108.**

Key Equations

Equation 17.1 (page 579) Water ionization constant.

$$K_w = [H_3O^+][OH^-] = 1.0 \times 10^{-14} \text{ at 25 °C}$$

Equation 17.2 (page 583) Definition of pH (see also Equation 4.3).

$$pH = -\log[H_3O^+]$$

Equation 17.3 (page 583) Definition of pOH.

$$pOH = -\log[OH^-]$$

Equation 17.4 (page 583) Definition of pK_w = pH + pOH (= 14.00 at 25 °C).

$$pK_w = 14.00 = pH + pOH$$

Equation 17.5 (page 585) Equilibrium expression for a general acid, HA, in water.

$$K_a = \frac{[H_3O^+][A^-]}{[HA]}$$

Equation 17.6 (page 585) Equilibrium expression for a general base, B, in water.

$$K_b = \frac{[BH^+][OH^-]}{[B]}$$

Equation 17.7 (page 588) Definition of pK_a.

$$pK_a = -\log K_a$$

Equation 17.8 (page 588) Relationship of K_a, K_b, and K_w, where K_a and K_b are for a conjugate acid–base pair.

$$K_a \times K_b = K_w$$

⚇WL Questions and problems for this chapter are available in OWL. Use the OWL access card that came with this textbook to access assigned questions and problems for this chapter.

18 Principles of Chemical Reactivity: Other Aspects of Aqueous Equilibria

© Cengage Learning/Charles D. Winters

Nature's Acids Many people grow rhubarb in gardens because the stalks of the plant, when stewed with sugar, make a wonderful dessert or filling for a pie or tart. But the leaves can make you sick! Why?

Rhubarb leaves are a source of at least seven organic acids, among them acetic and citric acids. But the most abundant is oxalic acid, $H_2C_2O_4$.

$$H_2O(\ell) + H_2C_2O_4(aq) \rightleftharpoons H_3O^+(aq) + HC_2O_4^-(aq)$$

$$H_2O(\ell) + HC_2O_4^-(aq) \rightleftharpoons H_3O^+(aq) + C_2O_4^{2-}(aq)$$

Rhubarb leaves contain between 0.1% and 1.4% by weight oxalic acid, while other leafy vegetables such as cabbage, spinach, and beet tops have lesser amounts.

While oxalic acid and other acids give the tart taste many of us enjoy with rhubarb, there is a problem with ingesting it: Oxalic acid interferes with essential elements in the body such as iron, magnesium, and, especially, calcium. The Ca^{2+} ion and oxalic acid react to form insoluble calcium oxalate, CaC_2O_4.

$$Ca^{2+}(aq) + H_2C_2O_4(aq) + 2\ H_2O(\ell) \rightarrow CaC_2O_4(s) + 2\ H_3O^+(aq)$$

Not only does this effectively remove calcium ions from the body, but calcium oxalate crystals can also grow into painful kidney and bladder stones. Because of this, people susceptible to kidney stones are put on a low-oxalic acid diet. Such people also have to be careful of taking too much vitamin C, a compound that can be turned into oxalic acid in the body. People have died from accidentally drinking antifreeze because the ethylene glycol in the antifreeze is converted to oxalic acid in the body. Symptoms of oxalic acid poisoning include nausea, vomiting, abdominal pain, and hemorrhages.

Because oxalic acid also occurs in other edible substances, including cocoa, peanuts, and tea, the average person consumes about 150 mg of oxalic acid a day. But will it kill you? For a person weighing about 145 pounds (65.7 kg), the lethal dose is around 24 grams of pure oxalic acid. You would have to eat a field of rhubarb leaves or drink an ocean of tea to come close to a fatal dose of oxalic acid. What would happen first, however, is that you may have severe diarrhea. Your gut recognizes oxalic acid as something to be eliminated and is stimulated to get rid of it.

In spite of the minor health risk from eating too much rhubarb, this plant and others have been cultivated for thousands of years for their healthful properties. The Chinese particularly have used rhubarb in traditional medicine for centuries. Indeed, it was so important that emperors of China in the 18th and 19th centuries forbade its export. Rhubarb was also cultivated in Russia and later in England. It made first its appearance in the United States in about 1800.

When you enjoy a slice of rhubarb pie, think about the topics in this chapter: the chemistry of acids and bases and of insoluble substances.

Questions:

1. Suppose you eat 28 grams of rhubarb leaves with an oxalic acid content of 1.2% by weight.
   (a) What volume of 0.25 M NaOH is required to titrate completely the oxalic acid in the leaves?
   (b) What mass of calcium oxalate could be formed from the oxalic acid in these leaves?
2. The solubility product constant of calcium oxalate is estimated to be 4×10^{-9}. What is its solubility in grams per liter?

Answers to these questions are available in Appendix N.

CHAPTER OUTLINE

18.1 The Common Ion Effect

18.2 Controlling pH: Buffer Solutions

18.3 Acid–Base Titrations

18.4 Solubility of Salts

18.5 Precipitation Reactions

18.6 Equilibria Involving Complex Ions

18.7 Solubility and Complex Ions

CHAPTER GOALS

See Chapter Goals Revisited (page 658) for Study Questions keyed to these goals.

- Understand the common ion effect.

- Understand the control of pH in aqueous solutions with buffers.

- Evaluate the pH in the course of acid–base titrations.

- Apply chemical equilibrium concepts to the solubility of ionic compounds.

OWL

Sign in to OWL at **www.cengage.com/owl** to view tutorials and simulations, develop problem-solving skills, and complete online homework assigned by your professor.

go Chemistry

Download mini lecture videos for key concept review and exam prep from OWL or purchase them from **www.cengagebrain.com**

In Chapter 3, we described four fundamental types of chemical reactions: acid–base reactions, precipitation reactions, gas-forming reactions, and oxidation-reduction reactions. In the present chapter, we want to apply the principles of chemical equilibria to a further understanding the first two kinds of reactions.

With regard to acid–base reactions, we are looking for answers to the following questions:

- How can we control the pH in a solution?
- What happens when an acid and base are mixed in any amount?

Precipitation reactions can also be understood in terms of chemical equilibria. The following questions are discussed in this chapter:

- If aqueous solutions of two ionic compounds are mixed, will precipitation occur?
- To what extent does an "insoluble" substance actually dissolve?
- What chemical reactions can be used to dissolve a precipitate?

18.1 The Common Ion Effect

In the previous chapter, we looked at the behavior of weak acids and bases in aqueous solution. But what happens to the pH of an acetic acid solution to which you added a significant concentration of its conjugate base, the acetate ion? The answer is that the pH of such a solution—weak acid/conjugate base or weak base/conjugate acid—is different than the pH of a solution of the weak acid or base by itself. The effect on the pH that is produced by having a significant concentration of acetate ion in an acetic acid solution, for example, is called the **common ion effect**. The name comes from the fact that the added acetate ion is "common" to the ionization equilibrium reaction of acetic acid.

$$CH_3CO_2H(aq) + H_2O(\ell) \rightleftharpoons H_3O^+(aq) + CH_3CO_2^-(aq)$$

We want to explore this aspect of acid–base chemistry here because it will be important in understanding *buffer solutions*, as you shall see in Section 18.2.

How does the common ion effect work (Figure 18.1)? If 1.0 L of a 0.25 M acetic acid solution has a pH of 2.67, what is the pH after adding 0.10 mol of sodium acetate? Sodium acetate, $NaCH_3CO_2$, is 100% dissociated into its ions, Na^+ and

FIGURE 18.1 The common ion effect. Approximately equal amounts of acetic acid (left flask, pH about 2.7) and sodium acetate, a base (right flask, pH about 9), were mixed in the beaker. The pH meter shows that the resulting solution in the beaker has a lower hydronium ion concentration (a pH of about 5) than the acetic acid solution, owing to the presence of acetate ion, the conjugate base of the acid and an ion common to the ionization reaction of the acid. (The acid and base solutions each had about the same concentration. Each solution contains universal indicator. This dye is red in low pH, yellow in slightly acidic media, and green in neutral to weakly basic media.)

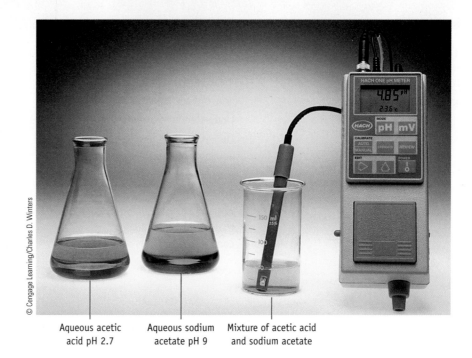

Aqueous acetic acid pH 2.7   Aqueous sodium acetate pH 9   Mixture of acetic acid and sodium acetate

$CH_3CO_2^-$, in water. Sodium ion has no effect on the pH of a solution (◀ Table 17.4 and Example 17.2). Thus, the important components of the solution are the weak acid (CH_3CO_2H) and its conjugate base ($CH_3CO_2^-$). Assume the acid ionizes to give H_3O^+ and $CH_3CO_2^-$, both in the amount x. This means that, relative to their initial concentrations, CH_3CO_2H decreases in concentration slightly (by an amount x) and $CH_3CO_2^-$ increases slightly (by an amount x).

- **The Common Ion Effect** In this ICE table, the first row (Initial) reflects the assumption that no ionization of the acid (or hydrolysis of the conjugate base) has yet occurred. Ionization of the acid in the presence of the conjugate base then produces x mol/L of hydronium ion and x mol/L more of the conjugate base.

Equation	CH_3CO_2H	+	H_2O	⇌	H_3O^+	+	$CH_3CO_2^-$
Initial (M)	0.25				0		0.10
Change (M)	$-x$				$+x$		$+x$
Equilibrium (M)	$(0.25 - x)$				x		$0.10 + x$

Because we have been able to define the equilibrium concentrations of acid and conjugate base and we know K_a, the hydronium ion concentration ($= x$) can be calculated from the usual equilibrium constant expression.

$$K_a = 1.8 \times 10^{-5} = \frac{[H_3O^+][CH_3CO_2^-]}{[CH_3CO_2H]} = \frac{(x)(0.10 + x)}{0.25 - x}$$

Now, because acetic acid is a weak acid and because it is ionizing in the presence of a significant concentration of its conjugate base, let us assume x is quite small. That is, it is reasonable to assume that $(0.10 + x)M \approx 0.10$ M and that $(0.25 - x)M \approx 0.25$ M. This leads to the "approximate" expression.

$$K_a = 1.8 \times 10^{-5} = \frac{[H_3O^+][CH_3CO_2^-]}{[CH_3CO_2H]} = \frac{(x)(0.10)}{0.25}$$

Solving this, we find that $x = [H_3O^+] = 4.5 \times 10^{-5}$ M and the pH is 4.35.

- **Equilibrium Constants and Temperature** Unless specified otherwise, all equilibrium constants and all calculations in this chapter are at 25 °C.

Without added $NaCH_3CO_2$, which provides the "common ion" $CH_3CO_2^-$, you can calculate that ionization of 0.25 M acetic acid will produce H_3O^+ and $CH_3CO_2^-$ ions in a concentration of 0.0021 M (to give a pH of 2.67). Le Chatelier's principle, however, predicts that the added common ion causes the reaction to proceed less far to the right. Hence, as we have found, $x = [H_3O^+]$ is less than 0.0021 M in the presence of added acetate ion. (Finally, note that the simplifying assumptions were valid.)

ⓌWL   INTERACTIVE EXAMPLE 18.1   Reaction of Lactic Acid with a Deficiency of Sodium Hydroxide

Problem What is the pH of the solution that results from adding 25.0 mL of 0.0500 M NaOH to 25.0 mL of 0.100 M lactic acid? (K_a for lactic acid $= 1.4 \times 10^{-4}$)

lactic acid ($HC_3H_5O_3$)
$K_a = 1.4 \times 10^{-4}$

lactate ion ($C_3H_5O_3^-$)

What Do You Know? You know the amounts of the acid and base (from the volume and concentration of each), and you know the ionization constant for the weak acid.

Strategy There are two parts to this problem: a stoichiometry problem followed by an equilibrium problem.

Part 1: *Stoichiometry Problem*

(a) Work a limiting reactant problem (in moles) to determine the amounts of lactic acid and lactate remaining after the reaction between lactic acid and sodium hydroxide.

(b) Calculate the concentrations of lactic acid and lactate ion that are present following the reaction of lactic acid with NaOH.

Part 2: *Equilibrium Problem*

(a) Set up an ICE table where the initial acid and conjugate base concentrations are known and the changes in the amount of acid consumed and amount of conjugate base produced are defined as x.

(b) Insert the equilibrium concentrations in the K_a expression for lactic acid and solve for x.

(c) Calculate the pH from the equilibrium hydronium ion concentration ($= x$).

Solution

Part 1: *Stoichiometry Problem*

(a) Amounts of NaOH and lactic acid used in the reaction

$$(0.0250 \text{ L NaOH})(0.0500 \text{ mol/L}) = 1.25 \times 10^{-3} \text{ mol NaOH}$$

$$(0.0250 \text{ L lactic acid})(0.100 \text{ mol/L}) = 2.50 \times 10^{-3} \text{ mol lactic acid}$$

(b) Amount of lactate ion produced by the acid–base reaction

Recognizing that NaOH is the limiting reactant, we have

$$(1.25 \times 10^{-3} \text{ mol NaOH})\left(\frac{1 \text{ mol lactate ion}}{1 \text{ mol NaOH}}\right) = 1.25 \times 10^{-3} \text{ mol lactate ion produced}$$

(c) Amount of lactic acid consumed

$$(1.25 \times 10^{-3} \text{ mol NaOH})\left(\frac{1 \text{ mol lactic acid}}{1 \text{ mol NaOH}}\right) = 1.25 \times 10^{-3} \text{ mol lactic acid consumed}$$

(d) Amount of lactic acid remaining when reaction is complete.

2.50×10^{-3} mol lactic acid available $- \ 1.25 \times 10^{-3}$ mol lactic acid consumed

$$= 1.25 \times 10^{-3} \text{ mol lactic acid remaining}$$

(e) Concentrations of lactic acid and lactate ion after reaction. Note that the total solution volume after reaction is 50.0 mL or 0.0500 L.

$$[\text{Lactic acid}] = \frac{1.25 \times 10^{-3} \text{ mol lactic acid}}{0.0500 \text{ L}} = 2.50 \times 10^{-2} \text{ M}$$

Strategy Map 18.1

PROBLEM

Calculate **pH** after combining solutions of **NaOH** and **lactic acid**.

↓

DATA/INFORMATION KNOWN

• Solution concentrations
• Solution volumes
• Value of acid K_a

> **STEP 1.** Stoichiometry problem.

Concentrations of **lactic acid** and **lactate ion** after acid–base reaction

> **STEP 2.** Enter **equilibrium concentrations** in ICE table.

At equilibrium:
$[H_3O^+] = x$
$[\text{Lactic acid}] = $ original conc. $- \ x$
$[\text{Lactate ion}] = $ original conc. $+ \ x$

> **STEP 3.** Enter **equilibrium concentrations** in K_a.

K_a **expression** with equilibrium concentrations in terms of **x**

> **STEP 4.** Solve K_a **expression** for **x**.

Value of $[H_3O^+]$

> **STEP 5.** Convert $[H_3O^+]$ to **pH**.

pH of solution

Because the amount of lactic acid remaining is the same as the amount of lactate ion produced, we have

$$[\text{Lactic acid}] = [\text{lactate ion}] = 2.50 \times 10^{-2} \text{ M}$$

Part 2: *Equilibrium Problem*

The concentrations just determined are used as the initial concentrations in an ICE table.

Equilibrium	$HC_3H_5O_3 + H_2O$	$\rightleftharpoons$	H_3O^+	+	$C_3H_5O_3^-$
Initial (M)	0.0250		0		0.0250
Change (M)	$-x$		$+x$		$+x$
Equilibrium (M)	$(0.0250 - x)$		x		$(0.0250 + x)$

Substituting the concentrations into the equilibrium expression, we have

$$K_a \text{ (lactic acid)} = 1.4 \times 10^{-4} = \frac{[H_3O^+][C_3H_5O_2^-]}{[HC_3H_5O_2]} = \frac{(x)(0.0250 + x)}{0.0250 - x}$$

Making the assumption that x is small with respect to 0.0250 M, we see that

$$K_a = 1.4 \times 10^{-4} \text{ M} = x = [H_3O^+], \text{ which gives a pH of } 3.85.$$

Think about Your Answer There are several final points to be made:

- Our assumption that $x \ll 0.0250$ is valid.

- By a separate calculation, we can show that a 0.025 M lactic acid solution has a pH of 2.73. The presence of 0.025 M lactate ion (a Brønsted base) in this solution has resulted in an increase of the pH to 3.85

- Here a base was added to the acidic solution. A useful observation is that *adding base to any solution will always result in an increase in pH.*

Check Your Understanding

You have a 0.30 M solution of formic acid (HCO_2H) and have added enough sodium formate ($NaHCO_2$) to make the solution 0.10 M in the salt. Calculate the pH of the formic acid solution before and after adding solid sodium formate.

> **REVIEW & CHECK FOR SECTION 18.1**
>
> What is the pH of the solution that results from adding 30.0 mL of 0.100 M NaOH to 45.0 mL of 0.100 M acetic acid?
>
> (a) 2.87   (b) 5.05   (c) 7.00

go Chemistry

Module 23: Understanding Acid–Base Buffers covers concepts in this section.

18.2 Controlling pH: Buffer Solutions

The addition of a small quantity of strong acid or base, say 0.010 mol, to a liter of human blood leads to a change in pH of only about 0.1 pH units from the normal pH of about 7.4. In comparison, if you add 0.010 mol of HCl to 1.0 L of pure water, the pH drops from 7 to 2. Addition of 0.010 mol of NaOH to pure water increases the pH from 7 to 12. Blood, and many other body fluids, are said to be buffered. A **buffer** causes solutions to be resistant to a change in pH when a strong acid or base is added (Figure 18.2).

There are two requirements for a buffer:

- Two substances are needed: an acid capable of reacting with added OH^- ions and a base that can consume added H_3O^+ ions.
- The acid and base must not react with each another.

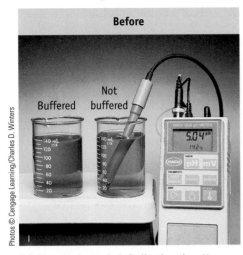 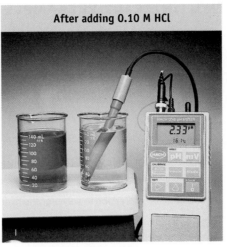

FIGURE 18.2 Buffer solutions.

(a) The pH electrode is indicating the pH of water that contains a trace of acid (and bromphenol blue acid–base indicator). The solution at the left is a buffer solution with a pH of about 7. (It also contains bromphenol blue dye.)

(b) When 5 mL of 0.10 M HCl is added to each solution, the pH of the water drops several units, whereas the pH of the buffer stays essentially constant, as implied by the fact that the indicator color does not change.

These requirements mean a buffer is usually prepared from a conjugate acid–base pair: (1) a weak acid and its conjugate base (acetic acid and acetate ion, for example), or (2) a weak base and its conjugate acid (ammonia and ammonium ion, for example). Some buffers commonly used in the laboratory are given in Table 18.1.

To see how a buffer works, consider an acetic acid/acetate ion buffer. Acetic acid, a weak acid, is needed to consume any added hydroxide ions.

$$CH_3CO_2H(aq) + OH^-(aq) \rightleftharpoons CH_3CO_2^-(aq) + H_2O(\ell) \qquad K = 1.8 \times 10^9$$

The equilibrium constant for the reaction is very large because OH^- is a much stronger base than acetate ion, $CH_3CO_2^-$ (◀ Section 17.6 and Table 17.3). This means that any OH^- entering the solution from an outside source is consumed completely. In a similar way, any hydronium ion added to the solution reacts with the acetate ion present in the buffer.

$$H_3O^+(aq) + CH_3CO_2^-(aq) \rightleftharpoons H_2O(\ell) + CH_3CO_2H(aq) \qquad K = 5.6 \times 10^4$$

The equilibrium constant for this reaction is also quite large because H_3O^+ is a much stronger acid than CH_3CO_2H.

The next several examples illustrate how to calculate the pH of a buffer solution, how to prepare a buffer, and how a buffer can control the pH of a solution.

● **Buffers and the Common Ion Effect** The common ion effect is observed for an acid (or base) ionizing in the presence of its conjugate base (or acid). A buffer is a solution of an acid, for example, and its conjugate base.

Table 18.1 Some Commonly Used Buffer Systems

Weak Acid	Conjugate Base	Acid K_a (pK_a)	Useful pH Range
Phthalic acid, $C_6H_4(CO_2H)_2$	Hydrogen phthalate ion, $C_6H_4(CO_2H)(CO_2)^-$	1.3×10^{-3} (2.89)	1.9–3.9
Acetic acid, CH_3CO_2H	Acetate ion, $CH_3CO_2^-$	1.8×10^{-5} (4.74)	3.7–5.8
Dihydrogen phosphate ion, $H_2PO_4^-$	Hydrogen phosphate ion, HPO_4^{2-}	6.2×10^{-8} (7.21)	6.2–8.2
Hydrogen phosphate ion, HPO_4^{2-}	Phosphate ion, PO_4^{3-}	3.6×10^{-13} (12.44)	11.4–13.4

Strategy Map 18.2

PROBLEM
Calculate **pH** of a **buffer solution**.

↓

DATA/INFORMATION KNOWN
• Concentration of **weak acid**   • Concentration of **conjugate base**   • Value of acid K_a

> STEP 1. Write **balanced equation** and K_a **expression**, and set up **ICE table**.

K_a expression and ICE table

> STEP 2. Enter **equilibrium concentrations** in ICE table.

At equilibrium:
$[H_3O^+] = x$
$[Weak\ acid] = $ original conc. $- x$
$[Conjugate\ base] = $
original conc. $+ x$

> STEP 3. Enter **equilibrium concentrations** in K_a.

K_a **expression** with equilibrium concentrations in terms of x

> STEP 4. Solve K_a **expression** for x.

Value of $[H_3O^+]$

> STEP 5. Convert $[H_3O^+]$ to **pH**.

pH of solution

 WL **INTERACTIVE EXAMPLE 18.2** pH of a Buffer Solution

Problem What is the pH of an acetic acid/sodium acetate buffer with $[CH_3CO_2H] = 0.700$ M and $[CH_3CO_2^-] = 0.600$ M?

What Do You Know? The hydronium ion concentration (and so the pH) of a buffer can be calculated if you know concentrations of the weak acid, its conjugate base, and K_a, as is the case here.

Strategy

- Write the balanced equation and the equilibrium constant expression.

- Set up an ICE table where the initial concentrations of acid and conjugate base are known, and the change in their concentrations is designated by x.

- Use the K_a expression to solve for the hydronium ion concentration (as in Example 17.5).

Solution

Equilibrium	CH_3CO_2H + H_2O	$\rightleftharpoons$	H_3O^+ +	$CH_3CO_2^-$
Initial (M)	0.700		0	0.600
Change (M)	$-x$		$+x$	$+x$
Equilibrium (M)	$0.700 - x$		x	$0.600 + x$

The appropriate equilibrium constant expression is

$$K_a = 1.8 \times 10^{-5} = \frac{[H_3O^+][CH_3CO_2^-]}{[CH_3CO_2H]} = \frac{(x)(0.600 + x)}{0.700 - x}$$

As explained on page 628, the value of x will be very small with respect to 0.700 or 0.600, so we can use the "approximate expression" to find x, the hydronium ion concentration.

$$K_a = 1.8 \times 10^{-5} = \frac{[H_3O^+][CH_3CO_2^-]}{[CH_3CO_2H]} = \frac{(x)(0.600)}{0.700}$$

$$x = 2.1 \times 10^{-5} \text{ M}$$

$$pH = -\log (2.1 \times 10^{-5}) = \boxed{4.68}$$

Think about Your Answer The pH of the buffer has a value between the pH of 0.700 M acetic acid (2.45) and 0.600 M sodium acetate (9.26).

Check Your Understanding

What is the pH of a buffer solution composed of 0.50 M formic acid (HCO_2H) and 0.70 M sodium formate ($NaHCO_2$)?

General Expressions for Buffer Solutions

In Example 18.2, we found the hydronium ion concentration of the acetic acid/acetate ion buffer solution by solving for x in the equation

$$K_a = 1.8 \times 10^{-5} = \frac{[H_3O^+][CH_3CO_2^-]}{[CH_3CO_2H]} = \frac{(x)(0.600)}{0.700}$$

If this equation is rearranged, we have a useful equation that can help you better understand how a buffer works.

$$[H_3O^+] = \frac{[CH_3CO_2H]}{[CH_3CO_2^-]} \times K_a$$

That is, the hydronium ion concentration in the acetic acid/acetate ion buffer is given by the ratio of the acid and conjugate base concentrations times the acid ionization constant. Indeed, this is true for all solutions of *a weak acid and its conjugate base*.

$$[H_3O^+] = \frac{[acid]}{[conjugate\ base]} \times K_a \tag{18.1}$$

It is often convenient to use Equation 18.1 in a different form. If we take the negative logarithm of each side of the equation, we have

$$-\log[H_3O^+] = \left\{-\log\frac{[acid]}{[conjugate\ base]}\right\} + (-\log K_a)$$

You know that $-\log[H_3O^+]$ is defined as pH, and $-\log K_a$ is defined as pK_a (◀ Sections 17.3 and 17.4). Furthermore, because

$$-\log\frac{[acid]}{[conjugate\ base]} = +\log\frac{[conjugate\ base]}{[acid]}$$

the preceding equation can be rewritten as

$$pH = pK_a + \log\frac{[conjugate\ base]}{[acid]} \qquad (18.2)$$

This equation is known as the **Henderson–Hasselbalch equation.**

Both Equations 18.1 and 18.2 show that *the pH of a buffer solution is controlled by two factors:*

- the strength of the acid (as expressed by K_a or pK_a)
- and the relative amounts of acid and conjugate base

The solution pH is established primarily by the value of K_a or pK_a, and the pH is fine-tuned by adjusting the acid-to-conjugate base ratio.

When the concentrations of conjugate base and acid are the same in a solution, the ratio [conjugate base]/[acid] is 1. The log of 1 is zero, so $pH = pK_a$ under these circumstances. If there is more of the conjugate base in the solution than acid, for example, then $pH > pK_a$. Conversely, if there is more acid than conjugate base in solution, then $pH < pK_a$.

● **Buffer Solutions** You will find it generally useful to consider all buffer solutions as composed of a weak acid and its conjugate base. Suppose, for example, a buffer is composed of the weak base ammonia and its conjugate acid ammonium ion. The hydronium ion concentration can be found from Equation 18.1 or 18.2 by assuming the buffer is composed of the weak acid NH_4^+ and its conjugate base, NH_3.

● **pK_a and Using The Henderson–Hasselbalch Equation** Many handbooks of chemistry list acid ionization constants in terms of pK_a values, so the approximate pH values of possible buffer solutions are readily apparent.

EXAMPLE 18.3   Using the Henderson–Hasselbalch Equation

Problem Benzoic acid ($C_6H_5CO_2H$, 2.00 g) and sodium benzoate ($NaC_6H_5CO_2$, 2.00 g) are dissolved in enough water to make 1.00 L of solution. Calculate the pH of the solution using the Henderson–Hasselbalch equation.

What Do You Know? You know the masses of weak acid and conjugate base and the volume of solution. (From these you can calculate the concentrations of these reagents, which you need for the Henderson–Hasselbalch equation.) You also know the K_a for the weak acid (Table 17.3 or Appendix H), from which you can calculate the required pK_a for the acid.

Strategy Calculate the acid and conjugate base concentrations and substitute these into the Henderson–Hasselbalch equation along with the value of pK_a for the acid.

Solution

K_a for benzoic acid is 6.3×10^{-5}. Therefore, $pK_a = -\log(6.3 \times 10^{-5}) = 4.20$

Next, you need the concentrations of the acid (benzoic acid) and its conjugate base (benzoate ion).

$$2.00\ g\ benzoic\ acid \left(\frac{1\ mol}{122.1\ g}\right) = 0.0164\ mol\ benzoic\ acid$$

$$2.00\ g\ sodium\ benzoate \left(\frac{1\ mol}{144.1\ g}\right) = 0.0139\ mol\ sodium\ benzoate$$

Because the solution volume is 1.00 L, the concentrations are [benzoic acid] = 0.0164 M and [sodium benzoate] = 0.0139 M. Therefore, using Equation 18.2, you have

$$pH = 4.20 + \log\frac{0.0139}{0.0164} = 4.20 + \log(0.848) = \boxed{4.13}$$

Think about Your Answer Notice that the pH is less than the pK_a because the concentration of acid is greater than the concentration of the conjugate base (and so the ratio of conjugate base to acid concentration is less than 1).

Check Your Understanding

Use the Henderson–Hasselbalch equation to calculate the pH of 1.00 L of a buffer solution containing 15.0 g of $NaHCO_3$ and 18.0 g of Na_2CO_3. (Consider this buffer as a solution of the weak acid HCO_3^- with CO_3^{2-} as its conjugate base.)

Preparing Buffer Solutions

To be useful, a buffer solution must have two characteristics:

- *pH Control:* It should control the pH at the desired value. The Henderson–Hasselbalch equation shows us how this can be done.

$$pH = pK_a + \log\frac{[\text{conjugate base}]}{[\text{acid}]}$$

First, an acid is chosen whose pK_a (or K_a) is near the intended value of pH (or $[H_3O^+]$). Second, the exact value of pH (or $[H_3O^+]$) is then achieved by adjusting the acid-to-conjugate base ratio. (Example 18.4 illustrates this approach.)

- *Buffer capacity:* The buffer should have the ability to keep the pH approximately constant after the addition of reasonable amounts of acid and base. For example, the concentration of acetic acid in an acetic acid/acetate ion buffer must be sufficient to consume all the hydroxide ion that may be added and still control the pH (see Example 18.4). Buffers are usually prepared as 0.10 M to 1.0 M solutions of reagents. However, any buffer will lose its ability to moderate changes if too much strong acid or base is added.

EXAMPLE 18.4 Preparing a Buffer Solution

Problem You wish to prepare 1.0 L of a buffer solution with a pH of 4.30. A list of possible acids (and their conjugate bases) follows:

Acid	Conjugate Base	K_a	pK_a
HSO_4^-	SO_4^{2-}	1.2×10^{-2}	1.92
CH_3CO_2H	$CH_3CO_2^-$	1.8×10^{-5}	4.74
HCO_3^-	CO_3^{2-}	4.8×10^{-11}	10.32

Which combination should be selected, and what should be the ratio of acid to conjugate base?

What Do You Know? You know the desired pH of the buffer, and you have a list of possible acid/conjugate base combinations.

Strategy Use either the general equation for a buffer (Equation 18.1) or the Henderson–Hasselbalch equation (Equation 18.2). Equation 18.1 informs you that $[H_3O^+]$ should be close to the acid K_a value, and Equation 18.2 tells you that pH should be close to the acid pK_a value. Either will establish which acid you will use. Having decided which acid to use, convert pH to $[H_3O^+]$ to use Equation 18.1. If you use Equation 18.2, use the pK_a value in the table. Finally, calculate the ratio of acid to conjugate base.

Solution The hydronium ion concentration for the buffer is found from the targeted pH.

$$pH = 4.30, \text{ so } [H_3O^+] = 10^{-pH} = 10^{-4.30} = 5.0 \times 10^{-5} \text{ M}$$

Of the acids given, only acetic acid (CH_3CO_2H) has a K_a value close to that of the desired $[H_3O^+]$ (or a pK_a close to a pH of 4.30). Now you need only to adjust the ratio $[CH_3CO_2H]/[CH_3CO_2^-]$ to achieve the desired hydronium ion concentration.

$$[H_3O^+] = 5.0 \times 10^{-5} \text{ M} = \frac{[CH_3CO_2H]}{[CH_3CO_2^-]}(1.8 \times 10^{-5})$$

Rearrange this equation to find the ratio $[CH_3CO_2H]/[CH_3CO_2^-]$.

$$\frac{[CH_3CO_2H]}{[CH_3CO_2^-]} = \frac{[H_3O^+]}{K_a} = \frac{5.0 \times 10^{-5}}{1.8 \times 10^{-5}} = \frac{2.8 \text{ mol/L}}{1.0 \text{ mol/L}}$$

Therefore, if you add 0.28 mol of acetic acid and 0.10 mol of sodium acetate (or any other pair of molar quantities in the ratio 2.8/1) to enough water to prepare 1.0 L of solution, the buffer solution will have a pH of 4.30.

Think about Your Answer If you prefer to use the Henderson–Hasselbalch equation, you would have

$$pH = 4.30 = 4.74 + \log\frac{[CH_3CO_2^-]}{[CH_3CO_2H]}$$

$$\log\frac{[CH_3CO_2^-]}{[CH_3CO_2H]} = 4.30 - 4.74 = -0.44$$

$$\frac{[CH_3CO_2^-]}{[CH_3CO_2H]} = 10^{-0.44} = 0.36$$

The ratio of conjugate base to acid, $[CH_3CO_2^-]/[CH_3CO_2H]$, is 0.36. The reciprocal of this ratio $\{= [CH_3CO_2H]/[CH_3CO_2^-] = 1/0.36\}$ is 2.8/1. This is the same result obtained previously using Equation 18.1.

Check Your Understanding

Using an acetic acid/sodium acetate buffer solution, what ratio of acid to conjugate base will you need to maintain the pH at 5.00? Describe how you would prepare such a solution.

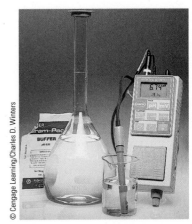

FIGURE 18.3 A commercial buffer solution. The solid acid and conjugate base in the packet are mixed with water to give a solution with the indicated pH. The quantity of water used does not matter because the ratio [acid]/[conjugate base] does not depend on the solution volume. (However, if too much water is added, the acid and conjugate base concentrations will be too low, and the buffer capacity could be exceeded. Again, buffer solutions usually have solute concentrations around 0.1 M to 1.0 M.)

Example 18.4 illustrates several important points concerning buffer solutions. The hydronium ion concentration depends not only on the K_a value of the acid but also on the ratio of acid and conjugate base concentrations. However, even though we write these ratios in terms of reagent concentrations, it is *the relative number of moles of acid and conjugate base that is important in determining the pH of a buffer solution.* Because both reagents are dissolved in the same solution, their concentrations depend on the same solution volume. In Example 18.4, the ratio 2.8/1 for acetic acid and sodium acetate implies that 2.8 times as many moles of acid were dissolved per liter as moles of sodium acetate.

$$\frac{[CH_3CO_2H]}{[CH_3CO_2^-]} = \frac{2.8 \text{ mol } CH_3CO_2H/L}{1.0 \text{ mol } CH_3CO_2^-/L} = \frac{2.8 \text{ mol } CH_3CO_2H}{1.0 \text{ mol } CH_3CO_2^-}$$

Notice that on dividing one concentration by the other, the volumes "cancel." This means that we only need to ensure that the ratio of moles of acid to moles of conjugate base is 2.8 to 1 in this example. The acid and its conjugate base could have been dissolved in any reasonable amount of water. This also means that *diluting a buffer solution will not change its pH.* Commercially available buffer solutions are often sold as premixed, dry reagents. To use them, you only need to mix the reagents in some volume of pure water (Figure 18.3).

Buffer Solutions

The following is a summary of important aspects of buffer solutions.
- A buffer resists changes in pH on adding small quantities of acid or base.
- A buffer consists of a weak acid and its conjugate base.
- The hydronium ion concentration of a buffer solution can be calculated from Equation 18.1,

$$[H_3O^+] = \frac{[acid]}{[conjugate\ base]} \times K_a$$

or the pH can be calculated from the Henderson–Hasselbalch equation (Equation 18.2).

$$pH = pK_a + log\frac{[conjugate\ base]}{[acid]}$$

- The pH depends primarily on the K_a of the weak acid and secondarily on the relative amounts of acid and conjugate base.
- The function of the weak acid of a buffer is to consume added base, and the

function of the conjugate base is to consume added acid. Such reactions affect the relative quantities of weak acid and its conjugate base. Because this ratio of acid to its conjugate base has only a secondary effect on the pH, the pH can be maintained relatively constant.
- The buffer must have sufficient capacity to react with reasonable quantities of added acid or base.

How Does a Buffer Maintain pH?

Now let us explore quantitatively how a given buffer solution can maintain the pH of a solution on adding a small amount of strong acid.

EXAMPLE 18.5   How Does a Buffer Maintain a Constant pH?

Problem What is the change in pH when 1.00 mL of 1.00 M HCl is added to (1) 1.000 L of pure water and to (2) 1.000 L of acetic acid/sodium acetate buffer with $[CH_3CO_2H] = 0.700$ M and $[CH_3CO_2^-] = 0.600$ M? (The pH of this acetic acid/acetate ion buffer is 4.68. See Example 18.2.)

What Do You Know? You know HCl is a strong acid and ionizes completely to supply H_3O^+ ions. In the first problem, when HCl is added to pure water, the concentration of HCl in the solution determines the pH. In the second problem, the conjugate base of the weak acid will react with the added HCl. Therefore, you need to perform a stoichiometry calculation to determine the new weak acid/conjugate base concentrations before you can calculate the pH.

Strategy

Part 1: *involves two steps:*

- Find the H_3O^+ concentration when adding 1.00 mL of acid to 1.000 L of pure water.

- Convert the value of $[H_3O^+]$ for the dilute solution to pH.

Part 2: *involves three steps:*

- A stoichiometry calculation to find how the concentrations of acid and conjugate base change on adding H_3O^+

- An equilibrium calculation to find $[H_3O^+]$ for a buffer solution where the concentrations of CH_3CO_2H and $CH_3CO_2^-$ are slightly altered owing to the reaction of $CH_3CO_2^-$ with added H_3O^+

- Conversion of $[H_3O^+]$ to pH

Solution

Part 1: *Adding Acid to Pure Water*

1.00 mL of 1.00 M HCl represents 0.00100 mol of acid. If this is added to 1.000 L of pure water, the H_3O^+ concentration of the water changes from 10^{-7} to almost 10^{-3},

$$c_1 \times V_1 = c_2 \times V_2$$

$$(1.00\ M)(0.00100\ L) = c_2 \times (1.001\ L)$$

$$c_2 = [H_3O^+]\ in\ diluted\ solution = 9.99 \times 10^{-4}\ M$$

and so the pH falls from 7.00 to 3.00.

Part 2: *Adding Acid to an Acetic Acid/Acetate Buffer Solution*

HCl is a strong acid that is 100% ionized in water and supplies H_3O^+, which reacts completely with the base (acetate ion) in the buffer solution according to the following equation:

$$H_3O^+(aq) + CH_3CO_2^-(aq) \rightleftharpoons H_2O(\ell) + CH_3CO_2H(aq)$$

	H_3O^+ from Added HCl	$CH_3CO_2^-$ from Buffer	CH_3CO_2H from Buffer
Initial amount of acid or base (mol $= c \times V$)	0.00100	0.600	0.700
Change (mol)	−0.00100	−0.00100	+0.00100
After reaction (mol)	0	0.599	0.701
Concentrations after reaction ($c =$ mol/V)	0	0.598	0.700

Because the added HCl reacts completely with acetate ion to produce acetic acid, the solution after this reaction (with $V = 1.001$ L) is once again a buffer containing only the weak acid and its salt. Now you only need to use Equation 18.1 (or the Henderson–Hasselbalch equation) to find $[H_3O^+]$ and the pH in the buffer solution as in Examples 18.2 and 18.3.

Equilibrium	CH_3CO_2H	+	H_2O	$\rightleftharpoons$	H_3O^+	+	$CH_3CO_2^-$
Initial (M)	0.700				0		0.598
Change (M)	−x				+x		+x
Equilibrium (M)	0.700 − x				x		0.598 +x

As usual, you can make the approximation that x, the concentration of H_3O^+ formed by ionizing acetic acid in the presence of acetate ion, is very small compared with 0.700 M or 0.598 M. Using Equation 18.1, you should get a pH of 4.68.

$$[H_3O^+] = x = \frac{[CH_3CO_2H]}{[CH_3CO_2^-]} \times K_a = \left(\frac{0.700 \text{ mol}}{0.598 \text{ mol}} \right)(1.8 \times 10^{-5}) = 2.1 \times 10^{-5} \text{ M}$$

$$pH = 4.68$$

Think about Your Answer Within the number of significant figures allowed, the pH of the buffer solution does not change after adding this amount of HCl. The buffer solution contains the conjugate base of the weak acid, and the base consumed the added HCl. In contrast, the pH changed by 4 units when the same amount of HCl was added to 1.0 L of pure water.

Check Your Understanding

Calculate the pH of 0.500 L of a buffer solution composed of 0.50 M formic acid (HCO_2H) and 0.70 M sodium formate ($NaHCO_2$) before and after adding 10.0 mL of 1.0 M HCl.

REVIEW & CHECK FOR SECTION 18.2

1. Which choice would be a good buffer solution?

   (a) 0.20 M HCN and 0.10 M KCN

   (b) 0.20 M HCl and 0.10 M KOH

   (c) 0.20 M CH_3CO_2H and 0.10 M HCO_2H

   (d) 0.10 HCl and 0.010 M KCl

2. If an acetic acid/sodium acetate buffer solution is prepared from 100. mL of 0.10 M acetic acid, what volume of 0.10 M sodium acetate must be added to have a pH of 4.00.

   (a) 100. mL   (b) 50. mL   (c) 36 mL   (d) 18 mL

3. What is the pH of a buffer composed of 100. mL of 0.20 M NH_4Cl and 200. mL of 0.10 M NH_3?

   (a) 4.85   (c) 7.00   (e) 10.05

   (b) 9.25   (d) 5.65

4. To prepare a buffer containing CH_3CO_2H and $NaCH_3CO_2$ and having a pH of 5, the ratio of the concentration of CH_3CO_2H to that of $NaCH_3CO_2$ should be about

    (a) 1/1   (b) 1.8/1   (c) 1/1.8   (d) 5.0/1

18.3 Acid–Base Titrations

A titration is one of the most useful ways of determining accurately the quantity of an acid, a base, or some other substance in a mixture or of ascertaining the purity of a substance. You learned how to perform the stoichiometry calculations involved in titrations in Chapter 4 (◄ Section 4.7). In Chapter 17, we described the following points regarding acid–base reactions (◄ Section 17.7):

- The pH at the equivalence point of a strong acid–strong base titration is 7. The solution at the equivalence point is truly "neutral" *only* when a strong acid is titrated with a strong base and vice versa.

- • **Equivalence Point** The equivalence point for a reaction is the point at which one reactant has been completely consumed by addition of another reactant. See page 159.

- If the substance being titrated is a weak acid or base, then the pH at the equivalence point is not 7 (◄ Table 17.5).
  - (a) A weak acid titrated with strong base leads to pH > 7 at the equivalence point due to the conjugate base of the weak acid.
  - (b) A weak base titrated with strong acid leads to pH < 7 at the equivalence point due to the conjugate acid of the weak base.

A knowledge of buffer solutions and how they work will now allow us to understand more fully how the pH changes in the course of an acid–base reaction.

Titration of a Strong Acid with a Strong Base

• **Weak Acid–Weak Base Titrations** Titrations combining a weak acid and weak base are generally not done because the equivalence point often cannot be accurately judged.

Figure 18.4 illustrates what happens to the pH as 0.100 M NaOH is slowly added to 50.0 mL of 0.100 M HCl.

$$HCl(aq) + NaOH(aq) \rightarrow NaCl(aq) + H_2O(\ell)$$

$$\text{Net ionic equation: } H_3O^+(aq) + OH^-(aq) \rightarrow 2\ H_2O(\ell)$$

Let us focus on four regions on this plot.

- pH of the initial solution
- pH as NaOH is added to the HCl solution before the equivalence point
- pH at the equivalence point
- pH after the equivalence point

Before beginning the titration, the 0.100 M solution of HCl has a pH of 1.00. As NaOH is added to the acid solution, the amount of HCl declines, and the acid remaining is dissolved in an ever-increasing volume of solution. Thus, $[H_3O^+]$ decreases,

FIGURE 18.4 The change in pH as a strong acid is titrated with a strong base. In this case, 50.0 mL of 0.100 M HCl is titrated with 0.100 M NaOH. The pH at the equivalence point is 7.0 for the reaction of a strong acid with a strong base.

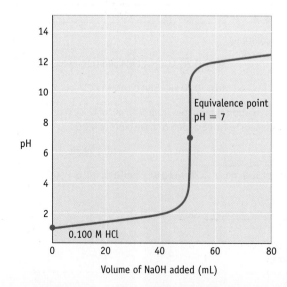

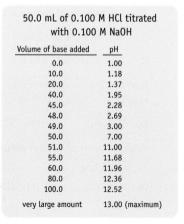

| 50.0 mL of 0.100 M HCl titrated with 0.100 M NaOH | |
Volume of base added	pH
0.0	1.00
10.0	1.18
20.0	1.37
40.0	1.95
45.0	2.28
48.0	2.69
49.0	3.00
50.0	7.00
51.0	11.00
55.0	11.68
60.0	11.96
80.0	12.36
100.0	12.52
very large amount	13.00 (maximum)

CASE STUDY

Take a Deep Breath

Maintenance of pH is vital to the cells of all living organisms because enzyme activity is influenced by pH. The primary protection against harmful pH changes in cells is provided by buffer systems, which maintain the intracellular pH of most cells between 6.9 and 7.4. Two important biological buffer systems control pH in this range: the bicarbonate/carbonic acid system (HCO_3^-/H_2CO_3) and the phosphate system ($HPO_4^{2-}/H_2PO_4^-$).

The bicarbonate/carbonic acid buffer is important in blood plasma, where three equilibria are important.

$$CO_2(g) \rightleftharpoons CO_2(\text{dissolved})$$

$$CO_2(\text{dissolved}) + H_2O(\ell) \rightleftharpoons H_2CO_3(aq)$$

$$H_2CO_3(aq) + H_2O(\ell) \rightleftharpoons$$
$$H_3O^+(aq) + HCO_3^-(aq)$$

The overall equilibrium for the second and third steps has $pK_{overall} = 6.3$ at 37 °C, the temperature of the human body. Thus,

$$7.4 = 6.3 + \log \frac{[HCO_3^-]}{[CO_2(\text{dissolved})]}$$

Although the value of $pK_{overall}$ is about 1 pH unit away from the blood pH, the natural partial pressure of CO_2 in the alveoli of the lungs (about 40 mm Hg) is sufficient to keep $[CO_2(\text{dissolved})]$ at about 1.2×10^{-3} M and $[HCO_3^-]$ at about 1.5×10^{-2} M as required to maintain this pH.

If blood pH rises above about 7.45, you can suffer from a condition called *alkalosis*. *Respiratory alkalosis* can arise from hyperventilation when a person breathes quickly to expel CO_2 from the lungs. This has the effect of lowering the CO_2 concentration, which in turn leads to a lower H_3O^+ concentration and a higher pH. This same condition can also arise from severe anxiety or from an oxygen deficiency at high altitude. It can

© Cengage Learning/Charles D. Winters

ultimately lead to overexcitability of the central nervous system, muscle spasms, convulsions, and death. One way to treat acute respiratory alkalosis is to breathe into a paper bag. The CO_2 you exhale is recycled. This raises the blood CO_2 level and causes the equilibria above to shift to the right, thus raising the hydronium ion concentration and lowering the pH.

Metabolic alkalosis can occur if you take large amounts of sodium bicarbonate to treat stomach acid (which is mostly HCl at a pH of about 1 to 2). It also commonly occurs when a person vomits profusely. This depletes the body of hydrogen ions, which leads to an increase in bicarbonate ion concentration.

Athletes can use the H_2CO_3/HCO_3^- equilibrium to enhance their performance. Strenuous activity produces high levels of lactic acid, and this can lower blood pH and cause muscle cramps. To counteract this, athletes will prepare before a race by hyperventilating for some seconds to raise blood

pH, thereby helping to neutralize the acidity from the lactic acid.

Acidosis is the opposite of alkalosis. There was a case of a toddler who came to the hospital with viral gastroenteritis and *metabolic acidosis*. He had severe diarrhea, was dehydrated, and had a high rate of respiration. One function of the bicarbonate ion is to neutralize stomach acid in the intestines. However, because of his diarrhea, the toddler was losing bicarbonate ions in his stool, and his blood pH was too low. To compensate, the toddler was breathing rapidly and blowing off CO_2 through the lungs (the effect of which is to lower $[H_3O^+]$ and raise the pH).

Respiratory acidosis results from a buildup of CO_2 in the body. This can be caused by pulmonary problems, by head injuries, or by drugs such as anesthetics and sedatives. It can be reversed by breathing rapidly and deeply. Doubling the breathing rate increases the blood pH by about 0.23 units.

Questions:

Phosphate ions are abundant in cells, both as the ions themselves and as important substituents on organic molecules. Most importantly, the pK_a for the $H_2PO_4^-$ ion is 7.20, which is very close to the high end of the normal pH range in the body.

$$H_2PO_4^-(aq) + H_2O(\ell) \rightleftharpoons$$
$$H_3O^+(aq) + HPO_4^{2-}(aq)$$

1. What should the ratio $[HPO_4^{2-}]/[H_2PO_4^-]$ be to control the pH at 7.4?
2. A typical total phosphate concentration in a cell, $[HPO_4^{2-}] + [H_2PO_4^-]$, is 2.0×10^{-2} M. What are the concentrations of HPO_4^{2-} and $H_2PO_4^-$ at pH 7.4?

Answers to these questions are available in Appendix N.

---

and the pH slowly increases. As an example, let us find the pH of the solution after 10.0 mL of 0.100 M NaOH has been added to 50.0 mL of 0.100 M HCl. Here, we set up a table to list the amounts of acid and base before reaction, the changes in those amounts, and the amounts remaining after reaction. Be sure to notice that the volume of the solution after reaction is the sum of the combined volumes of NaOH and HCl (60.0 mL or 0.0600 L in this case).

	$H_3O^+(aq)$	+	$OH^-(aq)$	→	$2 H_2O(\ell)$
Initial amount (mol = $c \times V$)	0.00500		0.00100		
Change (mol)	−0.00100		−0.00100		
After reaction (mol)	0.00400		0		
After reaction (c = mol/V)	0.00400 mol/0.0600 L		0		
	= 0.0667 M				

● **Titrations, Stoichiometry, and Limiting Reactants** In the titration of an acid with a base (as in Figure 18.4) the limiting reactant is the base before the equivalence point. After the equivalence point it will be the acid.

After addition of 10.0 mL of NaOH, the final solution has a hydronium ion concentration of 0.0667 M, and so the pH is

$$pH = -\log[H_3O^+] = -\log(0.0667) = 1.176$$

After 49.5 mL of base has been added—that is, just before the equivalence point—we can use the same approach to show that the pH is 3.3. The solution being titrated is still quite acidic, even very close to the equivalence point.

The pH of the equivalence point in an acid–base titration is the value at the steepest point in the vertical portion of the pH versus volume of titrant curve. (The **titrant** is the substance being added during the titration.) In the HCl/NaOH titration illustrated in Figure 18.4, you see that the pH increases very rapidly near the equivalence point. In fact, in this case the pH rises 7 units (the H_3O^+ concentration decreases by a factor of 10 million!) when only a drop or two of the NaOH solution is added, and the steepest point of the vertical portion of the curve is at a pH of 7.00.

> The pH of the solution at the equivalence point in a monoprotic strong acid–strong base reaction is always 7.00 (at 25 °C) because the solution contains a neutral salt.

After all of the HCl has been consumed and the slightest excess of NaOH has been added, the solution will be basic, and the pH will continue to increase as more NaOH is added (and the solution volume increases). For example, if we calculate the pH of the solution after 55.0 mL of 0.100 M NaOH has been added to 50.0 mL of 0.100 M HCl, we find

	$H_3O^+(aq)$	+	$OH^-(aq)$	→	$2 H_2O(\ell)$
Initial amount (mol = $c \times V$)	0.00500		0.00550		
Change (mol)	−0.00500		−0.00500		
After reaction (mol)	0		0.00050		
After reaction (c = mol/V)	0		0.00050 mol/0.1050 L = 0.0048 M		

At this point, the solution has a hydroxide ion concentration of 0.0048 M. Calculate the pOH from this value, then use this to calculate pH.

$$pOH = -\log[OH^-] = -\log(0.0048) = 2.32$$

$$pH = 14.00 - pOH = 11.68$$

Titration of a Weak Acid with a Strong Base

The titration of a weak acid with a strong base is somewhat different from the strong acid–strong base titration. Look carefully at the curve for the titration of 100.0 mL of 0.100 M acetic acid with 0.100 M NaOH (Figure 18.5),

$$CH_3CO_2H(aq) + NaOH(aq) \rightarrow NaCH_3CO_2(aq) + H_2O(\ell)$$

and focus on three points on this curve:

- *The pH before titration begins.* The pH before any base is added can be calculated from the weak acid K_a value and the acid concentration (◄ Example 17.5).
- *The pH at the equivalence point.* At the equivalence point, the solution contains only sodium acetate, the CH_3CO_2H and NaOH having been consumed. The pH is controlled by the acetate ion, the conjugate base of acetic acid.
- *The pH at the halfway point (half-equivalence point) of the titration.* Here, the pH is equal to the pK_a of the weak acid, a conclusion discussed in more detail in following paragraphs.

As NaOH is added to acetic acid, the base is consumed and sodium acetate is produced. Thus, at every point between the beginning of the titration (when only acetic acid is present) and the equivalence point (when only sodium acetate is present), the solution contains both acetic acid and its salt, sodium acetate. These are

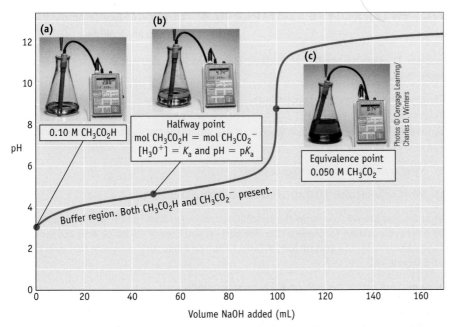

FIGURE 18.5 **The change in pH during the titration of a weak acid with a strong base.**
Here, 100.0 mL of 0.100 M acetic acid is titrated with 0.100 M NaOH. Note especially the following:
(a) Before titration. A 0.100 M solution of acetic acid has a pH of 2.87. **(b) Halfway point.** The pH at
the point at which half the acid has reacted with base is equal to the pK_a for the acid (pH = pK_a = 4.74).
(c) Equivalence point. The solution contains the acetate ion, a weak base. Therefore, the solution is
basic, with a pH of 8.72.

the components of a *buffer solution,* and the hydronium ion concentration can be
calculated from Equation 18.3 or 18.4.

$$[H_3O^+] = \frac{[\text{weak acid remaining}]}{[\text{conjugate base produced}]} \times K_a \qquad (18.3)$$

or

$$pH = pK_a + \log \frac{[\text{conjugate base produced}]}{[\text{weak acid remaining}]} \qquad (18.4)$$

The fact that a buffer is present for most of the region between the beginning of the
titration and the equivalence point is the reason that the pH of the solution only
rises slowly after a few milliliters of titrant has been added.

What happens to the pH when *exactly* half of the acid has been consumed by
base? Half of the acid (CH_3CO_2H) has been converted to the conjugate base
($CH_3CO_2^-$), and half remains. Therefore, the concentration of weak acid remaining is equal to the concentration of conjugate base produced ($[CH_3CO_2H]$ =
$[CH_3CO_2^-]$). Using Equations 18.3 or 18.4, we see that

$$[H_3O^+] = (1) \times K_a \qquad \text{or} \qquad pH = pK_a + \log(1)$$

Because $\log(1) = 0$, we come to the following general conclusion:

> At the halfway point in the titration of a weak acid with a strong base
>
> $$[H_3O^+] = K_a \text{ and } pH = pK_a$$
> $\qquad (18.5)$

In the particular case of the titration of acetic acid with a strong base, $[H_3O^+]$ =
1.8×10^{-5} M at the halfway point, and so the pH is 4.74. This is equal to the pK_a of
acetic acid.

• **Acid–Base Titrations** See Problem Solving Tip 18.2 on page 638: Calculating the pH at Various Stages in an Acid–Base Titration.

ⓄWL **INTERACTIVE EXAMPLE 18.6** Titration of Acetic Acid with Sodium Hydroxide

Problem Consider the titration of 100.0 mL of 0.100 M acetic acid with 0.100 M NaOH (see Figure 18.5).

$$CH_3CO_2H(aq) + OH^-(aq) \rightarrow CH_3CO_2^-(aq) + H_2O(\ell)$$

(a) What is the pH of the solution when 90.0 mL of 0.100 M NaOH has been added to 100.0 mL of 0.100 M acetic acid?

(b) What is the pH at the equivalence point?

(c) What is the pH after 110.0 mL of NaOH has been added?

What Do You Know? You know the concentrations and volumes of the acid and base and the balanced equation for the reaction. For part (a) you know K_a of acetic acid and for part (b) you know K_b for the acetate ion (◀ Table 17.3). You might also recognize that part (a) is like Example 18.1.

Strategy Each part of the problem has at least two steps:

(1) The first step in each part involves a limiting-reactant stoichiometry calculation to determine the amounts of acetic acid, sodium hydroxide, and acetate ion present after the acid–base reaction has gone to completion.

(2) **Part (a):** Before the equivalence point, do an equilibrium calculation to find $[H_3O^+]$ for a buffer solution where the amounts of CH_3CO_2H and $CH_3CO_2^-$ are known from the stoichiometry calculation. (See Example 18.1.)

Part (b): Only the acetate ion remains at the equivalence point. (See Example 17.7.) Determine the concentration of this ion after the acid–base reaction goes to completion. Using the balanced equation for the reaction of the acetate ion with water and its corresponding K_b, determine the equilibrium hydroxide ion concentration. Convert this to the equilibrium hydronium ion concentration and calculate the pH.

Part (c): After the equivalence point, the solution contains both the acetate ion and excess NaOH, but the latter controls the pH. Determine the concentration of excess NaOH (and $[OH^-]$), and then convert $[OH^-]$ to pH.

Solution

Part (a): pH before the Equivalence Point

First calculate the amounts of reactants before reaction (= concentration × volume) and then use the principles of stoichiometry to calculate the amounts of reactants and products after reaction. The limiting reactant is NaOH, so some CH_3CO_2H remains along with the product, $CH_3CO_2^-$.

Equation	CH_3CO_2H	+	OH^-	⇌	$CH_3CO_2^-$	+	H_2O
Initial (mol)	0.0100		0.00900		0		
Change (mol)	−0.00900		−0.00900		+0.00900		
After reaction (mol)	0.0010		0		0.00900		

The ratio of amounts (moles) of acid to conjugate base is the same as the ratio of their concentrations. Therefore, we can use the amounts of weak acid remaining and conjugate base formed to find the pH from Equation 18.3.

$$[H_3O^+] = \frac{mol\ CH_3CO_2H}{mol\ CH_3CO_2^-} \times K_a = \left(\frac{0.0010\ mol}{0.0090\ mol}\right)(1.8 \times 10^{-5}) = 2.0 \times 10^{-6}\ M$$

$$pH = -\log(2.0 \times 10^{-6}) = \boxed{5.70}$$

The pH is 5.70, in agreement with Figure 18.5. Notice that this pH is appropriate for a point after the halfway point (4.74) but before the equivalence point (8.72).

Part (b): *pH at the Equivalence Point*

To reach the equivalence point, 0.0100 mol of NaOH has been added to 0.0100 mol of CH_3CO_2H and 0.0100 mol of $CH_3CO_2^-$ has been formed.

Equation	CH_3CO_2H	+	OH^-	⇌	$CH_3CO_2^-$	+	H_2O
Initial (mol)	0.0100		0.0100		0		
Change (mol)	−0.0100		−0.0100		+0.0100		
After reaction (mol)	0		0		0.0100		

Because two solutions, each with a volume of 100.0 mL, have been combined, the concentration of $CH_3CO_2^-$ at the equivalence point is (0.0100 mol/0.200 L) = 0.0500 M. Next, set up an ICE table for the reaction of this weak base with water,

Equation	$CH_3CO_2^-$	+	H_2O	⇌	CH_3CO_2H	+	OH^-
Initial (M)	0.0500				0		0
Change (M)	−x				+x		+x
After reaction (M)	0.0500 − x				x		x

and calculate the concentration of OH^- ion using K_b for the weak base.

$$K_b \text{ for } CH_3CO_2^- = 5.6 \times 10^{-10} = \frac{[CH_3CO_2H][OH^-]}{[CH_3CO_2^-]} = \frac{(x)(x)}{0.0500 - x}$$

Making the usual assumption that x is small with respect to 0.0500 M,

$$x = [OH^-] = 5.3 \times 10^{-6} \text{ M and so the pOH} = 5.28$$

$$pH = 14.00 - 5.28 = 8.72$$

Part (c): *pH after the Equivalence Point*

Now the limiting reactant is CH_3CO_2H, and the solution contains excess OH^- ion from the unused NaOH as well as from the hydrolysis of $CH_3CO_2^-$.

Equation	CH_3CO_2H	+	OH^-	⇌	$CH_3CO_2^-$	+	H_2O
Initial (mol)	0.0100		0.0110		0		
Change (mol)	−0.0100		−0.0100		+0.0100		
After reaction (mol)	0		0.0010		0.0100		

The amount of OH^- produced by $CH_3CO_2^-$ hydrolysis is very small [see part (b)], so the pH of the solution after the equivalence point is determined by the excess NaOH (in 210 mL of solution).

$$[OH^-] = 0.0010 \text{ mol/0.210 L} = 4.8 \times 10^{-3} \text{ M (pOH} = 2.32)$$

$$pH = 14.00 - 2.32 = 11.68$$

Think about Your Answer The pH at the equivalence point (8.72) indicates the solution is mildly basic, as expected for a solution of the conjugate base of a weak acid. As more NaOH is added past the equivalence point the solution becomes substantially more basic (pH = 11.68).

Check Your Understanding

The titration of 0.100 M acetic acid with 0.100 M NaOH is described in the text. What is the pH of the solution when 35.0 mL of the base has been added to 100.0 mL of 0.100 M acetic acid?

Strategy Map 18.6

PROBLEM — Part (b)
Calculate **pH** at the *equivalence point* for a titration of **acetic acid** with **NaOH.**

DATA/INFORMATION KNOWN
• Solution concentrations
• Solution volumes
• Value of conjugate base K_b

STEP 1. Stoichiometry problem

Concentration of **acetate ion** at *equivalence point* of acid-base titration

STEP 2. Enter **equilibrium concentrations** in ICE table.

At equilibrium:
$[OH^-]$ = [Acetic acid] = x
[Acetate ion] = original conc. − x

STEP 3. Enter **equilibrium concentrations** in K_b.

K_b **expression** with equilibrium concentrations in terms of **x**

STEP 4. Solve K_b **expression** for **x.**

Value of $[OH^-]$

STEP 5. Convert $[OH^-]$ to **pH.**

pH of solution

FIGURE 18.6 Titration curve for a diprotic acid. The curve for the titration of 100.0 mL of 0.100 M oxalic acid ($H_2C_2O_4$, a weak diprotic acid) with 0.100 M NaOH. The first equivalence point (at 100 mL) occurs when the first hydrogen ion of $H_2C_2O_4$ is titrated, and the second (at 200 mL) occurs at the completion of the reaction. The curve for pH versus volume of NaOH added shows an initial rise at the first equivalence point and then another rise at the second equivalence point.

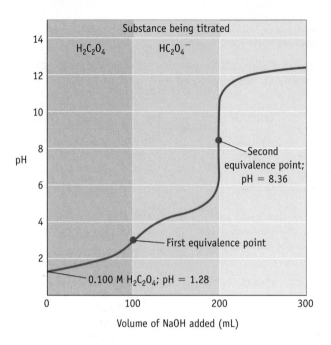

Titration of Weak Polyprotic Acids

The titrations illustrated thus far have been for the reaction of a monoprotic acid (HA) with a base such as NaOH. It is possible to extend the discussion of titrations to polyprotic acids such as oxalic acid, $H_2C_2O_4$.

$$H_2C_2O_4(aq) + H_2O(\ell) \rightleftharpoons HC_2O_4^-(aq) + H_3O^+(aq) \qquad K_{a1} = 5.9 \times 10^{-2}$$

$$HC_2O_4^-(aq) + H_2O(\ell) \rightleftharpoons C_2O_4^-(aq) + H_3O^+(aq) \qquad K_{a2} = 6.4 \times 10^{-5}$$

Figure 18.6 illustrates the curve for the titration of 100 mL of 0.100 M oxalic acid with 0.100 M NaOH. The first significant rise in pH is experienced when 100 mL of base has been added, indicating that the first proton of the acid has been titrated.

$$H_2C_2O_4(aq) + OH^-(aq) \rightleftharpoons HC_2O_4^-(aq) + H_2O(aq)$$

When the second proton of oxalic acid is titrated, the pH again rises significantly.

$$HC_2O_4^-(aq) + OH^-(aq) \rightleftharpoons C_2O_4^{2-}(aq) + H_2O(\ell)$$

The pH at this second equivalence point is controlled by the oxalate ion, $C_2O_4^{2-}$.

$$C_2O_4^{2-}(aq) + H_2O(\ell) \rightleftharpoons HC_2O_4^-(aq) + OH^-(aq)$$

$$K_b = K_w/K_{a2} = 1.6 \times 10^{-10}$$

Calculation of the pH at the equivalence point indicates that it should be about 8.4, as observed.

Titration of a Weak Base with a Strong Acid

Finally, it is useful to consider the titration of a weak base with a strong acid. Figure 18.7 illustrates the pH curve for the titration of 100.0 mL of 0.100 M NH_3 with 0.100 M HCl.

$$NH_3(aq) + H_3O^+(aq) \rightleftharpoons NH_4^+(aq) + H_2O(\ell)$$

The initial pH for a 0.100 M NH_3 solution is 11.13. As the titration progresses, the important species in solution are the weak acid NH_4^+ and its conjugate base, NH_3.

$$NH_4^+(aq) + H_2O(\ell) \rightleftharpoons NH_3(aq) + H_3O^+(aq) \qquad K_a = 5.6 \times 10^{-10}$$

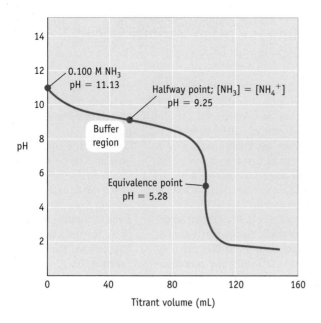

FIGURE 18.7 Titration of a weak base with a strong acid. The change in pH during the titration of a weak base (100.0 mL of 0.100 M NH_3) with a strong acid (0.100 M HCl). The pH at the half-neutralization point is equal to the pK_a for the conjugate acid (NH_4^+) of the weak base (NH_3) ($pH = pK_a = 9.25$). At the equivalence point, the solution contains the NH_4^+ ion, a weak acid, so the pH is about 5.

At the halfway point, the concentrations of NH_4^+ and NH_3 are the same, so

$$[H_3O^+] = \frac{[NH_4^+]}{[NH_3]} \times K_a = 5.6 \times 10^{-10}$$

$$[H_3O^+] = K_a$$

$$pH = pK_a = -\log(5.6 \times 10^{-10}) = 9.25$$

As the addition of HCl to NH_3 continues, the pH declines slowly because of the buffering action of the NH_3/NH_4^+ combination. Near the equivalence point, however, the pH drops rapidly. At the equivalence point, the solution contains only ammonium chloride, a weak Brønsted acid, and the solution is weakly acidic.

EXAMPLE 18.7   Titration of Ammonia with HCl

Problem What is the pH of the solution at the equivalence point in the titration of 100.0 mL of 0.100 M ammonia with 0.100 M HCl (see Figure 18.7)?

What Do You Know? This problem resembles part (b) of Example 18.6, although here you are titrating a weak base with a strong acid. In both examples you know the quantities of reagents and want to know the pH at the equivalence point. Here you need to know the K_a value for NH_4^+, the conjugate acid of the weak base, NH_3.

Strategy As in Example 18.6, this problem has two steps: (a) a stoichiometry calculation to find the concentration of NH_4^+ at the equivalence point, and (b) an equilibrium calculation to find $[H_3O^+]$ for a solution of the weak acid NH_4^+.

Solution

Part 1: Stoichiometry Problem

Here, we are titrating 0.0100 mol of NH_3 ($= c \times V$), so 0.0100 mol of HCl is required. Thus, 100.0 mL of 0.100 M HCl ($= 0.0100$ mol HCl) must be used in the titration.

Equation	NH_3	+	H_3O^+	$\rightleftharpoons$	NH_4^+	+	H_2O
Initial (mol = $c \times V$)	0.0100		0.0100		0		
Change on reaction (mol)	−0.0100		−0.0100		+0.0100		
After reaction (mol)	0		0		0.0100		
Concentration (M)	0		0		0.0100 mol (in 0.200L) = 0.0500 M		

PROBLEM SOLVING TIP 18.2

Calculating the pH at Various Stages in an Acid–Base Titration

Finding the pH at or before the equivalence point for an acid–base reaction always involves several calculation steps. There are no shortcuts. Consider the *titration of a weak base, B, with a strong acid* as in Example 18.7. (The same principles apply to other acid–base reactions.)

$$H_3O^+(aq) + B(aq) \rightleftharpoons BH^+(aq) + H_2O(\ell)$$

A: Determining the pH before the equivalence point.

Step 1. *Solve the stoichiometry problem.* Up to the equivalence point, acid is consumed completely to leave a solution containing some base (B) and its conjugate acid (BH⁺).

Use the principles of stoichiometry to calculate (a) the amount of acid added, (b) the amount of base remaining, and (c) the amount of conjugate acid (BH⁺) formed.

Step 2. Calculate the concentrations of B and BH⁺. Recognize that the volume of the solution at any point is the sum of the original volume of base solution plus the volume of acid solution added.

Step 3. *Calculate the pH.* At any point before the equivalence point, the solution is a buffer solution because both B and BH⁺ are present. Calculate [H₃O⁺] using the concentra-

tions of Step 2 and the value of K_a for the conjugate acid of the weak base.

B: Determining the pH at the equivalence point.

Calculate the concentration of the conjugate acid using the procedure of Part A. Use the value of K_a for the conjugate acid of the weak base and the procedure outlined in Example 18.7. (For a titration of a weak acid with a strong base, use the value of K_b for the conjugate base of the acid and follow the procedure outlined in Example 18.6.)

Roses are a natural indicator.

(a) The pigment in red rose petals was extracted with ethanol; the extract was a faint red. **(b)** After adding one drop of 6 M HCl, the color changed to a vivid red. **(c)** Adding two drops of 6 M NH₃ produced a green color, and **(d)** adding 1 drop each of HCl and NH₃ (to give a buffer solution) gave a blue solution. **(e)** Finally, adding a few milligrams of Al(NO₃)₃ turned the solution deep purple. (The deep purple color with aluminum ions was so intense that the solution had to be diluted significantly to take the photo.)

Part 2: Equilibrium Problem

When the equivalence point is reached, the solution consists of 0.0500 M NH₄⁺. The pH is determined by the hydrolysis of this weak acid.

Equation	NH_4^+	+	H_2O	$\rightleftharpoons$	NH_3	+	H_3O^+
Initial (M)	0.0500				0		0
Change (M)	$-x$				$+x$		$+x$
Equilibrium (M)	$0.0500 - x$				x		x

Using K_a for the weak acid NH₄⁺, we have

$$K_a = 5.6 \times 10^{-10} = \frac{[NH_3][H_3O^+]}{[NH_4^+]} = \frac{x^2}{0.0500 - x}$$

$$\text{Simplifying, } x = [H_3O^+] = \sqrt{(5.6 \times 10^{-10})(0.0500)} = 5.3 \times 10^{-6} \text{ M}$$

$$pH = 5.28$$

Think about Your Answer The pH at the equivalence point (5.28) indicates the solution is mildly acidic, as expected for a solution of the conjugate acid of a weak base.

Check Your Understanding

Calculate the pH after 75.0 mL of 0.100 M HCl has been added to 100.0 mL of 0.100 M NH₃. See Figure 18.7.

pH Indicators

Many organic compounds, both natural and synthetic, have a color that changes with pH (Figure 18.8). Not only does this add beauty and variety to our world, but it is also a useful property in chemistry.

You may have done an acid–base titration in the laboratory, and, before starting the titration, you added an **indicator**. The acid–base indicator is usually an organic compound that is itself a weak acid or weak base (as are many compounds that give flowers their colors). In aqueous solution, the acid form is in equilibrium with its conjugate base. Abbreviating the indicator's acid formula as HInd and the formula of its conjugate base as Ind⁻, we can write the equilibrium equation

$$HInd(aq) + H_2O(\ell) \rightleftharpoons H_3O^+(aq) + Ind^-(aq)$$

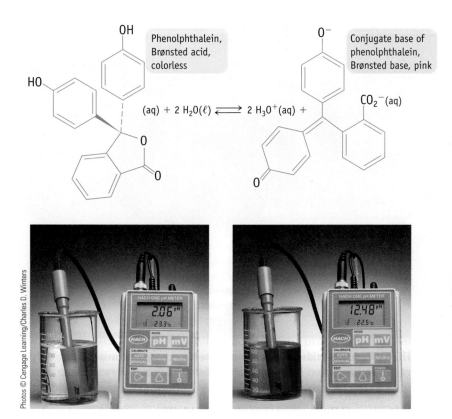

OH
Phenolphthalein,
Brønsted acid,
colorless

O⁻
Conjugate base of
phenolphthalein,
Brønsted base, pink

HO

$(aq) + 2\,H_2O(\ell) \rightleftharpoons 2\,H_3O^+(aq) +$

$CO_2^-(aq)$

FIGURE 18.8 Phenolphthalein, a common acid–base indicator. Phenolphthalein, a weak acid, is colorless. As the pH increases, the pink conjugate base form predominates, and the color of the solution changes. The change in color is most noticeable around pH 9. The dye is commonly used for strong acid + strong base or weak acid + strong base titrations because the pH changes from about 3 to about 10 in these cases. For other suitable indicator dyes, see Figure 18.10.

Photos © Cengage Learning/Charles D. Winters

The important characteristic of acid–base indicators is that the acid form of the compound (HInd) has one color and the conjugate base (Ind⁻) has another. To see how such compounds can be used as equivalence point indicators, let us write the usual equilibrium constant expressions for the dependence of hydronium ion concentration or pH on the indicator's ionization constant (K_a) and on the relative quantities of the acid and conjugate base.

$$[H_3O^+] = \frac{[HInd]}{[Ind^-]} \times K_a \quad \text{or} \quad pH = pK_a + \log\frac{[Ind^-]}{[HInd]}$$

These equations inform us that

- when the hydronium ion concentration is equivalent to the value of K_a (or when pH = pK_a), then [HInd] = [Ind⁻]
- when $[H_3O^+] > K_a$ (or pH < pK_a), then [HInd] > [Ind⁻]
- when $[H_3O^+] < K_a$ (or pH > pK_a), then [HInd] < [Ind⁻]

Now let us apply these conclusions to, for example, the titration of an acid with a base using an indicator whose pK_a value is nearly the same as the pH at the equivalence point (Figure 18.9). At the beginning of the titration, the pH is low and $[H_3O^+]$ is high; the acid form of the indicator (HInd) predominates, so its color is the one observed. As the titration progresses and the pH increases ($[H_3O^+]$ decreases), less of the acid HInd and more of its conjugate base exist in solution. Finally, just after we reach the equivalence point, [Ind⁻] is much larger than [HInd], and the color of [Ind⁻] is observed.

Several obvious questions remain to be answered. If you are trying to analyze for an acid and add an indicator that is a weak acid, won't this affect the analysis? Recall that you use only a tiny amount of an indicator in a titration. Although the acidic indicator molecules also react with the base as the titration progresses, so little indicator is present that any error is not significant.

Another question is whether you could accurately determine the pH by observing the color change of an indicator. In practice, your eyes are not quite that good. Usually, you see the color of HInd when [HInd]/[Ind⁻] is about 10/1, and the

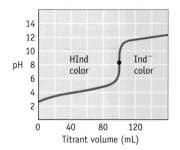

FIGURE 18.9 Indicator color changes in the course of a titration when the pK_a of the indicator HInd is about 8.

Table 18.2 Some Common Insoluble Compounds and Their K_{sp} Values*

Formula	Name	K_{sp} (25 °C)	Common Names/Uses
$CaCO_3$	Calcium carbonate	3.4×10^{-9}	Calcite, iceland spar
$MnCO_3$	Manganese(II) carbonate	2.3×10^{-11}	Rhodochrosite (forms rose-colored crystals)
$FeCO_3$	Iron(II) carbonate	3.1×10^{-11}	Siderite
CaF_2	Calcium fluoride	5.3×10^{-11}	Fluorite (source of HF and other inorganic fluorides)
$AgCl$	Silver chloride	1.8×10^{-10}	Chlorargyrite
$AgBr$	Silver bromide	5.4×10^{-13}	Used in photographic film
$CaSO_4$	Calcium sulfate	4.9×10^{-5}	Hydrated form is commonly called *gypsum*
$BaSO_4$	Barium sulfate	1.1×10^{-10}	Barite (used in "drilling mud" and as a component of paints)
$SrSO_4$	Strontium sulfate	3.4×10^{-7}	Celestite
$Ca(OH)_2$	Calcium hydroxide	5.5×10^{-5}	Slaked lime

* The values in this table were taken from *Lange's Handbook of Chemistry*, 15th edition, McGraw-Hill Publishers, New York, NY (1999). Additional K_{sp} values are given in Appendix J.

Do not confuse the *solubility* of a compound with its *solubility product constant*. The *solubility* of a salt is the quantity present in some volume of a saturated solution, expressed in moles per liter, grams per 100 mL, or other units. The *solubility product constant* is an equilibrium constant. Nonetheless, there is a connection between them: If one is known, the other can, in principle, be calculated.

Relating Solubility and K_{sp}

Solubility product constants are determined by careful laboratory measurements of the concentrations of ions in solution.

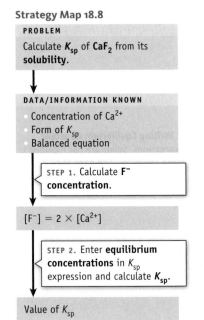

Strategy Map 18.8

PROBLEM

Calculate K_{sp} of CaF_2 from its **solubility.**

DATA/INFORMATION KNOWN

- Concentration of Ca^{2+}
- Form of K_{sp}
- Balanced equation

STEP 1. Calculate F^- **concentration.**

$[F^-] = 2 \times [Ca^{2+}]$

STEP 2. Enter **equilibrium concentrations** in K_{sp} expression and calculate K_{sp}.

Value of K_{sp}

⬤WL INTERACTIVE EXAMPLE 18.8 K_{sp} from Solubility Measurements

Problem Calcium fluoride, the main component of the mineral fluorite, dissolves to a slight extent in water.

$$CaF_2(s) \rightleftharpoons Ca^{2+}(aq) + 2\ F^-(aq) \qquad K_{sp} = [Ca^{2+}][F^-]^2$$

Calculate the K_{sp} value for CaF_2 if the calcium ion concentration has been found to be 2.3×10^{-4} mol/L.

What Do You Know? You know the balanced equation for the process, Ca^{2+} concentration, and the K_{sp} expression.

Strategy

- Calculate the fluoride ion concentration from $[Ca^{2+}]$ and stoichiometry.

- Insert the values for $[F^-]$ and $[Ca^{2+}]$ into the K_{sp} expression and calculate the value of K_{sp}.

Solution When CaF_2 dissolves in water, the balanced equation shows that the concentration of F^- ion must be twice the Ca^{2+} ion concentration.

If $[Ca^{2+}] = 2.3 \times 10^{-4}$ M, then $[F^-] = 2 \times [Ca^{2+}] = 4.6 \times 10^{-4}$ M

This means the solubility product constant is

$$K_{sp} = [Ca^{2+}][F^-]^2 = (2.3 \times 10^{-4})(4.6 \times 10^{-4})^2 = 4.9 \times 10^{-11}$$

Think about Your Answer A common student error is to forget the reaction stoichiometry. Be sure to notice that for every Ca^{2+} ion in solution there are two F^- ions.

Check Your Understanding

The barium ion concentration, $[Ba^{2+}]$, in a saturated solution of barium fluoride is 3.6×10^{-3} M. Calculate the value of the K_{sp} for BaF_2.

$$BaF_2(s) \rightleftharpoons Ba^{2+}(aq) + 2\ F^-(aq)$$

Minerals and Gems—The Importance of Solubility

Minerals and gems are among nature's most beautiful creations. Many, such as rubies, are metal oxides, and the various types of quartz are based on silicon dioxide. Another large class of gemstones consists largely of metal silicates. These include emerald, topaz, aquamarine, and tourmaline.

Yet another large class of minerals and of a few gemstones is carbonates. Rhodochrosite, one of the most beautiful red stones, is manganese(II) carbonate. And one of the most abundant minerals on earth is limestone, calcium carbonate, which is also a major component of sea shells and corals.

Hydroxides are represented by azurite, which is a mixed carbonate/hydroxide with the formula $Cu_3(OH)_2(CO_3)_2$. Turquoise is a mixed hydroxide/phosphate based on copper(II), the source of the blue color of turquoise.

Among the most common minerals are sulfides such as golden iron pyrite (FeS_2), black stibnite (Sb_2S_3), red cinnabar (HgS), and yellow orpiment (As_2S_3).

Other smaller classes of minerals exist; one of the smallest is the class based on the halides, and the best example is fluorite. Fluorite, CaF_2, exhibits a wide range of colors from purple to green to yellow.

What do all of these minerals and gems have in common? They are all insoluble or poorly soluble in water. If they were more soluble, they would be dissolved in the world's lakes and oceans.

Mineral samples (clockwise from the top center): red rhodochrosite, yellow orpiment, golden iron pyrite, green-blue turquoise, black stibnite, purple fluorite, and blue azurite. Formulas are in the text. See also Figure 18.11.

K_{sp} values for insoluble salts can be used to estimate the solubility of a solid salt or to determine whether a solid will precipitate when solutions of its anion and cation are mixed. Let us first look at an example of the estimation of the solubility of a salt from its K_{sp} value.

WL   INTERACTIVE EXAMPLE 18.9   Solubility from K_{sp}

Problem The K_{sp} for the mineral barite ($BaSO_4$, Figure 18.12) is 1.1×10^{-10} at 25 °C. Calculate the solubility of barium sulfate in pure water in (a) moles per liter and (b) grams per liter.

What Do You Know? You know the formula for the mineral and its K_{sp} value.

Strategy When $BaSO_4$ dissolves, equimolar amounts of Ba^{2+} ions and SO_4^{2-} ions are produced. Thus, the solubility of $BaSO_4$ can be estimated by calculating the equilibrium concentration of either Ba^{2+} or SO_4^{2-} from the solubility product constant.

- Write the balanced equation and the K_{sp} expression.

- Set up an ICE table, designating the unknown concentrations of Ba^{2+} and SO_4^{2-} in solution as x.

- Using x for $[Ba^{2+}]$ and $[SO_4^{2-}]$ in the K_{sp} expression, solve for x.

Solution The equation for the solubility of $BaSO_4$ and its K_{sp} expression are

$$BaSO_4(s) \rightleftharpoons Ba^{2+}(aq) + SO_4^{2-}(aq) \qquad K_{sp} = [Ba^{2+}][SO_4^{2-}] = 1.1 \times 10^{-10}$$

Denote the solubility of $BaSO_4$ (in mol/L) by x; that is, x moles of $BaSO_4$ dissolve per liter. Therefore, both $[Ba^{2+}]$ and $[SO_4^{2-}]$ must also equal x at equilibrium.

Equation	$BaSO_4(s)$	$\rightleftharpoons$	$Ba^{2+}(aq)$	+	$SO_4^{2-}(aq)$
Initial (M)			0		0
Change (M)			$+x$		$+x$
Equilibrium (M)			x		x

Because K_{sp} is the product of the barium ion and sulfate ion concentrations, K_{sp} is the square of the solubility, x,

$$K_{sp} = [Ba^{2+}][SO_4^{2-}] = 1.1 \times 10^{-10} = (x)(x) = x^2$$

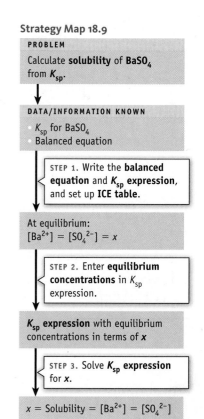

Strategy Map 18.9

PROBLEM
Calculate **solubility** of $BaSO_4$ from K_{sp}.

DATA/INFORMATION KNOWN
- K_{sp} for $BaSO_4$
- Balanced equation

STEP 1. Write the **balanced equation** and K_{sp} **expression**, and set up **ICE table**.

At equilibrium:
$[Ba^{2+}] = [SO_4^{2-}] = x$

STEP 2. Enter **equilibrium concentrations** in K_{sp} expression.

K_{sp} **expression** with equilibrium concentrations in terms of x

STEP 3. Solve K_{sp} **expression** for x.

x = Solubility = $[Ba^{2+}] = [SO_4^{2-}]$

(a) A sample of the mineral barite, which is mostly barium sulfate. "Drilling mud," used in drilling oil wells, consists of clay, barite, and calcium carbonate.

(b) Barium sulfate is opaque to x-rays, so it is used by physicians to examine the digestive tract. A patient drinks a "cocktail" containing $BaSO_4$, and the progress of the $BaSO_4$ through the digestive organs can be followed by x-ray analysis. This photo is an x ray of the gastrointestinal tract after a person ingested barium sulfate.

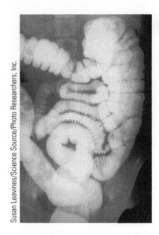

FIGURE 18.12 Barium sulfate. Barium sulfate, a white solid, is quite insoluble in water ($K_{sp} = 1.1 \times 10^{-10}$) (see Example 18.9).

and so the value of x is

$$x = [Ba^{2+}] = [SO_4^{2-}] = \sqrt{K_{sp}} = \sqrt{1.1 \times 10^{-10}} = 1.0 \times 10^{-5} \text{ M}$$

The solubility of $BaSO_4$ in pure water is 1.0×10^{-5} mol/L. To find its solubility in g/L, we only need to multiply by the molar mass of $BaSO_4$.

$$\text{Solubility in g/L} = (1.0 \times 10^{-5} \text{ mol/L})(233 \text{ g/mol}) = 0.0024 \text{ g/L}$$

Think about Your Answer As noted in Figure 18.12, $BaSO_4$ is used to investigate the digestive track. It is fortunate the compound is so insoluble, because water- and acid-soluble barium salts are toxic.

Check Your Understanding

Calculate the solubility of AgCN in moles per liter and grams per liter. K_{sp} for AgCN is 6.0×10^{-17}.

EXAMPLE 18.10 Solubility from K_{sp}

Problem Knowing that the K_{sp} value for MgF_2 is 5.2×10^{-11}, calculate the solubility of the salt in (a) moles per liter and (b) grams per liter.

What Do You Know? You know the formula for magnesium fluoride and its K_{sp} value.

Strategy The solubility must be defined in terms that will allow you to solve the K_{sp} expression for this value. From stoichiometry, we can say that if x mol of MgF_2 dissolves, then x mol of Mg^{2+} and $2x$ mol of F^- appear in solution. This means the MgF_2 solubility (in moles dissolved per liter) is equivalent to the concentration of Mg^{2+} ions in the solution.

- Write the balanced equation and the K_{sp} expression.

- Set up an ICE table, designating the unknown concentrations of Mg^{2+} and F^- in solution as x and $2x$, respectively.

- Using x for $[Mg^{2+}]$ and $2x$ for $[F^-]$ in the K_{sp} expression, solve for x.

Solution

$$MgF_2(s) \rightleftharpoons Mg^{2+}(aq) + 2\,F^-(aq) \qquad K_{sp} = [Mg^{2+}][F^-]^2 = 5.2 \times 10^{-11}$$

Equation	$MgF_2(s)$	$\rightleftharpoons$	$Mg^{2+}(aq)$	+	$2\,F^-(aq)$
Initial (M)			0		0
Change (M)			$+x$		$+2x$
Equilibrium (M)			x		$2x$

A CLOSER LOOK

Solubility Calculations

The K_{sp} value reported for lead(II) chloride, $PbCl_2$, is 1.7×10^{-5}. If we assume the appropriate equilibrium in solution is

$$PbCl_2(s) \rightleftharpoons Pb^{2+}(aq) + 2\ Cl^-(aq)$$

the calculated solubility of $PbCl_2$ is 0.016 M. The experimental value for the solubility of the salt, however, is 0.036 M, more than twice the calculated value! What is the problem? One is that the chemical behavior is sometimes much more complicated than the equation defining K_{sp}, as illustrated below.

The main problem in the lead(II) chloride case, and in many others, is that the compound dissolves but is not 100% dissociated into its constituent ions. Instead, it dissolves as the undissociated salt or forms ion pairs.

Other problems that lead to discrepancies between calculated and experimental solubilities are the reactions of ions (particularly anions) with water and complex ion formation. An example of the former effect is the product-favored reaction of sulfide ion with water, that is, *hydrolysis*.

$$S^{2-}(aq) + H_2O(\ell) \rightleftharpoons HS^-(aq) + OH^-(aq)$$

This means that the solubility of a metal sulfide is better described by a chemical equation such as

$$NiS(s) + H_2O(\ell) \rightleftharpoons$$
$$Ni^{2+}(aq) + HS^-(aq) + OH^-(aq)$$

Complex ion formation is illustrated by the fact that lead chloride is more soluble in the presence of excess chloride ion, owing to the formation of the complex ion $PbCl_4^{2-}$.

$$PbCl_2(s) + 2\ Cl^-(aq) \rightleftharpoons PbCl_4^{2-}(aq)$$

Hydrolysis and complex ion formation are discussed further on pages 649–650 and 655–658, respectively.

References:
- L. Meites, J. S. F. Pode, and H. C. Thomas: *Journal of Chemical Education*, Vol. 43, pp. 667–672, 1966.
- S. J. Hawkes: *Journal of Chemical Education*, Vol. 75, pp. 1179–1181, 1998.
- R. W. Clark and J. M. Bonicamp: *Journal of Chemical Education*, Vol. 75, pp. 1182–1185, 1998.
- R. J. Myers: *Journal of Chemical Education*, Vol. 63, pp. 687–690, 1986.

$$
\begin{array}{ccc}
PbCl_2(aq) & \xrightarrow{\ K = 0.63\ } & PbCl^+(aq) + Cl^-(aq) \\
\text{undissociated salt} & & \text{ion pairs} \\
\text{dissolved in water} & & \\
\Big\Updownarrow K = 0.0011 & & \Big\Updownarrow K = 0.026 \\
PbCl_2(s) & \xrightleftharpoons{\ K_{sp} = 1.7 \times 10^{-5}\ } & Pb^{2+}(aq) + 2\ Cl^-(aq) \\
\text{slightly soluble} & & \text{100\% dissociated} \\
\text{salt} & & \text{into ions}
\end{array}
$$

Substituting the equilibrium concentrations for $[Mg^{2+}]$ and $[F^-]$ into the K_{sp} expression, you find

$$K_{sp} = [Mg^{2+}][F^-]^2 = (x)(2x)^2 = 4x^3$$

Solving the equation for x,

$$x = \sqrt[3]{\frac{K_{sp}}{4}} = \sqrt[3]{\frac{5.2 \times 10^{-11}}{4}} = 2.4 \times 10^{-4}$$

you find that 2.4×10^{-4} moles of MgF_2 dissolve per liter. The solubility of MgF_2 in grams per liter is

$$(2.4 \times 10^{-4}\ \text{mol/L})(62.3\ \text{g/mol}) = 0.015\ \text{g } MgF_2/L$$

Think about Your Answer Problems like this one might provoke our students to ask the question, "Aren't you counting things twice when you multiply x by 2 and then square it as well?" in the expression $K_{sp} = (x)(2x)^2$. The answer is no. The 2 in the $2x$ term is based on the stoichiometry of the compound. The exponent of 2 on the F^- ion concentration arises from the rules for writing equilibrium expressions.

Check Your Understanding

Calculate the solubility of $Ca(OH)_2$ in moles per liter and grams per liter using the value of K_{sp} in Appendix J.

The *relative* solubilities of salts can often be deduced by comparing values of solubility product constants, but you must be careful! For example, the K_{sp} for silver chloride is

$$AgCl(s) \rightleftharpoons Ag^+(aq) + Cl^-(aq) \qquad K_{sp} = 1.8 \times 10^{-10}$$

whereas that for silver chromate is

$$Ag_2CrO_4(s) \rightleftharpoons 2\ Ag^+(aq) + CrO_4^{2-}(aq) \qquad K_{sp} = 9.0 \times 10^{-12}$$

In spite of the fact that Ag_2CrO_4 has a smaller numerical K_{sp} value than AgCl, the chromate salt is about 10 times *more* soluble than the chloride salt. If you determine solubilities from K_{sp} values as in the examples above, you would find the solubility of AgCl is 1.3×10^{-5} mol/L, whereas that of Ag_2CrO_4 is 1.3×10^{-4} mol/L. From this example and countless others, we conclude that

> Direct comparisons of the solubility of two salts on the basis of their K_{sp} values can be made only for salts having the same cation-to-anion ratio.

This means, for example, that you can directly compare solubilities of 1:1 salts such as the silver halides by comparing their K_{sp} values.

$$\text{AgI } (K_{sp} = 8.5 \times 10^{-17}) < \text{AgBr } (K_{sp} = 5.4 \times 10^{-13}) < \text{AgCl } (K_{sp} = 1.8 \times 10^{-10})$$
$$\xrightarrow{\text{increasing } K_{sp} \text{ and increasing solubility}}$$

Similarly, you could compare 1:2 salts such as the lead halides,

$$\text{PbI}_2 (K_{sp} = 9.8 \times 10^{-9}) < \text{PbBr}_2 (K_{sp} = 6.6 \times 10^{-6}) < \text{PbCl}_2 (K_{sp} = 1.7 \times 10^{-5})$$
$$\xrightarrow{\text{increasing } K_{sp} \text{ and increasing solubility}}$$

but you cannot directly compare a 1:1 salt (AgCl) with a 2:1 salt (Ag_2CrO_4).

Solubility and the Common Ion Effect

The test tube on the left in Figure 18.13 contains a precipitate of silver acetate, $AgCH_3CO_2$, in water. The solution is saturated, and the silver ions and acetate ions in the solution are in equilibrium with solid silver acetate.

$$AgCH_3CO_2(s) \rightleftharpoons Ag^+(aq) + CH_3CO_2^-(aq)$$

But what would happen if the silver ion concentration is increased, say by adding silver nitrate? Le Chatelier's principle (◄ Section 16.6) suggests—and we observe —that more silver acetate precipitate should form because a product ion has been added, causing the equilibrium to shift to form more silver acetate.

The ionization of weak acids and bases is affected by the presence of an ion common to the equilibrium process (Section 18.1), and the effect of adding silver ions to a saturated silver acetate solution is another example of the common ion effect. Adding a common ion to a saturated solution of a salt will lower the salt solubility (unless a complex ion can form; see *A Closer Look*, page 645).

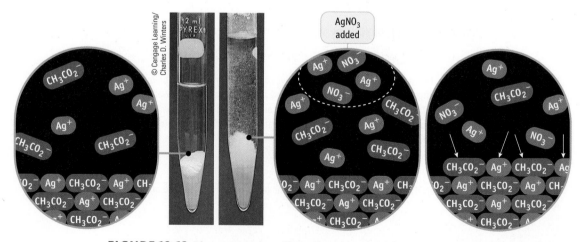

FIGURE 18.13 The common ion effect. The tube at the *left* contains a saturated solution of silver acetate, $AgCH_3CO_2$. When 1.0 M $AgNO_3$ is added to the tube *(right)*, more solid silver acetate forms.

EXAMPLE 18.11   The Common Ion Effect and Salt Solubility

Problem If solid AgCl is placed in 1.00 L of 0.55 M NaCl, what mass of AgCl will dissolve?

What Do You Know? You know the formula of the insoluble compound, its K_{sp} (Table 18.2 and Appendix J), and its molar mass. You also know that the presence of an ion common to the equilibrium (Cl^-) suppresses the solubility of AgCl.

Strategy To determine the solubility of AgCl in the presence of excess Cl^-, calculate the concentration of the Ag^+ ion.

- Write the balanced equation and the K_{sp} expression for AgCl.

- Set up an ICE table, and enter 0.55 M as the initial concentration of the common ion, Cl^-.

- Define the solubility of AgCl as x. This is the increase in the concentration of Ag^+ and Cl^- in solution as AgCl dissolves. Enter this information on the change (C) line in the ICE table.

- Fill in the last line (E) of the ICE table with x for [Ag^+] and $(0.55 + x)$ for [Cl^-]. Solve the K_{sp} expression for x.

Solution Set up an ICE table to show the concentrations of Ag^+ and Cl^- when equilibrium is attained.

Equation	AgCl(s)	$\rightleftharpoons$	Ag^+(aq)	+	Cl^-(aq)
Initial (M)			0		0.55
Change (M)			$+x$		$+x$
Equilibrium (M)			x		$0.55 + x$

The equilibrium concentrations from the table are substituted into the K_{sp} expression,

$$K_{sp} = 1.8 \times 10^{-10} = [Ag^+][Cl^-] = (x)(0.55 + x)$$

This is a quadratic equation and can be solved by the methods in Appendix A. An easier approach, however, is to make the approximation that x is very small with respect to 0.55 [and so $(0.55 + x) \approx 0.55$]. This is a reasonable assumption because we know that the solubility is very small without the common ion Cl^- and that it will be even smaller in the presence of added Cl^-. Therefore,

$$K_{sp} = 1.8 \times 10^{-10} = (x)(0.55)$$

$$x = [Ag^+] = 3.3 \times 10^{-10} \text{ M}$$

The solubility in grams per liter is then

$$(3.3 \times 10^{-10} \text{ mol/L})(143 \text{ g/mol}) = \boxed{4.7 \times 10^{-8} \text{ g/L}}$$

As predicted by Le Chatelier's principle, the solubility of AgCl in the presence of added Cl^- is less (3.3×10^{-10} M) than in pure water (1.3×10^{-5} M).

Think about Your Answer The approximation we made here is similar to the approximations we make in acid–base equilibrium problems. However, as a final step, we should check its validity by substituting the calculated value of x into the exact expression $K_{sp} = (x)(0.55 + x)$. If the product $(x)(0.55 + x)$ is the same as the given value of K_{sp}, the approximation is valid.

$$K_{sp} = (x)(0.55 + x) = (3.3 \times 10^{-10})(0.55 + 3.3 \times 10^{-10}) = 1.8 \times 10^{-10}$$

Check Your Understanding

Calculate the solubility of $BaSO_4$ (a) in pure water and (b) in the presence of 0.010 M $Ba(NO_3)_2$. K_{sp} for $BaSO_4$ is 1.1×10^{-10}.

EXAMPLE 18.12 The Common Ion Effect and Salt Solubility

Problem Calculate the solubility of silver chromate, Ag_2CrO_4, at 25 °C in the presence of 0.0050 M K_2CrO_4 solution.

$$Ag_2CrO_4(s) \rightleftharpoons 2\ Ag^+(aq) + CrO_4^{2-}(aq)$$

$$K_{sp} = [Ag^+]^2[CrO_4^{2-}] = 9.0 \times 10^{-12}$$

For comparison, the solubility of Ag_2CrO_4 in pure water is 1.3×10^{-4} mol/L.

What Do You Know? You know the formula and K_{sp} value for Ag_2CrO_4.

Strategy In the presence of chromate ion from the water-soluble salt K_2CrO_4, the concentration of Ag^+ ions produced by Ag_2CrO_4 will be less than in pure water. Assume the solubility of Ag_2CrO_4 is x mol/L. This means the concentration of Ag^+ ions will be $2x$ mol/L, whereas the concentration of CrO_4^{2-} ions will be x mol/L plus the concentration of CrO_4^{2-} already in the solution.

- Write the balanced equation and the K_{sp} expression for Ag_2CrO_4.

- Set up an ICE table, and enter 0.0050 M as the initial concentration of the common ion, CrO_4^{2-}.

- Designate the increase in the concentration of the CrO_4^{2-} ion in solution as x and the increase in $[Ag^+]$ as $2x$.

- Using $2x$ for $[Ag^+]$ and $(0.0050 + x)$ for $[CrO_4^{2-}]$ in the K_{sp} expression, solve for x.

Solution

Equation	$Ag_2CrO_4(s)$	$\rightleftharpoons$	$2\ Ag^+(aq)$	$+$	$CrO_4^{2-}(aq)$
Initial (M)			0		0.0050
Change (M)			$+2x$		$+x$
Equilibrium (M)			$2x$		$0.0050 + x$

Substituting the equilibrium concentrations into the K_{sp} expression, we have

$$K_{sp} = 9.0 \times 10^{-12} = [Ag^+]^2[CrO_4^{2-}]$$

$$K_{sp} = (2x)^2(0.0050 + x)$$

As in Example 18.11, you can make the approximation that x is very small with respect to 0.0050, and so $(0.0050 + x) \approx 0.0050$. Therefore, the approximate expression is

$$K_{sp} = 9.0 \times 10^{-12} = [Ag^+]^2[CrO_4^{2-}] \approx (2x)^2(0.0050)$$

Solving, we find x, the solubility of silver chromate in the presence of excess chromate ion, is

$$x = \text{Solubility of } Ag_2CrO_4 = \boxed{2.1 \times 10^{-5} \text{ M}}$$

Think about Your Answer The silver ion concentration in the presence of the common ion is

$$[Ag^+] = 2x = 4.2 \times 10^{-5} \text{ M}$$

This silver ion concentration is indeed less than its value in pure water (2.6×10^{-4} M), owing to the presence of an ion "common" to the equilibrium.

Check Your Understanding

Calculate the solubility of $Zn(CN)_2$ at 25 °C (a) in pure water and (b) in the presence of 0.10 M $Zn(NO_3)_2$. K_{sp} for $Zn(CN)_2$ is 8.0×10^{-12}.

There are two important general ideas from Examples 18.11 and 18.12:

- The solubility of a salt will be reduced by the presence of a common ion, in accordance with Le Chatelier's principle.
- We made the approximation that the amount of common ion added to the solution was very large in comparison with the amount of that ion coming from the insoluble salt, and this allowed us to simplify our calculations. This is almost always the case, but you should check to be sure.

The Effect of Basic Anions on Salt Solubility

The next time you are tempted to wash a supposedly insoluble salt down the kitchen or laboratory drain, stop and consider the consequences. Many metal ions such as lead, chromium, and mercury are toxic in the environment. Even if a so-called insoluble salt of one of these cations does not appear to dissolve, its solubility in water may be greater than you think, in part owing to the possibility that the anion of the salt is a weak base or the cation is a weak acid.

Lead(II) sulfide, PbS, which is found in nature as the mineral galena (Figure 18.14), provides an example of the effect of the acid–base properties of an ion on salt solubility. When placed in water, a trace amount dissolves,

$$PbS(s) \rightleftharpoons Pb^{2+}(aq) + S^{2-}(aq)$$

and one product of the reaction is the sulfide ion, which is itself a strong base.

$$S^{2-}(aq) + H_2O(\ell) \rightleftharpoons HS^-(aq) + OH^-(aq) \qquad K_b = 1 \times 10^5$$

This ion undergoes extensive hydrolysis (reaction with water) (◀ Table 17.3), which decreases its concentration, and the equilibrium process for dissolving PbS shifts to the right. Thus, the lead ion concentration in solution is greater than expected from the simple dissociation of the salt.

The lead(II) sulfide example leads to the following general observation:

> Any salt containing an anion that is the conjugate base of a weak acid will dissolve in water to a greater extent than given by K_{sp}.

This means that salts of phosphate, acetate, carbonate, and cyanide, as well as sulfide, can be affected, because all of these anions undergo the general hydrolysis reaction

$$X^-(aq) + H_2O(\ell) \rightleftharpoons HX(aq) + OH^-(aq)$$

The observation that ions from insoluble salts can undergo hydrolysis is related to another useful, general conclusion:

> Insoluble salts in which the anion is the conjugate base of a weak acid dissolve in strong acids.

Insoluble salts containing anions such as acetate, carbonate, hydroxide, phosphate, and sulfide dissolve in strong acids. For example, you know that if a strong acid is added to a water-insoluble metal carbonate such as $CaCO_3$, the salt dissolves (◀ Section 3.7).

$$CaCO_3(s) + 2 H_3O^+(aq) \rightarrow Ca^{2+}(aq) + 3 H_2O(\ell) + CO_2(g)$$

You can think of this as the result of a series of reactions.

$CaCO_3(s) \rightleftharpoons Ca^{2+}(aq) + CO_3^{2-}(aq)$ $K_{sp} = 3.4 \times 10^{-9}$

$CO_3^{2-}(aq) + H_3O^+(aq) \rightleftharpoons HCO_3^-(aq) + H_2O(\ell)$ $1/K_{a2} = 1/4.8 \times 10^{-11} = 2.1 \times 10^{10}$

$HCO_3^-(aq) + H_3O^+(aq) \rightleftharpoons H_2CO_3(aq) + H_2O(\ell)$ $1/K_{a1} = 1/4.2 \times 10^{-7} = 2.4 \times 10^6$

Overall: $CaCO_3(s) + 2 H_3O^+(aq) \rightleftharpoons Ca^{2+}(aq) + 2 H_2O(\ell) + H_2CO_3(aq)$

$$K_{net} = (K_{sp})(1/K_{a2})(1/K_{a1}) = 1.7 \times 10^8$$

Carbonic acid, a product of this reaction, is not stable,

$$H_2CO_3(aq) \rightleftharpoons CO_2(g) + H_2O(\ell) \qquad K \approx 10^5$$

and you see CO_2 bubbling out of the solution, a process that moves the $CaCO_3 + H_3O^+$ equilibrium even further to the right. Calcium carbonate dissolves completely in strong acid!

Many metal sulfides are also soluble in strong acids

$$FeS(s) + 2 H_3O^+(aq) \rightleftharpoons Fe^{2+}(aq) + H_2S(aq) + 2 H_2O(\ell)$$

FIGURE 18.14 Lead(II) sulfide (galena). This and other metal sulfides dissolve in water to a greater extent than expected because the sulfide ion reacts with water to form the very stable species HS⁻ and OH⁻.

$PbS(s) + H_2O(\ell) \rightleftharpoons$
    $Pb^{2+}(aq) + HS^-(aq) + OH^-(aq)$

The model of PbS shows that the unit cell is cubic, a feature reflected by the cubic crystals of the mineral galena.

● **Metal Sulfide Solubility** The true solubility of a metal sulfide is better represented by a modified solubility product constant, K_{spa}, which is defined as follows:

$MS(s) \rightleftharpoons M^{2+}(aq) + S^{2-}(aq)$
       $K_{sp} = [M^{2+}][S^{2-}]$

$S^{2-}(aq) + H_2O(\ell) \rightleftharpoons$
    $HS^-(aq) + OH^-(aq)$
       $K_b = [HS^-][OH^-]/[S^{2-}]$

Net reaction:

$MS(s) + H_2O(\ell) \rightleftharpoons$
    $HS^-(aq) + M^{2+}(aq) + OH^-(aq)$
   $K_{spa} = [M^{2+}][HS^-][OH^-] = K_{sp} \times K_b$

Values for K_{spa} for several metal sulfides are included in Appendix J.

FIGURE 18.15 The effect of the anion on salt solubility in acid. *(left)* A precipitate of AgCl (white) and Ag₃PO₄ (yellow). *(right)* Adding a strong acid (HNO₃) dissolves Ag₃PO₄ (and leaves insoluble AgCl). The basic anion PO_4^{3-} reacts with acid to give H_3PO_4, whereas Cl^- is too weakly basic to form HCl.

Add strong acid.

precipitate of
AgCl and Ag₃PO₄

precipitate of
AgCl

Photos © Cengage Learning/Charles D. Winters

as are metal phosphates (Figure 18.15),

$$Ag_3PO_4(s) + 3\ H_3O^+(aq) \rightleftharpoons 3\ Ag^+(aq) + H_3PO_4(aq) + 3\ H_2O(\ell)$$

and metal hydroxides.

$$Mg(OH)_2(s) + 2\ H_3O^+(aq) \rightleftharpoons Mg^{2+}(aq) + 4\ H_2O(\ell)$$

In general, the solubility of a salt containing the conjugate base of a weak acid is increased by addition of a stronger acid to the solution. In contrast, salts are not soluble in strong acid if the anion is the conjugate base of a strong acid. For example, AgCl is not soluble in strong acid

$$AgCl(s) \rightleftharpoons Ag^+(aq) + Cl^-(aq) \qquad\qquad K_{sp} = 1.8 \times 10^{-10}$$

$$H_3O^+(aq) + Cl^-(aq) \rightleftharpoons HCl(aq) + H_2O(\ell) \qquad K \ll 1$$

because Cl^- is a very weak base (◄ Table 17.3), and so its concentration is not lowered by a reaction with the strong acid H_3O^+ (Figure 18.15). This same conclusion would also apply to insoluble salts of Br^- and I^-.

REVIEW & CHECK FOR SECTION 18.4

1. What is the K_{sp} expression for silver carbonate?

   (a) $K_{sp} = [Ag^+][CO_3^{2-}]$ (c) $K_{sp} = [Ag^+][CO_3^{2-}]^2$

   (b) $K_{sp} = [Ag^+]^2[CO_3^{2-}]$

2. Using K_{sp} values, predict which salt in each pair is more soluble in water.

   (a) AgCl ($K_{sp} = 1.8 \times 10^{-10}$) or AgCN ($K_{sp} = 6.0 \times 10^{-17}$)

   (b) Mg(OH)₂ ($K_{sp} = 5.6 \times 10^{-12}$) or Ca(OH)₂ ($K_{sp} = 5.5 \times 10^{-5}$)

   (c) Ca(OH)₂ ($K_{sp} = 5.5 \times 10^{-5}$) or CaSO₄ ($K_{sp} = 4.9 \times 10^{-5}$)

3. What is the solubility of PbSO₄ in water at 25 °C? (K_{sp} for PbSO₄ = 2.5×10^{-8})

   (a) 2.5×10^{-8} M   (b) 1.6×10^{-4} M   (c) 6.3×10^{-16} M

4. What is the solubility of PbSO₄ in water at 25 °C if the solution already contains 0.25 M Na₂SO₄? (K_{sp} for PbSO₄ = 2.5×10^{-8})

   (a) 1.0×10^{-7} M   (b) 1.6×10^{-4} M   (c) 6.3×10^{-9} M

5. Which compound should be more soluble in 0.1 M HCl?

   (a) FeCO₃ ($K_{sp} = 3.1 \times 10^{-11}$)   (b) AgCl ($K_{sp} = 1.8 \times 10^{-10}$)

18.5 Precipitation Reactions

Metal-bearing ores contain the metal in the form of an insoluble salt (Figure 18.16), and, to complicate matters, ores often contain several such metal salts. Many industrial methods for separating metals from their ores involve dissolving metal salts to obtain the metal ion or ions in solution. The solution is then concentrated in some manner, and a precipitating agent is added to precipitate selectively only one type of metal ion as an insoluble salt. In the case of nickel, for example, the Ni^{2+} ion can be precipitated as insoluble nickel(II) sulfide or nickel(II) carbonate.

$$Ni^{2+}(aq) + HS^-(aq) + H_2O(\ell) \rightleftharpoons NiS(s) + H_3O^+(aq) \qquad K = 1.7 \times 10^{18}$$

$$Ni^{2+}(aq) + CO_3^{2-}(aq) \rightleftharpoons NiCO_3(s) \qquad K = 7.1 \times 10^{6}$$

After separating the insoluble salt, the final step in obtaining the metal itself is to reduce the metal cation to the metal either chemically or electrochemically (► Chapter 20).

Our immediate goal is to work out methods to determine whether a precipitate will form under a given set of conditions. For example, if Ag^+ and Cl^- are present at some given concentrations, will AgCl precipitate from the solution?

FIGURE 18.16 Minerals. Minerals are insoluble salts. The minerals shown here are light purple fluorite (calcium fluoride), black hematite [iron(III) oxide], and rust brown goethite, a mixture of iron(III) oxide and iron(III) hydroxide.

K_{sp} and the Reaction Quotient, Q

Silver chloride, like silver bromide, is used in photographic films. It dissolves to a very small extent in water and has a correspondingly small value of K_{sp}.

$$AgCl(s) \rightleftharpoons Ag^+(aq) + Cl^-(aq) \qquad K_{sp} = [Ag^+][Cl^-] = 1.8 \times 10^{-10}$$

But look at the problem from the other direction: If a solution contains Ag^+ and Cl^- ions at some concentration, will AgCl precipitate from solution? This is the same question we asked in Section 16.2 when we wanted to know if a given mixture of reactants and products was an equilibrium mixture, if the reactants continued to form products, or if products would revert to reactants. The procedure there was to calculate the reaction quotient, Q.

For silver chloride, the expression for the reaction quotient, Q, is

$$Q = [Ag^+][Cl^-]$$

Recall that *the difference between Q and K is that the concentrations in the reaction quotient expression, Q, may or may not be those at equilibrium.* For the case of a slightly soluble salt such as AgCl, we can reach the following conclusions (◄ Section 16.2).

1. If $Q = K_{sp}$, the solution is saturated.

When $Q = K_{sp}$ the ion concentrations have reached their maximum value.

2. If $Q < K_{sp}$, the solution is not saturated.

This can mean either of two things: (i) If solid AgCl is present, more will dissolve until equilibrium is achieved (when $Q = K_{sp}$). (ii) If solid AgCl is not already present, more $Ag^+(aq)$ or more $Cl^-(aq)$ (or both) could be added to the solution until precipitation of solid AgCl begins (when $Q > K_{sp}$).

3. If $Q > K_{sp}$, the solution is oversaturated and precipitation will occur.

The concentrations of Ag^+ and Cl^- in solution are too high, and AgCl will precipitate until $Q = K_{sp}$.

CASE STUD

Although there is cor
debate about the is
global warming, the m
CO_2 in Earth's atmo
There has been a sig
the beginning of the I
the 18th century. Alor
atmospheric CO_2, ther
mented increase in
absorbed by the world
the last 200 years the
about 550 billion to
about 22 million tons
a recent paper in *Natu*
took all of the man-ma
and released it into th
els of CO_2 would be
oceans had not absor
earth's temperature is
warmer, a point at whi
change would occur. T
by the oceans has preve
from entering the atm
evidence that the ocear
efficient sink. And tl
problem.

When CO_2 dissolv
forms carbonic acid, a
rium with hydronium i
ions (as well as a sm
carbonate ions).

$CO_2(g) + H_2O(\ell)$

$H_2CO_3(aq) + H_2O(\ell)$

$HCO_3^-(aq) + H_2O(\ell)$

Strategy Map 18.14

PROBLEM

What concentration of SO_4^{2-} is required to begin precipitation of $BaSO_4$?

↓

DATA/INFORMATION KNOWN

- K_{sp} value for $BaSO_4$
- Initial $[Ba^{2+}]$

↓

STEP 1. Write **expression** for K_{sp}.

$K_{sp} = [Ba^{2+}][SO_4^{2-}]$

STEP 2. Enter **Ba^{2+} concentration** and solve for $[SO_4^{2-}]$.

↓

$BaSO_4$ begins to precipitate when $[SO_4^{2-}]$ exceeds calculated value.

Strategy

- Write the balanced equation for dissolving $BaSO_4$ and the K_{sp} expression.

- **Part (a):** Use the K_{sp} expression to calculate $[SO_4^{2-}]$ when $[Ba^{2+}] = 0.010$ M.

- **Part (b):** Use the K_{sp} expression to calculate $[Ba^{2+}]$ when $[SO_4^{2-}] = 0.015$ M.

Solution

$$BaSO_4(s) \rightleftharpoons Ba^{2+}(aq) + SO_4^{2-}(aq) \qquad K_{sp} = [Ba^{2+}][SO_4^{2-}] = 1.1 \times 10^{-10}$$

(a) When the product of the ion concentrations exceeds the K_{sp} ($= 1.1 \times 10^{-10}$)—that is, when $Q > K_{sp}$—precipitation will occur. The Ba^{2+} ion concentration is known (0.010 M), so the SO_4^{2-} ion concentration necessary for precipitation can be calculated.

$$[SO_4^{2-}] = \frac{K_{sp}}{[Ba^{2+}]} = \frac{1.1 \times 10^{-10}}{0.010} = 1.1 \times 10^{-8} \text{ M}$$

The result tells us that if the sulfate ion concentration is just slightly greater than 1.1×10^{-8} M, $BaSO_4$ will begin to precipitate.

(b) If the sulfate ion concentration is increased to 0.015 M, the maximum concentration of Ba^{2+} ion that can exist in solution (in equilibrium with $BaSO_4$) is

$$[Ba^{2+}] = \frac{K_{sp}}{[SO_4^{2-}]} = \frac{1.1 \times 10^{-10}}{0.015} = 7.3 \times 10^{-9} \text{ M}$$

Think about Your Answer The fact that the barium ion concentration is so small when $[SO_4^{2-}] = 0.015$ M means that the Ba^{2+} ion has been essentially removed from solution. (It began at 0.010 M and has decreased by a factor of about 1 million.)

Check Your Understanding

What is the minimum concentration of I^- that can cause precipitation of PbI_2 from a 0.050 M solution of $Pb(NO_3)_2$? K_{sp} for PbI_2 is 9.8×10^{-9}. What concentration of Pb^{2+} ions remains in solution when the concentration of I^- is 0.0015 M?

EXAMPLE 18.15 K_{sp} and Precipitations

Problem Suppose you mix 100.0 mL of 0.0200 M $BaCl_2$ with 50.0 mL of 0.0300 M Na_2SO_4. Will $BaSO_4$ ($K_{sp} = 1.1 \times 10^{-10}$) precipitate?

What Do You Know? You know the concentration of two different compounds in solutions of differing volumes and so you can calculate the amount (moles) of each substance. You also know that $BaCl_2$ and Na_2SO_4 will combine to give a precipitate, $BaSO_4$, if they are mixed in sufficient concentration. You also know the K_{sp} value for $BaSO_4$.

Strategy Here, you mix two solutions, one containing Ba^{2+} ions and the other SO_4^{2-} ions, both with known concentration and volume, and insoluble $BaSO_4$ may be formed.

- Find the concentration of each of these ions after mixing.

- Knowing the ion concentrations in the combined solution, calculate Q and compare it with the K_{sp} value for $BaSO_4$ to decide if $BaSO_4$ will precipitate under these circumstances.

Solution First, use the equation $c_1V_1 = c_2V_2$ (◀ Section 4.5) to calculate c_2, the concentration of the Ba^{2+} and SO_4^{2-} ions after mixing, to give a new solution with a volume of 150.0 mL ($= V_2$).

$$[Ba^{2+}] \text{ after mixing} = \frac{(0.0200 \text{ mol/L})(0.1000 \text{ L})}{0.1500 \text{ L}} = 0.0133 \text{ M}$$

$$[SO_4^{2-}] \text{ after mixing} = \frac{(0.0300 \text{ mol/L})(0.0500 \text{ L})}{0.1500 \text{ L}} = 0.0100 \text{ M}$$

The equilibrium governing the reaction that can occur is

$$BaSO_4(s) \rightleftharpoons Ba^{2+}(aq) + SO_4^{2-}(aq) \qquad K_{sp} = [Ba^{2+}][SO_4^{2-}] = 1.1 \times 10^{-10}$$

Now the reaction quotient can be calculated.

$$Q = [Ba^{2+}][SO_4^{2-}] = (0.0133)(0.0100) = 1.33 \times 10^{-4}$$

Q is much larger than K_{sp}, so $BaSO_4$ precipitates.

Think about Your Answer The K_{sp} value for $BaSO_4$ is very small, so mixing solutions with Ba^{2+} and SO_4^{2-} ions, even in very low concentration, can lead to the precipitation of $BaSO_4$.

Check Your Understanding

You have 100.0 mL of 0.0010 M silver nitrate. Will AgCl precipitate if you add 5.0 mL of 0.025 M HCl?

REVIEW & CHECK FOR SECTION 18.5

Will $SrSO_4$ precipitate from a solution containing 2.5×10^{-4} M strontium ion, Sr^{2+}, if enough of the soluble salt Na_2SO_4 is added to make the solution 2.5×10^{-4} M in SO_4^{2-}? K_{sp} for $SrSO_4$ is 3.4×10^{-7}.

(a) yes   (b) no   (c) can't decide

Dimethylglyoximate complex of Ni^{2+} ion

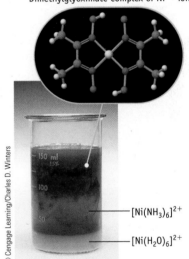

18.6 Equilibria Involving Complex Ions

Metal ions exist in aqueous solution as complex ions (◄ Section 17.11) (Figure 18.17). Complex ions consist of the metal ion and other molecules or ions bound into a single entity. In water, metal ions are always surrounded by water molecules, with the negative end of the polar water molecule, the oxygen atom, attracted to the positive metal ion. In the case of Ni^{2+}, the ion exists as $[Ni(H_2O)_6]^{2+}$ in water. On adding ammonia, water molecules are displaced successively, and in the presence of a high enough concentration of ammonia, the complex ion $[Ni(NH_3)_6]^{2+}$ exists. Many organic molecules also form complex ions with metal ions, one example being the complex with the dimethylglyoximate ion in Figure 18.17.

The molecules or ions that bind to metal ions are called **ligands** (► Chapter 22). In aqueous solution, metal ions and ligands exist in equilibrium, and the equilibrium constants for these reactions are referred to as **formation constants**, K_f (Appendix K). For example,

$$Cu^{2+}(aq) + NH_3(aq) \rightleftharpoons [Cu(NH_3)]^{2+}(aq) \qquad K_{f1} = 2.0 \times 10^4$$

$$[Cu(NH_3)]^{2+}(aq) + NH_3(aq) \rightleftharpoons [Cu(NH_3)_2]^{2+}(aq) \qquad K_{f2} = 4.7 \times 10^3$$

$$[Cu(NH_3)_2]^{2+}(aq) + NH_3(aq) \rightleftharpoons [Cu(NH_3)_3]^{2+}(aq) \qquad K_{f3} = 1.1 \times 10^3$$

$$[Cu(NH_3)_3]^{2+}(aq) + NH_3(aq) \rightleftharpoons [Cu(NH_3)_4]^{2+}(aq) \qquad K_{f4} = 2.0 \times 10^2$$

In these reactions, Cu^{2+} begins as $[Cu(H_2O)_4]^{2+}$, but ammonia successively displaces the water molecules. Overall, the formation of the tetraammine copper(II) complex ion has an equilibrium constant of 2.1×10^{13} ($= K_{f1} \times K_{f2} \times K_{f3} \times K_{f4}$).

$$Cu^{2+}(aq) + 4 NH_3(aq) \rightleftharpoons [Cu(NH_3)_4]^{2+}(aq) \qquad K_f = 2.1 \times 10^{13}$$

FIGURE 18.17 Complex ions. The green solution contains soluble $Ni(H_2O)_6^{2+}$ ions in which water molecules are bound to Ni^{2+} ions by ion–dipole forces. This complex ion gives the solution its green color. The Ni^{2+}-ammonia complex ion is purple. The red, insoluble solid is the dimethylglyoximate complex of the Ni^{2+} ion $[Ni(C_4H_7O_2N_2)_2]$ (model at top). Formation of this beautiful red insoluble compound is the classical test for the presence of the aqueous Ni^{2+} ion.

© Cengage Learning/Charles D. Winters

$[Ni(NH_3)_6]^{2+}$

$[Ni(H_2O)_6]^{2+}$

EXAMPLE 18.16 Complex Ion Equilibria

Problem What is the concentration of Cu^{2+} ions in a solution prepared by adding 0.00100 mol of $Cu(NO_3)_2$ to 1.00 L of 1.50 M NH_3? K_f for the copper-ammonia complex ion $[Cu(NH_3)_4]^{2+}$ is 2.1×10^{13}.

What Do You Know? Here you know the concentration of the species that form the complex ion (Cu^{2+} and NH_3), and you know the formation constant for the ion.

Strategy The formation constant for the complex ion is very large, so you start with the assumption that all of the Cu^{2+} ions react with NH_3 to form $[Cu(NH_3)_4]^{2+}$. That is, the initial concentration of the complex ion, $[Cu(NH_3)_4]^{2+}$ is 0.00100 M. This cation then dissociates to produce Cu^{2+} ions and additional NH_3 in solution. The equilibrium constant for the dissociation of $[Cu(NH_3)_4]^{2+}$ is the reciprocal of K_f because the dissociation of the ion is the reverse of its formation.

- Write a balanced equation for the dissociation of the complex ion that formed in solution and set up an ICE table.

- Assume that all of the Cu^{2+} ions in the solution are in the form of $[Cu(NH_3)_4]^{2+}$ (0.00100 M). This means that $[NH_3]$ = original concentration $- 4 \times 0.00100$ M.

- Assume the concentration of complex ion dissociated at equilibrium is x, so x mol/L of Cu^{2+} are released to solution as are $4x$ mol/L of NH_3.

- Use the equilibrium concentrations of the ions in the expression for $K_{dissociation}$ ($= 1/K_f$) and solve for x (which is the concentration of Cu^{2+} at equilibrium).

Solution Set up an ICE table for the dissociation of $[Cu(NH_3)_4]^{2+}$.

Equation	$[Cu(NH_3)_4]^{2+}(aq)$	$\rightleftharpoons$	$Cu^{2+}(aq)$	+	$4\,NH_3(aq)$
Initial (M)	0.00100		0		1.50 − 0.00400 M
Change (M)	−x		+x		+4x
Equilibrium (M)	0.00100 − x ≈ 0.00100		x		1.50 − 0.00400 + 4x ≈ 1.50

Here we assume that x is so small that the concentration of the complex ion is very nearly 0.00100 M and that the NH_3 concentration at equilibrium is essentially what was there originally.

$$K_{dissociation} = \frac{1}{K_f} = \frac{1}{2.1 \times 10^{13}} = \frac{[Cu^{2+}][NH_3]^4}{\{[Cu(NH_3)_4]^{2+}\}} = \frac{(x)(1.50)^4}{0.00100}$$

$$x = [Cu^{2+}] = 9.4 \times 10^{-18} \text{ M}$$

Think about Your Answer Make sure to test your assumption that x is so small it can neglected in determining the equilibrium concentrations of $[Cu(NH_3)_4]^{2+}$ and NH_3. It certainly is in this case.

Check Your Understanding

Silver nitrate (0.0050 mol) is added to 1.00 L of 1.00 M NH_3. What is the concentration of Ag^+ ions at equilibrium?

$$Ag^+(aq) + 2\,NH_3(aq) \rightleftharpoons [Ag(NH_3)_2]^+(aq) \qquad K_f = 1.1 \times 10^7$$

REVIEW & CHECK FOR SECTION 18.6

Iron(II) chloride (0.0025 mol) is added to 1.00 L of 0.500 M NaCN. What is the concentration of Fe^{2+} ions at equilibrium? K_f for $[Fe(CN)_6]^{4-}$ is 1.0×10^{35}.

(a) 1.0×10^{-35} M   (b) 1.9×10^{-36} M   (c) 5.2×10^{-38} M

18.7 Solubility and Complex Ions

Silver chloride does not dissolve either in water or in strong acid, but it does dissolve in ammonia because it forms a water-soluble complex ion, $[Ag(NH_3)_2]^+$ (Figure 18.18).

$$AgCl(s) + 2\,NH_3(aq) \rightleftharpoons [Ag(NH_3)_2]^+(aq) + Cl^-(aq)$$

We can view dissolving AgCl(s) as a two-step process. First, AgCl dissolves minimally in water, giving $Ag^+(aq)$ and $Cl^-(aq)$ ion. Then, the $Ag^+(aq)$ ion combines with NH_3 to give the ammonia complex. Lowering the $Ag^+(aq)$ concentration through complexation with NH_3 shifts the solubility equilibrium to the right, and more solid AgCl dissolves.

$$AgCl(s) \rightleftharpoons Ag^+(aq) + Cl^-(aq) \qquad\qquad K_{sp} = 1.8 \times 10^{-10}$$

$$Ag^+(aq) + 2\ NH_3(aq) \rightleftharpoons [Ag(NH_3)_2]^+(aq) \qquad K_f = 1.1 \times 10^7$$

This is an example of combining or "coupling" two (or more) equilibria where one is a product-favored reaction and the other is reactant-favored.

The large value of the formation constant for $[Ag(NH_3)_2]^+$ means that the equilibrium lies well to the right, and AgCl can dissolve in the presence of NH_3. If you combine K_f with K_{sp}, you obtain the net equilibrium constant for the interaction of AgCl and aqueous ammonia.

$$K_{net} = K_{sp} \times K_f = (1.8 \times 10^{-10})(1.1 \times 10^7) = 2.0 \times 10^{-3}$$

$$K_{net} = 2.0 \times 10^{-3} = \frac{\{[Ag(NH_3)_2]^+\}[Cl^-]}{[NH_3]^2}$$

Even though the value of K_{net} seems small, if you use a large concentration of NH_3, the concentration of $[Ag(NH_3)_2]^+$ in solution can be appreciable. Silver chloride is thus much more soluble in the presence of ammonia than in pure water.

The stabilities of various complex ions involving silver(I) can be compared by comparing values of their formation constants.

- **Complex Ions** Complex ions are prevalent in chemistry and are the basis of such biologically important substances as hemoglobin and vitamin B_{12}. They are described in more detail in Chapter 22.

Formation Equilibrium	K_f
$Ag^+(aq) + 2\ Cl^-(aq) \rightleftharpoons [AgCl_2]^-(aq)$	1.1×10^5
$Ag^+(aq) + 2\ S_2O_3{}^{2-}(aq) \rightleftharpoons [Ag(S_2O_3)_2]^{3-}(aq)$	2.9×10^{13}
$Ag^+(aq) + 2\ CN^-(aq) \rightleftharpoons [Ag(CN)_2]^-(aq)$	1.3×10^{21}

The formation of all three silver complexes is strongly product-favored. The cyanide complex ion $[Ag(CN)_2]^-$ is the most stable of the three.

Figure 18.18 shows what happens as complex ions are formed. Beginning with a precipitate of AgCl, adding aqueous ammonia dissolves the precipitate to give the

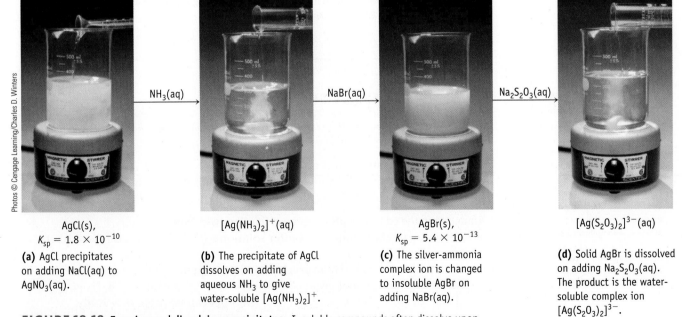

Photos © Cengage Learning/Charles D. Winters

$$NH_3(aq) \qquad NaBr(aq) \qquad Na_2S_2O_3(aq)$$

AgCl(s),
$K_{sp} = 1.8 \times 10^{-10}$

(a) AgCl precipitates on adding NaCl(aq) to $AgNO_3(aq)$.

$[Ag(NH_3)_2]^+(aq)$

(b) The precipitate of AgCl dissolves on adding aqueous NH_3 to give water-soluble $[Ag(NH_3)_2]^+$.

AgBr(s),
$K_{sp} = 5.4 \times 10^{-13}$

(c) The silver-ammonia complex ion is changed to insoluble AgBr on adding NaBr(aq).

$[Ag(S_2O_3)_2]^{3-}(aq)$

(d) Solid AgBr is dissolved on adding $Na_2S_2O_3(aq)$. The product is the water-soluble complex ion $[Ag(S_2O_3)_2]^{3-}$.

FIGURE 18.18 Forming and dissolving precipitates. Insoluble compounds often dissolve upon addition of a complexing agent.

soluble complex ion $[Ag(NH_3)_2]^+$. Silver bromide is even more stable than $[Ag(NH_3)_2]^+$, so AgBr ($K_{sp} = 5.4 \times 10^{-13}$) forms in preference to the complex ion on adding bromide ion. If thiosulfate ion, $S_2O_3^{2-}$, is then added, AgBr dissolves due to the formation of $[Ag(S_2O_3)_2]^{3-}$, a complex ion with a large formation constant (2.9×10^{13}).

EXAMPLE 18.17 Complex Ions and Solubility

Problem What is the value of the equilibrium constant, K_{net}, for dissolving AgBr in a solution containing the thiosulfate ion, $S_2O_3^{2-}$ (Figure 18.18)? Does AgBr dissolve readily on adding aqueous sodium thiosulfate to the solid?

What Do You Know? There are two equilibria here. One is for dissolving AgBr in water to give Ag^+ and Br^- ions, and its equilibrium constant is K_{sp}. The other is the formation of $[Ag(S_2O_3)_2]^{3-}$ ions from Ag^+ and $S_2O_3^{2-}$ ions; its equilibrium constant is K_f.

Strategy Summing several equilibrium processes gives the net chemical equation. K_{net} is the product of the values of K of the summed chemical equations (◀ Section 16.5).

Solution The overall reaction for dissolving AgBr in the presence of the thiosulfate anion is the sum of two equilibrium processes.

$$AgBr(s) \rightleftharpoons Ag^+(aq) + Br^-(aq) \qquad\qquad K_{sp} = 5.0 \times 10^{-13}$$

$$Ag^+(aq) + 2\ S_2O_3^{2-}(aq) \rightleftharpoons [Ag(S_2O_3)_2]^{3-}(aq) \qquad K_f = 2.9 \times 10^{13}$$

Net chemical equation:

$$AgBr(aq) + 2\ S_2O_3^{2-}(aq) \rightleftharpoons [Ag(S_2O_3)_2]^{3-}(aq) + Br^-(aq) \qquad K_{net} = K_{sp} \times K_f = 15$$

AgBr is predicted to dissolve readily in aqueous $Na_2S_2O_3$, as observed (Figure 18.18).

Think about Your Answer The value of K_{net} is greater than 1, indicating a product-favored reaction at equilibrium.

Check Your Understanding

Calculate the value of the equilibrium constant, K_{net}, for dissolving $Cu(OH)_2$ in aqueous ammonia (to form the complex ion $[Cu(NH_3)_4]^{2+}$) (◀ Figure 17.10).

REVIEW & CHECK FOR SECTION 18.7

What is the equilibrium constant for the process of dissolving AgI by the addition of aqueous NaCN? K_{sp} for AgI is 8.5×10^{-17}, and K_f for $[Ag(CN)_2]^-$ is 1.3×10^{21}.

(a) 6.5×10^{-4} (b) 8.7×10^4 (c) 1.1×10^5

---

 and

Sign in at **www.cengage.com/owl** to:
- View tutorials and simulations, develop problem-solving skills, and complete online homework assigned by your professor.
- For quick review and exam prep, download Go Chemistry mini lecture modules from OWL (or purchase them at **www.cengagebrain.com**)

 Access **How Do I Solve It?** tutorials on how to approach problem solving using concepts in this chapter.

CHAPTER GOALS REVISITED

Now that you have studied this chapter, you should ask whether you have met the chapter goals. In particular, you should be able to:

Understand the common ion effect
a. Predict the effect of the addition of a "common ion" on the pH of the solution of a weak acid or base (Section 18.1). **Study Questions: 1–4, 109.**

Understand the control of pH in aqueous solutions with buffers
a. Describe the functioning of buffer solutions. (Section 18.2). **Go Chemistry Module 23.**
b. Use the Henderson–Hasselbalch equation (Equation 18.2) to calculate the pH of a buffer solution of given composition. **Study Questions: 5–8, 11–14.**
c. Describe how a buffer solution of a given pH can be prepared. **Study Questions: 9, 10, 15–18, 83, 90, 101, 102, 109.**
d. Calculate the pH of a buffer solution before and after adding acid or base. **Study Questions: 19–22.**

Evaluate the pH in the course of acid–base titrations

a. Predict the pH of an acid–base reaction at its equivalence point (Section 18.3; see also Sections 17.6 and 17.7). **Study Questions: 24–30, 74, 76, 98, 108.**

Acid	Base	pH at Equivalence Point
Strong	Strong	= 7 (neutral)
Strong	Weak	< 7 (acidic)
Weak	Strong	> 7 (basic)

b. Understand the differences between the titration curves for a strong acid–strong base titration and titrations in which one of the substances is weak.
c. Describe how an indicator functions in an acid–base titration. **Study Questions: 31, 97–99.**

Apply chemical equilibrium concepts to the solubility of ionic compounds

a. Write the equilibrium constant expression—relating concentrations of ions in solutions to K_{sp}—for any insoluble salt (Section 18.4). **Study Questions: 37, 38.**
b. Calculate K_{sp} values from experimental data (Section 18.4). **Study Questions: 39–44.**
c. Estimate the solubility of a salt from the value of K_{sp} (Section 18.4). **Study Questions: 45–48, 96, 105.**
d. Calculate the solubility of a salt in the presence of a common ion (Section 18.4). **Study Questions: 53–56.**
e. Understand how hydrolysis of basic anions affects the solubility of a salt (Section 18.4). **Study Questions: 57, 58, 112.**
f. Decide if a precipitate will form when the ion concentrations are known Section 18.5). **Study Questions: 59–62, 73, 80, 88, 89.**
g. Calculate the ion concentrations that are required to begin the precipitation of an insoluble salt (Section 18.5). **Study Questions: 63, 78, 87–89.**
h. Understand that the formation of a complex ion can increase the solubility of an insoluble salt (Sections 18.6 and 18.7). **Study Questions: 65–70, 86, 92.**

Key Equations

Equation 18.1 (page 624) Hydronium ion concentration in a buffer solution composed of a weak acid and its conjugate base.

$$[H_3O^+] = \frac{[acid]}{[conjugate\ base]} \times K_a$$

Equation 18.2 (page 625) Henderson–Hasselbalch equation. To calculate the pH of a buffer solution composed of a weak acid and its conjugate base.

$$pH = pK_a + \log\frac{[conjugate\ base]}{[acid]}$$

Equation 18.3 (page 633) Equation to calculate the hydronium ion concentration before the equivalence point in the titration of a weak acid with a strong base. See also Equation 18.4 for the version of the equation based on the Henderson–Hasselbalch equation.

$$[H_3O^+] = \frac{[weak\ acid\ remaining]}{[conjugate\ base\ produced]} \times K_a$$

Equation 18.5 (page 633) The relationship between the pH of the solution and the pK_a of the weak acid (or $[H_3O^+]$ and K_a) at the halfway or half-neutralization point in the titration of a weak acid with a strong base (or of a weak base with a strong acid).

$$[H_3O^+] = K_a \text{ and } pH = pK_a$$

Equation 18.6 (page 641) The general equilibrium constant expression, K_{sp} (solubility constant) for dissolving a poorly soluble salt, A_xB_y.

$$A_xB_y(s) \rightleftharpoons x\ A^{y+}(aq) + y\ B^{x-}(aq) \qquad K_{sp} = [A^{y+}]^x[B^{x-}]^y$$

OWL Questions and problems for this chapter are available in OWL. Use the OWL access card that came with this textbook to access assigned questions and problems for this chapter.

19 Principles of Chemical Reactivity: Entropy and Free Energy

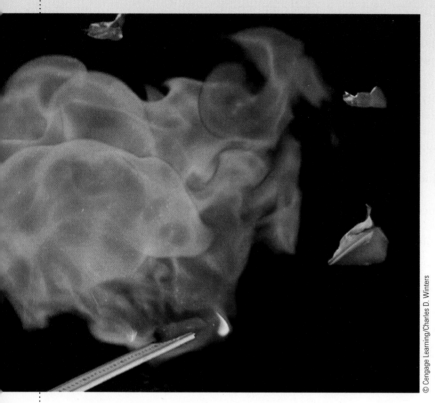

© Cengage Learning/Charles D. Winters

The Combustion of Hydrogen Gas. The exothermic reaction of hydrogen with oxygen to form water. (A hydrogen-filled balloon was ignited with a candle.)

Hydrogen for the Future?

In Chapter 5, you learned that the first law of thermodynamics states that energy is conserved in the universe. In this chapter, you will learn about the second law of thermodynamics. Two of the key concepts involved in the second law are (1) energy goes from being more concentrated to being more dispersed in the universe and (2) it is not possible to have 100% of the energy in a process be transferred as work. Energy is at the forefront of discussions of public policy today, and these fundamental laws of thermodynamics are key to understanding these issues.

One of the best energy sources for the future could be hydrogen. One could imagine a cycle in which water is decomposed by running an electric current through it to produce hydrogen and oxygen, and then the hydrogen and oxygen are recombined in a very exothermic process to yield water. Based on the first law of thermodynamics, however,

the best that we could hope for is to get the same exact amount of energy back from the process as we had to put in to decompose the water. The second law of thermodynamics makes things even worse. Because we cannot transform 100% of the energy released into useful work, the energy that we could use from the combustion step would be less than what would be needed to generate the hydrogen in the first place.

Is there a way around this problem? One widely used way to produce hydrogen is the reaction of carbon or methane with steam, for which there is a net release of energy when we account for both the energy needed to obtain the hydrogen and the energy released when the hydrogen is burned. Another potential solution is to use a renewable energy source such as solar or wind energy to decompose water. The laws of thermodynamics still apply but it is not as important in such a case that less energy is obtained than was needed because the input of energy came from an abundant and renewable energy resource.

Scientists are actively seeking ways to make hydrogen a useful, economical, and safe fuel for the future. Underlying this search, chemistry and the laws of thermodynamics play essential roles.

Questions:

1. Calculate $\Delta_r H°$, $\Delta_r S°$, and $\Delta_r G°$ for the reaction (at 298.15 K)
   $$2\ H_2O(\ell) \rightarrow 2\ H_2(g) + O_2(g)$$
2. Calculate $\Delta_r H°$, $\Delta_r S°$ and $\Delta_r G°$ for the reaction (at 298.15 K)
   $$CH_4(g) + H_2O(g) \rightarrow 3\ H_2(g) + CO(g)$$

Answers to these questions are available in Appendix N.

CHAPTER OUTLINE

19.1 Spontaneity and Energy Transfer as Heat

19.2 Dispersal of Energy: Entropy

19.3 Entropy: A Microscopic Understanding

19.4 Entropy Measurement and Values

19.5 Entropy Changes and Spontaneity

19.6 Gibbs Free Energy go **Chemistry**

19.7 Calculating and Using Free Energy

CHAPTER GOALS

See Chapter Goals Revisited (page 686) for Study Questions keyed to these goals.

- Understand the concept of entropy and its relationship to reaction spontaneity.

- Calculate the change in entropy for a system, its surroundings, and the universe to determine whether a process is spontaneous under standard conditions.

- Understand and use the Gibbs free energy.

Change is central to chemistry, so it is important to understand the factors that determine whether a change will occur. In chemistry, we encounter many examples of chemical change (chemical reactions) and physical change (the formation of mixtures, expansion of gases, and changes of state, to name a few). Chemists use the term **spontaneous** to represent a change that occurs without outside intervention. *Spontaneous changes occur only in the direction that leads to equilibrium.* Whether or not the process is spontaneous does not tell us anything about the rate of the change or the extent to which a process will occur before equilibrium is reached. It says only that the change will occur in a specific direction (toward equilibrium) and will occur naturally and unaided.

If a piece of hot metal is placed in a beaker of cold water, energy is transferred as heat spontaneously from the hot metal to the cooler water (Figure 19.1), and energy transfer will continue until the two objects are at the same temperature and thermal equilibrium is attained. Similarly, chemical reactions proceed spontaneously until equilibrium is reached, regardless of whether the position of the equilibrium favors products or reactants. We readily recognize that, starting with pure reactants, all product-favored reactions must be spontaneous, such as the formation of water from gaseous hydrogen and oxygen and the neutralization of $H_3O^+(aq)$ and $OH^-(aq)$. Be aware, though, that reactant-favored reactions are also spontaneous until equilibrium is achieved. Even though the dissolution of $CaCO_3$ is reactant-favored at equilibrium, if you place a handful of $CaCO_3$ in water, the process of dissolving will proceed spontaneously until equilibrium is reached.

Systems never change spontaneously in a direction that takes them farther from equilibrium. Given two objects at the same temperature, in contact but thermally isolated from their surroundings, it will never happen that one will heat up while the other becomes colder. Gas molecules will never spontaneously congregate at one end of a flask. Similarly, once equilibrium is established, the small amount of dissolved $CaCO_3$ in equilibrium with solid $CaCO_3$ will not spontaneously precipitate from solution, nor will a greater amount of $CaCO_3$ spontaneously dissolve.

The factors that determine the directionality and extent of change are among the topics of this chapter.

19.1 Spontaneity and Energy Transfer as Heat

We can readily recognize many chemical reactions that are spontaneous, such as hydrogen and oxygen combining to form water, methane burning to give CO_2 and H_2O, Na and Cl_2 reacting to form NaCl, and HCl(aq) and NaOH(aq) reacting to

© Cengage Learning/Charles D. Winters

FIGURE 19.1 A spontaneous process. The heated metal cylinder is placed in water. Energy transfers as heat spontaneously from the metal to water, that is, from the hotter object to the cooler object.

To understand the thermodynamic concepts introduced in this chapter, be sure to review the ideas of Chapter 5.

System: The part of the universe under study.

Surroundings: The rest of the universe exclusive of the system, capable of exchanging energy and/or matter with the system.

Exothermic: Energy transfers as heat from the system to the surroundings.

Endothermic: Energy transfers as heat from the surroundings to the system.

A Review of Concepts of Thermodynamics

First law of thermodynamics: The law of conservation of energy; energy cannot be created or destroyed. The change in internal energy of a system is the sum of energy transferred into or out of the system as heat and/or as work, $\Delta U = q + w$.

Enthalpy change: The energy transferred as heat under conditions of constant pressure.

State function: A quantity whose value depends only on the state of the system;

changes in a state function can be calculated by taking into account the initial and final states of a system.

Standard conditions: Pressure of 1 bar (1 bar = 0.98692 atm) and solution concentration of 1 m.

Standard enthalpy of formation, $\Delta_f H°$: The enthalpy change occurring when 1 mol of a compound is formed from its elements in their standard states.

● **Spontaneous Processes** A spontaneous physical or chemical change proceeds to equilibrium without outside intervention. Such a process may or may not be product-favored at equilibrium.

form H_2O and $NaCl(aq)$. A common feature of these reactions is that they are exothermic, so it would be tempting to conclude that evolution of energy as heat is the criterion that determines whether a reaction or process is spontaneous. Further inspection, however, reveals significant flaws in this reasoning. This is especially evident with the inclusion of some common spontaneous changes that are endothermic or energy neutral:

- *Dissolving NH_4NO_3.* The ionic compound NH_4NO_3 dissolves spontaneously in water. The process is endothermic ($\Delta_r H° = +25.7$ kJ/mol).
- *Expansion of a gas into a vacuum.* A system is set up with two flasks connected by a valve (Figure 19.2). One flask is filled with a gas, and the other is evacuated. When the valve is opened, the gas will flow spontaneously from one flask to the other until the pressure is the same throughout. The expansion of an ideal gas is energy neutral (although expansion of most real gases is endothermic).
- *Phase changes.* Melting of ice is an endothermic process. Above 0 °C, the melting of ice is spontaneous. Below 0 °C, melting of ice is not spontaneous. At 0 °C, no net change will occur; liquid water and ice coexist at equilibrium. This example illustrates that temperature can have a role in determining spontaneity and that equilibrium is somehow an important aspect of the problem.
- *Energy transfer as heat.* The temperature of a cold soft drink sitting in a warm environment will rise until the beverage reaches the ambient temperature. The energy required for this process comes from the surroundings. Energy transfer as heat from a hotter object (the surroundings) to a cooler object (the soft drink) is spontaneous.
- The reaction of H_2 and I_2 to form HI is endothermic, and the reverse reaction, the decomposition of HI to form H_2 and I_2, is exothermic. If $H_2(g)$ and $I_2(g)$ are mixed, a reaction forming HI will occur [$H_2(g) + I_2(g) \rightleftharpoons 2\,HI(g)$] until equilibrium is reached. Furthermore, if HI(g) is placed in a container, there will also be a reaction, but in the reverse direction, until equilibrium is achieved. Notice that approach to equilibrium occurs spontaneously from either direction.

On further reflection, it is logical to conclude that evolution of heat cannot be a sufficient criterion in determining spontaneity. The first law of thermodynamics tells us that in any process energy must be conserved. If energy is transferred out of a system, then the same amount of energy must be transferred to the surroundings. Exothermicity of the system is always accompanied by an endothermic change in the surroundings. If energy evolution were the only factor determining whether a change is spontaneous, then for every spontaneous process there would be a corresponding nonspontaneous change in the surroundings. We must search further than the first law of thermodynamics to determine whether a change is spontaneous.

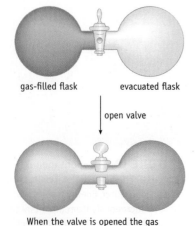

gas-filled flask   evacuated flask

open valve

When the valve is opened the gas expands irreversibly to fill both flasks.

FIGURE 19.2 Spontaneous expansion of a gas.

19.2 Dispersal of Energy: Entropy

We have shown that we cannot use energy itself as an indicator of spontaneity because energy is conserved in any process; we always end up with the same amount of energy as we had at the beginning. Imagine dropping this book on the floor. (But don't actually drop it!) It would fall spontaneously. The initial potential energy it has from being a certain distance above the floor is converted to kinetic energy. When the book hits the floor, the kinetic energy of the book is converted into other forms of kinetic energy: acoustic energy and thermal energy of the book, floor, and air, since these are all heated up somewhat. This example leads to two important points. First, the total energy is conserved. Second, there is a directionality to this process. The book will spontaneously fall to the ground, but we will not observe a book on its own jump spontaneously from the floor up to a desk. Is there a way to predict this directionality?

Let us consider the initial and final states for this process. Initially, the energy is concentrated in the book—the book has a certain potential energy as it is held above the floor. At the end of the process, this energy has been dispersed to the air, floor, and book. The energy has gone from being concentrated to being more dispersed. This is the indicator for which we have been searching. *In a spontaneous process, energy goes from being more concentrated to being more dispersed.*

There is a state function (◄ Section 5.4) called **entropy (S)** that allows us to quantify this. The **second law of thermodynamics** states that *a spontaneous process is one that results in an increase of the entropy of the universe.* In a spontaneous process ΔS(universe) is greater than zero; this corresponds to energy being dispersed in the process.

Because thermal energy is caused by the random motion of particles, potential energy is dispersed when it is converted to thermal energy. This conversion occurs when energy is transferred as heat, q. It is therefore not surprising that q is a part of the mathematical definition of ΔS. In addition, the effect of a given quantity of energy transferred as heat on energy dispersal is different at different temperatures. It turns out that a given q has a greater effect on ΔS at a lower temperature than at a higher temperature; that is, the extent of energy dispersal is inversely proportional to the temperature.

Our proposed definition for ΔS is thus q/T, but this is still not quite correct. We must be a little more specific about q. The value of q used in the calculation of an entropy change must be the energy transferred as heat under what are called *reversible conditions*, which we symbolize as q_{rev} (see *A Closer Look: Reversible and Irreversible Processes,* page 664). Our mathematical definition of ΔS is therefore q_{rev} divided by the absolute (Kelvin) temperature:

$$\Delta S = \frac{q_{rev}}{T}$$ (19.1)

As expected from this equation, the units for entropy are J/K.

• **Entropy** For a more complete discussion of entropy, see the papers by F. L. Lambert, such as "Entropy Is Simple, Qualitatively," *Journal of Chemical Education*, Vol. 79, pp. 1241–1246, 2002, and references therein. See also Lambert's site: **entropysite.oxy.edu**

• **Second Law of Thermodynamics** For a spontaneous process, ΔS (universe) > 0.

A CLOSER LOOK

Reversible and Irreversible Processes

To determine entropy changes experimentally, the energy transferred by heating and cooling must be measured for a reversible process. But what is a reversible process?

The test for reversibility is that after carrying out a change along a given path (for example, energy added as heat), it must be possible to return to the starting point by the same path (energy taken away as heat) without altering the surroundings. Melting of ice and freezing of water at 0 °C are examples of reversible processes. Given a mixture of ice and water at equilibrium, adding energy as heat in small increments will convert ice to water; removing energy as heat in small increments will convert water back to ice.

Reversibility is closely associated with equilibrium. Assume that we have a system at equilibrium. Reversible changes can be made by very slightly perturbing the equilibrium and letting the system readjust.

Spontaneous processes are often not reversible. Suppose a gas is allowed to expand into a vacuum. No work is done in this process because there is no force to resist expansion. To return the system to its original state, it is necessary to compress the gas. Doing so means doing work on the system, however, because the system will not return to its original state on its own. In this process, the energy content of the surroundings decreases by the amount of work expended by the surroundings. The system can be restored to its original state, but the surroundings will be altered in the process.

In summary, there are two important points concerning reversibility:

- At every step along a reversible pathway between two states, the system remains at equilibrium.
- Spontaneous processes often follow irreversible pathways and involve nonequilibrium conditions.

To determine the entropy change for a process, it is necessary to identify a reversible pathway. Only then can an entropy change for the process be calculated from q_{rev} and the Kelvin temperature.

REVIEW & CHECK FOR SECTION 19.2

1. In a spontaneous process, ΔS(universe) is

    (a) < 0 (b) $= 0$ (c) > 0

2. Which of the following is true for a spontaneous process but not for a nonspontaneous process? Energy in the universe is

    (a) concentrated   (b) conserved   (c) dispersed   (d) not conserved

3. For a particular system, q_{rev} at 25 °C is equal to + 950. J. What is its value of ΔS?

    (a) 38 J/K   (b) 3.19 J/K   (c) −3.19 J/K   (d) −38 J/K

19.3 Entropy: A Microscopic Understanding

Entropy is a measure of the extent of energy dispersal. In all spontaneous physical and chemical processes, energy changes from being localized or concentrated to being more dispersed or spread out. In a spontaneous process, the change in entropy, ΔS, of the universe indicates the extent to which energy is dispersed in a process carried out at a temperature T. So far, however, we have not explained why dispersal of energy occurs, nor have we given an equation for entropy itself. In order to do this, we will need to consider energy in its quantized form and matter on the atomic level.

Dispersal of Energy

We can explore the dispersal of energy using a simple example: energy being transferred as heat between hot and cold gaseous atoms. Consider an experiment involving two containers, one holding hot atoms and the other with cold atoms. Because they have translational energy, the atoms move randomly in each container and collide with the walls. When the containers are in contact, energy is transferred through the container walls. Eventually, both containers will be at the same temperature; the energy originally localized in the hotter atoms is distributed over a greater number of atoms; and the atoms in each container will have the same distribution of energies.

For further insight, we shall use a statistical explanation to show why energy is dispersed in a system. With statistical arguments, systems must include large

Possible distribution of energy packets

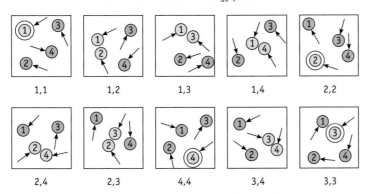

1,1   1,2   1,3   1,4   2,2

2,4   2,3   4,4   3,4   3,3

FIGURE 19.3 Energy dispersal. Possible ways of distributing two packets of energy among four atoms. To keep our analysis simple, we assume that initially there is one atom with two quanta of energy (1) and three atoms (2, 3 and 4) with no energy. There are 10 different ways to distribute the two quanta of energy among the four atoms.

numbers of particles for the arguments to be accurate. It will be easiest, however, if we look first at simple examples to understand the underlying concept and then extrapolate our conclusions to larger systems.

Consider a simple system in which, initially, there is one atom (1) with two discrete packets, or quanta, of energy and three other atoms (2, 3, and 4) with no energy (Figure 19.3). When these four atoms are brought together, the total energy in the system is 2 quanta. Collisions among the atoms allow energy to be transferred so that, over time, all distributions of the two packets of energy over the four atoms are seen. There are 10 different ways to distribute these 2 quanta of energy over the four atoms. Each of these 10 different ways to distribute the energy is called a **microstate**. In only one of these microstates do the 2 quanta remain on atom 1. In fact, only in 4 of the 10 microstates [1,1; 2,2; 3,3; and 4,4] is the energy concentrated on a single atom. In the majority of cases, 6 out of 10, the energy is distributed to two different atoms. Even in this small sample (four atoms) with only two packets of energy, it is more likely that at any given time the energy will be distributed to two atoms rather than concentrated on a single atom. There is a distinct preference that the energy will be dispersed over a greater number of atoms.

Let us now add more atoms to our system. We again begin with one atom (1) having 2 quanta of energy but now have five other atoms (2, 3, 4, 5, and 6) with no energy. Collisions let the energy be transferred between the atoms, and we now find there are 21 possible microstates (Figure 19.4). There are six microstates in which the energy is concentrated on one atom, including one in which the energy is still on atom 1, but there are now 15 out of 21 microstates (or 71.4%) in which the energy is present on two different atoms. As the number of particles increases, therefore, the number of microstates available increases dramatically, and the fraction of microstates in which the energy is concentrated rather than dispersed goes down dramatically. It is much more likely that the energy will be dispersed rather than concentrated.

Number of Microstates	Distribution of 2 Quanta of Energy Among Six Atoms					
6	1:1	2:2	3:3	4:4	5:5	6:6
5	1:2	2:3	3:4	4:5	5:6	
4	1:3	2:4	3:5	4:6		
3	1:4	2:5	3:6			
2	1:5	2:6				
1	1:6					

FIGURE 19.4 Distributing 2 quanta of energy among six atoms. (Atoms are labeled 1 to 6.) There are 21 ways—21 microstates—of distributing 2 quanta of energy among six atoms.

Now let us return to an example using a total of four atoms but increase the quantity of energy from 2 quanta to 6 quanta. Assume that we start with two atoms having 3 quanta of energy each. The other two atoms initially have zero energy (Figure 19.5). Through collisions, energy can be transferred to achieve different distributions of energy among the four atoms. In all, there are 84 microstates, falling into nine basic patterns. For example, one possible arrangement has one atom with 3 quanta of energy, and three atoms with 1 quantum each. There are four

FIGURE 19.5 Energy dispersal. Possible ways of distributing 6 quanta of energy among four atoms. A total of 84 microstates is possible.

Number of different ways to achieve this arrangement

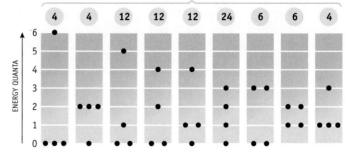

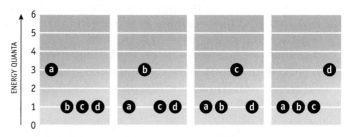

(a) Initially, four particles are separated from each other. Two particles each have 3 quanta of energy, and the other two have none. A total of 6 quanta of energy will be distributed once the four particles interact.

(b) Once the particles begin to interact, there are nine ways to distribute the 6 available quanta. Each of these arrangements will have multiple ways of distributing the energy among the four atoms. Part (c) shows how the arrangement on the right can be achieved four ways.

(c) There are four different ways to have four particles (a, b, c, and d) such that one particle has 3 quanta of energy and the other three each have 1 quantum of energy.

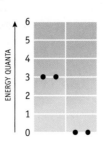

Ludwig Boltzmann (1844–1906). Engraved on his tombstone in Vienna, Austria, is his equation defining entropy. The constant *k* is now known as Boltzmann's constant.

microstates in which this is true (Figure 19.5c). Increasing the number of quanta from 2 to 6 with the same number of atoms increased the number of possible microstates from 10 to 84. In this case, increasing the amount of energy that is dispersed resulted in an increase in the number of microstates.

Statistical analyses for larger aggregates of atoms and energy quanta become increasingly complex, but the conclusions are even more compelling. As the number of particles and/or quanta increases, the number of energy microstates grows rapidly. A larger number of microstates allows the extent to which the energy is dispersed and/or the amount of the energy dispersed to increase. Ludwig Boltzmann proposed that the entropy of a system (the dispersal of energy at a given temperature) results from the number of microstates available. *As the number of microstates increases, so does the entropy of the system.* He expressed this idea in the equation

$$S = k \ln W \tag{19.2}$$

● **How Many Microstates?** To give you a sense of the number of microstates available to a substance, consider a mole of ice at 273 K, where $S° = 41.3$ J/K · mol. Using Boltzmann's equation, we find that $W = 10^{1,299,000,000,000,000,000,000,000}$. That is, there are many, many more microstates for 1 mol of ice than there are atoms in the universe (about 10^{80}).

which states that the entropy of a system, *S*, is proportional to the natural logarithm of the number of accessible microstates, *W*, that belong to a given energy of a system or substance. (The proportionality constant, *k*, is now known as **Boltzmann's constant** and has a value of 1.381×10^{-23} J/K.) Within these microstates, it turns out that those states that disperse energy over the largest number of atoms are vastly more probable than the others.

Dispersal of Matter: Dispersal of Energy Revisited

In many processes, it appears that the dispersal of matter also contributes to spontaneity. We shall see, however, that these effects can also be explained in terms of energy dispersal. Let us examine a specific case. Matter dispersal was illustrated in Figure 19.2 by the expansion of a gas into a vacuum. How is this spontaneous expansion of a gas related to energy dispersal and entropy?

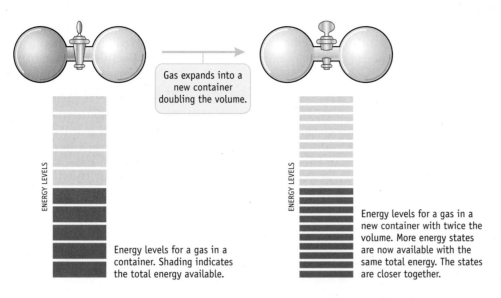

Gas expands into a new container doubling the volume.

ENERGY LEVELS

ENERGY LEVELS

Energy levels for a gas in a container. Shading indicates the total energy available.

Energy levels for a gas in a new container with twice the volume. More energy states are now available with the same total energy. The states are closer together.

FIGURE 19.6 Energy (and matter) dispersal. As the size of the container for the chemical or physical change increases, the number of microstates accessible to the atoms or molecules of the system increases, and the density of states increases. A consequence of the distribution of molecules over a greater number of microstates is an increase in entropy.

Note that for a gas in a container of the size likely to be found in a laboratory, the energy levels are so closely spaced that we do not usually think in terms of quantization of energy levels. For most purposes, the system can be regarded as having a continuum of energy levels.

We begin with the premise that all energy is quantized and that this applies to any system, including gas molecules in a room or in a reaction flask. You know from the previous discussion of kinetic-molecular theory that the molecules in a gas sample have a distribution of energies (◄ Figure 11.14) (often referred to as a *Boltzmann distribution*). The molecules are assigned to (or "occupy") quantized microstates, most of them in states near the average energy of the system, but fewer of them in states of high or low energy. (For a gas in a laboratory-sized container, the energy levels are so closely spaced that, for most purposes, there is a continuum of energy states.)

When the gas expands to fill a larger container, the average energy of the sample and the energy for the particles in a given energy range are constant. However, quantum mechanics shows (for now, you will have to take our word for it) that as a consequence of having a larger volume in which the molecules can move in the expanded state, there is an increase in the number of microstates and that those microstates are even more closely spaced than before (Figure 19.6). The result of this greater density of microstates is that the number of microstates available to the gas particles increases when the gas expands. Gas expansion, a dispersal of matter, leads to the dispersal of energy over a larger number of microstates and thus to an increase in entropy.

The logic applied to the expansion of a gas into a vacuum can be used to rationalize the mixing of two gases, the mixing of two liquids, or the dissolution of a solid in a liquid (Figure 19.7). For example, if flasks containing O_2 and N_2 are connected (in an experimental setup like that in Figure 19.6), the two gases diffuse together, eventually leading to a mixture in which O_2 and N_2 molecules are evenly distributed throughout the total volume. A mixture of O_2 and N_2 will never separate into samples of each component of its own accord. The gases spontaneously move toward a situation in which each gas and its energy are maximally dispersed. The energy of the system is dispersed over a larger number of microstates, and the entropy of the

● **Statistical Thermodynamics** The arguments presented here come from a branch of chemistry called *statistical thermodynamics*. See H. Jungermann, *Journal of Chemical Education*, Vol. 83, pp. 1686–1694, 2006.

Photos © Cengage Learning/Charles D. Winters

Time

FIGURE 19.7 Dissolving KMnO₄ in water. A small quantity of solid, purple $KMnO_4$ is added to water *(left)*. With time, the solid dissolves, and the highly colored MnO_4^- ions (and the K^+ ions) become dispersed throughout the solution. Entropy makes a large contribution to the mixing of liquids and solutions (page 477).

system increases. Indeed, this is a large part of the explanation for the fact that similar liquids (such as oil and gasoline or water and ethanol) will readily form homogeneous solutions. Recall the rule of thumb that "like dissolves like" (◀ Chapter 14).

A Summary: Entropy, Entropy Change, and Energy Dispersal

According to Boltzmann's equation (Equation 19.2), entropy is proportional to the number of ways that energy can be dispersed in a substance, that is, to the number of microstates available to the system (W). Thus, there will be a change in entropy, ΔS, if there is a change in the number of microstates over which energy can be dispersed.

$$\Delta S = S_{final} - S_{initial} = k \, (\ln W_{final} - \ln W_{initial}) = k \, \ln(W_{final}/W_{initial})$$

Our focus as chemists is on ΔS, and we shall be mainly concerned with the dispersion of energy in systems and surroundings during a physical or chemical change.

> **REVIEW & CHECK FOR SECTION 19.3**
>
> 1. As the number of microstates over which energy can be distributed in a system increases, its entropy
>
>   (a) decreases (b) increases (c) remains constant
>
> 2. Calculate the change in entropy for a system in going from a condition with 5 accessible microstates to 30 accessible microstates.
>
>   (a) -2.5×10^{-23} J/K (b) 9.4×10^{-24} J/K (c) 2.5×10^{-23} J/K (d) 8.3×10^{-23} J/K

19.4 Entropy Measurement and Values

For any substance under a given set of conditions, a numerical value for entropy can be determined. The greater the dispersal of energy, the greater the entropy and the larger the value of S. The point of reference for entropy values is established by the **third law of thermodynamics**. Defined by Ludwig Boltzmann, the third law states that *a perfect crystal at 0 K has zero entropy; that is, S = 0*. The entropy of an element or compound under any other set of conditions is the entropy gained by converting the substance from 0 K to those conditions. To determine the value of S, it is necessary to measure the energy transferred as heat under reversible conditions for the conversion from 0 K to the defined conditions and then to use Equation 19.1 ($\Delta S = q_{rev}/T$). Because it is necessary to add energy as heat to raise the temperature, *all substances have positive entropy values at temperatures above 0 K*. Negative values of entropy cannot occur. Recognizing that entropy is directly related to energy added as heat allows us to predict several general features of entropy values:

- Raising the temperature of a substance corresponds to adding energy as heat. Thus, the entropy of a substance will increase with an increase in temperature.
- Conversions from solid to liquid and from liquid to gas typically require large inputs of energy as heat. Consequently, there is a large increase in entropy in conversions involving changes of state (Figure 19.8).

Standard Entropy Values, S°

We introduced the concept of standard states into the earlier discussion of enthalpy (◀ Chapter 5), and we can similarly define the entropy of any substance in its standard state. The **standard molar entropy**, S°, of a substance is the entropy gained by converting 1 mol of it from a perfect crystal at 0 K to standard state conditions (1 bar, 1 molal for a solution) at the specified temperature. The units for standard molar entropy values are J/K · mol. Generally, values of S° found in tables of data refer to a temperature of 298 K. Appendix L lists many standard molar entropies at 298 K. More extensive lists of S° values can be found in standard reference sources such as the NIST tables (**webbook.nist.gov**).

● **Entropy Change on Gas Expansion** The entropy change for a gas expansion can be calculated from

$$\Delta S = nR\ln(V_{final}/V_{initial})$$

At a given temperature, V is proportional to the number of microstates, so the equation is related to $k \ln(W_{final}/W_{initial})$.

● **Negative Entropy Values** A glance at thermodynamic tables indicates that ions in aqueous solution can and do have negative entropy values listed. However, these are not absolute entropies. For ions, the entropy of H^+(aq) is arbitrarily assigned a standard entropy of zero.

(a)

(b)

The entropy of liquid bromine, $Br_2(\ell)$, is 152.2 J/K · mol, and that for bromine vapor is 245.47 J/K · mol.

The entropy of ice, which has a highly ordered molecular arrangement, is smaller than the entropy of liquid water.

FIGURE 19.8 Entropy and states of matter.

Scanning a list of standard entropies (such as those in Appendix L) will show that *large molecules generally have larger entropies than small molecules.* With a larger molecule, there are more ways for the molecule to rotate and vibrate, which provides a larger number of energy microstates over which energy can be distributed. As an example, consider the standard entropies for methane (CH_4), ethane (C_2H_6), and propane (C_3H_8), whose values are 186.3, 229.2, and 270.3 J/K · mol, respectively. Also, *molecules with more complex structures have larger entropies than molecules with simpler structures.* The effect of molecular structure can also be seen when comparing atoms or molecules of similar molar mass: Gaseous argon, CO_2, and C_3H_8 have entropies of 154.9, 213.7, and 270.3 J/K · mol, respectively.

Tables of entropy values also show that *entropies of gases are larger than those for liquids, and entropies of liquids are larger than those for solids.* In a solid, the particles have fixed positions in the solid lattice. When a solid melts, these particles have more freedom to assume different positions, resulting in an increase in the number of microstates available and an increase in entropy. When a liquid evaporates, constraints due to forces between the particles nearly disappear, the volume increases greatly, and a large entropy increase occurs. For example, the standard entropies of $I_2(s)$, $Br_2(\ell)$, and $Cl_2(g)$ are 116.1, 152.2, and 223.1 J/K · mol, respectively.

Finally, as illustrated in Figure 19.8, *for a given substance, a large increase in entropy accompanies changes of state,* reflecting the relatively large energy transfer as heat required to carry out these processes (as well as the dispersion of energy over a larger number of available microstates). For example, the entropies of liquid and gaseous water are 65.95 and 188.84 J/K · mol.

Entropy Values at 25 °C for some Hydrocarbons

	$S°$ (J/K · mol)
methane	186.3
ethane	229.2
propane	270.3

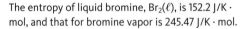

EXAMPLE 19.1   Entropy Comparisons

Problem Which substance has the higher entropy under standard conditions at 25 °C? Explain your reasoning. Check your answer against data in Appendix L.

(a) $NO_2(g)$ or $N_2O_4(g)$

(b) $I_2(g)$ or $I_2(s)$

What Do You Know? Entropy decreases in the order gas > liquid > solid, and larger molecules of related substances have greater entropies than smaller molecules.

Strategy For each part, identify the difference between the two substances and relate this to the general rules for entropy given above.

Solution

(a) Both NO_2 and N_2O_4 are gases. N_2O_4 is a larger molecule than NO_2 and so is expected to have the higher standard entropy.

(b) For a given substance, gases have higher entropies than solids, so $I_2(g)$ is expected to have the greater standard entropy.

Think about Your Answer $S°$ values in Appendix L confirm these predictions. At 25 °C, $S°$ for $NO_2(g)$ is 240.04 J/K · mol, and $S°$ for $N_2O_4(g)$ is 304.38 J/K · mol. $S°$ for $I_2(g)$ is 260.69 J/K · mol; $S°$ for $I_2(s)$ is 116.135 J/K · mol.

Check Your Understanding

Predict which substance in each par has the higher entropy and explain your reasoning.

(a) $O_2(g)$ or $O_3(g)$

(b) $SnCl_4(\ell)$ or $SnCl_4(g)$

Determining Entropy Changes in Physical and Chemical Processes

It is possible to use standard molar entropy values quantitatively to calculate the change in entropy that occurs in various processes under standard conditions. The standard entropy change for a reaction ($\Delta_r S°$) is the sum of the standard molar entropies of the products, each multiplied by its stoichiometric coefficient, minus the sum of the standard molar entropies of the reactants, each multiplied by its stoichiometric coefficient.

$$\Delta_r S° = \Sigma n S°(\text{products}) - \Sigma n S°(\text{reactants}) \qquad (19.3)$$

This equation allows us to calculate entropy changes for a *system* in which reactants are completely converted to products, under standard conditions. To illustrate, let us calculate $\Delta_r S°$ for the oxidation of NO with O_2.

$$2 \text{ NO(g)} + O_2(g) \rightarrow 2 \text{ NO}_2(g)$$

$\Delta_r S° = (2 \text{ mol NO}_2/\text{mol-rxn}) \, S°[\text{NO}_2(g)] -$
$\qquad \{(2 \text{ mol NO(g)/mol-rxn}) \, S°[\text{NO(g)}] + (1 \text{ mol } O_2/\text{mol-rxn}) \, S°[O_2(g)]\}$

$\qquad = (2 \text{ mol NO}_2/\text{mol-rxn})(240.0 \text{ J/K} \cdot \text{mol}) -$
$\qquad\qquad [(2 \text{ mol NO(g)/mol-rxn})(210.8 \text{ J/K} \cdot \text{mol}) + (1 \text{ mol } O_2/\text{mol-rxn})(205.1 \text{ J/K} \cdot \text{mol})]$

$\qquad = -146.7 \text{ J/K} \cdot \text{mol-rxn}$

The entropy of the system decreases, as is generally observed when some number of gaseous reactants has been converted to fewer molecules of gaseous products.

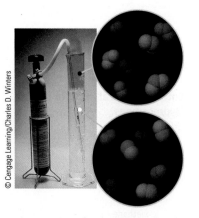

The reaction of NO with O_2. The entropy of the system decreases when two molecules of gas are produced from three molecules of gaseous reactants.

© Cengage Learning/Charles D. Winters

Strategy Map 19.2

PROBLEM

Calculate $\Delta_r S°$ for a reaction.

↓

DATA/INFORMATION

• Balanced chemical equation
• $S°$ values (Appendix L)

↓

STEP 1. Calculate $\Delta_r S°$ using Equation 19.3.

$\Delta_r S°$ for the reaction

OWL **INTERACTIVE EXAMPLE 19.2** Predicting and Calculating $\Delta_r S°$ for a Reaction

Problem Calculate the standard entropy changes for the following processes.

(a) Evaporation of 1.00 mol of liquid ethanol to ethanol vapor:

$$C_2H_5OH(\ell) \rightarrow C_2H_5OH(g)$$

(b) Formation of ammonia from hydrogen and nitrogen based on the following equation:

$$N_2(g) + 3 \text{ H}_2(g) \rightarrow 2 \text{ NH}_3(g)$$

What Do You Know? For each part, you are given a balanced chemical equation and asked to determine the standard entropy change for the reaction ($\Delta_r S°$). Values of standard molar entropies for the substances can be found in Appendix L.

Strategy Entropy changes for each system can be calculated from values of standard entropies (Appendix L) using Equation 19.3.

Solution

(a) Evaporation of ethanol

$\Delta_rS° = \Sigma nS°(\text{products}) - \Sigma nS°(\text{reactants})$

$= (1 \text{ mol } C_2H_5OH(g)/\text{mol-rxn}) \, S°[C_2H_5OH(g)] - (1 \text{ mol } C_2H_5OH(\ell)/\text{mol-rxn}) \, S°[C_2H_5OH(\ell)]$

$= (1 \text{ mol } C_2H_5OH(g)/\text{mol-rxn})(282.70 \text{ J/K} \cdot \text{mol}) - (1 \text{ mol } C_2H_5OH(\ell)/\text{mol-rxn})(160.7 \text{ J/K} \cdot \text{mol})$

$= +122.0 \text{ J/K} \cdot \text{mol-rxn}$

(b) Formation of ammonia

$\Delta_rS° = \Sigma nS°(\text{products}) - \Sigma nS°(\text{reactants})$

$= (2 \text{ mol } NH_3(g)/\text{mol-rxn}) \, S°[NH_3(g)] -$
$\{(1 \text{ mol } N_2(g)/\text{mol-rxn}) \, S°[N_2(g)] + (3 \text{ mol } H_2(g)/\text{mol-rxn}) \, S°[H_2(g)]\}$

$= (2 \text{ mol } NH_3(g)/\text{mol-rxn})(192.77 \text{ J/K} \cdot \text{mol}) -$
$[(1 \text{ mol } N_2(g)/\text{mol-rxn})(191.56 \text{ J/K} \cdot \text{mol}) + (3 \text{ mol } H_2(g)/\text{mol-rxn})(130.7 \text{ J/K} \cdot \text{mol})]$

$= -198.12 \text{ J/K} \cdot \text{mol-rxn}$

Think about Your Answer Predictions for the signs of these entropy changes can be made using the guidelines given in the text. In part (a), a large positive value for the entropy change is expected because the process converts ethanol from a liquid to a vapor. In part (b), a decrease in entropy is predicted because the number of moles of gases decreases from four to two.

● **Amount of Substance and Thermodynamic Calculations** In the calculation here and in all others in this chapter, when we write, for example,

$282.70 \text{ J/K} \cdot \text{mol}$

for the standard entropy of ethanol at 298 K, we mean

$282.70 \text{ J/K} \cdot \text{mol } C_2H_5OH(\ell)$

The identifying formula has been left off for the sake of simplicity.

Check Your Understanding

Calculate the standard entropy changes for the following processes using the entropy values in Appendix L. Are the signs of the calculated values of $\Delta_rS°$ in accord with predictions?

(a) Dissolving 1 mol of $NH_4Cl(s)$ in water: $NH_4Cl(s) \rightarrow NH_4Cl(aq)$

(b) Oxidation of ethanol: $C_2H_5OH(g) + 3 \, O_2(g) \rightarrow 2 \, CO_2(g) + 3 \, H_2O(g)$

REVIEW & CHECK FOR SECTION 19.4

1. Without doing any calculations, identify which of the following lists the materials in order of increasing entropy.

   (a) $H_2O(\ell) < NaCl(s) < NH_3(g)$ (c) $NaCl(s) < H_2O(\ell) < NH_3(g)$

   (b) $H_2O(\ell) < NH_3(g) < NaCl(s)$ (d) $NH_3(g) < H_2O(\ell) < NaCl(s)$

2. Without doing any calculations, predict the sign of $\Delta_rS°$ for the following reaction:

   $$Zn(s) + 2 \, HCl(aq) \rightarrow ZnCl_2(aq) + H_2(g)$$

   (a) $\Delta_rS° < 0$ (b) $\Delta_rS° = 0$ (c) $\Delta_rS° > 0$

3. Calculate $\Delta_rS°$ for the following reaction at 25 °C.

   $$2 \, H_2(g) + O_2(g) \rightarrow 2 \, H_2O(\ell)$$

   (a) $-326.6 \text{ J/K} \cdot \text{mol-rxn}$ (c) $139.9 \text{ J/K} \cdot \text{mol-rxn}$

   (b) $-139.9 \text{ J/K} \cdot \text{mol-rxn}$ (d) $326.6 \text{ J/K} \cdot \text{mol-rxn}$

19.5 Entropy Changes and Spontaneity

As illustrated by Example 19.2, the standard entropy change *for the system* in a physical or chemical change can be either positive (evaporation of ethanol) or negative (synthesis of ammonia from nitrogen and hydrogen). How does this information contribute to determining the spontaneity of the process?

A List of Common Entropy-Favored Processes

The discussion to this point allows the listing of several general principles involving entropy changes:

- The entropy of a substance will increase in going from a solid to a liquid to a gas.
- The entropy of any substance increases as the temperature is raised. Energy

must be added to a system to increase its temperature (that is, $q > 0$), so q_{rev}/T is necessarily positive.

- The entropy of a gas increases with an increase in volume. A larger volume pro-

vides a larger number of energy states over which to disperse energy.

- Reactions that increase the number of moles of gases in a system are accompanied by an increase in entropy.

As discussed previously (page 663), spontaneity is determined by the *second law of thermodynamics*, which states that *a spontaneous process is one that results in an increase of entropy in the universe.* The universe has two parts: the system and its surroundings (◄ Section 5.1), and so it makes sense that the entropy change for the universe is the sum of the entropy changes for the system and the surroundings. Under standard conditions, the entropy change for the universe, $\Delta S°$(universe) is

$$\Delta S°(\text{universe}) = \Delta S°(\text{system}) + \Delta S°(\text{surroundings}) \tag{19.4}$$

The calculation in Example 19.2 gave us the entropy change under standard conditions for a system, only half of the information needed. Now we will have to determine how the change being studied affects the entropy of the surroundings so that we can then find the entropy change for the universe.

The value of $\Delta S°$(universe) calculated from Equation 19.4 is the entropy change when reactants are converted *completely* to products, with all species at standard conditions. *A process is spontaneous under standard conditions if $\Delta S°$ (universe) is greater than zero.* As an example of the determination of reaction spontaneity, let us calculate $\Delta S°$(universe) for the reaction currently used to manufacture methanol, CH_3OH.

$$CO(g) + 2\ H_2(g) \rightarrow CH_3OH(\ell)$$

- **Using $\Delta S°$(universe)** For a process that is spontaneous under standard conditions:

$$\Delta S°(\text{universe}) > 0$$

For a process at equilibrium under standard conditions:

$$\Delta S°(\text{universe}) = 0$$

For a process that is not spontaneous under standard conditions:

$$\Delta S°(\text{universe}) < 0$$

Note that these conclusions refer to the *complete* conversion of reactants to products.

If $\Delta S°$(universe) is positive, the conversion of 1 mol of $CO(g)$ and 2 mol of $H_2(g)$ to 1 mol of $CH_3OH(\ell)$ will be spontaneous under standard conditions.

Calculating $\Delta S°$(system) To calculate $\Delta S°$(system), we start by defining the system to include the reactants and products. This means that $\Delta S°$(system) corresponds to the entropy change for the reaction, $\Delta_r S°$. Calculation of this entropy change follows the procedure given in Example 19.2.

$$\Delta S°(\text{system}) = \Delta_r S° = \Sigma n S°(\text{products}) - \Sigma n S°(\text{reactants})$$

$$= (1\ \text{mol}\ CH_3OH(\ell)/\text{mol-rxn})\ S°[CH_3OH(\ell)] - \\ \{(1\ \text{mol}\ CO(g)/\text{mol-rxn})\ S°[CO(g)] + (2\ \text{mol}\ H_2(g)/\text{mol-rxn})\ S°[H_2(g)]\}$$

$$= (1\ \text{mol}\ CH_3OH(\ell)/\text{mol-rxn})(127.2\ \text{J/K} \cdot \text{mol}) - \\ [(1\ \text{mol}\ CO(g)/\text{mol-rxn})(197.7\ \text{J/K} \cdot \text{mol}) + \\ (2\ \text{mol}\ H_2(g)/\text{mol-rxn})(130.7\ \text{J/K} \cdot \text{mol})]$$

$$= -331.9\ \text{J/K} \cdot \text{mol-rxn}$$

A decrease in entropy for the system is expected because three moles of gaseous reactants are converted to one mole of a liquid product.

Calculating $\Delta S°$(surroundings) We now need to calculate the entropy change for the surroundings. Recall from Equation 19.1 that for a reversible change, ΔS is equal to q_{rev}/T. Under constant pressure conditions and assuming a reversible process, the entropy change in the surroundings results from the fact that the enthalpy change for the reaction ($q_{rev} = \Delta_r H$) affects the surroundings. That is, the energy associated with an exothermic chemical reaction is dispersed into the surroundings. Recognizing that $\Delta H°$(surroundings) $= -\Delta_r H°$(system), the entropy change for the surroundings can be calculated by the equation

$$\Delta S°(\text{surroundings}) = \Delta H°(\text{surroundings})/T = -\Delta_r H°(\text{system})/T$$

For the synthesis of methanol by the reaction given, the enthalpy change can be calculated from enthalpy of formation data using Equation 5.6 (page 198).

$$\Delta H°(\text{system}) = \Sigma n\Delta_f H°(\text{products}) - \Sigma n\Delta_f H°(\text{reactants})$$

$$= (1 \text{ mol } CH_3OH(\ell)/\text{mol-rxn}) \, \Delta_f H°[CH_3OH(\ell)] - \{(1 \text{ mol } CO(g)/\text{mol-rxn}) \, \Delta_f H°[CO(g)] + (2 \text{ mol } H_2(g)/\text{mol-rxn}) \, \Delta_f H°[H_2(g)]\}$$

$$= (1 \text{ mol } CH_3OH(\ell)/\text{mol-rxn})(-238.4 \text{ kJ/mol}) - [(1 \text{ mol } CO(g)/\text{mol-rxn})(-110.5 \text{ kJ/mol}) + (2 \text{ mol } H_2(g)/\text{mol-rxn})(0 \text{ kJ/ mol})]$$

$$= -127.9 \text{ kJ/mol-rxn}$$

Assuming that the process is reversible and occurs at a constant temperature and pressure, the entropy change for the surroundings in the methanol synthesis is +429.2 J/K · mol-rxn, calculated as follows.

$$\Delta S°(\text{surroundings}) = -\Delta_r H°(\text{system})/T$$

$$= -[(-127.9 \text{ kJ/mol-rxn})/298 \text{ K}](1000 \text{ J/kJ})$$

$$= +429.2 \text{ J/K} \cdot \text{mol-rxn}$$

Calculating $\Delta S°$(universe), the Entropy Change for the System and Surroundings

The pieces are now in place to calculate the entropy change in the universe. For the formation of $CH_3OH(\ell)$ from $CO(g)$ and $H_2(g)$, $\Delta S°$(universe) is

$$\Delta S°(\text{universe}) = \Delta S°(\text{system}) + \Delta S°(\text{surroundings})$$

$$= -331.9 \text{ J/K} \cdot \text{mol-rxn} + 429.2 \text{ J/K} \cdot \text{mol-rxn}$$

$$= +97.3 \text{ J/K} \cdot \text{mol-rxn}$$

The positive value indicates an increase in the entropy of the universe. It follows from the second law of thermodynamics that this reaction is spontaneous.

ⓌWL INTERACTIVE EXAMPLE 19.3 Determining Whether a Process Is Spontaneous

Problem Calculate $\Delta S°$(universe) for the process of dissolving NaCl in water at 298 K.

What Do You Know? The process occurring is $NaCl(s) \rightarrow NaCl(aq)$. $\Delta S°$(universe) is equal to the sum of $\Delta S°$(system) and $\Delta S°$(surroundings). Values of $S°$ and $\Delta_f H°$ for NaCl(s) and NaCl(aq) are given in Appendix L.

Strategy The entropy change for the system, $\Delta S°$(system), can be calculated from values of $S°$ for the two species using Equation 19.3. $\Delta_r H°$ can be calculated from values of $\Delta_f H°$ for the two species using Equation 5.6 (page 234). $\Delta S°$(surroundings) is determined by dividing $-\Delta_r H°$ for the process by the Kelvin temperature. The sum of $\Delta S°$(system) and $\Delta S°$(surroundings) is $\Delta S°$(universe). If this value is positive, then the process is spontaneous under standard conditions.

Solution

Calculate $\Delta S°$(system)

$$\Delta S°(\text{system}) = \Sigma n S°(\text{products}) - \Sigma n S°(\text{reactants})$$

$$= (1 \text{ mol } NaCl(aq)/\text{mol-rxn}) \, S°[NaCl(aq)] - (1 \text{ mol } NaCl(s)/\text{mol-rxn}) \, S°[NaCl(s)]$$

$$= (1 \text{ mol } NaCl(aq)/\text{mol-rxn})(115.5 \text{ J/K} \cdot \text{mol}) - (1 \text{ mol } NaCl(s)/\text{mol-rxn})(72.11 \text{ J/K} \cdot \text{mol})$$

$$= +43.4 \text{ J/K} \cdot \text{mol-rxn}$$

Calculate $\Delta S°$(surroundings)

$$\Delta_r H°(\text{system}) = \Sigma n\Delta_f H°(\text{products}) - \Sigma n\Delta_f H°(\text{reactants})$$

$$= (1 \text{ mol } NaCl(aq)/\text{mol-rxn}) \, \Delta_f H°[NaCl(aq)] - (1 \text{ mol } NaCl(s)/\text{mol-rxn}) \, \Delta_f H°[NaCl(s)]$$

$$= (1 \text{ mol } NaCl(aq)/\text{mol-rxn})(-407.27 \text{ kJ/mol}) - (1 \text{ mol } NaCl(s)/\text{mol-rxn})(-411.12 \text{ kJ/mol})$$

$$= +3.85 \text{ kJ/mol-rxn}$$

Strategy Map 19.3

PROBLEM

Determine the value of $\Delta S°$(universe) for a chemical reaction.

KNOWN DATA/INFORMATION

- Balanced chemical equation
- Temperature
- Values of $S°$, $\Delta_f H°$ (Appendix L)

STEP 1. Calculate $\Delta_r S°$ using Equation 19.3.

$\Delta S°$(system)

STEP 2. Calculate $\Delta_r H°$ using Equation 5.6.
Change the sign of $\Delta_r H°$ and divide this by T (in K).

$\Delta S°$(surroundings)

STEP 3. Add $\Delta S°$(system) and $\Delta S°$(surroundings).

$\Delta S°$(universe)

The entropy change of the surroundings is determined by dividing $-\Delta_r H°(\text{system})$ by the Kelvin temperature.

$$\Delta S°(\text{surroundings}) = -\Delta_r H°(\text{system})/T$$

$$= [(-3.85 \text{ kJ/mol-rxn}/298 \text{ K})](1000 \text{ J/1 kJ})$$

$$= -12.9 \text{ J/K} \cdot \text{mol-rxn}$$

Calculate $\Delta S°$(universe)

The overall entropy change—the change of entropy in the universe—is the sum of the values for the system and the surroundings.

$$\Delta S°(\text{universe}) = \Delta S°(\text{system}) + \Delta S°(\text{surroundings})$$

$$= (+43.4 \text{ J/K} \cdot \text{mol-rxn}) + (-12.9 \text{ J/K} \cdot \text{mol-rxn})$$

$$= +30.5 \text{ J/K} \cdot \text{mol-rxn}$$

Think about Your Answer The sum of the two entropy quantities is positive, indicating that the entropy in the universe increases; thus, the process is spontaneous under standard conditions. Notice that the spontaneity of the process results from $\Delta S°$(system) and not from $\Delta S°$(surroundings).

Check Your Understanding

Is the reaction of hydrogen and chlorine to give hydrogen chloride gas predicted to be spontaneous under standard conditions (at 298 K)?

$$H_2(g) + Cl_2(g) \rightarrow 2 \text{ HCl}(g)$$

Predict the spontaneity of the reaction based on $\Delta_r H°$ and $\Delta_r S°$, and then calculate $\Delta S°$(universe) to verify your prediction.

● **$\Delta S°$(universe), Spontaneity, and Standard Conditions** It is important to reiterate that $\Delta H°$ and $\Delta S°$ values for a reaction are for the *complete* conversion of reactants to products under standard conditions. If $\Delta S°$(universe) is > 0, the reaction as written is spontaneous *under standard conditions*. However, one can calculate values for ΔS(universe) (without the superscript zero) for *nonstandard* conditions. If ΔS(universe) is > 0, the reaction is spontaneous under those conditions. However, it is possible that this same reaction under standard conditions is not spontaneous [$\Delta S°$(universe) < 0]. We will return to this point on pages 677–679.

In Summary: Spontaneous or Not?

In the preceding examples, predictions about the spontaneity of a process under standard conditions were made using values of $\Delta S°$(system) and $\Delta H°$(system) calculated from tables of thermodynamic data. It will be useful to look at all possibilities that result from the interplay of these two quantities. There are four possible outcomes when these two quantities are paired (Table 19.1). In two, $\Delta H°$(system) and $\Delta S°$(system) work in concert (Types 1 and 4 in Table 19.1). In the other two, the two quantities are opposed (Types 2 and 3).

Processes in which both the standard enthalpy and entropy changes favor energy dispersal (Type 1) are always spontaneous under standard conditions. Processes disfavored by both their standard enthalpy and entropy changes in the system (Type 4) can *never* be spontaneous under standard conditions. Let us consider examples that illustrate each situation.

Table 19.1 Predicting Whether a Reaction Will Be Spontaneous Under Standard Conditions

Reaction Type	$\Delta H°$(system)	$\Delta S°$(system)	Spontaneous Process? (Standard Conditions)
1	Exothermic, < 0	Positive, > 0	Spontaneous at all temperatures. $\Delta S°$(universe) > 0.
2	Exothermic, < 0	Negative, < 0	Depends on relative magnitudes of $\Delta H°$ and $\Delta S°$. Spontaneous at lower temperatures.
3	Endothermic, > 0	Positive, > 0	Depends on relative magnitudes of $\Delta H°$ and $\Delta S°$. Spontaneous at higher temperatures.
4	Endothermic, > 0	Negative, < 0	Not spontaneous at any temperature. $\Delta S°$(universe) < 0.

Combustion reactions are always exothermic and often produce a larger number of gaseous product molecules from a few reactant molecules. They are Type 1 reactions. The equation for the combustion of butane is an example.

$$2 \, C_4H_{10}(g) + 13 \, O_2(g) \rightarrow 8 \, CO_2(g) + 10 \, H_2O(g)$$

For this reaction, $\Delta_rH° = -5315.1$ kJ/mol · rxn, and $\Delta_rS° = 312.4$ J/K · mol-rxn. Both contribute to this reaction being spontaneous under standard conditions.

Hydrazine, N_2H_4, is used as a high-energy rocket fuel. Synthesis of N_2H_4 from gaseous N_2 and H_2 would be attractive because these reactants are inexpensive.

$$N_2(g) + 2 \, H_2(g) \rightarrow N_2H_4(\ell)$$

However, this reaction fits into the Type 4 category. The reaction is endothermic ($\Delta_rH° = +50.63$ kJ/mol-rxn), and the entropy change is negative ($\Delta_rS° = -331.4$ J/K·mol-rxn) (1 mol of liquid is produced from 3 mol of gases), so the reaction is not spontaneous under standard conditions, and complete conversion of reactants to products will not occur without outside intervention.

In the two other possible outcomes, entropy and enthalpy changes oppose each other. A process could be favored by the enthalpy change but disfavored by the entropy change (Type 2), or vice versa (Type 3). In either instance, whether a process is spontaneous depends on which factor is more important.

Temperature also influences the value of $\Delta S°$(universe). Because the enthalpy change for the surroundings is divided by the temperature to obtain $\Delta S°$(surroundings), the numerical value of $\Delta S°$(surroundings) will be smaller (either less positive or less negative) at higher temperatures. In contrast, $\Delta S°$(system) and $\Delta H°$(system) do not vary much with temperature. Thus, the effect of $\Delta S°$(surroundings) relative to $\Delta S°$(system) is diminished at higher temperature. Stated another way, at higher temperature, the enthalpy change becomes a less important factor in determining the overall entropy change. Consider the two cases where $\Delta H°$(system) and $\Delta S°$(system) are in opposition (Table 19.1):

- Type 2: Exothermic processes with $\Delta S°$(system) < 0. Such processes become less favorable with an increase in temperature.
- Type 3: Endothermic processes with $\Delta S°$(system) > 0. These processes become more favorable as the temperature increases.

The effect of temperature is illustrated by two examples. The first is the reaction of N_2 and H_2 to form NH_3. The reaction is exothermic, and thus it is favored by energy dispersal to the surroundings. The entropy change for the system is unfavorable, however, because the reaction, $N_2(g) + 3 \, H_2(g) \rightarrow 2 \, NH_3(g)$, converts four moles of gaseous reactants to two moles of gaseous products. The favorable enthalpy effect [$\Delta_rS°$(surroundings) $= -\Delta_rH°$(system)/T] becomes less important at higher temperatures. It is therefore reasonable to expect that the reaction will not be spontaneous if the temperature is high enough.

The second example considers the thermal decomposition of NH_4Cl (Figure 19.9). At room temperature, NH_4Cl is a stable, white, crystalline salt. When heated strongly, it decomposes to $NH_3(g)$ and $HCl(g)$. The reaction is endothermic (enthalpy-disfavored) but entropy-favored because of the formation of two moles of gas from one mole of a solid reactant. The reaction is increasingly favored at higher temperatures.

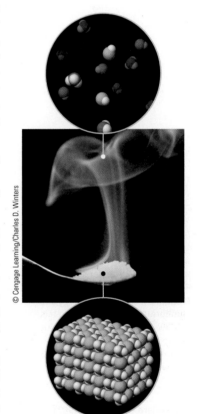

© Cengage Learning/Charles D. Winters

FIGURE 19.9 Thermal decomposition of $NH_4Cl(s)$. White, solid ammonium chloride, $NH_4Cl(s)$, is heated in a spoon. At high temperatures, decomposition to form $NH_3(g)$ and $HCl(g)$ is spontaneous. At lower temperatures, the reverse reaction, forming $NH_4Cl(s)$, is spontaneous. As gaseous $HCl(g)$ and $NH_3(g)$ cool, they recombine to form solid NH_4Cl, the white "smoke" seen in this photo.

REVIEW & CHECK FOR SECTION 19.5

1. Calculate $\Delta_rS°$ for the following reaction at 25 °C:

$$N_2(g) + 2 \, O_2(g) \rightarrow 2 \, NO_2(g)$$

(a) −480.1 J/K · mol-rxn

(b) −121.6 J/K · mol-rxn

(c) 121.6 J/K · mol-rxn

(d) 480.1 J/K · mol-rxn

2. Calculate $\Delta S°$(universe) for the following reaction at 25.0 °C:

$$C(graphite) + O_2(g) \rightarrow CO_2(g)$$

(a) -1317 J/K · mol-rxn

(c) 4.4 J/K · mol-rxn

(b) 3.1 J/K · mol-rxn

(d) 1320 J/K · mol-rxn

3. If $\Delta_r H° = +467.9$ kJ/mol-rxn and $\Delta_r S° = +560.7$ J/K · mol-rxn for the following reaction

$$2\ Fe_2O_3(s) + 3\ C(graphite) \rightarrow 4\ Fe(s) + 3\ CO_2(g)$$

then, under standard conditions, this reaction will be spontaneous

(a) at all temperatures

(c) at lower temperatures

(b) at higher temperatures

(d) at no temperature

19.6 Gibbs Free Energy

The method used so far to determine whether a process is spontaneous required evaluation of two quantities, $\Delta S°$(system) and $\Delta S°$(surroundings). Wouldn't it be convenient to have a single thermodynamic function that serves the same purpose? A function associated with the system only—one that does not require assessment of the surroundings—would be even better. Such a function exists. It is called the **Gibbs free energy**, with the name honoring J. Willard Gibbs (1839–1903). Gibbs free energy, G, often referred to simply as "free energy," is defined mathematically as

$$G = H - TS$$

where H is enthalpy, T is the Kelvin temperature, and S is entropy. In this equation, G, H, and S all refer to the system. Because enthalpy and entropy are state functions (◀ Section 5.4), free energy is also a state function.

Every substance possesses free energy, but the actual quantity is seldom known. Instead, just as with enthalpy (H) and internal energy (U), we are concerned with *changes* in free energy, ΔG, that occur in chemical and physical processes.

Let us first see how to use free energy as a way to determine whether a reaction is spontaneous. We can then ask further questions about the meaning of the term "free energy" and its use in deciding whether a reaction is product- or reactant-favored.

The Change in the Gibbs Free Energy, ΔG

Module 24: Gibbs Free Energy and Equilibrium covers concepts in this section.

Recall the equation defining the entropy change for the universe:

$$\Delta S(\text{universe}) = \Delta S(\text{surroundings}) + \Delta S(\text{system})$$

The entropy change of the surroundings equals the negative of the change in enthalpy of the system divided by T. Thus,

$$\Delta S(\text{universe}) = -\Delta H(\text{system})/T + \Delta S(\text{system})$$

Multiplying through this equation by $-T$, gives the equation

$$-T\Delta S(\text{universe}) = \Delta H(\text{system}) - T\Delta S(\text{system})$$

● **J. Willard Gibbs (1839–1903)** Gibbs received a Ph.D. from Yale University in 1863. His was the first Ph.D. in science awarded from an American university.

Gibbs defined the free energy function so that ΔG(system) $= -T\Delta S$(universe). Thus, the general expression relating changes in free energy to the enthalpy and entropy changes in the system is the following.

$$\Delta G = \Delta H - T\Delta S$$

Under standard conditions, we can rewrite this, the Gibbs free energy equation, as

$$\Delta G° = \Delta H° - T\Delta S° \qquad (19.5)$$

Gibbs Free Energy, Spontaneity, and Chemical Equilibrium

Because $\Delta_r G^\circ$ is related directly to ΔS°(universe), the Gibbs free energy can be used as a criterion of spontaneity for physical and chemical changes. As shown earlier, the signs of $\Delta_r G^\circ$ and ΔS°(universe) will be opposites [$\Delta_r G^\circ$(system) $= -T\Delta S^\circ$(universe)]. Therefore, we find the following relationships:

$\Delta_r G^\circ < 0$ The process is spontaneous in the direction written under standard conditions.

$\Delta_r G^\circ = 0$ The process is at equilibrium under standard conditions.

$\Delta_r G^\circ > 0$ The process is not spontaneous in the direction written under standard conditions.

To better understand the Gibbs function, let us examine the diagrams in Figure 19.10. The free energy of pure reactants is plotted on the left, and the free energy of the pure products on the right. The extent of reaction, plotted on the x-axis, goes from zero (pure reactants) to one (pure products). In both cases in Figure 19.10, the free energy initially declines as reactants begin to form products; it reaches a minimum at equilibrium and then increases again as we move from the equilibrium position to pure products. *The free energy at equilibrium, where there is a mixture of reactants and products, is always lower than the free energy of the pure reactants and of the pure products. A reaction proceeds spontaneously toward the minimum in free energy, which corresponds to equilibrium.*

$\Delta_r G^\circ$ *is the change in free energy accompanying the chemical reaction in which the reactants are converted completely to the products under standard conditions.* Mathematically, it is the difference in free energy between the products and the reactants under standard conditions. If the free energy of the products is less than that of the reactants, then $\Delta_r G^\circ < 0$, and the reaction is spontaneous under standard conditions (Figure 19.10a). Conversely, if the free energy of the products is greater than that of the reactants, then $\Delta_r G^\circ$ is positive ($\Delta_r G^\circ > 0$), and the reaction is not spontaneous under standard conditions (Figure 19.10b).

Notice that in Figure 19.10a, the equilibrium position occurs closer to the product side than to the reactant side. This is a product-favored reaction at equilibrium. In Figure 19.10b, we find the opposite. The reaction is reactant-favored at equilibrium. It is no accident that the reaction with a negative $\Delta_r G^\circ$ is product-favored, whereas the one with the positive $\Delta_r G^\circ$ is reactant-favored at equilibrium. It turns out that this is always true as the discussion below will show.

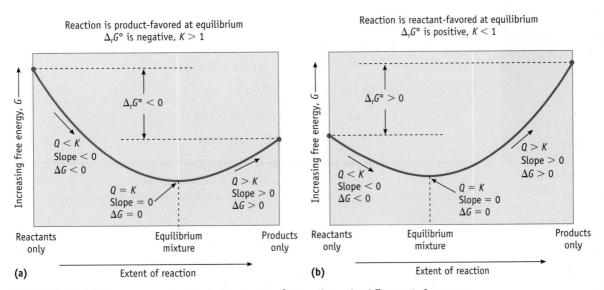

FIGURE 19.10 Free energy changes in the course of a reaction. The difference in free energy between the pure reactants in their standard states and the pure products in their standard states is $\Delta_r G^\circ$. Here, Q is the reaction quotient, and K is the equilibrium constant.

Now let us consider what happens to the instantaneous slope of the curve in Figure 19.10 as the reaction proceeds. Initially, this slope is negative, corresponding to a negative ΔG in moving from point to point. Eventually, however, the free energy reaches a minimum. At this point, the instantaneous slope of the graph is zero ($\Delta G = 0$) and the reaction has reached equilibrium. If we move past the equilibrium point, the instantaneous slope is positive ($\Delta G > 0$); proceeding further to products is not spontaneous. In fact, the reverse reaction will occur spontaneously; the reaction will once again proceed toward equilibrium.

The relationship of $\Delta_r G°$ (the value of $\Delta_r G$ under standard conditions) and the value of $\Delta_r G$ under nonstandard conditions is given by Equation 19.6:

$$\Delta_r G = \Delta_r G° + RT \ln Q \tag{19.6}$$

where R is the universal gas constant, T is the temperature in kelvins, and Q is the reaction quotient (◀ Section 16.2).

$$Q = \frac{[C]^c[D]^d}{[A]^a[B]^b} \quad \text{for } aA + bB \rightarrow cC + dD$$

Equation 19.6 informs us that, at a given temperature, $\Delta_r G$ is determined by the values of $\Delta_r G°$ and Q. When the system reaches equilibrium, no further net change in concentration of reactants and products will occur, and $\Delta_r G = 0$ and $Q = K$. Substituting these values into Equation 19.6 gives

$$0 = \Delta_r G° + RT \ln K \text{ (at equilibrium)}$$

Rearranging this equation leads to a useful relationship between the standard free energy change for a reaction and the equilibrium constant, K, Equation 19.7:

$$\Delta_r G° = -RT \ln K \tag{19.7}$$

From this equation, we learn that, when $\Delta_r G°$ is negative, K is greater than 1, and we say the reaction is *product-favored at equilibrium*. The more negative the value of $\Delta_r G°$, the larger the equilibrium constant. This makes sense because, as described in Chapter 16, large equilibrium constants are associated with product-favored reactions. The converse is also true: For *reactant-favored* reactions, $\Delta_r G°$ is positive, and K is less than 1. Finally, if $K = 1$ (a special set of conditions), then $\Delta_r G° = 0$.

Let us now see that Equation 19.6 can yield the relationsips between Q and K that we have been using since Chapter 16.

$$\Delta_r G = \Delta_r G° + RT \ln Q$$

Substituting $-RT \ln K$ for $\Delta_r G°$ (Equation 19.7) gives

$$\Delta_r G = -RT \ln K + RT \ln Q$$

This equation can be rearranged as follows:

$$\Delta_r G = RT (\ln Q - \ln K)$$

$$\Delta_r G/RT = \ln Q - \ln K$$

Furthermore, as long as the reaction is "descending" from the free energy of the pure reactants to the equilibrium position, $\Delta_r G$ is negative, implying that the reaction is spontaneous in the forward direction. Because $\Delta_r G$ is negative, we can write:

When $\Delta_r G < 0$, then $\ln Q - \ln K < 0$

Therefore, $\ln Q < \ln K$

and $Q < K$

This means that for a reaction to be spontaneous, $\Delta_r G$ must be negative (a spontaneous reaction), and Q must be less than K ($Q < K$), just as has been stated earlier. A similar analysis shows that if $\Delta_r G$ is positive, then $Q > K$.

A reactant-favored process. If a sample of yellow lead(II) iodide is placed in pure water, a small amount of the compound will dissolve spontaneously ($\Delta_r G < 0$ and $Q < K$) until equilibrium is reached. Because PbI_2 is quite insoluble ($K_{sp} = 9.8 \times 10^{-9}$), however, the process of dissolving the compound is reactant-favored at equilibrium. We may conclude, therefore, that the value of $\Delta_r G°$ is positive.

A Summary: Gibbs Free Energy ($\Delta_r G$ and $\Delta_r G°$), the Reaction Quotient (Q) and Equilibrium Constant (K), and Reaction Favorability

Let us summarize the relationships among $\Delta_r G°$, $\Delta_r G$, Q, and K.

- In Figure 19.10, you see that free energy decreases to a minimum as a system approaches equilibrium. The free energy of the mixture of reactants and products at equilibrium is always lower than the free energy of the pure reactants or of the pure products.
- When $\Delta_r G < 0$, the reaction is proceeding spontaneously toward equilibrium and $Q < K$.
- When $\Delta_r G > 0$, the reaction is beyond the equilibrium point and is not spontaneous in the forward direction. It will be spontaneous in the reverse direction; $Q > K$.
- When $\Delta_r G = 0$, the reaction is at equilibrium; $Q = K$.
- When $\Delta_r G° < 0$, the reaction is *spontaneous under standard conditions*. The system will proceed to an equilibrium position at which point the products will dominate in the reaction mixture because $K > 1$. That is, the reaction is *product-favored* at equilibrium.
- When $\Delta_r G° > 0$, the reaction is *not spontaneous under standard conditions*. The system will proceed to the equilibrium position at which point the reactants will dominate in the equilibrium mixture because $K < 1$. That is, the reaction is *reactant-favored* at equilibrium.
- For the special condition where $\Delta_r G° = 0$, the reaction is at equilibrium at standard conditions, with $K = 1$.

What Is "Free" Energy?

The term *free energy* was not arbitrarily chosen. In any given process, the free energy represents the maximum energy available to do useful work (mathematically, $\Delta G = w_{max}$). In this context, the word *free* means "available."

To illustrate the reasoning behind this relationship, consider a reaction carried out under standard conditions and in which energy is evolved as heat ($\Delta_r H° < 0$) and entropy decreases ($\Delta_r S° < 0$).

$$2 \text{ H}_2(g) + O_2(g) \rightarrow 2 \text{ H}_2O(g)$$

$$\Delta_r H° = -483.6 \text{ kJ/mol-rxn and } \Delta_r S° = -88.8 \text{ J/K} \cdot \text{mol-rxn}$$

$$\Delta_r G° = -483.6 \text{ kJ/mol-rxn} - (298 \text{ K})(-0.0888 \text{ kJ/mol-rxn}) = -457.2 \text{ kJ/mol-rxn}$$

At first glance, it might seem reasonable that all the energy released as heat (-483.6 kJ/mol-rxn) would be available. This energy could be transferred to the surroundings and would thus be available to do work. This is not the case, however. A negative entropy change in this reaction means that energy is less dispersed in the products than in the reactants. A portion of the energy released from the reaction must be used to reverse energy dispersal in the system; that is, to concentrate energy in the product. The energy left over is "free," or available to perform work. Here, the free energy change amounts to -457.2 kJ/mol-rxn.

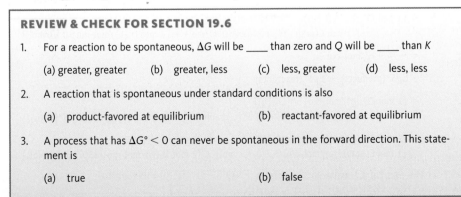

REVIEW & CHECK FOR SECTION 19.6

1. For a reaction to be spontaneous, ΔG will be ____ than zero and Q will be ____ than K

   (a) greater, greater   (b) greater, less   (c) less, greater   (d) less, less

2. A reaction that is spontaneous under standard conditions is also

   (a) product-favored at equilibrium   (b) reactant-favored at equilibrium

3. A process that has $\Delta G° < 0$ can never be spontaneous in the forward direction. This statement is

   (a) true   (b) false

19.7 Calculating and Using Free Energy

Standard Free Energy of Formation

The standard free energy of formation of a compound, $\Delta_f G°$, is the free energy change that occurs to form one mole of the compound from the component elements, with products and reactants in their standard states. By defining $\Delta_f G°$ in this way, *the free energy of formation of an element in its standard state is zero.*

Just as the standard enthalpy or entropy change for a reaction can be calculated using values of $\Delta_f H°$ (Equation 5.6) or $S°$ (Equation 19.3), the standard free energy change for a reaction can be calculated from values of $\Delta_f G°$ using a similar equation, where n represents the stoichiometric coefficient of the material in the balanced chemical equation under consideration:

$$\Delta_r G° = \Sigma n \Delta_f G°(\text{products}) - \Sigma n \Delta_f G°(\text{reactants}) \tag{19.8}$$

Calculating $\Delta_r G°$, the Free Energy Change for a Reaction Under Standard Conditions

The free energy change for a reaction under standard conditions can be calculated from thermodynamic data in two ways, either from standard enthalpy and entropy changes using values of $\Delta_f H°$ and $S°$ (as we did on the previous page for the formation of H_2O) or directly from values of $\Delta_f G°$ found in tables. These calculations are illustrated in the following two examples.

EXAMPLE 19.4 Calculating $\Delta_r G°$ from $\Delta_r H°$ and $\Delta_r S°$

Problem Calculate the standard free energy change, $\Delta_r G°$, for the formation of methane from carbon and hydrogen at 298 K, using tabulated values of $\Delta_f H°$ and $S°$. Is the reaction spontaneous under standard conditions? Is it product-favored or reactant-favored at equilibrium?

$$C(\text{graphite}) + 2\ H_2(g) \rightarrow CH_4(g)$$

What Do You Know? You are given a balanced chemical equation and asked to determine the standard free energy change for the reaction ($\Delta_r G°$). Values of standard molar enthalpies of formation and standard molar entropies for the substances can be found in Appendix L.

Strategy The values of $\Delta_f H°$ *and* $S°$ are first combined to find $\Delta_r H°$ *and* $\Delta_r S°$. With these values known, $\Delta_r G°$ can be calculated using Equation 19.5. When doing the calculation, remember that $S°$ values are given in units of J/K · mol, whereas $\Delta_f H°$ values are given in units of kJ/mol.

Solution

	C(graphite)	+	2 H₂(g)	→	CH₄(g)
$\Delta_f H°$ (kJ/mol)	0		0		−74.9
$S°$ (J/K·mol)	+5.6		+130.7		+186.3

From these values, we can find both $\Delta_r H°$ and $\Delta_r S°$ for the reaction:

$\Delta_r H° = \Sigma n \Delta_f H°(\text{products}) - \Sigma n \Delta_f H°(\text{reactants})$

$= (1\ \text{mol } CH_4(g)/\text{mol-rxn})\ \Delta_f H°[CH_4(g)] - \{(1\ \text{mol } C(\text{graphite})/\text{mol-rxn})\ \Delta_f H°[C(\text{graphite})] + (2\ \text{mol } H_2(g)/\text{mol-rxn})\ \Delta_f H°[H_2(g)]\}$

$= (1\ \text{mol } CH_4(g)/\text{mol-rxn})(-74.9\ \text{kJ/mol}) - [(1\ \text{mol } C(\text{graphite})/\text{mol-rxn})(0\ \text{kJ/mol}) + (2\ \text{mol } H_2(g)/\text{mol-rxn})(0\ \text{kJ/mol})]$

$= -74.9\ \text{kJ/mol-rxn}$

$\Delta_r S° = \Sigma n S°(\text{products}) - \Sigma n S°(\text{reactants})$

$= (1\ \text{mol } CH_4(g)/\text{mol-rxn})\ S°[CH_4(g)] - \{(1\ \text{mol } C(\text{graphite})/\text{mol-rxn})\ S°[C(\text{graphite})] + (2\ \text{mol } H_2(g)/\text{mol-rxn})\ S°[H_2(g)]\}$

$= (1\ \text{mol } CH_4(g)/\text{mol-rxn})(186.3\ \text{J/K} \cdot \text{mol}) - [1\ \text{mol } C(\text{graphite})/\text{mol-rxn}](5.6\ \text{J/K} \cdot \text{mol}) + (2\ \text{mol } H_2(g)/\text{mol-rxn})(130.7\ \text{J/K} \cdot \text{mol})]$

$= -80.7\ \text{J/K} \cdot \text{mol-rxn}$

Combining the values of $\Delta_r H°$ and $\Delta_r S°$ using Equation 19.5 gives $\Delta_r G°$.

$$\Delta_r G° = \Delta_r H° - T\Delta_r S°$$

$$= -74.9 \text{ kJ/mol-rxn} - [(298 \text{ K})(-80.7 \text{ J/K} \cdot \text{mol-rxn})](1 \text{ kJ/1000 J})$$

$$= -50.9 \text{ kJ/mol-rxn}$$

$\Delta_r G°$ is negative at 298 K, so the reaction is predicted to be spontaneous under standard conditions at this temperature. It is also predicted to be product-favored at equilibrium.

Think about Your Answer One of the most common mistakes students make is to forget to make sure that $\Delta_r H$ and $\Delta_r S$ are both in terms of the same energy unit (in this case, kJ). In this example, the product $T\Delta_r S°$ is negative (-24.0 J/mol-rxn) and disfavors the reaction. However, the entropy change is relatively small, and $\Delta_r H° = -74.9$ kJ/mol-rxn is the dominant term. Chemists call this an *enthalpy-driven reaction*.

Check Your Understanding

Using values of $\Delta_f H°$ and $S°$ to find $\Delta_r H°$ and $\Delta_r S°$, calculate the free energy change, $\Delta_r G°$, for the formation of 2 mol of $NH_3(g)$ from the elements at standard conditions and 25 °C.

$$N_2(g) + 3 \text{ H}_2(g) \rightarrow 2 \text{ NH}_3(g)$$

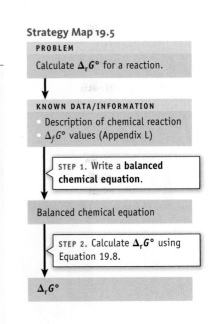

Strategy Map 19.5

PROBLEM

Calculate $\Delta_r G°$ for a reaction.

KNOWN DATA/INFORMATION

- Description of chemical reaction
- $\Delta_f G°$ values (Appendix L)

STEP 1. Write a **balanced chemical equation**.

Balanced chemical equation

STEP 2. Calculate $\Delta_r G°$ using Equation 19.8.

$\Delta_r G°$

⬛WL   **INTERACTIVE EXAMPLE 19.5**   Calculating $\Delta_r G°$ Using Free Energies of Formation

Problem Calculate the standard free energy change for the combustion of one mole of methane using values for standard free energies of formation of the products and reactants. Is the reaction spontaneous under standard conditions? Is it product-favored or reactant-favored at equilibrium?

What Do You Know? You are asked to determine the standard free energy change for a reaction ($\Delta_r G°$). Values of standard molar free energies of formation for the substances involved in the reaction can be found in Appendix L.

Strategy Write a balanced equation for the reaction. Then, use Equation 19.8 with values of $\Delta_f G°$ obtained from Appendix L.

Solution The balanced equation and values of $\Delta_f G°$ for each reactant and product are

$$CH_4(g) + 2 \text{ O}_2(g) \rightarrow 2 \text{ H}_2O(g) + CO_2(g)$$

$\Delta_f G°$(kJ/mol) -50.8 0 -228.6 -394.4

These values can then be substituted into Equation 19.8.

$$\Delta_r G° = \Sigma n\Delta_f G°(\text{products}) - \Sigma n\Delta_f G°(\text{reactants})$$

$$= \{(2 \text{ mol H}_2O(g)/\text{mol-rxn}) \, \Delta_f G°[\text{H}_2O(g)] + (1 \text{ mol CO}_2(g)/\text{mol-rxn}) \, \Delta_f G°[\text{CO}_2(g)]\}$$
$$- \{(1 \text{ mol CH}_4(g)/\text{mol-rxn}) \, \Delta_f G°[\text{CH}_4(g)] + (2 \text{ mol O}_2(g)/\text{mol-rxn}) \, \Delta_f G°[\text{O}_2(g)]\}$$

$$= [(2 \text{ mol H}_2O(g)/\text{mol-rxn})(-228.6 \text{ kJ/mol}) + (1 \text{ mol CO}_2(g)/\text{mol-rxn})(-394.4 \text{ kJ/mol})]$$
$$- [(1 \text{ mol CH}_4(g)/\text{mol-rxn})(-50.8 \text{ kJ/mol}) + (2 \text{ mol O}_2(g)/\text{mol-rxn})(0 \text{ kJ/mol})]$$

$$= -800.8 \text{ kJ/mol-rxn}$$

The large negative value of $\Delta_r G°$ indicates that the reaction is spontaneous under standard conditions and that it is product-favored at equilibrium.

Think about Your Answer Common errors made by students in this calculation are (1) overlooking the stoichiometric coefficients in the equation and (2) confusing the signs for the terms when using Equation 19.8.

Check Your Understanding

Calculate the standard free energy change for the oxidation of 1.00 mol of $SO_2(g)$ to form $SO_3(g)$ using values of $\Delta_f G°$.

Free Energy and Temperature

The definition of free energy, $G = H - TS$, informs us that free energy is a function of temperature, so $\Delta_r G°$ will change as the temperature changes (Figure 19.11). A consequence of this dependence on temperature is that, in certain instances, reactions can be product-favored at equilibrium at one temperature and reactant-favored at another. Those instances arise when the $\Delta_r H°$ and $T\Delta_r S°$ terms work in opposite directions:

- Processes that are entropy-favored ($\Delta_r S° > 0$) and enthalpy-disfavored ($\Delta_r H° > 0$)
- Processes that are enthalpy-favored ($\Delta_r H° < 0$) and entropy-disfavored ($\Delta_r S° < 0$)

Let us explore the relationship of $\Delta G°$ and T further and illustrate how it can be used to advantage.

Calcium carbonate is the primary component of limestone, marble, and seashells. Heating $CaCO_3$ produces lime, CaO, an important chemical, along with gaseous CO_2. The data below from Appendix L are at 298 K (25 °C).

	$CaCO_3(s)$	$\rightarrow$	$CaO(s)$	$+$	$CO_2(g)$
$\Delta_f G°$ (kJ/mol)	−1129.16		−603.42		−394.36
$\Delta_f H°$ (kJ/mol)	−1207.6		−635.09		−393.51
$S°$ (J/K · mol)	91.7		38.2		213.74

For the conversion of 1 mol of $CaCO_3(s)$ to 1 mol of $CaO(s)$ under standard conditions, $\Delta_r G° = +131.38$ kJ, $\Delta_r H° = +179.0$ kJ, and $\Delta_r S° = +160.2$ J/K. Although the reaction is entropy-favored, the large positive and unfavorable enthalpy change dominates at 298 K. Thus, the standard free energy change is positive at 298 K and 1 bar, indicating that the reaction is not spontaneous under standard conditions and is therefore reactant-favored at equilibrium.

The temperature dependence of $\Delta_r G°$ provides a means to turn the $CaCO_3$ decomposition into a product-favored reaction. Notice that the entropy change in the

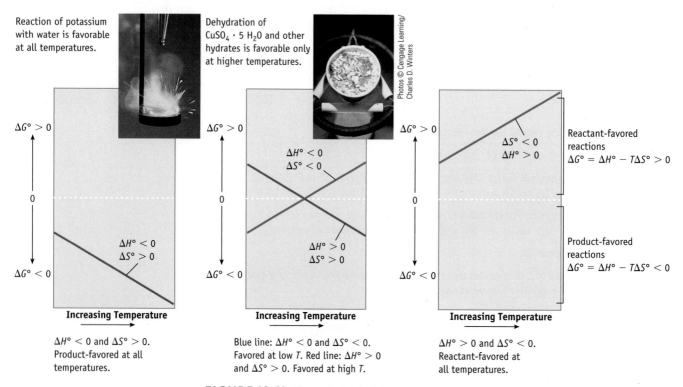

Reaction of potassium with water is favorable at all temperatures.

Dehydration of $CuSO_4 \cdot 5 H_2O$ and other hydrates is favorable only at higher temperatures.

Photos © Cengage Learning/ Charles D. Winters

$\Delta G° > 0$

$\Delta G° > 0$

$\Delta G° > 0$

$\Delta S° < 0$
$\Delta H° > 0$

Reactant-favored reactions
$\Delta G° = \Delta H° - T\Delta S° > 0$

0

0

0

$\Delta H° < 0$
$\Delta S° > 0$

$\Delta H° < 0$
$\Delta S° < 0$

$\Delta H° > 0$
$\Delta S° > 0$

Product-favored reactions
$\Delta G° = \Delta H° - T\Delta S° < 0$

$\Delta G° < 0$

$\Delta G° < 0$

$\Delta G° < 0$

Increasing Temperature

Increasing Temperature

Increasing Temperature

$\Delta H° < 0$ and $\Delta S° > 0$. Product-favored at all temperatures.

Blue line: $\Delta H° < 0$ and $\Delta S° < 0$. Favored at low T. Red line: $\Delta H° > 0$ and $\Delta S° > 0$. Favored at high T.

$\Delta H° > 0$ and $\Delta S° < 0$. Reactant-favored at all temperatures.

FIGURE 19.11 The variation in $\Delta_r G°$ with temperature.

CASE STUDY

Thermodynamics and Living Things

John C. Kotz

The laws of thermodynamics apply to all chemical reactions. It should come as no surprise, therefore, that issues of spontaneity and calculations involving ΔG also arise in studies of biochemical reactions. For biochemical processes, however, a different standard state is often used. Most of the usual definition is retained: 1 bar pressure for gases and 1 m concentration for aqueous solutes with the exception of one very important solute. Rather than using a standard state of 1 molal for hydronium ions (corresponding to a pH of about 0), biochemists use a hydronium concentration of 1×10^{-7} M, corresponding to a pH of 7. This pH is much more useful for biochemical reactions. When biochemists use this as the standard state, they write the symbol ′ next to the thermodynamic function. For example, they would write $\Delta G°'$ (pronounced *delta G zero prime*).

Living things require energy to perform their many functions. One of the main reactions involved in providing this energy is that of adenosine triphosphate (ATP) with water, a reaction for which $\Delta_r G°' = -30.5$ kJ/mol–rxn (◀ *The Chemistry of Life: Biochemistry*).

ATP, adenosine triphosphate

One of the key functions of the process of respiration is to produce molecules of ATP for our bodies to use. ATP is produced in the reaction of adenosine diphosphate (ADP) with hydrogen phosphate ($HP_i = HPO_4^{2-}$),

$$ADP + HP_i + H^+ \rightarrow ATP + H_2O$$
$$\Delta_r G°' = +30.5 \text{ kJ/mol-rxn}$$

a reaction that is not spontaneous under standard conditions and thus reactant-favored at equilibrium. How then do our bodies get this reaction to occur? The answer is to couple the production of ATP with another reaction that is even more product-favored than ATP production is reactant-favored. For example, organisms carry out the oxidation of carbohydrates in a multistep process, producing energy. One of the compounds produced in the process called *glycolysis* is phosphoenolpyruvate (PEP).

PEP, phosphoenolpyruvate

Its reaction with water is product-favored at equilibrium

$$PEP + H_2O \rightarrow Pyruvate + HP_i$$
$$\Delta_r G°' = -61.9 \text{ kJ/mol-rxn}$$

This reaction and ATP formation are linked through the HP_i that is produced in the PEP reaction. If both reactions are carried out, we obtain the following:

$$PEP + H_2O \rightarrow Pyruvate + HP_i$$
$$\Delta_r G°' = -61.9 \text{ kJ/mol-rxn}$$

$$ADP + HP_i + H^+ \rightarrow ATP + H_2O$$
$$\Delta_r G°' = +30.5 \text{ kJ/mol-rxn}$$

$$PEP + ADP + H^+ \rightarrow Pyruvate + ATP$$
$$\Delta_r G°' = -31.4 \text{ kJ/mol-rxn}$$

The overall reaction has a negative value for $\Delta_r G°'$ and thus is product-favored at equilibrium. ATP is formed in this process.

The coupling of reactions to produce a system that is product-favored is used in a multitude of reactions that occur in our bodies.

Questions:

1. Consider the hydrolysis reactions of creatine phosphate and adenosine-5′-monophosphate.

Creatine Phosphate + H_2O
$$\rightarrow \text{Creatine} + HP_i$$

$$\Delta_r G°' = -43.3 \text{ kJ/mol-rxn}$$

Adenosine-5′-Monophosphate + H_2O
$$\rightarrow \text{Adenosine} + HP_i$$

$$\Delta_r G°' = -9.2 \text{ kJ/mol-rxn}$$

Which of the following combinations produces a reaction that is product-favored at equilibrium: for creatine phosphate to transfer phosphate to adenosine or for adenosine-5′-monophosphate to transfer phosphate to creatine?

2. Assume the reaction A(aq) + B(aq) → C(aq) + H_3O^+(aq) produces one hydronium ion. What is the mathematical relationship between $\Delta G°'$ and $\Delta G°$ at 25 °C? (*Hint*: Use the equation $\Delta G = \Delta G° + RT \ln Q$ and substitute $\Delta G°'$ for ΔG.)

Answers to these questions are available in Appendix N.

reaction is positive as a result of the formation of CO_2 gas in the reaction. Thus, raising the temperature results in the value of $T\Delta_r S°$ becoming increasingly large. At a high enough temperature, $T\Delta_r S°$ will outweigh the enthalpy effect, and the process will become product-favored at equilibrium.

How high must the temperature be for this reaction to become product-favored? An estimate of the temperature can be obtained using Equation 19.5, by calculating the temperature at which $\Delta_r G° = 0$. Above that temperature, $\Delta_r G°$ will have a negative value.

$$\Delta_r G° = \Delta_r H° - T\Delta_r S°$$

$$0 = (179.0 \text{ kJ/mol-rxn})(1000 \text{ J/kJ}) - T(160.2 \text{ J/K} \cdot \text{mol-rxn})$$

$$T = 1117 \text{ K (or 844 °C)}$$

How accurate is this result? As noted earlier, this answer is only an estimate of the temperature needed. One source of error is the assumption that $\Delta_r H°$ and $\Delta S°$ do not vary with temperature, an assumption that is not strictly true. There is always a small variation in these values when the temperature changes—not large enough to be important if the temperature range is narrow, but potentially a problem over wider temperature ranges such as seen in this example. As an estimate, however, a temperature in the range of 850 °C for this reaction is reasonable.

● **CaCO₃ Decomposition** Experiments show that the pressure of CO_2 in an equilibrium system [$CaCO_3(s) \rightleftharpoons CaO(s) + CO_2(g)$] is 1 bar at about 900 °C ($K = 1$ and $\Delta G° = 0$), close to our estimated temperature.

ⓦWL INTERACTIVE EXAMPLE 19.6   Effect of Temperature on $\Delta_r G°$

Problem The decomposition of liquid $Ni(CO)_4$ to produce nickel metal and carbon monoxide has a $\Delta_r G°$ value of 40 kJ/mol-rxn at 25 °C.

$$Ni(CO)_4(\ell) \rightarrow Ni(s) + 4 \text{ CO}(g)$$

Use values of $\Delta_f H°$ and $S°$ for the reactant and products to estimate the temperature at which the reaction becomes product-favored at equilibrium.

What Do You Know? You are given a balanced chemical equation. Values of standard molar enthalpies of formation and standard molar entropies for the substances can be found in the chemical literature. You also know that the key temperature to determine is that at which $\Delta_r G°$ is zero.

Strategy The reaction is reactant-favored at equilibrium at 298 K. However, if the entropy change is positive for the reaction and the reaction is endothermic (with a positive value of $\Delta_r H°$), then a higher temperature may allow the reaction to become product-favored at equilibrium. Therefore, we first find $\Delta_r H°$ and $\Delta_r S°$ to see if their values meet our criteria for spontaneity at a higher temperature, and then we calculate the temperature at which the following condition is met: $0 = \Delta_r H° - T\Delta_r S°$.

Solution Values for $\Delta_f H°$ and $S°$ are obtained from the chemical literature for the substances involved.

	$Ni(CO)_4(\ell)$	$\rightarrow$	Ni(s)	+	4 CO(g)
$\Delta_f H°$(kJ/mol)	−632.0		0		−110.525
$S°$(J/K · mol)	320.1		29.87		197.67

For a process in which 1 mol of liquid $Ni(CO)_4$ is converted to 1 mol of Ni(s) and 4 mol of CO(g), we find

$$\Delta_r H° = +189.9 \text{ kJ/mol-rxn}$$

$$\Delta_r S° = +500.5 \text{ J/K mol-rxn}$$

We use these values of $\Delta_r H°$ and $\Delta_r S°$ to find the temperature at which $\Delta_r G° = 0$.

$$\Delta_r G° = \Delta_r H° - T\Delta_r S°$$

$$0 = (189.9 \text{ kJ/mol-rxn})(1000 \text{ J/kJ}) - T(500.5 \text{ J/K} \cdot \text{mol-rxn})$$

$$T = \boxed{379.4 \text{ K (or 106.2 °C)}}$$

Think about Your Answer At 298 K, the reaction is reactant-favored at equilibrium largely because it is quite endothermic. However, the positive entropy change allows the reaction to be product-favored at equilibrium at a higher temperature.

Check Your Understanding

Oxygen was first prepared by Joseph Priestley (1733–1804) by heating HgO. Use data in Appendix L to estimate the temperature required to decompose HgO(s) into Hg(ℓ) and $O_2(g)$.

Strategy Map 19.6

PROBLEM

Determine the **temperature** above which a chemical reaction is **product-favored** at **equilibrium**.

↓

KNOWN DATA/INFORMATION

● Balanced chemical equation
● Temperature
● Values of $\Delta_f H°$ and $S°$

↓

STEP 1. Calculate $\Delta_r H°$ using Equation 5.6.

$\Delta_r H°$

↓

STEP 2. Calculate $\Delta_r S°$ using Equation 19.3.

$\Delta_r S°$

↓

STEP 3. Use the equation $\Delta_r G° = \Delta_r H° - T\Delta_r S°$ to determine the temperature at which $\Delta_r G° = 0$.

↓

Temperature above which the reaction is **product-favored** at **equilibrium**.

Using the Relationship between $\Delta_r G°$ and K

Equation 19.7 provides a direct route to determine the standard free energy change from experimentally determined equilibrium constants. Alternatively, it allows calculation of an equilibrium constant from thermochemical data contained in tables or obtained from an experiment.

ⓌWL   INTERACTIVE EXAMPLE 19.7   Calculating K_p from $\Delta_r G°$

Problem Determine the standard free energy change, $\Delta_r G°$, for the formation of 1.00 mol of $NH_3(g)$ from nitrogen and hydrogen, and use this value to calculate the equilibrium constant for this reaction at 25 °C.

What Do You Know? You are given enough information so you can write a balanced chemical equation for the desired chemical reaction. Values of $\Delta_f G°$ are given in Appendix L.

Strategy The free energy of formation of ammonia represents the free energy change to form 1.00 mol of $NH_3(g)$ from the elements. The equilibrium constant for this reaction is calculated from $\Delta_r G°$ using Equation 19.7. Because the reactants and products are gases, the calculated value will be K_p.

Solution Begin by specifying a balanced equation for the chemical reaction under investigation.

$$\tfrac{1}{2} N_2(g) + \tfrac{3}{2} H_2(g) \rightleftharpoons NH_3(g)$$

The free energy change for this reaction is -16.37 kJ/mol-rxn ($\Delta_r G° = \Delta_f G°$ for $NH_3(g)$; Appendix L). In a calculation of K_p using Equation 19.7, we will need consistent units. If we use the gas constant, R, in units of 8.3145 J/K · mol, then the value of $\Delta_r G°$ must be in J/mol-rxn (and not kJ/mol-rxn). The temperature is 25 °C (298.15 K).

$$\Delta_r G° = -RT \ln K$$

$$-16{,}370 \text{ J/mol-rxn} = (-8.3145 \text{ J/K} \cdot \text{mol-rxn})(298.15 \text{ K}) \ln K_p$$

$$\ln K_p = 6.604$$

$$\boxed{K_p = 738}$$

Think about Your Answer The value of $\Delta_r G°$ is less than zero, indicating that this reaction is spontaneous under standard conditions and therefore product-favored at equilibrium. The value of K_p calculated is greater than 1 as it should be for such a process. This example illustrates how to calculate equilibrium constants from thermodynamic data. In fact, many equilibrium constants you find in the chemical literature are not experimentally determined but are instead calculated from thermodynamic data in this way.

Check Your Understanding

Determine the value of the equilibrium constant, K_p, for the decomposition of calcium carbonate to calcium oxide and carbon dioxide gas at 298.15 K.

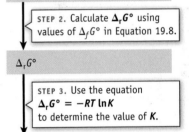

Strategy Map 19.7

PROBLEM

Determine the **equilibrium constant** for a reaction using **thermodynamic data**.

↓

KNOWN DATA/INFORMATION
- Information to write a balanced chemical equation
- The values of $\Delta_f G°$
- Temperature

STEP 1. Write the **balanced chemical equation**.

Balanced chemical equation

STEP 2. Calculate $\Delta_r G°$ using values of $\Delta_f G°$ in Equation 19.8.

$\Delta_r G°$

STEP 3. Use the equation $\Delta_r G° = -RT \ln K$ to determine the value of K.

Equilibrium constant, K

EXAMPLE 19.8   Calculating $\Delta_r G°$ from K_{sp} for an Insoluble Solid

Problem The value of K_{sp} for $AgCl(s)$ at 25 °C is 1.8×10^{-10}. Determine $\Delta_r G°$ for the process $Ag^+(aq) + Cl^-(aq) \rightleftharpoons AgCl(s)$ at 298.15 K.

What Do You Know You are given the value of K_{sp}. The process of interest is the reverse of the chemical equation for K_{sp}. You also know the temperature.

Strategy The chemical equation given is the opposite of the equation used to define K_{sp}; therefore, the equilibrium constant for this reaction is $1/K_{sp}$. This value is used in Equation 19.7 to calculate $\Delta_r G°$.

Solution

For $Ag^+(aq) + Cl^-(aq) \rightleftharpoons AgCl(s)$,

$$K = 1/K_{sp} = 1/1.8 \times 10^{-10} = 5.6 \times 10^9$$

$$\Delta_r G° = -RT \ln K = -(8.3145 \text{ J/K} \cdot \text{mol-rxn})(298.15 \text{ K}) \ln(5.6 \times 10^9)$$

$$= \boxed{-56 \text{ kJ/mol-rxn}}$$

Think about Your Answer The negative value of $\Delta_r G°$ indicates that the precipitation of AgCl from $Ag^+(aq)$ and $Cl^-(aq)$ is product-favored at equilibrium.

Check Your Understanding

Determine the value of $\Delta_r G°$ for the reaction $C(s) + CO_2(g) \rightleftharpoons 2\,CO(g)$ from data in Appendix L. Use this result to calculate the equilibrium constant.

REVIEW & CHECK FOR SECTION 19.7

1. Given that $\Delta_r H° = -2219$ kJ/mol-rxn and that $\Delta_r S° = -216$ J/K · mol-rxn at 25 °C, determine the value of $\Delta_r G°$ at 25 °C for the reaction

$$C_3H_8(g) + 5\,O_2(g) \rightarrow 3\,CO_2(g) + 4\,H_2O(\ell)$$

(a) −2283 kJ/mol-rxn
(b) −2214 kJ/mol-rxn
(c) −2155 kJ/mol-rxn
(d) 6.218×10^4 kJ/mol-rxn

2. Using values of $\Delta_f G°$, determine the value of $\Delta_r G°$ at 25 °C for the reaction

$$2\,KClO_3(s) \rightarrow 2\,KCl(s) + 3\,O_2(g)$$

(a) −225 kJ/mol-rxn
(b) −112 kJ/mol-rxn
(c) 112 kJ/mol-rxn
(d) 225 kJ/mol-rxn

3. The value of K_p for the following reaction at 425 °C is 0.018. What is the value of $\Delta_r G°$ at this temperature?

$$2\,HI(g) \rightleftharpoons H_2(g) + I_2(g)$$

(a) 1.0×10^1 kJ/mol-rxn
(b) −14 kJ/mol-rxn
(c) 23 kJ/mol-rxn
(d) 240 kJ/mol-rxn

OWL and go Chemistry

Sign in at **www.cengage.com/owl** to:

- View tutorials and simulations, develop problem-solving skills, and complete online homework assigned by your professor.
- For quick review and exam prep, download Go Chemistry mini lecture modules from OWL (or purchase them at **www.cengagebrain.com**)

Access **How Do I Solve It?** tutorials on how to approach problem solving using concepts in this chapter.

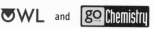

CHAPTER GOALS REVISITED

Now that you have studied this chapter, you should ask whether you have met the chapter goals. In particular, you should be able to:

Understand the concept of entropy and its relationship to reaction spontaneity

a. Understand that entropy is a measure of energy dispersal (Section 19.2). **Study Questions: 1, 2.**
b. Recognize that an entropy change is the energy transferred as heat for a reversible process divided by the Kelvin temperature. (*A Closer Look*, Section 19.2 and Equation 19.1) **Study Questions: 40–42.**
c. Identify common processes that are entropy-favored (Section 19.4). **Study Question: 3.**

Calculate the change in entropy for a system, its surroundings, and the universe to determine whether a process is spontaneous under standard conditions

a. Calculate entropy changes from tables of standard entropy values (Section 19.4). **Study Questions: 4–8.**
b. Use standard entropy and enthalpy changes to predict whether a reaction will be spontaneous under standard conditions (Section 19.5 and Table 19.1). **Study Questions: 9–12.**
c. Recognize how temperature influences whether a reaction is spontaneous (Section 19.5). **Study Questions: 13, 14, 23–26.**

Understand and use the Gibbs free energy

a.   Understand the connection between enthalpy and entropy changes and the Gibbs free energy change for a process (Section 19.6). **Study Questions: 15, 16.**

b.   Understand the relationship of Δ_rG, $\Delta_rG°$, Q, K, reaction spontaneity, and product- or reactant-favorability (Section 19.6).

Q	ΔG	Spontaneous?
$Q < K$	$\Delta G < 0$	Spontaneous to the right as the equation is written
$Q = K$	$\Delta G = 0$	Reaction is at equilibrium
$Q > K$	$\Delta G > 0$	Not spontaneous to the right; spontaneous to the left

K	ΔG°	Reactant-Favored or Product-Favored at Equilibrium?	Spontaneous Under Standard Conditions?
$K \gg 1$	$\Delta G° < 0$	Product-favored	Spontaneous under standard conditions
$K = 1$	$\Delta G° = 0$	$[C]^c[D]^d = [A]^a[B]^b$ at equilibrium	At equilibrium under standard conditions
$K \ll 1$	$\Delta G° > 0$	Reactant-favored	Not spontaneous under standard conditions

c.   Describe and use the relationship between the free energy change under standard conditions and equilibrium constants, and calculate K from $\Delta_rG°$ (Sections 19.6 and 19.7). **Study Questions: 27–30, 44, 48, 50, 61, and Go Chemistry Module 24.**

d.   Calculate the change in free energy at standard conditions for a reaction from the enthalpy and entropy changes under standard conditions or from the standard free energy of formation of reactants and products ($\Delta_fG°$) (Section 19.7). **Study Questions: 15–20, 46, 55, 56, 75, 79.**

e.   Know how free energy changes with temperature (Section 19.7). **Study Questions: 23–26, 57, 59, 70.**

Key Equations

Equation 19.1 (page 663) Calculate the entropy change from the energy transferred as heat for a reversible process and the temperature at which it occurs.

$$\Delta S = \frac{q_{rev}}{T}$$

Equation 19.2 (page 666) The Boltzmann equation: The entropy of a system, S, is proportional to the number of accessible microstates, W, belonging to a given energy of a system or substance.

$$S = k \ln W$$

Equation 19.3 (page 670) Calculate the standard entropy change under standard conditions for a process from the tabulated entropies of the products and reactants and the stoichiometric coefficients (n) of the substances in the balanced chemical equation for the reaction.

$$\Delta_rS° = \Sigma nS°(\text{products}) - \Sigma nS°(\text{reactants})$$

Equation 19.4 (page 672) Calculate the total entropy change for a system and its surroundings, to determine whether a process is spontaneous under standard conditions.

$$\Delta S°(\text{universe}) = \Delta S°(\text{system}) + \Delta S°(\text{surroundings})$$

Equation 19.5 (page 676) Calculate the free energy change for a process from enthalpy and entropy changes.

$$\Delta G° = \Delta H° - T\Delta S°$$

Equation 19.6 (page 678) Relates the free energy change under nonstandard conditions (Δ_rG) to the standard free energy change ($\Delta_rG°$) and the reaction quotient Q.

$$\Delta_rG = \Delta_rG° + RT \ln Q$$

Equation 19.7 (page 678): Relates the standard free energy change for a reaction and its equilibrium constant.

$$\Delta_r G° = -RT \ln K$$

Equation 19.8 (page 680): Calculate the standard free energy change for a reaction using tabulated values of $\Delta_f G°$ and the stoichiometric coefficients (n) of the substances in the balanced chemical equation for the reaction.

$$\Delta_r G° = \Sigma n \Delta_f G°(\text{products}) - \Sigma n \Delta_f G°(\text{reactants})$$

❂WL Questions and problems for this chapter are available in OWL. Use the OWL access card that came with this textbook to access assigned questions and problems for this chapter.

20 Principles of Chemical Reactivity: Electron Transfer Reactions

Courtesy General Motors/Orlando Sentinel/MCT

Battery Power

The 20th century was the century of the use and abuse of petroleum. The 21st century has to be the century of the discovery and use of alternative energy sources. Many of these will involve electrochemistry, the use of chemical reactions to produce electricity (batteries), or the use of electricity to produce chemical change (electrolysis).

The lithium-ion battery in your laptop computer or your phone, or the nickel-metal hydride battery in your hybrid car, works because redox reactions release chemical energy. (When you recharge the battery, you use electric energy to rejuvenate the chemicals in the battery.) But don't for a moment think that everything is understood about batteries or how to make them smaller, cheaper, or more efficient. One of the most important areas of research today is in battery technology, especially lithium-based batteries. Companies and universities around the world are searching for new battery designs that are lightweight, inexpensive, safe, environmentally kind, and that produce sufficient energy.

Electric vehicles. The forthcoming Chevrolet Volt will use lithium-ion batteries for propulsion. (For more on lithium-ion batteries see page 444.)

Questions:

1. The simplest way to write the reaction for discharge in a lithium-ion battery is

$$\text{Li(on carbon)(s)} + CoO_2(s) \rightarrow 6\ C(s) + LiCoO_2(s)$$

   (a) What are the oxidation numbers for cobalt in the two substances in the battery?
   (b) In such a battery, what reaction occurs at the cathode? At the anode?
   (c) An electrolyte is needed for ion conduction within the battery. From what you know about lithium chemistry (◄ page 256, for example), can the electrolyte in the battery be dissolved in water?

2. A lithium-ion camera battery is rated at 7500 mAh. That is, it can deliver 7500 milliamps (mA) or 7.5 amps of steady current for an hour.

   (a) How many moles of electrons can the battery deliver in one hour?
   (b) What mass of lithium is oxidized under these conditions in 1 hour?

Answers to these questions are available in Appendix N.

CHAPTER OUTLINE

20.1 Oxidation–Reduction Reactions **go Chemistry**

20.2 Simple Voltaic Cells

20.3 Commercial Voltaic Cells

20.4 Standard Electrochemical Potentials

20.5 Electrochemical Cells under Nonstandard Conditions

20.6 Electrochemistry and Thermodynamics

20.7 Electrolysis: Chemical Change Using Electrical Energy

20.8 Counting Electrons

CHAPTER GOALS

See Chapter Goals Revisited (page 731) for Study Questions keyed to these goals.

- Balance equations for oxidation–reduction reactions in acidic or basic solutions using the half-reaction approach.

- Understand the principles underlying voltaic cells.

- Understand how to use electrochemical potentials.

- Explore electrolysis, the use of electrical energy to produce chemical change.

Let us introduce you to electrochemistry and electron transfer reactions with a simple experiment. Place a piece of copper in an aqueous solution of silver nitrate. After a short time, metallic silver deposits on the copper, and the solution takes on the blue color typical of aqueous Cu^{2+} ions (Figure 20.1). The following oxidation–reduction (redox) reaction has occurred:

$$Cu(s) + 2\ Ag^+(aq) \rightarrow Cu^{2+}(aq) + 2\ Ag(s)$$

At the particulate level, Ag^+ ions in solution come into direct contact with the copper surface where the transfer of electrons occurs. Two electrons are transferred from a Cu atom to two Ag^+ ions. Copper ions, Cu^{2+}, enter the solution, and silver

OWL

Sign in to OWL at **www.cengage.com/owl** to view tutorials and simulations, develop problem-solving skills, and complete online homework assigned by your professor.

go Chemistry

Download mini lecture videos for key concept review and exam prep from OWL or purchase them from **www.cengagebrain.com**

A clean piece of copper wire will be placed in a solution of silver nitrate, $AgNO_3$.

Add $AgNO_3(aq)$

With time, the copper reduces Ag^+ ions to silver metal crystals, and the copper metal is oxidized to copper ions, Cu^{2+}.

After several days

The blue color of the solution is due to the presence of aqueous copper(II) ions.

Silver ions in solution

Surface of copper wire

Cu^{2+}

Photos © Cengage Learning/Charles D. Winters

FIGURE 20.1 The oxidation of copper by silver ions. Note that water molecules are not shown for clarity.

Step 2. *Separate the Process into Half-Reactions.*

Oxidation half-reaction: $Al(s) \rightarrow [Al(OH)_4]^-(aq)$
(Al oxidation number increased from 0 to +3.)

Reduction half-reaction: $H_2O(\ell) \rightarrow H_2(g)$
(H oxidation number decreased from +1 to 0.)

Step 3. *Balance the Half-Reactions for Mass.*

Addition of OH^- or OH^- and H_2O is required for mass balance in both half-reactions. In the case of the aluminum half-reaction, we simply add OH^- ions to the left side.

Oxidation half-reaction: $Al(s) + 4\ OH^-(aq) \rightarrow [Al(OH)_4]^-(aq)$

To balance the half-reaction for water reduction, notice that an oxygen-containing species must appear on the right side of the equation. Because H_2O is a reactant, we use OH^-, which is present in this basic solution, as the other product.

Reduction half-reaction: $2\ H_2O(\ell) \rightarrow H_2(g) + 2\ OH^-(aq)$

Step 4. *Balance the Half-Reactions for Charge.*

Electrons are added to balance charge.

Oxidation half-reaction: $Al(s) + 4\ OH^-(aq) \rightarrow [Al(OH)_4]^-(aq) + 3\ e^-$

Reduction half-reaction: $2\ H_2O(\ell) + 2\ e^- \rightarrow H_2(g) + 2\ OH^-(aq)$

Step 5. *Multiply the Half-Reactions by Appropriate Factors so That the Reducing Agent Donates as Many Electrons as the Oxidizing Agent Consumes.*

Here, electron balance is achieved by using 2 mol of Al to provide 6 mol of e^-, which are then acquired by 6 mol of H_2O.

Oxidation half-reaction: $2[Al(s) + 4\ OH^-(aq) \rightarrow [Al(OH)_4]^-(aq) + 3\ e^-]$

Reduction half-reaction: $3[2\ H_2O(\ell) + 2\ e^- \rightarrow H_2(g) + 2\ OH^-(aq)]$

Step 6. *Add the Half-Reactions.*

$$2\ Al(s) + 8\ OH^-(aq) \rightarrow 2\ [Al(OH)_4]^-(aq) + 6\ e^-$$

$$6\ H_2O(\ell) + 6\ e^- \rightarrow 3\ H_2(g) + 6\ OH^-(aq)$$

Net equation: $2\ Al(s) + 8\ OH^-(aq) + 6\ H_2O(\ell) \rightarrow 2\ [Al(OH)_4]^-(aq) + 3\ H_2(g) + 6\ OH^-(aq)$

Step 7. *Simplify by Eliminating Reactants and Products That Appear on Both Sides.*

Six OH^- ions can be canceled from the two sides of the equation:

$$2\ Al(s) + 2\ OH^-(aq) + 6\ H_2O(\ell) \rightarrow 2\ [Al(OH)_4]^-(aq) + 3\ H_2(g)$$

Think about Your Answer The final equation is balanced for mass and charge.

Mass balance: 2 Al, 14 H, and 8 O

Charge balance: There is a net −2 charge on each side.

Check Your Understanding

Voltaic cells based on the reduction of sulfur are under development. One such cell involves the reaction of sulfur with aluminum under basic conditions.

$$Al(s) + S(s) \rightarrow Al(OH)_3(s) + HS^-(aq)$$

(a)   Balance this equation, showing each balanced half-reaction.

(b)   Identify the oxidizing and reducing agents, the substance oxidized, and the substance reduced.

- Hydrogen balance can be achieved only with H^+/H_2O (in acid) or OH^-/H_2O (in base). Never add H or H_2 to balance hydrogen.
- Use H_2O or OH^- as appropriate to balance oxygen. Never add O atoms, O^{2-} ions, or O_2 for O balance.
- Never include $H^+(aq)$ and $OH^-(aq)$ in the same equation. A solution can be either acidic or basic, never both.

Balancing Oxidation–Reduction Equations—A Summary

- The number of electrons in a half-reaction reflects the change in oxidation number of the element being oxidized or reduced.
- Electrons are always a component of half-reactions but should never appear in the overall equation.

- Include charges in the formulas for ions. Omitting the charge, or writing the charge incorrectly, is one of the most common errors seen on student papers.
- The best way to become competent in balancing redox equations is to practice, practice, practice.

REVIEW & CHECK FOR SECTION 20.1

1. Copper(II) sulfide reacts with nitric acid according to the balanced equation:

$$3\ CuS(s) + 8\ H^+(aq) + 2\ NO_3^-(aq) \rightarrow 3\ Cu^{2+}(aq) + 3\ S(s) + 4\ H_2O(\ell) + 2\ NO(g)$$

   The substance oxidized is

   (a) CuS   (b) H^+ (c) NO_3^-

2. The balanced half-reaction for $NO_3^- \rightarrow NO$ in acid solution is

   (a) $NO_3^- \rightarrow NO + e^-$ (c) $NO_3^- \rightarrow NO + O^{2-} + e^-$

   (b) $2\ H^+ + e^- + NO_3^- \rightarrow NO + 2\ H_2O$ (d) $4\ H^+ + 3\ e^- + NO_3^- \rightarrow NO + 2\ H_2O$

3. The balanced half-reaction for $Br_2 \rightarrow BrO_3^-$ in basic solution is

   (a) $3\ OH^- + Br_2 \rightarrow 2\ BrO_3^- + H_2O + e^-$ (c) $e^- + OH^- + Br_2 \rightarrow 2\ BrO_3^- + H_2O$

   (b) $12\ OH^- + Br_2 \rightarrow 2\ BrO_3^- + 6\ H_2O + 10\ e^-$ (d) $10\ e^- + Br_2 + 6\ H_2O \rightarrow 2\ BrO_3^- + 6\ OH^-$

20.2 Simple Voltaic Cells

Let us use the reaction of copper metal and silver ions (Figure 20.1) as the basis of a voltaic cell. To do so, we place the components of the two half-reactions in separate compartments (Figure 20.5). This prevents the copper metal from transferring electrons directly to silver ions. Instead, electrons will be transferred through an external circuit, and useful work can potentially be done.

An Alternative Method of Balancing Equations in Basic Solution

Balancing redox equations in basic solution, which may require you to use OH^- ions and H_2O, can sometimes be more challenging than balancing equations in acidic solution. Rather than learn a separate method for balancing equations in basic solution, some find it easier to first balance an equation in basic solution as if it were in acidic solution and then add enough OH^- ions to both sides of the equation so that H^+ ions are converted to water. Let us show how this works for balancing the reaction of hypochlorite ion with manganese(IV) oxide to form chloride ion and permanganate ion.

(a) Balance the equation as if the reaction were carried out in acid by following Steps 1–7 in Example 20.2:

$$3\ ClO^-(aq) + H_2O(\ell) + 2\ MnO_2(s) \rightarrow$$
$$3\ Cl^-(aq) + 2\ MnO_4^-(aq) + 2\ H^+(aq)$$

(b) To adjust for the fact that this reaction is carried out in base rather than in acid, we will add the same number of hydroxide ions to both sides of the equation as there are $H^+(aq)$ ions. On the side that

had the $H^+(aq)$ ions, the $H^+(aq)$ ions and $OH^-(aq)$ ions will form $H_2O(\ell)$.

$$3\ ClO^-(aq) + H_2O(\ell) + 2\ MnO_2(s) + 2\ OH^-(aq) \rightarrow$$
$$3\ Cl^-(aq) + 2\ MnO_4^-(aq) + 2\ H^+(aq) + 2\ OH^-(aq)$$

$$3\ ClO^-(aq) + H_2O(\ell) + 2\ MnO_2(s) + 2\ OH^-(aq) \rightarrow$$
$$3\ Cl^-(aq) + 2\ MnO_4^-(aq) + 2\ H_2O(\ell)$$

(c) Simplify

$$3\ ClO^-(aq) + 2\ MnO_2(s) + 2\ OH^-(aq) \rightarrow$$
$$3\ Cl^-(aq) + 2\ MnO_4^-(aq) + H_2O(\ell)$$

The copper half-cell (on the left in Figure 20.5) consists of copper metal that serves as one electrode and a solution containing copper(II) ions. The half-cell on the right uses a silver electrode and a solution containing silver(I) ions. Important features of this simple cell are as follows:

- *The two half-cells are connected with a* **salt bridge** *that allows cations and anions to move between the two half-cells.* The electrolyte chosen for the salt bridge should contain ions that will not react with chemical reagents in both half-cells. In the example in Figure 20.5, $NaNO_3$ is used.
- *In all electrochemical cells, the* **anode** *is the electrode at which oxidation occurs. The electrode at which reduction occurs is always the* **cathode.** (In Figure 20.5, the copper electrode is the anode, and the silver electrode is the cathode.)
- *A negative sign can be assigned to the anode in a voltaic cell, and the cathode is marked with a positive sign.* The chemical oxidation occurring at the anode, which produces electrons, gives it a negative charge.
- *In all electrochemical cells, electrons flow in the external circuit from the anode to the cathode.* Electric current in the external circuit of a voltaic cell consists of electrons moving from the negative to the positive electrode.

The chemistry occurring in the cell pictured in Figure 20.5 is summarized by the following half-reactions and net ionic equation:

Cathode, reduction:	$2\ Ag^+(aq) + 2\ e^- \rightarrow 2\ Ag(s)$
Anode, oxidation:	$Cu(s) \rightarrow Cu^{2+}(aq) + 2\ e^-$
Net ionic equation:	$Cu(s) + 2\ Ag^+(aq) \rightarrow Cu^{2+}(aq) + 2\ Ag(s)$

• **Salt Bridges** A simple salt bridge can be made by adding gelatin to a solution of an electrolyte. Gelatin makes the contents semi-rigid so that the salt bridge is easier to handle. Porous glass disks and ion-permeable membranes are alternatives to a salt bridge. These devices allow ions to traverse from one half-cell to the other while keeping the two solutions from mixing.

The salt bridge is required in a voltaic cell for the reaction to proceed. In the Cu/Ag^+ voltaic cell, anions move in the salt bridge toward the copper half-cell, and cations move toward the silver half-cell (Figure 20.5). As $Cu^{2+}(aq)$ ions are formed in the copper half-cell by oxidation of copper metal, negative ions enter that cell from the salt bridge (and positive ions leave the cell), so that the numbers of positive and negative charges in the half-cell compartment remain in balance. The silver half-cell originally contained $AgNO_3$, so, as $Ag^+(aq)$ ions are

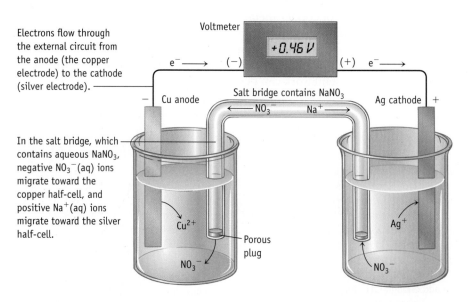

Net reaction: $Cu(s) + 2\ Ag^+(aq) \longrightarrow Cu^{2+}(aq) + 2\ Ag(s)$

FIGURE 20.5 A voltaic cell using Cu(s)|Cu²⁺(aq) and Ag(s)|Ag⁺(aq) half-cells. The cell is set up using 1.0 M Cu(NO₃)₂(aq) and 1.0 M AgNO₃(aq) solutions. Under these conditions, the cell will generate 0.46 volts at 25 °C.

Voltmeter

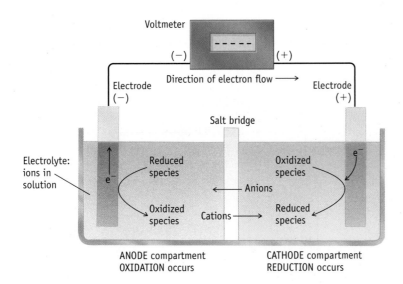

reduced to silver metal, negative ions (NO_3^-) move out of the half-cell into the salt bridge, and positive ions move into the cell. A complete circuit is required for current to flow. If the salt bridge is removed, reactions at the electrodes will cease.

In Figure 20.5, the electrodes are connected by wires to a voltmeter. In an alternative setup, the connections might be to a light bulb or other device that uses electricity. Electrons are produced by oxidation of copper, and Cu^{2+}(aq) ions enter the solution. The electrons traverse the external circuit to the silver electrode, where they reduce Ag^+(aq) ions to silver metal. To balance the extent of oxidation and reduction, two Ag^+(aq) ions are reduced for every Cu^{2+}(aq) ion formed. The main features of this and of all other voltaic cells are summarized in Figure 20.6.

EXAMPLE 20.4 Electrochemical Cells

Problem Describe how to set up a voltaic cell to generate an electric current using the reaction

$$Fe(s) + Cu^{2+}(aq) \rightarrow Cu(s) + Fe^{2+}(aq)$$

Which electrode is the anode, and which is the cathode? In which direction do electrons flow in the external circuit? In which direction do the positive and negative ions flow in the salt bridge? Write equations for the half-reactions that occur at each electrode.

What Do You Know? You are told the reaction proceeds in the direction written. You also know that the half-reaction occurring in one half-cell will involve an iron electrode and a solution of an iron(II) salt such as $Fe(NO_3)_2$. The other half-cell contains a copper electrode and a soluble copper(II) salt such as $Cu(NO_3)_2$.

Strategy First, identify the two different half-cells that make up the cell. These two half-cells are linked together with a salt bridge and an external circuit, as diagrammed in Figure 20.5. The description of the cell (anode, cathode), its chemistry (oxidizing and reducing agents, substances oxidized and reduced), and the movement of electrons in the external circuit and ions through the salt bridge will be similar to the features of the Cu/Cu^{2+} and Ag^+/Ag voltaic cell discussed in the text.

Solution This voltaic cell is similar to the one diagrammed in Figure 20.5. Here we assume the two half-cells are linked with a salt bridge containing KNO_3 as the electrolyte.

Iron is oxidized, so the iron electrode is the anode:

Oxidation, anode: $Fe(s) \rightarrow Fe^{2+}(aq) + 2 e^-$

Because copper(II) ions are reduced, the copper electrode is the cathode. The cathodic half-reaction is

Reduction, cathode: $Cu^{2+}(aq) + 2\ e^- \rightarrow Cu(s)$

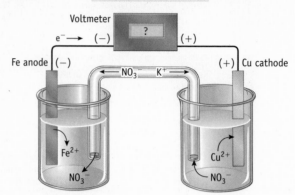

In the external circuit, electrons flow from the iron electrode (anode) to the copper electrode (cathode). In the salt bridge, negative ions flow toward the iron/iron(II) half-cell, and positive ions flow in the opposite direction.

Think about Your Answer A good strategy is to focus on the movement of the negative species in the cell (electron flow in the external circuit and anion flow in the salt bridge). These species always move in a circular fashion: electrons from anode to cathode, and anions in the salt bridge in the opposite direction. If you establish which reactant is oxidized (provides electrons), you then know which electrode is the anode, in which direction electrons move in the external circuit, and in which direction anions flow in the salt bridge. (That said, don't ignore the fact that cations flow in the opposite direction to the anions in the salt bridge.)

Check Your Understanding

Describe how to set up a voltaic cell using the following half-reactions:

Reduction half-reaction: $Ag^+(aq) + e^- \rightarrow Ag(s)$

Oxidation half-reaction: $Ni(s) \rightarrow Ni^{2+}(aq) + 2\ e^-$

Which is the anode, and which is the cathode? What is the overall cell reaction? What is the direction of electron flow in an external wire connecting the two electrodes? Describe the ion flow in a salt bridge (with $NaNO_3$) connecting the cell compartments.

Voltaic Cells with Inert Electrodes

In the half-cells described so far, the metal used as an electrode is also a reactant or a product in the redox reaction. Not all half-reactions involve a metal as a reactant or product, however. With the exception of carbon in the form of graphite, most nonmetals are unsuitable as electrode materials because they do not conduct electricity. It is not possible to make an electrode from a gas, a liquid (except mercury), or a solution. Ionic solids do not make satisfactory electrodes because the ions are locked tightly in a crystal lattice, and these materials do not conduct electricity.

In situations where reactants and products cannot serve as the electrode material, an inert or chemically unreactive electrode must be used. Such electrodes are made of materials that conduct an electric current but that are neither oxidized nor reduced in the cell.

Consider constructing a voltaic cell to accommodate the following product-favored reaction:

$$2\ Fe^{3+}(aq) + H_2(g) \rightarrow 2\ Fe^{2+}(aq) + 2\ H^+(aq)$$

Reduction half-reaction: $Fe^{3+}(aq) + e^- \rightarrow Fe^{2+}(aq)$

Oxidation half-reaction: $H_2(g) \rightarrow 2\ H^+(aq) + 2\ e^-$

Neither the reactants nor the products can be used as an electrode material. Therefore, the two half-cells are set up so that the reactants and products come in contact with an electrode where they can accept or give up electrons. Graphite is a commonly used electrode material: It is a conductor of electricity, and it is inexpensive (essential in commercial cells) and not readily oxidized under the conditions encountered in most cells. Mercury is used in certain types of cells. Platinum and gold are also commonly used because both are chemically inert under most circumstances, but they are generally too costly for commercial cells.

The *hydrogen electrode* is particularly important in the field of electrochemistry because it is used as a reference in assigning cell voltages (see Section 20.4) (Figure 20.7). The electrode material is platinum, chosen because hydrogen adsorbs on the metal's surface. In this half-cell's operation, hydrogen is bubbled over the electrode, and a large surface area maximizes the contact of the gas and the electrode. The aqueous solution contains $H^+(aq)$. The half-reactions involving $H^+(aq)$ and $H_2(g)$

$$2\ H^+(aq) + 2\ e^- \rightarrow H_2(g) \qquad \text{or} \qquad H_2(g) \rightarrow 2\ H^+(aq) + 2\ e^-$$

take place at the electrode surface, and the electrons involved in the reaction are conducted to or from the reaction site by the metal electrode.

A half-cell using the reduction of $Fe^{3+}(aq)$ to $Fe^{2+}(aq)$ can also be set up with a platinum electrode. In this case, the solution surrounding the electrode contains iron ions in two different oxidation states. Transfer of electrons to or from the reactant occurs at the electrode surface.

A voltaic cell involving the reduction of $Fe^{3+}(aq, 1.0\ M)$ to $Fe^{2+}(aq, 1.0\ M)$ with H_2 gas is illustrated in Figure 20.8. In this cell, the hydrogen electrode is the anode (H_2 is oxidized to H^+), and the iron-containing compartment is the cathode (Fe^{3+} is reduced to Fe^{2+}). The cell produces 0.77 V.

Electrochemical Cell Notations

Chemists often use a shorthand notation to simplify cell descriptions. For example, the cell involving the reduction of silver ion with copper metal is written as

$$Cu(s)\,|\,Cu^{2+}(aq, 1.0\ M)\,||\,Ag^+(aq, 1.0\ M)\,|\,Ag(s)$$
$$\underbrace{\text{Anode information}} \qquad \underbrace{\text{Cathode information}}$$

FIGURE 20.7 Hydrogen electrode. Hydrogen gas is bubbled over a platinum electrode in a solution containing H^+ ions. Such electrodes function best if they have a large surface area. Often, platinum wires are woven into a gauze, or the metal surface is roughened either by abrasion or by chemical treatment to increase the surface area.

© Cengage Learning/Charles D. Winters

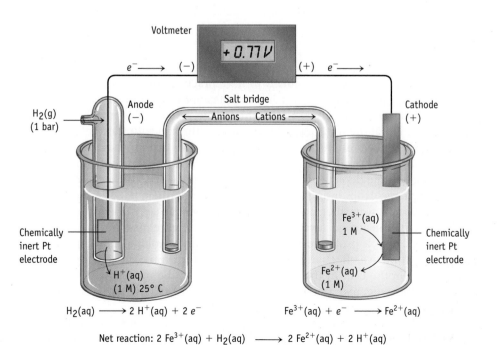

FIGURE 20.8 A voltaic cell with a hydrogen electrode. This cell has $Fe^{2+}(aq, 1.0\ M)$ and $Fe^{3+}(aq, 1.0\ M)$ in the cathode compartment and $H_2(g)$ and $H^+(aq, 1.0\ M)$ in the anode compartment. At 25 °C, the cell generates 0.77 V. (The hydrogen electrode in the anode compartment in this illustration is a simplified schematic. See Figure 20.7 for a photograph of a commercial hydrogen electrode.)

Net reaction: $2\ Fe^{3+}(aq) + H_2(aq) \longrightarrow 2\ Fe^{2+}(aq) + 2\ H^+(aq)$

The cell using H_2 gas to reduce Fe^{3+} ions is written as

$$Pt\,|\,H_2(P = 1\ bar)\,|\,H^+(aq,\ 1.0\ M)\,||\,Fe^{3+}(aq,\ 1.0\ M),\ Fe^{2+}(aq,\ 1.0\ M)\,|\,Pt$$

Anode information   Cathode information

By convention, on the left we write the anode and information with respect to the solution with which it is in contact. A single vertical line (|) indicates a phase boundary, and double vertical lines (||) indicate a salt bridge.

REVIEW & CHECK FOR SECTION 20.2

Answer the questions based on the cell below in which the following reaction occurs:

$$Ni^{2+}(aq) + Cd(s) \rightarrow Ni(s) + Cd^{2+}(aq)$$

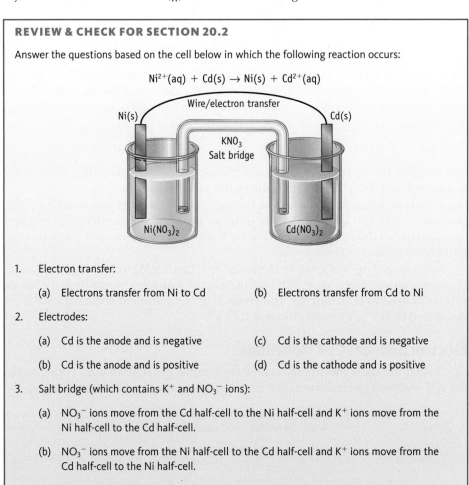

1. Electron transfer:

    (a) Electrons transfer from Ni to Cd   (b) Electrons transfer from Cd to Ni

2. Electrodes:

    (a) Cd is the anode and is negative   (c) Cd is the cathode and is negative

    (b) Cd is the anode and is positive   (d) Cd is the cathode and is positive

3. Salt bridge (which contains K^+ and NO_3^- ions):

    (a) NO_3^- ions move from the Cd half-cell to the Ni half-cell and K^+ ions move from the Ni half-cell to the Cd half-cell.

    (b) NO_3^- ions move from the Ni half-cell to the Cd half-cell and K^+ ions move from the Cd half-cell to the Ni half-cell.

20.3 Commercial Voltaic Cells

The cells described so far are unlikely to have practical use. They are neither compact nor robust, high priorities for most applications. In most situations, it is also important that the cell produce a constant voltage, but a problem with the cells described so far is that the voltage produced varies as the concentrations of ions in solution change (see Section 20.5). Also, the current production is low. Attempting to draw a large current results in a drop in voltage because the ion concentrations near the cathode become depleted if current is drawn too quickly. Furthermore, the negative anode charge drops owing to a build up of cations, the oxidation product.

The electrical work that can be drawn from a voltaic cell depends on the quantity of reagents consumed. A voltaic cell must have a large mass of reactants to produce current over a prolonged period. In addition, a voltaic cell that can be recharged is attractive. Recharging a cell means returning the reagents to their original sites in the cell. In the cells described so far, the movement of ions in the cell mixes the reagents, and they cannot be "unmixed" after the cell has been running.

Batteries can be classified as primary and secondary. **Primary batteries** cannot be returned to their original state by recharging, so when the reactants are

FIGURE 20.9 Some commercial voltaic cells. Commercial voltaic cells provide energy for a wide range of devices, come in a myriad of sizes and shapes, and produce different voltages. Some are rechargeable; others are discarded after use. One might think that there is nothing further to learn about batteries, but this is not true. Research on these devices is actively pursued in the chemical community.

consumed, the battery is "dead" and must be discarded. **Secondary batteries** are often called **storage batteries** or **rechargeable batteries**. The reactions in these batteries can be reversed, so the batteries can be recharged.

Years of development have led to many different commercial voltaic cells to meet specific needs (Figure 20.9), and several common ones are described below. All adhere to the principles that have been developed in earlier discussions.

Primary Batteries: Dry Cells and Alkaline Batteries

If you buy an inexpensive flashlight battery or dry cell battery, it might be a modern version of a voltaic cell invented by George LeClanché in 1866 (Figure 20.10). Zinc serves as the anode, and the cathode is a graphite rod placed down the center of the device. These cells are often called "dry cells" because there is no visible liquid phase. However, the cell contains a moist paste of NH_4Cl, $ZnCl_2$, and MnO_2, so water is present. The moisture is necessary because the ions present must be able to migrate from one electrode to the other. The cell generates a potential of 1.5 V using the following half-reactions:

Cathode, reduction: $2\ NH_4^+(aq) + 2\ e^- \rightarrow 2\ NH_3(g) + H_2(g)$

Anode, oxidation: $Zn(s) \rightarrow Zn^{2+}(aq) + 2\ e^-$

The products at the cathode are gases, which could build up pressure in the cell and cause it to rupture. The problem is avoided, however, by two other reactions that take place in the cell. Ammonia molecules bind to Zn^{2+} ions, and hydrogen gas is oxidized by MnO_2 to water.

$$Zn^{2+}(aq) + 2\ NH_3(g) + 2\ Cl^-(aq) \rightarrow Zn(NH_3)_2Cl_2(s)$$

$$2\ MnO_2(s) + H_2(g) \rightarrow Mn_2O_3(s) + H_2O(\ell)$$

Zinc-carbon dry cells were widely used because of their low cost, but they have several disadvantages. If current is drawn from the battery rapidly, the gaseous products cannot be consumed rapidly enough, so the cell resistance rises, and the voltage drops. In addition, the zinc electrode and ammonium ions are in contact in the cell, and these chemicals react slowly. Recall that zinc reacts with acid to form hydrogen. The ammonium ion, $NH_4^+(aq)$, is a weak Brønsted acid and reacts slowly with zinc. Because of this reaction, these voltaic cells cannot be stored indefinitely, a fact you may have learned from experience. When the zinc outer shell deteriorates, the battery can leak acid and perhaps damage the appliance in which it is contained.

At the present time, you are more likely to use **alkaline batteries** in your camera or flashlight because they generate current up to 50% longer than a dry cell of the

● **Batteries** The word *battery* has become part of our common language, designating any self-contained device that generates an electric current. The term battery has a more precise scientific meaning, however. It refers to a collection of two or more voltaic cells. For example, the 12-volt battery used in automobiles is made up of six voltaic cells. Each voltaic cell develops a voltage of 2 volts. Six cells connected in series produce 12 volts.

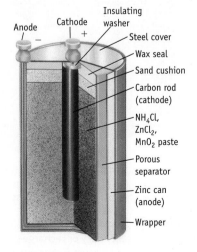

FIGURE 20.10 The common dry cell battery. Sometimes called the *zinc-carbon battery*.

same size. The chemistry of alkaline cells is quite similar to that in a zinc-carbon dry cell, except that the material inside the cell is basic (alkaline). Alkaline cells use the oxidation of zinc and the reduction of MnO_2 to generate a current, but NaOH or KOH is used in the cell instead of the acidic salt NH_4Cl.

Cathode, reduction: $2\ MnO_2(s) + H_2O(\ell) + 2\ e^- \rightarrow Mn_2O_3(s) + 2\ OH^-(aq)$

Anode, oxidation: $Zn(s) + 2\ OH^-(aq) \rightarrow ZnO(s) + H_2O(\ell) + 2\ e^-$

Alkaline cells, which produce 1.54 V (approximately the same voltage as the zinc-carbon dry cell), have the further advantage that the cell potential does not decline under high current loads because no gases are formed.

Prior to 2000, mercury-containing batteries were widely used in calculators, cameras, watches, heart pacemakers, and other devices. However, these small batteries were banned in the United States in the 1990s because of environmental problems. They have been replaced by several other types of batteries, such as silver oxide batteries and zinc-oxygen batteries. Both operate under alkaline conditions, and both have zinc anodes. In the silver oxide battery, which produces a voltage of about 1.5 V, the cell reactions are

Cathode, reduction: $Ag_2O(s) + H_2O(\ell) + 2\ e^- \rightarrow 2\ Ag(s) + 2\ OH^-(aq)$

Anode, oxidation: $Zn(s) + 2\ OH^-(aq) \rightarrow ZnO(s) + H_2O(\ell) + 2\ e^-$

The zinc-oxygen battery, which produces about 1.15–1.35 V, is unique in that atmospheric oxygen and not a metal oxide is the oxidizing agent.

Cathode, reduction: $O_2(g) + 2\ H_2O(\ell) + 4\ e^- \rightarrow 4\ OH^-(aq)$

Anode, oxidation: $Zn(s) + 2\ OH^-(aq) \rightarrow ZnO(s) + H_2O(\ell) + 2\ e^-$

These batteries are used in hearing aids, pagers, and medical devices.

Secondary or Rechargeable Batteries

When a zinc-carbon dry cell or an alkaline cell ceases to produce a usable electric current, it is discarded. In contrast, some types of cells can be recharged, often hundreds of times. Recharging requires applying an electric current from an external source to restore the cell to its original state. There are a number of well known types of secondary batteries in common use.

Lead Storage Battery

An automobile battery—the lead storage battery—is probably the best-known rechargeable battery (Figure 20.11). The 12-V version of this battery contains six voltaic cells, each generating about 2 V. The lead storage battery can produce a large initial current, an essential feature when starting an automobile engine.

The anode of a lead storage battery is metallic lead. The cathode is also made of lead, but it is covered with a layer of compressed, insoluble lead(IV) oxide, PbO_2. The electrodes, arranged alternately in a stack and separated by thin fiberglass sheets, are immersed in aqueous sulfuric acid. When the cell supplies electrical energy, the lead anode is oxidized to lead(II) sulfate, an insoluble substance that adheres to the electrode surface. The two electrons produced per lead atom move through the external circuit to the cathode, where PbO_2 is reduced to Pb^{2+} ions that, in the presence of H_2SO_4, also form lead(II) sulfate.

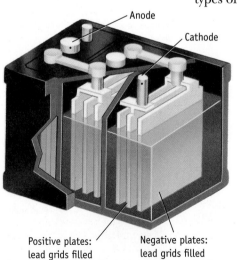

Anode

Cathode

Positive plates: lead grids filled with PbO_2

Negative plates: lead grids filled with spongy lead

FIGURE 20.11 Lead storage battery, a secondary or rechargeable battery. Each cell of the battery generates 2 V.

Cathode, reduction: $PbO_2(s) + 4\ H^+(aq) + SO_4{}^{2-}(aq) + 2\ e^- \rightarrow PbSO_4(s) + 2\ H_2O(\ell)$

Anode, oxidation: $Pb(s) + SO_4{}^{2-}(aq) \rightarrow PbSO_4(s) + 2\ e^-$

Net ionic equation: $Pb(s) + PbO_2(s) + 2\ H_2SO_4(aq) \rightarrow 2\ PbSO_4(s) + 2\ H_2O(\ell)$

When current is generated, sulfuric acid is consumed and water is formed. Because water is less dense than sulfuric acid, the density of the solution decreases during this process. Therefore, one way to determine whether a lead storage battery needs to be recharged is to measure the density of the solution.

A lead storage battery is recharged by supplying electrical energy. The $PbSO_4$ coating the surfaces of the electrodes is converted back to metallic lead and PbO_2, and sulfuric acid is regenerated. Recharging this battery is possible because the reactants and products remain attached to the electrode surface. The lifetime of a lead storage battery is limited, however, because, with time, the coatings of PbO_2 and $PbSO_4$ flake off of the surface and fall to the bottom of the battery case.

Scientists and engineers would like to find an alternative to lead storage batteries, especially for use in cars. Lead storage batteries have the disadvantage of being large and heavy. In addition, lead and its compounds are toxic and their disposal adds a further complication. Nevertheless, at this time, the advantages of lead storage batteries often outweigh their disadvantages.

Tesla Motors Roadster, an all-electric car. This limited edition sports car (cost is over $100,000) is powered by a rechargeable nickel-metal hydride battery that will last for over 200 miles of driving.

Nickel-Cadmium Battries

Nickel-cadmium ("Ni-cad") batteries, used in a variety of cordless appliances such as telephones, video camcorders, and cordless power tools, are lightweight and rechargeable. The chemistry of the cell is based on the oxidation of cadmium and the reduction of nickel(III) oxide under basic conditions. As with the lead storage battery, the reactants and products formed when producing a current are solids that adhere to the electrodes.

Cathode, reduction: $NiO(OH)(s) + H_2O(\ell) + e^- \rightarrow Ni(OH)_2(s) + OH^-(aq)$

Anode, oxidation: $Cd(s) + 2\ OH^-(aq) \rightarrow Cd(OH)_2(s) + 2\ e^-$

Ni-cad batteries produce a nearly constant voltage. However, their cost is relatively high, and there are restrictions on their disposal because cadmium compounds are toxic and present an environmental hazard.

Nickel-Metal Hydride

Rechargeable nickel-metal hydride batteries (NiMH) are now in common use. You may have them in your camera or portable music player. Their most important use, however, is in electric or hybrid cars.

In the nickel-metal hydride battery, electrons are generated when H atoms interact with OH^- ions at the metal alloy anode.

$$Alloy(H) + OH^- \rightarrow Alloy + H_2O + e^-$$

The reaction at the cathode is the same as in Ni-cad batteries.

$$NiO(OH) + H_2O + e^- \rightarrow Ni(OH)_2 + OH^-$$

The "alloy" in NiMH batteries is generally a mixture of a rare earth metal such as lanthanum, cerium, or neodymium and another metal such as nickel, cobalt, manganese or aluminum.

Lithium Batteries

As we introduced on page 690 (and in Chapter 13, page 444), much of the battery research now focuses on lithium batteries, and there are two types in common use: lithium-ion and a variant, lithium-ion polymer batteries. The chemistry of both is the same.

Cathode, reduction: $Li_{1-x}CoO_2(s) + x\ Li^+ + x\ e^- \rightarrow LiCoO_2(s)$

Anode, oxidation: $Li_xC_6(s) \rightarrow x\ Li^+ + x\ e^- + 6\ C(s)$

Lithium-ion polymer batteries are popular because they can be made in any shape and so fit into products such as Amazon's Kindle and Apple's portable

● Be Alert: Children Can Ingest Lithium Batteries! Very small lithium batteries are used in many devices such as toys, musical greeting cards, and appliances. Unfortunately, because they are small and shiny, children put them in their mouths and can swallow them. In 2010 the *New York Times* (June 1, 2010, page D5) reported that an increasing number of children have died from reactions induced by the battery's current when the battery was swallowed.

devices. They also deliver a higher current density than lithium-ion batteries and can be charged and discharged through many cycles.

Finally, an important reason for the development of lithium-based batteries is their light weight. Lithium has a much lower density (0.53 g/cm³) than cadmium (8.69 g/cm³), nickel (8.91 g/cm³), or lead (11.34 g/cm³). This is important not only in portable devices where people do not want to carry around heavy objects but also in the automobile industry. A lighter weight for a battery in a car results in more of the energy going into moving the rest of the car and less being used simply to transport the battery.

The importance of lithium-based batteries in our 21st-century economy cannot be overstated. Most of the current development and manufacture is being done in Japan, South Korea, and China, but much of the basic technology of lithium batteries was developed in the United States, and investment is increasing.

Fuel Cells

An advantage of voltaic cells is that they are small and portable, but their size is also a limitation. The amount of electric current produced is limited by the quantity of reagents contained in the cell. When one of the reactants is completely consumed, the cell will no longer generate a current. Fuel cells avoid this limitation because the reactants (fuel and oxidant) can be supplied continuously to the cell from an external reservoir.

Although the first fuel cells were first constructed more than 150 years ago, little was done to develop this technology until the space program rekindled interest in these devices. Hydrogen-oxygen fuel cells have been used in NASA's Gemini, Apollo, and Space Shuttle programs. Not only are they lightweight and efficient, but they also have the added benefit that they generate drinking water for the ship's crew.

In a hydrogen-oxygen fuel cell (Figure 20.12), hydrogen is pumped onto the anode of the cell, and O_2 (or air) is directed to the cathode where the following reactions occur:

Cathode, reduction: $O_2(g) + 2\ H_2O(\ell) + 4\ e^- \rightarrow 4\ OH^-(aq)$

Anode, oxidation: $H_2(g) \rightarrow 2\ H^+(aq) + 2\ e^-$

The two halves of the cell are separated by a special material called a proton exchange membrane (PEM). Protons, $H^+(aq)$, formed at the anode traverse the PEM and react with the hydroxide ions produced at the cathode, forming water. The net reaction in the cell is thus the formation of water from H_2 and O_2. Cells currently in use run at temperatures of 70–140 °C and produce about 0.9 V.

Hydrogen-oxygen fuel cells operate at 40–60% efficiency and meet most of the requirements for use in automobiles: They operate at room temperature or slightly

● **Energy for Automobiles** Energy available from systems that can be used to power an automobile.

Chemical System	W · h/kg* (1 W · h = 3600 J)
Lead-acid battery	18–56
Nickel-cadmium battery	33–70
Sodium-sulfur battery	80–140
Lithium polymer battery	150
Gasoline-air combustion engine	12,200

* watt-hour/kilogram

FIGURE 20.12 Fuel cell design. Hydrogen gas is oxidized to $H^+(aq)$ at the anode surface. On the other side of the proton exchange membrane (PEM), oxygen gas is reduced to $OH^-(aq)$. The $H^+(aq)$ ions travel through the PEM and combine with $OH^-(aq)$, forming water.

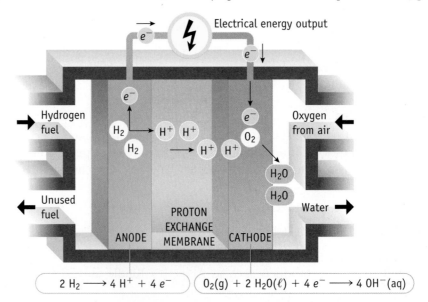

above, start rapidly, and develop a high current density. However, cost is a serious problem as is the absence of a carbon-free way of making hydrogen. These faults, and the cost and difficulty of building a nationwide infrastructure for distributing hydrogen, remain major stumbling blocks that would need to be overcome before hydrogen fuel cells could be adopted for widespread use in automobiles.

REVIEW & CHECK FOR SECTION 20.3

1. The following half-reaction occurs in an automobile lead-acid battery:

$$PbO_2(s) + 4\ H^+(aq) + SO_4^{2-}(aq) + 2\ e^- \rightarrow PbSO_4(s) + 2\ H_2O(\ell)$$

   (a) Lead(IV) oxide is oxidized.   (c) Hydrogen ions are oxidized.

   (b) Lead(IV) oxide is reduced.   (d) Hydrogen ions are reduced.

2. You can determine if a lead-based battery is no longer working by checking the density of the solution in the battery. If the density is low, then the battery is "dead." This works because

   (a) PbO_2 is consumed.   (c) Water evaporates.

   (b) Sulfuric acid is consumed.   (d) Lead is consumed.

20.4 Standard Electrochemical Potentials

Different electrochemical cells produce different potentials: 1.5 V for alkaline cells, about 3.7 V for a lithium-ion battery, and about 2.0 V for the individual cells in a lead storage battery. In this section, we want to identify some of the factors affecting cell voltages and develop ways to calculate the expected potential of a cell based on the chemistry in the cell and the conditions used.

Electromotive Force

Electrons generated at the anode of an electrochemical cell move through the external circuit toward the cathode, and the force needed to move the electrons arises from a difference in the potential energy of electrons at the two electrodes. This difference in potential energy per electrical charge is called the **electromotive force** or **emf**, for which the literal meaning is "force causing electrons to move." Emf has units of volts (V); 1 volt is the potential difference needed to impart one joule of energy to an electric charge of one coulomb ($1\ J = 1\ V \times 1\ C$). *One coulomb is the quantity of charge that passes a point in an electric circuit when a current of 1 ampere flows for 1 second ($1\ C = 1\ A \times 1\ s$).*

● **Electrochemical Units**
• The coulomb (abbreviated C) is the standard (SI) unit of electrical charge (Appendix C, Table 6).
• 1 joule = 1 volt × 1 coulomb.
• 1 coulomb = 1 ampere × 1 second.

Measuring Standard Potentials

Imagine you planned to study cell voltages in a laboratory with two objectives: (1) to understand the factors that affect these values and (2) to be able to predict the potential of a voltaic cell. You might construct a number of different half-cells, link them together in various combinations to form voltaic cells (as in Figure 20.13), and measure the cell potentials. After a few experiments, it would become apparent that cell potentials depend on a number of factors: the half-cells used (i.e., the reaction in each half-cell and the overall or net reaction in the cell), the concentrations of reactants and products in solution, the pressure of gaseous reactants, and the temperature.

To simplify matters for now, let us consider only cell potentials measured under **standard conditions**:

• Reactants and products are present in their standard states.
• Solutes in aqueous solution have a concentration of 1.0 M.
• Gaseous reactants or products have a pressure of 1.0 bar.

A cell potential measured under these conditions is called the **standard potential** and is denoted by $E°_{cell}$. Unless otherwise specified, all values of $E°_{cell}$ refer to measurements at 298 K (25 °C).

FIGURE 20.13 A voltaic cell using Zn|Zn²⁺(aq, 1.0 M) and H₂|H⁺(aq, 1.0 M) half-cells.

- When zinc and acid are combined in an electrochemical cell, the cell generates a potential of 0.76 V under standard conditions.
- The electrode in the H₂|H⁺(aq, 1.0 M) half-cell is the cathode, and the Zn electrode is the anode.
- Electrons flow in the external circuit to the hydrogen half-cell from the zinc half-cell.
- The positive sign of the measured voltage indicates that the hydrogen electrode is the cathode or positive electrode.

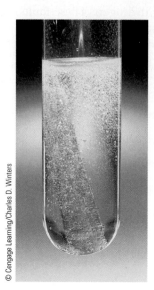

© Cengage Learning/Charles D. Winters

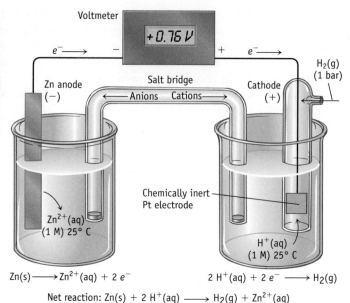

$$Zn(s) \longrightarrow Zn^{2+}(aq) + 2\ e^-$$

$$2\ H^+(aq) + 2\ e^- \longrightarrow H_2(g)$$

Net reaction: $Zn(s) + 2\ H^+(aq) \longrightarrow H_2(g) + Zn^{2+}(aq)$

(a) The product-favored reaction of aqueous hydrogen ions with zinc produces H₂(g) and Zn²⁺ ions.

(b) A voltaic cell that makes use of the product-favored reduction of aqueous hydrogen ions by zinc.

Suppose you set up a number of standard half-cells and connect each in turn to a **standard hydrogen electrode (SHE)**. Your apparatus would look like the voltaic cell in Figure 20.13b. For now, we will concentrate on three aspects of this cell:

1. *The reaction that occurs.* The reaction occurring in the cell pictured in Figure 20.13 could be *either* the reduction of Zn²⁺ ions with H₂ gas

$$Zn^{2+}(aq) + H_2(g) \rightarrow Zn(s) + 2\ H^+(aq)$$

Zn²⁺(aq) is the oxidizing agent, and H₂ is the reducing agent.
Standard hydrogen electrode would be the anode (negative electrode).

or the reduction of H⁺(aq) ions by Zn(s).

$$Zn(s) + 2\ H^+(aq) \rightarrow Zn^{2+}(aq) + H_2(g)$$

Zn is the reducing agent, and H⁺(aq) is the oxidizing agent.
Standard hydrogen electrode would be the cathode (positive electrode).

The reaction that actually occurs is the one that is product-favored at equilibrium and thus is spontaneous under standard conditions. In this case we know the product-favored reaction is the reduction of H⁺ ions by zinc metal (Figure 20.13a).

2. *Direction of electron flow in the external circuit.* In a voltaic cell, electrons always flow from the anode (negative electrode) to the cathode (positive electrode). We can tell the direction of electron movement—and therefore which is the anode and which is the cathode—by placing a voltmeter in the circuit. A positive potential is observed if the voltmeter terminal with a plus sign (+) is connected to the positive electrode or cathode [and the terminal with the minus sign (−) is connected to the negative electrode or anode]. Connected in the opposite way (plus to minus and minus to plus), the voltmeter will indicate a negative potential.

3. *Cell potential.* In Figure 20.13, the voltmeter is hooked up with its positive terminal connected to the hydrogen half-cell, and a reading of +0.76 V is observed. The hydrogen electrode is thus the positive electrode or cathode, and the reactions occurring in this cell must be

Reduction, cathode:	$2\ H^+(aq) + 2\ e^- \rightarrow H_2(g)$
Oxidation, anode:	$Zn(s) \rightarrow Zn^{2+}(aq) + 2\ e^-$
Net cell reaction:	$Zn(s) + 2\ H^+(aq) \rightarrow Zn^{2+}(aq) + H_2(g)$

EMF, Cell Potential, and Voltage

Electromotive force (emf) and cell potential (E_{cell}) are often used synonymously, but the two are subtly different. E_{cell} is a measured quantity, so its value is affected by how the measurement is made. To understand this point, consider as an analogy water in a pipe under pressure. Water pressure can be viewed as analogous to emf; it represents a force that will cause water in the pipe to move. If we open a faucet, water will flow. Opening the faucet will, however, decrease the pressure in the system.

Emf is the potential difference when no current flows. To measure E_{cell}, a voltmeter is placed in the external circuit. Although voltmeters have high internal resistance to minimize current flow, a small current flows nonetheless. As a result, the value of E_{cell} will be slightly different than the emf.

Finally, there is a difference between a *potential* and a *voltage*. The voltage of a cell has a magnitude but no sign. In contrast, the potential of a half-reaction or a cell has a sign (+ or −) and a magnitude.

This is an important result. All of the species are present in the cell at standard conditions, so the reaction can, in principle, proceed in either direction. The product-favored reaction, however, proceeds from left to right as written. This shows us that, of the two reducing agents present, zinc is better than H_2, and H^+ ions are a better oxidizing agent than zinc ions.

The potential of $+0.76$ V measured for the oxidation of zinc with hydrogen ions also reflects the difference in potential energy of an electron at each electrode. From the direction of flow of electrons in the external circuit (Zn electrode → H_2 electrode), we conclude that the potential energy of an electron at the zinc electrode is higher than the potential energy of the electron at the hydrogen electrode.

Hundreds of electrochemical cells like that shown in Figure 20.13 can be set up, allowing us to determine the relative oxidizing or reducing ability of various chemical species and to determine the electrical potential generated by the reaction under standard conditions. A few results are given in Figure 20.14, where half-reactions are listed as reductions. That is, the chemical species on the left are oxidizing agents and are listed in descending oxidizing ability.

● **$E°$ Values** An extensive listing of $E°$ values is found in Appendix M, and still larger tables of data can be found in chemistry reference books. A common convention, used in Appendix M, lists standard reduction potentials in two groups, one for acid and neutral solutions and the other for basic solutions.

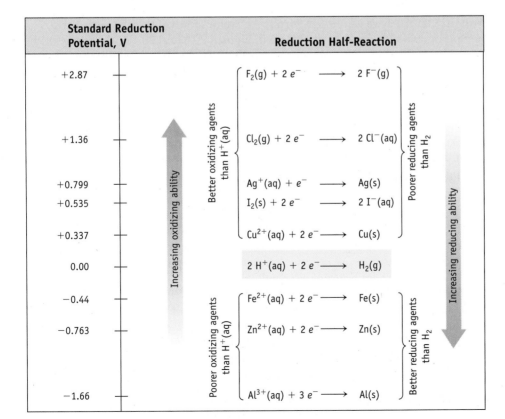

Standard Reduction Potential, V	Reduction Half-Reaction
+2.87	$F_2(g) + 2\,e^- \longrightarrow 2\,F^-(g)$
+1.36	$Cl_2(g) + 2\,e^- \longrightarrow 2\,Cl^-(aq)$
+0.799	$Ag^+(aq) + e^- \longrightarrow Ag(s)$
+0.535	$I_2(s) + 2\,e^- \longrightarrow 2\,I^-(aq)$
+0.337	$Cu^{2+}(aq) + 2\,e^- \longrightarrow Cu(s)$
0.00	$2\,H^+(aq) + 2\,e^- \longrightarrow H_2(g)$
−0.44	$Fe^{2+}(aq) + 2\,e^- \longrightarrow Fe(s)$
−0.763	$Zn^{2+}(aq) + 2\,e^- \longrightarrow Zn(s)$
−1.66	$Al^{3+}(aq) + 3\,e^- \longrightarrow Al(s)$

FIGURE 20.14 A potential ladder for reduction half-reactions.
- The position of a half-reaction on this potential ladder reflects the relative ability of the species at the left to act as an oxidizing agent.
- The higher the compound or ion is in the list, the better it is as an oxidizing agent. Conversely, the atoms or ions on the right are reducing agents. The lower they are in the list, the better they are as a reducing agent.
- The potential for each half-reaction is given with its standard reduction potential, $E°_{reduction}$.
- For more information see J. R. Runo and D. G. Peters, *Journal of Chemical Education*, Vol 70, p. 708, 1993.

Standard Reduction Potentials

By doing experiments such as that illustrated by Figure 20.13, we not only have a notion of the relative oxidizing and reducing abilities of various chemical species, but we can also rank them quantitatively.

If $E°_{cell}$ is a measure of the standard potential for the cell, then $E°_{cathode}$ and $E°_{anode}$ can be taken as measures of electrode potential. Because $E°_{cell}$ reflects the difference in electrode potentials, $E°_{cell}$ must be the difference between $E°_{cathode}$ and $E°_{anode}$.

● **Equation 20.1** Equation 20.1 is another example of calculating a change from $X_{final} - X_{initial}$. Electrons move to the cathode (the "final" state) from the anode (the "initial" state). Thus, Equation 20.1 resembles equations you have seen previously in this book (such as Equations 5.6 and 6.5).

$$E°_{cell} = E°_{cathode} - E°_{anode} \tag{20.1}$$

Here, $E°_{cathode}$ and $E°_{anode}$ are the standard *reduction* potentials for the half-cell reactions that occur at the cathode and anode, respectively. Equation 20.1 is important for three reasons:

- If you have values for $E°_{cathode}$ and $E°_{anode}$, you can calculate the standard potential, $E°_{cell}$, for a voltaic cell.
- *When the calculated value of $E°_{cell}$ is positive, the reaction as written is predicted to be product-favored at equilibrium.* Conversely, if the calculated value of $E°_{cell}$ is negative, the reaction is predicted to be reactant-favored at equilibrium. Such a reaction will be product-favored at equilibrium in a direction opposite to the way it is written.
- If you measure $E°_{cell}$ and know either $E°_{cathode}$ or $E°_{anode}$, you can calculate the other value. This value would tell you how one half-cell reaction compares with others in terms of relative oxidizing or reducing ability.

But here is a dilemma. You cannot measure individual half-cell potentials. Just as values for $\Delta_f H°$ and $\Delta_f G°$ were established by choosing a reference point (the elements in their standard states), scientists have selected a reference point for half-reactions. We assign a potential of exactly 0 V to the half-reaction that occurs at a standard hydrogen electrode (SHE).

$$2\ H^+(aq,\ 1\ M) + 2\ e^- \rightarrow H_2(g,\ 1\ bar) \quad E° = 0.00\ V$$

With this standard, we can now determine $E°$ values for half-cells by measuring $E°_{cell}$ in experiments such as those described in Figures 20.8 and 20.13, where one of the electrodes is the standard hydrogen electrode. We can then quantify the information with reduction potential tables such as Figure 20.14 and use these values to make predictions about $E°_{cell}$ for new voltaic cells.

Tables of Standard Reduction Potentials

The experimental approach just described leads to lists of $E°$ values such as seen in Figure 20.14, Table 20.1, and Appendix M. Let us list some important points concerning these tables and then illustrate them in the discussion and examples that follow.

1. The half-reactions are written as "oxidized form + electrons → reduced form." The species on the left side of the reaction arrow is an oxidizing agent, and the species on the right side of the reaction arrow is a reducing agent. Therefore, *all potentials are for reduction reactions*, and the potentials (in volts vs. SHE) are called *standard reduction potentials*.
2. The more positive the value of $E°$ for the reactions in Figure 20.14, Table 20.1, and similar tables, the better the oxidizing ability of the ion or compound on the left side of the reaction. This means *$F_2(g)$ is the best oxidizing agent in the table*. Lithium ion at the lower-left corner of Table 20.1 is the poorest oxidizing agent because its $E°$ value is the most negative.
3. The more negative the value of the reduction potential, $E°$, the less likely the half-reaction will occur as a reduction, and the more likely the reverse

half-reaction will occur (as an oxidation). Thus, Li(s) is the strongest reducing agent in the table, and F⁻ is the weakest reducing agent. The reducing agents in the table (the ions, elements, and compounds at the right) increase in strength from the top to the bottom.

4. The reaction between any substance on the left in this table (an oxidizing agent) with any substance *lower* than it on the right (a reducing agent) is product-favored at equilibrium. This has been called the *northwest–southeast rule*.

Table 20.1 Standard Reduction Potentials in Aqueous Solution at 25 °C*

Reduction Half-Reaction		$E°$ (V)
$F_2(g) + 2\,e^-$	$\rightarrow 2\,F^-(aq)$	+2.87
$H_2O_2(aq) + 2\,H^+(aq) + 2\,e^-$	$\rightarrow 2\,H_2O(\ell)$	+1.77
$PbO_2(s) + SO_4{}^{2-}(aq) + 4\,H^+(aq) + 2\,e^-$	$\rightarrow PbSO_4(s) + 2\,H_2O(\ell)$	+1.685
$MnO_4{}^-(aq) + 8\,H^+(aq) + 5\,e^-$	$\rightarrow Mn^{2+}(aq) + 4\,H_2O(\ell)$	+1.51
$Au^{3+}(aq) + 3\,e^-$	$\rightarrow Au(s)$	+1.50
$Cl_2(g) + 2\,e^-$	$\rightarrow 2\,Cl^-(aq)$	+1.36
$Cr_2O_7{}^{2-}(aq) + 14\,H^+(aq) + 6\,e^-$	$\rightarrow 2\,Cr^{3+}(aq) + 7\,H_2O(\ell)$	+1.33
$O_2(g) + 4\,H^+(aq) + 4\,e^-$	$\rightarrow 2\,H_2O(\ell)$	+1.229
$Br_2(\ell) + 2\,e^-$	$\rightarrow 2\,Br^-(aq)$	+1.08
$NO_3{}^-(aq) + 4\,H^+(aq) + 3\,e^-$	$\rightarrow NO(g) + 2\,H_2O(\ell)$	+0.96
$OCl^-(aq) + H_2O(\ell) + 2\,e^-$	$\rightarrow Cl^-(aq) + 2\,OH^-(aq)$	+0.89
$Hg^{2+}(aq) + 2\,e^-$	$\rightarrow Hg(\ell)$	+0.855
$Ag^+(aq) + e^-$	$\rightarrow Ag(s)$	+0.799
$Hg_2{}^{2+}(aq) + 2\,e^-$	$\rightarrow 2\,Hg(\ell)$	+0.789
$Fe^{3+}(aq) + e^-$	$\rightarrow Fe^{2+}(aq)$	+0.771
$I_2(s) + 2\,e^-$	$\rightarrow 2\,I^-(aq)$	+0.535
$O_2(g) + 2\,H_2O(\ell) + 4\,e^-$	$\rightarrow 4\,OH^-(aq)$	+0.40
$Cu^{2+}(aq) + 2\,e^-$	$\rightarrow Cu(s)$	+0.337
$Sn^{4+}(aq) + 2\,e^-$	$\rightarrow Sn^{2+}(aq)$	+0.15
$2\,H^+(aq) + 2\,e^-$	$\rightarrow H_2(g)$	0.00
$Sn^{2+}(aq) + 2\,e^-$	$\rightarrow Sn(s)$	−0.14
$Ni^{2+}(aq) + 2\,e^-$	$\rightarrow Ni(s)$	−0.25
$V^{3+}(aq) + e^-$	$\rightarrow V^{2+}(aq)$	−0.255
$PbSO_4(s) + 2\,e^-$	$\rightarrow Pb(s) + SO_4{}^{2-}(aq)$	−0.356
$Cd^{2+}(aq) + 2\,e^-$	$\rightarrow Cd(s)$	−0.40
$Fe^{2+}(aq) + 2\,e^-$	$\rightarrow Fe(s)$	−0.44
$Zn^{2+}(aq) + 2\,e^-$	$\rightarrow Zn(s)$	−0.763
$2\,H_2O(\ell) + 2\,e^-$	$\rightarrow H_2(g) + 2\,OH^-(aq)$	−0.8277
$Al^{3+}(aq) + 3\,e^-$	$\rightarrow Al(s)$	−1.66
$Mg^{2+}(aq) + 2\,e^-$	$\rightarrow Mg(s)$	−2.37
$Na^+(aq) + e^-$	$\rightarrow Na(s)$	−2.714
$K^+(aq) + e^-$	$\rightarrow K(s)$	−2.925
$Li^+(aq) + e^-$	$\rightarrow Li(s)$	−3.045

Increasing strength of oxidizing agents (left margin, bottom to top)

Increasing strength of reducing agents (right margin, top to bottom)

* In volts (V) versus the standard hydrogen electrode.

Product-favored reactions will always involve a reducing agent that is "southeast" of the proposed oxidizing agent.

Reduction Half-Reaction

$$I_2(s) + 2\ e^- \longrightarrow 2\ I^-(aq)$$
$$Cu^{2+}(aq) + 2\ e^- \longrightarrow Cu(s)$$
$$2\ H^+(aq) + 2\ e^- \longrightarrow H_2(g)$$
$$Fe^{2+}(aq) + 2\ e^- \longrightarrow Fe(s)$$
$$Zn^{2+}(aq) + 2\ e^- \longrightarrow Zn(s)$$

The northwest–southeast rule: The reducing agent always lies to the southeast of the oxidizing agent in a product-favored reaction.

- **Northwest–Southeast Rule** This guideline is a reflection of the idea of moving down a potential energy "ladder" in a product-favored reaction.

For example, Zn can reduce Fe^{2+}, H^+, Cu^{2+}, and I_2, but, of the species on this list, Cu can reduce only I_2.

5. The algebraic sign of the half-reaction reduction potential is the sign of the electrode when it is attached to the H_2/H^+ standard cell (see Figures 20.8 and 20.13).

6. Electrochemical potentials depend on the nature of the reactants and products and their concentrations, not on the quantities of material used. Therefore, changing the stoichiometric coefficients for a half-reaction does not change the value of $E°$. For example, the reduction of Fe^{3+} has an $E°$ of $+0.771$ V, whether the reaction is written as

- **Changing Stoichiometric Coefficients** The volt is defined as "energy/charge" (V = J/C). Multiplying a reaction by some number causes both the energy and the charge to be multiplied by that number. Thus, the ratio "energy/charge = volt" does not change.

$$Fe^{3+}(aq,\ 1\ M) + e^- \rightarrow Fe^{2+}(aq,\ 1\ M) \qquad E° = +0.771\ V$$

or as

$$2\ Fe^{3+}(aq,\ 1\ M) + 2\ e^- \rightarrow 2\ Fe^{2+}(aq,\ 1\ M) \qquad E° = +0.771\ V$$

Using Tables of Standard Reduction Potentials

Tables or "ladders" of standard reduction potentials are very useful. They allow you to predict the potential of a new voltaic cell, provide information that can be used to balance redox equations, and help predict which redox reactions are product-favored.

Calculating Standard Cell Potentials, $E°_{cell}$

The standard reduction potentials for half-reactions were obtained by measuring cell potentials. It makes sense, therefore, that these values can be combined to give the potential of some new cell.

The net reaction occurring in a voltaic cell using silver and copper half-cells is

$$2\ Ag^+(aq) + Cu(s) \rightarrow 2\ Ag(s) + Cu^{2+}(aq)$$

The silver electrode is the cathode, and the copper electrode is the anode. We know this because silver ion is reduced (to silver metal) and copper metal is oxidized (to Cu^{2+} ions). (Recall that oxidations always occur at the anode and reductions at the cathode.) Also notice that the $Cu^{2+}|Cu$ half-reaction is "southeast" of the $Ag^+|Ag$ half-reaction in the potential ladder (Figure 20.14 and Table 20.1).

$$E°_{cathode} = +0.799\ V \qquad Ag^+(aq) + e^- \longrightarrow Ag(s)$$

"Distance" from $E°_{cathode}$ to $E°_{anode}$ is 0.799 V − 0.337 V = 0.462 V.

Cu is "southeast" of Ag^+

$$E°_{anode} = +0.337\ V \qquad Cu^{2+}(aq) + 2\ e^- \longrightarrow Cu(s)$$

The potential for the voltaic cell is the difference between the standard reduction potentials for the two half-reactions.

$$E°_{cell} = E°_{cathode} - E°_{anode}$$

$$E°_{cell} = (+0.799\ V) - (+0.337\ V)$$

$$E°_{cell} = +0.462\ V$$

Notice that the value of $E°_{cell}$ is related to the "distance" between the cathode and anode reactions on the potential ladder. The products have a lower potential energy than the reactants, and the cell potential, $E°_{cell}$, has a positive value.

A positive potential calculated for the $Ag^+|Ag$ and $Cu^{2+}|Cu$ cell ($E°_{cell} = +0.462$ V) confirms that the reduction of silver ions in water with copper metal is product-favored at equilibrium (Figure 20.1). We might ask, however, about the value of $E°_{cell}$ if a reactant-favored equation had been selected. For example, what is $E°_{cell}$ for the reduction of copper(II) ions with silver metal?

Cathode, reduction: $Cu^{2+}(aq) + 2\ e^- \rightarrow Cu(s)$

Anode, oxidation: $\underline{2\ Ag(s) \rightarrow 2\ Ag^+(aq) + 2\ e^-}$

Net ionic equation: $2\ Ag(s) + Cu^{2+}(aq) \rightarrow 2\ Ag^+(aq) + Cu(s)$

Cell Voltage Calculation

$$E°_{cathode} = +0.337 \text{ V and } E°_{anode} = +0.799 \text{ V}$$

$$E°_{cell} = E°_{cathode} - E°_{anode} = (+0.337 \text{ V}) - (0.799 \text{ V})$$

$$E°_{cell} = -0.462 \text{ V}$$

The negative sign for $E°_{cell}$ indicates that the reaction as written is reactant-favored at equilibrium. The products of the reaction (Ag^+ and Cu) have a higher potential energy than the reactants (Ag and Cu^{2+}). For the indicated reaction to occur, a potential of at least 0.462 V would have to be imposed on the system by an external source of electricity (see Section 20.7).

Relative Strengths of Oxidizing and Reducing Agents

Five half-reactions, selected from Appendix M, are arranged from the half-reaction with the highest (most positive) $E°$ value to the one with the lowest (most negative) value.

$E°$, V		Reduction Half-Reaction
+1.36		$Cl_2(g) + 2\ e^- \longrightarrow 2\ Cl^-(aq)$
+0.80		$Ag^+(aq) + e^- \longrightarrow Ag(s)$
+0.00	Increasing strength as oxidizing agents	$2\ H^+(aq) + 2\ e^- \longrightarrow H_2(g)$
−0.25		$Ni^{2+}(aq) + 2\ e^- \longrightarrow Ni(s)$
−0.76		$Zn^{2+}(aq) + 2\ e^- \longrightarrow Zn(s)$

- The list on the left is headed by Cl_2, an element that is a strong oxidizing agent and thus is easily reduced. At the bottom of the list is $Zn^{2+}(aq)$, an ion not easily reduced and thus a poor oxidizing agent.
- On the right, the list is headed by $Cl^-(aq)$, an ion that can be oxidized to Cl_2 only with difficulty. It is a very poor reducing agent. At the bottom of the list is zinc metal, which is readily oxidized and is a good reducing agent.

By arranging these half-reactions based on $E°$ values, we have also arranged the chemical species on the two sides of the equation in order of their strengths as oxidizing or reducing agents. In this list, from strongest to weakest, the order is

Oxidizing agents: $Cl_2 > Ag^+ > H^+ > Ni^{2+} > Zn^{2+}$

strong ⟶ weak

Reducing agents: $Zn > Ni > H_2 > Ag > Cl^-$

strong ⟶ weak

Finally, notice that the value of $E°_{cell}$ is greater the farther apart the oxidizing and reducing agents are on the potential ladder. For example,

$$Zn(s) + Cl_2(g) \rightarrow Zn^{2+}(aq) + 2\ Cl^-(aq) \qquad E° = +2.12 \text{ V}$$

Manganese in the Oceans

Manganese is a key component of some oxidation–reduction cycles in the oceans. According to an article in the journal *Science*, it "can perform this role because it exists in multiple oxidation states and is recycled rapidly between these states by bacterial processes."

Figure A shows how this cycle is thought to work. Manganese(II) ions in subsurface water are oxidized to form manganese(IV) oxide, MnO_2. Particles of this insoluble solid sink toward the ocean floor. However, some encounter hydrogen sulfide, which is produced in the ocean depths, rising toward the surface. Another redox reaction occurs, producing sulfur and manganese(II) ions. The newly formed Mn^{2+} ions diffuse upward, where they are again oxidized.

The manganese cycle had been thought to involve only the +2 and +4 oxidation states of manganese, and analyses of water samples assumed the dissolved manganese existed only as Mn^{2+} ions. One reason for this is that the intermediate oxidation state, Mn^{3+}, is not predicted to be stable in water. It should disproportionate to the +2 and +4 states.

$$2\ Mn^{3+}(aq) + 2\ H_2O(\ell) \longrightarrow$$
$$Mn^{2+}(aq) + MnO_2(s) + 4\ H^+(aq)$$

It is known, however, that Mn^{3+} can exist when complexed with species such as pyrophosphate ions, $P_2O_7^{4-}$.

Several years ago, geochemists suggested that Mn^{3+} ions could exist in natural water. They could be produced by bacterial action and stabilized by phosphate from ATP or ADP. They speculated that the Mn^{3+} ion could play an important part in the natural manganese cycle.

Now, other researchers have indeed discovered that, in oxygen-poor waters, the manganese(III) ion, Mn^{3+}, can persist. These ions were found in anoxic zones (zones without dissolved oxygen) below 100 m in the Black Sea and below about 15 m in the Chesapeake Bay. It is now clear that Mn^{3+} ions, which had previously been known only in the laboratory, can exist in natural waters under the right circumstances and that the manganese cycle may have to be revised.

Questions:

1. Given the following reduction potentials, show that Mn^{3+} should disproportionate to Mn^{2+} and MnO_2 at standard conditions.

$$4\ H^+(aq) + MnO_2(s) + e^- \longrightarrow$$
$$Mn^{3+}(aq) + 2\ H_2O(\ell)$$

$E° = 0.95\ V$

$$Mn^{3+}(aq) + e^- \longrightarrow Mn^{2+}(aq)$$

$E° = 1.50\ V$

2. Balance the following equations in acid solution.
   (a) Reduction of MnO_2 with HS^- to Mn^{2+} and S
   (b) Oxidation of Mn^{2+} with O_2 to MnO_2
3. Calculate $E°$ for the oxidation of Mn^{2+} with O_2 to MnO_2.

Answers to these questions are available in Appendix N.

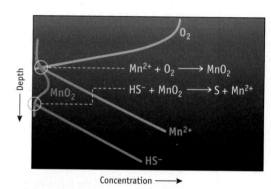

FIGURE A Manganese chemistry in the oceans. Relative concentrations of important species as a function of depth in the oceans. See K. S. Johnson, *Science*, Vol. 313, p. 1896, 2006 and R. E. Trouwborst, B. G. Clement, B. M. Tebor, B. T. Glazer, and G. W. Luther, III, *Science*, Vol. 313, pp. 1955–1957, 2006.

Substituting values for the constants in Equation 20.2, and using 298 K as the temperature, gives

$$E = E° - \frac{0.0257}{n}\ln Q \quad \text{at } 25\ °C \qquad (20.3)$$

or, in a commonly used form using base-10 logarithms,

$$E = E° - \frac{0.0592}{n}\log Q$$

In essence, the term $(RT/nF)\ln Q$ "corrects" the standard potential $E°$ for nonstandard conditions or concentrations.

⬛WL **INTERACTIVE EXAMPLE 20.6** Using the Nernst Equation

Problem A voltaic cell is set up at 25 °C with the half-cells: $Al^{3+}(0.0010\ M)\,|\,Al$ and $Ni^{2+}(0.50\ M)\,|\,Ni$. Write an equation for the reaction that occurs when the cell generates an electric current, and determine the cell potential.

What Do You Know? You know the temperature and the identity and concentrations of the reactants and products.

Strategy

- Determine which substance is oxidized (Al or Ni) by looking at the appropriate half-reactions in Table 20.1 and deciding which is the better reducing agent (Example 20.5).

- Add the half-reactions to determine the net ionic equation, and calculate $E°_{cell}$.

- Use the Nernst equation to calculate E, the potential under nonstandard conditions.

Solution Aluminum metal is a stronger reducing agent than Ni metal. (Conversely, Ni^{2+} is a better oxidizing agent than Al^{3+}.) Therefore, Al is oxidized, and the $Al^{3+}\,|\,Al$ compartment is the anode.

Cathode, reduction:	$3\,[Ni^{2+}(aq) + 2\,e^- \rightarrow Ni(s)]$
Anode, oxidation:	$2\,[Al(s) \rightarrow Al^{3+}(aq) + 3\,e^-]$
Net ionic equation:	$2\,Al(s) + 3\,Ni^{2+}(aq) \rightarrow 2\,Al^{3+}(aq) + 3\,Ni(s)$

$$E°_{cell} = E°_{cathode} - E°_{anode}$$

$$E°_{cell} = (-0.25\ V) - (-1.66\ V) = 1.41\ V$$

The expression for Q is written based on the cell reaction. In the net reaction, $Al^{3+}(aq)$ has a coefficient of 2, so this concentration is squared. Similarly, $[Ni^{2+}(aq)]$ is cubed. Solids are not included in the expression for Q (◀ Section 16.2).

$$Q = \frac{[Al^{3+}]^2}{[Ni^{2+}]^3}$$

The net equation requires transfer of six electrons from two Al atoms to three Ni^{2+} ions, so $n = 6$, and the Nernst equation gives

$$E_{cell} = E°_{cell} - \frac{0.0257}{n}\ln\frac{[Al^{3+}]^2}{[Ni^{2+}]^3}$$

$$= +1.41\ V - \frac{0.0257}{6}\ln\frac{[0.0010]^2}{[0.50]^3}$$

$$= +1.41\ V - 0.00428\,\ln(8.0 \times 10^{-6})$$

$$= +1.41\ V - 0.00428\,(-11.74)$$

$$= \boxed{1.46\ V}$$

Think about Your Answer The concentrations of Al^{3+} and Ni^{2+} both affect the cell potential. Analysis of the $\ln Q$ term in the Nernst equation shows that if $[Ni^{2+}] = 1\ M$ but $[Al^{3+}] < 1\ M$, then $E_{cell} > E°_{cell}$. The reaction is more product-favored in this situation. The reverse situation (with $[Ni^{2+}] < 1\ M$ and $[Al^{3+}] = 1\ M$) would lead to $E_{cell} < E°_{cell}$. In this example, the very low value of $[Al^{3+}]$ has the greater effect, and E_{cell} is greater than $E°_{cell}$.

Check Your Understanding

A voltaic cell is set up with an aluminum electrode in a 0.025 M $Al(NO_3)_3(aq)$ solution and an iron electrode in a 0.50 M $Fe(NO_3)_2(aq)$ solution. Determine the cell potential, E_{cell}, at 298 K.

Strategy Map 20.6

PROBLEM

Write an equation for the reaction involving **Al** and **Al³⁺** with **Ni** and **Ni²⁺**. Calculate E_{cell}.

KNOWN DATA/INFORMATION

- Concentrations of **Al³⁺** and **Ni²⁺**
- Look up $E°$ values for the half-reactions.

STEP 1. Decide which half-reactions occur at the **anode** and **cathode** based on $E°$ **values.**

Cathode reaction (reduction of Ni^{2+}) and **anode reaction** (oxidation of Al)

STEP 2. Add half-reactions to obtain **net cell reaction.**

Net cell reaction

STEP 3. Calculate $E°_{cell}$.

$E°_{cell} = E°_{cathode} - E°_{anode}$

STEP 4. Use **Nernst equation** with known ion concentrations, calculated $E°_{cell}$, and **n value** to calculate E_{cell}.

E_{cell} under nonstandard conditions

Example 20.6 demonstrates the calculation of a cell potential if concentrations are known. It is also useful to apply the Nernst equation in the opposite sense, using a measured cell potential to determine an unknown concentration. A device that does just this is the pH meter (Figure 20.16). In an electrochemical cell in which $H^+(aq)$ is a reactant or product, the cell voltage will vary predictably with the hydrogen ion concentration. The cell voltage is measured and the value used to calculate pH. Example 20.7 illustrates how E_{cell} varies with the hydrogen ion concentration in a simple cell.

FIGURE 20.16 Measuring pH.

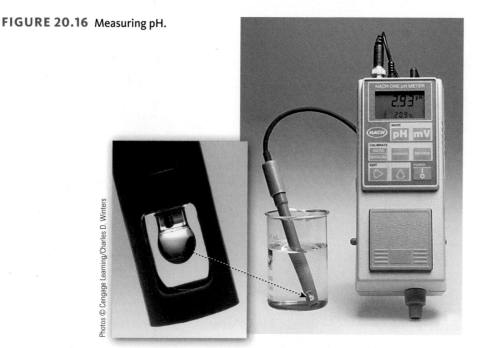

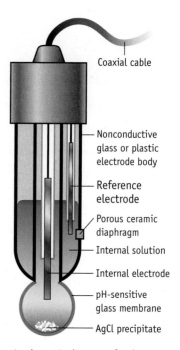

Coaxial cable

Nonconductive glass or plastic electrode body

Reference electrode

Porous ceramic diaphragm

Internal solution

Internal electrode

pH-sensitive glass membrane

AgCl precipitate

Photos © Cengage Learning/Charles D. Winters

Measuring pH with a portable pH meter that can be used in the field.

A schematic diagram of a glass electrode.

EXAMPLE 20.7   Variation of E_{cell} with Concentration

Problem A voltaic cell is set up with copper and hydrogen half-cells. Standard conditions are employed in the copper half-cell, Cu^{2+}(aq, 1.00 M) | Cu(s). The hydrogen gas pressure is 1.00 bar, and [H^+(aq)] in the hydrogen half-cell is the unknown. A value of 0.490 V is recorded for E_{cell} at 298 K. Determine the pH of the solution.

What Do You Know? You know the temperature, the identity of the reactants and products, the concentration of the Cu^{2+} ion, the partial pressure of H_2, and E_{cell}.

Strategy

• Determine which is the better oxidizing and reducing agent in order to decide what net reaction is occurring in the cell. Write a balanced equation for the reaction.

• Calculate $E°_{cell}$.

• Use the Nernst equation with the given Cu^{2+} ion concentration to calculate the hydrogen ion concentration.

• Calculate the pH.

Solution Based on their positions in a table of standard reduction potentials, hydrogen is a better reducing agent than copper metal, so Cu(s) | Cu^{2+}(aq, 1.00 M) is the cathode, and H_2(g, 1.00 bar) | H^+(aq, ? M) is the anode.

Cathode, reduction: Cu^{2+}(aq) + 2 e^- → Cu(s)

Anode, oxidation: H_2(g) → 2 H^+(aq) + 2 e^-

Net ionic equation: H_2(g) + Cu^{2+}(aq) → Cu(s) + 2 H^+(aq)

$$E°_{cell} = E°_{cathode} - E°_{anode}$$

$$E°_{cell} = (+0.337 \text{ V}) - (0.00 \text{ V}) = +0.337 \text{ V}$$

The reaction quotient, Q, is derived from the balanced net ionic equation.

$$Q = \frac{[H^+]^2}{[Cu^{2+}]P_{H_2}}$$

The net equation requires the transfer of two electrons, so $n = 2$. The value of $[Cu^{2+}]$ is 1.00 M, but $[H^+]$ is unknown. Substitute this information into the Nernst equation (and don't overlook the fact that $[H^+]$ is squared in the expression for Q).

$$E = E° - \frac{0.0257}{n} \ln \frac{[H^+]^2}{[Cu^{2+}]P_{H_2}}$$

$$0.490\ V = 0.337\ V - \frac{0.0257}{2} \ln \frac{[H^+]^2}{(1.00)(1.00)}$$

$$-11.9 = \ln[H^+]^2$$

$$[H^+] = 3 \times 10^{-3}\ M$$

$$pH = 2.6$$

Think about Your Answer Be sure to write the balanced equation. Without it you may not have the correct exponents in the term for Q in the Nernst equation.

Check Your Understanding

The half-cells Fe^{2+}(aq, 0.024 M) | Fe(s) and H^+(aq, 0.056 M) | H_2(1.0 bar) are linked by a salt bridge to create a voltaic cell. Determine the cell potential, E_{cell}, at 298 K.

In the real world, using a hydrogen electrode in a pH meter is not practical. The apparatus is clumsy; it is anything but robust; and platinum (for the electrode) is costly. Common pH meters use a glass electrode, so called because it contains a thin glass membrane separating the cell from the solution whose pH is to be measured (Figure 20.16). Inside the glass electrode is a silver wire coated with AgCl and a solution of HCl; outside is the solution of unknown pH to be evaluated. A Ag/AgCl or calomel electrode—the latter a common reference electrode using a mercury(I)–mercury redox couple (Hg_2Cl_2 | Hg)—serves as the second electrode of the cell. The potential across the glass membrane depends on $[H^+]$. Common pH meters give a direct readout of pH.

REVIEW & CHECK FOR SECTION 20.5

Calculate E_{cell} at 298 K for a cell involving Sn and Cu and their ions:

$$Sn(s) | Sn^{2+}(aq,\ 0.25\ M) || Cu^{2+}(aq,\ 0.10\ M) | Cu(s)$$

(a) 0.47 V   (b) 0.49 V   (c) 0.50 V

20.6 Electrochemistry and Thermodynamics
Work and Free Energy

The first law of thermodynamics states that the internal energy change in a system (ΔU) is related to two quantities, heat (q) and work (w): $\Delta U = q + w$ [Section 5.4]. This equation also applies to chemical changes that occur in a voltaic cell. As current flows, energy is transferred from the system (the voltaic cell) to the surroundings.

In a voltaic cell, the decrease in internal energy in the system will manifest itself ideally as electrical work done on the surroundings by the system. In practice, however, some energy is usually evolved as heat by the voltaic cell. The maximum work done by an electrochemical system (ideally, assuming no heat is generated) is proportional to the potential difference (volts) and the quantity of charge (coulombs):

$$w_{max} = nFE \qquad (20.4)$$

In this equation, E is the cell voltage, and nF is the quantity of electric charge transferred from anode to cathode.

The free energy change for a process is, by definition, the maximum amount of work that can be obtained [Section 19.6]. Because the maximum work and the cell potential are related, $E°$ and $\Delta_r G°$ can be related mathematically (taking care to assign signs correctly). The maximum work done on the surroundings when electricity is produced by a voltaic cell is $+nFE$, with the positive sign denoting an increase in energy in the surroundings. The energy content of the cell decreases by this amount. Thus, $\Delta_r G$ for the voltaic cell has the opposite sign.

$$\Delta_r G = -nFE \tag{20.5}$$

● **Units in Equation 20.6** n has units of mol e^-, and F has units of (C/mol e^-). Therefore, nF has units of coulombs (C). Because 1 J = 1 C · V, the product nFE will have units of energy (J).

Under standard conditions, the appropriate equation is

$$\Delta_r G° = -nFE° \tag{20.6}$$

This expression shows that, the more positive the value of $E°$, the more negative the value of $\Delta_r G°$ for the reaction. Also, because of the relationship between $\Delta_r G°$ and K, the farther apart the half-reactions are on the potential ladder, the more strongly product-favored the reaction is at equilibrium.

EXAMPLE 20.8   Relating $E°$ and $\Delta_r G°$

Problem The standard cell potential, $E°_{cell}$, for the reduction of silver ions with copper metal (Figure 20.5) is +0.462 V at 25 °C. Calculate $\Delta_r G°$ for this reaction.

What Do You Know? You know the cell potential under standard conditions and therefore know to use Equation 20.6 to calculate the change in free energy. In this equation F is a known constant, but n has to be determined. However, you can write the balanced equation for the cell reaction, and so obtain n from that.

Strategy Use Equation 20.6 where $E°_{cell}$ and F are known. The value of n, the number of moles of electrons transferred between copper metal and silver ions, comes from the balanced equation.

Solution In this cell, copper is the anode, and silver is the cathode. The overall cell reaction is

$$Cu(s) + 2\ Ag^+(aq) \rightarrow Cu^{2+}(aq) + 2\ Ag(s)$$

which means that each mole of copper transfers 2 mol of electrons to 2 mol of Ag^+ ions. That is, $n = 2$. Now use Equation 20.6.

$$\Delta_r G° = -nFE° = -(2\text{ mol }e^-)(96{,}485\text{ C/mol }e^-)(0.462\text{ V}) = -89{,}200\text{ C} \cdot \text{V}$$

Because 1 C · V = 1 J, we have

$$\Delta_r G° = -89{,}200\text{ J or } -89.2\text{ kJ}$$

Think about Your Answer This example demonstrates an effective method of obtaining thermodynamic values from electrochemical experiments. Keep in mind that a positive $E°$ implies a negative $\Delta_r G°$.

Check Your Understanding

The following reaction has an $E°$ value of −0.76 V:

$$H_2(g) + Zn^{2+}(aq) \rightarrow Zn(s) + 2\ H^+(aq)$$

Calculate $\Delta_r G°$ for this reaction. Is the reaction product- or reactant-favored at equilibrium?

$E°$ and the Equilibrium Constant

When a voltaic cell produces an electric current, the reactant concentrations decrease, and the product concentrations increase. The cell voltage also changes. As reactants are converted to products, the value of E_{cell} decreases and the cell potential eventually reaches zero; no further net reaction occurs, and equilibrium is achieved.

This situation can be analyzed using the Nernst equation. When $E_{cell} = 0$, the reactants and products are at equilibrium, and the reaction quotient, Q, is equal to the equilibrium constant, K. Substituting the appropriate symbols and values into the Nernst equation,

$$E = 0 = E° - \frac{0.0257}{n} \ln K$$

and collecting terms gives an equation that relates the cell potential and equilibrium constant:

$$\ln K = \frac{nE°}{0.0257} \quad \text{at 25 °C (298 K)} \qquad \textbf{(20.7)}$$

Equation 20.7 can be used to determine values for equilibrium constants, as illustrated in Example 20.9.

⦿WL INTERACTIVE EXAMPLE 20.9 *E° and Equilibrium Constants*

Problem Calculate the equilibrium constant for the reaction at 298 K:

$$Fe(s) + Cd^{2+}(aq) \rightleftharpoons Fe^{2+}(aq) + Cd(s)$$

What Do You Know? You have the balanced chemical equation and know that Equation 20.7 is required. You need to know $E°$ and n, the number of electrons transferred.

Strategy First, determine $E°_{cell}$ from $E°$ values for the two half-reactions (see Example 20.5) and from those the value of n, the other parameter required in Equation 20.7.

Solution The half-reactions and $E°$ values are

Cathode, reduction: $Cd^{2+}(aq) + 2\ e^- \rightarrow Cd(s)$

Anode, oxidation: $Fe(s) \rightarrow Fe^{2+}(aq) + 2\ e^-$

Net ionic equation: $Fe(s) + Cd^{2+}(aq) \rightleftharpoons Fe^{2+}(aq) + Cd(s)$

$$E°_{cell} = E°_{cathode} - E°_{anode}$$

$$E°_{cell} = (-0.40\ V) - (-0.44\ V) = +0.04\ V$$

Now substitute $n = 2$ and $E°_{cell}$ into Equation 20.7.

$$\ln K = \frac{nE°}{0.0257} = \frac{(2)(0.04\ V)}{0.0257} = 3.1$$

$$K = 20$$

Think about Your Answer The relatively small positive voltage (0.04 V) for the cell indicates that the cell reaction is only mildly product-favored. A value of 20 for the equilibrium constant is in accord with this observation.

Check Your Understanding

Calculate the equilibrium constant at 25 °C for the reaction

$$2\ Ag^+(aq) + Hg(\ell) \rightleftharpoons 2\ Ag(s) + Hg^{2+}(aq)$$

The relationships between $E°$, K, and $\Delta_r G°$, which are summarized in Table 20.2. Values of $E°$ can be used to obtain equilibrium constants for many different chemical systems. One example is the determination of solubility product constants, K_{sp}. Let us begin with an electrode in which an insoluble ionic compound, AgCl, is a component of a half-cell. Figure 20.17 illustrates how the potential for the reduction of AgCl in the presence of Cl^- ion (1.00 M) can be determined.

$$AgCl(s) + e^- \rightarrow Ag(s) + Cl^-(aq) \qquad E° = +0.222\ V$$

Strategy Map 20.9

PROBLEM

Calculate an **equilibrium constant** using **electrochemical data**.

↓

KNOWN DATA/INFORMATION

- Balanced equation

STEP 1. Decide which half-reactions occur at **anode** and **cathode** based on $E°$ **values** and calculate $E°_{cell}$.

$$E°_{cell} = E°_{cathode} - E°_{anode}$$

STEP 2. Use Equation 20.7 with calculated $E°_{cell}$, and n **value** to calculate $\ln K$.

Calculated $\ln K$

STEP 3. Convert $\ln K$ to K.

Equilibrium constant, K

Table 20.2 Summary of the Relationship of K, $\Delta_rG°$, and $E°$

K	$\Delta_rG°$	$E°$	Reactant-Favored or Product-Favored at Equilibrium?	Spontaneous under Standard Conditions?
$K \gg 1$	$\Delta_rG° < 0$	$E° > 0$	Product-favored	Spontaneous under standard conditions
$K = 1$	$\Delta_rG° = 0$	$E° = 0$	$[C]^c[D]^d = [A]^a[B]^b$ at equilibrium	At equilibrium under standard conditions
$K \ll 1$	$\Delta_rG° > 0$	$E° < 0$	Reactant-favored	Not spontaneous under standard conditions

When paired with the standard hydrogen electrode, the standard reduction potential for the AgCl|Ag half-cell is +0.222 V. If this half-reaction is then paired with that for a standard silver electrode in a hypothetical voltaic cell, the cell reactions could be written as

Cathode, reduction: $\quad AgCl(s) + e^- \rightarrow Ag(s) + Cl^-(aq)$

Anode, oxidation: $\quad\quad Ag(s) \rightarrow Ag^+(aq) + e^-$

Net ionic equation: $\quad\overline{AgCl(s) \rightarrow Ag^+(aq) + Cl^-(aq)}$

The equation for the net reaction represents the equilibrium of solid AgCl and its ions. The cell potential is negative,

$$E°_{cell} = E°_{cathode} - E°_{anode} = (+0.222 \text{ V}) - (+0.799 \text{ V}) = -0.577 \text{ V}$$

indicating a reactant-favored process, as would be expected based on the low solubility of AgCl. Using Equation 20.7, the value of K_{sp} can then be obtained from $E°_{cell}$.

$$\ln K = \frac{nE°}{0.0257} = \frac{(1)(-0.577 \text{ V})}{0.0257} = -22.5$$

$$K_{sp} = e^{-22.5} = 2 \times 10^{-10}$$

FIGURE 20.17 Measurement of the standard electrode potential for the Ag|AgCl electrode.

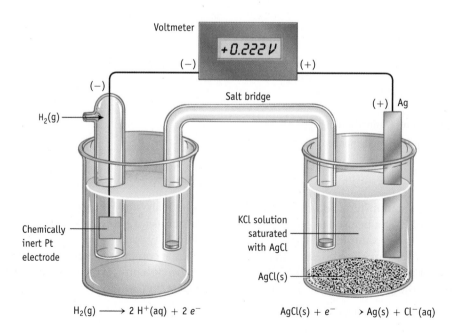

REVIEW & CHECK FOR SECTION 20.6

In Appendix M, the following standard reduction potential is reported:

$$[Zn(CN)_4]^{2-}(aq) + 2\ e^- \rightarrow Zn(s) + 4\ CN^-(aq) \qquad E° = -1.26\ V$$

Use this information, along with the data on the $Zn^{2+}(aq)\,|\,Zn$ half-cell, to calculate the equilibrium constant for the reaction

$$Zn^{2+}(aq) + 4\ CN^-(aq) \rightarrow [Zn(CN)_4]^{2-}(aq)$$

What is the value for the formation constant for this complex ion at 25 °C?

(a) 2.8×10^8 (b) 6.3×10^{16} (c) 1.9×10^{68}

20.7 Electrolysis: Chemical Change Using Electrical Energy

Thus far, we have described electrochemical cells that use product-favored redox reactions to generate an electric current. Equally important, however, is the opposite process, **electrolysis**, the use of electrical energy to bring about chemical change.

Electrolysis of water is a classic chemistry experiment, and the electroplating of metals is another example of electrolysis (Figure 20.18). In electroplating, an electric current is passed through a solution containing a salt of the metal to be plated. The object to be plated is the cathode. When metal ions in solution are reduced, the metal deposits on the object's surface.

Electrolysis is also important because it is widely used in the refining of metals such as aluminum and in the production of chemicals such as chlorine.

Electrolysis of Molten Salts

All electrolysis experiments are set up in a similar manner. The material to be electrolyzed, either a molten salt or a solution, is contained in an electrolysis cell. As was the case with voltaic cells, ions must be present in the liquid or solution for a current

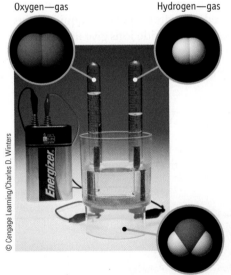

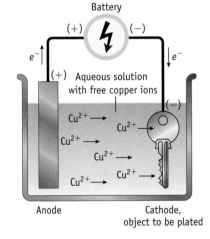

(a) Electrolysis of water produces hydrogen (at the cathode) and oxygen gas (at the anode).

(b) Electroplating adds a layer of metal to the surface of an object *(left)*, either to protect the object from corrosion or to improve its physical appearance. The procedure uses an electrolysis cell, set up with the object to be plated as the cathode and a solution containing a salt of the metal to be plated *(right)*.

FIGURE 20.18 Electrolysis.

FIGURE 20.19 The preparation of sodium and chlorine by the electrolysis of molten NaCl. In the molten state, sodium ions migrate to the negative cathode, where they are reduced to sodium metal. Chloride ions migrate to the positive anode, where they are oxidized to elemental chlorine.

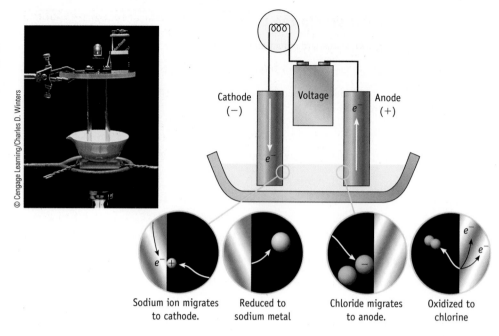

© Cengage Learning/Charles D. Winters

Sodium ion migrates to cathode.

Reduced to sodium metal

Chloride migrates to anode.

Oxidized to chlorine

to flow as it is the movement of ions within the cell that constitutes the electric current. The cell has two electrodes that are connected to a source of DC (direct-current) voltage. If the applied voltage is high enough, chemical reactions occur at the two electrodes. Reduction occurs at the negatively charged cathode, with electrons being transferred from that electrode to a chemical species in the cell. Oxidation occurs at the positive anode, with electrons from a chemical species being transferred to that electrode.

Let us focus our attention first on the chemical reactions that occur at each electrode in the electrolysis of a molten salt. Sodium chloride melts at about 800 °C, and in the molten state sodium ions (Na^+) and chloride ions (Cl^-) are freed from their rigid arrangement in the crystalline lattice. Therefore, if a potential is applied to the electrodes, sodium ions are attracted to the negative electrode, and chloride ions are attracted to the positive electrode (Figure 20.19). If the potential is high enough, chemical reactions occur at each electrode. At the negative cathode, Na^+ ions accept electrons and are reduced to sodium metal (a liquid at this temperature). Simultaneously, at the positive anode, chloride ions give up electrons and form elemental chlorine.

Cathode ($-$), reduction:	$2\ Na^+ + 2\ e^- \rightarrow 2\ Na(\ell)$
Anode ($+$), oxidation:	$2\ Cl^- \rightarrow Cl_2(g) + 2\ e^-$
Net ionic equation:	$2\ Na^+ + 2\ Cl^- \rightarrow 2\ Na(\ell) + Cl_2(g)$

Electrons move through the external circuit under the force exerted by the applied potential, and the movement of positive and negative ions in the molten salt

PROBLEM SOLVING TIP 20.3

Whether you are describing a voltaic cell or an electrolysis cell, the terms *anode* and *cathode* always refer to the electrodes at which oxidation and reduction occur, respectively. The polarity of the electrodes is reversed, however.

Electrochemical Conventions: Voltaic Cells and Electrolysis Cells

Type of Cell	Electrode	Function	Polarity
Voltaic	Anode	Oxidation	$-$
	Cathode	Reduction	$+$
Electrolysis	Anode	Oxidation	$+$
	Cathode	Reduction	$-$

constitutes the current within the cell. Finally, it is important to recognize that the reaction is not spontaneous. The energy required for this reaction to occur has been provided by the electric current.

Electrolysis of Aqueous Solutions

Sodium ions (Na^+) and chloride ions (Cl^-) are the primary species present in molten NaCl. Only chloride ions can be oxidized, and only sodium ions can be reduced. Electrolysis of a substance in aqueous solution is more complicated than the electrolysis of a molten salt, however, because water is now present. Water is an *electroactive* substance; that is, it can be oxidized or reduced in an electrochemical process.

Consider the electrolysis of aqueous sodium iodide (Figure 20.20). In this experiment, the electrolysis cell contains $Na^+(aq)$, $I^-(aq)$, and H_2O molecules. Possible *reduction reactions* at the *negative cathode* include

$$Na^+(aq) + e^- \rightarrow Na(s)$$

$$2\ H_2O(\ell) + 2\ e^- \rightarrow H_2(g) + 2\ OH^-(aq)$$

Possible *oxidation reactions* at the *positive anode* are

$$2\ I^-(aq) \rightarrow I_2(aq) + 2\ e^-$$

$$2\ H_2O(\ell) \rightarrow O_2(g) + 4\ H^+(aq) + 4\ e^-$$

In the electrolysis of aqueous NaI, $H_2(g)$ and $OH^-(aq)$ are formed by water reduction at the cathode, and iodine is formed at the anode. Thus, the overall cell process can be summarized by the following equations:

Cathode ($-$), reduction: $2\ H_2O(\ell) + 2\ e^- \rightarrow H_2(g) + 2\ OH^-(aq)$

Anode ($+$), oxidation: $2\ I^-(aq) \rightarrow I_2(aq) + 2\ e^-$

Net ionic equation: $2\ H_2O(\ell) + 2\ I^-(aq) \rightarrow H_2(g) + 2\ OH^-(aq) + I_2(aq)$

where $E°_{cell}$ has a negative value.

$$E°_{cell} = E°_{cathode} - E°_{anode} = (-0.8277\ V) - (+0.621\ V) = -1.449\ V$$

This process is not spontaneous under standard conditions, and a potential of *at least* 1.45 V must be *applied* to the cell for these reactions to occur. If the process had involved the oxidation of water instead of iodide ion at the anode, the required

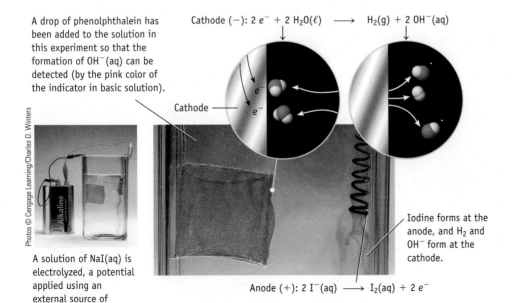

A drop of phenolphthalein has been added to the solution in this experiment so that the formation of $OH^-(aq)$ can be detected (by the pink color of the indicator in basic solution).

Cathode ($-$): $2\ e^- + 2\ H_2O(\ell) \longrightarrow H_2(g) + 2\ OH^-(aq)$

Cathode

A solution of NaI(aq) is electrolyzed, a potential applied using an external source of electricity.

Anode ($+$): $2\ I^-(aq) \longrightarrow I_2(aq) + 2\ e^-$

Iodine forms at the anode, and H_2 and OH^- form at the cathode.

FIGURE 20.20 Electrolysis of aqueous NaI.

Photos © Cengage Learning/Charles D. Winters

potential would be -2.057 V $[E^\circ_{cathode} - E^\circ_{anode} = (-0.8277\,\text{V}) - (+1.229\,\text{V})]$, and if the reaction involving the reduction of Na^+ and the oxidation of I^- had occurred, the required potential would be -3.335 V $[E^\circ_{cathode} - E^\circ_{anode} = (-2.714$ V$) - (+0.621\,\text{V})]$. The reaction occurring is the one requiring the smaller applied potential, so the net cell reaction in the electrolysis of NaI(aq) is the oxidation of iodide and reduction of water.

What happens if an aqueous solution of some other metal halide such as $SnCl_2$ is electrolyzed? Consult Appendix M, and consider all possible half-reactions. In this case, aqueous Sn^{2+} ion is much more easily reduced ($E^\circ = -0.14$ V) than water ($E^\circ = -0.83$ V) at the cathode, so tin metal is produced. At the anode, two oxidations are possible: Cl^-(aq) to Cl_2(g) or H_2O to O_2(g). Experiments show that chloride ion is oxidized in preference to water, so the reactions occurring on electrolysis of aqueous tin(II) chloride are (Figure 20.21)

Cathode $(-)$, reduction: $Sn^{2+}(aq) + 2\,e^- \rightarrow Sn(s)$

Anode $(+)$, oxidation: $\underline{2\,Cl^-(aq) \rightarrow Cl_2(g) + 2\,e^-}$

Net ionic equation: $Sn^{2+}(aq) + 2\,Cl^-(aq) \rightarrow Sn(s) + Cl_2(g)$

$$E^\circ_{cell} = E^\circ_{cathode} - E^\circ_{anode} = (-0.14\,\text{V}) - (+1.36\,\text{V}) = -1.50\,\text{V}$$

Formation of Cl_2 at the anode in the electrolysis of $SnCl_2$(aq) is contrary to a prediction based on E° values. If the electrode reactions were

Cathode $(-)$, reduction: $Sn^{2+}(aq) + 2\,e^- \rightarrow Sn(s)$

Anode $(+)$, oxidation: $2\,H_2O(\ell) \rightarrow O_2(g) + 4\,H^+(aq) + 4\,e^-$

$$E^\circ_{cell} = (-0.14\,\text{V}) - (+1.23\,\text{V}) = -1.37\,\text{V}$$

a smaller applied potential would seemingly be required. To explain the formation of chlorine instead of oxygen, we must take into account rates of reaction. The oxidation of Cl^-(aq) is much more rapid than the oxidation of H_2O.

● **Overvoltage** Voltages higher than the minimum are typically used to speed up reactions that would otherwise be slow. The term *overvoltage* is often used and refers to the voltage needed to make a reaction occur at a reasonable rate.

This "problem" is used to advantage in the commercially important electrolysis of aqueous NaCl, where a voltage high enough to oxidize both Cl^- and H_2O is used. However, because chloride ion is oxidized much faster than H_2O, Cl_2 is the major product in this electrolysis. This is the predominant means by which chlorine is produced for commercial use.

Another instance in which rates are important concerns electrode materials. Graphite, commonly used to make inert electrodes, can be oxidized. For the half-reaction $CO_2(g) + 4\,H^+(aq) + 4\,e^- \rightarrow C(s) + 2\,H_2O(\ell)$, E° is $+0.20$ V, indicating that carbon is slightly easier to oxidize than copper ($E^\circ = +0.34$ V). Based on this value, oxidation of a graphite electrode might reasonably be expected to occur during an electrolysis. And indeed it does, albeit slowly; graphite electrodes used in electrolysis cells slowly deteriorate and eventually have to be replaced.

FIGURE 20.21 Electrolysis of aqueous tin(II) chloride. Tin metal collects at the negative cathode. Chlorine gas is formed at the positive anode. Elemental chlorine is formed in the cell, in spite of the fact that the potential for the oxidation of Cl^- is more negative than that for oxidation of water. (That is, chlorine should be less easily oxidized than water.) This is the result of chemical kinetics and illustrates the complexity of some aqueous electrochemistry.

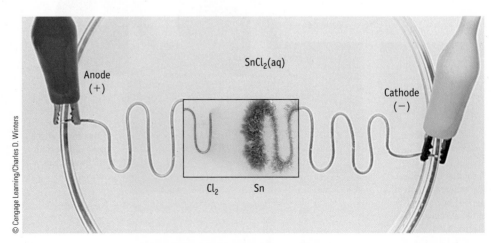

One other factor—the concentration of electroactive species in solution—must be taken into account when discussing electrolyses. As shown in Section 20.6, the potential at which a species in solution is oxidized or reduced depends on concentration. Unless standard conditions are used, predictions based on $E°$ values are merely qualitative. In addition, the rate of a half-reaction depends on the concentration of the electroactive substance at the electrode surface. At a very low concentration, the rate of the redox reaction may depend on the rate at which an ion diffuses from the solution to the electrode surface.

EXAMPLE 20.10 Electrolysis of Aqueous Solutions

Problem Predict how products of the electrolysis of aqueous solutions of NaF, NaCl, NaBr, and NaI are likely to be different and predict $E°_{cell}$ for each electrolysis. (The electrolysis of NaI is illustrated in Figure 20.20.)

What Do You Know? You know the identity of the compounds to be electrolyzed, but you will need to know the $E°$ values for their half-reactions and for water electrolysis.

Strategy The main criterion used to predict the chemistry in an electrolytic cell should be the ease of oxidation and reduction, an assessment based on $E°$ values.

Solution The cathode reaction presents no problem: Water is reduced to hydroxide ion and H_2 gas in preference to reduction of $Na^+(aq)$ (as in the electrolysis of aqueous NaI). Thus, the primary cathode reaction in all cases is

$$2 H_2O(\ell) + 2 e^- \rightarrow H_2(g) + 2 OH^-(aq)$$

$$E°_{cathode} = -0.83 V$$

At the anode, you need to assess the ease of oxidation of the halide ions relative to water. Based on $E°$ values, this should be $I^-(aq) > Br^-(aq) > Cl^-(aq) >> F^-(aq)$. Fluoride ion is much more difficult to oxidize than water, and electrolysis of an aqueous solution containing this ion results exclusively in O_2 formation. That is, the primary anode reaction for NaF(aq) is

$$2 H_2O(\ell) \rightarrow O_2(g) + 4 H^+(aq) + 4 e^-$$

$$E°_{anode} = +1.23 V$$

Therefore, in this case,

$$E°_{cell} = (-0.83 V) - (+1.23 V) = -2.06 V$$

Recall that chlorine is the primary product at the anode in the electrolysis of aqueous solutions of chloride salts (as in Figure 20.20). Therefore, the primary anode reaction in NaCl(ag) is

$$2 Cl^-(aq) \rightarrow Cl_2(g) + 2 e^-$$

$$E°_{cell} = (-0.83 V) - (+1.36 V) = -2.19 V$$

Bromide ions are considerably easier to oxidize than chloride ions, so Br_2 may be expected as the primary product in the electrolysis of aqueous NaBr. For NaBr(aq), the primary anode reaction is

$$2 Br^-(aq) \rightarrow Br_2(\ell) + 2 e^-$$

so $E°_{cell}$ is

$$E°_{cell} = (-0.83 V) - (+1.08 V) = -1.91 V$$

Thus, the electrolysis of NaBr resembles that of NaI (Figure 20.19) in producing the halogen, hydrogen gas, and hydroxide ion. The half-reactions and the cell potential for aqueous NaI were given on page 727.

Think about Your Answer As described above, you would predict from $E°$ values the ease of oxidation of halide ions is $I^-(aq) > Br^-(aq) > Cl^-(aq) >> F^-(aq)$. This is confirmed by the results.

Check Your Understanding

Predict the chemical reactions that will occur at the two electrodes in the electrolysis of an aqueous sodium hydroxide solution. What is the minimum voltage needed to cause this reaction to occur?

A CLOSER LOOK

Electrochemistry and Michael Faraday

The terms *anion, cation, electrode,* and *electrolyte* originated with Michael Faraday (1791–1867), one of the most influential people in the history of chemistry. Faraday was apprenticed to a bookbinder in London when he was 13. This situation suited him perfectly, as he enjoyed reading the books sent to the shop for binding. By chance, one of these volumes was a small book on chemistry, which whetted his appetite for science, and he began performing experiments on electricity. In 1812, a patron of the shop invited Faraday to accompany him to the Royal Institute to attend a lecture by one of the most famous chemists of the day, Sir Humphry Davy. Faraday was intrigued by Davy's lecture and wrote to ask Davy for a position as an assistant. He was

accepted and began work in 1813. Faraday was so talented that his work proved extraordinarily fruitful, and only 12 years later he was made the director of the laboratory of the Royal Institute.

It has been said that Faraday's contributions were so enormous that, had there been Nobel Prizes when he was alive, he would have received at least six. These could have been awarded for discoveries such as the following:

• Electromagnetic induction, which led to the first transformer and electric motor
• The laws of electrolysis (the effect of electric current on chemicals)
• The magnetic properties of matter

• Benzene and other organic chemicals (which led to important chemical industries)
• The "Faraday effect" (the rotation of the plane of polarized light by a magnetic field)
• The introduction of the concept of electric and magnetic fields

Michael Faraday (1791–1867)

© Oesper Collection in the History of Chemistry/

In addition to making discoveries that had profound effects on science, Faraday was an educator. He wrote and spoke about his work in memorable ways, especially in lectures to the general public that helped to popularize science.

REVIEW & CHECK FOR SECTION 20.7

You have a solution containing several metal ions, K^+, Fe^{2+}, Al^{3+}, Ag^+. Which ion will require the lowest potential to plate out on the cathode?

(a) K^+ (b) Fe^{2+} (c) Al^{3+} (d) Ag^+

20.8 Counting Electrons

Metallic silver is produced at the cathode in the electrolysis of aqueous $AgNO_3$ in which one mole of electrons is required to produce one mole of silver. In contrast, two moles of electrons are required to produce one mole of tin (Figure 20.21):

$$Sn^{2+}(aq) + 2\ e^- \rightarrow Sn(s)$$

It follows that if the number of moles of electrons flowing through the electrolysis cell could be measured, the number of moles of silver or tin produced could be calculated. Conversely, if the amount of silver or tin produced is known, then the number of moles of electrons moving through the circuit could be calculated.

The number of moles of electrons consumed or produced in an electron transfer reaction is obtained by measuring the current flowing in the external electric circuit in a given time. The **current** flowing in an electrical circuit is the amount of charge (in units of coulombs, C) per unit time, and the usual unit for current is the ampere (A). One ampere equals the passage of one coulomb of charge per second.

$$\text{Current (amperes, A)} = \frac{\text{electric charge (coulombs, C)}}{\text{time, } t \text{ (seconds, s)}} \qquad (20.8)$$

● **Faraday Constant** The Faraday constant is the charge carried by 1 mol of electrons: 9.6485338×10^4 C/mol e^-.

The current passing through an electrochemical cell and the time for which the current flows are easily measured quantities. Therefore, the charge (in coulombs) that passes through a cell can be obtained by multiplying the current (in amperes) by the time (in seconds). Knowing the charge and using the Faraday constant as a conversion factor, we can calculate the number of moles of electrons that passed through an electrochemical cell. In turn, we can use this quantity to calculate the quantities of reactants and products. The following example illustrates this type of calculation.

⬙WL   INTERACTIVE EXAMPLE 20.11 Using the Faraday Constant

Problem A current of 2.40 A is passed through a solution containing $Cu^{2+}(aq)$ for 30.0 minutes, with copper metal being deposited at the cathode. What mass of copper, in grams, is deposited?

What Do You Know? You know the current passed through the cell and the time over which it was passed.

Strategy The current and time can be used to calculate the amount of charge that passed through the cell. The Faraday constant can then be used to relate this to the amount (moles) of electrons that were used. This in turn can be related to the amount of copper metal deposited and finally to the mass of copper.

Solution

1. Calculate the charge (number of coulombs) passing through the cell in 30.0 minutes.

$$\text{Charge (C)} = \text{current (A)} \times \text{time (s)}$$

$$= (2.40 \text{ A})(30.0 \text{ min})(60.0 \text{ s/min})$$

$$= 4.32 \times 10^3 \text{ C}$$

2. Calculate the number of moles of electrons (i.e., the number of Faradays of electricity).

$$(4.32 \times 10^3 \text{ C})\left(\frac{1 \text{ mol } e^-}{96{,}485 \text{ C}}\right) = 4.48 \times 10^{-2} \text{ mol } e^-$$

3. Calculate the amount of copper and, from this, the mass of copper.

$$\text{mass of copper} = (4.48 \times 10^{-2} \text{ mol } e^-)\left(\frac{1 \text{ mol Cu}}{2 \text{ mol } e^-}\right)\left(\frac{63.55 \text{ g Cu}}{1 \text{ mol Cu}}\right) = 1.42 \text{ g}$$

Think about Your Answer The key relation in this calculation is "current = charge/time." Most situations will involve knowing two of these three quantities from experiment and calculating the third.

Check Your Understanding

1. Calculate the mass of O_2 produced in the electrolysis of water, using a current of 0.445 A for a period of 45 minutes.

2. In the commercial production of sodium by electrolysis, the cell operates at 7.0 V and a current of 25×10^3 A. What mass of sodium can be produced in 1 hour?

Strategy Map 20.11

PROBLEM

Calculate the **mass of copper** deposited in electrolysis cell.

↓

KNOWN DATA/INFORMATION
- Current in cell
- Time

> STEP 1. Use **current × time** to calculate **charge**.

Charge passing through cell (in **coulombs**)

> STEP 2. Use **Faraday constant** to calculate **moles of electrons** passed.

Moles of electrons passed in given **time**

> STEP 3. Relate moles of **electrons** to moles of **metal**.

Amount of metal **(mol)** produced

> STEP 4. Convert **amount** to **mass**.

Mass of metal **(g)** produced

REVIEW & CHECK FOR SECTION 20.8

If you wish to convert 0.0100 mol of $Au^{3+}(aq)$ ions into Au(s) in a "gold-plating" process, how long must you electrolyze a solution if the current passing through the circuit is 2.00 amps?

(a) 483 seconds

(b) 4.83×10^4 seconds

(c) 965 seconds

(d) 1450 seconds

CHAPTER GOALS REVISITED

Now that you have studied this chapter, you should ask whether you have met the chapter goals. In particular, you should be able to:

Balance equations for oxidation–reduction reactions in acidic or basic solutions using the half-reaction approach Study Questions: 1–6, 15, 16, 49, 50, 84, 85, and Go Chemistry Module 25.

Understand the principles underlying voltaic cells

a. In a voltaic cell, identify the half-reactions occurring at the anode and the cathode, the polarity of the electrodes, the direction of electron flow in the external circuit, and the direction of ion flow in the salt bridge (Section 20.2). Study Questions: 7–10, 51.

 and

Sign in at **www.cengage.com/owl** to:
- View tutorials and simulations, develop problem-solving skills, and complete online homework assigned by your professor.
- For quick review and exam prep, download Go Chemistry mini lecture modules from OWL (or purchase them at **www.cengagebrain.com**)

❓ Access **How Do I Solve It?** tutorials on how to approach problem solving using concepts in this chapter.

b. Appreciate the chemistry and advantages and disadvantages of dry cells, alkaline batteries, lead storage batteries, lithium batteries, and Ni-cad batteries (Section 20.3). **Study Questions: 11, 12.**

c. Understand how fuel cells work, and recognize the difference between batteries and fuel cells (Section 20.3). **Study Question: 101.**

Understand how to use electrochemical potentials

a. Understand the process by which standard reduction potentials are determined, and identify standard conditions as applied to electrochemistry (Section 20.4).

b. Describe the standard hydrogen electrode ($E° = 0.00$ V), and explain how it is used as the standard to determine the standard potentials of half-reactions (Section 20.4).

c. Know how to use standard reduction potentials to determine cell voltages for cells under standard conditions (Equation 20.1). **Study Questions: 13–16.**

d. Know how to use a table of standard reduction potentials (Table 20.1 and Appendix M) to rank the strengths of oxidizing and reducing agents, to predict which substances can reduce or oxidize another species, and to predict whether redox reactions will be product-favored or reactant-favored (Section 20.4). **Study Questions: 17–24, 55, 56.**

e. Use the Nernst equation (Equations 20.2 and 20.3) to calculate the cell potential under nonstandard conditions (Section 20.5). **Study Questions: 25–28, 75–79.**

f. Explain how cell voltage allows the determination of pH (Section 20.5) and other ion concentrations. **Study Questions: 29, 30, 77.**

g. Use the relationships between cell voltage ($E°_{cell}$) and free energy ($\Delta_r G°$) (Equations 20.5 and 20.6) and between $E°_{cell}$ and an equilibrium constant for the cell reaction (Equation 20.7) (Section 20.6 and Table 20.2). **Study Questions: 31–36, 60, 61, 63, 78, 82.**

Explore electrolysis, the use of electrical energy to produce chemical change

a. Describe the chemical processes occurring in an electrolysis. Recognize the factors that determine which substances are oxidized and reduced at the electrodes (Section 20.7). **Study Questions: 37–42, 74.**

b. Relate the amount of a substance oxidized or reduced to the amount of current and the time the current flows (Section 20.8). **Study Questions: 43–48, 59, 64–66.**

Key Equations

Equation 20.1 (page 711) Calculating a standard cell potential, $E°_{cell}$, from standard half-cell potentials.

$$E°_{cell} = E°_{cathode} - E°_{anode}$$

Equation 20.2 (page 717) The Nernst equation, the relationship of the cell potential under nonstandard conditions (E) to that under standard conditions ($E°$). R is the gas constant (8.314472 J/K · mol); T is the temperature (K); and n is the number of moles of electrons transferred between oxidizing and reducing agents. F is the **Faraday constant** (9.6485338×10^4 C/mol of e^-), and Q is the reaction quotient.

$$E = E° - (RT/nF) \ln Q$$

Equation 20.3 (page 718) Nernst equation (at 298 K).

$$E = E° - \frac{0.0257}{n} \ln Q$$

Equation 20.4 (page 721) The amount of work done (w) by an electrochemical system.

$$w_{max} = nFE$$

Equations 20.5 and 20.6 (page 722) Relationship between free energy change and the cell potential under nonstandard or standard conditions, respectively.

$$\Delta G = -nFE \text{ or } \Delta G° = -nFE°$$

Equation 20.7 (page 723) Relationship between the equilibrium constant and the standard cell potential for a reaction (at 298 K).

$$\ln K = \frac{nE°}{0.0257}$$

Equation 20.8 (page 730) Relationship between current, electric charge, and time.

$$\text{Current (amperes, A)} = \frac{\text{electric charge (coulombs, C)}}{\text{time, } t \text{ (seconds, s)}}$$

OWL Questions and problems for this chapter are available in OWL. Use the OWL access card that came with this textbook to access assigned questions and problems for this chapter.

21 The Chemistry of the Main Group Elements

Quartz (SiO$_2$), dry ice (solid CO$_2$), and pure silicon.

© Cengage Learning/Charles D. Winters

Carbon and Silicon

Carbon and silicon are both significant elements in the earth's crust. Mendeleev placed them in the same periodic group based on the similarity in stoichiometry of their simple compounds such as their oxides (CO$_2$ and SiO$_2$). We know now that carbon is the backbone of millions of organic compounds. As we will see in this chapter, silicon, the second most abundant element in the earth's crust, is vitally important in the structures of a large number of minerals.

Probe a bit further into silicon chemistry, however, and what emerges are the surprising differences between these elements. We get a hint of this when we compare the two oxides, CO$_2$, a gas above −78 °C and 1 atm pressure, and SiO$_2$, a rock hard solid with a very high melting point. Another interesting contrast can be made between the reactions of the two simple hydrogen compounds, CH$_4$ and SiH$_4$, with water. The reaction of methane with water is reactant-favored, whereas silane will explode on contact with water.

In this chapter, we will evaluate and summarize the inorganic chemistry of the main group elements and explore the similarities and differences between elements in each periodic group.

Questions:

1. Write balanced chemical equations for the reactions of H$_2$O(ℓ) with CH$_4$ (forming CO$_2$ and H$_2$) and SiH$_4$ (forming SiO$_2$ and H$_2$).
2. Using thermodynamic data, calculate the standard free energy change for the reactions in Question 1. Is either reaction product-favored at equilibrium?
3. Look up the electronegativities of carbon, silicon, and hydrogen. What conclusion can you draw concerning the polarity of C—H and Si—H bonds?
4. Carbon and silicon compounds with the formulas (CH$_3$)$_2$CO (acetone) and [(CH$_3$)$_2$SiO]$_n$ (a silicone polymer) also have quite different structures (Chapter 10). Draw Lewis structures for these species. This difference, along with the difference between structures of CO$_2$ and SiO$_2$, suggests a general observation about silicon compounds. Based on that observation, do you expect that a silicon compound with a structure similar to ethene (C$_2$H$_4$) exists?

Answers to these questions are available in Appendix N.

CHAPTER OUTLINE

21.1 Element Abundances

21.2 The Periodic Table: A Guide to the Elements

21.3 Hydrogen

21.4 The Alkali Metals, Group 1A

21.5 The Alkaline Earth Elements, Group 2A

21.6 Boron, Aluminum, and the Group 3A Elements

21.7 Silicon and the Group 4A Elements

21.8 Nitrogen, Phosphorus, and the Group 5A Elements

21.9 Oxygen, Sulfur, and the Group 6A Elements

21.10 The Halogens, Group 7A

CHAPTER GOALS

See Chapter Goals Revisited (page 780) for Study Questions keyed to these goals.

- Relate the formulas and properties of compounds to the periodic table.

- Describe the chemistry of the main group or A-Group elements, particularly H; Na and K; Mg and Ca; B and Al; Si; N and P; O and S; and F and Cl.

- Apply the principles of stoichiometry, thermodynamics, and electrochemistry to the chemistry of the main group elements.

The main group or A-Group elements occupy an important place in the world of chemistry. Eight of the 10 most abundant elements on the earth are in these groups. Likewise, the top 10 chemicals produced by the U.S. chemical industry are all main group elements or their compounds.

Because main group elements and their compounds are economically important—and because they have interesting chemistries—we devote this chapter to a brief survey of these elements.

21.1 Element Abundances

The abundance of the first 18 elements in the solar system is plotted against their atomic numbers in Figure 21.1. Hydrogen and helium are the most abundant by a wide margin because most of the mass of the solar system resides in the sun, and these elements are the sun's primary components. Lithium, beryllium, and boron

OWL

Sign in to OWL at **www.cengage.com/owl** to view tutorials and simulations, develop problem-solving skills, and complete online homework assigned by your professor.

go Chemistry

Download mini lecture videos for key concept review and exam prep from OWL or purchase them from **www.cengagebrain.com**

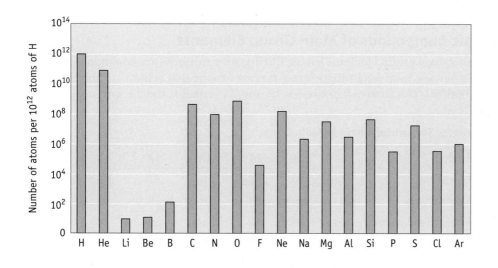

FIGURE 21.1 Abundance of elements 1–18 in the solar system. Li, Be, and B have relatively low abundances because they are circumvented when elements are made in stars. The common elements such as C, O, and Ne are made in stars by the accretion of alpha particles (helium nuclei). Helium has an atomic number of 2. If three He atoms combine, they produce an atom with atomic number 6 (carbon). Adding yet another He atom gives an atom with atomic number 8 (oxygen), and so on. (Notice that the vertical axis uses a logarithmic scale. This means, for example, there are 10^{12} H atoms for every 100 B atoms.)

Table 21.1 The 10 Most Abundant Elements in Earth's Crust

Rank	Element	Abundance (ppm)*
1	Oxygen	474,000
2	Silicon	277,000
3	Aluminum	82,000
4	Iron	56,300
5	Calcium	41,000
6	Sodium	23,600
7	Magnesium	23,300
8	Potassium	21,000
9	Titanium	5600
10	Hydrogen	1520

*ppm = g per 1000 kg. Most abundance data taken from J. Emsley: *The Elements*, New York, Oxford University Press, 3rd edition, 1998.

are low in abundance, but carbon's abundance is very high. From this point on, with the exception of iron and nickel, elemental abundances gradually decline as the atomic number increases.

Ten elements account for 99% of the aggregate mass of the earth's crust (Table 21.1), and oxygen, silicon, and aluminum represent more than 80% of this mass. Oxygen and nitrogen are the primary components of the atmosphere, and oxygen-containing water is highly abundant on the surface, underground, and as a vapor in the atmosphere. Many common minerals also contain these elements, including limestone ($CaCO_3$) and quartz or sand (SiO_2, ◄ Figure 2.8). Aluminum and silicon occur together in many minerals; among the more common ones are feldspar, granite, and clay.

21.2 The Periodic Table: A Guide to the Elements

The similarities in the properties of certain elements guided Mendeleev when he created the first periodic table (◄ page 51). He placed elements in groups based partly on the composition of their common compounds with oxygen and hydrogen (Table 21.2). We now understand that the elements are grouped according to the arrangements of their valence electrons.

Recall that the metallic character of the elements declines on moving from left to right in the periodic table. Elements in Group 1A, the alkali metals, are the most metallic elements in the periodic table. Elements on the far right are nonmetals, and in between are the metalloids. Metallic character also increases from the top of a group to the bottom. This is especially well illustrated by Group 4A. Carbon, at the top of the group, is a nonmetal; silicon and germanium are metalloids; and tin and lead are metals (Figure 21.2). The significance of metallic character in a discussion of the chemistry of the elements is readily apparent; metals typically form ionic compounds, whereas compounds composed only of nonmetals are covalent. Typically, ionic compounds are crystalline solids that have high melting points and conduct electricity in the molten state. Covalent compounds, on the other hand, can be gases, liquids, or solids and have low melting and boiling points.

Valence Electrons

The *ns* and *np* electrons are the valence electrons for main group elements (where *n* is the period in which the element is found) (◄ Section 7.3). The chemical behavior of an element is determined by the valence electrons.

When considering electronic structure, a useful reference point is the noble gases (Group 8A). Helium has an electron configuration of $1s^2$; the other noble gases have ns^2np^6 valence electron configurations. The dominant characteristic of the noble gases is their lack of reactivity. Indeed, the first two elements in the group do not form any compounds that can be isolated. The other four elements are now known to have limited chemistry, however, and the discovery of xenon compounds in the 1960s ranks as one of the most interesting developments in modern chemistry.

Ionic Compounds of Main Group Elements

Ions of main group elements having filled *s* and *p* subshells are common—justifying the often-seen statement that elements react in ways that achieve a "noble gas configuration." The elements in Groups 1A and 2A form 1+ and 2+ ions with electron

FIGURE 21.2 Group 4A elements. A nonmetal, carbon (graphite crucible); a metalloid, silicon (round, lustrous bar); and metals tin (chips of metal) and lead (a bullet, a toy, and a sphere).

Table 21.2 Similarities within Periodic Groups*

Group	1A	2A	3A	4A	5A	6A	7A
Common oxide	M_2O	MO	M_2O_3	EO_2	E_4O_{10}	EO_3	E_2O_7
Common hydride	MH	MH_2	MH_3	EH_4	EH_3	EH_2	EH
Highest oxidation state	+1	+2	+3	+4	+5	+6	+7
Common oxoanion			BO_3^{3-}	CO_3^{2-}	NO_3^{-}	SO_4^{2-}	ClO_4^{-}
				SiO_4^{4-}	PO_4^{3-}		

*M denotes a metal and E denotes a nonmetal or metalloid.

configurations that are the same as those for the previous noble gases. All common compounds of these elements (e.g., NaCl, CaCO₃) are ionic. The metallic elements in Group 3A (aluminum, gallium, indium, and thallium, but not the metalloid boron) form compounds containing 3+ ions.

Elements of Groups 6A and 7A can achieve a noble gas configuration by adding electrons. The Group 7A elements (halogens) form anions with a 1− charge (the halide ions, F^-, Cl^-, Br^-, I^-), and the Group 6A elements form anions with a 2− charge (O^{2-}, S^{2-}, Se^{2-}, Te^{2-}). In Group 5A chemistry, 3− ions with a noble gas configuration (such as the nitride ion, N^{3-}) are also known. The energy required to form highly charged anions is large, however, which means that other types of chemical behavior will usually take precedence.

EXAMPLE 21.1   Reactions of Group 1A–3A Elements

Problem Give the formula and name for the product in each of the following reactions. Write a balanced chemical equation for the reaction.

(a) $Ca(s) + S_8(s)$

(b) $Rb(s) + I_2(s)$

(c)   lithium and chlorine

(d)   aluminum and oxygen

What Do You Know? The reactions listed are between metals in periodic groups 1A, 2A, and 3A and nonmetals in periodic groups 6A and 7A. The products formed will be ionic compounds made up of cations formed from metals by loss of electrons and anions formed from nonmetals by the gain of electrons. Remember that equations must be balanced.

Strategy Predictions are based on the assumption that ions are formed with the electron configuration of the nearest noble gas. Group 1A elements form 1+ ions; Group 2A elements form 2+ ions; and metals in Group 3A form 3+ ions. In their reactions with metals, halogen atoms typically add a single electron to give anions with a 1− charge; Group 6A elements add two electrons to form anions with a 2− charge. For names of products, refer to the nomenclature discussion on page 67.

Solution

Balanced Equation		Product Name
(a)	$8\ Ca(s) + S_8(s) \rightarrow 8\ CaS(s)$	Calcium sulfide
(b)	$2\ Rb(s) + I_2(s) \rightarrow 2\ RbI(s)$	Rubidium iodide
(c)	$2\ Li(s) + Cl_2(g) \rightarrow 2\ LiCl(s)$	Lithium chloride
(d)	$4\ Al(s) + 3\ O_2(g) \rightarrow 2\ Al_2O_3(s)$	Aluminum oxide

Think about Your Answer It will be useful to review the formulas and names of common ions. These were given earlier in the text (Section 2.7).

Check Your Understanding

Write balanced chemical equations for the reactions forming the following compounds from the elements.

(a)   NaBr   (b)   CaSe   (c)   PbO   (d) $AlCl_3$

Molecular Compounds of Main Group Elements

Many avenues of reactivity are open to nonmetallic main group elements. Reactions with metals generally result in formation of ionic compounds, whereas compounds containing only metalloids and nonmetallic elements are generally molecular in nature.

Molecular compounds are encountered with the Group 3A element boron (Figure 21.3), and the chemistry of carbon in Group 4A is dominated by molecular

© Cengage Learning/Charles D. Winters

FIGURE 21.3 Boron halides. Liquid BBr₃ (left) and solid BI₃ (right). Formed from a metalloid and a nonmetal, both are molecular compounds. Both are sealed in glass ampules to prevent these boron compounds from reacting with H₂O in the air.

Table 21.3 Fluorine Compounds Formed by Main Group Elements

Group	Compound	Bonding
1A	NaF	Ionic
2A	MgF_2	Ionic
3A	AlF_3	Ionic
4A	SiF_4	Covalent
5A	PF_5	Covalent
6A	SF_6	Covalent
7A	IF_7	Covalent
8A	XeF_4	Covalent

compounds with covalent bonds (◄ Chapters 8 and 10). Similarly, nitrogen chemistry is dominated by molecular compounds. Consider ammonia, NH_3; the various nitrogen oxides (such as "laughing gas," N_2O); and nitric acid, HNO_3. In each of these species, nitrogen bonds covalently to another nonmetallic element. Also in Group 5A, phosphorus reacts with chlorine to produce the molecular compound PCl_3 (◄ page 93).

The valence electron configuration of an element determines the composition of its molecular compounds. Involving all the valence electrons in the formation of a compound is a frequent occurrence in main group element chemistry. We should not be surprised to discover compounds in which the central element has the highest possible oxidation number (such as P in PF_5). The highest oxidation number is readily predictable: it equals the group number. Thus, the highest (and only) oxidation number of sodium in its compounds is +1; the highest oxidation number of C is +4; and the highest oxidation number of phosphorus is +5 (Tables 21.2 and 21.3).

EXAMPLE 21.2 Predicting Formulas for Compounds of Main Group Elements

Problem Predict the formula for each of the following:

(a) The product of the reaction between germanium and excess oxygen

(b) The product of the reaction of arsenic and excess fluorine

(c) A compound formed from phosphorus and excess chlorine

(d) The fully deprotonated anion of selenic acid

What Do You Know? You know the symbol and location of each of the elements in the periodic table.

Strategy You can predict that in each reaction the element bonded to the halogen or oxygen in the product will achieve its most positive oxidation number, a value equal to the number of its periodic group.

Solution

(a) The Group 4A element germanium should have a maximum oxidation number of +4. Thus, its oxide has the formula GeO_2.

(b) Arsenic, in Group 5A, reacts vigorously with fluorine to form AsF_5, in which arsenic has an oxidation number of +5.

(c) PCl_5 is formed when the Group 5A element phosphorus reacts with excess chlorine.

(d) The chemistries of S and Se are similar. Sulfur, in Group 6A, has a maximum oxidation number of +6, so it forms SO_3 and sulfuric acid, H_2SO_4. Selenium, also in Group 6A, has analogous chemistry, forming SeO_3 and selenic acid, H_2SeO_4. The fully deprotonated anion of this acid is the selenate ion, SeO_4^{2-}.

Think about Your Answer Notice that in several questions the use of excess oxidizing agent was specified. This is important because, for most of the nonmetals, several different products were possible. For example, the oxidation of carbon can produce either CO_2 or CO depending on whether there is an excess or a deficiency of oxygen. You will learn more about this in later sections of this chapter.

Check Your Understanding

Write the formula for each of the following:

(a) hydrogen telluride

(b) sodium arsenate

(c) selenium hexachloride

(d) perbromic acid

There are many similarities among elements in the same periodic table group. This means you can use compounds of more common elements as examples when you encounter compounds of elements with which you are not familiar. For example, water, H_2O, is the simplest hydrogen compound of oxygen. You can reasonably expect the hydrogen compounds of other Group 6A elements to be H_2S, H_2Se, and H_2Te; all are well known.

EXAMPLE 21.3   Predicting Formulas

Problem Predict the formula for each of the following:

(a)   A compound of hydrogen and phosphorus

(b)   The hypobromite ion

(c)   Germane (the simplest hydrogen compound of germanium)

(d)   Two oxides of tellurium

What Do You Know? We expect the elements in a periodic group to behave similarly in their chemistry, in most instances. Thus, we can use known examples from the more common elements found early in the periodic table and extrapolate to predict the behavior of the less common elements found in the same group.

Strategy Recall as examples some of the compounds of lighter elements in a group, and then assume other elements in that group will form analogous compounds.

Solution

(a)   Phosphine, PH_3, has a composition analogous to ammonia, NH_3.

(b)   Hypobromite ion, BrO^-, is similar to the hypochlorite ion, ClO^-, the anion of hypochlorous acid ($HClO$).

(c) GeH_4 is analogous to CH_4 and SiH_4, other Group 4A hydrogen compounds.

(d)   Te and S are in Group 6A. TeO_2 and TeO_3 are analogs of the oxides of sulfur, SO_2 and SO_3.

Think about Your Answer Keep in mind that there can be differences between elements within a periodic group. One difference is that second-period elements generally obey the octet rule, whereas in the third period and below, compounds with expanded octets can form (Section 8.5). For example, in reactions with fluorine, nitrogen forms NF_3 whereas phosphorus forms both PF_3 and PF_5.

Check Your Understanding

Identify a compound or ion based on a second-period element that has a formula and Lewis structure analogous to each of the following:

(a) PH_4^+ (b) S_2^{2-} (c) P_2H_4 (d) PF_3

EXAMPLE 21.4   Recognizing Incorrect Formulas

Problem One formula is incorrect in each of the following groups. Pick out the incorrect formula, and indicate why it is incorrect.

(a) $CsSO_4$, KCl, $NaNO_3$, Li_2O

(b) MgO, CaI_2, $BaPO_4$, $CaCO_3$

(c) CO, CO_2, CO_3

(d) PF_5, PF_4^+, PF_7, PF_6^-

What Do You Know? One formula in each set of four is wrong. You can determine which by applying the rules for oxidation states set forth in this section.

Strategy Look for errors such as incorrect charges on ions or an oxidation number exceeding the maximum possible for the periodic group.

Solution

(a) CsSO$_4$. Sulfate ion has a 2− charge, so this formula would require a Cs^{2+} ion. Cesium, in Group 1A, forms only 1+ ions. The formula of cesium sulfate is Cs$_2$SO$_4$.

(b) BaPO$_4$. This formula implies a Ba^{3+} ion (because the phosphate anion is PO$_4^{3-}$). The cation charge does not equal the group number. The formula of barium phosphate is Ba$_3$(PO$_4$)$_2$.

(c) CO$_3$. Given that O has an oxidation number of −2, carbon would have to have an oxidation number of +6. Carbon is in Group 4A, however, and can have a maximum oxidation number of +4.

(d) PF$_7$. The formula implies P has an oxidation state of +7, whereas +5 is expected for the maximum possible P oxidation state.

Think about Your Answer There are a lot of facts to learn in chemistry. Fortunately, many can be organized in ways that make learning easier. Using the periodic table to predict the chemistry of each element, and recognizing similarities among elements in groups, is one of the essential aspects of chemistry. This has been the goal of this section of the chapter.

Check Your Understanding

Explain why compounds with the following formulas would not be expected to exist: Na$_2$Cl, CaCH$_3$CO$_2$, Mg$_2$O.

REVIEW & CHECK FOR SECTION 21.2

1. Which of the following formulas is incorrect?

    (a) CaH$_2$ (b) CaI$_2$ (c) CaS   (d) Ca$_2$O$_3$

2. What is the name of the product of the reaction between elemental phosphorus and excess oxygen, P$_4$O$_{10}$?

    (a) phosphorus oxide   (c) phosphoric acid

    (b) phosphorus decaoxide   (d) tetraphosphorus decaoxide

3. Like sulfur, selenium forms compounds in several different oxidation states. Which of the following is not likely to be an oxidation state of selenium in its compounds?

    (a) −2   (b) +3   (c) +6   (d) +4

4. What is the highest oxidation state that antimony can have in its compounds?

    (a) 0   (b) +1   (c) +3   (d) +5

21.3 Hydrogen

Chemical and Physical Properties of Hydrogen

Hydrogen has three isotopes, two of them stable (protium and deuterium) and one radioactive (tritium).

Isotopes of Hydrogen

Isotope Mass (u)	Symbol	Name
1.0078	^{1}H (H)	Hydrogen (protium)
2.0141	^{2}H (D)	Deuterium
3.0160	^{3}H (T)	Tritium

Of the three isotopes, only H and D are found in nature in significant quantities. Tritium, which is produced by cosmic ray bombardment of nitrogen in the atmosphere, is found to the extent of 1 atom per 10^{18} atoms of ordinary hydrogen. Tritium has a half-life of 12.26 years.

A CLOSER LOOK

Hydrogen, Helium, and Balloons

In 1783, Jacques Charles first used hydrogen to fill a balloon large enough to float above the French countryside (◄ page 394). In World War I, hydrogen-filled observation balloons were used. The Graf Zeppelin, a passenger-carrying dirigible built in Germany in 1928, was also filled with hydrogen. It carried more than 13,000 people between Germany and the United States until 1937, when it was replaced by the Hindenburg. The Hindenburg was designed to be filled with helium. At that time, World War II was approaching, and the United States, which has much of the world's supply of helium, would not sell the gas to Germany. As a consequence, the Hindenburg had to use hydrogen. The Hindenburg exploded and burned when landing in Lakehurst, New Jersey, in May 1937. Of the 62 people on board, only about half escaped uninjured. As a result of this disaster, hydrogen has acquired a reputation as being a very dangerous substance. Actually, it is as safe to handle as many other fuels.

The Hindenburg. This hydrogen-filled dirigible crashed in Lakehurst, New Jersey, in May 1937. Some have speculated that the aluminum paint coating the skin of the dirigible was involved in sparking the fire.

Under standard conditions, hydrogen is a colorless gas. Its very low boiling point, 20.7 K, reflects its nonpolar character and low molar mass. As the least dense gas known, it is ideal for filling lighter-than-air craft.

Deuterium compounds have been the subject of much research. One important observation is that, because D has twice the mass of H, reactions involving D atom transfer are slightly slower than those involving H atoms. This knowledge led to a way to produce D_2O or "heavy water." Hydrogen can be produced, albeit expensively, by electrolysis of water (Figure 21.4).

$$2\ H_2O(\ell) + \text{electrical energy} \rightarrow 2\ H_2(g) + O_2(g)$$

Any sample of natural water always contains a tiny concentration of D_2O. When electrolyzed, H_2O is electrolyzed more rapidly than D_2O. Thus, as the electrolysis proceeds, the liquid remaining is enriched in D_2O. Repeating this process many times will eventually give pure D_2O, often called "heavy water." Large amounts of D_2O are now produced because this compound is used as a moderator in some nuclear reactors that are used for power generation.

Hydrogen combines chemically with virtually every other element, except the noble gases. There are three different types of binary hydrogen-containing compounds.

Ionic metal hydrides are formed in the reaction of H_2 with a Group 1A or 2A metal.

$$2\ Na(s) + H_2(g) \rightarrow 2\ NaH(s)$$

$$Ca(s) + H_2(g) \rightarrow CaH_2(s)$$

These compounds contain the hydride ion, H^-, in which hydrogen has a -1 oxidation number.

Molecular compounds (such as H_2O, HF, and NH_3) are generally formed by direct combination of hydrogen with nonmetallic elements (Figure 21.5). The oxidation number of the hydrogen atom in these compounds is $+1$, but covalent bonds to hydrogen are the rule.

$$N_2(g) + 3\ H_2(g) \rightarrow 2\ NH_3(g)$$

$$F_2(g) + H_2(g) \rightarrow 2\ HF(g)$$

FIGURE 21.4 Electrolysis of water. Electrolysis of water (containing dilute H_2SO_4 as an electrolyte) gives O_2 *(left)* and H_2 *(right)*.

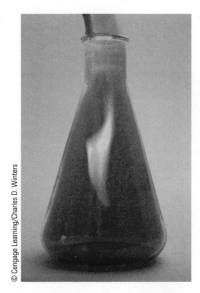

FIGURE 21.5 The reaction of H_2 and Br_2. Hydrogen gas burns in an atmosphere of bromine vapor to give hydrogen bromide.

Hydrogen is absorbed by many metals to form *interstitial hydrides*, the third general class of hydrogen compounds. This name refers to the structures of these species, in which the hydrogen atoms reside in the spaces between the metal atoms (called *interstices*) in the crystal lattice. Palladium metal, for example, can soak up 1000 times its volume of hydrogen (at STP). Most interstitial hydrides are non-stoichiometric; that is, the ratio of metal and hydrogen is not a whole number. When interstitial metal hydrides are heated, H_2 is driven out. This phenomenon allows these materials to store H_2, just as a sponge can store water. It suggests one way to store hydrogen for use in automobiles (◄ page 708).

Preparation of Hydrogen

About 300 billion liters (STP) of hydrogen gas are produced annually worldwide, and virtually all is used immediately in the manufacture of ammonia (► Section 21.8), methanol (◄ Section 10.4), or other chemicals.

Some hydrogen is made from coal and steam, a reaction that has been used for more than 100 years.

$$C(s) + H_2O(g) \longrightarrow \underbrace{H_2(g) + CO(g)}_{\text{water gas or synthesis gas}}$$

$$\Delta_rH° = +131 \text{ kJ/mol-rxn}$$

The reaction is carried out by injecting water into a bed of red-hot coke. The mixture of gases produced, called *water gas* or *synthesis gas*, was used until about 1950 as a fuel for cooking, heating, and lighting. However, it has serious drawbacks. It produces only about half as much heat as an equal amount of methane does, and the flame is nearly invisible. Moreover, because it contains carbon monoxide, water gas is toxic.

The largest quantity of hydrogen is now produced by the *catalytic steam reformation of hydrocarbons* such as methane in natural gas (Figure 21.6). Methane reacts with steam at high temperature to give H_2 and CO.

$$CH_4(g) + H_2O(g) \rightarrow 3\ H_2(g) + CO(g) \qquad \Delta_rH° = +206 \text{ kJ/mol-rxn}$$

The reaction is rapid at 900–1000 °C and goes nearly to completion. More hydrogen can be obtained in a second step in which the CO formed in the first step reacts with more water. This so-called *water gas shift reaction* is run at 400–500 °C and is slightly exothermic.

$$H_2O(g) + CO(g) \rightarrow H_2(g) + CO_2(g) \qquad \Delta_rH° = -41 \text{ kJ/mol-rxn}$$

The CO_2 formed in the process is removed by reaction with CaO (to give solid $CaCO_3$), leaving fairly pure hydrogen.

Perhaps the cleanest way to make hydrogen on a relatively large scale is the electrolysis of water (Figure 21.4). This approach provides not only hydrogen gas but also high-purity O_2. Because electricity is quite expensive, however, this method is not generally used commercially.

Table 21.4 and Figure 21.7 give examples of reactions used to produce H_2 gas in the laboratory. The most often used method is the reaction of a metal with an acid. Alternatively, the reaction of aluminum with aqueous NaOH (Figure 21.7b) also generates hydrogen. During World War II, this reaction was used to obtain hydrogen to inflate small balloons for weather observation and to raise radio antennas. Metallic aluminum was plentiful at the time because it came from damaged aircraft.

The combination of a metal hydride and water (Figure 21.7c) is an efficient but expensive way to synthesize H_2 in the laboratory. The reaction is commonly used in laboratories to dry organic solvents because the metal hydride reacts with traces of water present in the solvent.

FIGURE 21.6 Production of water gas. Water gas, also called *synthesis gas*, is a mixture of CO and H_2. It is produced by treating coal, coke, or a hydrocarbon like methane with steam at high temperatures in plants such as that pictured here. Methane has the advantage that it gives more total H_2 per gram than other hydrocarbons, and the ratio of the by-product CO_2 to H_2 is lower.

Table 21.4 Methods for Preparing H₂ in the Laboratory

1.	Metal + Acid → metal salt + H₂
	$Mg(s) + 2 HCl(aq) \rightarrow MgCl_2(aq) + H_2(g)$
2.	Metal + H₂O → metal hydroxide or oxide + H₂
	$2 Na(s) + 2 H_2O(\ell) \rightarrow 2 NaOH(aq) + H_2(g)$
	$2 Fe(s) + 3 H_2O(\ell) \rightarrow Fe_2O_3(s) + 3 H_2(g)$
	$2 Al(s) + 2 KOH(aq) + 6 H_2O(\ell) \rightarrow 2 K[Al(OH)_4](aq) + 3 H_2(g)$
3.	Metal hydride + H₂O → metal hydroxide + H₂
	$CaH_2(s) + 2 H_2O(\ell) \rightarrow Ca(OH)_2(s) + 2 H_2(g)$

REVIEW & CHECK FOR SECTION 21.3

1. Which of the following elements does not react with hydrogen?

    (a) neon   (b) nitrogen   (c) potassium   (d) fluorine

2. Which of the methods below is the most suitable for the preparation of large quantities of hydrogen (such as the amounts needed as a reagent for the synthesis of compounds such as ammonia)?

    (a) Electrolysis of water

    (b) The reaction of metal hydrides with water

    (c) The high temperature reaction of methane and water

    (d) The reaction of zinc and hydrochloric acid

21.4 The Alkali Metals, Group 1A

Sodium and potassium are, respectively, the sixth and eighth most abundant elements in the earth's crust by mass. In contrast, lithium is relatively rare, as are rubidium and cesium. Only traces of radioactive francium occur in nature. Its longest-lived isotope (²²³Fr) has a half-life of only 22 minutes.

The Group 1A elements are metals, and all are highly reactive with oxygen, water, and other oxidizing agents (◄ Figure 7.13, page 256). In all cases, compounds

Group 1A
Alkali metals

Lithium
3
Li
20 ppm
Sodium
11
Na
23,600 ppm
Potassium
19
K
21,000 ppm
Rubidium
37
Rb
90 ppm
Cesium
55
Cs
0.0003 ppm
Francium
87
Fr
trace

Element abundances are in **parts per million** *in the earth's crust.*

FIGURE 21.7
Producing hydrogen gas.

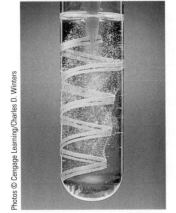

(a) The reaction of magnesium and acid. The products are hydrogen gas and a magnesium salt.

(b) The reaction of aluminum and NaOH. The products of this reaction are hydrogen gas and a solution of Na[Al(OH)₄].

(c) The reaction of CaH₂ and water. The products are hydrogen gas and Ca(OH)₂.

FIGURE 21.8 The importance of salt. All animals, including humans, need a certain amount of salt in their diet. Sodium ions are important in maintaining electrolyte balance and in regulating osmotic pressure. For an interesting account of the importance of salt in society, culture, history, and economy, see *Salt, A World History*, by M. Kurlansky, New York, Penguin Books, 2003.

of the Group 1A metals contain the element as a 1+ ion. The free metal is never found in nature. Most sodium and potassium compounds are water-soluble (◄ solubility guidelines, Figure 3.10), so it is not surprising that sodium and potassium compounds are found either in the oceans or in underground deposits that are the residue of ancient seas. To a much smaller extent, these elements are also found in minerals, such as Chile saltpeter ($NaNO_3$).

Despite the fact that sodium is only slightly more abundant than potassium on the earth, sea water contains significantly more sodium than potassium (2.8% NaCl versus 0.8% KCl). Why the great difference? Most compounds of both elements are water-soluble, so why didn't rain dissolve Na- and K-containing minerals over the centuries and carry them down to the sea, so that they appear in the same proportions in the oceans as on land? The answer lies in the fact that potassium is an important factor in plant growth. Most plants contain four to six times as much combined potassium as sodium. Thus, most of the potassium ions in groundwater from dissolved minerals are taken up preferentially by plants, whereas sodium salts continue on to the oceans. (Because plants require potassium, commercial fertilizers usually contain a significant amount of potassium salts.)

Some NaCl is essential in the diet of humans and other animals because many biological functions are controlled by the concentrations of Na^+ and Cl^- ions (Figure 21.8). The fact that salt has long been recognized as important is evident in surprising ways. For example, we are paid a "salary" for work done. This word is derived from the Latin *salarium*, which meant "salt money" because Roman soldiers were paid in salt.

Preparation of Sodium and Potassium

Sodium is produced by reducing sodium ions in sodium salts. However, because common chemical reducing agents are not powerful enough to convert sodium ions to sodium metal, the metal is usually prepared by electrolysis.

The English chemist Sir Humphry Davy first isolated sodium in 1807 by the electrolysis of molten sodium carbonate. However, the element remained a laboratory curiosity until 1824, when it was found sodium could be used to reduce aluminum chloride to aluminum metal. At that time, metallic aluminum was rare and very valuable, so this discovery inspired considerable interest in manufacturing sodium. By 1886, a practical method of sodium production had been devised (the reduction of NaOH with carbon). Unfortunately for sodium producers, in this same year Charles Hall and Paul Heroult invented the electrolytic method for aluminum production (see page 755), thereby eliminating this market for sodium.

Sodium is currently produced by the electrolysis of molten NaCl (◄ Section 20.7). The Downs cell for the electrolysis of molten NaCl operates at 7 to 8 V with currents of 25,000 to 40,000 amps (Figure 21.9). The cell is filled with a mixture of dry NaCl, $CaCl_2$, and $BaCl_2$. Adding other salts to NaCl lowers the melting point from that of pure NaCl (800.7 °C) to about 600 °C. [Recall that solutions have lower melting points than pure solvents (Chapter 14).] Sodium is produced at a copper or iron cathode that surrounds a circular graphite anode. Directly over the cathode is an inverted trough in which the low-density, molten sodium (melting point, 97.8 °C) collects. Chlorine, a valuable by-product, collects at the anode.

Potassium can also be made by electrolysis. Molten potassium is soluble in molten KCl, however, making separation of the metal difficult. The preferred method for preparation of potassium uses the reaction of sodium vapor with molten KCl, with potassium being continually removed from the equilibrium mixture.

$$Na(g) + KCl(\ell) \rightleftharpoons K(g) + NaCl(\ell)$$

Properties of Sodium and Potassium

Sodium and potassium are silvery metals that are soft and easily cut with a knife (◄ Figure 2.6). They are just a bit less dense than water. Their melting points are quite low, 97.8 °C for sodium and 63.7 °C for potassium.

● **Alkali Metals React with Water—Thermodynamics and Kinetics** The alkali metals all react vigorously with water, and it is easily observed that the violence of the reaction increases with atomic number. This seems counter to the argument presented in *A Closer Look* (page 746) that lithium is the best reducing agent. However, the reducing ability of a metal is a thermodynamic property, whereas the violence of the reaction is mainly a consequence of reaction rate.

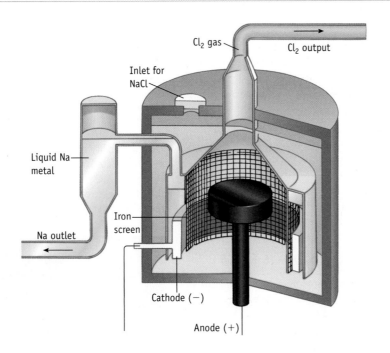

FIGURE 21.9 A Downs cell for **preparing sodium.** A circular iron cathode is separated from the graphite anode by an iron screen. At the temperature of the electrolysis, about 600 °C, sodium is a liquid. It rises to the top and is drawn off periodically. Chlorine gas is produced at the anode and collected inside the inverted cone in the center of the cell.

All of the alkali metals are highly reactive. When exposed to moist air, the metal surface quickly becomes coated with a film of oxide or hydroxide. Consequently, the metals must be stored in a way that avoids contact with air, typically by placing them in kerosene or mineral oil.

The high reactivity of Group 1A metals is exemplified by their reaction with water, which generates an aqueous solution of the metal hydroxide and hydrogen gas (◄ Figure 7.13, page 256),

$$2 \text{ Na(s)} + 2 \text{ H}_2\text{O}(\ell) \rightarrow 2 \text{ Na}^+(\text{aq}) + 2 \text{ OH}^-(\text{aq}) + \text{H}_2(\text{g})$$

and their reaction with any of the halogens to yield a metal halide (◄ Figure 1.2),

$$2 \text{ Na(s)} + \text{Cl}_2(\text{g}) \rightarrow 2 \text{ NaCl(s)}$$

$$2 \text{ K(s)} + \text{Br}_2(\ell) \rightarrow 2 \text{ KBr(s)}$$

Chemistry often produces surprises. Group 1A metal oxides, M_2O, are known, but they are not the principal products of reactions between the Group 1A elements and oxygen. Instead, the primary product of the reaction of sodium and oxygen is sodium *peroxide*, Na_2O_2, whereas the principal product from the reaction of potassium and oxygen is KO_2, potassium *superoxide*.

$$2 \text{ Na(s)} + \text{O}_2(\text{g}) \rightarrow \text{Na}_2\text{O}_2(\text{s})$$

$$\text{K(s)} + \text{O}_2(\text{g}) \rightarrow \text{KO}_2(\text{s})$$

Both Na_2O_2 and KO_2 are ionic compounds. The Group 1A cation is paired with either the peroxide ion (O_2^{2-}) or the superoxide ion (O_2^-). These compounds are not merely laboratory curiosities. They are used in oxygen-generation devices in places where people are confined, such as submarines, aircraft, and spacecraft, or when an emergency supply is needed. When a person breathes, 0.82 L of CO_2 is exhaled for every 1 L of O_2 inhaled. Thus, a requirement of an O_2-generation system is that it should produce a larger volume of O_2 than the volume of CO_2 taken in. This requirement is met with superoxides (Figure 21.10). With KO_2 the reaction is

$$4 \text{ KO}_2(\text{s}) + 2 \text{ CO}_2(\text{g}) \rightarrow 2 \text{ K}_2\text{CO}_3(\text{s}) + 3 \text{ O}_2(\text{g})$$

Courtesy of Mine Safety Appliances Company

FIGURE 21.10 A closed-circuit breathing apparatus that generates its own oxygen. One source of oxygen is potassium superoxide (KO_2). Both carbon dioxide and moisture exhaled by the wearer into the breathing tube react with the KO_2 to generate oxygen. Because the rate of the chemical reaction is determined by the quantity of moisture and carbon dioxide exhaled, the production of oxygen is regulated automatically. With each exhalation, more oxygen is produced by volume than is required by the user.

The Reducing Ability of the Alkali Metals

The uses of the Group 1A metals depend on their reducing ability. The values of $E°$ reveal that Li is the best reducing agent in the group, whereas Na is the poorest; the remainder of these metals have roughly comparable reducing abilities.

Reduction Potential

Element	$E°$ (V)
$Li^+(aq) + e^- \rightarrow Li(s)$	−3.045
$Na^+(aq) + e^- \rightarrow Na(s)$	−2.714
$K^+(aq) + e^- \rightarrow K(s)$	−2.925
$Rb^+(aq) + e^- \rightarrow Rb(s)$	−2.925
$Cs^+(aq) + e^- \rightarrow Cs(s)$	−2.92

Analysis of $E°$ is a thermodynamic problem, and to understand it better we can break the process of metal oxidation, $M(s) \rightarrow M^+(aq) + e^-$, into a series of steps. Here, we imagine that the metal sublimes to vapor, an electron is removed to form the gaseous cation, and the cation is hydrated. The first two steps require energy, but the last is exothermic. From Hess's law (page 194), we

know that the overall energy change should be

$$\Delta_rH_{net} = \Delta_{sub}H + IE + \Delta_{hyd}H$$

The element that is the best reducing agent should have the most negative (or least positive) value of Δ_rH. Because it is exothermic, the hydration enthalpy serves to offset the energy of the endothermic steps ($\Delta_{sub}H$ and IE). Thus, a more exothermic hydration enthalpy leads to a more negative Δ_rH. For the alkali metals, enthalpies of hydration range from −506 kJ/mol for Li^+ to −180 kJ/mol for Cs^+. The fact that $\Delta_{hyd}H$ is so much more exothermic for Li^+ than for Cs^+ largely

accounts for the difference in reducing ability.

While this analysis of the problem gives us a reasonable explanation for the great reducing ability of lithium, recall that $E°$ is directly related to $\Delta_rG°$ and not to $\Delta H°$. However, $\Delta_rG°$ is largely determined by $\Delta_rH°$ for these reactions, so it is possible to relate variations in $E°$ to variations in $\Delta_rH°$.

Potassium is a very good reducing agent and reacts vigorously with water.

Important Lithium, Sodium, and Potassium Compounds

Electrolysis of aqueous sodium chloride (*brine*) is the basis of one of the largest chemical industries in the United States.

$$2\ NaCl(aq) + 2\ H_2O(\ell) \rightarrow Cl_2(g) + 2\ NaOH(aq) + H_2(g)$$

Two of the products from this process—chlorine and sodium hydroxide—give the industry its name: the *chlor-alkali industry*. More than 10 billion kilograms of Cl_2 and NaOH is produced annually in the United States.

Sodium carbonate, Na_2CO_3, is another commercially important compound of sodium. It is also known by two common names, *soda ash* and *washing soda*. In the past, it was largely manufactured by combining NaCl, ammonia, and CO_2 in the *Solvay process* (which remains the method of choice in many countries). In the United States, however, sodium carbonate is obtained from naturally occurring deposits of the mineral *trona*, $Na_2CO_3 \cdot NaHCO_3 \cdot 2\ H_2O$ (Figure 21.11).

The chlor-alkali process has long used electrolysis cells with mercury cathodes. However, because of the environmental dangers of mercury, this process is largely being phased out, and there is considerable interest in the manufacture of sodium hydroxide by other methods. One approach is a revival of the old "soda-lime process," which produces NaOH from inexpensive lime (CaO) and soda ash (Na_2CO_3).

$$Na_2CO_3(aq) + CaO(s) + H_2O(\ell) \rightarrow 2\ NaOH(aq) + CaCO_3(s)$$

The insoluble calcium carbonate by-product is filtered off and then heated (calcining) to convert it to lime, which is recycled into the reaction system.

$$CaCO_3(s) \rightarrow CaO(s) + CO_2(g)$$

Sodium bicarbonate, $NaHCO_3$, also known as *baking soda*, is another common compound of sodium. Not only is $NaHCO_3$ used in cooking, but it is also added in

(a) *(above)* A mine in California. The mineral trona is taken from a mine 1600 feet deep.

(b) *(right)* Blocks of trona are cut from the face of the mine.

FIGURE 21.11 Producing soda ash. Trona mined in Wyoming and California is processed into soda ash (Na_2CO_3) and other sodium-based chemicals. Soda ash is the ninth most widely used chemical in the United States. Domestically, about half of all soda ash production is used in making glass. The remainder goes to make chemicals such as sodium silicate, sodium phosphate, and sodium cyanide. Some is also used to make detergents, in the pulp and paper industry, and in water treatment.

small amounts to table salt. NaCl is often contaminated with small amounts of $MgCl_2$. The magnesium salt is hygroscopic; that is, it picks water up from the air and, in doing so, causes the NaCl to clump. Adding $NaHCO_3$ converts $MgCl_2$ to magnesium carbonate, a nonhygroscopic salt.

$$MgCl_2(s) + 2\,NaHCO_3(s) \rightarrow MgCO_3(s) + 2\,NaCl(s) + H_2O(\ell) + CO_2(g)$$

Large deposits of sodium nitrate, $NaNO_3$, are found in Chile, which explains its common name of "Chile saltpeter." These deposits are thought to have formed by bacterial action on organisms in shallow seas. The initial product was ammonia, which was subsequently oxidized to nitrate ion; combination with sea salt led to sodium nitrate. Because nitrates in general, and alkali metal nitrates in particular, are highly water-soluble, deposits of $NaNO_3$ are found only in areas with very little rainfall.

Sodium nitrate is important because it can be converted to potassium nitrate by an exchange reaction.

$$NaNO_3(aq) + KCl(aq) \rightleftharpoons KNO_3(aq) + NaCl(s)$$

Equilibrium favors the products here because, of the four salts involved in this reaction, NaCl is least soluble in hot water. Sodium chloride precipitates, and the KNO_3 that remains in solution can be recovered by evaporating the water.

Potassium nitrate has been used for centuries as the oxidizing agent in gunpowder. A mixture of KNO_3, charcoal, and sulfur will react when ignited.

$$2\,KNO_3(s) + 4\,C(s) \rightarrow K_2CO_3(s) + 3\,CO(g) + N_2(g)$$

$$2\,KNO_3(s) + 2\,S(s) \rightarrow K_2SO_4(s) + SO_2(g) + N_2(g)$$

Notice that both reactions (which are doubtless more complex than those written here) produce gases. These gases propel the bullet from a gun or cause a firecracker to explode.

In the past several years lithium has become a significant commodity because of its use in lithium ion batteries. In fact, there is some concern that shortages of lithium will arise if the use of lithium ion batteries in cars becomes widespread (◀ page 707). Currently, most lithium is obtained from concentrated brines in Chile and Bolivia, from which Li_2CO_3 is extracted. There are also ores from which lithium can be obtained. Similar to the other alkali metals, lithium metal is obtained from its compounds by molten salt electrolysis.

● **Gunpowder** Gunpowder was developed well over 1000 years ago by the Chinese. One of the three ingredients is KNO_3, commonly called *saltpeter*. Although there are places in the world where it is prevalent and can be mined, for centuries it was obtained by mixing manure or human waste with wood ashes. Animal and human waste contain ammonia, which is converted by bacterial oxidation to nitrates. Wood ashes contain potash, potassium carbonate, the source of potassium in KNO_3. Sodium nitrate will also work in gunpowder, but it is hygroscopic and the powder is not effective when damp. See *Gunpowder* by J. Kelly, Basic Books, 2004.

Group **2A**
Alkaline earths

Beryllium
4
Be
2.6 ppm

Magnesium
12
Mg
23,300 ppm

Calcium
20
Ca
41,000 ppm

Strontium
38
Sr
370 ppm

Barium
56
Ba
500 ppm

Radium
88
Ra
6×10^{-7} ppm

Element abundances are in **parts per million** *in the earth's crust.*

21.5 The Alkaline Earth Elements, Group 2A

The "earth" part of the name alkaline earth dates back to the days of medieval alchemy. To alchemists, any solid that did not melt and was not changed by fire into another substance was called an "earth." Compounds of the Group 2A elements, such as CaO, were alkaline according to experimental tests conducted by the alchemists: They had a bitter taste and neutralized acids. With very high melting points, these compounds were unaffected by fire.

Calcium and magnesium rank fifth and eighth, respectively, in abundance on the earth. Both elements form many commercially important compounds, and we shall focus our attention on these species.

Like the Group 1A elements, the Group 2A elements are very reactive, so they are found in nature as compounds. Unlike most of the compounds of the Group 1A metals, however, many compounds of the Group 2A elements have low water-solubility, which explains their occurrence in various minerals (Figure 21.12). Calcium minerals include limestone ($CaCO_3$), gypsum ($CaSO_4 \cdot 2\,H_2O$), and fluorite (CaF_2). Magnesite ($MgCO_3$), talc or soapstone ($3\,MgO \cdot 4\,SiO_2 \cdot H_2O$), and asbestos ($3\,MgO \cdot 4\,SiO_2 \cdot 2\,H_2O$) are common magnesium-containing minerals. The mineral dolomite, $MgCa(CO_3)_2$, contains both magnesium and calcium.

Limestone, a sedimentary rock, is found widely on the earth's surface. Many of these deposits contain the fossilized remains of marine life. Other forms of calcium carbonate include marble and Icelandic spar, the latter occurring as large, clear crystals with the interesting optical property of birefringence (Figure 21.12).

Common minerals of Group 2A elements.

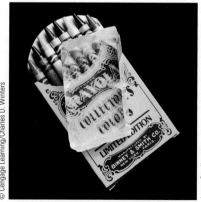

Icelandic spar. This mineral, one of a number of crystalline forms of $CaCO_3$, displays birefringence, a property in which a double image is formed when light passes through the crystal.

The walls of Arizona's Grand Canyon are largely limestone or dolomite.

FIGURE 21.12 Various minerals containing calcium and magnesium.

Properties of Calcium and Magnesium

Calcium and magnesium are fairly high-melting, silvery metals. The chemical properties of these elements present few surprises. They are oxidized by a wide range of oxidizing agents to form ionic compounds that contain the M^{2+} ion. For example, these elements combine with halogens to form MX_2, with oxygen or sulfur to form MO or MS, and with water to form hydrogen and the metal hydroxide, $M(OH)_2$ (Figure 21.13). With acids, hydrogen is evolved (Figure 21.7), and a salt of the metal cation and the anion of the acid results.

Metallurgy of Magnesium

Several hundred thousand tons of magnesium are produced annually, largely for use in lightweight alloys. (Magnesium has a very low density, 1.74 g/cm³.) Most aluminum used today contains about 5% magnesium to improve its mechanical properties and to make it more resistant to corrosion. Other alloys having more magnesium than aluminum are used when a high strength-to-weight ratio is needed and when corrosion resistance is important, such as in aircraft and automotive parts and in lightweight tools.

Interestingly, magnesium-containing minerals are not the source of this element. Most magnesium is obtained from sea water, in which Mg^{2+} ion is present in a concentration of about 0.05 M. To obtain magnesium metal, magnesium ions in sea water are first precipitated (Figure 21.14) as the relatively insoluble hydroxide [K_{sp} for $Mg(OH)_2 = 5.6 \times 10^{-12}$]. Calcium hydroxide, the source of OH^- in this reaction, is prepared in a sequence of reactions beginning with $CaCO_3$, which may be in the form of seashells. Heating $CaCO_3$ gives CO_2 and CaO, and addition of water to CaO gives calcium hydroxide. When $Ca(OH)_2$ is added to sea water, $Mg(OH)_2$ precipitates:

$$Mg^{2+}(aq) + Ca(OH)_2(s) \rightleftharpoons Mg(OH)_2(s) + Ca^{2+}(aq)$$

Magnesium hydroxide is isolated by filtration and then converted to magnesium chloride by reaction with hydrochloric acid.

$$Mg(OH)_2(s) + 2\ HCl(aq) \rightarrow MgCl_2(aq) + 2\ H_2O(\ell)$$

After evaporating the water, anhydrous magnesium chloride remains. Solid $MgCl_2$ melts at 714 °C, and the molten salt is electrolyzed to give the metal and chlorine.

$$MgCl_2(\ell) \rightarrow Mg(s) + Cl_2(g)$$

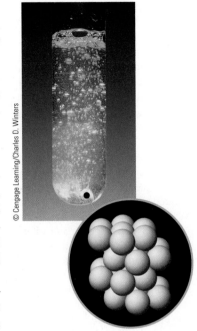

FIGURE 21.13 The reaction of calcium and warm water. Hydrogen bubbles are seen rising from the metal surface. The other reaction product is $Ca(OH)_2$. The inset is a model of hexagonal close-packed calcium metal (◀ page 448).

© Cengage Learning/Charles D. Winters

FIGURE 21.14 The process used to produce magnesium metal from the magnesium in sea water.

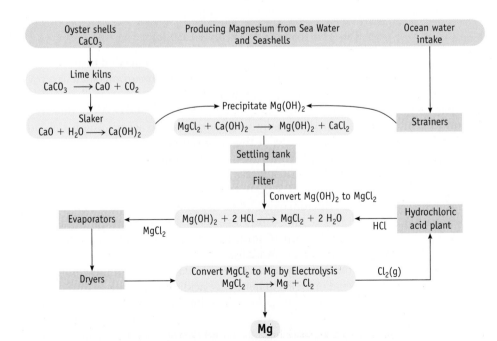

Alkaline Earth Metals and Biology

Plants and animals derive energy from the oxidation of a sugar, glucose, with oxygen. Plants are unique, however, in being able to synthesize glucose from CO_2 and H_2O using sunlight as an energy source. This process is initiated by *chlorophyll*, a very large, magnesium-based molecule.

In your body, the alkaline earth metal ions Mg^{2+} and Ca^{2+} serve regulatory functions. Although the two metal ions are required by living systems, the other Group 2A elements are toxic. Beryllium compounds are carcinogenic, and soluble barium salts are poisons. You may be concerned if your physician asks you to drink a "barium cocktail" to check the condition of your digestive tract. Don't be afraid, because the "cocktail" contains very insoluble $BaSO_4$ ($K_{sp} = 1.1 \times 10^{-10}$), so it passes through your digestive tract without a significant amount being absorbed. Barium sulfate is opaque to x-rays, so its path through your organs appears on the developed x-ray image.

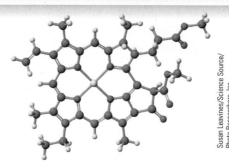

A molecule of chlorophyll. Magnesium is its central element.

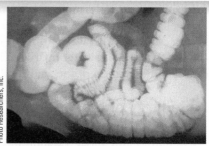

X-ray of a gastrointestinal tract using $BaSO_4$ to make the organs visible.

The calcium-containing compound *hydroxyapatite* is the main component of tooth enamel. Cavities in your teeth form when acids (such as soft drinks) decompose the weakly basic hydroxyapatite coating.

$$Ca_5(OH)(PO_4)_3(s) + 4\ H_3O^+(aq) \rightarrow$$
$$5\ Ca^{2+}(aq) + 3\ HPO_4^{2-}(aq) + 5\ H_2O(\ell)$$

This reaction can be prevented by converting hydroxyapatite to the much more acid-resistant coating of fluoroapatite.

$$Ca_5(OH)(PO_4)_3(s) + F^-(aq) \rightarrow$$
$$Ca_5F(PO_4)_3(s) + OH^-(aq)$$

The source of the fluoride ion can be sodium fluoride or sodium monofluorophosphate (Na_2FPO_3, commonly known as MFP) in your toothpaste.

Calcium Minerals and Their Applications

The most common calcium minerals are the fluoride, phosphate, and carbonate salts of the element. Fluorite, CaF_2, and fluoroapatite, $Ca_5F(PO_4)_3$, are important as commercial sources of fluorine. Almost half of the CaF_2 mined is used in the steel industry, where it is added to the mixture of materials that is melted to make crude iron. The CaF_2 acts to remove some impurities and improves the separation of molten metal from silicates and other by-products resulting from the reduction of iron ore to the metal (Chapter 22). A second major use of fluorite is in the manufacture of hydrofluoric acid by a reaction of the mineral with concentrated sulfuric acid.

$$CaF_2(s) + H_2SO_4(\ell) \rightarrow 2\ HF(g) + CaSO_4(s)$$

Hydrofluoric acid is used to make cryolite, Na_3AlF_6, a material needed in aluminum production (▶ Section 21.6) and in the manufacture of fluorocarbons such as tetrafluoroethylene, the precursor to Teflon (Table 10.12).

Apatites have the general formula $Ca_5X(PO_4)_3$ (X = F, Cl, OH). More than 100 million tons of apatite is mined annually, with Florida alone accounting for about one third of the world's output. Most of this material is converted to phosphoric acid by reaction with sulfuric acid. Phosphoric acid is needed in the manufacture of a multitude of products, including fertilizers and detergents, baking powder, and various food products (▶ Section 21.8.).

Calcium carbonate and calcium oxide (*lime*) are of special interest. The thermal decomposition of $CaCO_3$ to give lime (and CO_2) is one of the oldest chemical reactions known. Lime is one of the top 10 industrial chemicals produced today, with about 20 billion kilograms produced annually.

Limestone, which consists mostly of calcium carbonate, has been used in agriculture for centuries. It is spread on fields to neutralize acidic compounds in the soil and to supply Ca^{2+}, an essential nutrient. Because magnesium carbonate is often present in limestone, "liming" a field also supplies Mg^{2+}, another important nutrient for plants.

For several thousand years, lime has been used in *mortar* (a lime, sand, and water paste) to secure stones to one another in building houses, walls, and roads. The

Apatite. The mineral has the general formula of $Ca_5X(PO_4)_3$ (X = F, Cl, OH). (The apatite is the elongated crystal in the center of a matrix of other rock.)

A CLOSER LOOK

Of Romans, Limestone, and Champagne

The stones of the Appian Way in Italy, a road conceived by the Roman senate in about 310 BC, are cemented with mortar made from limestone. The Appian Way was intended to serve as a military road linking Rome to seaports from which soldiers could embark to Greece and other Mediterranean countries. The road stretches 560 kilometers (350 miles) from Rome to Brindisi on the Adriatic Sea (at the heel of the Italian "boot"). It took almost 200 years to construct. The road had a standard width of 14 Roman feet, approximately 6 meters, large enough to allow two chariots to pass, and featured two sidewalks of 1.2 m each. Every 16 km or so, there were horse-changing stations with taverns, shops, and *latrinae*, the famous Roman restrooms.

All over the Roman Empire, buildings, temples, and aqueducts were constructed of blocks of limestone and marble, and the mortar to cement the blocks was made by heating the chips from stone cutting (to give CaO). In central France, the Romans dug chalk (also $CaCO_3$) from the ground for cementing sandstone blocks. This activity created huge caves that remain to this day and are used for aging and storing champagne.

The Appian Way in Italy.

Champagne in a limestone cave in France.

Chinese used it to set stones in the Great Wall. The Romans perfected its use, and the fact that many of their constructions still stand today is testament both to their skill and to the usefulness of lime. The famous Roman road, the Appian Way, used lime mortar between several layers of its stones.

The utility of mortar depends on some simple chemistry. Mortar consists of one part lime to three parts sand, with water added to make a thick paste. The first reaction, referred to as *slaking*, occurs after the solids are mixed with water. This produces a slurry containing calcium hydroxide, which is known as *slaked lime*.

$$CaO(s) + H_2O(\ell) \rightleftharpoons Ca(OH)_2(s)$$

When the wet mortar mix is placed between bricks or stone blocks, it slowly reacts with CO_2 from the air, and the slaked lime is converted to calcium carbonate.

$$Ca(OH)_2(s) + CO_2(g) \rightleftharpoons CaCO_3(s) + H_2O(\ell)$$

The sand grains are bound together by the particles of calcium carbonate.

REVIEW & CHECK FOR SECTION 21.5

1. Which of the following insoluble calcium compounds does not dissolve in hydrochloric acid?

    (a) limestone, $CaCO_3$

    (b) slaked lime, $Ca(OH)_2$

    (c) gypsum, $CaSO_4 \cdot 2H_2O$

    (d) hydroxyapatite, $Ca_5(OH)(PO_4)_3$

2. Calcium minerals are the raw materials for a variety of large-scale industrial processes. Which of the following is not an industrial process?

    (a) Converting limestone, $CaCO_3$, to lime

    (b) Converting fluorite, CaF_2, to HF

    (c) Converting slaked lime, $Ca(OH)_2$, to lime

    (d) Converting apatite minerals to phosphate fertilizers

Gallium. Gallium, with a melting point of 29.8 °C, is one of the few metals that can be a liquid at or near room temperature. (Others are Hg and Cs.)

● **Boron, a Strange Element**
Within 1 week in 1808 three chemists claimed to have discovered boron: Sir Humphry Davy, J. Gay-Lussac, and L-J. Thenard. They had not; all were impure samples. One hundred years would pass before pure boron was finally isolated. But then the mystery began. What is its structure? Until recently three forms were known, but a fourth was discovered in 2009. The molecular structure of each of the forms has an icosahedron of twelve boron atoms (Figure 21.15).

Isolation of pure, elemental boron from boron-containing minerals is extremely difficult and is done in small quantities. Like most metals and metalloids, boron can be obtained by chemically or electrolytically reducing an oxide or halide. Magnesium has often been used for chemical reductions, but the product of this reaction is a noncrystalline boron of low purity.

$$B_2O_3(s) + 3\ Mg(s) \rightarrow 2\ B(s) + 3\ MgO(s)$$

Boron has several allotropes, all characterized by having an icosahedron of boron atoms as one structural element (Figure 21.15c). Partly as a result of extended covalent bonding, elemental boron is a very hard and refractory (resistant to heat) semiconductor. In this regard, it differs from the other Group 3A elements; Al, Ga, In, and Tl are all relatively low-melting, rather soft metals with high electrical conductivity.

Metallic Aluminum and Its Production

The low cost of aluminum and the excellent characteristics of its alloys with other metals (low density, strength, ease of handling in fabrication, and inertness toward corrosion, among others), have led to its widespread use. You know it best in the form of aluminum foil, aluminum cans, and parts of aircraft.

Pure aluminum is soft and weak; moreover, it loses strength rapidly at temperatures higher than 300 °C. What we call "aluminum" is actually aluminum alloyed with small amounts of other elements to strengthen the metal and improve its properties. A typical alloy may contain about 4% copper with smaller amounts of silicon, magnesium, and manganese. Softer, more corrosion-resistant alloys for window frames, furniture, highway signs, and cooking utensils may include only manganese.

The standard reduction potential of aluminum [$Al^{3+}(aq) + 3\ e^- \rightarrow Al(s)$; $E° = -1.66\ V$] tells you that aluminum is easily oxidized. From this, we might expect aluminum to be highly susceptible to corrosion but, in fact, it is quite resistant. Aluminum's corrosion resistance is due to the formation of a thin, tough, and transparent skin of Al_2O_3 that adheres to the metal surface. An important feature of the protective oxide layer is that it rapidly self-repairs. If you penetrate the surface coating by scratching it or using some chemical agent, the exposed metal surface immediately reacts with oxygen (or another oxidizing agent) to form a new layer of oxide over the damaged area (Figure 21.16).

Aluminum was first prepared by reducing $AlCl_3$ using sodium or potassium. This was a costly process, and, in the 19th century, aluminum was a precious metal.

FIGURE 21.16 Corrosion of aluminum.

(a) A ball of aluminum foil is added to a solution of copper(II) nitrate and sodium chloride. Normally, the coating of chemically inert Al_2O_3 on the surface of aluminum protects the metal from further oxidation.

(b) In the presence of the Cl^- ion, the coating of Al_2O_3 is breached, and aluminum reduces copper(II) ions to copper metal. The reaction is rapid and so exothermic that the water can boil on the surface of the foil. [The blue color of aqueous copper(II) ions will fade as these ions are consumed in the reaction.]

At the 1855 Paris Exposition, in fact, a sample of aluminum was exhibited along with the crown jewels of France. In an interesting coincidence, in 1886 Frenchman Paul Heroult (1863–1914) and American Charles Hall (1863–1914) simultaneously and independently conceived of the electrochemical method used today. The Hall–Heroult method bears the names of the two discoverers.

Aluminum is found in nature as aluminosilicates, minerals such as clay that are based on aluminum, silicon, and oxygen. As these minerals weather, they break down to various forms of hydrated aluminum oxide, $Al_2O_3 \cdot n\, H_2O$, called *bauxite*. Mined in huge quantities, bauxite is the raw material from which aluminum is obtained. The first step is to purify the ore, separating Al_2O_3 from iron and silicon oxides. This is done by the *Bayer process*, which relies on the amphoteric, basic, or acidic nature of the various oxides. (Silica, SiO_2, is an acidic oxide, whereas Al_2O_3 is amphoteric, and Fe_2O_3 is a basic oxide.) Silica and Al_2O_3 dissolve in a hot concentrated solution of caustic soda (NaOH), leaving insoluble Fe_2O_3 to be filtered out.

$$Al_2O_3(s) + 2\,NaOH(aq) + 3\,H_2O(\ell) \rightarrow 2\,Na[Al(OH)_4](aq)$$

$$SiO_2(s) + 2\,NaOH(aq) + 2\,H_2O(\ell) \rightarrow Na_2[Si(OH)_6](aq)$$

If a solution containing aluminate and silicate anions is treated with CO_2, Al_2O_3 precipitates, and the silicate ion remains in solution. Recall that CO_2 is an acidic oxide that forms the weak acid H_2CO_3 in water, so the Al_2O_3 precipitation in this step is an acid–base reaction.

$$H_2CO_3(aq) + 2\,Na[Al(OH)_4](aq) \rightarrow Na_2CO_3(aq) + Al_2O_3(s) + 5\,H_2O(\ell)$$

Metallic aluminum is obtained from purified bauxite by electrolysis (Figure 21.17). Bauxite is first mixed with cryolite, Na_3AlF_6, to give a lower-melting mixture (melting temperature = 980 °C) that is electrolyzed in a cell with graphite electrodes. The cell operates at a relatively low voltage (4.0–5.5 V) but with an extremely high current (50,000–150,000 A). Aluminum is produced at the cathode and oxygen at the anode. To produce 1 kg of aluminum requires 13 to 16 kilowatt-hours of energy plus the energy required to maintain the high temperature.

● **Charles Martin Hall (1863–1914)** Hall was only 22 years old when he worked out the electrolytic process for extracting aluminum from Al_2O_3 in a woodshed behind the family home in Oberlin, Ohio. He went on to found a company that eventually became ALCOA, the Aluminum Corporation of America.

Oesper Collection in the History of Chemistry/ University of Cincinnati

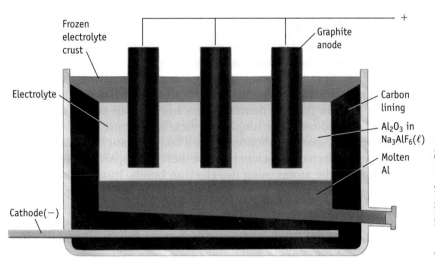

(a) Purified aluminum-containing ore (bauxite) is essentially Al_2O_3. In the Hall–Heroult process this is mixed with cryolite (Na_3AlF_6) and other fluorides such as AlF_3. (The additives serve to lower the melting point of the mixture and increase the conductivity). The aluminum-containing substances are reduced at a graphite cathode to give molten aluminum. Oxygen is produced at a graphite anode, and the gas reacts slowly with the carbon to give CO_2, leading to eventual destruction of the electrode.

(b) Molten aluminum alloy, produced from recycled metal, at 760 °C, in 1.6×10^4-kg capacity crucibles.

Courtesy, Allied Metal Company, Chicago

FIGURE 21.17 Industrial production of aluminum.

Aluminum chloride can react with a chloride ion to form the anion $[AlCl_4]^-$. Aluminum fluoride, in contrast, can accommodate three additional F^- ions to form an octahedral $[AlF_6]^{3-}$ ion. This anion is found in cryolite, Na_3AlF_6, the compound added to aluminum oxide in the electrolytic production of aluminum metal. Apparently, the Al^{3+} ion can bind to six of the smaller F^- ions, whereas only four of the larger Cl^-, Br^-, or I^- ions can surround an Al^{3+} ion.

REVIEW & CHECK FOR SECTION 21.6

1. In terms of abundance of the elements in the earth's crust, aluminum ranks

   (a) first   (b) second   (c) third   (d) fourth

2. The element below aluminum in Group 3A is gallium, and there are numerous similarities in the chemistry of these two elements. For example, the hydroxides of both elements are amphoteric. A consequence of this is that both gallium hydroxide and aluminum hydroxide

   (a) are insoluble   (c) dissolve only in base

   (b) dissolve only in acid   (d) dissolve in acid and in base

---

21.7 Silicon and the Group 4A Elements

Carbon is a nonmetal; silicon and germanium are classified as metalloids; and tin and lead are metals. As a result, the elements of Group 4A have a broad range of chemical behavior.

The Group 4A elements are characterized by half-filled valence shells with two electrons in the ns orbital and two electrons in np orbitals. The bonding in carbon and silicon compounds is largely covalent and involves sharing four electron pairs with neighboring atoms. In germanium compounds, the +4 oxidation state is common (GeO_2 and $GeCl_4$), but some +2 oxidation state compounds exist (GeI_2). Oxidation numbers of both +2 and +4 are common in compounds of tin and lead (such as $SnCl_2$, $SnCl_4$, PbO, and PbO_2). Oxidation numbers two units less than the group number are often encountered for heavier elements in Groups 3A–7A.

Silicon

Silicon is second after oxygen in abundance in the earth's crust, so it is not surprising that we are surrounded by silicon-containing materials: bricks, pottery, porcelain, lubricants, sealants, computer chips, and solar cells. The computer revolution is based on the semiconducting properties of silicon.

Reasonably pure silicon can be made in large quantities by heating pure silica sand with purified coke to approximately 3000 °C in an electric furnace.

$$SiO_2(s) + 2\ C(s) \rightarrow Si(\ell) + 2\ CO(g)$$

The molten silicon is drawn off the bottom of the furnace and allowed to cool to a shiny blue-gray solid. Because extremely high-purity silicon is needed for the electronics industry, purifying raw silicon requires several steps. First, the silicon in the impure sample is allowed to react with chlorine to convert the silicon to liquid silicon tetrachloride.

$$Si(s) + 2\ Cl_2(g) \rightarrow SiCl_4(\ell)$$

Silicon tetrachloride (boiling point of 57.6 °C) is carefully purified by distillation and then reduced to silicon using magnesium.

$$SiCl_4(g) + 2\ Mg(s) \rightarrow 2\ MgCl_2(s) + Si(s)$$

The magnesium chloride is washed out with water, and the silicon is remelted and cast into bars. A final purification is carried out by zone refining, a process in which a special heating device is used to melt a narrow segment of the silicon rod. The heater is moved slowly down the rod. Impurities contained in the silicon tend to

Group 4A

Carbon
6
C
480 ppm

Silicon
14
Si
277,000 ppm

Germanium
32
Ge
1.8 ppm

Tin
50
Sn
2.2 ppm

Lead
82
Pb
14 ppm

Element abundances are in **parts per million** *in the earth's crust.*

remain in the liquid phase because the melting point of a mixture is lower than that of the pure element (Chapter 14). The silicon that crystallizes above the heated zone is therefore of a higher purity (Figure 21.19).

Silicon Dioxide

The simplest oxide of silicon is SiO_2, commonly called *silica*, a constituent of many rocks such as granite and sandstone. Quartz is a pure crystalline form of silica, but impurities in quartz produce gemstones such as amethyst (Figure 21.20).

Silica and CO_2 are oxides of two elements in the same chemical group, so similarities between them might be expected. In fact, SiO_2 is a high-melting solid (quartz melts at 1610 °C), whereas CO_2 is a gas at room temperature and 1 atm. This great disparity arises from the different structures of the two oxides. Carbon dioxide is a molecular compound, with the carbon atom linked to each oxygen atom by a double bond. In contrast, SiO_2 is a network solid, which is the preferred structure because the bond energy of two Si=O double bonds is much less than the bond energy of four Si—O single bonds. The contrast between SiO_2 and CO_2 is an example of a more general phenomenon. Multiple bonds, often encountered between second-period elements, are rare among elements in the third period and beyond. There are many compounds with multiple bonds to carbon but very few compounds featuring multiple bonds to silicon.

Quartz crystals are used to control the frequency of radio and television transmissions. Because these and related applications use so much quartz, there is not enough natural quartz to fulfill demand, and quartz is therefore synthesized. Noncrystalline, or vitreous, quartz, made by melting pure silica sand, is placed in a steel "bomb," and dilute aqueous NaOH is added. A "seed" crystal is placed in the mixture, just as you might use a seed crystal in a hot sugar solution to grow rock candy. When the mixture is heated above the critical temperature of water (above 400 °C and 1700 atm) over a period of days, pure quartz crystallizes.

Silicon dioxide is resistant to attack by all acids except HF, with which it reacts to give SiF_4 and H_2O.

$$SiO_2(s) + 4\ HF(\ell) \rightarrow SiF_4(g) + 2\ H_2O(\ell)$$

Silicon dioxide also dissolves slowly in hot, molten NaOH or Na_2CO_3 to give Na_4SiO_4, sodium silicate.

$$SiO_2(s) + 2\ Na_2CO_3(\ell) \rightarrow Na_4SiO_4(s) + 2\ CO_2(g)$$

After the molten mixture has cooled, hot water under pressure is added. This partially dissolves the material to give a solution of sodium silicate. After filtering off insoluble sand or glass, the solvent is evaporated to leave sodium silicate, called *water glass*. The biggest single use of this material is in household and industrial detergents, in which it is included because a sodium silicate solution maintains pH by its

FIGURE 21.19 Pure silicon. The manufacture of very pure silicon begins with producing the volatile liquid silanes $SiCl_4$ or $SiHCl_3$. After carefully purifying these by distillation, they are reduced to elemental silicon with extremely pure Mg or Zn. The resulting spongy silicon is purified by zone refining. The end result is a cylindrical rod of ultrapure silicon such as those seen in this photograph. Thin wafers of silicon are cut from the bars and are the basis for the semiconducting chips in computers and other devices.

Synthetic quartz. These crystals were grown from silica in sodium hydroxide. The colors come from added Co^{2+} ions (blue) or Fe^{2+} ions (brown).

FIGURE 21.20 Various forms of quartz.

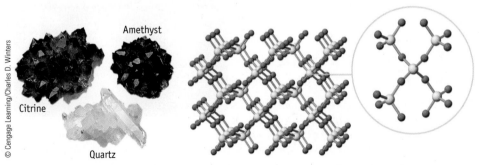

Amethyst

Citrine

Quartz

(a) Pure quartz is colorless, but the presence of small amounts of impurities adds color. Purple amethyst and brown citrine crystals are quartz with iron impurities.

(b) Quartz is a network solid in which each Si atom is bound tetrahedrally to four O atoms, each O atom linked to another Si atom.

Silica gel. Silica gel is solid, noncrystalline SiO_2. Packages of the material are often used to keep electronic equipment dry when stored. Silica gel is also used to clarify beer; passing beer through a bed of silica gel removes minute particles that would otherwise make the brew cloudy. Yet another use is in kitty litter.

buffering ability. Additionally, sodium silicate is used in various adhesives and binders, especially for gluing corrugated cardboard boxes.

If sodium silicate is treated with acid, a gelatinous precipitate of SiO_2 called *silica gel* is obtained. Washed and dried, silica gel is a highly porous material with many uses. It is a drying agent, readily absorbing up to 40% of its own weight of water. Small packets of silica gel are often placed in packing boxes of merchandise during storage. The material is frequently stained with $(NH_4)_2CoCl_4$, a humidity detector that is pink when hydrated and blue when dry.

Silicate Minerals with Chain and Ribbon Structures

The structure and chemistry of silicate minerals is an enormous topic in geology and chemistry. Although all silicates are built from tetrahedral SiO_4 units, they have different properties and a wide variety of structures because of the way these tetrahedral SiO_4 units link together.

The simplest silicates, *orthosilicates*, contain SiO_4^{4-} anions. The $4-$ charge of the anion is balanced by four M^+ ions, two M^{2+} ions, or a combination of ions. Olivine, an important mineral in the earth's mantle, contains Mg^{2+} and Fe^{2+}, with the Fe^{2+} ion giving the mineral its characteristic olive color, and gem-like zircons are $ZrSiO_4$. Calcium orthosilicate, Ca_2SiO_4, is a component of Portland cement, the most common type of cement used in many parts of the world. (It consists mostly of a mixture of CaO and SiO_2 with the remainder largely aluminum and iron oxides.)

A group of minerals called *pyroxenes* have as their basic structural unit a chain of SiO_4 tetrahedra.

If two such chains are linked together by sharing oxygen atoms, the result is an *amphibole*, of which the asbestos minerals are one example. The molecular chain results in asbestos being a fibrous material.

Silicates with Sheet Structures and Aluminosilicates

Linking many silicate chains together produces a sheet of SiO_4 tetrahedra (Figure 21.21). This sheet is the basic structural feature of some of the earth's most important minerals, particularly the clay minerals (such as china clay), mica, talc, and

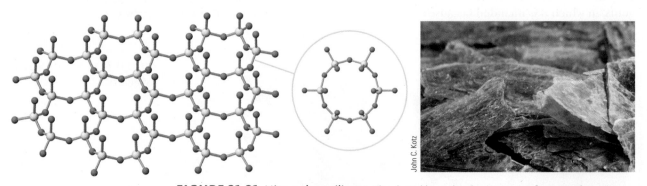

FIGURE 21.21 Mica, a sheet silicate. The sheet-like molecular structure of mica explains its physical appearance. As in the pyroxenes, each silicon is bonded to four oxygen atoms, but the Si and O atoms form a sheet of six-member rings of Si atoms with O atoms in each edge. The ratio of Si to O in this structure is 1 to 2.5. A formula of $SiO_{2.5}$ requires a positive ion, such as Na^+, to counterbalance the charge. Thus, mica and other sheet silicates, and aluminosilicates such as talc and many clays, have positive ions between the sheets. The sheet structure leads to the characteristic feature of mica, that it is often found as "books" of thin, silicate sheets. Mica is used in furnace windows and as insulation, and flecks of mica give the glitter to "metallic" paints.

the chrysotile form of asbestos. However, these minerals do not contain just silicon and oxygen. Rather, they are often referred to as *aluminosilicates* because they frequently have Al^{3+} ions in place of Si^{4+} (which means that other positive ions such as Na^+, K^+, and Mg^{2+} must also be present in the lattice to balance the net negative and positive charges). In kaolinite clay, for example, the sheet of SiO_4 tetrahedra is bonded to a sheet of AlO_6 octahedra. In addition, some Si^{4+} ions can be replaced by Al^{3+} atoms. Another example is muscovite, a form of mica. Aluminum ions have replaced some Si^{4+} ions, and there are charge-balancing K^+ ions, so it is best represented by the formula $KAl_2(OH)_2(Si_3AlO_{10})$.

There are some interesting uses of clays, one being in medicine (Figure 21.22). In certain cultures, clay is eaten for medicinal purposes. Several remedies for the relief of upset stomach contain highly purified clays that absorb excess stomach acid as well as potentially harmful bacteria and their toxins by exchanging the intersheet cations in the clays for the toxins, which are often organic cations.

Other aluminosilicates include the feldspars, common minerals that make up about 60% of the earth's crust, and zeolites (Figure 21.22). Both materials are composed of SiO_4 tetrahedra in which some of the Si atoms have been replaced by Al atoms, along with alkali and alkaline earth metal ions for charge balance. The main feature of zeolite structures is their regularly shaped tunnels and cavities. Hole diameters are between 300 and 1000 pm, and small molecules such as water can fit into the cavities of the zeolite structure. As a result, zeolites can be used as drying agents to selectively absorb water from air or a solvent. Small amounts of zeolites are often sealed into multipane windows to keep the air dry between the panes.

Zeolites are also used as catalysts. ExxonMobil, for example, has patented a process in which methanol, CH_3OH, is converted to gasoline in the presence of specially tailored zeolites. In addition, zeolites are added to detergents, where they function as water-softening agents because the sodium ions of the zeolite can be exchanged for Ca^{2+} ions in hard water, effectively removing Ca^{2+} ions from the water.

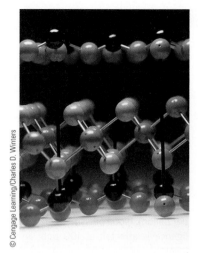

Kaolinite clay. The basic structural feature of many clays, and kaolinite in particular, is a sheet of SiO_4 tetrahedra (black and red spheres) bonded to a sheet of AlO_6 octahedra (gray and green spheres).

Silicone Polymers

Silicon and chloromethane (CH_3Cl) react at 300 °C in the presence of a catalyst, Cu powder. The primary product of this reaction is $(CH_3)_2SiCl_2$.

$$Si(s) + 2\ CH_3Cl(g) \rightarrow (CH_3)_2SiCl_2(\ell)$$

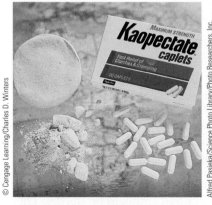

(a) Remedies for stomach upset. One of the ingredients in Kaopectate is kaolin, one form of clay. The off-white objects are pieces of clay purchased in a market in Ghana, West Africa. This clay was made to be eaten as a remedy for stomach ailments. Eating clay is widespread among the world's different cultures.

(b) The stucture of a zeolite. Zeolites, which have Si, Al, and O linked in a polyhedral framework, are often portrayed in drawings like this. Each edge consists of an Si—O—Si, Al—O—Si, or Al—O—Al bond. The channels in the framework can selectively capture small molecules or ions or act as catalytic sites.

(c) Apophyllite, a crystalline zeolite.

(d) Consumer products that remove odor-causing molecules from the air often contain zeolites.

FIGURE 21.22. Aluminosilicates.

CASE STUDY

Lead, Beethoven, and a Mystery Solved

Lead anchors the bottom of Group 4A. One of a handful of elements known since ancient times, it has a variety of modern uses. It ranks fifth among metals in usage behind iron, copper, aluminum, and zinc. The major uses of the metal and its compounds are in storage batteries (page 706), pigments, ammunition, solders, plumbing, and bearings.

Unfortunately, lead and its compounds are cumulative poisons, particularly in children. At a blood level as low as 50 ppb (parts per billion), blood pressure is elevated; intelligence is affected at 100 ppb; and blood levels higher than 800 ppb can lead to coma and possible death. Health experts believe that more than 200,000 children become ill from lead poisoning annually, a problem caused chiefly by children eating paint containing lead-based pigments. Older homes often contain lead-based paint because white lead [2 $PbCO_3 \cdot Pb(OH)_2$] was the pigment used in white paint until about 40 years ago, when it was replaced by TiO_2. Lead salts have a sweet taste, which may contribute to the tendency of children to chew on painted objects.

The symptoms of lead poisoning include nausea, abdominal pain, irritability, headaches, and excess lethargy or hyperactivity. Indeed, these describe some of the symptoms of the illness that affected Ludwig van Beethoven. As a child, he was recognized as a musical prodigy and was thought to be the greatest pianist in Europe by the time he was 19. But then he fell ill, and, by the time he was 29, he wrote to his brother to say he

Ludwig van Beethoven (1770–1827).

Erich Lessing/Art Resource, NY

was considering suicide. By the time he died in 1827 at the age of 56, his belly, arms, and legs were swollen, and he complained constantly of pain in his joints and in his big toe. It is said he wandered the streets of Vienna with long, uncombed hair, dressed in a top hat and long coat, and scribbling in a notebook.

An autopsy at the time showed he died of kidney failure. Kidney stones had destroyed his kidneys, stones that presumably came from gout, the buildup of uric acid in his body. (Gout leads to joint pain, among other things.) But why did he have gout?

It was well known in the time of the Roman Empire that lead and its salts are toxic. The Romans drank wine sweetened with a very concentrated grape juice syrup that was prepared by boiling the juice in a

lead kettle. The resulting syrup, called *Sapa*, had a very high concentration of lead, and many Romans contracted gout. So, if Beethoven enjoyed drinking wine, which was often kept in lead-glass decanters, he could have contracted gout and lead poisoning. One scientist has also noted that he may have been one of a small number of people who have a "metal metabolism disorder," a condition that prevents the excretion of toxic metals like lead.

In 2005, scientists at Argonne National Laboratory examined fragments of Beethoven's hair and skull and found both were extremely high in lead. The hair sample, for example, had 60 ppm lead, about 100 times higher than normal.

The mystery of what caused Beethoven's death has been solved. But what remains a mystery is how he contracted lead poisoning.

Questions:

1. If blood contains 50. ppb lead, how many atoms of lead are there in 1.0 L of blood? (Assume d(blood) = 1.0 g/mL.)

2. Research has found that port wine stored for a year in lead-glass decanters contains 2000 ppm lead. If the decanter contains 750 mL of wine (d = 1.0 g/mL), what mass of lead has been extracted into the wine?

Answers to these questions are available in Appendix N.

Halides of Group 4A elements other than carbon hydrolyze readily. Thus, the reaction of $(CH_3)_2SiCl_2$ with water initially produces $(CH_3)_2Si(OH)_2$. On standing, these molecules combine to form a condensation polymer by eliminating water. The polymer is called polydimethylsiloxane, a member of the *silicone* family of polymers.

$$(CH_3)_2SiCl_2 + 2\ H_2O \rightarrow (CH_3)_2Si(OH)_2 + 2\ HCl$$

$$n\ (CH_3)_2Si(OH)_2 \rightarrow [-(CH_3)_2SiO-]_n + n\ H_2O$$

Silicone polymers are nontoxic and have good stability to heat, light, and oxygen; they are chemically inert and have valuable antistick and antifoam properties. They can take the form of oils, greases, and resins. Some have rubber-like properties ("Silly Putty," for example, is a silicone polymer). More than 1 million tons of silicone polymers is made worldwide annually. These materials are used in a wide variety of products: lubricants, peel-off labels, lipstick, suntan lotion, car polish, and building caulk.

© Cengage Learning/Charles D. Winters

Silicone. Some examples of products containing silicones, polymers with repeating —R_2Si—O— units.

21.8 Nitrogen, Phosphorus, and the Group 5A Elements

Group 5A elements are characterized by the ns^2np^3 configuration with its half-filled np subshell. In compounds of the Group 5A elements, the primary oxidation numbers are +3 and +5, although common nitrogen compounds display a range of oxidation numbers from −3 to +5. Once again, as in Groups 3A and 4A, the most positive oxidation number is less common for the heavier elements. In many arsenic, antimony, and bismuth compounds, the element has an oxidation number of +3. Not surprisingly, compounds of these elements with oxidation numbers of +5 are powerful oxidizing agents.

This part of our tour of the main group elements will concentrate on the chemistries of nitrogen and phosphorus. Nitrogen is found primarily as N_2 in the atmosphere, where it constitutes 78.1% by volume (75.5% by weight). In contrast, phosphorus occurs in the earth's crust in solids. More than 200 different phosphorus-containing minerals are known; all contain the tetrahedral phosphate ion, PO_4^{3-}, or a derivative of this ion. By far, the most abundant phosphorus-containing minerals are apatites (see page 750).

Nitrogen and its compounds play a key role in our economy, with ammonia making a particularly notable contribution. Phosphoric acid is an important commodity chemical, and it finds its greatest use in producing fertilizers.

Both phosphorus and nitrogen are part of every living organism. Phosphorus is contained in nucleic acids and phospholipids, and nitrogen occurs in proteins and nucleic acids.

Properties of Nitrogen and Phosphorus

Nitrogen (N_2) is a colorless gas that liquifies at 77 K (−196 °C) (◄ Figure 12.1, page 418). Its most notable feature is its reluctance to react with other elements or compounds because the N≡N triple bond has a large bond dissociation enthalpy (945 kJ/mol) and because the molecule is nonpolar. Nitrogen does, however, react with hydrogen to give ammonia in the presence of a catalyst (◄ Case Study, page 574) and with a few metals (notably lithium and magnesium) to give metal nitrides, compounds containing the N^{3-} ion.

$$3 \ Mg(s) + N_2(g) \longrightarrow Mg_3N_2(s)$$
magnesium nitride

Elemental nitrogen is a very useful material. Because of its lack of reactivity, it is used to provide a nonoxidizing atmosphere for packaged foods and wine and to pressurize electric cables and telephone wires. Liquid nitrogen is valuable as a coolant in freezing biological samples such as blood and semen, in freeze-drying food, and for other applications that require extremely low temperatures.

Group 5A

Nitrogen 7 **N** 25 ppm	
Phosphorus 15 **P** 1000 ppm	
Arsenic 33 **As** 1.5 ppm	
Antimony 51 **Sb** 0.2 ppm	
Bismuth 83 **Bi** 0.048 ppm	

Element abundances are in **parts per million** *in the earth's crust.*

Liquid nitrogen. Biological samples—such as embryos or semen from animals or humans—can be stored in liquid nitrogen (at −196 °C) for long periods of time.

Simon Fraser/MRC Unit, Newcastle General Hospital/Science Photo Library/Photo Researchers, Inc.

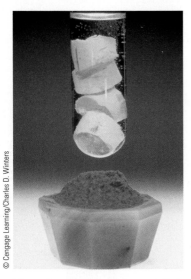

The red and white allotropes of phosphorus.

Elemental phosphorus was first derived from human waste (see *A Closer Look: Making Phosphorus*, page 767), but it is now produced by the reduction of phosphate minerals in an electric furnace.

$$2\ Ca_3(PO_4)_2(s) + 10\ C(s) + 6\ SiO_2(s) \rightarrow P_4(g) + 6\ CaSiO_3(s) + 10\ CO(g)$$

The phosphorus vapor can then be cooled under water, preventing its spontaneous combustion, and eventually yielding the solid forms of phosphorus present at room temperature. Waxy white phosphorus is the most common allotrope of phosphorus, but paradoxically it is the least stable thermodynamically. Rather than occurring as a diatomic molecule with a triple bond, like its second-period relative nitrogen (N_2), phosphorus is made up of tetrahedral P_4 molecules in which each P atom is joined to three others via single bonds. Red phosphorus is a polymer of P_4 units.

white phosphorus, P_4 polymeric red phosphorus

Nitrogen Compounds

A notable feature of the chemistry of nitrogen is the wide diversity of its compounds. Compounds with nitrogen in all oxidation numbers between -3 and $+5$ are known (Figure 21.23).

Hydrogen Compounds of Nitrogen: Ammonia and Hydrazine

Ammonia is a gas at room temperature and pressure. It has a very penetrating odor and condenses to a liquid at $-33\ ^\circ C$ under 1 atm of pressure. Solutions in water, often referred to as ammonium hydroxide, are basic due to the reaction of ammonia with water (◄ Section 17.4 and Figure 3.14).

$$NH_3(aq) + H_2O(\ell) \rightleftharpoons NH_4^+(aq) + OH^-(aq)\quad K_b = 1.8 \times 10^{-5}\ \text{at}\ 25\ ^\circ C$$

Ammonia is a major industrial chemical and is prepared by the Haber process (◄ page 574), largely for use as a fertilizer.

Hydrazine, N_2H_4, is a colorless, fuming liquid with an ammonia-like odor (mp, 2.0 °C; bp, 113.5 °C). Almost 1 million kilograms of hydrazine is produced annually by the Raschig process—the oxidation of ammonia with alkaline sodium hypochlorite in the presence of gelatin (which is added to suppress metal-catalyzed side reactions that lower the yield of hydrazine).

$$2\ NH_3(aq) + NaClO(aq) \rightleftharpoons N_2H_4(aq) + NaCl(aq) + H_2O(\ell)$$

Hydrazine, like ammonia, is also a base,

$$N_2H_4(aq) + H_2O(\ell) \rightleftharpoons N_2H_5^+(aq) + OH^-(aq) \qquad K_b = 8.5 \times 10^{-7}$$

and it is a strong reducing agent, as reflected in the reduction potential for the following half-reaction in basic solution:

$$N_2(g) + 4\ H_2O(\ell) + 4\ e^- \rightarrow N_2H_4(aq) + 4\ OH^-(aq) \qquad E^\circ = -1.15\ V$$

Hydrazine's reducing ability is exploited in its use in wastewater treatment for chemical plants. It removes oxidizing ions such as CrO_4^{2-} by reducing them, thus preventing them from entering the environment. A related use is the treatment of water boilers in large electric-generating plants. Oxygen dissolved in the water presents a serious problem in these plants because the dissolved gas can oxidize (corrode) the metal of the boiler and pipes. Hydrazine reduces the amount of dissolved oxygen in water.

$$N_2H_4(aq) + O_2(g) \rightarrow N_2(g) + 2\ H_2O(\ell)$$

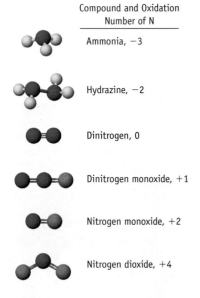

Compound and Oxidation Number of N

Ammonia, -3

Hydrazine, -2

Dinitrogen, 0

Dinitrogen monoxide, $+1$

Nitrogen monoxide, $+2$

Nitrogen dioxide, $+4$

Nitric acid, $+5$

FIGURE 21.23 Compounds and oxidation numbers for nitrogen. In its compounds, the N atom can have oxidation states ranging from -3 to $+5$.

A Healthy Saltwater Aquarium and the Nitrogen Cycle

by Jeffrey Keaffaber, University of Florida

Large saltwater aquariums like those at Sea World in Florida, the Shedd Aquarium in Chicago, and the new Georgia Aquarium in Atlanta, are a continual source of enjoyment. So are smaller aquariums in your home. Maintaining these facilities is not trivial, however; a healthy environment for its marine inhabitants is essential. For this, chemistry plays an important role.

A key part of aquarium maintenance involves control of the concentrations of various dissolved nitrogen-containing species, including ammonia, nitrite ion, and nitrate ion, all of which are stressful to fish at low concentrations and toxic in higher concentrations. The chemistry that relates to maintaining proper balance among these species is called the *nitrogen cycle*.

© Avico Ltd/Alamy

Nitrification

The nitrogen cycle begins with the production of ammonia (and, in acid solution, its conjugate acid, the ammonium ion, NH_4^+), a fundamental waste product of protein metabolism in an aquarium habitat. Unless removed, the ammonia concentration will build up over time. To remove it, the aquarium water is cycled through sand filters infused with aerobic, oxygen-loving bacteria. These bacteria utilize enzymes that catalyze the oxidation of ammonia and ammonium ion by O_2 to form first nitrite ion and then nitrate ion. The overall process is called *nitrification*, and the saltwater bacteria that mediate each oxidation step are *Nitrosococcus sp.* and *Nitrococcus sp.*, respectively. Half-reactions representing this chemistry are as follows:

Oxidation half-reactions

$$NH_4^+(aq) + 8\ OH^-(aq) \rightarrow$$
$$NO_2^-(aq) + 6\ H_2O(\ell) + 6\ e^-$$

$$NO_2^-(aq) + 2\ OH^-(aq) \rightarrow$$
$$NO_3^-(aq) + H_2O(\ell) + 2\ e^-$$

Reduction half-reaction

$$O_2(aq) + 2\ H_2O(\ell) + 4\ e^- \rightarrow 4\ OH^-(aq)$$

When setting up an aquarium at home, it is appropriate to monitor the concentrations of various nitrogen species. Initially,

the NH_3/NH_4^+ concentration rises, but then it begins to fall as oxidation occurs. With time, the nitrite ion concentration builds up, peaks, and then decreases, with an accompanying increase in nitrate ion concentration. To reach a stable situation, as much as 6 weeks may be required.

Denitrification

Nitrate is much less toxic than ammonia and the nitrite ion, but its buildup must also be limited. In a small aquarium, nitrate ion concentration can be controlled by partial exchange of water. However, because of environmental restrictions this is not possible for large aquariums; they must use a closed water treatment process.

To remedy the buildup of the nitrate ion, another biologically catalyzed process is used that reduces the nitrate ion to nitrogen gas, N_2. A reducing agent is required, and early designs of denitrifying filters utilized methanol, CH_3OH, as the reducing agent. Among naturally occurring saltwater bacteria capable of nitrate reduction under low oxygen (anoxic) conditions, *Pseudomonas sp.* is commonly used. The bacteria, utilizing enzymes to catalyze reaction of nitrate and methanol to form N_2 and CO_2, are introduced to sand filters where methanol is added.

A stable pH is also important to the health of aquarium fish. Therefore, the

marine environment in a saltwater aquarium is maintained at a relatively constant pH of 8.0–8.2. To aid in this, the CO_2 produced by methanol oxidation remains dissolved in solution, and increases the buffer capacity of the seawater.

Questions:

1. Write a balanced net ionic equation for the oxidation of NH_4^+ by O_2 to produce H_2O and NO_2^-.

2. Write half-reactions for the reduction of NO_3^- to N_2 and for the oxidation of CH_3OH to CO_2 in basic solution. Then, combine these half-reactions to obtain the balanced equation for the reduction of NO_3^- by CH_3OH.

3. Consider the carbon-containing species CO_2, H_2CO_3, HCO_3^-, and CO_3^{2-}. Which one is present in largest concentration at the pH conditions of the aquarium? Give a short explanation. K_a of $H_2CO_3 = 4.2 \times 10^{-7}$ and K_a of $HCO_3^- = 4.8 \times 10^{-11}$. (See Study Question 18.110.)

4. A large, 2.2×10^7 L, aquarium contains 1.7×10^4 kg of dissolved NO_3^-. Calculate concentrations in ppm (mg/L) N, ppm NO_3^-, and the molar concentration of NO_3^-.

Answers to these questions are available in Appendix N.

Oxides and Oxoacids of Nitrogen

Nitrogen is unique among all elements in the number of binary oxides it forms (Table 21.5). All are thermodynamically unstable with respect to decomposition to N_2 and O_2; that is, all have positive $\Delta_f G°$ values. Most are slow to decompose, however, and so are described as kinetically stable.

Dinitrogen monoxide, N_2O, commonly called *nitrous oxide*, is a nontoxic, odorless, and tasteless gas in which nitrogen has the lowest oxidation number (+1) among the nitrogen oxides. It can be made by the careful decomposition of ammonium nitrate at 250 °C.

$$NH_4NO_3(s) \rightarrow N_2O(g) + 2\ H_2O(g)$$

It is used as an anesthetic in minor surgery and has been called "laughing gas" because of its euphoriant effects. Because it is soluble in vegetable fats, the largest commercial use of N_2O is as a propellant and aerating agent in cans of whipped cream.

Nitrogen monoxide, NO, is an odd-electron molecule. It has 11 valence electrons, giving it one unpaired electron and making it a free radical. The compound has recently been the subject of intense research because it has been found to be important in a number of biochemical processes.

Nitrogen dioxide, NO_2, is the brown gas you see when a bottle of nitric acid is allowed to stand in the sunlight.

$$2\ HNO_3(aq) \rightarrow 2\ NO_2(g) + H_2O(\ell) + \tfrac{1}{2}\ O_2(g)$$

Nitrogen dioxide is also a culprit in air pollution. Nitrogen monoxide forms when atmospheric nitrogen and oxygen are heated in internal combustion engines. Released into the atmosphere, NO rapidly reacts with O_2 to form NO_2.

$$2\ NO(g) + O_2(g) \rightarrow 2\ NO_2(g)$$

Table 21.5 **Some Oxides of Nitrogen**

Formula	Name	Structure	Nitrogen Oxidation Number	Description
N_2O	Dinitrogen monoxide (nitrous oxide)	:N≡N—O: (linear)	+1	Colorless gas (laughing gas)
NO	Nitrogen monoxide (nitric oxide)	*	+2	Colorless gas; odd-electron molecule (paramagnetic)
N_2O_3	Dinitrogen trioxide	O=N—N(=O)—O: (planar)	+3	Blue solid (mp, −100.7 °C); reversibly dissociates to NO and NO_2 above its mp.
NO_2	Nitrogen dioxide	O—N—O	+4	Brown, paramagnetic gas; odd-electron molecule
N_2O_4	Dinitrogen tetraoxide	O=N(—O)—N(=O)—O: (planar)	+4	Colorless liquid/gas; dissociates to NO_2 (◄ Figure 16.8)
N_2O_5	Dinitrogen pentaoxide	O=N(—O:)—O—N(=O)—O:	+5	Colorless solid

*It is not possible to draw a Lewis structure that accurately represents the electronic structure of NO.
◄ Chapter 8. Also note that only one resonance structure is shown for each structure.

A CLOSER LOOK

Making Phosphorus

He stoked his small furnace with more charcoal and pumped the bellows until his retort glowed red hot. Suddenly something strange began to happen. Glowing fumes filled the vessel and from the end of the retort dripped a shining liquid that burst into flames.

J. Emsley: *The 13th Element*, p. 5. New York, John Wiley, 2000.

John Emsley begins his story of phosphorus, its discovery, and its uses, by imagining what the German alchemist Hennig Brandt must have seen in his laboratory that day in 1669. He was in search of the philosopher's stone, the magic elixir that would turn the crudest substance into gold. (Some may recall that the first Harry Potter novel was titled *Harry Potter and the Philosopher's Stone* when it was published in Great Britain.)

Brandt was experimenting with urine, which had served as the source of useful chemicals since Roman times. It is not surprising that phosphorus could be extracted from this source. Humans consume much more phosphorus, in the form of phosphate, than they require, and the excess phosphorus (about 1.4 g per day) is excreted in the urine. It is nonetheless extraordinary that Brandt was able to isolate the element. According to an 18th-century chemistry book, about 30 g of phosphorus could be obtained from 60 gallons of urine. And the process was not simple. Another 18th-century recipe states that "50 or 60 pails full" of urine was to be used. "Let it lie steeping . . . till it putrefy and breed worms." The chemist was then to reduce the whole to a paste and finally to heat the paste very strongly in a retort. After some days, phosphorus distilled from the mixture and was collected in water. (We know now that carbon from the organic compounds in the urine would have reduced the phosphate to phosphorus.) Phosphorus was made in this manner for more than 100 years.

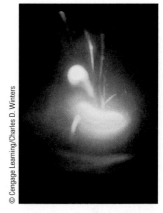

The glow of phosphorus burning in air.

Nitrogen dioxide has 17 valence electrons, so it is also an odd-electron molecule. Because the odd electron largely resides on the N atom, two NO_2 molecules can combine, forming an N—N bond and producing N_2O_4, *dinitrogen tetraoxide*.

$$2\ NO_2(g) \longrightarrow N_2O_4(g)$$

deep brown   colorless
gas   (mp, $-11.2\ °C$)

Solid N_2O_4 is colorless and consists entirely of N_2O_4 molecules. However, as the solid melts and the temperature increases to the boiling point, the color darkens as N_2O_4 dissociates to form brown NO_2. At the normal boiling point (21.5 °C), the distinctly brown gas consists of 15.9% NO_2 and 84.1% N_2O_4.

When NO_2 is bubbled into water, nitric acid and nitrous acid form.

$$2\ NO_2(g) + H_2O(\ell) \rightarrow HNO_3(aq) + HNO_2(aq)$$

nitric acid   nitrous acid

Nitric acid has been known for centuries and has become an important compound in our modern economy. The oldest way to make the acid is to treat $NaNO_3$ with sulfuric acid (Figure 21.24).

$$2\ NaNO_3(s) + H_2SO_4(\ell) \rightarrow 2\ HNO_3(\ell) + Na_2SO_4(s)$$

Enormous quantities of nitric acid are now produced industrially by the oxidation of ammonia in the multistep *Ostwald process*. The acid has many applications, but by far the greatest amount is turned into ammonium nitrate (for use as a fertilizer) by the reaction of nitric acid and ammonia.

Nitric acid is a powerful oxidizing agent, as the large, positive $E°$ values for the following half-reactions illustrate:

$$NO_3^-(aq) + 4\ H_3O^+(aq) + 3\ e^- \rightarrow NO(g) + 6\ H_2O(\ell) \qquad E° = +0.96\ V$$

$$NO_3^-(aq) + 2\ H_3O^+(aq) + e^- \rightarrow NO_2(g) + 3\ H_2O(\ell) \qquad E° = +0.80\ V$$

Nitrous oxide, N_2O. This oxide readily dissolves in fats, so the gas is added, under pressure, to cans of cream. When the valve is opened, the gas expands, whipping the cream. N_2O is also an anesthetic and is considered safe for medical uses. However, significant dangers arise from using it as a recreational drug. Long-term use can induce nerve damage and cause problems such as weakness and loss of feeling.

FIGURE 21.24 The preparation and properties of nitric acid.

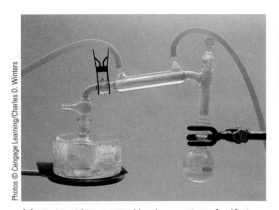

(a) Nitric acid is prepared by the reaction of sulfuric acid and sodium nitrate. (Here the reactants were heated in the flask on the right, and HNO₃ was distilled and collected in the flask cooled with ice on the left.) Pure HNO₃ is colorless, but samples of the acid are often brown because of NO₂ formed by decomposition of the acid. In this photo this gas fills the apparatus and colors the liquid in the distillation flask.

(b) When concentrated nitric acid reacts with copper, the metal is oxidized to copper(II) ions, and NO₂ gas is a reaction product.

Concentrated nitric acid attacks and oxidizes most metals. (Aluminum is an exception; see page 757.) In this process, the nitrate ion is reduced to one of the nitrogen oxides. Which oxide is formed depends on the metal and on reaction conditions. In the case of copper, for example, either NO or NO₂ is produced, depending on the concentration of the acid (Figure 21.24b).

In dilute acid:

$$3\ Cu(s) + 8\ H_3O^+(aq) + 2\ NO_3^-(aq) \rightarrow 3\ Cu^{2+}(aq) + 12\ H_2O(\ell) + 2\ NO(g)$$

In concentrated acid:

$$Cu(s) + 4\ H_3O^+(aq) + 2\ NO_3^-(aq) \rightarrow Cu^{2+}(aq) + 6\ H_2O(\ell) + 2\ NO_2(g)$$

Four metals (Au, Pt, Rh, and Ir) that are not attacked by nitric acid are often described as the "noble metals." The alchemists of the 14th century, however, knew that if they mixed HNO₃ with HCl in a ratio of about 1:3, this aqua regia, or "kingly water," would attack even gold, the noblest of metals.

$$10\ Au(s) + 6\ NO_3^-(aq) + 40\ Cl^-(aq) + 36\ H_3O^+(aq) \rightarrow$$
$$10\ [AuCl_4]^-(aq) + 3\ N_2(g) + 54\ H_2O(\ell)$$

Hydrogen Compounds of Phosphorus and Other Group 5A Elements

The phosphorus analog of ammonia, phosphine (PH₃), is a poisonous, highly reactive gas with a faint garlic-like odor. Industrially, it is made by the reaction of white phosphorus and aqueous NaOH.

$$P_4(s) + 3\ KOH(aq) + 3\ H_2O(\ell) \rightarrow PH_3(g) + 3\ KH_2PO_2(aq)$$

The other hydrides of the heavier Group 5A elements are also toxic and become more unstable as the atomic number of the element increases. Nonetheless, arsine (AsH₃) is used in the semiconductor industry as a starting material in the preparation of gallium arsenide (GaAs) semiconductors.

Phosphorus Oxides and Sulfides

The most important compounds of phosphorus are those with oxygen, and there are at least six simple binary compounds containing just phosphorus and oxygen. All of them can be thought of as being derived structurally from the P₄ tetrahedron

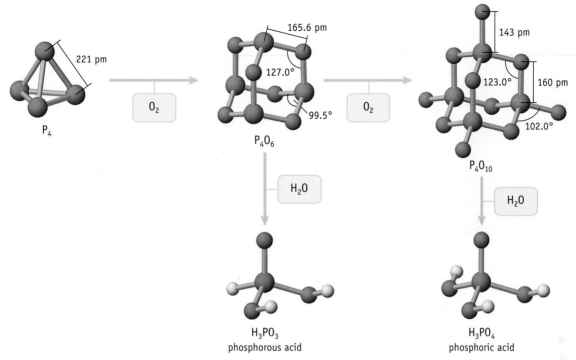

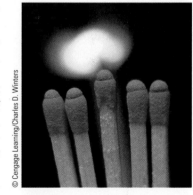

FIGURE 21.25 Phosphorus oxides. Other binary P—O compounds have formulas between P_4O_6 and P_4O_{10}. They are formed by starting with P_4O_6 and adding O atoms successively to the P atom vertices.

of white phosphorus. For example, if P_4 is carefully oxidized, P_4O_6 is formed; an O atom has been placed into each P—P bond in the tetrahedron (Figure 21.25).

The most common and important phosphorus oxide is P_4O_{10}, a fine white powder commonly called *phosphorus pentaoxide* because its empirical formula is P_2O_5. In P_4O_{10}, each phosphorus atom is surrounded tetrahedrally by O atoms.

Phosphorus also forms a series of compounds with sulfur. Of these, the most important is P_4S_3. In this phosphorus sulfide, S atoms are placed into only three of the P—P bonds. The principal use of P_4S_3 is in "strike anywhere" matches, the kind that light when you rub the head against a rough object. The active ingredients are P_4S_3 and the powerful oxidizing agent potassium chlorate, $KClO_3$. The "safety match" is now more common than the "strike anywhere" match. In safety matches, the head is predominantly $KClO_3$, and the material on the match book is red phosphorus (about 50%), Sb_2S_3, Fe_2O_3, and glue.

Matches. The head of a "strike anywhere" match contains P_4S_3 and the oxidizing agent $KClO_3$. (Other components are ground glass, Fe_2O_3, ZnO, and glue.) Safety matches have sulfur (3–5%) and $KClO_3$ (45–55%) in the match head and red phosphorus in the striking strip.

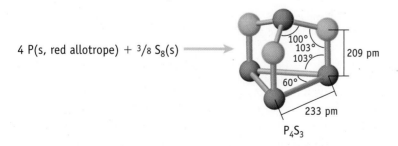

Enormous quantities of phosphorus compounds are used around the world, and most of this begins with phosphate rock, which is largely $Ca_3(PO_4)_2$ or apatite. As outlined in Figure 21.26, phosphate rock is converted to impure phosphoric acid and then to other products, or to elemental phosphorus, from which pure acid and other products are made.

(a) Mining phosphate rock. Phosphate rock is primarily $Ca_3(PO_4)_2$, and most mined in the United States comes from Florida.

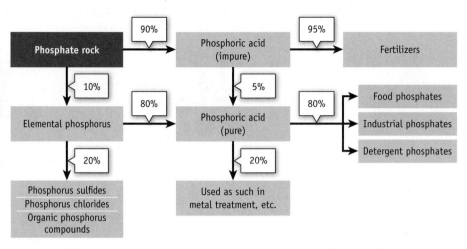

(b) Uses of phosphorus and phosphoric acid.

FIGURE 21.26 Uses of phosphate rock, phosphorus, and phosphoric acid.

Phosphorus Oxoacids and Their Salts

A few of the many known phosphorus oxoacids are illustrated in Table 21.6. Indeed, there are so many acids and their salts in this category that structural principles have been developed to organize and understand them.

(a) All P atoms in the oxoacids and their anions (conjugate bases) are four-coordinate and tetrahedral.

(b) All the P atoms in the acids have at least one P—OH group (and this occurs often in the anions as well). In every case, the H atom is ionizable as H^+.

(c) Some oxoacids have one or more P—H bonds. These H atoms are not ionizable as H^+.

(d) Polymerization can occur by P—O—P bond formation to give both linear and cyclic species. Two P atoms are never joined by more than one P—O—P bridge.

Table 21.6 Phosphorus Oxoacids

Formula	Name	Structure	pK_a
H_3PO_4	Orthophosphoric acid		2.21, 7.21, 12.67
$H_4P_2O_7$	Pyrophosphoric acid (diphosphoric acid)		0.85, 1.49, 5.77, 8.22
$(HPO_3)_3$	Metaphosphoric acid		
H_3PO_3	Phosphorous acid (phosphonic acid)		2.00, 6.59
H_3PO_2	Hypophosphorous acid (phosphinic acid)		1.24

(e) When a P atom is surrounded only by O atoms (as in H_3PO_4), its oxidation number is +5. For each P—OH that is replaced by P—H, the oxidation number drops by 2 (because P is considered more electronegative than H). For example, the oxidation number of P in H_3PO_2 is +1.

Orthophosphoric acid, H_3PO_4, and its salts are far more important commercially than other P—O acids. Millions of tons of phosphoric acid are made annually, some using white phosphorus as the starting material. The element is burned in oxygen to give P_4O_{10}, and the oxide reacts with water to produce the acid (Figure 21.27).

$$P_4O_{10}(s) + 6\ H_2O(\ell) \rightarrow 4\ H_3PO_4(aq)$$

This approach gives a pure product, so it is used to make phosphoric acid for use in food products in particular. The acid is nontoxic, and it gives the tart or sour taste to carbonated "soft drinks," such as various colas (about 0.05% H_3PO_4) or root beer (about 0.01% H_3PO_4).

A major use for phosphoric acid is to impart corrosion resistance to metal objects such as nuts and bolts, tools, and car-engine parts by plunging the object into a hot acid bath. Car bodies are similarly treated with phosphoric acid containing metal ions such as Zn^{2+}, and aluminum trim is "polished" by treating it with the acid.

The reaction of H_3PO_4 with strong bases produces salts such as NaH_2PO_4, Na_2HPO_4, and Na_3PO_4. In industry, the monosodium and disodium salts are produced using Na_2CO_3 as the base, but an excess of the stronger (and more expensive) base NaOH is required to remove the third proton to give Na_3PO_4.

Sodium phosphate (Na_3PO_4) is used in scouring powders and paint strippers because the anion PO_4^{3-} is a relatively strong base in water ($K_b = 2.8 \times 10^{-2}$). Sodium monohydrogen phosphate, Na_2HPO_4, which has a less basic anion than PO_4^{3-}, is widely used in food products. Kraft has patented a process using the salt in the manufacture of pasteurized cheese, for example. Thousands of tons of Na_2HPO_4 are still used for this purpose, even though the function of the salt in this process is not completely understood. In addition, a small amount of Na_2HPO_4 in pudding mixes enables the mix to gel in cold water, and the basic anion raises the pH of cereals to provide "quick-cooking" breakfast cereal. (The OH^- ion from HPO_4^{2-} hydrolysis accelerates the breakdown of the cellulose material in the cereal.)

Calcium phosphates are used in a broad spectrum of products. For example, the weak acid $Ca(H_2PO_4)_2 \cdot H_2O$ is used as the acid leavening agent in baking powder. A typical baking powder contains (along with inert ingredients) 28% $NaHCO_3$, 10.7% $Ca(H_2PO_4)_2 \cdot H_2O$, and 21.4% $NaAl(SO_4)_2$ (also a weak acid). The weak acids react with sodium bicarbonate to produce CO_2 gas. For example,

$$Ca(H_2PO_4)_2 \cdot H_2O(s) + 2\ NaHCO_3(aq) \rightarrow 2\ CO_2(g) + 3\ H_2O(\ell) + Na_2HPO_4(aq) + CaHPO_4(aq)$$

Finally, calcium monohydrogen phosphate, $CaHPO_4$, is used as an abrasive and polishing agent in toothpaste.

FIGURE 21.27 Reaction of P_4O_{10} and water. The white solid oxide reacts vigorously with water to give orthophosphoric acid, H_3PO_4. (The heat generated vaporizes the water, so steam is visible.)

© Cengage Learning/Charles D. Winters

REVIEW & CHECK FOR SECTION 21.8

1. Construct Lewis structures for the several resonance forms of N_2O. What is the predicted N—N bond order?

   (a) 1   (b) 2   (c) between 2 and 3   (d) 3

2. Which statement about ammonia is *not correct*?

   (a) Ammonia can be made by a direct reaction of the elements.

   (b) Solutions of ammonia are acidic.

   (c) Ammonia is a gas at room temperature and atmospheric pressure.

   (d) Ammonia is used as a reactant in the synthesis of nitric acid.

3. What is the oxidation state of phosphorus in phosphorous acid, $HPO(OH)_2$?

   (a) 0   (b) +1   (c) +3   (d) +5

Common household products containing sulfur or sulfur-based compounds.

Alternatively, the oxide can be reduced to elemental lead by combining it with fresh lead sulfide.

$$2\ PbO(s) + PbS(s) \rightarrow 3\ Pb(s) + SO_2(g)$$

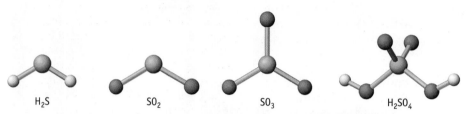

Models of some common sulfur-containing molecules: H_2S, SO_2, SO_3, and H_2SO_4.

Sulfur dioxide (SO_2), a colorless, toxic gas with a sharp odor, is produced on an enormous scale by the combustion of sulfur and by roasting sulfide ores in air. The combustion of sulfur in sulfur-containing coal and fuel oil creates particularly large environmental problems. It has been estimated that about 2.0×10^8 tons of sulfur oxides (primarily SO_2) are released into the atmosphere each year by human activities; this is more than half of the total emitted by all other natural sources of sulfur in the environment.

Sulfur dioxide readily dissolves in water. The most important reaction of this gas is its oxidation to SO_3.

$$SO_2(g) + \tfrac{1}{2}\ O_2(g) \rightarrow SO_3(g) \qquad \Delta_r H^\circ = -98.9\ \text{kJ/mol-rxn}$$

Sulfur trioxide is almost never isolated but is converted directly to sulfuric acid by reaction with water in the "contact process."

The largest use of sulfur is the production of sulfuric acid, H_2SO_4, the compound produced in largest quantity by the chemical industry (◄ page 115). In the United States, roughly 70% of the acid is used to manufacture superphosphate fertilizer from phosphate rock. Plants need a soluble form of phosphorus for growth, but calcium phosphate and apatite [$Ca_5X(PO_4)_3$, X = F, OH, Cl] are insoluble. Treating phosphate-containing minerals with sulfuric acid produces a mixture of soluble phosphates. The balanced equation for the reaction of excess sulfuric acid and calcium phosphate, for example, is

$$Ca_3(PO_4)_2(s) + 3\ H_2SO_4(\ell) \rightarrow 2\ H_3PO_4(\ell) + 3\ CaSO_4(s)$$

but it does not tell the whole story. Concentrated superphosphate fertilizer is actually mostly $CaHPO_4$ or $Ca(H_2PO_4)_2$ plus some H_3PO_4 and $CaSO_4$. (Notice that the chemical principle behind this reaction is that sulfuric acid is a stronger acid than H_3PO_4 (◄ Table 17.3), so the $PO_4{}^{3-}$ ion is protonated by sulfuric acid.)

Smaller amounts of sulfuric acid are used in the conversion of ilmenite, a titanium-bearing ore, to TiO_2, which is then used as a white pigment in paint, plastics, and paper. The acid is also used to manufacture iron and steel as well as petroleum products, synthetic polymers, and paper.

REVIEW & CHECK FOR SECTION 21.9

1. Which of the following is not a common oxidation number for sulfur in its compounds?

   (a) −2   (b) +6   (c) +3   (d) +4

2. Which statement about oxygen is *not* true?

   (a) Liquid oxygen is attracted to a magnet.

   (b) The allotropes of oxygen are O_2 and O_3.

   (c) Oxygen is the most abundant element in the earth's crust.

   (d) All electrons in O_2 are paired.

Snot-tites and Sulfur Chemistry

Sulfur chemistry can be important in cave formation, as a spectacular example in the jungles of southern Mexico amply demonstrates. Toxic hydrogen sulfide gas spews from the Cueva de Villa Luz along with water that is milky white with suspended sulfur particles. The cave can be followed downward to a large underground stream and a maze of actively enlarging cave passages. Water rises into the cave from underlying sulfur-bearing strata, releasing hydrogen sulfide at concentrations up to 150 ppm. Yellow sulfur crystallizes on the cave walls around the inlets. The sulfur and sulfuric acid are produced by the following reactions:

$$2\ H_2S(g) + O_2(g) \rightarrow 2\ S(s) + 2\ H_2O(\ell)$$
$$2\ S(s) + 2\ H_2O(\ell) + 3\ O_2(g) \rightarrow$$
$$2\ H_2SO_4(aq)$$

The cave atmosphere is poisonous to humans, so gas masks are essential for would-be explorers. But surprisingly, the cave is teeming with life. Several species of bacteria thrive on sulfur compounds in acidic environments. The chemical energy released in their metabolism is used to obtain carbon for their bodies from calcium carbonate and carbon dioxide, both of which are abundant in the cave. One result is that bacterial filaments hang from the walls and ceilings in bundles. Because the filaments look like something coming from a runny nose, cave explorers refer to them as "snot-tites." Other microbes feed on the bacteria, and so on up the food chain—which includes spiders, gnats, and pygmy snails—all the way to sardine-like fish that swim in the cave stream. This entire ecosystem is supported by reactions involving sulfur within the cave.

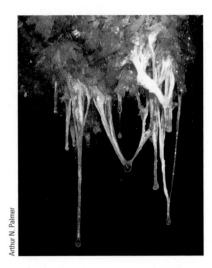

Snot-tites. Filaments of sulfur-oxidizing bacteria (dubbed "snot-tites") hang from the ceiling of a Mexican cave containing an atmosphere rich in hydrogen sulfide. The bacteria thrive on the energy released by oxidation of the hydrogen sulfide, forming the base of a complex food chain. Droplets of sulfuric acid on the filaments have an average pH of 1.4, with some as low as zero! Drops that landed on explorers in the cave burned their skin and disintegrated their clothing.

Arthur N. Palmer

21.10 The Halogens, Group 7A

Fluorine and chlorine are the most abundant halogens in the earth's crust, with fluorine somewhat more abundant than chlorine. If their abundance in sea water is measured, however, the situation is quite different. Chlorine has an abundance in sea water of 18,000 ppm, whereas the abundance of fluorine in the same source is only 1.3 ppm. This variation is a result of the differences in the solubility of their salts and plays a role in the methods used to recover the elements themselves.

Preparation of the Elements

Fluorine

The water-insoluble mineral fluorspar (calcium fluoride, CaF_2) is one of the many sources of fluorine. Because the mineral was originally used as a flux in metalworking, its name comes from the Latin word meaning "to flow." In the 17th century, it was discovered that solid CaF_2 would emit light when heated, and the phenomenon was called *fluorescence*. In the early 1800s, when it was recognized that a new element was contained in fluorspar, A. M. Ampère (1775–1836) suggested that the element be called fluorine.

Although fluorine was recognized as an element by 1812, it was not until 1886 that it was isolated by the French chemist Henri Moisson (1852–1907) in elemental form as a very pale yellow gas by the electrolysis of KF dissolved in anhydrous HF. Indeed, because F_2 is such a powerful oxidizing agent, chemical oxidation of F^- to F_2 is not feasible, and electrolysis is the only practical way to obtain gaseous F_2 (Figure 21.31).

Fluorine is still prepared by the Moisson method, but the preparation is difficult because F_2 is so reactive. It corrodes the equipment and reacts violently with traces of grease or other contaminants. Furthermore, the products of electrolysis,

Group **7A**
Halogens
Fluorine
9
F
950 ppm
Chlorine
17
Cl
130 ppm
Bromine
35
Br
0.37 ppm
Iodine
53
I
0.14 ppm
Astatine
85
At
trace

Element abundances are in **parts per million** *in the earth's crust.*

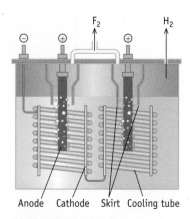

FIGURE 21.31 Schematic of an electrolysis cell for producing fluorine.

F_2 and H_2, can recombine explosively, so they must not be allowed to come into contact with each other. (Compare with the reaction of H_2 and Br_2 in Figure 21.5.) Current U.S. production of fluorine is approximately 5000 metric tons per year.

Chlorine

Chlorine is a strong oxidizing agent, and to prepare this element from chloride ion by a chemical reaction requires an even stronger oxidizing agent. In the laboratory permanganate ions or dichromate ions in acid solution will serve this purpose (Figure 21.32). Elemental chlorine was first made by the Swedish chemist Karl Wilhelm Scheele (1742–1786) in 1774, who combined sodium chloride with an oxidizing agent in an acidic solution.

Industrially, chlorine is made by electrolysis of brine (concentrated aqueous NaCl). The other product of the electrolysis, NaOH, is also a valuable industrial chemical. About 80% of the chlorine produced is made using an electrochemical cell similar to the one depicted in Figure 21.33. Oxidation of chloride ion to Cl_2 gas occurs at the anode and reduction of water occurs at the cathode.

Anode reaction (oxidation): $\qquad 2\,Cl^-(aq) \rightarrow Cl_2(g) + 2\,e^-$

Cathode reaction (reduction): $\qquad 2\,H_2O(\ell) + 2\,e^- \rightarrow H_2(g) + 2\,OH^-(aq)$

Activated titanium is used for the anode, and stainless steel or nickel is preferred for the cathode. The anode and cathode compartments are separated by a membrane that is not permeable to water but allows Na^+ ions to pass to maintain the charge balance. Thus, the membrane functions as a "salt" bridge between the anode and cathode compartments. The energy consumption of these cells is in the range of 2000–2500 kWh per ton of NaOH produced.

Bromine

The standard reduction potentials of the halogens indicate that their strength as oxidizing agents decreases going from F_2 to I_2.

Half-Reaction	Reduction Potential ($E°$, V)
$F_2(g) + 2\,e^- \rightarrow 2\,F^-(aq)$	2.87
$Cl_2(g) + 2\,e^- \rightarrow 2\,Cl^-(aq)$	1.36
$Br_2(\ell) + 2\,e^- \rightarrow 2\,Br^-(aq)$	1.08
$I_2(s) + 2\,e^- \rightarrow 2\,I^-(aq)$	0.535

This means that Cl_2 will oxidize Br^- ions to Br_2 in aqueous solution, for example.

$$Cl_2(aq) + 2\,Br^-(aq) \rightarrow 2\,Cl^-(aq) + Br_2(aq)$$

$$E°_{net} = E°_{cathode} - E°_{anode} = 1.36\ V - (1.08\ V) = +0.28\ V$$

In fact, this is the commercial method of preparing bromine when NaBr is obtained from natural brine wells in Arkansas and Michigan.

Iodine

Iodine is a lustrous, purple-black solid, easily sublimed at room temperature and atmospheric pressure (◄ Figure 13.20). The element was first isolated in 1811 from seaweed and kelp, extracts of which had long been used for treatment of goiter, the enlargement of the thyroid gland. It is now known that the thyroid gland produces a growth-regulating hormone (thyroxine) that contains iodine. Consequently, most table salt in the United States has 0.01% NaI added to provide the necessary iodine in the diet.

A laboratory method for preparing I_2 is illustrated in Figure 21.34. The commercial preparation depends on the source of I^- and its concentration. One method is interesting because it involves some chemistry described earlier in this book. Iodide ions are first precipitated with silver ions to give insoluble AgI.

$$I^-(aq) + Ag^+(aq) \rightarrow AgI(s)$$

FIGURE 21.32 **Chlorine preparation.** Chlorine is prepared by oxidation of chloride ion using a strong oxidizing agent. Here, oxidation of NaCl is accomplished using $K_2Cr_2O_7$ in H_2SO_4. (The Cl_2 gas is bubbled into water in a receiving flask.)

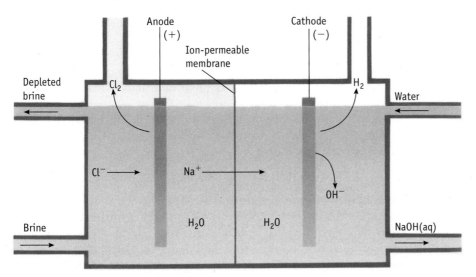

FIGURE 21.33 A membrane cell for the production of NaOH and Cl$_2$ gas from a saturated, aqueous solution of NaCl (brine). Here, the anode and cathode compartments are separated by a water-impermeable but ion-conducting membrane. A widely used membrane is made of Nafion, a fluorine-containing polymer that is a relative of polytetrafluoroethylene (Teflon). Brine is fed into the anode compartment and dilute sodium hydroxide or water into the cathode compartment. Overflow pipes carry the evolved gases and NaOH away from the chambers of the electrolysis cell.

This is reduced by clean scrap iron to give iron(II) iodide and metallic silver.

$$2\ AgI(s) + Fe(s) \rightarrow FeI_2(aq) + 2\ Ag(s)$$

The silver is recycled by oxidizing it with nitric acid (forming silver nitrate) which is then reused. Finally, iodide ion from water-soluble FeI$_2$ is oxidized to iodine with chlorine [with iron(III) chloride as a by-product].

$$2\ FeI_2(aq) + 3\ Cl_2(aq) \rightarrow 2\ I_2(s) + 2\ FeCl_3(aq)$$

Fluorine Compounds

Fluorine is the most reactive of all of the elements, forming compounds with every element except He and Ne. In most cases, the elements combine directly, and some reactions can be so vigorous as to be explosive. This reactivity can be explained by at least two features of fluorine chemistry: the relatively weak F—F bond compared with chlorine and bromine, and, in particular, the relatively strong bonds formed by

FIGURE 21.34 The preparation of iodine. A mixture of sodium iodide and manganese(IV) oxide was placed in the flask *(left).* On adding concentrated sulfuric acid *(right),* brown iodine vapor is evolved.

$$2\ NaI(s) + 2\ H_2SO_4(aq) + MnO_2(s) \rightarrow$$
$$Na_2SO_4(aq) + MnSO_4(aq)$$
$$+ 2\ H_2O(\ell) + I_2(g)$$

Bond Dissociation Enthalpies of Some Halogen Compounds (kJ/mol)

X	X—X	H—X	C—X (in CX_4)
F	155	565	485
Cl	242	432	339
Br	193	366	285
I	151	299	213

fluorine to other elements. This is illustrated by the table of bond dissociation enthalpies in the margin.

In addition to its oxidizing ability, another notable characteristic of fluorine is its small size. These properties lead to the formation of compounds where a number of F atoms can be bonded to a central element in a high oxidation state. Examples include PtF_6, UF_6, IF_7, and XeF_4.

Hydrogen fluoride is an important industrial chemical. More than 1 million tons of hydrogen fluoride is produced annually worldwide, almost all by the action of concentrated sulfuric acid on fluorspar.

$$CaF_2(s) + H_2SO_4(\ell) \rightarrow CaSO_4(s) + 2\ HF(g)$$

The U.S. capacity for HF production is approximately 210,000 metric tons, but demand often exceeds supply for this chemical. Anhydrous HF is used in a broad range of industries: in the production of refrigerants, herbicides, pharmaceuticals, high-octane gasoline, aluminum, plastics, electrical components, and fluorescent lightbulbs.

The fluorspar used to produce HF must be very pure and free of SiO_2 because HF reacts readily with silicon dioxide.

$$SiO_2(s) + 4\ HF(aq) \rightarrow SiF_4(g) + 2\ H_2O(\ell)$$

$$SiF_4(g) + 2\ HF(aq) \rightarrow H_2SiF_6(aq)$$

This series of reactions explains why HF can be used to etch or frost glass (such as the inside of fluorescent light bulbs). It also explains why HF is not shipped in glass containers (unlike HCl, for example).

The aluminum industry consumes about 10–40 kg of cryolite, Na_3AlF_6, per metric ton of aluminum produced. The reason is that cryolite is added to aluminum oxide to produce a lower-melting mixture that can be electrolyzed. Cryolite is found in only small quantities in nature, so it is made in various ways, among them the following reaction:

$$6\ HF(aq) + Al(OH)_3(s) + 3\ NaOH(aq) \rightarrow Na_3AlF_6(s) + 6\ H_2O(\ell)$$

About 3% of the hydrofluoric acid produced is used in uranium fuel production. To separate uranium isotopes in a gas centrifuge (▶ page 841), the uranium must be in the form of a volatile compound. Naturally occurring uranium is processed to give UO_2. This oxide is treated with hydrogen fluoride to give UF_4, which is then reacted with F_2 to produce the volatile solid UF_6.

$$UO_2(s) + 4\ HF(aq) \rightarrow UF_4(s) + 2\ H_2O(\ell)$$

$$UF_4(s) + F_2(g) \rightarrow UF_6(s)$$

This last step consumes 70%–80% of the fluorine produced annually.

Chlorine Compounds

Hydrogen Chloride

Hydrochloric acid, an aqueous solution of hydrogen chloride, is a valuable industrial chemical. Hydrogen chloride gas can be prepared by the reaction of hydrogen and chlorine, but the rapid, exothermic reaction is difficult to control. The classical method of making HCl in the laboratory uses the reaction of NaCl and sulfuric acid, a procedure that takes advantage of the facts that HCl is a gas and that H_2SO_4 will not oxidize the chloride ion.

$$2\ NaCl(s) + H_2SO_4(\ell) \rightarrow Na_2SO_4(s) + 2\ HCl(g)$$

Hydrogen chloride gas has a sharp, irritating odor. Both gaseous and aqueous HCl react with metals and metal oxides to give metal chlorides and, depending on the reactant, hydrogen or water.

$$Mg(s) + 2\ HCl(aq) \rightarrow MgCl_2(aq) + H_2(g)$$

$$ZnO(s) + 2\ HCl(aq) \rightarrow ZnCl_2(aq) + H_2O(\ell)$$

Oxoacids of Chlorine

Oxoacids of chlorine range from HClO, in which chlorine has an oxidation number of +1, to $HClO_4$, in which the oxidation number is equal to the group number, +7. All are strong oxidizing agents.

Oxoacids of Chlorine

Acid	Name	Anion	Name
HClO	Hypochlorous	ClO^-	Hypochlorite
$HClO_2$	Chlorous	ClO_2^-	Chlorite
$HClO_3$	Chloric	ClO_3^-	Chlorate
$HClO_4$	Perchloric	ClO_4^-	Perchlorate

Hypochlorous acid, HClO, forms when chlorine dissolves in water. In this reaction, half of the chlorine is oxidized to hypochlorite ion and half is reduced to chloride ion in a **disproportionation reaction**.

$$Cl_2(g) + 2 H_2O(\ell) \rightleftharpoons H_3O^+(aq) + HClO(aq) + Cl^-(aq)$$

If Cl_2 is dissolved in cold aqueous NaOH instead of in pure water, hypochlorite ion and chloride ion form.

$$Cl_2(g) + 2 OH^-(aq) \rightleftharpoons ClO^-(aq) + Cl^-(aq) + H_2O(\ell)$$

Under basic conditions, the equilibrium lies far to the right. The resulting alkaline solution is the "liquid bleach" used in home laundries. The bleaching action of this solution is a result of the oxidizing ability of ClO^-. Most dyes are colored organic compounds, and hypochlorite ion oxidizes dyes to colorless products.

When calcium hydroxide is combined with Cl_2, solid $Ca(ClO)_2$ is the product. This compound is easily handled and is the "chlorine" that is sold for swimming pool disinfection.

When a basic solution of hypochlorite ion is heated, another disproportionation occurs, forming chlorate ion and chloride ion:

$$3 ClO^-(aq) \rightarrow ClO_3^-(aq) + 2 Cl^-(aq)$$

Sodium and potassium chlorates are made in large quantities this way. The sodium salt can be reduced to ClO_2, a compound used for bleaching paper pulp. Some $NaClO_3$ is also converted to potassium chlorate, $KClO_3$, the preferred oxidizing agent in fireworks and a component of safety matches.

Perchlorates, salts containing ClO_4^-, are powerful oxidants. Pure perchloric acid, $HClO_4$, is a colorless liquid that explodes if shocked. It explosively oxidizes organic materials and rapidly oxidizes silver and gold. Dilute aqueous solutions of the acid are safe to handle, however.

Perchlorate salts of most metals are usually relatively stable, albeit unpredictable. Great care should be used when handling any perchlorate salt. Ammonium perchlorate, for example, bursts into flame if heated above 200 °C.

$$2 NH_4ClO_4(s) \rightarrow N_2(g) + Cl_2(g) + 2 O_2(g) + 4 H_2O(g)$$

The strong oxidizing ability of the ammonium salt accounts for its use as the oxidizer in the solid booster rockets for the Space Shuttle. The solid propellant in these rockets is largely NH_4ClO_4, the remainder being the reducing agent, powdered aluminum. Each launch requires about 750 tons of ammonium perchlorate, and more than half of the sodium perchlorate currently manufactured is converted to the ammonium salt. The process for making this conversion is an exchange reaction that takes advantage of the fact that ammonium perchlorate is less soluble in water than sodium perchlorate:

$$NaClO_4(aq) + NH_4Cl(aq) \rightleftharpoons NaCl(aq) + NH_4ClO_4(s)$$

• **Disproportionation** A reaction in which an element or compound is simultaneously oxidized and reduced is called a disproportionation reaction. Here, Cl_2 is oxidized to ClO^- and reduced to Cl^-.

Use of a perchlorate salt. The solid-fuel booster rockets of the Space Shuttle utilize a mixture of NH_4ClO_4 (oxidizing agent) and Al powder (reducing agent).

REVIEW & CHECK FOR SECTION 21.10

1. Which halogen has the highest bond dissociation enthalpy?

   (a) F_2 (b) Cl_2 (c) Br_2 (d) I_2

2. Which of the following statements is *not* correct?

   (a) The ease of oxidation of the halide ions is $F^- < Cl^- < Br^- < I^-$.

   (b) Fluorine is the most abundant halogen in the earth's crust.

   (c) F_2 is prepared industrially by electrolysis of aqueous NaF.

   (d) HF is used to etch glass.

 and

Sign in at **www.cengage.com/owl** to:
- View tutorials and simulations, develop problem-solving skills, and complete online homework assigned by your professor.
- For quick review and exam prep, download Go Chemistry mini lecture modules from OWL (or purchase them at **www.cengagebrain.com**)

? Access **How Do I Solve It?** tutorials on how to approach problem solving using concepts in this chapter.

CHAPTER GOALS REVISITED

Now that you have studied this chapter, you should ask whether you have met the chapter goals. In particular, you should be able to:

Relate the formulas and properties of compounds to the periodic table

a. Predict several chemical reactions of the Group A elements (Section 21.2). **Study Questions: 13, 14.**

b. Predict similarities and differences among the elements in a given group, based on periodic properties (Section 21.2). **Study Questions: 22, 64, 65.**

c. Know which reactions produce ionic compounds, and predict formulas for common ions and common ionic compounds based on electron configurations (Section 21.2). **Study Questions: 5, 6, 16.**

d. Recognize when a formula is incorrectly written, based on general principles governing electron configurations (Section 21.2).

Describe the chemistry of the main group or A-Group elements, particularly H; Na and K; Mg and Ca; B and Al; Si; N and P; O and S; and F and Cl

a. Identify the most abundant elements, know how they are obtained, and list some of their common chemical and physical properties.

b. Be able to summarize briefly a series of facts about the most common compounds of main group elements (ionic or covalent bonding, color, solubility, simple reaction chemistry) (Sections 21.3–21.10).

c. Identify uses of common elements and compounds, and understand the chemistry that relates to their usage (Sections 21.3–21.10).

Apply the principles of stoichiometry, thermodynamics, and electrochemistry to the chemistry of the main group elements
Study Questions: 18, 24, 29, 32, 34, 43, 48, 50, 54, 58, 61, 75, 92, 94.

22 The Chemistry of the Transition Elements

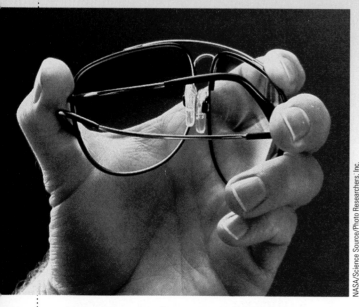

NASA/Science Source/Photo Researchers, Inc.

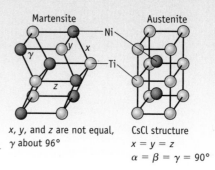

x, y, and z are not equal, γ about 96°

CsCl structure
$x = y = z$
$\alpha = \beta = \gamma = 90°$

Two phases of nitinol. The austenite form has a structure like CsCl.

Nitinol frames for glasses. Can you do this with your eye glasses? If you can, it is likely they are made of nitinol, a nickel–titanium alloy. They snap back to the proper fit even after they are twisted like a pretzel.

Memory Metal
In the 1960s, William J. Buehler, an engineer at the Naval Ordnance Laboratory in Maryland, was looking for a material for use in the nose cone of a Navy missile. It was important that the material be impact- and heat-resistant and not lose these properties when it was bent and shaped. He prepared long, thin strips of an alloy of nickel and titanium to demonstrate that they could be folded and unfolded many times without breaking. At a meeting to discuss this material, one of his associates held a cigarette lighter to a folded-up piece of metal and was amazed to observe that the metal strip immediately unfolded and assumed its original shape. Thus, memory metal was discovered. This unusual alloy is now called *nitinol*, a name constructed out of "nickel," "titanium," and "Naval Ordnance Laboratory."

Memory metal is an alloy with roughly the same number of Ni and Ti atoms. When the atoms are arranged in the highly symmetrical austenite phase, the alloy is relatively rigid. In this phase a specific shape is established that will be "remembered." If the alloy is cooled below its "phase transition temperature," it enters a less symmetrical but flexible phase (martensite). Below this transition temperature, the metal is fairly soft and may be bent and twisted out of shape. When warmed above the phase transition temperature, nitinol returns to its original shape. The temperature at which the change in shape occurs varies with small differences in the nickel-to-titanium ratio.

Besides eye glasses frames, nitinol is now used in stents to reinforce blood vessels and in orthodontics.

Questions:
1. What are the dimensions of the austenite unit cell? Assume the Ti and Ni atoms are just touching along the unit cell diagonal. (Atom radii: Ti = 145 pm; Ni = 125 pm.)
2. Calculate the density of nitinol based on the austenite unit cell parameters. Does the calculated density of the austenite unit cell agree with the reported density of 6.5 g/cm³?
3. Are Ti and Ni atoms paramagnetic or diamagnetic?

Answers to these questions are available in Appendix N.

CHAPTER OUTLINE

22.1 Properties of the Transition Elements

22.2 Metallurgy

22.3 Coordination Compounds

22.4 Structures of Coordination Compounds

22.5 Bonding in Coordination Compounds

22.6 Colors of Coordination Compounds

22.7 Organometallic Chemistry: Compounds with Metal–Carbon Bonds

CHAPTER GOALS

See Chapter Goals Revisited (page 817) for Study Questions keyed to these goals.

- Identify and explain the chemical and physical properties of the transition elements.

- Understand the composition, structure, and bonding in coordination compounds.

- Relate ligand field theory to the magnetic and spectroscopic properties of the complexes.

- Apply the effective atomic number (EAN) rule to simple organometallic compounds of the transition metals.

The transition elements are the large block of elements in the central portion of the periodic table. All are metals and bridge the *s*-block elements at the left and the *p*-block elements on the right (Figure 22.1). The transition elements are often divided into two groups, depending on the valence electrons involved in their chemistry. The first group are the ***d*-block elements**, because their occurrence in the periodic table coincides with the filling of the *d* orbitals. The second group are the ***f*-block elements**, characterized by filling of the *f* orbitals. Contained within this group of elements are two subgroups: the *lanthanides*, elements that occur between La and Hf, and the *actinides*, elements that occur between Ac and Rf.

This chapter focuses primarily on the *d*-block elements, and within this group we concentrate mainly on the elements in the fourth period, that is, the elements of the first transition series, scandium to zinc.

OWL

Sign in to OWL at **www.cengage.com/owl** to view tutorials and simulations, develop problem-solving skills, and complete online homework assigned by your professor.

go Chemistry

Download mini lecture videos for key concept review and exam prep from OWL or purchase them from **www.cengagebrain.com**

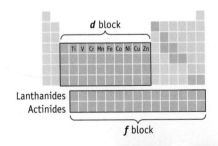

Group **4B:**
Titanium (Ti)

Group **5B:**
Vanadium (V)

Group **6B:**
Chromium (Cr)

Group **7B:**
Manganese (Mn)

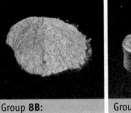

FIGURE 22.1
The transition metals. The *d*-block elements and *f*-block elements (the transition elements) are highlighted in a darker shade of blue.

Group **8B:**
Iron (Fe)

Group **8B:**
Cobalt (Co)

Group **8B:**
Nickel (Ni)

Group **1B:**
Copper (Cu)

Group **2B:**
Zinc (Zn)

Photos © Cengage Learning/Charles D. Winters

FIGURE 22.3 Typical reactions of transition metals. These metals react with oxygen, with halogens, and with acids under appropriate conditions.

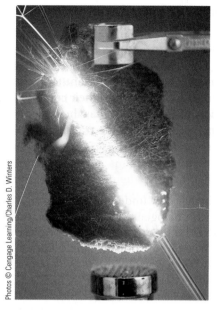

Photos © Cengage Learning/Charles D. Winters

(a) Steel wool reacts with O_2.

(b) Steel wool reacts with chlorine gas, Cl_2.

(c) Iron chips react with aqueous HCl.

Higher oxidation numbers are more common in compounds of the elements in the second and third transition series. For example, the naturally occurring sources of molybdenum and tungsten are the ores molybdenite (MoS_2) and wolframite (WO_3). This general trend is carried over in the *f*-block. The lanthanides form primarily 3+ ions. In contrast, actinide elements usually have higher oxidation numbers in their compounds; +4 and even +6 are typical. For example, UO_3 is a common oxide of uranium, and UF_6 is a compound important in processing uranium fuel for nuclear reactors [▶ Section 23.6].

Periodic Trends in the *d*-Block: Size, Density, Melting Point

The periodic table is the most useful single reference source for a chemist. Not only does it provide data that have everyday use, but it also organizes the elements with respect to their chemical and physical properties. Let us look at three physical properties of the transition elements that vary periodically: atomic radii, density, and melting point.

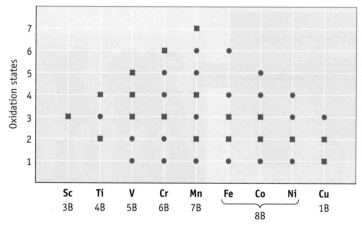

(a) The most common oxidation states are indicated with *red squares;* less common oxidation states are indicated with *blue dots.*

© Cengage Learning/Charles D. Winters

(b) Aqueous solutions of chromium compounds with two different oxidation numbers: +3 in $Cr(NO_3)_3$ (violet) and $CrCl_3$ (green), and +6 in K_2CrO_4 (yellow) and $K_2Cr_2O_7$ (orange). The two Cr(III) species have different colors in solution because there are different complex ions in solution. The complex ion in the purple solution is $[Cr(H_2O)_6]^{3+}$; the complex ion in the green solution is $[Cr(H_2O)_4Cl_2]^+$.

FIGURE 22.4 Oxidation states of the transition elements in the first transition series.

A CLOSER LOOK

Corrosion of Iron

It is hard not to be aware of corrosion. Those of us who live in the northern part of the United States are well aware of the problems of rust on our automobiles. It is estimated that 20% of iron production each year goes solely to replace iron that has rusted away.

Qualitatively, we describe corrosion as the deterioration of metals by a product-favored oxidation reaction. The corrosion of iron, for example, converts iron metal to red-brown rust, which is hydrated iron(III) oxide, $Fe_2O_3 \cdot H_2O$. This process requires both air and water, and it is enhanced if the water contains dissolved ions and if the metal is stressed (e.g., if it has dents, cuts, and scrapes on the surface).

The corrosion process occurs in what is essentially a small electrochemical cell. There is an anode and a cathode, an electrical connection between the two (the metal itself), and an electrolyte in contact with both anode and cathode. When a metal corrodes, the metal is oxidized on anodic areas of the metal surface.

Anode, oxidation $M(s) \longrightarrow M^{n+} + n\ e^-$

The electrons are consumed by several possible half-reactions in cathodic areas.

Cathode, reduction

$$2\ H_3O^+(aq) + 2\ e^- \longrightarrow H_2(g) + 2\ H_2O(\ell)$$

$$2\ H_2O(\ell) + 2\ e^- \longrightarrow H_2(g) + 2\ OH^-(aq)$$

$$O_2(g) + 2\ H_2O(\ell) + 4\ e^- \longrightarrow 4\ OH^-(aq)$$

The rate of iron corrosion is controlled by the rate of the cathodic process. Of the three possible cathodic reactions, the one that is fastest is determined by acidity and the amount of oxygen present. If little or no oxygen is present—as when a piece of iron is buried in soil such as moist clay—hydronium ions or water are reduced, and $H_2(g)$ and hydroxide ions are the products. Iron(II) hydroxide is relatively insoluble and will precipitate on the metal surface, inhibiting the further formation of Fe^{2+}.

Anode	$Fe(s) \longrightarrow Fe^{2+}(aq) + 2\ e^-$
Cathode	$2\ H_2O(\ell) + 2\ e^- \longrightarrow H_2(g) + 2\ OH^-(aq)$
Precipitation	$Fe^{2+}(aq) + 2\ OH^-(aq) \longrightarrow Fe(OH)_2(s)$
Net reaction	$Fe(s) + 2\ H_2O(\ell) \longrightarrow H_2(g) + Fe(OH)_2(s)$

If both water and O_2 are present, the chemistry of iron corrosion is somewhat different, and the corrosion reaction is about 100 times faster than without oxygen.

Anode	$2\ Fe(s) \longrightarrow 2\ Fe^{2+}(aq) + 4\ e^-$
Cathode	$O_2(g) + 2\ H_2O(\ell) + 4\ e^- \longrightarrow 4\ OH^-(aq)$
Precipitation	$2\ Fe^{2+}(aq) + 4\ OH^-(aq) \longrightarrow 2\ Fe(OH)_2(s)$
Net reaction	$2\ Fe(s) + 2\ H_2O(\ell) + O_2(g) \longrightarrow 2\ Fe(OH)_2(s)$

If oxygen is present but not in excess, further oxidation of the iron(II) hydroxide leads to the formation of magnetic iron oxide Fe_3O_4, (which can be thought of as a mixed oxide of Fe_2O_3 and FeO).

$$6\ Fe(OH)_2(s) + O_2(g) \longrightarrow 2\ Fe_3O_4 \cdot H_2O(s) + 4\ H_2O(\ell)$$
green hydrated magnetite

$$Fe_3O_4 \cdot H_2O(s) \longrightarrow H_2O(\ell) + Fe_3O_4(s)$$
black magnetite

It is the black magnetite that you find coating an iron object that has corroded by resting in moist soil.

If the iron object has free access to oxygen and water, as in the open or in flowing water, red-brown iron(III) oxide will form.

$$4\ Fe(OH)_2(s) + O_2(g) \longrightarrow 2\ Fe_2O_3 \cdot H_2O(s) + 2\ H_2O(\ell)$$
red-brown

This is the familiar rust you see on cars and buildings and the substance that colors the water red in some mountain streams or in your home.

The corrosion or rusting of iron results in major economic loss.

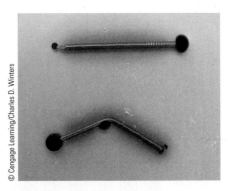

Anode and cathode reactions in iron corrosion. Two iron nails were placed in an agar gel that contains phenolphthalein and $K_3[Fe(CN)_6]$. Iron(II) ion, formed at the tip and where the nail is bent, reacts with $[Fe(CN)_6]^{3-}$ to form blue-green $Fe_4[Fe(CN)_6]_3 \cdot 14\ H_2O$ (Prussian blue). Hydrogen and $OH^-(aq)$ are formed at the other parts of the surface of the nail, the latter being detected by the pink color of the acid–base indicator. In this electrochemical cell, regions of stress—the ends and the bent region of the nail—act as anodes, and the remainder of the surface serves as the cathode.

Titanium, a versatile element. Titanium takes it name from the Titans, a race of powerful gods in Greek mythology. It is the least dense metal in the *d*-block elements and is resistant to corrosion. For this reason it is used in aircraft and jet engines, not to mention golf clubs. Because it is also nonallergenic, it is used in jewelry and in artificial hip joints (shown here). It also forms many compounds, such as TiO_2. Titanium(IV) oxide is in the paper on which this book is printed. It prevents the type and figures on one side of the paper from being seen on the other side.

Metal Atom Radii

The variation in atomic radii for the transition elements in the fourth, fifth, and sixth periods is illustrated in Figure 7.9. The radii of the transition elements vary over a fairly narrow range, with a small decrease to a minimum being observed around the middle of this group of elements. This similarity of radii can be understood based on electron configurations. Atom size is determined by the electrons in the outermost orbital, which for these elements is the *ns* orbital ($n = 4$, 5, or 6). Progressing from left to right in the periodic table, the size decline expected from increasing the number of protons in the nucleus is mostly canceled out by an opposing effect, repulsion from additional electrons in the $(n - 1)d$ orbitals.

The radii of the *d*-block elements in the fifth and sixth periods in each group are almost identical. The reason is that the lanthanide elements immediately precede the third series of *d*-block elements. The filling of 4*f* orbitals is accompanied by a steady contraction in size, consistent with the general trend of decreasing size from left to right in the periodic table. At the point where the 5*d* orbitals begin to fill again, the radii have decreased to a size similar to that of elements in the previous period. The decrease in size that results from the filling of the 4*f* orbitals is called the **lanthanide contraction**.

The similar sizes of the second and third series *d*-block elements have significant consequences for their chemistry. For example, the "platinum group metals" (Ru, Os, Rh, Ir, Pd, and Pt) form similar compounds. Thus, it is not surprising that minerals containing these metals are found in the same geological zones on earth. Nor is it surprising that it is difficult to separate these elements from one another.

Density

The variation in metal radii causes the densities of the transition elements to first increase and then decrease across a period (Figure 22.5). Although the overall change in radii among these elements is small, the effect is magnified because the volume is actually changing with the cube of the radius [$V = (4/3)\pi r^3$].

The lanthanide contraction explains why elements in the sixth period have the highest density. The relatively small radii of sixth-period transition metals, combined with the fact that their atomic masses are considerably larger than their counterparts in the fifth period, causes sixth-period metal densities to be very large.

Melting Point

The melting point of any substance reflects the forces of attraction between the atoms, molecules, or ions that compose the solid. With transition elements, the

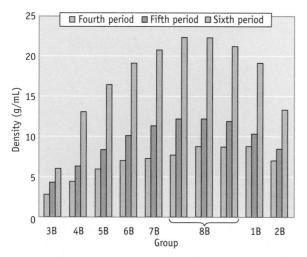

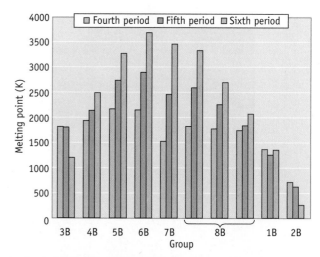

FIGURE 22.5 Periodic properties in the transition series. Density *(left)* and melting point *(right)* of the *d*-block elements.

melting points rise to a maximum around the middle of the series (Figure 22.5), then descend. Again, these elements' electron configurations provide us with an explanation. The variation in melting point indicates that the strongest metallic bonds occur when the *d* subshell is about half filled. This is also the point at which the largest number of electrons occupy the bonding molecular orbitals in the metal. (See the discussion of bonding in metals in Section 13.3.)

REVIEW & CHECK FOR SECTION 22.1

1. What is the maximum oxidation number that manganese can exhibit in its compounds?

   (a) +3   (b) +4   (c) +5   (d) +6   (e) +7

2. The "lanthanide contraction" is often given as an explanation for the fact that the sixth period transition elements have

   (a) densities smaller than that of the fifth period transition elements

   (b) atomic radii that are similar to the fifth period transition elements

   (c) melting points that are lower than the fifth period transition elements

---

22.2 **Metallurgy**

A few metals occur in nature as the free elements. This group includes copper (Figure 22.6), silver, and gold. Most metals, however, are found as oxides, sulfides, halides, carbonates, or other ionic compounds (Figure 22.7). Some metal-containing mineral deposits have little economic value, either because the concentration of the metal is too low or because the metal is difficult to separate from impurities. The relatively few minerals from which elements can be obtained profitably are called *ores* (Figure 22.7). **Metallurgy** is the general name given to the process of obtaining metals from their ores.

Very few ores are chemically pure substances. Instead, the desired mineral is usually mixed with large quantities of impurities such as sand and clay, called **gangue** (pronounced "gang"). Generally, the first step in a metallurgical process is to separate the mineral from the gangue. Then the ore is converted to the metal, a reduction process. Pyrometallurgy and hydrometallurgy are two methods of recovering metals from their ores. As the names imply, **pyrometallurgy** involves high temperatures and **hydrometallurgy** uses aqueous solutions (and thus is limited to the relatively low temperatures at which water is a liquid). Iron and copper metallurgy illustrate these two methods of metal production.

FIGURE 22.6 Naturally occurring copper. Copper occurs as the metal (native copper) and as minerals such as blue azurite [2 $CuCO_3 \cdot Cu(OH)_2$] and green malachite [$CuCO_3 \cdot Cu(OH)_2$].

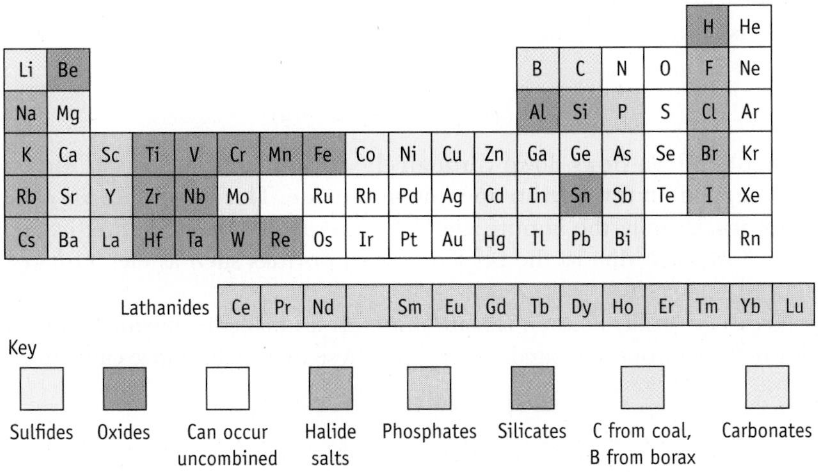

																H	He
Li	Be											B	C	N	O	F	Ne
Na	Mg											Al	Si	P	S	Cl	Ar
K	Ca	Sc	Ti	V	Cr	Mn	Fe	Co	Ni	Cu	Zn	Ga	Ge	As	Se	Br	Kr
Rb	Sr	Y	Zr	Nb	Mo		Ru	Rh	Pd	Ag	Cd	In	Sn	Sb	Te	I	Xe
Cs	Ba	La	Hf	Ta	W	Re	Os	Ir	Pt	Au	Hg	Tl	Pb	Bi			Rn

Lathanides | Ce | Pr | Nd | | Sm | Eu | Gd | Tb | Dy | Ho | Er | Tm | Yb | Lu |

Key

Sulfides   Oxides   Can occur uncombined   Halide salts   Phosphates   Silicates   C from coal, B from borax   Carbonates

FIGURE 22.7 Sources of the elements. A few transition metals, such as copper and gold, occur naturally as the metal. Most other elements are found naturally as oxides, sulfides, or other salts.

FIGURE 22.8 A blast furnace.
The largest modern furnaces have
hearths 14 meters in diameter. They
can produce as much as 10,000 tons
of iron per day.

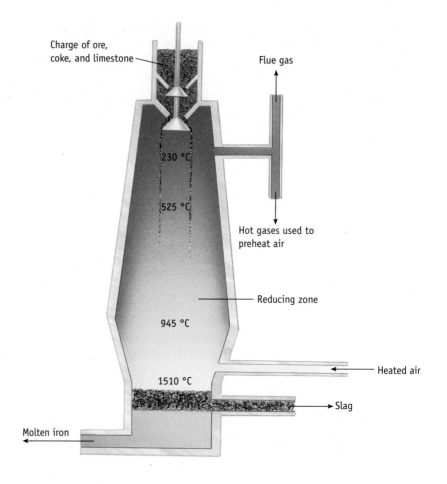

Pyrometallurgy: Iron Production

● **Coke: A Reducing Agent** Coke is
made by heating coal in a tall, narrow
oven that is sealed to keep out oxygen.
Heating drives off volatile chemicals,
including benzene and ammonia. What
remains is nearly pure carbon.

The production of iron from its ores is carried out in a blast furnace (Figure 22.8).
The furnace is charged with a mixture of ore (usually hematite, Fe_2O_3), coke (which
is primarily carbon), and limestone ($CaCO_3$). A blast of hot air forced in at the bottom of the furnace causes the coke to burn with such an intense heat that the temperature at the bottom is almost 1500 °C. The quantity of air input is controlled so
that carbon monoxide is the primary product. Both carbon and carbon monoxide
participate in the reduction of iron(III) oxide to give impure metal.

$$Fe_2O_3(s) + 3\ C(s) \longrightarrow 2\ Fe(\ell) + 3\ CO(g)$$

$$Fe_2O_3(s) + 3\ CO(g) \longrightarrow 2\ Fe(\ell) + 3\ CO_2(g)$$

Much of the carbon dioxide formed in the reduction process (and from heating the
limestone) is reduced on contact with unburned coke and produces more reducing
agent.

$$CO_2(g) + C(s) \longrightarrow 2\ CO(g)$$

The molten iron flows down through the furnace and collects at the bottom,
where it is tapped off through an opening in the side. This impure iron is called *cast
iron* or *pig iron*. Usually, the impure metal is either brittle or soft (undesirable properties for most uses) due to the presence of impurities such as elemental carbon,
phosphorus, and sulfur.

Iron ores generally contain silicate minerals and silicon dioxide. Lime (CaO),
formed when limestone is heated, reacts with these materials to give calcium silicate.

$$SiO_2(s) + CaO(s) \longrightarrow CaSiO_3(\ell)$$

This is an acid–base reaction because CaO is a basic oxide and SiO_2 is an acidic oxide. The calcium silicate, molten at the temperature of the blast furnace and less

dense than molten iron, floats on the iron. Other nonmetal oxides dissolve in this layer and the mixture, called *slag*, is easily removed.

Pig iron from the blast furnace may contain as much as 4.5% carbon, 0.3% phosphorus, 0.04% sulfur, 1.5% silicon, and some other elements as well. The impure iron must be purified to remove these nonmetal components. Several processes are available to accomplish this task, but the most important uses the *basic oxygen furnace* (Figure 22.9). The process in the furnace removes much of the carbon and all of the phosphorus, sulfur, and silicon. Pure oxygen is blown into the molten pig iron and oxidizes phosphorus to P_4O_{10}, sulfur to SO_2, and carbon to CO_2. These nonmetal oxides either escape as gases or react with basic oxides such as CaO that are added or are used to line the furnace. For example,

$$P_4O_{10}(g) + 6\ CaO(s) \longrightarrow 2\ Ca_3(PO_4)_2(\ell)$$

The result is ordinary *carbon steel*. Almost any degree of flexibility, hardness, strength, and malleability can be achieved in carbon steel by reheating and cooling in a process called *tempering*. (See *Case Study: High Strength Steels and Unit Cells*, page 454.) The resulting material can then be used in a wide variety of applications. The major disadvantages of carbon steel are that it corrodes easily and that it loses its properties when heated strongly.

Other transition metals, such as chromium, manganese, and nickel, can be added during the steel-making process, giving *alloys* (solid solutions of two or more metals) that have specific physical, chemical, and mechanical properties. One well-known alloy is stainless steel, which contains 18% to 20% Cr and 8% to 12% Ni. Stainless steel is much more resistant to corrosion than carbon steel. Another alloy of iron is alnico V. Used in loudspeaker magnets because of its permanent magnetism, it contains five elements: Al (8%), Ni (14%), Co (24%), Cu (3%), and Fe (51%).

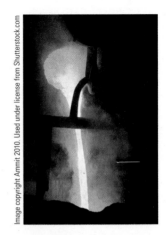

FIGURE 22.9 Molten iron being poured from a basic oxygen furnace.

Hydrometallurgy: Copper Production

In contrast to iron ores, which are mostly oxides, most copper minerals are sulfides. Copper-bearing minerals include chalcopyrite ($CuFeS_2$), chalcocite (Cu_2S), and covellite (CuS). Because ores containing these minerals generally have a very low percentage of copper, enrichment is necessary. This step is carried out by a process known as *flotation*. First, the ore is finely powdered. Next, oil is added and the mixture is agitated with soapy water in a large tank (Figure 22.10). At the same time, compressed air is forced through the mixture, so that the lightweight, oil-covered copper sulfide particles are carried to the top as a frothy mixture. The heavier gangue settles to the bottom of the tank, and the copper-laden froth is skimmed off.

Hydrometallurgy can be used to obtain copper from an enriched ore. In one method, enriched chalcopyrite ore is treated with a solution of copper(II) chloride. A reaction ensues that leaves copper in the form of solid, insoluble CuCl, which is easily separated from the iron that remains in solution as aqueous $FeCl_2$.

$$CuFeS_2(s) + 3\ CuCl_2(aq) \longrightarrow 4\ CuCl(s) + FeCl_2(aq) + 2\ S(s)$$

Aqueous NaCl is then added, and CuCl dissolves because of the formation of the complex ion $[CuCl_2]^-$.

$$CuCl(s) + Cl^-(aq) \longrightarrow [CuCl_2]^-(aq)$$

Copper(I) compounds in solution are unstable with respect to Cu(0) and Cu(II). Thus, $[CuCl_2]^-$ disproportionates to the metal and $CuCl_2$, and the latter is used to treat further ore.

$$2\ [CuCl_2]^-(aq) \longrightarrow Cu(s) + Cu^{2+}(aq) + 4\ Cl^-(aq)$$

Approximately 10% of the copper produced in the United States is obtained with the aid of bacteria. Acidified water is sprayed onto copper-mining wastes that contain low levels of copper. As the water trickles down through the crushed rock, the bacterium *Thiobacillus ferrooxidans* breaks down the iron sulfides in the rock and

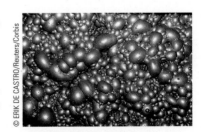

FIGURE 22.10 Enriching copper ore by the flotation process. The less dense particles of Cu_2S are trapped in the soap bubbles and float. The denser gangue settles to the bottom.

FIGURE 22.11 Electrolytic refining of copper. (a) Slabs of impure copper, called *blister copper*, form the anode, and pure copper is deposited at the cathode. **(b)** The electrolysis cells at a copper refinery.

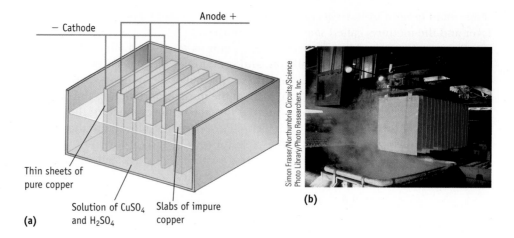

Thin sheets of pure copper

Solution of $CuSO_4$ and H_2SO_4

Slabs of impure copper

(a)

(b)

converts iron(II) to iron(III). Iron(III) ions oxidize the sulfide ion of copper sulfide to sulfate ions, leaving copper(II) ions in solution. Then the copper(II) ion is reduced to metallic copper by reaction with iron.

$$Cu^{2+}(aq) + Fe(s) \longrightarrow Cu(s) + Fe^{2+}(aq)$$

The purity of the copper obtained via these metallurgical processes is about 99%, but this is not acceptable because even traces of impurities greatly diminish the electrical conductivity of the metal. Consequently, a further purification step is needed—one involving electrolysis (Figure 22.11). Thin sheets of pure copper metal and slabs of impure copper are immersed in a solution containing $CuSO_4$ and H_2SO_4. The pure copper sheets serve as the cathode of an electrolysis cell, and the impure slabs are the anode. Copper in the impure sample is oxidized to copper(II) ions at the anode, and copper(II) ions in solution are reduced to pure copper at the cathode.

REVIEW & CHECK FOR SECTION 22.2

1. In the reaction between $CuFeS_2(s)$ and $CuCl_2(aq)$, what element is reduced and what element is oxidized?

    (a) Sulfur is reduced, iron is oxidized.   (c) Sulfur is reduced, copper is oxidized.

    (b) Copper is reduced, sulfur is oxidized.   (d) This is not a redox reaction.

2. What is the most common form in which iron is found in the earth's crust?

    (a) Iron is found in elemental form   (c) Iron sulfide

    (b) Iron oxide   (d) Iron silicate

22.3 Coordination Compounds

When metal salts dissolve, water molecules cluster around the ions (page 419). The negative end of each polar water molecule is attracted to the positively charged metal ion, and the positive end of the water molecule is attracted to the anion. As noted earlier (◄ Section 12.2), the energy of the ion–solvent interaction (solvation energy) is an important aspect of the solution process. But there is much more to this story.

Complexes and Ligands

A green solution formed by dissolving nickel(II) chloride in water contains $Ni^{2+}(aq)$ and $Cl^-(aq)$ ions (Figure 22.12). If the solvent is removed, a green crystalline solid is obtained. The formula of this solid is often written as $NiCl_2 \cdot 6\,H_2O$, and the compound is called nickel(II) chloride hexahydrate. Addition of ammonia to

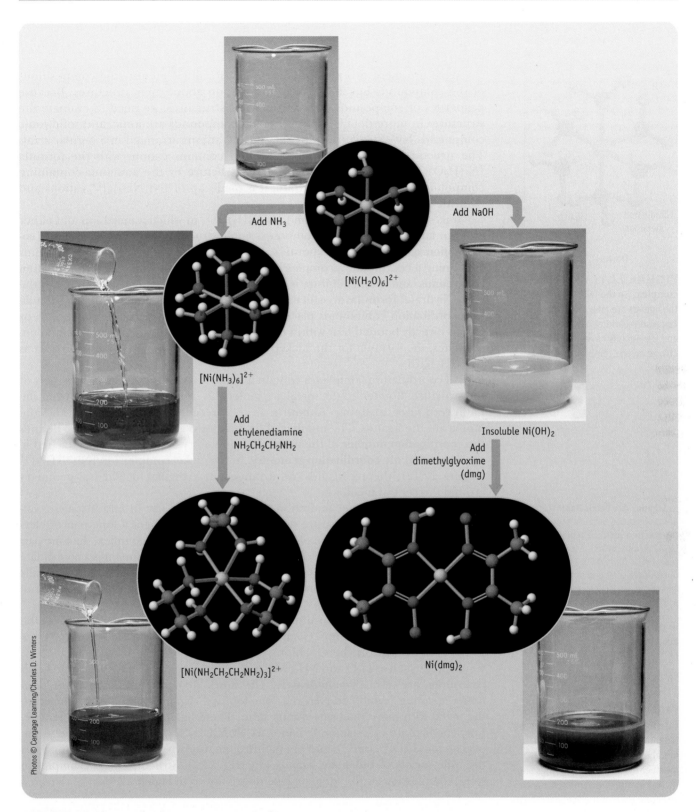

Photos © Cengage Learning/Charles D. Winters

Add NH₃

[Ni(H₂O)₆]²⁺

Add NaOH

[Ni(NH₃)₆]²⁺

Add
ethylenediamine
NH₂CH₂CH₂NH₂

Insoluble Ni(OH)₂

Add
dimethylglyoxime
(dmg)

[Ni(NH₂CH₂CH₂NH₂)₃]²⁺

Ni(dmg)₂

FIGURE 22.12 Coordination compounds of Ni²⁺ ion. The transition metals and their ions form a wide range of compounds, often with beautiful colors and interesting structures. One purpose of this chapter is to explore some commonly observed structures and explain how these compounds can be so colorful.

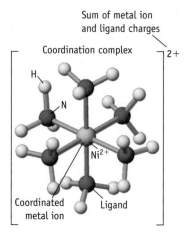

Sum of metal ion
and ligand charges

Coordination complex

Coordinated
metal ion

Ligand

$[Ni(NH_3)_6]^{2+}$

FIGURE 22.13 **A coordination complex.** In the $[Ni(NH_3)_6]^{2+}$ ion, the ligands are NH_3 molecules. Because the metal has a 2+ charge and the ligands have no charge, the charge on the complex ion is 2+.

● **Ligands Are Lewis Bases** Ligands are Lewis bases because they furnish the electron pair; the metal ion is a Lewis acid because it accepts electron pairs (◄ Section 17.11). Thus, the coordinate covalent bond between ligand and metal can be viewed as a Lewis acid–Lewis base interaction.

● **Bidentate Ligands** All common bidentate ligands bind to *adjacent* sites on the metal.

FIGURE 22.14 **Common bidentate ligands.** Coordination of these bidentate ligands to a transition metal ion results in five- or six-member metal-containing rings and no ring strain. (Only one resonance structure each of the acetylacetonate ion and phenanthroline are shown.)

the aqueous nickel(II) chloride solution gives a lilac-colored solution from which another compound, $NiCl_2 \cdot 6\ NH_3$, can be isolated. This formula looks very similar to the formula for the hydrate, with ammonia substituted for water.

What are these two nickel species? The formulas identify the compositions of the compounds but fail to give information about their structures. Because properties of compounds derive from their structures, we need to evaluate the structures in more detail. Typically, metal compounds are ionic, and solid ionic compounds have structures with cations and anions arranged in a regular array. The structure of hydrated nickel chloride contains cations with the formula $[Ni(H_2O)_6]^{2+}$ and chloride anions. The structure of the ammonia-containing compound is similar to the hydrate; it is made up of $[Ni(NH_3)_6]^{2+}$ cations and chloride anions.

Ions such as $[Ni(H_2O)_6]^{2+}$ and $[Ni(NH_3)_6]^{2+}$, in which a metal ion and either water or ammonia molecules compose a single structural unit, are examples of **coordination complexes**, also known as **complex ions** (Figure 22.13). Compounds containing a coordination complex as part of the structure are called **coordination compounds**, and their chemistry is known as **coordination chemistry**. Although the older "hydrate" formulas are still used, the preferred method of writing the formula for coordination compounds places the metal atom or ion and the molecules or anions directly bonded to it within brackets to show that it is a single structural entity. Thus, the formula for the nickel(II)–ammonia compound is better written as $[Ni(NH_3)_6]Cl_2$.

All coordination complexes contain a metal atom or ion as the central part of the structure. Bonded to the metal are molecules or ions called **ligands** (from the Latin verb *ligare*, meaning "to bind"). In the preceding examples, water and ammonia are the ligands. The number of ligand atoms attached to the metal defines the **coordination number** of the metal. The geometry described by the attached ligands is called the **coordination geometry**. In the nickel complex ion $[Ni(NH_3)_6]^{2+}$ (Figure 22.12), the six ligands are arranged in a regular octahedral geometry around the central metal ion.

Ligands can be either neutral molecules or anions (or, in rare instances, cations). The characteristic feature of a ligand is that it contains a lone pair of electrons. In the classic description of bonding in a coordination complex, the lone pair of electrons on a ligand is shared with the metal ion. The attachment is a coordinate covalent bond (◄ Section 8.5), because the electron pair being shared was originally on the ligand. The name "coordination complex" derives from the name given to this kind of bonding.

The net charge on a coordination complex is the sum of the charges on the metal and its attached groups. Complexes can be cations (as in the two nickel complexes used as examples here), anions, or uncharged.

Ligands such as H_2O and NH_3, which coordinate to the metal via a single Lewis base atom, are termed **monodentate**. The word "dentate" comes from the Latin *dentis*, meaning "tooth," so NH_3 is a "one-toothed" ligand. Some ligands attach to the metal with more than one donor atom. These ligands are called **polydentate**. Ethylenediamine (1,2-diaminoethane), $H_2NCH_2CH_2NH_2$, often abbreviated as en; oxalate ion, $C_2O_4^{2-}$ (ox^{2-}); and phenanthroline, $C_{12}H_8N_2$ (phen), are examples of the wide variety of **bidentate** ligands (Figure 22.14). Structures and examples of some complex ions with bidentate ligands are shown in Figure 22.15.

(a) $H_2NCH_2CH_2NH_2$, en
ethylenediamine

(b) $C_2O_4^{2-}$, ox
oxalate ion

(c) $CH_3COCHCOCH_3^-$, acac$^-$
acetylacetonate ion

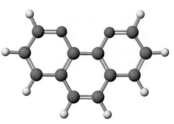

(d) $C_{12}H_8N_2$, phen
phenanthroline

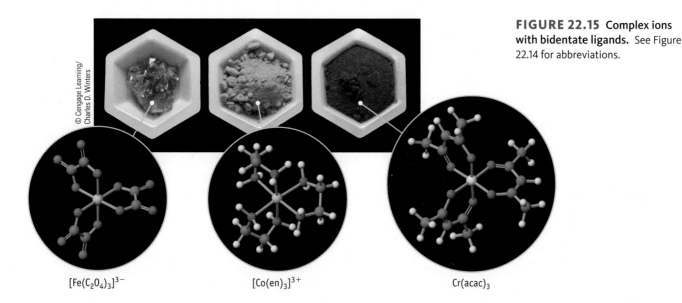

FIGURE 22.15 Complex ions with bidentate ligands. See Figure 22.14 for abbreviations.

$[Fe(C_2O_4)_3]^{3-}$ $[Co(en)_3]^{3+}$ $Cr(acac)_3$

Polydentate ligands are also called **chelating ligands**, or just *chelates* (pronounced "key-lates"). The name derives from the Greek *chele*, meaning "claw." Because two or more bonds are broken to separate the ligand from the metal, complexes with chelated ligands often have greater stability than those with monodentate ligands. Chelated complexes are important in everyday life. One way to clean the rust out of water-cooled automobile engines and steam boilers is to add a solution of oxalic acid. Iron(III) oxide reacts with oxalic acid to give a water-soluble iron oxalate complex ion:

$$3\ H_2O(\ell) + Fe_2O_3(s) + 6\ H_2C_2O_4(aq) \longrightarrow 2\ [Fe(C_2O_4)_3]^{3-}(aq) + 6\ H_3O^+(aq)$$

Ethylenediaminetetraacetate ion ($EDTA^{4-}$), a hexadentate ligand, is an excellent chelating ligand (Figure 22.16). It can wrap around a metal ion, encapsulating it. Salts of this anion are often added to commercial salad dressings to remove traces of free metal ions from solution; otherwise, these metal ions can act as catalysts for the oxidation of the oils in the dressing. Without $EDTA^{4-}$, the dressing would quickly become rancid. Another use is in bathroom cleansers. The $EDTA^{4-}$ ion removes deposits of $CaCO_3$ and $MgCO_3$ left by hard water by coordinating to Ca^{2+} or Mg^{2+} to create soluble complex ions.

Complexes with polydentate ligands play particularly important roles in biochemistry, one example of which is described in *A Closer Look: Hemoglobin*, page 798.

Formulas of Coordination Compounds

It is useful to be able to predict the formula of a coordination complex, given the metal ion and ligands, and to derive the oxidation number of the coordinated metal ion, given the formula of a coordination compound. The following examples explore these issues.

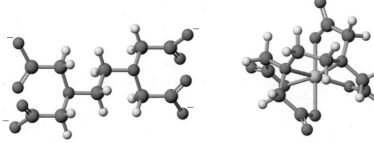

(a) ethylenediaminetetraacetate, $EDTA^{4-}$ **(b)** $[Co(EDTA)]^-$

FIGURE 22.16 $EDTA^{4-}$, a hexadentate ligand. **(a)** Ethylenediaminetetraacetate, $EDTA^{4-}$. **(b)** $[Co(EDTA)]^-$. Notice the five-member rings created when this ligand bonds to the metal.

EXAMPLE 22.1   Formulas of Coordination Complexes

Problem Give the formulas of the following coordination complexes:

(a)   A Ni^{2+} ion is bound to two water molecules and two bidentate oxalate ions.

(b)   A Co^{3+} ion is bound to one Cl^- ion, one ammonia molecule, and two bidentate ethylenedi-amine (en) molecules.

What Do You Know? The composition of the coordination complex is given for each part of the question.

Strategy The problem requires determining the net charge, which equals the sum of the charges of the various component parts of the complex ion. With that information, the metal and ligands can be assembled in the formula, which is placed in brackets, and the net charge indicated.

Solution

(a)   This complex ion is constructed from two neutral H_2O molecules, two $C_2O_4^{2-}$ ions, and one Ni^{2+} ion, so the net charge on the complex is $2-$. The formula for the complex ion is

$$[Ni(C_2O_4)_2(H_2O)_2]^{2-}$$

(b)   This cobalt(III) complex ion combines two en molecules and one NH_3 molecule, both having no charge, as well as one Cl^- ion and a Co^{3+} ion. The net charge is $2+$. The formula for this complex (writing out the entire formula for ethylenediamine) is

$$[Co(H_2NCH_2CH_2NH_2)_2(NH_3)Cl]^{2+}$$

Think about Your Answer Enclosing the formula of a complex in brackets is done to designate that this is a single unit. Species shown outside the brackets are counterions (when shown).

Check Your Understanding

(a)   What is the formula of a complex ion composed of one Co^{3+} ion, three ammonia molecules, and three Cl^- ions?

(b)   What is the formula for the coordination complex of iron (II) having two ethylenediamine ligands and two bromide ion ligands?

EXAMPLE 22.2   Coordination Compounds

Problem In each of the following coordination complexes, determine the metal's oxidation number and coordination number.

(a) $[Co(en)_2(NO_2)_2]Cl$ (c) $Pt(NH_3)_2Cl_4$

(b) $Pt(NH_3)_2(C_2O_4)$ (d) $[Co(NH_3)_5Cl]SO_4$

What Do You Know? You are given the formulas of compounds involving complex ions. The oxidation number is the charge that must be present on the metal. The coordination number is the number of ligand attachments to the metal. Remember that bidentate ligands coordinate in two positions.

Strategy Each formula consists of a complex ion or molecule made up of the metal ion, neutral, and anionic ligands. Counterions (shown outside the square brackets) may also be present. The oxidation number of the metal is the charge necessary to balance the sum of the negative charges associated with the anionic ligands and counterion. The coordination number is the number of donor atoms in the ligands that are bonded to the metal. Remember that the bidentate ligands in these examples (en, oxalate ion) attach to the metal at two sites and that any counterions present are not part of the complex ion—that is, they are not ligands.

Solution

(a) A chloride ion with a 1− charge is outside the brackets, showing that the charge on the complex ion must be 1+. There are two nitrite ions (NO_2^-) and two neutral bidentate ethylenediamine ligands in the complex. To give a 1+ charge on the complex ion, the cobalt ion must have a charge of 3+; that is, the sum of 2− (two nitrites), 0 (two en ligands), and 3+ (the cobalt ion) equals 1+. Each en ligand fills two coordination positions, and the two nitrite ions fill two more positions. The coordination number of the metal is 6.

(b) There is an oxalate ion ($C_2O_4^{2-}$) and two neutral ammonia ligands. To balance the charge on the oxalate ion, platinum must have a 2+ charge; that is, it has an oxidation number of +2. The coordination number is 4, with an oxalate ligand filling two coordination positions and each ammonia molecule filling one.

(c) There are four chloride ions (Cl^-) and two neutral ammonia ligands. In this complex, the oxidation number of the metal is +4, and the coordination number is 6.

(d) A sulfate ion with a 2– charge is outside the brackets, showing that the charge on the complex ion must be 2+. There is one chloride ion (Cl^-) and five neutral ammonia ligands in the complex ion. To give a 2+ charge on the complex ion, the cobalt must be 3+; the sum of 1− (one chloride), zero (five NH_3 ligands), and 3+ (the cobalt ion) equals 2+. The oxidation number of the metal is +3 and the coordination number is 6 (sulfate is not coordinated to the metal).

Think about Your Answer Knowing the oxidation state of the metal in a coordination compound is important. From this, you can determine the electron configuration of the metal. As we shall see in Section 22.5, this information will be needed to determine important properties of a complex such as magnetism and color.

Check Your Understanding

(a) Determine the metal's oxidation number and coordination number in (i) $K_3[Co(NO_2)_6]$ and in (ii) $Mn(NH_3)_4Cl_2$.

(b) Determine the oxidation number and coordination number of cobalt in the complex $NH_4[Co(edta)]$.

Naming Coordination Compounds

Coordination compounds are named according to an established system. The three compounds below are named according to the rules that follow.

Compound	Systematic Name
$[Ni(H_2O)_6]SO_4$	Hexaaquanickel(II) sulfate
$[Cr(en)_2(CN)_2]Cl$	Dicyanobis(ethylenediamine)chromium(III) chloride
$K[Pt(NH_3)Cl_3]$	Potassium amminetrichloroplatinate(II)

1. In naming a coordination compound that is a salt, name the cation first and then the anion. (This is how all salts are commonly named.)
2. When giving the name of the complex ion or molecule, name the ligands first, in alphabetical order, followed by the name of the metal. (When determining alphabetical order, any prefix is not considered part of the name.)
3. Ligands and their names:
   (a) If a ligand is an anion whose name ends in *-ite* or *-ate*, the final *e* is changed to *o* (sulfate ⟶ sulfato or nitrite ⟶ nitrito).
   (b) If the ligand is an anion whose name ends in *-ide*, the ending is changed to *o* (chloride ⟶ chloro, cyanide ⟶ cyano).
   (c) If the ligand is a neutral molecule, its common name is usually used with several important exceptions: Water as a ligand is referred to as *aqua;* ammonia is called *ammine,* and CO is called *carbonyl.*

Common Monodentate Ligands and Their Names

Ligand	Name
F^-	fluoro
Cl^-	chloro
Br^-	bromo
I^-	iodo
H_2O	aqua
NH_3	ammine
CO	carbonyl
CN^-	cyano
OH^-	hydroxo
$-NO_2^-$	nitro

A CLOSER LOOK

Hemoglobin

Metal-containing coordination compounds figure prominently in many biochemical reactions. Perhaps the best-known example is hemoglobin, the chemical in the blood responsible for O_2 transport. It is also one of the most thoroughly studied bioinorganic compounds.

Hemoglobin (Hb) is a large iron-containing protein that includes four polypeptide segments, each containing an iron(II) ion locked inside a porphyrin ring system and coordinated to a nitrogen atom from another part of the protein. A sixth site is available to attach to oxygen.

One segment of the hemoglobin molecule resembles the myoglobin structure shown below. (Myoglobin, the oxygen-storage protein in muscle, has only one polypeptide chain with an enclosed heme group.) In myoglobin, the heme group (the name given to the metal and porphyrin ring system) is enclosed within a polypeptide chain. The iron ion in the porphyrin ring is shown in Figure B. The first and sixth coordination positions are taken up by nitrogen atoms from amino acids of the polypeptide chain.

Hemoglobin functions by reversibly adding oxygen to the sixth coordination position of each iron, giving a complex called *oxyhemoglobin*.

Because hemoglobin features four iron centers, a maximum of four molecules of oxygen can bind to the molecule. The binding to oxygen is cooperative; that is, binding one molecule enhances the tendency to bind the second, third, and fourth molecules. Formation of the oxygenated complex is favored, but not too highly, because oxygen must also be released by the molecule to body tissues. Interestingly, an increase in acidity leads to a decrease in the stability of the oxygenated complex. This phenomenon is known as the *Bohr effect*, named for Christian Bohr, Niels Bohr's father. Release of oxygen in tissues is facilitated by an increase in acidity that results from the presence of CO_2 formed by metabolism.

Among the notable properties of hemoglobin is its ability to form a complex with carbon monoxide. This complex is very stable, with the equilibrium constant for the following reaction being about 200 (where Hb is hemoglobin):

$$HbO_2(aq) + CO(g) \rightleftharpoons HbCO(aq) + O_2(g)$$

When CO complexes with iron, the oxygen-carrying capacity of hemoglobin is lost. Consequently, CO (and the isoelectronic CN^- ion) is highly toxic to humans. Exposure to even small amounts greatly reduces the capacity of the blood to transport oxygen.

Hemoglobin abnormalities are well known. One of the most common abnormalities causes sickle cell anemia.

FIGURE A Porphyrin ring of the heme group. The tetradentate ligand surrounding the iron(II) ion in hemoglobin is a dianion of a substituted molecule called a *porphyrin*. Because of the double bonds in this structure, all of the carbon and nitrogen atoms in the dianion of the porphyrin lie in a plane. In addition, the nitrogen lone pairs are directed toward the center of the molecule, and the molecular dimensions are such that a metal ion may fit nicely into the cavity.

porphyrin

$\downarrow$ $-2\,H^+$

porphyrin^{2-}

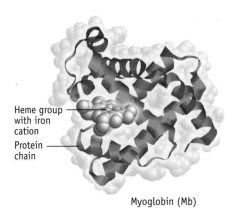

Heme group with iron cation

Protein chain

Myoglobin (Mb)

FIGURE B The heme group in myoglobin. This protein is a close relative of hemoglobin. The heme group with its iron ion is shown.

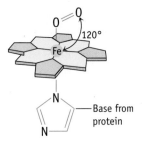

120°

Base from protein

FIGURE C Oxygen binding. Oxygen binds to the iron of the heme group in oxyhemoglobin (and in myoglobin). Interestingly, the Fe—O—O angle is bent.

(d) When there is more than one of a particular monodentate ligand with a simple name, the number of ligands is designated by the appropriate prefix: *di, tri, tetra, penta,* or *hexa.* If the ligand name is complicated, the prefix changes to *bis, tris, tetrakis, pentakis,* or *hexakis,* followed by the ligand name in parentheses.

4. If the coordination complex is an anion, the suffix *-ate* is added to the metal name.

5. Following the name of the metal, the oxidation number of the metal is given in Roman numerals.

EXAMPLE 22.3   Naming Coordination Compounds

Problem Name the following compounds:

(a) $[Cu(NH_3)_4]SO_4$

(c) $Co(phen)_2Cl_2$

(b) $K_2[CoCl_4]$

(d) $[Co(en)_2(H_2O)Cl]Cl_2$

What Do You Know? Formulas of each compound are given.

Strategy Apply the rules for nomenclature given above.

Solution

(a) The complex ion (in square brackets) is composed of four NH_3 molecules (named *ammine* in a complex) and the copper ion. To balance the $2-$ charge on the sulfate counterion, copper must have a $2+$ charge. The compound's name is

> tetraamminecopper(II) sulfate

(b) The complex ion $[CoCl_4]^{2-}$ has a $2-$ charge. With four Cl^- ligands, the cobalt ion must have a $2+$ charge, so the sum of charges is $2-$. The name of the compound is

> potassium tetrachlorocobaltate(II)

(c) This is a neutral coordination compound. The ligands include two Cl^- ions and two neutral bidentate *phen* (phenanthroline) ligands (page 794). The metal ion must have a $2+$ charge (Co^{2+}). The name, listing ligands in alphabetical order, is

> dichlorobis(phenanthroline)cobalt(II)

(d) The complex ion has a $2+$ charge because it is paired with two uncoordinated Cl^- ions. The cobalt ion is Co^{3+} because it is bonded to two neutral en ligands, one neutral water, and one Cl^-. The name is

> aquachlorobis(ethylenediamine)cobalt(III) chloride

Think about Your Answer The earliest studies of coordination compounds created a colorful array of names, and a few of these are still used (Prussian Blue, mentioned earlier, and Zeise's salt, $K[Pt(C_2H_4)Cl_3]$, are two examples). However, a systematic nomenclature is essential to communication among scientists working in the area. The rules of nomenclature make it possible for any scientist to derive a formula from the name, or to name a compound if given its formula.

Check Your Understanding

Name the following coordination compounds.

(a) $[Ni(H_2O)_6]SO_4$ (b) $[Cr(en)_2(CN)_2]Cl$ (c) $K[Pt(NH_3)Cl_3]$ (d) $K[CuCl_2]$

REVIEW & CHECK FOR SECTION 22.3

1. What is the coordination number for the metal in the compound tetraaquaoxalatoiron(III) chloride?

    (a) 4   (b) 6   (c) 7   (d) 2

2. What is the oxidation number of the metal in $(NH_4)_3[Fe(CN)_6]$?

    (a) 0   (b) $+1$ (c) $+2$ (d) $+3$

3. Which of the following is a complex ion based on an iron(II) ion?

    (a) $[Fe(H_2O)_4Cl_2]ClO_4$

    (c) $Fe(acac)_3$

    (b) $Na_4[Fe(C_2O_4)_3]$

    (d) $[Fe(NH_3)_5Cl]SO_4$

4. In which of the following complexes is the metal 6-coordinate?

   (a) lithium tetrachlorocobaltate(II)

   (b) pentaamminehydroxochromium(III) chloride

   (c) bis(ethylenediamine)silver(I) chloride

   (d) diamminedichlorozinc(II)

22.4 Structures of Coordination Compounds

Common Coordination Geometries

The geometry of a coordination complex is defined by the arrangement of donor atoms of the ligands around the central metal ion. Metal ions in coordination compounds can have coordination numbers ranging from 2 to 12. Only complexes with coordination numbers of 2, 4, and 6 are common, however, so we will concentrate on species such as $[ML_2]^{n\pm}$, $[ML_4]^{n\pm}$, and $[ML_6]^{n\pm}$, where M is the metal ion and L is a *monodentate* ligand. Within these stoichiometries, the following geometries are encountered:

- All $[ML_2]^{n\pm}$ complexes are linear. The two ligands are on opposite sides of the metal, and the L—M—L bond angle is 180°. Common examples include $[Ag(NH_3)_2]^+$ and $[CuCl_2]^-$.
- Tetrahedral geometry occurs in many $[ML_4]^{n\pm}$ complexes. Examples include $TiCl_4$, $[CoCl_4]^{2-}$, $[NiCl_4]^{2-}$, and $[Zn(NH_3)_4]^{2+}$.
- Some $[ML_4]^{n\pm}$ complexes can have square planar geometry. This geometry is most often seen with metal ions that have eight *d* electrons. Examples include $Pt(NH_3)_2Cl_2$, $[Ni(CN)_4]^{2-}$, and the nickel complex with the dimethylglyoximate (dmg$^-$) ligand in Figure 22.12.
- Octahedral geometry is found in complexes with the stoichiometry $[ML_6]^{n\pm}$ (Figure 22.12).

Isomerism

Isomerism is one of the most interesting aspects of molecular structure. Recall that the chemistry of organic compounds is greatly enlivened by the multitude of isomeric compounds that are known.

Isomers are classified as follows:

- *Structural isomers* have the same molecular formula but different bonding arrangements of atoms.
- *Stereoisomers* have the same atom-to-atom bonding sequence, but the atoms differ in their arrangement in space. There are two types of stereoisomers: geometric isomers (such as *cis* and *trans* alkenes, page 349) and optical isomers (nonsuperimposable mirror images that have the unique property that they rotate the plane of plane-polarized light, page 342).

All three types of isomerism—structural, geometric, and optical—are encountered in coordination chemistry.

Structural Isomerism

The two most important types of structural isomerism in coordination chemistry are coordination isomerism and linkage isomerism. **Coordination isomerism** occurs when it is possible to exchange a coordinated ligand and the uncoordinated counterion. For example, dark violet $[Co(NH_3)_5Br]SO_4$ and red $[Co(NH_3)_5SO_4]Br$ are coordination isomers. In the first compound, bromide ion is a ligand and sulfate is a counterion; in the second, sulfate is a ligand and bromide is the counterion. A diagnostic test for this kind of isomer is often made based on chemical reactions. For example, these two compounds can be distinguished by precipitation reactions.

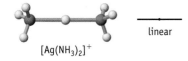

linear

$[Ag(NH_3)_2]^+$

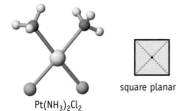

square planar

$Pt(NH_3)_2Cl_2$

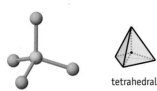

tetrahedral

$[NiCl_4]^{2-}$

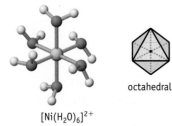

octahedral

$[Ni(H_2O)_6]^{2+}$

Common Coordination Geometries

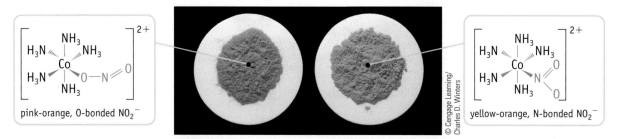

FIGURE 22.17 Linkage isomers, [Co(NH₃)₅ONO]²⁺ and [Co(NH₃)₅NO₂]²⁺. These complex ions, whose systematic names are pentaamminenitritocobalt(III) and pentaamminenitrocobalt(III), respectively, were the first known examples of this type of isomerism. (Each compound has a Cl⁻ as a counterion.)

Addition of $Ba^{2+}(aq)$ to a solution of $[Co(NH_3)_5Br]SO_4$ gives a precipitate of $BaSO_4$, indicating the presence of sulfate ion in solution. In contrast, no reaction occurs if $Ba^{2+}(aq)$ is added to a solution of $[Co(NH_3)_5SO_4]Br$. In this complex, sulfate ion is attached to Co^{3+} and is not a free ion in solution.

$$[Co(NH_3)_5Br]SO_4 + Ba^{2+}(aq) \longrightarrow BaSO_4(s) + [Co(NH_3)_5Br]^{2+}(aq)$$

$$[Co(NH_3)_5SO_4]Br + Ba^{2+}(aq) \longrightarrow \text{no reaction}$$

Linkage isomerism occurs when it is possible to attach a ligand to the metal through different atoms. The two most common ligands with which linkage isomerism arises are thiocyanate, SCN^-, and nitrite, NO_2^-. The Lewis structure of the thiocyanate ion shows that there are lone pairs of electrons on sulfur and nitrogen. The ligand can attach to a metal either through sulfur (called *S-bonded thiocyanate*) or through nitrogen (called *N-bonded thiocyanate*). Nitrite ion can attach either at oxygen or at nitrogen. The former are called nitrito complexes; the latter are nitro complexes (Figure 22.17).

Ligands forming linkage isomers

Geometric Isomerism

Geometric isomerism results when the atoms bonded directly to the metal have a different spatial arrangement. The simplest example of geometric isomerism in coordination chemistry is *cis–trans* isomerism, which occurs in both square-planar and octahedral complexes. An example of *cis–trans* isomerism is seen in the square-planar complex Pt(NH₃)₂Cl₂ (Figure 22.18a). In this complex, the two Cl⁻ ions can be either adjacent to each other (*cis*) or on opposite sides of the metal (*trans*). The

● **Cis–Trans Isomerism** *Cis–trans* isomerism is not possible for tetrahedral complexes. All L—M—L angles are 109.5°, and all positions are equivalent in this three-dimensional structure.

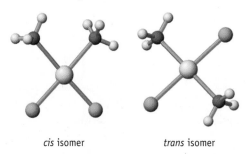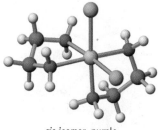

cis isomer   *trans* isomer   *cis* isomer, purple   *trans* isomer, green

(a) The square-planar complex Pt(NH₃)₂Cl₂ can exist in two geometries, *cis* and *trans*.

(b) *Cis* and *trans* octahedral isomers are possible for [Co(en)₂Cl₂]⁺.

FIGURE 22.18 *Cis–trans* isomers.

fac isomer

mer isomer

FIGURE 22.19 *Fac* and *mer* isomers of Cr(NH₃)₃Cl₃. In the *fac* isomer, the three chloride ligands (and the three ammonia ligands) are arranged at the corners of a triangular face. In the *mer* isomer, the three similar ligands follow a meridian.

cis isomer is effective in the treatment of testicular, ovarian, bladder, and osteogenic sarcoma cancers, but the *trans* isomer has no effect on these diseases.

Cis–trans isomerism in an octahedral complex is illustrated by a complex ion with two bidentate ethylenediamine ligands and two chloride ligands, $[Co(H_2NCH_2CH_2NH_2)_2Cl_2]^+$. In this complex, the two Cl^- ions occupy positions that are either adjacent (the purple *cis* isomer) or opposite (the green *trans* isomer) (Figure 22.18b).

Another common type of geometric isomerism occurs for octahedral complexes with the general formula MX_3Y_3. A *fac* isomer has three identical ligands lying at the corners of a triangular face of an octahedron defined by the ligands (*fac* = facial), whereas the ligands follow a meridian in the *mer* isomer (*mer* = meridional). *Fac* and *mer* isomers of $Cr(NH_3)_3Cl_3$ are shown in Figure 22.19.

Optical Isomerism

Optical isomerism (chirality) occurs for octahedral complexes when the metal ion coordinates to three bidentate ligands or when the metal ion coordinates to two bidentate ligands and two monodentate ligands in a *cis* position. The complexes $[Co(en)_3]^{3+}$ and *cis*-$[Co(en)_2Cl_2]^+$, illustrated in Figure 22.20, are examples of chiral complexes. The diagnostic test for chirality is met with both species: Mirror images of these molecules are not superimposable (page 342). Solutions of the optical isomers rotate plane-polarized light in opposite directions.

EXAMPLE 22.4 Isomerism in Coordination Chemistry

Problem For which of the following compounds or complex ions do isomers exist? If isomers are possible, identify the type of isomerism (structural, geometric, or optical).

(a) $[Co(NH_3)_4Cl_2]^+$

(b) $Pt(NH_3)_2(CN)_2$ (square-planar)

(c) $Co(NH_3)_3Cl_3$

(d) $K_3[Fe(C_2O_4)_3]$

(e) $Zn(NH_3)_2Cl_2$ (tetrahedral)

(f) $[Co(NH_3)_5SCN]^{2+}$

What Do You Know? From the formulas, which are given, you can infer coordination number and geometry. Structural isomers are those for which there are different attachments of atoms; geometric isomers are those where the spatial arrangement of atoms is different. Complexes that are optically active have nonsuperimposable mirror images.

Strategy Determine the number of ligands attached to the metal and decide whether the ligands are monodentate or bidentate. Knowing how many donor atoms are coordinated to the metal (the coordination number) will allow you to establish the metal geometry. At that point, it is necessary to recall the possible types of isomers for each geometry. The only isomerism possible for square-planar complexes is geometric (*cis* and *trans*). Tetrahedral complexes do not have geometric or optical isomers. Six-coordinate complexes of the formula MA_4B_2 can be either *cis* or *trans*. *Mer* and *fac* isomers are possible with a stoichiometry of MA_3B_3. Optical activity arises for metal complexes of the formula *cis*-$M(bidentate)_2X_2$ and $M(bidentate)_3$ (among others). Drawing pictures of the molecules will help you visualize the isomers.

$[Co(en)_3]^{3+}$

$[Co(en)_3]^{3+}$ mirror image

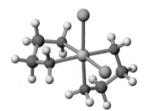

cis-$[Co(en)_2Cl_2]^+$

cis-$[Co(en)_2Cl_2]^+$ mirror image

FIGURE 22.20 Chiral metal complexes. Both $[Co(en)_3]^{3+}$ and *cis*-$[Co(en)_2Cl_2]^+$ are chiral. Notice that the mirror images of the two compounds are not superimposable.

Solution

(a) Two geometric isomers can be drawn for octahedral complexes with a formula of MA_4B_2. One isomer has two Cl^- ions in *cis* positions (adjacent positions, at a 90° angle), and the other isomer has the Cl^- ligands in *trans* positions (with a 180° angle between the ligands). Optical isomers are not possible.

cis isomer   *trans* isomer

(b) In this square-planar complex, the two NH_3 ligands (and the two CN^- ligands) can be either *cis* or *trans*. These are geometric isomers. Optical isomers are not possible.

cis isomer   *trans* isomer

● **Optical Isomerism** Square-planar complexes are incapable of optical isomerism based at the metal center; mirror images are always superimposable. Chiral tetrahedral complexes are possible, but examples of complexes with a metal bonded tetrahedrally to four different monodentate ligands are rare.

(c) Two geometric isomers of this octahedral complex, with chloride ligands either *fac* or *mer*, are possible. In the *fac* isomer, the three Cl^- ligands are all at 90° to each other; in the *mer* isomer, two Cl^- ligands are at 180°, and the third is 90° from the other two. Optical isomers are not possible.

fac isomer   *mer* isomer

(d) Ignore the counterions, K^+. The anion is an octahedral complex. Remember that the bidentate oxalate ion occupies two coordination sites of the metal, and that three oxalate ligands means that the metal has a coordination number of 6. Mirror images of complexes of the stoichiometry M(bidentate)$_3$ are not superimposable; therefore, two optical isomers are possible. (Here the ligands, $C_2O_4^{2-}$, are drawn abbreviated as O—O.)

nonsuperimposable mirror images of $[Fe(ox)_3]^{3-}$

(e) Only a single structure is possible for tetrahedral complexes such as $Zn(NH_3)_2Cl_2$.

(f) Only linkage isomerism (structural isomerism) is possible for this octahedral cobalt complex. Either the sulfur or the nitrogen of the SCN^- anion can be attached to the cobalt(III) ion in this complex.

S-bonded SCN^- N-bonded SCN^-

Think about Your Answer More complex situations occur when a complex has three or more different ligands. In those instances it is important to be able to visualize the three dimensional molecule and to organize your approach.

Check Your Understanding

What types of isomers are possible for the following compounds or complex ions?

(a) $K[Co(NH_3)_2Cl_4]$

(c) $[Co(NH_3)_5Cl]^{2+}$

(e) $Na_2[MnCl_4]$ (tetrahedral)

(b) $Pt(en)Cl_2$ (square-planar)

(d) $[Ru(phen)_3]Cl_3$

(f) $[Co(NH_3)_5NO_2]^{2+}$

REVIEW & CHECK FOR SECTION 22.4

1. Which of the following complexes cannot exist as *cis–trans* isomers?

    (a) $Fe(NH_3)_4Cl_2$

    (c) Square-planar $[PtCl_2(CN)_2]^{2-}$

    (b) $Ni(H_2O)_4(C_2O_4)$ $(C_2O_4^{2-}$ = oxalate)

    (d) $Mn(en)_2Cl_2$ (en = ethylenediamine)

2. Which of the following complexes can exist as optical isomers?

    (a) Square-planar $Pt(NH_3)(H_2O)(CN)Cl$

    (c) $Fe(NH_3)_4Cl_2$

    (b) $Fe(en)_2(CN)_2$ (en = ethylenediamine)

    (d) $Cr(NH_3)_3Cl_3$

3. Which statement best describes the isomers for the complex $Fe(en)_2Cl_2$? (en = ethylenediamine)

    (a) Only a single structure can be drawn.

    (b) There are two geometric isomers, one of which has an optical isomer.

    (c) The compound exists as a pair of optical isomers.

    (d) This compound exists as two structural isomers.

4. How many geometric isomers are possible for $Fe(en)(NH_3)_2Cl_2$? (To answer this question, try to draw out possible structures in an organized manner. Start by placing the bidentate ethylenediamine ligand in two coordination sites, then add chloride and ammonia ligands into the remaining positions in various ways.)

    (a) 0

    (b) 1

    (c) 2

    (d) 3

22.5 Bonding in Coordination Compounds

Metal–ligand bonding in a coordination complex was described earlier in this chapter as being covalent, resulting from the sharing of an electron pair between the metal and the ligand donor atom. Although frequently used, this description is not capable of explaining the color and magnetic behavior of these complexes. As a consequence, the covalent bonding picture has now largely been superseded by two other bonding models: molecular orbital theory and ligand field theory.

The bonding model based on molecular orbital theory assumes that the metal and the ligand bond through the molecular orbitals formed by atomic orbital overlap between metal and ligand. The **ligand field model**, in contrast, focuses on repulsion (and destabilization) of electrons in the metal coordination sphere. The ligand field model also assumes that the positive metal ion and the negative ligand lone pair are attracted electrostatically; that is, the bond arises when a positively charged metal ion attracts a negative ion or the negative end of a polar molecule. For the most part, the molecular orbital and ligand field models predict similar, qualitative results regarding color and magnetic behavior. Here, we will focus on the ligand field approach and illustrate how it explains color and magnetism of transition metal complexes.

The *d* Orbitals: Ligand Field Theory

To understand ligand field theory, it is necessary to look at the *d* orbitals, particularly with regard to their orientation relative to the positions of ligands in a metal complex.

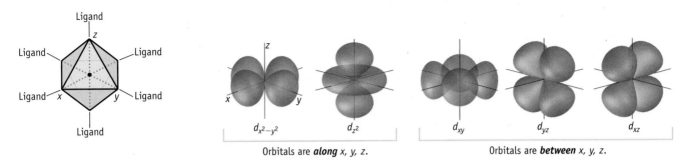

FIGURE 22.21 **The d orbitals.** The five d orbitals and their spatial relation to the ligands on the x-, y-, and z-axes.

We look first at octahedral complexes. Assume the ligands in an octahedral complex lie along the x-, y-, and z-axes. This results in the five d orbitals (Figure 22.21) being subdivided into two sets: the $d_{x^2-y^2}$ and d_{z^2} orbitals in one set and the d_{xy}, d_{xz}, and d_{yz} orbitals in the second. The $d_{x^2-y^2}$ and d_{z^2} orbitals are directed along the x-, y-, and z-axes, whereas the orbitals of the second group are aligned between these axes.

In an isolated atom or ion, the five d orbitals have the same energy. For a metal atom or ion in a coordination complex, however, the d orbitals have different energies. According to the ligand field model, repulsion between d electrons on the metal and electron pairs of the ligands destabilizes electrons that reside in the d orbitals; that is, it causes their energy to increase. Electrons in the various d orbitals are not affected equally, however, because of their different orientations in space relative to the position of the ligand lone pairs (Figure 22.22). Electrons in the $d_{x^2-y^2}$ and d_{z^2} orbitals experience a larger repulsion because these orbitals point directly at the incoming ligand electron pairs. A smaller repulsive effect is experienced by electrons in the d_{xy}, d_{xz}, and d_{yz} orbitals. The difference in degree of repulsion means that an energy difference exists between the two sets of orbitals in an octahedral complex. This difference, called the **ligand field splitting** and denoted by the symbol Δ_0, is a function of the metal and the ligands and varies predictably from one complex to another.

A different splitting pattern is encountered with square-planar complexes (Figure 22.23). Assume that the four ligands are along the x- and y-axes. The $d_{x^2-y^2}$ orbital also points along these axes, so it has the highest energy. The d_{xy} orbital (which also lies in the xy-plane, but does not point at the ligands) is next highest in energy, followed by the d_{z^2} orbital. The d_{xz} and d_{yz} orbitals, both of which partially point in the z-direction, have the lowest energy.

The d-orbital splitting pattern for a tetrahedral complex is the reverse of the pattern observed for octahedral complexes. Three orbitals (d_{xz}, d_{xy}, d_{yz}) are higher in energy, whereas the $d_{x^2-y^2}$ and d_{z^2} orbitals are below them in energy (Figure 22.23).

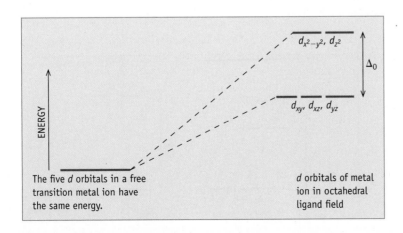

The five d orbitals in a free transition metal ion have the same energy.

d orbitals of metal ion in octahedral ligand field

FIGURE 22.22 **Ligand field splitting for an octahedral complex.** The d-orbital energies increase as the ligands approach the metal along the x-, y-, and z-axes. The d_{xy}, d_{xz}, and d_{yz} orbitals, not pointed toward the ligands, are less destabilized than the $d_{x^2-y^2}$ and d_{z^2} orbitals. Thus, the d_{xy}, d_{xz}, and d_{yz} orbitals are at lower energy. (Δ_0 stands for the splitting in an octahedral ligand field.)

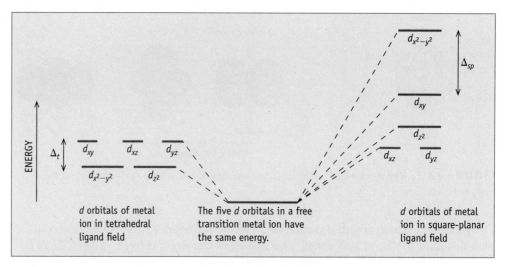

FIGURE 22.23 Splitting of the *d* orbitals in *(left)* tetrahedral and *(right)* square-planar geometries. (Δ_t and Δ_{sp} are, respectively, the splitting in tetrahedral and square-planar ligand fields.)

Electron Configurations and Magnetic Properties

The *d*-orbital splitting in coordination complexes provides the means to explain both the magnetic behavior and the color of these complexes. To understand this explanation, however, we must first understand how to assign electrons to the various orbitals in each geometry.

A gaseous Cr^{2+} ion has the electron configuration $[Ar]3d^4$. The term *gaseous* in this context is used to denote a single, isolated atom or ion with all other particles located an infinite distance away. In this situation, the five $3d$ orbitals have the same energy. The four electrons reside singly in different *d* orbitals, according to Hund's rule, and the Cr^{2+} ion has four unpaired electrons.

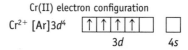

When the Cr^{2+} ion is part of an octahedral complex, the five *d* orbitals do not have identical energies. As illustrated in Figure 22.22, these orbitals divide into two sets, with the d_{xy}, d_{xz}, and d_{yz} orbitals having a lower energy than the $d_{x^2-y^2}$ and d_{z^2} orbitals. Having two sets of orbitals means that two different electron configurations are possible (Figure 22.24). Three of the four *d* electrons in Cr^{2+} are assigned to the lower-energy d_{xy}, d_{xz}, and d_{yz} orbitals. The fourth electron either can be assigned to an orbital in the higher-energy $d_{x^2-y^2}$ and d_{z^2} set or can pair up with an electron already in the lower-energy set. The first arrangement is called **high spin**, because it has the maximum number of unpaired electrons, four in the case of Cr^{2+}. The second arrangement is called **low spin**, because it has the minimum number of unpaired electrons possible.

FIGURE 22.24 High- and low-spin cases for an octahedral chromium(II) complex. *(left, high spin)* If the ligand field splitting (Δ_0) is smaller than the pairing energy (P), the electrons are placed in different orbitals, and the complex has four unpaired electrons. *(right, low spin)* If the splitting is larger than the pairing energy, all four electrons will be in the lower-energy orbital set. This requires pairing two electrons in one of the orbitals, so the complex will have two unpaired electrons.

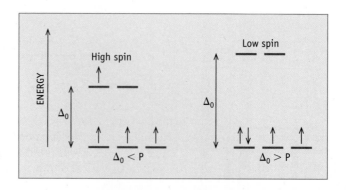

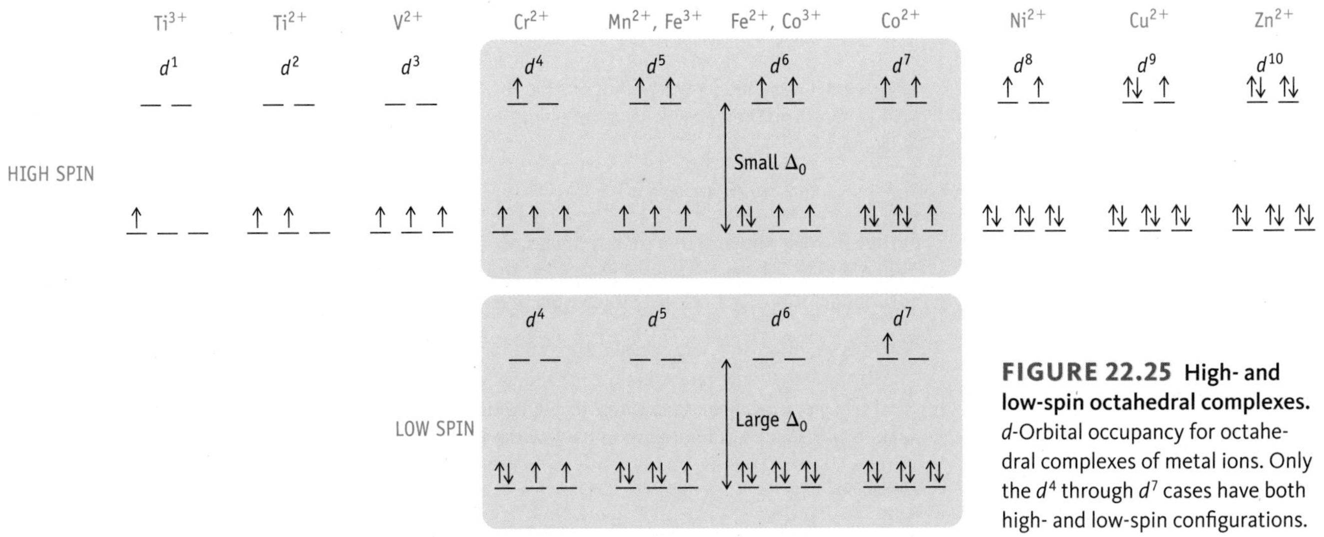

| | Ti^{3+} | Ti^{2+} | V^{2+} | Cr^{2+} | Mn^{2+}, Fe^{3+} | Fe^{2+}, Co^{3+} | Co^{2+} | Ni^{2+} | Cu^{2+} | Zn^{2+} |

FIGURE 22.25 High- and low-spin octahedral complexes. d-Orbital occupancy for octahedral complexes of metal ions. Only the d^4 through d^7 cases have both high- and low-spin configurations.

At first glance, a high-spin configuration appears to contradict conventional thinking. It seems logical that the most stable situation would occur when electrons occupy the lowest-energy orbitals. A second factor intervenes, however. Because electrons are negatively charged, repulsion increases when they are assigned to the same orbital. This destabilizing effect bears the name **pairing energy (P)**. The preference for an electron to be in the lowest-energy orbital and the pairing energy have opposing effects (Figure 22.24). Low-spin complexes arise when the splitting of the d orbitals by the ligand field is large—that is, when Δ_0 has a large value. The energy gained by putting all of the electrons in the lowest-energy level is the dominant effect when $\Delta_0 > P$. In contrast, high-spin complexes occur if the value of Δ_0 is smaller than the energy required to pair electrons ($\Delta_0 < P$).

For octahedral complexes, high- and low-spin complexes can occur only for configurations d^4 through d^7 (Figure 22.25). Complexes of the d^6 metal ion, Fe^{2+}, for example, can have either high spin or low spin. The complex formed when the Fe^{2+} ion is placed in water, $[Fe(H_2O)_6]^{2+}$, is high spin, whereas the $[Fe(CN)_6]^{4-}$ complex ion is low spin.

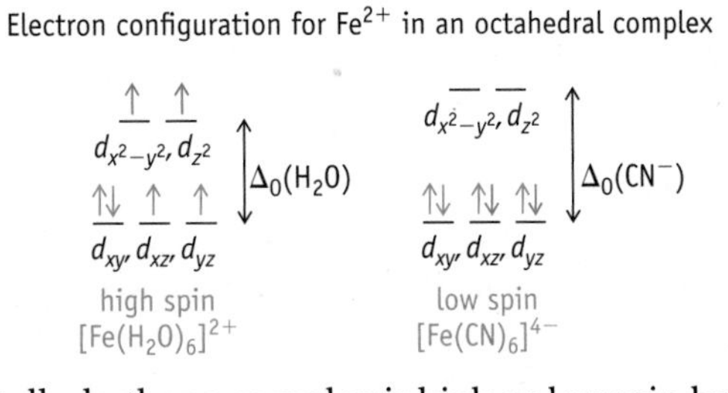

Electron configuration for Fe^{2+} in an octahedral complex

It is possible to tell whether a complex is high or low spin by examining its magnetic behavior. The high-spin complex $[Fe(H_2O)_6]^{2+}$ has four unpaired electrons and is *paramagnetic* (attracted by a magnet), whereas the low-spin $[Fe(CN)_6]^{4-}$ complex has no unpaired electrons and is *diamagnetic* (repelled by a magnet) (◄ page 227).

Most complexes of Pd^{2+} and Pt^{2+} ions are square-planar, the electron configuration of these metals being [noble gas]$(n - 1)d^8$. In a square-planar complex, there are four sets of orbitals (Figure 22.23). For square-planar d^8 complexes, all except the highest-energy orbital are filled, and all electrons are paired, resulting in diamagnetic (low-spin) complexes.

Nickel, which is found above palladium in the periodic table, forms both square-planar and tetrahedral complexes (as well as octahedral complexes). For example, the complex ion $[Ni(CN)_4]^{2-}$ is square-planar, whereas the $[NiCl_4]^{2-}$ ion is tetrahedral. Magnetism allows us to differentiate between these two geometries. Based on the ligand field splitting pattern, the cyanide complex is expected to be diamagnetic, whereas the chloride complex is paramagnetic with two unpaired electrons.

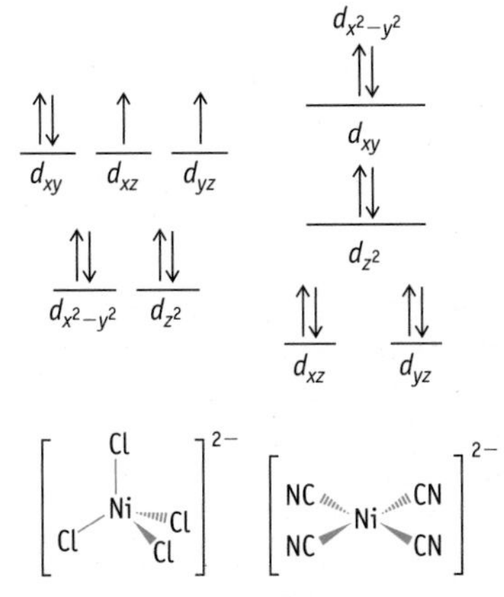

Nickel(II) complexes and magnetism. The anion $[NiCl_4]^{2-}$ is a paramagnetic tetrahedral complex. In contrast, $[Ni(CN)_4]^{2-}$ is a diamagnetic square-planar complex.

EXAMPLE 22.5   High- and Low-Spin Complexes and Magnetism

Problem Give the electron configuration for the metal ion in each of the following complexes. How many unpaired electrons are present in each? Are the complexes paramagnetic or diamagnetic?

(a)   Low-spin $[Co(NH_3)_6]^{3+}$

(b)   High-spin $[CoF_6]^{3-}$

What Do You Know? From their formulas, we know that these have six ligands and we can presume that they have octahedral geometry. Both are complexes of cobalt(III), a metal ion with a d^6 configuration. One complex is low spin, the other high spin.

Strategy Set up an energy-level diagram for an octahedral complex. In low-spin complexes, the electrons are added preferentially to the lower-energy set of orbitals. In high-spin complexes, the first five electrons are added singly to each of the five orbitals, then additional electrons are paired with electrons in orbitals in the lower-energy set.

Solution

(a)   The six electrons of the Co^{3+} ion fill the lower-energy set of orbitals entirely. This d^6 complex ion has no unpaired electrons and is diamagnetic.

(b)   To obtain the electron configuration in high-spin $[CoF_6]^{3-}$, place one electron in each of the five d orbitals, and then place the sixth electron in one of the lower-energy orbitals. The complex has four unpaired electrons and is paramagnetic.

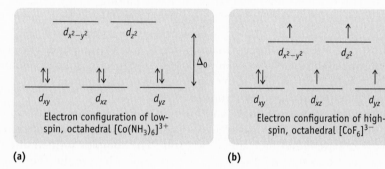

(a) Electron configuration of low-spin, octahedral $[Co(NH_3)_6]^{3+}$

(b) Electron configuration of high-spin, octahedral $[CoF_6]^{3-}$

Think about Your Answer The two complexes have different magnetic properties; low-spin $[Co(NH_3)_6]^{3+}$ is diamagnetic and high-spin $[CoF_6]^{3-}$ is paramagnetic. As expected, the low-spin complex has fewer unpaired electrons than the high-spin complex.

Check Your Understanding

For each of the following complex ions, give the oxidation number of the metal, depict possible low- and high-spin configurations, give the number of unpaired electrons in each configuration, and tell whether each is paramagnetic or diamagnetic.

(a) $[Ru(H_2O)_6]^{2+}$

(b) $[Ni(NH_3)_6]^{2+}$

REVIEW & CHECK FOR SECTION 22.5

1.   Which of the following complexes has the greatest number of unpaired electrons?

(a)   Low-spin $[Fe(CN)_6]^{4-}$

(b)   High-spin $[Mn(NH_3)_6]^{2+}$

(c) $[V(H_2O)_6]^{3+}$

(d) $[Ni(en)_3]^{2+}$ (en = ethylenediamine)

2.   Which of the following complexes is diamagnetic?

(a)   Square-planar $[PtCl_4]^{2-}$

(b)   Tetrahedral $[NiCl_4]^{2-}$

(c) $[Fe(H_2O)_6]^{3+}$

(d)   High-spin $[CoF_6]^{3-}$

Fe³⁺(aq)   Co²⁺(aq)   Ni²⁺(aq)   Cu²⁺(aq)   Zn²⁺(aq)

© Cengage Learning/Charles D. Winters

FIGURE 22.26 Aqueous solutions of some transition metal ions. Compounds of transition metal elements are often colored, whereas those of main group metals are usually colorless. Pictured here, from *left* to *right*, are solutions of the nitrate salts of Fe^{3+}, Co^{2+}, Ni^{2+}, Cu^{2+}, and Zn^{2+}.

22.6 Colors of Coordination Compounds

The range of colors observed for compounds of the transition elements is one of their most interesting features (Figure 22.26). In contrast, compounds of main group metals are usually colorless. The underlying reason for this difference is that the color of transition metal compounds results from *d*-orbital splitting. Before discussing how *d*-orbital splitting is involved, however, we need to look more closely at what we mean by color.

Color

Visible light, consisting of radiation with wavelengths from 400 to 700 nm (◄ Section 6.1), represents a very small portion of the electromagnetic spectrum. Within this region are all the colors you see when white light passes through a prism: red, orange, yellow, green, blue, indigo, and violet (ROY G BIV). Each color is identified with a portion of the wavelength range.

Isaac Newton did experiments with light and established that the mind's perception of color requires only three colors! When we see white light, we are seeing a mixture of all of the colors—in other words, the superposition of red, green, and blue. If one or more of these colors is absent, the light of the other colors that reaches your eyes is interpreted by your mind as color.

Figure 22.27 will help you in analyzing perceived colors. This color wheel shows the three primary colors—red, green, and blue—as overlapping disks arranged in a triangle. The secondary colors—cyan, magenta, and yellow—appear where two disks overlap. The overlap of all three disks in the center produces white light.

The colors we perceive are determined as follows:

- Light of a single primary color is perceived as that color: Red light is perceived as red, green light as green, blue light as blue.
- Light made up of two primary colors is perceived as the color shown where the disks in Figure 22.27 overlap: Red and green light together appear yellow, green and blue light together are perceived as cyan; and red and blue light are perceived as magenta.
- Light made up of the three primary colors is white (colorless).

In discussing the color of a substance such as a coordination complex *in solution*, these guidelines are turned around because the color we see is the color of light *not* absorbed by the solution.

- Red color is the result of the absence of green and blue light from white light.
- Green color results if red and blue light are absent from white light.
- Blue color results if red and green light are absent.

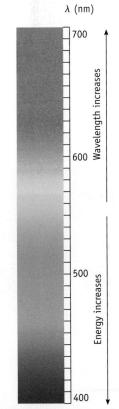

λ (nm)

700

600

500

400

Wavelength increases

Energy increases

The ROY G BIV spectrum of colors of visible light. The colors used in printing this book are cyan, magenta, yellow, and black. The blue in ROY G BIV is actually cyan, according to color industry standards. Magenta doesn't have its own wavelength region. Rather, it is a mixture of blue and red.

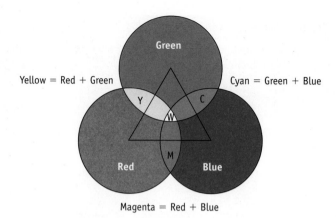

FIGURE 22.27 Using color disks to analyze colors. The three primary colors are red, green, and blue. Adding light of two primary colors gives the secondary colors yellow (= red + green), cyan (= green + blue), and magenta (= red + blue). Adding all three primary colors results in white light.

The secondary colors are rationalized similarly. Absorption of blue light gives yellow (the color across from it in Figure 22.27); absorption of red light results in cyan; and absorption of green light results in magenta.

Now we can apply these ideas to explain colors in transition metal complexes. Focus on what kind of light is *absorbed*. A solution of $[Ni(H_2O)_6]^{2+}$ is green. Green light is the result of removing red and blue light from white light. As white light passes through an aqueous solution of Ni^{2+}, red and blue light are absorbed, and green light is allowed to pass (Figure 22.28). Similarly, the $[Co(NH_3)_6]^{3+}$ ion is yellow because blue light has been absorbed and red and green light pass through.

The Spectrochemical Series

Recall that atomic spectra are obtained when electrons are excited from one energy level to another (◄ Section 6.3). The energy of the light absorbed or emitted is related to the energy levels of the atom or ion under study. The concept that light is absorbed when electrons move from lower to higher energy levels applies to all substances, not just atoms. It is the basic premise for the absorption of light for transition metal coordination complexes.

In coordination complexes, the splitting between d orbitals often corresponds to the energy of visible light, so light in the visible region of the spectrum is absorbed when electrons move from a lower-energy d orbital to a higher-energy d orbital. This change, as an electron moves between two orbitals having different energies in a complex, is called a **d-to-d transition**. Qualitatively, such a transition for $[Co(NH_3)_6]^{3+}$ might be represented using an energy-level diagram such as that shown here.

FIGURE 22.28 Light absorption and color. The color of a solution is due to the color of the light *not* absorbed by the solution. Here, a solution of Ni^{2+} ion in water absorbs red and blue light and so appears green. See also Figures 4.15–4.19.

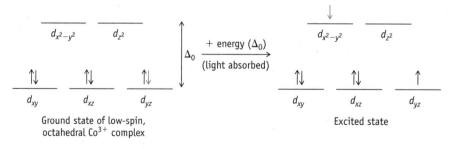

Experiments with coordination complexes reveal that, for a given metal ion, some ligands cause a small energy separation of the d orbitals, whereas others cause a large separation. In other words, some ligands create a small ligand field, and

Table 22.3 The Colors of Some Co^{3+} Complexes*

Complex Ion	Wavelength of Light Absorbed (nm)	Color of Light Absorbed	Color of Complex
$[CoF_6]^{3-}$	700	Red	Blue
$[Co(C_2O_4)_3]^{3-}$	600, 420	Yellow, violet	Dark green
$[Co(H_2O)_6]^{3+}$	600, 400	Yellow, violet	Blue-green
$[Co(NH_3)_6]^{3+}$	475, 340	Blue, ultraviolet	Yellow-orange
$[Co(en)_3]^{3+}$	470, 340	Blue, ultraviolet	Yellow-orange
$[Co(CN)_6]^{3-}$	310	Ultraviolet	Pale yellow

*The complex with fluoride ion, $[CoF_6]^{3-}$, is high spin and has one absorption band. The other complexes are low spin and have two absorption bands. In all but one case, one of these absorptions occurs in the visible region of the spectrum. The wavelengths are measured at the top of that absorption band.

others create a large one. An example is seen in the spectroscopic data for several cobalt(III) complexes presented in Table 22.3.

- Both $[Co(NH_3)_6]^{3+}$ and $[Co(en)_3]^{3+}$ are yellow-orange, because they absorb light in the blue portion of the visible spectrum. These compounds have very similar spectra, to be expected because both have six amine-type donor atoms (H—NH_2 or R—NH_2).
- Although $[Co(CN)_6]^{3-}$ does not have an absorption band whose maximum falls in the visible region, it is pale yellow. The peak of the absorption band occurs in the ultraviolet region, but the absorption band is broad and extends minimally into the visible (blue) region.
- $[Co(C_2O_4)_3]^{3-}$ and $[Co(H_2O)_6]^{3+}$ have similar absorptions, in the yellow and violet regions. Their colors are shades of green with a small difference due to the relative amount of light of each color being absorbed.

The wavelength of maximum absorption among the listed complexes ranges from 700 nm for $[CoF_6]^{3-}$ to 310 nm for $[Co(CN)_6]^{3-}$. The ligands change from member to member of this series, and we can conclude that the energy of the light absorbed by the complex is related to the different ligand field splittings, Δ_0, caused by the different ligands. Fluoride ion causes the smallest splitting of the d orbitals among the complexes listed in Table 22.3, whereas cyanide causes the largest splitting.

Spectra of complexes of other metals provide similar results. Based on this information, ligands can be listed in order of their ability to split the d orbitals. This list is called the **spectrochemical series** because it was determined by spectroscopy. A short list, with some of the more common ligands, follows:

$$F^-, Cl^-, Br^-, I^- < C_2O_4^{2-} < H_2O < NH_3 = en < phen < CN^-$$

small orbital splitting   large orbital splitting
small Δ_0 large Δ_0

The spectrochemical series is applicable to a wide range of metal complexes. Indeed, the ability of ligand field theory to explain the differences in the colors of the transition metal complexes is one of the strengths of this theory.

Based on the relative position of a ligand in the series, predictions can be made about a compound's magnetic behavior. Recall that d^4, d^5, d^6, and d^7 complexes can be high or low spin, depending on the ligand field splitting, Δ_0. Complexes formed with ligands near the left end of the spectrochemical series are expected to have small Δ_0 values and, therefore, are likely to be high spin. In contrast, complexes with ligands near the right end are expected to have large Δ_0 values and low-spin configurations. The complex $[CoF_6]^{3-}$ is high spin, whereas $[Co(NH_3)_6]^{3+}$ and the other complexes in Table 22.3 are low spin.

PROBLEM SOLVING TIP 22.1

Ligand Field Theory

This summary of the concepts of ligand field theory may help you to keep the broader picture in mind.

- Ligand–metal bonding results from the electrostatic attraction between a metal cation and either an anion or a polar molecule.
- The ligands define a coordination geometry. Common geometries are linear (coordination number = 2), tetrahedral and square-planar (coordination number = 4), and octahedral (coordination number = 6).
- The placement of the ligands around the metal causes the d orbitals on the metal

to have different energies. In an octahedral complex, for example, the d orbitals divide into two groups: a higher-energy group ($d_{x^2-y^2}$ and d_{z^2}) and a lower-energy group (d_{xy}, d_{xz}, and d_{yz}).
- Electrons are placed in the metal d orbitals in a manner that leads to the lowest total energy. Two competing features determine the placement: the relative energy of the sets of orbitals and the electron pairing energy.
- For the electron configurations of d^4, d^5, d^6, and d^7 in octahedral complexes, two electron configurations are possible:

high spin, which occurs when the orbital splitting is small, and low spin, which occurs with a large orbital splitting. To determine whether a complex is high spin or low spin, measure its magnetism to determine the number of unpaired electrons.
- The d-orbital splitting (the energy difference between the metal d-orbital energies) often corresponds to the energy associated with visible light. As a consequence, many metal complexes absorb visible light and thus are colored.

EXAMPLE 22.6   Spectrochemical Series

Problem An aqueous solution of $[Fe(H_2O)_6]^{2+}$ is light blue-green. Do you expect the d^6 Fe^{2+} ion in this complex to have a high- or low-spin configuration? How could you test your prediction experimentally?

What Do You Know? This is an octahedral complex of Fe^{2+}, an ion with a d^6 configuration. The blue-green color is the result of the absorption of visible light due to a d-to-d transition. The complex could be either low spin (no unpaired electrons) or high spin (four unpaired electrons).

Strategy Use the color wheel in Figure 22.27 to determine what color light is transmitted and therefore what color of light has been absorbed. Based on the energy of the light absorbed, determine if Δ_0 is high or low. If it is low, then the complex is high spin. If it is high, then the complex is low spin.

Solution In this case, the blue-green color of the solution indicates that blue and green light is transmitted. This implies that red light was absorbed. The low energy of the light absorbed suggests that $[Fe(H_2O)_6]^{2+}$ is likely to be a high-spin complex. If the complex is high spin, it will have four unpaired electrons and be paramagnetic; if, contrary to our prediction, it is low spin, it will have no unpaired electrons and be diamagnetic. Thus, identifying the presence of four unpaired electrons by measuring the compound's magnetism can be used to verify the high-spin configuration experimentally.

Think about Your Answer The spectrochemical series was generated by measuring the absorptions for a series of complexes of a given metal, varying the ligands. As in the example here, the spectrochemical series is useful because it allows you to predict whether a complex is likely to be high or low spin. The more definitive evidence for high or low spin comes from information on the magnetism of the complex.

Check Your Understanding

1. The complex $[Co(NH_3)_5OH]^{2+}$ has an absorption maximum at 500 nm. What color of light is absorbed and what color is the complex?

2. Figure 22.17 shows a sample of yellow $[Co(NH_3)_5NO_2]^{2+}$. What color of light is absorbed by this complex? Predict whether this is a high-spin or low-spin complex.

REVIEW & CHECK FOR SECTION 22.6

Solutions of each of the following complexes absorb light in the visible region of the spectrum. Predict which complex will absorb light of the longest wavelength.

(a) $[Cr(H_2O)_6]^{3+}$ (b) $[Cr(NH_3)_6]^{3+}$ (c) $[CrF_6]^{3-}$ (d) $[Cr(CN)_6]^{3-}$

CASE STUDY

Accidental Discovery of a Chemotherapy Agent

There are many naturally occurring metal-based molecules such as heme, vitamin B_{12}, and the enzyme involved in fixing nitrogen (nitrogenase). Chemists have also synthesized various metal-based compounds for medical purposes. One of these, *cisplatin* [$PtCl_2(NH_3)_2$], was known for many years, but it was discovered serendipitously to be effective in treatment of certain kinds of cancers.

In 1965, Barnett Rosenberg, a biophysicist at Michigan State University, set out to study the effect of electric fields on living cells, but the results of his experiments were very different from his expectations. He and his students had placed an aqueous suspension of live *Escherichia coli* bacteria in an electric field between supposedly inert platinum electrodes. Much to their surprise, they found that cell growth was significantly affected. After careful experimentation, the effect on cell division was found to be due to a trace of a complex of platinum, ammonia, and chloride ion formed by an electrolytic process involving the platinum electrode in the presence of ammonia in the growth medium.

To follow up on this interesting discovery, Rosenberg and his students tested the effect of *cis-* and *trans-*$PtCl_2(NH_3)_2$ on cell growth and found that only the *cis* isomer

was effective. This led Rosenberg and others to study the effect of so-called *cisplatin* on cancer cell growth, and the result is that cisplatin and similar compounds are now used to treat genitourinary tumors. In fact, testicular cancer in particular is now considered largely curable because of cisplatin chemotherapy.

The chemistry of cisplatin has now been thoroughly studied and illustrates many of the principles of transition metal coordination chemistry. It has been found that cisplatin has a half-life of 2.5 hours for the replacement of a Cl^- ligand by water at 310 K (in a first order reaction) and that the replacement of a second Cl^- ligand by water is slightly faster.

The aqua species are acidic and damaging to the kidneys, so cisplatin is generally used in a saline solution to prevent the hydrolysis reactions. It has been found that, in blood plasma at pH 7.4 and with a Cl^- ion concentration of about 1.04×10^{-5} M, $PtCl_2(NH_3)_2$

and $PtCl(OH)(NH_3)_2$ are the dominant species. In the cell nucleus, however, the Cl^- ion concentration is lower, and the aqua species are present in higher concentration.

Question:

If a patient is given 10.0 mg of cisplatin, what quantity remains as cisplatin at 24 hours?

The answer to this question is available in Appendix N.

22.7 Organometallic Chemistry: Compounds with Metal–Carbon Bonds

One of the largest and most active areas of chemistry over the past half-century has been the field of organometallic chemistry, the study of molecules having metal–carbon bonds. Thousands of such compounds have been made and characterized, and much of the activity has involved transition metals. Many have unique bonding modes and structures. In recent years, organometallic compounds have also found particularly widespread use as reagents in organic synthesis and as catalysts for economically important chemical reactions.

Carbon Monoxide Complexes of Metals

In the earlier discussion of hemoglobin (page 798), you learned that the iron in this biologically important complex binds not only to O_2 but also to CO. Our understanding of metal–CO complexes, however, emerged from the solution to a problem in industrial chemistry at the end of the 19th century.

The synthesis of sodium carbonate from NaCl, CO_2, NH_3, and H_2O by the Solvay process was an important chemical industry in the late 19th century (and remains so today in some parts of the world; page 746). In the late 1800s, a Solvay plant in England had a problem: The valves used to conduct the gaseous reactants and products rapidly corroded. A German chemist, Ludwig Mond (1839–1909), traced this problem to a small quantity of CO in the gas stream. Carbon monoxide gas reacted with nickel metal in the valves to form $Ni(CO)_4$ (Figure 22.29), tetracarbonyl nickel, a volatile liquid with a low boiling point (bp 47 °C). The

FIGURE 22.29 Metals carbonyls. $Ni(CO)_4$ is a tetrahedral molecule. Pentacarbonyliron(0) $[Fe(CO)_5]$ is trigonal bipyramidal, whereas $Mo(CO)_6$ is octahedral, and the geometry around the manganese atom in $Mn_2(CO)_{10}$ is likewise octahedral (with the sixth position being an Mn—Mn bond).

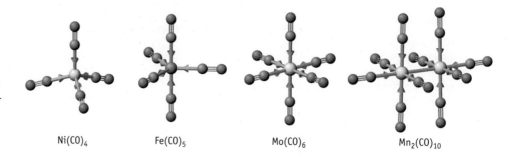

$Ni(CO)_4$ $Fe(CO)_5$ $Mo(CO)_6$ $Mn_2(CO)_{10}$

deterioration of the valves occurred when the gaseous $Ni(CO)_4$ that formed was carried away in the effluent gas stream.

$$Ni(s) + 4\ CO(g) \rightleftharpoons Ni(CO)_4(g)$$

Mond determined that the reaction forming $Ni(CO)_4$ is reversible; at moderate temperatures or at low pressures, CO is liberated, and nickel metal re-forms. He then exploited the formation and decomposition reactions in a process to obtain pure nickel. Nickel and cobalt are generally found together in nature. Because of this, the two metals are obtained together when ores are refined, and they are difficult to separate. However, if a mixture of the metals is exposed to CO, nickel is converted to $Ni(CO)_4$ whereas cobalt is unchanged. Volatile $Ni(CO)_4$ is easily separated from solid cobalt metal. The $Ni(CO)_4$ can then be decomposed to give pure nickel. The Mond process, as it is now known, was the preferred procedure to obtain pure nickel for the first half of the 20th century.

Nickel is unique among metals in its facile reaction with CO, but many other transition metal carbonyl compounds are now known and can be synthesized by a variety of procedures. One of the most common is reductive carbonylation, in which a metal salt is reduced in the presence of CO, usually under high pressure. In effect, reduction of the metal salt gives metal atoms. Before these very reactive atoms aggregate to form the unreactive bulk metal, however, they react with CO. Metal carbonyls of most of the transition metals can be made by this route. Simple examples include hexacarbonyls of the Group 6B metals [$Cr(CO)_6$, $Mo(CO)_6$, and $W(CO)_6$,] as well as $Fe(CO)_5$ and $Mn_2(CO)_{10}$ (Figure 22.29).

The Effective Atomic Number Rule and Bonding in Organometallic Compounds

An important observation has guided researchers studying these metal–CO complexes and other organometallic compounds. The **effective atomic number (EAN) rule**, now often referred to as the **18-electron rule**, states that compounds in which the sum of the metal valence electrons plus the electrons donated by the ligand groups totals 18 are likely to be stable. Thus, the 18-electron rule predicts the stoichiometry of a number of compounds. Iron(0), for example, which has eight valence electrons, would be expected to coordinate to five CO ligands. Each CO donates two electrons for a total of 10; adding these to the eight valence electrons on the metal gives 18 electrons around the Fe atom (Figure 22.30). We now recognize that the 18-electron rule is similar to the octet rule in main group chemistry, in that it defines the likely number of bonds to an element.

The bonding in traditional coordination compounds is described as being due to attractive forces between a positively charged metal ion and a polar molecule or an anion, and the properties of these species are in accord with substantial ionic character to the metal–ligand bond. In metal carbonyls, the zerovalent metal lacks a charge, and CO is only slightly polar. Both features argue against ionic bonding. The best model for these species instead describes the bonding between the metal and CO as covalent. Each CO donates a C-atom lone pair to the metal atom to form a sigma bond (Figure 22.31). However, carbon monoxide is a very poor donor, and this alone would not lead to these species being stable. Rather, in conjunction with

● **Terminology in Organometallic Chemistry** The standard terminology of coordination chemistry applies to metal carbonyls. In tetracarbonylnickel(0), for example, the metal is formally zerovalent, and CO molecules are ligands. The metal has a coordination number of 4 and a tetrahedral coordination geometry.

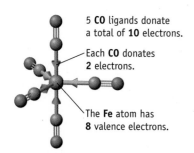

5 **CO** ligands donate a total of **10** electrons.

Each **CO** donates **2** electrons.

The **Fe** atom has **8** valence electrons.

FIGURE 22.30 The EAN rule and $Fe(CO)_5$. The EAN or 18-electron rule states that stable organometallic compounds frequently have 18 valence electrons around the central metal. (There are also many 16-electron molecules, particularly of the heavier transition metals. See Study Question 22.34.) Here, the zerovalent Fe atom has the configuration $[Ar]3d^64s^2$, so eight valence electrons are available for bonding.

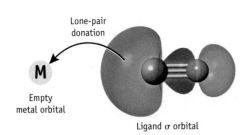
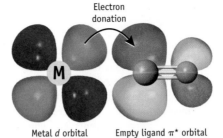

Ligand to metal sigma bonding.
Donation of **CO** lone pair to empty orbital on **M**.

Metal to ligand pi bonding.
Donation of electrons from filled **M** *d* orbital to empty π* antibonding orbital on **CO**.

FIGURE 22.31 Bonding in metal carbonyls. The current understanding is that the CO ligand donates a lone pair of electrons to the low-valent metal to form a sigma bond. The electron-rich metal then donates electrons from a *d* orbital to the antibonding π* orbital of the CO. There is a "synergistic" effect; the sigma and pi bonds complement each other.

the sigma bond, the electron-rich metal donates a pair of electrons to form a π bond formed by overlap of a *d* orbital of the metal and a π* antibonding orbital of CO. The latter interaction is described as *d*π–*p*π bonding. Based on this model of bonding, carbon monoxide in these compounds is best described as a σ donor and π acceptor ligand.

Ligands in Organometallic Compounds

The bonding model for metal–CO complexes leads to important conclusions: Only metals of low charge and with filled or partially filled *d* orbitals can form stable bonds to CO, and only ligands capable of forming π bonds with the metal are capable of forming low-valent metal compounds. Thus, this area is dominated by low-valent metals and special types of ligands that are capable of π bonding.

Some of the most common ligands in coordination chemistry cannot engage in π bonding and thus cannot form low-valent compounds with transition metals. For example, low-valent metal complexes with NH_3 or amines (such as $N(CH_3)_3$) do not exist, but phosphine complexes of zerovalent metals are well known. Phosphines, such as $P(CH_3)_3$, are the phosphorus analogs of amines and so have a lone pair of electrons on the P atom. Thus, phosphines can donate this lone pair to a metal to form a sigma bond. In addition, phosphorus atoms have empty 3*d* orbitals, and these can overlap with the filled *d* orbitals of a low-valent metal. Ammonia, with no empty valence orbitals to overlap with a filled metal *d* orbitals cannot form such π bonds.

Yet another class of molecules that can serve as ligands in organometallic compounds are organic species such as ethylene (C_2H_4) and benzene (C_6H_6). For example, in the anion of Zeise's salt, ethylene binds to a Pt^{2+} ion through donation of the two π electrons of the double bond (Figure 22.32). As in metal carbonyls, there is also a metal-to-ligand π bond formed by overlap of filled metal *d* orbitals with the antibonding π orbitals from the ligand. This combination of bonding modes strengthens the metal–ligand bond.

Benzene can be thought of as a tridentate ligand capable of donating three π electron pairs to a metal atom (which then donates *d* electrons back to the ligand in a π-type interaction). Such molecules also obey the 18-electron rule. For example, in dibenzenechromium (Figure 22.33), the Cr(0) atom with six valence electrons is bound to two ligands, each donating six π electrons.

Dibenzenechromium is one of many organometallic compounds often referred to as "sandwich" compounds. (See *Case Study: Ferrocene—The Beginning of a Chemical Revolution*, page 817.) These are molecular compounds in which a low-valent metal atom is "sandwiched" between two organic ligands. In addition, they are often referred to as π complexes because the π electrons of the ligand are involved in bonding. Modern terminology uses the Greek letter eta (η) to indicate this type of attachment, so dibenzenechromium is properly symbolized as $Cr(\eta^6\text{-}C_6H_6)_2$ (where the superscript 6 indicates the number of carbon atoms involved in bonding on each ligand).

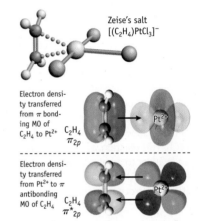

Zeise's salt $[(C_2H_4)PtCl_3]^-$

Electron density transferred from π bonding MO of C_2H_4 to Pt^{2+} C_2H_4 π_{2p} Pt^{2+}

Electron density transferred from Pt^{2+} to π antibonding MO of C_2H_4 C_2H_4 π^*_{2p} Pt^{2+}

FIGURE 22.32 Bonding in the anion of Zeise's salt.

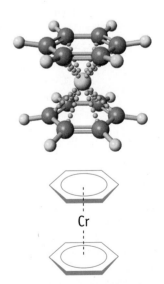

Cr

FIGURE 22.33 Dibenzenechromium. Two views of the molecule. (*top*): A computer-generated model. (*bottom*): A line drawing typically used by chemists.

Ferrocene—The Beginning of a Chemical Revolution

Unexpected discoveries open up new areas of science. One example was the synthesis of compounds of Xe (page 308), a result that destroyed the myth that the noble gases were unreactive and led to a rich chemistry of these elements. At approximately the same time, another discovery, the synthesis of ferrocene, $Fe(C_5H_5)_2$, set the stage for the rapid and exciting growth of organometallic chemistry, the chemistry of compounds containing metal–carbon bonds. This destroyed another common myth of the time, that metal–carbon bonds are inherently unstable. Xenon compounds and ferrocene spearheaded a renaissance in inorganic chemistry that continues to this day.

The synthesis of ferrocene in two quite dissimilar research laboratories was accidental and unexpected. The first report, from an academic laboratory, described the reaction of the cyclopentadienide anion, $[C_5H_5]^-$ with $FeCl_2$. This reaction was intended to provide a precursor to an elusive organic compound fulvalene, $(C_5H_4)_2$ but a completely new substance—ferrocene—was obtained instead.

$2\ C_5H_5MgCl + FeCl_2 \rightarrow Fe(\eta^5\text{-}C_5H_5)_2 + 2\ MgCl_2$
(in diethyl ether)

The second report, from an industrial laboratory, described a high-temperature process in which cyclopentadiene was passed over a catalyst that contained, among other things, iron(II) oxide.

$2\ C_5H_6(g) + FeO(s) \rightarrow$
$\qquad Fe(\eta^5\text{-}C_5H_5)_2(s) + H_2O(g)$
(at high temperature)

The properties of ferrocene were unexpected. Ferrocene is a diamagnetic orange solid, has a relatively low melting point (mp 173 °C), and is soluble in organic solvents but not in water. Most striking is its thermal and oxidative stability. Ferrocene is unaffected by oxygen, water, aqueous bases and nonoxidizing acids under ambient conditions, and can be heated to over 450 °C without decomposing. These observations contradicted conventional wisdom that held that metal–carbon bonds were weak and unstable, hard to make, and reactive.

The structure of ferrocene was quickly established by x-ray crystallography. Like dibenzenechromium (Figure 22.33), ferrocene is a molecular "sandwich" compound, containing an iron atom sandwiched between two planar hydrocarbon rings (Figure). Iron(II), a d^6 metal ion, is in its low-spin configuration. The cyclopentadienyl anions can be thought of as tridentate ligands with the six π electrons (three pairs) of each organic ring being donated to the metal ion.

Once the structure of ferrocene was known, the race was on to make other complexes with unsaturated hydrocarbon ligands. Synthesis of cyclopentadienyl compounds of other metals soon followed, and now hundreds are known in combination with CO, benzene, ethylene, and many other carbon-containing compounds. The discovery of ferrocene was one of the keys that unlocked the field of organometallic chemistry.

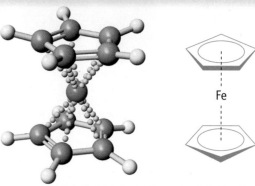

The structure of ferrocene, $Fe(\eta^5\text{-}C_5H_5)_2$

Questions:
1. Rationalize ferrocene's diamagnetism.
2. Dibenzenechromium is also a "sandwich" compound. What is the oxidation state of chromium in this compound? Is it diamagnetic or paramagnetic?
3. Do ferrocene and dibenzenechromium obey the 18-electron rule (page 814)?
4. Ferrocene can be oxidized to form the ferrocenium cation, $[Fe(\eta\text{-}C_5H_5)_2]^+$. The standard reduction potential for the ferrocenium–ferrocene half-reaction is 0.400 V. Identify several oxidizing agents in Appendix M that are capable of oxidizing ferrocene. Is elemental chlorine, Cl_2, a sufficiently strong oxidizing agent to carry out this oxidation?
5. Write an equation for one way to synthesize nickelocene, $Ni(\eta^5\text{-}C_5H_5)_2$. Nickelocene is paramagnetic. Predict the number of unpaired electrons in this compound.

Answers to these questions are available in Appendix N.

EXAMPLE 22.7   The Effective Atomic Number Rule

Problem Show that each of the following molecules or ions satisfies the EAN rule.

(a) $[Fe(CO)_2(\eta^5-C_5H_5)]^-$ (b) $[Mn(CO)_5]^-$ (c) $Co(C_2H_4)_2(\eta^5-C_5H_5)$

What Do You Know? The formulas of the several complexes are given. Recognize that CO and C_2H_4 are both two-electron donors and that the $C_5H_5^-$ ion is a six-electron donor.

Strategy A complex obeys the EAN rule if the total number of electrons around the metal (valence electrons from the metal plus electrons donated by the ligands) equals 18. For the metal center, take its total number of valence electrons, and add or subtract electrons as necessary to adjust for negative or positive charges.

Solution

(a) The overall charge is 1−, which is equal to the charge on the $C_5H_5^-$ group. Thus, the Fe atom must have no charge.

> 6 electrons for $C_5H_5^-$ + 8 electrons for Fe + 4 electrons for 2 CO groups = 18 electrons

(b) Because CO is neutral the Mn center must have a negative charge. This means the manganese center has effectively eight valence electrons. Together with the five CO groups, each donating two electrons, the total is 18 electrons.

(c) This is a neutral molecule, so the negative charge on the $C_5H_5^-$ group must be balanced by a positive charge on Co. The Co^+ ion has eight valence electrons. Each C_2H_4 molecule donates two electrons.

> 6 electrons for $C_5H_5^-$ + 8 electrons for Co^+ + 4 electrons for 2 C_2H_4 molecules = 18 electrons

Think about Your Answer The octet rule and the 18-electron rule have similar rationales. According to the octet rule, 8 electrons (4 pairs) are required to fill all the bonding and nonbonding orbitals for main group elements. By analogy, complexes obeying the 18-electron rule will have 18 electrons (9 pairs) filling all the bonding and nonbonding orbitals for *d*-block elements.

Check Your Understanding

What must the charge of the metal in each species be for the species to satisfy the EAN rule?

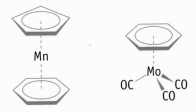

REVIEW & CHECK FOR SECTION 22.7

Which of the following formulas does not obey the 18-electron rule?

(a) $Mn(CO)_5H$

(b) $[V(CO)_6]^-$

(c) $[Co(\eta^5-C_5H_5)_2]^+$

(d) The anion in Zeise's salt, $[Pt(\eta^2-C_2H_4)Cl_3]^-$

CHAPTER GOALS REVISITED

Now that you have studied this chapter, you should ask whether you have met the chapter goals. In particular, you should be able to:

Identify and explain the chemical and physical properties of the transition elements
a. Identify the general classes of transition elements (Section 22.1).
b. Identify the transition metals from their symbols and positions in the periodic table, and recall some physical and chemical properties (Section 22.1).
c. Understand the electrochemical nature of corrosion (Section 22.1).
d. Describe the metallurgy of iron and copper (Section 22.2).

Understand the composition, structure, and bonding in coordination compounds
a. Given the formula for a coordination complex, identify the metal and its oxidation state, the ligands, the coordination number and coordination geometry, and the overall charge on the complex (Section 22.3). Relate names and formulas of complexes. **Study Questions: 9–18, 48, 49, 57.**
b. Given the formula for a complex, be able to recognize whether isomers will exist, and draw their structures (Section 22.4). **Study Questions: 41, 44, 45, 47, 48.**

c. Describe the bonding in coordination complexes (Section 22.5). **Study Question: 42.**

d. Apply the principles of stoichiometry, thermodynamics, and equilibrium to transition metal compounds. **Study Questions: 64, 67, 68, 70.**

Relate ligand field theory to the magnetic and spectroscopic properties of the complexes

a. Understand why substances are colored (Section 22.6). **Study Questions: 31, 32, 43.**

b. Understand the relationship between the ligand field splitting, magnetism, and color of complexes (Sections 22.5–22.6). **Study Questions: 23–30, 36, 37, 44, 45, 52, 55, 56, 60, 63.**

Apply the effective atomic number (EAN) rule to simple organometallic compounds of the transition metals

a. Apply the EAN rule to molecules containing a low-valent metal and ligands such as C_6H_6, C_2H_4, and CO (Section 22.7). **Study Questions: 33, 34, 68, 71.**

◉WL Questions and problems for this chapter are available in OWL. Use the OWL access card that came with this textbook to access assigned questions and problems for this chapter.

23 Nuclear Chemistry

A Primordial Nuclear Reactor

The natural nuclear reactors in West Africa have been called "one of the greatest natural phenomena that ever occurred." In 1972, a French scientist noticed that the uranium taken from a mine in Oklo, Gabon, was strangely deficient in ^{235}U. Uranium exists in nature as two principal isotopes, ^{238}U (99.275% abundant) and ^{235}U (0.72% abundant). It is ^{235}U that most readily undergoes nuclear fission and is used to fuel nuclear power plants around the world. But the uranium found in the Oklo mines had an isotope ratio like that of the spent fuel that comes from modern reactors. Based on this and other evidence, scientists concluded that ^{235}U at the Oklo mine was once about 3% of the total and that a "natural" fission process occurred in the bed of uranium ore nearly 2 billion years ago.

But intriguing questions remained: Why could fission occur to a significant extent in this natural deposit of uranium and why didn't the "reactor" explode? Apparently, there must have been a moderator of neutron energy and a regulation mechanism. In a modern nuclear power reactor, control rods slow down the neutrons from nuclear fission so that they can induce fission in other ^{235}U nuclei; that is, they moderate the neutron energy. Without a moderator, the neutrons just fly off. In the Oklo reactors, water seeping into the bed of uranium ore could have acted as a moderator.

The reason the Oklo reactor did not explode is that water could have also been the regulator. As the fission process heated the water, it boiled off as steam. This caused the fission to stop, but it began again when more water seeped in.

Courtesy of Francois Gauthier-Lafaye

The natural nuclear reactor in Oklo, Gabon (West Africa). Nearly 2 billion years ago, a natural formation containing uranium oxide (the yellow material) underwent fission that started and stopped over a period of a million years.

Scientists now believe this natural reactor would turn on for about 30 minutes and then shut down for several hours before turning on again. There is evidence these natural reactors functioned intermittently for about 1 million years, until the concentration of uranium isotopes was too low to keep the reaction going.

Questions:

1. How many protons and neutrons are there in the ^{235}U and ^{238}U nuclei?

2. Although plutonium does not occur naturally on earth now, it is thought to have been produced in the Oklo reactor (and then decayed). Write balanced nuclear equations for:
   (a) the reaction of ^{238}U and a neutron to give ^{239}U
   (b) the decay of ^{239}U to ^{239}Np and then to ^{239}Pu by beta emission
   (c) the decay of ^{239}Pu to ^{235}U by alpha emission

Answers to these questions are available in Appendix N.

CHAPTER OUTLINE

23.1   Natural Radioactivity
23.2   Nuclear Reactions and Radioactive Decay
23.3   Stability of Atomic Nuclei
23.4   Rates of Nuclear Decay
23.5   Artificial Nuclear Reactions
23.6   Nuclear Fission
23.7   Nuclear Fusion
23.8   Radiation Health and Safety
23.9   Applications of Nuclear Chemistry

CHAPTER GOALS

See Chapter Goals Revisited (page 850) for Study Questions keyed to these goals.

- Identify radioactive elements, and describe natural and artificial nuclear reactions.

- Calculate the binding energy and binding energy per nucleon for a particular isotope.

- Understand rates of radioactive decay.

- Understand artificial nuclear reactions.

- Understand issues of health and safety with respect to radioactivity.

- Be aware of some uses of radioactive isotopes in science and medicine.

History of science scholars cite areas of study out of which modern chemistry arose: technology, medicine, and alchemy. The third of these pillars, alchemy, was pursued in many cultures on three continents for well over 1000 years. Simply stated, the goal of the ancient alchemists was to turn less valuable materials into gold. We now recognize the futility of these efforts, because this goal is not reachable by chemical processes. But we now also know that their dream of transmuting one element into another can be achieved by means of nuclear reactions. This happens naturally in the decomposition of uranium and other radioactive elements, and scientists can intentionally carry out such reactions in the laboratory. The goal is no longer to make gold, however. Far more important and valuable products of nuclear reactions are possible.

Nuclear chemistry encompasses a wide range of topics that share one thing in common: they involve changes in the nucleus of an atom. While nuclear "chemistry" is a major focus in this chapter, this subject cuts across many areas of science and modern society. Radioactive isotopes are used in medicine. Nuclear power provides a sizable fraction of energy for modern society. And, then there are nuclear weapons. . . .

23.1 Natural Radioactivity

In the late 19th century, while studying radiation emanating from uranium and thorium, Ernest Rutherford (1871–1937) stated, "There are present at least two distinct types of radiation—one that is readily absorbed, which will be termed for convenience **α (alpha) radiation**, and the other of a more penetrative character, which will be termed **β (beta) radiation**." Subsequently, charge-to-mass ratio measurements showed that α radiation is composed of helium nuclei (He^{2+}) and β radiation is composed of electrons (e^-) (Table 23.1).

Table 23.1 Characteristics of α, β, and γ Radiation

Name	Symbols	Charge	Mass (g/particle)
Alpha	^4_2He, $^4_2\alpha$	2+	6.65×10^{-24}
Beta	$^0_{-1}e$, $^0_{-1}\beta$	1−	9.11×10^{-28}
Gamma	γ	0	0

OWL

Sign in to OWL at **www.cengage.com/owl** to view tutorials and simulations, develop problem-solving skills, and complete online homework assigned by your professor.

go Chemistry

Download mini lecture videos for key concept review and exam prep from OWL or purchase them from **www.cengagebrain.com**

FIGURE 23.1 The relative penetrating ability of α, β, and γ radiation. Highly charged α particles interact strongly with matter and are stopped by a piece of paper. Beta particles, with less mass and a lower charge, interact to a lesser extent with matter and thus can penetrate farther. Gamma radiation is the most penetrating.

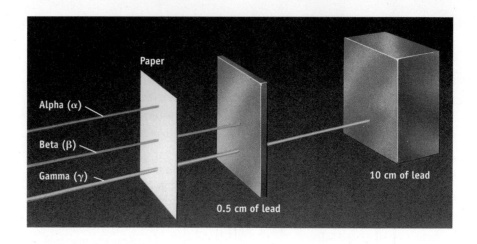

● **Common Symbols: α and β** Symbols used to represent alpha and beta particles do not include a plus or minus superscript, respectively, to show that they have a charge.

Rutherford hedged his bet when he said at least two types of radiation existed. A third type was later discovered by the French scientist Paul Villard (1860–1934); he named it γ **(gamma) radiation**, using the third letter in the Greek alphabet in keeping with Rutherford's scheme. Unlike α and β radiation, γ radiation is not affected by electric and magnetic fields. Rather, it is a form of electromagnetic radiation like x-rays but even more energetic.

Early studies measured the penetrating power of the three types of radiation (Figure 23.1). Alpha radiation is the least penetrating; it can be stopped by several sheets of ordinary paper or clothing. Beta particles can penetrate several millimeters of living bone or tissue, but about 0.5 cm of lead will stop the particles. Gamma radiation is the most penetrating. Thick layers of lead or concrete are required to shield the body from this radiation, and γ-rays can pass completely through the human body.

Alpha and β particles typically possess high kinetic energies. The energy of γ radiation is similarly very high. The energy associated with this radiation is transferred to any material used to stop the particle or absorb the radiation. This fact is important because the damage caused by radiation is related to the energy absorbed (see Section 23.8).

23.2 Nuclear Reactions and Radioactive Decay

Equations for Nuclear Reactions

In 1903, Rutherford and Frederick Soddy (1877–1956) proposed that radioactivity is the result of a natural change of an isotope of one element into an isotope of a different element. Such processes are called **nuclear reactions**.

Consider a reaction in which radium-226 (the isotope of radium with mass number 226) emits an α particle to form radon-222. The equation for this reaction is

$$^{226}_{88}\text{Ra} \rightarrow {}^{4}_{2}\alpha + {}^{222}_{86}\text{Rn}$$

● **Symbols Used in Nuclear Equations** The mass number is included as a superscript, and the atomic number is included as a subscript preceding the symbols for reactants and products. This is done to facilitate balancing these equations.

In a nuclear reaction, the sum of the mass numbers of reacting particles must equal the sum of the mass numbers of products. Furthermore, to maintain nuclear charge balance, the sum of the atomic numbers of the products must equal the sum of the atomic numbers of the reactants. These principles are illustrated using the preceding nuclear equation:

	$^{226}_{88}\text{Ra}$	$\rightarrow$	$^{4}_{2}\alpha$	+	$^{222}_{86}\text{Rn}$
	radium-226	$\rightarrow$	α particle	+	radon-222
Mass number: (protons + neutrons)	226	=	4	+	222
Atomic number: (protons)	88	=	2	+	86

Alpha-particle emission causes a decrease of two units in atomic number and four units in the mass number.

Similarly, nuclear mass and nuclear charge balance accompany β-particle emission, as illustrated by the decomposition of uranium-239:

	$^{239}_{92}\text{U}$	$\rightarrow$	$^{0}_{-1}\beta$	+	$^{239}_{93}\text{Np}$
	uranium-239	$\rightarrow$	β particle	+	neptunium-239
Mass number: (protons + neutrons)	239	=	0	+	239
Atomic number: (protons)	92	=	−1	+	93

The β particle has a charge of $1-$. Charge balance requires that the atomic number of the product be one unit greater than the atomic number of the reacting nucleus. The mass number does not change in this process.

How does a nucleus, composed of protons and neutrons, eject an electron? It is a complex process, but the net result is the conversion within the nucleus of a neutron to a proton and an electron.

$$^{1}_{0}\text{n} \longrightarrow {}^{0}_{-1}\text{e} + {}^{1}_{1}\text{p}$$

neutron   electron   proton

Notice that the mass and charge numbers balance in this equation.

What is the origin of the gamma radiation that accompanies many nuclear reactions? Recall that a photon of visible light is emitted when an atom undergoes a transition from an excited electronic state to a lower energy state (◄ Section 6.3). Gamma radiation originates from transitions between nuclear energy levels. Nuclear reactions often result in the formation of a product nucleus in an excited nuclear state. One option is to return to the ground state by emitting a photon. The energy of γ radiation is a measure of the large energy difference between the energy levels in the nucleus.

Radioactive Decay Series

Several naturally occurring radioactive isotopes decay to form a product that is also radioactive. When this happens, the initial nuclear reaction is followed by a second nuclear reaction; if the situation is repeated, a third and a fourth nuclear reaction occur; and so on. Eventually, a nonradioactive isotope is formed to end the series. Such a sequence of nuclear reactions is called a **radioactive decay series**. In each step of this sequence, the reactant nucleus is called the *parent*, and the product is called the *daughter*.

Uranium-238, the most abundant of three naturally occurring uranium isotopes, heads one of four radioactive decay series. This series begins with the loss of an α particle from $^{238}_{92}\text{U}$ to form radioactive $^{234}_{90}\text{Th}$. Thorium-234 then decomposes by β emission to $^{234}_{91}\text{Pa}$, which emits a β particle to give $^{234}_{92}\text{U}$. Uranium-234 is an α emitter, forming $^{230}_{90}\text{Th}$. Further α and β emissions follow, until the series ends with formation of the stable, nonradioactive isotope, $^{206}_{82}\text{Pb}$. In all, this radioactive decay series converting $^{238}_{92}\text{U}$ to $^{206}_{82}\text{Pb}$ is made up of 14 reactions, with eight α and six β particles being emitted. The series is portrayed graphically by plotting atomic number versus mass number (Figure 23.2). An equation can be written for each step in the sequence. Equations for the first four steps in the uranium-238 radioactive decay series are as follows:

Step 1. $^{238}_{92}\text{U} \rightarrow {}^{234}_{90}\text{Th} + {}^{4}_{2}\alpha$

Step 2. $^{234}_{90}\text{Th} \rightarrow {}^{234}_{91}\text{Pa} + {}^{0}_{-1}\beta$

Step 3. $^{234}_{91}\text{Pa} \rightarrow {}^{234}_{92}\text{U} + {}^{0}_{-1}\beta$

Step 4. $^{234}_{92}\text{U} \rightarrow {}^{230}_{90}\text{Th} + {}^{4}_{2}\alpha$

Uranium ore contains trace quantities of the radioactive elements formed in the radioactive decay series. A significant development in nuclear chemistry was Marie Curie's discovery in 1898 of radium and polonium as trace components of

FIGURE 23.2 The uranium-238 radioactive decay series. The steps in this radioactive decay series are shown graphically in this plot of mass number versus atomic number. Each α-decay step lowers the atomic number by two units and the mass number by four units. Beta-particle emission does not change the mass but raises the atomic number by one unit. Half-lives of the isotopes are included on the chart. Notice that several of the isotopes in this series decompose by two different pathways.

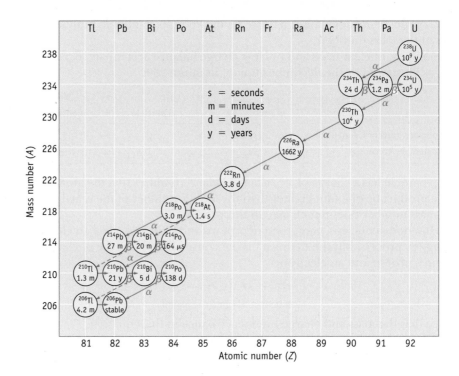

pitchblende, a uranium ore. The amount of each of these elements is small because the isotopes of these elements have short half-lives. Marie Curie isolated only a single gram of radium from 7 tons of ore. It is a credit to her skills as a chemist that she extracted sufficient amounts of radium and polonium from uranium ore to identify these elements.

The uranium-238 radioactive decay series is also the source of the environmental hazard radon. Trace quantities of uranium are often present naturally in the soil and rocks, and radon-222 is being continuously formed. Because radon is chemically inert, it is not trapped by chemical processes in soil or water and is free to seep into mines or into homes through pores in cement block walls, through cracks in the basement floor or walls, or around pipes. Because it is more dense than air, radon tends to collect in low spots, and its concentration can build up in a basement if steps are not taken to remove it.

The major health hazard from radon, when it is inhaled by humans, arises not from radon itself but from its decomposition product, polonium.

$$^{222}_{86}\text{Rn} \rightarrow {}^{218}_{84}\text{Po} + {}^{4}_{2}\alpha \qquad t_{1/2} = 3.82 \text{ days}$$

$$^{218}_{84}\text{Po} \rightarrow {}^{214}_{82}\text{Pb} + {}^{4}_{2}\alpha \qquad t_{1/2} = 3.04 \text{ minutes}$$

Radon does not undergo chemical reactions or form compounds that can be taken up in the body. Polonium, however, is not chemically inert. Polonium-218 can lodge in body tissues, where it undergoes α decay to give lead-214, another radioactive isotope. The range of an α particle in body tissue is quite small, perhaps 0.7 mm. This is approximately the thickness of the epithelial cells of the lungs, however, so α-particle radiation can cause serious damage to lung tissues.

Virtually every home in the United States has some level of radon, and kits can be purchased to test for the presence of this gas. If a significant amount of radon gas is detected in your home, you should take corrective actions such as sealing cracks around the foundation and in the basement. It may be reassuring to know that the health risks associated with radon are low. The likelihood of getting lung cancer from exposure to radon is about the same as the likelihood of dying in an accident in your home.

Radon detector. This kit is intended for use in the home to detect radon gas. The small device is placed in the home's basement for a given time period and is then sent to a laboratory to measure the amount of radon that might be present.

EXAMPLE 23.1   Radioactive Decay Series

Problem A second radioactive decay series begins with $^{235}_{92}U$ and ends with $^{207}_{82}Pb$.

(a) How many α and β particles are emitted in this series?

(b) The first three steps of this series are (in order) α, β, and α emission. Write an equation for each of these steps.

What Do You Know? You are given the starting (long-lived radioactive) isotope and the final (stable) isotope. A series of alpha and beta reactions is the link between these two species.

Strategy First, find the total change in atomic number and mass number. A combination of α and β particles is required that will decrease the total nuclear mass by 28 (235 − 207) and at the same time decrease the atomic number by 10 (92 − 82). Each equation must give symbols for the parent and daughter nuclei and the emitted particle. In the equations, the sums of the atomic numbers and mass numbers for reactants and products must be equal.

Solution

(a) Mass declines by 28 mass units (235 − 207). Because a decrease of 4 mass units occurs with each α emission, 7 α particles must be emitted. Also, for each α emission, the atomic number decreases by 2. Emission of 7 α particles would cause the atomic number to decrease by 14, but the actual decrease in atomic number is 10 (92 − 82). This means that 4 β particles must also have been emitted because each β emission increases the atomic number of the product by one unit. Thus, the radioactive decay sequence involves emission of 7 α and 4 β particles.

(b) Step 1. $^{235}_{92}U \rightarrow {}^{231}_{90}Th + {}^{4}_{2}\alpha$

Step 2. $^{231}_{90}Th \rightarrow {}^{231}_{91}Pa + {}^{0}_{-1}\beta$

Step 3. $^{231}_{91}Pa \rightarrow {}^{227}_{89}Ac + {}^{4}_{2}\alpha$

Think about Your Answer

Because the mass number changes in these series only when an alpha particle is lost, all of the mass numbers in a given decay series are some multiple of four less than the first isotope in the series. For the series beginning with U-238, the mass numbers are in the series 238, 234, 230, . . . , 206. This series is sometimes called the $4n + 2$ *series* because each mass number (M) fits the equation $4n + 2 = M$, where n is an integer (n is 59 for the first member of this series and 51 for the last member). For the series headed by $^{235}_{92}U$, the mass numbers are 235, 231, 227, . . . , 207; this is the $4n + 3$ *series*.

Two other decay series are possible. One, called the $4n$ *series* and beginning with ^{232}Th, is found in nature; the other, the $4n + 1$ series, is not. No member of this series has a very long half-life. During the 4.5 billion years since this planet was formed, all radioactive members of this series have completely decayed to the end product of this series, ^{209}Bi.

Check Your Understanding

(a) Six α and four β particles are emitted in the thorium-232 radioactive decay series before a stable isotope is reached. What is the final product in this series?

(b) The first three steps in the thorium-232 decay series (in order) are α, β, and β emission. Write an equation for each step.

Other Types of Radioactive Decay

Most naturally occurring radioactive elements decay by emission of α, β, and γ radiation. Other nuclear decay processes became known, however, when new radioactive elements were synthesized by artificial means. These include **positron ($^{0}_{+1}\beta$) emission** and **electron capture**.

Positrons ($^{0}_{+1}\beta$) and electrons have the same mass but opposite charge. The positron is the antimatter analog to an electron. *Positron emission* by polonium-207, for example, results in the formation of bismuth-207.

● **Positrons** Positrons were discovered by Carl Anderson (1905–1991) in 1932. The positron is one of a group of particles that are known as *antimatter*. If matter and antimatter particles collide, mutual annihilation occurs, with energy being emitted.

	$^{207}_{84}Po$	$\rightarrow$	$^{0}_{+1}\beta$	$+$	$^{207}_{83}Bi$
	polonium-207	$\rightarrow$	positron	$+$	bismuth-207
Mass number: (protons + neutrons)	207	$=$	0	$+$	207
Atomic number: (protons)	84	$=$	1	$+$	83

To retain charge balance, positron decay results in a decrease in the atomic number.

In *electron capture,* an extranuclear electron is captured by the nucleus. The mass number is unchanged, and the atomic number is reduced by 1. (In an old nomenclature, the innermost electron shell was called the *K shell,* and electron capture was called *K capture.*)

• **Neutrinos and Antineutrinos** Beta particles having a wide range of energies are emitted. To balance the energy associated with β decay, it is necessary to postulate the concurrent emission of another particle, the *antineutrino.* Similarly, neutrino emission accompanies positron emission. Much study has gone into detecting neutrinos and antineutrinos. These massless, chargeless particles are not included when writing nuclear equations.

	$^{7}_{4}Be$	$+$	$^{0}_{-1}e$	$\rightarrow$	$^{7}_{3}Li$
	beryllium-7	$+$	electron	$\rightarrow$	lithium-7
Mass number: (protons + neutrons)	7	$+$	0	$=$	7
Atomic number: (protons)	4	$+$	-1	$=$	3

In summary, most unstable nuclei decay by one of four paths: α or β decay, positron emission, or electron capture. Gamma radiation often accompanies these processes. Section 23.6 introduces a fifth way that nuclei decompose, *fission.*

EXAMPLE 23.2 Nuclear Reactions

Problem Complete the following equations. Give the symbol, mass number, and atomic number of the product species.

(a) $^{37}_{18}Ar + ^{0}_{-1}e \rightarrow ?$ (c) $^{35}_{16}S \rightarrow ^{35}_{17}Cl + ?$

(b) $^{11}_{6}C \rightarrow ^{11}_{5}B + ?$ (d) $^{30}_{15}P \rightarrow ^{0}_{+1}\beta + ?$

What Do You Know? You are given the reactant and one of two products formed in these nuclear reactions. You know that both the masses and charges must balance in a balanced nuclear equation.

Strategy The missing product in each reaction can be determined by recognizing that the sums of mass numbers and atomic numbers for products and reactants must be equal. When you know the nuclear mass and nuclear charge of the product, you can identify it with the appropriate symbol.

Solution

(a) This is an electron capture reaction. The product has a mass number of $37 + 0 = 37$ and an atomic number of $18 - 1 = 17$. Therefore, the symbol for the product is $^{37}_{17}Cl$.

(b) This missing particle has a mass of zero and a charge of 1+; these are the characteristics of a positron, $^{0}_{+1}\beta$. If this particle is included in the equation, the sums of the atomic numbers ($6 = 5 + 1$) and the mass numbers (11) on either side of the equation are equal.

(c) A beta particle, $^{0}_{-1}\beta$, is required to balance the mass numbers (35) and atomic numbers ($16 = 17 - 1$) in the equation.

(d) The product nucleus has mass number 30 and atomic number 14. This identifies the unknown as $^{30}_{14}Si$.

Think about Your Answer The product in each reaction is a stable isotope. As we will see in the next section, there is a narrow range of neutron-to-proton ratios for stable products. Spontaneous nuclear reactions always occur in a way that moves either into or closer to the narrow range of stable isotopes.

Check Your Understanding

Indicate the symbol, the mass number, and the atomic number of the missing product in each of the following nuclear reactions.

(a) $^{13}_{7}N \rightarrow ^{13}_{6}C + ?$ (b) $^{41}_{20}Ca + ^{0}_{-1}e \rightarrow ?$ (c) $^{90}_{38}Sr \rightarrow ^{90}_{39}Y + ?$ (d) $^{22}_{11}Na \rightarrow ? + ^{0}_{+1}\beta$

23.3 Stability of Atomic Nuclei

We can learn something about nuclear stability from Figure 23.3. In this plot, the horizontal axis represents the number of protons, and the vertical axis gives the number of neutrons for known isotopes. Each circle represents an isotope identified by the number of neutrons and protons contained in its nucleus. The black circles represent stable (nonradioactive) isotopes, about 300 in number, and the red circles represent some of the known radioactive isotopes. For example, the three isotopes of hydrogen are $^{1}_{1}H$ and $^{2}_{1}H$ (neither is radioactive) and $^{3}_{1}H$ (tritium, radioactive). For lithium, the third element, isotopes with mass numbers 4, 5, 6, and 7 are known. The isotopes with masses of 6 and 7 (shown in black) are stable, whereas the other two isotopes (in red) are radioactive.

Figure 23.3 contains the following information about nuclear stability:

- Stable isotopes fall in a very narrow range called the **band of stability**. It is remarkable how few isotopes are stable.
- Only two stable isotopes ($^{1}_{1}H$ and $^{3}_{2}He$) have more protons than neutrons.
- Up to calcium ($Z = 20$), stable isotopes often have equal numbers of protons and neutrons or only one or two more neutrons than protons.
- Beyond calcium, the neutron-to-proton ratio is always greater than 1. As the mass increases, the band of stable isotopes deviates more and more from a 1:1 neutron-to-proton ratio (the line in Figure 23.3 for which $N/Z = 1$).
- Beyond bismuth (83 protons and 126 neutrons), all isotopes are unstable and radioactive. There is apparently no nuclear "superglue" strong enough to hold heavy nuclei together.
- Isotopes that fall farther from the band of stability tend to have shorter half-lives than do unstable isotopes nearer to the band of stability.
- Elements of even atomic number have more stable isotopes than do those of odd atomic number. More stable isotopes have an even number of neutrons than have an odd number. Roughly 200 isotopes have an even number of neutrons and an even number of protons, whereas only about 120 isotopes have an odd number of either protons or neutrons. Only five stable isotopes ($^{2}_{1}H$, $^{6}_{3}Li$, $^{10}_{5}B$, $^{14}_{7}N$, and $^{180}_{73}Ta$) have odd numbers of both protons and neutrons.

The Band of Stability and Radioactive Decay

Besides being a criterion for stability, the neutron-to-proton ratio can assist in predicting what type of radioactive decay will be observed. Unstable nuclei decay in a manner that brings them toward a stable neutron-to-proton ratio—that is, toward the band of stability.

FIGURE 23.3 Stable and unstable isotopes. A graph of the number of neutrons (*N*) versus the number of protons (*Z*) for stable (*black circles*) and radioactive (*red circles*) isotopes from hydrogen to bismuth. This graph is used to assess criteria for nuclear stability and to predict modes of decay for unstable nuclei.

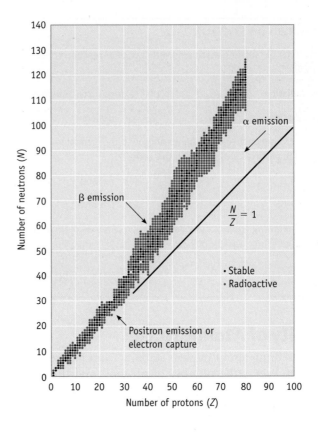

- All elements beyond bismuth ($Z = 83$) are unstable. To reach the band of stability starting with these elements, a process that decreases the atomic number is needed. Alpha emission is an effective way to lower Z, the atomic number, because each emission decreases the atomic number by 2. For example, americium, the radioactive element used in smoke detectors, decays by α emission:

$$^{243}_{95}\text{Am} \rightarrow {}^{4}_{2}\alpha + {}^{239}_{93}\text{Np}$$

- Beta emission occurs for isotopes that have a high neutron-to-proton ratio—that is, isotopes above the band of stability. With β decay, the atomic number increases by 1, and the mass number remains constant, resulting in a lower neutron-to-proton ratio:

$$^{60}_{27}\text{Co} \rightarrow {}^{0}_{-1}\beta + {}^{60}_{28}\text{Ni}$$

- Isotopes with a low neutron-to-proton ratio, below the band of stability, decay by positron emission or by electron capture. Both processes lead to product nuclei with a lower atomic number and the same mass number and move the product closer to the band of stability:

$$^{13}_{7}\text{N} \rightarrow {}^{0}_{+1}\beta + {}^{13}_{6}\text{C}$$

$$^{41}_{20}\text{Ca} + {}^{0}_{-1}\text{e} \rightarrow {}^{41}_{19}\text{K}$$

EXAMPLE 23.3   Predicting Modes of Radioactive Decay

Problem Identify probable mode(s) of decay for each isotope and write an equation for the decay process.

(a) oxygen-15, $^{15}_{8}\text{O}$

(b) uranium-234, $^{234}_{92}\text{U}$

(c) fluorine-20, $^{20}_{9}\text{F}$

(d) manganese-56, $^{56}_{25}\text{Mn}$

What Do You Know? You are given the formulas for four radioactive isotopes. The possible modes of decomposition are α, β, or positron emission and electron capture. The preferred mode will give a more stable isotope and create a nucleus that is closer to the band of stability.

Strategy In parts (a), (c), and (d), compare the mass number with the atomic mass. If the mass number of the isotope is higher than the atomic weight, then there are too many neutrons, and β emission is likely. If the mass number is lower than the atomic weight, then there are too few neutrons, and positron emission or electron capture is the more likely process. It is not possible to choose between the latter two modes of decay without further information. For part (b), note that isotopes with atomic number greater than 83 are likely to be α emitters.

Solution

(a) Oxygen-15 has 7 neutrons and 8 protons, so the neutron-to-proton (n/p) ratio is less than 1—too low for ^{15}O to be stable. Nuclei with too few neutrons are expected to decay by either positron emission or electron capture. In this instance, the process is $^{0}_{+1}\beta$ emission, and the equation is $^{15}_{8}\text{O} \rightarrow {}^{0}_{+1}\beta + {}^{15}_{7}\text{N}$.

(b) Alpha emission is a common mode of decay for isotopes of elements with atomic numbers higher than 83. The decay of uranium-234 is one example:

$$^{234}_{92}\text{U} \rightarrow {}^{230}_{90}\text{Th} + {}^{4}_{2}\alpha$$

(c) Fluorine-20 has 11 neutrons and 9 protons, a high n/p ratio. The ratio is lowered by β emission:

$$^{20}_{9}\text{F} \rightarrow {}^{0}_{-1}\beta + {}^{20}_{10}\text{Ne}$$

(d) The atomic weight of manganese is 54.85. The higher mass number, 56, suggests that this radioactive isotope has an excess of neutrons, in which case it would be expected to decay by β emission:

$$^{56}_{25}\text{Mn} \rightarrow {}^{0}_{-1}\beta + {}^{56}_{26}\text{Fe}$$

Think about Your Answer Decomposition can occur by four possible paths, α and β emission, positron emission, and electron capture. Whether the decay follows beta emission or positron emission (or electron capture) is predicted by the neutron/proton ratio, relative to the n/p ratio in stable isotopes.

Check Your Understanding

Write an equation for the probable mode of decay for each of the following unstable isotopes, and write an equation for that nuclear reaction.

(a) silicon-32, $^{32}_{14}\text{Si}$

(b) titanium-45, $^{45}_{22}\text{Ti}$

(c) plutonium-239, $^{239}_{94}\text{Pu}$

(d) potassium-42, $^{42}_{19}\text{K}$

Nuclear Binding Energy

An atomic nucleus can contain as many as 83 protons and still be stable. For stability, nuclear binding (attractive) forces must be greater than the electrostatic repulsive forces between the closely packed protons in the nucleus. **Nuclear binding energy**, E_b, is defined as the energy required to separate the nucleus of an atom into protons and neutrons. For example, the nuclear binding energy for deuterium is the energy required to convert one mole of deuterium ($^{2}_{1}\text{H}$) nuclei into one mole of protons and one mole of neutrons.

$$^{2}_{1}\text{H} \rightarrow {}^{1}_{1}\text{p} + {}^{1}_{0}\text{n} \qquad E_b = 2.15 \times 10^8 \text{ kJ/mol}$$

The positive sign for E_b indicates that energy is required for this process. A deuterium nucleus is more stable than an isolated proton and an isolated neutron, just as the H_2 molecule is more stable than two isolated H atoms. Recall, however, that the H—H bond energy is only 436 kJ/mol. The energy holding a proton and a neutron together in a deuterium nucleus, 2.15×10^8 kJ/mol, is about 500,000 times larger than the typical covalent bond energies.

To further understand nuclear binding energy, we turn to an experimental observation and a theory. The experimental observation is that the mass of a nucleus is always less than the sum of the masses of its constituent protons and neutrons. The theory is that the "missing mass," called the **mass defect**, is equated with the energy that holds the nuclear particles together.

The mass defect for deuterium is the difference between the mass of a deuterium nucleus and the sum of the masses of a proton and a neutron. Mass spectrometric measurements (◄ Section 2.3) give the masses of these particles to a high level of accuracy, providing the numbers needed to carry out calculations of mass defects.

Masses of atomic nuclei are not generally listed in reference tables, but masses of atoms are. Calculation of the mass defect can be carried out using masses of atoms instead of masses of nuclei. By using atomic masses, we are including in this calculation the masses of extranuclear electrons in the reactants and the products. Because the same number of extranuclear electrons appears in products and reactants, this does not affect the result. Thus, for one mole of deuterium nuclei, the mass defect is found as follows:

$$\begin{array}{cccc}
{}^2_1\text{H} & \rightarrow & {}^1_1\text{H} & + & {}^1_0\text{n} \\
2.01410 \text{ g/mol} & & 1.007825 \text{ g/mol} & & 1.008665 \text{ g/mol}
\end{array}$$

$$\text{Mass defect} = \Delta m = \text{mass of products} - \text{mass of reactants}$$

$$= [1.007825 \text{ g/mol} + 1.008665 \text{ g/mol}] - 2.01410 \text{ g/mol}$$

$$= 0.00239 \text{ g/mol}$$

The relationship between mass and energy is contained in Albert Einstein's 1905 theory of special relativity, which holds that mass and energy are different manifestations of the same quantity. Einstein defined the energy–mass relationship: Energy is equivalent to mass times the square of the speed of light; that is, $E = mc^2$. In the case of atomic nuclei, it is assumed that the missing mass (the mass defect, Δm) is equated with the binding energy holding the nucleus together.

$$E_b = (\Delta m)c^2 \tag{23.1}$$

If Δm is given in kilograms and the speed of light is given in meters per second, E_b will have units of joules (because $1 \text{ J} = 1 \text{ kg} \cdot \text{m}^2/\text{s}^2$). For the decomposition of one mole of deuterium nuclei to one mole of protons and one mole of neutrons, we have

$$E_b = (2.39 \times 10^{-6} \text{ kg/mol})(2.998 \times 10^8 \text{ m/s})^2$$

$$= 2.15 \times 10^{11} \text{ J/mol of } {}^2_1\text{H nuclei } (= 2.15 \times 10^8 \text{ kJ/mol of } {}^2_1\text{H nuclei})$$

The nuclear stabilities of different elements are compared using the **binding energy per mole of nucleons**. (**Nucleon** is the general name given to nuclear particles—that is, protons and neutrons.) A deuterium nucleus contains two nucleons, so the binding energy per mole of nucleons, E_b/n, is 2.15×10^8 kJ/mol divided by 2, or 1.08×10^8 kJ/mol nucleon.

$$E_b/n = \left(\frac{2.15 \times 10^8 \text{ kJ}}{\text{mol } {}^2_1\text{H nuclei}} \right) \left(\frac{1 \text{ mol } {}^2_1\text{H nuclei}}{2 \text{ mol nucleons}} \right)$$

$$E_b/n = 1.08 \times 10^8 \text{ kJ/mol nucleons}$$

The binding energy per nucleon can be calculated for any atom whose mass is known. Then, to compare nuclear stabilities, we can plot binding energies per nucleon as a function of mass number (Figure 23.4). The greater the binding energy per nucleon, the greater the stability of the nucleus. From the graph in Figure 23.4, the point of maximum nuclear stability occurs at a mass of 56 (i.e., at iron in the periodic table).

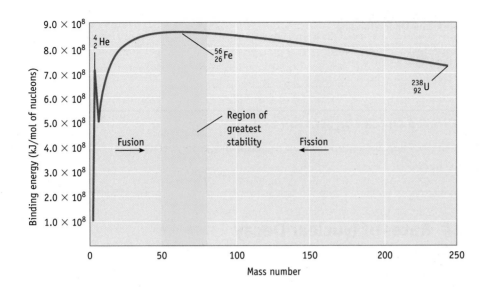

FIGURE 23.4 Relative stability of nuclei. Binding energy per nucleon for the most stable isotope of elements between hydrogen and uranium is plotted as a function of mass number. (Fission and fusion are discussed on pages 840–843.)

EXAMPLE 23.4   Nuclear Binding Energy

Problem Calculate the binding energy, E_b (in kJ/mol), and the binding energy per nucleon, E_b/n (in kJ/mol nucleon), for carbon-12.

What Do You Know? The mass of carbon-12 is, by definition, exactly 12 g/mol. You will need the molar masses of hydrogen atoms ($_1^1H$) and neutrons (1.007825 g/mol and 1.008665 g/mol respectively, to determine the mass defect.

Strategy The mass defect is the difference between the mass of carbon-12 and the masses of 6 protons, 6 neutrons, and 6 electrons. The mass of 1 mol of protons and 1 mol of electrons can be taken into account by using the molar mass of $_1^1H$. Use the values given to calculate the mass defect (in g/mol). The binding energy is calculated from the mass defect using Equation 23.1.

Solution The mass of $_1^1H$ is 1.007825 g/mol, and the mass of $_0^1n$ is 1.008665 g/mol. Carbon-12, $_6^{12}C$, is the standard for the atomic masses in the periodic table, and its mass is defined as exactly 12 g/mol

$$\Delta m = [(6 \times \text{mass } _1^1H) + (6 \times \text{mass } _0^1n)] - \text{mass } _6^{12}C$$

$$= [(6 \times 1.007825 \text{ g/mol}) + (6 \times 1.008665 \text{ g/mol})] - 12.000000 \text{ g/mol}$$

$$= 9.8940 \times 10^{-2} \text{ g/mol nuclei}$$

The binding energy is calculated using Equation 23.1. Using the mass in kilograms and the speed of light in meters per second gives the binding energy in joules:

$$E_b = (\Delta m)c^2$$

$$= (9.8940 \times 10^{-5} \text{ kg/mol})(2.99792 \times 10^8 \text{ m/s})^2$$

$$= 8.8923 \times 10^{12} \text{ J/mol nuclei} (= 8.8923 \times 10^9 \text{ kJ/mol nuclei})$$

The binding energy per nucleon, E_b/n, is determined by dividing the binding energy by 12 (the number of nucleons)

$$\frac{E_b}{n} = \frac{8.8923 \times 10^9 \text{ kJ/mol nuclei}}{12 \text{ mol nucleons/mol nuclei}}$$

$$= 7.4102 \times 10^8 \text{ kJ/mol nucleons}$$

Think about Your Answer Be sure that the units of mass and the speed of light are in kg and m/s, respectively, in this calculation. (Note that $1 \text{ kg} \cdot \text{m}^2/\text{s}^2 = 1 \text{ J}$.) The binding energy is a very large quantity of energy compared to those of ordinary chemical reactions. Compare binding energy to the very exothermic reaction of hydrogen and oxygen to form water vapor, for which $\Delta_r H°$ is only −242 kJ per mol of water vapor formed at 25 °C.

Check Your Understanding

Calculate the binding energy per nucleon, in kilojoules per mole, for lithium-6. The molar mass of ^6_3Li is 6.015125 g/mol.

REVIEW & CHECK FOR SECTION 23.3

Among the listed species, which has the highest binding energy per nucleon?

(a) ^2_1D (b) $^{20}_{10}Ne$ (c) $^{56}_{26}Fe$ (d) $^{238}_{92}U$

23.4 Rates of Nuclear Decay

Half-Life

When a new radioactive isotope is identified, its *half-life* is usually measured. Half-life ($t_{1/2}$) is used in nuclear chemistry in the same way it is used when discussing the kinetics of first-order chemical reactions (◄ Section 15.4): It is the time required for half of a sample to decay to products (Figure 23.5). Recall that for first-order kinetics the half-life is independent of the amount of sample.

Half-lives for radioactive isotopes cover a wide range of values. Uranium-238 has one of the longer half-lives, 4.47×10^9 years, a length of time close to the age of the earth (estimated at 4.5–4.6×10^9 years). Thus, roughly half of the uranium-238 present when the planet was formed is still around. At the other end of the range of half-lives are isotopes such as the 277 isotope of element 112 (recently named copernicium, ^{277}Cn), which has a half-life of 240 microseconds ($1 \, \mu s = 1 \times 10^{-6}$ s).

Half-life provides an easy way to estimate the time required before a radioactive element is no longer a health hazard. Strontium-90, for example, is a β emitter with a half-life of 29.1 years. Significant quantities of strontium-90 were dispersed into the environment in atmospheric nuclear bomb tests in the 1960s and 1970s, and, from the half-life, we know that almost one half is still around. The health problems associated with strontium-90 arise because calcium and strontium have similar chemical properties. Strontium-90 is taken into the body and deposited in bone, taking the place of calcium. Radiation damage by strontium-90 (a β emitter) in bone has been directly linked to bone-related cancers.

FIGURE 23.5 Decay of 20.0 mg of oxygen-15. After each half-life period of 2.0 minutes, the mass of oxygen-15 decreases by one half. (Oxygen-15 decays by positron emission.)

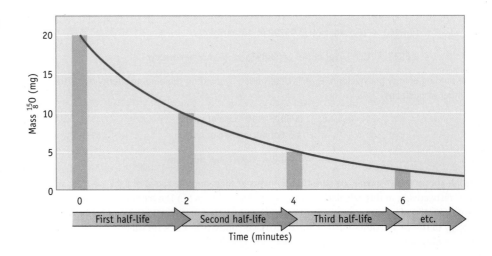

EXAMPLE 23.5   Using Half-Life

Problem Radioactive iodine-131, used to treat hyperthyroidism, has a half-life of 8.04 days.

(a) If you have 8.8 μg (micrograms) of this isotope, what mass remains after 32.2 days?

(b) How long will it take for a sample of iodine-131 to decay to one eighth of its activity?

(c) Estimate the length of time necessary for the sample to decay to 10% of its original activity.

What Do You Know? The half-life of ^{131}I, 8.04 days, is given in the problem. Estimates of the masses remaining are made based on this value.

Strategy This problem asks you to use half-life to qualitatively assess the rate of decay. After one half-life, half of the sample remains. After another half-life, the amount of sample is again decreased by half to one fourth of its original value. To answer these questions, assess the number of half-lives that have elapsed and use this information to determine the amount of sample remaining.

Solution

(a) The time elapsed, 32.2 days, is 4 half-lives (32.2/8.04 = 4). The amount of iodine-131 has decreased to 1/16 of the original amount [1/2 $\times$ 1/2 $\times$ 1/2 $\times$ 1/2 = (1/2)^4 = 1/16]. The amount of iodine remaining is 8.8 μg $\times$ (1/2)^4 or 0.55 μg.

(b) After 3 half-lives, the amount of iodine-131 remaining is 1/8 [= (1/2)^3] of the original amount. The amount remaining is 8.8 μg $\times$ (1/2)^3 = 1.1 μg.

(c) After 3 half-lives, 1/8 (12.5%) of the sample remains; after 4 half-lives, 1/16 (6.25%) remains. It will take between 3 and 4 half-lives, between 24.15 and 32.2 days, to decrease the amount of sample to 10% of its original value.

Think about Your Answer You will find it useful to make approximations as we have done in (c). An exact time can be calculated from the first-order rate law (page 519 and Equation 23.5).

Check Your Understanding

Tritium ($^3_1$H), a radioactive isotope of hydrogen, has a half-life of 12.3 years.

(a) Starting with 1.5 mg of this isotope, what mass (mg) remains after 49.2 years?

(b) How long will it take for a sample of tritium to decay to one eighth of its activity?

(c) Estimate the length of time necessary for the sample to decay to 1% of its original activity.

Kinetics of Nuclear Decay

The rate of nuclear decay is determined from measurements of the **activity** (A) of a sample. Activity refers to the number of disintegrations observed per unit of time, a quantity that can be measured readily with devices such as a Geiger–Müller counter (Figure 23.6). *Activity is proportional to the number of radioactive atoms present (N).*

$$A \propto N \tag{23.2}$$

If the number of radioactive nuclei N is reduced by half, the activity of the sample will be half as large. Doubling N will double the activity. This evidence indicates that the rate of decomposition is first order with respect to N. Consequently, the equations describing rates of radioactive decay are the same as those used to describe the kinetics of first-order chemical reactions; the change in the number of radioactive atoms N per unit of time is proportional to N:

$$\frac{\Delta N}{\Delta t} = -kN \tag{23.3}$$

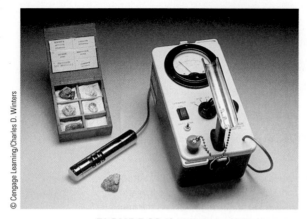
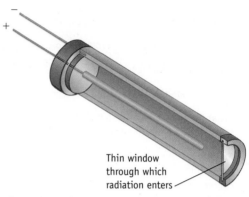

Thin window through which radiation enters

FIGURE 23.6 A Geiger–Müller counter. A charged particle (an α or β particle) enters the gas-filled tube (diagram at the *right*) and ionizes the gas. The gaseous ions migrate to electrically charged plates and are recorded as a pulse of electric current. The current is amplified and used to operate a counter. A sample of carnotite, a mineral containing uranium oxide, is also shown in the photograph.

The integrated rate equation can be written in two ways depending on the data used:

$$\ln\left(\frac{N}{N_0}\right) = -kt \tag{23.4}$$

or

$$\ln\left(\frac{A}{A_0}\right) = -kt \tag{23.5}$$

Here, N_0 and A_0 are the number of atoms and the activity of the sample initially, respectively, and N and A are the number of atoms and the activity of the sample after time t, respectively. Thus, N/N_0 is the fraction of atoms remaining after a given time (t), and A/A_0 is the fraction of the activity remaining after the same period. In these equations, k is the rate constant (decay constant) for the isotope in question. The relationship between half-life and the first-order rate constant is the same as seen with chemical kinetics (Equation 15.4, page 524):

$$t_{1/2} = \frac{0.693}{k} \tag{23.6}$$

Equations 23.3–23.6 are useful in several ways:

- If the activity (A) or the number of radioactive nuclei (N) is measured in the laboratory over some period t, then k can be calculated. The decay constant k can then be used to determine the half-life of the sample.
- If k is known, the fraction of a radioactive sample (N/N_0) still present after some time t has elapsed can be calculated.
- If k is known, the time required for that isotope to decay to a fraction of the original activity (A/A_0) can be calculated.

EXAMPLE 23.6   Kinetics of Radioactive Decay

Problem A sample of radon-222 has an initial α-particle activity (A_0) of 7.0×10^4 dps (disintegrations per second). After 6.6 days, its activity (A) is 2.1×10^4 dps. What is the half-life of radon-222?

What Do You Know? You are given the initial and final activities of the sample of ^{222}Rn and the time elapsed.

Strategy Values for A, A_0, and t are given. The problem can be solved using Equation 23.5 with k as the unknown. Once k is found, the half-life can be calculated using Equation 23.6.

Solution

$$\ln(2.1 \times 10^4 \text{ dps}/7.0 \times 10^4 \text{ dps}) = -k \text{ (6.6 day)}$$

$$\ln(0.30) = -k(6.6 \text{ day})$$

$$k = 0.18 \text{ days}^{-1}$$

From k we obtain $t_{1/2}$:

$$t_{1/2} = 0.693/0.18 \text{ days}^{-1} = 3.8 \text{ days}$$

Think about Your Answer Notice that the activity decreased to between one half and one fourth of its original value. The 6.6 days of elapsed time represents one full half-life and part of another half-life.

Check Your Understanding

(a) A sample of $Ca_3(PO_4)_2$ containing phosphorus-32 has an activity of 3.35×10^3 dpm. Exactly 2 days later, the activity is 3.18×10^3 dpm. Calculate the half-life of phosphorus-32.

(b) A highly radioactive sample of nuclear waste products with a half-life of 200. years is stored in an underground tank. How long will it take for the activity to diminish from an initial activity of 6.50×10^{12} dpm to a fairly harmless activity of 3.00×10^3 dpm?

Radiocarbon Dating

In certain situations, the age of a material can be determined based on the rate of decay of a radioactive isotope. The best-known example of this procedure is the use of carbon-14 to date historical artifacts.

Naturally occurring carbon is primarily carbon-12 and carbon-13 with isotopic abundances of 98.9% and 1.1%, respectively. In addition, traces of a third isotope, carbon-14, are present to the extent of about 1 in 10^{12} atoms in atmospheric CO_2 and in living materials. Carbon-14 is a β emitter with a half-life of 5730 years. A 1-gram sample of carbon from living material will show about 14 disintegrations per minute, not a lot of radioactivity but nevertheless detectable by modern methods.

Carbon-14 is formed in the upper atmosphere by nuclear reactions initiated by neutrons in cosmic radiation:

$$^{14}_{7}\text{N} + ^{1}_{0}\text{n} \rightarrow ^{14}_{6}\text{C} + ^{1}_{1}\text{H}$$

Once formed, carbon-14 is oxidized to $^{14}CO_2$. This product enters the carbon cycle, circulating through the atmosphere, oceans, and biosphere.

The usefulness of carbon-14 for dating comes about in the following way. Plants absorb CO_2 and convert it to organic compounds, thereby incorporating carbon-14 into living tissue. As long as a plant remains alive, this process will continue, and the percentage of carbon that is carbon-14 in the plant will equal the percentage in the atmosphere. When the plant dies, carbon-14 will no longer be taken up. Radioactive decay continues, however, with the carbon-14 activity decreasing over time. After 5730 years, the activity will be 7 dpm/g; after 11,460 years, it will be 3.5 dpm/g; and so on. By measuring the activity of a sample, and knowing the half-life of carbon-14, it is possible to calculate when a plant (or an animal that was eating plants) died.

As with all experimental procedures, carbon-14 dating has limitations. Although the procedure assumes that the amount of carbon-14 in the atmosphere hundreds or thousands of years ago is the same as it is now, in fact the percentage has varied by as much as 10% (Figure 23.7). Furthermore, it is not possible to use carbon-14 to date an object that is less than about 100 years old; the radiation level

FIGURE 23.7 **Variation of atmospheric carbon-14 activity.** The amount of carbon-14 has changed with the variation in cosmic ray activity. To obtain the data for the pre-1990 part of the curve shown in this graph, scientists carried out carbon-14 dating of artifacts for which the age was accurately known (often through written records). Data for this figure were obtained using carbon-14 dating of tree rings.

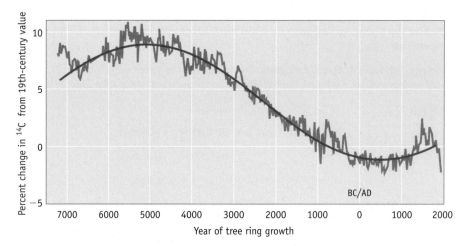

Source: Hans E. Suess, La Jolla Radiocarbon Laboratory

● **Willard Libby (1908–1980)** Libby received the 1960 Nobel Prize in chemistry for developing carbon-14 dating techniques. Carbon-14 dating is widely used in fields such as anthropology.

Oesper Collection in the History of Chemistry, University of Cincinnati

from carbon-14 will not change enough in this short time period to permit accurate detection of a difference from the initial value. In most instances, the accuracy of the measurement is, in fact, only about ±100 years. Finally, it is not possible to determine ages of objects much older than about 40,000 years. By then, after nearly seven half-lives, the radioactivity will have decreased virtually to zero. But for the span of time between 100 and 40,000 years, this technique has provided important information (Figure 23.8).

EXAMPLE 23.7   Radiochemical Dating

Problem To test the concept of carbon-14 dating, J. R. Arnold and W. F. Libby applied this technique to analyze samples of acacia and cyprus wood whose ages were already known. (The acacia wood, which was supplied by the Metropolitan Museum of Art in New York, came from the tomb of Zoser, the first Egyptian pharaoh to be entombed in a pyramid. The cyprus wood was from the tomb of Sneferu.) The average activity based on five determinations on one of these wood samples was 7.04 dpm per gram of carbon. Assume (as Arnold and Libby did) that the original activity of carbon-14, A_0, was 12.6 dpm per gram of carbon. Calculate the approximate age of the sample.

What Do You Know? You are given the initial and final ^{14}C activity. The first-order rate constant can be calculated from the half-life.

Strategy First, determine the rate constant for the decay of carbon-14 from its half-life ($t_{1/2}$ for ^{14}C is 5.73×10^3 years). Then, use Equation 23.5.

Solution

$$k = 0.693/t_{1/2} = 0.693/5730 \text{ yr}$$

$$= 1.21 \times 10^{-4} \text{ yr}^{-1}$$

$$\ln(A/A_0) = -kt$$

$$\ln\left(\frac{7.04 \text{ dpm/g}}{12.6 \text{ dpm/g}}\right) = (-1.21 \times 10^{-4} \text{ yr}^{-1})t$$

$$t = 4.8 \times 10^3 \text{ yr}$$

The wood is about 4800 years old.

Think about Your Answer This problem uses real data from an early research paper in which the carbon-14 dating method was being tested. The age of the wood was known to be 4750 ± 250 years. (See J. R. Arnold and W. F. Libby: *Science*, Vol. 110, p. 678, 1949.)

© JEAN LOUIS PRADELS/PHOTOPQR/ LA DEPECHE DU MIDI/NEWSCOM

FIGURE 23.8 **The Iceman.** The world's oldest preserved human remains were discovered in the ice of a glacier high in the Alps. Carbon-14 dating techniques allowed scientists to determine that he lived about 5300 years ago. See pages 50 and 1 for more information on the Iceman.

Check Your Understanding

A sample of the inner part of a redwood tree felled in 1874 was shown to have ^{14}C activity of 9.32 dpm/g. Calculate the approximate age of the tree when it was cut down. Compare this age with that obtained from tree ring data, which estimated that the tree began to grow in 979 $\pm$ 52 BC. Use 13.4 dpm/g for the value of A_0.

REVIEW & CHECK FOR SECTION 23.4

1. The age of the earth is thought to be 4.5 billion years. Approximately what percentage of ^{232}Th has decomposed during this period of time? The half-life of ^{232}Th is 1.4×10^{10} years.

   (a) 20%   (b) 40%   (c) 60%   (d) 80%

2. The half-life of ^{32}P is 14.3 days. How much ^{32}P in a sample will remain after 71.5 days have elapsed?

   (a) 1/4   (b) 1/8   (c) 1/16   (d) 1/32

23.5 Artificial Nuclear Reactions

How many different isotopes are found on earth? All of the stable isotopes occur naturally. A few unstable (radioactive) isotopes that have long half-lives are found in nature; the best-known examples are uranium-235, uranium-238, and thorium-232. Trace quantities of other radioactive isotopes with short half-lives are present because they are being formed continuously by nuclear reactions. They include isotopes of radium, polonium, and radon, along with other elements produced in various radioactive decay series, and carbon-14, formed in a nuclear reaction initiated by cosmic radiation.

Naturally occurring isotopes account for only a very small fraction of the currently known radioactive isotopes, however. The rest—several thousand—have been synthesized via artificial nuclear reactions, sometimes referred to as transmutation.

The first artificial nuclear reaction was identified by Rutherford about 90 years ago. Recall the classic experiment that led to the nuclear model of the atom in which gold foil was bombarded with α particles. In the years following that experiment, Rutherford and his coworkers bombarded many other elements with α particles. In 1919, one of these experiments led to an unexpected result: When nitrogen atoms were bombarded with α particles, protons were detected among the products. Rutherford correctly concluded that a nuclear reaction had occurred. Nitrogen had undergone a *transmutation* to oxygen:

$$\,_2^4He + \,_7^{14}N \rightarrow \,_8^{17}O + \,_1^1H$$

During the next decade, other nuclear reactions were discovered by bombarding other elements with α particles. Progress was slow, however, because in most cases α particles are simply scattered by target nuclei. The bombarding particles cannot get close enough to the nucleus to react because of the strong repulsive forces between the positively charged α particle and the positively charged atomic nucleus.

Two advances were made in 1932 that greatly extended nuclear reaction chemistry. The first involved the use of particle accelerators to create high-energy particles as projectiles. The second was the use of neutrons as the bombarding particles.

The α particles used in the early studies on nuclear reactions came from naturally radioactive materials such as uranium and had relatively low energies, at least by today's standards. Particles with higher energy were needed, so J. D. Cockcroft (1897–1967) and E. T. S. Walton (1903–1995), working in Rutherford's laboratory in Cambridge, England, turned to protons. Protons are formed when hydrogen atoms ionize in a cathode-ray tube, and it was known that they could be accelerated to

● **Discovery of Neutrons** Neutrons had been predicted to exist for more than a decade before they were identified in 1932 by James Chadwick (1891–1974). Chadwick produced neutrons in a nuclear reaction between α particles and beryllium: $\,_2^4\alpha + \,_4^9Be \rightarrow \,_6^{12}C + \,_0^1n$.

● **Glenn T. Seaborg (1912–1999)** Seaborg figured out that thorium and the elements that followed it fit under the lanthanides in the periodic table. For this insight, he and Edwin McMillan shared the 1951 Nobel Prize in chemistry. Over a 21-year period, Seaborg and his colleagues synthesized 10 new transuranium elements. To honor Seaborg's scientific contributions, the name "seaborgium" was assigned to element 106. It marked the first time an element was named for a living person.

Lawrence Berkeley Laboratory

The Search for New Elements

By 1936, guided first by Mendeleev's predictions and later by atomic theory, chemists had identified all but two of the elements with atomic numbers between 1 and 92. From this point onward, all new elements to be discovered came from artificial nuclear reactions. Two gaps in the periodic table were filled when radioactive technetium and promethium, the last two elements with atomic numbers less than 92, were identified in 1937 and 1942, respectively. The first success in the search for elements with atomic numbers higher than 92 came with the 1940 discovery of neptunium and plutonium.

Since 1950, laboratories in the United States (Lawrence Berkeley National Laboratory), Russia (Joint Institute for Nuclear Research at Dubna, near Moscow), and Europe (Institute for Heavy Ion Research at Darmstadt, Germany) have competed to make new elements. Syntheses of new transuranium elements use a standard methodology. An element of fairly high atomic number is bombarded with a beam of high-energy particles. Initially, neutrons were used; later, helium nuclei and then larger nuclei such as ^{11}B and ^{12}C were used; and, more recently, highly charged ions of elements such as calcium, chromium, cobalt, and zinc have been chosen. The bombarding particle fuses with the nucleus of the target atom, forming a new nucleus that lasts for a short time before decomposing. New elements are detected by their decomposition products, a signature of particles with specific masses and energies.

By using bigger particles and higher energies, the list of known elements reached 106 by the end of the 1970s. To further extend the search, Russian scientists used a new idea, matching precisely the energy of the bombarding particle with the energy required to fuse the nuclei. This technique enabled the synthesis of elements 107, 108, and 109 in Darmstadt in the early 1980s, and the synthesis of elements 110, 111, and 112 in the following decade. Lifetimes of these elements were in the millisecond range; copernicium-277, $^{277}_{112}Cn$, for example, has a half-life of 240 μs.

Yet another breakthrough was needed to extend the list further. Scientists have long known that isotopes with specific so-called *magic numbers* of neutrons and protons are more stable. Elements with 2, 8, 20, 50, and 82 protons are members of this category, as are elements with 126 neutrons. The magic numbers correspond to filled shells in the nucleus. Their significance is analogous to the significance of filled shells for electronic structure. Theory had predicted that the next magic numbers would be 114 protons and 184 neutrons. Using this information, researchers discovered element 114 in early 1999. The Dubna group reporting this discovery found that the mass 289 isotope had an exceptionally long half-life, about 20 seconds.

Work along these lines is continuing and has resulted in the discovery of elements 113, 115, 116 and 118. Then, as we were preparing this edition of the book, the discovery of element 117, Uus, was announced (on April 19, 2010). An international team working at the Flerov Laboratories in Dubna, Russia, described the formation of 6 atoms of this element by bombarding ^{249}Bk with ^{48}Ca atoms. (The isotopes ^{293}Uus and ^{294}Uus were detected.) Assuming confirmation of this element from other labs will follow, the periodic table now includes all the elements from 1 to 118!

Footnote: Until an official name is chosen for element 117, it will be known by a placeholder name, ununseptium, and the corresponding three-letter symbol Uus. This name simply stands for "one-one-seven," because the atomic number of the element is 117.

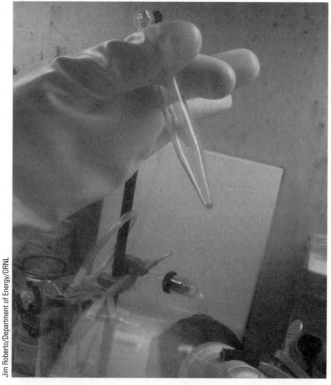

Jim Roberto/Department of Energy/ORNL

Discovery of element 117. Scientists at the Flerov Laboratories in Dubna, Russia, used radioactive berkelium-249 as the starting material to make two isotopes of element 117. Berkelium-249 is contained in the greenish fluid in the tip of the vial. It was made in the research reactor at Oak Ridge National Laboratory.

higher energy by applying a high voltage. Cockcroft and Walton found that when energetic protons struck a lithium target, the following reaction occurs:

$$^{7}_{3}Li + ^{1}_{1}p \rightarrow 2\ ^{4}_{2}He$$

This was the first example of a reaction initiated by a particle that had been artificially accelerated to high energy. Since this experiment was conducted, the technique has been developed much further, and the use of particle accelerators in nuclear chemistry is now commonplace. Particle accelerators operate on the principle that a charged particle placed between charged plates will be accelerated to a high speed and high energy. Modern examples of this process are seen in the synthesis of the transuranium elements, several of which are described in more detail in *A Closer Look: The Search for New Elements*, on the previous page.

Experiments using neutrons as bombarding particles were first carried out in both the United States and Great Britain in 1932. Nitrogen, oxygen, fluorine, and neon were bombarded with energetic neutrons, and α particles were detected among the products. Using neutrons made sense: Because neutrons have no charge, it was reasoned that these particles would not be repelled by the positively charged nuclear particles. Thus, neutrons did not need high energies to react.

In 1934, Enrico Fermi (1901–1954) and his coworkers showed that nuclear reactions using neutrons are more favorable if the neutrons have low energy. A low-energy neutron is simply captured by the nucleus, giving a product in which the mass number is increased by one unit. Because of the low energy of the bombarding particle, the product nucleus does not have sufficient energy to fragment in these reactions. The new nucleus is produced in an excited state, however; when the nucleus returns to the ground state, a γ-ray is emitted. Reactions in which a neutron is captured and a γ-ray is emitted are called **(n, γ) reactions**.

The (n, γ) reactions are the source of many of the radioisotopes used in medicine and chemistry. An example is radioactive phosphorus, $^{32}_{15}P$, which is used in chemical studies such as tracing the uptake of phosphorus in the body.

$$^{31}_{15}P + ^{1}_{0}n \rightarrow ^{32}_{15}P + \gamma$$

Transuranium elements, elements with an atomic number greater than 92, were first made in a nuclear reaction sequence beginning with an (n, γ) reaction. Scientists at the University of California at Berkeley bombarded uranium-238 with neutrons. Among the products identified were neptunium-239 and plutonium-239. These new elements were formed when ^{239}U decayed by β radiation.

$$^{238}_{92}U + ^{1}_{0}n \rightarrow ^{239}_{92}U$$

$$^{239}_{92}U \rightarrow ^{239}_{93}Np + ^{0}_{-1}\beta$$

$$^{239}_{93}Np \rightarrow ^{239}_{94}Pu + ^{0}_{-1}\beta$$

Four years later, a similar reaction sequence was used to make americium-241. Plutonium-239 was found to add two neutrons to form plutonium-241, which decays by β emission to give americium-241.

● **Transuranium Elements in Nature** Neptunium, plutonium, and americium were unknown prior to their preparation via these nuclear reactions. Later, these elements were found to be present in trace quantities in uranium ores.

EXAMPLE 23.8 Nuclear Reactions

Problem Write equations for the nuclear reactions described below.

(a) Fluorine-19 undergoes an (n, γ) reaction to give a radioactive product that decays by β emission. (Write equations for both nuclear reactions.)

(b) When an atom of beryllium-9 (the only stable isotope of beryllium) reacts with an alpha particle emitted by a plutonium atom, a neutron is ejected.

What Do You Know? Reactants and one of two products are given for each reaction.

Strategy The equations are written so that both mass and charge are balanced.

Solution

(a) $^{19}_{9}F + ^{1}_{0}n \rightarrow ^{20}_{9}F + \gamma$

$^{20}_{9}F \rightarrow ^{20}_{10}Ne + ^{0}_{-1}\beta$

(b) $^{239}_{94}Pu \rightarrow ^{235}_{92}U + ^{4}_{2}\alpha$

$^{4}_{2}\alpha + ^{9}_{4}Be \rightarrow ^{12}_{6}C + ^{1}_{0}n$

Think about Your Answer Both answers deal with neutrons. The equation given for part (a), an (n, γ) reaction, illustrates a process that is easy to carry out. In a lab, the neutrons might be produced in a small device called a *Pu-Be neutron source*. The answer to part (b) of this question describes the nuclear reactions in such a device.

Check Your Understanding

Technetium is one of the two elements with atomic numbers less than 83 for which there is no stable isotope (promethium, element 61, is the other). Nevertheless, technetium is a very important element because of its extensive use in medical imaging (page 847). It is produced in a two-step process. First, ^{98}Mo undergoes an (n, γ) reaction; then, the resulting unstable isotope decomposes to ^{99}Tc. Write equations for these two reactions.

REVIEW & CHECK FOR SECTION 23.5

Which of the following is seldom used as a bombarding particle to initiate a nuclear reaction?

(a) protons   (b) neutrons   (c) electrons   (d) alpha particles

---

23.6 Nuclear Fission

In 1938, two chemists, Otto Hahn (1879–1968) and Fritz Strassman (1902–1980), isolated and identified barium in a sample of uranium that had been bombarded with neutrons. How was barium formed? The answer to that question explained one of the most significant scientific discoveries of the 20th century. The uranium nucleus had split into smaller pieces in the process we now call **nuclear fission**.

The details of nuclear fission were unraveled through the work of a number of scientists. They determined that a uranium-235 nucleus initially captured a neutron to form uranium-236. This isotope underwent nuclear fission to produce two new nuclei, one with a mass number around 140 and the other with a mass around 90, along with several neutrons (Figure 23.9). The nuclear reactions that led to formation of barium when a sample of ^{235}U was bombarded with neutrons are

$$^{235}_{92}\text{U} + {}^{1}_{0}\text{n} \rightarrow {}^{236}_{92}\text{U}$$

$$^{236}_{92}\text{U} \rightarrow {}^{141}_{56}\text{Ba} + {}^{92}_{36}\text{Kr} + 3\,{}^{1}_{0}\text{n}$$

An important aspect of fission reactions is that they produce more neutrons than are used to initiate the process. Under the right circumstances, these neutrons then serve to continue the reaction. If one or more of these neutrons are captured by another ^{235}U nucleus, then a further reaction can occur, releasing still more neutrons. This sequence repeats over and over. Such a mechanism, in which each step generates a reactant to continue the reaction, is called a **chain reaction**.

● **The Atomic Bomb** In an atomic bomb, each nuclear fission step produces 3 neutrons, which leads to about 3 more fissions and 9 more neutrons, which leads to 9 more fission steps and 27 more neutrons, and so on. The rate of fission depends on the number of neutrons, so the nuclear reaction occurs faster and faster as more and more neutrons are formed, leading to an enormous output of energy in a short time span.

FIGURE 23.9 Nuclear fission. Neutron capture by $^{235}_{92}$U produces $^{236}_{92}$U. This isotope undergoes fission, which yields several fragments along with several neutrons. These neutrons initiate further nuclear reactions by adding to other $^{235}_{92}$U nuclei. The process is highly exothermic, producing about 2×10^{10} kJ/mol.

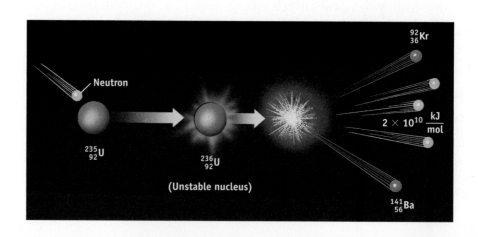

A nuclear fission chain reaction has three general steps:

1. *Initiation.* The reaction of a single atom is needed to start the chain. Fission of ^{235}U is initiated by the absorption of a neutron.
2. *Propagation.* This part of the process repeats itself over and over, with each step yielding more product. The fission of ^{236}U releases neutrons that initiate the fission of other uranium atoms.
3. *Termination.* Eventually, the chain will end. Termination could occur if the reactant (^{235}U) is used up, or if the neutrons that continue the chain escape from the sample without being captured by ^{235}U.

To harness the energy produced in a nuclear reaction, it is necessary to control the rate at which a fission reaction occurs. This is managed by balancing the propagation and termination steps by limiting the number of neutrons available. In a nuclear reactor, this balance is accomplished by using cadmium rods to absorb neutrons. By withdrawing or inserting the rods, the number of neutrons available to propagate the chain can be changed, and the rate of the fission reaction (and the rate of energy production) can be increased or decreased.

Uranium-235 and plutonium-239 are the fissionable isotopes most commonly used in power reactors. Natural uranium contains only 0.72% of uranium-235; more than 99% of the natural element is uranium-238. The percentage of uranium-235 in natural uranium is too small to sustain a chain reaction, however, so the uranium used for nuclear fuel must be enriched in this isotope. One way to do so is by gaseous centrifugation (Figure 23.10). Plutonium, on the other hand, occurs naturally in only trace quantities and so must be made by a nuclear reaction. The raw material for this nuclear synthesis is the more abundant uranium isotope, ^{238}U. Addition of a neutron to ^{238}U gives ^{239}U, which, as noted earlier, undergoes two β emissions to form ^{239}Pu.

Currently, there are over 100 operating nuclear power plants in the United States and more than 400 worldwide. About 20% of this country's electricity (and 17% of the world's energy) comes from nuclear power (Figure 23.11). Although one might imagine that nuclear energy would be called upon to meet the ever-increasing needs of society, no new nuclear power plants are under construction in the United States. Among other things, the disasters at Chernobyl (in the former Soviet Union) in 1986 and Three Mile Island (in Pennsylvania) in 1979 have sensitized the public to the issue of safety. The cost to construct a nuclear power plant (measured in terms of dollars per kilowatt-hour of power) is considerably more than the cost for a natural gas–powered facility, and there are severe regulatory restrictions for nuclear power. Disposal of highly radioactive nuclear waste is another

● **Lise Meitner (1878–1968)**
Meitner's greatest contribution to 20th-century science was her explanation of the process of nuclear fission. She and her nephew, Otto Frisch, also a physicist, published a paper in 1939 that was the first to use the term *nuclear fission.* Element number 109 is named meitnerium to honor Meitner's contributions. The leader of the team that discovered this element said that "She should be honored as the most significant woman scientist of [the 20th] century."

AIP-Emilio Segré Visual Archives, Herzfeld Collection

Oak Ridge National Laboratory

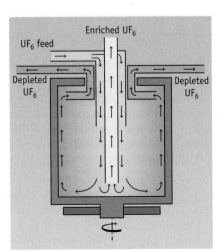

FIGURE 23.10 Isotope separation. *(left)* Separation of uranium isotopes for use in atomic weaponry or in nuclear power plants is done with gas centrifuges. *(right)* UF_6 gas is injected into the centrifuge from a tube passing down through the center of a tall, spinning cylinder. The heavier $^{238}UF_6$ molecules experience more centrifugal force and move to the outer wall of the cylinder; the lighter $^{235}UF_6$ molecules stay closer to the center. A temperature difference inside the rotor causes the $^{235}UF_6$ molecules to move to the top of the cylinder and the $^{238}UF_6$ molecules to move to the bottom. (See the *New York Times,* page F1, March 23, 2004.)

FIGURE 23.11 Percentage of electricity generation by nuclear power in various countries.
This varies from 76.2% in France, to 19.7% in the United States, to a little more than 2% in China and India. (Data from "Energy, Electricity and Nuclear Power Estimates for the Period up to 2030" from the International Atomic Energy Agency, 2009 edition.)

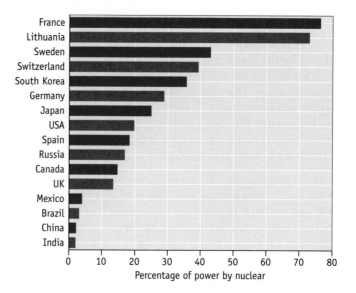

thorny problem, with 20 metric tons of waste being generated per year at each reactor.

In addition to technical problems, nuclear energy production brings with it significant geopolitical security concerns. The process for enriching uranium for use in a reactor is the same process used for generating weapons-grade uranium. Also, some nuclear reactors are designed so that one by-product of their operation is the isotope plutonium-239, which can be removed and used in a nuclear weapon. Despite these problems, nuclear fission is an important part of the energy profile in a number of countries. For example, 76.2% of power production in France and 24.9% in Japan is nuclear generated.

REVIEW & CHECK FOR SECTION 23.6

Identify the other element generated in this fission reaction: $^{235}_{92}U + ^{1}_{0}n \rightarrow ^{141}_{55}Cs + ? + 2^{1}_{0}n$

(a) $^{93}_{37}Rb$ (b) $^{141}_{56}Ba$ (c) $^{92}_{36}Kr$

23.7 Nuclear Fusion

In a **nuclear fusion** reaction, several small nuclei react to form a larger nucleus. Tremendous amounts of energy can be generated by such reactions. An example is the fusion of deuterium and tritium nuclei to form $^{4}_{2}He$ and a neutron:

$$^{2}_{1}H + ^{3}_{1}H \rightarrow ^{4}_{2}He + ^{1}_{0}n \qquad \Delta E = -1.7 \times 10^{9} \text{ kJ/mol}$$

Fusion reactions provide the energy of our sun and other stars. Scientists have long dreamed of being able to harness fusion to provide power. To do so, a temperature of 10^6 to 10^7 K, like that in the interior of the sun, would be required to bring the positively charged nuclei together with enough energy to overcome nuclear repulsions. At the very high temperatures needed for a fusion reaction, matter does not exist as atoms or molecules; instead, matter is in the form of a *plasma* made up of unbound nuclei and electrons.

Three critical requirements must be met before nuclear fusion could represent a viable energy source. First, the temperature must be high enough for fusion to occur. The fusion of deuterium and tritium, for example, requires a temperature of 10^7 K or more. Second, the plasma must be confined long enough to release a net output of energy. Third, the energy must be recovered in some usable form.

Harnessing a nuclear fusion reaction for a peaceful use has not yet been achieved. Nevertheless, many attractive features encourage continuing research in this field. The hydrogen used as "fuel" is cheap and available in almost unlimited amounts. As a further benefit, most radioisotopes produced by fusion have short half-lives, so they remain a radiation hazard for only a short time.

23.8 Radiation Health and Safety
Units for Measuring Radiation

Several units of measurement are used to describe levels and doses of radioactivity. Generally the units used in the United States are not the same as the SI units of measurement.

In the United States, the degree of radioactivity is often measured in **curies** (Ci). Less commonly used in the United States is the SI unit, the **becquerel** (Bq). Both units measure the number of disintegrations per second; 1 Ci is 3.7×10^{10} dps (disintegrations per second), while 1 Bq represents 1 dps. The curie and the becquerel are used to report the amount of radioactivity when multiple kinds of unstable nuclei are decaying and to report amounts necessary for medical purposes.

By itself, the degree of radioactivity does not provide a good measure of the amount of energy in the radiation or the amount of damage that the radiation can cause to living tissue. Two additional kinds of information are necessary. The first is the amount of energy absorbed; the second is the effectiveness of the particular kind of radiation in causing tissue damage. The amount of energy absorbed by living tissue is measured in **rads**. *Rad* is an acronym for "radiation absorbed dose." One rad represents 0.01 J of energy absorbed per kilogram of tissue. Its SI equivalent is the **gray** (Gy); 1 Gy denotes the absorption of 1 J per kilogram of tissue.

Different forms of radiation cause different amounts of biological damage. The amount of damage depends on how strongly a form of radiation interacts with matter. Alpha particles cannot penetrate the body any farther than the outer layer of skin. If α particles are emitted within the body, however, they will cause between 10 and 20 times the amount of damage done by γ-rays, which can go entirely through a human body without being stopped. In determining the amount of biological damage to living tissue, differences in damaging power are accounted for using a "quality factor." This quality factor has been set at 1 for β and γ radiation, 5 for low-energy protons and neutrons, and 20 for α particles or high-energy protons and neutrons.

Biological damage is quantified in a unit called the **rem** (an acronym for "roentgen equivalent man"). A dose of radiation in rem is determined by multiplying the energy absorbed in rads by the quality factor for that kind of radiation. The rad and the rem are very large in comparison to normal exposures to radiation, so it is more common to express exposures in millirems (mrem). The SI equivalent of the rem is the **sievert** (Sv), determined by multiplying the dose in grays by the quality factor.

Radiation: Doses and Effects

Exposure to a small amount of radiation is unavoidable. Earth is constantly being bombarded with radioactive particles from outer space. There is also some exposure to radioactive elements that occur naturally on earth, including ^{14}C, ^{40}K (a radioactive isotope that occurs naturally in 0.0117% abundance), ^{238}U, and ^{232}Th. Radioactive elements in the environment that were created artificially (in the fallout from nuclear bomb tests, for example) also contribute to this exposure. For some people, medical procedures using radioisotopes are a major contributor.

The average dose of background radioactivity to which a person in the United States is exposed is about 200 mrem per year (Table 23.2). Well over half of that amount comes from natural sources over which we have no control. Of the 60–70 mrem per year exposure that comes from artificial sources, nearly 90% is delivered in medical procedures such as x-ray examinations and radiation therapy. Considering the controversy surrounding nuclear power, it is interesting to note that less than 0.5% of the total annual background dose of radiation that the average person receives can be attributed to the nuclear power industry.

● **The Roentgen** The roentgen (R) is an older unit of radiation exposure. It is defined as the amount of x-rays or γ radiation that will produce 2.08×10^9 ions in 1 cm³ of dry air. The roentgen and the rad are similar in size. Wilhelm Roentgen (1845–1923) first produced and detected x radiation. Element 111 has been named roentgenium in his honor.

Table 23.2 Radiation Exposure of an Individual for One Year from Natural and Artificial Sources

	Millirem/Year	Percentage
Natural Sources		
Cosmic radiation	50.0	25.8
The earth	47.0	24.2
Building materials	3.0	1.5
Inhaled from the air	5.0	2.6
Elements found naturally in human tissue	21.0	10.8
Subtotal	**126.0**	**64.9**
Medical Sources		
Diagnostic x-rays	50.0	25.8
Radiotherapy	10.0	5.2
Internal diagnosis	1.0	0.5
Subtotal	**61.0**	**31.5**
Other Artificial Sources		
Nuclear power industry	0.85	0.4
Luminous watch dials, TV tubes	2.0	1.0
Fallout from nuclear tests	4.0	2.1
Subtotal	**6.9**	**3.5**
Total	**193.9**	**99.9**

Describing the biological effects of a dose of radiation precisely is not a simple matter. The amount of damage done depends not only on the kind of radiation and the amount of energy absorbed, but also on the particular tissues exposed and the rate at which the dose builds up. A great deal has been learned about the effects of radiation on the human body by studying the survivors of the bombs dropped over Japan in World War II and the workers exposed to radiation from the reactor disaster at Chernobyl. From studies of the health of these survivors, we have learned that the effects of radiation are not generally observable below a single dose of 25 rem. At the other extreme, a single dose of >200 rem will be fatal to about half the population (Table 23.3).

Our information is more accurate when dealing with single, large doses than it is for the effects of chronic, smaller doses of radiation. One current issue of debate in the scientific community is how to judge the effects of multiple smaller doses or long-term exposure (see *A Closer Look: What Is A Safe Exposure?*).

Table 23.3 Effects of a Single Dose of Radiation

Dose (rem)	Effect
0–25	No effect observed
26–50	Small decrease in white blood cell count
51–100	Significant decrease in white blood cell count, lesions
101–200	Loss of hair, nausea
201–500	Hemorrhaging, ulcers, death in 50% of population
500	Death

A CLOSER LOOK

What Is a Safe Exposure?

Is the exposure to natural background radiation totally without effect? Can you equate the effect of a single dose and the effect of cumulative, smaller doses that are spread out over a long period of time? The assumption generally made is that no "safe maximum dose," or level below which absolutely no damage will occur, exists. However, the accuracy of this assumption has come into question. These issues are not testable with human subjects, and tests based on animal studies are not completely reliable because of the uncertainty of species-to-species variations.

The model used by government regulators to set exposure limits assumes that the relationship between exposure to radiation and incidence of radiation-induced problems, such as cancer, anemia, and immune system problems, is linear. Under this assumption, if a dose of 2*x* rem causes damage in 20% of the population, then a dose of *x* rem will cause damage in 10% of the population. But is this true? Cells do possess mechanisms for repairing damage. Many scientists believe that this self-repair mechanism renders the human body less susceptible to damage from smaller doses of radiation, because the damage will be repaired as part of the normal course of events. They argue that, at extremely low doses of radiation, the self-repair response results in less damage.

The bottom line is that much still remains to be learned in this area. And the stakes are significant.

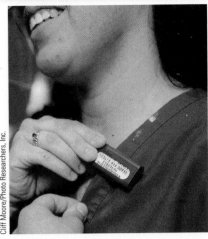

The film badge. These badges, worn by scientists using radioactive materials, are used to monitor cumulative exposure to radiation.

Cliff Moore/Photo Researchers, Inc.

23.9 Applications of Nuclear Chemistry

We tend to think about nuclear chemistry in terms of power plants and bombs. In truth, radioactive elements are now used in many areas of science and medicine, and they are of ever-increasing importance to our lives. Because describing all of their uses would take several books, we have selected just a few examples to illustrate the diversity of applications of radioactivity.

Nuclear Medicine: Medical Imaging

Diagnostic procedures using nuclear chemistry are essential in medical imaging, which entails the creation of images of specific parts of the body. There are three principal components to constructing a radioisotope-based image:

- A radioactive isotope, administered as the element or incorporated into a compound, that concentrates the radioactive isotope in the tissue to be imaged
- A method of detecting the type of radiation involved
- A computer to assemble the information from the detector into a meaningful image

The choice of a radioisotope and the manner in which it is administered are determined by the tissue in question. A compound containing the isotope must be absorbed more by the target tissue than by the rest of the body. Table 23.4 lists radioisotopes that are commonly used in nuclear imaging processes, their half-lives, and

Table 23.4 Radioisotopes Used in Medical Diagnostic Procedures

Radioisotope	Half-Life (h)	Imaging
^{99m}Tc	6.0	Thyroid, brain, kidneys
^{201}Tl	73.0	Heart
^{123}I	13.2	Thyroid
^{67}Ga	78.2	Various tumors and abscesses
^{18}F	1.8	Brain, sites of metabolic activity

A CLOSER LOOK

Technetium-99m

Technetium was the first new element to be made artificially. One might think that this element would be a chemical rarity, but this is not so: Its importance in medical imaging has brought a great deal of attention to it. Although all the technetium in the world must be synthesized by nuclear reactions, the element is readily available and even inexpensive; its price is about $60 per gram, only a bit more than gold (in 2010).

Technetium-99m is formed when molybdenum-99 decays by β emission. Technetium-99m then decays to its ground state with a half-life of 6.01 hours, giving off a 140-KeV γ-ray in the process. (Technetium-99 is itself unstable, decaying to stable ^{99}Ru with a half-life of 2.1×10^5 years.)

Technetium-99m is produced in hospitals using a molybdenum–technetium generator. Sheathed in lead shielding, the generator contains the artificially synthesized isotope ^{99}Mo in the form of molybdate ion, MoO_4^{2-}, adsorbed on a column of alumina, Al_2O_3. The MoO_4^{2-} is continually being converted into the pertechnate ion, $^{99m}TcO_4^-$ by β emission. When it is needed, the $^{99m}TcO_4^-$ is washed from the column using a saline solution. Technetium-99m may be used directly as the pertechnate ion or converted into other compounds. The pertechnate ion or radiopharmaceuticals made from it are administered intravenously to the patient. Such small quantities are needed that 1 μg (microgram) of technetium-99m is sufficient for the average hospital's daily imaging needs.

One use of ^{99m}Tc is for imaging the thyroid gland. Because I^-(aq) and TcO_4^-(aq) ions have very similar sizes, the thyroid will (mistakenly) take up TcO_4^-(aq) along with iodide ion. This uptake concentrates ^{99m}Tc in the thyroid and allows a physician to obtain images such as the one shown in the accompanying Figure.

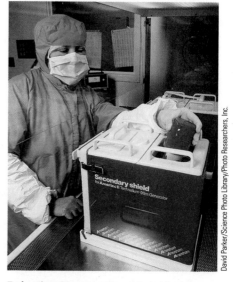

Technetium-99m generator. A technician is handling a sample containing the MoO_4^{2-} that will be converted to $^{99m}TcO_4^-$.

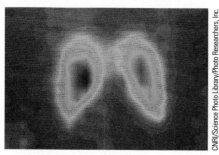

(a) Healthy human thyroid gland.

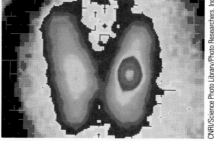

(b) Thyroid gland showing effect of hyperthyroidism.

Thyroid imaging. Technetium-99m concentrates in sites of high activity. Images of this gland, which is located at the base of the neck, were obtained by recording γ-ray emission after the patient was given radioactive technetium-99m. Current technology creates a computer color-enhanced scan.

the tissues they are used to image. All of the isotopes in Table 23.4 are γ emitters; γ radiation is preferred for imaging because it is less damaging to the body in small doses than either α or β radiation.

Technetium-99m is used in more than 85% of the diagnostic scans done in hospitals each year (see *A Closer Look: Technetium-99m.* The "m" stands for *metastable*, a term used to identify an unstable state that exists for a finite period of time. Recall that atoms in excited electronic states emit visible, infrared, and ultraviolet radiation (◄ Chapter 6). Similarly, a nucleus in an excited state gives up its excess energy, but in this case a much higher energy is involved, and the emission occurs as γ radiation. The γ-rays given off by ^{99m}Tc are detected to produce the image (Figure 23.12).

Another medical imaging technique based on nuclear chemistry is *positron emission tomography* (PET). In PET, an isotope that decays by positron emission is incorporated into a carrier compound and given to the patient. When emitted, the positron travels no more than a few millimeters before undergoing matter–antimatter annihilation.

$$^{0}_{+1}\beta + ^{0}_{-1}e \rightarrow 2\gamma$$

The two emitted γ-rays travel in opposite directions. By determining where high numbers of γ-rays are being emitted, one can construct a map showing where the positron emitter is located in the body.

An isotope often used in PET is ^{15}O. A patient is given gaseous O_2 that contains ^{15}O. This isotope travels throughout the body in the bloodstream, allowing images of the brain and bloodstream to be obtained (Figure 23.13). Because positron emitters are typically very short-lived, PET facilities must be located near a cyclotron where the radioactive nuclei are prepared and then immediately incorporated into a carrier compound.

Nuclear Medicine: Radiation Therapy

To treat most cancers, it is necessary to use radiation that can penetrate the body to the location of the tumor. Gamma radiation from a cobalt-60 source is commonly used. Unfortunately, the penetrating ability of γ-rays makes it virtually impossible to destroy diseased tissue without also damaging healthy tissue in the process. Nevertheless, this technique is a regularly sanctioned procedure, and its successes are well known.

Analytical Methods: The Use of Radioactive Isotopes as Tracers

Radioactive isotopes can be used to help determine the fate of compounds in the body or in the environment. These studies begin with a compound that contains a radioactive isotope of one of its component elements. In biology, for example, scientists can use radioactive isotopes to measure the uptake of nutrients. Plants take up phosphorus-containing compounds from the soil through their roots. By adding a small amount of radioactive ^{32}P, a β emitter with a half-life of 14.3 days, to fertilizer and then measuring the rate at which the radioactivity appears in the leaves, plant biologists can determine the rate at which phosphorus is taken up. The outcome can assist scientists in identifying hybrid strains of plants that can absorb phosphorus quickly, resulting in faster-maturing crops, better yields per acre, and more food or fiber at less expense.

To measure pesticide levels, a pesticide can be tagged with a radioisotope and then applied to a test field. By counting the disintegrations of the radioactive tracer, information can be obtained about how much pesticide accumulates in the soil, is taken up by the plant, and is carried off in runoff surface water. After these tests are completed, the radioactive isotope decays to harmless levels in a few days or a few weeks because of the short half-lives of the isotopes used.

Analytical Methods: Isotope Dilution

Imagine, for the moment, that you wanted to estimate the volume of blood in an animal subject. How might you do this? Obviously, draining the blood and measuring its volume in volumetric glassware is not a desirable option.

One technique uses a method called *isotope dilution*. In this process, a small amount of radioactive isotope is injected into the bloodstream. After a period of time to allow the isotope to become distributed throughout the body, a blood sample is taken and its radioactivity measured. The calculation used to determine the total blood volume is illustrated in the next example.

EXAMPLE 23.9   Analysis Using Radioisotopes

Problem A 1.00-mL solution containing 0.240 μCi of tritium is injected into a dog's bloodstream. After a period of time to allow the isotope to be dispersed, a 1.00-mL sample of blood is drawn. The radioactivity of this sample is found to be 4.3×10^{-4} μCi/mL. What is the total volume of blood in the dog?

What Do You Know? You know the concentration (activity) and volume of a concentrated solution and you then measure the concentration of the dilute solution. The unknown is the volume of the dilute solution.

Strategy In this problem, we relate the activity of the sample (in Ci) to the amount of the radioisotope present. The total amount of solute is 0.240 μCi, and the concentration (measured on the small sample of blood) is 4.3×10^{-4} μCi/mL. The unknown is the total volume of blood, V.

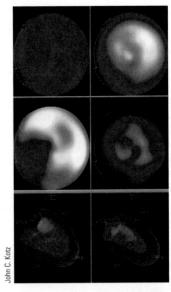

John C. Kotz

FIGURE 23.12 Heart imaging with technetium-99m. The radioactive element technetium-99m, a γ emitter, is injected into a patient's vein in the form of the pertechnetate ion (TcO_4^-) or as a complex ion with an organic ligand. A series of scans of the γ emissions of the isotope are made while the patient is resting and then again after strenuous exercise. Bright areas in the scans indicate that the isotope is binding to the tissue in that area. The scans in this figure show a normal heart function for one of the authors of this book.

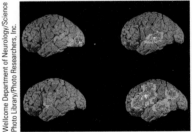

Wellcome Department of Neurology/Science Photo Library/Photo Researchers, Inc.

FIGURE 23.13 PET scans of the brain. These scans show the left side of the brain; *red* indicates an area of highest activity. *(upper left) Sight* activates the visual area in the occipital cortex at the back of the brain. *(upper right) Hearing* activates the auditory area in the superior temporal cortex of the brain. *(lower left) Speaking* activates the speech centers in the insula and motor cortex. *(lower right) Thinking* about verbs, and speaking them, generates high activity, including in the hearing, speaking, temporal, and parietal areas.

● **Boron Neutron Capture Therapy (BNCT).** To avoid the side effects associated with more traditional forms of radiation therapy, BNCT is an experimental treatment that has been explored in the last 10 to 15 years. BNCT is unusual in that boron-10, the isotope of boron used as part of the treatment, is not radioactive. This isotope is highly effective in capturing neutrons, however: 2500 times better than boron-11 and eight times better than uranium-235. When the nucleus of a boron-10 atom captures a neutron, the resulting boron-11 nucleus has so much energy that it fragments to form an α particle and a lithium-7 atom. Although the α particles do a great deal of damage, because their penetrating power is so low, the damage remains confined to an area not much larger than one or two cells in diameter.

Solution The blood contains a total of 0.240 μCi of radioactive material. Exactly 1.00 mL containing this amount was injected. After dilution in the bloodstream, 1.00 mL of blood, representative of the total volume V, is found to have an activity of 4.3×10^{-4} μCi/mL.

$$(0.240 \ \mu\text{Ci/mL})(1.00 \ \text{mL}) = (4.3 \times 10^{-4} \ \mu\text{Ci/mL})(V)$$

$$V = 560 \ \text{mL}$$

Think about Your Answer This is solved as a classic dilution problem, by equating $C_{\text{dil}} \times V_{\text{dil}} = C_{\text{conc}} \times V_{\text{conc}}$

Check Your Understanding

To measure the solubility for $PbCrO_4$ you mix a small amount of a lead(II) salt having radioactive ^{212}Pb with a lead salt containing 0.0100 g of lead. To this mixture you add enough K_2CrO_4 to completely precipitate the lead(II) ions as $PbCrO_4$. The supernatant solution still contains a trace of lead, of course, and when you evaporate 10.00 mL of this solution to solid $PbCrO_4$, you find a radioactivity that is 4.17×10^{-5} of what it is for the pure ^{212}Pb salt. Calculate the solubility of $PbCrO_4$ in mol/L. (Adapted from C. E. Housecraft and A. G. Sharpe, *Inorganic Chemistry*, Pearson, 3rd edition, 2008, p. 84.)

Space Science: Neutron Activation Analysis and the Moon Rocks

The first manned space mission to the moon brought back a number of samples of soil and rock—a treasure trove for scientists. One of their first tasks was to analyze these samples to determine their identity and composition. Most analytical methods require chemical reactions using at least a small amount of material; however, this was not a desirable option, considering that the moon rocks were at the time the most valuable rocks in the world.

A few scientists got a chance to work on this unique project, and one of the analytical tools they used was **neutron activation analysis**. In this nondestructive process, a sample is irradiated with neutrons. Most isotopes add a neutron to form a new isotope that is one mass unit higher in an excited nuclear state. When the nucleus decays to its ground state, it emits a γ-ray. The energy of the γ-ray identifies the element, and the number of γ-rays can be counted to determine the amount of the element in the sample. Using neutron activation analysis, it is possible to analyze for a number of elements in a single experiment (Table 23.5).

Neutron activation analysis has many other uses. This analytical procedure yields a kind of fingerprint that can be used to identify a substance. For example, this technique has been applied in determining whether an art work is real or fraudulent. Analysis of the pigments in paints on a painting can be carried out without damaging the painting to determine whether the composition resembles modern paints or paints used hundreds of years ago.

Food Science: Food Irradiation

Refrigeration, canning, and chemical additives provide significant protection in terms of food preservation, but in some parts of the world these procedures are unavailable, and stored-food spoilage may claim as much as 50% of the food crop. Irradiation with γ-rays from sources such as ^{60}Co and ^{137}Cs is an option for prolonging the shelf life of foods. Relatively low levels of radiation may retard the growth of organisms, such as bacteria, molds, and yeasts, that can cause food spoilage. After irradiation, milk in a sealed container has a minimum shelf life of 3 months without refrigeration. Chicken normally has a 3-day refrigerated shelf life; after irradiation, it may have a 3-week refrigerated shelf life.

Higher levels of radiation, in the 1- to 5-Mrad (1 Mrad = 1×10^6 rad) range, will kill every living organism. Foods irradiated at these levels will keep indefinitely when sealed in plastic or aluminum-foil packages. Ham, beef, turkey, and corned beef

Table 23.5 Rare Earth Analysis of Moon Rock Sample 10022 (Fine-Grain Igneous Rock)

Element	Concentration (ppm)
La	26.4
Ce	68
Nd	66
Sm	21.2
Eu	2.04
Gd	25
Tb	4.7
Dy	31.2
Ho	5.5
Er	16
Yb	17.7
Lu	2.55

Source: L. A. Haskin, P. A, Helmke, and R. O. Allen: *Science,* Vol. 167, p. 487, 1970. The concentrations of rare earths in moon rocks were quite similar to the values in terrestrial rocks except that the europium concentration is much depleted. The January 30, 1970, issue of *Science* was devoted to analysis of moon rocks.

CASE STUDY

Nuclear Medicine and Hyperthyroidism

The primary function of the thyroid gland, located in your neck, is the production of thyroxine (3,5,3′,5′-tetraiodothyronine) and 3,5,3′-triiodothyronine. These chemical compounds are hormones that help to regulate the rate of metabolism, a term that refers to all of the chemical reactions that take place in the body. In particular, the thyroid hormones play an important role in the processes that release energy from food.

Abnormally low levels of thyroxine result in a condition known as *hypothyroidism*. Its symptoms include lethargy and feeling cold much of the time. The remedy for this condition is medication, consisting of thyroxine pills. The opposite condition, hyperthyroidism, also occurs in some people. In this condition, the body produces too much of the hormone. Hyperthyroidism is diagnosed by symptoms such as nervousness, heat intolerance, increased appetite, and muscle weakness and fatigue when blood sugar is too rapidly depleted. The standard remedy for hyperthyroidism is to destroy part of the thyroid gland, and one way to do so is to use a compound containing radioactive iodine-123 or iodine-131.

To understand this procedure, you need to know something about iodine in the body. Iodine is an essential element. Some diets provide iodine naturally (seaweed, for example, is a good source of iodine), but in the Western world most iodine taken up by the body comes from iodized salt, NaCl containing a few percent of NaI. An adult man or woman of average size should take in about 150 μg (micrograms) of iodine

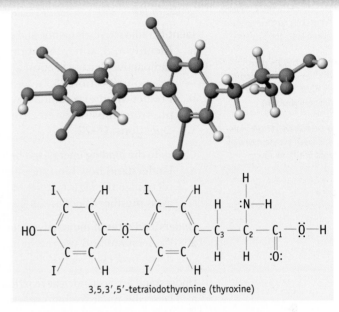

Thyroxine. The hormone 3,5,3′,5′-tetraiodothyronine (thyroxine) exerts a stimulating effect on metabolism.

3,5,3′,5′-tetraiodothyronine (thyroxine)

(1μg $= 10^{-6}$ g) in the daily diet. In the body, iodide ion is transported to the thyroid, where it serves as one of the raw materials in making thyroxine.

The fact that iodine concentrates in the thyroid tissue is essential to the procedure for using radioiodine therapy as a treatment for hyperthyroidism. Typically, an aqueous NaI solution is used in which a small fraction of iodide is the radioactive isotope iodine-131 or iodine-123, and the rest is nonradioactive iodine-127. The radioactivity destroys thyroid tissue, resulting in a decrease in thyroid activity.

Questions:

1. ^{131}I decays by beta emission. What is the product of this decay? Write a balanced nuclear equation for this process.

2. The half-life of ^{131}I is 8.04 days, and that of ^{123}I is 13.3 hours. If you begin with an equal number of moles of each isotope, at the end of 7 days what are their relative amounts?

Answers to these questions are available in Appendix N.

---

sterilized by radiation have been used on many Space Shuttle flights, for example. An astronaut said, "The beautiful thing was that it didn't disturb the taste, which made the meals much better than the freeze-dried and other types of foods we had."

These procedures are not without their opponents, and the public has not fully embraced irradiation of foods. An interesting argument favoring this technique is that radiation is less harmful than other methodologies for food preservation. This type of sterilization offers greater safety to food workers because it lessens chances of exposure to harmful chemicals, and it protects the environment by avoiding contamination of water supplies with toxic chemicals.

Food irradiation is commonly used in European countries, Canada, and Mexico. Its use in the United States is currently regulated by the U.S. Food and Drug Administration (FDA) and Department of Agriculture (USDA). In 1997, the FDA approved the irradiation of refrigerated and frozen uncooked meat to control pathogens and extend shelf-life, and in 2000 the USDA approved the irradiation of eggs to control *Salmonella* infection.

CHAPTER GOALS REVISITED

Now that you have studied this chapter, you should ask whether you have met the chapter goals. In particular, you should be able to:

Identify radioactive elements, and describe natural and artificial nuclear reactions

a. Identify α, β, and γ radiation, the three major types of radiation in natural radioactive decay (Section 23.1), and write balanced equations for nuclear reactions (Section 23.2). **Study Questions: 11–16, 32, 41.**

b. Predict whether a radioactive isotope will decay by α or β emission, or by positron emission or electron capture (Sections 23.2 and 23.3). **Study Questions: 17–22.**

Calculate the binding energy and binding energy per nucleon for a particular isotope

a. Understand how binding energy per nucleon is defined (Section 23.3) and recognize the significance of a graph of binding energy per nucleon versus mass number (Section 23.3). **Study Questions: 23–28, 52.**

Understand rates of radioactive decay

a. Understand and use mathematical equations that characterize the radioactive decay process (Section 23.4). **Study Questions: 29–38, 47, 50, 54, 58.**

Understand artificial nuclear reactions

a. Describe nuclear chain reactions, nuclear fission, and nuclear fusion (Sections 23.6 and 23.7). **Study Questions: 39–46.**

Understand issues of health and safety with respect to radioactivity

a. Describe the units used to measure intensity, and understand how they pertain to health issues (Sections 23.8 and 23.9).

Be aware of some uses of radioactive isotopes in science and medicine

Key Equations

Equation 23.1 (page 830): The equation relating interconversion of mass (m) and energy (E). This equation is applied in the calculation of binding energy (E_b) for nuclei.

$$E_b = (\Delta m)c^2$$

Equation 23.2 (page 833): The activity of a radioactive sample (A) is proportional to the number of radioactive atoms (N).

$$A \propto N$$

Equation 23.3 (page 833): The change in the number of radioactive elements with time is equal to the product of the rate constant (k, decay constant) and number of atoms present (N).

$$\Delta N/\Delta t = -kN$$

Equation 23.4 (page 834): The rate law for nuclear decay based on number of radioactive atoms initially present (N_0) and the number N after time t.

$$\ln(N/N_0) = -kt$$

Equation 23.5 (page 834): The rate law for nuclear decay based on the measured activity of a sample (A).

$$\ln(A/A_0) = -kt$$

Equation 23.6 (page 834): The relationship between the half-life and the rate constant for a nuclear decay process.

$$t_{1/2} = 0.693/k$$

LIST OF APPENDICES

A Using Logarithms and Solving Quadratic Equations   A-2

B Some Important Physical Concepts   A-6

C Abbreviations and Useful Conversion Factors   A-9

D Physical Constants   A-13

E A Brief Guide to Naming Organic Compounds   A-15

F Values for the Ionization Energies and Electron Attachment Enthalpies of the Elements   A-18

G Vapor Pressure of Water at Various Temperatures   A-19

H Ionization Constants for Aqueous Weak Acids at 25 °C   A-20

I Ionization Constants for Aqueous Weak Bases at 25 °C   A-22

J Solubility Product Constants for Some Inorganic Compounds at 25 °C   A-23

K Formation Constants for Some Complex Ions in Aqueous Solution at 25 °C   A-25

L Selected Thermodynamic Values   A-26

M Standard Reduction Potentials in Aqueous Solution at 25 °C   A-32

N Answers to Chapter Opening Questions and Case Study Questions   A-36

O Answers to Check Your Understanding Questions   A-47

P Answers to Review & Check Questions   A-63

Q Answers to Selected Interchapter Study Questions   A-72

R Answers to Selected Study Questions   A-75

A Using Logarithms and Solving Quadratic Equations

An introductory chemistry course requires basic algebra plus a knowledge of (1) exponential (or scientific) notation, (2) logarithms, and (3) quadratic equations. The use of exponential notation was reviewed on pages 33–35, and this appendix reviews the last two topics.

A.1 Logarithms

Two types of logarithms are used in this text: (1) common logarithms (abbreviated log) whose base is 10 and (2) natural logarithms (abbreviated ln) whose base is e (= 2.71828):

$$\log x = n, \text{ where } x = 10^n$$
$$\ln x = m, \text{ where } x = e^m$$

Most equations in chemistry and physics were developed in natural, or base e, logarithms, and we follow this practice in this text. The relation between log and ln is

$$\ln x = 2.303 \log x$$

Despite the different bases of the two logarithms, they are used in the same manner. What follows is largely a description of the use of common logarithms.

A common logarithm is the power to which you must raise 10 to obtain the number. For example, the log of 100 is 2, since you must raise 10 to the second power to obtain 100. Other examples are

$$\log 1000 = \log (10^3) = 3$$
$$\log 10 = \log (10^1) = 1$$
$$\log 1 = \log (10^0) = 0$$
$$\log 0.1 = \log (10^{-1}) = -1$$
$$\log 0.0001 = \log (10^{-4}) = -4$$

To obtain the common logarithm of a number other than a simple power of 10, you must resort to a log table or an electronic calculator. For example,

$$\log 2.10 = 0.322, \text{ which means that } 10^{0.322} = 2.10$$
$$\log 5.16 = 0.713, \text{ which means that } 10^{0.713} = 5.16$$
$$\log 3.125 = 0.4949, \text{ which means that } 10^{0.4949} = 3.125$$

To check this on many calculators, enter the number, and then press the "log" key. You should make sure that you understand how to use your particular calculator.

To obtain the natural logarithm ln of the numbers shown here, use a calculator having this function. Enter each number, and press "ln:"

$$\ln 2.10 = 0.742, \text{ which means that } e^{0.742} = 2.10$$
$$\ln 5.16 = 1.641, \text{ which means that } e^{1.641} = 5.16$$

To find the common logarithm of a number greater than 10 or less than 1 with a log table, first express the number in scientific notation. Then find the log of each part of the number and add the logs. For example,

$$\log 241 = \log (2.41 \times 10^2) = \log 2.41 + \log 10^2$$
$$= 0.382 + 2 = 2.382$$
$$\log 0.00573 = \log (5.73 \times 10^{-3}) = \log 5.73 + \log 10^{-3}$$
$$= 0.758 + (-3) = -2.242$$

Significant Figures and Logarithms

Notice that the mantissa has as many significant figures as the number whose log was found.

Obtaining Antilogarithms

If you are given the logarithm of a number, and find the number from it, you have obtained the "antilogarithm," or "antilog," of the number. Two common procedures used by electronic calculators to do this are as follows:

Procedure A	Procedure B
1. Enter the value of the log or ln.	1. Enter value of the log or ln.
2. Press 2ndF.	2. Press INV.
3. Press 10^x or e^x.	3. Press log or ln x.

Make sure you can properly perform this operation on your calculator by working the following examples:

1. Find the number whose log is 5.234:
   Recall that log $x = n$, where $x = 10^n$. In this case, $n = 5.234$. Find the value of 10^n, the antilog. In this case,

$$10^{5.234} = 10^{0.234} \times 10^5 = 1.71 \times 10^5$$

Notice that the characteristic (5) sets the decimal point; it is the power of 10 in the exponential form. The mantissa (0.234) gives the value of the number x, 1.71 in this case.

2. Find the number whose log is -3.456:

$$10^{-3.456} = 10^{0.544} \times 10^{-4} = 3.50 \times 10^{-4}$$

Notice here that -3.456 is expressed as the sum of -4 and $+0.544$.

Mathematical Operations Using Logarithms

Because logarithms are exponents, operations involving them follow the same rules used for exponents. Thus, multiplying two numbers can be done by adding logarithms:

$$\log xy = \log x + \log y$$

For example, we multiply 563 by 125 by adding their logarithms and finding the antilogarithm of the result:

$$\log 563 = 2.751$$
$$\log 125 = \underline{2.097}$$
$$\log xy = 4.848$$
$$xy = 10^{4.848} = 10^{0.848} \times 10^4 = 7.05 \times 10^4$$

One number (x) can be divided by another (y) by subtraction of their logarithms:

$$\log \frac{x}{y} = \log x - \log y$$

For example, to divide 125 by 742,

$$\log 125 = 2.097$$
$$-\log 742 = \underline{2.870}$$
$$\log \frac{x}{y} = -0.773$$

$$\frac{x}{y} = 10^{-0.773} = 10^{0.227} \times 10^{-1} = 1.68 \times 10^{-1}$$

● **Logarithms and Nomenclature** The number to the left of the decimal in a logarithm is called the **characteristic**, and the number to the right of the decimal is the **mantissa**.

Similarly, powers and roots of numbers can be found using logarithms.

$$\log x^y = y(\log x)$$

$$\log \sqrt[y]{x} = \log x^{1/y} = \frac{1}{y} \log x$$

As an example, find the fourth power of 5.23. We first find the log of 5.23 and then multiply it by 4. The result, 2.874, is the log of the answer. Therefore, we find the antilog of 2.874:

$$(5.23)^4 = ?$$
$$\log (5.23)^4 = 4 \log 5.23 = 4(0.719) = 2.874$$
$$(5.23)^4 = 10^{2.874} = 748$$

As another example, find the fifth root of 1.89×10^{-9}:

$$\sqrt[5]{1.89 \times 10^{-9}} = (1.89 \times 10^{-9})^{1/5} = ?$$

$$\log(1.89 \times 10^{-9})^{1/5} = \frac{1}{5}\log(1.89 \times 10^{-9}) = \frac{1}{5}(-8.724) = -1.745$$

The answer is the antilog of -1.745:

$$\left(1.89 \times 10^{-9}\right)^{1/5} = 10^{-1.745} = 1.80 \times 10^{-2}$$

A.2 Quadratic Equations

Algebraic equations of the form $ax^2 + bx + c = 0$ are called **quadratic equations**. The coefficients a, b, and c may be either positive or negative. The two roots of the equation may be found using the *quadratic formula:*

$$x = \frac{-b \pm \sqrt{b^2 - 4ac}}{2a}$$

As an example, solve the equation $5x^2 - 3x - 2 = 0$. Here $a = 5$, $b = -3$, and $c = -2$. Therefore,

$$x = \frac{3 \pm \sqrt{(-3)^2 - 4(5)(-2)}}{2(5)}$$

$$= \frac{3 \pm \sqrt{9 - (-40)}}{10} = \frac{3 \pm \sqrt{49}}{10} = \frac{3 \pm 7}{10}$$

$$= 1 \text{ and } -0.4$$

How do you know which of the two roots is the correct answer? Mathematically, both roots are possible, but in chemistry problems you have to decide in each case which root has physical significance. It is *usually* true in this course, however, that negative values are not significant.

When you have solved a quadratic expression, you should always check your values by substitution into the original equation. In the previous example, we find that $5(1)^2 - 3(1) - 2 = 0$ and that $5(-0.4)^2 - 3(-0.4) - 2 = 0$.

The most likely place you will encounter quadratic equations is in the chapters on chemical equilibria, particularly in Chapters 16 through 18. Here, you will often be faced with solving an equation such as

$$1.8 \times 10^{-4} = \frac{x^2}{0.0010 - x}$$

This equation can certainly be solved using the quadratic formula (to give $x = 3.4 \times 10^{-4}$). You may find the **method of successive approximations** to be especially convenient, however. Here we begin by making a reasonable approximation of x. This approximate value is substituted into the original equation, which is then solved to give what is hoped to be a more correct value of x. This process is repeated until the answer converges on a particular value of x—that is, until the value of x derived from two successive approximations is the same.

Step 1: First, assume that x is so small that $(0.0010 - x) \approx 0.0010$. This means that

$$x^2 = 1.8 \times 10^{-4}\,(0.0010)$$
$$x = 4.2 \times 10^{-4} \text{ (to 2 significant figures)}$$

Step 2: Substitute the value of x from Step 1 into the denominator of the original equation, and again solve for x:

$$x^2 = 1.8 \times 10^{-4}(0.0010 - 0.00042)$$
$$x = 3.2 \times 10^{-4}$$

Step 3: Repeat Step 2 using the value of x found in that step:

$$x = \sqrt{1.8 \times 10^{-4}(0.0010 - 0.00032)} = 3.5 \times 10^{-4}$$

Step 4: Continue repeating the calculation, using the value of x found in the previous step:

$$x = \sqrt{1.8 \times 10^{-4}(0.0010 - 0.00035)} = 3.4 \times 10^{-4}$$

Step 5: $\qquad x = \sqrt{1.8 \times 10^{-4}(0.0010 - 0.00034)} = 3.4 \times 10^{-4}$

Here, we find that iterations after the fourth step give the same value for x, indicating that we have arrived at a valid answer (and the same one obtained from the quadratic formula).

Here are some final thoughts on using the method of successive approximations. First, in some cases the method does not work. Successive steps may give answers that are random or that diverge from the correct value. In Chapters 16 through 18, you confront quadratic equations of the form $K = x^2/(C - x)$. The method of successive approximations works as long as $K < 4C$ (assuming one begins with $x = 0$ as the first guess, that is, $K \approx x^2/C$). This is always going to be true for weak acids and bases (the topic of Chapters 17 and 18), but it may *not* be the case for problems involving gas phase equilibria (Chapter 16), where K can be quite large.

Second, values of K in the equation $K = x^2/(C - x)$ are usually known only to two significant figures. We are therefore justified in carrying out successive steps until two answers are the same to two significant figures.

Finally, we highly recommend this method of solving quadratic equations, especially those in Chapters 17 and 18. If your calculator has a memory function, successive approximations can be carried out easily and rapidly.

B* Some Important Physical Concepts

B.1 Matter

The tendency to maintain a constant velocity is called *inertia*. Thus, unless acted on by an unbalanced force, a body at rest remains at rest, and a body in motion remains in motion with uniform velocity. Matter is anything that exhibits inertia; the quantity of matter is its mass.

B.2 Motion

Motion is the change of position or location in space. Objects can have the following classes of motion:

- Translation occurs when the center of mass of an object changes its location. Example: a car moving on the highway.
- Rotation occurs when each point of a moving object moves in a circle about an axis through the center of mass. Examples: a spinning top, a rotating molecule.
- Vibration is a periodic distortion and then recovery of original shape. Examples: a struck tuning fork, a vibrating molecule.

B.3 Force and Weight

Force is that which changes the velocity of a body; it is defined as

$$\text{Force} = \text{mass} \times \text{acceleration}$$

The SI unit of force is the **newton**, N, whose dimensions are kilograms times meter per second squared ($kg \cdot m/s^2$). A newton is therefore the force needed to change the velocity of a mass of 1 kilogram by 1 meter per second in a time of 1 second.

Because the earth's gravity is not the same everywhere, the weight (a force) corresponding to a given mass is not a constant. At any given spot on earth, gravity is constant, however, and therefore weight is proportional to mass. When a balance tells us that a given sample (the "unknown") has the same weight as another sample (the "weights," as given by a scale reading or by a total of counterweights), it also tells us that the two masses are equal. The balance is therefore a valid instrument for measuring the mass of an object independently of slight variations in the force of gravity.

B.4 Pressure†

Pressure is force per unit area. The SI unit, called the *pascal*, Pa, is

$$1 \text{ pascal} = \frac{1 \text{ newton}}{m^2} = \frac{1 \text{ kg} \cdot m/s^2}{m^2} = \frac{1 \text{ kg}}{m \cdot s^2}$$

The International System of Units also recognizes the bar, which is 10^5 Pa and which is close to standard atmospheric pressure (Table 1).

Chemists also express pressure in terms of the heights of liquid columns, especially water and mercury. This usage is not completely satisfactory because the

---

*Adapted from F. Brescia, J. Arents, H. Meislich, et al.: *General Chemistry,* 5th ed. Philadelphia: Harcourt Brace, 1988.
†See Section 11.1.

Table 1 Pressure Conversions

From	To	Multiply By
atmosphere	mm Hg	760 mm Hg/atm (exactly)
atmosphere	lb/in²	14.6960 lb/(in² · atm)
atmosphere	kPa	101.325 kPa/atm
bar	Pa	10^5 Pa/bar (exactly)
bar	lb/in²	14.5038 lb/(in² · bar)
mm Hg	torr	1 torr/mm Hg (exactly)

pressure exerted by a given column of a given liquid is not a constant but depends on the temperature (which influences the density of the liquid) and the location (which influences the magnitude of the force exerted by gravity). Such units are therefore not part of the SI, and their use is now discouraged. The older units are still used in books and journals, however, and chemists must be familiar with them.

The pressure of a liquid or a gas depends only on the depth (or height) and is exerted equally in all directions. At sea level, the pressure exerted by the earth's atmosphere supports a column of mercury about 0.76 m (76 cm, or 760 mm) high.

One **standard atmosphere** (atm) is the pressure exerted by exactly 76 cm of mercury at 0 °C (density, 13.5951 g/cm³) and at standard gravity, 9.80665 m/s². The **bar** is equivalent to 0.9869 atm. One **torr** is the pressure exerted by exactly 1 mm of mercury at 0 °C and standard gravity.

B.5   Energy and Power

The SI unit of energy is the product of the units of force and distance, or kilograms times meter per second squared (kg · m/s²) times meters (× m), which is kg · m²/s²; this unit is called the **joule**, J. The joule is thus the work done when a force of 1 newton acts through a distance of 1 meter.

Work may also be done by moving an electric charge in an electric field. When the charge being moved is 1 coulomb (C) and the potential difference between its initial and final positions is 1 volt (V), the work is 1 joule. Thus,

$$1 \text{ joule} = 1 \text{ coulomb volt (CV)}$$

Another unit of electric work that is not part of the International System of Units but is still in use is the **electron volt**, eV, which is the work required to move an electron against a potential difference of 1 volt. (It is also the kinetic energy acquired by an electron when it is accelerated by a potential difference of 1 volt.) Because the charge on an electron is 1.602×10^{-19} C, we have

$$1 \text{ eV} = 1.602 \times 10^{-19} \text{ CV} \times \frac{1 \text{ J}}{1 \text{ CV}} = 1.602 \times 10^{-19} \text{ J}$$

If this value is multiplied by Avogadro's number, we obtain the energy involved in moving 1 mol of electron charges (1 faraday) in a field produced by a potential difference of 1 volt:

$$1\frac{\text{eV}}{\text{particle}} = \frac{1.602 \times 10^{-19} \text{ J}}{\text{particle}} \times \frac{6.022 \times 10^{23} \text{ particles}}{\text{mol}} \cdot \frac{1 \text{ kJ}}{1000 \text{ J}} = 96.49 \text{ kJ/mol}$$

Power is the amount of energy delivered per unit time. The SI unit is the watt, W, which is 1 joule per second. One kilowatt, kW, is 1000 W. Watt-hours and kilowatt-hours are therefore units of energy (Table 2). For example, 1000 watt-hours, or 1 kilowatt-hour, is

$$1.0 \times 10^3 \text{ W} \cdot \text{h} \times \frac{1 \text{ J}}{1 \text{ W} \cdot \text{s}} \times \frac{3.6 \times 10^3 \text{ s}}{1 \text{ h}} = 3.6 \times 10^6 \text{ J}$$

Table 2 Energy Conversions

From	To	Multiply By
calorie (cal)	joule	4.184 J/cal (exactly)
kilocalorie (kcal)	cal	10^3 cal/kcal (exactly)
kilocalorie	joule	4.184×10^3 J/kcal (exactly)
liter atmosphere (L · atm)	joule	101.325 J/L · atm
electron volt (eV)	joule	1.60218×10^{-19} J/eV
electron volt per particle	kilojoules per mole	96.485 kJ · particle/eV · mol
coulomb volt (CV)	joule	1 CV/J (exactly)
kilowatt-hour (kW-h)	kcal	860.4 kcal/kW-h
kilowatt-hour	joule	3.6×10^6 J/kW-h (exactly)
British thermal unit (BTU)	calorie	252 cal/BTU

C Abbreviations and Useful Conversion Factors

Table 3 Some Common Abbreviations and Standard Symbols

Term	Abbreviation	Term	Abbreviation
Activation energy	E_a	Equilibrium constant	K
Ampere	A	Concentration basis	K_c
Aqueous solution	aq	Pressure basis	K_p
Atmosphere, unit of pressure	atm	Ionization weak acid	K_a
Atomic mass unit	u	Ionization weak base	K_b
Avogadro's constant	N	Solubility product	K_{sp}
Bar, unit of pressure	bar	Formation constant	K_f
Body-centered cubic	bcc	Ethylenediamine	en
Bohr radius	a_0	Face-centered cubic	fcc
Boiling point	bp	Faraday constant	F
Celsius temperature	°C	Gas constant	R
Charge number of an ion	z	Gibbs free energy	G
Coulomb, electric charge	C	Standard free energy	$G°$
Curie, radioactivity	Ci	Standard free energy of formation	$\Delta_f G°$
Cycles per second, hertz	Hz	Standard free energy change for reaction	$\Delta_r G°$
Debye, unit of electric dipole	D	Half-life	$t_{1/2}$
Electron	e^-	Heat	q
Electron volt	eV	Hertz	Hz
Electronegativity	χ	Hour	h
Energy	E	Joule	J
Enthalpy	H	Kelvin	K
Standard enthalpy	$H°$	Kilocalorie	kcal
Standard enthalpy of formation	$\Delta_f H°$	Liquid	ℓ
Standard enthalpy of reaction	$\Delta_r H°$	Logarithm, base 10	log
Entropy	S	Logarithm, base e	ln
Standard entropy	$S°$	Millimeters of mercury, unit of pressure	mm Hg
Entropy change for reaction	$\Delta_r S°$	Minute	min

(continued)

Table 3 Some Common Abbreviations and Standard Symbols (continued)

Term	Abbreviation	Term	Abbreviation
Molar	M	Pressure	P
Molar mass	M	Proton number	Z
Mole	mol	Rate constant	k
Osmotic pressure	Π	Standard temperature and pressure	STP
Pascal, unit of pressure	Pa	Temperature	T
Planck's constant	h	Volt	V
Pound	lb	Watt	W
Primitive cubic (unit cell)	pc	Wavelength	λ

C.1   Fundamental Units of the SI System

The metric system was begun by the French National Assembly in 1790 and has undergone many modifications. The International System of Units or *Système International* (SI), which represents an extension of the metric system, was adopted by the 11th General Conference of Weights and Measures in 1960. It is constructed from seven base units, each of which represents a particular physical quantity (Table 4).

Table 4 SI Fundamental Units

Physical Quantity	Name of Unit	Symbol
Length	meter	m
Mass	kilogram	kg
Time	second	s
Temperature	kelvin	K
Amount of substance	mole	mol
Electric current	ampere	A
Luminous intensity	candela	cd

The first five units listed in Table 4 are particularly useful in general chemistry and are defined as follows:

1. The *meter* was redefined in 1960 to be equal to 1,650,763.73 wavelengths of a certain line in the emission spectrum of krypton-86.
2. The *kilogram* represents the mass of a platinum–iridium block kept at the International Bureau of Weights and Measures at Sèvres, France.
3. The *second* was redefined in 1967 as the duration of 9,192,631,770 periods of a certain line in the microwave spectrum of cesium-133.
4. The *kelvin* is 1/273.16 of the temperature interval between absolute zero and the triple point of water.
5. The *mole* is the amount of substance that contains as many entities as there are atoms in exactly 0.012 kg of carbon-12 (12 g of ^{12}C atoms).

C.2 Prefixes Used with Traditional Metric Units and SI Units

Decimal fractions and multiples of metric and SI units are designated by using the prefixes listed in Table 5. Those most commonly used in general chemistry appear in italics.

Table 5 Traditional Metric and SI Prefixes

Factor	Prefix	Symbol	Factor	Prefix	Symbol
10^{12}	tera	T	10^{-1}	*deci*	d
10^9	giga	G	10^{-2}	*centi*	c
10^6	mega	M	10^{-3}	*milli*	m
10^3	*kilo*	k	10^{-6}	micro	μ
10^2	hecto	h	10^{-9}	*nano*	n
10^1	deka	da	10^{-12}	*pico*	p
			10^{-15}	femto	f
			10^{-18}	atto	a

C.3 Derived SI Units

In the International System of Units, all physical quantities are represented by appropriate combinations of the base units listed in Table 4. A list of the derived units frequently used in general chemistry is given in Table 6.

Table 6 Derived SI Units

Physical Quantity	Name of Unit	Symbol	Definition
Area	square meter	m^2	
Volume	cubic meter	m^3	
Density	kilogram per cubic meter	kg/m^3	
Force	newton	N	$kg \cdot m/s^2$
Pressure	pascal	Pa	N/m^2
Energy	joule	J	$kg \cdot m^2/s^2$
Electric charge	coulomb	C	$A \cdot s$
Electric potential difference	volt	V	$J/(A \cdot s)$

Table 7 Common Units of Mass and Weight

1 pound = 453.39 grams
1 kilogram = 1000 grams = 2.205 pounds
1 gram = 1000 milligrams
1 gram = 6.022×10^{23} atomic mass units
1 atomic mass unit = 1.6605×10^{-24} gram
1 short ton = 2000 pounds = 907.2 kilograms
1 long ton = 2240 pounds
1 metric tonne = 1000 kilograms = 2205 pounds

Table 8 **Common Units of Length**

1 inch = 2.54 centimeters (exactly)
1 mile = 5280 feet = 1.609 kilometers
1 yard = 36 inches = 0.9144 meter
1 meter = 100 centimeters = 39.37 inches = 3.281 feet = 1.094 yards
1 kilometer = 1000 meters = 1094 yards = 0.6215 mile
1 Ångstrom = 1.0×10^{-8} centimeter = 0.10 nanometer = 100 picometers $= 1.0 \times 10^{-10}$ meter = 3.937×10^{-9} inch

Table 9 **Common Units of Volume**

1 quart = 0.9463 liter   **1 liter = 1.0567 quarts**
1 liter = 1 cubic decimeter = 1000 cubic centimeters = 0.001 cubic meter
1 milliliter = 1 cubic centimeter = 0.001 liter = 1.056×10^{-3} quart
1 cubic foot = 28.316 liters = 29.924 quarts = 7.481 gallons

D Physical Constants

Table 10 Physical Constants

Quantity	Symbol	Traditional Units	SI Units
Acceleration of gravity	g	980.6 cm/s^2	9.806 m/s^2
Atomic mass unit (1/12 the mass of ^{12}C atom)	u	1.6605×10^{-24} g	1.6605×10^{-27} kg
Avogadro's number	N	$6.02214179 \times 10^{23}$ particles/mol	$6.02214179 \times 10^{23}$ particles/mol
Bohr radius	a_0	0.052918 nm 5.2918×10^{-9} cm	5.2918×10^{-11} m
Boltzmann constant	k	1.3807×10^{-16} erg/K	1.3807×10^{-23} J/K
Charge-to-mass ratio of electron	e/m	1.7588×10^{8} C/g	1.7588×10^{11} C/kg
Electron rest mass	m_e	9.1094×10^{-28} g 0.00054858 u	9.1094×10^{-31} kg
Electronic charge	e	1.6022×10^{-19} C 4.8033×10^{-10} esu	1.6022×10^{-19} C
Faraday constant	F	96,485 C/mol e$^-$ 23.06 kcal/V $\cdot$ mol e$^-$	96,485 C/mol e$^-$ 96,485 J/V $\cdot$ mol e$^-$
Gas constant	R	$0.082057 \dfrac{\text{L} \cdot \text{atm}}{\text{mol} \cdot \text{K}}$ $1.987 \dfrac{\text{cal}}{\text{mol} \cdot \text{K}}$	$8.3145 \dfrac{\text{Pa} \cdot \text{dm}^3}{\text{mol} \cdot \text{K}}$ 8.3145 J/mol $\cdot$ K
Molar volume (STP)	V_m	22.414 L/mol	22.414×10^{-3} m^3/mol 22.414 dm^3/mol
Neutron rest mass	m_n	1.67493×10^{-24} g 1.008665 u	1.67493×10^{-27} kg
Planck's constant	h	6.6261×10^{-27} erg $\cdot$ s	$6.6260693 \times 10^{-34}$ J $\cdot$ s
Proton rest mass	m_p	1.6726×10^{-24} g 1.007276 u	1.6726×10^{-27} kg
Rydberg constant	R Rhc	—	1.0974×10^{7} m^{-1} 2.1799×10^{-18} J
Velocity of light (in a vacuum)	c	2.9979×10^{10} cm/s (186,282 miles/s)	2.9979×10^{8} m/s
$\pi = 3.1416$			
$e = 2.7183$			
$\ln X = 2.303 \log X$			

Table 11 Specific Heats and Heat Capacities for Some Common Substances at 25 °C

Substance	Specific Heat (J/g · K)	Molar Heat Capacity (J/mol · K)
Al(s)	0.897	24.2
Ca(s)	0.646	25.9
Cu(s)	0.385	24.5
Fe(s)	0.449	25.1
Hg(ℓ)	0.140	28.0
H_2O(s), ice	2.06	37.1
H_2O(ℓ), water	4.184	75.4
H_2O(g), steam	1.86	33.6
C_6H_6(ℓ), benzene	1.74	136
C_6H_6(g), benzene	1.06	82.4
C_2H_5OH(ℓ), ethanol	2.44	112.3
C_2H_5OH(g), ethanol	1.41	65.4
$(C_2H_5)_2O$(ℓ), diethyl ether	2.33	172.6
$(C_2H_5)_2O$(g), diethyl ether	1.61	119.5

Table 12 Heats of Transformation and Transformation Temperatures of Several Substances

Substance	MP (°C)	Heat of Fusion		BP (°C)	Heat of Vaporization	
		J/g	kJ/mol		J/g	kJ/mol
Elements*						
Al	660	395	10.7	2518	12083	294
Ca	842	212	8.5	1484	3767	155
Cu	1085	209	13.3	2567	4720	300
Fe	1535	267	13.8	2861	6088	340
Hg	−38.8	11	2.29	357	295	59.1
Compounds						
H_2O	0.00	333	6.01	100.0	2260	40.7
CH_4	−182.5	58.6	0.94	−161.5	511	8.2
C_2H_5OH	−114	109	5.02	78.3	838	38.6
C_6H_6	5.48	127.4	9.95	80.0	393	30.7
$(C_2H_5)_2O$	−116.3	98.1	7.27	34.6	357	26.5

*Data for the elements are taken from J. A. Dean: *Lange's Handbook of Chemistry,* 15th ed. New York: McGraw-Hill Publishers, 1999.

E A Brief Guide to Naming Organic Compounds

It seems a daunting task to devise a systematic procedure that gives each organic compound a unique name, but that is what has been done. A set of rules was developed to name organic compounds by the International Union of Pure and Applied Chemistry (IUPAC). The IUPAC nomenclature allows chemists to write a name for any compound based on its structure or to identify the formula and structure for a compound from its name. In this book, we have generally used the IUPAC nomenclature scheme when naming compounds.

In addition to the systematic names, many compounds have common names. The common names came into existence before the nomenclature rules were developed, and they have continued in use. For some compounds, these names are so well entrenched that they are used most of the time. One such compound is acetic acid, which is almost always referred to by that name and not by its systematic name, ethanoic acid.

The general procedure for systematic naming of organic compounds begins with the nomenclature for hydrocarbons. Other organic compounds are then named as derivatives of hydrocarbons. Nomenclature rules for simple organic compounds are given in the following section.

E.1 Hydrocarbons

Alkanes

The names of alkanes end in "-ane." When naming a specific alkane, the root of the name identifies the longest carbon chain in the compound. Specific substituent groups attached to this carbon chain are identified by name and position.

Alkanes with chains of one to ten carbon atoms are given in Table 10.2. After the first four compounds, the names derive from Greek and Latin numbers—pentane, hexane, heptane, octane, nonane, decane—and this regular naming continues for higher alkanes. For substituted alkanes, the substituent groups on a hydrocarbon chain must be identified both by a name and by the position of substitution; this information precedes the root of the name. The position is indicated by a number that refers to the carbon atom to which the substituent is attached. (Numbering of the carbon atoms in a chain should begin at the end of the carbon chain that allows the substituent groups to have the lowest numbers.)

Names of hydrocarbon substituents are derived from the name of the hydrocarbon. The group —CH_3, derived by taking a hydrogen from methane, is called the *methyl group;* the —C_2H_5 group is the *ethyl group.* The nomenclature scheme is easily extended to derivatives of hydrocarbons with other substituent groups such as —Cl (chloro), —NO_2 (nitro), —CN (cyano), —D (deuterio), and so on (Table 13). If two or more of the same substituent groups occur, the prefixes "di-," "tri-," and "tetra-" are added. When different substituent groups are present, they are generally listed in alphabetical order.

Table 13 **Names of Common Substituent Groups**

Formula	Name	Formula	Name
—CH_3	methyl	—D	deuterio
—C_2H_5	ethyl	—Cl	chloro
—$CH_2CH_2CH_3$	propyl (*n*-propyl)	—Br	bromo
—$CH(CH_3)_2$	1-methylethyl (isopropyl)	—F	fluoro
—$CH{=}CH_2$	ethenyl (vinyl)	—CN	cyano
—C_6H_5	phenyl	—NO_2	nitro
—OH	hydroxo		
—NH_2	amino		

Example:

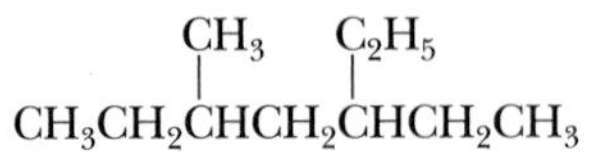

$$\underset{CH_3CH_2\overset{\displaystyle CH_3}{\overset{|}{C}}HCH_2\overset{\displaystyle C_2H_5}{\overset{|}{C}}HCH_2CH_3}{}$$

Step	Information to Include	Contribution to Name
1	An alkane	Name will end in "-ane"
2	Longest chain is 7 carbons	Name as a *heptane*
3	—CH_3 group at carbon 3	3-*methyl*
4	—C_2H_5 group at carbon 5	5-*ethyl*
Name:		5-ethyl-3-methylheptane

Cycloalkanes are named based on the ring size and by adding the prefix "cyclo"; for example, the cycloalkane with a six-member ring of carbons is called *cyclohexane.*

Alkenes

Alkenes have names ending in "-ene." The name of an alkene must specify the length of the carbon chain and the position of the double bond (and when appropriate, the configuration, either *cis* or *trans*). As with alkanes, both identity and position of substituent groups must be given. The carbon chain is numbered from the end that gives the double bond the lowest number.

Compounds with two double bonds are called *dienes,* and they are named similarly—specifying the positions of the double bonds and the name and position of any substituent groups.

For example, the compound $H_2C{=}C(CH_3)CH(CH_3)CH_2CH_3$ has a five-carbon chain with a double bond between carbon atoms 1 and 2 and methyl groups on carbon atoms 2 and 3. Its name using IUPAC nomenclature is **2,3-dimethyl-1-pentene**. The compound $CH_3CH{=}CHCCl_3$ with a *cis* configuration around the double bond is named **1,1,1-trichloro-*cis*-2-butene**. The compound $H_2C{=}C(Cl)CH{=}CH_2$ is **2-chloro-1,3-butadiene**.

Alkynes

The naming of alkynes is similar to the naming of alkenes, except that *cis–trans* isomerism isn't a factor. The ending "-yne" on a name identifies a compound as an alkyne.

Benzene Derivatives

The carbon atoms in the six-member ring are numbered 1 through 6, and the name and position of substituent groups are given. The two examples shown here are **1-ethyl-3-methylbenzene** and **1,4-diaminobenzene**.

1-ethyl-3-methylbenzene   1,4-diaminobenzene

E.2 Derivatives of Hydrocarbons

The names for alcohols, aldehydes, ketones, and acids are based on the name of the hydrocarbon with an appropriate suffix to denote the class of compound, as follows:

- **Alcohols:** Substitute "-ol" for the final "-e" in the name of the hydrocarbon, and designate the position of the —OH group by the number of the carbon atom. For example, $CH_3CH_2CHOHCH_3$ is named as a derivative of the 4-carbon hydrocarbon butane. The —OH group is attached to the second carbon, so the name is **2-butanol**.
- **Aldehydes:** Substitute "-al" for the final "-e" in the name of the hydrocarbon. The carbon atom of an aldehyde is, by definition, carbon-1 in the hydrocarbon chain. For example, the compound $CH_3CH(CH_3)CH_2CH_2CHO$ contains a 5-carbon chain with the aldehyde functional group being carbon-1 and the —CH_3 group at position 4; thus, the name is **4-methylpentanal**.
- **Ketones:** Substitute "-one" for the final "-e" in the name of the hydrocarbon. The position of the ketone functional group (the carbonyl group) is indicated by the number of the carbon atom. For example, the compound $CH_3COCH_2CH(C_2H_5)CH_2CH_3$ has the carbonyl group at the 2 position and an ethyl group at the 4 position of a 6-carbon chain; its name is **4-ethyl-2-hexanone**.
- **Carboxylic acids (organic acids):** Substitute "-oic" for the final "-e" in the name of the hydrocarbon. The carbon atoms in the longest chain are counted beginning with the carboxylic carbon atom. For example, *trans*-$CH_3CH{=}CHCH_2CO_2H$ is named as a derivative of *trans*-3-pentene—that is, ***trans*-3-pentenoic acid**.

An **ester** is named as a derivative of the alcohol and acid from which it is made. The name of an ester is obtained by splitting the formula RCO_2R' into two parts, the RCO_2— portion and the —R' portion. The —R' portion comes from the alcohol and is identified by the hydrocarbon group name; derivatives of ethanol, for example, are called *ethyl* esters. The acid part of the compound is named by dropping the "-oic" ending for the acid and replacing it by "-oate." The compound $CH_3CH_2CO_2CH_3$ is named **methyl propanoate**.

Notice that an anion derived from a carboxylic acid by loss of the acidic proton is named the same way. Thus, $CH_3CH_2CO_2^-$ is the **propanoate anion**, and the sodium salt of this anion, $Na(CH_3CH_2CO_2)$, is **sodium propanoate**.

F Values for the Ionization Energies and Electron Attachment Enthalpies of the Elements

First Ionization Energies for Some Elements (kJ/mol)

1A (1)													3A (13)	4A (14)	5A (15)	6A (16)	7A (17)	8 (18)
H 1312	2A (2)																	He 2371
Li 520	Be 899	3B (3)	4B (4)	5B (5)	6B (6)	7B (7)	8B (8,9,10)			1B (11)	2B (12)		B 801	C 1086	N 1402	O 1314	F 1681	Ne 2081
Na 496	Mg 738												Al 578	Si 786	P 1012	S 1000	Cl 1251	Ar 1521
K 419	Ca 599	Sc 631	Ti 658	V 650	Cr 652	Mn 717	Fe 759	Co 758	Ni 757	Cu 745	Zn 906		Ga 579	Ge 762	As 947	Se 941	Br 1140	Kr 1351
Rb 403	Sr 550	Y 617	Zr 661	Nb 664	Mo 685	Tc 702	Ru 711	Rh 720	Pd 804	Ag 731	Cd 868		In 558	Sn 709	Sb 834	Te 869	I 1008	Xe 1170
Cs 377	Ba 503	La 538	Hf 681	Ta 761	W 770	Re 760	Os 840	Ir 880	Pt 870	Au 890	Hg 1007		Tl 589	Pb 715	Bi 703	Po 812	At 890	Rn 1037

Table 14 Electron Attachment Enthalpy Values for Some Elements (kJ/mol)*

H −72.77						
Li −59.63	Be 0†	B −26.7	C −121.85	N 0	O −140.98	F −328.0
Na −52.87	Mg 0	Al −42.6	Si −133.6	P −72.07	S −200.41	Cl −349.0
K −48.39	Ca 0	Ga −30	Ge −120	As −78	Se −194.97	Br −324.7
Rb −46.89	Sr 0	In −30	Sn −120	Sb −103	Te −190.16	I −295.16
Cs −45.51	Ba 0	Tl −20	Pb −35.1	Bi −91.3	Po −180	At −270

*Derived from data taken from H. Hotop and W. C. Lineberger: *Journal of Physical Chemistry, Reference Data*, Vol. 14, p. 731, 1985. (This paper also includes data for the transition metals.) Some values are known to more than two decimal places. See also: http://en.wikipedia.org/wiki/Electron_affinity_(data_page)
†Elements with an electron attachment enthalpy of zero indicate that a stable anion A⁻ of the element does not exist in the gas phase.

G Vapor Pressure of Water at Various Temperatures

Table 15 Vapor Pressure of Water at Various Temperatures

Temperature (°C)	Vapor Pressure (torr)	Temperature (°C)	Vapor Pressure (torr)	Temperature (°C)	Vapor Pressure (torr)	Temperature (°C)	Vapor Pressure (torr)
−10	2.1	21	18.7	51	97.2	81	369.7
−9	2.3	22	19.8	52	102.1	82	384.9
−8	2.5	23	21.1	53	107.2	83	400.6
−7	2.7	24	22.4	54	112.5	84	416.8
−6	2.9	25	23.8	55	118.0	85	433.6
−5	3.2	26	25.2	56	123.8	86	450.9
−4	3.4	27	26.7	57	129.8	87	468.7
−3	3.7	28	28.3	58	136.1	88	487.1
−2	4.0	29	30.0	59	142.6	89	506.1
−1	4.3	30	31.8	60	149.4	90	525.8
0	4.6	31	33.7	61	156.4	91	546.1
1	4.9	32	35.7	62	163.8	92	567.0
2	5.3	33	37.7	63	171.4	93	588.6
3	5.7	34	39.9	64	179.3	94	610.9
4	6.1	35	42.2	65	187.5	95	633.9
5	6.5	36	44.6	66	196.1	96	657.6
6	7.0	37	47.1	67	205.0	97	682.1
7	7.5	38	49.7	68	214.2	98	707.3
8	8.0	39	52.4	69	223.7	99	733.2
9	8.6	40	55.3	70	233.7	100	760.0
10	9.2	41	58.3	71	243.9	101	787.6
11	9.8	42	61.5	72	254.6	102	815.9
12	10.5	43	64.8	73	265.7	103	845.1
13	11.2	44	68.3	74	277.2	104	875.1
14	12.0	45	71.9	75	289.1	105	906.1
15	12.8	46	75.7	76	301.4	106	937.9
16	13.6	47	79.6	77	314.1	107	970.6
17	14.5	48	83.7	78	327.3	108	1004.4
18	15.5	49	88.0	79	341.0	109	1038.9
19	16.5	50	92.5	80	355.1	110	1074.6
20	17.5						

H Ionization Constants for Aqueous Weak Acids at 25 °C

Table 16 Ionization Constants for Aqueous Weak Acids at 25 °C

Acid	Formula and Ionization Equation	K_a
Acetic	$CH_3CO_2H \rightleftharpoons H^+ + CH_3CO_2^-$	1.8×10^{-5}
Arsenic	$H_3AsO_4 \rightleftharpoons H^+ + H_2AsO_4^-$ $H_2AsO_4^- \rightleftharpoons H^+ + HAsO_4^{2-}$ $HAsO_4^{2-} \rightleftharpoons H^+ + AsO_4^{3-}$	$K_1 = 5.8 \times 10^{-3}$ $K_2 = 1.1 \times 10^{-7}$ $K_3 = 3.2 \times 10^{-12}$
Arsenous	$H_3AsO_3 \rightleftharpoons H^+ + H_2AsO_3^-$ $H_2AsO_3^- \rightleftharpoons H^+ + HAsO_3^{2-}$	$K_1 = 6.0 \times 10^{-10}$ $K_2 = 3.0 \times 10^{-14}$
Benzoic	$C_6H_5CO_2H \rightleftharpoons H^+ + C_6H_5CO_2^-$	6.3×10^{-5}
Boric	$H_3BO_3 \rightleftharpoons H^+ + H_2BO_3^-$ $H_2BO_3^- \rightleftharpoons H^+ + HBO_3^{2-}$ $HBO_3^{2-} \rightleftharpoons H^+ + BO_3^{3-}$	$K_1 = 7.3 \times 10^{-10}$ $K_2 = 1.8 \times 10^{-13}$ $K_3 = 1.6 \times 10^{-14}$
Carbonic	$H_2CO_3 \rightleftharpoons H^+ + HCO_3^-$ $HCO_3^- \rightleftharpoons H^+ + CO_3^{2-}$	$K_1 = 4.2 \times 10^{-7}$ $K_2 = 4.8 \times 10^{-11}$
Citric	$H_3C_6H_5O_7 \rightleftharpoons H^+ + H_2C_6H_5O_7^-$ $H_2C_6H_5O_7^- \rightleftharpoons H^+ + HC_6H_5O_7^{2-}$ $HC_6H_5O_7^{2-} \rightleftharpoons H^+ + C_6H_5O_7^{3-}$	$K_1 = 7.4 \times 10^{-3}$ $K_2 = 1.7 \times 10^{-5}$ $K_3 = 4.0 \times 10^{-7}$
Cyanic	$HOCN \rightleftharpoons H^+ + OCN^-$	3.5×10^{-4}
Formic	$HCO_2H \rightleftharpoons H^+ + HCO_2^-$	1.8×10^{-4}
Hydrazoic	$HN_3 \rightleftharpoons H^+ + N_3^-$	1.9×10^{-5}
Hydrocyanic	$HCN \rightleftharpoons H^+ + CN^-$	4.0×10^{-10}
Hydrofluoric	$HF \rightleftharpoons H^+ + F^-$	7.2×10^{-4}
Hydrogen peroxide	$H_2O_2 \rightleftharpoons H^+ + HO_2^-$	2.4×10^{-12}
Hydrosulfuric	$H_2S \rightleftharpoons H^+ + HS^-$ $HS^- \rightleftharpoons H^+ + S^{2-}$	$K_1 = 1 \times 10^{-7}$ $K_2 = 1 \times 10^{-19}$
Hypobromous	$HOBr \rightleftharpoons H^+ + OBr^-$	2.5×10^{-9}
Hypochlorous	$HOCl \rightleftharpoons H^+ + OCl^-$	3.5×10^{-8}
Nitrous	$HNO_2 \rightleftharpoons H^+ + NO_2^-$	4.5×10^{-4}
Oxalic	$H_2C_2O_4 \rightleftharpoons H^+ + HC_2O_4^-$ $HC_2O_4^- \rightleftharpoons H^+ + C_2O_4^{2-}$	$K_1 = 5.9 \times 10^{-2}$ $K_2 = 6.4 \times 10^{-5}$
Phenol	$C_6H_5OH \rightleftharpoons H^+ + C_6H_5O^-$	1.3×10^{-10}
Phosphoric	$H_3PO_4 \rightleftharpoons H^+ + H_2PO_4^-$ $H_2PO_4^- \rightleftharpoons H^+ + HPO_4^{2-}$ $HPO_4^{2-} \rightleftharpoons H^+ + PO_4^{3-}$	$K_1 = 7.5 \times 10^{-3}$ $K_2 = 6.2 \times 10^{-8}$ $K_3 = 3.6 \times 10^{-13}$
Phosphorous	$H_3PO_3 \rightleftharpoons H^+ + H_2PO_3^-$ $H_2PO_3^- \rightleftharpoons H^+ + HPO_3^{2-}$	$K_1 = 1.6 \times 10^{-2}$ $K_2 = 7.0 \times 10^{-7}$
Selenic	$H_2SeO_4 \rightleftharpoons H^+ + HSeO_4^-$ $HSeO_4^- \rightleftharpoons H^+ + SeO_4^{2-}$	$K_1 = $ very large $K_2 = 1.2 \times 10^{-2}$

Table 16 Ionization Constants for Aqueous Weak Acids at 25 °C (continued)

Acid	Formula and Ionization Equation	K_a
Selenous	$H_2SeO_3 \rightleftarrows H^+ + HSeO_3^-$ $HSeO_3^- \rightleftarrows H^+ + SeO_3^{2-}$	$K_1 = 2.7 \times 10^{-3}$ $K_2 = 2.5 \times 10^{-7}$
Sulfuric	$H_2SO_4 \rightleftarrows H^+ + HSO_4^-$ $HSO_4^- \rightleftarrows H^+ + SO_4^{2-}$	$K_1 = $ very large $K_2 = 1.2 \times 10^{-2}$
Sulfurous	$H_2SO_3 \rightleftarrows H^+ + HSO_3^-$ $HSO_3^- \rightleftarrows H^+ + SO_3^{2-}$	$K_1 = 1.2 \times 10^{-2}$ $K_2 = 6.2 \times 10^{-8}$
Tellurous	$H_2TeO_3 \rightleftarrows H^+ + HTeO_3^-$ $HTeO_3^- \rightleftarrows H^+ + TeO_3^{2-}$	$K_1 = 2 \times 10^{-3}$ $K_2 = 1 \times 10^{-8}$

I   Ionization Constants for Aqueous Weak Bases at 25 °C

Table 17   Ionization Constants for Aqueous Weak Bases at 25 °C

Base	Formula and Ionization Equation	K_b
Ammonia	$NH_3 + H_2O \rightleftharpoons NH_4^+ + OH^-$	1.8×10^{-5}
Aniline	$C_6H_5NH_2 + H_2O \rightleftharpoons C_6H_5NH_3^+ + OH^-$	4.0×10^{-10}
Dimethylamine	$(CH_3)_2NH + H_2O \rightleftharpoons (CH_3)_2NH_2^+ + OH^-$	7.4×10^{-4}
Ethylamine	$C_2H_5NH_2 + H_2O \rightleftharpoons C_2H_5NH_3^+ + OH^-$	4.3×10^{-4}
Ethylenediamine	$H_2NCH_2CH_2NH_2 + H_2O \rightleftharpoons H_2NCH_2CH_2NH_3^+ + OH^-$ $H_2NCH_2CH_2NH_3^+ + H_2O \rightleftharpoons H_3NCH_2CH_2NH_3^{2+} + OH^-$	$K_1 = 8.5 \times 10^{-5}$ $K_2 = 2.7 \times 10^{-8}$
Hydrazine	$N_2H_4 + H_2O \rightleftharpoons N_2H_5^+ + OH^-$ $N_2H_5^+ + H_2O \rightleftharpoons N_2H_6^{2+} + OH^-$	$K_1 = 8.5 \times 10^{-7}$ $K_2 = 8.9 \times 10^{-16}$
Hydroxylamine	$NH_2OH + H_2O \rightleftharpoons NH_3OH^+ + OH^-$	6.6×10^{-9}
Methylamine	$CH_3NH_2 + H_2O \rightleftharpoons CH_3NH_3^+ + OH^-$	5.0×10^{-4}
Pyridine	$C_5H_5N + H_2O \rightleftharpoons C_5H_5NH^+ + OH^-$	1.5×10^{-9}
Trimethylamine	$(CH_3)_3N + H_2O \rightleftharpoons (CH_3)_3NH^+ + OH^-$	7.4×10^{-5}

J Solubility Product Constants for Some Inorganic Compounds at 25 °C

Table 18A Solubility Product Constants at 25 °C

Cation	Compound	K_{sp}	Cation	Compound	K_{sp}
Ba^{2+}	*$BaCrO_4$	1.2×10^{-10}	Hg_2^{2+}	*Hg_2Br_2	6.4×10^{-23}
	$BaCO_3$	2.6×10^{-9}		Hg_2Cl_2	1.4×10^{-18}
	BaF_2	1.8×10^{-7}		*Hg_2I_2	2.9×10^{-29}
	*$BaSO_4$	1.1×10^{-10}		Hg_2SO_4	6.5×10^{-7}
Ca^{2+}	$CaCO_3$ (calcite)	3.4×10^{-9}	Ni^{2+}	$NiCO_3$	1.4×10^{-7}
	*CaF_2	5.3×10^{-11}		$Ni(OH)_2$	5.5×10^{-16}
	*$Ca(OH)_2$	5.5×10^{-5}	Ag^+	*$AgBr$	5.4×10^{-13}
	$CaSO_4$	4.9×10^{-5}		*$AgBrO_3$	5.4×10^{-5}
Cu^+, Cu^{2+}	$CuBr$	6.3×10^{-9}		$AgCH_3CO_2$	1.9×10^{-3}
	CuI	1.3×10^{-12}		$AgCN$	6.0×10^{-17}
	$Cu(OH)_2$	2.2×10^{-20}		Ag_2CO_3	8.5×10^{-12}
	$CuSCN$	1.8×10^{-13}		*$Ag_2C_2O_4$	5.4×10^{-12}
Au^+	$AuCl$	2.0×10^{-13}		*$AgCl$	1.8×10^{-10}
				Ag_2CrO_4	1.1×10^{-12}
Fe^{2+}	$FeCO_3$	3.1×10^{-11}		*AgI	8.5×10^{-17}
	$Fe(OH)_2$	4.9×10^{-17}		$AgSCN$	1.0×10^{-12}
Pb^{2+}	$PbBr_2$	6.6×10^{-6}		*Ag_2SO_4	1.2×10^{-5}
	$PbCO_3$	7.4×10^{-14}	Sr^{2+}	$SrCO_3$	5.6×10^{-10}
	$PbCl_2$	1.7×10^{-5}		SrF_2	4.3×10^{-9}
	$PbCrO_4$	2.8×10^{-13}		$SrSO_4$	3.4×10^{-7}
	PbF_2	3.3×10^{-8}	Tl^+	$TlBr$	3.7×10^{-6}
	PbI_2	9.8×10^{-9}		$TlCl$	1.9×10^{-4}
	$Pb(OH)_2$	1.4×10^{-15}		TlI	5.5×10^{-8}
	$PbSO_4$	2.5×10^{-8}	Zn^{2+}	$Zn(OH)_2$	3×10^{-17}
Mg^{2+}	$MgCO_3$	6.8×10^{-6}		$Zn(CN)_2$	8.0×10^{-12}
	MgF_2	5.2×10^{-11}			
	$Mg(OH)_2$	5.6×10^{-12}			
Mn^{2+}	$MnCO_3$	2.3×10^{-11}			
	*$Mn(OH)_2$	1.9×10^{-13}			

The values reported in this table were taken from J. A. Dean: *Lange's Handbook of Chemistry*, 15th ed. New York: McGraw-Hill Publishers, 1999. Values have been rounded off to two significant figures.

*Calculated solubility from these K_{sp} values will match experimental solubility for this compound within a factor of 2. Experimental values for solubilities are given in R. W. Clark and J. M. Bonicamp: *Journal of Chemical Education*, Vol. 75, p. 1182, 1998.

Table 18B K_{spa} **Values* for Some Metal Sulfides at 25 °C**

Substance	K_{spa}
HgS (red)	4×10^{-54}
HgS (black)	2×10^{-53}
Ag_2S	6×10^{-51}
CuS	6×10^{-37}
PbS	3×10^{-28}
CdS	8×10^{-28}
SnS	1×10^{-26}
FeS	6×10^{-19}

*The equilibrium constant value K_{spa} for metal sulfides refers to the equilibrium $MS(s) + H_2O(\ell) \rightleftharpoons M^{2+}(aq) + OH^-(aq) + HS^-(aq)$; see R. J. Myers, *Journal of Chemical Education*, Vol. 63, p. 687, 1986.

K Formation Constants for Some Complex Ions in Aqueous Solution at 25 °C

Table 19 Formation Constants for Some Complex Ions in Aqueous Solution at 25 °C*

Formation Equilibrium	K
$Ag^+ + 2\,Br^- \rightleftharpoons [AgBr_2]^-$	2.1×10^7
$Ag^+ + 2\,Cl^- \rightleftharpoons [AgCl_2]^-$	1.1×10^5
$Ag^+ + 2\,CN^- \rightleftharpoons [Ag(CN)_2]^-$	1.3×10^{21}
$Ag^+ + 2\,S_2O_3^{2-} \rightleftharpoons [Ag(S_2O_3)_2]^{3-}$	2.9×10^{13}
$Ag^+ + 2\,NH_3 \rightleftharpoons [Ag(NH_3)_2]^+$	1.1×10^7
$Al^{3+} + 6\,F^- \rightleftharpoons [AlF_6]^{3-}$	6.9×10^{19}
$Al^{3+} + 4\,OH^- \rightleftharpoons [Al(OH)_4]^-$	1.1×10^{33}
$Au^+ + 2\,CN^- \rightleftharpoons [Au(CN)_2]^-$	2.0×10^{38}
$Cd^{2+} + 4\,CN^- \rightleftharpoons [Cd(CN)_4]^{2-}$	6.0×10^{18}
$Cd^{2+} + 4\,NH_3 \rightleftharpoons [Cd(NH_3)_4]^{2+}$	1.3×10^7
$Co^{2+} + 6\,NH_3 \rightleftharpoons [Co(NH_3)_6]^{2+}$	1.3×10^5
$Cu^+ + 2\,CN^- \rightleftharpoons [Cu(CN)_2]^-$	1.0×10^{24}
$Cu^+ + 2\,Cl^- \rightleftharpoons [CuCl_2]^-$	3.2×10^5
$Cu^{2+} + 4\,NH_3 \rightleftharpoons [Cu(NH_3)_4]^{2+}$	2.1×10^{13}
$Fe^{2+} + 6\,CN^- \rightleftharpoons [Fe(CN)_6]^{4-}$	1.0×10^{35}
$Hg^{2+} + 4\,Cl^- \rightleftharpoons [HgCl_4]^{2-}$	1.2×10^{15}
$Ni^{2+} + 4\,CN^- \rightleftharpoons [Ni(CN)_4]^{2-}$	2.0×10^{31}
$Ni^{2+} + 6\,NH_3 \rightleftharpoons [Ni(NH_3)_6]^{2+}$	5.5×10^8
$Zn^{2+} + 4\,OH^- \rightleftharpoons [Zn(OH)_4]^{2-}$	4.6×10^{17}
$Zn^{2+} + 4\,NH_3 \rightleftharpoons [Zn(NH_3)_4]^{2+}$	2.9×10^9

*Data reported in this table are taken from J. A. Dean: *Lange's Handbook of Chemistry,* 15th ed. New York: McGraw-Hill Publishers, 1999.

L Selected Thermodynamic Values

Table 20 **Selected Thermodynamic Values***

Species	$\Delta_f H°$ (298.15 K) (kJ/mol)	$S°$ (298.15 K) (J/K · mol)	$\Delta_f G°$ (298.15 K) (kJ/mol)
Aluminum			
Al(s)	0	28.3	0
AlCl$_3$(s)	−705.63	109.29	−630.0
Al$_2$O$_3$(s)	−1675.7	50.92	−1582.3
Barium			
BaCl$_2$(s)	−858.6	123.68	−810.4
BaCO$_3$(s)	−1213	112.1	−1134.41
BaO(s)	−548.1	72.05	−520.38
BaSO$_4$(s)	−1473.2	132.2	−1362.2
Beryllium			
Be(s)	0	9.5	0
Be(OH)$_2$(s)	−902.5	51.9	−815.0
Boron			
BCl$_3$(g)	−402.96	290.17	−387.95
Bromine			
Br(g)	111.884	175.022	82.396
Br$_2$(ℓ)	0	152.2	0
Br$_2$(g)	30.91	245.47	3.12
BrF$_3$(g)	−255.60	292.53	−229.43
HBr(g)	−36.29	198.70	−53.45
Calcium			
Ca(s)	0	41.59	0
Ca(g)	178.2	158.884	144.3
Ca^{2+}(g)	1925.90	—	—
CaC$_2$(s)	−59.8	70.	−64.93
CaCO$_3$(s, calcite)	−1207.6	91.7	−1129.16
CaCl$_2$(s)	−795.8	104.6	−748.1
CaF$_2$(s)	−1219.6	68.87	−1167.3
CaH$_2$(s)	−186.2	42	−147.2
CaO(s)	−635.09	38.2	−603.42
CaS(s)	−482.4	56.5	−477.4
Ca(OH)$_2$(s)	−986.09	83.39	−898.43
Ca(OH)$_2$(aq)	−1002.82	—	−868.07
CaSO$_4$(s)	−1434.52	106.5	−1322.02

*Most thermodynamic data are taken from the NIST Chemistry WebBook at http://webbook.nist.gov.

Table 20 Selected Thermodynamic Values (continued)

Species	$\Delta_f H°$ (298.15 K) (kJ/mol)	$S°$ (298.15 K) (J/K · mol)	$\Delta_f G°$ (298.15 K) (kJ/mol)
Carbon			
C(s, graphite)	0	5.6	0
C(s, diamond)	1.8	2.377	2.900
C(g)	716.67	158.1	671.2
$CCl_4(\ell)$	−128.4	214.39	−57.63
$CCl_4(g)$	−95.98	309.65	−53.61
$CHCl_3(\ell)$	−134.47	201.7	−73.66
$CHCl_3(g)$	−103.18	295.61	−70.4
$CH_4(g, methane)$	−74.87	186.26	−50.8
$C_2H_2(g, ethyne)$	226.73	200.94	209.20
$C_2H_4(g, ethene)$	52.47	219.36	68.35
$C_2H_6(g, ethane)$	−83.85	229.2	−31.89
$C_3H_8(g, propane)$	−104.7	270.3	−24.4
$C_6H_6(\ell, benzene)$	48.95	173.26	124.21
$CH_3OH(\ell, methanol)$	−238.4	127.19	−166.14
$CH_3OH(g, methanol)$	−201.0	239.7	−162.5
$C_2H_5OH(\ell, ethanol)$	−277.0	160.7	−174.7
$C_2H_5OH(g, ethanol)$	−235.3	282.70	−168.49
CO(g)	−110.525	197.674	−137.168
$CO_2(g)$	−393.509	213.74	−394.359
$CS_2(\ell)$	89.41	151	65.2
$CS_2(g)$	116.7	237.8	66.61
$COCl_2(g)$	−218.8	283.53	−204.6
Cesium			
Cs(s)	0	85.23	0
$Cs^+(g)$	457.964	—	—
CsCl(s)	−443.04	101.17	−414.53
Chlorine			
Cl(g)	121.3	165.19	105.3
$Cl^-(g)$	−233.13	—	—
$Cl_2(g)$	0	223.08	0
HCl(g)	−92.31	186.2	−95.09
HCl(aq)	−167.159	56.5	−131.26
Chromium			
Cr(s)	0	23.62	0
$Cr_2O_3(s)$	−1134.7	80.65	−1052.95
$CrCl_3(s)$	−556.5	123.0	−486.1

(continued)

Table 20 Selected Thermodynamic Values (continued)

Species	$\Delta_f H°$ (298.15 K) (kJ/mol)	$S°$ (298.15 K) (J/K · mol)	$\Delta_f G°$ (298.15 K) (kJ/mol)
Copper			
Cu(s)	0	33.17	0
CuO(s)	−156.06	42.59	−128.3
CuCl$_2$(s)	−220.1	108.07	−175.7
CuSO$_4$(s)	−769.98	109.05	−660.75
Fluorine			
F$_2$(g)	0	202.8	0
F(g)	78.99	158.754	61.91
F$^-$(g)	−255.39	—	—
F$^-$(aq)	−332.63	—	−278.79
HF(g)	−273.3	173.779	−273.2
HF(aq)	−332.63	88.7	−278.79
Hydrogen			
H$_2$(g)	0	130.7	0
H(g)	217.965	114.713	203.247
H$^+$(g)	1536.202	—	—
H$_2$O(ℓ)	−285.83	69.95	−237.15
H$_2$O(g)	−241.83	188.84	−228.59
H$_2$O$_2$(ℓ)	−187.78	109.6	−120.35
Iodine			
I$_2$(s)	0	116.135	0
I$_2$(g)	62.438	260.69	19.327
I(g)	106.838	180.791	70.250
I$^-$(g)	−197	—	—
ICl(g)	17.51	247.56	−5.73
Iron			
Fe(s)	0	27.78	0
FeO(s)	−272	—	—
Fe$_2$O$_3$(s, hematite)	−825.5	87.40	−742.2
Fe$_3$O$_4$(s, magnetite)	−1118.4	146.4	−1015.4
FeCl$_2$(s)	−341.79	117.95	−302.30
FeCl$_3$(s)	−399.49	142.3	−344.00
FeS$_2$(s, pyrite)	−178.2	52.93	−166.9
Fe(CO)$_5$(ℓ)	−774.0	338.1	−705.3
Lead			
Pb(s)	0	64.81	0
PbCl$_2$(s)	−359.41	136.0	−314.10
PbO(s, yellow)	−219	66.5	−196
PbO$_2$(s)	−277.4	68.6	−217.39
PbS(s)	−100.4	91.2	−98.7

Table 20 **Selected Thermodynamic Values (continued)**

Species	$\Delta_f H°$ (298.15 K) (kJ/mol)	$S°$ (298.15 K) (J/K · mol)	$\Delta_f G°$ (298.15 K) (kJ/mol)
Lithium			
Li(s)	0	29.12	0
Li$^+$(g)	685.783	—	—
LiOH(s)	−484.93	42.81	−438.96
LiOH(aq)	−508.48	2.80	−450.58
LiCl(s)	−408.701	59.33	−384.37
Magnesium			
Mg(s)	0	32.67	0
MgCl$_2$(s)	−641.62	89.62	−592.09
MgCO$_3$(s)	−1111.69	65.84	−1028.2
MgO(s)	−601.24	26.85	−568.93
Mg(OH)$_2$(s)	−924.54	63.18	−833.51
MgS(s)	−346.0	50.33	−341.8
Mercury			
Hg(ℓ)	0	76.02	0
HgCl$_2$(s)	−224.3	146.0	−178.6
HgO(s, red)	−90.83	70.29	−58.539
HgS(s, red)	−58.2	82.4	−50.6
Nickel			
Ni(s)	0	29.87	0
NiO(s)	−239.7	37.99	−211.7
NiCl$_2$(s)	−305.332	97.65	−259.032
Nitrogen			
N$_2$(g)	0	191.56	0
N(g)	472.704	153.298	455.563
NH$_3$(g)	−45.90	192.77	−16.37
N$_2$H$_4$(ℓ)	50.63	121.52	149.45
NH$_4$Cl(s)	−314.55	94.85	−203.08
NH$_4$Cl(aq)	−299.66	169.9	−210.57
NH$_4$NO$_3$(s)	−365.56	151.08	−183.84
NH$_4$NO$_3$(aq)	−339.87	259.8	−190.57
NO(g)	90.29	210.76	86.58
NO$_2$(g)	33.1	240.04	51.23
N$_2$O(g)	82.05	219.85	104.20
N$_2$O$_4$(g)	9.08	304.38	97.73
NOCl(g)	51.71	261.8	66.08
HNO$_3$(ℓ)	−174.10	155.60	−80.71
HNO$_3$(g)	−135.06	266.38	−74.72
HNO$_3$(aq)	−207.36	146.4	−111.25

(continued)

Table 20 **Selected Thermodynamic Values (continued)**

Species	$\Delta_f H°$ (298.15 K) (kJ/mol)	$S°$ (298.15 K) (J/K · mol)	$\Delta_f G°$ (298.15 K) (kJ/mol)
Oxygen			
$O_2(g)$	0	205.07	0
$O(g)$	249.170	161.055	231.731
$O_3(g)$	142.67	238.92	163.2
Phosphorus			
$P_4(s, white)$	0	41.1	0
$P_4(s, red)$	−17.6	22.80	−12.1
$P(g)$	314.64	163.193	278.25
$PH_3(g)$	5.47	210.24	6.64
$PCl_3(g)$	−287.0	311.78	−267.8
$P_4O_{10}(s)$	−2984.0	228.86	−2697.7
$H_3PO_4(\ell)$	−1279.0	110.5	−1119.1
Potassium			
$K(s)$	0	64.63	0
$KCl(s)$	−436.68	82.56	−408.77
$KClO_3(s)$	−397.73	143.1	−296.25
$KI(s)$	−327.90	106.32	−324.892
$KOH(s)$	−424.72	78.9	−378.92
$KOH(aq)$	−482.37	91.6	−440.50
Silicon			
$Si(s)$	0	18.82	0
$SiBr_4(\ell)$	−457.3	277.8	−443.9
$SiC(s)$	−65.3	16.61	−62.8
$SiCl_4(g)$	−662.75	330.86	−622.76
$SiH_4(g)$	34.31	204.65	56.84
$SiF_4(g)$	−1614.94	282.49	−1572.65
$SiO_2(s, quartz)$	−910.86	41.46	−856.97
Silver			
$Ag(s)$	0	42.55	0
$Ag_2O(s)$	−31.1	121.3	−11.32
$AgCl(s)$	−127.01	96.25	−109.76
$AgNO_3(s)$	−124.39	140.92	−33.41
Sodium			
$Na(s)$	0	51.21	0
$Na(g)$	107.3	153.765	76.83
$Na^+(g)$	609.358	—	—
$NaBr(s)$	−361.02	86.82	−348.983
$NaCl(s)$	−411.12	72.11	−384.04
$NaCl(g)$	−181.42	229.79	−201.33
$NaCl(aq)$	−407.27	115.5	−393.133

Table 20 Selected Thermodynamic Values (continued)

Species	$\Delta_f H°$ (298.15 K) (kJ/mol)	$S°$ (298.15 K) (J/K · mol)	$\Delta_f G°$ (298.15 K) (kJ/mol)
Sodium (continued)			
NaOH(s)	−425.93	64.46	−379.75
NaOH(aq)	−469.15	48.1	−418.09
Na_2CO_3(s)	−1130.77	134.79	−1048.08
Sulfur			
S(s, rhombic)	0	32.1	0
S(g)	278.98	167.83	236.51
S_2Cl_2(g)	−18.4	331.5	−31.8
SF_6(g)	−1209	291.82	−1105.3
H_2S(g)	−20.63	205.79	−33.56
SO_2(g)	−296.84	248.21	−300.13
SO_3(g)	−395.77	256.77	−371.04
$SOCl_2$(g)	−212.5	309.77	−198.3
H_2SO_4(ℓ)	−814	156.9	−689.96
H_2SO_4(aq)	−909.27	20.1	−744.53
Tin			
Sn(s, white)	0	51.08	0
Sn(s, gray)	−2.09	44.14	0.13
$SnCl_4$(ℓ)	−511.3	258.6	−440.15
$SnCl_4$(g)	−471.5	365.8	−432.31
SnO_2(s)	−577.63	49.04	−515.88
Titanium			
Ti(s)	0	30.72	0
$TiCl_4$(ℓ)	−804.2	252.34	−737.2
$TiCl_4$(g)	−763.16	354.84	−726.7
TiO_2(s)	−939.7	49.92	−884.5
Zinc			
Zn(s)	0	41.63	0
$ZnCl_2$(s)	−415.05	111.46	−369.398
ZnO(s)	−348.28	43.64	−318.30
ZnS(s, sphalerite)	−205.98	57.7	−201.29

M Standard Reduction Potentials in Aqueous Solution at 25 °C

Table 21 Standard Reduction Potentials in Aqueous Solution at 25 °C

Acidic Solution	Standard Reduction Potential $E°$ (volts)
$F_2(g) + 2\,e^- \longrightarrow 2\,F^-(aq)$	2.87
$Co^{3+}(aq) + e^- \longrightarrow Co^{2+}(aq)$	1.82
$Pb^{4+}(aq) + 2\,e^- \longrightarrow Pb^{2+}(aq)$	1.8
$H_2O_2(aq) + 2\,H^+(aq) + 2\,e^- \longrightarrow 2\,H_2O$	1.77
$NiO_2(s) + 4\,H^+(aq) + 2\,e^- \longrightarrow Ni^{2+}(aq) + 2\,H_2O$	1.7
$PbO_2(s) + SO_4{}^{2-}(aq) + 4\,H^+(aq) + 2\,e^- \longrightarrow PbSO_4(s) + 2\,H_2O$	1.685
$Au^+(aq) + e^- \longrightarrow Au(s)$	1.68
$2\,HClO(aq) + 2\,H^+(aq) + 2\,e^- \longrightarrow Cl_2(g) + 2\,H_2O$	1.63
$Ce^{4+}(aq) + e^- \longrightarrow Ce^{3+}(aq)$	1.61
$NaBiO_3(s) + 6\,H^+(aq) + 2\,e^- \longrightarrow Bi^{3+}(aq) + Na^+(aq) + 3\,H_2O$	≈1.6
$MnO_4{}^-(aq) + 8\,H^+(aq) + 5\,e^- \longrightarrow Mn^{2+}(aq) + 4\,H_2O$	1.51
$Au^{3+}(aq) + 3\,e^- \longrightarrow Au(s)$	1.50
$ClO_3{}^-(aq) + 6\,H^+(aq) + 5\,e^- \longrightarrow \frac{1}{2}\,Cl_2(g) + 3\,H_2O$	1.47
$BrO_3{}^-(aq) + 6\,H^+(aq) + 6\,e^- \longrightarrow Br^-(aq) + 3\,H_2O$	1.44
$Cl_2(g) + 2\,e^- \longrightarrow 2\,Cl^-(aq)$	1.36
$Cr_2O_7{}^{2-}(aq) + 14\,H^+(aq) + 6\,e^- \longrightarrow 2\,Cr^{3+}(aq) + 7\,H_2O$	1.33
$N_2H_5{}^+(aq) + 3\,H^+(aq) + 2\,e^- \longrightarrow 2\,NH_4{}^+(aq)$	1.24
$MnO_2(s) + 4\,H^+(aq) + 2\,e^- \longrightarrow Mn^{2+}(aq) + 2\,H_2O$	1.23
$O_2(g) + 4\,H^+(aq) + 4\,e^- \longrightarrow 2\,H_2O$	1.229
$Pt^{2+}(aq) + 2\,e^- \longrightarrow Pt(s)$	1.2
$IO_3{}^-(aq) + 6\,H^+(aq) + 5\,e^- \longrightarrow \frac{1}{2}\,I_2(aq) + 3\,H_2O$	1.195
$ClO_4{}^-(aq) + 2\,H^+(aq) + 2\,e^- \longrightarrow ClO_3{}^-(aq) + H_2O$	1.19
$Br_2(\ell) + 2\,e^- \longrightarrow 2\,Br^-(aq)$	1.08
$AuCl_4{}^-(aq) + 3\,e^- \longrightarrow Au(s) + 4\,Cl^-(aq)$	1.00
$Pd^{2+}(aq) + 2\,e^- \longrightarrow Pd(s)$	0.987
$NO_3{}^-(aq) + 4\,H^+(aq) + 3\,e^- \longrightarrow NO(g) + 2\,H_2O$	0.96
$NO_3{}^-(aq) + 3\,H^+(aq) + 2\,e^- \longrightarrow HNO_2(aq) + H_2O$	0.94
$2\,Hg^{2+}(aq) + 2\,e^- \longrightarrow Hg_2{}^{2+}(aq)$	0.920
$Hg^{2+}(aq) + 2\,e^- \longrightarrow Hg(\ell)$	0.855
$Ag^+(aq) + e^- \longrightarrow Ag(s)$	0.7994
$Hg_2{}^{2+}(aq) + 2\,e^- \longrightarrow 2\,Hg(\ell)$	0.789
$Fe^{3+}(aq) + e^- \longrightarrow Fe^{2+}(aq)$	0.771

Table 21 Standard Reduction Potentials in Aqueous Solution at 25 °C (continued)

Acidic Solution	Standard Reduction Potential $E°$ (volts)
$SbCl_6^-(aq) + 2\,e^- \longrightarrow SbCl_4^-(aq) + 2\,Cl^-(aq)$	0.75
$[PtCl_4]^{2-}(aq) + 2\,e^- \longrightarrow Pt(s) + 4\,Cl^-(aq)$	0.73
$O_2(g) + 2\,H^+(aq) + 2\,e^- \longrightarrow H_2O_2(aq)$	0.682
$[PtCl_6]^{2-}(aq) + 2\,e^- \longrightarrow [PtCl_4]^{2-}(aq) + 2\,Cl^-(aq)$	0.68
$I_2(aq) + 2\,e^- \longrightarrow 2\,I^-(aq)$	0.621
$H_3AsO_4(aq) + 2\,H^+(aq) + 2\,e^- \longrightarrow H_3AsO_3(aq) + H_2O$	0.58
$I_2(s) + 2\,e^- \longrightarrow 2\,I^-(aq)$	0.535
$TeO_2(s) + 4\,H^+(aq) + 4\,e^- \longrightarrow Te(s) + 2\,H_2O$	0.529
$Cu^+(aq) + e^- \longrightarrow Cu(s)$	0.521
$[RhCl_6]^{3-}(aq) + 3\,e^- \longrightarrow Rh(s) + 6\,Cl^-(aq)$	0.44
$Cu^{2+}(aq) + 2\,e^- \longrightarrow Cu(s)$	0.337
$Hg_2Cl_2(s) + 2\,e^- \longrightarrow 2\,Hg(\ell) + 2\,Cl^-(aq)$	0.27
$AgCl(s) + e^- \longrightarrow Ag(s) + Cl^-(aq)$	0.222
$SO_4^{2-}(aq) + 4\,H^+(aq) + 2\,e^- \longrightarrow SO_2(g) + 2\,H_2O$	0.20
$SO_4^{2-}(aq) + 4\,H^+(aq) + 2\,e^- \longrightarrow H_2SO_3(aq) + H_2O$	0.17
$Cu^{2+}(aq) + e^- \longrightarrow Cu^+(aq)$	0.153
$Sn^{4+}(aq) + 2\,e^- \longrightarrow Sn^{2+}(aq)$	0.15
$S(s) + 2\,H^+ + 2\,e^- \longrightarrow H_2S(aq)$	0.14
$AgBr(s) + e^- \longrightarrow Ag(s) + Br^-(aq)$	0.0713
$2\,H^+(aq) + 2\,e^- \longrightarrow H_2(g)\,(\text{reference electrode})$	0.0000
$N_2O(g) + 6\,H^+(aq) + H_2O + 4\,e^- \longrightarrow 2\,NH_3OH^+(aq)$	−0.05
$Pb^{2+}(aq) + 2\,e^- \longrightarrow Pb(s)$	−0.126
$Sn^{2+}(aq) + 2\,e^- \longrightarrow Sn(s)$	−0.14
$AgI(s) + e^- \longrightarrow Ag(s) + I^-(aq)$	−0.15
$[SnF_6]^{2-}(aq) + 4\,e^- \longrightarrow Sn(s) + 6\,F^-(aq)$	−0.25
$Ni^{2+}(aq) + 2\,e^- \longrightarrow Ni(s)$	−0.25
$Co^{2+}(aq) + 2\,e^- \longrightarrow Co(s)$	−0.28
$Tl^+(aq) + e^- \longrightarrow Tl(s)$	−0.34
$PbSO_4(s) + 2\,e^- \longrightarrow Pb(s) + SO_4^{2-}(aq)$	−0.356
$Se(s) + 2\,H^+(aq) + 2\,e^- \longrightarrow H_2Se(aq)$	−0.40
$Cd^{2+}(aq) + 2\,e^- \longrightarrow Cd(s)$	−0.403
$Cr^{3+}(aq) + e^- \longrightarrow Cr^{2+}(aq)$	−0.41
$Fe^{2+}(aq) + 2\,e^- \longrightarrow Fe(s)$	−0.44
$2\,CO_2(g) + 2\,H^+(aq) + 2\,e^- \longrightarrow H_2C_2O_4(aq)$	−0.49
$Ga^{3+}(aq) + 3\,e^- \longrightarrow Ga(s)$	−0.53
$HgS(s) + 2\,H^+(aq) + 2\,e^- \longrightarrow Hg(\ell) + H_2S(g)$	−0.72
$Cr^{3+}(aq) + 3\,e^- \longrightarrow Cr(s)$	−0.74
$Zn^{2+}(aq) + 2\,e^- \longrightarrow Zn(s)$	−0.763
$Cr^{2+}(aq) + 2\,e^- \longrightarrow Cr(s)$	−0.91

(continued)

Table 21 Standard Reduction Potentials in Aqueous Solution at 25 °C (continued)

Acidic Solution	Standard Reduction Potential $E°$ (volts)
$FeS(s) + 2\,e^- \longrightarrow Fe(s) + S^{2-}(aq)$	−1.01
$Mn^{2+}(aq) + 2\,e^- \longrightarrow Mn(s)$	−1.18
$V^{2+}(aq) + 2\,e^- \longrightarrow V(s)$	−1.18
$CdS(s) + 2\,e^- \longrightarrow Cd(s) + S^{2-}(aq)$	−1.21
$ZnS(s) + 2\,e^- \longrightarrow Zn(s) + S^{2-}(aq)$	−1.44
$Zr^{4+}(aq) + 4\,e^- \longrightarrow Zr(s)$	−1.53
$Al^{3+}(aq) + 3\,e^- \longrightarrow Al(s)$	−1.66
$Mg^{2+}(aq) + 2\,e^- \longrightarrow Mg(s)$	−2.37
$Na^+(aq) + e^- \longrightarrow Na(s)$	−2.714
$Ca^{2+}(aq) + 2\,e^- \longrightarrow Ca(s)$	−2.87
$Sr^{2+}(aq) + 2\,e^- \longrightarrow Sr(s)$	−2.89
$Ba^{2+}(aq) + 2\,e^- \longrightarrow Ba(s)$	−2.90
$Rb^+(aq) + e^- \longrightarrow Rb(s)$	−2.925
$K^+(aq) + e^- \longrightarrow K(s)$	−2.925
$Li^+(aq) + e^- \longrightarrow Li(s)$	−3.045

Basic Solution	Standard Reduction Potential $E°$ (volts)
$ClO^-(aq) + H_2O + 2\,e^- \longrightarrow Cl^-(aq) + 2\,OH^-(aq)$	0.89
$OOH^-(aq) + H_2O + 2\,e^- \longrightarrow 3\,OH^-(aq)$	0.88
$2\,NH_2OH(aq) + 2\,e^- \longrightarrow N_2H_4(aq) + 2\,OH^-(aq)$	0.74
$ClO_3^-(aq) + 3\,H_2O + 6\,e^- \longrightarrow Cl^-(aq) + 6\,OH^-(aq)$	0.62
$MnO_4^-(aq) + 2\,H_2O + 3\,e^- \longrightarrow MnO_2(s) + 4\,OH^-(aq)$	0.588
$MnO_4^-(aq) + e^- \longrightarrow MnO_4^{2-}(aq)$	0.564
$NiO_2(s) + 2\,H_2O + 2\,e^- \longrightarrow Ni(OH)_2(s) + 2\,OH^-(aq)$	0.49
$Ag_2CrO_4(s) + 2\,e^- \longrightarrow 2\,Ag(s) + CrO_4^{2-}(aq)$	0.446
$O_2(g) + 2\,H_2O + 4\,e^- \longrightarrow 4\,OH^-(aq)$	0.40
$ClO_4^-(aq) + H_2O + 2\,e^- \longrightarrow ClO_3^-(aq) + 2\,OH^-(aq)$	0.36
$Ag_2O(s) + H_2O + 2\,e^- \longrightarrow 2\,Ag(s) + 2\,OH^-(aq)$	0.34
$2\,NO_2^-(aq) + 3\,H_2O + 4\,e^- \longrightarrow N_2O(g) + 6\,OH^-(aq)$	0.15
$N_2H_4(aq) + 2\,H_2O + 2\,e^- \longrightarrow 2\,NH_3(aq) + 2\,OH^-(aq)$	0.10
$[Co(NH_3)_6]^{3+}(aq) + e^- \longrightarrow [Co(NH_3)_6]^{2+}(aq)$	0.10
$HgO(s) + H_2O + 2\,e^- \longrightarrow Hg(\ell) + 2\,OH^-(aq)$	0.0984
$O_2(g) + H_2O + 2\,e^- \longrightarrow OOH^-(aq) + OH^-(aq)$	0.076
$NO_3^-(aq) + H_2O + 2\,e^- \longrightarrow NO_2^-(aq) + 2\,OH^-(aq)$	0.01
$MnO_2(s) + 2\,H_2O + 2\,e^- \longrightarrow Mn(OH)_2(s) + 2\,OH^-(aq)$	−0.05
$CrO_4^{2-}(aq) + 4\,H_2O + 3\,e^- \longrightarrow Cr(OH)_3(s) + 5\,OH^-(aq)$	−0.12
$Cu(OH)_2(s) + 2\,e^- \longrightarrow Cu(s) + 2\,OH^-(aq)$	−0.36
$S(s) + 2\,e^- \longrightarrow S^{2-}(aq)$	−0.48
$Fe(OH)_3(s) + e^- \longrightarrow Fe(OH)_2(s) + OH^-(aq)$	−0.56
$2\,H_2O + 2\,e^- \longrightarrow H_2(g) + 2\,OH^-(aq)$	−0.8277

Table 21 Standard Reduction Potentials in Aqueous Solution at 25 °C (continued)

Basic Solution	Standard Reduction Potential $E°$ (volts)
$2\ NO_3^-(aq) + 2\ H_2O + 2\ e^- \longrightarrow N_2O_4(g) + 4\ OH^-(aq)$	−0.85
$Fe(OH)_2(s) + 2\ e^- \longrightarrow Fe(s) + 2\ OH^-(aq)$	−0.877
$SO_4^{2-}(aq) + H_2O + 2\ e^- \longrightarrow SO_3^{2-}(aq) + 2\ OH^-(aq)$	−0.93
$N_2(g) + 4\ H_2O + 4\ e^- \longrightarrow N_2H_4(aq) + 4\ OH^-(aq)$	−1.15
$[Zn(OH)_4]^{2-}(aq) + 2\ e^- \longrightarrow Zn(s) + 4\ OH^-(aq)$	−1.22
$Zn(OH)_2(s) + 2\ e^- \longrightarrow Zn(s) + 2\ OH^-(aq)$	−1.245
$[Zn(CN)_4]^{2-}(aq) + 2\ e^- \longrightarrow Zn(s) + 4\ CN^-(aq)$	−1.26
$Cr(OH)_3(s) + 3\ e^- \longrightarrow Cr(s) + 3\ OH^-(aq)$	−1.30
$SiO_3^{2-}(aq) + 3\ H_2O + 4\ e^- \longrightarrow Si(s) + 6\ OH^-(aq)$	−1.70

N Answers to Chapter Opening Questions and Case Study Questions

Chapter 1

Chapter Opening: Gold!

1. 5.2×10^5 g Au; $19 million

2. Gold = Au; copper = Cu; zinc = Zn

3. 1337 K

4. (a) No, gold is not the most dense element.

   (b) 0.217 cm^3 Au

Case Study: CO$_2$ in the Oceans

1. Carbon dioxide

2. Calcium: Ca; copper: Cu; manganese: Mn; iron: Fe

3. Most dense: copper, least dense: calcium

4. Elements: calcium, carbon, oxygen

   Compound name: calcium carbonate

Let's Review

Case Study: Out of Gas!

1. Fuel density in kg/L: (1.77 lb/L) (0.4536 kg/lb) = 0.803 kg/L

2. Mass of fuel already in tank: 7682 L (0.803 kg/L) = 6170 kg

   Mass of fuel needed: 22,300 kg − 6,170 kg = 16,100 kg (answer to three significant figures)

   Volume of fuel needed: 16,130 kg (1 L/0.803 kg) = 20,100 L

Chapter 2

Chapter Opening: The Periodic Table, the Central Icon of Chemistry

1. Eka-silicon is germanium. Its atomic weight is 72.61 (predicted 72), and its density is 5.32 g/cm^3 (predicted value 5.5 g/cm^3).

2. Other elements missing from Mendeleev's periodic table include Sc, Ga, the noble gases (He, Ne, Ar, Kr, Xe), and all of the radioactive elements except Th and U.

Case Study: Using Isotopes: Ötzi, the Iceman of the Alps

1. ^{18}O: 18 − 8 = 10 neutrons

   ^{204}Pb: 204 − 82 = 122 neutrons

   ^{206}Pb: 206 − 82 = 124 neutrons

2. ^{14}C: 14 − 6 = 8 neutrons

3. ^{87}Sr/^{86}Sr = 0.710

   The ratio found in the Iceman is slightly higher.

Case Study: Mummies, Bangladesh, and the Formula of Compound 606

1. Amount As: 19.024 g As × (1 mol As/74.9216 g As) = 0.25392 mol As

   Amount Cu: 48.407 g Cu × (1 mol Cu/63.546 g Cu) = 0.76176 mol Cu

   Amount S: 32.569 g S × (1 mol S/32.066 g S) = 1.0157 mol S

   Mole ratio Cu/As: 0.76176 mol Cu/0.25392 mol As = 3.0000 mol Cu/1 mol As

   Mole ratio S/As: 1.0157 mol S/0.25392 mol As = 4.0001 mol S/1 mol As

   Empirical formula: Cu$_3$AsS$_4$

2. Amount C: 39.37 g C × (1 mol C/12.011 g C) = 3.278 mol C

   Amount H: 3.304 g H × (1 mol H/1.0079 g H) = 3.278 mol H

   Amount O: 8.741 g O × (1 mol O/15.999 g O) = 0.5463 mol O

   Amount N: 7.652 g N × (1 mol N/14.007 g N) = 0.5463 mol N

   Amount As: 40.932 g As × (1 mol As/74.9216 g As) = 0.54633 mol As

   Mole ratio C/O and H/O: 3.278 mol C/0.5463 mol O = 6.000 mol C/1 mol O. A similar calculation yields 6.000 mol H/1 mol O.

Mole ratio N/O and As/O: 0.5463 mol N/ 0.5463 mol O = 1.000 mol N/1 mol O. A similar calculation yields 1.000 mol As/1 mol O.

Empirical formula: C_6H_6AsNO

Molar mass of C_6H_6AsNO: 183.0 g/mol

Compound 1: 549/183.0 = 3.00; therefore, the molecular formula is $C_{18}H_{18}As_3N_3O_3$.

Compound 2: 915/183.0 = 5.00; therefore, the molecular formula is $C_{30}H_{30}As_5N_5O_5$.

Chapter 3

Chapter Opening: Black Smokers and Volcanoes

$Fe^{2+}(aq) + H_2S(aq) + 2 H_2O(\ell) \rightarrow$
$FeS(s) + 2 H_3O^+(aq)$

$2 Bi^{3+}(aq) + 3 H_2S(aq) + 6 H_2O(\ell) \rightarrow$
$Bi_2S_3(s) + 6 H_3O^+(aq)$

$Ca^{2+}(aq) + SO_4^{2-}(aq) \rightarrow CaSO_4(s)$

Case Study: Killing Bacteria with Silver

1. 100×10^{15} Ag^+ ions (1 mol/6.022×10^{23} ions) = 2×10^{-7} mol Ag^+

2. 2×10^{-7} mol Ag^+ (107.9 g Ag^+/1 mol Ag^+) = 2×10^{-5} g Ag^+ ions

Chapter 4

Chapter Opening: The Chemistry of Pyrotechnics

1. 1.0 g Fe (1 mol Fe/55.85 g Fe)(2 mol Fe_2O_3/ 4 mol Fe)(159.7 g Fe_2O_3/1 mol Fe_2O_3) = 1.4 g Fe_2O_3

2. 10.0 g Fe_2O_3 (1 mol Fe_2O_3/159.7 g Fe_2O_3) (2 mol Al/1 mol Fe_2O_3)(26.98 g Al/1 mol Al) = 3.38 g Al

3. 20 g Al (1 mol Al/26.98 g Al) = 0.7 mol Al
10 g Fe_2O_3 (1 mol Fe_2O_3/159.7 g Fe_2O_3) = 0.06 mol Fe_2O_3

Mole ratio from balanced equation: 2 mol Al/ 1 mol Fe_2O_3

Mole ratio actually present: 0.7 mol Al/0.06 mol Fe_2O_3 = 1×10^1 mol Al/1 mol Fe_2O_3

There is an excess of Al compared to what the balanced equation requires, so Fe_2O_3 is the limiting reactant.

Case Study: Green Chemistry and Atom Economy

Reactant molecules contain 5 C, 10 H, 3 O. Combined molar mass = 118.1 g/mol

Desired product (methyl methacrylate) contains 5 C, 8 H, 2 O. Molar mass = 100.1 g/mol

% Atom economy = $(100.1/118.1) \times 100$ = 84.75%

Case Study: How Much Salt Is There in Seawater?

1. Step 1: Calculate the amount of Cl^- in the diluted solution from titration data.

   Mol Cl^- in 50 mL of dilute solution = mol Ag^+ = (0.100 mol/L)(0.02625 L) = 2.63×10^{-3} mol Cl^-

   Step 2: Calculate the concentration of Cl^- in the dilute solution.

   Concentration of Cl^- in dilute solution = 2.63×10^{-3} mol/0.0500 L = 5.25×10^{-2} M

   Step 3: Calculate the concentration of Cl^- in seawater.

   Seawater was initially diluted to one hundredth its original concentration. Thus, the concentration of Cl^- in seawater (undiluted) = 5.25 M

Case Study: Forensic Chemistry: Titrations and Food Tampering

1. Step 1: Calculate the amount of I_2 in solution from titration data.

   Amount I_2 = (0.0425 mol $S_2O_3^{2-}$/L)(0.0253 L) (1 mol I_2/2 mol $S_2O_3^{2-}$) = 5.38×10^{-4} mol I_2

   Step 2: Calculate the amount of NaClO present based on the amount of I_2 formed, and from that value calculate the mass of NaClO.

   Mass NaClO = 5.38×10^{-4} mol I_2 (1 mol HClO/ 1 mol I_2)(1 mol NaClO/1 mol HClO)(74.44 g NaClO/1 mol NaClO) = 0.0400 g NaClO

Chapter 5

Chapter Opening: Energy and Your Diet

1. Orange, all commercial varieties (197 kJ) < apple, raw with skin (218 kJ) < potato, flesh and skin, raw (321 kJ) < rice, brown, long grain, uncooked (1548 kJ) < popcorn, unpopped kernels (1568 kJ) < glazed doughnut, yeast-leavened, unenriched (1686 kJ) < peanut M&Ms (2156 kJ) < mixed nuts, dry roasted, with peanuts, with salt (2485 kJ)

2. In the following, we assume water vapor, $H_2O(g)$, is formed upon oxidation.

   Balanced Equation: $C_2H_5OH(\ell) + 3 O_2(g) \rightarrow 2 CO_2(g) + 3 H_2O(g)$

   $\Delta_r H° = (2$ mol CO_2/1 mol-rxn$)[\Delta_f H° CO_2(g)]$ $+ (3$ mol H_2O/1 mol-rxn$)[\Delta_f H° H_2O(g)] -$ $\{(1$ mol C_2H_5OH/1 mol-rxn$)[\Delta_f H° C_2H_5OH(\ell)]$ $+ (3$ mol O_2/1 mol-rxn$)[\Delta_f H° O_2(g)]\}$

$\Delta_r H° = (2 \text{ mol } CO_2/1 \text{ mol-rxn})[-393.5 \text{ kJ}/ \text{mol } CO_2] + (3 \text{ mol } H_2O/1 \text{ mol-rxn})[-241.8 \text{ kJ}/ \text{mol } H_2O] - \{(1 \text{ mol } C_2H_5OH/1 \text{ mol-rxn}) [-277.0 \text{ kJ/mol } C_2H_5OH] + (3 \text{ mol } O_2/ 1 \text{ mol-rxn})[0 \text{ kJ/mol } O_2]\} = -1235.4 \text{ kJ/mol-rxn}$

$100. \text{ g } C_2H_5OH (1 \text{ mol } C_2H_5OH/46.07 \text{ g} C_2H_5OH)(1 \text{ mol-rxn}/1 \text{ mol } C_2H_5OH) (-1235.4 \text{ kJ/mol-rxn}) = -2680 \text{ kJ}$

The energy evolved as heat is 2680 kJ. This is more energy than all of the foods pictured except for the mixed nuts.

Case Study: The Fuel Controversy–Alcohol and Gasoline

In the following, we assume water vapor, $H_2O(g)$, is formed upon oxidation.

1. Burning ethanol: $C_2H_5OH(\ell) + 3 O_2(g) \rightarrow 2 CO_2(g) + 3 H_2O(g)$

   $\Delta_r H° = (2 \text{ mol } CO_2/\text{mol-rxn})[\Delta_f H°(CO_2)] + (3 \text{ mol } H_2O/\text{mol-rxn})[\Delta_f H°(H_2O)] - (1 \text{ mol } C_2H_5OH/\text{mol-rxn})[\Delta_f H°(C_2H_5OH)]$

   $\Delta_r H° = (2 \text{ mol } CO_2/\text{mol-rxn})[-393.5 \text{ kJ/mol} CO_2] + (3 \text{ mol } H_2O/\text{mol-rxn})[-241.8 \text{ kJ/mol} H_2O] - (1 \text{ mol } C_2H_5OH/\text{mol-rxn})[-277.0 \text{ kJ}/ \text{mol } C_2H_5OH] = -1235.4 \text{ kJ/mol-rxn}$

   1 mol ethanol per 1 mol-rxn; therefore, q per mol is -1235.4 kJ/mol

   q per gram: $-1235.4 \text{ kJ/mol} (1 \text{ mol } C_2H_5OH/ 46.07 \text{ g } C_2H_5OH) = -26.80 \text{ kJ/g } C_2H_5OH$

   Burning octane: $C_8H_{18}(\ell) + 12.5 O_2(g) \rightarrow 8 CO_2(g) + 9 H_2O(g)$

   $\Delta_r H° = (8 \text{ mol } CO_2/\text{mol-rxn})[\Delta_f H°(CO_2)] + (9 \text{ mol } H_2O/\text{mol-rxn})[\Delta_f H°(H_2O)] - (1 \text{ mol } C_8H_{18}/\text{mol-rxn})[\Delta_f H°(C_8H_{18})]$

   $\Delta_r H° = (8 \text{ mol } CO_2/\text{mol-rxn})[-393.5 \text{ kJ/mol} CO_2] + (9 \text{ mol } H_2O/\text{mol-rxn})[-241.8 \text{ kJ/mol} H_2O] - (1 \text{ mol } C_8H_{18}/\text{mol-rxn})[-250.1 \text{ kJ/mol} C_8H_{18}] = -5074.1 \text{ kJ/mol-rxn}$

   1 mol octane per mol-rxn; therefore, q per mol is -5074.1 kJ/mol

   q per gram: $-5074.1 \text{ kJ/1 mol } C_8H_{18} (1 \text{ mol} C_8H_{18}/114.2 \text{ g } C_8H_{18}) = -44.43 \text{ kJ/g } C_8H_{18}$

2. For ethanol, per liter: $q = -26.80 \text{ kJ/g} (785 \text{ g/L}) = -2.10 \times 10^4 \text{ kJ/L}$

   For octane, per liter: $q = -44.43 \text{ kJ/g} (699 \text{ g/L}) = -3.11 \times 10^4 \text{ kJ/L}$

   Octane produces almost 50% more energy per liter of fuel.

3. Mass of CO_2 per liter of ethanol = 1.000 L $(785 \text{ g } C_2H_5OH/L)(1 \text{ mol } C_2H_5OH/46.07 \text{ g} C_2H_5OH)(2 \text{ mol } CO_2/1 \text{ mol } C_2H_5OH)(44.01 \text{ g} CO_2/1 \text{ mol } CO_2) = 1.50 \times 10^3 \text{ g } CO_2$

   Mass of CO_2 per liter of octane = 1.000 L $(699 \text{ g } C_8H_{18}/L)(1 \text{ mol } C_8H_{18} /114.2 \text{ g } C_8H_{18}) (8 \text{ mol } CO_2/1 \text{ mol } C_8H_{18})(44.01 \text{ g } CO_2/1 \text{ mol} CO_2) = 2.16 \times 10^3 \text{ g } CO_2$

4. Volume of ethanol needed to obtain 3.11×10^4 kJ of energy from oxidation: $2.10 \times 10^4 \text{ kJ/L} C_2H_5OH)(x) = 3.11 \times 10^4 \text{ kJ}$ (where x is volume of ethanol)

   Volume of ethanol = $x = 1.48$ L

   Mass of CO_2 produced by burning 1.48 L of ethanol = $(1.50 \times 10^3 \text{ g } CO_2/L C_2H_5OH) (1.48 \text{ L } C_2H_5OH) = 2.22 \times 10^3 \text{ g } CO_2$

   To obtain the same amount of energy, slightly more CO_2 is produced by burning ethanol than by burning octane.

5. Your car will travel about 50% farther on a liter of octane, and it will produce slightly less CO_2 emissions, than if you burned 1.0 L of ethanol.

Chapter 6
Chapter Opening: Fireworks

1. Yellow light has a longer wavelength than blue light.

2. Blue light has a greater energy per photon than yellow light.

3. The energy of light emitted by atoms is determined by the energy levels of the electrons in an atom. See discussion in the text, Section 6.3.

Case Study: What Makes the Colors in Fireworks?

1. Yellow light is from the 589 and 590 nm emissions.

2. Primary emission for Sr is red. This has a longer wavelength than yellow light.

3. $4 Mg(s) + KClO_4(s) \rightarrow KCl(s) + 4 MgO(s)$

Chapter 7
Chapter Opening: Rubies and Sapphires— Pretty Stones

1. Cr: $[Ar]3d^5 4s^1$; Cr^{3+}: $[Ar]3d^3$

2. Cr^{2+} is paramagnetic with four unpaired electrons. Cr^{3+} is paramagnetic with three unpaired electrons.

3. The radius of Al^{3+} is 57 pm. The radius of Cr^{3+} is thus only 7 pm larger than that of Al^{3+}.

4. Fe^{2+}: $[Ar]3d^6$; Ti^{4+}: $[Ar]$

Case Study: Metals in Biochemistry and Medicine

1. Fe: $[Ar]3d^64s^2$; Fe^{2+}: $[Ar]3d^6$; Fe^{3+}: $[Ar]3d^5$

2. Both iron ions are paramagnetic.

3. Cu: $[Ar]3d^{10}4s^1$; Cu^+: $[Ar]3d^{10}$; Cu^{2+}: $[Ar]3d^9$; Cu^{2+} is paramagnetic; Cu^+ is diamagnetic.

4. The slightly larger size of Cu compared to Fe is related to greater electron–electron repulsions.

5. Fe^{2+} is larger than Fe^{3+} and will fit less well into the structure. As a result, some distortion of the ring structure from planarity occurs.

Chapter 8

Chapter Opening: Chemical Bonding in DNA

1. Carbon and phosphorus (in phosphate) achieve the noble gas configuration by forming four bonds.

2. In each instance, there are four bonds to the element; VSEPR predicts that these atoms will have tetrahedral geometry with 109.5° angles.

3. Bond angles in these rings are 120°. To achieve this preferred bond angle, the rings must be planar.

4. Thymine and cytosine are polar molecules.

Case Study: Hydroxyl Radicals, Atmospheric Chemistry, and Hair Dyes

1. 18 valence electrons

$$H-\overset{..}{\underset{..}{O}}-\overset{..}{N}=\overset{..}{\underset{..}{O}}$$

All atoms have a formal charge of zero. There are no equivalent resonance structures.

2. 30 valence electrons

$$\left[\begin{array}{c}\overset{Formal\ charge = -1}{\underset{..}{\overset{..}{O}}:} \\ \overset{..}{\underset{..}{O}}=\overset{}{C}-\overset{..}{\underset{..}{O}}-\overset{..}{\underset{..}{O}}-H\end{array}\right]^{-}\longleftrightarrow\left[\begin{array}{c}:\overset{..}{\underset{..}{O}}: \\ \parallel \\ :\overset{..}{\underset{..}{O}}-C-\overset{..}{\underset{..}{O}}-\overset{..}{\underset{..}{O}}-H \\ \underset{Formal\ charge = -1}{}\end{array}\right]^{-}$$

All atoms have zero formal charge except for those indicated.

Case Study: Ibuprofen, A Study in Green Chemistry

1. The Lewis structures for the key portions of the molecules where the reaction takes place are the following:

Bond broken: 1 C—O and 1 C≡O

$$(1\ mol\ C\!-\!O/1\ mol\text{-rxn})(358\ kJ/mol\ C\!-\!O) +$$
$$(1\ mol\ C\!\equiv\!O/1\ mol\text{-rxn})(1046\ kJ/mol\ C\!\equiv\!O)$$
$$= 1404\ kJ/mol\text{-rxn}$$

Bonds formed: 1 C—C, 1 C=C, and 1 C—O

$$(1\ mol\ C\!-\!C/1\ mol\text{-rxn})(346\ kJ/mol\ C\!-\!C) +$$
$$(1\ mol\ C\!=\!O/1\ mol\text{-rxn})(745\ kJ/mol\ C\!=\!O) +$$
$$(1\ mol\ C\!-\!O/1\ mol\text{-rxn})(358\ kJ/mol\ C\!-\!O)$$
$$= 1449\ kJ/mol\text{-rxn}$$

$$\Delta_r H° \approx 1404\ kJ/mol\text{-rxn} - 1449\ kJ/mol\text{-rxn}$$
$$= -45\ kJ/mol\text{-rxn}$$

The reaction is thus predicted to be exothermic.

2. All of the atoms in ibuprofen have a formal charge of zero.

3. The most polar bond in the molecule is the O—H bond.

4. The molecule is not symmetrical and so is polar.

5. The shortest bond in the molecule is the O—H bond.

6. The C=O bond has the highest bond order. The C=C bonds in the ring have an order of 1.5.

7. Yes, there are 120° bond angles present: the bond angles around the C atoms in the ring and those around the C atom in the $-CO_2H$ group are all 120°. There are no 180° angles in this molecule.

8. There is one acid group in ibuprofen $(-CO_2H)$, therefore 1 mol of ibuprofen will react with 1 mol of NaOH.

   200. mg ibuprofen (1 g/1000 mg)(1 mol ibuprofen/206.3 g ibuprofen)(1 mol NaOH/ 1 mol ibuprofen)(1 L/0.0259 mol NaOH) (1000 mL/1 L) = 37.4 mL

   Therefore 37.4 mL of the NaOH solution would be required.

Chapter 9

Chapter Opening: The Noble Gases: Not So Inert

1. XeF_2 is linear. The electron-pair geometry is trigonal-bipyramidal. Three lone pairs are located in the equatorial plane, and the two F atoms are located in the axial positions. This symmetrical structure will not have a dipole.

2. The Xe atom is sp^3d hybridized. Xe—F bonds: overlap of Xe sp^3d orbitals with F $2p$ orbital. 3 lone pairs in Xe sp^3d orbitals.

3. This 36-electron molecule has a bent molecular structure.

Case Study: Green Chemistry, Safe Dyes, and Molecular Orbitals

1. C_8H_4BrNO

2. The energy per photon is inversely proportional to wavelength. The light absorbed by butter yellow has a smaller wavelength than that absorbed by nitrated butter yellow; therefore butter yellow absorbs higher energy light than does nitrated butter yellow.

3. Tyrian purple: 9

Nitrated butter yellow: 8

Chapter 10
Chapter Opening: The Food of the Gods

1. The only structural difference between theobromine and caffeine occurs on the N in the six-member ring that is between the two C=O groups. In theobromine, there is an H atom attached to this N. In caffeine, there is a CH_3 group attached.

2. 5.00 g sample (2.16 g theobromine/100 g sample) = 0.108 g theobromine

Case Study: An Awakening with L-DOPA

1. L-DOPA is chiral. Its chiral center is indicated in the following structural formula.

2. Dopamine is not chiral. Epinephrine is chiral. Its chiral center is indicated in the following structural formula.

3. 5.0 g L-DOPA (1 mol L-DOPA/197.2 g L-DOPA) = 0.025 mol L-DOPA

Case Study: Green Adhesives

1. Structural formulas:

phenol   urea   formaldehyde

2. The electron-pair and molecular geometries are both trigonal-planar. The C—H σ bonds are each formed by the overlap of an sp^2 hybrid orbital on the C atom with the $1s$ orbital of the H atom. The σ bond between the C and O is formed by the overlap of an sp^2 hybrid orbital on the C with an sp^2 hybrid orbital on the O. The π bond between the C and O is formed by the overlap of a $2p$ orbital on C with a $2p$ orbital on O.

3. Similarity: Both nylon-6,6 and proteins are polyamides.

Differences:

1. In proteins there is one direction for the amide linkage: CONH. In nylon-6,6 two orientations are present: CONH and NHCO.

2. In proteins there is only one C between the amide linkages. In nylon-6,6 there are four or six carbons between amide linkages.

3. Proteins have numerous R groups that can be attached to the carbon in between the amide groups, whereas nylon-6,6 has only hydrogen atoms attached to the carbons in between the amide groups.

4. Proteins are chiral, whereas nylon-6,6 is not.

Chapter 11
Chapter Opening: The Atmosphere and Altitude Sickness

1. $P(O_2)$ at 3000 m is 70% of 0.21 atm, the value of $P(O_2)$ at sea level. Thus, $P(O_2)$ at 3000 m = 0.21 atm $\times$ 0.70 = 0.15 atm (110 mm Hg). At the top of Everest, $P(O_2)$ = 0.21 atm $\times$ 0.29 = 0.061 atm (46 mm Hg).

2. Blood saturation levels (estimated from table): at 3000 m, >95%; at top of Everest, 75%.

Case Study: What to Do with All of that CO_2? More on Green Chemistry

1. 9.1×10^9 g C (1 mol C/12.0 g C) (1 mol CO_2/1 mol C)(44.0 g CO_2/1 mol CO_2) = 3.3×10^{10} g CO_2

2. 1.0×10^6 g CO_2 (1 mol CO_2/44.0 g CO_2) (1 mol $(NH_4)_2CO_3$/1 mol CO_2) (96.1 g $(NH_4)_2CO_3$/1 mol $(NH_4)_2CO_3$) = 2.2×10^6 g $(NH_4)_2CO_3$

Chapter 12

Case Study: Methane Hydrates

1. One CH_4 molecule is contained in a clathrate cage that has 12 faces, 20 apexes, and 30 edges; each edge is an O—H—O unit, so there are 30 hydrogen bonds.

2. 164 m^3 = 1.64×10^5 L of CH_4. Use the ideal gas law to calculate n ($n = 7.32 \times 10^3$ mol CH_4). Using $\Delta_f H°$ values, calculate the enthalpy change for the combustion of 1.00 mole of CH_4 to form $CO_2(g)$ and $H_2O(g)$ ($\Delta_r H° = -802$ kJ/mol). $\Delta_r H°$ for 164 m^3 of CH_4 = $(-802$ kJ/mol) $(7.32 \times 10^3$ mol) = -5.87×10^6 kJ.

Case Study: A Pet Food Catastrophe

1. % N in melamine = $84/126 \times 100 = 67\%$

   % N in cyanuric acid = $42/129 \times 100 = 33\%$

   Both of these compounds have a greater percentage of nitrogen than the average protein.

2. 454 g sample (0.14 g melamine/1,000,000 g sample) = 6.4×10^{-5} g melamine = 0.064 mg melamine = 64 μg melamine

Chapter 13

Chapter Opening: Lithium and "Green Cars"

1. 73,000,000 metric tons Li_2CO_3 (13.88 metric tons Li/73.89 metric tons Li_2CO_3) = 1.4×10^7 metric tons Li (= 14 million metric tons Li)

2. The unit cell for lithium metal is body-centered cubic. (There are atoms at each of the corners of a cube and one atom embedded in the middle of the cube.)

3. 351 pm (1 m/10^{12} pm)(100 cm/1 m) = 3.51×10^{-8} cm

   V of unit cell = $(3.51 \times 10^{-8}$ cm$)^3$ = 4.32×10^{-23} cm^3

   Mass of unit cell = 6.941 g Li/mol (1 mol/6.022×10^{23} atoms)(2 atoms/unit cell) = 2.305×10^{-23} g/unit cell

   Density = mass/volume = 2.305×10^{-23} g/4.32×10^{-23} cm^3 = 0.533 g/cm^3

4. One method is to carry out the following gas-forming reaction:

   $Li_2CO_3(aq) + 2\ HCl(aq) \rightarrow 2\ LiCl(aq) + H_2O(\ell) + CO_2(g)$

Case Study: High-Strength Steel and Unit Cells

1. Ferrite has a body-centered cubic unit cell.

2. Austenite has a face-centered cubic unit unit cell of iron atoms with a carbon atom embedded in the body of the cell.

3. The face-centered unit cell of austenite transforms into a body-centered unit cell, but this unit cell is not cubic because it is longer in one dimension than the other two (this is called a *body-centered tetragonal unit cell*). The carbon atoms remain embedded in the lattice.

Case Study: Graphene—The Hottest New Network Solid

1. We want to know the distance x in the diagram of the hexagon. You know the interior angles in a hexagon are all 120°. To find x, we can find the distance y in the figure and then double it ($x = 2y$). The angle bounded by the side of the hexagon and y is 30°, and, from geometry, $\cos 30° = y/$(hexagon side) so $y = (\cos 30°)$ (139 pm) = (0.866)(139 pm) = 120. pm. Finally, $x = 241$ pm.

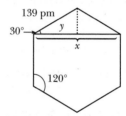

2. 1.0 μm (1 m/10^6 μm)(10^{12} pm/1 m)(1 C_6-ring/241 pm) = 4.1×10^3 C_6-rings

3. The thickness would be approximately 150 pm. This corresponds to the diameter of a carbon atom, found by multiplying the radius of a carbon atom (given in Figure 7.6) by 2 and rounding off to two significant figures.

Chapter 14

Chapter Opening: Survival at Sea

$\Pi = cRT = (1.15$ mol/L)(0.082057 L atm/mol · K)(298 K) = 28.1 atm

Case Study: Exploding Lakes and Diet Coke

1. $PV = nRT$

   4.0 atm(0.025 L) = n(0.08206 L · atm/mol · K) 298 K; $n = 4.1 \times 10^{-3}$ mol

2. $P_1V_1 = P_2V_2$

   4.0 atm(0.025 L) = 3.7×10^{-4} atm (V_2); $V_2 = 270$ L

   The gas expanded by a factor of 11,000 (= 270 L/0.025 L).

3. Solubility of $CO_2 = k_H P_g = 0.034$ mol/kg · bar $(3.7 \times 10^{-4}$ bar$) = 1.3 \times 10^{-5}$ mol/kg

4. Before opening: Solubility of CO_2 before opening $= k_H P_g = 0.034$ mol/kg · bar $(4.0$ bar$) = 0.14$ mol/kg

   Assume the density of the solution is 1.0 g/cm³, therefore 1.0 L corresponds to 1.0 kg. The amount of CO_2 dissolved is 0.14 mol.

   After opening: Solubility of $CO_2 = 1.3 \times 10^{-5}$ mol/kg as calculated in part 3 above. The amount of CO_2 in 1.0 kg is 1.3×10^{-5} mol.

   Mass of CO_2 released $= (0.14$ mol $CO_2 - 1.3 \times 10^{-5}$ mol $CO_2)(44.01$ g $CO_2/1$ mol $CO_2) = 6.0$ g CO_2

Chapter 15

Chapter Opening: Where Did the Indicator Go?

See the answer to Study Question 75 for this chapter in Appendix R.

Case Study: Enzymes—Nature's Catalysts

1. To decompose an equivalent amount of H_2O_2 catalytically would take 1.0×10^{-7} years; this is equivalent to 3.2 seconds.

2.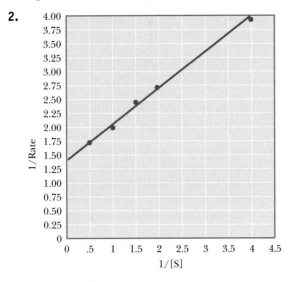

[S]	1/[S]	Rate	1/Rate
2.50	0.400	0.588	1.70
1.00	1.00	0.500	2.00
0.714	1.40	0.417	2.40
0.526	1.90	0.370	2.70
0.250	4.00	0.256	3.91

From the graph, we obtain a value of 1/Rate = 1.47 when 1/[S] = 0. From this, $R_{max} = 0.68$ mmol/min.

Chapter 16

Chapter Opening: Dynamic and Reversible

1. Endothermic. Raising the temperature (adding energy as heat) leads to conversion of reactants to products.

2. Apply LeChatelier's Principle: Increasing the concentration of Cl^- by adding HCl results in a shift to the right to produce more $[CoCl_4]^{2-}$. Decreasing the concentration of Cl^- by addition of water results in a shift to the left and the formation of more $[Co(H_2O)_6]^{2+}$.

3. The various stresses applied have caused the system to adjust in either direction. (Better evidence: Show that when heating and cooling the system is *repeated several times* the system cycles back and forth between the two colors.)

Case Study: Applying Equilibrium Concepts— The Haber–Bosch Ammonia Process

1. (a) Oxidize part of the NH_3 to HNO_3, then react NH_3 and HNO_3 (an acid–base reaction) to form NH_4NO_3.

   $4\ NH_3 + 5O_2 \rightarrow 4\ NO_2 + 6\ H_2O$

   $2\ NO_2 + H_2O \rightarrow HNO_3 + HNO_2$

   $HNO_3 + NH_3 \rightarrow NH_4NO_3$

   (b) $\Delta_r H° = (1$ mol $(NH_2)_2CO/$mol-rxn$)$ $[\Delta_f H°\{(NH_2)_2CO\}] + (1$ mol $H_2O/$mol-rxn$)$ $[\Delta_f H°(H_2O)] - (2$ mol $NH_3/$mol-rxn$)$ $[\Delta_f H°(NH_3)] - (1$ mol $CO_2/$mol-rxn$)$ $[\Delta_f H°(CO_2)]$

   $\Delta_r H° = (1$ mol $(NH_2)_2CO/$mol-rxn$)$ $(-333.1$ kJ/mol$) + (1$ mol $H_2O/$mol-rxn$)$ $(-241.8$ kJ/mol$) - (2$ mol $NH_3/$mol-rxn$)$ $(-45.90$ kJ/mol$) - (1$ mol $CO_2/$mol-rxn$)$ $(-393.5$ kJ/mol$)$

   $\Delta_r H° = -89.6$ kJ/mol-rxn.

   The reaction as written is exothermic, so the equilibrium will be more favorable for product formation at a low temperature. The reaction converts three moles of gaseous reactants to one mole of gaseous products; thus, high pressure will be more favorable to product formation.

2. (a) For $CH_4(g) + H_2O(g) \rightarrow CO(g) + 3\ H_2(g)$

   $\Delta_r H° = (1$ mol $CO/$mol-rxn$)[\Delta_f H°(CO)] - (1$ mol $CH_4/$mol-rxn$)[\Delta_f H°(CH_4)] - (1$ mol $H_2O/$mol-rxn$)[\Delta_f H°(H_2O)]$

   $\Delta_r H° = (1$ mol $CO/$mol-rxn$)(-110.5$ kJ/mol$) - (1$ mol $CH_4/$mol-rxn$)(-74.87$ kJ/mol$) - (1$ mol $H_2O/$mol-rxn$)(-241.8$ kJ/mol$) = 206.2$ kJ/mol-rxn (endothermic)

For $CO(g) + H_2O(g) \rightarrow CO_2(g) + H_2(g)$

$\Delta_r H° = (1 \text{ mol } CO_2/\text{mol-rxn})[\Delta_f H°(CO_2)] -$ (1 mol CO/mol-rxn)$[\Delta_f H°(CO)] -$ (1 mol H_2O/mol-rxn)$[\Delta_f H°(H_2O)]$

$\Delta_r H° = (1 \text{ mol } CO_2/\text{mol-rxn})(-393.5 \text{ kJ/mol})$ $- (1 \text{ mol CO/mol-rxn})(-110.5 \text{ kJ/mol}) -$ (1 mol H_2O/mol-rxn)$(-241.8 \text{ kJ/mol}) =$ -41.2 kJ/mol-rxn (exothermic)

(b) (15 billion kg $= 1.5 \times 10^{13}$ g)

Add the two equations:
$CH_4(g) + 2 H_2O(g) \rightarrow CO_2(g) + 4 H_2(g)$

CH_4 required $= (1.5 \times 10^{13}$ g $NH_3)(1$ mol $NH_3/17.03$ g $NH_3)(3$ mol $H_2/2$ mol $NH_3)$ (1 mol $CH_4/4$ mol $H_2)(16.04$ g $CH_4/1$ mol $CH_4) = 5.3 \times 10^{12}$ g CH_4

CO_2 formed $= (1.5 \times 10^{13}$ g $NH_3)(1$ mol $NH_3/17.03$ g $NH_3)(3$ mol $H_2/2$ mol $NH_3)$ (1 mol $CO_2/4$ mol $H_2)(44.01$ g $CO_2/1$ mol $CO_2) = 1.5 \times 10^{13}$ g CO_2

Chapter 17
Chapter Opening: Aspirin is More Than 100 Years Old!

1. Aspirin, with a smaller pK_a, is a stronger acid than acetic acid.

2. The acidic hydrogen is the H on the $-CO_2H$ (carboxylic acid) functional group.

3. $C_6H_4(CO_2CH_3)CO_2H + H_2O \rightarrow$ $C_6H_4(CO_2CH_3)CO_2^- + H_3O^+$

Case Study: Would You Like Some Belladonna Juice in Your Drink?

1. (100. mg $C_{17}H_{23}NO_3)(1$ g/1000 mg)(1 mol $C_{17}H_{23}NO_3/289.4$ g $C_{17}H_{23}NO_3) = 3.46 \times 10^{-4}$ mol $C_{17}H_{23}NO_3$

2. The proton will attach to the N.

3. The pK_a of protonated atropine (4.35) is less than that of the ammonium ($pK_a = 9.26$), methylammonium ($pK_a = 10.70$), and anilinium ($pK_a = 4.60$) ions.

4. The site of chirality is a carbon atom with four different groups attached; this is the carbon atom that has the $-CH_2OH$ group attached.

Chapter 18
Chapter Opening: Nature's Acids

1. (a) $H_2C_2O_4(aq) + 2 NaOH(aq) \rightarrow$ $Na_2C_2O_4(aq) + 2 H_2O(\ell)$

28 g rhubarb (1.2 g $H_2C_2O_4$/100 g rhubarb) (1 mol $H_2C_2O_4/90.04$ g $H_2C_2O_4)(2$ mol NaOH/1 mol $H_2C_2O_4)(1$ L/0.25 mol NaOH)(1000 mL/1 L) $= 30.$ mL

(b) $Ca^{2+}(aq) + H_2C_2O_4(aq) + 2 H_2O(\ell) \rightarrow$ $CaC_2O_4(s) + 2 H_3O^+(aq)$

28 g rhubarb (1.2 g $H_2C_2O_4$/100 g rhubarb) (1 mol $H_2C_2O_4/90.04$ g $H_2C_2O_4)(1$ mol $CaC_2O_4/1$ mol $H_2C_2O_4)(128.1$ $CaC_2O_4/$ 1 mol $CaC_2O_4) = 0.48$ g CaC_2O_4

2. $CaC_2O_4(s) \rightleftharpoons Ca^{2+}(aq) + C_2O_4^{2-}(aq)$

$K_{sp} = 4 \times 10^{-9} = [Ca^{2+}][C_2O_4^{2-}]$

Let x = solubility of CaC_2O_4

$4 \times 10^{-9} = x^2$

$x = 6 \times 10^{-5}$ mol/L

Solubility in g/L $= 6 \times 10^{-5}$ mol/L (128.1 g $CaC_2O_4/1$ mol $CaC_2O_4) = 0.008$ g/L

Case Study: Take a Deep Breath!

1. pH $= pK_a + \log[HPO_4^{2-}]/[H_2PO_4^-]$

$7.4 = 7.20 + \log[HPO_4^{2-}]/[H_2PO_4^-]$

$[HPO_4^{2-}]/[H_2PO_4^-] = 1.6$

2. Assign $x = [HPO_4^{2-}]$, then $[H_2PO_4^-] = (0.020 - x)$

$1.6 = x/(0.020 - x); x = 0.012$

$[HPO_4^{2-}] = x = 0.012$ mol/L

$[H_2PO_4^-] = 0.020 - x = 0.008$ mol/L

Case Study: Chemical Equilibria in the Oceans

For both of these compounds, the chemical equilibrium is:

$CaCO_3(s) \rightleftharpoons Ca^{2+}(aq) + CO_3^{2-}(aq)$

$K_{sp} = [Ca^{2+}][CO_3^{2-}]$ and the solubility (in mol/L) is equal to $(K_{sp})^{1/2}$.

Aragonite:

Solubility (in g/L) $= [(6.0 \times 10^{-9})^{1/2}$ mol $CaCO_3/L](100.1$ g $CaCO_3/1$ mol $CaCO_3) = 7.8 \times 10^{-3}$ g $CaCO_3/L$

Calcite:

Solubility (in g/L) $= [(3.4 \times 10^{-9})^{1/2}$ mol $CaCO_3/L](100.1$ g $CaCO_3/1$ mol $CaCO_3) = 5.8 \times 10^{-3}$ g $CaCO_3/L$

Aragonite is 7.8×10^{-3} g/L/5.8×10^{-3} g/L $= 1.3$ times more soluble than calcite.

Chapter 19

Chapter Opening: Hydrogen for the Future?

1. $\Delta_r H° = (2 \text{ mol } H_2/\text{mol-rxn})\Delta_f H°[H_2(g)] +$
$(1 \text{ mol } O_2/\text{mol-rxn})\Delta_f H°[O_2(g)] -$
$(2 \text{ mol } H_2O/\text{mol-rxn})\Delta_f H°[H_2O(\ell)]$
$= (2 \text{ mol } H_2/\text{mol-rxn})[0 \text{ kJ/mol } H_2] +$
$(1 \text{ mol } O_2/\text{mol-rxn})[0 \text{ kJ/mol } O_2] -$
$(2 \text{ mol } H_2O/\text{mol-rxn})[-285.83 \text{ kJ/mol } H_2O]$
$= +571.66 \text{ kJ/mol-rxn}$

$\Delta_r S° = (2 \text{ mol } H_2/\text{mol-rxn})S°[H_2(g)] +$
$(1 \text{ mol } O_2/\text{mol-rxn})S°[O_2(g)] -$
$(2 \text{ mol } H_2O/\text{mol-rxn})S°[H_2O(\ell)]$
$= (2 \text{ mol } H_2/\text{mol-rxn})[130.7 \text{ J/K} \cdot \text{mol } H_2] +$
$(1 \text{ mol } O_2/\text{mol-rxn})[205.07 \text{ J/K} \cdot \text{mol } O_2] -$
$(2 \text{ mol } H_2O/\text{mol-rxn})[69.95 \text{ J/K} \cdot \text{mol } H_2O]$
$= +326.6 \text{ J/K} \cdot \text{mol-rxn} =$
$+0.3266 \text{ kJ/K} \cdot \text{mol-rxn}$

$\Delta_r G° = \Delta_r H° - T \Delta_r S°$
$= 571.66 \text{ kJ/mol-rxn} - 298.15 \text{ K}(0.3266 \text{ kJ/K} \cdot$
$\text{mol-rxn}) = +474.29 \text{ kJ/mol-rxn}$

2. $\Delta_r H° = (3 \text{ mol } H_2/\text{mol-rxn})\Delta_f H°[H_2(g)] +$
$(1 \text{ mol } CO/\text{mol-rxn})\Delta_f H°[CO(g)] -$
$\{(1 \text{ mol } CH_4/\text{mol-rxn})\Delta_f H°[CH_4(g)] +$
$(1 \text{ mol } H_2O/\text{mol-rxn})\Delta_f H°[H_2O(g)]\}$
$= (3 \text{ mol } H_2/\text{mol-rxn})[0 \text{ kJ/mol } H_2] +$
$(1 \text{ mol } CO/\text{mol-rxn})[-110.525 \text{ kJ/mol } CO] -$
$\{(1 \text{ mol } CH_4/\text{mol-rxn})[-74.87 \text{ kJ/mol } CH_4] +$
$(1 \text{ mol } H_2O/\text{mol-rxn})[-241.83 \text{ kJ/mol } H_2O]\}$
$= +206.18 \text{ kJ/mol-rxn}$

$\Delta_r S° = (3 \text{ mol } H_2/\text{mol-rxn})S°[H_2(g)] +$
$(1 \text{ mol } CO/\text{mol-rxn})S°[CO(g)] -$
$\{(1 \text{ mol } CH_4/\text{mol-rxn})S°[CH_4(g)] +$
$(1 \text{ mol } H_2O/\text{mol-rxn})S°[H_2O(g)]\}$
$= (3 \text{ mol } H_2/\text{mol-rxn})[130.7 \text{ J/K} \cdot \text{mol } H_2] +$
$(1 \text{ mol } CO/\text{mol-rxn})[197.674 \text{ J/K} \cdot \text{mol } CO] -$
$\{(1 \text{ mol } CH_4/\text{mol-rxn})[186.26 \text{ J/K} \cdot \text{mol } CH_4] +$
$(1 \text{ mol } H_2O/\text{mol-rxn})[188.84 \text{ J/K} \cdot \text{mol } H_2O]\}$
$= +214.7 \text{ J/K} \cdot \text{mol-rxn} =$
$+0.2147 \text{ kJ/K} \cdot \text{mol-rxn}$

$\Delta_r G° = \Delta_r H° - T \Delta_r S° = 206.18 \text{ kJ/mol-rxn} -$
$298.15 \text{ K}(0.2147 \text{ kJ/K} \cdot \text{mol-rxn})$
$= +142.17 \text{ kJ/mol-rxn}$

Case Study: Thermodynamics and Living Things

1. Creatine phosphate + $H_2O \rightarrow$ creatine + HP_i
$$\Delta G° = -43.3 \text{ kJ/mol}$$

Adenosine + $HP_i \rightarrow$ adenosine monophosphate + H_2O $\Delta G° = +9.2 \text{ kJ/mol}$

Net reaction (sum of the two reactions):

Creatine phosphate + adenosine $\rightarrow$
creatine + adenosine monophosphate

For this, $\Delta G° = -43.3 \text{ kJ/mol} + 9.2 \text{ kJ/mol} = -34.1 \text{ kJ/mol}$. The negative value indicates that the transfer of phosphate from creatine phosphate to adenosine is product-favored.

2. $\Delta G°' = \Delta G° + RT \ln[C][H_3O^+]/[A][B] =$
$\Delta G° + (8.31 \times 10^{-3} \text{ kJ/mol} \cdot \text{K})(298 \text{ K})$
$\ln[1][1 \times 10^{-7}]/[1][1]$

$\Delta G°' = \Delta G° - 39.9 \text{ kJ/mol}$

Chapter 20

Chapter Opening: Battery Power

1. (a) The oxidation number of cobalt in CoO_2 is $+4$ and in $LiCoO_2$ is $+3$.

(b) Cobalt(IV) is reduced in this reaction, so CoO_2 is the cathode, and Li is the anode.

(c) Lithium in the battery is elemental Li in a carbon matrix. Elemental Li reacts readily with water (◄ page 324).

2. (a) (1 hour) (3600 s/1 hour)(7.5 C/s)(1 mol e^-/96,485 C) = 0.28 mol e^-

(b) (0.28 mol e^-) (1 mol Li/1 mol e^-)(6.941 g Li/1 mol Li) = 1.9 g Li

Case Study: Manganese in the Oceans

1. Cathode reaction: $Mn^{3+} + e^- \rightarrow Mn^{2+}$

Anode reaction: $Mn^{3+} + 2 H_2O \rightarrow MnO_2 + 4 H^+ + e^-$

Net reaction: $2 Mn^{3+} + 2 H_2O \rightarrow MnO_2 + Mn^{2+} + 4 H^+$

$E°_{cell} = E°(\text{cathode}) - E°(\text{anode}) = 1.50 \text{ V} - 0.95 \text{ V} = 0.55 \text{ V}$

The positive value associated with disproportionation (the net reaction) is positive, indicating a product-favored reaction.

2. (a) $MnO_2 + HS^- + 3 H^+ \rightarrow Mn^{2+} + S + 2 H_2O$

(b) $2 Mn^{2+} + O_2 + 2 H_2O \rightarrow 2 MnO_2 + 4 H^+$

3. Cathode reaction: $O_2 + 4 H^+ + 4 e^- \rightarrow 2 H_2O$ $(E° = 1.229 \text{ V})$

Anode reaction: $Mn^{2+} + 2 H_2O \rightarrow MnO_2 + 4 H^+ + 2 e^-$ $(E° = 1.23 \text{ V}, \text{ from Appendix M})$

$E°_{cell} = E°(\text{cathode}) - E°(\text{anode}) = 1.229 \text{ V} - 1.23 \text{ V} = 0 \text{ V}.$

Chapter 21

Chapter Opening: Carbon and Silicon

1. $CH_4(g) + 2\,H_2O(\ell) \rightarrow CO_2(g) + 4\,H_2(g)$

$SiH_4(g) + 2\,H_2O(\ell) \rightarrow SiO_2(s) + 4\,H_2(g)$

2. For CH_4:

$\Delta_rG° = (1\text{ mol }CO_2/\text{mol-rxn})\Delta_fG°[CO_2(g)] + (4\text{ mol }H_2/\text{mol-rxn})\Delta_fG°[H_2(g)] - \{(1\text{ mol }CH_4/\text{mol-rxn})\Delta_fG°[CH_4(g)] + (2\text{ mol }H_2O/\text{mol-rxn})\Delta_fG°[H_2O(\ell)]\}$

$\Delta_rG° = (1\text{ mol }CO_2/\text{mol-rxn})(-394.359\text{ kJ/mol }CO_2) + (4\text{ mol }H_2/\text{mol-rxn})(0\text{ kJ/mol }H_2) - \{(1\text{ mol }CH_4/\text{mol-rxn})(-50.8\text{ kJ/mol }CH_4) + (2\text{ mol }H_2O/\text{mol-rxn})(-237.15\text{ kJ/mol }H_2O)\} = 130.7\text{ kJ/mol-rxn}$

For SiH_4:

$\Delta_rG° = (1\text{ mol }SiO_2/\text{mol-rxn})\Delta_fG°[SiO_2(s)] + (4\text{ mol }H_2/\text{mol-rxn})\Delta_fG°[H_2(g)] - \{(1\text{ mol }SiH_4/\text{mol-rxn})\Delta_fG°[SiH_4(g)] + (2\text{ mol }H_2O/\text{mol-rxn})\Delta_fG°[H_2O(\ell)]\}$

$\Delta_rG° = (1\text{ mol }SiO_2/\text{mol-rxn})(-856.97\text{ kJ/mol }SiO_2) + (4\text{ mol }H_2/\text{mol-rxn})(0\text{ kJ/mol }H_2) - \{(1\text{ mol }SiH_4/\text{mol-rxn})(56.84\text{ kJ/mol }SiH_4) + (2\text{ mol }H_2O/\text{mol-rxn})(-237.15\text{ kJ/mol }H_2O)\} = -439.51\text{ kJ/mol-rxn}$

The reaction involving SiH_4 is product-favored at equilibrium, whereas the reaction involving CH_4 is reactant-favored at equilibrium.

3. Electronegativities: C 2.5, Si 1.9, H 2.2. From this we conclude that the polarities of C—H and Si—H bonds are in the opposite directions. In SiH_4, the H has a slight negative charge, and in CH_4 the H has a slight positive charge.

4. General observation from these examples: Carbon often bonds to other atoms via double bonds, whereas Si does not. We would not expect $H_2Si{=}SiH_2$ to exist as a molecular species; instead a polymeric structure $-[SiH_2SiH_2]_n-$ is predicted.

$$H_3C-\overset{\overset{\displaystyle :O:}{\|}}{C}-CH_3 \qquad \left(-\overset{\overset{\displaystyle CH_3}{|}}{\underset{\underset{\displaystyle CH_3}{|}}{Si}}-\overset{..}{\underset{..}{O}}-\right)_n$$

Case Study: Hard Water

1. For Mg^{2+}: $(50.\text{ mg})(1\text{ mmol }Mg^{2+}/24.31\text{ mg})(1\text{ mmol }CaO/\text{mmol }Mg^{2+})(56.08\text{ mg }CaO/1\text{ mmol }CaO) = 120\text{ mg }CaO$

For Ca^{2+}: $(150\text{ mg})(1\text{ mmol }Ca^{2+}/40.08\text{ mg})(1\text{ mmol }CaO/\text{mmol }Ca^{2+})(56.08\text{ mg }CaO/1\text{ mmol }CaO) = 210\text{ mg }CaO$

Total $CaO = 120\text{ mg} + 210\text{ mg} = 330\text{ mg}$ (two significant figures)

We get 2 mol $CaCO_3$ per mole Ca^{2+} and 1 mol each of $CaCO_3$ and $MgCO_3$ per mole Mg^{2+}

$CaCO_3$ from Ca^{2+} reaction: $(0.15\text{ g }Ca^{2+})(1\text{ mol}/40.08\text{ g }Ca^{2+})(2\text{ mol }CaCO_3/1\text{ mol }Ca^{2+})(100.1\text{ g }CaCO_3/1\text{ mol }CaCO_3) = 0.75\text{ g}$

$CaCO_3$ from Mg^{2+} reaction: $(0.050\text{ g }Mg^{2+})(1\text{ mol}/24.31\text{ g }Mg^{2+})(1\text{ mol }CaCO_3/1\text{ mol }Mg^{2+})(100.1\text{ g }CaCO_3/1\text{ mol }CaCO_3) = 0.21\text{ g}$

$MgCO_3$ from Mg^{2+} reaction: $(0.050\text{ g }Mg^{2+})(1\text{ mol}/24.31\text{ g }Mg^{2+})(1\text{ mol }MgCO_3/1\text{ mol }Mg^{2+})(84.31\text{ g }MgCO_3/1\text{ mol }MgCO_3) = 0.17\text{ g}$

Total mass of solids $= 0.75\text{ g} + 0.21\text{ g} + 0.17\text{ g} = 1.1\text{ g}$ (two significant figures)

2. $CaCO_3(s) + 2\,CH_3CO_2H(aq) \rightarrow Ca(CH_3CO_2H)_2(aq) + H_2O(\ell) + CO_2(g)$

This is a gas-forming reaction.

Case Study: Lead, Beethoven, and a Mystery Solved

1. 50 ppb is 50 g in 1×10^9 g of blood. Assume the density of blood is 1.0 g/mL. In 1.0×10^3 mL (i.e., 1.0 L) of blood, there will be 50×10^{-6} g of Pb. From this:

$(50 \times 10^{-6}\text{ g})(1\text{ mol }Pb/207.2\text{ g }Pb)(6.022 \times 10^{23}\text{ atoms }Pb/\text{mol }Pb) = 1.5 \times 10^{17}\text{ atoms Pb}$

2. $(750\text{ mL wine})(1.0\text{ g wine/mL wine})(2000\text{ g }Pb/1,000,000\text{ g wine}) = 1.5\text{ g Pb}$

Case Study: A Healthy Saltwater Aquarium and the Nitrogen Cycle

1. $2\,NH_4^+(aq) + 4\,OH^-(aq) + 3\,O_2(aq) \rightarrow 2\,NO_2^-(aq) + 6\,H_2O(\ell)$

2. Reduction half-reaction: $2\,NO_3^-(aq) + 6\,H_2O(\ell) + 10\,e^- \rightarrow N_2(g) + 12\,OH^-(aq)$

Oxidation half-reaction: $CH_3OH(aq) + 6\,OH^-(aq) \rightarrow CO_2(aq) + 5\,H_2O(\ell) + 6\,e^-$

Net: $6\,NO_3^-(aq) + 5\,CH_3OH \rightarrow 3\,N_2(g) + 5\,CO_2(aq) + 6\,OH^-(aq) + 7\,H_2O(\ell)$

3. HCO_3^- is the predominant species. Recall that when acid and base concentrations are equal, $pH = pK_a$. If H_2CO_3 and HCO_3^- are present in equal concentrations, the pH would be about 6.4. If HCO_3^- and CO_3^{2-} are present in equal concentrations, the pH would be 10.2. For the pH to be about 8 (in a salt water aquarium), $[HCO_3^-]$ would have to be higher than either of the other carbonate species.

4. Concentration of N in ppm (mg N/L) =
$[(1.7 \times 10^4 \text{ kg NO}_3^-)(10^6 \text{ mg NO}_3^-/\text{kg NO}_3^-)$
$(14.0 \text{ mg N}/62.0 \text{ mg NO}_3^-)]/(2.2 \times 10^7 \text{ L}) =$
$1.7 \times 10^2 \text{ mg/L} = 1.7 \times 10^2 \text{ ppm}$

Concentration of NO_3^- in ppm (mg/L) =
$(1.7 \times 10^4 \text{ kg})(10^6 \text{ mg/kg})/(2.2 \times 10^7 \text{ L}) =$
770 mg/L

Concentration of NO_3^- in mol/L =
$[(1.7 \times 10^4 \text{ kg})(10^3 \text{ g/kg})(1 \text{ mol}/62.0 \text{ g})]/$
$(2.2 \times 10^7 \text{ L}) = 0.012 \text{ mol/L}$

Chapter 22
Chapter Opening: Memory Metal

1. Define length of the side of the cube as x, then the length of the diagonal across the cube is $x\sqrt{3}$. This is set equal to:
$2 r_{Ti} + 2 r_{Ni}$, i.e., $x\sqrt{3} = 2 r_{Ti} + 2 r_{Ni} = 540$ pm; $x = 312$ pm ($a = b = c = 3.12 \times 10^{-8}$ cm)

2. Calculated density:

Mass of one unit cell is the mass of one Ti and one Ni atom = $(47.87 \text{ g/mol})(1 \text{ mol}/6.022 \times 10^{23} \text{ atoms Ti}) + (58.69 \text{ g/mol})(1 \text{ mol}/6.022 \times 10^{23} \text{ atoms Ti}) = 1.77 \times 10^{-22}$ g

Volume of the unit cell is $x^3 = (3.12 \times 10^{-8} \text{ cm})^3 = 3.04 \times 10^{-23}$ cm^3

Calculated density = 1.77×10^{-22} g/3.04×10^{-23} cm$^3 = 5.82$ g/cm^3

The agreement is not very good, probably because atoms don't pack together as tightly as is assumed.

3. As free atoms, both Ti and Ni are paramagnetic.

Case Study: Accidental Discovery of a Chemotherapy Drug

1. First-order kinetics: $\ln[x/x_o] = - kt$

$\ln[x/10 \text{ mg}] = - 7.6 \times 10^{-5} \text{ s}^{-1}[24 \text{ h} \times 3600 \text{ s/h}]$

$x/10 \text{ mg} = 1.4 \times 10^{-3}$; $x = 1.4 \times 10^{-2}$ mg remain

Case Study: Ferrocene—The Beginning of a Chemical Revolution

1. Fe^{2+} in ferrocene has an electron configuration [Ar] $3d^6$ and is present in the low-spin state.

2. Cr(0). Cr(0) in this compound is assumed to have an electron configuration [Ar] $3d^6$, is present in the low-spin state, and so is diamagnetic.

3. Both are in accordance with the 18-electron rule.

4. Select oxidizing agents from Table 20.1 (page 917) based on the northeast–southwest rule (above $E° = 0.400$ v). Common oxidizing agents include the halogens H_2O_2 and MnO_4^-. Cl_2 is a sufficiently strong oxidizing agent to carry out this oxidation.

5. $NiCl_2 + 2 \text{ Na}[C_5H_5] \rightarrow Ni(\eta\text{-}C_5H_5)_2 + 2 \text{ NaCl}$. Nickelocene is predicted to have two unpaired electrons (Ni^{2+}, with a d^8 configuration, in an octahedral environment.)

Chapter 23
Chapter Opening: A Primordial Nuclear Reactor

1. ^{235}U: 92 protons, $235 - 92 = 143$ neutrons

^{238}U: 92 protons, 146 neutrons

2. (a) $^{238}\text{U} + {}_0^1\text{n} \rightarrow {}^{239}\text{U}$

(b) $^{239}\text{U} \rightarrow {}^{239}\text{Np} + {}_{-1}^0\beta$

$^{239}\text{Np} \rightarrow {}^{239}\text{Pu} + {}_{-1}^0\beta$

(c) $^{239}\text{Pu} \rightarrow {}^{235}\text{U} + {}_2^4\alpha$

Case Study: Nuclear Medicine and Hyperthyroidism

1. $^{131}\text{I} \rightarrow {}^{131}\text{Xe} + {}_{-1}^0\beta$

2. Calculate the fraction ($= f$) of each remaining after 7 days

For ^{123}I: $k = 0.693/t_{1/2} = 0.693/13.3 \text{ h} = 0.0521 \text{ h}^{-1}$

$\ln(f) = - 0.0521 \text{ h}^{-1}[7 \text{ d}(24 \text{ h/d})]$

$f = 1.6 \times 10^{-4}$

For ^{131}I: $k = 0.693/t_{1/2} = 0.693/8.04 \text{ d} = 0.0862 \text{ d}^{-1}$

$\ln(f) = - 0.0862 \text{ d}^{-1}(7 \text{ d})$

$f = 0.55$

Ratio of amounts remaining, $[^{131}\text{I}]/[^{123}\text{I}]$ = $0.55/(1.6 \times 10^{-4}) = 3400$

The amount of ^{131}I isotope is 3400 times greater than the amount of ^{123}I.

O Answers to Check Your Understanding Questions

Let's Review

LR 1 0.154 nm (1 m/10^9 nm) (10^{12} pm/1 m) = 154 pm

0.154 nm (1 m/10^9 nm) (100 cm/1 m) = 1.54×10^{-8} cm

LR 2 Student A: average = -0.1 °C; percent error
= {[(-0.1 + 273.15) K $-$ 273.15 K]/273.15 K} $\times$ 100%
= -0.04%

Student B: average = $+0.01$ °C; percent error
= {[(0.01 + 273.15) K $-$ 273.15 K]/273.15 K} $\times$ 100%
= 0.004%

Student B has the smaller error.

LR 3 $x = 3.9 \times 10^5$. The difference between 110.7 and 64 is
47. Dividing 47 by 0.056 and 0.00216 gives an answer
with two significant figures.

LR 4 (19,320 kg/m^3)(10^3 g/1 kg)(1 m^3/10^6 cm^3)
= 19.32 g/cm^3

LR 5 Change all dimensions to centimeters: 7.6 m
= 760 cm; 2.74 m = 274 cm; 0.13 mm = 0.013 cm.

Volume of paint = (760 cm)(274 cm)(0.013 cm)
= 2.7×10^3 cm^3

Volume (L) = (2.7×10^3 cm^3)(1 L/10^3 cm^3) = 2.7 L

Mass = (2.7×10^3 cm^3)(0.914 g/cm^3) = 2.5×10^3 g

Chapter 2

2.1 (1) Mass number with 26 protons and 30 neutrons is 56
(2) (59.930788 u)(1.661 $\times 10^{-24}$ g/u) = 9.955 $\times 10^{-23}$ g
(3) ^{64}Zn has 30 protons, 30 electrons, and (64 $-$ 30)
= 34 neutrons.

2.2 Use Equation 2.2 for the calculation.

Atomic mass = (34.96885)(75.77/100) + (36.96590)
(24.23/100) = 35.45. (Accuracy is limited by the value
of the percent abundance to four significant figures.)

2.3 Use Equation 2.2 for the calculation. Let x = percent
abundance of ^{20}Ne and y = percent abundance of
^{22}Ne.

20.1797 u = (x/100)(19.992435 u) + (0.27/100)
(20.993843 u) + (y/100)(21.991383 u)

Because all the percent abundances must sum to
100%, y = 100 $-$ x $-$ 0.27 = 99.73 $-$ x.

20.1797 u = (x/100)(19.992435 u) + 0.27/100)
(20.993843 u) + [(99.73 $-$ x)/100](21.991383 u)

x = 90.5; therefore the percent abundance of ^{20}Ne =
90.5% and the percent abundance of ^{22}Ne = 9.2%.

2.4 (a) (1) NaF: 1 Na^+ and 1 F^- ion. (2) $Cu(NO_3)_2$:
1 Cu^{2+} and 2 NO_3^- ions. (3) $NaCH_3CO_2$: 1 Na^+
and 1 $CH_3CO_2^-$ ion.
(b) $FeCl_2$, $FeCl_3$

2.5 Na_2S, Na_3PO_4, BaS, $Ba_3(PO_4)_2$

2.6 (2.6×10^{24} atoms)(1.000 mol/6.022×10^{23}
atoms)(197.0 g Au/1.000 mol) = 850 g Au

Volume = (850 g Au)(1.00 cm^3/19.32 g) = 44 cm^3

Volume = 44 cm^3 = (thickness)(area) = (0.10 cm)(area)

Area = 440 cm^2

Length = width = $\sqrt{440 \text{ cm}^2}$ = 21 cm

2.7 Molar mass $H_3C_6H_5O_7$ = 8(1.01) + 6(12.01) +
7(16.00) = 192.14 g/mol

(454 g $H_3C_6H_5O_7$)(1 mol $H_3C_6H_5O_7$/192.14 g
$H_3C_6H_5O_7$) = 2.36 mol $H_3C_6H_5O_7$

(2.36 mol $H_3C_6H_5O_7$)(6.022×10^{23} molecules/1 mol)
= 1.42×10^{24} molecules $H_3C_6H_5O_7$

(1.42×10^{24} molecules $H_3C_6H_5O_7$)(6 atoms C/1 molecule $H_3C_6H_5O_7$) = 8.54×10^{24} atoms C

2.8 (1) 1.00 mol $(NH_4)_2CO_3$ (molar mass 96.09 g/mol)
has 28.0 g of N (29.2%), 8.06 g of H (8.39%),
12.0 g of C (12.5%), and 48.0 g of O (50.0%)
(2) 454 g C_8H_{18} (1 mol C_8H_{18}/114.2 g)(8 mol C/1 mol
C_8H_{18})(12.01 g C/1 mol C) = 382 g C

2.9 (1) C_5H_4
(2) $C_2H_4O_2$
(3) (88.17 g C)(1 mol C/12.011 g C) = 7.341 mol C

(11.83 g H)(1 mol H/1.008 g H) = 11.74 mol H

11.74 mol H/7.341 mol C = 1.6 mol H/1 mol C
= (8/5); (mol H/1 mol C) = 8 mol H/5 mol C

The empirical formula is C_5H_8. The molar mass,
68.11 g/mol, closely matches this formula, so C_5H_8
is also the molecular formula.
(4) (78.90 g C)(1 mol C/12.011 g C) = 6.569 mol C

(10.59 g H)(1 mol H/1.008 g H) = 10.51 mol H

(10.51 g O)(1 mol O/16.00 g O) = 0.6569 mol O

10.51 mol H/0.6569 mol O = 16 mol H/1 mol O

6.569 mol C/0.6569 mol O = 10 mol C/1 mol O

The empirical formula is $C_{10}H_{16}O$.

2.10 $(1.25 \text{ g Ga})(1 \text{ mol Ga}/69.72 \text{ g Ga}) = 0.0179 \text{ mol Ga}$

$1.68 \text{ g product} - 1.25 \text{ g Ga} = 0.43 \text{ g O}$

$(0.43 \text{ g O})(1 \text{ mol O}/16.00 \text{ g O}) = 0.027 \text{ mol O}$

Mole ratio $= 0.027 \text{ mol O}/0.0179 \text{ mol Ga}$
$= 1.5 \text{ mol O}/1.0 \text{ mol Ga} = 3.0 \text{ mol O}/2.0 \text{ mol Ga}$

Empirical formula $= \text{Ga}_2\text{O}_3$

2.11 $(0.586 \text{ g K})(1 \text{ mol K}/39.10 \text{ g K}) = 0.0150 \text{ mol K}$

$(0.480 \text{ g O})(1 \text{ mol O}/16.00 \text{ g O}) = 0.0300 \text{ mol O}$

The ratio of moles K to moles O atoms is $1:2$; the empirical formula is KO_2.

2.12 Mass of water lost on heating is $0.235 \text{ g} - 0.128 \text{ g}$
$= 0.107 \text{ g}$; 0.128 g NiCl_2 remain

$(0.107 \text{ g H}_2\text{O})(1 \text{ mol H}_2\text{O}/18.016 \text{ g H}_2\text{O})$
$= 0.00594 \text{ mol H}_2\text{O}$

$(0.128 \text{ g NiCl}_2)(1 \text{ mol NiCl}_2/129.6 \text{ g NiCl}_2)$
$= 0.000988 \text{ mol NiCl}_2$

Mole ratio $= 0.00594 \text{ mol H}_2\text{O}/0.000988 \text{ mol}$
$\text{NiCl}_2 = 6.01$: Therefore $x = 6$

The formula for the hydrate is $\text{NiCl}_2 \cdot 6 \text{ H}_2\text{O}$.

Chapter 3

3.1 (a) $2 \text{ C}_4\text{H}_{10}(g) + 13 \text{ O}_2(g) \longrightarrow$
$$8 \text{ CO}_2(g) + 10 \text{ H}_2\text{O}(g)$$
(b) $2 \text{ Pb}(\text{C}_2\text{H}_5)_4(\ell) + 27 \text{ O}_2(g) \longrightarrow$
$$2 \text{ PbO}(s) + 16 \text{ CO}_2(g) + 20 \text{ H}_2\text{O}(g)$$

3.2 (a) LiNO_3 is soluble and gives $\text{Li}^+(aq)$ and $\text{NO}_3^-(aq)$
ions.
(b) CaCl_2 is soluble and gives $\text{Ca}^{2+}(aq)$ and $\text{Cl}^-(aq)$
ions.
(c) CuO is not water soluble.
(d) NaCH_3CO_2 is soluble and gives $\text{Na}^+(aq)$ and
$\text{CH}_3\text{CO}_2^-(aq)$ ions.

3.3 (a) $\text{Na}_2\text{CO}_3(aq) + \text{CuCl}_2(aq) \longrightarrow$
$$2 \text{ NaCl}(aq) + \text{CuCO}_3(s)$$
(b) No reaction; no insoluble compound is produced.
(c) $\text{NiCl}_2(aq) + 2 \text{ KOH}(aq) \longrightarrow$
$$\text{Ni}(\text{OH})_2(s) + 2 \text{ KCl}(aq)$$

3.4 (a) $3 \text{ CaCl}_2(aq) + 2 \text{ Na}_3\text{PO}_4(aq) \longrightarrow$
$$\text{Ca}_3(\text{PO}_4)_2(s) + 6 \text{ NaCl}(aq)$$
$3 \text{ Ca}^{2+}(aq) + 2 \text{ PO}_4^{3-}(aq) \longrightarrow \text{Ca}_3(\text{PO}_4)_2(s)$
(b) $\text{FeCl}_3(aq) + 3 \text{ KOH}(aq) \longrightarrow$
$$\text{Fe}(\text{OH})_3(s) + 3 \text{ KCl}(aq)$$
$\text{Fe}^{3+}(aq) + 3 \text{ OH}^-(aq) \longrightarrow \text{Fe}(\text{OH})_3(s)$
(c) $\text{Pb}(\text{NO}_3)_2(aq) + 2 \text{ KCl}(aq) \longrightarrow$
$$\text{PbCl}_2(s) + 2 \text{ KNO}_3(aq)$$
$\text{Pb}^{2+}(aq) + 2 \text{ Cl}^-(aq) \longrightarrow \text{PbCl}_2(s)$

3.5 (a) $\text{H}_3\text{PO}_4(aq) + \text{H}_2\text{O}(\ell) \rightleftharpoons$
$$\text{H}_3\text{O}^+(aq) + \text{H}_2\text{PO}_4^-(aq)$$
(b) Acting as an acid:
$\text{H}_2\text{PO}_4^-(aq) + \text{H}_2\text{O}(\ell) \rightleftharpoons$
$$\text{HPO}_4^{2-}(aq) + \text{H}_3\text{O}^+(\ell)$$
Acting as a base:
$\text{H}_2\text{PO}_4^-(aq) + \text{H}_2\text{O}(\ell) \rightleftharpoons$
$$\text{H}_3\text{PO}_4(aq) + \text{OH}^-(aq)$$
Because $\text{H}_2\text{PO}_4^-(aq)$ can react as a Brønsted acid
and as a base, it is said to be amphiprotic.

3.6 $\text{Mg}(\text{OH})_2(s) + 2 \text{ HCl}(aq) \longrightarrow$
$$\text{MgCl}_2(aq) + 2 \text{ H}_2\text{O}(\ell)$$
Net ionic equation: $\text{Mg}(\text{OH})_2(s) + 2 \text{ H}_3\text{O}^+(aq)$
$$\longrightarrow \text{Mg}^{2+}(aq) + 4 \text{ H}_2\text{O}(\ell)$$

3.7 $\text{BaCO}_3(s) + 2 \text{ HNO}_3(aq) \longrightarrow$
$$\text{Ba}(\text{NO}_3)_2(aq) + \text{CO}_2(g) + \text{H}_2\text{O}(\ell)$$
Barium carbonate and nitric acid produce barium nitrate, carbon dioxide, and water.

3.8 (a) Fe in Fe_2O_3, $+3$; (b) S in H_2SO_4, $+6$;
(c) C in CO_3^{2-}, $+4$; (d) N in NO_2^+, $+5$

3.9 Dichromate ion is the oxidizing agent and is reduced.
(Cr with a $+6$ oxidation number is reduced to Cr^{3+}
with a $+3$ oxidation number.) Ethanol is the reducing
agent and is oxidized. (The C atoms in ethanol have
an oxidation number of -2. The oxidation number is
0 in acetic acid.)

3.10 (a) Gas-forming reaction:
$\text{CuCO}_3(s) + \text{H}_2\text{SO}_4(aq) \longrightarrow$
$$\text{CuSO}_4(aq) + \text{H}_2\text{O}(\ell) + \text{CO}_2(g)$$
Net ionic equation:
$\text{CuCO}_3(s) + 2 \text{ H}_3\text{O}^+(aq) \longrightarrow$
$$\text{Cu}^{2+}(aq) + 3 \text{ H}_2\text{O}(\ell) + \text{CO}_2(g)$$
(b) Oxidation–reduction: $4 \text{ Ga}(s) + 3 \text{ O}_2(g) \longrightarrow$
$$2 \text{ Ga}_2\text{O}_3(s)$$
(c) Acid–base reaction:
$\text{Ba}(\text{OH})_2(aq) + 2 \text{ HNO}_3(aq) \longrightarrow$
$$\text{Ba}(\text{NO}_3)_2(aq) + 2 \text{ H}_2\text{O}(\ell)$$
Net ionic equation:
$\text{OH}^-(aq) + \text{H}_3\text{O}^+(aq) \longrightarrow 2 \text{ H}_2\text{O}(\ell)$
(d) Precipitation reaction:
$\text{CuCl}_2(aq) + (\text{NH}_4)_2\text{S}(aq) \longrightarrow$
$$\text{CuS}(s) + 2 \text{ NH}_4\text{Cl}(aq)$$
Net ionic equation:
$\text{Cu}^{2+}(aq) + \text{S}^{2-}(aq) \longrightarrow \text{CuS}(s)$

Chapter 4

4.1 $(454 \text{ g C}_3\text{H}_8)(1 \text{ mol C}_3\text{H}_8/44.10 \text{ g C}_3\text{H}_8)$
$= 10.3 \text{ mol C}_3\text{H}_8$

$10.3 \text{ mol C}_3\text{H}_8 (5 \text{ mol O}_2/1 \text{ mol C}_3\text{H}_8)$
$(32.00 \text{ g O}_2/1 \text{ mol O}_2) = 1650 \text{ g O}_2$

$(10.3 \text{ mol C}_3\text{H}_8)(3 \text{ mol CO}_2/1 \text{ mol C}_3\text{H}_8)$
$(44.01 \text{ g CO}_2/1 \text{ mol CO}_2) = 1360 \text{ g CO}_2$

$(10.3 \text{ mol C}_3\text{H}_8)(4 \text{ mol H}_2\text{O}/1 \text{ mol C}_3\text{H}_8)$
$(18.02 \text{ g H}_2\text{O}/1 \text{ mol H}_2\text{O}) = 742 \text{ g H}_2\text{O}$

4.2 (a) Amount Al $= (50.0 \text{ g Al})(1 \text{ mol Al}/26.98 \text{ g Al})$
$= 1.85 \text{ mol Al}$

Amount $\text{Fe}_2\text{O}_3 = (50.0 \text{ g Fe}_2\text{O}_3)(1 \text{ mol Fe}_2\text{O}_3/$
$159.7 \text{ g Fe}_2\text{O}_3) = 0.313 \text{ mol Fe}_2\text{O}_3$

Mol Al/mol $\text{Fe}_2\text{O}_3 = 1.85/0.313 = 5.92$

This is more than the $2:1$ ratio required, so the
limiting reactant is Fe_2O_3.
(b) Mass Fe $= (0.313 \text{ mol Fe}_2\text{O}_3)(2 \text{ mol Fe}/$
$1 \text{ mol Fe}_2\text{O}_3) (55.85 \text{ g Fe}/1 \text{ mol Fe}) = 35.0 \text{ g Fe}$

4.3 $(0.143 \text{ g O}_2)(1 \text{ mol O}_2/32.00 \text{ g O}_2)(3 \text{ mol TiO}_2/$
$3 \text{ mol O}_2)(79.87 \text{ g TiO}_2/1 \text{ mol TiO}_2) = 0.357 \text{ g TiO}_2$

Percent TiO_2 in sample $= (0.357 \text{ g}/2.367 \text{ g})(100\%)$
$= 15.1\%$

4.4 $(1.612 \text{ g CO}_2)(1 \text{ mol CO}_2/44.010 \text{ g CO}_2)(1 \text{ mol C}/1 \text{ mol CO}_2) = 0.03663 \text{ mol C}$

$(0.7425 \text{ g H}_2\text{O})(1 \text{ mol H}_2\text{O}/18.015 \text{ g H}_2\text{O})(2 \text{ mol H}/1 \text{ mol H}_2\text{O}) = 0.08243 \text{ mol H}$

$0.08243 \text{ mol H}/0.03663 \text{ mol} = 2.250 \text{ H}/1 \text{ C} = 9 \text{ H}/4 \text{ C}$

The empirical formula is C_4H_9, which has a molar mass of 57 g/mol. This is one half of the measured value of molar mass, so the molecular formula is C_8H_{18}.

4.5 $(0.240 \text{ g CO}_2)(1 \text{ mol CO}_2/44.01 \text{ g CO}_2)(1 \text{ mol C}/1 \text{ mol CO}_2)(12.01 \text{ g C}/1 \text{ mol C}) = 0.06549 \text{ g C}$

$(0.0982 \text{ g H}_2\text{O})(1 \text{ mol H}_2\text{O}/18.02 \text{ g H}_2\text{O})(2 \text{ mol H}/1 \text{ mol H}_2\text{O})(1.008 \text{ g H}/1 \text{ mol H}) = 0.01099 \text{ g H}$

Mass O (by difference) =
$0.1342 \text{ g} - 0.06549 \text{ g} - 0.01099 \text{ g} = 0.0577 \text{ g}$

Amount C = $0.06549 \text{ g}(1 \text{ mol C}/12.01 \text{ g C})$
$= 0.00545 \text{ mol C}$

Amount H = $0.01099 \text{ g H}(1 \text{ mol H}/1.008 \text{ g H})$
$= 0.01090 \text{ mol H}$

Amount O = $0.0577 \text{ g O}(1 \text{ mol O}/16.00 \text{ g O})$
$= 0.00361 \text{ mol O}$

To find a whole-number ratio, divide each value by 0.00361; this gives 1.51 mol C : 3.02 mol H : 1 mol O. Multiply each value by 2, and round off to 3 mol C : 6 mol H : 2 mol O. The empirical formula is $C_3H_6O_2$; given the molar mass of 74.1, this is also the molecular formula.

4.6 $(26.3 \text{ g})(1 \text{ mol NaHCO}_3/84.01 \text{ g NaHCO}_3)$
$= 0.313 \text{ mol NaHCO}_3$

$0.313 \text{ mol NaHCO}_3/0.200 \text{ L} = 1.57 \text{ M}$

Ion concentrations: $[\text{Na}^+] = [\text{HCO}_3^-] = 1.57 \text{ M}$

4.7 $(2.00 \text{ M})(V_{\text{conc}}) = (1.00 \text{ M})(0.250 \text{ L})$; $V_{\text{conc}} = 0.125 \text{ L}$

To prepare the solution, measure accurately 125 mL of 2.00 M NaOH into a 250-mL volumetric flask, and add water to give a total volume of 250 mL.

4.8 (a) $\text{pH} = -\log(2.6 \times 10^{-2}) = 1.59$
(b) $-\log[\text{H}^+] = 3.80$; $[\text{H}^+] = 1.6 \times 10^{-4} \text{ M}$

4.9 HCl is the limiting reagent.

$(0.350 \text{ mol HCl}/1 \text{ L})(0.0750 \text{ L})(1 \text{ mol CO}_2/2 \text{ mol HCl})(44.01 \text{ g CO}_2/1 \text{ mol CO}_2) = 0.578 \text{ g CO}_2$

4.10 $(0.953 \text{ mol NaOH}/1 \text{ L})(0.02833 \text{ L NaOH})$
$= 0.0270 \text{ mol NaOH}$

$(0.0270 \text{ mol NaOH})(1 \text{ mol CH}_3\text{CO}_2\text{H}/1 \text{ mol NaOH})$
$= 0.0270 \text{ mol CH}_3\text{CO}_2\text{H}$

$(0.0270 \text{ mol CH}_3\text{CO}_2\text{H})(60.05 \text{ g/mol})$
$= 1.62 \text{ g CH}_3\text{CO}_2\text{H}$

$0.0270 \text{ mol CH}_3\text{CO}_2\text{H}/0.0250 \text{ L} = 1.08 \text{ M}$

4.11 $(0.100 \text{ mol HCl}/1 \text{ L})(0.02967 \text{ L}) = 0.00297 \text{ mol HCl}$

$(0.00297 \text{ mol HCl})(1 \text{ mol NaOH}/1 \text{ mol HCl})$
$= 0.00297 \text{ mol NaOH}$

$0.00297 \text{ mol NaOH}/0.02500 \text{ L} = 0.119 \text{ M NaOH}$

4.12 Moles acid = moles base = $(0.323 \text{ mol/L})(0.03008 \text{ L})$
$= 9.716 \times 10^{-3} \text{ mol}$

Molar mass = $0.856 \text{ g acid}/9.716 \times 10^{-3} \text{ mol acid}$
$= 88.1 \text{ g/mol}$

4.13 $(0.196 \text{ mol Na}_2\text{S}_2\text{O}_3/1 \text{ L})(0.02030 \text{ L})$
$= 0.00398 \text{ mol Na}_2\text{S}_2\text{O}_3$

$(0.00398 \text{ mol Na}_2\text{S}_2\text{O}_3)(1 \text{ mol I}_2/2 \text{ mol Na}_2\text{S}_2\text{O}_3)$
$= 0.00199 \text{ mol I}_2$

0.00199 mol I_2 is in excess and was not used in the reaction with ascorbic acid.

I_2 originally added = $(0.0520 \text{ mol I}_2/1 \text{ L})(0.05000 \text{ L})$
$= 0.00260 \text{ mol I}_2$

I_2 used in reaction with ascorbic acid
$= 0.00260 \text{ mol} - 0.00199 \text{ mol} = 6.1 \times 10^{-4} \text{ mol I}_2$

$(6.1 \times 10^{-4} \text{ mol I}_2)(1 \text{ mol C}_6\text{H}_8\text{O}_6/1 \text{ mol I}_2)$
$(176.1 \text{ g}/1 \text{ mol}) = 0.11 \text{ g C}_6\text{H}_8\text{O}_6$

4.14 A Beer's Law plot of the calibration data was constructed by plotting concentration of Cu^{2+} along the x-axis and the absorbance along the y-axis. The equation for the best-fit line is $y = 13.0x + 0.011$.

Substituting the absorbance for the solution of unknown concentration yields

$0.418 = 13.0x + 0.011$

$x = 0.0315$

Thus the concentration of Cu^{2+} in the solution of unknown concentration is 0.0315 M.

Chapter 5

5.1 $C = 59.8 \text{ J}/[(25.0 \text{ g})(1.00 \text{ K})] = 2.39 \text{ J/g} \cdot \text{K}$

5.2 $(15.5 \text{ g})(C_{\text{metal}})(18.9 \text{ °C} - 100.0 \text{ °C}) + (55.5 \text{ g})(4.184 \text{ J/g} \cdot \text{K})(18.9 \text{ °C} - 16.5 \text{ °C}) = 0$

$C_{\text{metal}} = 0.44 \text{ J/g} \cdot \text{K}$

5.3 $1.00 \text{ L}(1000 \text{ mL}/1 \text{ L})(0.7849 \text{ g/cm}^3) = 785 \text{ g}$

Heat liquid from 25.0 °C to 78.3 °C.

$\Delta T = 78.3 \text{ °C} - 25.0 \text{ °C} = 53.3 \text{ °C} = 53.3 \text{ K}$
$q = (2.44 \text{ J/g} \cdot \text{K})(785 \text{ g})(53.3 \text{ K})$
$= 1.02 \times 10^5 \text{ J} = 102 \text{ kJ}$

Boil the liquid.
$q = 38.56 \text{ kJ/mol}(785 \text{ g})(1 \text{ mol}/46.08 \text{ g})$
$= 657 \text{ kJ}$

Total = 102 kJ + 657 kJ = 759 kJ

5.4 Energy transferred as heat from tea + energy as heat expended to melt ice = 0.

$(250 \text{ g})(4.2 \text{ J/g} \cdot \text{K})(273.2 \text{ K} - 291.4 \text{ K}) +$
$x \text{ g}(333 \text{ J/g}) = 0$

$x = 57 \text{ g}$

57 g of ice melts with energy supplied by cooling 250 g of tea from 18.2 °C (291.4 K) to 0 °C (273.2 K).

Mass of ice remaining = mass of ice initially − mass of ice melted

Mass of ice remaining = 75 g − 57 g = 18 g

5.5 $(15.0 \text{ g C}_2\text{H}_6)(1 \text{ mol C}_2\text{H}_6/30.07 \text{ g C}_2\text{H}_6) = 0.499 \text{ mol C}_2\text{H}_6$

$\Delta H° = 0.499 \text{ mol C}_2\text{H}_6(1 \text{ mol-rxn}/2 \text{ mol C}_2\text{H}_6)$
$(-2857.3 \text{ kJ/mol-rxn})$

$= -713 \text{ kJ}$

5.6 Mass of final solution = 400. g

$\Delta T = 27.78 \ ^\circ C - 25.10 \ ^\circ C = 2.68 \ ^\circ C = 2.68 \ K$

Amount of HCl used = amount of NaOH used =
$C \times V = (0.400 \ mol/L) \times 0.200 \ L = 0.0800 \ mol$

Energy transferred as heat by acid–base reaction +
energy gained as heat to warm solution = 0

$q_{rxn} + (4.20 \ J/g \cdot K)(400. \ g)(2.68 \ K) = 0$

$q_{rxn} = -4.50 \times 10^3 \ J$

This represents the energy transferred as heat in the
reaction of 0.0800 mol HCl.

Energy transferred as heat per mole = $\Delta_r H$ =
$-4.50 \ kJ/0.0800 \ mol \ HCl = -56.3 \ kJ/mol \ HCl$

5.7 (a) Energy evolved as heat in reaction + energy as heat
absorbed by H_2O + energy as heat absorbed by
bomb = 0

$q_{rxn} + (1.50 \times 10^3 \ g)(4.20 \ J/g \cdot K)(27.32 \ ^\circ C - 25.00 \ ^\circ C) + (837 \ J/K)(27.32 \ ^\circ C - 25.00 \ ^\circ C) = 0$

$q_{rxn} = -16,600 \ J$ (energy as heat evolved in burning 1.0 g sucrose)

(b) Energy evolved as heat per mole = $(-16.6 \ kJ/g$
sucrose$)(342.3 \ g \ sucrose/1 \ mol \ sucrose) =$
$-5670 \ kJ/mol \ sucrose$

5.8 $C(s) + O_2(g) \longrightarrow CO_2(g) \qquad \Delta_r H_1^\circ = -393.5 \ kJ$

$2 \ [S(s) + O_2(g) \longrightarrow SO_2(g)]$
$\qquad \Delta_r H^\circ = 2 \ \Delta_r H_2^\circ = 2(-296.8) = -593.6 \ kJ$

$CO_2(g) + 2 \ SO_2(g) \longrightarrow CS_2(g) + 3 \ O_2(g)$
$\qquad \Delta_r H^\circ = -\Delta_r H_3^\circ = +1103.9 \ kJ$

Net: $C(s) + 2 \ S(s) \longrightarrow CS_2(g)$
$\qquad \Delta_r H^\circ_{net} = -393.5 \ kJ + -593.6 \ kJ + 1103.9 \ kJ$
$\qquad = +116.8 \ kJ$

5.9 $\Delta_r H^\circ = (6 \ mol \ CO_2/mol\text{-}rxn)\Delta_f H^\circ[CO_2(g)] +$
$(3 \ mol \ H_2O/mol\text{-}rxn)\Delta_f H^\circ[H_2O(\ell)] - \{(1 \ mol \ C_6H_6/1 \ mol\text{-}rxn)\Delta_f H^\circ[C_6H_6(\ell)] + (15\!\!/\!\!2 \ mol \ O_2/mol\text{-}rxn)\Delta_f H^\circ[O_2(g)]\}$

$= (6 \ mol/mol\text{-}rxn)(-393.5 \ kJ/mol) + (3 \ mol/mol\text{-}rxn)(-285.8 \ J/mol) - (1 \ mol/mol\text{-}rxn)(+49.0 \ kJ/mol) - 0$

$= -3267.4 \ kJ/mol\text{-}rxn$

Chapter 6

6.1 (a) Highest frequency, violet; lowest frequency, red
(b) The frequency of a microwave oven, 2.45 GHz, is
higher than the FM radio frequency, 91.7 MHz.
(c) The wavelength of x-rays is shorter than the wavelength of ultraviolet light.

6.2 (a) E (per atom) $= -Rhc/n^2$
$= (-2.179 \times 10^{-18})/(3^2) \ J/atom$
$= -2.421 \times 10^{-19} \ J/atom$
(b) E (per mol) $= (-2.421 \times 10^{-19} \ J/atom)$
$(6.022 \times 10^{23} \ atoms/mol)$
$(1 \ kJ/10^3 \ J)$
$= -145.8 \ kJ/mol$

6.3 The least energetic line is from the electron transition
from $n = 2$ to $n = 1$.

$\Delta E = -Rhc[1/1^2 - 1/2^2]$
$= -(2.179 \times 10^{-18} \ J/atom)(3/4)$
$= -1.634 \times 10^{-18} \ J/atom$

The photon emitted thus has $E_{photon} = 1.634 \times 10^{-18} \ J$

$\nu = E_{photon}/h$
$= (1.634 \times 10^{-18} \ J)/(6.626 \times 10^{-34} \ J \cdot s)$
$= 2.466 \times 10^{15} \ s^{-1}$

$\lambda = c/\nu = (2.998 \times 10^8 \ m/s^{-1})/(2.466 \times 10^{15} \ s^{-1})$
$= 1.216 \times 10^{-7} \ m$ (or 121.6 nm)

6.4 First, calculate the velocity of the neutron:
$v = [2E/m]^{1/2} = [2(6.21 \times 10^{-21} \ kg \cdot m^2 \ s^{-2})/(1.675 \times 10^{-27} \ kg)]^{1/2}$
$= 2720 \ m \cdot s^{-1}$

Use this value in the de Broglie equation:
$\lambda = h/mv = (6.626 \times 10^{-34} \ kg \cdot m^2 \ s^{-2} \cdot s)/(1.675 \times 10^{-27} \ kg)(2720 \ m \ s^{-1})$
$= 1.45 \times 10^{-10} \ m = 0.145 \ nm$

Chapter 7

7.1 (a) chlorine (Cl)
(b) $1s^2 2s^2 2p^6 3s^2 3p^3$

7.2 Calcium has two valence electrons in the $4s$ subshell.
Quantum numbers for these two electrons are $n = 4$,
$\ell = 0$, $m_\ell = 0$, and $m_s = \pm 1/2$

7.3 Obtain the answers from Table 7.3.

7.4

All three ions are paramagnetic with three, two, and
four unpaired electrons, respectively.

7.5 (a) Increasing atomic radius: C < B < Al
(b) Increasing ionization energy: Al < B < C
(c) Carbon is predicted to have the more negative
electron attachment enthalpy.

Chapter 8

8.1

8.2

methanol   hydroxylamine

8.3 (a) CN^- : formal charge on C is -1; formal charge on N is 0.

(b) SO_3^{2-}: formal charge on S is $+1$; formal charge on each O is -1.

8.4 Resonance structures for the HCO_3^- ion:

$$\left[\overset{..}{\underset{..}{O}}=C-\overset{..}{\underset{..}{O}}:\right]^- \longleftrightarrow \left[:\overset{..}{\underset{..}{O}}-C=\overset{..}{\underset{..}{O}}\right]^-$$
$$\quad\quad\quad :\overset{..}{O}-H \quad\quad\quad\quad\quad :\overset{..}{O}-H$$

(a) No. Three resonance structures are needed in the description of CO_3^{2-}; only two are needed to describe HCO_3^-.

(b) In each resonance structure: carbon's formal charge is 0; the oxygen of the $-OH$ group and the double-bonded oxygen have a formal charge of zero; the singly bonded oxygen has a formal charge of -1. The average formal charge on the latter two oxygen atoms is $-\frac{1}{2}$. In the carbonate ion, each of the three oxygen atoms has an average formal charge of $-\frac{2}{3}$.

(c) H^+ would be expected to add to one of the oxygens with a negative formal charge; that is, one of the oxygens with formal charge of $-\frac{1}{2}$ in this structure.

8.5 $\left[:\overset{..}{F}-\overset{..}{Cl}-\overset{..}{F}:\right]^+$ ClF_2^+, 2 bond pairs and 2 lone pairs.

$\left[:\overset{..}{F}-\overset{..}{\underset{..}{Cl}}-\overset{..}{F}:\right]^-$ ClF_2^-, 2 bond pairs and 3 lone pairs.

8.6 Tetrahedral geometry around carbon. The $Cl-C-Cl$ bond angle will be close to $109.5°$.

8.7 For each species, the electron-pair geometry and the molecular shape are the same. BF_3: trigonal-planar; BF_4^-: tetrahedral. Adding F^- to BF_3 adds an electron pair to the central atom and changes the shape.

8.8 The electron-pair geometry around I is trigonal-bipyramidal. The molecular geometry of the ion is linear.

$$\left[\begin{array}{c} :\overset{..}{C}l: \\ | \\ :-I \\ | \\ :\overset{..}{C}l: \end{array}\right]^-$$

8.9 (a) In PO_4^{3-}, there is tetrahedral electron-pair geometry. The molecular geometry is also tetrahedral.

$$\left[\begin{array}{c} :\overset{..}{O}: \\ | \\ :\overset{..}{O}-P-\overset{..}{O}: \\ | \\ :\overset{..}{O}: \end{array}\right]^{3-}$$

(b) In SO_3^{2-}, there is tetrahedral electron-pair geometry. The molecular geometry is trigonal-pyramidal.

$$\left[\begin{array}{c} :\overset{..}{O}-S-\overset{..}{O}: \\ | \\ :\overset{..}{O}: \end{array}\right]^{2-}$$

(c) In IF_5, there is octahedral electron-pair geometry. The molecular geometry is square-pyramidal.

8.10 (a) The H atom is the positive pole in each case. $H-F$ ($\Delta\chi = 1.8$) is more polar than $H-I$ ($\Delta\chi = 0.5$).

(b) The B atom is the positive pole in each case. $B-F$ ($\Delta\chi = 2.0$) is more polar than $B-C$ ($\Delta\chi = 0.5$).

(c) $C-Si$ ($\Delta\chi = 0.6$) is more polar than $C-S$ ($\Delta\chi = 0.1$). In $C-Si$, C is the negative pole, and Si is the positive pole. In $C-S$, S is the negative pole, and C the positive pole.

8.11 16 valence electrons

$$\left[:S\equiv C-\overset{..}{N}:\right]^- \longleftrightarrow \left[:\overset{..}{S}=C=\overset{..}{N}:\right]^- \longleftrightarrow \left[:\overset{..}{S}-C\equiv N:\right]^-$$

Formal charges $\quad + \quad 0 \quad 2- \quad\quad\quad 0 \quad 0 \quad - \quad\quad\quad - \quad 0 \quad 0$

Formal charge considerations favor the middle structure because it has less formal charge than the left structure and, unlike the right structure, it has the negative formal charge on the most electronegative atom in the ion.

Bond polarity: For the C–N bond, $\Delta\chi = 0.5$, so this bond is polar and should have a partially positive C and a partially negative N. For the C–S bond, $\Delta\chi = 0.1$, so this bond should be only slightly polar with a partially positive C and a partially negative S. Comparison of formal charge and bond polarity: The bonding in SCN^- will be closest to the middle resonance structure with a smaller contribution of the resonance structure on the right. From this we conclude that both N and S will have a negative formal charge, with N having the more negative value. The polarities of the $C-N$ and $C-S$ bonds match this description, with N and S being the negative end of each polar bond.

8.12 (a) $BFCl_2$, polar, negative side is the F atom because F is the most electronegative atom in the molecule.

(b) NH_2Cl, polar, negative side is the Cl atom.

(c) SCl_2, polar, Cl atoms are on the negative side.

8.13 (a)

$$:\ddot{O}:$$
$$:\ddot{Cl}-\overset{}{S}-\ddot{Cl}:$$

Formal charges: S = +1, O = −1, Cl = 0

(b) Geometry: trigonal-pyramidal. The molecule is polar. The positive charge is on sulfur, the negative charge on oxygen.

8.14 $CH_4(g) + 2\ O_2(g) \longrightarrow CO_2(g) + 2\ H_2O(g)$

Break 4 C—H bonds and 2 O=O bonds:
(4 mol)(413 kJ/mol) + (2 mol)(498 kJ/mol) = 2648 kJ

Make 2 C=O bonds and 4 H—O bonds:
(2 mol)(745 kJ/mol) + (4 mol)(463 kJ/mol) = 3342 kJ

$\Delta_r H° = 2648\ kJ − 3342\ kJ = −694\ kJ/mol\text{-rxn}$
(value calculated using enthalpies of formation = −802 kJ/mol-rxn)

Chapter 9

9.1 This molecule has a tetrahedral electron-pair geometry and also a tetrahedral molecular shape. The carbon is sp^3 hybridized. The C—H bond arises from overlap of a carbon sp^3 orbital and the hydrogen 1s orbital. The C—Cl bonds each arise from overlap of a carbon sp^3 orbital and a chlorine 3p orbital.

9.2 The carbon and nitrogen atoms in CH_3NH_2 are sp^3 hybridized. The C—H bonds arise from overlap of carbon sp^3 orbitals and hydrogen 1s orbitals. The bond between C and N is formed by overlap of sp^3 orbitals from these atoms. Overlap of nitrogen sp^3 and hydrogen 1s orbitals gives the two N—H bonds, and there is a lone pair in the remaining sp^3 orbital on nitrogen.

9.3 This molecule has a trigonal-bipyramidal electron-pair geometry and a linear molecular shape. The xenon is sp^3d hybridized. The Xe—F bonds each arise from the overlap of one of the axial sp^3d hybrid orbitals on Xe with a 2p orbital on the F. The three lone pairs around Xe are in the equatorial sp^3d hybrid orbitals.

9.4 (a) BH_4^-, tetrahedral electron-pair geometry, sp^3
(b) SF_5^-, octahedral electron-pair geometry, sp^3d^2
(c) OSF_4, trigonal-bipyramidal electron-pair geometry, sp^3d
(d) ClF_3, trigonal-bipyramidal electron-pair geometry, sp^3d
(e) BCl_3, trigonal-planar electron-pair geometry, sp^2
(f) XeO_6^{4-}, octahedral electron-pair geometry, sp^3d^2

9.5 The two CH_3 carbon atoms are sp^3 hybridized, and the center carbon atom is sp^2 hybridized. For each of the carbon atoms in the methyl groups, the sp^3 orbitals overlap with hydrogen 1s orbitals to form the three C—H bonds, and the fourth sp^3 orbital overlaps with an sp^2 orbital on the central carbon atom, forming a carbon–carbon sigma bond. Overlap of an sp^2 orbital on the central carbon and an oxygen sp^2 orbital gives the sigma bond between these elements. The π bond between carbon and oxygen arises by overlap of a p orbital from each element.

9.6 H_2^+: $(\sigma_{1s})^1$ The ion has a bond order of ½ and is expected to exist. A bond order of ½ is predicted for He_2^+ and H_2^-, both of which are predicted to have electron configurations $(\sigma_{1s})^2(\sigma^*_{1s})^1$.

9.7 Li_2^- is predicted to have an electron configuration $(\sigma_{1s})^2(\sigma^*_{1s})^2(\sigma_{2s})^2(\sigma^*_{2s})^1$ and a bond order of ½, the positive value implying that the ion might exist.

9.8 O_2^+: [core electrons] $(\sigma_{2s})^2(\sigma^*_{2s})^2(\sigma_{2p})^2(\pi_{2p})^4(\pi^*_{2p})^1$. The bond order is 2.5. The ion is paramagnetic with one unpaired electron.

Chapter 10

10.1 (a) Isomers of C_7H_{16}

$$CH_3CH_2CH_2CH_2CH_2CH_2CH_3 \qquad \text{heptane}$$

$$\begin{array}{c} CH_3 \\ | \\ CH_3CH_2CH_2CH_2CHCH_3 \end{array} \qquad \text{2-methylhexane}$$

$$\begin{array}{c} CH_3 \\ | \\ CH_3CH_2CH_2CHCH_2CH_3 \end{array} \qquad \text{3-methylhexane}$$

$$\begin{array}{c} CH_3 \\ | \\ CH_3CH_2CHCHCH_3 \\ | \\ CH_3 \end{array} \qquad \text{2,3-dimethylpentane}$$

$$\begin{array}{c} CH_3 \\ | \\ CH_3CH_2CH_2CCH_3 \\ | \\ CH_3 \end{array} \qquad \text{2,2-dimethylpentane}$$

$$\begin{array}{c} CH_3 \\ | \\ CH_3CH_2CCH_2CH_3 \\ | \\ CH_3 \end{array} \qquad \text{3,3-dimethylpentane}$$

$$\begin{array}{c} CH_3 \\ | \\ CH_3CHCH_2CHCH_3 \\ | \\ CH_3 \end{array} \qquad \text{2,4-dimethylpentane}$$

3-Ethylpentane is pictured on page 446.

$$\begin{array}{c} H_3C \quad CH_3 \\ | \quad\quad | \\ CH_3C-CHCH_3 \\ | \\ CH_3 \end{array} \qquad \text{2,2,3-trimethylbutane}$$

(b) Two isomers, 3-methylhexane and 2,3-dimethylpentane, are chiral.

10.2 The names accompany the structures in the answer to Check Your Understanding 10.1.

10.3 Isomers of C_6H_{12} in which the longest chain has six C atoms:

Names (in order, top to bottom): 1-hexene, *cis*-2-hexene, *trans*-2-hexene, *cis*-3-hexene, *trans*-3-hexene. None of these isomers is chiral.

10.4 (a)

bromoethane

(b)

2,3-dibromobutane

10.5 1,4-diaminobenzene

10.6

$CH_3CH_2CH_2CH_2OH$ 1-butanol

$CH_3CH_2CHCH_3$ 2-butanol

CH_3CHCH_2OH 2-methyl-1-propanol

CH_3

CH_3CCH_3 2-methyl-2-propanol

CH_3

10.7 (a) $CH_3CH_2CH_2OH$: 1-propanol, has an alcohol (—OH) group

CH_3CO_2H: ethanoic acid (acetic acid), has a carboxylic acid (—CO_2H) group

$CH_3CH_2NH_2$: ethylamine, has an amino (—NH_2) group

(b) 1-propyl ethanoate (propyl acetate)

(c) Oxidation of this primary alcohol first gives propanal, CH_3CH_2CHO. Further oxidation gives propanoic acid, $CH_3CH_2CO_2H$.

(d) *N*-ethylacetamide, $CH_3CONHCH_2CH_3$

(e) The amine is protonated by hydrochloric acid, forming ethylammonium chloride, $[CH_3CH_2NH_3]Cl$.

10.8 Kevlar is a condensation polymer, prepared by the reaction of terephthalic acid and 1,4-diaminobenzene.

$$n\, H_2NC_6H_4NH_2 + n\, HO_2CC_6H_4CO_2H \rightarrow \;(\text{HNC}_6\text{H}_4\text{NHCOC}_6\text{H}_4\text{CO})_n + 2n\, H_2O$$

Chapter 11

11.1 0.83 bar (0.82 atm) > 75 kPa (0.74 atm) > 0.63 atm > 250 mm Hg (0.33 atm)

11.2 $V_2 = P_1V_1/P_2 = (745 \text{ mm Hg})(65.0 \text{ L})/$
$[0.70(745 \text{ mm Hg})] = 93 \text{ L}$

11.3 $V_1 = 45 \text{ L}$ and $T_1 = 298 \text{ K}$; $V_2 = ?$ and $T_2 = 263 \text{ K}$
$V_2 = V_1(T_2/T_1) = (45 \text{ L})(263 \text{ K}/298 \text{ K}) = 40. \text{ L}$

11.4 $V_2 = V_1(P_1/P_2)(T_2/T_1)$
$= (22 \text{ L})(150 \text{ atm}/0.993 \text{ atm})(295 \text{ K}/304 \text{ K})$
$= 3200 \text{ L}$

At 5.0 L per balloon, there is sufficient He to fill 640 balloons.

11.5 44.8 L of O_2 is required; 44.8 L of $H_2O(g)$ and 22.4 L of $CO_2(g)$ are produced.

11.6 $PV = nRT$

$(750/760 \text{ atm})(V) =$
$(1300 \text{ mol})(0.08206 \text{ L} \cdot \text{atm/mol} \cdot \text{K})(296 \text{ K})$

$V = 3.2 \times 10^4 \text{ L}$

11.7 $d = PM/RT$; $M = dRT/P$

$M = (5.02 \text{ g/L})(0.082057 \text{ L} \cdot \text{atm/mol} \cdot \text{K})$
$(288.2 \text{ K})/(745/760 \text{ atm}) = 121 \text{ g/mol}$

11.8 $PV = (m/M)RT$; $M = mRT/PV$

$M = (0.105 \text{ g})(0.082057 \text{ L} \cdot \text{atm/mol} \cdot \text{K})$
$(296.2 \text{ K})/[(561/760) \text{ atm} (0.125 \text{ L})] = 27.7 \text{ g/mol}$

11.9 $n(N_2) = PV/RT = (1.10 \text{ atm})(25.0 \text{ L})/$
$(0.082057 \text{ L} \cdot \text{atm/mol} \cdot \text{K})(298.2 \text{ K}) = 1.12 \text{ mol } N_2$

1.12 mol N_2 (2 mol Na/3 mol N_2)
$(22.99 \text{ g Na}/1 \text{ mol Na}) = 17.2 \text{ g Na}$

11.10 $n(H_2) = PV/RT$
$= (542/760 \text{ atm})(355 \text{ L})/$
$(0.08206 \text{ L} \cdot \text{atm/mol} \cdot \text{K})(298.2 \text{ K})$

$n(H_2) = 10.3 \text{ mol}$
$n(NH_3) = (10.3 \text{ mol } H_2)(2 \text{ mol } NH_3/3 \text{ mol } H_2)$
$= 6.90 \text{ mol } NH_3$
$P (125 \text{ L}) = (6.90 \text{ mol})(0.082057 \text{ L} \cdot \text{atm/mol} \cdot \text{K})$
(298.2 K)

$P(NH_3) = 1.35 \text{ atm}$

11.11 $P_{halothane}$ (5.00 L)
  = (0.0760 mol)(0.08206 L · atm/mol · K) (298.2 K)
  $P_{halothane}$ = 0.372 atm (or 283 mm Hg)
  P_{oxygen} (5.00 L)
  = (0.734 mol)(0.08206 L · atm/mol · K)(298.2 K)
  P_{oxygen} = 3.59 atm (or 2730 mm Hg)
  $P_{total} = P_{halothane} + P_{oxygen}$
  = 283 mm Hg + 2730 mm Hg = 3010 mm Hg

11.12 For He: Use Equation 11.9, with $M = 4.00 \times 10^{-3}$ kg/mol, $T = 298$ K, and $R = 8.314$ J/mol · K to calculate the rms speed of 1360 m/s. A similar calculation for N_2, with $M = 28.01 \times 10^{-3}$ kg/mol, gives an rms speed of 515 m/s.

11.13 The molar mass of CH_4 is 16.0 g/mol.

$$\frac{\text{Rate for CH}_4}{\text{Rate for unknown}} = \frac{n \text{ molecules}/1.50 \text{ min}}{n \text{ molecules}/4.73 \text{ min}} = \sqrt{\frac{M_{unknown}}{16.0}}$$

$M_{unknown}$ = 159 g/mol

Chapter 12

12.1 Because F^- is the smaller ion, water molecules can approach more closely and interact more strongly. Thus, F^- should have the more negative enthalpy of hydration.

12.2
$$H_3C-O\cdots H-O$$

with H above O and CH_3 below the second O

Hydrogen bonding in methanol entails the attraction of the hydrogen atom bearing a partial positive charge (δ^+) on one molecule to the oxygen atom bearing a partial negative charge (δ^-) on a second molecule. The strong attractive force of hydrogen bonding will cause the boiling point and the enthalpy of vaporization of methanol to be quite high.

12.3 Water is a polar solvent, while hexane and CCl_4 are nonpolar. London dispersion forces are the primary forces of attraction between all pairs of dissimilar solvents. For mixtures of water with the other solvents, dipole–induced dipole forces will also be important.

12.4 (a) O_2: induced dipole–induced dipole forces only.
(b) CH_3OH: strong hydrogen bonding (dipole–dipole forces) as well as induced dipole–induced dipole forces.
(c) Forces between water molecules: strong hydrogen bonding and induced dipole–induced dipole forces. Between N_2 and H_2O: dipole-induced dipole forces and induced dipole–induced dipole forces.
      Relative strengths: a < forces between N_2 and H_2O in c < b < forces between water molecules in c.

12.5 $(1.00 \times 10^3$ g) (1 mol/32.04 g) (35.2 kJ/mol)
  $= 1.10 \times 10^3$ kJ

12.6 (a) At 40 °C, the vapor pressure of ethanol is about 120 mm Hg.
(b) The equilibrium vapor pressure of ethanol at 60 °C is about 320 mm Hg. At 60 °C and 600 mm Hg, ethanol is a liquid. If vapor is present, it will condense to a liquid.

Chapter 13

13.1 (a) The strategy to solve this problem is given in Example 13.1.

Step 1. Mass of the unit cell
  = (197.0 g/mol)(1 mol/6.022×10^{23} atom)
      (4 atoms/unit cell)
  = 1.309×10^{-21} g/unit cell

Step 2. Volume of unit cell
  = (1.309×10^{-21} g/unit cell)(1 cm³/19.32 g)
  = 6.773×10^{-23} cm³/unit cell

Step 3. Length of side of unit cell
  = [6.773×10^{-23} cm³/unit cell]$^{1/3}$
  = 4.076×10^{-8} cm

Step 4. Calculate the radius from the edge dimension.

Diagonal distance = 4.076×10^{-8} cm $(2^{1/2})$ = 4 (r_{Au})

r_{Au} = 1.441×10^{-8} cm (= 144.1 pm)

(b) To verify a body-centered cubic structure, calculate the mass contained in the unit cell. If the structure is bcc, then the mass will be the mass of 2 Fe atoms. (Other possibilities: fcc − mass of 4 Fe; primitive cubic − mass of 1 Fe atom). This calculation uses the four steps from the previous exercise in reverse order.

Step 1. Use radius of Fe to calculate cell dimensions. In a body-centered cube, atoms touch across the diagonal of the cube.

Diagonal distance = side dimension ($\sqrt{3}$) = 4 r_{Fe}

Side dimension of cube = 4 (1.26×10^{-8} cm)/ ($\sqrt{3}$) = 2.910×10^{-8} cm

Step 2. Calculate unit cell volume

Unit cell volume = $(2.910 \times 10^{-8}$ cm$)^3$ = 2.464×10^{-23} cm³

Step 3. Combine unit cell volume and density to find the mass of the unit cell.

Mass of unit cell = 2.464×10^{-23} cm³(7.8740 g/cm³) = 1.940×10^{-22} g

Step 4. Calculate the mass of 2 Fe atoms, and compare this to the answer from step 3.

Mass of 2 Fe atoms
  = 55.85 g/mol (1 mol/6.022×10^{23} atoms)
      (2 atoms)
  = 1.85×10^{-22} g.

This is a fairly good match, and clearly much better than the two other possibilities, primitive and fcc.

13.2 M_2X. In a face-centered cubic unit cell, there are four anions and eight tetrahedral holes in which to place metal ions. All of the tetrahedral holes are inside the unit cell, so the ratio of atoms in the unit cell is 2 : 1.

13.3 We need to calculate the mass and volume of the unit cell from the information given. The density of KCl will then be mass/volume. Select units so the density is calculated as g/cm^3

Step 1. Mass: The unit cell contains 4 K^+ ions and 4 Cl^- ions

Unit cell mass = $(39.10 \text{ g/mol})(1 \text{ mol}/6.022 \times 10^{23} \text{ K}^+ \text{ ions})(4 \text{ K}^+ \text{ ions}) + (35.45 \text{ g/mol})(1 \text{ mol}/6.022 \times 10^{23} \text{ Cl}^- \text{ ions})(4 \text{ Cl}^- \text{ ions})$

$= 2.355 \times 10^{-22} \text{ g} + 2.597 \times 10^{-22} \text{ g}$
$= 4.952 \times 10^{-22} \text{ g}$

Step 2. Volume: Assuming K^+ and Cl^- ions touch along one edge of the cube, the side dimension = $2\ r_{K^+} + 2\ r_{Cl^-}$. The volume of the cube is the cube of this value. (Convert the ionic radius from pm to cm.)

$V = [2(1.33 \times 10^{-8} \text{ cm}) + 2(1.81 \times 10^{-8} \text{ cm})]^3$
$= 2.477 \times 10^{-22} \text{ cm}^3$

Step 3. Density = mass/volume
$= 4.952 \times 10^{-22} \text{ g}/2.477 \times 10^{-22} \text{ cm}^3)$
$= 2.00 \text{ g/cm}^3$

Chapter 14

14.1 (a) 10.0 g sucrose = 0.0292 mol;
250. g H_2O = 13.9 mol

$X_{sucrose} = (0.0292 \text{ mol})/(0.0292 \text{ mol} + 13.9 \text{ mol})$
$= 0.00210$

$c_{sucrose} = (0.0292 \text{ mol sucrose})/(0.250 \text{ kg solvent})$
$= 0.117\ m$

Weight % sucrose =
$(10.0 \text{ g sucrose}/260 \text{ g soln})(100\%) = 3.85\%$

(b) 1.08×10^4 ppm =
1.08×10^4 mg Na^+ per 1000 g soln

$= (1.08 \times 10^4 \text{ mg Na}^+/1000 \text{ g soln})$
$(1050 \text{ g soln}/1 \text{ L})$

$= 1.13 \times 10^4$ mg Na^+/L
$= 11.3$ g Na^+/L

$(11.3 \text{ g Na}^+/\text{L})(58.44 \text{ g NaCl}/23.0 \text{ g Na}^+) =$
28.8 g NaCl/L

14.2 $\Delta_{soln}H° = \Delta_f H°[\text{NaOH(aq)}] - \Delta_f H°[\text{NaOH(s)}]$
$= -469.2 \text{ kJ/mol} - (-425.9 \text{ kJ/mol})$
$= -43.3 \text{ kJ/mol}$

14.3 Solubility of $CO_2 = k_H P_g = 0.034 \text{ mol/kg} \cdot \text{bar} \times$
$0.33 \text{ bar} = 1.1 \times 10^{-2} \text{ mol/kg}$

14.4 The solution contains sucrose $[(10.0 \text{ g})(1 \text{ mol}/342.3 \text{ g})$
$= 0.0292 \text{ mol}]$ in water $[(225 \text{ g})(1 \text{ mol}/18.02 \text{ g})$
$= 12.5 \text{ mol}]$.

$X_{water} = (12.5 \text{ mol H}_2\text{O})/(12.5 \text{ mol} + 0.0292 \text{ mol})$
$= 0.998$

$P_{water} = X_{water}P°_{water} = 0.998(149.4 \text{ mm Hg})$
$= 149 \text{ mm Hg}$

14.5 $c_{glycol} = \Delta T_{bp}/K_{bp} = 1.0 \text{ °C}/(0.512 \text{ °C}/m)$
$= 2.0\ m = 2.0 \text{ mol/kg}$

$mass_{glycol} = (2.0 \text{ mol/kg})(0.125 \text{ kg})(62.07 \text{ g/mol}) = 15 \text{ g}$

14.6 $c_{glycol} = (525 \text{ g})(1 \text{ mol}/62.07 \text{ g})/(3.00 \text{ kg}) = 2.82\ m$

$\Delta T_{fp} = K_{fp} \times m = (-1.86 \text{ °C}/m)(2.82\ m) = -5.24 \text{ °C}$

You will be protected only to about -5 °C and not to -25 °C.

14.7 c (mol/L) = 0.033 g bradykinin (1 mol bradykinin/1060 g bradykinin)/0.0500 L = 6.2×10^{-4} M

$\Pi = cRT$
$= (6.2 \times 10^{-4} \text{ mol/L})(0.082057 \text{ L·atm/mol·K})(293 \text{ K})$
$= 0.015 \text{ atm}$

14.8 $\Delta T_{fp} = 5.265 \text{ °C} - 5.50 \text{ °C} = -0.24 \text{ °C}$

$\Delta T_{fp} = K_{fp} \times m_{solute}$

$m_{solute} = -0.24 \text{ °C}/-5.12 \text{ °C}/m = 0.046 \text{ mol/kg}$

$n_{solute} = (0.046 \text{ mol solute/kg benzene})$
$(0.02346 \text{ kg benzene}) = 0.0011 \text{ mol solute}$

$M = 0.448 \text{ g}/0.0011 \text{ mol} = 4.2 \times 10^2 \text{ g/mol}$

$M(\text{compound})/M(\text{empirical formula})$
$= 4.2 \times 10^2 \text{ g/mol}/104.1 \text{ g/mol} = 4.0$

Molecular formula = $(C_2H_5)_8Al_4F_4$

14.9 c (mol/L) = Π/RT = $[(1.86 \text{ mm Hg})(1 \text{ atm}/760 \text{ mm Hg})]/[(0.08206 \text{ L} \cdot \text{atm/mol} \cdot \text{K})(298 \text{ K})]$
$= 1.00 \times 10^{-4}$ M

$(1.00 \times 10^{-4} \text{ mol/L})(0.100 \text{ L}) = 1.00 \times 10^{-5} \text{ mol}$

Molar mass = $1.40 \text{ g}/1.00 \times 10^{-5} \text{ mol}$
$= 1.4 \times 10^5 \text{ g/mol}$

(Assuming the polymer is composed of CH_2 units, the polymer is about 10,000 units long.)

14.10 $c_{NaCl} = (25.0 \text{ g NaCl})(1 \text{ mol}/58.44 \text{ g})/(0.525 \text{ kg})$
$= 0.815\ m$

$\Delta T_{fp} = K_{fp} \times m \times i = (-1.86 \text{ °C}/m)(0.815\ m)(1.85)$
$= -2.80 \text{ °C}$

Chapter 15

15.1 $-\frac{1}{2}(\Delta[\text{NOCl}]/\Delta t) = \frac{1}{2}(\Delta[\text{NO}]/\Delta t) = \Delta[\text{Cl}_2]/\Delta t$

15.2 For the first 2 hours:
$-\Delta[\text{sucrose}]/\Delta t = -[(0.033 - 0.050) \text{ mol/L}]/(2.0 \text{ h})$
$= 0.0085 \text{ mol/L} \cdot \text{h}$

For the last 2 hours:
$-\Delta[\text{sucrose}]/\Delta t = -[(0.010 - 0.015) \text{ mol/L}]/(2.0 \text{ h})$
$= 0.0025 \text{ mol/L} \cdot \text{h}$

Instantaneous rate at 4 hours = $0.0045 \text{ mol/L} \cdot \text{h}$. (Calculated from the slope of a line tangent to the curve at the defined concentration.)

15.3 Compare experiments 1 and 2: Doubling $[O_2]$ causes the rate to double, so the rate is first order in $[O_2]$. Compare experiments 2 and 4: Doubling $[NO]$ causes the rate to increase by a factor of 4, so the rate is second order in $[NO]$. Thus, the rate law is

Rate $= -(1/2)(\Delta[NO]/\Delta t) = k[NO]^2[O_2]$

Notice that stoichiometry requires the factor of $(1/2)$ in the reaction rate expression.

Using the data in experiment 1 to determine k:

$(1/2)0.028$ mol/L $\cdot$ s $= k[0.020$ mol/L$]^2[0.010$ mol/L$]$

$k = 3.5 \times 10^3$ L^2/mol$^2 \cdot$ s

15.4 Rate $= k[Pt(NH_3)_2Cl_2] = (0.27$ h$^{-1})(0.020$ mol/L$)$

$= 0.0054$ mol/L $\cdot$ h

15.5 ln ([sucrose]/[sucrose]$_o$) $= -kt$

ln ([sucrose]/[0.010]) $= -(0.21$ h$^{-1})(5.0$ h$)$

[sucrose] $= 0.0035$ mol/L

15.6 (a) The fraction remaining is $[CH_3N_2CH_3]/[CH_3N_2CH_3]_o$.

ln $([CH_3N_2CH_3]/[CH_3N_2CH_3]_o)$
$= -(3.6 \times 10^{-4}$ s$^{-1})(150$ s$)$

$[CH_3N_2CH_3]/[CH_3N_2CH_3]_o = 0.95$

(b) After the reaction is 99% complete $[CH_3N_2CH_3]/[CH_3N_2CH_3]_o = 0.010$

ln $(0.010) = -(3.6 \times 10^{-4}$ s$^{-1})(t)$

$t = 1.3 \times 10^4$ s $(210$ min$)$

15.7 $1/[HI] - 1/[HI]_o = kt$

$1/[HI] - 1/[0.010$ M$] = (30.$ L/mol $\cdot$ min$)(12$ min$)$

[HI] $= 0.0022$ M

15.8 $(0.060$ M/0.24 M$) = 0.25$; thus ¼ of the original material remains and two half-lives have transpired.
$t_{1/2} = 141$ min.

$k = \ln2/t_{1/2} = (\ln 2)/141$ min $= 4.9 \times 10^{-3}$ min^{-1}

Initial rate $= k[H_2O_2]_0$
$= 4.9 \times 10^{-3}$ min^{-1} $(0.24$ mol/L$)$
$= 1.2 \times 10^{-3}$ mol/L $\cdot$ min

15.9 (a) For ^{241}Am, $t_{1/2} = 0.693/k = 0.693/(0.0016$ y$^{-1})$
$= 430$ y

For ^{125}I, $t_{1/2} = 0.693/(0.011$ d$^{-1}) = 63$ d

(b) ^{125}I decays much faster.

(c) ln $[(n)/(1.6 \times 10^{15}$ atoms$)] = -(0.011$ d$^{-1})(2.0$ d$)$

$n/1.6 \times 10^{15}$ atoms $= 0.978$; $n = 1.57 \times 10^{15}$ atoms

Since the answer should have two significant figures, we should round this off to 1.6×10^{15} atoms. The approximately 2% that has decayed is not discernable within the limits of accuracy of the data presented.

15.10 An Arrhenius plot was constructed by plotting ln k on the y-axis and $1/T$ on the x-axis. Using Microsoft Excel, the equation of the best-fit line is $y = -22336x + 27.304$.

$E_a = -R \cdot$ (slope)
$= -(0.0083145$ kJ/mol $\cdot$ K$)(-22336$ K$)$
$= 1.9 \times 10^2$ kJ/mol

15.11 ln $(k_2/k_1) = (-E_a/R)(1/T_2 - 1/T_1)$

ln $[(1.00 \times 10^4)/(4.5 \times 10^3)]$
$= -(E_a/8.3145 \times 10^{-3}$ kJ/mol $\cdot$ K$)(1/283$ K $- 1/274$ K$)$

$E_a = 57$ kJ/mol

15.12 All three steps are bimolecular.

For step 3: Rate $= k[N_2O][H_2]$.

There are two intermediates, $N_2O_2(g)$ and $N_2O(g)$.

When the three equations are added, N_2O_2 (a product in the first step and a reactant in the second step) and N_2O (a product in the second step and a reactant in the third step) cancel, leaving the net equation:

2 NO$(g) + 2$ H$_2(g) \longrightarrow$ N$_2(g) + 2$ H$_2$O(g).

15.13 (a) 2 NH$_3$(aq) $+$ OCl$^-$(aq) $\longrightarrow$
$\qquad$ N$_2$H$_4$(aq) $+$ Cl$^-$(aq) $+$ H$_2$O(ℓ)
(b) The second step is the rate-determining step.
(c) Rate $= k[NH_2Cl][NH_3]$
(d) NH$_2$Cl, N$_2$H$_5^+$, and OH$^-$ are intermediates.

15.14 Overall reaction: 2 NO$_2$Cl$(g) \longrightarrow 2$ NO$_2(g) +$ Cl$_2(g)$

Rate $= k'[NO_2Cl]^2/[NO_2]$ (where $k' = k_1k_2/k_{-1}$)

Increasing $[NO_2]$ causes the reaction rate to decrease.

Chapter 16

16.1 (a) $K = [CO]^2/[CO_2]$
(b) $K = [Cu^{2+}][NH_3]^4/[Cu(NH_3)_4^{2+}]$
(c) $K = [H_3O^+][CH_3CO_2^-]/[CH_3CO_2H]$

16.2 (a) $Q = [2.18]/[0.97] = 2.2$. The system is not at equilibrium; $Q < K$. To reach equilibrium, [isobutane] will increase and [butane] will decrease.

(b) $Q = [2.60]/[0.75] = 3.5$. The system is not at equilibrium; $Q > K$. To reach equilibrium, [butane] will increase and [isobutane] will decrease.

16.3 (a)

Equation	$C_6H_{10}I_2$ $\rightleftharpoons$	C_6H_{10}	$+$ I_2
Initial (M)	0.050	0	0
Change (M)	−0.035	+0.035	+0.035
Equilibrium (M)	0.015	0.035	0.035

(b) $K = (0.035)(0.035)/(0.015) = 0.082$

16.4

Equation	$PCl_5(g)$ $\rightleftharpoons$	$PCl_3(g)$	$+$ $Cl_2(g)$
Initial (atm)	2.000	0	0
Change (atm)	−x	+x	+x
Equilibrium (atm)	2.000 − x	x	x

The partial pressure of Cl_2 at equilibrium is 0.814 atm, so $x = 0.814$ atm.

$P(PCl_5) = 2.000$ atm $- 0.814$ atm $= 1.186$ atm

$P(PCl_3) = P(Cl_2) = 0.814$ atm

$K_p = (0.814)(0.814)/1.186 = 0.559$

16.5

Equation	H_2	+	I_2	$\rightleftharpoons$	2 HI
Initial (M)	6.00×10^{-3}		6.00×10^{-3}		0
Change (M)	$-x$		$-x$		$+2x$
Equilibrium (M)	$0.00600 - x$		$0.00600 - x$		$+2x$

$$K_c = 33 = \frac{(2x)^2}{(0.00600 - x)^2}$$

$x = 0.0045$ M, so $[H_2] = [I_2] = 0.0015$ M and $[HI] = 0.0090$ M.

16.6

Equation	$PCl_5(g)$	$\rightleftharpoons$	$PCl_3(g)$	+	$Cl_2(g)$
Initial (M)	0.1000		0		0
Change (M)	$-x$		$+x$		$+x$
Equilibrium (M)	$0.1000 - x$		x		x

$$K_c = [PCl_3][Cl_2]/[PCl_5]$$

$$33.3 = x^2/0.1000 - x$$

We cannot use the simplifying assumption in this case (K is > 1 and $100 \cdot K > 0.1000$), so we must solve using the quadratic formula.

$$x^2 + 33.3x - 3.33 = 0$$

Using the quadratic formula, $x = 0.0997$ (the other root, $x = -33.4$, is not possible because it leads to negative concentrations).

$$[PCl_3] = [Cl_2] = 0.0997 \text{ M}$$

$$[PCl_5] = 0.1000 \text{ M} - 0.0997 \text{ M} = 0.0003 \text{ M}$$

16.7 (a) $K' = K^2 = (2.5 \times 10^{-29})^2 = 6.3 \times 10^{-58}$

(b) $K'' = 1/K^2 = 1/(6.3 \times 10^{-58}) = 1.6 \times 10^{57}$

16.8

Equation	Butane	$\rightleftharpoons$	Isobutane
Initial (M)	0.20		0.50
After adding 2.0 M more isobutane	0.20		2.00 + 0.50
Change (M)	$+x$		$-x$
Equilibrium (M)	$0.20 + x$		$2.50 - x$

$$K = \frac{[\text{isobutane}]}{[\text{butane}]} = \frac{(2.50 - x)}{(0.20 + x)} = 2.50$$

Solving for x gives $x = 0.57$ M. Therefore, [isobutane] = $2.50 - 0.57 = 1.93$ M and [butane] = $0.20 + 0.57 = 0.77$ M.

Chapter 17

17.1 $[H_3O^+] = 4.0 \times 10^{-3}$ M; $[OH^-] = K_w/[H_3O^+] = 2.5 \times 10^{-12}$ M

17.2 (a) pH = 7

(b) pH < 7 (NH_4^+ is an acid)

(c) pH < 7 $[Al(H_2O)_6]^{3+}$ is an acid

(d) pH > 7 (HPO_4^{2-} is a stronger base than it is an acid)

17.3 (a) NH_4^+ is a stronger acid than HCO_3^-. CO_3^{2-}, the conjugate base of HCO_3^-, is a stronger base than NH_3, the conjugate base of NH_4^+.

(b) Reactant-favored; the reactants are the weaker acid and base.

(c) Reactant-favored; the reactants are the weaker acid and base, so reaction lies to the left.

17.4 From the pH, we can calculate $[H_3O^+] = 1.9 \times 10^{-3}$ M. Also, $[\text{butanoate}^-] = [H_3O^+] = 1.9 \times 10^{-3}$ M. Use these values along with [butanoic acid] to calculate K_a.

$K_a = [1.9 \times 10^{-3}][1.9 \times 10^{-3}]/(0.055 - 1.9 \times 10^{-3})$
$= 6.8 \times 10^{-5}$

17.5 $K_a = 1.8 \times 10^{-5} = [x][x]/(0.10 - x)$

$x = [H_3O^+] = [CH_3CO_2^-] = 1.3 \times 10^{-3}$ M; $[CH_3CO_2H] = 0.099$ M; pH = 2.87

17.6 $HF(aq) + H_2O(\ell) \rightleftharpoons H_3O^+(aq) + F^-(aq)$

$K_a = 7.2 \times 10^{-4} = [x][x]/(0.00150 - x)$

The x in the denominator cannot be dropped. This equation must be solved with the quadratic formula or by successive approximations. Because this problem follows the section dealing with the method of successive approximations, that method will be used here.

$7.2 \times 10^{-4} = x^2/(0.00150)$; $x = 1.0 \times 10^{-3}$

$7.2 \times 10^{-4} = x^2/(0.00150 - 1.0 \times 10^{-3})$; $x = 6.0 \times 10^{-4}$

$7.2 \times 10^{-4} = x^2/(0.00150 - 6.0 \times 10^{-4})$; $x = 8.0 \times 10^{-4}$

$7.2 \times 10^{-4} = x^2/(0.00150 - 8.0 \times 10^{-4})$; $x = 7.1 \times 10^{-4}$

$7.2 \times 10^{-4} = x^2/(0.00150 - 7.1 \times 10^{-4})$; $x = 7.5 \times 10^{-4}$

$7.2 \times 10^{-4} = x^2/(0.00150 - 7.5 \times 10^{-4})$; $x = 7.3 \times 10^{-4}$

$7.2 \times 10^{-4} = x^2/(0.00150 - 7.3 \times 10^{-4})$; $x = 7.4 \times 10^{-4}$

$7.2 \times 10^{-4} = x^2/(0.00150 - 7.4 \times 10^{-4})$; $x = 7.4 \times 10^{-4}$

The result has converged to two significant figures.

$[H_3O^+] = [F^-] = 7.4 \times 10^{-4}$ M

$[HF] = 0.00150$ M $- 7.4\ 10^{-4}$ M $= 7.6 \times 10^{-4}$ M

17.7 $OCl^-(aq) + H_2O(\ell) \rightleftharpoons HOCl(aq) + OH^-(aq)$

$K_b = 2.9 \times 10^{-7} = [x][x]/(0.015 - x)$

$x = [OH^-] = [HOCl] = 6.6 \times 10^{-5}$ M

pOH = 4.18; pH = 9.82

17.8 Equivalent amounts of acid and base react to form water, $CH_3CO_2^-$ and Na^+. Acetate ion hydrolyzes to a small extent, giving CH_3CO_2H and OH^-. We need to determine $[CH_3CO_2^-]$ and then solve a weak base equilibrium problem to determine $[OH^-]$.

Amount $CH_3CO_2^-$ = moles base = 0.12 mol/L $\times$ 0.015 L $= 1.8 \times 10^{-3}$ mol

Total volume = 0.030 L, so $[CH_3CO_2^-] = (1.8 \times 10^{-3}$ mol$)/0.030$ L $= 0.060$ M

$CH_3CO_2^-(aq) + H_2O(\ell) \rightleftharpoons CH_3CO_2H(aq) + OH^-(aq)$

$K_b = 5.6 \times 10^{-10} = [x][x]/(0.060 - x)$

$x = [OH^-] = [CH_3CO_2H] = 5.8 \times 10^{-6}$ M

pOH = 5.24; pH = 8.76

17.9 $H_2C_2O_4(aq) + H_2O(\ell) \rightleftharpoons$
$$H_3O^+(aq) + HC_2O_4^-(aq)$$

$K_{a1} = 5.9 \times 10^{-2} = [x][x]/(0.10 - x)$

The x in the denominator cannot be dropped. This equation must be solved with the quadratic formula or by successive approximations.

$x = [H_3O^+] = [HC_2O_4^-] = 5.3 \times 10^{-2}$ M

pH = 1.28

$K_{a2} = [H_3O^+][C_2O_4^{2-}]/[HC_2O_4^-]$; because $[H_3O^+] = [HC_2O_4^-]$

$[C_2O_4^{2-}] = K_{a2} = 6.4 \times 10^{-5}$ M

Chapter 18

18.1 pH of 0.30 M HCO_2H:

$K_a = [H_3O^+][HCO_2^-]/[HCO_2H]$

$1.8 \times 10^{-4} = [x][x]/[0.30 - x]$

$x = 7.3 \times 10^{-3}$ M; pH = 2.13

pH of 0.30 M formic acid + 0.10 M $NaHCO_2$

$K_a = [H_3O^+][HCO_2^-]/[HCO_2H]$

$1.8 \times 10^{-4} = [x][0.10 + x]/(0.30 - x)$

$x = 5.4 \times 10^{-4}$ M; pH = 3.27

18.2

Equation	$HCO_2H + H_2O$	$\rightleftharpoons$	H_3O^+	$+$	HCO_2^-
Initial (M)	0.50		0		0.70
Change (M)	$-x$		$+x$		$+x$
Equilibrium (M)	$0.50 - x$		x		$0.70 + x$

$K_a = 1.8 \times 10^{-4} = (x)(0.70 + x)/(0.50 - x)$

The value of x will be insignificant compared to 0.50 M and 0.70 M.

$1.8 \times 10^{-4} = (x)(0.70)/(0.50)$

$x = [H_3O^+] = 1.3 \times 10^{-4}$ M

pH $= -\log[H_3O^+] = 3.89$

18.3 $(15.0$ g $NaHCO_3)(1$ mol$/84.01$ g$) = 0.179$ mol $NaHCO_3$, and $(18.0$ g $Na_2CO_3)(1$ mol$/106.0$ g$) = 0.170$ mol Na_2CO_3

pH = pK_a + log {[base]/[acid]}

pH $= -\log (4.8 \times 10^{-11}) + \log \{[0.170]/[0.179]\}$

pH $= 10.32 - 0.02 = 10.30$

18.4 pH = pK_a + log {[base]/[acid]}

$5.00 = -\log (1.8 \times 10^{-5}) + \log \{[\text{base}]/[\text{acid}]\}$

$5.00 = 4.74 + \log \{[\text{base}]/[\text{acid}]\}$

[base]/[acid] = 1.8

To prepare this buffer solution, the ratio [base]/[acid] must equal 1.8. For example, you can dissolve 1.8 mol (148 g) of $NaCH_3CO_2$ and 1.0 mol (60.05 g) of CH_3CO_2H in enough water to make 1.0 L of solution.

18.5 Initial pH (before adding acid):

pH = pK_a + log {[base]/[acid]}
$= -\log (1.8 \times 10^{-4}) + \log \{[0.70]/[0.50]\}$
$= 3.74 + 0.15 = 3.89$

After adding acid, the added HCl will react with the weak base (formate ion) and form more formic acid. The net effect is to change the ratio of [base]/[acid] in the buffer solution.

Initial amount $HCO_2H = 0.50$ mol/L $\times$ 0.500 L
$= 0.25$ mol

Initial amount $HCO_2^- = 0.70$ mol/L $\times$ 0.50 L
$= 0.35$ mol

Amount HCl added = 1.0 mol/L $\times$ 0.010 L
$= 0.010$ mol

Amount HCO_2H after HCl addition
$= 0.25$ mol $+ 0.010$ mol $= 0.26$ mol

Amount HCO_2^- after HCl addition
$= 0.35$ mol $- 0.010$ mol $= 0.34$ mol

pH = pK_a + log {[base]/[acid]}

pH $= -\log (1.8 \times 10^{-4}) + \log \{[0.34]/[0.26]\}$

pH $= 3.74 + 0.12 = 3.86$

18.6 35.0 mL base will partially neutralize the acid.

Initial amount CH_3CO_2H
$= (0.100$ mol/L$)(0.1000$ L$) = 0.0100$ mol

Amount NaOH added
$= (0.100$ mol/L$)(0.0350$ L$) = 0.00350$ mol

Amount CH_3CO_2H after reaction
$= 0.0100 - 0.00350 = 0.0065$ mol

Amount $CH_3CO_2^-$ after reaction $= 0.0035$ mol

$[CH_3CO_2H]$ after reaction = 0.0065 mol/0.1350 L
$= 0.048$ M

$[CH_3CO_2^-]$ after reaction = 0.00350 mol/0.1350 L
$= 0.0259$ M

$K_a = [H_3O^+][CH_3CO_2^-]/[CH_3CO_2H]$

$1.8 \times 10^{-5} = [x][0.0259 + x]/[0.048 - x]$

$x = [H_3O^+] = 3.3 \times 10^{-5}$ M; pH = 4.48

18.7 75.0 mL acid will partially neutralize the base.

Initial amount $NH_3 = (0.100$ mol/L$)(0.1000$ L$)$
$= 0.0100$ mol

Amount HCl added $= (0.100$ mol/L$)(0.0750$ L$)$
$= 0.00750$ mol

Amount NH_3 after reaction $= 0.0100 - 0.00750$
$= 0.0025$ mol

Amount NH_4^+ after reaction $= 0.00750$ mol

Solve using the Henderson–Hasselbach equation; use K_a for the weak acid NH_4^+:

pH = pK_a + log {[base]/[acid]}

pH $= -\log (5.6 \times 10^{-10}) + \log \{[0.0025]/[0.00750]\}$

pH $= 9.25 - 0.48 = 8.77$

18.8 $BaF_2(s) \rightleftharpoons Ba^{2+}(aq) + 2\,F^-(aq)$

$[F^-] = 2[Ba^{2+}] = 2(3.6 \times 10^{-3}\ M) = 7.2 \times 10^{-3}\ M$

$K_{sp} = [Ba^{2+}][F^-]^2 = (3.6 \times 10^{-3})(7.2 \times 10^{-3})^2$
$= 1.9 \times 10^{-7}$

18.9 $AgCN(s) \rightleftharpoons Ag^+(aq) + CN^-(aq)$

$K_{sp} = [Ag^+][CN^-]$

Let x = solubility of AgCN in mol/L

$6.0 \times 10^{-17} = x^2$

$x = 7.7 \times 10^{-9}$ mol/L

$(7.7 \times 10^{-9}$ mol AgCN/L$)(133.9$ g AgCN/1 mol AgCN$) = 1.0 \times 10^{-6}$ g/L

18.10 $Ca(OH)_2(s) \rightleftharpoons Ca^{2+}(aq) + 2\,OH^-(aq)$

$K_{sp} = [Ca^{2+}][OH^-]^2;\ K_{sp} = 5.5 \times 10^{-5}$

$5.5 \times 10^{-5} = [x][2x]^2$ (where x = solubility in mol/L)

$x = 2.4 \times 10^{-2}$ mol/L

Solubility in g/L =
$(2.4 \times 10^{-2}$ mol/L$)(74.1$ g/mol$) = 1.8$ g/L

18.11 (a) In pure water:

$K_{sp} = [Ba^{2+}][SO_4^{2-}];\ 1.1 \times 10^{-10} = [x][x];$

$x = 1.0 \times 10^{-5}$ mol/L

(b) In 0.010 M $Ba(NO_3)_2$, which furnishes 0.010 M Ba^{2+} in solution:

$K_{sp} = [Ba^{2+}][SO_4^{2-}]$
$1.1 \times 10^{-10} = [0.010 + x][x]$
$x = 1.1 \times 10^{-8}$ mol/L

18.12 (a) In pure water:

$K_{sp} = [Zn^{2+}][CN^-]^2;\ 8.0 \times 10^{-12} = [x][2x]^2 = 4x^3$

Solubility = $x = 1.3 \times 10^{-4}$ mol/L

(b) In 0.10 M $Zn(NO_3)_2$, which furnishes 0.10 M Zn^{2+} in solution:

$K_{sp} = [Zn^{2+}][CN^-]^2;\ 8.0 \times 10^{-12} = [0.10 + x][2x]^2$

Solubility = $x = 4.5 \times 10^{-6}$ mol/L

18.13 When $[Pb^{2+}] = 1.1 \times 10^{-3}$ M, $[I^-] = 2.2 \times 10^{-3}$ M.

$Q = [Pb^{2+}][I^-]^2 = [1.1 \times 10^{-3}][2.2 \times 10^{-3}]^2$
$= 5.3 \times 10^{-9}$

This value is less than K_{sp}, which means that the system has not yet reached equilibrium and more PbI_2 will dissolve.

18.14 $K_{sp} = [Pb^{2+}][I^-]^2$. Let x be the concentration of I^- required at equilibrium.

$9.8 \times 10^{-9} = [0.050][x]^2$

$x = [I^-] = 4.4 \times 10^{-5}$ mol/L. A concentration greater than this value will result in precipitation of PbI_2.

Let x be the concentration of Pb^{2+} in solution, in equilibrium with 0.0015 M I^-.

$9.8 \times 10^{-9} = [x][1.5 \times 10^{-3}]^2$

$x = [Pb^{2+}] = 4.4 \times 10^{-3}$ M

18.15 First, determine the concentrations of Ag^+ and Cl^-; then calculate Q, and see whether it is greater than or less than K_{sp}. Concentrations are calculated using the final volume, 105.0 mL, in the equation $C_{dil} \times V_{dil} = C_{conc} \times V_{conc}$.

$[Ag^+](0.1050\ L) = (0.0010\ mol/L)(0.1000\ L)$

$[Ag^+] = 9.5 \times 10^{-4}$ M

$[Cl^-](0.1050\ L) = (0.025\ M)(0.0050\ L)$

$[Cl^-] = 1.2 \times 10^{-3}$ M

$Q = [Ag^+][Cl^-] = [9.5 \times 10^{-4}][1.2 \times 10^{-3}]$
$= 1.1 \times 10^{-6}$

Because $Q > K_{sp}$, precipitation occurs.

18.16

Equation	$Ag(NH_3)_2^+$ $\rightleftharpoons$	Ag^+ +	$2\,NH_3$
Initial (M)	0.0050	0	$1.00 - 2(0.0050)$
Change	$-x$	$+x$	$+2x$
Equilibrium (M)	$0.0050 - x$	x	$0.99 + 2x$

$K = 1/K_f = 1/1.1 \times 10^7 = [x][0.99]^2/0.0050$
$x = [Ag^+] = 4.6 \times 10^{-10}$ mol/L

18.17 $Cu(OH)_2(s) \rightleftharpoons Cu^{2+}(aq) + 2\,OH^-(aq)$

$K_{sp} = [Cu^{2+}][OH^-]^2$

$Cu^{2+}(aq) + 4\,NH_3(aq) \rightleftharpoons Cu(NH_3)_4^{2+}(aq)$

$K_{form} = [Cu(NH_3)_4^{2+}]/[Cu^{2+}][NH_3]^4$

Net: $Cu(OH)_2(s) + 4\,NH_3(aq) \rightleftharpoons$
$Cu(NH_3)_4^{2+}(aq) + 2\,OH^-(aq)$

$K_{net} = K_{sp} \times K_f = (2.2 \times 10^{-20})(2.1 \times 10^{13})$
$= 4.6 \times 10^{-7}$

Chapter 19

19.1 (a) O_3; larger molecules generally have higher entropies than smaller molecules.
(b) $SnCl_4(g)$; gases have higher entropies than liquids.

19.2 (a) $\Delta_r S° = \Sigma n S°(products) - \Sigma n S°(reactants)$

$\Delta_r S° = (1\ mol/mol\text{-}rxn)\ S°[NH_4Cl(aq)]$
$- (1\ mol/mol\text{-}rxn)\ S°[NH_4Cl(s)]$

$\Delta_r S° = (1\ mol/mol\text{-}rxn)(169.9\ J/mol \cdot K)$
$- (1\ mol/mol\text{-}rxn)(94.85\ J/mol \cdot K)$

$= 75.1\ J/K \cdot mol\text{-}rxn$

A gain in entropy for the formation of a mixture (solution) is expected.

(b) $\Delta_r S° = (2\ mol\ CO_2/mol\text{-}rxn)\ S°(CO_2) + (3\ mol\ H_2O/mol\text{-}rxn)\ S°(H_2O) - [(1\ mol\ C_2H_5OH/mol\text{-}rxn)\ S°(C_2H_5OH) + (3\ mol\ O_2/mol\text{-}rxn)\ S°(O_2)]$

$\Delta_r S° = (2\ mol/mol\text{-}rxn)(213.74\ J/mol \cdot K)$
$+ (3\ mol/mol\text{-}rxn)(188.84\ J/mol \cdot K)$
$- [(1\ mol/mol\text{-}rxn)(282.70\ J/mol \cdot K)$
$+ (3\ mol/mol\text{-}rxn)(205.07\ J/mol \cdot K)]$

$\Delta_r S° = +96.09\ J/K \cdot mol\text{-}rxn$

An increase in entropy is expected because there is a increase in the number of moles of gases.

19.3 $\Delta S°(\text{system}) = \Delta_r S°$
$$= \Sigma n S°(\text{products}) - \Sigma n S°(\text{reactants})$$

$\Delta_r S° = (2 \text{ mol HCl/mol-rxn}) \, S°[\text{HCl(g)}]$
$- \{(1 \text{ mol H}_2/\text{mol-rxn}) \, S°[\text{H}_2\text{(g)}]$
$+ (1 \text{ mol Cl}_2/\text{mol-rxn}) \, S°[\text{Cl}_2\text{(g)}]\}$

$= (2 \text{ mol HCl/mol-rxn})(186.2 \text{ J/K} \cdot \text{mol HCl})$
$- \{(1 \text{ mol H}_2/\text{mol-rxn})(130.7 \text{ J/K} \cdot \text{mol H}_2)$
$+ (1 \text{ mol Cl}_2/\text{mol-rxn}) \, (223.08 \text{ J/K} \cdot \text{mol Cl}_2\text{(g)}\}$

$= 18.6 \text{ J/K} \cdot \text{mol-rxn}$

$\Delta_r H° = \Sigma n \Delta_f H°(\text{products}) - \Sigma n \Delta_f H°(\text{reactants})$

$\Delta_r H° = (2 \text{ mol HCl/mol-rxn}) \, \Delta_f H°[\text{HCl(g)}]$
$- \{(1 \text{ mol H}_2/\text{mol-rxn}) \, \Delta_f H°[\text{H}_2\text{(g)}]$
$+ (1 \text{ mol Cl}_2/\text{mol-rxn}) \, \Delta_f H°[\text{Cl}_2\text{(g)}]\}$

$= (2 \text{ mol HCl/mol-rxn})(-92.31 \text{ kJ/mol HCl})$
$- \{(1 \text{ mol H}_2/\text{mol-rxn})(0 \text{ kJ/mol H}_2)$
$+ (1 \text{ mol Cl}_2/\text{mol-rxn}) \, (0 \text{ kJ/mol Cl}_2\text{(g)}\}$

$= -184.62 \text{ kJ/mol-rxn}$

Both $\Delta_r H° \, (< 0)$ and $\Delta_r S° \, (> 0)$ are favorable, so this reaction is predicted to be spontaneous under standard conditions.

$\Delta S°(\text{surroundings}) = -\Delta_r H°/T$
$= -(-184.62 \text{ kJ/mol-rxn})/298.15 \text{ K}$
$= 0.61922 \text{ kJ/K} \cdot \text{mol-rxn} = 619.22 \text{ J/K} \cdot \text{mol-rxn}$

$\Delta S°(\text{universe}) = \Delta S°(\text{system}) + \Delta S°(\text{surroundings})$
$= 18.6 \text{ J/K} \cdot \text{mol-rxn} + 619.22 \text{ J/K} \cdot \text{mol-rxn}$
$= 637.8 \text{ J/K} \cdot \text{mol-rxn}$

$\Delta S°(\text{universe}) > 0$, so the reaction is spontaneous under standard conditions.

19.4 For the reaction $N_2\text{(g)} + 3 \, H_2\text{(g)} \longrightarrow 2 \, NH_3\text{(g)}$:

$\Delta_r H° = (2 \text{ mol/mol-rxn}) \, \Delta_f H° \text{ for NH}_3\text{(g)}$
$= (2 \text{ mol/mol-rxn})(-45.90 \text{ kJ/mol})$
$= -91.80 \text{ kJ/mol-rxn}$

$\Delta_r S° = (2 \text{ mol/mol-rxn}) \, S°(\text{NH}_3) - [(1 \text{ mol/mol-rxn})$
$S°(\text{N}_2) + (3 \text{ mol/mol-rxn}) \, S°(\text{H}_2)]$

$\Delta_r S° = (2 \text{ mol/mol-rxn})(192.77 \text{ J/ mol} \cdot \text{K})$
$- [(1 \text{ mol/mol-rxn})(191.56 \text{ J/mol} \cdot \text{K})$
$+ (3 \text{ mol/mol-rxn})(130.7 \text{ J/mol} \cdot \text{K})]$

$\Delta_r S° = -198.1 \text{ J/K} \cdot \text{mol-rxn} \, (= 0.1981 \text{ kJ/K} \cdot \text{mol-rxn})$

$\Delta_r G° = \Delta_r H° - T\Delta_r S° =$
$-91.80 \text{ kJ/mol-rxn} - (298 \text{ K})(-0.1981 \text{ kJ/K} \cdot \text{mol-rxn})$

$\Delta_r G° = -32.8 \text{ kJ/mol-rxn}$

19.5 $SO_2\text{(g)} + \frac{1}{2} \, O_2\text{(g)} \longrightarrow SO_3\text{(g)}$

$\Delta_r G° = \Sigma n \Delta_f G°(\text{products}) - \Sigma n \Delta_f G°(\text{reactants})$

$\Delta_r G° = (1 \text{ mol/mol-rxn}) \Delta_f G°[\text{SO}_3\text{(g)}]$
$- \{(1 \text{ mol/mol-rxn}) \Delta_f G°[\text{SO}_2\text{(g)}]$
$+ (0.5 \text{ mol/mol-rxn}) \Delta_f G°[\text{O}_2\text{(g)}]\}$

$\Delta_r G° = -371.04 \text{ kJ/mol-rxn}$
$- (-300.13 \text{ kJ/mol-rxn} + 0 \text{ kJ/mol-rxn})$

$= -70.91 \text{ kJ/mol-rxn}$

19.6 $HgO\text{(s)} \longrightarrow Hg\text{(}\ell\text{)} + \frac{1}{2} \, O_2\text{(g)}$; determine the temperature at which $\Delta_r G° = \Delta_r H° - T\Delta_r S° = 0$. T is the unknown in this problem.

$\Delta_r H° = [-\Delta_f H° \text{ for HgO(s)}] = 90.83 \text{ kJ/mol-rxn}$

$\Delta_r S° = S°[\text{Hg}(\ell)] + \frac{1}{2} \, S°(\text{O}_2) - S°[\text{HgO(s)}]$

$\Delta_r S° = (1 \text{ mol/mol-rxn})(76.02 \text{ J/mol} \cdot \text{K})$
$+ [(0.5 \text{ mol/mol-rxn})(205.07 \text{ J/mol} \cdot \text{K})$
$- (1 \text{ mol/mol-rxn})(70.29 \text{ J/mol} \cdot \text{K})]$
$= 108.26 \text{ J/K} \cdot \text{mol-rxn}$

$\Delta_r H° - T(\Delta_r S°) = 0$
$= 90,830 \text{ J/mol-rxn} - T(108.26 \text{ J/K} \cdot \text{mol-rxn})$

$T = 839 \text{ K} \, (566 \, °\text{C})$

19.7 $CaCO_3\text{(s)} \rightleftharpoons CaO\text{(s)} + CO_2\text{(g)}$

$\Delta_r G° = \Sigma \, n\Delta_f G°(\text{products}) - \Sigma \, n\Delta_f G°(\text{reactants})$

$\Delta_r G° = \Delta_f G°(\text{CaO}) + \Delta_f G°(\text{CO}_2) - \Delta_f G°(\text{CaCO}_3)$

$\Delta_r G° = (1 \text{ mol CaO/mol-rxn})(-603.42 \text{ kJ/mol CaO})$
$+ (1 \text{ mol CO}_2/\text{mol-rxn})(-394.359 \text{ kJ/mol CO}_2)$
$- (1 \text{ mol CaCO}_3/\text{mol-rxn})$
$(-1129.16 \text{ kJ/mol CaCO}_3)$

$\Delta_r G° = 131.38 \text{ kJ/mol-rxn}$

$\Delta_r G° = -RT \ln K$

$131,380 \text{ J/mol-rxn}$
$= -(8.3145 \text{ J/mol-rxn} \cdot \text{K})(298 \text{ K})(\ln K)$

$K_p = 9.37 \times 10^{-24}$

19.8 $C\text{(s)} + CO_2\text{(g)} \rightleftharpoons 2 \, CO\text{(g)}$

$\Delta_r G° = \Sigma \, n\Delta_f G°(\text{products}) - \Sigma \, n\Delta_f G°(\text{reactants})$

$\Delta_r G° = 2 \, \Delta_f G°(\text{CO}) - \Delta_f G°(\text{CO}_2)$

$\Delta_r G° = (2 \text{ mol/mol-rxn})(-137.17 \text{ kJ/mol})$
$- (1 \text{ mol/mol-rxn})(-394.36 \text{ kJ/mol})$

$\Delta_r G° = 120.02 \text{ kJ/mol-rxn}$

$\Delta_r G° = -RT \ln K$

$120,020 \text{ J/mol-rxn}$
$= -(8.3145 \text{ J/mol-rxn} \cdot \text{K})(298 \text{ K})(\ln K)$

$K = 9.18 \times 10^{-22}$

Chapter 20

20.1 Oxidation half-reaction: $Al\text{(s)} \longrightarrow Al^{3+}\text{(aq)} + 3 \, e^-$
Reduction half-reaction: $2 \, H^+\text{(aq)} + 2 \, e^- \longrightarrow H_2\text{(g)}$
Overall reaction: $2 \, Al\text{(s)} + 6 \, H^+\text{(aq)} \longrightarrow$
$2 \, Al^{3+}\text{(aq)} + 3 \, H_2\text{(g)}$

Al is the reducing agent and is oxidized; $H^+\text{(aq)}$ is the oxidizing agent and is reduced.

20.2 (1) $2 \, VO^{2+}\text{(aq)} + Zn\text{(s)} + 4 \, H^+\text{(aq)} \longrightarrow$
$Zn^{2+}\text{(aq)} + 2 \, V^{3+}\text{(aq)} + 2 \, H_2O(\ell)$
$2 \, V^{3+}\text{(aq)} + Zn\text{(s)} \longrightarrow 2 \, V^{2+}\text{(aq)} + Zn^{2+}\text{(aq)}$
(2) Oxidation (Fe^{2+}, the reducing agent, is oxidized):
$Fe^{2+}\text{(aq)} \longrightarrow Fe^{3+}\text{(aq)} + e^-$
Reduction (MnO_4^-, the oxidizing agent, is reduced)
$MnO_4^-\text{(aq)} + 8 \, H^+\text{(aq)} + 5 \, e^- \longrightarrow$
$Mn^{2+}\text{(aq)} + 4 \, H_2O(\ell)$

Overall reaction:

$MnO_4^-\text{(aq)} + 8 \, H^+\text{(aq)} + 5 \, Fe^{2+}\text{(aq)} \longrightarrow$
$Mn^{2+}\text{(aq)} + 5 \, Fe^{3+}\text{(aq)} + 4 \, H_2O(\ell)$

20.3 (a) Oxidation half-reaction:

$$Al(s) + 3\ OH^-(aq) \longrightarrow Al(OH)_3(s) + 3\ e^-$$

Reduction half-reaction:

$$S(s) + H_2O(\ell) + 2\ e^- \longrightarrow HS^-(aq) + OH^-(aq)$$

Overall reaction:

$$2\ Al(s) + 3\ S(s) + 3\ H_2O(\ell) + 3\ OH^-(aq) \longrightarrow$$
$$2\ Al(OH)_3(s) + 3\ HS^-(aq)$$

(b) Aluminum is the reducing agent and is oxidized; sulfur is the oxidizing agent and is reduced.

20.4 Construct two half-cells, the first with a silver electrode and a solution containing $Ag^+(aq)$, and the second with a nickel electrode and a solution containing $Ni^{2+}(aq)$. Connect the two half-cells with a salt bridge. When the electrodes are connected through an external circuit, electrons will flow from the anode (the nickel electrode) to the cathode (the silver electrode). The overall cell reaction is $Ni(s) + 2\ Ag^+(aq) \longrightarrow Ni^{2+}(aq) + 2\ Ag(s)$. To maintain electrical neutrality in the two half-cells, negative ions will flow from the $Ag\ |\ Ag^+$ half-cell to the $Ni\ |\ Ni^{2+}$ half-cell, and positive ions will flow in the opposite direction.

20.5 (a) Using Appendix M, the order determined for these metals from least strong reducing agent to strongest reducing agent is $Hg < Pb < Sn$. (The further down the table, the stronger the metal is as a reducing agent.)

(b) F_2, Cl_2, and Br_2 all can oxidize mercury to mercury(II); Hg is located "southeast" of them on the table. I_2 cannot oxidize mercury to mercury(II); mercury is located "northeast" rather than "southeast" of it.

20.6 Overall reaction: $2\ Al(s) + 3\ Fe^{2+}(aq) \longrightarrow$
$$2\ Al^{3+}(aq) + 3\ Fe(s)$$

$(E^\circ{}_{cell} = 1.22\ V,\ n = 6)$

$$E_{cell} = E^\circ{}_{cell} - (0.0257/n)\ \ln\ \{[Al^{3+}]^2/[Fe^{2+}]^3\}$$
$$= 1.22 - (0.0257/6)\ \ln\ \{[0.025]^2/[0.50]^3\}$$
$$= 1.22\ V - (-0.023)\ V = 1.24\ V$$

20.7 Overall reaction: $Fe(s) + 2\ H^+(aq) \longrightarrow$
$$Fe^{2+}(aq) + H_2(g)$$

$(E^\circ{}_{cell} = 0.44\ V,\ n = 2)$

$$E_{cell} = E^\circ{}_{cell} - (0.0257/n)\ \ln\ \{[Fe^{2+}]P_{H_2}/[H^+]^2\}$$
$$= 0.44 - (0.0257/2)\ \ln\ \{[0.024]1.0/[0.056]^2\}$$
$$= 0.44\ V - 0.026\ V = 0.41\ V$$

20.8 $\Delta_rG^\circ = -nFE^\circ$
$$= -(2\ mol\ e^-)(96,500\ C/mol\ e^-)$$
$$(-0.76\ V)(1\ J/1\ C \cdot V)$$
$$= 146,680\ J = 150\ kJ$$

The negative value of E° and the positive value of ΔG° both indicate a reactant-favored reaction.

20.9 $E^\circ{}_{cell} = E^\circ{}_{cathode} - E^\circ{}_{anode} = 0.799\ V - 0.855\ V$
$$= -0.056\ V;\ n = 2$$

$$E^\circ = (0.0257/n)\ \ln\ K$$

$$-0.056 = (0.0257/2)\ \ln\ K$$

$$K = 0.013$$

20.10 Cathode: $2\ H_2O(\ell) + 2\ e^- \longrightarrow 2\ OH^-(aq) + H_2(g)$

$E^\circ{}_{cathode} = -0.83\ V$

Anode: $4\ OH^-(aq) \longrightarrow O_2(g) + 2\ H_2O(\ell) + 4\ e^-$

$E^\circ{}_{anode} = 0.40\ V$

Overall: $2\ H_2O(\ell) \longrightarrow 2\ H_2(g) + O_2(g)$

$E^\circ{}_{cell} = E^\circ{}_{cathode} - E^\circ{}_{anode} = -0.83\ V - 0.40\ V = -1.23\ V$

The minimum voltage needed to cause this reaction to occur is 1.23 V.

20.11 (1) O_2 is formed at the anode, by the reaction

$$2\ H_2O(\ell) \longrightarrow 4\ H^+(aq) + O_2(g) + 4\ e^-.$$

$(0.445\ A)(45\ min)(60\ s/min)(1\ C/1\ A \cdot s)(1\ mol\ e^-/96,500\ C)(1\ mol\ O_2/4\ mol\ e^-)(32\ g\ O_2/1\ mol\ O_2) = 0.10\ g\ O_2$

(2) The cathode reaction (electrolysis of molten NaCl) is

$$Na^+(melt) + e^- \longrightarrow Na(\ell).$$

$(25 \times 10^3\ A)(60\ min)(60\ s/min)(1\ C/1\ A \cdot s)(1\ mol\ e^-/96,500\ C)(1\ mol\ Na/mol\ e^-)(23\ g\ Na/1\ mol\ Na) = 21,450\ g\ Na = 21\ kg$

Chapter 21

21.1 (a) $2\ Na(s) + Br_2(\ell) \longrightarrow 2\ NaBr(s)$
(b) $Ca(s) + Se(s) \longrightarrow CaSe(s)$
(c) $2\ Pb(s) + O_2(g) \longrightarrow 2\ PbO(s)$

Lead(II) oxide, a red compound commonly called *litharge*, is the most widely used inorganic lead compound. Maroon-colored lead(IV) oxide is the product of lead oxidation in lead-acid storage batteries (Chapter 20). Other oxides such as Pb_3O_4 also exist.

(d) $2\ Al(s) + 3\ Cl_2(g) \longrightarrow 2\ AlCl_3(s)$

21.2 (a) H_2Te
(b) Na_3AsO_4
(c) $SeCl_6$
(d) $HBrO_4$

21.3 (a) NH_4^+ (ammonium ion)
(b) O_2^{2-} (peroxide ion)
(c) N_2H_4 (hydrazine)
(d) NF_3 (nitrogen trifluoride)

21.4 (a) In Na_2Cl, chlorine would have the unlikely charge of $2-$ (to balance the two positive charges of the two Na^+ ions).

(b) This compound would require either the calcium ion to have the formula Ca^+ or the acetate ion to have the formula $CH_3CO_2^{2-}$. In all of its compounds, calcium occurs as the Ca^{2+} ion. The acetate ion, formed from acetic acid by loss of H^+, has a $1-$ charge.

(c) In Mg_2O, the magnesium ions would need to have the incorrect charge of $1+$ to balance charge of the O^{2-} ion or else the oxygen would need to have the incorrect charge of $4-$ to balance the charge of the two Mg^{2+} ions. Neither of these possiblities is acceptable.

Chapter 22

22.1 (a) $Co(NH_3)_3Cl_3$
(b) $Fe(H_2NCH_2CH_2NH_2)_2Br_2 = Fe(en)_2Br_2$

22.2 (a) (i) $K_3[Co(NO_2)_6]$: a complex of cobalt(III) with a coordination number of 6
(ii) $Mn(NH_3)_4Cl_2$: a complex of manganese(II) with a coordination number of 6
(b) $NH_4[Co(edta)]$: a complex of cobalt(III) with a co-ordination number of 6

22.3 (a) hexaaquanickel(II) sulfate
(b) dicyanobis(ethylenediamine)chromium(III) chloride
(c) potassium amminetrichloroplatinate(II)
(d) potassium dichlorocuprate(I)

22.4 (a) Geometric isomers are possible (with the NH_3 ligands in *cis* and *trans* positions).
(b) Only a single structure is possible.
(c) Only a single structure is possible.
(d) This compound is chiral; there are two optical isomers.
(e) Only a single structure is possible.
(f) Two structural isomers are possible based on coordination of the NO_2^- ligand through oxygen or nitrogen.

22.5 (a) $[Ru(H_2O)_6]^{2+}$: an octahedral complex of ruthenium(II) (d^6). A low-spin complex has no unpaired electrons and is diamagnetic. A high-spin complex has four unpaired electrons and is paramagnetic.

high-spin Ru^{2+} low-spin Ru^{2+}

(b) $[Ni(NH_3)_6]^{2+}$: an octahedral complex of nickel(II) (d^8). Only one electron configuration is possible; it has two unpaired electrons and is paramagnetic.

Ni^{2+} ion (d^8)

22.6 (1) A wavelength of 500 nm corresponds to green light being absorbed. The complex ion will appear magenta.
(2) The complex appears yellow because blue light is being absorbed. The high energy of blue light indicates that Δ_o is large and the complex is therefore low spin.

22.7 (a) The $C_5H_5^-$ ligand is an anion (6 π electrons), C_6H_6 is a neutral ligand (6 π electrons), so Mn must be +1 (6 valence electrons). There is a total of 18 valence electrons.
(b) The ligands in this complex all are neutral, so the Mo atom must have no charge. The C_6H_6 ligand contributes six electrons, each CO contributes two electrons for a total of six, and Mo has six valence electrons. The total is 18 electrons.

Chapter 23

23.1 (a) Emission of six α particles leads to a decrease of 24 in the mass number and a decrease of 12 in the atomic number. Emission of four β particles increases the atomic number by 4 but doesn't affect the mass. The final product of this process has a mass number of $232 - 24 = 208$ and an atomic number of $90 - 12 + 4 = 82$, identifying it as $^{208}_{82}Pb$.

(b) Step 1: $^{232}_{90}Th \longrightarrow {}^{228}_{88}Ra + {}^{4}_{2}\alpha$
Step 2: $^{228}_{88}Ra \longrightarrow {}^{228}_{89}Ac + {}^{0}_{-1}\beta$
Step 3: $^{228}_{89}Ac \longrightarrow {}^{228}_{90}Th + {}^{0}_{-1}\beta$

23.2 (a) $^{0}_{+1}\beta$ (b) $^{41}_{19}K$ (c) $^{0}_{-1}\beta$ (d) $^{22}_{10}Ne$

23.3 (a) $^{32}_{14}Si \longrightarrow {}^{32}_{15}P + {}^{0}_{-1}\beta$

(b) $^{45}_{22}Ti \longrightarrow {}^{45}_{21}Sc + {}^{0}_{+1}\beta$ or
$^{45}_{22}Ti + {}^{0}_{-1}e \longrightarrow {}^{45}_{21}Sc$

(c) $^{239}_{94}Pu \longrightarrow \alpha + {}^{235}_{92}U$

(d) $^{42}_{19}K \longrightarrow {}^{42}_{20}Ca + {}^{0}_{-1}\beta$

23.4 $\Delta m = 0.03435$ g/mol
$\Delta E = (3.435 \times 10^{-5}$ kg/mol$)(2.998 \times 10^8$ m/s$)^2$
$= 3.087 \times 10^{12}$ J/mol $(= 3.087 \times 10^9$ kJ/mol$)$
$E_b = 5.146 \times 10^8$ kJ/mol nucleons

23.5 (a) 49.2 years is exactly four half-lives; quantity remaining $= 1.5$ mg$(1/2)^4 = 0.094$ mg
(b) Three half-lives, 36.9 years
(c) 1% is between six half-lives, 73.8 years (1/64 remains), and seven half-lives, 86.1 years (1/128 remains). [Using the integrated first-order rate equation with $[R]/[R]_0 = 0.010$ and $k = (\ln 2)/t_{1/2} = 0.0564$ y^{-1}, the amount of time is calculated to be 81.7 years.]

23.6 (a) $\ln ([A]/[A_o]) = -kt$
$\ln ([3.18 \times 10^3]/[3.35 \times 10^3]) = -k(2.00$ d$)$
$k = 0.0260$ d^{-1}
$t_{1/2} = 0.693/k = 0.693/(0.0260$ d$^{-1}) = 26.7$ d
(b) $k = 0.693/t_{1/2} = 0.693/200$ y $= 3.47 \times 10^{-3}$ y^{-1}
$\ln ([A]/[A_o]) = -kt$
$\ln ([3.00 \times 10^3]/[6.50 \times 10^{12}]) =$
$\qquad\qquad\qquad -(3.47 \times 10^{-3}$ y$^{-1})t$
$\ln (4.62 \times 10^{-10}) = -(3.47 \times 10^{-3}$ y$^{-1})t$
$t = 6190$ y

23.7 $\ln ([A]/[A_o]) = -kt$
$\ln ([9.32]/[13.4]) = -(1.21 \times 10^{-4}$ y$^{-1})t$
$t = 3.00 \times 10^3$ y

This compares quite well with the estimated date.

23.8 $^{98}_{42}Mo + {}^{1}_{0}n \rightarrow {}^{99}_{42}Mo + \gamma$
$^{99}_{42}Mo \rightarrow {}^{99}_{43}Tc + {}^{0}_{-1}\beta$

23.9 $4.17 \times 10^{-5}(0.0100$ g $Pb^{2+}) = 4.17 \times 10^{-7}$ g Pb^{2+}

Solubility $= [4.17 \times 10^{-7}$ g Pb^{2+} (1 mol Pb^{2+}/207.2 g Pb)(1 mol $PbCrO_4$/1 mol Pb^{2+})]/0.01000 L $= 2.01 \times 10^{-7}$ mol $PbCrO_4$/L

P   Answers to Review & Check Questions

Chapter 1

Section 1.4

1. (b) Na
2. (d) silicon

Section 1.5

1. (d) water
2. (c) potassium

Section 1.6

1. (c) gasoline burns in air
2. (b) polypropylene < water < soda bottle plastic
3. (d) volume = 13.6 cm^3

Section 1.7

(b) The campfire wood burns.

Section 1.8

1. (d) 1.0 g of water vapor at 100 °C
2. (a) A mixture of H_2 and O_2 has lower chemical potential energy than H_2O.

Let's Review

Section 1

1. (b) −196 °C
2. (c) 0.16 cm^3
3. (a) 0.750 L
4. (c) 1.9 L
5. (a) 5.59 × 10^3 mg
6. (d) 2 × 10^{-8} g/L
7. (c) 1050 kJ

Section 2

1. (c) 0.03 g
2. (b) Student B

Section 3

1. (c) 3
2. (a) 1
3. (b) 10.32 g
4. (c) 0.65 cm^2

Section 4

1. (a) 23.3 km^2
2. (c) 2.8 × 10^{17} km

Section 5

1. (a) 2.4
2. (d) 78 g

Section 6

(c) 130 m

Chapter 2

Section 2.2

1. (b) 30 neutrons
2. (c) Na

Section 2.3

(d) $^{109}_{47}$Ag, 48.161%

Section 2.4

1. (a) ^{63}Cu
2. (b) 40%

Section 2.5

1.   (a) Ge

2.   (a) Si

3.   (c) O

4.   (c) allotropes

Section 2.6

(b) $C_3H_7NO_2S$

Section 2.7

1.   (b) 2+

2.   (a) loses 3 electrons

3.   (b) ammonium sulfide

4.   (a) $Ba(CH_3CO_2)_2$

5.   (a) vanadium(III) oxide

6.   (a) MgO

Section 2.8

1.   (a) O_2F_2

2.   (b) nitrogen pentaoxide, N_2O_5

3.   (c) dinitrogen tetrafluoride

4.   (b) tetraphosphorus decaoxide

Section 2.9

1.   (c) 164.09 g/mol

2.   (b) 4.0 g H_2

3.   (c) 0.40 mol Ca

4.   (d) 30 g HF

5.   (a) 6.0×10^{23} atoms O

Section 2.10

1.   (d) butane, C_4H_{10}

2.   (a) benzene, C_6H_6

3.   (b) $C_6H_{12}O_6$

4.   (b) C_5H_6O, $C_{10}H_{12}O_2$

Section 2.11

(d) $MgSO_4 \cdot 3H_2O$

Chapter 3

Section 3.1

1.   (b) 2, 3, 1

2.   (c) 12,000

Section 3.2

(b) 9 O_2, 8 H_2O

Section 3.3

(d) All chemical equilibria are product-favored.

Section 3.4

1.   (a) $Ba(NO_3)_2$

2.   (b) Ca^{2+} and 2 NO_3^-

3.   (b) acetic acid

Section 3.5

1.   (b) Ag_2S

2.   (d) 1, 2, 3, and 4

3.   (c) $2\,Ag^+(aq) + CO_3^{2-}(aq) \rightarrow Ag_2CO_3(s)$

Section 3.6

1.   (d) H_2SO_4

2.   (d) PO_4^{3-}

3.   (c) $CH_3CO_2H(aq) + OH^-(aq) \rightarrow$ $H_2O(\ell) + CH_3CO_2^-(aq)$

4.   (b) BaO

Section 3.7

(b) Na_2CO_3

Section 3.8

1.   (c) +6

2.   (c) 2 and 4

3.   (c) $KMnO_4$

Section 3.9

(c) acid–base and precipitation

Chapter 4

Section 4.1

1.   (b) 0.60 moles Cl_2

2.   (a) 0.60 g Mg

Section 4.2

(b) O_2

Section 4.3

1.   (c) 41.8 g
2.   (c) 32.5%

Section 4.4

(c) C_4H_4O

Section 4.5

1.   (c) 0.080 M
2.   (b) 0.010 M
3.   (b) 15 mL of 0.20 M KOH

Section 4.6

1.   (a) 0.10 M HCl
2.   (c) 1.301

Section 4.7

1.   (a) acidic
2.   (a) 80.0 mL

Section 4.8

(b) A is reduced to half the original value.

Chapter 5

Section 5.1

(a) melting of ice at 0 °C

Section 5.2

1.   (b) warming 2.0 g of copper by 10 °C
2.   (c) 20 g of water at 20 °C + 10 g of water at 30 °C

Section 5.3

1.   (a) Raising the temperature of 100 g of water by 1.0 °C
2.   (b) All of the ice melts; liquid water is at a temperature between 0 °C and 10 °C.

Section 5.4

1.   (b) 2 and 4
2.   (c) 2 C(s) + O_2(g) → 2 CO(g)

Section 5.5

1.   (c) 90.8 kJ
2.   (b) −425 kJ

Section 5.6

1.   (b) endothermic
2.   (b) The calculated value of $\Delta_{solution}H°$ will be too small.

Section 5.7

1.   (d) Al(s) + 3/2 Cl_2(g) → $AlCl_3$(s)
2.   (c) $1/2\ \Delta_r H_1° - \Delta_r H_2° + 1/2\ \Delta_r H_3°$

Chapter 6

Section 6.1

(d) microwaves from a microwave oven

Section 6.2

1.   (a) 3.9×10^5 J/mol
2.   (a) UVB

Section 6.3

1.   (d) from $n = 6$ to $n = 2$
2.   (b) from $n = 4$ to $n = 2$

Section 6.4

(b) Wavelength will be shorter.

Section 6.5

1.   (b) $4p$
2.   (d) 16
3.   (c) both n and ℓ

Section 6.6

1.   (d) $3f$
2.   (b) $n = 4$, $\ell = 1$, $m_\ell = 0$
3.   (c) 2

Chapter 7

Section 7.1

(d) 98

Section 7.2

(a) $4s$

(b) $6s$

(c) $5s$

Section 13.4

1.   (d) NaF

2.   (c) $Na^+(g) + Cl^-(g) \rightarrow NaCl(s)$

3.   (c) -287 kJ/mol

Section 13.5

1.   (c) buckyballs (C_{60})

2.   (b) an amorphous solid

Section 13.6

1.   (a) NaCl

2.   (b) 15 g

Section 13.7

1.   (d) Supercritical CO_2 can be obtained at 25 °C if a high enough pressure is applied.

2.   (b) At 66 K and 0.50 atm N_2 is a gas.

Chapter 14

Section 14.1

1.   (b) 0.063

     (b) 3.7

2.   (a) weight percent

Section 14.2

     (c) -23.55 kJ/mol

Section 14.3

     (b) 0.097 mol

Section 14.4

1.   (b) $Na_2SO_4 <$ sugar $< KBr < C_2H_4(OH)_2$

2.   (d) naphthalene, $C_{10}H_8$

3.   (c) 1.0 g propylene glycol, $C_3H_6(OH)_2$

4.   (c) 122 g/mol

Section 14.5

     (b) 40 million atoms

Chapter 15

Section 15.1

1.   (b) The rate of disappearance of $NO(g)$ is twice the rate of disappearance of $O_2(g)$.

2.   (c) 0.6×10^{-5} mol/L · min

Section 15.2

     (e) all of the above

Section 15.3

     (b) 1.24×10^{-5} mol/L·s

Section 15.4

1.   (b) between 6 and 7

2.   (d) A graph of $\ln[SO_2Cl_2]$ vs. time gives a straight line.

3.   (b) second order

Section 15.5

1.   (c) $\ln k$ vs. $1/T$

2.   (b) A higher proportion of reactant molecules exceeds the activation energy.

Section 15.6

1.   (b) This reaction might occur in a single elementary step.

2.   (c) Rate $= k[A]^2[B]$

Chapter 16

Section 16.1

     (a) More $CaCO_3$ would precipitate.

Section 16.2

1.   (b) $K = [SO_2]^2[O_2]/[SO_3]^2$

2.   (b) No, it is not at equilibrium, and the reaction proceeds further to the right.

Section 16.3

     (b) 6.1×10^{-4}

Section 16.4

     (b) 0.011 M

Section 16.5

(c) $K_p = 2.7 \times 10^{33}$

Section 16.6

1. (b) shift right
2. (a) shift left
3. (a) exothermic

Chapter 17

Section 17.1

(c) H_2CO_3, CH_3CO_2H, H_3PO_4

Section 17.2

1. (c) amphiprotic
2. (b) base
3. (b) acid = HNO_3 and conjugate base = NO_3^-
   (a) base = NH_3 and conjugate acid = NH_4^+
4. (c) HF/F^- and $CH_3CO_2H/CH_3CO_2^-$

Section 17.3

1. (b) 11.08
2. (a) 4.8×10^{-5} M
3. (d) 1.4×10^{-4} M

Section 17.4

1. (a) HF
2. (c) HCN
3. (a) $C_6H_5CO_2H$
4. (b) 9.25
5. (c) 10.15

Section 17.5

1. (b) decrease
2. (a) increase

Section 17.6

1. (b) product-favored
2. (b) right

Section 17.7

1. (a) acidic
2. (a) acidic
3. (b) basic

Section 17.8

1. (b) 6.3×10^{-6} M
2. (a) 2.3×10^{-5}
3. (b) pH = 8.37, $[Na^+]$ = 0.10 M, $[CHO_2^-]$ = 0.10 M, $[OH^-] = 2.4 \times 10^{-6}$ M
4. (c) greater than 7

Section 17.9

(c) 10.16

Section 17.10

1. (a) H_2SeO_4
2. (b) $[Fe(H_2O)_6]^{3+}$
3. (a) HOCl

Section 17.11

1. (b) BCl_3
2. (c) Brønsted base and (d) Lewis base

Chapter 18

Section 18.1

(b) 5.05

Section 18.2

1. (a) 0.20 M HCN and 0.10 M KCN
2. (d) 18 mL
3. (b) 9.25
4. (c) 1/1.8

Section 18.3

1. (b) 1.48
2. (d) 11.29
3. (c) methyl red

Section 18.4

1. (b) $K_{sp} = [Ag^+]^2[CO_3^{2-}]$
2. (a) AgCl
   (b) $Ca(OH)_2$
   (c) $Ca(OH)_2$
3. (b) 1.6×10^{-4} M
4. (a) 1.0×10^{-7} M
5. (a) $FeCO_3$

Section 18.5

(b) no

Section 18.6

(b) 1.9×10^{-36} M

Section 18.7

(c) 1.1×10^{5}

Chapter 19

Section 19.1

1. (b) toward equilibrium
2. (b) false
3. (b) sometimes spontaneous

Section 19.2

1. (c) > 0
2. (c) dispersed
3. (b) 3.19 J/K

Section 19.3

1. (b) increases
2. (c) 2.5×10^{-23} J/K

Section 19.4

1. (c) $NaCl(s) < H_2O(\ell) < NH_3(g)$
2. (c) $\Delta_r S° > 0$
3. (a) -326.6 J/K·mol-rxn

Section 19.5

1. (b) -121.6 J/K·mol-rxn
2. (d) 1320 J/K·mol-rxn
3. (b) at higher temperatures

Section 19.6

1. (d) less, less
2. (a) product-favored at equilibrium
3. (b) false

Section 19.7

1. (c) -2155 kJ/mol-rxn
2. (a) -225 kJ/mol-rxn
3. (c) 23 kJ/mol-rxn

Chapter 20

Section 20.1

1. (a) CuS
2. (d) $4 H^+ + 3 e^- + NO_3^- \rightarrow NO + 2 H_2O$
3. (b) $12 OH^- + Br_2 \rightarrow 2 BrO_3^- + 6 H_2O + 10 e^-$

Section 20.2

1. (b) Electrons transfer from Cd to Ni.
2. (a) Cd is the anode and is negative.
3. (b) NO_3^- ions move from the Ni half-cell to the Cd half-cell, and K^+ ions move from the Cd half-cell to the Ni half-cell.

Section 20.3

1. (b) Lead(IV) oxide is reduced.
2. (b) Sulfuric acid is consumed.

Section 20.4

1. (a) 1.562 V
2. (d) Mg
3. (a) ii and iv

Section 20.5

(a) 0.47 V

Section 20.6

(b) 6.3×10^{16}

Section 20.7

(d) Ag^+

Section 20.8

(d) 1450 seconds

Chapter 21

Section 21.2

1. (d) Ca_2O_3
2. (d) tetraphosphorus decaoxide
3. (b) $+3$
4. (d) $+5$

Section 21.3

1. (a) neon

2. (c) the high temperature reaction of methane and water

Section 21.4

1. (c) high melting point (>400 °C)

2. (a) Li^+

3. (c) two Na^+ ions and one O_2^{2-} ion

Section 21.5

1. (c) gypsum, $CaSO_4 \cdot 2H_2O$

2. (c) converting slaked lime, $Ca(OH)_2$, to lime

Section 21.6

1. (c) third

2. (d) dissolve in acid and in base

Section 21.7

1. (c) SiO_2

2. (c) +4

Section 21.8

1. (c) between 2 and 3

2. (b) Solutions of ammonia are acidic.

3. (c) +3

Section 21.9

1. (c) +3

2. (d) All electrons in O_2 are paired.

Section 21.10

1. (b) Cl_2

2. (c) F_2 is prepared industrially by electrolysis of aqueous NaF.

Chapter 22

Section 22.1

1. (e) +7

2. (b) Atomic radii that are similar to the 5th period transition elements.

Section 22.2

1. (b) Copper is reduced; sulfur is oxidized.

2. (b) iron oxide

Section 22.3

1. (b) 6

2. (d) +3

3. (b) $Na_4[Fe(C_2O_4)_3]$

4. (b) pentaamminehydroxochromium(III) chloride

Section 22.4

1. (b) $Ni(H_2O)_4(C_2O_4)$

2. (b) $Fe(en)_2(CN)_2$

3. (b) There are two geometric isomers, one of which has an optical isomer.

4. (d) 3

Section 22.5

1. (b) high-spin $[Mn(NH_3)_6]^{2+}$

2. (a) square-planar $[PtCl_4]^{2-}$

Section 22.6

(c) $[CrF_6]^{3-}$

Section 22.7

(d) the anion in Zeise's salt, $[Pt(\eta^2\text{-}C_2H_4)Cl_3]^-$

Chapter 23

Section 23.2

1. (a) alpha radiation

2. (b) 7 α and 4 β

3. (c) $^{90}_{38}Sr \rightarrow {}^{90}_{39}Y + \beta$

Section 23.3

(c) $^{56}_{26}Fe$

Section 23.4

1. (a) 20%

2. (d) 1/32

Section 23.5

(c) electrons

Section 23.6

(a) $^{93}_{37}Rb$

Index/Glossary

Italicized page numbers indicate pages containing illustrations, and those followed by "t" indicate tables. Glossary terms, printed in boldface, are defined here as well as in the text.

abbreviations, A-9
ABS plastic, 378
absolute temperature scale. *See* **Kelvin temperature scale.**
absolute zero The lowest possible temperature, equivalent to −273.15 °C, used as the zero point of the Kelvin scale, 22, 390
    zero entropy at, 668
absorbance The negative logarithm of the transmittance, 166
absorption spectrum A plot of the intensity of light absorbed by a sample as a function of the wavelength of the light, 167, 810
    excited states and, *213*
absorptivity, molar, 167
abundance(s), of elements in Earth's crust, 54t
    of isotopes, 46, 48t
acceptor level, in semiconductor, 459
accuracy The agreement between the measured quantity and the accepted value, 26
    of acid–base indicators, 639
acetaldehyde, 366t
acetaminophen, structure of, 339
acetate ion, buffer solution of, 623t
    reaction with phosphoric acid, 592
acetic acid, 368t
    buffer solution of, 623t
    decomposition product of aspirin, 576
    hydrogen bonding in, 424
    ionization of, 619
    orbital hybridization in, 321
    production of, 365
    quantitative analysis of, 143
    reaction with ammonia, 595
    reaction with sodium bicarbonate, 593
    reaction with sodium hydroxide, 115, 369
    reaction with water, 112
    structure of, 340, 365
    synthesis of, 144
    titration with sodium hydroxide, 632, 634
    as weak acid, 100, 584, 587
    as weak electrolyte, 103
acetic anhydride, 142
acetone, 366t
    hydrogenation of, 302
    structure of, *322*, 365
acetylacetonate ion, as ligand, 794
acetylene, orbital hybridization in, 322
acetylsalicylic acid. *See* aspirin.

acid(s) A substance that, when dissolved in pure water, increases the concentration of hydrogen ions, 110–119. *See also* **Brønsted acid(s), Lewis acid(s).**
    Arrhenius definition of, 111
    bases and, 576–617. *See also* **acid–base reaction(s).**
    Brønsted definition, 577
    Brønsted–Lowry definition, 112–114
    carboxylic. *See* **carboxylic acid(s).**
    common, 111t
    Lewis definition of, 611–615
    reaction with bases, 114–118
    strengths of, 585, 587
        direction of reaction and, 591
        molecular structure and, 608
        strong. *See* **strong acid.**
        weak. *See* **weak acid.**
acid ionization constant (K_a) The equilibrium constant (or the ionization of an add in aqueous solution, 585, 586t
    relation to conjugate base ionization constant, 588
    values of, A-20
acid rain, 117
acid–base adduct, 611
acid–base indicator(s), 638–640
acid–base pairs, conjugate, 580, 581t
acid–base reaction(s) An exchange reaction between an acid and a base producing a salt and water, 114–118, 127
    characteristics of, 594t
    equivalence point of, 159, 630
    pH after, 602
    titration using, 158–160, 630–640
acidic oxide(s) An oxide of a nonmetal that acts as an add, 117
acidic solution A solution in which the concentration of hydronium ions is greater than the concentration of hydroxide ion, 582
acidosis, 631
acoustic energy The energy of compression and expansion of spaces between molecules, 16
Acrilan, 375t
acrylonitrile, in ABS plastic, 378
actinide(s) The series of elements between actinium and rutherfordium in the periodic table, 58, 243, 783
activation energy (E_a) The minimum amount of energy that must be absorbed by a system to cause it to react, 690
    experimental determination, 531–533
    reduction by catalyst, 534
activity (*A*) A measure of the rate of nuclear decay, the number of disintegrations observed in a sample per unit time, 833

activity, thermodynamic, 553
actual yield The measured amount of product obtained from a chemical reaction, 141
acute mountain sickness, 384
addition polymer(s) A synthetic organic polymer formed by directly joining monomer units, 373–377
    production from ethylene derivatives, 373
addition reaction(s), of alkenes and alkynes, 352
adduct, acid–base, 611
adenine, 304
    hydrogen bonding to thymine, 429
    structure of, *304*
adenosine 5'-triphosphate (ATP), structure of, 683
adhesive force A force of attraction between molecules of two different substances, 440
adhesives, chemistry of, 381
    on geckos' toes, 416
adipoyl chloride, 379
adrenaline, 364
aerobic fermentation, 358
aerosol, 500t
air, components of, 404t
    density of, 396
    fractional distillation of, 772
air bags, *385*, 392
albite, dissolved by rain water, 163
alchemy, 821
alcohol(s) Any of a class of organic compounds characterized by the presence of a hydroxyl group bonded to a saturated carbon atom, 357–363
    energy content of, 25
    naming of, A-17
    oxidation to carbonyl compounds, 365
    solubility in water, 361
aldehyde(s) Any of a class of organic compounds characterized by the presence of a carbonyl group, in which the carbon atom is bonded to at least one hydrogen atom, 365–366
    naming of, A-17
alkali metal(s) The metals in Group 1A of the periodic table, 54
    electron configuration of, 239
    ions, enthalpy of hydration, 419
    reaction with oxygen, 745
    reaction with water, 54, 743, 744
    reduction potentials of, 746
alkaline battery, 705
alkaline earth metal(s) The elements in Group 2A of the periodic table, 54, 748–751
    biological uses of, 750
    electron configuration of, 239

alkalosis, 631

alkane(s) Any of a class of hydrocarbons in which each carbon atom is bonded to four other atoms, 343–348

  derivatives of, 359t

  general formula of, 344t

  naming of, A-15

  properties of, 348

  reaction with chlorine, 348

  reaction with oxygen, 348

  standard enthalpies of vaporization of, 435t

Alka-Seltzer®, *126*

  composition of, 576

alkene(s) Any of a class of hydrocarbons in which there is at least one carbon–carbon double bond, 349–353

  general formula of, 344t

  hydrogenation of, 353

  naming of, A-16

alkyl groups, 347

alkylation, of benzene, 356

alkyne(s) Any of a class of hydrocarbons in which there is at least one carbon–carbon triple bond, 351, 352t

  general formula of, 344t

  naming of, A-16

allene, structure of, *340*

allotrope(s) Different forms of the same element that exist in the same physical state under the same conditions of temperature and pressure, 55

  boron, 753, 754

  carbon, 55

  oxygen, 772. *See also* ozone.

  phosphorus, 56, *763*, 764

  sulfur, 57, 772

alloy(s) A mixture of a metal with one or more other elements that retains metallic characteristics

  aluminum, 754

  atom substitution in, 248

  iron, 791

  magnesium in, 749

  memory metal, 782

alnico V, 791

  ferromagnetism of, 227

alpha particle(s) A helium nucleus ejected from certain radioactive substances, 821

  bombardment with, 837

  predicting emission of, 828

alpha ray(s). *See* **alpha particle(s).**

alpha-hydroxy acid(s), *605*

altitude sickness, 384

aluminosilicates, 761

  separation of, 755

aluminum, abundance of, 55, 753

  chemistry of, 757

  oxidation of, 137

  production of, 754

  reaction with bromine, 58, *59*

  reaction with copper ions, 693

  reaction with iron(III) oxide, 124

  reaction with sodium hydroxide, 742, *743*

  reaction with water, 697

  recycling of, 205

  reduction by sodium, 744

  structure of, *446*

aluminum bromide, dimerization of, 757

aluminum carbide, reaction with water, 142

aluminum hydroxide, amphoterism of, 612

aluminum oxide, 754

  amphoterism of, 755

  in gemstones, 232

amalgam, mercury, 717

americium, 839

  radioactive half-life of, 526

amide(s) Any of a class of organic compounds characterized by the presence of an amido (-NRCO-) group, 365, 370

amide link, 379

amine(s) A derivative of ammonia in which one or more of the hydrogen atoms are replaced by organic groups, 362

  as acids and bases, 615

amino group A functional group related to ammonia, in which some or all of the hydrogen atoms are replaced by organic groups, 365, 370

ammonia, aqueous, equilibrium constant expression for, 552

  bond angles in, 282

  combustion of, balanced equation for, 97

  decomposition of, 515, *524*

  as Lewis base, 614

  as ligand, 794

  molecular polarity of, 294

  orbital hybridization in, 314

  oxidation of, 137, *138*, 141

  percent composition of, 78

  pH of, 154

  production by Haber process, 574

  production of, as equilibrium process, 99

    equilibrium constant for, 567

    spontaneity of, 675

    stoichiometry of, 398

  reaction with acetic acid, 595

  reaction with boron trifluoride, 576

  reaction with hydrochloric acid, 595

  reaction with hydrogen chloride, 116, 403

  reaction with sodium hypochlorite, 764

  reaction with water, 113

  relation to amines, 362

  synthesis of, equilibrium constant, 685

  titration with hydrochloric acid, 637

  waste product of fish metabolism, 765

  as weak base, 584, 587

ammonium carbonate, reaction with carbon dioxide, 412

ammonium chloride, decomposition of, 675

  in dry cell battery, 705

ammonium ion, 65, 66

  in Lewis adduct, 612

ammonium nitrate, decomposition of, 766

  dissolution of, 662

  enthalpy of solution, 479

ammonium perchlorate, in rocket fuel, 779

amorphous solid(s) A solid that lacks long-range regular structure and displays a melting range instead of a specific melting point, 463

amount, of pure substance, 72. *See also* **mole (mol).**

amounts table, in solving stoichiometry problems, 135. *See also* ICE table.

ampere (A) The unit of electric current, 730, A-10

Ampere, André Marie, 775

amphibole, 760

amphiprotic substance A substance that can behave as either a Brønsted acid or a Brønsted base, 113, 579

amphoteric substance A substance, such as a metal hydroxide, that can behave as either an acid or base, 612, 613t

  aluminum oxide, 755

amplitude The maximum height of a wave, as measured from the axis of propagation, *206*

analysis, chemical. *See* **chemical analysis.**

  spectrophotometric, 167

Anderson, Carl, 825

Angstrom unit (Å), *24*

angular momentum quantum number, 220

  number of nodal surfaces and, 224

anhydrous compound The substance remaining after the water has been removed (usually by heating) from a hydrated compound, 87

aniline, reaction with sulfuric acid, 363

  structure of, 355

  as weak base, 587

anilinium hydrogen sulfate, 363

anion(s) An ion with a negative electric charge, 62

  as Brønsted acids and bases, 578, 611

  naming, 67

  noble gas electron configuration in, 257

  sizes of, 254

anode The electrode of an electrochemical cell at which oxidation occurs, 699

  in corrosion, 787

antibonding molecular orbital A molecular orbital in which the energy of the electrons is higher than that of the parent orbital electrons, 326

antifreeze, 489, 490

  ethylene glycol in, 475

antilogarithms, A-3

antimatter, 825

antimony, abundance of, 763

  isotopic abundance of, 49

antineutrino, 826

apatite(s), 750, 769

apophyllite, *761*

Appian Way, mortar in, 751

approximations, successive, method of, 563, A-4

aqua regia, 768

aquarium, nitrogen cycle in, 765

aqueous solution A solution in which the solvent is water, 101–105

  balancing redox equations in, 695–698

  electrolysis in, 727

  equilibrium constant expression for, 552

aragonite, 652

argon, isotope ratios, 50

argyria, 126

Arnold, James R., 836

aromatic compound(s) Any of a class of hydrocarbons characterized by the presence of a benzene ring or related structure, 323, 353–356

  general formula of, 344t

  naming of, A-16

  properties of, 356

Arrhenius, Svante, 110

Arrhenius equation A mathematical expression that relates reaction rate to activation energy, collision frequency, and molecular orientation, 531

arsenic, abundance of, 763

  poisoning by, 84

arsine, 768

asbestos, 748, 760

ascorbic acid, reaction with iodine, 512

  structure of, 86

  titration of, 164

aspirin, 370

  history of, 576

  melting point of, *14*

  molar mass of, 75

  structure of, 355

  synthesis of, 148

astatine, abundance of, 775

astronomical unit, *29*

atmosphere. *See also* air.

  carbon-14 isotope in, 835, *836*

  composition of, 404

  pressure–temperature profile of, 404

standard, 386. *See also* **standard atmosphere (atm).**

atom(s) The smallest particle of an element that retains the characteristic chemical properties of that element, 10
Bohr model of, 211–216
composition of, 45
electron configurations of, 237–245. *See also* electron configuration(s).
mass of, 44
quantization of energy in, 212, 219
size of, 44, 247. *See also* atomic radius.
structure of, 43

atom economy, green chemistry and, 144

atomic bomb, 840

atomic mass The experimentally determined mass of an atom of one isotope, 47–49. *See also* **atomic weight.**

atomic mass unit (u) The unit of a scale of relative atomic masses of the elements; 1 u = 1/12 of the mass of a carbon atom with six protons and six neutrons, 44
equivalent in grams, 44

atomic number (Z) The number of protons in the nucleus of an atom of an element, 44
chemical periodicity and, 53
even versus odd, and nuclear stability, 827
in nuclear symbol, 822

atomic orbital(s) The matter wave for an allowed energy state of an electron in an atom, 220–222
assignment of electrons to, 235–238
energies of, and electron assignments, 235–238
number of electrons in, 234t
order of energies in, 235, 244
orientations of, 224
overlapping of, in valence bond theory, 310
penetration of, 236
quantum numbers of, 220
shapes of, 222–226

atomic radius, bond length and, 298
determining, 450
effective nuclear charge and, 249
periodicity, 247
in transition elements, 788

atomic reactor, 841

atomic theory of matter A theory that describes the structure and behavior of substances in terms of ultimate chemical particles called atoms and molecules, 43

atomic weight The average mass of an atom in a natural sample of the element, 47

atropine, 604

Atwater system, for energy content of foods, 25

Aufbau principle, 235

austenite, 454, 782

autoionization of water Proton transfer between two water molecules to produce a hydronium ion and a hydroxide ion, 581

automobile, electric, 444, *707*
energy available from power sources, 708t

Autumn, Kellar, 416

average reaction rate, 510

Avogadro, Amedeo, 72, 73, 392

Avogadro's hypothesis Equal volumes of gases under the same conditions of temperature and pressure have equal numbers of particles, 392–393
kinetic-molecular theory and, 407

Avogadro's number The number of particles in one mole of any substance (6.0221415×10^{23}), 72

axial position, in cyclohexane structure, 349
in trigonal-bipyramidal molecular geometry, 284

azo dye, 334

azomethane, decomposition of, 522

azurite, *641, 643, 784, 789*

background radiation, 843, 844t

bacteria, copper production by, 791

baking powder, *595,* 771

baking soda, 118, 746
reaction with vinegar, 593

balance, laboratory, precision of, *31*

balanced chemical equation A chemical equation showing the relative amounts of reactants and products, 96–98
enthalpy and, 189
equilibrium constant and, 566–568

ball-and-stick models, 60, 341

balloon, helium, 741
hot-air, *396*
hydrogen, 394, 741
models of electron pair geometries, 280
weather, 394, *395*

Balmer, Johann, 210

Balmer equation, 211

Balmer series, 211, 214

band gap, 458

band of stability, nuclear, 827

band theory, of metallic bonding, 457
of semiconductors, 458

bar A unit of pressure; 1 bar = 100 kPa, 386, A-6

barium, abundance of, 748

barium chloride, as strong electrolyte, 102
reaction with sodium sulfate, 109, 654

barium nitrate, in fireworks, 217
as strong electrolyte, 103

barium sulfate, formation by precipitation, 109
precipitation of, 653
solubility of, 643
as x-ray contrast agent, 750

barometer An apparatus used to measure atmospheric pressure, 386
mercury, 386

base(s) A substance that, when dissolved in pure water, increases the concentration of hydroxide ions, 110–119. *See also* **Brønsted base(s), Lewis base(s).**
acids and, 576–617. *See also* **acid–base reaction(s).**
Arrhenius definition of, 111
Brønsted–Lowry definition of, 112–114, 577
common, 111t
strong. *See* **strong base.**
weak. *See* **weak base.**
in DNA, 304
Lewis definition of, 611–615
of logarithms, A-2
reaction with acids, 114–118
strengths of, 585, 587
direction of reaction and, 591

base ionization constant (K_b) The equilibrium constant for the ionization of a base in aqueous solution, 585, 586t
relation to conjugate acid ionization constant, 588

base pairing, hydrogen bonding and, 429

base units, SI, 21t, A-10

basic oxide(s) An oxide of a metal that acts as a base, 117

basic oxygen furnace, 791

basic solution A solution in which the concentration of hydronium ions is less than the concentration of hydroxide ions, 582

battery A device consisting of two or more electrochemical cells, 704
electrochemical, 690
energy per kilogram, 708t

bauxite, 755

Bayer process, 755

becquerel The SI unit of radioactivity, 1 decomposition per second, 843

Beer–Lambert law The absorbance of a sample is proportional to the path length and the concentration, 167

Beethoven, Ludwig van, 762

belladonna, atropine in, 604

bends, 482

benzaldehyde, structure of, 366

benzene, boiling point elevation and freezing point depression constants for, 489t
bonding in, resonance structures of, 272, 323
derivatives of, 355, A-16
liquid and solid volumes, *418*
molecular orbital configuration of, 333
in organometallic compounds, 815
reactions of, 356
standard enthalpy of formation of, 199
structure of, 354
vapor pressure of, 488

benzoic acid, 368t
buffer solution of, 625
pH in aqueous solution, 598
structure of, 355

benzonitrile, structure of, 340

benzyl acetate, 370

benzyl butanoate, 370t

beryllium, abundance of, 748

beryllium dichloride, orbital hybridization in, 317

beta particle(s) An electron ejected at high speed from certain radioactive substances, 821
predicting emission of, 828

beta rays. *See* **beta particle(s).**

bicarbonate ion. *See also* hydrogen carbonate ion.
in biological buffer system, 631

bidentate ligands, 794

bimolecular process, 536

binary compound(s) A compound formed from two elements, 60

binding energy The energy required to separate a nucleus into individual protons and neutrons, 829–832
per nucleon, 830

biochemistry, thermodynamics and, 683

biological effects of radiation, 844, 845

birefringence, 748

bismuth, abundance of, 763

bismuth subsalicylate, in Pepto-Bismol, *107*

bisphenol-A, structure of, 377

black powder, 217

black smokers, metal sulfides from, 92

black tongue, Pepto-Bismol and, *107*

blackbody radiation, 207

blast furnace, 790

bleach, detection in food tampering, 165
hypochlorite ion in, 779
sodium hypochlorite in, 475

blood, buffers in, 622
oxygen saturation of, 384
pH of, 154, 631

blue vitriol, 87

boat form, 349

body-centered cubic (bcc) unit cell, 447
Bohr, Christian, 798
Bohr, Niels, 211
Bohr effect, in hemoglobin, 798
boiling point The temperature at which the vapor pressure of a liquid is equal to the external pressure on the liquid, 439, A-14t
    for common compounds, 435t
    hydrogen bonding and, 422, *423*
    intermolecular forces and, 420, 422t
boiling point elevation, 487
boiling point elevation constant (K_{bp}), 488
Boltzmann, Ludwig, 405, 666
Boltzmann distribution curves. *See* Maxwell–Boltzmann distribution curves.
bomb calorimeter, 192
bombardier beetle, *513*
bond(s) An interaction between two or more atoms that holds them together by reducing the potential energy of their electrons, 261. *See also* bonding.
    coordinate covalent, 276, 611, 794
    covalent, 262
    formation of, 261
    ionic, 261
    multiple, 264
        molecular geometry and, 284–286
    polar, 287–291
    properties of, 297–303
    sigma, 311
    structural formulas showing, 49
    wedge representation of, *50*, 341
bond angle The angle between two atoms bonded to a central atom, 280
    effect of lone pairs on, 282
    in strained hydrocarbons, 348
bond dissociation energy. *See* bond energy.
Bond dissociation enthalpy The enthalpy change for breaking a bond in a molecule, with the reactants and products in the gas phase at standard conditions, 299–303
    average, 300t
    bond order and, 299
    of halogen compounds, 778t
bond energy, acid strength and, 607
    of carbon–carbon bonds, 342
    compared to nuclear binding energy, 829
    in network solids, 759
bond length The distance between the nuclei of two bonded atoms, 298, 299t
    atomic radius and, 298
    in benzene, 323
    bond order and, 298
bond order The number of bonding electron pairs shared by two atoms in a molecule, 297
    bond dissociation enthalpy and, 299
    bond length and, 298
    fractional, 298, 327
    molecular orbitals and, 327
bond pair(s) Two electrons, shared by two atoms, that contribute to the bonding attraction between the atoms, 264
    angles between, 280
    in formal charge equation, 270
    molecular polarity and, 291–297, 306t
bond polarity, electronegativity and, 287–291
    formal charge and, 289
bond strength. *See* bond energy.
bonding, in carbon compounds, 338–382
    in coordination compounds, 804–808
        ligand field theory of, 804–808
    in metal carbonyls, 814, *815*
    metallic, band theory of, 457
    molecular orbital theory of, 309, 324–335, 804

molecular structure and, 260–307
    multiple, 264, 319–324
    valence bond theory of, 309–324
bonding molecular orbital A molecular orbital in which the energy of the electrons is lower than that of the parent orbital electrons, 326
boranes, 756
borate ion, structure of, 756
borax, 60, 662, 753, 756
boric acid, 662, 756
    reaction with glycerin, 578
    in slime, 376
Born, Max, 219
Born–Haber cycle, 461, *462*
boron, abundance of, 753
    atomic weight of, 47
    chemistry of, 753, 756
    coordinate covalent bonds to, 662
    halides of, *737*
    similarity to silicon, 751
boron hydrides, 756
boron neutron capture therapy (BNCT), 848
boron tetrafluoride ion, formal charges in, 680
boron trifluoride, molecular polarity of, 679
    orbital hybridization in, 317
    reaction with ammonia, 276
    structure of, 290
borosilicate glass, 756
Bosch, Carl, 574
Boyle, Robert, 387, 394
Boyle's law, 387
    kinetic-molecular theory and, 407
Brandt, Hennig, 302, 767
breathing apparatus, closed-circuit, 745
brine, electrolysis of, 776
British thermal unit (Btu), A-8t
bromine, abundance of, 775
    atomic weight of, 48
    oxides of, 89
    physical states of, 7
    production of, 776
    reaction with aluminum, 58, *59*
    reaction with iodide ion, *716*
    reaction with nitrogen monoxide, 536
bromobenzene, mass spectrum of, *85*
bromphenol blue, *623*
Brønsted, Johannes N., 110
Brønsted–Lowry acid(s) A proton donor, 112–114, 577
Brønsted–Lowry base(s) A proton acceptor, 112–114, 577
bronze, 20
bubble gum, rubber in, 377
buckminsterfullerene ("buckyball"), *55, 56*
Buehler, William J., 782
buffer solution(s) A solution that resists a change in pH when hydroxide or hydronium ions are added, 622–629
    biological, 631
    capacity of, 626
    common, 623t
    constant pH of, 628
    general expressions for, 624
    preparation of, 626
buret, *159*
1,3-butadiene, as copolymer, 376
    structure of, 351, 377
butane, combustion of, balanced equation for, 97
    conversion to isobutane, 556, 569
    structural isomers of, 344

butanone, 372
1-butene, structure of, 340, 349
2-butene, *cis-trans* isomers of, 341, 349
    iodine-catalyzed isomerization, 533
butenes, isomers of, 349, 350t
butter yellow dye, 334
butyl butanoate, 370t
butyric acid, 367, 368t

cabbage, reaction with acid and base, *110*
cacao, 338
cadaverine, 363
cadmium selenide, in quantum dots, 773
cadmium, in nickel-cadmium battery, 707
    in nuclear reactor, 841
cadmium sulfide, as pigment, 784
cadmium-ammonia complex, formation constant of, 656
caffeine, extraction with supercritical carbon dioxide, 439, 442
    structure of, 338, *588*
calcining, 746
calcite, 652
    ions in, 61
calcium, abundance of, 748
    chemistry of, 748–751
    reaction with oxygen, *262*
    reaction with water, 749
calcium carbonate, decomposition of, 198, 200, 202
    decomposition of, temperature and spontaneity, 682
    equilibrium with carbon dioxide in solution, 549
    in iron production, 790
    in limestone, 98, 748
    precipitation from hard water, 752
    reaction with hydrochloric acid, 118, 157
    in shellfish, 652
    solubility of, 640
    uses of, 750
calcium dihydrogen phosphate, *595*
calcium fluoride, 775
    in fluorite, 748
    solubility of, 642
calcium hypochlorite, 779
calcium ion, reaction with oxalic acid, 618
calcium orthosilicate, 760
calcium oxalate, solubility of, 618
calcium oxide, 117
    uses of, 750
calcium phosphate, 768, 769, 771
calcium silicate, in blast furnace, 790
calcium sulfate, in gypsum, 86, 748
calculation, significant figures in, 33
calculator, logarithms on, A-2
    pH and, 155
    scientific notation on, 30
calibration plot, for spectrophotometric analysis, 168
calomel electrode, 721
caloric fluid, 175
calorie (cal) The quantity of energy required to raise the temperature of 1.00 g of pure liquid water from 14.5 °C to 15.5 °C, 26, A-8t
calorimetry The experimental determination of the energy changes of reactions, 190–194
camphor, boiling point elevation and freezing point depression constants for, 494t
candela (cd), A-10
Cannizzaro, Stanislao, 51
capacity, of buffer solution, 626
capillary action, 440

carbohydrates, biological oxidation of, 683
  energy content of, 25
  structure of, 367
carbon, abundance of, 758
  allotropes of, 55, 463
  atomic mass unit relative to, 44
  binding energy per nucleon, 831
  Lewis structures involving, 266–267
  nanotubes, 465
  organic compounds of, 338–382
  oxidation of, 566
  radioactive isotopes of, 835
  as reducing agent, 124t
  similarity to silicon, 734
  in steel crystal lattice, 454
carbon dioxide, bond order in, 298
  bonding in, 264
  density of, 396
  dissolved in Lake Nyos, 485
  enthalpy of formation, 194
  as greenhouse gas, 412
  Henry's law constant, 482t
  as Lewis acid, 614
  molecular geometry of, 285
  molecular polarity of, 293
  in oceans, 17, 652
  phase diagram of, 468, *469*
  reaction with water, 116
  resonance structures, 290
  sublimation of, 183, 468
  supercritical, 439, 442, 469
carbon monoxide, bond order in, 298
  calculating enthalpy of formation, 194
  as ligand in metal complexes, 813
  molecular orbital configuration of, 331, *332*
  reaction with hemoglobin, 798
  reaction with iron(III) oxide, 119
  reaction with methanol, 315
  reaction with nitrogen dioxide, 517, 541
  in water gas, 704
carbon steel, 454, 791
carbon tetrachloride, structure of, 268
carbonate ion, 64
  as polyprotic base, 605
  bond order in, 299
  molecular geometry, 241
  resonance structures of, 273
carbonate radical, 235
carbonates, solubility in strong acids, 649
carbonic acid, 116
  in biological buffer system, 631
  in oceans, 652
  as polyprotic acid, 579t
carbonic anhydrase, 540
carbonyl group The functional group that characterizes aldehydes and ketones, consisting of a carbon atom doubly bonded to an oxygen atom, 364
carbonyls, 814
carboxyl group The functional group that consists of a carbonyl group bonded to a hydroxyl group, 365
  in weak acids, *111*
carboxylic acid(s) Any of a class of organic compounds characterized by the presence of a carboxyl group, 365, 367
  acid strengths of, 609
  naming of, A-17
carboxylate ion, resonance in, 610
β-carotene, *351*, 495
Carothers, Wallace, 379
cast iron, 790
catalase, 513

catalyst(s) A substance that increases the rate of a reaction while not being consumed in the reaction, 353, 513
  effect on reaction rates, 87–89
  in Haber–Bosch process, 568
  homogeneous and heterogeneous, 528
  in rate equation, 62
  zeolites as, 761
catalytic steam reformation, hydrogen production by, 742
cathode The electrode of an electrochemical cell at which reduction occurs, 699
  in corrosion, 787
cation(s) An ion with a positive electrical charge, 61
  as Brønsted acids and bases, 578
  as Lewis acids, 612
  naming, 66
  noble gas electron configuration in, 257
  sizes of, 254
caves, sulfur-oxidizing bacteria in, 775
cell(s), electrochemical, 699–704
  galvanic, 692
  unit, 446
  voltaic, 699–704
cell potential, 709–717
Celsius temperature scale A scale defined by the freezing and boiling points of pure water, defined as 0 °C and 100 °C, 22
cesium, abundance of, 743
  melting point of, *55*
cesium chloride, structure of, 452
Chadwick James, 837
chain reaction, 540
chair form, 349
chalcocite, 791
chalcogens, 56
chalcopyrite, 791
chalk, mining of, 751
champagne, storage of, 751
characteristic The part of a logarithm to the left of the decimal point, A-3
charge, balanced in chemical equation, 108
  balancing in ionic compounds, 65
  conservation of, 693, 774
  in Coulomb's law, 62
charge density, 423
charge distribution, in covalent compounds, 270
  in molecules, 289
Charles, Jacques Alexandre César, 389, 394, 741
Charles's law, 390
  kinetic-molecular theory and, 407
chelating ligand A ligand that forms more than one coordinate covalent bond with the central metal ion in a complex, 795
chemical analysis The determination of the amounts or identities of the components of a mixture, 77, 142–148
chemical bonds. *See* **bond(s)**, bonding.
chemical change(s) A change that involves the transformation of one or more substances into one or more different substances, 16. *See also* **reaction(s)**.
chemical compound(s). *See* **compound(s)**.
chemical energy The energy stored in chemical compounds, 16
chemical equation(s) A written representation of a chemical reaction, showing the reactants and products, their physical states, and the direction in which the reaction proceeds, 16, 93–95
  balancing, 96–98, 692–699
  manipulating, equilibrium constant and, 566–568

chemical equilibrium, factors affecting, 568–573
chemical kinetics The study of the rates of chemical reactions under various conditions and of reaction mechanisms, 506–546
chemical property, 16
chemical reaction(s). *See* reaction(s).
chemistry, green. *See* green chemistry.
chemocline, 485
china clay, 760
chiral compound A molecule that is not superimposable on its mirror image, 341, 802. *See also* **enantiomers.**
  center of chirality of, 340
  optical activity of, 341
chlor-alkali industry, 746
chlorate ion, formal charges in, 271
  Lewis structure of, 266
chlorine, abundance of, 775
  coordinate covalent bonds to, 277
  formation by aqueous electrolysis, 728
  oxoacids of, 779
  production of, 776
  reaction with alkanes, 348
  reaction with iron, 95
  reaction with nitrogen monoxide, 514
  reaction with phosphorus, 93, *94*, 133
  reaction with sodium, *3*, 123, 261, *262*
  from sodium chloride electrolysis, 744
chlorine difluoride ion, molecular geometry of, 283
chlorobenzene, structure of, 355
chloroform, boiling point elevation and freezing point depression constants of, 29t
chloromethane, reaction with silicon, 761
chlorophyll, magnesium in, 744
chocolate, 338
chromium, line spectrum of, 228
chromium(III) ion, in ruby, 232
cinnabar, 2, 772
cinnamaldehyde, structure of, 314
circle, area of, 39
cisplatin, atomic distances in, 36
  discovery of, 813
  rate of substitution reaction, 69
  structure of, 91
cis-trans isomers, 87, 322, 341
  cisplatin, 813
  in coordination compounds, 801
citric acid, 368t
  reaction with sodium hydrogen carbonate, 126
  structure of, *588*
Clapeyron, Emile, 438
Clausius, Rudolf, 438
Clausius–Clapeyron equation, 438
clay(s), 754
cleavage, of crystalline solids, 463
clock reaction, iodine, 512
close packing, in crystal lattice, 449
clown fish, 17
coagulation, of colloids, 500
coal, carbon dioxide from, 412
coal tar, aromatic compounds from, 354t
cobalt, colors of complexes of, 811t
cobalt(II) chloride, reaction with hydrochloric acid, 548
cobalt(II) chloride hexahydrate, 86, 87, *421*
Cockcroft, J. D., 837
coefficient(s), stoichiometric, 95, 551
coffee, decaffeination with supercritical carbon dioxide, 439, 442
coffee-cup calorimeter, 190

cohesive force A force of attraction between molecules of a single substance, 440

coke, in iron production, 790
   water gas from, 742

colligative properties The properties of a solution that depend only on the number of solute particles per solvent molecule and not on the nature of the solute or solvent, 473, 486–498
   of solutions of ionic compounds, 496–498

collision theory A theory of reaction rates that assumes that molecules must collide in order to react, 527–531

colloid(s) A state of matter intermediate between a solution and a suspension, in which solute particles are large enough to scatter light but too small to settle out, 499–502
   types of, 500t

color(s), of acid–base indicators, *640*
   of coordination compounds, 809–812
   of fireworks, 204, 217
   of gemstones, 232
   of glass, 784
   pi bonding and, 334
   of transition metal compounds, 784
   visible light, 206, 809

combined gas law. *See* general gas law.

combustion analysis, determining empirical formula by, 145–148

combustion calorimeter, 192

combustion reaction The reaction of a compound with molecular oxygen to form products in which all elements are combined with oxygen, 96

common ion effect The limiting of acid (or base) ionization caused by addition of its conjugate base (or conjugate acid), 619–622
   solubility and, 646–649

common logarithms, A-2

common names, 347, A-15
   of binary compounds, 71

compact disc player, light energy in, 209

completion, reaction going to, 555

complex(es), 612. *See also* **coordination compound(s).**
   in enzyme-catalyzed reaction, 540
   formation constants of, A-25
   solubility and, 655–658

complex ion formation, solubility and, 645

compound(s) Matter that is composed of two or more kinds of atoms chemically combined in definite proportions, 12
   binary, naming, 71
   coordination. *See* **coordination compound(s).**
   covalent, 262
   determining formulas of, 77–86
   hydrated, 86, 792
   ionic, 61–70
   molecular, 70–71
   naming, 67
   odd-electron, 278, 330
   specific heat capacity of, 176t
   standard molar enthalpy of formation of, 197

compressibility The change in volume with change in pressure, 387

concentration(s) The amount of solute dissolved in a given amount of solution, 149
   in collision theory, 527
   effect on equilibrium of changing, 569
   in equilibrium constant expressions, 550
   graph of, determining reaction rate from, 508, 509
   of ions in solution, 149–153
   known, preparation of, 151–153
   measurement by osmotic pressure, 493

partial pressures as, 552, 554
   rate of change, 507–511
   reaction rate and, 512–519
   relation to absorbance, 167
   units of, 474

condensation The movement of molecules from the gas to the liquid phase, 434

condensation polymer(s) A synthetic organic polymer formed by combining monomer units in such a way that a small molecule, usually water, is split out, 370, 377–380
   silicone, 762

condensed formula A variation of a molecular formula that shows groups of atoms, 59, 341

condition(s), standard. *See* **standard state.**

conduction band, 458

conductor(s), band theory of, 457

conjugate acid–base pair(s) A pair of compounds or ions that differs by the presence of one hydrogen ion, 580, 581t
   in buffer solutions, 623
   ionization constants of, 588
   strengths of, 585

conservation law, charge, 693
   energy, 18, 173
   mass, 5, 134, 693
   matter, 93

constant(s), acid and base ionization, 585, 586t
   Boltzmann, 666
   equilibrium, 542, 550–558
   Faraday, 717, 730
   formation, 655
   gas, 394, 492
   Henry's law, 482t
   physical, A-13t
   Planck's, 208
   rate, 514, 515
   Rydberg, 211
   significant figures in, 32
   solubility product, 641
   van der Waals, 411
   water ionization, 581

continuous spectrum The spectrum of white light emitted by a heated object, consisting of light of all wavelengths, 210

conversion factor(s) A multiplier that relates the desired unit to the starting unit, 24, 35, A-9
   in mass/mole problems, 72

coordinate covalent bond(s) Interatomic attraction resulting from the sharing of a lone pair of electrons from one atom with another atom, 276, 611, 794

coordination complex(es), 612

coordination compound(s) A compound in which a metal ion or atom is bonded to one or more molecules or anions to define a structural unit, 792–799
   bonding in, 804–808
   colors of, 809–812
   formulas of, 795–797
   magnetic properties of, 807
   naming of, 797
   spectrochemical series of, 810
   structures of, 800–804

coordination geometry The arrangement in space of the central metal ion and the ligands attached to it, 794

coordination isomers Two or more complexes in which a coordinated ligand and a noncoordinated counterion are exchanged, 800

coordination number The number of ligands attached to the central metal ion in a coordination compound, 794
   geometry and, 800

Copernicus, element 112, 11

Copernicus, Nicolaus, 11

copolymer A polymer formed by combining two or more different monomers, 376

copper, 20
   biochemistry of, 257
   electrolytic refining, 792
   ores of, *789*
   production of, 791
   reaction with nitric acid, 124
   reaction with silver ions, 120, 691–693

copper(II) carbonate, *641*

copper(I) chloride, in fireworks, 217

copper(II) chloride, dissociation of, 150

copper(II) ion, complexes of, 612

copper(II) sulfate pentahydrate, 87

coral, calcium carbonate in, *110*

core electrons The electrons in an atom's completed set of shells, 239, 262
   molecular orbitals containing, 328

corrosion The deterioration of metals by oxidation–reduction reactions, 785, 787
   of aluminum, 754

corundum, 762

cosmic radiation, 843

coulomb (C) The quantity of charge that passes a point in an electric circuit when a current of 1 ampere flows for 1 second, 709, 730, A-7, A-11

Coulomb's law The force of attraction between the oppositely charged ions of an ionic compound is directly proportional to their charges and inversely proportional to the square of the distance between them, 68, 418
   lattice energy and, 460

covalent bond(s) An interatomic attraction resulting from the sharing of electrons between the atoms, 262
   polar and nonpolar, 287–291
   valence bond theory of, 309–318

covalent compound(s) A compound formed by atoms that are covalently bonded to each other, 262

covalent radius, 249

covellite, 791

cracking, in petroleum refining, 358

Crick, Francis, 387

critical point The upper end of the curve of vapor pressure versus temperature, 439

critical pressure The pressure at the critical point, 439

critical temperature The temperature at the critical point; above this temperature the vapor cannot be liquefied at any pressure, 439
   of superconductor, 467

crocoite, *641*

cross-linked polyethylene (CLPE), 374

cross-linking, in vulcanized rubber, 376

cryolite, 750
   in fireworks, 217
   in Hall–Heroult process, 755, 778

crystal lattice A solid, regular array of positive and negative ions, 20, 68, 446

crystallites, in steel, 454

crystallization, colloids and, 499
   temperature change and, 484

cubic centimeter, 25

cubic close-packed (ccp), 449

cubic unit cell A unit cell having eight identical points at the corners of a cube, 447

curie A unit of radioactivity, 843

Curie, Marie and Pierre, 65, 772, 823

Cusumano, James, 6

cyanate ion, resonance structures, 676

cyanide ion, reaction with water, 113
cyanuric acid, hydrogen bonding with melamine, 433
cycloalkanes, 348
  general formula of, 344t
  naming of, A-16
cycloalkenes, 351
cyclobutane, structure of, 348
cyclohexane, structure of, 348, 349
cyclohexene, structure of, 351
cyclopentadiene ion, in ferrocene, 817
cyclopentane, structure of, 340, 348
cyclopropane, conversion to propene, 520
  structure of, 348
cysteine, molecular geometry of, 286
  structure of, 61
cytosine, 304
  hydrogen bonding to guanine, 429
  structure of, 260

d orbital(s). *See also* **atomic orbital(s).**
  in bonding of third period elements, 277
  in coordination compounds, 804
  hybrid orbitals and, 335
d-block elements, properties of, 783
d-to-*d* transition, 810
Dalton, John, 52, 401
Dalton's law of partial pressures The total pressure of a mixture of gases is the sum of the pressures of the components of the mixture, 401
data, graphing of, 37
dating, radiocarbon, 840
Davisson, C. J., 216
Davy, Humphry, 730, 744
de Broglie, Louis Victor, 216
Debye, Peter, *292*
debye unit, 292
decay constant, for radioactivity, 834
decay series, radioactive, 823–836
deciliter, 25
decomposition, determining molecular formula by, 81
decompression sickness, 482
deep-sea diving, gas laws and, 410
Deepwater Horizon fire, methane and, 426
defined quantity, significant figures in, 78
delocalization, molecular orbital, 458
delta (Δ), symbol for change, 176
delta (δ), symbol for partial charge, 287
denitrification, by bacteria, 765
density The ratio of the mass of an object to its volume, 13, 19
  of air, 396
  balloons and, *396*
  of gas, calculation from ideal gas law, 396
  of sulfuric acid in lead storage battery, 707
  of transition elements, 788
  units of, 13
dental amalgam, 717
deoxyribonucleic acid, bonding in, 260
  hydrogen bonding in, 429
  structure of, *24*
2-deoxyribose, 367
  structure of, 260, *304*
derived units, SI, A-11
desalination, by reverse osmosis, 472, 493
detergent, 502
deuterium, 46
  binding energy of, 829
  fusion of, 842
  preparation of, 399, 741
diagonal relationship, in periodic table, 753

diamagnetism The physical property of being repelled by a magnetic field, 227, 807
diamminedichloroplatinum(II), isomers of, 801
diamond, as insulator, 458
  interatomic distances in, *24*
  structure of, 55
  synthesis of, 463
diapers, synthetic polymers in, *381*
diatomic molecule(s), atomic radii in, 248, *249*
  of elements, 56
  heteronuclear, 331
  homonuclear, 329
dibenzenechromium, 815
diberyllium cation, 328
diborane, 756
  hybridization in, 335
dibromine pentaoxide, 82
1,2-dichloroethylene, isomers of, 323
  molecular polarity of, 296
dichromate ion, as oxidizing agent, 124t
  reaction with ethanol, 125
diene(s) A hydrocarbon containing two double bonds, 351
  naming of, A-16
dietary Calorie, 26
diethyl ether, 360
  vapor pressure curves for, *437*
diethyl ketone, 366t
diffraction, of electrons, 216
  of x-rays by crystals, *450*
diffusion The gradual mixing of the molecules of two or more substances by random molecular motion, 408
  probability and, 666–668
dihelium, molecular orbital energy level diagram of, 326
dihydrogen phosphate ion, amphiprotic nature of, 579
  buffer solution of, 623t
L-3,4-dihydroxyphenylalanine (L-DOPA), 364
diiodocyclohexane, dissociation of, 560
dilithium, molecular orbital energy level diagram of, 328
dilution, buffer pH and, 627
  isotope, 847
  preparation of solutions by, 152
  serial, 154
dimensional analysis A general problem-solving approach that uses the dimensions or units of each value to guide you through calculations, 24, 35
dimethyl ether, structure of, *59,* 340, 424
dimethyl sulfide, 773
dimethylbutane, isomers of, 346
2,3-dimethylbutane, structure of, *147*
dimethylglyoximate ion, *655*
dimethylglyoxime (DMG), reaction with nickel(II) ion, 43
dimethylphthalate (DMT), in recycling PET plastic, 379
2,2-dimethylpropane, structure of, 347
dinitrogen, bonding in, 348
dinitrogen monoxide, 766
dinitrogen oxide, 766
dinitrogen pentaoxide, 766
  decomposition of, 62
  mechanism, 91
  rate equation, 67
dinitrogen tetraoxide, 766, 767
  decomposition of, 571
dinitrogen trioxide, 766

diode, semiconductor, 51
dioxovanadium (V) ion, reaction with zinc, 695
dioxygen. *See* oxygen.
dipolar bond. *See* **polar covalent bond.**
dipole(s), induced, 427
dipole moment (μ) The product of the magnitude of the partial charges in a molecule and the distance by which they are separated, 292
dipole-induced dipole interaction The electrostatic force between two neutral molecules, one having a perm anent dipole and the other having an induced dipole, 427
dipole–dipole interaction The electrostatic force between two neutral molecules that have permanent dipole moments, 420
diprotic acid, 112
disaccharides, 315
dispersion(s), colloidal, 499
dispersion forces Intermolecular attractions involving induced dipoles, 428–430
disproportionation reaction, 718, 779
distillation, in petroleum refining, 358
disulfur dichloride, preparation of, 166
DNA. *See* deoxyribonucleic acid.
dolomite, 748
domain, ferromagnetic, 227
donor level, in semiconductor, 459
dopamine, 312
dopant, in semiconductor, 459
double bond A bond formed by sharing two pairs of electrons, one pair in a sigma bond and the other in a pi bond, 264
  in alkenes, 349
  valence bond theory of, 319–321
double displacement reaction(s). *See* **exchange reaction(s).**
Downs cell, for producing sodium, 744, *745*
dry cell battery, 705
dry ice, 183, *184,* 469
dye(s), green chemistry of, 328
  pH indicating, 155
  rate of reaction with bleach, 64, 508
dynamic equilibrium, molecular description of, 100
  vapor pressure and, 436
dynamite, *360*

earth, alchemical meaning of, 748
effective atomic number (EAN) rule. *See* **18-electron rule.**
effective nuclear charge (Z*) The nuclear charge experienced by an electron in am multielectron atom, as modified by the other electrons, 236, 237t, 244
  atomic radius and, 249
effusion The movement of gas molecules through a membrane or other porous barrier by random molecular motion, 408
Ehrlich, Paul, 84
18-electron rule Organometallic compounds in which the number of metal valence electrons plus the number of electrons donated by the ligand groups totals 18 are likely to be stable, 814
Einstein, Albert, 209, 830
eka-silicon, 42, 51
elastic collision, 411
elastomer(s) A synthetic organic polymer with very high elasticity, 376
electric current, unit of, 730
electric field, polar molecules aligned in, *292*
electrical energy The energy due to the motion of electrons in a conductor, 16

electrochemical cell(s) A device that produces an electric current as a result of an electron transfer reaction, 699–704
commercial, 704–709
corrosion and, 787
nonstandard conditions for, 717–421
notation for, 703
potential of, 709–717
work done by, 916
electrochemistry, 690–733
electrode(s) A device such as a metal plate or wire for conducting electrons into and out of solutions in electrochemical cells, 101, 699
hydrogen, 703
inert, 702
pH, 155
standard hydrogen, 710
terminology for, 726t
electrolysis The use of electrical energy to produce chemical change, 692, 725–729
aluminum production by, 755
of aqueous a solutions, 727
electrodes in, 726t
fluorine production by, 775
hydrogen produced by, 741
of a odium chloride, 398, 726, 744
of water, *18*
electrolyte(s) A substance that ionizes in water or on melting to form an electrically conducting solution, 101
electromagnetic radiation Radiation that consists of wave-like electric and magnetic fields, including light, microwaves, radio signals, and x-rays, 205–207
gamma rays as, 822
electromotive force (emf), 709
electron(s) (e⁻) A negatively charged subatomic particle found in the space about the nucleus, 43
assignment to atomic orbitals, 235–237
as beta particle, 821
bond pair, 264
configuration. See electron configuration(s).
core, 239, 262
   molecular orbitals containing, 328
counting, 730
delocalization of, 452
diffraction of, *216*
in electrochemical cell, direction of flow, 710
lone pair of, 264
nuclear charge experience by, 236
octet of, 263, 264
pairing, magnetic properties and, 227
quantization of potential energy, 212, 219
shells and subshells, 220, 234t
spin. *See* electron spin transfer in oxidation–reduction reactions, 120
valence, 239, 261–263. *See also* **bond pair(s)**, **lone pair(s)**.
   of main group elements, 736
   repulsions of, 280
in voltaic cell, direction of flow, 699
wave properties of, 216
wavelength of, 218
electron affinity The negative of the internal energy change occurring when an atom of the element in the gas phase gains an electron, 252
electron attachment enthalpy (ₑₐH) The enthalpy change occurring when an atom of the element in the gas phase gains an electron, 252
acid strength and, 607
central atom in Lewis structure and, 265
values of, A-18

electron capture A nuclear process in which an inner-shell electron is captured, 825
predicting, 774
electron cloud pictures, 223
electron configuration(s), in coordination compounds, 806
of elements, 238t
of heteronuclear diatomic molecules, 331
of homonuclear diatomic molecules, 329–331
of ions, 245–247
Lewis notation for, 263
main group, 239–241
noble gas notation for, 239
orbital box notation for, 233
spdf notation for, 239
transition elements, 242, 246, 785
electron density The probability of finding an atomic electron within a given region of space, related to the square of the electron's wavefunction, 220
in molecules, 294
electron spin, pairing of, 227
quantization of, 226
electron spin quantum number, 226
electron transfer reaction(s). *See* **oxidation–reduction reaction(s)**.
electron volt (eV), A-7
electron-deficient molecule, 335, 756
electronegativity (χ) A measure of the ability of an atom in a molecule to attract electrons to itself, 288
hydrogen bonding and, 423
electroneutrality, principle of, 289
electron-pair geometry The geometry determined by all the bond pairs and lone pairs in the valence shell of the central atom, 282
orbital hybridization and, 312
electroplating, 725
electrostatic energy The energy due to the separation of electrical charges, 16
electrostatic force(s) Forces of attraction or repulsion caused by electric charges, 68
electrostatic potential surface, 680
element(s) Matter that is composed of only one kind of atom, 10
abundances of, 735, 7369t
   in Earth's crust, 54t
   atomic number of, 48
atomic weight of, 51
attachment enthalpy values of, A-18
d-block, 783
diatomic molecules of, *56*
f-block, 783
ionization energies of, A-18
isotopes of, 46–47
main group, *52*, 53
   chemistry of, 734–780
molar mass of, 72
monatomic ions of, charges on, 62
names of, 10–11
oxidation number of zero, 121
p-block, 240
   molecular orbitals involving, 329, 331
physical states of, 417
s-block, 239
sources of, *789*
specific heat capacity of, 176t
standard enthalpy of formation of, 198
standard enthalpy of vaporization of, 435t
symbol for, 44–45
synthesis of, 838
transition, *52*, 53, 58. *See also* **transition elements.**
transuranium, 838, 839

elementary step A simple event in which some chemical transformation occurs; one of a sequence of events that form the reaction mechanism, 536
rate equation for, 537
elephants, frontalin in, 343
empirical formula A molecular formula showing the simplest possible ratio of atoms in a molecule, 79
determination by combustion analysis, 145–148
relation to molecular formula, 80
emulsifying agent, 501
emulsion A colloidal dispersion of one liquid in another, 500t
enantiomers A stereoisomeric pair consisting of a chiral compound and its mirror image isomer, 341
end point. *See* **equivalence point.**
endothermic process A thermodynamic process in which energy as heat flows into a system from its surroundings, 175, 662
enthalpy change of, 188
energy The capacity to do work and transfer heat, 173–175, A-7. *See also* **enthalpy** and **heat** entries.
activation. *See* **activation energy.**
binding, 829–832
color of photons and, 210
conversion of forms of, *17*, 18
density, in batteries vs. gasoline, 708t
direction of transfer, 174
dispersal of, 663, 664
in food, 25, 172
internal, 184
ionization. *See* **ionization energy.**
lattice, 461
law of conservation of, 18, 173
levels in hydrogen atom, 212
mass equivalence of, 830
nuclear, 841
quantization of, 212, 219, 667
relation to frequency of radiation, 208
sources for human activity, 173
state changes and, 180–183
units of, A-7
energy level diagram, *195*
energy resources and usage, 204–205
enthalpy (H) The sum of the internal energy of the system and the product of its pressure and volume, 186
bond dissociation, 299–303
effect on acid strength, 608
enthalpy change (ΔH) The energy as heat transferred at constant pressure, 186, 662
for chemical reactions, 188–190
sign conventions for, 186
as state function, 186
enthalpy of formation, standard molar, 197
values of, A-26
enthalpy of fusion (fusH) The energy as heat required to convert one mole of a substance from a solid to a liquid at constant temperature, 464, 466t, A-14t
enthalpy of hydration, 419
of alkali metals, 746
enthalpy of solution (solnH) The amount of energy as heat involved in the process of solution formation, 479–482
measurement of, 481
enthalpy of solvation, 419
enthalpy of sublimation (sublimationH) The energy as heat required to convert one mole of a substance from a solid to a gas, 467
enthalpy of vaporization (vapH) The quantity of energy as heat required to convert 1 mol of a liquid to a gas at constant temperature, 434, 435t, A-14t

intermolecular forces and, 420, 422t
of nonpolar substances, 428t, 435t
relation to vapor pressure, 438
entropy (S) A measure of the dispersal of energy in a system, 663
effect on acid strength, 608
molecular structure and, 669
physical state and, 669
second law of thermodynamics and, 672
solution process and, 477–478
standard molar, 668
statistical basis of, 664–666
entropy change (ΔS), for universe, system, and surroundings, 672
of reaction, 670
enzyme(s) A biological catalyst, 513
catalysis by, 540
ephedrine, structure of, 97
epichlorohydrin, structure of, 377
epinephrine, 364
epoxy copolymer, 377
Epsom salt, 88
equation(s), Arrhenius, 531
Balmer, 211
Beer–Lambert law, 167
Bohr, 212
boiling point elevation, 488
Boltzmann, 666
bond order, 298, 327
Boyle's law, 388
buffer solution pH, 624
Celsius–Kelvin scale conversion, 23
Charles's law, 390
chemical, 16, 93–95
Clausius–Clapeyron, 438
Coulomb's law, 68
Dalton's law, 401
de Broglie, 216
dilution, 152
Einstein's, 830
enthalpy of formation, 198
entropy, 663
entropy change of reaction, 670
equilibrium constant expression, 551
equilibrium constant of electrochemical cell, 723
first law of thermodynamics, 184
formal charge, 270
free energy change at nonequilibrium conditions, 678
freezing point depression, 490
general gas law, 391
Gibbs free energy, 676
Graham's law, 409
half-life, 524
for radioactive decay, 834
heat and temperature change, 176
Henderson–Hasselbalch, 625
Henry's law, 482
Hess's law, 194
ideal gas law, 394
integrated rate, 519–527
ion pair energy, 460
ionization constants, for acids and bases, 585
for water, 581
K_a and K_b, 588
kinetic energy, 405
molarity, 149
mole fraction, 474
Nernst, 717
net ionic, 108
of strong acid–strong base reactions, 114
nuclear reactions, 822

osmotic pressure, 492
pH, 154, 583
of buffer solution, 624
pK_a, 588
Planck's, 207–210
pressure–volume work, 185
quadratic, 563, A-4
Raoult's law, 486
rate, 507, 514. *See also* **rate equation(s).**
rms speed, 406
Schrödinger, Erwin, 219
second law of thermodynamics, 672
speed of a wave, 205
standard free energy change of reaction, 678
standard potential, 711, 714
straight line, 67
van der Waals, 411
equatorial position, in cyclohexane structure, 349
in trigonal-bipyramidal molecular geometry, 284
equilibrium A condition in which the forward and reverse reaction rates in a physical or chemical system are equal, 98–101
chemical, 547–575. *See also* chemical equilibrium.
dynamic, 100
factors affecting, 568–573
Le Chatelier's principle and, 483, 568–573
in osmosis, 492
in reaction mechanism, 542
reversibility and, 664
solution process as, 476
successive, 655
thermal, 174
equilibrium constant (K) The constant in the equilibrium constant expression, 542, 550–558
calculating from initial concentrations and pH, 596
calculating from standard potential, 722
calculations with, 561–565
concentration vs. partial pressure, 552, 554
determining, 558–861
Gibbs free energy change and, 678
meaning of, 554
for product-favored vs. reactant-favored reactions, 100, 554
relation to reaction quotient, 555–556
simplifying assumption in, 563–564, 597, A-5
unitless, 552, 553
values of, 554t
for weak acid and base (K_a and K_b), 584–589
equilibrium constant expression A mathematical expression that relates the concentrations of the reactants and products at equilibrium at a particular temperature to a numerical constant, 551
for gases, 552, 554, 560
reverse reaction, 566
stoichiometric multipliers and, 566–568
sum of reactions, 569
equilibrium vapor pressure The pressure of the vapor of a substance at equilibrium in contact with its liquid or solid phase in a sealed container, 437
in phase diagram, 468
equivalence point The point in a titration at which one reactant has been exactly consumed by addition of the other reactant, 159
of acid–base reaction, 630
error The difference between the measured quantity and the accepted value, 27

ester(s) Any of a class of organic compounds structurally related to carboxylic acids, but in which the hydrogen atom of the carboxyl group is replaced by a hydrocarbon group, 365, 369
hydrolysis of, 369
naming of, A-17
esterification reaction A reaction between a carboxylic add and an alcohol in which a molecule of water is formed, 369
ethane, orbital hybridization in, 315
1,2-ethanediol, 359t, 360
ethanol, 358, 359t
as fuel, 201
hydrogen bonding in, 424
mass spectrum of, 84, *85*
miscibility with water, 477
NMR spectrum of, *229*
as nonelectrolyte, 103
oxidation to acetic acid, 365
reaction with dichromate ion, 125
standard enthalpy of formation of, 197
structure of, *59, 340*
vapor pressure curves for, *437*
ethene, 349
ether(s) Any of a class of organic compounds characterized by the presence of an oxygen atom singly bonded to two carbon atoms, 360
ethyl acetate, 369
ethylene glycol, 359t, 360
as antifreeze, *361*, 475, 489, 490
as nonelectrolyte, 103
specific heat capacity of, 176t
ethylene, 349
derivatives of, as monomers, 375t
in organometallic compounds, 808
orbital hybridization in, 319
reaction with water, 359
1,2-ethylenediamine (en), as ligand, 794
reaction with succinic acid, 380
ethylenediaminetetraacetate ion (EDTA^{4-}), as ligand, 795
eugenol, 488
europium, isotopes of, 91
evaporation. *See* **vaporization.**
exact number, significant figures in, 62
exchange energy, Hund's rule and, 240
exchange reaction(s) A chemical reaction that proceeds by the interchange of reactant cation–anion partners, 105
excited state The state of an atom in which at least one electron is not in the lowest possible energy level, 512
nuclear, 823
exclusion principle. *See* **Pauli exclusion principle.**
exothermic process A thermodynamic process in which energy as heat flows from a system to its surroundings, 175, 662
enthalpy change of, 188
expansion molding, polystyrene, 375
exponent, 29
exponential notation, A-2. *See also* scientific notation.
extensive properties Physical properties that depend on the amount of matter present, 14
extrinsic semiconductor, 459

f orbital(s). *See* **atomic orbital(s).**
face-centered cubic (fcc) unit cell, 447, 449
factor-label method. *See* **dimensional analysis.**
Fahrenheit temperature scale A scale defined by the freezing and boiling points of pure water, defined as 32 °F and 212 °F, 22

family, in periodic table. *See* **group(s).**

Faraday, Michael, 354, 730

Faraday constant (F) The proportionality constant that relates standard free energy of reaction to standard potential; the charge carried by one mole of electrons, 717, 730

fat(s), energy content of, 25

   unsaturated, *352*

feldspar, 761

Fermi, Enrico, 839

Fermi level The highest filled electron energy level in a metal at absolute zero temperature, 457

ferrite, unit cell of, 454

ferrocene, 817

ferromagnetism A form of paramagnetism, seen in some metals and their alloys, in which the magnetic effect is greatly enhanced, 227

filling order, of electron subshells in atoms, 235

film badge, radiation monitoring, 845

filtration, *10*

fire extinguisher, carbon dioxide, 396

fire retardant, boric acid as, 756

fireworks, colors of, 204, 217

   metals in, 132, *137*

first law of thermodynamics The total energy of the universe is constant, 183–187, 662

   half-life of, 524

   integrated rate equation, 519

   nuclear, 833

fission The highly exothermic process by which very heavy nuclei split to form lighter nuclei, 820, 1840

fixed notation, 29

flotation, for ore treatment, 791

fluid, 7

   supercritical, 439, 469

fluorapatite, 750

fluorescence, 775

fluorine, abundance of, 775

   bonding in, 264

   chemistry of, 777–778

   compounds of, hydrogen bonding in, 422

      with main group elements, 738t

   production of, 775

   reaction with nitrogen dioxide, 539

   sigma bond in, 311

fluorite, *241,* 642, *643, 651,* 748

   ions in, 61

fluorspar, *241,* 775

flux, borax as, 756

   cryolite as, 755

   fluorspar as, 775

foam, 500t

food, energy content of, 25, 146

   irradiation, 848

   tampering, titration for detecting, 165

fool's gold. *See* iron pyrite.

force(s), A-6

   intermolecular. *See* **intermolecular forces.**

forensic chemistry, 4

formal charge The charge on an atom in a molecule or ion calculated by assuming equal sharing of the bonding electrons, 270

   bond polarity and, 289

   relation to acid strength, 609

formaldehyde, 366t

   Lewis structure of, 265

   orbital hybridization in, 320, *321*

   released by synthetic adhesives, 381

   structure of, 365

formation, standard enthalpy change of, 197

   standard molar free energy of, 680

formation constant An equilibrium constant for the formation of a complex ion, 655

   values of, A-25

formic acid, 367, 368t

   pH in aqueous solution, 599

   reaction with sodium hydroxide, 594

   in water, equilibrium constant expression for, 566

   as weak acid, 587

formula(s), chemical, 12

   condensed, 341

   empirical, 79, 145

   general, of hydrocarbons, 344t

   of ionic compounds, 65

      structures and, 452–456

   molecular. *See* **molecular formula.**

   perspective, 341

   predicting, 738–739

   structural, 341, 347. *See also* **structural formula.**

formula unit The simplest ratio of ions in an ionic compound, similar to the molecular formula of a molecular compound, 75

formula weight, 75

fractional abundance, 49

fragment ion, in mass spectra, *85*

francium, abundance of, 743

Franklin, Rosalind, 303, 429

free energy. *See* **Gibbs free energy (G).**

free energy change (ΔG), 676

   equilibrium constant and, 678

free radical(s) A neutral atom or molecule containing an unpaired electron, 278

freezing point depression, 490

   for ionic solutions, 497t, 498

freezing point depression constant (K_{fp}), 490

frequency (v) The number of complete waves passing a point in a given amount of time, 205

   relation to energy of radiation, 208

frequency factor, in Arrhenius equation, 531

Friedel–Crafts reaction, 614

Frisch, Otto, *841*

frontalin, 343

fructose, 367

   fuel, density of, 36

   ethanol as, 201

   hydrogen as, 660

   methanol as, *139*

fuel cell A voltaic cell in which reactants are continuously added, 708

Fuller, R. Buckminster, 56

fulvalene, 817

functional group A structural fragment found in all member of a class of compounds, 357, 359t

fusion The state change from solid to liquid, 180

   enthalpy of, 464, 466t, A-14t

   heat of, 180

fusion, nuclear The highly exothermic process by which comparatively light nuclei combine to form heavier nuclei, 842

galena, 649, 772

gallium, *754*

   abundance of, 753

   isotopes of, 91

   melting point of, *55*

gallium arsenide, 460, 768

gallium oxide, formula of, 82

Galvani, Luigi, 692

galvanic cell(s), 692

gamma ray(s) High energy electromagnetic radiation, 206, 822

gangue A mixture of sand and clay in which a desired mineral is usually found, 789

gas(es) The phase of matter in which a substance has no definite shape and a volume defined only by the size of its container, 7

   compressibility of, 387, 418

   density, calculation from ideal gas law, 396

   diffusion of, 408, 667

   dissolution in liquids, 482

   in equilibrium constant expression, 552, 554, 560

   expansion as spontaneous process, 662

   ideal, 394

   kinetic-molecular theory of, 403–407, 417

   laws governing, 387–393, 407

   mixtures of, partial pressures in, 400–403

   noble. *See* **noble gas(es).**

   nonideal, 410

   pressure of, 386

   properties of, 384–414

   solubility in water, 428t

   speeds of molecules in, 405

   standard molar volume, 395

   volume effects on equilibria of, 571

gas centrifuge, isotope separation by, 841

gas constant (R) The proportionality constant in the ideal gas law, 0.082057 L · atm/mol · K or 8314510 J/mol · K, 394

   in Arrhenius equation, 531

   in kinetic energy–temperature relation, 405

   in Nernst equation, 717

   in nonequilibrium free energy change, 678

   in osmotic pressure equation, 492

gas-forming reaction(s), 118–119, 127

gasoline, energy per kilogram, 708t

Gay-Lussac, Joseph, 392

GC-MS. *See* mass spectrometers.

gecko, wall-climbing ability of, 416

Geiger–Muller counter, *834*

gel A colloidal dispersion with a structure that prevents it from flowing, 500t

gems, solubility of, 643

general gas law An equation that allows calculation of pressure, temperature, and volume when a given amount of gas undergoes a change in conditions, 391

geological sequestration of carbon dioxide, 412

geometric isomers Isomers in which the atoms of the molecule are arranged in different geometric relationships, 341, 801

   of alkenes, 349

Gerlach, Walther, 226

germanium, 51

   abundance of, 758

   compounds of, 758

   as semiconductor, 459

Germer, L. H., 216

Gibbs, J. Willard, 676

Gibbs free energy (G) A thermodynamic state function relating enthalpy, temperature, and entropy, 676–679

   cell potential and, 722

   work and, 679

Gimli Glider, 36

glass, colors of, 784

   etching by hydrogen fluoride, 778

   structure of, 464

glass electrode, 721

glassware, laboratory, *14, 25, 149,* 151

   significant figures and, *32*

global warming, 412

glucose, combustion of, stoichiometry of, 135

   structure and isomers of, 367

glue, chemistry of, 381
glycerol, 359t, 360
glycine, protein chain from, 381
    structure of, 615
glycolysis, 683
goethite, *651*
gold, density of, 14
    oxidation by fluorine, 716
    uses of, 1
Goodyear, Charles, 376
gout, lead poisoning and, 762
Graham, Thomas, 408, 499
Graham's law, 409
gram (g), 25
graph(s), analysis of, 37
graphene, 465
graphite, 465
    structure of, 55
graphite electrode, 703
    oxidation of, 728
gravitational energy The energy due to the attraction between masses, 16
gray The SI unit of radiation dosage, 843
green chemistry, 442
    of adhesives, 381
    atom economy and, 144
    of dyes and pigments, 334
    principles of, 5–6
ground state The state of an atom in which all electrons are in the lowest possible energy levels, 212
group(s) The vertical columns in the periodic table of the elements, 53
    ion charge related to, 62–63
    similarities within, 736t
Group 1A elements, 54. *See also* **alkali metal(s)**.
    chemistry of, 743–752
Group 2A elements, 54. *See also* **alkaline earth metal(s)**.
    chemistry of, 748–751
Group 3A elements, 60
    chemistry of, 753–758
Group 4A elements, 60
    chemistry of, 758–762
    hydrogen compounds of, 422
Group 5A elements, 56
    chemistry of, 763–771
Group 6A elements, 56
    chemistry of, 772–774
Group 7A elements, 57. *See also* **halogens**.
    chemistry of, 775–779
Group 8A elements, 57. *See also* **noble gas(es)**.
guanine, 304
    hydrogen bonding to cytosine, 429
guidelines, assigning oxidation numbers, 121
    solubility of ionic compounds in water, *104*
Gummi Bear, *189*
gunpowder, 747
gypsum, 86, 748, 772
    ions in, 61

Haber, Fritz, 574
Haber–Bosch process, 574
Hahn, Otto, 840
hair dye, hydroxyl radicals in, 279
half-cell A compartment of an electrochemical cell in which a half-reaction occurs, 699
half-life (t$_{1/2}$) The time required for the concentration of one of the reactants to reach half of its initial value, 523
    for radioactive decay, 834
half-reaction method A systematic procedure for balancing oxidation–reduction reactions, 692

half-reactions The two chemical equations into which the equation for an oxidation–reduction reaction can be divided, one representing the oxidation process and the other the reduction process, 121, 693
    sign of standard reduction potential for, 712
    standard potentials for, 711, *712*
halide ions Ions of the elements of Group 7A, 67
halides, as strong electrolytes, 103
    compounds with aluminum, 757
Hall, Charles Martin, 755
Hall–Heroult process, aluminum production by, 755
halogenation, of benzene, 356
halogens The elements in Group 7A of the periodic table, 57
    chemistry of, 775–779
    electron attachment enthalpies of, 253
    electron configuration of, 241
    oxidation number of, 122
    as oxidizing agents, 124t
    ranked by oxidizing ability, 716
    reaction with akenes and alkynes, 352
    reaction with alkali metals, 746
halothane, 402
hard water, 752
    detergents and, 502
heat, a form of energy, 175
    as reactant or product, 572
    sign conventions for, 176, 185t
    temperature change and, 176
    transfer, as spontaneous process, 662
        calculations, 178
        during phase change, 180
heat capacity, 176
heat of fusion The quantity of energy as heat required to convert a solid to a liquid at constant temperature, 180. *See also* **enthalpy of fusion**.
heat of solution. *See* **enthalpy of solution**.
heat of vaporization The quantity of energy as heat requited to convert 1 mol of a liquid to a gas at constant temperature, 180. *See also* **enthalpy of vaporization**.
heat pack, supersaturated solution in, 478
heavy water, 46
Heisenberg, Werner, 219
Heisenberg's uncertainty principle It is impossible to determine both the position and the momentum of an electron in an atom simultaneously with great certainty, 220
helium, in balloons, 741
    discovery of, 57
    nucleus as alpha particle, 821
    orbital box diagram, 233
    use in deep-sea diving, 410
hematite, *651*, 790
    ions in, 61
heme group, 798
hemoglobin, 257, 798
    carbonic anhydrase and, 540
Henderson–Hasselbalch equation, 625
Henry's law The concentration of a gas dissolved in a liquid at a given temperature is directly proportional to the partial pressure of the gas above the liquid, 482
Herculon, 375t
Heroult, Paul, 755
hertz The unit of frequency, or cycles per second; 1 Hz = 1^{s-1}, 205
Hertz, Heinrich, 205
Hess's law If a reaction is the sum of two or more other reactions, the enthalpy change for the overall process is the sum of the enthalpy changes for the constituent reactions, 194–200

heterogeneous mixture A mixture in which the properties in one region or sample are different from those in another region or sample, 9–10
heteronuclear diatomic molecule(s) A molecule composed of two atoms of different elements, 331
hexachloroethane, in fireworks, 217
hexadentate ligands, 795
hexagonal close-packed (hcp) unit cell, *448*, 449
hexamethylenediamine, 379
hexane, structural isomers of, 345
    structure of, *147*
hexose, 367
high-density polyethylene (HDPE), 374
high spin configuration The electron configuration for a coordination complex with the maximum number of unpaired electrons, 806
highest occupied molecular orbital (HOMO), 329
Hindenburg, 741
Hippocrates, 576
hippuric acid, formula of, 80, *81*
hole(s), in crystal lattice, 452
    in metals, 458
    in semiconductors, 459
homogeneous catalyst A catalyst that is in the same phase as the reaction mixture, 534
homogeneous mixture A mixture in which the properties are the same throughout, regardless of the optical resolution used to examine it, 9–10
homonuclear diatomic molecule(s) A molecule composed of two identical atoms, 329
    electron configurations of, 329–331
Hund's rule The most stable arrangement of electrons is that with the maximum number of unpaired electrons, all with the same spin direction, 240
    hybrid orbitals and, 314
    molecular orbitals and, 325
hybrid, resonance, 272
hybrid orbital(s) An orbital formed by mixing two or more atomic orbitals, 312–324
    in benzene, 323
    geometries of, *313*
hydrated compound A compound in which molecules of water are associated with ions, 86, 792
    formula unit of, *75*
hydration, enthalpy of, 419
    of ions, 480
hydrazine, 71
    formula of, 80
    production by Raschig reaction, 541, 764
    reaction with water, 764
    synthesis of, spontaneity of, 675
hydrides, 741
    boron, 756
    of Group 3A elements, 753
    reaction with water, 742, *743*
hydrocarbon(s) A compound that contains only carbon and hydrogen, 343–357
    catalytic swam reformation of, 704
    combustion analysis of, 145–148
    derivatives of, naming of, A-17
    Lewis structures of, 267
    naming of, 71, A-15
    strained, 348
    types of, 344t
hydrochloric acid, 778. *See also* hydrogen chloride.
    reaction with ammonia, 595
    reaction with calcium carbonate, 112
    titration with ammonia, 637
    titration with sodium hydroxide, 630

hydrofluoric acid, production of, 744
hydrogen, in balloons, 703
  binary compounds of, 71
  bonding in, 264
  bridging, 335
  chemistry of, 696–705
  compounds of, 422
    Lewis structures of, 267
    with nitrogen, 764
  electron configuration of, 239
  as fuel, 660
  in fuel cell, 708
  fusion of, 842
  ionization energy of, 228
  ions formed by, 63
  laboratory preparation of, 705t
  line emission spectrum, *211*
    explanation of, 213–216
  molecular orbital energy level diagram, 326
  orbital box diagram, 233
  oxidation number of, 122
  in oxoanions, 67
  potential energy during bond formation, 310
  reaction with iodine, 561
  reaction with nitrogen, 398
  reaction with oxygen, 10
  as reducing agent, 124t
hydrogen bonding Attraction between a hydrogen atom and a very electronegative atom to produce an unusually strong dipole–dipole attraction, 361, 422–427
  in DNA, 305
  in polyamides, 380
hydrogen chloride, as strong electrolyte, 103
  emitted by volcanoes, 163
  production of, 778
  reaction with ammonia, 116, 403
  reaction with magnesium, 191
  reaction with 2-methylpropene, 353
  reaction with sodium hydroxide, 114
hydrogen electrode, 703
  as pH meter, 719, *720*
  standard, 710
hydrogen fluoride, electrostatic potential map of, 294
  production of, 778
  reaction with silica, 759
  reaction with silicon dioxide, 778
  sigma bond in, 311
hydrogen halides, acidity and structure of, 607
  standard enthalpies of formation of, 198t
  standard enthalpies of vaporization of, 435t
hydrogen iodide, decomposition of, 75
  equilibrium with hydrogen and iodine, 662
hydrogen ion. *See* hydronium ion.
hydrogen peroxide, catalyzed decomposition of, 513
  decomposition of, 664
  in hair dye, 552
hydrogen phosphate ion, buffer solution of, 623t
hydrogen phthalate ion, buffer solution of, 623t
hydrogen sulfate ion, structure of, 269t
hydrogen sulfide, as polyprotic acid, 579t
  dissociation of, 560
  properties of, 773
  sulfur-oxidizing bacteria and, 775
hydrogenation An addition reaction in which the reagent is molecular hydrogen, 301, 353
  of benzene, 356

hydrolysis reaction A reaction with water in which a bond to oxygen is broken, 369
  of anions of insoluble salt, 649
  solubility and, 645
hydrometallurgy Recovery of metals from their ores by reactions in aqueous solution, 789, 791
hydronium ion, $H_3O^+(aq)$, 112, 113
  concentration expressed as pH, 154
  as Lewis adduct, 611
hydrophilic colloids, 500
hydrophobic colloids, 500
hydroxide ion, $OH^-(aq)$, 112
  formal charges in, 271
  as Lewis base, 612
hydroxides, solubility in strong acids, 650
hydroxyapatite, 750
hydroxyl ion, molecular geometry of, 283
hydroxyl radical, 279
p-hydroxyphenyl-2-butanone, 366
hygroscopic salt, 747
hyperthyroidism, 849
hypertonic solution, 496
hypochlorite ion, formal charges in, 270
  self oxidation–reduction, 537
  structure of, 269t
hypochlorous acid, 779
  structure of, 269t
hypophosphoric acid, 764t
hypothesis A tentative explanation or prediction bated on experimental observations, 3
hypothyroidism, 849
hypotonic solution, 496
hypoxia, 384

ibuprofen, synthesis of, 144, 301
ice, density of, 13
  hydrogen bonding in, 424
  melting of, 180–182
  slipperiness of, 468
  structure of, 60, 425
ICE table, 551
  calculating K_a value from, 596
  for common ion effect, 620
Iceland spar, 748
Iceman, 1, 50, 84
  radiochemical dating of, *836*
ideal gas A simplification of real gases in which it is assumed that there are no forces between the molecules and that the molecules occupy no volume, 394
ideal gas law A law that relates pressure, volume, number of moles, and temperature for an ideal gas, 394–398
  departures from, 410, 417
  osmotic pressure equation and, 492
  stoichiometry and, 398–400
ideal solution A solution that obeys Raoult's law, 486
ilmenite, 774
imaging, medical, 845
indicator(s) A substance used to signal the equivalence point of a titration by a change in some physical property such as color, 159
  acid–base, 155, 506, 638–640
indium, abundance of, 753
induced dipole(s) Separation of charge in a normally nonpolar molecule, caused by the approach of a polar molecule, 427
induced dipole-induced dipole force The electrostatic force between two neutral molecules, both having induced dipoles, 428–430
inert gas(es). *See* **noble gas(es).**
inertia, A-6
infant formula, adulteration of, 433

infrared (IR) radiation, 206
initial rate The instantaneous reaction rate at the start of the reaction, 516
insoluble compound(s), 640
  solubility product constants of, 642t
instantaneous reaction rate, 510
integrated rate equation, 519–527
  for nuclear decay, 834
integrity, in science, 5
intensive properties Physical properties that do not depend on the amount of matter present, 14
intercept, of straight-line graph, 37, 522
intermediate. *See* **reaction intermediate.**
  in rate law, 542
intermolecular forces Interactions between molecules, between ions, or between molecules and ions, 361, 411, 416–443
  determining types of, 430, 431
  energies of, 418, 421t
internal energy The sum of the potential and kinetic energies of the particle in the system, 184
  measurement of, 192
  relation to enthalpy change, 186
International Union of Pure and Applied Chemistry (IUPAC), 347, A-15
interstitial hydrides, 742
intravenous solution(s), tonicity of, 496
intrinsic semiconductor, 459
iodide ion, reaction with bromine, *716*
  reaction with iron(III) in, 559
iodine, abundance of, 775
  as catalyst, 533
  clock reaction, 512
  dissociation of, 563
  production of, 776, 777
  radioactive half-life of, 526
  reaction with hydrogen, 561
  reaction with sodium thiosulfate, 165
  solubility in liquids, 430
  solubility in polar and nonpolar solvents, 478
iodine-131, treatment of hyperthyroidism, 849
iodine tetrafluoride ion, molecular geometry of, 284
ion(s) An atom or group of atoms that has lost or gained one or more electrons so that it is no longer electrically neutral, 12, 61. *See also* **anion(s); cation(s).**
  acid–base properties of, 590t
  in aqueous solution, 101
  complex. See **coordination compound(s).**
  concentrations of, 150
  direction of flow in voltaic cells, 700
  electron configurations of, 245–247
  energy of interactions between, 460
  formation by metals and nonmetals, 62
  hydration of, 419, 480
  monatomic, 62
  noble gas electron configuration in, 257
  polyatomic, 64
  sizes of, 254
  spectator, 108
ion–dipole attraction The electrostatic force between an ion and a neutral molecule that has a permanent dipole moment, 418
ion exchange, in water softener, 752
ionic bond(s) The attraction between a positive ion and a negative ion resulting from the complete (or nearly complete) transfer of one or more electrons from one atom to another, 261
ionic compound(s) A compound formed by the combination of positive and negative ions, 61–70
  balancing charges in, 65
  bonding in, 460

colligative properties of solutions of, 496–498
crystal cleavage, 69
formula weight of, 75
formulas of, 65–66
lattice energy of, 460, 461t
of main group elements, 736
melting point of, 466t, 467
naming, 67
properties of, 68
solubility in water, 104, 479, *481*
temperature and, *484*
ionic radius, lattice energy and, 467
periodicity of, 254
solubility and, 481
ionic solid(s) A solid formed by the condensation of anions and cations, 452–456
ionization constant(s) The equilibrium constant for an ionization reaction, 581
acid and base, 585, 586t, A-20, A-22
water, 581
ionization energy The energy change required to remove an electron from an atom or ion in the gas phase, 249, 250t
periodicity of, 251
values of, A-18
iridium, density of, 784
iron, biochemistry of, 257
combustion of, 202
corrosion of, 787
meteorite, *248*
most stable isotope, 830, *831*
production of, 790
reaction with chlorine, 95
reaction with copper ions, 701
reaction with oxygen, 96
unit cell of, 454
iron(III) hydroxide, formation by precipitation, 107
in hemoglobin, 798
oxidation–reduction titration of, 162
reaction with permanganate ion, 125
reaction with potassium permanganate, 697
in sapphire, 232
iron(III) ion, paramagnetism of, 246, *247*
reaction with iodide ion, 559
iron(III) nitrate, dilution of, 153
iron(III) oxide, formation by corrosion, 787
reaction with aluminum, 124
reaction with carbon monoxide, 119
reduction of, 790
iron pyrite, 12, *643*, 772
irreversible process A process that involves nonequilibrium conditions, 664
isobutane, conversion to butane, 556, 569
isoelectronic ions Ions that have the same number of electrons but different numbers of protons, 255
isoelectronic species Molecules or ions that have the same number of valence electrons and similar Lewis structures, 269
isomer(s) Two or more compounds with the same molecular formula but different arrangements of atoms, 323
cis-trans. See *cis-trans* isomers.
geometric, 341, 349, 801
Markovnikov's rule and, 352
mer-fac, 802
number of, 344
optical, 341, 802
of organic compounds, 340–342
structural. See **structural isomers.**
isomerization, *cis-trans*, 533
in petroleum refining, 358
in gasoline, 205, 358

isoprene, in rubber, 376
isopropyl alcohol, 359t
isostructural species, 269
isotonic solution, 496
isotope(s) Atoms with the same atomic number but different mass numbers because of a difference in the number of neutrons, 46–48
hydrogen, 740
in mass spectra, 85
metastable, 846
percent abundance of, 46, 48t
radioactive, as tracers, 847
radioactive decay of, 823
separation of, 841
stable and unstable, 827, *828*
isotope dilution, volume measurement by, 847
isotope labeling, 369
isotope ratios, geographic variations of, 50

jasmine, oil of, 370
JELL-O®, 499
joule (J) The SI unit of energy, 25, A-7, A-11
Joule, James P., 175

K capture. *See* **electron capture.**
kaolin, 761
Kekulé, August, 323, 354
kelvin (K), 23, 390, A-10
in heat calculations, 179
Kelvin, Lord (William Thomson), *22,* 390
Kelvin temperature scale A scale in which the unit is the same size as the Celsius degree but the zero point is the lowest possible temperature, 22. *See also* **absolute zero.**
ketone(s) Any of a class of organic compounds characterized by the presence of a carbonyl group, in which the carbon atom is bonded to two other carbon atoms, 365–366
naming of, A-17
Kevlar, structure of, 380
Kidney stones, oxalic acid and, 618
Kilcoin, Caitlin, 4
kilocalorie (kcal), 26, A-8t
kilogram (kg) The SI base unit of mass, 25, A-10
kilojoule (kJ), 25
kilopascal (kPa), 386
kinetic energy The energy of a moving object, dependent on its mass and velocity, 7, 16
of alpha and beta particles, 822
distribution in gas, 528, *529*
temperature and, 403
distribution in liquid, *434*
total, 404
kinetic stability, of organic compounds, 342
kinetic-molecular theory A theory of the behavior of matter at the molecular level, 7
departures from assumptions of, 410
gas laws and, 407
of gases, 403–407
physical states and, 417
kinetics. *See* **chemical kinetics.**
Kohlrausch, Friedrich, 581

lactic acid, 368t
acid ionization constant of, 590
ionization of, 621
optical isomers of, 341, *342*
Lake Nyos, 25
Lake Otsego, 29, 29t
lakes, freezing of, 425, 427
lanthanide contraction The decrease in ionic radius that results from the filling of the 4*f* orbitals, 788

lanthanide(s) The series of elements between lanthanum and hafnium in the periodic table, 58, 243, 783
laser, synthetic ruby in, 232
lattice energy (lattice*U*) The energy of formation of one mole of a solid crystalline ionic compound from ions in the gas phase, 461
ionic radius and, 467
relation to solubility, 480
lattice point(s) The corners of the unit cell in a crystal lattice, 446
laughing gas, 766
Lavoisier, Antoine Laurent, 44, 93, 94
law A concise verbal or mathematical statement of a relation that is always the same under the same conditions, 4
Beer–Lambert, 167
Boyle's, 387
Charles's, 391
of chemical periodicity, 53
of conservation of energy, 18, 173
of conservation of mass, 5, 134
of conservation of matter, 93
Coulomb's, 68, 418
Dalton's, 401
general gas, 491
Graham's, 409
Henry's, 482
Hess's, 194
ideal gas, 394–398
Raoult's, 486
rate. *See* **rate equation(s).**
of thermodynamics, first, 183–187, 662
second, 663
third, 668
Le Chatelier's principle A change in any of the factors determining an equilibrium will cause the system to adjust to reduce the effect of the change, 483, 569
common ion effect and, 620
lead, abundance of, 758
density of, 13
isotope ratios, 50
pollution by, 762
lead(II) chloride, solubility of, 645
lead(II) chromate, *641*
formation by precipitation, 106, *107*
lead(II) iodide, dissolution of, *678*
lead(IV) oxide, in lead storage battery, 706
lead(II) sulfide, formation by precipitation, 106, *107*
roasting of, 773
solubility of, 649
least-squares analysis, 37
lecithin, 501
Leclanché, Georges, 705
Lego bricks, ABS plastic in, 378
lemon juice, pH of, 156
length, measurement of, 23
levo enantiomer, 364
Lewis, Gilbert Newton, 263, 611
Lewis acid(s) A substance that can accept a pair of electrons to form a new bond, 611–615
molecular, 614
Lewis base(s) A substance that can donate a pair of electrons to form a new bond, 611–615
ligands as, 794
molecular, 614
Lewis electron dot symbol/structure(s) A notation for the electron configuration of an atom or molecule, 263
predicting, 267–270
procedure for constructing, 265
Li, Kaichang, 381

Libby, Willard, *836*

ligand(s) The molecules or anions bonded to the central metal atom in a coordination compound, 655, 794

as Lewis bases, 794

naming of, 797

in organometallic compounds, 815

spectrochemical series of, 810

ligand field splitting (Δ₀) The difference in potential energy between sets of *d* orbitals in a metal atom or ion surrounded by ligands, 805

spectrochemical series and, 810

ligand field theory A theory of metal-ligand bonding in coordination compounds, 804–805

light. *See also* **electromagnetic radiation.**

absorption and reemission by metals, 458

absorption in molecules with pi bond frameworks, 334

plane-polarized, 341, 342

scattering by colloids, 499

speed of, 206, 830

visible, 206, 809

light-emitting diode (LED), 460

lime, 750, 751

reaction with water, 117

in soda-lime process, 746

in water softener, 752

limestone, 748

dissolving in vinegar, *118*

in iron production, 790

in stalactites and stalagmites, 98

limiting reactant The reactant present in limited supply that determines the amount of product formed, 137–141, 400

line emission spectrum The spectrum of light emitted by excited atoms in the gas phase, consisting of discrete wavelengths, 210, 211

linear electron-pair geometry, orbital hybridization and, *313*

linear molecular geometry, 281, 284, *285*, 800

in carbon compounds, 339

linear regression analysis, 37

linkage isomers Two or more complexes in which a ligand is attached to the metal atom through different atoms, 801

liquid(s) The phase of matter in which a substance has no definite shape but a definite volume, 7

compressibility of, 418

miscible and immiscible, 477

properties of, 432–442

liter (L) A unit of volume convenient for laboratory use; 1 L = 1000 cm³, 25

lithium, abundance of, 743

effective nuclear charge in, 236

reaction with water, 399

transmutation to helium, 838

uses of, 444

lithium aluminum hydride, as reducing catalyst, 366

lithium carbonate, 747

production of, 444

lithium-ion battery, 690, 707

litmus, 155

logarithms, 154, A-2

operations with, A-3

London dispersion forces, 428–430

lone pair(s) Pairs of valence electrons that do not contribute to bonding in a covalent molecule, 264

effect on electron-pair geometry, 282

in formal charge equation, 270

in Lewis base, 612

in ligands, 794

valence bond theory and, 312

low-density polyethylene (LDPE), 374

low spin configuration The electron configuration for a coordination complex with the minimum number of unpaired electrons, 806

lowest unoccupied molecular orbital (LUMO), 329

Lowry, Thomas M., 110

Lucite, 375t

lycopodium powder, *513*

Lyman series, 214

Mackintosh, Charles, 376

macroscopic level Processes and properties on a scale large enough to be observed directly, 8

magic numbers, of protons and neutrons, 838

magnesite, 748

magnesium, abundance of, 748

chemistry of, 748–751

combustion of, 120

in fireworks, 217

production of, 749

reaction with hydrogen chloride, 191

reaction with nitrogen, 763

magnesium carbonate, in magnesite, 748

magnesium chloride, in table salt, 747

magnesium fluoride, solubility of, 644

magnesium(II) hydroxide, precipitation of, 653

magnesium oxide, structure of, 455

magnesium perchlorate, water absorption by, 146

magnetic quantum number, 221

magnetic resonance imaging (MRI), 229

magnetism, atomic basis of, 227

magnetite, 787

main group element(s), *52, 53*

atomic radii, *248*

chemistry of, 734–780

electron configurations, 239–241

ionic compounds of, 736

ionization energies, 251

molecular compounds of, 737

malachite, *641, 789*

malic acid, 368t, *605*

manganese, oxidation–reduction cycle in sea water, 718

manganese(II) carbonate, *641*

manganese(II) dioxide, in dry cell battery, 705

manometer, U-tube, 387

mantissa The part of a logarithm to the right of the decimal point, A-3

Markovnikov, Vladimir, 352

Markovnikov's rule, 352

martensite, 454, 782

mass A measure of the quantity of matter in a body, 25

conservation of, 693, 822

energy equivalence of, 830

law of conservation of, 5, 134

weight and, A-6

mass balance, 134

mass defect, 47, 830

mass number (A) The sum of the number of protons and neutrons in the nucleus of an atom of an element, 44

in nuclear symbol, 822

mass percent. *See* **percent composition.**

mass spectrometer, 46, *47*

determining formula with, 84

matches, phosphorus sulfide in, 769

matter Anything that has mass and occupies space, 7, A-6

classification of, 6–10

dispersal of, 666

law of conservation of, 93

states of, 7, 417

matter wave, 216

mauveine, 334

Maxwell, James Clerk, 205, 405

Maxwell–Boltzmann distribution curves, 405, 528, *529,* 667

McMillan, Edwin, *837*

mean square speed, of gas molecules, 405

measured quantity, significant figures in, 32

measurement(s), units of, 21–26, 386, A-10–A-12

mechanical energy The energy due to the motion of macroscopic objects, 16

mechanism, reaction. *See* **reaction mechanism.**

Meitner, Lise, *841*

melamine, 4

pet food adulterated with, 433

melting point The temperature at which the crystal lattice of a solid collapses and solid is converted to liquid, 466t, 467, A-14t

identifying compounds by, 14

of ionic solids, 69

of transition elements, 788

membrane, semipermeable, 491

membrane cell, chlorine production by, 776, *777*

Mendeleev, Dmitri Ivanovitch, 42, 50–51

meniscus, *440, 441*

Menten, Maud L., 540

Mentos, reaction with Diet Coke, 485

mercury, from cinnabar, *2*

line emission spectrum, *211*

melting point of, 784

in pressure measurement, 386

mercury battery, 706

mercury(II) oxide, decomposition of, 94

mer-fac isomers, 802

meta position, 355

metal(s) An element characterized by a tendency to give up electrons and by good thermal and electrical conductivity, 53

band theory of, 457

biochemistry of, 257

cations formed by, 62–63

coordination compounds of, 792

electronegativity of, 288

enthalpy of fusion of, *466*

hydrated cations as Brønsted acids, 579, 610

memory, 782

plating by electrolysis, 725

reaction with nitric acid, 768

as reducing agents, 124t

sulfides, in black smokers, 92

transition. *See* **transition elements.**

unit cell types of, *448*

uses of, 444

metal sulfides, solubility of, 649

solubility product constants of, A-24

metallic character, periodicity of, 736

metalloid(s) An element with properties of both metals and nonmetals, 54

electronegativity of, 288

metallurgy, 789–792

metaphosphoric acid, 770t

metastable isotope, 846

metathesis reaction(s). *See* **exchange reaction(s).**

meteorite, iron, *248*

meter (m) The SI base unit of length, 23

definition of, A-10

methane, bond angles in, 282

bond order in, 298

combustion analysis of, 145

combustion of, standard free energy change, 681

enthalpy of formation, 195

as greenhouse gas, 426

hybrid orbitals in, 312, 314

hydrogen produced from, 742

standard free energy of formation of, 680

structure of, 60

methane hydrate, 426

methanol, 358, 359t

hydrogen bonding in, 432

in denitrification, 765

as fuel, 139

orbital hybridization in, 316

reaction with carbon monoxide, 367

reaction with halide ions, 530

spontaneity of formation reaction, 672

synthesis of, 139

methyl acetate, reaction with sodium hydroxide, 516

methyl chloride, reaction with halide ions, 530

methyl ethyl ketone, 366t

methyl mercaptan, 773

methyl methacrylate, synthesis of, 144

methyl salicylate, 670, 494

methylacetsmide, structure of, 371

methylamine, 316

electrostatic potential map of, 294

as weak base, 587

methylamines, 362

2-methyl-I,3-butadiene. *See* isoprene.

3-methylbutyl acetate, 370t

2-methylpentane, structure of, 345

2-methylpropene, reaction with hydrogen chloride, 353

structure of, 340, 349

methylsiloxane, 761

metric system A decimal system for recording and reporting scientific measurements, in which all units are expressed as powers of I0 times some basic unit, 21

mica, structure of, 760

Michaelis, Leonor, 540

microstate, 665

microwave radiation, 206

milk, adulteration of, 433

millerite, 144

milligram (mg), 25

milliliter (mL) A unit of volume equivalent to one thousandth of a liter; 1 mL = 1cm³, 25

millimeter of mercury (mm Hg) A common unit of pressure, defined as the pressure that can support a 1-millimeter column of mercury; 760mm Hg = 1 atm, 386, A-7t

mineral oil, density of, 39

minerals, analysis of, 143

clay, 761

silicate, 760

solubility of, 643

mischmetal, 58

miscibility, 477

mixture(s) A combination of two or more substances in which each substance retains its identity, 9–10

analysis of, 142–148

gaseous, partial pressures in, 400–403

models, molecular, 60

moderator, nuclear, 820

Mohr method, 163

Moisson, Henri, 775

molal boiling point elevation constant (K_{bp}), 488

molality (m) The number of moles of solute per kilogram of solvent, 474

molar absorptivity, 167

molar enthalpy of vaporization ($_{vap}H$), relation to molar enthalpy of condensation, 434

molar heat capacity, 176t, 177

values of, A-14t

molar mass (M) The mass in grams of one mole of particles of any substance, 72

calculation from colligative properties, 493

determination by titration, 161

effusion rate and, 408

enthalpy of vaporization and, 435

from ideal gas law, 397

molecular speed and, 406

polarizability and, 427

molar volume, standard, 395

molarity (M) The number of moles of solute per liter of solution, 149, 474

mole (mol) The SI base unit for amount of substance, 72, A-10

conversion to mass units, 72

of reaction, 141, 188

mole fraction (X) The ratio of the number of moles of one substance to the total number of moles in a mixture of substances, 402, 474

in Raoult's law, 486

molecular compound(s) A compound formed by the combination of atoms without significant ionic character, 61, 70–71. *See also* **covalent compound(s).**

as Brønsted acids and bases, 578

as Lewis acids, 614

of main group elements, 737

as nonelectrolytes, 103

molecular formula A written formula that expresses the number of atoms of each type within one molecule of a compound, 58

determining, 77–86

empirical formula and, 79

relation to empirical formula, 80

molecular geometry The arrangement in space of the central atom and the atoms directly attached to it, 282

hybrid orbitals and, 313

molecular polarity and, 291–297, 306t

multiple bonds and, 284–286

molecular models, 60

molecular orbital(s), bonding and anti-bonding, 326

from atomic p orbitals, 329

highest occupied (HOMO), 329

lowest unoccupied (LUMO), 329

molecular orbital theory A model of bonding in which pure atomic orbitals combine to produce molecular orbitals that are delocalized over two or more atoms, 309, 324–335, 804

for metals and semiconductors, 456–460

resonance and, 332

molecular polarity, 291–297, 306t

intermolecular forces and, 418

miscibility and, 477

of surfactants, 501

molecular solid(s) A solid formed by the condensation of covalently bonded molecules, 463

solubilities of, 478

molecular structure, acid–base properties and, 607–611

bonding and, 260–306

entropy and, 669

VSEPR model of, 280–286

molecular weight. *See* **molar mass.**

molecularity The number of particles colliding in an elementary step, 536

reaction order and, 537

molecule(s) The smallest unit of a compound that retains the composition and properties of that compound, 12, 58

calculating mass of, 76

collisions of, reaction rate and, 527–531

molar mass of, 74

nonpolar, interactions of, 427–430

polar, interactions of, 420

shapes of, 280–286

speeds in gases, 405

volume of, 410

molybdenite, 786

molybdenum, generation of technetium from, 846

monatomic ion(s) An ion consisting of one atom bearing an electric charge, 62

naming, 66–67

Mond, Ludwig, 813

Mond process, 814

monodentate ligands, 794

monomer(s) The small units from which a polymer is constructed, 373

monoprotic acid(s) A Brønsted acid that can donate one proton, 579

monosaccharides, 367

moon, rock samples analyzed, 848

mortar, lime in, 750, 751

Moseley, Henry C. J., 52

Mulliken, Robert S., 309

multiple bonds, 264

molecular geometry and, 284–286

in resonance structures, 272

valence bond theory of, 319–324

mummy, 1, 84

mussels, glue from, 381

Mylar, 378

myoglobin, 798

n-type semiconductor, 460

naming, of alcohols, 359t

of alcohols, A-17

of aldehydes and ketones, 366t, A-17

of alkanes, 344t, 346, A-15

of alkenes, 349, A-16

of alkynes, 352t, A-16

of anions and cations, 66–67

of aromatic compounds, A-16

of benzene derivatives, A-16

of binary nonmetal compounds, 71

of carboxylic acids, 368, A-17

of coordination compounds, 797

of esters, 369, 370t, A-17

of ionic compounds, 67

of substituted alkanes, 352

nanometer, 23

naphthalene, melting point, 14

solubility in benzene, 478

structure of, 354

National Institute of Standards and Technology (NIST), 26, 197

natural logarithms, A-2

neon, line emission spectrum of, 211

mass spectrum of, 47

neptunium, 839

Nernst equation A mathematical expression that relates the potential of an electrochemical cell to the concentrations of the cell reactants and products, 717

Nestlé, Henry, 338

net ionic equation(s) A chemical equation involving only those substances undergoing chemical changes in the course of the reaction, 108

of strong acid–strong base reactions, 114

network solid(s) A solid composed of a network of covalently bonded atoms, 463
  bonding in, 458
  silicon dioxide, 759
  solubilities of, 479
neurotransmitter, 364
neutral solution A solution in which the concentrations of hydronium ion and hydroxide ion are equal, 581
neutralization reaction(s) An add–bate reaction that produces a neutral solution of a salt and water, 115, 594
neutrino(s) A massless, chargeless particle emitted by some nuclear reactions, 826
neutron(s) An electrically neutral subatomic particle found in the nucleus, 43
  bombardment with, 839
  conversion to electron and proton, 823
  discovery of, 837
  in nuclear reactor, 820
  nuclear stability and, 827
neutron activation analysis, 848
neutron capture reactions, 839
newton (N) The SI unit of force, $1 N = 1 kg \cdot m/s^2$, A-6, A-11
nickel, as catalyst in diamond synthesis, 694
  coordination complex with ammonia, 794
  in memory metal, 782
nickel-cadmium (Ni-cad) battery, 707
nickel(II) carbonate, reaction with sulfuric acid, 163
nickel carbonyl, 813
  decomposition of, temperature and spontaneity, 684
nickel(II) chloride hexahydrate, 88, 792, 793
nickel(II) complexes, stability, 655
nickel(II) formate, 682
nickel(II) ion, light absorption by, 164
nickel-metal hydride battery, 690
nickel(II) oxide, reaction with chlorine trifluoride, 416
nickel(II) sulfide, quantitative analysis of, 144
nicotine, structure of, 368, 615
nicotinic acid, structure of, 620
nitinol, 782
nitrate ion, concentration in aquarium, 765
  molecular geometry of, 286
  resonance structures of, 609
  structure of, 660t
nitration, of benzene, 356
nitric acid, 772
  as oxidizing agent, 124t
  pH of, 156
  production by Ostwald process, 767
  reaction with aluminum, 757
  reaction with copper, 124
  reaction with metals, 768
  strength of, 607
  structure of, 268, 269t
nitric oxide. See nitrogen monoxide.
nitride(s), 763
nitrification, by bacteria, 765
nitrite ion, concentration in aquarium, 765
  linkage isomers containing, 801
  molecular geometry of, 286
  resonance structures of, 274
nitrito complex, 801
nitro complex, 801
nitrogen, abundance of, 763
  bond enthalpy of triple bond, 763
  bond order in, 298
  chemistry of, 763–768

compounds of, hydrogen bonding in, 422
  with hydrogen, 764
fixation of, 56
Henry's law constant, 482t
Lewis structures involving, 266–267
liquid, 389, 763
liquid and gas volumes, 418
molecular compounds of, 738
molecular orbital configuration of, 329
oxidation numbers of, 764
oxides of, 766
reaction with hydrogen, 398
reaction with oxygen, 564, 572
in testing food for protein content, 433
transmutation to oxygen, 837
nitrogen dioxide, 766
  dimerization of, 279, 280, 557, 571, 767
  free radical, 278
  production of hydroxyl radicals by, 279
  reaction with carbon monoxide, 217, 541
  reaction with fluorine, 539
  reaction with water, 117
nitrogen monoxide, 766
  biological roles of, 279
  free radical, 278
  reaction with bromine, 236
  reaction with chlorine, 514
  reaction with oxygen, mechanism of, 542–544
  reaction with ozone, 555
nitrogen narcosis, 410
nitrogen trifluoride, molecular polarity of, 296
  structure of, 268
nitrogenase, 784
nitroglycerin, 360
  decomposition of, 199
nitronium ion, Lewis structure of, 266
nitrosyl bromide, formation of, 536
nitrosyl chloride, rate of decomposition of, 510
nitrous acid, 767
  strength of, 607
nitrous oxide. See dinitrogen oxide.
Nobel, Alfred, 360
noble gas(es) The elements in Group 8A of the periodic table, 57
  compounds of, 277, 308
  electron configuration of, 63, 225, 263, 736
  in ions, 257
noble gas notation An abbreviated form of spdf notation that replaces the completed electron shells with the symbol of the corresponding noble gas in brackets, 239
noble metals, 768
nodal surface A surface on which there is zero probability of finding an electron, 223, 224
node(s) A point of zero amplitude of a wave, 206
nonbonding electrons. See lone pair(s).
nonelectrolyte A substance that dissolves in water to form an electrically nonconducting solution, 102, 103
nonequilibrium conditions, reaction quotient at, 555
  reaction quotient at, 717
nonideal gas, 410
nonideal solution, 486
nonmetal(s) An element characterized by a lack of metallic properties, 53
  anions formed by, 63
  binary compounds of, 71
  electronegativity of, 288
nonpolar covalent bond A covalent bond in which there is equal sharing of the bonding electron pair, 287

nonpolar molecules, 294
  interactions of, 427–430
nonspontaneous reaction, 661. See also **reactant-favored reaction(s).**
normal boiling point The boiling point when the external pressure is 1 atm, 439
  for common compounds, 435t
northwest–southeast rule A product-favored reaction involves a reduction agent below and to the right of the oxidizing agent in the table of standard reduction potentials, 713
nuclear binding energy, 829–832
nuclear charge, effective, 236, 237t, 244
nuclear chemistry, 820–850
nuclear energy, 841
nuclear fission, 840
nuclear fusion, 842
nuclear magnetic resonance (NMR), 229
nuclear magnetic resonance (NMR) spectrometer, 142
nuclear medicine, 845
nuclear reaction(s) A reaction involving one or more atomic nuclei, resulting in a change in the identities of the isotopes, 822–826
  artificial, 837–840
  predicting types of, 828
  rates of, 833–837
nuclear reactor A container in which a controlled nuclear reaction occurs, 841
  natural, 820
nuclear spin, quantization of, 229
nucleation, in bubble formation, 485
nucleon A nuclear particle, either a neutron or a proton, 830
nucleus The core of an atom, made up of protons and neutrons, 43
  stability of, 827–833
Nutrient Data Laboratory site, 25
nutrition label, energy content on, 25
nylon, 379

octahedral electron-pair geometry, orbital hybridization and, 313, 318
octahedral holes, 452
octahedral molecular geometry, 281, 800
octane, combustion of, 96
  heat of combustion, 193
octet, of electrons, 263
octet rule When forming bonds, atoms of main group elements gain, lose, or share electrons to achieve a stable configuration having eight valence electrons, 264
  exceptions to, 264, 276–279
odd-electron compounds, 278, 330, 766, 767
oil(s), soaps and, 501
Oklo, natural nuclear rector at, 820
oleic acid, 368t
olivine, 760
optical isomers Isomers that are nonsuperimposable mirror images of each other, 341, 802
orbital(s) The matter wave for an allowed energy state of an electron in an atom or molecule, 220
  atomic. See **atomic orbital(s).**
  molecular. See molecular orbital(s).
orbital box diagram A notation for the electron configuration of an atom in which each orbital is shown as a box and the number and spin direction of the electrons are shown by arrows, 233
orbital hybridization The combination of atomic orbitals to form a set of equivalent hybrid orbitals that matches the electron pair geometry of the compound, 312–324

orbital overlap Partial occupation of the same region of space by orbitals from two atoms, 310
order, bond. *See* **bond order.**
reaction. *See* **reaction order.**
ore(s) A sample of matter containing a desired mineral or element, usually with large quantities of impurities, 789
insoluble salts in, 651
organic compounds, bonding in, 338–382
naming of, 344t, 346, A-15
organometallic chemistry, 813–817
orientation of reactants, effect on reaction rate, 530
Orlon, 375t
orpiment, *61, 84, 643*
ions in, 61
ortho position, 355
orthophosphoric acid, 770t, 771
orthorhombic sulfur, 772
orthosilicates, 760
osmium, density of, 784
osmosis The movement of solvent molecules through a semipermeable membrane from a region of lower solute concentration to a region of higher solute concentration, 491
reverse, 493
osmotic pressure (Π) The pressure exerted by osmosis in a solution system at equilibrium, 492
Ostwald, Wilhelm, 72
Ostwald process, 767
Ötzi the Iceman, 1, 50, 84, 836
overlap, orbital, 310
overvoltage, 728
oxalate ion, as ligand, 794
oxalic acid, 368t
molar mass of, 76
as polyprotic acid, 579t
in rhubarb, 618
titration of, 158, 636
oxidation The loss of electrons by an atom, ion, or molecule, 120
of transition metals, 785
oxidation number(s) A number assigned to each element in a compound in order to keep track of the electrons during a reaction, 121
formal charges and, 271
of Group 3A elements, 753
of Group 4A elements, 758
of Group 5A elements, 763, 764
of phosphorus, 771
in redox reaction, 692
relation to group number, 738
of transition elements, 785, *786*
oxidation reaction(s), of alcohols, 365
oxidation–reduction reaction(s) A reaction involving the transfer of one or more electrons from one species to another, 119–125, 690–733
in acidic and basic solutions, 695–698
balancing equations for, 695–699
recognizing, 123
titration using, 162
oxides, as acids and bases, 116
oxidizing agent(s) The substance that accepts electrons and is reduced in an oxidation–reduction reaction, 120, 692
relative strengths of, 711, *712*, 715
oxoacid(s), acid strengths of, 607
Lewis structures of, 268, 269t
of chlorine, 779
oxoacids, phosphorus, 770
oxoanion(s) Polyatomic anions containing oxygen, 67
as Brønsted bases, 611
Lewis structures of, 268, 269t

oxy-acetylene torch, *352*
oxygen, abundance of, 772
allotropes of, 57, 772
compounds of, hydrogen bonding in, 422
with nitrogen, 766
with phosphorus, 768, *769*
corrosion and, 787
deprivation and sickness, 384
discovery of, 94
dissolving in water, 427
in fuel cell, 708
Henry's law constant, 482t
in iron production, 790
isotope ratios, 50
Lewis structures involving, 266–267
molecular orbital configuration of, 329
oxidation number of, 122
as oxidizing agent, 120, 124t
paramagnetism of, 227, 228, 325, 329
partial pressure and altitude, 404
from photosynthesis, 772
reaction with alkali metals, 745
reaction with alkanes, 348
reaction with calcium, *262*
reaction with hemoglobin, 798
reaction with hydrogen, 16
reaction with nitrogen, 564, 572
reaction with nitrogen monoxide, mechanism of, 542–544
toxicity in deep-sea diving, 410
oxygen-15, in PET imaging, 847
oxyhemoglobin, 798
ozone, 57, 772
decomposition of, 536
fractional bond order of, 298
molecular orbital configuration of, 332
reaction with nitrogen monoxide, 555
resonance structures of, 272
solar radiation absorbed by, 404

p-block elements, 240
molecular orbitals involving, 329, 331
p orbital(s). *See* **atomic orbital(s).**
p-type semiconductor, 459
packing, in crystal lattice, 449
paint, transition metal pigments in, 784
white lead in, 762
pairing, of electron spins, 227
pairing energy The additional potential energy due to the electrostatic repulsion between two electrons in the same orbital, 807
palladium, hydrogen absorption by, 205
para position, 355
paramagnetism The physical property of being attracted by a magnetic field, 227, 329, 807
of transition metal ions, 246, *247*, 785
parent ion, in mass spectra, 85
Parkinson's disease, 364
parsec, *29*
partial charge(s) The charge on an atom in a molecule or ion calculated by assuming sharing of the bonding electrons proportional to the electronegativity of the atom, 287, 291
partial pressure(s) The pressure exerted by one gas in a mixture of gases, 401
in equilibrium constant expression, 552, 554, 560
particle accelerator, 838
particulate level Representations of chemical phenomena in terms of atoms and molecules. Also called submicroscopic level, 8
parts per million (ppm), 475

pascal (Pa) The SI unit of pressure; 1 Pa = 1 N/m², 386, A-6, A-11
path length, light absorption and, *166*, 167
Pauli, Wolfgang, 233
Pauli exclusion principle No two electrons in an atom can have the same set of four quantum numbers, 233
molecular orbitals and, 325
Pauling, Linus, *288*
and electronegativity, 288
and theory of resonance, 272
and valence bond theory, 309, 312
pentane, structural isomers of, 344
pentenes, isomers of, 350
pentose, 367
peptide linkage, 371
Pepto-Bismol, black tongue and, *107*
percent abundance The percentage of the atoms of a natural sample of the pure element represented by a particular isotope, 46
percent composition The percentage of the mass of a compound represented by each of its constituent elements, 77
percent error The difference between the measured quantity and the accepted value, expressed as a percentage of the accepted value, 27
percent yield The actual yield of a chemical reaction as a percentage of its theoretical yield, 142
perchlorate ion, structure of, 269t
perchlorates, 779
perchloric acid, structure of, 269t
perchboroethylene, 442
perhydroxyl ion, 279
periodic table of the elements, 10, 42, 50–58, 736–740
electron configurations and, *237*
historical development of, 50–51
ion charges and, 63–64
periodicity, of atomic radii, 247
of chemical properties, 50, 255–258
of electron attachment enthalpies, 252
of electronegativity, 288
of ionic radius, 254
of ionization energy, 251
periods The horizontal rows in the periodic table of the elements, 53
Perkin, William, 334
permanganate ion, as oxidizing agent, 124t
reaction with iron(II) ion, 125, 162
perovskite, structure of, 455
peroxides, 745
oxidation number of oxygen in, 122
peroxymonocarbonate ion, 279
perspective formula, 341
pertechnate ion, 846
pet food, adulteration of, 433
Peter, Daniel, 338
chemistry of, 358
pH The negative of the base-10 logarithm of the hydrogen ion concentration; a measure of acidity, 154–156, 580–583
in aquarium, 765
in buffer solutions, 622–629
of blood, 631
calculating equilibrium constant from, 596
calculating from equilibrium constant, 598–603
change in, during acid–base titration, 630
common ion effect and, 618–621
pH meter, 155, 721
phase change, as spontaneous process, 662
condensation, 434
heat transfer in, 180
vaporization, 433

phase diagram A graph showing which phases of a substance exist at various temperatures and pressures, 468–469

phase transition temperature, 782

phenanthroline, as ligand, 794

phenol, structure of, 355

phenolphthalein, structure of, 506, *639*

Philosopher's Stone, 767

phosgene, molecular polarity of, 293

phosphate ion, buffer solution of, 623t
  in biological buffer system, 631
  resonance structures of, 275
  structure of, 269t

phosphate rock, 769, 770

phosphates, solubility in strong acids, 650

phosphine, 71, 768
  rate of decomposition of, 510

phosphines, in organometallic compounds, 815

phosphoenolpyruvate (PEP), 683

phosphoric acid, 771
  as polyprotic acid, 579t, 587
  reaction with acetate ion, 592
  structure of, 269t

phosphorus, abundance of, 763
  allotropes of, 56, *763, 764*
  chemistry of, 768–771
  coordinate covalent bonds to, 277
  discovery of, 767
  molecular compounds of, 738
  oxides of, 768, *769*
  reaction with chlorine, 93, *94, 133*
  reaction with oxygen, 96
  sulfides of, 769

phosphorous acid, 770t

phosphorus oxoacids, 770

phosphorus pentachloride, decomposition of, 562

phosphorus pentafluoride, orbital hybridization in, 318

photocell, *209*

photoelectric effect The ejection of electrons from a metal bombarded with light of at least a minimum frequency, 209

photon(s) A "particle" of electromagnetic radiation having zero mass and an energy given by Planck's law, 209

photovoltaic cell, 204

phthalic acid, buffer solution of, 623t

physical change A change that involves only physical properties, 15

physical properties Properties of a substance that can be observed and measured without changing the composition of the substance, 13–15
  temperature dependence of, 13

pi (π) bond(s) The second (and third, if present) bond in a multiple bond; results from sideways overlap of p atomic orbitals, 320–324
  in benzene, 333
  molecular orbital view of, 329
  in organometallic compounds, 815
  in ozone, 332

picometer, 23

pie filling, specific heat capacity of, 177

pig iron, 790

pigments, green chemistry of, 334
  transition metals in, 784

Pins, Raffaele, 576

pitchblende, 824

pKₐ The negative of the base-10 logarithm of the acid ionization constant, 588
  at midpoint of acid–base titration, 633
  pH of buffer solution and, 625

planar node. *See* **atomic orbital(s).**

Planck, Max, 208

Planck's constant (*h*) The proportionality constant that relates the frequency of radiation to its energy, 208

Planck's equation, 207–210

plasma A gas-like phase of matter that consists of charged particles, 842

plaster of Paris, 86

plastic(s). *See also* **polymer(s).**

plastic sulfur, 772

plating, by electrolysis, 725

platinum, in cisplatin, 813
  in oxidation of ammonia, *138*

platinum electrode, 703

platinum group metals, 788

Plexiglas, 375t

plotting. *See* graph(s).

plutonium-239, fission of, 841

pOH The negative of the base-10 logarithm of the hydroxide ion concentration; a measure of basicity, 583

poisoning, arsenic, 84
  carbon monoxide, 798
  hydrogen sulfide, 773
  lead, 762

polar covalent bond A covalent bond in which there is unequal sharing of the bonding electron pair, 287

polarity, bond, 287–291
  molecular, 291–297, 306t
    intermolecular forces and, 418
    solubility of alcohols and, 361
    solubility of carboxylic acids and, 368

polarizability The extent to which the electron cloud of an atom or molecule can be distorted by an external electric charge, 427

polarized light, rotation by optically active compounds, 341, *342*

polonium, 57
  abundance of, 772
  from decay of uranium, 824

polyacrylate polymer, in disposable diapers, *381*

polyacrylonitrile, 375t

polyamide(s) A condensation polymer formed by linking monomers by amine groups, 379

polyatomic ion(s) An ion consisting of more than one atom, 64
  names and formulas of, 64t, 67
  oxidation numbers in, 122

polybutadiene, in ABS plastic, 378

polydentate ligands, 794

polydimethylsiloxane, 762

polyester(s) A condensation polymer formed by linking monomers by ester groups, 377–379

polyethylene, 373, 374, 375t
  in disposable diapers, 381
  structure of, *446*

polyethylene terephthalate (PET), 378, 379

polyisoprene, 376

polylactic acid (PLA), 378

polymer(s) A large molecule composed of many smaller repeating units, usually arranged in a chain, 373–381
  addition, 373–377
  classification of, 373
  condensation, 370, 377–380
  osmotic pressure of, 492, 493
  silicone, 761–762

polymethyl methacrylate, 375t

polypropylene, 375t
  in disposable diapers, *381*

polyprotic acid(s) A Brønsted acid that can donate more than one proton, 579
  ionization constants of, 588

pH of, 605
  titration of, 636

polyprotic base(s) A Brønsted base that can accept more than one proton, 579
  pH of, 605

polystyrene, 374, 375t, 377

polytetrafluoroethylene, 375t

polyvinyl acetate (PVA), 374, 375t

polyvinyl alcohol, 375

polyvinyl chloride (PVC), 375t

popcorn, percent yield of, *142*

porphyrin, 798

Portland cement, 760

positron(s) A particle having the same mass as an electron but a positive charge, 825
  emitters of, 846
  predicting emission of, 828

positron emission tomography (PET), 846

potassium, abundance of, 743
  preparation of, 744
  reaction with water, *54*

potassium chlorate, decomposition of, 772
  in fireworks, 217
  in matches, 769

potassium dichromate, 152
  oxidation of alcohol by, *365*

potassium dihydrogen phosphate, crystallization of, *484*

potassium fluoride, dissolution of, 479
  electrolysis of, 775

potassium nitrate, 747
  in fireworks, 217

potassium perchlorate, 779
  in fireworks, 217

potassium permanganate, 150
  absorption spectrum of, *168*
  dissolution of, *667*
  oxidation of alcohol by, *365*
  reaction with iron(II) ion, 697
  in redox titration, 162

potassium superoxide, 745
  molecular orbital configuration of, 330

potential, of electrochemical cell, 709–717

potential energy The energy that results from an object's position, 16
  bond formation and, 310
  of electron in hydrogen atom, 212

potential ladder, *712*

pounds per square inch (psi), 387, A-7

power The amount of energy delivered per unit time, A-7

powers, calculating with logarithms, A-4
  on calculator, 31

precipitate A water-insoluble solid product of a reaction, usually of water-soluble reactants, 106
  gelatinous, 499

precipitation reaction(s) An exchange reaction that produces an insoluble salt, or precipitate, from soluble reactants, 105–110, 126, 640–650
  solubility product constant and, 651–655

precision The agreement of repeated measurements of a quantity with one another, 26

prefixes, for ligands, 798
  for SI units, 21, 22t, A-10

pressure The force exerted on an object divided by the area over which the force is exerted, 386, A-6
  atmospheric, altitude and, 384
  constant, calorimetry at, 190
  critical, 439
  effect on solubility, 482
  gas, volume and, 388
  partial. *See* partial pressure.
  relation to boiling point, 439

standard, 395
units of, 386, A-6
vapor. *See* **vapor pressure.**
pressure–volume work, 185
Priestley, Joseph, 94
primary alcohols, 365
primary battery, 704
primary colors, 809
primary standard A pure, solid add or base that can be accurately weighed for preparation of a titrating reagent, 160
primitive cubic (pc) unit cell, 447
principal quantum number, 212, 220
probability, diffusion and, 666–668
in quantum mechanics, 219
problem-solving strategies, 38
Problem Solving Tip, aqueous solutions of salts, 591
balanced equations and equilibrium constants, 567
balancing equations in basic solution, 699
balancing oxidation–reduction equations, 699
buffer solutions, 628
calculating *T*, 179
common entropy-favored processes, 672
concepts of thermodynamics, 662
determining ionic compounds, 69
determining strong and weak acids, 585
drawing Lewis electron dot structures, 271
drawing structural formulas, 347
electrochemical conventions for voltaic cells and electrolysis cells, 726
finding empirical and molecular formulas, 82
formulas for ions and ionic compounds, 68
ligand field theory, 812
pH during acid–base reaction, 638
pH of equal molar amounts of acid and base, 603
preparing a solution by dilution, 153
reactions with a limiting reactant, 141
recognizing gas-forming reactions, 119
relating rate equations and reaction mechanisms, 543
resonance structures, 278
stoichiometry calculations involving solutions, 157
stoichiometry calculations, 135
using a calculator, 31
using Hess's law, 197
using the quadratic formula, 564
writing net ionic equations, 108
product(s) A substance formed in a chemical reaction, 16, 93
effect of adding or removing, 569
in equilibrium constant expression, 551
heat as, 572
rate of concentration change, 509
product-favored reaction(s) A reaction in which reactants are completely or largely converted to products at equilibrium, 100
equilibrium constant for, 554
predicting, 678
propane, combustion of, balanced equation for, 97
enthalpy of combustion, 189
percent composition of, 78
structure of, 340
1,2,3-propanetriol, 359t, 360
propanoic acid, as weak acid, 587
1-propanol, 359t
2-propanol, 359t
structure of, 302

propene, 349, 520
hydrogenation of, 301
reaction with bromine, 353
propionic acid, 368t
proportionality constant, 388, 405
proportionality symbol, 514
propyl alcohol, 359t
propyl propanoate, 372
propylene, 349
protein(s) A polymer formed by condensation of amino adds, sometimes conjugated with other groups
energy content of, 25
as hydrophilic colloids, 501
testing food for, 433
proton(s) A positively charged subatomic particle found in the nucleus, 43
bombardment with, 837
donation by Brønsted acid, 112, 577
nuclear stability and, 827
proton exchange membrane (PEM), 708
Prussian blue, 784, *785*
purification, of mixtures, 10
putrescine, 363
pyridine, structure of, 615
pyrite, iron, 12
pyrometallurgy Recovery of metals from their ores by high-temperature processes, 789
pyrophosphate ion, manganese(III) ion and, 718
pyrophosphoric acid, 770t
pyroxenes, structure of, 760

quadratic equations, A-4
quadratic formula, use in concentration problems, 562
qualitative information Nonnumerical experimental observations, such as descriptive or comparative data, 3, 21
quantitative analysis, 143
quantitative information Numerical experimental data, such as measurements of changes in mass or volume, 3, 21
quantity, of pure substance, 72
quantization, of electron spin, 226
of electron's potential energy, 212, 219
of energy in macroscopic system, 667
of nuclear spin, 229
Planck's assumption of, 208
quantum dots, *773*
quantum mechanics A general theoretical approach to atomic behavior that describes the electron in an atom as a matter wave, 219–222
quantum number(s) A set of numbers with integer values that define the properties of an atomic orbital, 219
allowed values of, 220
angular momentum, 220
in macroscopic system, 667
magnetic, 221
Pauli exclusion principle and, 233
principal, 212, 220
spin, 226
quartz, 759
structure of, *463*
quenching, in steel production, 454

rad A unit of radiation dosage, 843
radial distribution plot, 222
radiation, background, 843
cancer treatment with, 847
cosmic, 843
electromagnetic, 205–207

health effects of, 843–845
safe exposure, 845
treatment of food with, 848
units of, 843
radiation absorbed dose (rad), 843
radioactive decay series A series of nuclear reactions by which a radioactive isotope decays to form a stable isotope, 823–826
radioactivity, natural, 821
radiocarbon dating, 835
radium, abundance of, 748
from decay of uranium, 824
radius, atomic, 247
covalent, 249
ionic, 254, 467, 487
radon, from decay of uranium, 834
radioactive half-life of, 526, 834
Ramsay, Sir William, 574
Raoult's law The vapor pressure of the solvent is proportional to the mole fraction of the solvent in a solution, 486
rare earth elements. *See* **lanthanides.**
rare gas(es). *See* **noble gas(es).**
Raschig reaction, 541, 764
rate. *See* **reaction rate(s).**
rate constant *(k)* The proportionality constant in the rate equation, 514, 515
Arrhenius equation for, 531
half-life and, 524
for radioactivity, 834
units of, 515, 516
rate equation(s) The mathematical relationship between reactant concentration and reaction rate, 514
determining, 516
for elementary step, 537
first-order, nuclear, 833
graphical determination of, 522
integrated, 519–527
for nuclear decay, 834
reaction mechanisms and, 538–542
rate law. *See* **rate equation(s).**
rate-determining step The slowest elementary step of a reaction mechanism, 538
reactant(s) A starting substance in a chemical reaction, 16, 93
concentration of, reaction rate and, 513–519
effect of adding or removing, 569
in equilibrium constant expression, 551
heat as, 572
rate of concentration change, 509
reactant-favored reaction(s) A reaction in which only a small amount of reactants is converted to products at equilibrium, 100
equilibrium constant for, 554, 555
predicting, 678
reaction(s) A process in which substances are changed into other substances by rearrangement, combination, or separation of atoms, 16. *See also under element, compound, or chemical group of interest.*
acid–base. *See* **acid–base reaction(s).**
addition, 352
in aqueous solution, 101
stoichiometry of, 157–158
types of, 126
autoionization, 581
chain, 840
condensation, 377
coupling of, 683
direction of, acid–base strength and, 591
reaction quotient and, 555
disproportionation, 779

electron transfer, 690–733. *See also* **oxidation–reduction reaction(s).**
enthalpy change for, 188–190
esterification, 369
exchange, 105, 126
free energy change for, 677
Friedel–Crafts, 614
gas laws and, 398–400
gas-forming, 118–119, 127
hydrogenation, 301
hydrolysis, 369
moles of, 141, 188
(n, γ), 839
neutralization, 115, 594
neutron capture, 839
nuclear, 822–826
  artificial, 837–840
  rates of, 833–837
order of. *See* **reaction order.**
oxidation of alcohols, 365
oxidation–reduction. *See* **oxidation–reduction reaction(s).**
in petroleum refining, 358
precipitation, 105–110, 126, 640–650
  solubility product constant and, 651–655
product-favored vs. reactant-favored, 100, 200–202, 554, 661
predicting, 678
rate of. *See* **reaction rate(s).**
reduction of aldehydes and ketones, 366, 369
reduction of carboxylic acids, 368
reductive carbonylation, 814
reverse, equilibrium constant expression for, 566
reversibility of, 98, 549
spontaneity of, predicting, 674, 677
standard enthalpy of, 188
standard reduction potentials of, 711, 712t
substitution, 356
sum of, equilibrium constant expression for, 567
thermite, 124, 132, 140
water gas, 742
reaction coordinate diagram, 529–530
reaction intermediate A species that is produced in one step of a reaction mechanism and completely consumed in a later step, 534
in rate law, 542
reaction mechanism(s) The sequence of events at the molecular level that controls the speed and outcome of a reaction, 507, 535–544
effect of catalyst on, 534
rate equation and, 538–544
rate-determining step of, 538
reaction order The exponent of a concentration term in the reaction's rate equation, 514
determining, 516
molecularity and, 537
reaction quotient (Q) The product of concentrations of products divided by the product of concentrations of reactants, each raised to the power of its stoichiometric coefficient in the chemical equation, 555–557. *See also* **equilibrium constant.**
Gibbs free energy change and, 678
relation to cell potential, 717
solubility product constant and, 651
reaction rate(s) The change in concentration of a reagent per unit time, 507–511
Arrhenius equation and, 531
average vs. instantaneous, 510
catalysts and, 533–535
collision theory of, 527–531
conditions affecting, 512

effect of temperature, 528
expression for. *See* **rate equation(s).**
initial, 516
radioactive disintegration, 833–835
stoichiometry and, 509, 510
rechargeable battery, 705
recipe, stoichiometry and, 134
redox reaction(s). *See* **oxidation–reduction reaction(s).**
reducing agent(s) The substance that donates electrons and is oxidized in an oxidation–reduction reaction, 120, 692
relative strengths of, 711, 712, 715
reduction The gain of electrons by an atom, ion, or molecule, 119
of transition metals, 785
reduction potential(s), standard, 711, 713t
standard, values of, A-32
reduction reaction(s), of aldehydes and ketones, 366, 369
of carboxylic acids, 368
reductive carbonylation reaction, 814
reformation, in petroleum refining, 358
relative atomic mass. *See* **atomic weight.**
rem A unit of radiation dosage to biological tissue, 843
resin, in ion exchanger, 752
resonance, molecular orbital theory and, 332
resonance stabilization, 356
resonance structure(s) The possible structures of a molecule for which more than one Lewis structure can be written, differing by the number of bond pairs between a given pair of atoms, 272–276
amides, 371
benzene, 272, 323, 354
carbonate ion, 273
effect on acid strength, 609
nitrate ion, 273
nitrite ion, 274
ozone, 272
respiration, production of ATP by, 683
reverse osmosis The application of pressure greater than the osmotic pressure of impure solvent to force solvent through a semipermeable membrane to the region of lower solute concentration, 493
water purification by, 472
reversibility, equilibrium and, 549
of chemical reactions, 98
reversible process A process for which it is possible to return to the starting conditions along the same path without altering the surroundings, 664
rhodochrosite, *641, 643*
rhubarb, oxalic acid in, 618
ribonucleic acid, hydrogen bonding in, 429
ring structure, in benzene, 323
RNA. *See* ribonucleic acid.
Roberts, Ainé, *394*
rock salt structure, 455
roentgen A unit of radiation dosage, 843
Roentgen, Wilhelm, 843
Roman numerals, in names of cations, 66, 67
root-mean-square (rms) speed The square root of the average of the squares of the speeds of the molecules ma sample, 406
roots, calculating with logarithms, A-4
  on calculator, 31
rose petals, pigment as indicator, *638*
Rosenberg, Barnett, 813
rotation, A-6
  around bonds in alkanes, 345
  of polarized light, 341, *342*
  around sigma and pi bonds, 323, 341
rounding off, 33

ROY G BIV, 809
rubber, isoprene in, 376
  natural and synthetic, 376
  styrene-butadiene, 376
  vulcanized, 376
rubidium, abundance of, 743
ruby, 232
  ion charges in, 65
  synthetic, 757
rust, 787. *See also* iron(III) oxide.
Rutherford, Ernest, 43, 821, 837
rutile, in rubies, 232
Rydberg, Johannes, 210
Rydberg constant, 211

s-block elements, 239
s orbital(s). *See* **atomic orbital(s).**
  structure of, 354
safety match, 769, 779
salad dressing, as emulsion, 501
salicylic acid, 142, 370. 576
salt(s) An ionic compound whose cation comes from a base and whose anion comes from an acid, 114–118
acid–base properties of, 589–591
calculating pH of aqueous solution, 601
concentration in seawater, 163
electrolysis of, 725
hydrated, 421
insoluble, precipitation of, 651–655
solubility of, 640–650
solubility product constants of, 642t
salt bridge A device for maintaining the balance of ion charges in the compartments of an electrochemical cell, 699
saltpeter, 744, 747
salvarsan, in syphilis treatment, 84
sandwich compounds, 815
Sapa, 762
sapphire, 232, 757
sarin, 604
saturated compound(s) A hydrocarbon containing only single bonds, 344. *See also* **alkanes.**
saturated solution(s) A stable solution in which the maximum amount of solute has been dissolved, 476, 641
reaction quotient in, 651
scanning electron microscopy (SEM), *23*
Scheele, Carl Wilhelm, 94, 776
Schrödinger, Erwin, 281
science, goals of, 5
  methods of, 2–5
scientific notation, 29–31, A-2
  operations in, 30
screening, of nuclear charge, 236
scrubber, flue gas, 412
SCUBA diving, gas laws and, 410
  Henry's law and, 482
sea urchin, 17
sea water, density of, 35
  halogens in, 775
  magnesium in, 749
  pH of, 156
  reverse osmosis of, 493
  salt concentration in, 163
  sodium and potassium ions in, 744
Seaborg, Glenn T., *837*
second, definition of, A-10
second-order reaction, 514
  half-life of, 524
  integrated rate equation, 521
second law of thermodynamics The total entropy of the universe is continually increasing, 519

secondary alcohols, 365

secondary battery, 705

secondary colors, 809

seesaw molecular geometry, 284, *285*

selenium, abundance of, 772

uses of, 772, 773

semiconductor(s) Substances that can conduct small quantities of electric current, 458–460

band theory of, 458

semimetals. *See* **metalloid(s).**

semipermeable membrane A thin sheet of material through which only certain types of molecules can pass, 491

sequestration, geological, of carbon dioxide, 412

serial dilution, 154

setae, on geckos' toes, 416

SI Abbreviation for Système International d'Unités, a uniform system of measurement units in which a single base unit is used for each measured physical quantity, 21, A-10

sievert The SI unit of radiation dosage to biological tissue, 843

sigma (σ) bond(s) A bond formed by the overlap of orbitals head to head, and with bonding electron density concentrated along the axis of the bond, 311

sign conventions, for electron affinity, 252

for energy calculations, 185t

for voltaic cells, *701*

significant figure(s) The digits in a measured quantity that are known exactly, plus one digit that is inexact to the extent of ±1, 31–34

in atomic masses, 73

in calculating empirical formulas, 82

logarithms and, A-3

silane, comparison to methane, 734

silica, 759

silica gel, 760

silicates, minerals containing, 760

structure of, 463

silicon, abundance of, 758

bond energy compared to carbon, 342

chemistry of, 758–762

compounds of, *56*

purification of, 758

as semiconductor, 459

similarity to boron, 751

similarity to carbon, 734

structure of, *446*

silicon carbide, structure of, 456

silicon dioxide, 759

comparison to carbon dioxide, 734

reaction with hydrogen fluoride, 778

silicon tetrachloride, 758

molecular geometry of, 281

silicone polymers, 761–762

Silly Putty, 762

silt, formation of, 501

silver, as bacteriocide, 126

silver acetate, solubility of, 646

silver bromide, solubility of, 641

in aqueous thiosulfate ion, 658

silver chloride, free energy change of dissolution, 685

K_{sp} of, 724

precipitation of, 106, 108

solubility of, 645, 647, 652

in aqueous ammonia, 566, 656

silver chromate, 163

formation by precipitation, 107

solubility of, 646, 648

silver nitrate, reaction with potassium chloride, 106, 108

silver oxide battery, 706

simple cubic (sc) unit cell, 447

single bond A bond formed by sharing one pair of electrons; a sigma bond, 311

slag, in blast furnace, 791

slaked lime, 117

slime, 376

slope, of straight line graph, 37, 522

snot-tires, 775

soap A salt produced by the hydrolysis of a fat or oil by a strong base, 501

hard water and, 752

soapstone, 748

soda ash, 746, *747. See* sodium carbonate.

soda lime process, 746

Soddy, Frederick, 822

sodium, abundance of, 743

in fireworks, 217

preparation of, 744

reaction with chlorine, 3, 123, 261, *262*

reaction with water, *5*

sodium acetate, calculating pH of aqueous solution, 601

in heat pack, 478

sodium azide, 398

in air bags, *385*, 392

sodium benzoate, 80

sodium bicarbonate, 746. *See also* sodium hydrogen carbonate.

reaction with acetic acid, 593

sodium borohydride, 756

as reducing catalyst, 366

sodium carbonate, 151

calculating pH of aqueous solution, 605

industrial uses, 746

primary standard for acid–base titration, 161

sodium chloride, as strong electrolyte, 101

composition of, *3, 12*

crystal lattice of, 69

electrolysis of, 398, 726, 744

entropy of solution process, 673

ion charges in, 65

lattice enthalpy calculation for, 462

melting ice and, 496

oxidation by dichromate ion, 776

reaction with sulfuric acid, 778

standard enthalpy of formation of, 197

structure of, *446*, 453

sodium fluoride, 43

sodium glycinate, 279

sodium halides, products of electrolysis of, 729

sodium hydrogen carbonate, reaction with citric acid, 126

reaction with tartaric acid, 118

sodium hydroxide, commercial preparation of, 746

enthalpy of solution, 479

production of, 776

reaction with acetic acid, 115

reaction with aluminum, 742, *743*

reaction with formic acid, 594

reaction with hydrogen chloride, 114

reaction with methyl acetate, 516

titration of acetic acid, 632, 634

titration with hydrochloric acid, 630

sodium hypochlorite, 165

in bleach, 475

reaction with ammonia, 764

sodium iodide, aqueous, electrolysis of, 727

sodium ions, in ion exchanger, 752

sodium laurylbenzenesulfonate, structure of, 502

sodium monohydrogen phosphate, 771

sodium nitrate, reaction with sulfuric acid, 767, *768*

sodium perchlorate, 779

sodium peroxide, 745

sodium phosphate, 771

sodium polyacrylate, in disposable diapers, *381*

sodium silicate, 759

sodium stearate, as soap, 501

sodium sulfate, quantitative analysis of, 143

reaction with barium chloride, 109

sodium thiosulfate, reaction with iodine, 165

sol A colloidal dispersion of a solid substance in a fluid medium, 500t

solid(s) The phase of matter in which a substance has both definite shape and definite volume, 7

amorphous, 463

chemistry of, 444–470

compressibility of, 418

concentration of, in equilibrium constant expression, 552

dissolution in liquids, 478

ionic, 452–456

molecular, 463

network, 463

types of, 445t

solubility The concentration of solute in equilibrium with undissolved solute in a saturated solution, 476

common ion effect and, 646–648

of complex ions, 645–648

estimating from solubility product constant, 642–646

factors affecting, 482–484

of gases in water, 428t

intermolecular forces and, 421

of ionic compounds in water, 104

of salts, 640–650

undissociated salt and, 645

solubility product constant (K_{sp},) An equilibrium constant relating the concentrations of the ionization products of a dissolved substance, 641

isotopic measurement of, 848

reaction quotient and, 651

standard potential and, 723, 724t

values of, A-23

solute The substance dissolved in a solvent to form a solution, 101, 473

solution(s) A homogeneous mixture in a single phase, 8–10, 472–504

acidic and basic, redox reactions in, 695–698

aqueous, balancing redox equations, 695–698

pH and pOH of, 154–156, 583

reactions in, 101

boiling process in, 488

buffer. *See* **buffer solution(s).**

concentrations in, 149–153

enthalpy of, 479–482

ideal, 486

osmosis in, 491

process of forming, 476–482

Raoult's law, 486

saturated, 476

solvation, effect on acid strength, 608

enthalpy of, 419

Solvay process, 746, 813

solvent The medium in which a solute is dissolved to form a solution, 101, 473

soy, adhesive based on, 381

space-filling models, 60, 341

spatulae, on geckos' toes, 416

spdf notation A notation for the electron configuration of an atom in which the number of electrons assigned to a subshell is shown as a superscript alter the subshell's symbol, 221, 239

specific heat, values of, A-14t

specific heat capacity (C) The quantity of heat required to raise the temperature of 1.00 g of a substance by 1.00 kelvin, 176
  determining, 178
  hydrogen bonding and, 427

spectator ion(s) An ion that is present in a solution in which a reaction takes place but that is not involved in the net process, 108

spectrochemical series An ordering of ligands by the magnitudes of the splitting energies they cause, 810–812

spectrometer, mass, 46, 47

spectrophotometry An analytical method based on the absorption and transmission of specific wavelengths of light, 164–169

spectrum, absorption, 164, 810
  electromagnetic, 207, 809
    continuous, 210
    of heated body, 207, 208
    line, 210, 211

speed(s), of gas molecules, 405
  of wave, 205

sponge, skeletal structure of, *23*

spontaneous reaction, 661. *See also* **product favored reaction(s).**
  effect of temperature on, 675
  Gibbs free energy change and, 677–679
  relation to enthalpy and entropy, 674t

square-planar molecular geometry, 284, 285, 800

square-pyramidal molecular geometry, 284, 285

stability, band of, 827
  of organic compounds, 342
  standard enthalpy of formation and, 198

stabilization, resonance, 356

stainless steel, 791

stalactites and stalagmites, 98

standard atmosphere (atm) A unit of pressure; 1 atm = 760mm Hg, 386, A-7t

standard conditions In an electrochemical cell, all reactants and products are pure liquids or solids, or 1.0 M aqueous solutions, or gases at a pressure of 1 bar, 709

standard deviation A measure of precision, calculated as the square root of the sum of the squares of the deviations for each measurement from the average divided by one less than the number of measurements, 28

standard free energy change of reaction ($_rG°$) The free energy change for a reaction in which all reactants and products are in their standard states, 676

standard hydrogen electrode (SHE), 710

standard molar enthalpy of formation ($_fH°$) The enthalpy change of a reaction for the formation of one mol of a compound directly from its elements, all in their standard states, 197, 662
  enthalpy of solution calculated from, 481
  values of, A-26

standard molar enthalpy of fusion ($_{fus}H°$) The quantity of energy as heat required to convert 1 mol of a solid to a liquid at 1 bar and constant temperature, 189

standard molar enthalpy of vaporization ($A_{vap}H°$) The quantity of energy as heat required to convert 1 mol of a liquid to a gas at 1 bar and constant temperature, 189, 428t, 434

standard molar entropy(S) The entropy of a substance in its most stable form at a pressure of 1 bar, 668
  values of, A-26

standard molar free energy of formation ($_fG°$) The free energy change for the formation of one mole of a compound from its elements, all in their standard states, 680
  values of, A-26

standard molar volume The volume occupied by 1 mol of gas at standard temperature and pressure; 22.414 L, 395

standard potential (E_{cell}) The potential of an electrochemical cell measured under standard conditions, 709
  of alkali metals, 746
  calculation of, 711, 714
  equilibrium constant calculated from, 722

standard reaction enthalpy ($\Delta_rH°$) The enthalpy change of a reaction that occurs with all reactants and products in their standard states, 188
  product-favored vs. reactant-favored reactions and, 200–202

standard reduction potential(s), 711, 713t
  of halogens, 776t
  values of, A-32

standard state The most stable form of an element or compound in the physical state in which it exists at 1 bar and the specified temperature, 188, 662

standard temperature and pressure (STP) A temperature of 0 °C and a pressure of exactly 1 atm, 395

standardization The accurate determination of the concentration of an add, base, or other reagent for use in a titration, 160

standing wave A single-frequency wave having fixed points of zero amplitude, 219

starch, 367

starch-iodide paper, 165

stars, elements formed in, 43

state(s), changes of, 180
  ground and excited, 212
  physical, of matter, 7, 417
  reaction enthalpy and, 189
  standard. *See* **standard state.**

state function A quantity whose value is determined only by the state of the system, 186, 662

stearic acid, 368t

steel, high-strength, 454
  production of, 791

stereoisomers Two or more compounds with the same molecular formula and the same atom-to-atom bonding but with different arrangements of the atoms in space, 340

sterilization, by irradiation, 848

Stern, Otto, 226

stibnite, *643*

stoichiometric coefficients The multiplying numbers assigned to the species in a chemical equation in order to balance the equation, 95
  electrochemical cell potential and, 714
  exponents in rate equation vs., 514
  fractional, 189
  in calculating enthalpy of formation, 198
  in equilibrium constant expression, 551

stoichiometric factor(s) A conversion factor relating moles of one species in a reaction to moles of another species in the same reaction, *134, 399*
  in solution stoichiometry, 157
  in titrations, 160

stoichiometry The study of the quantitative relations between amounts of reactants and products, 95
  ICE table and, 551
  ideal gas law and, 398–400
  mass relationships in, 133–136
  rate equation of elementary step and, 537
  reaction rates and, 509, 510

of reactions in aqueous solution, 157–164
  unit cell and, 446

storage battery, 705

STP. *See* **standard temperature and pressure (STP).**

strained hydrocarbons, 348

Strassman, Fritz, 840

strategies, problem-solving, 38

stratosphere, 528

strong acid(s) An add that ionizes completely in aqueous solution, 111, 584
  reaction with strong base, 584
  reaction with weak base, 585
  titration of, 630–632

strong base(s) A base that ionizes completely in aqueous solution, 111, 584

strong electrolyte A substance that dissolves in water to form a good conductor of electricity, 101, *548*

strontium, abundance of, 748
  in fireworks, 217
  isotopes of, ratios, 50

structural formula A variation of a molecular formula that expresses how the atoms in a compound are connected, 59, 341, 347

structural isomers Two or more compounds with the same molecular formula but with different atoms bonded to each other, 340, 800
  of alcohols, 359, 360
  of alkanes, 344
  of alkenes, 349

structure, molecular, 261

styrene, in ABS plastic, 378
  structure of, 355

styrene-butadiene rubber (SBR), 376

Styrofoam, 375t

Styron, 375t

subatomic particles A collective term for protons, neutrons, and electrons, 43
  properties of, 45t

sublimation The direct conversion of a solid to a gas, 183, 467

submicroscopic level Representations of chemical phenomena in terms of atoms and molecules; also oiled particulate level, 8
  entropy at, 664–666

subshells, labels for, 221
  number of electrons in, 234t
  order of energies of, 235

substance(s), pure A form of matter that cannot be separated into two different species by any physical technique and that has a unique set of properties, 8
  amount of, 72

substituent groups, common, A-16t

substitution reaction(s), of aromatic compounds, 356

substrate, in enzyme-catalyzed reaction, 540

successive approximations, method of, 563, 599

successive equilibria, 655

succinic acid, reaction with 1,2-ethylenediamine, 380

sucrose, as nonelectrolyte, 103
  enthalpy of combustion, 189
  half-life of decomposition, 525
  rate of decomposition of, 511
  structure of, 367

sugar, dietary Calories in, 189

sulfate ion, orbital hybridization in, 319
  resonance structures of, 275
  structure of, 269t

sulfide(s), in black smokers, 92
  roasting of, 773

sulfonate group, in detergents, 502

sulfur, abundance of, 772
  allotropes of, 57, 772, 773
  chemistry of, 773
  combustion of, 552
  compounds with phosphorus, 769
  in iron pyrite, 12
  mining of, 115
  natural deposits of, 772
  reaction with oxygen, 96
sulfur dioxide, 774
  as Lewis acid, 614
  reaction with oxygen, 558
  reaction with water, 117
  scrubbing from flue gas, 412
sulfur hexafluoride, 277
  molecular orbitals in, 335
sulfur tetrafluoride, molecular polarity of, 295
  orbital hybridization in, 319
sulfur trioxide, 774
sulfuric acid, 774
  in lead storage battery, 706
  as polyprotic acid, 579
  production of, from elemental sulfur, 775
  properties and uses of, 115
  reaction with nickel(II) carbonate, 119
  reaction with sodium chloride, 778
  reaction with sodium nitrate, 767, 768
  reaction with water, 112
  structure of, 269t
sunscreens, 210
supercritical fluid A substance at or above the critical
  temperature and pressure, 439, 469
superoxide ion, molecular orbital configuration
  of, 330
superoxides, 745
  oxidation number of oxygen in, 122
superphosphate fertilizer, 774
supersaturated solution(s) A solution that temporar-
  ily contains more than the saturation amount of
  solute, 478
  reaction quotient in, 651
surface area, of colloid, 499
  reaction rate and, 513
surface density plot, 222, *223*
surface tension The energy required to disrupt the
  surface of a liquid, 439–441
  detergents and, 501
surfactant(s) A substance that changes the properties of a surface, typically in a colloidal suspension,
  501–502
  supercritical carbon dioxide and, 442
surroundings Everything outside the system in a
  thermodynamic process, 174, 662
  entropy change for, 672
sustainability 5–6
sweat, cooling by, 436
symbol(s), in chemistry, 8, 10–11
  ionic charge in, 62
symmetry, chirality and, 341, 346
  molecular polarity and, 295
synthesis gas, 742
system The substance being evaluated (or energy
  content in a thermodynamic process, 174, 662
  entropy change for, 672
systematic names, 347, A-15
Système International d'Unités, 21, A-10

T-shaped molecular geometry, 284, *285*
talc, 748
tartaric acid, 368t
  as polyprotic acid, *579*
  reaction with sodium hydrogen carbonate,
    118

technetium, 840
technetium-99m, 846
Teflon, 375t
tellurium, abundance of, 772
temperature A physical property that determines the
  direction of heat flow in an object on contact with
  another object, 175
  change in, heat and, 176
  in collision theory, 528, 530
  constant during phase change, 180
  critical, 439
  effect on solubility, 483
  effect on spontaneity of processes, 675
  electromagnetic radiation emission and, 207,
    208
  energy and, 175
  equilibrium constant and, 572
  equilibrium vapor pressure and, 437
  free energy and, 682–684
  gas, volume and, 390
  ionization constant for water and, 581t
  molecular speed and, 406
  physical properties and, 13, *14*
  reaction rate and, 512
  scales for measuring, 21–23
  standard, 395
tempering, of steel, 791
terephthalic acid, structure of, 378
termolecular process, 536
tertiary alcohols, 365
tetracarbonyl nickel, 813
tetrachloromethane. *See* carbon tetrachloride.
tetraethyl lead, combustion of, balanced equation for, 97
  effusion of, 409
tetragonal unit cell, *447*
tetrahedral electron-pair geometry, orbital
  hybridization and, *313*, 314
tetrahedral holes, 453
tetrahedral molecular geometry, 281, 800
  in carbon compounds, 339, 340
  in DNA backbone, 304
3,5,3',5'-tetraiodothyronine, structure of, 849
thallium, abundance of, 753
Thenard, Louis, *143*
thenardite, 143
theobromine, structure of, 338
theoretical yield The maximum amount of product
  that can be obtained from the given amounts of
  reactants in a chemical reaction, 141
theory A unifying principle that explains a body of
  facts and the laws based on them, 5
  atomic. *See* **atomic theory of matter.**
  kinetic-molecular. *See* **kinetic-molecular theory.**
  molecular orbital. *See* **molecular orbital
    theory.**
  quantum. *See* **quantum mechanics.**
  valence bond. *See* **valence bond theory.**
  VSEPR. *See* **valence shell electron pair repul-
    sion (VSEPR) model.**
thermal energy The energy due to the motion of
  atoms, molecules, or ions, 16
thermal equilibrium A condition in which the system
  and its surroundings are at the same temperature,
  174
thermite reaction, *124*, 132, 140
thermodynamics The study of energy conversion as
  heat or work, 173, 662
  first law of, 183–187, 662
  second law of, 663
  third law of, 668
thermoplastic polymer(s) A polymer that softens but
  is unaltered on heating, 373

thermosetting polymer(s) A polymer that degrades
  or decomposes on heating 373
thermosphere, 404
Thiobacillus ferrooxidans, 791
thiocyanate ion, linkage isomers containing, 801
third law of thermodynamics The entropy of a pure,
  perfectly formed crystal at 0 K is zero, 668
Thompson, Benjamin (Count Rumford), 175
Thomson, Sir Joseph John, 51
Thomson, William. *See* Kelvin, Lord.
three-center bond, 335, 756
thymine, 304
  hydrogen bonding to adenine, 429
  structure of, 260
thyroid gland, imaging of, 846
  treatment of hyperthyroidism, 849
thyroxine, 776, 849
tin, abundance of, 758
tin(II) chloride, aqueous, electrolysis of, 728
tin(IV) iodide, formula of, 82
titanium, in memory metal, 782
  uses of, *788*
titanium(IV) ion, in sapphire, 232
titanium(IV) oxide, 762, 774
  as pigment, 784
  quantitative analysis of, 145
  in rubies, 232
titrant The substance being added during a titration,
  632
titration A procedure for quantitative analysis of a
  substance by an essentially complete reaction in
  solution with a measured quantity of a reagent of
  known concentration, 158–160
  acid–base, 158, 630–640
  curves for, *630,* 632, *633*
  oxidation–reduction, 162
toluene, structure of, 354
tonicity, 496
torr A unit of pressure equivalent to one millimeter of mercury, 386, A-7t
Torricelli, Evangelista, *386*
tracer, radioactive, 847
transition, *d*-to-*d*, 810
transition elements Some elements that lie in rows
  4 to 7 of the periodic table, comprising scandium
  through zinc, yttrium through cadmium, and lanthanum through mercury, 52, 53, 58, 242, 246
  atomic radii, 249, *250*
  cations formed by, 63
  chemistry of, 782–818
  commercial production of, 789–792
  electron configurations of, 785
  naming in ionic compounds, 67
  oxidation numbers of, 785, *786*
  properties of, 784–789
transition state The arrangement of reacting
  molecules and atoms at the point of maximum
  potential energy, 529
translation, A-6
transmittance (T) The ratio of the amount of light
  transmitted by the sample to the amount of
  incident light, 165
transmutation. *See* **nuclear reaction(s).**
transuranium elements, 838, 839
trendline, 37
triatomic molecules, of elements, 56
trichlorobenzene, isomers of, 355
triclinic unit cell, *447*
trigonal-pyramidal molecular geometry, 282
trigonal-bipyramidal electron-pair geometry,
  orbital hybridization and, 313, 318
trigonal-bipyramidal molecular geometry, 281
  axial and equatorial positions in, 284, *285*

trigonal-planar electron-pair geometry, in benzene, 324
  orbital hybridization and, *313, 316*
trigonal-planar molecular geometry, 281
  in carbon compounds, 339
  in DNA structure, 305
trigonal-pyramidal molecular geometry, orbital hybridization and, *313*
triiodide ion, orbital hybridization in, 319
trimethylamine, structure of, 615
triple bond A bond formed by sharing three pairs of electrons, one pair in a sigma bond and the other two in pi bonds, 264
  valence bond theory of, 322
triple point The temperature and pressure at which the solid, liquid, and vapor phases of a substance are in equilibrium, 468
tritium, 46, 740
  fusion of, 842
trona, 746
troposphere, 404
tungsten, melting point of, 784
turquoise, *643*
Tyndall effect, 499
Tyrian purple dye, 334

U.S. Food and Drug Administration (FDA), 165
ulexite, 753
ultraviolet catastrophe, 207
ultraviolet (UV) radiation, 206
  skin damage and, 210
uncertainty principle, 220
unimolecular process, 536
unit(s), of measurement, 21–26, 386
  SI, 21, A-10
unit cell(s) The smallest repeating unit in a crystal lattice, 446
  number of atoms in, 448
  shapes of, *447*
universe, entropy change for, 672
  total energy of, 18
unpaired electrons, in transition metal ions, 785
  paramagnetism of, 227
unsaturated compound(s) A hydrocarbon containing double or triple carbon–carbon bonds, 351
unsaturated solution(s), reaction quotient in, 651
ununseptium, discovery of, 838
uracil, 429
uranium, fission reaction of, 840
  isotopes of, 820
  isotopic enrichment, 841
  isotopic separation, 778
  radioactive series from, 823
uranium-235, fission of, 841
uranium hexafluoride, 778, 786, *841*
urine, phosphorus distilled from, 767

valence band, 458
valence bond theory A model of bonding in which a bond arises from the overlap of atomic orbitals on two atoms to gives bonding orbital with electrons localized between the atoms, 309–324
valence electron(s) The outermost and most reactive electrons of an atom, 239, 261–263
  Lewis symbols and, 263
  of main group elements, 736
valence shell electron pair repulsion (VSEPR) model A model for predicting the shapes of molecules in which structural electron pairs are arranged around each atom to minimize electron-electron repulsions, 280
valeric acid, 368t

van der Waals, Johannes, 411
van der Waals equation A mathematical expression that describes the behavior of nonideal gases, 411
van der Waals force(s), 416, 418, 431
van't Hoff, Jacobus Henrikus, 497
van't Hoff factor The ratio of the experimentally measured freezing point depression of a solution to the value calculated from the apparent molality, 497
vapor pressure The pressure of the vapor of a substance in contact with its liquid or solid phase in a sealed container, 436–439
  Raoult's law and, 486
  relation to enthalpy of vaporization, 438
  of water, A-19
vaporization The state change from liquid to gas, 180, 433
  enthalpy of, 435t, A-14t
    intermolecular forces and, 420, 422t
  heat of, 180
Vectra, 375t
velocity, of wave, 205
Verneuil, Auguste, 232
vibration, A-6
Villard, Paul, 822
vinegar, 365
  pH of, 154
  reaction with baking soda, 598
vinyl alcohol, structure of, 381
viscosity The resistance of a liquid to flow, 440
visible light, 206, 814
vitamin B₁₂, cobalt in, 784
vitamin C. *See* ascorbic acid.
volatile organic compounds (VOCs), 442
volatility The tendency of the molecules of a substance to escape from the liquid phase into the gas phase, 437
volcano, chloride ions emitted by, 163
  sulfur emitted by, 772
volt (V) The electric potential through which 1 coulomb of charge must pass in order to do 1 joule of work, 709, A-7, A-11
Volta, Alessandro, 692
voltage, cell potential vs., 711
voltaic cell(s), 692, 699–704
  commercial, 704–709
  electrodes in, 726t
volume, constant, calorimetry at, 192
  constant, heat transfer at, 186
  gas, pressure and, 388
    temperature and, 390
  measurement of, 25
  per molecule, 410
  physical state changes and, 417
  standard molar, 395
volumetric flask, *149, 151*

wallboard, gypsum in, 86
Walton, E. T. S., 837
washing soda, 746. *See also* sodium carbonate.
water, amphiprotic nature of, 113
  autoionization of, 581
  balancing redox equations with, 695–698
  boiling point elevation and freezing point depression constants of, 521t
  bond angles in, 282
  as Brønsted base, 112
  concentration of, in equilibrium constant expression, 552
  corrosion and, 787
  critical temperature, 439
  density of, temperature and, 13t, 425
  as electroactive substance, 759

electrolysis of, *11,* 772
electrostatic potential map of, 294
enthalpy of formation, 194
expansion on freezing, 7
formation by acid–base reactions, 158
generated by hydrogen–oxygen fuel cell, 708
hard, 752
heat of fusion, 180
heat of vaporization, 180
heavy, 46
in hydrated compounds, 86
interatomic distances in, 24
iodine solubility in, 430
ionization constant for (K_w), 581
molecular polarity of, 293
orbital hybridization in, 315
pH of, 154
phase diagram of, 468, *649*
purification of, *10*
  by reverse osmosis, 472
reaction with alkali metals, 54, 743, 744
reaction with aluminum, 697
reaction with hydrides, 742, *743*
reaction with insoluble salts, 649
reaction with lithium, 399
reaction with potassium, 54
reaction with sodium, *5*
relation to alcohols, 361
solubility of alcohols in, 361
solubility of gases in, 428t
solubility of ionic compounds in, 104, 479, *481*
solvent in aqueous solution, 101
specific heat capacity of, 176t, 177, 427
standard molar enthalpy of vaporization of, 435
treatment, methods of, with hydrazine, 764
triple point of, 468
vapor pressure of, A-19
  curves for, *437*
water gas reaction, 742
water glass, 759
water softener, 752
Watson, James D., 429
watt A unit of power, defined as 1 joule/second, A-7
wave, matter as, 216
wave mechanics. *See* **quantum mechanics.**
wave-particle duality The idea that the electron has properties of both a wave and a particle, 216–218
wavefunction(s) (Ψ) A set of equations that characterize the electron as a matter wave, 219
  addition and subtraction of, 326
  phases of, 329
  radial and angular components of, 225
wavelength (λ) The distance between successive crests (or troughs) in a wave, 205
  choice for spectrophotometric analysis, 167
  of moving mass, 216
weak acid(s) An acid that is only partially ionized in aqueous solution, 111, 584
  calculating pH of aqueous solution, 598
  in buffer solutions, 623
  ionization constants, A-20
  reaction with strong base, 594
  reaction with weak base, 595
  titration of, 632
weak base(s) A base that is only partially ionized in aqueous solution, 111, 584
  in buffer solutions, 623
  calculating pH of aqueous solution, 600
  ionization constants, A-22
  titration of, 636
weak electrolyte A substance that dissolves in water to form a poor conductor of electricity, 102, 103

weather, heat of vaporization of water and, 436
weight, mass and, A-6
weight percent, of solution, 475
Wilkins, Maurice, 303, 429
wintergreen, oil of, 370, 526
wolframite, 786
work Energy transfer that occurs as a mass is moved through a distance against an opposing force, 183
   in electrochemical cell, 721
   energy transferred by, 175
   Gibbs free energy and, 679
   pressure–volume, 185
   sign conventions for, 185t

xanthines, 338
xenon, compounds of, 308
xenon difluoride, *308,* 663
xenon oxytetrafluoride, molecular geometry of, 286
xenon tetrafluoride, 284, *285*
xerography, selenium in, 772
x ray(s), 206
x-ray crystallography, 303, 450

yeast, acetic acid produced by, *368*
yield, of product in a chemical reaction, 141

Zeise's salt, 815
zeolite(s), 761
   in ion exchanger, 752
zeroes, as significant figures, 32
zero-order reaction, 514, 515
   half-life of, 524
   integrated rate equation, 522
zinc, reaction with dioxovanadium(V) ion, 695
   reaction with hydrochloric acid, 110, 158
zinc blende, structure of, 453
zinc chloride, in dry cell battery, 705
zinc-oxygen battery, 706
zinc sulfide, structure of, 453
zone refining, 758, *759*

PHYSICAL AND CHEMICAL CONSTANTS

Avogadro's number $N = 6.0221415 \times 10^{23}$/mol
Electronic charge $e = 1.60217653 \times 10^{-19}$ C
Faraday's constant $F = 9.6485338 \times 10^4$ C/mol electrons
Gas constant $R = 8.314472$ J/K · mol
$= 0.082057$ L · atm/K · mol

π $\pi = 3.1415926536$
Planck's constant $h = 6.6260693 \times 10^{-34}$ J · sec
Speed of light $c = 2.99792458 \times 10^8$ m/sec
(in a vacuum)

USEFUL CONVERSION FACTORS AND RELATIONSHIPS

Length
SI unit: Meter (m)
1 kilometer $= 1000$ meters
$= 0.62137$ mile
1 meter $= 100$ centimeters
1 centimeter $= 10$ millimeters
1 nanometer $= 1.00 \times 10^{-9}$ meter
1 picometer $= 1.00 \times 10^{-12}$ meter
1 inch $= 2.54$ centimeter (exactly)
1 Ångstrom $= 1.00 \times 10^{-10}$ meter

Mass
SI unit: Kilogram (kg)
1 kilogram $= 1000$ grams
1 gram $= 1000$ milligrams
1 pound $= 453.59237$ grams $= 16$ ounces
1 ton $= 2000$ pounds

Volume
SI unit: Cubic meter (m^3)
1 liter (L) $= 1.00 \times 10^{-3}$ m^3
$= 1000$ cm^3
$= 1.056710$ quarts
1 gallon $= 4.00$ quarts

Energy
SI unit: Joule (J)
1 joule $= 1$ kg · m^2/s^2
$= 0.23901$ calorie
$= 1$ C $\times$ 1 V
1 calorie $= 4.184$ joules

Pressure
SI unit: Pascal (Pa)
1 pascal $= 1$ N/m^2
$= 1$ kg/m · s^2
1 atmosphere $= 101.325$ kilopascals
$= 760$ mm Hg $= 760$ torr
$= 14.70$ lb/in^2
$= 1.01325$ bar
1 bar $= 10^5$ Pa (exactly)

Temperature
SI unit: kelvin (K)
0 K $= -273.15$ °C
K $= $ °C $+ 273.15$°C
? °C $= (5$ °C/9 °F$)($°F $- 32$ °F$)$
? °F $= (9$ °F/5 °C$)($°C$) + 32$ °F

LOCATION OF USEFUL TABLES AND FIGURES

Atomic and Molecular Properties

Atomic electron configurations	Table 7.3
Atomic radii	Figures 7.6, 7.9
Bond dissociation enthalpies	Table 8.9
Bond lengths	Table 8.8
Electron Attachment Enthalpy	Figure 7.11, Appendix F
Electronegativity	Figure 8.11
Elements and their unit cells	Figure 13.5
Hybrid orbitals	Figure 9.3
Ionic radii	Figure 7.12
Ionization energies	Figure 7.10, Table 7.5

Thermodynamic Properties

Enthalpy, free energy, entropy	Appendix L
Lattice energies	Table 13.2
Specific heat capacities	Appendix D

Acids, Bases and Salts

Common acids and bases	Table 3.1
Formation constants	Appendix K
Ionization constants for weak acids and bases	Table 17.3, Appendix H, I
Names and composition of polyatomic ions	Table 2.4
Solubility guidelines	Figure 3.10
Solubility constants	Appendix J

Miscellaneous

Charges on common monatomic cations and anions	Figure 2.18
Common polymers	Table 10.12
Oxidizing and reducing agents	Table 3.3
Selected alkanes	Table 10.2
Standard reduction potentials	Table 20.1, Appendix M

STANDARD ATOMIC WEIGHTS OF THE ELEMENTS 2007 Based on relative atomic mass of $^{12}C = 12$, where ^{12}C is a neutral atom in its nuclear and electronic ground state.[†]

Name	Symbol	Atomic Number	Atomic Weight	Name	Symbol	Atomic Number	Atomic Weight
Actinium*	Ac	89	(227)	Molybdenum	Mo	42	95.96(2)
Aluminum	Al	13	26.9815386(2)	Neodymium	Nd	60	144.242(3)
Americium*	Am	95	(243)	Neon	Ne	10	20.1797(6)
Antimony	Sb	51	121.760(1)	Neptunium*	Np	93	(237)
Argon	Ar	18	39.948(1)	Nickel	Ni	28	58.6934(4)
Arsenic	As	33	74.92160(2)	Niobium	Nb	41	92.90638(2)
Astatine*	At	85	(210)	Nitrogen	N	7	14.0067(2)
Barium	Ba	56	137.327(7)	Nobelium*	No	102	(259)
Berkelium*	Bk	97	(247)	Osmium	Os	76	190.23(3)
Beryllium	Be	4	9.012182(3)	Oxygen	O	8	15.9994(3)
Bismuth	Bi	83	208.98040(1)	Palladium	Pd	46	106.42(1)
Bohrium	Bh	107	(272)	Phosphorus	P	15	30.973763(2)
Boron	B	5	10.811(7)	Platinum	Pt	78	195.084(9)
Bromine	Br	35	79.904(1)	Plutonium*	Pu	94	(244)
Cadmium	Cd	48	112.411(8)	Polonium*	Po	84	(209)
Cesium	Cs	55	132.9054519(2)	Potassium	K	19	39.0983(1)
Calcium	Ca	20	40.078(4)	Praseodymium	Pr	59	140.90765(2)
Californium*	Cf	98	(251)	Promethium*	Pm	61	(145)
Carbon	C	6	12.0107(8)	Protactinium*	Pa	91	231.03588(2)
Cerium	Ce	58	140.116(1)	Radium*	Ra	88	(226)
Chlorine	Cl	17	35.453(2)	Radon*	Rn	86	(222)
Chromium	Cr	24	51.9961(6)	Rhenium	Re	75	186.207(1)
Cobalt	Co	27	58.933195(5)	Rhodium	Rh	45	102.90550(2)
Copernicium	Cn	112	(285)	Roentgenium	Rg	111	(280)
Copper	Cu	29	63.546(3)	Rubidium	Rb	37	85.4678(3)
Curium*	Cm	96	(247)	Ruthenium	Ru	44	101.07(2)
Darmstadtium	Ds	110	(281)	Rutherfordium	Rf	104	(267)
Dubnium	Db	105	(268)	Samarium	Sm	62	150.36(2)
Dysprosium	Dy	66	162.500(1)	Scandium	Sc	21	44.955912(6)
Einsteinium*	Es	99	(252)	Seaborgium	Sg	106	(271)
Erbium	Er	68	167.259(3)	Selenium	Se	34	78.96(3)
Europium	Eu	63	151.964(1)	Silicon	Si	14	28.0855(3)
Fermium*	Fm	100	(257)	Silver	Ag	47	107.8682(2)
Fluorine	F	9	18.9984032(5)	Sodium	Na	11	22.98976928(2)
Francium*	Fr	87	(223)	Strontium	Sr	38	87.62(1)
Gadolinium	Gd	64	157.25(3)	Sulfur	S	16	32.065(5)
Gallium	Ga	31	69.723(1)	Tantalum	Ta	73	180.94788(2)
Germanium	Ge	32	72.64(1)	Technetium*	Tc	43	(98)
Gold	Au	79	196.966569(4)	Tellurium	Te	52	127.60(3)
Hafnium	Hf	72	178.49(2)	Terbium	Tb	65	158.92535(2)
Hassium	Hs	108	(270)	Thallium	Tl	81	204.3833(2)
Helium	He	2	4.002602(2)	Thorium*	Th	90	232.03806(2)
Holmium	Ho	67	164.93032(2)	Thulium	Tm	69	168.93421(2)
Hydrogen	H	1	1.00794(7)	Tin	Sn	50	118.710(7)
Indium	In	49	114.818(3)	Titanium	Ti	22	47.867(1)
Iodine	I	53	126.90447(3)	Tungsten	W	74	183.84(1)
Iridium	Ir	77	192.217(3)	Ununhexium	Uuh	116	(292)
Iron	Fe	26	55.845(2)	Ununoctium	Uuo	118	(294)
Krypton	Kr	36	83.798(2)	Ununpentium	Uup	115	(228)
Lanthanum	La	57	138.90547(7)	Ununquadium	Uuq	114	(289)
Lawrencium*	Lr	103	(262)	Ununseptium	Uus	117	(294)
Lead	Pb	82	207.2(1)	Ununtrium	Uut	113	(284)
Lithium	Li	3	6.941(2)	Uranium*	U	92	238.02891(3)
Lutetium	Lu	71	174.9668(1)	Vanadium	V	23	50.9415(1)
Magnesium	Mg	12	24.3050(6)	Xenon	Xe	54	131.293(6)
Manganese	Mn	25	54.938045(5)	Ytterbium	Yb	70	173.054(5)
Meitnerium	Mt	109	(276)	Yttrium	Y	39	88.90585(2)
Mendelevium*	Md	101	(258)	Zinc	Zn	30	65.38(2)
Mercury	Hg	80	200.59(2)	Zirconium	Zr	40	91.224(2)

[†]The atomic weights of many elements can vary depending on the origin and treatment of the sample. This is particularly true for Li; commercially available lithium-containing materials have Li atomic weights in the range of 6.939 and 6.996. The uncertainties in atomic weight values are given in parentheses following the last significant figure to which they are attributed.

*Elements with no stable nuclide; the value given in parentheses is the atomic mass number of the isotope of longest known half-life. However, three such elements (Th, Pa, and U) have a characteristic terrestrial isotopic composition, and the atomic weight is tabulated for these. **http://www.chem.qmw.ac.uk/iupac/AtWt/**